Ship Design and Construction

WRITTEN BY A GROUP OF AUTHORITIES

ROBERT TAGGART, EDITOR

1980

Published by
The Society of Naval Architects and Marine Engineers
One World Trade Center, Suite 1369
New York, N.Y. 10048

Library of Congress Catalog Card No. 80-52966
ISBN No. 0-9603048-0-0.
Printed in the United States of America
First Printing, October 1980
Second Printing, May 1986
All errata reported as of this date
has been incorporated in
this printing

Introduction

A Word from the President

The Society of Naval Architects and Marine Engineers embarked on a program of publishing significant treatises on the subject of naval architecture, marine engineering, and shipbuilding with the 1939 edition of *Principles of Naval Architecture.* This was followed in 1942 with the publication of *Marine Engineering* and in 1955 with *Design and Construction of Steel Merchant Ships. Principles of Naval Architecture* was revised and updated in 1967 and is currently undergoing another revision to be published within the next few years. *Marine Engineering* was also completely revised and a new edition published in 1971.

This present volume of *Ship Design and Construction* evolved from the 1955 *Design and Construction of Steel Merchant Ships* and a 1969 revision that bore the same title as the current volume. The subject matter of the new book, however, is almost totally different from what was contained in the 1969 edition. There is a change not only in the text material but also in the philosophy on which that material is based. This 1980 edition concerns itself with the practical aspects of ship design as they relate to the requirements of the owner and operator and as they relate to the characteristics of the mission that the ship is to perform. Producibility of the ship is stressed as a design goal as well as the need to investigate shipbuilding and ship operating economics to derive a cost effective vessel that will perform the task intended. In addition to the design aspects, coverage is provided of such subjects as shipbuilding contracts, government regulations, shipyard production techniques, launchings, trials, and guarantee surveys.

Through the cooperation of the Publications Committee of the Society and the Control Committees for this book and for the forthcoming edition of *Principles of Naval Architecture,* positive steps have been taken to develop a compatible series of volumes that will cover the gamut of technology from the basic concept of a vessel to the ultimate delivery to its owner and operator. This has proved to be an exacting task, particularly during a period when new intergovernmental regulations have been undergoing rapid development and when environmental demands have forced changes not only in the ships themselves but also in the facilities where they are constructed and the waterways in which they operate.

The Control Committee that formulated the overall pattern and tone of this edition of *Ship Design and Construction,* selected the authors, and guided the publication through the many rough spots inherent in the production of such a work was ably chaired by E. Scott Dillon. The other Committee members were Nicholas Bachko, Francis G. Bartlett, Hollinshead de Luce, Edward V. Lewis, John J. Nachtsheim, George C. Nickum, J. Randolph Paulling, Jr., Robert I. Price, Robert J. Tapscott, and Robert T. Young. The dedication of the Chairman and his Committee to this long and arduous task is greatly appreciated by the Society. It is my sincere belief that the product of their efforts will be of great benefit to the maritime industry.

LESTER ROSENBLATT
President

Foreword

With the passage of time since the 1969 edition of *Ship Design and Construction,* progress in the related arts and sciences has increasingly dictated the need for an updated version. Accordingly, in February, 1976 the Society's Executive Committee directed that the revision proceed promptly. In March 1976, President L. V. Honsinger appointed the Control Committee and in April 1976, the Editor was appointed.

The purpose of the book remains essentially the same as that of the prior editions; namely, a textbook "to assist students and others entering the field of shipbuilding towards a knowledge of how merchant ships are designed and constructed and to provide them with a good background for further study." Nevertheless, a number of considerations led the Committee to modify extensively the scope and organization of the book.

At the outset, the Committee recognized that within a few years the Society's book *Principles of Naval Architecture* would also be revised and that it contained material which more properly pertained to design and construction rather than theoretical naval architecture. Therefore it recommended, and the Publications Committee as well as the Executive Committee approved, the inclusion in *Ship Design and Construction* of new chapters on Load Lines, Tonnage, and Launching which would then be deleted from future editions of *Principles of Naval Architecture.* As a partial trade-off toward page reduction, the Committee eliminated the 1969 edition chapter on Submersibles because of its relatively narrow field of interest and the lack of major new developments for commercial operations.

In an effort to accord the subject matter more uniform treatment, the five chapter concentration on structure of the prior edition gave way to a shorter three chapter version. On the other hand, the Committee sensed a need for material which would give the student familiarity with a greater variety of important vessel types. Therefore, it enlarged the treatment of Basic Design into two chapters, the first to describe the basic design in its general application and the second to show how basic designs are developed for vessels with a wide variety of missions. Thus, a central theme is expounded in the early chapters which pervades the entire book and emphasizes the effects which the type of cargo and the vessel's mission have in developing markedly different configurations and basic designs. Because of the importance of cargo on design, more space is allocated to cargo handling with separate chapters devoted respectively to dry and liquid cargos.

As an overall guide to organization of text material, Chapters I and II constitute the Basic Design section, while Chapters III to XIV cover Final Design; the remainder pertains to various aspects of Ship Construction. Along the line of more even treatment to the overall subject matter, the Committee enlarged the section on Ship Construction by adding chapters on Contractual Arrangements and Trials as well as the chapter on Launching mentioned previously.

Significant strides in the application of computers to both design and construction since the 1969 edition prompted consideration of a special chapter devoted to the role of computers. However, difficulties of integrating such a chapter into the remainder of the text led to a decision calling upon the authors to include computer applications in each chapter as appropriate. Additionally, in keeping with the trend toward increased use of metrication both in the United States and abroad, the Committee received approval to accord primacy to measurements in accordance with the *Systeme International d'Unites* (*SI*) in the text and illustrations with English units retained only in secondary status as an aid to students learning the metric system.

After first drafts of the various chapters of *Ship Design and Construction* (*SDC*) had been prepared, the Executive Committee decided to proceed with a new edition of *Principles of Naval Architecture* (*PNA*). It then became more important to harmonize the contents of the two books with *PNA* containing the theoretical aspects of naval architecture with *SDC* applying that theory to practice. Fortunately, John J. Nachtsheim, Chairman of the Control Committee and Edward V. Lewis, Editor of *PNA* were members of the *SDC* Control Committee, greatly facilitating the integration process. This bore fruit especially with the treatment of the strength of ships and the design of principal structural members which had not achieved sufficient coordination in earlier editions.

We are indebted to the Editor and the members of the Control Committee who have painstakingly reviewed all of the chapters and made many valuable comments. In some cases they actually provided some of the text which the authors greatly appreciated. Special mention is due Past President Young, who in spite of the extra work and responsibility placed on his shoulders after his election to the Presidency, continued to serve as an active Committee member throughout his entire term. Additionally, we would like to express our sincere appreciation to the American Bureau of Shipping. Not only have five Bureau personnel served either as authors or Control Committee members, but the Bureau has consistently provided assistance and information to other authors and to the Society in the preparation of this volume.

As a result of the collaborative effort involved in its preparation the 1980 edition of *Ship Design and Construction* will better meet the needs of all naval architects. Because of its comprehensive treatment and the near impossibility for one person to retain specialized knowledge in every technical field covered by this edition, the book should be valued by practicing naval architects as well.

E. SCOTT DILLON
Chairman, CONTROL COMMITTEE

Preface

The 1980 edition of *Ship Design and Construction* is a descendant of the *Design and Construction of Steel Merchant Ships,* published by the Society in 1955, and the revision of that book entitled *Ship Design and Construction* published in 1969. Although its antecedents covered much of the same general subject matter, the present volume has been essentially completely rewritten and thus stands alone as a significantly different form of treatise on the subject.

The emphasis has been placed upon the design and construction of ships to fulfill specific missions; throughout the text the rationale for configuring the ship to do a specific job or a specified multiplicity of jobs is highlighted. As a result, few of the chapters contained herein are directly comparable to those found in the previous editions. Additionally chapters on Load Lines, Tonnage, and Launching, previously covered in the *Principles of Naval Architecture* are now more logically contained within this volume as well as chapters on Contracting Arrangements and Trials and Preparations for Delivery.

A general format has been adopted that leads the reader through the derivation of mission requirements, development of conceptual and preliminary designs, including hull form and arrangements, deriving acceptable load lines, and performing tonnage calculations. Ensuing chapters deal with the overall structural design, the design of structural components, and with the selection and connection of hull materials. With these basic elements decided upon, the more detailed aspects of design are treated including hull outfit and fittings, and cargo handling techniques and equipment for dry, liquid, and hazardous cargos. The final design aspects wind up with treatments of maneuvering, navigation, and motion control, techniques for controlling the interior environment of the ship, and methods and materials for preservation of the hull. In making the transition from design to construction the various stages of cost estimating, contracts, and governmental oversight are discussed followed by a detailed explanation of the equipment and techniques involved in ship construction. The various processes used in ship launching, including the most modern methods of transferring a vessel from the building site to a waterborne condition, are described and launching calculation techniques are delineated. The volume concludes with a discussion of ship trials and the final preparations required for delivery from the shipyard to the owner.

In this 1980 edition, the 1969 edition Glossary has been significantly expanded to cover all unfamiliar terms used in both design and construction of ships rather than only the construction terms defined previously. Acronyms, abbreviations, and symbols have been defined as they appear within the text instead of the previous practice of including them in separate tables. In general, the symbols used are in accordance with the 1963 International Towing Tank Conference Committee on the Presentation of Data.

UNITS OF MEASURE

The Metric Conversion Act of 1975 (P.L. 94-168) declared a national policy of coordinating the increasing use of metric systems of measurement and established the United States Metric Board to coordinate voluntary conversion to the International System of Units, *SI*. One of the major departures of *SI* from previous metric systems is the use of distinctly separate units for *mass* and *force*. In *SI*, the unit of *force*, the newton (N), instead of being related to gravity, is defined as being equal to the acceleration it imparts to a unit *mass*, the kilogram (kg). The *SI* unit for *mass* (not *force*) is the kilogram, used to specify the quantity of matter in a body. The *SI* unit for *force* is the newton. One newton applied to a mass of one kilogram gives an acceleration of one meter per second squared. *Weight* is sometimes defined as the force which, when applied to a body, would give it an acceleration equal to the local acceleration of free fall. However, this technical use of the term is generally disregarded in commercial and everyday use, when reference to the *weight* of a body is used to indicate its *mass*. Because of this conventional usage, it has not been possible to delete the dual use of the term *weight* as a quantity throughout the entire text nor to specify whether mass or force is intended. To this extent, the present volume must be considered as an initial step in the mental conversion process between past thinking and more precise engineering definition of terms of mensuration.

The practice followed throughout the book has been to present all dimensions in *SI* units followed by U.S. Customary units in parentheses. Occasionally, to avoid confusion, separate comparable tables or graphs are presented in the two sets of units. Also, on some illustrations, *SI* units only are given to eliminate unnecessary crowding. When expressing displacement, deadweight, buoyancy, or other vertical forces associated with gravitational acceleration the conventional use of long tons has been retained; furthermore, long tons and metric tons have been used interchangeably because of the small difference between these two measures. Similar treatment has been used in dealing with horsepower. For a complete listing of the *SI* unit terms and conversion factors used throughout the text, the reader is referred to the Glossary under *SI* Units.

THE INTERGOVERNMENTAL MARITIME CONSULTATIVE ORGANIZATION

The Intergovernmental Maritime Consultative Organization (IMCO) is a relatively new forum for the consideration of international maritime problems. It was created in 1958 and comprises a forum in which worldwide maritime problems, except those concerning rates and tariffs, are presented, evaluated, and solved. It is a standards-making body, a medium of exchange of information on shipping matters, and a means of promoting measures to facilitate the movement of ships and their cargo. IMCO has facilitated many international agreements on safety, pollution, and ship requirements and a mechanism has been established for keeping these agreements up to date. The organization does not possess direct regulatory powers. However, international agreements developed by IMCO on the subject of shipping and other sea-related questions, when brought into effect by assent of the required number of participating national governments, do become binding upon mariners of those nations through the respective national legislative processes. IMCO also functions as a source of information and counsel to other elements of the United Nations organization having an interest in maritime matters.

In its relatively brief existence, IMCO has dealt with a wide variety of problems related to the sea. The types of craft discussed range from conventional displacement ships with a variety of missions to offshore structures, hydrofoils, and air cushion vehicles together with their equipment and requirements for the personnel to operate them. Not only is operation of the ship considered, but the impact of the ship on the environment as well. The concepts of traffic separation and ship control disciplines have been considered as they relate to the Rules of the Road in various restricted areas of the world's sea lanes.

Some significant agreements which IMCO has evolved are: The International Convention of Safety of Life at Sea 1960; International Convention on Loadlines 1966; International Convention on Tonnage Measurement of Ships 1969; the International Convention on Facilitation of International Maritime Traffic 1965; the International Convention of Intervention on the High Seas in case of Oil Pollution Casualties 1969; International Convention on Civil Liability for Oil Pollution Damage 1969; International Convention for Prevention of Pollution of the Seas by Oil 1973; Revision of the Safety of Life at Sea Convention 1974; International Fire Safety Amendments of 1966 and 1967; Conventions on Containers in International Trade 1972; International Regulation for Preventing Collisions at Sea 1972; Code for the Construction of Chemical Ships; and Code for the Construction of Gas Carriers.

These various conventions and their effects on ship design and construction are mentioned in several chapters of this book. Additional details on how the IMCO actions have been transformed into rules and regulations for the building and operating of United States ships are given in Chapter XI "Design for Transport of Liquid and Hazardous Cargos."

ACKNOWLEDGMENTS

The authors of the chapters of this edition of *Ship Design and Construction* wish to extend their appreciation for the following contributions:

Mr. Kiss (Chapter I) is indebted to numerous individuals and organizations for suggestions, advice, photographs, and insights which led to the creation of this chapter. Mr. E. Scott Dillon, author of this chapter in the previous edition is deserving of the initial individual acknowledgment, since he provided an excellent basis on which to build and since he served as one of the author's principal mentors in the area of ship design. Special thanks are due to Sharon Bowers for her accurate typing and reproduction of numerous drafts of the text. In addition the following individuals provided essential assistance in gathering data, preparing illustrations, converting English units to Metric, and generally offering useful critical reviews of the text: Charles B. Cherrix, Thomas G. Connors, Alexander C. Landsburg, George H. Levine, Robert M. McNaull, Earl Schneider, Paul Speicher, Earl Taylor, Wesley Williams, and Warren B. Wilson.

Mr. Michel (Chapter II) extends his appreciation to numerous companies and individuals for the use of their data and illustrations in the preparation of this text. Gratitude goes to the following people who contributed directly to this chapter: George C. Nickum, of Nickum & Spaulding Associates, who prepared the sections on "Ferryboats and Fishing Vessels" and who further contributed to the section on "Tugboats"; William W. Hamilton, of Friede & Goldman, Ltd., who provided the section on "Towboats;" Ronald K. Kiss, MarAd, who contributed the section on "LNG Vessels"; and Calvin V. Norton, of Friede & Goldman, Ltd., who collected the data, prepared many of the illustrations, collated, and proofread the manuscript.

Mr. Tapscott (Chapter III) is grateful for the ship arrangement drawings supplied by Ronald K. Kiss and material used from Chapter XI of the previous edition authored by E. A. Maier and A. E. Stanford.

Mr. Ritola (Chapter IV) would like to thank James R. Graf of the American Bureau of Shipping; Mr. Graf provided immeasurable assistance in the technical aspects and provided the excellent sketches.

Chapter VI was initially authored by Dr. J. Randolph Paulling and a later draft was coauthored by Dr. Rolf Glasfeld. However, although the output of these two authors was of high technical caliber, the Control Committee decided that the material was more applicable to *Principles of Naval Architecture* than to *Ship Design and Construction*. As a result of this decision, and concurrence by the *PNA* Control Committee, this material will be readapted for that publication. The tremendous effort put forth by these authors in attempting to meet *SDC* deadlines is sincerely appreciated. The revised text of Chapter VI was prepared by David B. Bannerman and Hsien Y. Jan and is directed toward those aspects of structural design that are particularly applicable to the problems encountered by the shipyard naval architect in developing a structure that is not only technically adequate but is also in consonance with regulatory agency requirements. The present chapter incorporates material from the 1969 edition of *Ship Design and Construction*, specifically from Chapter III, by Henry A. Schade, and Chapter IV by David B. Bannerman and Robert S. Little. These coauthors would like to express their appreciation to the American Bureau of Shipping and particularly to Stanley Stiansen for making available the resources of that organization and to Drs. Paulling and Glasfeld for the material extracted from their earlier drafts. In addition they would like to acknowledge the assistance of Matias Wojnarowski, Hsao H. Chen, and Donald Liu in preparing the text, and Robert Curry in reviewing the text.

Mr. Stiansen (Chapter VII) wishes to acknowledge the contributions, in preparing and reviewing the text, of Matias Wojnarowski and Robert Curry, both of the American Bureau of Shipping, and of Antonios Latsis, formerly of ABS. Hugo Petrillo of ABS participated in the preparation of the figures. The present chapter incorporates material from the previous edition's chapter of the same title by Thomas M. Buermann.

Mr. Stern (Chapter VIII) expresses his appreciation to B. L. Alia and M. F. Wheatcroft for their helpful suggestions, to C. R. Herbert for his preparation of the figures, and to Mrs. A. Tessalone for her assistance in preparation of the manuscript.

Mr. Mallett (Chapter IX) acknowledges the contribution of MacGregor-Comarain, Inc., in providing illustrations for the text. The majority of the remaining illustrations were adaptations from the previous edition of *Ship Design and Construction*.

Mr. Boylston (Chapter X) appreciates the assistance in obtaining, and the permission given to publish the numerous illustrations for this chapter furnished by the following organizations: MacGregor-Comarain, Inc.; Line Fast Corp.; Friede & Goldman, Ltd.; Morgan Engineering Co.; Navire; Shipping World and Shipbuilder; Paceco, Inc.; Dravo Corp.; Siwertell; and A. B. Hagglund and Soner.

Vice Admiral Price (Chapter XI) is indebted to the following for their assistance in either the development of the text or the use of figures and tables from their own articles pertaining to the subject: the late J. L. Moss, Marcona Corp. concerning slurry; George E. Nickum, Nickum & Spaulding Associates, regarding fish; S. Fraser Sammis, National Cargo Bureau, on ore and grains; T. R. Farrell, Lloyd's Register of Shipping, on chemicals; George B. King, BP Tanker Co., Ltd., on inerting of tankers; Robert J. Lakey, Helge Rindal Inc., on gasses and chemicals; W. Hickman, National Steel and Shipbuilding Co., on tanker piping; Walter Neal, Keystone Shipping Co., on compatible cargo; William DuBarry Thomas, J. J. Henry Co. Inc., on LNG; William Kime, USCG, on damage criteria; William M. Benkert, American Institute of Merchant Shipping, on the overall outline of the chapter. The author's ability to meet the Control Committee's schedule is attributable to two dedicated secretaries, Pat Woolridge and Marilyn Poppe.

Mr. Hunley (Chapter XII) would like to note that although he is listed as a co-author of this chapter, his contributions were primarily the reorganization and consolidation of material furnished by Arthur Pitchersky and Abraham Taplin (Ship Control). These gentlemen prepared the multiple drafts involved. Additionally, George Prentiss is to be thanked for his assistance in preparing the section on anchors and anchoring and Robert Taggart for the material on environmental force calculations and on dynamic positioning.

Mr. Devoluy (Chapter XIV) appreciates the assistance of William Briggs and John Peart in supplying information on shop plate priming.

Messrs. Hoffmann and Bachko (Chapter XV) are grateful for the assistance and information supplied by the U. S. Coast Guard, the Maritime Administration, and the American Bureau of Shipping.

Prof. Bartlett (Chapter XVI) wishes to acknowledge the assistance rendered by Filippo Cali of Cali Associates, Inc. and by his former colleague, James F. Hallock. Additionally he is indebted to the personnel of Ingalls Shipbuilding Corporation for several of the illustrations provided.

Mr. Leavitt (Chapter XVII) wishes to state that the greater part of the end launching material is based on unpublished notes developed during his many years as Chief Naval Architect of the Ingalls Shipbuilding Corporation. Several of the figures, with modifications, have been taken from *Principles of Naval Architecture.* Side launching formulas are from *Static and Dynamics of the Ship, Theory of Buoyancy, Stability and Launching* by V. Semyonov-Tyan-Shansky, Peace Publishers with symbols changed for consistency. The Ingalls Shipbuilding Division of Litton Industries is thanked for making available time and office, typing, and reproduction facilities for the preparation of this chapter.

Mr. Jack (Chapter XVIII) expresses his appreciation to Joseph C. Czudak, former Construction Representative for MarAd, for his valuable contributions toward the development of the chapter and particularly in the preparation of the sections on "Testing" and "Delivery"; also to Michael W. Walsh, MarAd, for his guidance and assistance in the preparation of the section on "Stability Tests."

We must note with regret the death of one of the co-authors of Chapter IX, Norman J. Thompson. Although he had faithfully completed his obligations as an author prior to his death, and had completed his review of the galley proof, he did not have the opportunity to see the results of his efforts in published form.

This dedication on the part of Norman Thompson was characteristic of that exhibited by all of the Authors and the Control Committee who took part in the preparation of this book. The Editor was indeed fortunate to have been closely associated with all of these outstanding and highly competent individuals who gave unstintingly of their time and effort in bringing this publication to fruition.

Particularly worthy of note is the work performed by David B. Bannerman who had served the Society as Chairman of the Control Committee for the previous edition of *Ship Design and Construction*. When the selectee for preparation of the Glossary and the Index requested relief from that task midway through the preparation of the book, Mr. Bannerman cheerfully took over. Additionally, when problems arose with the text of Chapter VI, he again jumped into the breech and orchestrated a complete rewriting of that chapter between March 1980 and the publication date.

The Editor is very grateful to the people on the staff of Robert Taggart Incorporated who have suffered through the lengthy procedure of developing the text and illustrations of this book. Miss Evelyn Cerny kept careful track of the movements of the many chapters through the various stages of development, completely typed several of the chapters from handwritten drafts, made editorial corrections in all chapters for each of three submissions to the Control Committee, and reproduced more than 75,000 pages of text to meet the requirements for review by all concerned. Jeffrey Lown and Caren Cathers prepared the majority of the illustrations that are used throughout the text.

The staff at Society headquarters has done a masterful job of final editing, the correcting of the galley and page proofs, and the layout of the latter; Trevor Lewis-Jones is due specific credit for keeping these final phases of preparation on schedule despite the inevitable last minute problems that arose. The Society's Technical Coordinator, Philip Poullada, was of signficant help in the work, particularly many of the illustrations in Chapter III.

Finally, the Editor would like to express his appreciation to the Control Committee, and to its Chairman, E. Scott Dillon. Although many Committee members doubled as authors they all continued to lend full support to the Editor throughout more than four years of preparation of this edition of *Ship Design and Construction*. It is our sincere hope that the final product proves worthy of this dedication.

ROBERT TAGGART
Editor

Table of Contents

TABLE OF CONTENTS

Chapter VII STRUCTURAL COMPONENTS

STANLEY G. STIANSEN, Vice President, American Bureau of Shipping

Chapter VIII HULL MATERIALS AND WELDING

IRVING L. STERN, Assistant Chief Surveyor, American Bureau of Shipping

Chapter IX HULL OUTFIT AND FITTINGS

DANIEL T. MALLETT, Naval Architect & Assistant Head-Hull Department, George G. Sharp, Inc.*

NORMAN J. THOMPSON, Chief Naval Architect, George G. Sharp, Inc.**

NORMAN W. LEMLEY, Chief, Survival Systems Branch, U. S. Coast Guard

Chapter X CARGO HANDLING-DRY CARGO

JOHN W. BOYLSTON, Marine Manager, El Paso Marine Company

Chapter XI DESIGN FOR TRANSPORT OF LIQUID AND HAZARDOUS CARGOS

ROBERT I. PRICE, Commander, Atlantic Area and Third Coast Guard District, U. S. Coast Guard

TABLE OF CONTENTS

Chapter XII — SHIP MANEUVERING, NAVIGATION AND MOTION CONTROL

WILLIAM H. HUNLEY, Deputy Director Hull Group, Naval Sea Systems Command
NORMAN W. LEMLEY, Chief, Survival Systems Branch, U. S. Coast Guard

Chapter XIII — CONTROL OF THE SHIP'S INTERIOR ENVIRONMENT

VICTOR R. SCHELLENBERG, Engineering Section Manager, Newport News Shipbuilding

Chapter XIV — HULL PRESERVATION

RAYMOND P. DEVOLUY, Consultant
DAVID T. BLOODGOOD, Coatings and Materials Engineer, Bethlehem Steel Corporation

Chapter XV — SHIPBUILDING COSTING AND CONTRACT ARRANGEMENTS

NICHOLAS BACHKO, Vice President, United States Lines
LUDWIG HOFFMANN, Consultant

Chapter XVI — SHIP CONSTRUCTION

HAROLD G. ACKER, Engineer, Bethlehem Steel Corporation
FRANCIS G. BARTLETT, Lecturer, University of New South Wales

TABLE OF CONTENTS

Chapter XVII LAUNCHING

CLYDE M. LEAVITT, Consultant

Chapter XVIII TRIALS AND PREPARATION FOR DELIVERY

ROBERT L. JACK, Consultant

* Now Retired

** Deceased

Note: The office affiliations given are those at the time of writing the chapters

CHAPTER I

Ronald K. Kiss

Mission Analysis and Basic Design

Section 1
Introduction

1.1 Definition. The term *basic design* refers to determination of major ship characteristics affecting cost and performance. Thus, basic design includes the selection of ship dimensions, hull form, power (amount and type), preliminary arrangement of hull and machinery, and major structure. Proper selections assure the attainment of the mission requirements such as good seakeeping performance, maneuverability, the desired speed, endurance, cargo capacity, and deadweight. Furthermore, it includes checks and modifications for achievement of required cargo handling capability, quarters, hotel services, subdivision and stability standards, freeboard and tonnage measurement; all while considering the ship as part of a profitable transportation, industrial, or service system.

Section 2 describes the procedures for establishing the mission requirements before the basic design is undertaken. These requirements, such as the nature of the cargos and/or passengers to be carried, have a powerful influence on the design.

Basic design encompasses both concept design and preliminary design. It results in the determination of major ship characteristics, permitting the preparation of initial cost estimates. In the overall design process, basic design is followed by contract design and detail design. Contract design, as its name implies, develops plans and specifications suitable for shipyard bidding and contract award. Well prepared contract plans and specifications will be clear and in sufficient detail to avoid costly contingency items and protect bidders from obscure or inadequate description of requirements. Detail design is the shipyard's responsibility for further developing the contract plans as required to prepare shop drawings used for the actual construction of the vessel.

An understanding of the entire design sequence is essential to anyone seeking to develop a basic design. The four steps involved are illustrated in the Design Spiral, Evans (1959)[1] as an iterative process working from mission requirements to a detail design, Fig. 1. These steps are amplified further below:

a. Concept Design. The very first effort, concept design, translates the mission requirements into naval architectural and engineering characteristics. Essentially, it embodies technical feasibility studies to determine such fundamental elements of the proposed ship as length, beam, depth, draft, fullness, power, or alternative sets of characteristics, all of which meet the required speed, range, cargo cubic, and deadweight. It includes preliminary light-ship weight estimates usually derived from curves, formulas, or experience. Alternative designs are generally analyzed in parametric studies during this phase to determine the most economical design solution or whatever other controlling parameters are considered determinant. The selected concept design then is used as a talking paper for obtaining approximate construction costs, which often determine whether or not to initiate the next level of development, the preliminary design.

b. Preliminary Design. A ship's preliminary design further refines the major ship characteristics affecting cost and performance. Certain controlling factors such as length, beam, horsepower, and deadweight would not be expected to change upon completion of this phase. Its completion provides a precise definition of a vessel that will meet the mission requirements; this provides the basis for development of contract plans and specifications.

c. Contract Design. The contract design stage yields a set of plans and specifications which form an integral part of the shipbuilding contract document. It encompasses one or more loops around the design spiral, thereby further refining the preliminary design. This stage delineates more precisely such features as hull form based on a faired set of lines, powering based on model testing, seakeeping and maneuvering characteristics, the effect of number of propellers on hull form, structural details, use of different types of steel, spacing and type of frames. Paramount, among the contract design features, is a weight and center of gravity estimate taking into account the location and weight of each

[1] Complete references are listed at end of chapter.

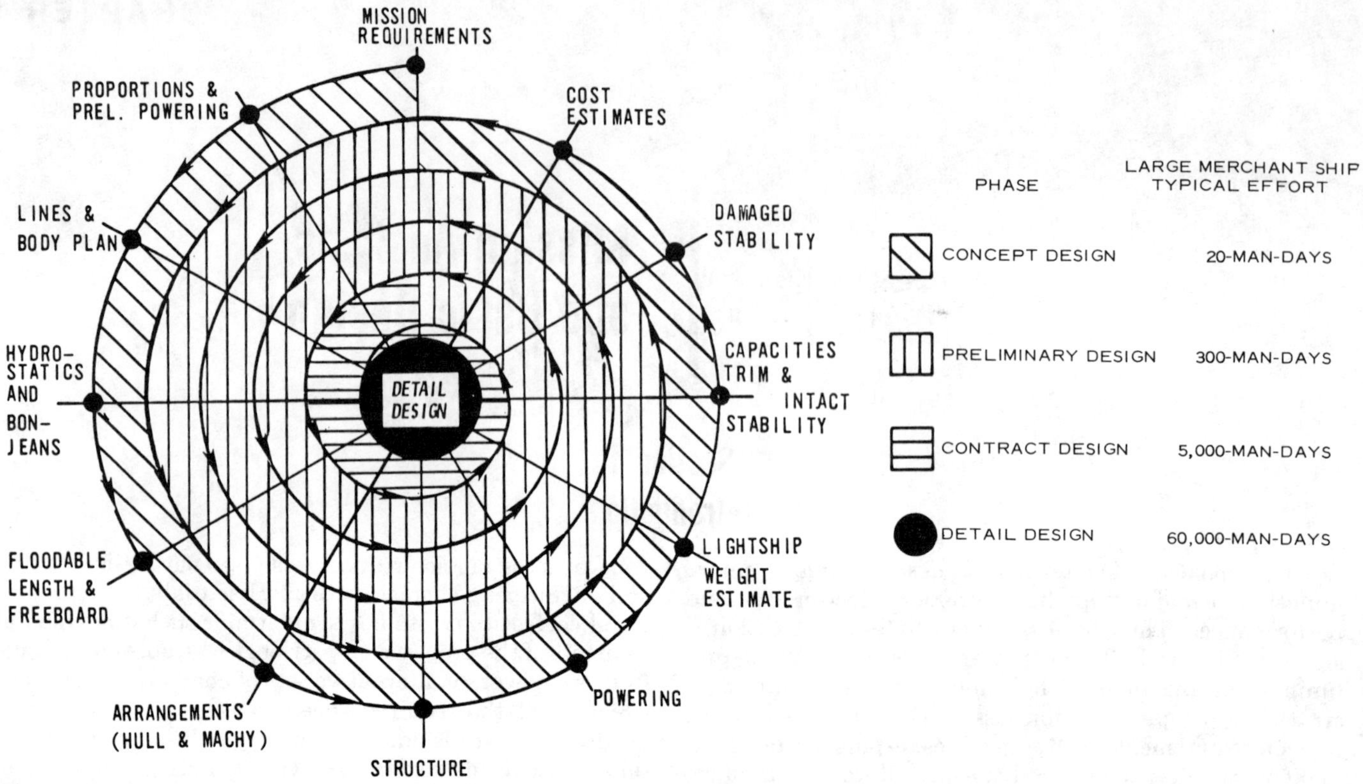

Fig. 1 Basic design spiral

major item in the ship. The final general arrangement is also developed during this stage. This fixes the overall volumes and areas of cargo, machinery, stores, fuel oil, fresh water, living and utility spaces and their interrelationship, as well as their relationship to other features such as cargo handling equipment, and machinery components.

The accompanying specifications delineate quality standards of hull and outfit and the anticipated performance for each item of machinery and equipment. They describe the tests and trials that shall be performed successfully in order that the vessel will be considered acceptable.

Table 1A shows a typical list of plans developed in the contract design of a major ship. Smaller, less complex vessels may not require every plan listed for adequate definition, but the list does provide an indication of the level of detail considered in contract design. Table 1B is a list of the typical sections covered in a commercial ship specification.

d. Detail Design. The final stage of ship design is the development of detailed working plans. These plans are the installation and construction instructions to the ship fitters, welders, outfitters, metal workers, machinery vendors, pipefitters, etc. As such, they are not considered to be a part of the basic design process. One unique element to consider in this stage of design is that up to this point, each phase of the design is passed from one engineering group to another. At this stage the interchange is from engineer to artisan, that is, the engineer's product at this point is no longer to be interpreted, adjusted, or corrected by any other engineer. This engineering product must unequivocally define the desired end result and be producible and operable.

In summary, this chapter considers basic design as that portion of the overall ship design process which commences with concept design and carries preliminary design to the point where there is reasonable assurance that the major features have been determined with sufficient dependability to allow the orderly development of contract plans and specifications. This development will form a basis to obtain shipyard prices within a predetermined price range that will result in an efficient ship with the requisite performance characteristics.

1.2 General Aspects. The late 1960's and 1970's saw a number of major new developments which in one way or another had an impact on the general basic design problem. Among the most significant was the computer. While the computer affects *how* basic design is performed, other changes have impacted on *what* constitutes the basic design problem. For example, one revolutionary development was the change from breakbulk to containerized cargos in the liner trades. Other developments in other ship types created similar new considerations. For tankers, size mushroomed; the increasing demand for petroleum and other raw materials by the industrialized nations of the world has necessitated ever larger tankers and bulk carriers to meet the enormous demand at acceptable costs.

Man is looking increasingly to the sea for all major resources; offshore drilling for oil and gas has burgeoned from a small industry located mainly in the shallow areas of the Gulf of Mexico to a worldwide colossus moving into deeper water and more severe sea conditions (Durfee et al, 1976). These developments have caused a revolution in the design of offshore drilling rigs/ships/units and the entire support fleet necessary for such a challenging undertaking. This includes crew boats, offshore supply boats, high powered towing vessels, pipe laying barges/ships, and countless other specialized craft. Future developments cannot be foretold, but it seems certain that other minerals will be sought from the sea necessitating entire new fleets of vessels designed for tasks not yet known.

Thus, the difficulty of basic ship design will vary with the degree of departure from past practice. Some ship operating companies are closely tied to successful previous designs, and they will permit little variation from these baselines in the development of replacement vessel designs. If the prospective mission appears to parallel existing operations, this may be a sound approach. Consequently, in such situations, basic design may be limited to examination of minor modifications to dimensions, powering, and arrangements.

At the other extreme, totally new seagoing missions, such as the ocean transportation of liquified natural gas (LNG), when first introduced, caused the designer to begin with a blank piece of paper and proceed through rational design engineering with crude assumptions subject to frequent and painstaking revision and development.

1.3 Ship Types. For convenience, Table 2 separates watercraft into three categories:

Table 1A—Typical Plans Developed During Contract Design Stage

Outboard Profile, General Arrangement
Inboard Profile, General Arrangement
General Arrangement of All Decks and Holds
Arrangement of Crew Quarters
Arrangement of Commissary Spaces
Lines
Midship Section
Steel Scantling Plan
Arrangement of Machinery—Plan Views
Arrangement of Machinery—Elevations
Arrangement of Machinery—Sections
Arrangement of Main Shafting
Power and Lighting System—One Line Diagram
Fire Control Diagram by Decks and Profile
Ventilation and Air Conditioning Diagram
Diagrammatic Arrangements of all Piping Systems
Heat Balance and Steam Flow Diagram—Normal Power at Normal Operating Conditions
Electric Load Analysis
Capacity Plan
Curves of Form
Floodable Length Curves
Preliminary Trim and Stability Booklet
Preliminary Damage Stability Calculations

1. *Commercial Vessels.* To transport cargo or passengers.

2. *Industrial Vessels.* To perform specialized marine functions; such as fishing or pipe laying, often using specialized personnel.

3. *Service Vessels.* To provide support capability to commercial ships and/or industrial vessels.

Table 2 is not intended to be all-inclusive. Moreover, there can be a wide variation of a design within a given type

Table 1B—Typical Sections in a Commercial Ship Specification

General
Structural Hull
Houses and Interior Bulkheads
Sideports, Doors, Hatches, Manholes
Hull Fittings
Deck Coverings
Insulation, Lining, and Battens
Kingposts, Booms, Masts, Davits
Rigging and Lines
Ground Tackle
Piping-Hull Systems
Air Conditioning, Heating, and Ventilation
Fire Detection and Extinguishing
Painting and Cementing
Navigating Equipment
Life Saving Equipment
Commissary Spaces
Utility Spaces and Workshops
Furniture and Furnishings
Plumbing Fixtures and Accessories
Hardware
Protection Covers
Miscellaneous Equipment and Storage
Name Plates, Notices, and Markings
Joiner Work and Interior Decoration
Stabilization Systems
Container Stowage and Handling
Main and Auxiliary Machinery
Main Turbines
Reduction Gears—Main Propulsion
Main Shafting, Bearings, and Propeller
Vacuum Equipment
Distilling Plant
Fuel Oil System
Lubricating Oil System
Sea Water System
Fresh Water System
Feed and Condensate Systems
Steam Generating Plant
Forced Draft System
Steam and Exhaust Systems
Machinery Space Ventilation
Air Conditioning Refrigeration Equipment
Ship's Service Refrigeration
Cargo Refrigeration—Direct Expansion System
Liquid Cargo System
Cargo Hold Dehumidification System
Pollution Abatement Systems and Equipment
Tank Level Indicators
Compressed Air Systems
Pumps
General Requirements for Machinery Pressure Piping Systems
Insulation—Lagging for Piping and Machinery
Emergency Generator Engine
Auxiliary Turbines
Tanks—Miscellaneous
Ladders, Gratings, Floor Plates, Platforms, and Walkways in Machinery Spaces
Engineers' and Electricians' Workshop, Stores and Repair Equipment
Hull Machinery
Instruments and Miscellaneous Gage Boards—Mechanical
Spares—Engineering
Electrical Systems, General
Generators
Switchboards
Electrical Distribution
Auxiliary Motors and Controls
Lighting
Radio Equipment
Navigation Equipment
Interior Communications
Storage Batteries
Test Equipment, Electrical
Centralized Engine Room and Bridge Control
Planning and Scheduling, Plans, Instruction Books, etc.
Tests and Trials
Deck, Engine, and Stewards' Equipment and Tools, Portable

Table 2—Representative Vessel Types

COMMERCIAL VESSELS	INDUSTRIAL VESSELS	SERVICE VESSELS
General Cargo Ships	Suction Dredges	Tugboats without barges
Containerships	Pipe-laying Vessels	Offshore Supply Boats
Tankers	Drilling Vessels	Crewboats
Liquefied Gas Carriers	Semi-Submersibles	Crane Support Ships
Bulk Carriers	Incinerator Vessels	Diving Support Ships
Ore/Bulk/Oil (OBO) Carriers	Hopper Dredges	Fire Boats
Integrated Tug/Barges	Fish Processing Vessels	Pilot Boats
Roll-on/Roll-off Ships	Fish Catching Vessels	Towboat without tow
Ferries	Fisheries Research Vessels	
Barge Carriers	Oceanographic Research Vessels	
Heavy-Lift Ships	Hydrographic Survey Vessels	
Chemical Tankers	Ocean Mining Vessels	
Lumber Carriers	Seismic Exploration Vessels	
Towboats with barges		
Passenger Ships		

of vessel. For example, the general cargo ship may range from: a small coaster tramping in the Mediterranean to a larger liner in the Transpacific trade; a ship with several 'tween decks to a design with deep holds and limited 'tween-deck area; a multipurpose ship with capacity for liquid bulk cargo and refrigerated cargo to an austere dry cargo ship.

Some representative vessels from the list on Table 2 are shown in Figs. 2 through 13 which illustrate a wide diversity in the size, shape, and overall configuration of these vessels. One may well ask, "Why? What causes this?"

The answer can be provided in one word—MISSION. For commercial ships their mission is to function as a system to carry cargo or passengers. The characteristics of the payload exert a powerful influence on the overall design. Designs for carrying passengers differ significantly from designs for carrying crude oil. People and their effects impose relatively light payload, and swift voyages are desired to permit adequate time in port. On the other hand, the requirement to ship crude oil in vast tonnages places a premium on ship deadweight capacity.

For example these contrasting requirements yield passenger ships, Fig. 9, with high freeboard, multiple decks, long superstructures, extensive hotel facilities, fine hull forms

Fig. 2 SS AFRICAN COMET—general cargo ship built in 1963 by the Ingalls Shipbuilding Corporation for Farrell Lines Inc.; Molded dimensions 174.3 m (572 ft) by 22.86 m (75 ft) by 13.0 m (42.5 ft)

Fig. 3 SS ROBERT E. LEE—LASH barge carrying ship built in 1974 by Avondale Shipyards, Inc. for Waterman Steamship Corporation; Molded dimensions 272.3 m (893.3 ft) by 30.48 m (100 ft) by 18.3 m (60 ft)

Fig. 4 SS LNG AQUARIUS—first Liquefied Natural Gas (LNG) tanker constructed in the United States at General Dynamics, Quincy Shipbuilding Division in 1977 for Energy Transportation Corporation; Molded dimensions 285.3 m (936 ft) by 43.7 m (143.5 ft) by 25 m (82 ft)

Fig. 5 SS SEA LAND GALLOWAY—high-speed containership built at A.G. Weser in 1972 for Sea-Land Service Inc.; Molded dimensions 288.4 m (946.1 ft) by 32.2 m (105.5 ft) by 20.9 m (68.5 ft)

for high speed, possibly multiple screws for high power with reduced risk of vibration, and probably some form of roll stabilizing device such as a passive tank or fin stabilizer. On the other hand the tanker, Fig. 7, will have low freeboard, a single deck, a relatively short deck house aft, small hotel load, full hull form for maximum deadweight capacity, usually a single screw, and no stabilizer.

Similarly, a Liquefied Natural Gas (LNG) tanker, Fig. 4, looks different from either a passenger liner or a crude oil tanker. Here again, the cargo drives the overall geometric development. LNG has a specific gravity of between 0.45 and 0.5 and is carried at −162°C (−260°F). A detailed code of construction requirements has been developed by the Intergovernmental Maritime Consultative Organization (IMCO), as well as by the U.S. Coast Guard (USCG) and the American Bureau of Shipping (ABS) for both LNG and Liquified Petroleum Gas (LPG). Among the many requirements, which are more fully described in Chapter XI, are double bottoms and double sides. The resultant ship generally has relatively shallow full-load draft, a double walled structure designed to support one of several specially designed, usually proprietary containment systems, extensive insulation requirements, a high, relatively short aft deckhouse, hull fullness somewhat less than an oil tanker, substantial horsepower to provide a high sustained speed since the cargo boils-off during the voyage, and large amounts of special steels or aluminum capable of containing extremely low temperature liquids without becoming brittle.

In the case of industrial and service vessels, they may or may not carry a cargo, but their mission or duty requirements cause them to also have distinctive configurations. Consider, for example, an offshore supply boat, Fig. 12, and an inland waterways towboat, Fig. 13. The supply boat operates in a seagoing environment and must carry a range of materials, from drilling mud and blowout plugs to drill pipe for offshore drilling rigs and platforms. In contrast, the towboat hull must carry a large amount of installed propulsion machinery and convert this high power to effective thrust within a restricted draft albeit normally in calm water.

The designs resulting from these differing missions lead to supply boats with moderate freeboard, small deckhouses very far forward, wide open deck aft, and moderate speed and power, whereas the towboat has low freeboard (primarily because of the restricted waters it plies), wide long deckhouse, shallow draft, high pilot house (to see over large barge flotillas lashed ahead of the towboat), multiple screws, and elaborate rudder systems to maximize the ahead and astern controllability of the boat and barge flotilla.

Similar explanations can be fashioned for all of the vessels listed in Table 2, but the point is that the cargo and its characteristics, or in the case of industrial and service vessels, the mission, dictates the final vessel configuration.

Fig. 6 MV JOHN HENRY—small heavy-lift ship built at Peterson Builders, Inc. in 1978. Molded dimensions 91.4 m (300 ft) by 16.76 m (55 ft) with a draft of 5.13 m (16.83 ft)

Fig. 7 SS AMERICAN INDEPENDENCE—Very large crude carrier built at the Bethlehem Steel Corporation, Sparrows Point Shipyard in 1977 for Gulf Trading and Transportation Company; Molded dimensions 335.3 m (1,100 ft) by 54.3 m (178 ft) by 26.2 m (86 ft)

Section 2
Mission Requirements

2.1 Systems Analysis. Before a naval architect can begin the basic design of a vessel, it is important to work closely with the owner to understand and define the mission. The study which should be undertaken will result in the definition of required size and speed and any other major mission requirements. In one sense the establishment of mission requirements overlaps the initial phase of concept design. Here, the naval architect must interact with the owner to ensure that the emerging missions are such that they can be fulfilled in terms of a practical and economical vessel. Parametric studies covering both ship characteristics and economic factors will be required in this process.

Owners usually contemplate one or a combination of the following alternatives:

- replacement or conversion of overage or obsolete vessels;
- expansion or modification of services on an existing route, in an effort to enlarge their participation;
- development of a new service or carrying a different kind of cargo on an existing route aimed at capturing an increased percentage of the trade;
- development of a vessel to undertake an old or new industrial operation at sea;
- development of a vessel to support commercial or industrial vessels engaged in ocean technology.

Table 3—Sample Required Freight Rate (RFR) Economic Analysis of a Potential Product Tanker Operation

Preliminary Ship Characteristics:	
Deadweight, tons	30,000
Cargo Deadweight, tons	28,200
Length, m (ft)	161.5 (530)
Beam, m (ft)	25.9 (85)
Draft, m (ft)	11.9 (39)
Shaft Horsepower	14,000
Gross Tonnage	15,000
Speed, knots	16
Crew	25
Fuel Consumption at sea, t/d	74.5
Fuel Consumption in port, t/d	12.6
Estimated Construction Cost	$42,400,000
Trade Route Characteristics:	
Round Trip Voyage Length	2,720 miles
Sea Days	7.1
Port Days	3.0
Days per Voyage	10.1
Operating Days per yr	350.0
Trips per Year	34.7
Cargo Delivered per yr	978,840 tons
Annual Operating Costs (Thousands of Dollars)	
Crew Wages	$1,414
Subsistence	48
Stores/Supplies/Equipment	89
Maintenance and Repair	358
Insurance	193
Other	33
Port Charges	673
Fuel	1,967
Subtotal	$4,775
Annual Capital Cost Construction Amortized over 20 yr at an interest rate of 10 percent— Thousands of Dollars	$4,980
Total Annual Cost	$9,755
Required Freight Rate = $\frac{\$9,755}{978,840 \text{ tons}}$ =	$9.97 per ton

In every situation the owner is confronted with decisions concerning the number of ships required, their type, size, and speed. This is true even in a relatively simple situation, for example a point-to-point operation, wherein a steel company hauls ore or taconite from a Lake Superior loading port to a discharge facility in Ohio. Even in this case the most economical ship is not necessarily the largest that can transit the limiting locks and channel depths. Krappinger (1966) has shown that this would only be the case if the cargo handling rate could be increased sufficiently to hold down port time to no more than that required by smaller vessels. The number of ships would depend upon the amount of ore to be moved per year and optimum ship speed and size. Examples of the contribution of basic design and computer-aided systems analysis to bulk carrier optimization for specific routes are furnished by Benford et al. (1962), Mack-Forlist and Hettena (1966), and Everett et al (1972).

Similarly, mission analysis must be done for industrial and service vessels, but often, only a single vessel is required and initial cost becomes the controlling factor.

At the other end of the scale of complexity is a service with multiple choices of itineraries (including entry into or the by-passing of certain ports because of their particular restrictions), alternative methods of cargo handling, frequency of sailing (an important factor in a vigorously competitive market), and alternatives as to desirable types of cargo excluding low-paying bulk or accepting moderately sized parcels of grain, for example. The problem of optimizing ship transport capability for such a service entails selection from an exceedingly large number of possibilities.

The shipowner's traditional approach is an attempt to analyze one or two permutations of the most significant variables. Inputs include known traffic statistics coupled with estimates of future trends. More specifically, the owner considers quantities of various commodities which the company could be expected to carry with differing sailing frequencies, ship types, and turnaround time. Pro forma economic analyses combine estimates of revenue, operating expense, capital charges, and overhead to predict a rate of return on investment or a required freight rate (RFR) for each permutation.

Table 3 is an example of a typical economic analysis which may be performed by a prospective tanker operator. In this example the RFR for the candidate ship was $9.97 per ton. If the trade required a total throughput of about 2,940,000

Fig. 8 ITB PRESQUE ISLE—integrated tug barge Great Lakes ore carrier built in 1973; tug built by Halter Marine Services, barge by Erie Marine, Inc. for Litton Great Lakes Corporation; Molded dimensions of integrated vessel 304.8 m (1,000 ft) by 31.9 m (104.6 ft) by 14.2 m (46.5 ft)

Fig. 9 SS QUEEN ELIZABETH 2—passenger liner built in 1969 at John Brown Shipbuilding Company for Cunard Line Ltd.; Molded dimensions 293.5 m (963 ft) by 32 m (105 ft) by 17.1 m (56 ft)

Fig. 10 GLOMAR PACIFIC—drilling ship built in 1977 by Levingston Shipbuilding Company, Inc. for Global Marine Inc.; Molded dimensions 137 m (449.5 ft) by 21.9 m (372 ft) by 10.7 m (35 ft)

Fig. 11 M/T DR. JACK—6,000 horsepower tug built in 1976 by Halter Marine Services, Inc. for Jackson Marine Corporation; Molded dimensions 45.1 m (148 ft) by 12.2 m (40 ft) by 6.7 m (22 ft)

tons per year, three of these ships would satisfy the requirement. However, it may be feasible to design a larger or faster vessel which does not exceed the physical constraints of length, beam, and draft or the storage capacity of shore facilities. If the vessel were sufficiently larger and/or faster only two ships might suffice. A similar RFR analysis would indicate which fleet would be most economical. The example in Table 3 has not been complicated by the introduction of varied financing alternatives for the construction cost, but it is possible to break the annual capital cost into components which may be financed out of owner equity, bonds, or other types of loans, each with a different interest rate. The advantage of the RFR calculation is the absence of a need to estimate revenue. However, in most trades it will be essential to determine whether sufficient cargo can be obtained to provide adequate income to exceed the total annual cost.

There are practical limits to the combinations and degree of refinement to which fleet simulation studies can be carried. The number of variables is large. Therefore, such items as fuel consumption and capital charges, which are dependent upon ship characteristics, can only be approximated. It is here that the basic designer makes his most important contribution to fleet simulation studies. In effect, he undertakes repeatedly first iterations of the design concept process described in Section 3.2 to quickly estimate capital and operating costs corresponding to speed, capacity, and endurance of the alternative ships used in the fleet simulation.

Computers offer an opportunity to increase the number of parameters considered in a given period of time. The type of computer program and its sophistication must vary depending upon the size and total cost of the transportation system or industrial project. Obviously, a shrimp boat or supply boat will not require the same amount of analysis to determine the basic design requirements as an LNG tanker or containership.

For a project as large as an LNG or containerized transportation system, it behooves the naval architect to consider the ship as part of the total system, since its size and speed can have costly impact on the shore terminals. Number of berths, number and size of gasification and liquefaction plants, and overall storage requirements are several of the more important nonmarine factors which should be optimized together with the ship. Obviously, a complex and sophisticated analysis is warranted, and a computer-aided analysis offers a very practical way to consider all of the important interactions. Schmitt (1976) provides a thorough overview of precisely the type of computer simulation deemed essential for a large LNG transportation project.

The selected analytical approach is up to the discretion of the designer based upon individual preference and the characteristics of the shipping system being developed. For example, a computer model could account for the varying number of containers, container cranes, tractors, shoreside storage area, and container handling equipment, all of which would vary with different size ships. These and similar considerations are discussed by Ericksen (1972) and Miller (1970) for containership systems. Benford (1967b) presents

Table 4—Mission Requirements

ECONOMIC PARAMETERS

1. Number of ships
2. Projected economic life—years
3. Itinerary and schedule of departures, including anticipated average loading conditions (basis for speed and power margins, cargo handling studies, and freshwater allowance for backup if only one evaporator is installed)
4. Dry bale cubic and deadweight. (In case of bulk carrier, owner may prefer maximum deadweight obtainable economically at specified draft.)
5. Refrigerated cargo cubic and deadweight, number of boxes, and temperature level for each box
6. Liquid cargo cubic and deadweight, type of deep tanks; i.e., conventional or cofferdammed, water-jacketed, flush-inside surface
7. Number of passengers and crew and habitability standard:
   - Area per person
   - Number of one, two, three, etc., person rooms
   - Number of spare berths beyond nominal passenger capacity for more flexibility in booking
   - Type and size of public spaces
   - Is passenger elevator to be provided?

   NOTE: These requirements are pertinent to cargo ships with 12 or fewer passengers. Relatively few new combination passenger-cargo and even fewer passenger transoceanic ships are likely to be built in the future. Thus, their design is becoming a lost art. However, a naval architect responsible for the designing of a passenger vessel would need to know in greater detail more about the owner's intended service, Sharp (1947).
8. Limitation on vessel acquisition cost
9. Dry bulk cubic and stowage factor
10. Special cargo lockers, cubic and deadweight
11. Number, weight, and size of vehicles to be carried
12. Special provisions for container stowage; that is, the number, type (i.e., general cargo, reefer vans, etc.), size, weight, and whether stowed in cellular structure or if on deck, the number per stack
13. Type of containment system for special cargo; such as, LNG, Ammonia, Chemicals, etc.
14. As to tankers only:
   - Number of segregated cargos
   - Cargo pumping rate
15. Type of stabilization (if any)
16. Location of bunkering ports, fishing grounds, or industrial projects to be serviced
17. Types of machinery plant owner is willing to consider with respect to future crewing and maintenance

RESTRICTIONS

1. Limiting lock width, length, draft, or similar limitations due to harbor channel or canal features
2. Spacing of fixed bulk unloading gear ashore
3. Limiting heights of bulk cargo handling equipment ashore or container cranes ashore
4. Tidal range at all ports
5. Dry dock facility limitations
6. Depth of water to be worked for industrial vessels
7. Geographic seaway location for input into vessel motion requirements and analyses
8. General plan for cargo handling, including data on port facilities, heavy lift requirements, and special problems (i.e., tidal variations, sideport requirements, etc.)
9. Subdivision standard
10. Tonnage limitations
11. Loadline rules
12. Coast Guard regulations
13. Classification society requirements
14. As to tankers only:
   - Classification by Coast Guard grade of most hazardous cargo to be carried
   - Limiting capacity of shoreside tankage

Fig. 12 M/V SABLE ISLAND—offshore supply boat built in 1977 by Halter Marine Services, Inc. for Arthur Levy, Inc.; Molded dimensions 54.9 m (180 ft) by 12.2 m (40 ft) by 4.3 m (14 ft)

Fig. 13 M/V ED RENSHAW—Western River tow boat built in 1977 by St. Louis Ship for United Barge Company; Molded dimensions 51.8 m (170 ft) by 13.7 m (45 ft) by 3.4 m (11 ft)

a basic computer-aided approach to the selection of ship size for cargo ships which can be adapted to other ship types. The use of computer programs for selecting optimal tanker dimensions is given by Roseman, et al (1974). A fundamental appreciation of computer applications for basic characteristic selection is given by Mentz (1975a) and (1975b). The place of the ship in the overall transportation system is well defined by Nachtsheim (1972). Finally, see Sato (1967) for a computer-aided analysis of the effect of principal dimensions on weight and cost of large tankers.

These systems analyses gradually define the full spectrum of design requirements, beginning with the determination of number of ships, size, speed, progressing to more specialized vessel requirements and restrictions, and concluding with detailed owner requirements such as number of crew and single or double staterooms. The importance of these studies cannot be over estimated, for no matter how well a ship design meets the established requirements, it is doomed to fail if the requirements were not correctly set. The summarized results are simply denoted as mission requirements of which Table 4 is an example.

The remainder of this chapter will concentrate on mission requirements and details of the design process, using containerships as a basis for discussion. A single type of ship has been selected for this purpose to focus more clearly on the principles involved in that process. However, it should be realized that the same process is utilized regardless of the ship type which must be specially suited to the mission it is to fulfill. Chapter II will provide design guidance on other classes of vessels along with some useful design data.

Section 3
Concept Design

3.1 General. Concept designs are required both for the foregoing studies (Section 2), wherein the mission is defined in engineering parameters, and for the next step, which is preliminary design, namely, the final ship proportions, arrangements, power plant type, and structural layout that will satisfy the mission requirements. For any particular set of requirements, there is an infinite number of combinations which give the transport or mission capability desired; i.e., for cargo ships—capacity, deadweight, speed, and endurance, Fig. 14. The problem is to arrive at an economical solution within the constraints of the mission requirements. Fortunately, there is evidence (Mandel and Leopold, 1966) which suggests that little variation in cost occurs when proportions of ships vary over a fairly wide range. This is graphically illustrated in Fig. 14 (Murphy et al, 1965). Therefore considerable deviation from the theoretical optimum will not necessarily incur a significant cost penalty, although such deviations may affect seakeeping and maneuvering performance.

The concept design process may be carried out in either of two ways. Until recently the designer relied upon accumulated experience and data for the type of vessel being designed. He focused on past practice and interpolated and extrapolated from similar existing vessels. Presumably the existing vessels represented optimum or near optimum designs and small deviations would not be economically unsatisfactory. A new and more powerful alternative is to conduct a systematic parametric analysis that can be used to derive optimum proportions. Although longhand techniques can be utilized in performing these parametric analyses, the advent of the modern computer has rendered the process not only more rapid but also permits a more thorough investigation of a wider range of parameters.

The empirical approach serves only as a rough guide, if a novel ship type is to be designed. However, this method does provide a quick and fairly reliable starting point when adequate data are available and plotted for ready comparison. In pursuing this approach, the concept design phase essentially involves the translation of the owner's requirements, or mission requirements, into a broad definition of an item of hardware that can be produced and operated in a manner that will satisfy the stated mission. In the design process it is essentially the first trip around the outer loop of the design spiral shown in Fig. 1. During this process the mission may be altered to meet cost, material, or engineering constraints imposed by the current economic situation or the current status of technology. On the other hand, the initial conceptual designs that have satisfied similar mission requirements in the past may be revised extensively to fulfill the stated mission more economically, more safely, or with greater efficiency. Whether the concept design phase is based on the empirical approach or on a series of parametric studies, the ultimate purpose is to define in general a vessel that can be built and will satisfy the mission.

The following references are sources of either graphical or tabulated ship data on the following types: cargo ships, tankers, bulk carriers, combination bulk carriers, containerships, LNG tankers, and drill rigs: (Benford, 1962); (Johnson, 1965); (Gilfillan, 1969); (Dorman, 1966); (Henry and Karsch, 1966); (Thomas and Schwendtner, 1971); (Danforth, 1977); (Dorman and de Koff, 1971); (Comstock, 1967).

3.2 Proportions Derived from Parametric Studies. Both computer-aided studies and longhand calculations will be discussed in this section.

a. Computer Aided Parametric Studies. Given the availability of suitable computer programs, wide variations in proportions can be examined subject only to the limits of the reliability of the underlying design equations incorporated in the program. Murphy et al (1968) covers one of the earliest programs initially developed in 1963 for breakbulk cargo ships which has been updated and modified to encompass containerships, tankers, and bulk carrier designs. Although the foundations of computerized basic design were

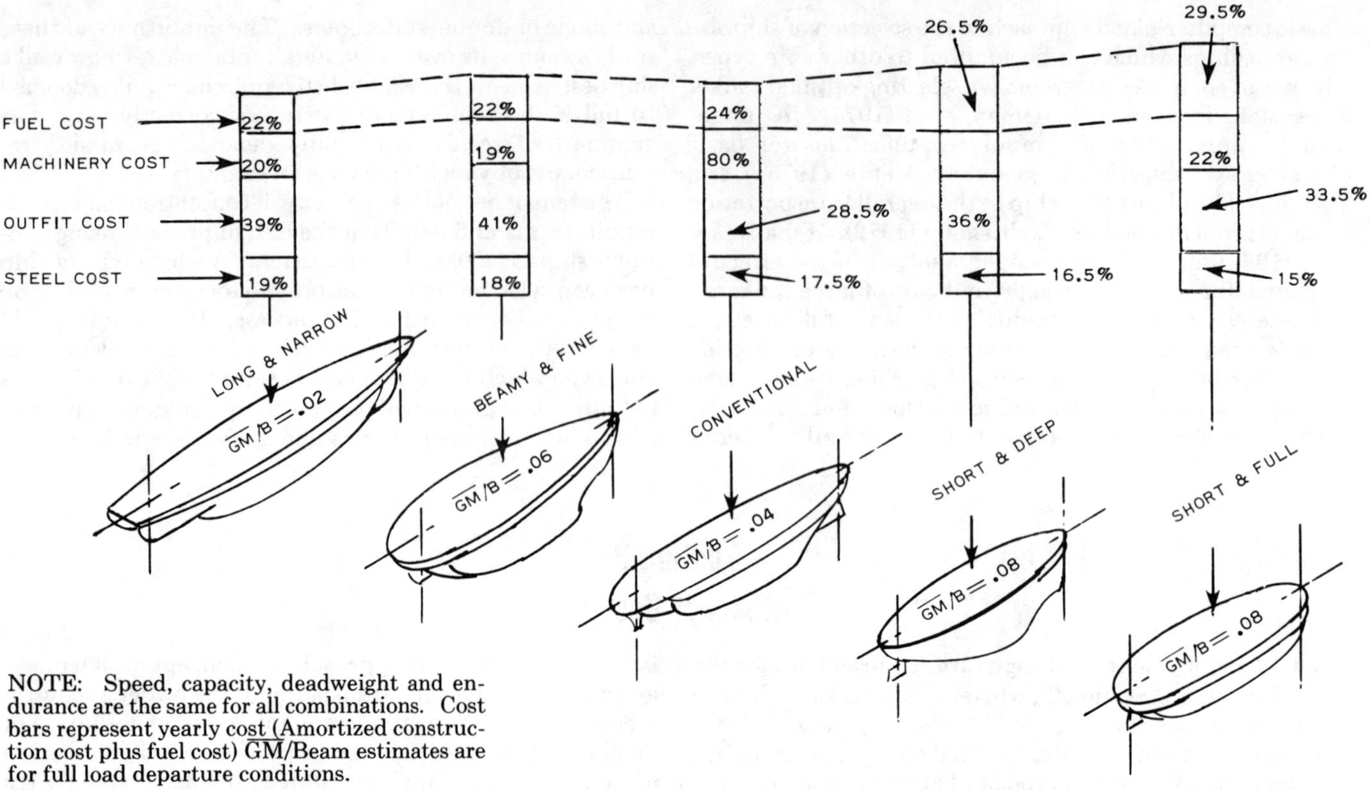

NOTE: Speed, capacity, deadweight and endurance are the same for all combinations. Cost bars represent yearly cost (Amortized construction cost plus fuel cost) $\overline{GM}$/Beam estimates are for full load departure conditions.

Fig. 14 Effect of form and dimensional combinations on cost

laid by the foregoing program and Mandel and Leopold (1966), development of such programs has been quite limited. Major published efforts include Eames and Drummond (1976), Gilfillan (1966–67), Fisher (1972), and Nowacki (1970).

The reason for the seemingly slow development of this tool apparently lies in the personal nature of concept design. Most of the programs are written for batch processing wherein data are input on cards or tape, calculations made by the computer according to fixed formulas with results displayed on a high speed line printer. The main effort of the designer becomes the preparation of input data and the analysis of computer output. Traditionally, the experience of the designer is directly applied at each step of the design process, in terms of methods, concepts, and most importantly the reasonableness of the answers obtained. He should be encouraged to accept the idea that there are an infinite number of technical solutions to the mission requirements, not just the one solution that emerges from trial and error. To exploit fully the potential of computers, the design procedure should cover as wide a range as possible of ship proportions.

Present trends are toward the development of interactive computer programs where the designer can interact with a computer through a remote typewriter type terminal and cathode ray tube (CRT) display. Ultimately, perhaps, the design estimating relationships and the judgment and experience of the naval architect can then be married to the speed and precision of a computer to yield the best of both worlds.

Finally, no matter what systems are adopted, the influence exerted by the proportions on displacement, powering, weights, costs, fuel consumption, stability, etc., must be properly expressed to yield optimum designs. Fortunately, as mentioned earlier for general cargo ships, the curve of measure of merit (Required Freight Rate, Net Present Value, etc.) tends to be relatively flat in the area of the optimum. This permits the use of less than perfect approximations with small effect on the final design characteristics. Furthermore, it permits the selection of ship proportions favorable to good seakeeping and maneuvering performance.

The programs now in existence are developed for different ship types and use different search techniques to arrive at optimum designs. Parsons (1975) gives a good discussion of these techniques. Fundamentally, the programs either: vary certain major parameters such as length or C_B and iterate around the design spiral making trial and error variations to seek an optimum; or systematically develop a large family of ships with ranges of displacement, L/D, B/T, $v/\sqrt{gL}$ and C_P selected to cover reasonable combinations or proportions for the owner's requirements. Fig. 15 (Murphy et al, 1965) shows the logical flow diagram for the latter type of program, which is the most powerful approach for exploring all possibilities.

All programs must have certain estimating relationships

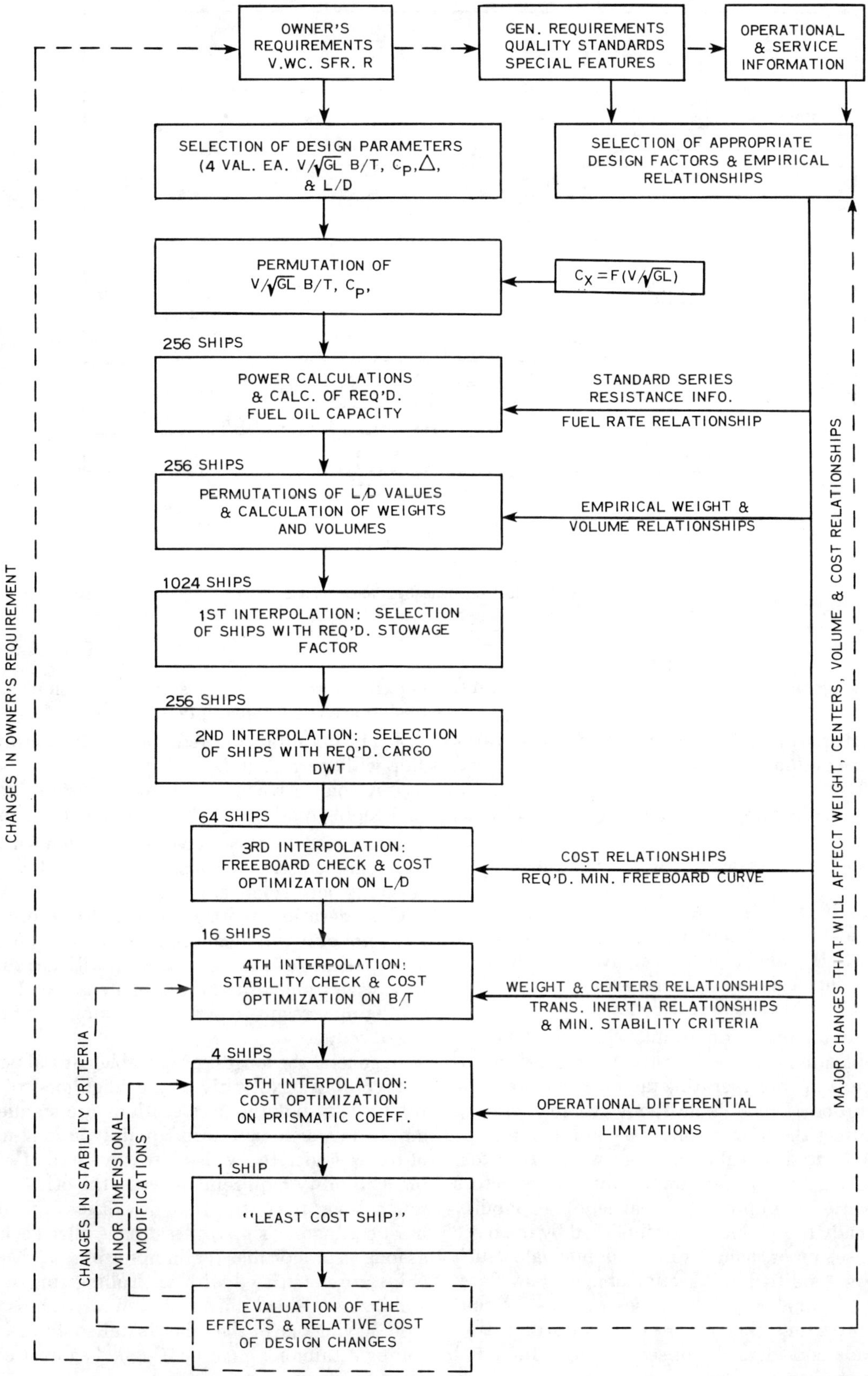

Fig. 15 Flow chart for least cost vessel

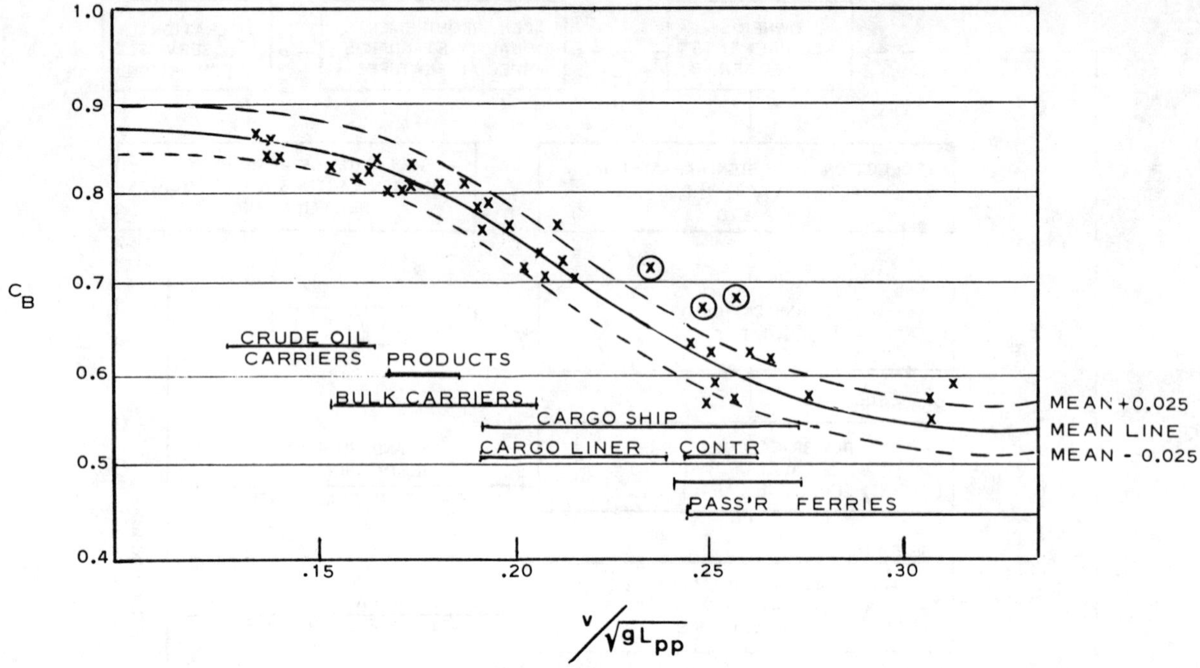

Fig. 16 Block coefficient vs Froude number

built into the program or available as source input data. These include:

1. Power estimating relationships such as Taylor's Standard Series residuary resistance values as reanalyzed by Gertler (1954).

2. Mathematical curves fitted to empirical data items, such as:

- Weights and costs of steel, outfit, and machinery, and;
- other ship characteristics as functions of governing parameters: midship coefficient, C_X vs $v/\sqrt{gL}$; machinery space volume vs SHP; number of 20 ft equivalent containers vs $L \times B \times D \times C_B$; freeboard vs L; transverse inertia coefficient, C_{IT}, vs C_P; fuel vs SHP; etc.

Often these are technical constraints which must be imposed before optimum cost configurations can be derived. In addition, a host of ship operating factors can also be affected by ship proportions. Table 5 shows some operating factors which affect the proportions of a containership.

Of the factors listed in Table 5, rough weather performance is becoming increasingly important. The factors which can be varied to improve the seakeeping capability of a ship are exactly those which are established by the basic design. Of course, minor changes can be adopted later, such as bulbous bows, modified hull section shapes, bow flare, freeboard forward, and cargo loading restrictions. Nonetheless, it is important that future design optimization programs include consideration of seakeeping. Included should be speed reduction in waves, vertical accelerations, seaway bending moments, roll angle predictions, deck wetness, propeller emergence, and slamming. Essential to the capability to include these considerations in preliminary design are the following factors:

- The definition or prediction of the seaway in which the ship will operate, and;
- A qualitative or quantitative criterion for the various seakeeping qualities of the ship to meet its mission.

Some guidelines can be offered which will assist the designer in selecting ship characteristics favorable to good seagoing performance.

Consideration may be given first to minimizing rolling, since not only is it objectionable in itself, but it may necessitate a change in course or speed, with a resulting increase in voyage time. Furthermore, a change of course to ease rolling may result in serious pitching and hence a drastic speed reduction.

In general the longest practicable natural period of rolling is favorable. Not only is the likelihood of synchronous rolling less, but the accelerations are smaller for a given amplitude, although unfortunately, a long natural period of roll is associated with a low value of $\overline{GM}$ which may not meet stability requirements. On the other hand, rolling is much easier to control than any other type of motion, and most oceangoing ships at least have bilge keels, as deep and as long as practicable, for damping of roll. Many passenger ships are now fitted with antirolling fins or passive tank stabilizers. Naval ships and even cargo ships would benefit from having one or the other installed, but the reduction in rolling attainable must be weighed against considerations of added weight, space, and cost.

The next consideration should be pitching and heaving motions and the associated vertical accelerations, shipping

Table 5—Ship Proportions with Possible Significant Effect on Ship Design Requirements

Ship Design Requirements	Length	Beam	Depth	Draft	$LWL/\nabla^{1/3}$	C_B	B/T	L/D
Port accessibility	x	x		x				
Berth accessibility	x	x	x	x				
Fees governed by admeasurement tonnage	x	x	x			x		
Pilotage and pier rental costs not based on tonnage	x			x				
Availability of dry docks	x	x		x				
Rough weather performance	x	x	x	x	x	x	x	x
Calm weather performance	x	x		x	x	x	x	
Fuel consumption and endurance	x	x	x	x	x	x	x	x
Effective use of overload horsepower	x				x	x		
Maneuvering and directional stability	x	x		x	x	x	x	
Longitudinal ship deflection	x		x					x
Propeller induced hull vibration	x	x	x	x	x	x	x	x
Capital charges to amortize first cost	x	x	x			x		x

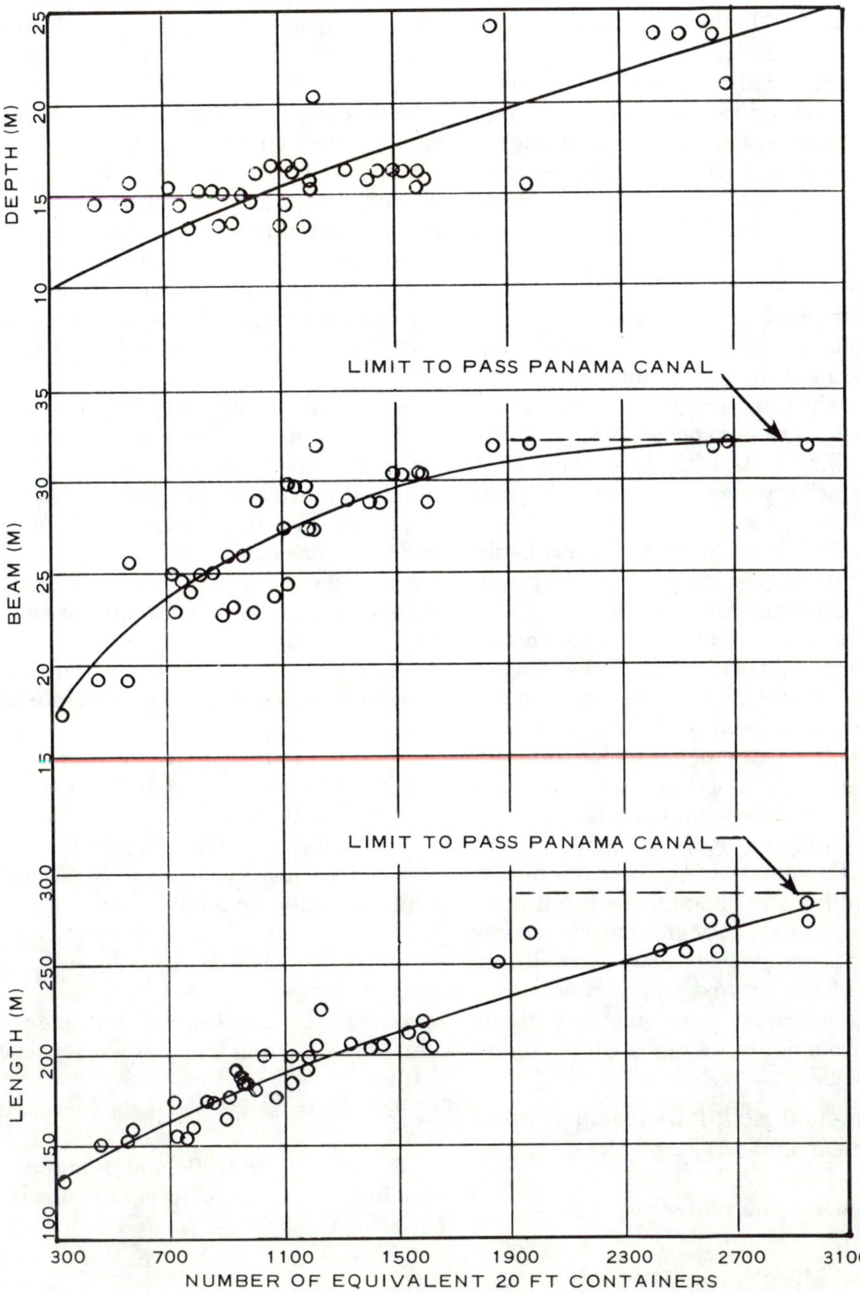

Fig. 17 Equivalent 20-ft containers versus Lpp, B, and D

water and slamming. Of first importance is attaining subcritical pitching and heaving behavior over as wide a range of conditions as possible in head seas, which generally give the worst pitching and heaving behavior. The period ratios for pitching and heaving have a vital effect on ship behavior and hence on attainable sea speed. The natural periods of pitching and heaving are usually quite close together, but attention should focus on the pitching period since it has the greater effect on wetness, slamming, and vertical accelerations. Increased length in relation to other dimensions and reduced longitudinal gyradius are both favorable to a lower period ratio.

The following references provide a basis for considering seakeeping early in the design: Lewis, E. V. (1967); Aertssen (1968); Chryssostomides (1972); Hadler and Sarchin (1973); Ochi and Bales (1977); Cleary (1977).

Seakeeping can be an important consideration when establishing mission requirements since the speed in a seaway will be less than the calm water speed. This may influence the final selection of proportions, necessitate a higher calm water design speed or dictate some other adjustment in owner's requirements. The Federal Power Commission (1977) gives a simplified treatment of a method to account for speed reductions in a seaway.

The selection of a measure of merit for any of the programs involves operational costs as well as construction costs. It then becomes important to account for the most important variable costs which happen to be construction and fuel consumption costs. Other expenses; such as, crew wages and container handling costs, while large, would not be expected to vary appreciably over a practical range of proportions.

Other effects of variations in proportions are probably small in relation to capital charges and fuel consumption. Accordingly, a practical approach for hand calculations is to find proportions producing the minimum sum of annual capital and fuel costs. Since this simplification will most likely produce satisfactory results, it is often used in computer assisted parametric studies as well. However, the computer program can easily and quickly include second order effects and hence provide a preliminary *pro forma* economic analysis along with the optimum design.

In summary, the validity of any computer generated results is dependent upon the quality of the basic empirical relationships, the assumptions made, and the input data used. From time to time, any such program needs recompiling to reflect updating of cost, weight, and other data, or to suit special features of the desired ship. However, a number of design constants can be manipulated without recompiling the program and merely require card punching as input data. These include:

1. adjustment to bare hull EHP for appendage resistance, propulsive coefficient, and service margins;
2. $\overline{KB}$ as percentage of draft;
3. $\overline{KG}$ of steel and outfit as percentage of depth;
4. cost of fuel; etc.

The program listing for Murphy et al (1965) and a representative set of empirical equations and design constants for general purpose dry cargo ships can be obtained from the Maritime Administration (MarAd), as can similar listings and data for containership, tanker, and bulk carrier design programs.

b. Longhand Calculations. Many of the considerations and factors discussed in Section 3.2*a* for computer-aided parametric studies are directly applicable to longhand calculations. Obviously, the time required to perform all but relatively limited or simplified parametric analyses by hand is prohibitive.

The exact procedure to follow is dependent upon the type of ship being designed. Whereas for a tanker or general cargo ship certain dimensions can be varied by a given percentage and new alternatives developed, for a containership any dimensional change in length should account for the discrete length of the specified container: 20 ft, 40 ft, etc.

Of course, if alternative types of power plants are considered, particularly nuclear propulsion, the effect of different initial costs, fuel costs, crew wages, machinery and fuel weight, and space, is to introduce variables that affect ship proportions appreciably. These variables must be treated separately (Benford, 1963).

Prior to making any adjustments in dimensions, it is prudent to complete at least one cycle around the design spiral of Fig. 1. The detailed steps are described in Section 4.

In general the method used is as follows:

1. Denote as a basis ship dimensions and coefficients with subscript 1. Assign subscript 2, 3, etc., to alternative ships with different lengths.

2. Choose an alternative ship with its new length (say five percent longer). It is similar to the basis ship in that it has the same general arrangement, speed, endurance, cargo capacity, and deadweight as the basis ship. Say $L_2 = 1.05L_1$

3. Both ships have the same stability standard, therefore, $B_1/D_1 = B_2/D_2$ (accurate enough for this purpose).

4. Cargo cubic is closely a function of $L \times B \times D \times (1 + 0.5C_{B_1})$; therefore, $L_1 \times B_1 \times D_1 \times (1 + 0.5C_{B_1}) = L_2 \times B_2 \times D_2 \times (1 + 0.5C_{B_2})$

5. Obtain C_{B_2} from Figure 16 for $v/\sqrt{gL_2}$

6. By substituting $B_2 = B_1 \times D_2/D_1$ and $L_2 = 1.05L_1$ in the equation under 4:

$$D_2 = D_1 \sqrt{\frac{(1 + 0.5C_{B_1})}{1.05\,(1 + 0.5C_{B_2})}} \tag{1}$$

and L_2, B_2, D_2, and C_{B_2} are thus quantified. Obtain a new light ship LS_2 from the appropriate design chart using these values.

7. $\Delta_2 = DWT_2 + LS_2$ (accurate enough for first trial)

8. Design draft T_2, which becomes a dependent variable in the displacement equation, may be determined by equating

$$\frac{\Delta_1}{\Delta_2} = \frac{L_1 B_1 T_1 C_{B_1}}{L_2 B_2 T_2 C_{B_2}} \tag{2}$$

or

$$T_2 = \frac{\Delta_2 L_1 B_1 T_1 C_{B1}}{\Delta_1 L_2 B_2 C_{B2}} \tag{3}$$

9. Check SHP and fuel oil. If necessary, modify LS_2, DWT_2, Δ_2, and T_2. These procedures can be carried out with varying degrees of refinement to suit accuracy desired, in accordance with the methods of Sections 4 and 5. Stability may also be checked in similar fashion if necessary.

As indicated previously, T_2 will ordinarily work out to a satisfactory value in the case of general purpose dry cargo ships or containerships. If it does not, then either B or C_B or a combination thereof would have to be modified, or the owners would have to accept less deadweight at the port with limiting draft.

10. Estimate yearly cost of ship and fuel oil.
11. Repeat process for as many different lengths as desired.
12. By plotting the sum of annual cost of fuel and capital recovery against ship length, an optimum length may be selected giving least cost. Expected profitability of the optimum solution is then calculated by considering annual revenue minus all annual cost including the investment cost of the least cost ship found in the foregoing. Profitability should be appraised in terms of capital recovery factor, CRF, or required freight rate, RFR, based on a realistic rate of return on invested capital in the shipping business.

$$\text{CRF} = \frac{i\,(1+i)^n}{(1+i)^n - 1} \tag{4}$$

where: i = interest rate per interest period
n = number of interest periods

The CRF multiplied by the investment gives the uniform end-of-year receipt required to recover the investment in n years with interest rate i.

During the pro forma feasibility stage, a rate of at least 10 percent after taxes appears necessary. For a treatment of principles and methods of profitability analysis, see Benford (1963).

For bulk carriers, as opposed to the containership example, it is usually easier to find suitable proportions, since parametric studies are simplified whenever dimensional limitations apply. This can occur, for example, when the vertical clearance under fixed, shoreside, ore-handling equipment is critical with the vessel in light condition, resulting in a ship which may be depth limited. Canal or other locks may govern length, beam, or draft, or at times, all three at once. In such cases, dimensional variables are eliminated from parametric studies, which are then reduced to variations in C_B.

However, the typical bulk carrier is a draft-limited ship. Thus, studies are confined to variations in L/D, L/B, and C_B with L/D equal to or less than the classification society limit.

3.3 Proportions Derived from Empirical Data. The majority of early containerships were the result of conversions from tankers or dry cargo ships to cellular container carriers. With the present well established and expanding container system, most ships are custom designed for economical operation on their intended liner trade.

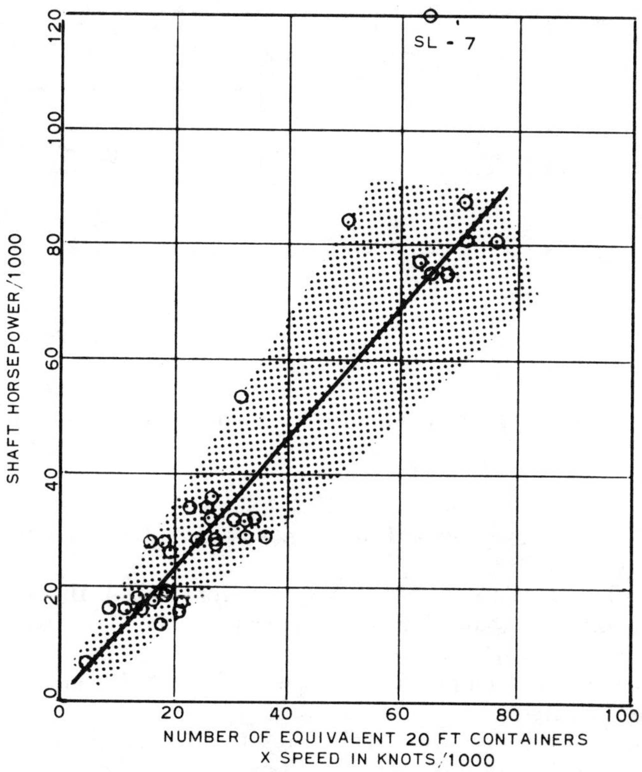

Fig. 18 Initial power estimate for containership

Henry and Karsch (1966) provide a comprehensive survey of containerships giving guidance towards selection of proportions and estimation of light ship weight for ships with large deck loads of containers. The concept design procedure involves working around the outer loop of the design spiral shown in Fig. 1. Specific steps are:

1. Select length, beam, and depth of the ship for the number of containers to be carried, Fig. 17. This plot and others of similar type are not intended to serve as substitutes for original design data assembled by the naval architect, but as illustrations of methods. Every design office should develop (and continually refine) its own set of curves. The curves of empirical data represent vessels with normal or average speed, endurance, ballast tankage, extent of superstructure, engine room length, evaporating capacity, container weight, freeboard, and L/D and L/B ratios. Caution should be exercised when deviating from normal parameters—particularly with regard to speed.
2. Obtain C_B from plot of $v/\sqrt{gL}$ vs C_B, Fig. 16.
3. Obtain initial power estimate, Fig. 18.
4. Tabulate steel, outfit, and machinery weights from Figs. 19, 20, and 21. Add weight margin of three percent of total steel, outfit, and machinery to obtain total light ship weight.
5. Obtain displacement (Δ) by adding estimated deadweight items to light ship weight.
6. Check required power, Fig. 22. If power is not confirmed, iterate last three steps until confirmed.

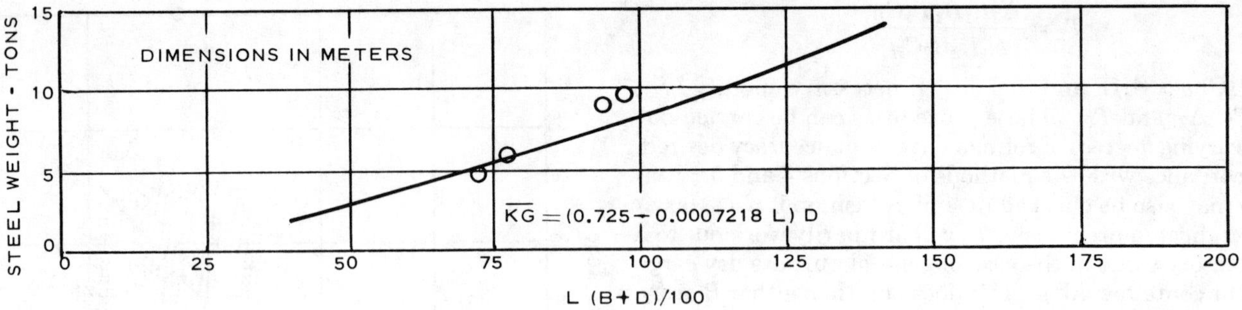

Fig. 19 Containership steel weight

7. Solve for saltwater draft in displacement equation

$$T = \frac{\nabla}{Lpp \cdot B \cdot C_B} \tag{5}$$

8. Check freeboard and iterate last five steps if necessary.

9. Tabulate the center of gravity for steel, outfit, and machinery with previously determined weights, Figs. 19, 20, and 21. Compute light ship center and add a margin, say 0.3 m (1 ft). Compute vertical center of gravity $\overline{KG}$ of total displacement.

10. Calculate $\overline{GM}$ and if it is less than 0.025B revise dimensions and iterate last seven steps.

There are numerous variations on the procedures outlined above. Light ship weight may be related to overall geometry and estimated as a single quantity offering a more simplified method. The question of vertical center of gravity and $\overline{GM}$ can be deferred for later review. Such short cuts should be dependent upon the experience and skill of the naval architect. Furthermore, other simplified procedures can and should be adapted for other ship types. Benford (1962) provides a deadweight procedure for dry cargo ships. Chapter II considers several other types of vessels.

An example of the use of these charts is as follows: Find preliminary dimensions of a 23 knot cellular containership capable of carrying 500 40-ft containers and 500 20-ft containers at an average weight of 20 tons and 12 tons respectively. Containers may be carried three high on deck and may be 8 ft or 8.5 ft high. Fuel oil is to be carried in wing tanks and segregated ballast tankage will be provided in the double bottom. Cruising radius is 10,000 miles, and the last ocean crossing before bunkering is 4,000 miles. Steam trubines with four stages of feed heating have been selected.

Solution:

Equivalent number of 20-ft containers = 500 × 2 + 500 = 1,500 T.E.U.

From Fig. 17: Lpp = 215 m (705 ft), B = 30.5 m (100 ft) D = 16.5 m (54 ft).

D was chosen at 16.5 m (54 ft) instead of the 17.5 m (59 ft) indicated in Fig. 17 because of the discrete height of containers, e.g., assume six high below deck = 14.6 m (48 ft) plus 1.8 m (6 ft) double bottom. The extra one meter (3 ft) needed for 8.5 ft containers can be accommodated in hatch coamings.

$$v = 23 \text{ knots} \times 0.5144 \text{ m/s/knot} = 11.8 \text{ m/s}$$

$$\text{and } v/\sqrt{gL_{pp}} = 11.8/\sqrt{9.8 \times 215} = 0.258$$

From Fig. 16: C_B ranges between 0.58 and 0.63 but since containerships have trended lower, and in some cases below the mean minus 0.025 line, say C_B = 0.59.

Select initial power estimate from Fig. 18: Assumed SHP = 40,000 SHP

Estimate Light ship weight from Figs. 19, 20, and 21:

	Weight	$\overline{KG}$
Steel	9,300 tons	9.4 m (30.5 ft)
Outfit	2,800 tons	14.1 m (46.3 ft)
Machinery	1,280 tons	7.8 m (25.4 ft)
Subtotal	13,380 tons	
3 percent Margin	400 tons	
Light ship	13,780 tons	

First trial for total:

Light ship	=	13,780 tons
20 ft Containers (500 × 12T.)	=	6,000 tons
40 ft Containers (500 × 20T.)	=	10,000 tons
Fuel Oil	=	?
Crew, effects, fresh water	=	400 tons
Total Displacement less fuel oil	=	30,180 tons

First approximation of fuel oil is obtained by using a fuel rate of 225g/SHP/hr (0.5 lb/SHP/hr). The following fuel weight estimate includes a 25 percent margin on the last 4,000 mile ocean crossing before bunkering.

Fuel Wt. = 40,000 SHP × (10,000 + 0.25 × 4,000) mi × 225g/SHP/hr ÷ (23 mi/hr × 1,000,000g/ton)

Fuel Wt. = 4,300 tons

The first trial Δ then is 34,480 tons.

Check required power, Fig. 22, at a displacement of 34,480 tons and a speed of 23 knots.

SHP = 27,000

Service factor 1.25, SHP = 27,000 × 1.25 = 33,750 SHP which is less than original estimate of 40,000 SHP.

The machinery weight is therefore revised by again using Fig. 21 and a second light ship is determined using a revised estimate of specific fuel rate from Fig. 23.

Revised Machinery Weight	=	1,180 tons
Revised Fuel Weight	=	3,515 tons
Revised Total	=	33,595 tons

Rechecking required power from Fig. 22 results in an SHP estimate of 32,625 which will again result in a change in fuel oil required.

Since Δ is to be adjusted downward, less SHP will be required. Hence, fuel oil excess will be more than the 118 tons arrived at by multiplying the fuel weight by the ratio of the new and the old powers; say 130 tons.

Recalculate using estimated fuel oil of 3,385 tons and

$$\Delta = 30{,}080 + 3{,}385 = 33{,}465 \text{ tons}$$

$$\text{SHP (Fig. 22)} = 26{,}000 \times 1.25 = 32{,}500 \text{ SHP, and}$$

$$FO = \frac{32{,}500}{32{,}625} \times 3{,}397 = 3{,}383 \text{ tons}$$

which is two tons less than used to obtain Δ (acceptable within range of accuracy of method). Therefore, say FO = 3385 tons

$$\text{Total displacement} = 30{,}080 + 3{,}385 = 33{,}465 \text{ tons}$$

$$T = \frac{33{,}465 \text{ tons}}{1.026 \dfrac{\text{tons}}{\text{m}^3} \times 215 \text{ m} \times 30.5 \text{ m} \times 0.59} = 8.4 \text{ m (27.5 ft)}$$

By inspection it is obvious, assuming normal shear, that 8.4 meters is less than the draft corresponding to minimum freeboard. This conclusion can be reached by comparison with other existing designs of similar dimensions. However, freeboard may be checked as shown in Chapter IV.

The resultant design draft is actually quite a bit lower than the actual draft of containerships of this size. However, since there has been a trend to increasing container weights, the 12 tons per 20-ft container and 20 tons per 40-ft container may be expected to increase in the future and such weight growth can be readily accepted by this design with some reduction in speed.

Finally, the preliminary transverse metacentric height ($\overline{GM}_T$) is examined.

The vertical center of gravity for steel, outfit, and machinery are estimated using Figs. 19, 20, and 21 to provide the $\overline{KG}$ of the light ship weight.

Net light ship weight $\overline{KG}$	= 10.2 m (33.5 ft)
Margin	= +0.3 m (1.0 ft)
Total light ship weight $\overline{KG}$	= 10.5 m (34.5 ft)

For three high deck stowage cargo container $\overline{KG}$ may be approximated as $0.6D + 4.3$ m = 14.2 m (46.6 ft), the $\overline{KG}$ of fuel at 6.1 m (20 ft) (assumes fuel in wing deep tanks) and the $\overline{KG}$ of crew, effects, freshwater at $0.5 \times D$ = 8.3 m (27.2 ft), the overall full load $\overline{KG}$ is determined to be:

	Weight (*tons*)	$\overline{KG}$ (*m*)	*Moment* (*m − tons*)
Light ship	13,680	10.2	139,540
Cargo	16,000	14.2	227,200
Fuel	3,385	6.1	20,650
Miscellaneous	400	8.3	3,320
Total	33,465	11.7	390,710

The $\overline{GM}_T$ may now be calculated as follows:

$I_T = C_{IT} \times L_{WL} \times B^3$ where

$C_{IT} = 0.937\, C_P - 0.0122$

Approximating $L_{WL} = 1.02 \times Lpp = 219.3$ m (719.3 ft) and

$$C_{PWL} = \frac{C_B}{C_x} \times \frac{Lpp}{L_{WL}} = \frac{0.59}{0.98} \times \frac{215}{219.3} = 0.59$$

then:

I_T = 268,100 m^4 (30,998,200 ft^4)

$\nabla = \Delta/1.026$ = 32,620 m^3 (1,151,900 ft^3)

$\overline{BM}_T = \dfrac{I_T}{\nabla}$ = 8.2 m (27.0 ft)

$\overline{KB}$ = 0.54T = 4.5 m (14.9 ft)

$\overline{KM}_T = \overline{KB} + \overline{BM}_T$ = 12.7 m (41.7 ft)

$\overline{GM}_T = \overline{KM}_T - \overline{KG}$ = 1.0 m (3.3 ft)

$\overline{GM}_T/B$ = 0.033

Based upon an analysis of existing containerships now in service the preliminary dimensions are acceptable if $\overline{GM}_T/B$ equals or exceeds 0.025. It is reasonable to assume that an adequate operational $\overline{GM}_T$ can be maintained by pressing up the segregated ballast double bottom tanks as bunkers are consumed. For dry cargo ships a higher $\overline{GM}_T/B$ is

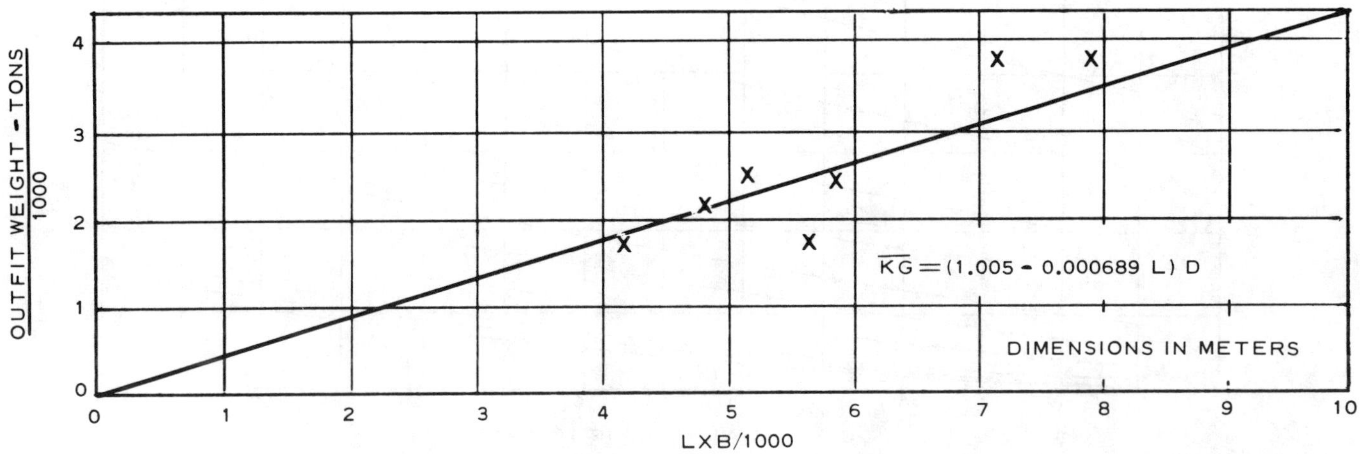

Fig. 20 Containership outfit weight

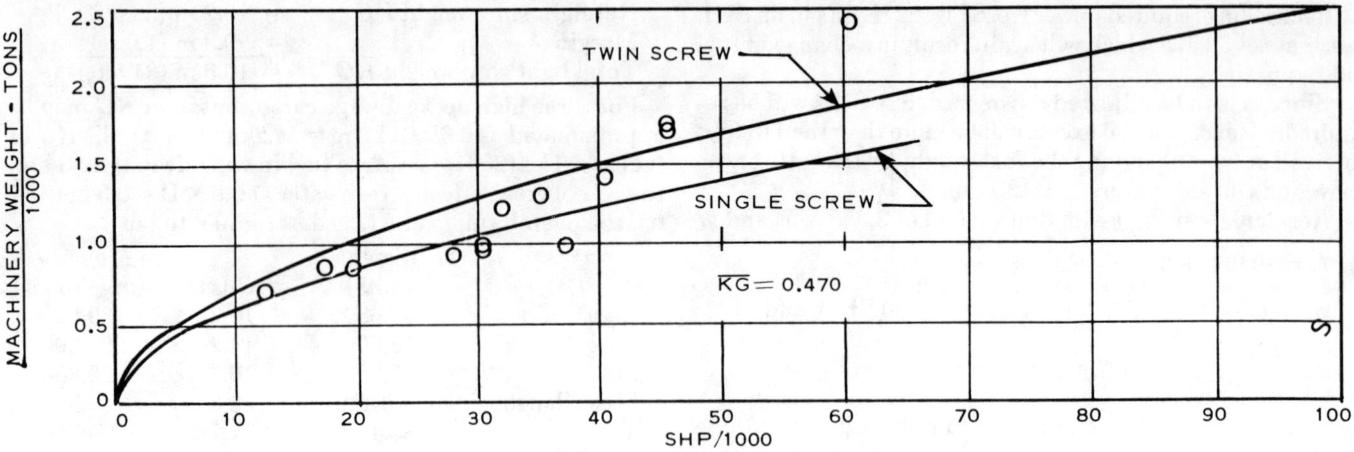

Fig. 21 Containership machinery steam plant

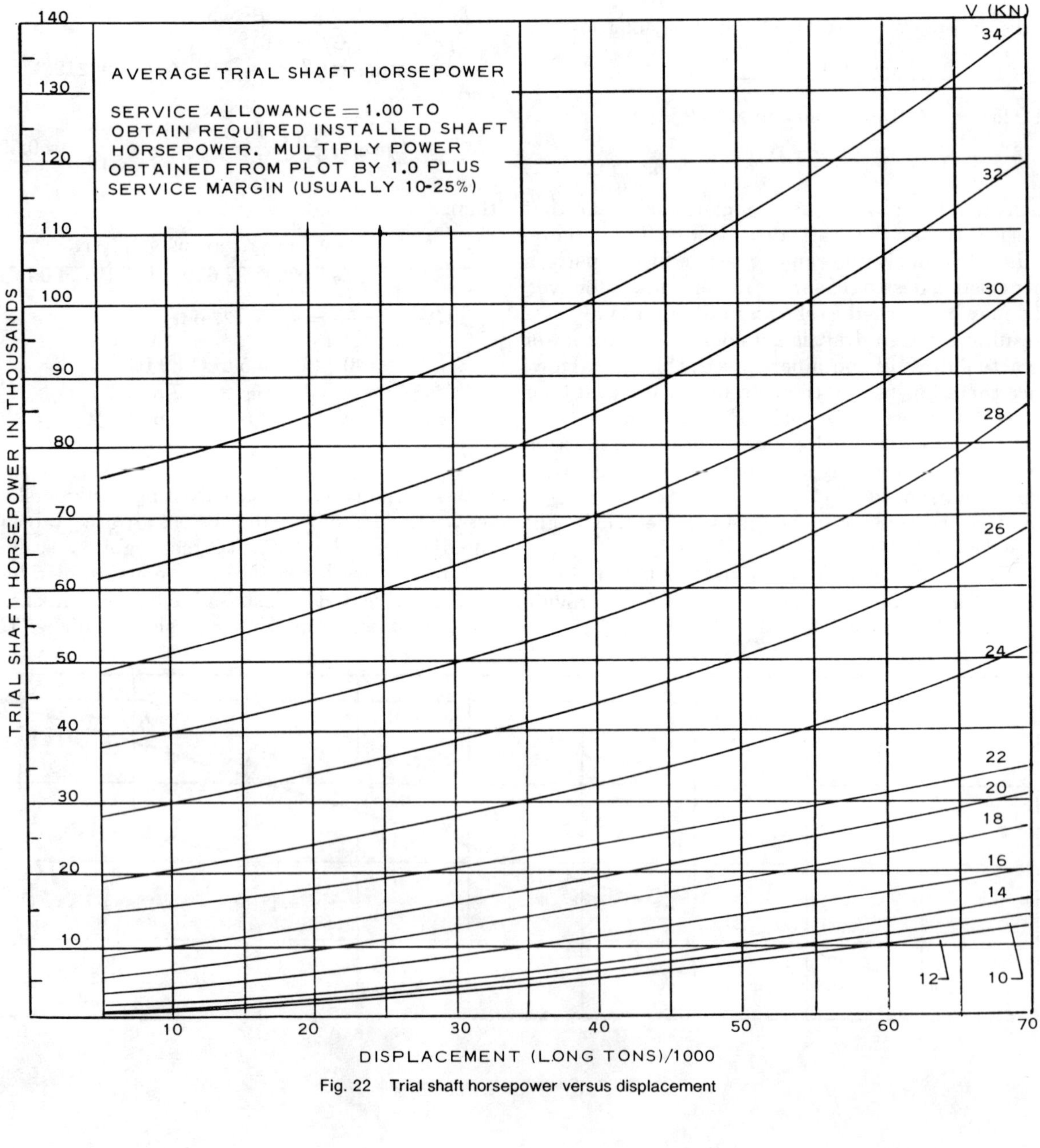

Fig. 22 Trial shaft horsepower versus displacement

usually desired, but the ease of loading containers with heavy containers stowed low permits a lower criterion for selection of basic dimensions.

Discussion of the various factors used in the foregoing example follows.

The number of containers and speed are the dominant requirements for containerships. The average weight per container is a very important factor and varies among different trade routes. Recent information would indicate that 12 tons for a 20-foot container and 20 tons for a 40-foot container are acceptable overall averages, but the trend has been towards increasing weight rather than toward lighter container loads. In consonance with the primary competitive requirements, containerships have tended to grow in size and speed and also show a trend toward lower block coefficients. Although hull forms with fine ends have less displacement than fuller hull forms at the same draft, expansion of length and beam to provide the required container cubic is usually sufficient to develop adequate displacement at a draft of 35 feet or less.

Accordingly, it is convenient to enter plots of ship dimensions versus number of 20-ft equivalent containers as shown in Fig. 17.

The plotted points in Fig. 16 show the relationship between C_B and Froude number for a number of actual ship types.

Fig. 18, used for an initial horsepower estimate is based on recent designs and is, at best, crude. It would be completely adequate simply to select the power installed in a ship of similar dimensions and speed as an initial guess. Empirical curves of light ship weights are dependent on speed, date of construction, endurance, type of machinery, cargo gear (if any), and reefer system. A breakdown into steel, outfit, and machinery permits the designer to use judgment in the appropriate area, especially if the subject design contains any unique requirements which will affect weight. A widely used definition of light ship, upon which Figs. 19, 20, and 21 are based, contained in Troost (1957), is as follows:

> "The light ship weight shall be the weight of the ship with all its equipment and outfit including permanent ballast (solid and liquid), spare parts, regardless of by whom furnished, water in boilers to normal steaming level, machinery in working condition, lubricating oil in all machinery, but not in storage tanks, water in piping systems in machinery spaces, water in fresh water systems and firemains, but without any items of consumable or variable load, as liquids in double bottom tanks, deep tanks, storage or reserve supply tanks, or any allowance for cargo, stores, passengers, crew and their effects."

Total deadweight at design displacement is made up of fuel oil and the owner's requirement for cargo, plus miscellaneous items, including fresh water, stores, passengers, crew, and their effects. Many modern containerships are equipped with sufficient seawater evaporating capacity to take care of all fresh water requirements. For preliminary estimating, nonrevenue deadweight of cargo ships exclusive of fuel oil may be taken at 400 tons. This weight is not greatly dependent on ship size since it represents crew, effects, and consumable supplies which vary little with ship size.

Fuel oil weight is dependent upon endurance with a predetermined margin, horsepower, and fuel consumption rate. Fig. 22 gives average trial condition SHP, i.e., clean bottom, calm water, no wind. The usual allowance for wind, weather, and foul bottom is 25 percent or, as often designated, a *service factor* of 1.25. However, larger high-speed vessels may require less, especially on relatively smooth-water routes, such as New York to Buenos Aires, in which case a service factor of about 1.15 would be appropriate.

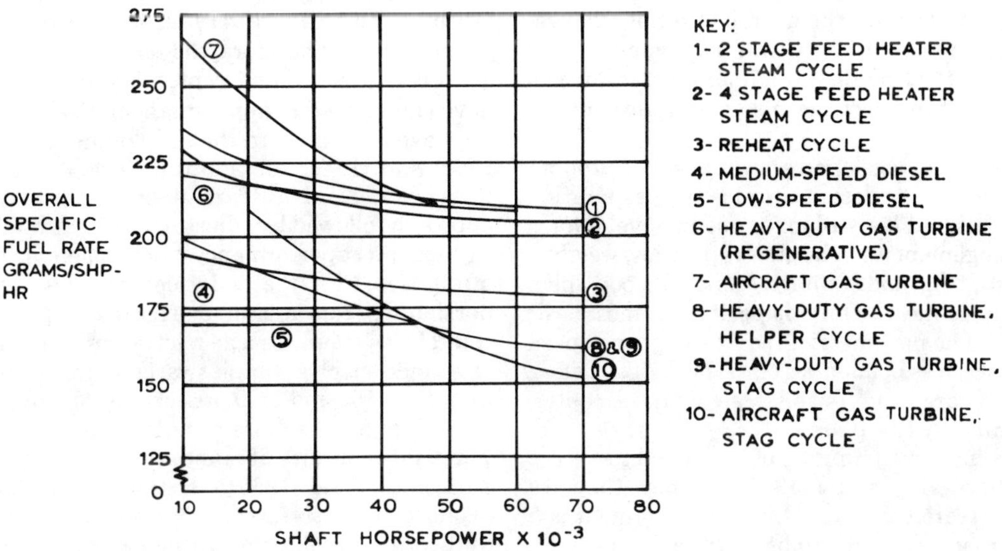

Fig. 23 All-purpose fuel rate

Conversely, smaller vessels in a rough-water route, such as the North Atlantic-European run, would require a service factor of about 1.35. Hawkins and Levine (1969) and Giblon (1975) discuss service margins and the detailed considerations involved. For other than steam plants Subsection 4.6d on engine ratings should be consulted.

As for fuel consumption, Femenia (1973) presents curves of all-purpose rates for steam turbine plants of various cycles and steam conditions. The curves (modified for the metric system) shown in Fig. 23 may be taken as representing a conservative prediction of attainable fuel rates in service with average auxiliary loads. By adding the deadweight items and combining total deadweight with light ship weight, the full load total displacement is found.

Having thus derived length, depth, beam, block coefficient and displacement empirically, saltwater molded draft is obtained from the displacement equation $T = \nabla/(L \times B \times C_B)$, in which ∇ is the molded volume of the underwater body corresponding to Δ, which is the sum of light ship (LS) weight and deadweight (DWT) items, not including an allowance for shell and appendages. A value of $0.005 \times \Delta$ is approximately correct for the latter which, because of its relative insignificance, may be neglected in the first stages of design.

The final step of checking the transverse stability utilizes an inertia coefficient to approximate the transverse inertia. This is the type of design data that should be developed as part of the naval architect's basic design file. The coefficient may be plotted as a function of C_P or C_W without introducing too much scatter, however, it is preferable to maintain such plots for similar ship types.

The $\overline{KB}$ may be estimated by comparison with other design or by use of the Morrish formula for C_X less than 0.90 (Attwood and Pengelly, 1960) and the Postdunine formula for C_X greater than 0.90.

$$\overline{KB} = T - \frac{T}{3}\left(0.5 + \frac{C_B}{C_W}\right) \quad \text{for } C_X \leq 0.90 \tag{6}$$

$$\overline{KB} = T\left(\frac{C_W}{C_B + C_W}\right) \quad \text{for } C_X \geq 0.90 \tag{7}$$

Thereafter the calculation of $\overline{GM}_T$ is straightforward.

The principal virtue of the data given herein lies in the case where proportions may be roughed out and the probability that their accuracy will be sufficient (barring abnormalities) to start laying out the ship for the first stage of design. Watson (1977) provides additional insight into the methods of containership design.

Section 4
Steps in the Preliminary Design Process

4.1 General. Once the initial proportions have been selected in the concept design as described in Section 3, more detailed analyses and calculations are required which could be characterized as additional iterations around the design spiral, Fig. 1. This is necessary to develop additional data and to check results and modify earlier assumptions. The preliminary profile and deck arrangements are usually roughed out at this point. A number of configurations can be considered, depending on the degree of innovation required by the mission. Often several arrangements are developed to the point where their profitability or utility can be compared in order to select the most economic configuration.

4.2 Lines. Having developed an initial set of proportions, the next step on the design spiral is the preparation of preliminary lines and body plan for use in developing a preliminary arrangement and check of capacities, weights, centers, freeboard, trim, and stability. Since the body plan almost completely defines the hull shape, it is the most basic of all ship plans. The maneuvering, seakeeping, and speed characteristics, as well as boundaries of the decks, capacities, VCG, and LCG of cargo and fuel oil depend on the configuration represented by this plan.

Delineation of lines and definition of associated terms are given by Owen and Niedermair (1967) but how they are developed is not treated therein. Lines development is a part of the design process and will be discussed here.

Thus far, hull fullness has been considered mainly in terms of block coefficient, C_B. For purposes of drawing the body plan, it is necessary to consider the maximum section coefficient, C_X and prismatic coefficient, C_P, wherein $C_B = C_X \times C_P$.

There are an infinite number of shapes satisfying the displacement equation for any set of values of L, B, T, C_P, and Δ. The challenge lies in developing an optimum hull form or, at least, one having acceptable performance. Resistance and SHP are very sensitive to even minor changes in hull form. Therefore, the selection of ship lines requires great care to avoid unacceptable results. Although experts have endeavored to apply mathematical theories developed for wave resistance to the development of hull lines (Inui, 1962) and (Pien, 1964), a fully acceptable procedure still does not exist. Naval architects prefer to use an empirical approach followed by adjustments to the hull form suggested by model test performance. Several empirical methods are often used. First, a body plan may be developed by proportioning from a similar existing ship form. This task could be as simple as a geometric variation in beam or draft or it could involve complex variations to account for changes in L, B, T, C_P, and LCB. Lackenby (1950) provides a useful procedure for the more complex variations. Alternatively, waterline contours of standard lines may be used, such as Series 60 (Todd, 1963). Extensive data are provided on displacement (SNAME, 1958) and planing (SNAME, 1963) hull forms, in addition to model test results. Finally, a rough, faired set of lines may be developed without any

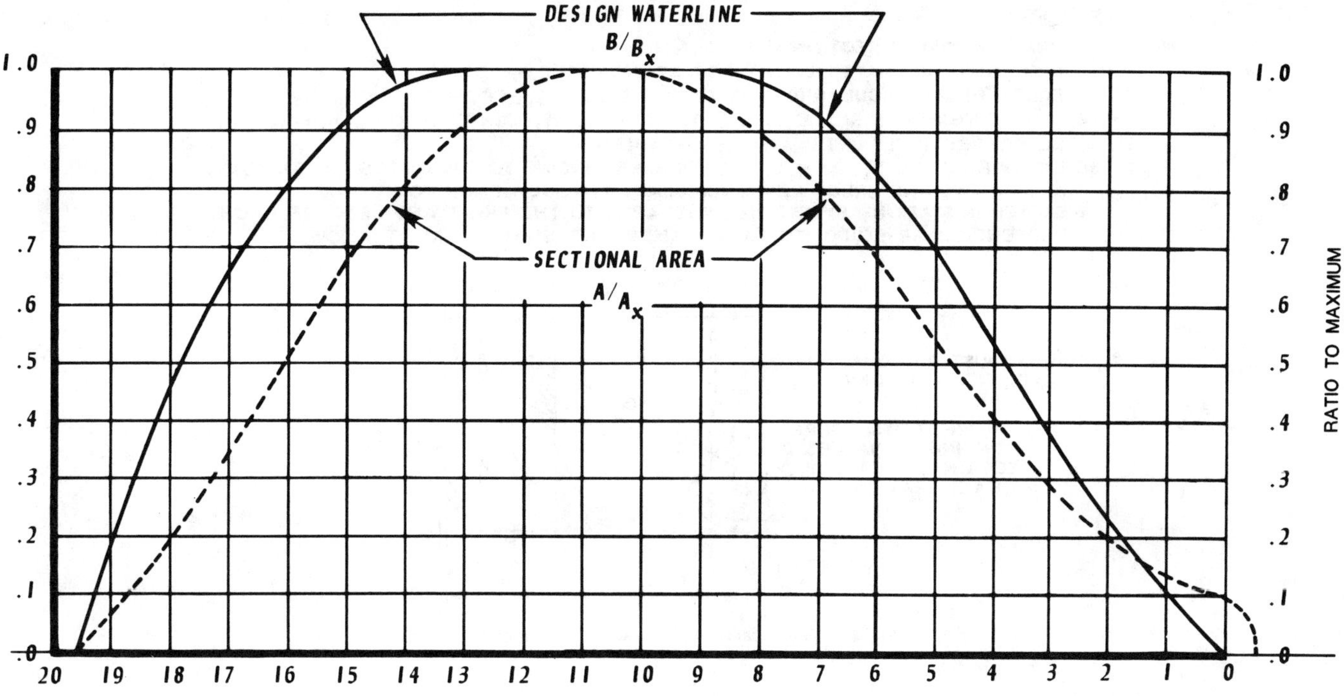

Fig. 24 Containership sectional area and design waterline curves

parent, relying solely on the eye and judgment of the designer.

Important controlling elements of any ship hull form are the sectional area and design waterline curves. Fig. 24 shows these curves for a typical containership. Ordinates of sectional area for any station represent the area under the design waterline at that station divided by the maximum section area under the design waterline. In effect, this curve represents longitudinal distribution of displacement. The DWL curve in Fig. 24 is not the same scale in length with respect to beam but is depicted this way for convenience. The figure also highlights two additional points. First, the design waterline curve indicates the design waterline is straight between stations 9 and 13 but, since the sectional area curve is not, the ship has no parallel middlebody, and, second, the fact that the sectional area curve is positive at station 0 and forward, indicates this ship has a bulb which is not apparent from the DWL curve.

The shape of transverse sections is largely controlled by the design waterline. When the separation between the sectional area curve and the design waterline curve is small, U-shape sections are indicated. Conversely, wide separation denotes V-shape sections. For convenience ordinates of B/B_X, are usually plotted with A/A_X as in Fig. 24.

The shapes of sectional area and design waterline curves affect LCB and C_{IT} and also influence EHP and SHP appreciably. Such elements on both curves as amount of hollowness or convexity of forebody entrance and afterbody exit, half angles of entrance and exit, and hardness of shoulders are all variables that affect performance. Also, the size, shape, and location of a bulb may be very significant.

Design guidelines are difficult to generalize. Optimum hull forms will vary considerably with Froude number, prismatic coefficient, and displacement-length ratio as demonstrated by Taylor (1943) and in the series 60 tests (Todd, 1953).

Other sources of systematic series data are contained in: (Baker, 1933); (Ridgely-Nevitt, 1963), (Ridgely-Nevitt, 1967), (Lasky and Campbell, 1966); (Kiss, 1972), (Linblad, 1961), (Saunders, 1957), (Hadler and Hubble, 1971), (Strom-Tejsen and Pien, 1968), (Lindgren and Williams, 1968), (Clement and Blount, 1963), (Muntjewerf, 1970), and (Mercier, 1973).

After the section area and design waterline curves are laid out to the designer's satisfaction, the rough body plan can then be completed by the procedure shown in Fig. 25. This should meet the requirements for the first preliminary design pass around the design spiral.

After completion of the rough body plan, refinement of other parameters shown on the design spiral may be considered. For a containership it is well to develop quickly a container stowage arrangement since the fixed shape of the containers may dictate a minor hull form adjustment which may permit additional cargo to be carried without seriously degrading the quality of the hull form. Once again, the significant influence of the cargo/mission can be readily observed.

a. Lines for Model Test. Subdivision, trim, stability, capacities, freeboard, and deadweight are first calculated on the basis of dimensions and hull form from a roughly faired preliminary body plan. Once these requirements are tentatively satisfied, it is then essential to refine the lines and conduct a model test for assured performance.

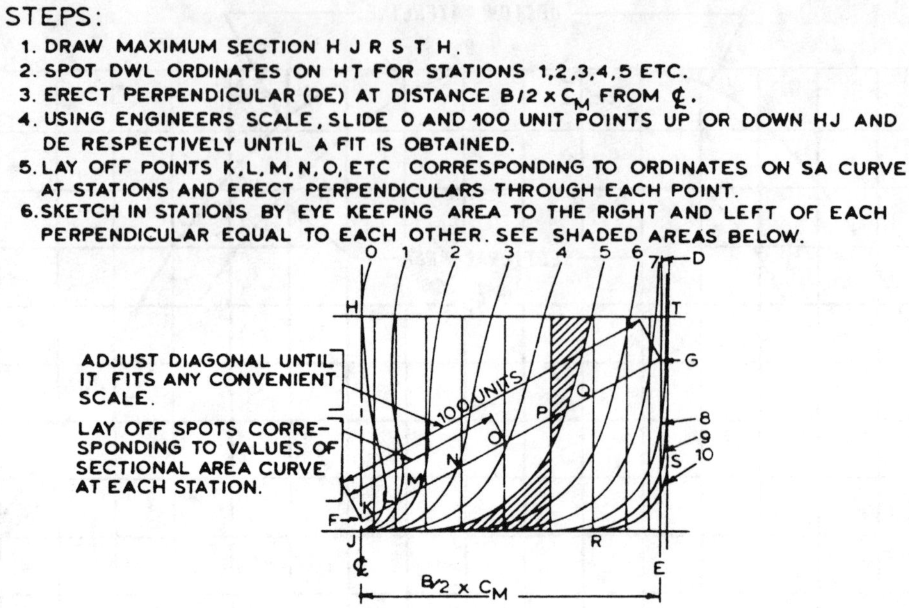

Fig. 25 Graphical aid to fair rough underwater body plan

Justification for model testing expenditures can be demonstrated by the following example: Tentatively assuming fuel oil at \$20 per bbl, interest rate of return after taxes at 10 percent, steaming time at 50 percent, fuel rate 220 grams per SHP-hr, 30,000 SHP, economic life of 20 years, taxes at 48 percent, the present worth P, of two percent savings over the life of the ship as a result of improved hull form is from, Benford (1970):

$$P = \frac{R}{CRF} \tag{8}$$

When CRF is capital recovery factor before taxes and CRF' is capital recovery factor after taxes, R is uniform annual amount of money, or annual savings.

$$CRF = \frac{i \quad (1+i)^n}{(1+i)^n - 1} \quad \text{See Benford (1963)} \tag{9}$$

and

$$CRF' = \frac{CRF - \dfrac{t}{n}}{1-t} = \frac{0.1175 - \dfrac{0.48}{20}}{1-0.48} = 0.180 \tag{10}$$

where:

n = economic life in years

t = tax rate

then:

$$R = \frac{\text{days}}{\text{yr}} \times \frac{\text{hr}}{\text{day}} \times \frac{\text{g}}{\text{SHP-hr}} \times \frac{\dfrac{\text{time at sea}}{\text{total time}} \times \dfrac{\$}{\text{bbl}} \times \text{SHP} \times \dfrac{\text{SHP saved}}{\text{SHP}}}{\dfrac{\text{g}}{\text{bbl}}}$$

$$R = \frac{365 \times 24 \times 220 \times 0.50 \times 20.00 \times 30{,}000 \times 0.02}{152{,}863}$$

and savings = \$75,700 per year

$$P = \frac{75{,}700}{0.180} \cong \$420{,}000 \text{ saved over the life of one ship}$$

If more than one ship is built to this one basic hull form, the present worth of the savings is a direct multiple of the number of ships involved. Accordingly, the economics are clear for any important ship and especially for multiple-ship programs, so that it is wise to model test thoroughly the proposed lines.

Stuntz (1963) provides a more complete explanation of how model basins can help the designer to develop the optimum hull form. Model tests should not be commenced until the naval architect is fully satisfied with the design and sufficient cycles around the design spiral have been made to assure that no surprises calling for lines changes will be required.

As noted earlier, the design to squeeze in another row of containers could have influenced the designer to introduce a small modification to an otherwise good parent design. Likewise waterlines may require broadening locally to fit the bull gear of a ship with machinery aft. Changes of this type, if adopted, usually alter the sectional area curve and can increase the power requirement five percent or more. If the naval architect decides to go ahead with such changes, it is all the more important to make comparative model tests which will then furnish a reliable basis for economic analysis of the tradeoffs.

When all the decisions concerning form modifications have been made, the lines should be carefully faired before sending them to the model basin. This process can be done by hand (in the traditional way) using wood battens and lead ducks, or by using mathematical procedures programmed for computers.

The concept of mathematical representation of ships' lines was probably first brought out in the United States by Admiral D. W. Taylor in his famous paper delivered before the International Engineering Congress in San Francisco in 1915, (Taylor, 1915). Although he envisioned, as have many others, the establishment of a mathematical relationship between ship form and wave making resistance, he had another fundamental reason for the adaptation of mathematical curves to ships' lines. He desired to run a series of model resistance tests of a ship form with a systematic variation of certain basic form parameters. Therefore he modified somewhat the lines of the armored cruiser, HMS *Leviathan,* and fitted to them a series of mathematical waterlines and station lines based upon those parameters he wished to vary. This resulted in Experimental Model Basin Model 632.

The waterlines were separated fore and aft, and a fifth degree curve was fitted to each half of each waterline. Before fitting the mathematical curve he reduced each waterline to nondimensional form; i.e., the ordinates of each point were divided by the maximum half-beam of the waterline and the abscissas of each point were divided by the entrance length of the ship.

Admiral Taylor carried this principle further to fit fourth degree parabolas and hyperbolas to the section curves. By systematic variation of the prismatic coefficient, displacement-length ratio, and beam-draft ratio he produced a series of mathematical forms which were constructed, and tested, and which resulted in the well-known Taylor's Standard Series, Taylor (1943).

It is a little known fact that during the period when Admiral Taylor served at the Experimental Model basin in Washington for fourteen years and later served as Chief Constructor of the U. S. Navy (1914–1922), many of the ship designs produced were faired mathematically using the foregoing techniques. Although the forebody/afterbody separation does not permit precise fairing from a mathematical standpoint, and although the abovewater body was faired by hand, this still represents a major accomplishment many years before computers simplified this approach to a routine exercise.

Still working with manual calculation techniques, Taggart (1955) extended Taylor's fifth degree waterline and sectional area curves to a sixth degree format to be utilized in the fairing of mold loft offsets. This permitted interaction between the drafting room and mold loft to create faired lines that were a considerable improvement over those previously developed using battens or splines for fairing.

As computers became available to reduce the labor involved, the path was opened for the application of many other techniques for the mathematical fairing of ship lines. Actually, the age-old methods employed by linesmen and mold loftsmen of fairing with battens or splines resulted in a novel form of mathematics called *spline theory* where the bending characteristics of a beam partially fixed at a few points are duplicated mathematically. This theory forms the basis of the U. S. Navy's hull fairing computer program HULDEF. The purpose of HULDEF, described by Laskey and Daidola (1977) and by Gebhardt and Thompson (1976) is to obtain a consistent hull surface definition for the various tasks involved in contract design. It allows the designer to start with a preliminary set of ship lines and generate a contract design lines plan and final mold loft offsets suitable for construction of a ship model for hydrodynamic testing. HULDEF thus provides an interactive tool allowing the designer to control the ship's hull surface definition in a conventional way, but with the speed and precision of a computer. Other proprietary programs also exist such as that described by Collatz and Seiffert (1976).

Work is continuing on the problem of fully relying on the computer to generate ship lines without first developing a rough body plan. One technique, extending the earlier work of Admiral Taylor, is described by Taggart and Magnusson (1967). Papers presented at the First International Symposium on Computer-Aided Surface Definition (1977) describe many of the most recent developments in this area.

b. Seakeeping. As noted in Section 3.2a, there are an increasing number of computer programs to aid in analyzing or predicting the seakeeping performance of a ship. All of these are based upon the original strip theory developed by Korvin-Kroukovsky (1961), or variations thereon. Although these programs have not yet been coupled with optimization programs, they are well suited for analysis of the preliminary design if applied before final hull proportions are fixed. As in the development of hull form for minimum resistance and propulsion, it is difficult to provide general guidelines for seakeeping. In fact, opinions of naval architects vary with respect to the effect on seakeeping of such parameters as $L_{WL}/\nabla^{1/3}$, C_P, U vs. V-shaped bow sections, and large bulbs. These factors and others are well discussed by Lewis (1959), and Lewis (1955).

The effects of some of the significant parameters on seakeeping performance are outlined below:

1. *Length.* This is possibly the most important factor in determining ship behavior. Longer is better from a seakeeping viewpoint. This is emphasized by the fact that in heavy seas, ships often reduce speed to prevent synchronism with long waves to avoid the associated severe motions. Longer ships require less speed reduction to accomplish this.

2. *Slenderness ratio.* Most seakeeping data indicate that a high slenderness ratio $L_{WL}/\nabla^{1/3}$ is desirable. At equal speeds, vertical bow accelerations tend to be lower and less green water appears to be shipped by a vessel with higher $L_{WL}/\nabla^{1/3}$.

3. *Draft.* The forward draft is very important in considering slamming. There should be sufficient ballast capacity to assure adequate draft forward during the light operations. Ochi (1971) indicates that the magnitude of slamming is drastically decreased, and its peak location moves forward as draft is increased. T/L should exceed 0.045 to permit maintaining a reasonable speed and still avoid severe slamming in a seaway. See also (Vossers et al, 1960) and (Ochi, 1973).

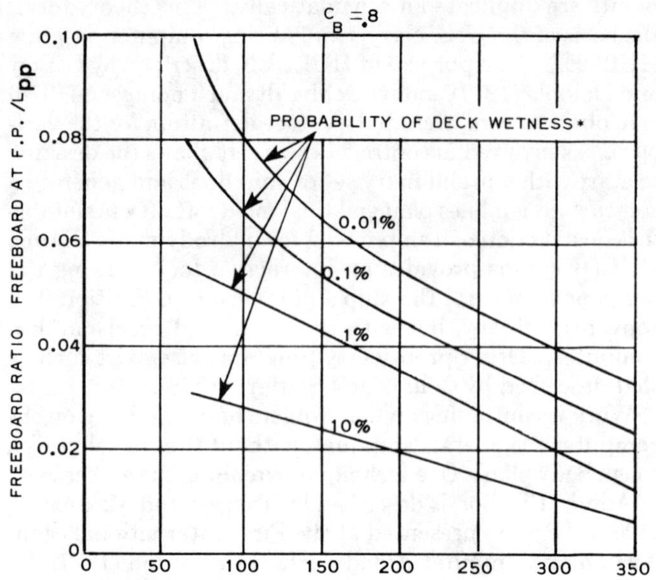

*PLOT IS BASED ON ASSUMED RELATIONSHIP BETWEEN WIND SPEED AND SIGNIFICANT WAVE HEIGHT FOR NORTH ATLANTIC SEA CONDITIONS

Fig. 26 Freeboard ratio versus ship length in meters

4. *Prismatic Coefficient.* Although deck wetness tends to decrease with decreasing C_P, the overall motion characteristics tend to increase with higher C_P. However, higher C_P results in greater speed loss at higher speeds but not at low Froude numbers.

5. *Waterplane Area Coefficient.* Both still-water and wave bending moments are proportional to C_W, i.e., as C_W increases so does the bending moment amidship (Swaan, 1979). However, the choice of C_W is usually based on other reasons.

6. *Freeboard and Flare.* Although variation in the above-water form has little effect on the motion amplitudes, freeboard and flare are extremely important in reducing the frequency of green water on deck. Lewis (1959) indicates that with increasing ship speed the ratio of bow freeboard to length should be increased. Goodrich (1964) provides further data relating the frequency of deck wetness to block coefficient, length, and the ratio of bow freeboard to length. Fig. 26, Ewing (1967) shows typical data. Finally, some flare is useful to deflect water outward as the bow moves downward in a wave, but extreme flare may produce damaging impact loads at the forecastle and cause greater speed loss.

7. *Shape of Forebody Sections.* Analytical studies (Ewing, 1967) indicate that over the full range of wave lengths, smaller amplitudes in heaving motion result when V-shaped sections are used. Pitch amplitudes are reduced by U-shaped sections only in long waves, whereas in short waves U-shaped sections increase pitching motion. Experimental studies have shown that V-shaped sections slam more frequently, but with a lower unit pressure per slam. Thus, a ship with V-shaped bow sections can maintain a higher sea speed than a U-shaped ship before reading the same level of impact pressure due to slamming. Hull damage can be minimized by avoiding large flat bottom areas in the forward quarter length. Conversely, U-shaped sections result in less resistance in waves and induce smaller wave bending moments.

8. *Longitudinal Radius of Gyration.* A decrease in the longitudinal radius of gyration reduces the pitching period and thus relative bow motions as well as bow accelerations while causing only slight increase in stern accelerations. Although a low longitudinal radius of gyration is desirable for good seakeeping performance, it is difficult for the designer to control unless there is flexibility in weight distribution.

c. Maneuverability. Ships with cutaway forefoot and minimum skeg aft answer the helm quickly with low tactical diameters. However, again there is a tradeoff in that such ships, especially full-ended low L/B ships, tend to require continual rudder adjustment to maintain a straight course. When such directional instability is suspected, it is advisable to check with a model test spiral maneuver and increase skeg area as required, while at the same time increasing rudder area to retain design tactical diameter. On twin-rudder ships, the same effect may be obtained by rudder toe-in in at sacrifice of increased resistance. For a complete treatment of this subject see Mandel (1967). Proceedings of the STAR Symposium (SNAME, 1975) provide the latest thinking concerning requirements and criteria on this important factor.

d. Bulbous Bows. A bulbous bow is intended to interact with the primary wave making characteristics of the ship in such a way as to develop an independent wave system which reduces the total wave system generated by the hull-bulb combination. Taylor (1943), Inui (1962), and Couch and Moss (1966) have clearly established the resistance improvements possible by using large bulbs. This has been especially so for full-ended bulk carriers. On these latter ships benefits are due to reduction of wave *breaking* and flow separation under the hull, not wave cancellation. Since bulbs do not significantly affect seakeeping (Dillon and Lewis, 1955), the efficiency of their use can frequently be determined during the resistance and propulsion model tests.

e. Transom Sterns. The trends toward unitizing cargo in containers and barges and toward roll-on/roll-off methods of transport has made the transom stern common for a wide range of single-screw ship types. Careful design of the hull is required to minimize some of the potential problems and deficiencies of this type stern. For example, the flow of water around the hull tends to follow the buttocks instead of diagonals into the propeller. As a result, less of the frictional boundary layer flows through the propeller disc, yielding lower wake fractions and hull efficiencies on ships with transom sterns.

f. Effect of Number of Screws on Hull Form. The transition from single to twin or multiple screws may be a very difficult decision for an owner. It is often a tradeoff between safety by redundancy, superior maneuverability and greater freedom from cavitation against better economy with a single screw. Strom-Tejsen (1972) presents the results of a set of comparative resistance and propulsion tests for contrarotating propellers, single propeller, overlapping propellers, and conventional twin screws of a large high-powered containership. Alterations to the after 35 percent of the hull form were introduced to change from twin to single screw; section shapes were changed from V to U-shaped to improve the flow of the viscous boundary layer through the propeller disc and yield a high propulsive coefficient. For this design, it is interesting to note that the required power for 25.5 knots is reduced by 14.9 percent when switching from twin screws to single to overlapping propellers, and by 23 percent if contrarotating propellers are chosen instead of twin screws. The SHP obtained during model tests on these forms were as follows:

Twin Screws	51,600
Single Screw	44,300
Overlapping Screw	44,000
Contrarotating Propellers	39,700

The decision whether to switch from single to twin screws is discussed more completely under Subsection 4.6c.

g. Vibration. Hull vibration can cause any of a number of problems for a ship, ranging from major crew discomfort to costly structural or machinery damage. The entire subject of ship vibrations has been important for years, but as installed power continues to rise, and dramatic increases have occurred in the last few years, the difficulties become magnified as pointed out by Lewis, F. M. (1967).

In the early stages of preliminary design, the designer can do little more than maximize the hull-propeller clearance and attempt to develop hull lines which will provide as much uniformity as possible in the wake field where the propeller will operate. Reed (1971) provides useful guidelines on propeller clearance.

As the design becomes better defined, it is possible to perform vibration analyses to provide an early indication of potential problems. As a first step in the prediction of the response of a ship system, the magnitude of the propeller-excited vibratory forces must be estimated. The propeller forces and moments exciting the ship hull are hydrodynamically induced by the non-uniform inflow into the propeller. The cyclic forces generated by the non-uniformity of the flow can be estimated from a wake survey, see Section 4.6b. However, in the absence of a survey an estimate can be made by using a survey from a similar ship. A preliminary estimate of propeller characteristics must also be made. With the knowledge of the longitudinal and tangential velocity distribution around the propeller blades, the thrust and torque fluctuations, and positions of the points of applications, the horizontal and vertical bearing forces in the propeller plane can be estimated by a refined two-dimensional quasi-steady airfoil method originated by Burrill (1943–4). Supplemental calculations may include a harmonic analysis of the circumferential distribution of the velocity vectors at various radii, and of the thrust and torque fluctuations, and of the bearing forces.

The harmonic content of the wake pattern and the dynamic response characteristics of the hull structure and machinery plant will influence the choice of the number of propeller blades. Alternating thrust and torque, thrust eccentricity and vertical and horizontal bearing forces may then be calculated.

The fundamental flexural and torsional frequencies of vibration of a ship's hull may be estimated by empirical methods developed by Burrill (1943–4), Todd (1961), and McGoldrick and Noonan (1966); however, those methods often yield inconsistent results. More refined predictions of hull natural frequencies and mode shapes can be obtained by the application of existing theories (McGoldrick et al, 1953) and related computer programs (Cuthill and Henderson, 1964).

After the preliminary sizing and arrangement of the machinery and shafting components, the vibratory characteristics of the proposed system are determined and evaluated. Except in the case of longitudinal shaft vibrations, it is normally feasible to obtain satisfactory vibration behavior of the propulsion system without compromising the vibration characteristics of the hull. In other words, longitudinal shafting vibrations are those most likely to excite hull response.

Given the machinery characteristics; number, weight, and location of the screws; SHP; RPM; speed, the longitudinal vibration of the proposed shaft arrangement should be studied. The fundamental frequencies and mode shapes may be found using the Holzer Method. Additional details on design considerations of the type discussed above are presented by Noonan (1971). Further design related information is given by Reed (1973), Reed (1971), Hagen and Hammer (1969), and Noonan (1976).

If problems are detected during the contract design development certain possibile corrective actions may still be taken. For example, the propeller forces can be reduced by increasing the clearance around the propeller or opting to use a highly skewed propeller which tends to reduce the unsteady propeller thrust and the unsteady pressure forces. Cumming et al (1972), Valentine and Dashnaw (1975), and a number of papers presented at the SNAME 1978 Vibration Symposium give some elaboration on the use of skewed propellers.

4.3 Floodable Length and Freeboard. After establishing the hull form, hydrostatic curves can be calculated (see Owen and Niedermair (1967). Presently, a host of computer programs exist to relieve the designer of the need either to approximate the geometric properties and the tedium of manual calculation. The preliminary spacing and location of main transverse watertight bulkheads should now be established and checked by constructing a floodable length curve for the full load draft using expected cargo permeabilities. For containerships, recent studies have shown that 0.74 represents an average permeability for the contents of a container hold. For many years, the target for good operating stability, used by MarAd for general purpose, dry cargo ships, containerships, and dry bulk carriers has been that ships have sufficient intact $\overline{GM}$ to withstand flooding of any one main compartment in certain loading conditions with assumed permeabilities. This is further discussed in Subsection 4.9b. A preliminary check of the damage stability early in the design process can assist in establishing

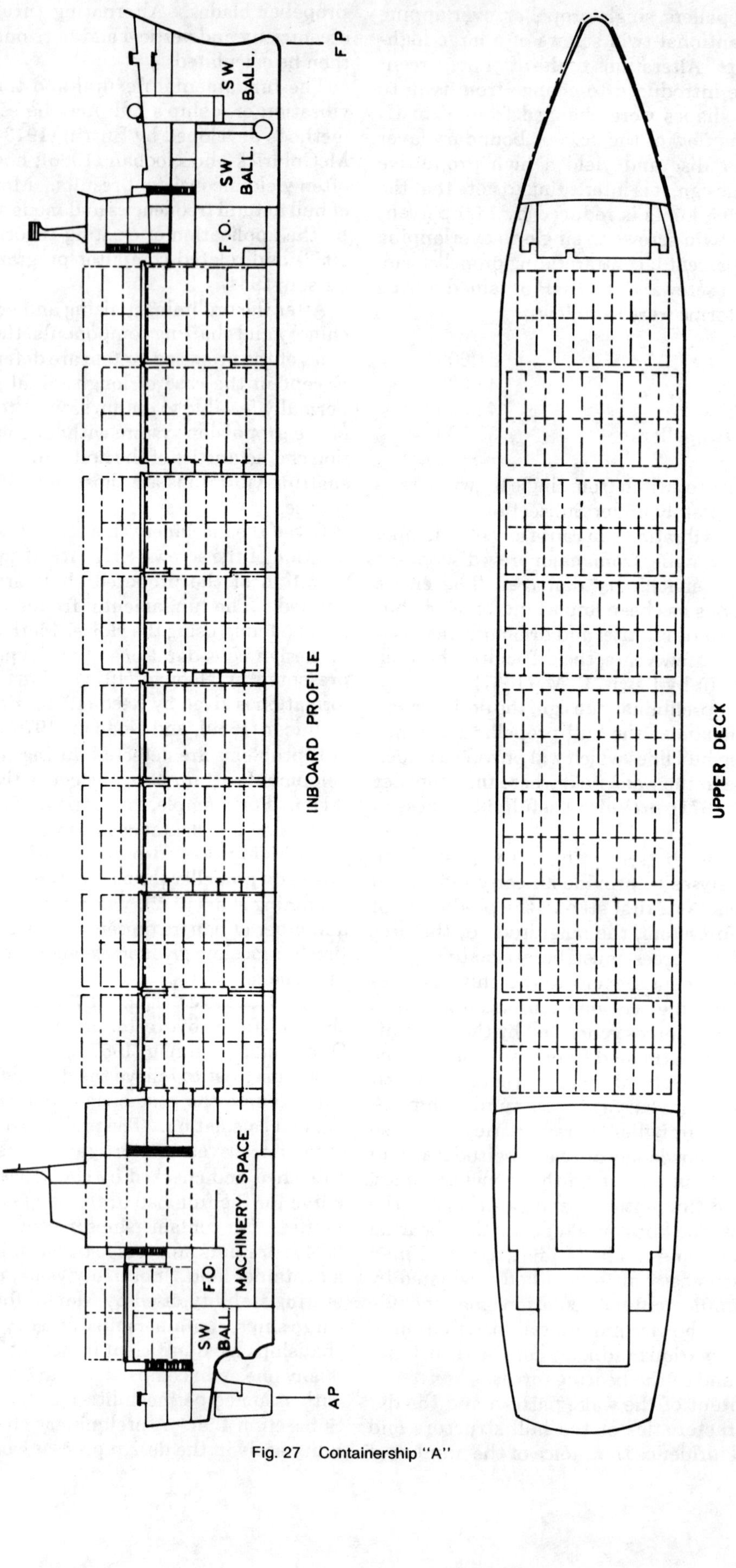

Fig. 27 Containership "A"

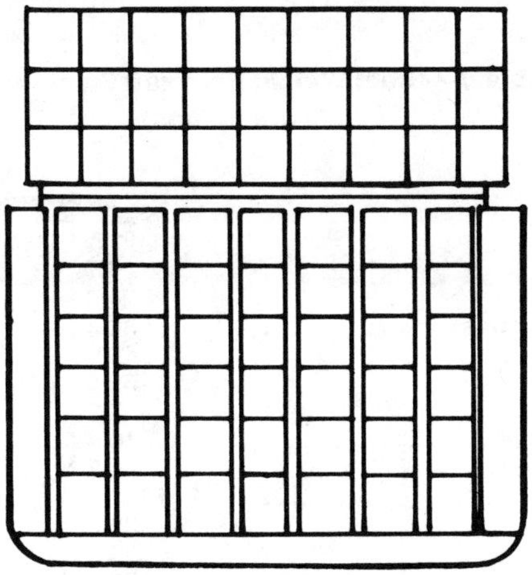

MIDSHIP OUTLINE

Fig. 27(a) Containership "A"

proper compartment lengths. Tankers and LNG carriers in some cases have more stringent requirements imposed upon them. Chapter XI discusses these requirements in greater depth.

Freeboard is determined by the International Load Line Convention (1966) as described in Chapter IV. Bulk carriers may obtain reduced freeboards by meeting certain requirements and having the capability of surviving one compartment flooding under specified conditions.

4.4 General Arrangement. Having verified the freeboard and established the potential bulkhead locations vis-a-vis floodable length, the designer can now develop additional details of the general arrangement. The sometimes conflicting requirements of compartmentation and cargo handling, machinery location, mooring, and accommodation and navigation arrangements must now be addressed. This subject is covered in detail in Chapter III. For information on arranging machinery spaces, see Harrington (1971).

The general arrangement drawing now becomes the focal point of all future design modifications and compromises. The functional efficiency of the design is initially established at this juncture. For all types of vessels the mission should be of paramount importance in developing the general arrangement. For containerships, this means carrying the required number of containers as efficiently as possible accounting for stability, overstow, cargo handling time, and flexibility. Sacrifices may be made in the arrangement of machinery, the hull form, and accommodations if they contribute to improved earning capacity.

As will be noted further on, the general arrangement and structural layout are interdependent. The general arrangement also serves in preliminary and final weight estimating and in locating centers of gravity, *VCG, LCG,* and *TCG*. It also fixes the location of tankage with a direct effect on damaged stability requirements, having especially adverse consequences when unsymmetrical flooding occurs in a bilged compartment or damage is sustained over a watertight flat.

4.5 Structure. Upon completion of the initial general arrangement, a preliminary midship section is prepared. Special factors and considerations in the design of midship sections, overall structural integrity of the main hull girder, the selection of scantlings and production details are provided in Chapters VI, VII, VIII, and XVI. It is at the preliminary design stage, however, where basic decisions are made such as the choice of framing system (transverse, longitudinal, or mixed), the type of hatch girder and support system, and principal structural materials such as high strength steel, ordinary strength steel, aluminum, fiberglass, prestressed concrete, or combinations of these.

The structural design itself can be a subject for a suboptimization design study. Evans and Khoushy (1963), provide a thorough treatment of structural design optimization. In addition, their examples provide general guidance to designers in making design decisions without detailed analytical studies. It appears that the structural design optimization should seek to minimize weight. This will reduce cost and minimize the loss of cargo deadweight due to structure. However, minimizing weight may not always result in minimum cost. If stiffener spacing is varied, web frames curved, many different shapes used, and plate stiffeners located unsymmetrically, the design may become difficult to fabricate and offset savings in material costs by increasing labor costs. Thus, the designer should usually strive to utilize only one stiffener spacing, to reduce the number of shapes and generally to simplify the structure as needed.

There are different philosophies regarding the influence of a hull's structure on the general arrangement. Some designers prefer that the structural design should be developed to implement the selected, optimal general arrangement. Others believe that the general arrangement should be modified if structural continuity of bulkheads, decks, and girders can be improved. In the latter case, the rationale is that improved structural continuity tends to result in a more sound, vibration free design.

Table 6—Steel Weight Itemized Summaries
(Units are in tons)

		CONTAINERSHIPS		
Code	**Item**	**A**	**B**	**C**
0–0	Stem casting	7.7	28.6	—
1	Stern frame casting	24.6	—	73.3
	Total Forgings and Castings	**32.3**	**28.6**	**73.3**
1–0	Flat plate keel	45.0	60.6	84.1
1	Shell plating	1253.0	1848.5	2304.3
2	Bulwarks	7.3	14.9	59.7
3	Bilge keels	10.8	6.2	—
5	Rubbing Strips and fenders	6.7	—	—
	Total Shell Plating	**1322.8**	**1930.3**	**2448.1**
2–0	Center vertical keel	32.5	60.0	50.0
1	Trans. framing in I. B.	199.3	328.2	178.8
2	Long. framing in I. B.	185.9	425.9	369.4
3	Trans. framing outside I. B.	227.2	294.7	148.7
4	Long. framing outside I. B.	175.2	282.9	175.8
5	Framing in peaks	54.2	191.7	165.3
6	Transom and cants	4.7	—	11.6
7	Web frames	170.5	454.9	52.4
	Total Framing	**1049.5**	**2038.3**	**1152.0**
3–0	Main Deck P_L and Beams	187.8	785.9	977.5
1	Second Deck P_L and Beams	87.5	319.2	641.7
2	Third Deck P_L and Beams	35.2	24.6	52.6
3	Foscle Deck P_L and Beams	—	80.9	46.6
4	Upper Deck P_L and Beams	326.4	—	—
	Total Deck Plating and Beams	**636.9**	**1210.6**	**1718.4**
4-0	Main trans. W. T. bhds.	322.6	295.4	1257.0
1	Trans. W. T. and O. T. bhds.	66.7	211.2	168.4
2	Long. W. T. and O. T. bhds.	48.6	832.3	515.7
3	Structural N.W.T. bhds.	156.6	18.0	94.9
4	Non-structural bhds.	32.6	53.5	39.2
5	Trunks—structural	16.4	56.3	43.4
7	Stair enclosures	3.0	—	0.3
8	Hatch coamings	49.7	176.4	89.5
9	Box Girders	—	391.7	—
	Total Bulkheads and Trunks	**696.2**	**2034.8**	**2208.4**
5-0	Pillars and girders	21.6	29.8	299.7
	Total Pillars and Girders	**21.6**	**29.8**	**299.7**
6-0	Inner bottom plating	243.4	386.6	396.7
1	Platform deck	—	—	224.5
4	Cofferdam flats	4.3	28.3	26.0
5	Cofferdam floors	4.1	18.7	55.4
6	Miscellaneous flats and floors	16.0	158.7	9.7
	Total Hull Miscellaneous	**267.8**	**592.3**	**712.3**
7-0	Main engine foundations	27.0	73.0	67.1
1	Boiler foundations	2.1	In I.B.	20.8
2	Aux. mach. foundations	28.3	8.0	45.4
3	Shaft stools foundations	4.9	1.0	5.8
4	Miscellaneous foundations	25.5	35.6	20.6
	Total Foundations	**87.8**	**117.6**	**159.7**
8-0	Main House	—		114.5
1	Upper House	—		92.5
2	Boat House	111.7		92.4
3	Cabin House	68.6		56.0
4	Sun House	56.4	401.6	49.1
5	Bridge House	33.6		14.1
6	Nav. Bridge House	34.9		—
7	House Top	12.8		—
8	Miscellaneous houses and structure	—	—	510.0
9	Stack enclosure	13.4	12.5	—
	Total Superstructures	**331.4**	**414.1**	**928.6**
9-0	Subtotal	**4446.3**	**8396.4**	**9700.5**
	Welding	111.1	228.0	200.0
	Total Welding	**111.1**	**228.0**	**200.0**
	TOTAL STEEL	**4557.4**	**8624.4**	**9900.5**

Of course, the ideal solution must depend on the specific design problem, but the need to harmonize structure with the general arrangement and ease of construction is clearly evident. It should be considered early in the design process.

4.6 Speed and Powering. Closely allied with the hull form design discussed in Subsection 4.2 is the prediction or estimation of required power to produce a given ship speed.

a. Power Estimation. A preliminary estimate has already been made, but as the designer refines the design by sequential steps around the design spiral increasingly refined power estimates should be made. The following list systematically presents the methods of increasingly improved power estimates:

1. Gross approximation from known ships of similar size and speed by Admiralty constant (SHP is proportional to $\Delta^{2/3}V^3$) or charts, such as those of Fig. 22.

2. Estimation of EHP from Taylor's Standard Series or Series 60, with

$$\text{SHP} = \text{EHP} \times \frac{1}{\eta_H} \times \frac{1}{\eta_O} \times \frac{1}{\eta_R} \times \frac{1}{\eta_S} \quad (11)$$

or, simply

$$\text{SHP} = \frac{\text{EHP}}{\eta_P} \quad (12)$$

where:

η_H = $(1 - t)/(1 - w)$ = hull efficiency
t = thrust deduction factor
w = Taylor's wake fraction
η_O = open water propeller efficiency
η_R = relative rotative efficiency
η_S = shaft transmission efficiency
η_P = propulsive coefficient (sometimes symbolized by PC or P_C

The values of t, w, η_O, η_R, and η_S may be estimated from ships of similar characteristics as described in (Todd, 1967) and (Troost, 1957).

Reduced assumed typical shaft lengths and the requisite number of bearings, resulting from an increasing number of designs with machinery aft, together with improved lubricated stern tube bearings, has reduced the frictional losses in transmission. For steam turbines it is suggested that η_S be taken as 0.99 for engines aft and 0.98 for others. Steam turbine reduction gear efficiency is on the order of 98 percent. An additional 3–4 percent loss may be appropriate for gearing losses in medium speed diesel installations as discussed in SNAME T & R Bulletin 3-27 (1975).

Gas turbines are arranged in a variety of ways to drive a ship, using non-reversing gears, reversing gears, and electric drives of various types. The total transmission efficiencies range between 0.97 for parallel shaft non-reverse gear to 0.83 for low power D.C. generator—D.C. motor. SNAME T & R Bulletin 3-28 (1976) provides a complete discussion of transmission efficiencies for gas turbines.

3. Estimation from Taylor's Standard Series TSS, (Taylor, 1943), (Gertler, 1954), or Series 60 (Todd, 1963), with correction needed to predict the power of the new ship from a known ship of similar size and speed.

$$\text{SHP new ship} = \text{SHP basis ship} \times \frac{\text{EHP new ship TSS}}{\text{EHP basis ship TSS}}$$

4. Same as above but with further corrections for:

- Bilge keels or antirolling tanks,
- addition of bulbous bow,
- effect of change in propeller rpm, diameter, blade thickness or speed of advance coefficient, etc.
- change in LCB or any form distortion,
- loss due to bow thruster opening interference
- drag of standard appendages (rudders, struts, bossings, shafting).

5. Same as above except that the EHP test results from a small model of the new ship are substituted for the EHP derived from Taylor's Standard Series.

6. SHP from self-propelled large model tests with appendages and stock propeller.

7. SHP from self-propelled large model tests with appendages and final designed propeller.

From the designer's point of view, it would be well to remember that steps 1. through 4. represent estimates of what the speed and power relationship should be, whereas steps 5., 6., and 7. progressively approach the actual performance obtained in the design. Standardization trials develop final full-scale performance. With the wealth of model basin data now available, it is a waste of time and money to model test a hull form merely to determine what the power requirement ought to be. On the other hand, for the reasons previously outlined in the discussion of lines and body plans, once a tentative hull form has been developed and checked out for satisfaction of other design features, model testing is essential for any important ship to corroborate estimated performance and/or look for ways to improve it.

Finally, all of the foregoing powers must be modified to determine the required installed power. Service margins involved in this determination are discussed in Section 3.3a.

b. Design Information from Model Tests. Depending upon the size, novelty, importance of the vessel being designed and the number to be built, any of a number of aspects should be investigated in the model tank.

Flow observations in a circulating flow channel can detect harmful separation at the stern, which may be corrected by adjustment of waterline exit angles. Likewise the alignment of chine lines or bow thruster openings can also be studied to minimize power losses. Flow observations are also essential for positioning of bilge keels and shaft struts. Observation of humps and hollows in the wave profile, when the model is running at design speed, may indicate useful modifications of shoulders, bulbs or entrance angles.

Wake surveys in the propeller disc provide important data for proper design of the propeller, as well as for wake analyses to consider propeller-induced hull vibration (Hadler and Cheng, 1965). A vastly uneven wake distribution over the propeller disc points out a need for improved stern design.

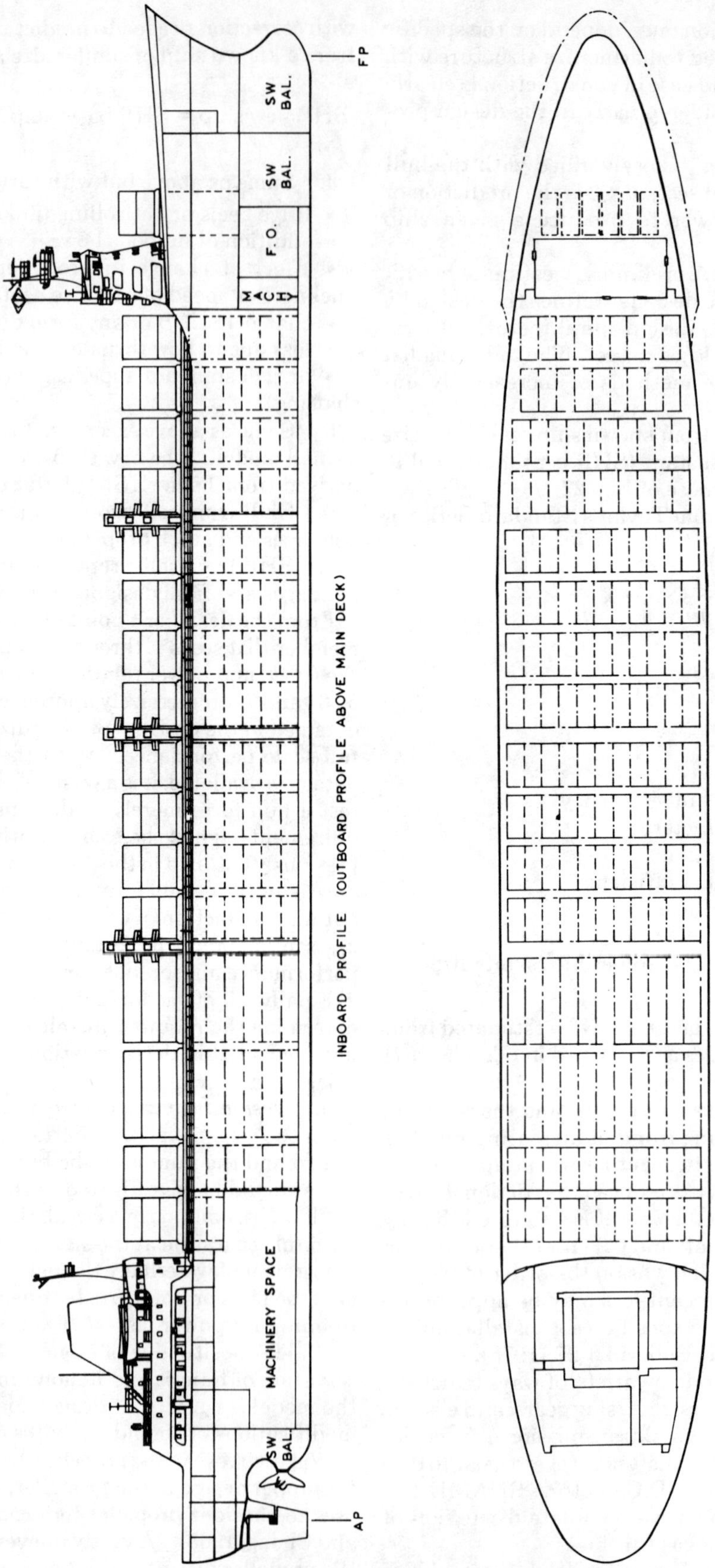

Fig. 28 Containership "B"

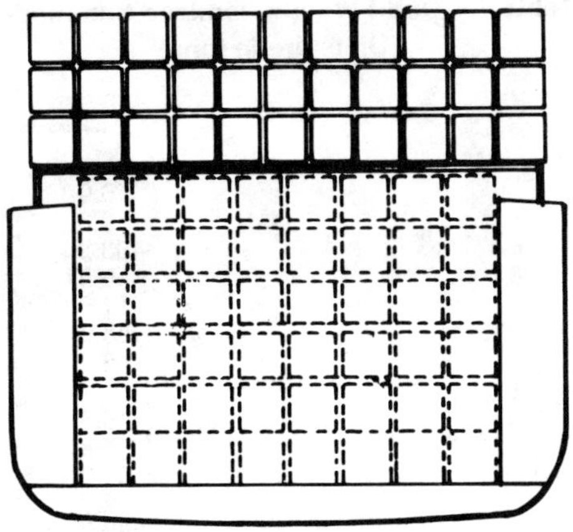

MIDSHIP OUTLINE

Fig. 28 (a) Containership "B"

Cavitation tunnel tests in simulated wakes are indicated when the propeller will be heavily loaded to confirm the adequacy of the final propeller design from the standpoint of blade erosion and power loss. Such a test can also give an indication of potential vibration problems. For a further check on the probability of this unfavorable occurrence, vibratory forces can be measured on the model using the techniques of Kuo (1966). The presence of cavitation on a propeller has an important effect on the propeller-induced pressure forces on the ship hull. In many cases these dynamic pressures are the main cause of excessive vibration on high-powered ships. The lower frequencies of the pressure field induced by the cavitating propeller are primarily responsible for hull vibrations, while the higher frequencies often cause excessive noise. It is now possible to measure these dynamic hull-pressure and shaft forces in a depressurized towing tank (Oosterveld and van Oossomen, 1975). The absolute necessity of tests in a depressurized towing tank is still the subject of debate among model test experts, but it is important to be aware of what tests are possible.

Free running self-propelled models may be used to determine tactical diameters by means of turning circle tests. Z-maneuvers demonstrate the ability to check a turn and yield some information on directional stability which may be supplemented by spiral tests. Astern runs at light and loaded displacement supply astern speed estimates needed for rudder torque calculations to permit sizing the steering gear.

Lastly, tests in various seaway situations are used to confirm motions of the ship or floating platform which must operate over a wide range of sea conditions to carry out its mission properly.

Considering the many aids to improved efficiency and power reduction available at the model basin, the design agent owes it to his client to inform him fully of the high dividends which a well-conceived model test program will yield even though a substantial investment may be required.

The total cost of even very elaborate model test programs can be controlled by using small models 1.5 to 2 m (5 to 6.5 ft) at a model basin which specializes in this work.

For the price of a single 6 m model in a large model basin, three or four small models can be built and run for EHP, maneuvering, and seakeeping tests. The best of these can then be run at larger model sizes, say 4 to 10 m (13 to 33 ft) for a final series of tests.

c. Choice of the Number and Type of Propellers. Single-screw propulsion predominates for most applications in merchant ships. It is simple, and consequently less costly to purchase and operate. Furthermore, for most merchant ships with normal hull forms, single screw propulsion also results in the highest possible hydrodynamic efficiency. This fact stems from two primary factors:

1. Greater recovery of energy dissipated by the ship in generating its frictional boundary layer which converges into the propeller disc on the centerline of the ship, thereby producing a higher wake fraction and better hull efficiency;
2. Minimum appendage resistance, i.e., no need for the protruding bossings or propeller struts required on multiple-screw ships.

Twin or multiple propellers are installed on a variety of vessels for very proper reasons. Included among these situations are:

- Power requirements in excess of acceptable practice for a single screw due to risk of cavitation or propeller induced vibration;
- A need for redundancy and increased reliability of two propulsion units, as on passenger or naval surface combat ships (the designer should however recognize that twin screws are generally in a more exposed and vulnerable location outboard of the ship's centerline);
- Slow speed maneuverability requirements;
- Hull form may be conductive to multiple propellers without high efficiency losses, as on shallow draft vessels with broad, flat sterns, and on full stern vessels.

The increasing power levels in modern ships discussed in

Table 7—Outfit Weight Itemized Summaries
(Units are in tons)

Code	Item	CONTAINERSHIPS A	B	C
10-0	Steel masts, kingposts, etc.	6.0	13.6	72.1
1	Steel booms	0.3	—	16.4
2	Steel hatch covers and beams	463.2	373.6	616.1
3	Steel stairways	6.5	—	1.8
4	Steel sheet metal work	17.9	11.5	7.2
	Total Struct. Steel in Outfit	**493.9**	**398.7**	**713.6**
11-0	Deck castings, mooring ftgs.	42.9	52.4	51.4
1	Mast and spar forgings	—	—	2.5
2	Rails and stanchions	5.8	9.1	4.0
3	Ladders	5.8	8.9	18.4
4	Miscellaneous hull fittings	39.2	130.6	32.2
5	Ratproofing	2.2	0.5	2.4
6	Container cell guides etc.	267.9	356.6	450.5
7	Gantry crane rails			57.4
	Total Hull Attachments	**363.8**	**558.1**	**618.8**
12-0	Sliding W.T. doors			
1	Hinged W.T. doors	3.7	1.5	5.9
2	Manholes and scuttles	4.8	3.7	4.4
3	Airports, windows and lights	3.9	5.4	6.0
4	Hatches and ports O.T. or W.T.	5.6	—	7.6
5	N.W.T. steel doors	3.8	6.0	2.2
6	Skylights and companions	—	—	1.4
	Total Lights, Doors, Hatches	**21.8**	**16.6**	**27.5**
13-2	Hold ceiling and sparring	—	—	25.4
3	Miscellaneous carpenter work	4.4	—	1.0
8	Miscellaneous deck coverings	9.5	20.9	26.8
9	Cement and tile coverings	3.0	9.3	4.9
	Total Carpenter Work and Decking	**16.9**	**30.2**	**58.1**
14-1	Interior joiner work	81.1	106.9	100.9
2	Furniture	12.8	7.7	20.5
5	Accommodation ladder	1.5	4.4	3.1
8	Insulation in quarters	9.7	43.1	39.5
	Total Joiner Work	**105.1**	**162.1**	**164.0**
15-0	Anchors, chains, lines	72.6	158.5	90.7
1	Boats and boat handling	9.5	23.6	14.5
2	Rigging and blocks	1.1	1.6	8.9
3	Canvas work	0.2	0.2	0.3
4	Miscellaneous deck outfit	4.9	16.3	23.0
8	Paint	63.4	114.0	134.0
9	Cement	14.7	—	20.2
	Total Deck Outfit	**166.4**	**314.2**	**291.6**
16-0	Galley and pantry equipment	5.7	4.1	5.7
1	Utility space equipment	8.6	2.9	4.0
2	Steward's outfit	4.5	1.9	3.0
	Total Steward's Outfit	**18.8**	**8.9**	**12.7**
17-0	Fire Det. and ext. system	13.0	37.7	14.8
1	Heating system	16.2	3.1	2.7
2	Ventilation—natural	18.5	6.8	12.7
3	Ventilation—mechanical	29.0	10.5	20.3
4	Refrigeration systems	39.7	26.0	1102.7
5	Plumbing fixtures and drains	28.7	17.1	25.7
6	Deck scuppers	1.0	21.4	15.3
	Total Hull Engineering	**146.1**	**122.6**	**1194.2**
18-0	Bilge and ballast system	37.9	111.1	38.2
1	Cargo oil system	—	18.2	22.1
3	Fire mains	21.5	22.2	31.9
4	San. and fresh water system	4.0	18.6	5.0
5	Fuel oil transfer system	52.7	43.8	48.9
6	Vents, sgd. and overflows	22.5	73.4	44.7
7	Valve operating gear	0.3	2.2	9.2
9	Cross connections	—	9.0	—
	Total Piping	**138.9**	**298.5**	**200.0**
19-0	Deck Machinery	38.9	122.7	121.1
1	Steer. gear and rudder	66.6	156.5	69.2
2	Communicating system	2.3	6.4	5.7
3	Electric plant	146.2	169.0	202.0
4	Dumb waiters and elevators	—	3.4	4.2
7	Bow thruster system	13.6	15.0	17.9
	Total Miscellaneous Machinery	**267.6**	**473.0**	**420.1**
	TOTAL OUTFIT	**1739.3**	**2382.9**	**3700.6**

Subsection 4.2g has focused attention on alternative propeller systems: Contrarotating propellers have demonstrated significant SHP reductions in model tests on high speed ships (Hadler et al, 1964). Furthermore, machinery systems have been developed to transmit efficiently the contrarotation (Steele, 1968), (Folenta et al, 1976). In another area, heavily loaded propellers on large tankers and certain tugboats, have been found to be suited for the application of ducted propellers which can reduce vibratory forces, increase thrust, or increase speed depending on the propulsive requirements.

d. Machinery Selection. A vital part of the design process is the selection of the main propulsion machinery. The designer is faced with a wide spectrum of options, ranging from small high speed diesels to large steam turbines to complex combined cycle propulsion plants such as COGAS (Combination Gas Turbine and Steam Turbine). The selection depends on the power requirements and the intended service.

All marine power plants now being used fall into one of the following four classes of thermodynamic cycles.

1. Brayton cycle—gas turbines,
2. Otto cycle—internal combustion, reciprocating gasoline engines,
3. Diesel cycle—high speed diesel engine, rpm $\gtrsim$ 750; medium speed diesel engine, 1000 $\gtrsim$ rpm $\gtrsim$ 200; slow speed diesel engine, rpm $\lesssim$ 200,
4. Rankine cycle—steam turbine.

These cycles are more fully described by Baumeister (1958) along with their various sub-classes; closed, open, regenerative, etc. The basic operating characteristics are common to each cycle and form the true foundation for selecting a system.

The major considerations used by the designer in the early stages of the design spiral are:

- Horsepower range,
- specific fuel consumption, (*SF*) g/hp-hr,
- specific weight, kg/hp.

Other performance factors often considered simultaneously with the above are:

- Initial and life-cycle costs,
- maintenance and repair requirements,
- lubricating oil consumption rates,
- maximum-to-continuous power ratio.

Considering all of these factors is in itself a difficult design trade-off problem subject to optimization procedures. The horsepower requirements are the most straightforward and for very high power and selection is very simple as can be seen from the following list of power ranges:

	Horsepower
Steam Turbine	8,000–100,000+
Gas Turbine	500–45,000
Diesel	15–45,000
Gasoline	10–500

Once the propulsion systems have been narrowed down to a list of definite candidates which satisfy such requirements as fuel consumption and weight, the life cycle cost usually becomes the deciding factor. Femenia (1973) provides an in-depth analysis of one approach to an overall cost based selection process.

e. Engine Ratings. Steam turbines are usually designed to operate at optimum efficiency at about 10 percent below rated maximum continuous shaft horsepower. This power level is defined in SNAME T & R Bulletin 3-11 (1973) as the shaft horsepower at which the ship is capable of operating most of her service life.

Diesel engines, prevalent outside the U. S., on the Great Lakes and Rivers and receiving increased application in recent U. S. ship designs, have very different ratings and performance characteristics when compared to steam turbines. Slow speed diesel engines are generally fitted for direct drive of fixed pitch propellers, although controllable pitch propellers are occasionally used. Medium speed diesel engines drive either fixed pitch or controllable pitch propellers through a reduction gear.

SNAME T & R Bulletin 3-27 (1975) defines power plant ratings:

> "*Maximum continuous shaft horsepower* is the shaft horsepower at which the power plant is designed to operate continuously and meet the requirements of the classification society. It may or may not coincide with the maximum continuous rating of the engine.
> "*Normal continuous service shaft horsepower* is the shaft horsepower at which the ship is intended to run most of its service life."

Unlike the steam plant the normal continuous service shaft horsepower usually used for diesel machinery is in the range of 85 percent of the maximum continuous shaft horsepower rating of the engine. Selecting a normal rating below the maximum holds engine maintenance costs to reasonable levels. Thus, the maximum continuous power rating is only utilized when it is necessary to recover time lost due to poor weather or port delays and the increased maintenance costs are offset by improved ship operating schedules.

The service margin discussed in Subsection 3.3a is normally applied to the normal continuous shaft horsepower of a marine diesel. In addition to accounting for wind, waves, and fouling the margin provides some protection against the inaccuracy often occurring in propeller design. If the propeller is overpitched, it may be impossible to bring the diesel up to its full RPM and power. This basic problem exists with any internal combustion engine due to its inflexible power—RPM (constant torque) relationship.

Gas turbine power plants have similarities to both diesel and steam plants. Suffice to note that like diesel engines the normal continuous shaft horsepower usually selected is 85 percent of the maximum continuous shaft horsepower.

4.7 Weights and Center of Gravity Estimate. Critically important throughout the design process of a ship is the continual estimation of the ship's weight and location of the center of gravity (vertically, longitudinally, and transversely). If the ship is symmetrical, transverse centers may be ignored; but many vessels have off center weights,

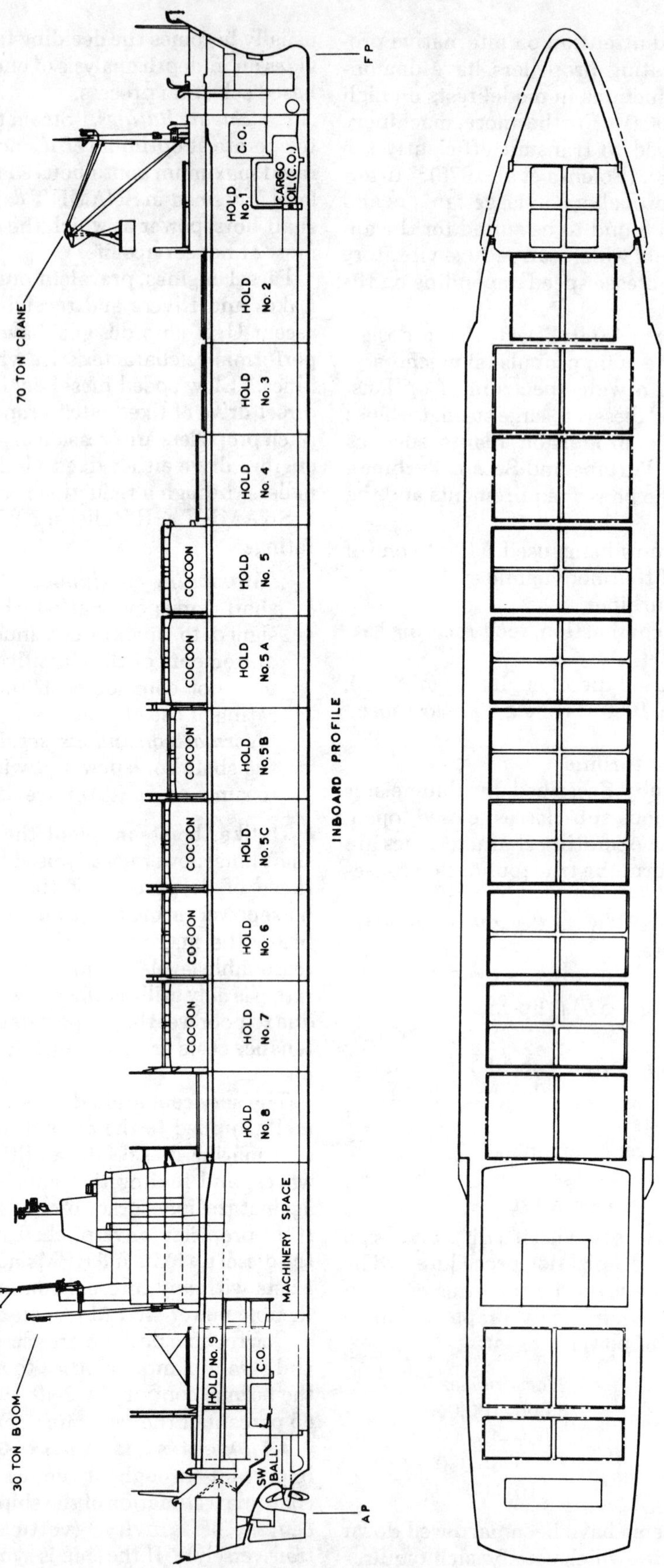

Fig. 29 Containership "C"

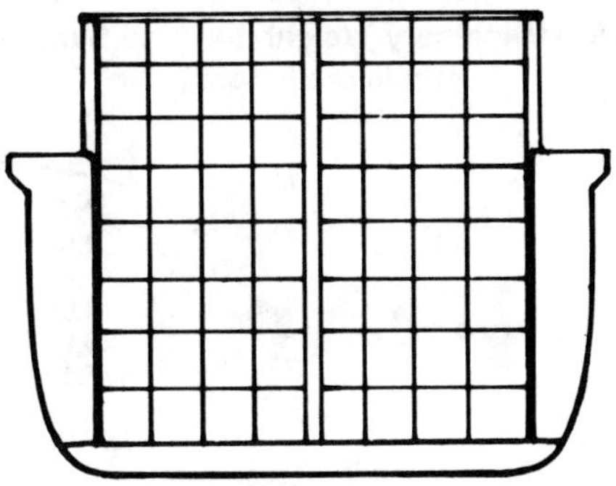

Fig. 29(a) Containership "C"

such as generators, heavy-lift cargo gear, or other heavy auxiliaries. If the designer fails to recognize such asymmetry, modify the transverse location or appropriately compensate for it, permanent list can result. The material take-offs which provide detailed weights are also the essential element of the cost estimate.

As described in Subsection 3.3a, preliminary weights are usually based on graphical presentations of previous designs. These estimates are adjusted for known major deviations from the norm; such as, refrigerated cargo holds with large insulation weights, unusual amounts of cargo gear, or special structural materials. Figs. 19 and 20 provide steel and outfit weights for containerships. Similar data for other classes of ships are provided in Chapter II. Other weight estimating techniques useful for preliminary weight estimates are described by Watson and Gilfillan (1977) and Lamb (1970).

An intermediate weight estimate is associated with a completed preliminary design. Weight take-offs are conducted on the basis of the design details normally available. This will usually include a general arrangement, midship section, shell expansion, and scantling plan. Various systems for classifying weights exist, but the most commonly used system in the United States is the MarAd standard classification (MarAd, 1962) described and illustrated by Moore (1967). Forms for use with the MarAd system are available from the Maritime Administration.

Once again, no substitute exists for experience and a complete file of data when a naval architect endeavors to estimate weights and centers for the numerous items not shown on the preliminary plans; such as furniture, joiner work, foundations, and hull piping. Despite the need to make approximations in many categories, the result is usually adequate and consistent with the accuracy of the preliminary design in general.

Weight summaries of three containerships are listed in Tables 6 through 9 for reference. Tables 10 and 11 provide principal characteristics for these ships, and arrangements are given in Fig. 27, 28, and 29. Ship A, Fig. 27, has fixed ballast; Ship B, Fig. 28, has no fixed ballast; Ship C, Fig. 29, is a combination containership with a high percentage of below deck refrigerated container capacity.

Final weight estimates are usually prepared by the contractor within 90 days after contract award. At that time, the complete contract design with guidance plans and diagrammatics are available for consultation. Supplementary sketches and drawings are also developed where additional information is still necessary.

The basic design function of weight and stability assessment does not end here. Ships tend to grow in the wrong direction during construction, i.e., to become heavier and with a higher center of gravity. There is a wide range in such trends, dependent upon how firm and well conceived the owner's requirements are in the first place. However, all too many cases have been reported in recent years of unwelcome operating restrictions following disappointing inclining experiments. They are usually caused by changes or developments under the shipbuilding contract which add unpredicted weights topside or can be the result of a poor initial weight estimate. Examples of unpredictable weight growths are: Overly thick underlayment for deck coverings (to smooth out wavy decks); use of stock materials heavier than needed or specified; standardization of masts and rigging on the heaviest of different lift capacities, and addition of superstructure quarters. As a means of controlling weight and center of gravity growth, the design agent should monitor working plans, check purchase orders, survey the ship under construction, and check light-ship weight at launching. In effect, a running check on weights and centers should be maintained. The shipbuilder, for his part, should be required to periodically report quantitative assessments of effect of changes under contract, or other departures from construction envisaged in the final weight estimate. The usual allowance in the original weight estimate for weight growth for cargo, cargo-passenger, and container ships is three percent and 0.3 m (1.0 ft) for *VCG* rise.

4.8 Capacities and Centers of Gravity of Deadweight Items. This subject is thoroughly covered by Owen and Niedermair (1967) and Moore (1967). It is necessary only to add a few thoughts on application to basic design.

a. Capacity and Deadweight. A modular cargo such as containers may cause some difficulty in providing sufficient underdeck capacity due to a slight interference which

Table 8—Machinery Weight Itemized Summaries
(Units are in tons)

Code	Item	CONTAINERSHIPS A	B	C
20-0	Main propulsion	39.9	50.8	51.7
1	Turbine drain and leak-off system	1.4	1.1	0.9
2	Main reduction gears	73.0	86.4	89.1
3	Main condenser	29.9	53.3	56.0
4	Main air ejector	3.1	1.2	1.0
5	Main circulating system	33.7	38.9	29.2
	Total Main Propulsion Units	**181.0**	**231.7**	**227.9**
21-0	Feed heaters	8.4	10.0	24.8
1	Feed and condensate system	30.5	31.3	20.6
	Total Feed and Condensate Equipment	**38.9**	**41.3**	**45.4**
22-0	Make-up feed	—	2.7	—
1	Contaminated system	3.7	6.1	8.8
2	Salt water evap. system	6.4	18.6	—
	Total Evaporator System	**10.1**	**27.4**	**8.8**
23-0	Shafting	63.4	65.6	92.4
1	Bearings and stern tube	36.1	33.6	28.0
2	Propellers	23.6	29.9	30.5
3	Miscellaneous shafting parts	7.6	4.4	3.3
4	Shafting and propeller spares	0.7	9.8	0.8
	Total Shafting and Propellers	**131.4**	**143.3**	**155.0**
24-0	Lub. oil system	30.5	35.6	17.1
1	Miscellaneous engine oil tanks	0.7	0.7	1.1
	Total Lubricating Oil System	**31.2**	**36.3**	**18.2**
25-0	Service compressed air serv. system	3.7	3.3	9.0
1	Starting air system	—	2.1	0.6
	Total Air Systems	**3.7**	**5.4**	**9.6**
26-0	Boilers	196.0	278.6	220.9
2	Soot blowers	1.2	1.2	4.3
3	Boiler draft system	11.0	20.5	15.2
4	Automatic combustion control	0.2	—	1.6
5	Stacks and uptakes	25.2	20.2	22.0
6	F.O. service system	8.1	10.1	6.6
	Total Boilers and F.O. System	**241.7**	**330.6**	**270.6**
27-0	Main steam piping	10.7	30.9	9.7
1	Aux. steam piping	10.4	47.5	12.0
2	Exhaust and escape piping	20.6	20.8	17.0
3	Steam drain system	5.0	9.5	2.8
4	Whistles	0.7	0.4	0.8
	Total Steam Piping	**47.4**	**109.1**	**42.3**
28-0	Access	34.5	21.9	25.6
1	Work shop	8.2	8.9	4.2
2	Lifting and handling gear	6.9	7.5	11.0
3	Machinery space ventilation	9.8	22.5	24.3
4	Machinery space fixtures	0.3	1.1	0.5
5	Spare parts	7.3	7.4	13.1
6	Miscellaneous instruments and gages	1.4	5.9	5.4
	Total Miscellaneous	**68.4**	**75.2**	**84.1**
29-0	Liquids in machinery (Groups 12 to 19)	7.7	32.8	40.8
1	Water (Groups 20–28)	45.8	86.0	108.4
2	Oil (Groups 20–28)	19.3	38.9	23.5
	Total Liquids in Machinery	**72.8**	**157.7**	**172.7**
	TOTAL MACHINERY	**826.6**	**1158.0**	**1034.6**

Table 9—Light Ship Weight Summaries, Tons

Item	CONTAINERSHIPS A	B	C
Steel	4557	8624	9901
Outfit	1739	2383	3701
Machinery	827	1158	1035
Subtotal	**7123**	**12165**	**14637**
*Margin	76	420	187
Fixed Ballast	3329	—	—
Light Ship	**10528**	**12585**	**14824**

* Margin adjusted to agree with ship inclining results. Margin for weight estimates may vary between 3–6 percent of subtotal.

would eliminate two or more containers. The freedom to carry containers on deck tends to offset this underdeck capacity limitation. Recently there has been a move to accommodate an increasing percentage of the containers below deck in order to provide the maximum security and protection from wind, weather, and waves. Further, the cell guide structure used for underdeck stowage provides for much simpler and more rapid container handling since the time consuming deck lashing problem is reduced. For general cargo ships earning ability is a function of capacity. Therefore early reliable capacity estimates are extremely important.

Dry bulk carriers and certain other ships use deadweight as the measure of earning ability. Owners are always delighted when the deadweight survey reveals that the final actual deadweight capacity exceeds the design requirements. Thus, the light ship estimate must be accomplished with care in the contract design stage and conservatively approximated (with adequate margins) during the preliminary design.

b. Centers of Gravity of Deadweight Items. Aside from their relationship to ship operating revenue, capacity calculations include locating centers of all spaces containing significant deadweight items. The weights and centers of these items are indispensable to stability and trim studies.

An operator may wish to adopt an operating stability standard determined from densities of different commodities and their distribution within the ship representing anticipated loadings in service. This is usually the case with container or barge carrying ships where, because of the high degree of selectivity of stowage locations with these types and the usual practice of weighing each container, it is practical in some cases to use a cargo center more favorable to stability than the homogeneous center. Furthermore, such ships are usually designed for on-deck stowage of several tiers of containers or lighters, which calls for careful checking of the cargo center of gravity for use in stability calculations.

c. Trim. The position of the longitudinal center of gravity is a dominant factor in the final trim of a ship. For general purpose, dry cargo ships, expected to carry a full cubic load of cargo, it is difficult to correct adverse trim by shifting *LCB,* since the *LCG* of cargo moves along with the center of buoyancy. It is easier to improve trim by adding to or subtracting from forecastle deck extensions over No. 1 and No. 2 holds or poop deck extensions forward. This is less of a problem on dry bulk carriers and tankers, as space usually exists within the hull to adjust *LCG* of cargo without disturbing *LCB.*

In the case of containerships and barge carriers, the operator once again has some flexibility in the longitudinal location of cargo. However, a design which has the minimum number of operational limitations is clearly preferable because the owner already must consider other factors; such as, shipper preference for below deck stowage, overstow problems due to varying port destinations, and simply late arrival of cargo at the ship.

Trim and adequate draft for ocean crossings, when in ballast (no cargo), is an important consideration for all types of cargo ships. The problem was especially serious during World War II, when return trips were usually in ballast, and in many cases there was insufficient tankage to provide proper draft for good seakeeping. As a result of operating experience at that time, the target for satisfactory ballast condition of general dry cargo ships was established as a draft corresponding to three and a half percent of *Lpp* forward and four and a half percent of *Lpp* aft. For bulk carriers, it is considered good ship operation to have a ballast capacity corresponding to at least 40 percent of the deadweight available to be used in case of bad weather. The 1973 International Convention for the Prevention of Marine Pollution as modified by the Tanker Safety and Pollution Prevention Protocols developed in 1978 requires that crude tankers over 20,000 dwt and product tankers over 30,000 dwt have sufficient segregated ballast capacity to provide a mean draft equal to 2 m (6.5 ft) + $0.02L$ with propeller immersion and a trim not exceeding $0.015L$. As noted previously, the large wing tanks on containerships generally provide ample segregated ballast capacity therein or in double bottom tanks normally carrying oil fuel now carried in the wing tanks.

d. Stability and Trim Sheet. Operating stability and trim for various loading conditions anticipated in service are recorded as shown in Fig. 30. This figure is a convenient summary of the intact condition of the combined effects of hull form, arrangements, free surface, transverse $\overline{GM}$ available; and weight and centers of light ship, cargo, fuel oil, water, stores, and so on.

By comparing the $\overline{GM}$ available from trim sheets against $\overline{GM}$ required to sustain damage, adequacy of the ship's stability characteristics is determined over the operating range of drafts. This test is valid for typical cargo and passenger ships, as these vessels inherently have forms producing large righting moments under the most severe rolling angles likely to be encountered in service (Moore, 1967). On the other hand, fishing boats with low freeboards, such as tuna clippers, have an unfortunate history of many losses due to capsizing. They are therefore in a totally different category in which dynamic stability requires careful investigation.

Tugs require still other factors to be considered such as the towline force which can exert a large overturning moment. In general the U. S. Coast Guard wind heel criteria have been generally accepted on a worldwide basis as being reasonable and providing for safe operation without jeopardizing the design. Garske et al (1973) provide an excellent overview of various stability criteria for various ship types.

4.9 Damaged Stability. In basic design, considerations of subdivision and damaged stability are becoming increasingly important, as a strong body of opinion is urging international adoption of a minimum of one-compartment standard for cargo ships through the deliberations of IMCO. This was proposed by the U. S. at the International SOLAS Conference of 1960, but was not adopted. Subsequent ef-

(continued on page 44)

Table 10—General Characteristics of Containerships

Item	A Fixed Ballast		B No Fixed Ballast		C Comb. Reefer	
Length Overall (LOA) m, (ft-in.)	185.9	(610-0)	219.6	(720-5)	247.9	(813-3)
Length Between Perpendiculars (Lpp), m (ft-in.)	177.1	(581-0)	206.3	(677-0)	234.4	(769-0)
Beam (B)	23.8	(78-0)	28.9	(95-0)	27.4	(90-0)
Depth to Strength Deck (D_S), m(ft-in.)	16.6	(54-6)	16.5	(54-0)	16.2	(53-0)
Draft (Maximum) Molded (T), m (ft.-in.)	9.6	(31-7)	10.4	(34-0)	10.1	(33-0)
Displacement, tons	22,253	(24,848)	38,881	(38,256)	43,036	(42,344)
Deadweight, tons	14,554	(14,320)	26,089	(25,670)	27,970	(27,520)
Deadweight/Displacement Ratio		0.579		0.682		0.638
Ratios						
Lpp/D_S		10.66		12.54		14.50
T/D_S		0.579		0.629		0.623
B/T		2.47		2.79		2.72
Lpp/B		7.45		7.13		8.54
Form						
Length for Coefficients, m (ft-in.)	177.1	(581-0)	206.3	(677-0)	234.4	(769-0)
Draft, Design, m (ft-in.)	8.2	(27-0)	9.5	(31-0)	8.8	(29-0)
Displacement, Molded tons (Long Tons)	22,300	(21,940)	34,000	(33,450)	36,790	(36,200)
Block coefficient		0.628		0.587		0.631
Prismatic coefficient		0.644		0.604		0.656
Maximum section coefficient		0.975		0.973		0.963
Lpp/(Vol)$^{1/3}$		6.34		6.42		7.11
Waterplane coefficient		0.740		0.722		0.760
Wetted surface, total, m^3 (ft^3)	5,180	(55,800)	6,970	(75,000)	8,320	(89,600)
$\overline{KM}$, m (ft)	9.9	(32-6½)	12.0	(39-5)	12.0	(39-4)
Moment to trim 1 cm, ton-m		782		465		624
Tons/cm		24.0		44.4		50.0
Tonnage						
Gross, U.S.		17,902		23,760		30,990
Net, U.S.		12,901		16,094		25,196

Table 11—General Characteristics of Containerships

Item	A Fixed Ballast			B No Fixed Ballast			C Comb. Reefer		
Weights(tons) and Centers(m)	*Wt tons*	*VCG meters*	*LCG meters*	*Wt. tons*	*VCG meters*	*LCG meters*	*Wt. tons*	*VCG meters*	*LCG meters*
Net steel	4557	9.0	94.6	8624	9.8	106.8	9901	10.5	127.1
Outfit	1739	13.9	94.3	2383	14.2	111.7	3701	14.6	134.2
Machinery	827	9.0	151.5	1158	8.4	174.2	1035	8.8	133.3
Margin	76	(+.1)	—	420	(+0.3)	—	187	(+0.3)	—
Fixed ballast	3329	2.8	80.9	—	—	—	—	—	—
Light ship, total	10528	7.9	94.6	12585	10.8	114.2	14824	11.7	133.3
Miscellaneous deadweight	—	—	—	152	11.6	189.8	250	4.4	158.3
Passengers, crew and stores	120	15.8	135.9	85	15.3	179.8	40	18.9	182.9
Fuel oil	2265	3.6	80.9	6727	8.0	93.0	4641	7.0	83.6
Fresh water	220	8.0	141.5	588	4.5	173.4	325	10.2	156.3
Salt water ballast	755	0.8	97.1	2470	1.0	97.9	132	0.9	176.1
Stabilizer tank	—	—		163	9.3	126.1	232	0.9	185.8
Container cargo	10960	11.9	90.2	15485	14.1	103.7	21900	12.3	114.0
Deadweight, total	14320	10.0	90.3	25670	11.0	102.8	27520	11.3	110.8
Full load, total	24848	9.2	92.8	38256	10.9	106.6	42344	11.4	118.6
Stability									
$\overline{GM}$, light ship, m, (ft.)		3.8	(12.5)		5.5	(18.2)		3.8	(12.6)
GM, loaded, m, (ft.)		0.5	(1.7)		0.9	(2.9)		0.4	(1.5)
Capacities									
No. of passengers		—				—		12	
No. of crew accom.		42			45			37	
20 Equivalent containers (above deck)		316			590			526	
20 Equivalent containers (below deck)		612			746			1186	
Machinery Particulars									
Type of machinery		steam			steam			steam	
SHP, maximum and RPM ABS		17500/97			32000/110			28500/106.5	
Number of boilers		2			2			2	
Steam conditions, kPa/°C (PSIG/°F)		6000/513° (870/955°)			6000/576° (870/960°)			6000/513° (870/955°)	
Speed									
On trial at 80 percent of normal SHP at maximum draft, knots		20.1			22.8			20.7	

FORM MA-335 (8-25-53)

U.S. DEPARTMENT OF COMMERCE
MARITIME ADMINISTRATION

STABILITY AND TRIM CALCULATIONS

SHIP: CONTAINERSHIP "A" | DATE: | BY: | SHEET:

CONDITION: FULL LOAD DEPARTURE

CONDITION	ITEM	DISP AND WEIGHT TONS	C.G. ABOVE BASE: LEVER FT.	C.G. ABOVE BASE: VERT. MOM. FT. TONS	C.G. FROM F.P.: FEET AFT	AFTER MOMENTS FT. TONS	FEET FOR'D.	FORWARD MOMENTS FT. TONS	FREE SURFACE
	LIGHT SHIP	10,528	26.2		315.4				
CONTAINERS ABOVE DECK	CONTAINER TIER 8	1,264	72.1		302.2				
" " "	7	1,264	63.6		302.2				
CONTAINERS BELOW DECK	6	1,392	51.8		296.8				
" " "	5	1,392	43.3		296.8				
" " "	4	1,296	34.9		308.5				
" " "	3	1,600	26.0		289.6				
" " "	2	1,536	17.5		283.8				
" " "	1	1,216	9.0		290.3				
" " "	TOTAL CONTAINER CAP.	10.960	39.1		295.8				
	FUEL OIL TANKS	2,265	11.9		265.4				9,270
	S.W. BALLAST TANKS	755	9.5		318.5				12,856
	FRESH WATER TANKS	220	26.4		464.3				1,042
	STORES	100	53.1		437.7				
	CREW & EFFECTS	20	47.0		486.2				
TOTAL - SHIP IN CONDITION		24,848	30.2		304.3				23,168

MEAN S.W. DRAFT	29.88 Feet	LCG	304.3 Feet	LCF	318.72 Feet
KM	32.84 Feet	LCB	300.6 Feet	DRAFT FWD.	28.38 Feet
KG	30.2 Feet	TRIM LEVER FWD, AFT	3.7 Feet	DRAFT AFT	30.91 Feet
GM	2.64 Feet	MOMENT TO TRIM 1"	2,750 FT Tons		
CORR. FOR F.S.	0.93 Feet	TRIM IN INCHES FWD, AFT	33.7 Inches		
GM AVAILABLE	1.71 Feet				
GM REQUIRED	1.09 Feet				

COMM-DC 40552

Fig. 30 Trim and stability of containership "A"

(*continued from page 41*)
forts by the U. S. Coast Guard to gain national agreement through rulemaking have likewise been resisted. At present, the U. S. Coast Guard has no subdivision regulations for cargo ships other than SOLAS 1974.

The Load Line Convention specifies a one-compartment flooding criterion in a condition of equilibrium with positive metacentric height for large bulk carriers, as a condition for reduced freeboard assignment. The probabilistic approach to damaged stability has been developed for application to passenger vessels. This subject is more fully covered by Robertson et al (1974). In addition, since 1936, U. S. cargo ships financed under terms of the Merchant Marine Act of 1936, as amended, have been built to a minimum of one-compartment standard of subdivision and damaged stability, hence many U. S. flag merchant ships must meet this standard.

It bears repeating that much can be done toward increased safety by eliminating possibilities of unsymmetrical flooding and flooding over flats. Whereas valve-controlled cross connections and reefer blowout plugs work in theory to equalize heeling moments, in practice there is always the danger of malfunction of such devices—which are never used until disaster strikes.

An equally important consideration is avoidance of oily ballast operation through basic design. If sufficient attention is paid in the early design stage, there is no reason why the superior standard of safety cannot be obtained at reasonable cost on ships with a modest cruising radius requirement.

Recent U. S. Coast Guard Regulations and IMCO conventions have extensively stiffened the design requirements relating to environmental matters. Chapter XI elaborates on these requirements which must be considered in the very early stages of the design process.

a. Criteria for Cargo Ships Generally. Robertson (1967) and Garske et al (1973)) thoroughly treat the subject of damaged stability, with reference to methods of calculation as applied to merchant ships. Criteria for its application to cargo ships, including containerships and bulk carriers, as developed by MarAd for ships financed under terms of the Merchant Marine Act of 1936 are that the ships shall be capable of surviving the final stage of flooding of any one compartment at any time during any voyage, generally in accordance with the following:

1. The assumed longitudinal extent of damage shall be equal to 10 ft plus three percent of the subdivision length of the vessel.
2. The transverse extent shall be equal to one-fifth of the maximum beam, measured inboard from the ship's side, at the level of the deepest subdivision load line.
3. The vertical extent shall be from baseline upward without limit.

Where any damage of a lesser extent than provided in 1., 2., and 3. above would result in a more severe condition regarding heel or loss of metacentric height, such damage shall also be considered in the calculations.

4. Permeabilities shall be as follows:

	Permeability
Assigned to cargo or stores	60
Occupied by machinery	80
Intended for liquids	0 or 95*

* Whichever results in the more severe requirement

5. The final conditions of the ship, after damage and after equalization measures have been taken, shall be as follows:

• The residual metacentric height shall be positive.
• In the case of unsymmetrical flooding, heel should be not greater than 15 degrees, nor shall it immerse the margin line.

Compliance with the above is usually demonstrated by the preparation of calculations for nine standard operating conditions, i.e., full cargo, half cargo, and no cargo; each cargo condition with full, half, and 10 percent consumables. General cargo is assumed at the homogeneous center of the volume used in determining the bale capacity of the vessel.

Where the cargo deep tanks are convertible for dry or liquid cargo, these are assumed to be loaded with dry cargo, and where deep tanks are provided exclusively for the carriage of liquid cargos, the full cargo, half cargo, and no cargo conditions shall include liquid cargos in the same proportions.

The proportions and arrangements of the vessel shall be such that the stability characteristics, when loaded as above, can be maintained in accordance with the damaged stability requirements for one-compartment flooding.

In very exceptional cases, where compliance with the above requirements would unduly penalize the design, and no other practical solution is available, MarAd will consider corrective measures; such as cross flooding fittings or operational restrictions for cargo and/or liquids, if it can be demonstrated that these remedies are practical from an operational viewpoint. In these exceptional cases, the operational restrictions would be subject to specific approval by the Administration, and the stability booklet must include pertinent instructions as required to maintain satisfactory operational stability.

b. Criteria for Container and Barge-Carrying Ships. These types pose special problems relative to permeability (μ) assumptions due to possible large voids outside the containers and barges and the probable slow rate of leakage into the sealed units. A solution often used is based on calculation of an average permeability of the compartment, taking $\mu = 0.95$ for voids and $\mu = 0.60$ for container or barge volume.

Recent studies have indicated that the permeability of loaded containers would be more nearly 0.74 and this is under consideration at IMCO as are general damage stability requirements for cargo ships. For barge carriers the same permeability of cargo spaces should be used for the contents of a barge.

c. Tankers and LNG Carriers. Stability criteria exist for these types of vessels in the U. S. Coast Guard requirements. Other requirements have been developed in cooperation with other maritime nations at IMCO. These are more fully described in Chapter XI.

Section 5
Summation and Adjustment

5.1 General. After the naval architect completes one preliminary design loop around the design spiral shown in Fig. 1, decisions must be made regarding insufficient or excess deadweight, capacity, available metacentric height, $\overline{GM}$, trim, freeboard, and power.

As each successive loop is completed the deviations between characteristics obtained and desired must be evaluated. To a large extent, these characteristics are interdependent. For example, if displacement must be increased to reach design deadweight, more power and fuel oil will be required to maintain design speed and endurance. Furthermore, if the vessel is draft-limited, the added displacement can only come from increased B, L, or C_B, or a combination thereof, which introduces cumulative effects on light-ship weight and power. As a general rule, it is often necessary to overcorrect deficiencies to offset cumulative adverse effects of the correction.

Corrections should not be attempted if the deviation is within the probable error of the approximations used for estimating capacities, centers, and powering. The best procedure at this point, if more accuracy is desired, is to go through a new cycle with more refined and detailed design work.

If deadweight and capacities are both too low, then a small linear increase of L, B, D, and T is indicated. On the other hand, if deadweight is too low and capacities too high, L, B, and D may be adjusted slightly downward, while draft is increased considerably.

Excessive trim usually requires rearrangement of spaces rather than shifting LCB as indicated previously, although with slow speed ships, LCB can be a distance from the optimum location without serious influence on resistance.

Insufficient or excessive $\overline{GM}$, in most cases, should be taken care of by a change in beam accompanied by an adjustment in depth to maintain capacity, and a variation in draft to maintain displacement. It should be noted that an increase in beam per se will affect light ship appreciably more than a change in depth, and that an increase in beam-to-draft ratio will result in higher $\overline{GM}$ required for damaged stability. When the vessel is beam-limited, a deficiency in stability may be offset by aluminum superstructures and higher strength steel kingposts as well as other lightweight topside outfit, such as aluminum boats. Use of extreme V-sections to increase waterplane inertia (usually at the expense of increased SHP) may be considered. There are expensive remedies which should be tested before adoption as to profitability of investment.

Probably the most unsettling discovery is an adverse model basin report in which the required EHP and (or) SHP is higher than predicted by Taylor's Standard Series or Series 60. Unless there is some peculiarity of the design inherently penalizing its hydrodynamic performance, there is no alternative in such cases but to study the model for clues as to the causes of the deficiencies and correct them or find a more suitable parent form and start over if correction is not possible.

It may be debatable whether cost estimating is basic design work but, in any event, a cost check is of overriding interest to the prospective owners. As soon as all adjustments have been made at the close of each design loop, the weights of steel and outfit and amount of power should be priced and combined with estimated shipyard overhead, labor, and profit to indicate a fair and reasonable bid price. To this figure must be added allowances for changes under contract, owner's engineering, inspection, and plan approval (See Chapter XV).

If the overall price is too high, then the owner is confronted with a choice between:

- abandoning the project,
- raising more capital,
- reducing the size of the project,
- reducing quality of outfitting.

With a good functional design and value-engineered specification for a cargo ship, there is usually not much opportunity for significant cost reduction except by reducing certain characteristics such as speed, cubic capacity, or deadweight, or eliminating major special features such as bow thrusters, constant-tension mooring winches, heavy lift gear, 'tween deck hydraulic hatch covers, and the like. However, it is the fond hope (sometimes realized) of every creative naval architect to develop new concepts of ship arrangements or improve conventional designs to such an extent that the economic feasibility of the proposed vessel will be assured even in the face of rising costs.

Section 6
Design Philosophy

6.1 General. In the field of ship design, for any type of ship, there are usually several approaches to the accomplishment of individual ship operating functions, sometimes without much difference in their relative merits. Experienced managements will hold tenaciously to particular concepts. For example, conservative ship owners will not install newly developed equipment unless and until it has been proved successful by more venturesome operators in service at sea. A partial list of basic design features subject to controversy of this type follows:

- Single vs twin boilers,
- machinery amidships vs part way aft vs all the way aft

vs forward of amidship,

- compact and crowded machinery space for large cargo cubic vs enlarged machinery space for better accessibility with attendant loss in cargo cubic,
- triple hatches vs twin hatches vs single hatch (transversely),
- V-shaped vs U-shaped bow sections,
- bulbous vs conventional vs cylindrical bows,
- higher strength steel vs ordinary strength steel vs aluminum alloys,
- minimum vs maximum cargo gear vs no cargo gear,
- diesel vs steam propulsion vs gas turbine vs combinations,
- single screw vs twin screw vs contrarotating propellers,
- break bulk vs unitized cargo handling,
- small vs large power margins for speed maintenance and engine maintenance,
- utility of bow thrusters,
- utility of antirolling tanks, or fin stabilizers,
- marginal operating stability with maximum cargo cubic vs improved operating stability and less cargo cubic,
- refrigerated container holds vs self contained container refrigeration,
- shallow vs deep containerships,
- unmanned vs 1-man vs 2-man engine rooms,
- type of LNG containment system,
- sluicing vs separate piping system for tanker discharge,
- bridge forward, midships or aft,
- deep holds vs 'tweendecks,
- heavy fuel vs light fuel, and
- conventional vs ducted propellers in heavily loaded applications.

Should the naval architect be convinced of the unsoundness of his client's partiality, he should nevertheless refrain from opposition until he can present a substantiating economic analysis containing effects of all the elements considered by the shipowner in the situation (Benford, 1963), (Benford, 1967a).

Consistently, the designer, striving for novel, more efficient, ship arrangements in unorthodox configurations, and relying on the equipment as yet undeveloped for marine use, needs to be doubly sure of the economic potential of his ideas before approaching his client.

The naval architect is continually being challenged by demands for new vessel applications, new missions, new cargo forms; many being conceived by engineers and businessmen outside the traditional merchant shipping community. The vessel must often fit into an existing capital intensive system which places important constraints on the design. Features which may be optimum for a ship in an unconstrained system may not be optimum given the shoreside interface and other demands of the total system.

This is not to discourage new ideas—far from it—but to caution the designer to master his knowledge of the operating environment, and particularly that of his client's services, keeping an open mind and flexible stance, and staying within the bound of practicality by meeting the hard test of economic feasibility, with due regard to actual operating limitations.

Intuitive decisions have all too frequently dominated the most important aspects of basic design, sometimes with unfortunate consequences. In spite of the greatly increased understanding of the economics of ship design, there is still a large tendency to fall back on intuition. This must be strongly resisted by the designers of the future, who will be dealing with vessels of ever-increasing cost and complexity requiring more not less analytical evaluations to effect the proper decisions during design.

References

Aertssen, G. (1968), "Laboring of Ships in Rough Seas with Special Emphasis on the Fast Ships," Spring Meeting, SNAME.

Attwood, Edward L. and H. S. Pengelly (1960), *Theoretical Naval Architecture,* revised by A. J. Sims, Longmans, Green and Company, Ltd., London.

Baker, G. S. (1933), *Ship Design, Resistance and Screw Propulsion,* Vol. 1, *The Design of Ship Forms and Their Resistance,* Charles Birchall and Sons Ltd., Liverpool, England.

Baumeister, T. (1958), *Mechanical Engineers' Handbook,* McGraw-Hill Book Company.

Benford, H., K. C. Thornton, and E. B. Williams (1962), "Current Trends in the Design of Iron Ore Ships," SNAME.

Benford, H. (1962), "General Cargo Ship Economics and Design," University of Michigan, College of Engineering, Research Administration Report on Project 04465, August.

Benford, H. (1963), "Principles of Engineering Economy in Ship Design," *Trans.* SNAME.

Benford, H. (1967a), "Practical Application of Economics to Merchant Ship Design," *Marine Technology,* January.

Benford, H. (1967b), "On the Rational Selection of Ship Size," *Trans.* SNAME.

Benford, H. (1970), "Measures of Merit for Ship Design," *Marine Technology,* October.

Burrill, L. C. (1943–44), "Calculation of Marine Propeller Performance Characteristics," NECIES, Vol. 60, pp.

Chryssostomides, C. (1972), "Seakeeping Considerations in a Total Design Methodology," 9th Office of Naval Research Symposium on Naval Hydrodynamics, Paris.

Cleary, W. A. (1977), "Practical Uses of Sea Wave Calculations in Ship Design," Fifth Pan American Naval Engineering Congress.

Clement, E. and D. L. Blount (1963), "Resistance Tests of a Systematic Series of Planing Hull Forms," *Trans.* SNAME.

Collatz, G. and E. Seifert (1977), "Interactive Fairing of Ship Lines—A Procedure Developed for the Model Manufacture at the Hamburg Model Basin," First International Symposium on Computer-Aided Hull-Surface Definition, September SNAME.

Couch, R. B. and J. Moss (1966), "Application of Large Protruding Bulbs to Ships of High Block Coefficient," *Trans.* SNAME.

Cumming, R., W. Morgan, and R. Boswell (1972), "Highly Skewed Propellers," *Trans.* SNAME.

Cuthill, E. H. and F. H. Henderson (1964), "Description and Usage of GBRC1—General Bending Response Code 1," Naval Ship Research and Development Report 1925, October.

Danforth, Leon J. (1977), "Environmental Constraints on Drill Rig Configurations," *Marine Technology,* July.

Dillon, E. S. and E. V. Lewis, (1955), "Ships with Bulbous Bows in Smooth Water and in Waves," *Trans.* SNAME.

Dorman, W. J. (1966), "Combination Bulk Carriers," *Marine Technology,* October.

Dorman, W. J. and D. DeKoff (1971), "Characteristics of Recent Large Containership Designs," *Marine Technology,* October.

Durfee, J., J. Lisnyk, and E. N. Castrinakis (1976), "U. S. Offshore Industry—Potential for the Future," SNAME, Chesapeake Section, Symposium on Marine Resource Development in the Middle Atlantic States, October.

Eames, M. C. and T. G. Drummond (1976), "Concept Exploration—An Approach to Small Warship Design," *Trans.* RINA.

Ericksen, Stian (1972), "Optimizing Containerships and Their Terminals," Spring Meeting SNAME.

Evans, J. Harvey and D. Khoushy (1963), "Optimized Design of Midship Section Structure," *Trans.* SNAME.

Evans, J. Harvey (1959), "Basic Design Concepts," *Naval Engineers Journal,* November.

Everett, John L., A. C. Hax, V. A. Lewinson, and D. Nudds (1976), "Optimization of a Fleet of Large Tankers and Bulkers: A Linear Programming Approach," *Marine Technology,* January.

Ewing, J. A. (1967), "The Effect of Speed, Forebody Shape and Weight Distribution on Ship Motions," *Trans.* RINA.

Federal Power Commission (1977), "El Paso—Computerized Financial Documentation and Shipping Fleet Simulation," Bureau of Natural Gas, Systems Analysis Branch, May.

Femenia, Jose (1973), "Economic Comparisons of Various Marine Power Plants," *Trans.* SNAME.

First International Symposium on Computer-Aided Surface Definition, SNAME, 1977.

Barnhill, R. E., and R. F. Riesenfeld, "Surface Representation."

Hoffman, D., and T. Zielinski, "The Use of Conformal Mapping Techniques for Hull Surface Definition."

Nowacki, H., G. Creutz, and F. C. Munchmeyer, "Ship Lines Creation by Computer-Objectives, Methods and Results."

Rawat, P., "Mathematical Ship Lines—What Do They Mean."

Fisher, K. W. (1972), "Economic Optimization Procedures in Preliminary Ship Design," *Trans.* RINA.

Folenta, D. J., J. Motzer, F. X. Critelli (1976), "Design and Development of High Horsepower Marine Planetary Gears," Gas Turbine and Fluids Engineering Conference, ASME.

Garske, W., Jr., R. E. Johnson, and A. C. Landsburg (1973), "Trim Stability and Loading Manuals: A Survey of Needs, Uses, and Improvements," *Trans.* SNAME.

Gebhardt, John C., and R. B. Thompson (1976), "Implementing the U. S. Navy's Hull Definition Program in U. S. Shipyards," REAPS Technical Symposium, CADCOM Report 76-18.

Gertler, M. (1954), "A Reanalysis of the Original Test Data for the Taylor Standard Series," DTMB Report No. 806, March.

Giblon, Robert P. (1975), "Service Margin and Power Plant Selection," *STAR Symposium, SNAME.*

Gilfillan, A. W. (1966–1967), "Preliminary Design by Computer," *Trans.* IESS, Vol. 110.

Gilfillan, A. W. (1969), "The Economic Design of Bulk Cargo Carriers," *Trans.* RINA.

Goodrich, G. J. (1964), "The Influence of Freeboard on Wetness," Fifth Symposium on Naval Hydrodynamics.

Hadler, J. B., W. B. Morgan, and K. A. Meyer (1964), "Advanced Propeller Propulsion for High Powered Single Screw Ships," *Trans.* SNAME.

Hadler, J. B., and H. M. Cheng (1965), "Analysis of Experimental Wake Data in Way of Propeller Plane of Single and Twin Screw Ship Models," *Trans.* SNAME.

Hadler, J. B., and N. Hubble (1971), "Prediction of the Power Performance of the Series 62 Planing Hull Forms," *Trans.* SNAME.

Hadler, J. B., and T. H. Sarchin (1973), "Seakeeping Criteria and Specifications," *Seakeeping Symposium,* SNAME.

Hagen, A., and N. Hammer (1969), "Shipboard Noise and Vibration from a Habitability Viewpoint," *Marine Technology,* January.

Hammer, N. (1970), "Ship Vibration—Recent Experience and Prospects for the Future," N.Y. Met Section, SNAME, February.

Harrington, R. L., Editor (1971), *Marine Engineering,* SNAME, New York.

Hattori, Y., and Y. Matida (1977), "Some Problems in Practical Improvement of Mathematic Fairing," SNAME, First International Symposium on Computer-Aided Hull-Surface Definition, September.

Hawkins, Seth, and George Levine (1969), "Comments on Service Margins for Ships," Chesapeake Section, SNAME, May.

Henry, J. J., and H. J. Karsch (1966), "Container Ships," *Trans.* SNAME.

Inui, T. (1962), "Wave-making Resistance of Ships," *Trans.* SNAME.

Johnson, R. P., and H. P. Rumble (1965), "Weight, Cost and Design Characteristics of Tankers and Dry Cargo Ships," *Marine Technology,* April.

Kiss, Ronald K. (1972), "Aspects of Simplified Hull Forms—Past, Present and Future," *Marine Technology,* October.

Korvin-Kroukovsky, J. B. (1961), *Theory of Seakeeping,* SNAME.

Krappinger, Odo (1966),"Great Lakes Ore Carrier Economics and Preliminary Design," SNAME, Great Lakes and Great Rivers Section, May.

Kuo, C. (1966), "A Method of Analyzing Propeller Excited Vibratory Forces on a Model," DTMB Report No. 2266, December.

Lackenby, H. (1950), "On the Systematic Geometric Variation of Ship Forms," *Trans.* RINA.

Lamb, T. (1970), "Merchant Ship Weight Estimation," Society of Aeronautical Weight Engineers, May.

Lamb, T. (1970), "A Ship Design Procedure," *Marine Technology,* October.

Laskey, N. V., and G. T. R. Campbell (1966), "The Design of Hull Forms for Supertankers and Bulk Carriers," *Marine Technology,* January.

Lasky, Marc P., and John C. Daidola (1977), "Design Experience with Hull Form Definition During Pre-Detail Design," SNAME Proceedings, First International Symposium on Computer-Aided Hull Surface Definition.

Lewis, E. V. (1955), "Ship Speeds in Irregular Seas," *Trans.* SNAME.

Lewis, E. V. (1959), "Increasing the Sea Speed of Merchant Ships," *Trans.* SNAME.

Lewis, E. V. (1967), *Principles of Naval Architecture,* John P. Comstock, Editor, SNAME.

Lewis, Frank M. (1967), *Principles of Naval Architecture,* John P. Comstock, Editor, SNAME.

Lindblad, A. F. (1961), "On the Design of Lines for Merchant Ships," *Transactions of Chalmers University of Technology,* Gothenburg, Sweden, No. 240.

Lindgren, H., and A. Williams (1968). "Systematic Tests with Small, Fast Displacement Vessels," SNAME Spring Meeting.

Loukakis, T. A., and C. Chryssostomidis (1975), "Seakeeping Standard Series for Cruiser—Stern Ships," *Trans.* SNAME.

Mack-Forlist, D. M., and R. Hettena (1966),"An Economic Feasibility Study of United States Bulk Carriers," *Marine Technology,* April.

Mandel, P., and R. Leopold (1966), "Optimization Methods Applied to Ship Design," *Trans.* SNAME.

Mandel, Philip (1967), *Principles of Naval Architecture,* John P. Comstock, Editor, SNAME.

MarAd, "Classification of Weights—Standard Grouping for Merchant Ships," U. S. Department of Commerce, Maritime Administration, Superintendent of Documents, Clearinghouse for Federal Scientific and Technical Information, Springfield, VA, 22151, Revised July 1962.

McGoldrick, R. T., et al (1953), "Recent Developments in the Theory of Ship Vibration," Naval Ship Research and Development Report 739.

McGoldrick, R. T., and E. F. Noonan (1966), *Vibration Characteristics of Ship Hulls, Applied Mechanics Surveys,* Spartan Books.

Mentz, Paul B. (1975a), "What Price Speed?—In International Marine Transportation," SNAME, Chesapeake Section, September.

Mentz, Paul B. (1975b), "Water Transportation Productivity—A Selective Assessment," SNAME, Chesapeake Section, September.

Mercier, John (1973), "Index of Methodical Series Ship Model Resistance Tests," SNAME T & R Bulletin No. 1-31.

Miller, Dave S. (1970), "The Economics of the Containership Subsystem," *Marine Technology,* April.

Moore, Charles S. (1967), *Principles of Naval Architecture,* John P. Comstock, Editor, SNAME.

Muntjewerf, J. J. (1970), "Methodical Series Experiments on Cylindrical Bows," *Trans.* RINA.

Murphy, R. D., D. J. Sabat, and R. J. Taylor (1965), "Least Cost Ship Characteristics by Computer Techniques," *Marine Technology,* April.

Nachtsheim, John J. (1972), "Ship System Economics—Less the Ship," SNAME Spring Meeting.

Narita, H., Y. Kunitake, and H. Yagi (1974), "Application and Development of a Large Ducted Propeller for the 280,000 DWT Tanker M. S. *Thorsage," Trans.* SNAME.

Noonan, E. F. (1971), "Design Considerations for Shipboard Vibrations," *Marine Technology,* January.

Noonan, E. F. (1976), "An Assessment of Current Shipboard Vibration Technology," Ship Structures Symposium 1975, SNAME.

Nowacki, Horst (1970), "Tanker Preliminary Design—An Optimization Problem with Constraints," *Trans.* SNAME.

Ochi, M. K. (1971), "A Method to Estimate Slamming Characteristics for Ship Design," *Marine Technology,* April.

Ochi, M. K. (1973), "Prediction of Slamming Characteristics and Hull Responses for Ship Design," *Trans.* SNAME.

Ochi, M. K., and S. L. Bales (1977), "Effect of Various Spectral Formulations in Predicting Responses of Marine Vehicles and Ocean Structures," Offshore Technology Conference.

Oosterveld, M. W. C., and P. van Oossomen (1975), "Recent Results of Tests in the NSMB Depressurized Towing Tank," *STAR Symposium,* SNAME.

Owen, W. Selkirk and John C. Niedermair (1967), *Principles of Naval Architecture,* John P. Comstock, Editor, SNAME.

Parson, M. G. (1975), "Optimization Methods for Use in Computer-Aided Ship Design," *STAR Symposium,* SNAME.

Pien, P. C. (1964), "The Application of Wave Making Resistance Theory to the Design of Ship Hulls with Low Total Resistance," Fifth Symposium of Naval Hydrodynamics, Bergen, Norway. Also, DTMB Report No. 2144.

Reed, E. (1971), "The Design of Ships to Avoid Propeller Excited Vibrations," *Trans.* SNAME.

Reed, E. (1973), "Acceptable Levels of Vibration on Ships," *Marine Technology,* April.

Ridgeley-Nevitt, C. (1963), "The Development of Parent Hulls for a High Displacement Length Series of Trawler Forms," SNAME.

Ridgeley-Nevitt, C. (1967), "The Resistance of a High Displacement Length Ratio Trawler Series," *Trans.* SNAME.

Robertson, James B. Jr. (1967), *Principles of Naval Architecture,* John P. Comstock, Editor, SNAME.

Robertson, J. B., G. W. Nickum, R. I. Price, and E. Middleton (1974), "The New Equivalent International Regulations on Subdivision and Stability of Passenger Ships," *Trans.* SNAME.

Roseman, Donald P., Morton Gertler, and Robert E. Kohl (1974), "Characteristics of Bulk Products Carriers for Restricted Draft Service," *Trans.* SNAME.

Sato, S. (1967), "Effects of Principal Dimensions on Weight and Cost of Large Ships," N. Y. Met Section, SNAME, February.

Saunders, H. E. (1957), *Hydrodynamics in Ship Design, Vols. I and II,* SNAME, New York.

Saunders, H. E. (1965), *Hydrodynamics in Ship Design, Vol. III,* SNAME, New York.

Schmitt, Ivan W. (1976), "The Trans-Alaska Gas Project Site Selection and System Optimization," *Marine Technology,* January.

Sharp, G. G. (1947), "Some Considerations in the Design of Modern Ships," *Trans.* SNAME.

SNAME (1958), "Index to Model and Expanded Resistance Data Sheets," No. 1-175, T & R Bulletin No. 1-14, and associated sheets.

SNAME (1963), "How to Use SNAME Small Craft Data Sheets for Design and Resistance Prediction," T & R Bulletin No. 1-23, and associated sheets.

SNAME (1973), "Marine Steam Power Plan Heat Balance Practices," T & R Bulletin 3-11.

SNAME (1975), "Proposed Procedures for Determining Ship Controllability Requirements and Capabilities," Panel H-10, *STAR Symposium.*

SNAME (1975), "Marine Diesel Power Plant Practices," T & R Bulletin 3-27.

SNAME (1976), "Marine Gas Turbine Power Plant Performance Practices," T & R Bulletin No. 3-28.

Steele, T. W. (1968), "Propulsion Machinery Considerations for Contrarotating Propeller Systems," N.Y. Met. Section, SNAME.

Strom-Tejsen, Jorgen, and Pao Pien (1968), "A Hull Form Design Procedure for High Speed Displacement Ships," Spring Meeting, SNAME.

Strom-Tejsen, J. (1972), "A Comparison of Contra-Rotating Propellers with Other Propulsion Systems: Results of Model Experiments," *Marine Technology,* January.

Stuntz, George (1963), "The Use of Model Basins in the Design Process," Chesapeake Section, SNAME, December.

Swaan, W. A. (1959), "Amidships Bending Moments for Ships in Waves," International Shipbuilding Progress, September.

Taggart, R. (1955), "Mathematical Fairing of Ship's Lines for Mold Loft Layout," *Naval Engineers Journal,* May.

Taggart, R., and E. F. Magnusson (1967), "Mathematical Derivation of Destroyer-Type Hulls," *Naval Engineers Journal,* February.

Taylor, D. W. (1915), "Calculations for Ships' Forms and the Light Thrown by Model Experiments Upon Resistance, Propulsion and Rolling of Ships," *Transactions of the International Engineering Congress.*

Taylor, D. W. (1943), *The Speed and Power of Ships,* 3rd Edition, U. S. Government Printing Office.

Thomas, W. DuBarry, and A. Schwendtner (1971), "LNG Carriers: The Current State of the Art," *Trans.* SNAME.

Todd, F. H. (1961), *Ship Hull Vibration,* Edward Arnold Ltd., London.

Todd, F. H. (1953), "Some Further Experiments on Single Screw Merchant Ship Forms—Series 60," *Trans.* SNAME. Also, DTMB Report No. 1712, "Series 60 Methodical Experiments with Models of Single Screw Merchant Ships," July.

Todd, F. H. (1967), *Principles of Naval Architecture,* John P. Comstock, Editor, SNAME.

Troost, L. (1957), "A Simplified Method for Preliminary Powering of Single Screw Merchant Ships," *Trans.* SNAME.

Valentine, D. T., and F. Dashnaw (1975), "Highly Skewed Propellers for San Clemente Class, Ore, Bulk, Oil Carrier—Design Considerations, Model and Full Scale Evaluation," *STAR Symposium,* SNAME.

Vossers, G., W. A. Swaan, and H. Rijken (1960), "Experiments with Series 60 Models in Waves," *Trans.* SNAME.

Watson, D. G. M., and A. W. Gilfallan (1977), "Some Ship Design Methods," *The Naval Architect,* July.

CHAPTER II

Walter H. Michel

Mission Impact on Vessel Design

Section 1
Introduction

1.1 General. The design procedures in going from the owner's basic concept and requirements through to the development of final major particulars and contract plans have been set forth in Chapter I, illustrating the use of basic principles of naval architecture in association with the specific characteristics and features that must be incorporated in the design if the vessel is to meet its mission requirements. It is the purpose of the present chapter to identify the major characteristics and features for specific types of commercial vessels, industrial vessels, and service vessels, with particular emphasis on the evolutionary developments that have occurred for each type. This material will serve to highlight the features that appear most important today and suggest the trends for the future.

Through the course of this chapter, representative data will be given to illustrate the relationships between size and weight (displacement to light ship or to deadweight capacity), proportions to speed and power, etc. appropriate to the particular type of vessel, from which one can project a first set of general characteristics of a new design for the particular mission under consideration. References to other sources of similar information will also be given. However, essential to producing a satisfactory design is the designer's ability to relate the mission requirements to those of comparative existing vessels, to incorporate those modern features proven to be effective, to discard those features that are outmoded, and to be alert to the possibility of adopting new ideas.

In this regard, it is important that the designer develop a sizable data bank of particulars and descriptions of other vessels, gleaned from technical publications, periodicals, or company brochures, from which he can select the most representative as a basis for the development of his own design with its own special requirements. This approach parallels the course that the owner normally takes in specifying the particular mission requirements that he wants, insofar as they reflect changes in capacity, power, behavior, or special features from an existing norm with which the owner is familiar. An early rapport between the owner and designer is more probable with this approach than if the designer has convoluted too many design spirals too soon in the design process.

Furthermore, unless one reviews the *state of the art* as exemplified by recent practice, one can overlook many important and necessary features that might otherwise not be apparent from the usual parametric relationships used in computer optimization of the major vessel characteristics. Some examples:

- High-speed containerships are of two types; i.e., shallow depth designs with as many as five tiers of on-deck containers and which requires an unusually high forecastle or forward bridge houses, as a measure to protect deck cargo from damage by green seas over the bow; and deep designs with most containers below deck and unusually high freeboard through the length.

- Ships that have their main deck almost entirely encumbered by cargo, or by extensive hatches and hatch gear, are benefited by having underdeck passageways running the greater length of the ship for personnel movement, piping, and electric cable runs.

- Service vessels such as tugs and supply vessels normally have minimum freeboards, not necessarily for the expected reason of easier line handling but rather to keep gross register tonnage at a low figure to escape more severe regulatory requirements on crew and equipment. For larger vessels engaged in full ocean service, where sufficiently low gross tonnage cannot be achieved, it is not surprising that freeboards are proportionately greater to insure greater stability and seakeeping.

Recognition and application of such special features for particular types of vessel are the substance of the art of ship design, the need for which has not diminished even though an increasingly greater scientific approach is demanded to establish the viability of the overall design.

This chapter focuses on such special features and trends for the typical commercial, industrial, or service vessel, noting that the more detailed and systematic considerations involved in the development of a design are covered in subsequent chapters, as well as in numerous standard references. However, in the section on offshore drilling units, design criteria and particulars are elaborated upon insofar as they differ significantly from the principal requirements of seafaring vessels and have little prior documentation.

Section 2
Commercial Ships

2.1 Introduction. Over the past quarter of a century there have been significant changes in the size, appearance and general characteristics of ships engaged in international commerce. The advance of design and construction technology through this period encouraged the development of ship types that could satisfy the growing economic demands for ships of greater capacity, propelled at faster speeds, and having the ability for faster turn-around in port, with the ultimate goal of increasing ton-miles per day at maximum profit.

Tankers and other bulk carriers have grown dramatically through this period, in sheer size and in installed power. Specialized carriers such as containerships, barge carriers, ro/ro's and liquefied gas carriers were developed and have already become dominant in size and performance (if not in numbers) over the long-established general cargo ship. General cargo ships themselves have experienced major changes, most notably in the incorporation of hatches that practically span the entire holds below and with fast-handling hatch covers and high capacity cargo gear, all directed toward the rapid handling of cargo.

The classical midship deckhouse has practically disappeared from the scene, either having been relocated to an after position along with the engine room below, or to a position well forward in the ship to suit particular purposes, Figs. 1 and 2. Such departures have been mandated by greater utilization of the optimal midship space of the ship for both hold and deck cargo, despite the disadvantages in navigability with deckhouse aft or in habitability with quarters forward.

The trend in mission requirements today is toward a consolidation and refinement of these advances, with less emphasis on the development of larger, faster, or newer types of carrier. Aside from world-wide political and economic upheavals that could forcibly modify directions in ship design and construction, there are existing factors that control this present trend.

1. *Port and Waterway Limitations.* In addition to the usual draft and berthing limitations of various major ports, along with restrictions on length and beam in waterways such as the Panama Canal, most trading ports, and the countries they serve throughout the world, have an inability or an unwillingness to handle and absorb through their systems any large volume of specially handled or packaged cargos.

2. *Fuel Costs.* The large increase in fuel cost that began in the mid 70's and which is likely to prevail, has significantly affected the economic equation that previously favored increasing ship speeds.

3. *Regulatory Requirements.* There is increased emphasis on greater compartmentation (as a function of increased size) as a preventative against massive oil spills in tankers, and against loss of all types of ships due to collision.

For a greater appreciation of how commercial ship designs evolved from changing mission requirements through the years (and in fact how the mission requirements themselves were influenced by various economic, political, and technological changes), the illuminating paper "Forty Years of Ship Designs Under the Merchant Marine Act, 1936–1976" (Dillon et al, 1976)[1] should be required reading. The pros and cons of *standardized* versus *optimized* designs, the sometimes radical shift in emphasis dictated by rapid economic and political changes, the pitfalls encountered in reaching out beyond existing practice, and the overview of the status of design and construction at the present time, all contribute to a better understanding and a necessary sense of humility for the proper approach to design for the future.

2.2 General Cargo Ships. The *general cargo* ship is so designated because of its ability to carry a variety of commodities in a variety of forms such as sacked, boxed, palletized, refrigerated, containerized, and with the possible accommodation for bulk materials such as grain in designated holds and special oils in tank compartments.

a. Types of General Cargo Ships. Cargo liner is the designation given to those general cargo ships engaged in international trade between specific ports and on regular schedules. Their cargo mix is fairly well established as to types and quantity, and the owner can usually so specify within close limits that the designer can readily establish the suitable vessel size and power through the design process detailed in Chapter I. Additionally, cargo space allocations along with optimum hatch sizes and appropriate cargo handling gear can be satisfactorily determined to meet the specified trade requirements.

Cargo ships that are intended for one major cargo may be even more definitively designed for a particular commodity (lumber, newsprint, livestock, etc.). However, consideration must be given to the vessel requirements for the return voyage, whether a specific return cargo is contemplated or whatever the market can offer in the way of breakbulk, along with adequate ballasting capacity so that the entire round trip operation can be accomplished with optimum economy, within schedules, and with adequate seakeeping ability and safety.

[1] Complete references are listed at end of chapter.

The *general purpose* cargo ship (or more commonly called the *tramp ship*) is the designation given to a general breakbulk cargo ship that has no set trade route or schedule, but plies between ports all over the world indiscriminately, following the dictates of each particular cargo consignment. Its economic justification lies in its ability to handle and deliver cargos of unusual size and/or to service ports not normally attended by the cargo liner. To these ends, the modern tramp ship is frequently equipped with heavy lift cargo gear, in addition to its normal cargo gear, to handle lifts of 100 tons or more, and perhaps with at least one extra long cargo hold and hatch for cargo of exceptional dimensions. It usually is proportioned with larger beam for stability in handling heavy lifts and for capability of topside loading, and with moderate draft for entry into a greater number of ports. Because high speed is not a requisite in this trade, the ship is fuller and of less installed power than a comparable cargo liner, with resulting lower costs per ton of cargo deadweight in both construction and operation.

A *standard design* type of general cargo ship attempts to embody the major features of size, speed, and cargo capability that are representative of a particular vessel type and service, whether it be cargo liner or tramp ship. Necessarily it will be a compromise, perhaps being less ship than desired for some services and more ship than required for others. However, its advantage for a prospective owner (or a number of owners engaged in similar trade) lies in the more certain knowledge of its capabilities and operating economies, its low construction cost and delivery—all particularly enhanced if it has demonstrated to be an excellent performer. A standard design may often be undertaken for a single owner as a fleet-replacement series, or may be developed by an individual shipyard as an offering to the industry with the economies of multiple construction.

Perhaps the most exemplary standard cargo ship of the past quarter-century has been the *Mariner*, the design developed by the Maritime Administration (MarAd) in its program of rejuvenation of the entire U. S. Merchant Marine. Over fifty of these vessels (with some modifications) were constructed throughout the 50s and 60s, in a number of shipyards and for numerous owners, and the soundness of the design is evident in that a large percentage of these are still gainfully employed.

Directed towards cargo liner service, the *Mariner* design had to be based on such competitive characteristics as high speed, effective cargo handling, and other modern features for its time, without being so specialized as to discourage its use in less demanding established trades. Typical of the considerations involved in finalizing the vessel particulars was that concerning hatch and hold sizes, as indicated by the following quote from Russo and Sullivan (1953):

> "It is well to agree on the fact that what constitutes an efficient layout from the point of view of cargo carrying and cargo handling is greatly dependent upon the special requirements of each trade. It is obvious that if a great preponderance of the cargo moving in a given trade route consists of long rails, piling, pipes, etc., the optimum ship for that trade should have extra long and deep holds, long hatches, centerline pillars, special rig to facilitate handling extra long drafts of cargo in and out of the holds, etc. Extra long holds result in very large bale cubic per cargo gear; extra long hatches result in a smaller number of hatches, and a smaller amount of cargo handling gear, per ship. A ship designed with the object of being optimum for handling and stowing long pieces of cargo would not be an optimum ship to carry general cargoes. The ability of the ship to survive damage involving flooding also may be lessened.
>
> An operator who wants to move general cargoes in and out of a large number of ports, on the other hand, has no special interest in extra long holds and

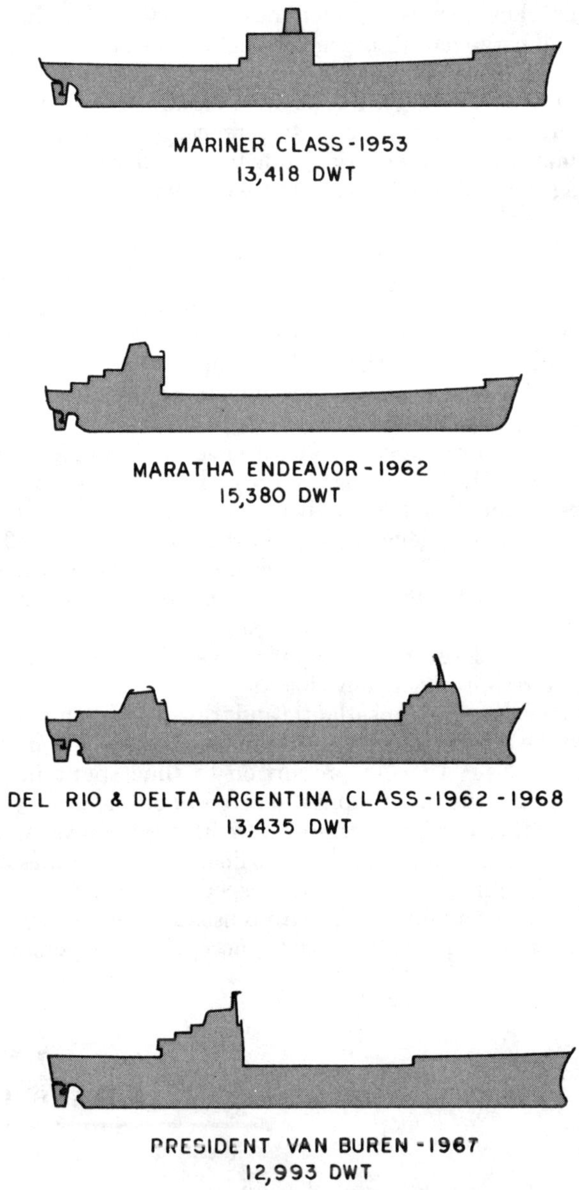

Fig. 1 Changing profiles—general cargo ships

hatches. He will prefer to have as many hatches, and cargo handling gears, as he can get satisfactorily in the ship, in order to facilitate reaching and moving the cargo that he must handle in each port, with a minimum of interference. The requirements of the majority of long trade routes will average somewhere between these extremes."

A comprehensive description of the design process and the design considerations that led to the *Mariner* is given in the same reference. While some of the particular considerations may not be valid today, the overall treatment of the subject wherein the various major factors are weighed in arriving at design decisions should be valuable reading for today's designers.

b. Size and Speed. There have been no significant changes in the basic range of size or speed of general cargo ships in the past twenty-five years. With a few notable exceptions, ship lengths (L_{pp}) have remained in the 137 to 168-m (450 to 550-ft) range, with displacements of from 15,000 to 25,000 tons. Speeds for cargo liners still center around 20 knots and for tramp ships about 16 knots, with variations of several knots either way, depending on the service requirements envisioned.

Typically, the transatlantic and transpacific cargo liners have the greater speed requirements, directed towards reducing at-sea time in proportion to time spent in port (seemingly the more rapid the cargo-handling ability, the greater the economic pressure to reduce sea-time even further). Thus, some modern cargo liners in these trades have been designed for 23–25 knot speeds (approaching the high-speed containerships) with consequent hull forms with finer lines and greater lengths, accepting the penalty of greater light ship weight and cost in proportion to the cargo deadweight assigned.

For shorter trade routes, such as the U. S. Eastern seaboard or Gulf of Mexico to South America, ship speed requirements are correspondingly reduced to the present range of about 17 to 19 knots, allowing somewhat fuller forms and shorter ship lengths for the same amount of cargo deadweight. Tramp ships by the nature of their service, as has been previously indicated, tend toward fuller forms for greater deadweight and toward slower speeds for greater operating economy, in view of their unscheduled operations.

It is of interest to note that the deadweight capacity of the cargo liners through this range of ship sizes is within a narrow band of 12,000 to 15,000 tons (for the tramp ship or special cargo carrier, it is about 15,000 to 20,000 tons), and it is apparent that the dictates of speed result in the variety of ship form necessary to accomplish the basic purpose of cargo transport. Future economic evaluations may alter the equation for optimum cargo ton-miles per hour, particularly in view of the increasing cost of energy, but the distinctive trends between services as illustrated above, would be expected to remain.

Table 1 gives a general indication of major characteristics of a number of notable general cargo ships that are seen to conform to the pattern as stated above.

c. Hatch and Hold Configuration. Probably the most significant advance in breakbulk general cargo ship design has been the increase in hatch openings through the main deck leading into the holds below, affording tremendous increase in cargo handling speed, stowage, loading selectivity and versatility, and in general a greater overall control of the cargo process. The classical arrangement through the years

Table 1—Cargo Ships

	BREAKBULK		LUMBER	
	Maratha Endeavor (Shelterdeck)	Golden Chalice	Columbialand	Tiha
Length O.A., m (ft)	158.5 (520)	145.46 (477.23)	175.21 (574.83)	160.02 (525)
Length B.P., m (ft)	150.27 (493)	136.95 (449.31)	167.64 (550)	152.38 (500)
Beam, m (ft)	21.52 (70.6)	22 (72.19)	26.52 (87)	21.95 (72)
Depth, m (ft)	12.04 (39.5)	12.4 (40.69)	13.72 (45)	12.23 (40.13)
Max Draft, m (ft)	8.87 (29.11)	9 (29.52)	9.86 (32.34)	8.63 (28.31)
Speed vs. Knots	15.0	14.9	15.5	15.0
Block Coefficient, C_B	0.715	0.733	0.780	0.759
Froude Number	0.201	0.209	0.197	0.200
Max SHP	6,130	6,470	11,400	7,700
Propulsion	Diesel	Diesel	Diesel	Diesel
Dry Cargo, Bale, m^3	22,114	21,509	33,946	22,512
(ft^3)	(780,941)	(759,600)	(1,198,795)	(795,000)
Reefer Cargo, m^3	0	0	0	0
(ft^3)	0	0	0	0
Light Ship, tons	5,500	4,382	8,302	5,860
Deadweight, tons	15,380	15,660	26,335	16,240
Displacement, tons	20,880	20,042	33,202	22,100
Length/Depth	12.48	11.04	12.22	12.46
Length/Beam	6.98	6.22	6.32	6.94
Heavy Lift, tons	160	None	None	None

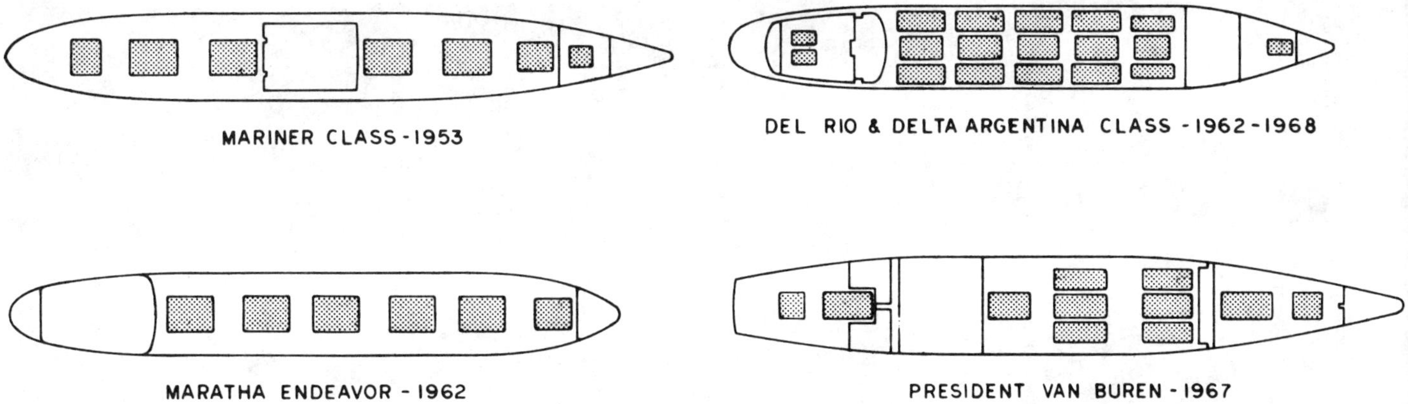

Fig. 2 Evolution in hatch arrangements—1953–1968

has been for a single line of relatively narrow hatches (less than half the ship's beam) and correspondingly short to allow for cargo winch gear and hatch cover stowage in between. With this arrangement the majority of cargo had to be man-handled beyond the hatch opening to the outer extremities of the 'tween deck or hold, with great expenditures of time and manpower. Even the *Mariner* class which boasted large hatches (for its day) had a *hatch square* of less than 30 percent of the deck area.

Revolutionary developments were manifested in the 1960s with the introduction of twin and triple-hatch arrangements which greatly enlarged the hatch square percentage and tremendously improved cargo handling ability. Noteworthy in this regard were the Delta Lines vessels, with triple-hatch arrangements throughout half the vessel's length, spanning about 75 percent of the ship's breadth and with a *hatch square* equivalent to about 60 percent of the deck area involved. The dramatic difference between the age-old arrangement and this new concept is illustrated in Fig. 2.

The ability to plumb virtually all of the cargo without secondary handling, and with a minimum of broken stowage or damage, has established the multiple-hatch arrangement, almost without exception, as the required configuration for the general cargo ship of today. The associated hatch cover problems are discussed in Chapters IX and X.

d. Major Design Considerations. With the overall size, speed, and capacity determined on the basis of the general service requirements outlined above and in accordance with the methodology of Chapter I, more specific consideration must be given to such items as primary ship structure, cargo handling and deck arrangements, compartmentation, and loading conditions, in order to bring the design to fruition.

The large hatch openings prevalent in today's cargo ship

Table 1—Cargo Ships (Continued)

LINERS

Mormacargo	Delta Argentina	Strathardle	President Van Buren	Mariner Class
167.87 (550.75)	159.11 (522)	171.38 (562.28)	174.93 (573.92)	171.8 (563.65)
158.5 (520)	148.44 (487)	160.16 (525.46)	165.96 (544.5)	161.13 (528.5)
22.86 (75)	21.34 (70)	24.23 (79.48)	24.99 (82)	23.16 (76)
12.95 (42.5)	13.72 (45)	13.97 (45.82)	13.87 (45.5)	13.56 (44.5)
9.63 (31.58)	9.45 (31)	9.15 (30)	9.32 (30.58)	9.1 (29.84)
24+	18.6	22.45	23	20
0.563	0.639	0.568	0.544	0.616
0.313	0.251	0.291	0.293	0.259
19,000	14,250	20,700	24,000	19,750
Steam	Steam	Diesel	Steam	Steam
18,839	16,978	19,809	22,538	20,862
(665,300)	(599,580)	(699,556)	(795,920)	(736,723)
1,133	1,339	637	1,390	857
(40,000)	(47,280)	(22,500)	(49,100)	(30,254)
7,700	5,850	7,804	8,237	7,675
12,100	13,435	12,552	12,993	13,418
19,800	19,285	20,356	21,230	21,093
12.24	10.82	11.47	11.97	11.88
6.93	6.96	6.61	6.64	6.95
75	75	30	70	2 @ 60

Fig. 3 Cargo liner DEL ORO

reduces the width of effective main deck structure to an extent that would require extra thick deck and sheer strake plating (if normal shipbuilding mild steel were used) in order to maintain the required hull girder strength. To reduce weight as well as cost of construction, high strength steels have become favored for this application, and in fact the standard low alloy, high strength steels were developed principally to meet needs of this nature.

However, it is not sufficient only to satisfy the requirements of effective midship section modulus in the usual sense. Consideration must be given to buckling of the thinner higher stressed plating, to high torsional stresses that may prevail in quartering seas and which may cause high stress concentrations at hatch corners, and to the need for adequate section modulus in a diagonal direction when the vessel is in a bending mode and concurrently heeled to a significant angle under wave forces. Chapter VI elaborates on the analyses necessary to establish primary structural scantlings, and the required classification standards to be met.

In addition, consideration must be given to the disposition of cargo along the ship's length (plus fuel and ballast as necessary to maintain proper trim in service) insofar as it may affect still-water bending moment. An arrangement that may appear optimum from a cargo distribution standpoint may incur high bending moments requiring excessive structure; or conversely a well-proportioned structure based on general ship bending guidelines may not allow effective cargo carrying or proper ballasting under some desired conditions. This consideration must be addressed early in the design process.

Further, early attention must be given to the amount of *usable* space to be contained within the ship's envelope as it affects the gross and net register tons that will be assigned to the vessel (see Chapter V). Where the intended cargo is foreseen to be light in weight, requiring large stowage underdeck volume, the *shelterdeck* type of cargo ship with its shallower draft and higher feedboard may be indicated, providing maximum cubic at minimum net register tonnage (for example, the ro/ro ship, as discussed in Section 2.4). However, the shelterdeck vessel suffers from having less deadweight capacity needed for heavier cargos than a comparable full scantling ship, and also from a generally higher light ship center of gravity which can pose a stability problem, particularly if deck cargos are contemplated. Tonnage regulations permit shelterdeck ships to be designed as full scantling vessels to achieve maximum deadweight, with special provisions to allow ready conversion to the lighter displacement, lower net tonnage shelterdecker, should the trade so dictate.

Today, however, the full scantling ship is more prevalent than the shelterdeck type, due largely to the flexibility afforded by carrying containers on deck. The ship's envelope can be optimized for deadweight cargo, and the possible need for greater cargo volume with lighter cargo can be met by deck containers, without increasing the register tonnage.

Finally, scrutiny should be given to the space allocated to dry cargo. While it may be impressive to produce a large cubic cargo capacity, it could be wasteful if not utilized properly. Thus, remote and inaccessible areas that cannot readily be stowed with other than dunnage (particularly in view of the prevalence of unitized cargos) should be re-allocated to other beneficial purposes such as tonnage-exempted ballast, revenue-producing cargo oils, or additional fuel for longer range potential.

In summary, it can be seen that considerations of cargo handling, distribution, and stowage can have marked influence on structural scantling requirements, subdivision, and register tonnage among other factors.

Thus, aside from the overall mission requirements that tend to establish the general particulars of size, speed and deadweight, it is ultimately the dictates of cargo that need to be addressed in order to finalize the design characteristics and ultimate vessel particulars. From the classic arrangement depicted in Fig. 2, Chapter I, to the highly cargo efficient requirement of the *Del Oro,* Fig. 3, the direction in which general cargo ship designs have proceeded to suit the demands of cargo is clearly illustrated.

2.3 Containerships. The general cargo ship designed to be capable of carrying all of its cargo in unitized containers is designated as the full *containership.* Prior to its advent in the 1960s, containers had been carried to some extent on deck or in selected holds as part of the cargo mix on typical general cargo ships, but it required the technological break-through of the all-hatch concept described above to provide full and effective stowage of containers throughout the vessel, and to make the containership a practical reality.

a. Container Characteristics. Containers come in a number of standard sizes (see Chapter X for detail dimensions and capacities), and all are basically selected to allow their use as trailer truck bodies or to be carried on railroad flat cars for the overland portions of their total voyage. The concept originating in the United States of an inter-modal carrier that could transport a valuable type of cargo, usually from door-to-door, with a minimum of handling but with speed and efficiency, captured the imagination of the transportation industry in the highly developed countries of Europe, and later Japan. The United States enjoyed a pivotal position in most of this trade; and the containership became the darling of the shipping world.

b. Size and speed. The containerships of the early 60s were little different in principal characteristics from the standard cargo ship of the time (Dillon et al, 1976). With the rapid increase in demand for containerized transport the requirements for high capacity and greater speed were mandated, and shortly led to ships of double the previous capacity and extending up to the exceptional speed of 33 knots for cargo ships (Boylston et al, 1974) as shown in Table 2. Within hardly more than a decade, the containership has largely lost resemblance to the general cargo ship in characteristics, if not in function, and in fact is today the successor to the bygone passenger liner in size, sleekness, and speed.

c. Configuration. The full containership embodies the concept of full cellular stowage within the holds, plumbing directly down thru a multiple array of hatches, in a guided arrangement necessary to secure the containers without dunnage against motions at sea. Additionally, most containerships are designed to carry containers on deck, stacked three to four high and secured by systems of lashing, purposefully to afford sufficient deadweight carrying capacity for what is normally a high-cubic, low weight cargo system. In order to protect this exposed cargo against the forces of the sea, particularly in regard to green seas taken over the bow, an unusally high forecastle is provided, along with a well-flared bow shape, Fig. 4. Alternately, the bridge house and quarters are located forward, which further provides for better visibility, without encumbrance from the container stack, as shown in Fig. 5, Chapter I.

However, a number of containerships have been designed to carry a higher percentage of containers below deck, to minimize the problem associated with exposed stowage. Such ships have great depths of hull, with high freeboard (and windage) and show a firm commitment to container or other high volume cargo, Table 2.

d. Structural Considerations. The operational demands of high speeds, efficient stowage and rapid handling of containers, results in a long, fine hull with a main deck devoted almost entirely to hatch openings throughout its length and breadth. The need for intensive structural analysis concerning hull girder bending, torsion, and slamming is imperative, in addition to the further demanding task of determining hull and hatch cover scantlings for cargo support, described by Boylston et al (1974).

Table 2—Containerships

	American Lancer	Rhodri Mawr	Liverpool Bay	Sealand SL-7	Seatrain Euroliner
Length B.P., m (ft)	200.2 (670.0)	100.0 (328.0)	274.3 (900.0)	268.4 (880.5)	223.4 (733.0)
Beam, m (ft)	25.9 (85.0)	16.8 (55.0)	32.3 (105.8)	32.2 (105.5)	30.0 (98.4)
Depth, m (ft)	15.4 (50.5)	7.9 (26.0)	24.6 (80.7)	19.5 (64.0)	19.2 (63.0)
Draft, m (ft)	8.5 (27.8)	4.1 (13.5)	11.0 (36.0)	10.6 (34.8)	9.9 (32.5)
Vs, Knots	21.5	14.0	26.0	33.0	26.4
C_B	0.709	—	0.595	0.540	0.534
Froude Number	0.247	0.230	0.258	0.331	0.290
Propulsion	Steam	Diesel (2)	Steam	Steam	Gas Turbine
No. of Equivalent 20 ft Container Units	1,178	184	2,300	1,974	1,632
Shaft Horsepower	27,300	4,200	80,000	120,000	59,240
Light Ship, tons	9,725	—	23,870	24,671	12,636
Deadweight, tons	22,275	2,645	34,020	27,140	23,382
Displacement, tons	32,000	—	57,890	51,811	36,018
Length/Depth	13.27	12.62	11.15	13.76	11.72
Length/Beam	7.88	5.96	8.50	8.35	7.50

Fig. 4 Containership HAWAIIAN ENTERPRISE

The American Bureau of Shipping (ABS) and the Ship Structure Committee (SSC), among others, have both conducted much research on this subject (Elbatouti et al, 1976). A detailed study of these reports is essential for any designer embarking on a design assignment of this type.

e. Future Trends. The high-speed containerships illustrated here were all conceived and built before the energy crisis of the mid-70s. Today, it is clearly indicated that these high speeds are no longer economically justified. Existing ships are being run at reduced power, and in some instances are having power plants and propulsion modified for even greater economy (Novak et al, 1977). It is highly likely that containerships, in the immediate future at least, will be of fuller form and moderate speed, easing back toward the proportions of the general cargo ship.

2.4 Roll-On Roll-Off Ships. The ro/ro was the forerunner of the full containership by nearly a decade, embodying the concept of transporting trailers, complete with undercarriage; thus the name *trailership* was alternately used for this type of vessel. The rapid handling of cargo on and off the vessel, in the form of wheeled vehicles, by means of stern or bow ramps and even sideports (for smaller vehicles) was of course a well-established practice before this time. The novel feature of the ro/ro was in the adaptation to a high-speed, transoceanic system featuring multi-deck ships. This occurred in an era of increasing demand for rapid international movement of high premium containerized cargo (National Academy of Sciences, 1946 and 1958).

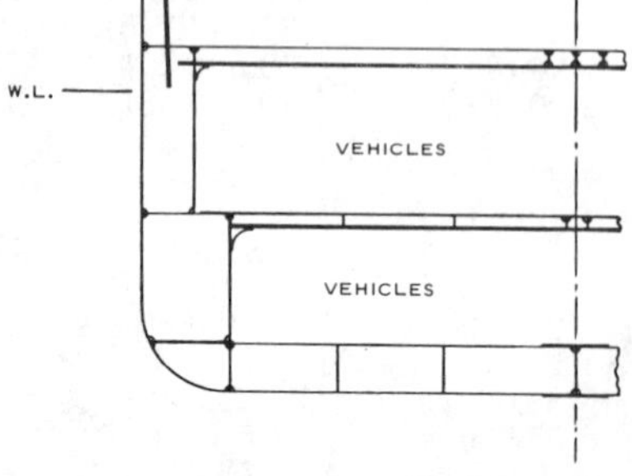

Fig. 5 Midsection ATLANTIC SAGA

The inherent disadvantage of the full trailership, on an equivalent cargo basis, is the waste of cubic capacity required for undercarriage and all-around clearances. Even when tailored to basic standard containers moved in on low-slung dollies, which affords some 40 percent increase in cargo capacity, it cannot meet the carrying capacity of the full containership with its more efficient vertical modular stowage.

Thus, during the containership surge, the ro/ro diminished in popularity although its use was sufficiently maintained (particularly in regard to the transport of the vehicles themselves as cargo) to enable ro/ro technology to advance through this period. It has evolved to the point of perhaps being the most favored high-speed cargo vessel today, in the form of a combination carrier wherein the upper deck and perhaps some selected holds are committed to containers as a significant portion of the cargo, with the remaining lower decks being assigned to wheeled vehicles ranging from trailers to minicars as the traffic dictates. Provision may also be made for the carriage of general cargo, incorporating the usual hold/hatch/cargo gear or simply by fork-lift operation.

Thus, the roll-on, roll-off ship of today is no longer the full trailership as conceived over two decades ago. It retains the name ro/ro to designate that it features the use of loading ramps for wheeled vehicles, even in such instances when the major cargo may be of other containment.

a. Configuration. The most prevalent arrangement of the present-day large trans-ocean ro/ro consigns most of the below-deck cargo cubic to wheeled vehicles and/or to general cargo that is wheeled into place by dolly or folklift through the loading ramps. Internal ramps lead from the loading deck to the other 'tween-deck spaces, and in some foreign designs the entire ship's length is devoid of transverse bulkheads from the stern to the forward quarter of the vessel. A typical sectional view is shown in Fig. 5.

The upper deck is consigned to containers, and is flush without hatches except possibly in the forward end for the stowage of additional containers in the narrow hold spaces below, not otherwise suitable for vehicle movement. Fuel and ballast (for $\overline{GM}$ control) are carried in double bottoms and narrow wing tanks, the latter extending up above the design waterline sufficiently to preclude loss due to incidental side damage. However, such arrangement does not meet requirements of one compartment subdivision (to 1/5th beam). In order for ships of this type to qualify as American flag cargo liners, full-depth transverse bulkheads need to be installed, with large access openings for vehicular passage, and equipped with mechanically operated watertight doors that resemble the modern deck hatch cover installed vertically.

A more versatile configuration embodies all of the general purpose cargo ship's features described in Section 2.2, including large deck hatches, container stowage on deck and in holds, general cargo gear and heavy lifts, etc. plus the ability to accommodate rolling stock through loading ramps. Ro/ro cargo may not necessarily be the main commodity, nevertheless as previously mentioned, the designation of ro/ro ship remains (as of today) to indicate that the feature is incorporated. A typical example of an American-flag ship of this arrangement is the *Mormacsea* whereby roll-on, roll-off cargo is confined to the second deck, which serves as the bulkhead deck to meet stability and floodable length requirements.

It should be noted that the typical ro/ro is a shelterdeck vessel, with freeboard assignment to its lower continuous deck, resulting in a considerable reduction in gross tonnage and related fees.

b. Size and Speed. Typically, the combination ro/ro ship of today has the same deadweight capacity, speed, and power as the general cargo ship that would otherwise be used in parallel service, Table 3. However, because of the contemplated low density cargo, the principal dimensions of the vessel are roughly 10 percent greater in length, beam, and depth, resulting in a lower draft and significantly higher freeboard to the upper deck than the comparable general cargo ship.

2.5 Barge Carrying Ships. The barge carrying ship is essentially a containership, the major differences being that its *containers* are in order of magnitude or more larger than the standard container and are handled to and from the water instead of the dock. For example, the standard 20-ft container has a cubic capacity of approximately 1200 ft^3, the standard LASH lighter capacity of 20,000 ft^3, and the standard SEABEE barge a capacity of 40,000 ft^3.

There are important ramifications in those differences, that makes the barge carrier highly competitive in many

Table 3—Ro/Ro Ships

	Atlantic Saga	Australian Enterprise	Atlantic Span	Ponce de Leon	Avon Forest
Length B.P., m (ft)	183.0 (600.4)	168.0 (551.2)	183.7 (602.7)	196.0 (643.0)	189.4 (621.5)
Beam, m (ft)	27.4 (90.0)	25.0 (82.0)	26.4 (86.6)	28.0 (92.0)	22.9 (75.0)
Depth, m (ft)	19.3 (63.3)	16.4 (53.8)	19.3 (63.3)	18.3 (60.1)	17.7 (58.0)
Draft, m (ft)	9.0 (29.5)	9.0 (29.5)	8.2 (27.0)	8.2 (27.0)	4.1 (13.5)
VS, Knots	20.75	21.0	21.5	25.0	19.5
C_B	0.604	0.581	—	0.528	0.664
Froude Number	0.252	0.266	0.261	0.293	0.233
SHP	19,800	22,000 (M)	20,700 (M)	32,000	17,000
Propulsion	Diesel (1)	Diesel (3)	Diesel (1)	Steam	Diesel (2)
Autos	1,200	147	1,100+	300*	1,900
40-Ft. containers	—	14	—	—	—
20-Ft. containers	500	583	405	—	—
Light Ship, tons	11,806	8,091	—	11,000	10,155
Deadweight, tons	15,725	14,082	14,370	13,100	1,807
Displacement, tons	27,531	22,173	—	24,100	11,962
Length/Depth	9.482	10.243	9.519	10.692	10.716
Length/Beam	6.671	6.720	6.959	6.989	8.287

* Also 260 40-ft Trailers

Fig. 6 SEABEE ship loading barges

trading areas throughout the world. For example:

- In heavily industrialized port areas, particularly where associated with a developed river system, there is great convenience and economy in having full barge loads of commodity move directly to waterfront plants or move to remote terminals upriver for discharge and further inland distribution. There is not the need for intermediate handling or transfer of cargo, and further benefit accrues from the cheaper river transportation compared to road or rail. Such port/river complexes as New Orleans/Mississippi and Rotterdam/Rhine are prime examples.
- In less developed (or antiquated) port areas throughout the world, where commerce is heavy but where facilities are inadequate for rapid cargo handling, warehousing, or distribution, barge-contained cargos can more readily be accommodated and dispatched.
- In all major port areas, the possibility of delays of several days to weeks confronted by typical cargo vessels in awaiting dock-side space for loading or unloading is circumvented by the barge-carrying ship, which can handle its barge cargos in the open roadsteads or at anchor in ports.
- By virtue of their size and construction (compared to standard containers), barges of the size presently used can carry a greater variety and unit size of cargo, and thus possess the potential for greater utility in established trade or for unusual cargo carrying. Further, they are far less likely to be lost, strayed, stolen, or pilfered.

While the principal characteristics of size, capacity and speed of the barge carrying ship will depend on the mission requirements, from large intercontinental liners to small local inter-port feeder vessels, the basic configuration is essentially dictated by the method in which the barges are handled on and off the vessel.

- *Elevator:* In this scheme, representative of the SEABEE class, Fig. 6, Table 4, barges are floated into position at the stern of the ship and lifted by powered means to a designated deck elevation, and then trolleyed forward into the ship's hold or upper deck. There are three deck levels including the main deck that can accommodate the barges, in a sort of *sliding file* arrangement, aft to forward. There are necessarily no transverse bulkheads through the cargo area, and the arrangement lends itself ideally to the alternate carrying of roll-on/roll-off cargo.
- *Traveling Crane:* In this arrangement typical of the *LASH* ships, Fig. 2, Chapter I, Table 4, barges are lifted at the stern by a deck-mounted gantry crane, which travels forward and deposits the barges into deep hold *cells*. The barges are specially designed to stack one above the other on corner posts built into the barge structure. Full transverse bulkheads at traditional spacing are provided, but because of the large *hatch* openings, main deck structure is limited to the side areas. Alternately, the ship can serve as a containership, in part or fully, with the addition of a container crane.

• *Float-on/float-off* is one of the simpler arrangements, requiring little in elaborate handling equipment. Similar to the naval *Landing Ship Dock* (LSD), the vessel has the features of a floating drydock, to ballast down for accepting or discharging barges from the stern. Because of generally limited depths of water in most ports, this arrangement is basically limited to one tier of barges (in some instances with a full main deck over, for containers, vehicles, or other cargo). Its simplicity along with its capacity restrictions make it most suitable as a feeder vessel, whereby it dispatches barges to and from main ports to adjacent areas.

Additional methods of barge handling have been considered, but the above represent those most commonly in use.

a. Arrangement. In consideration of the large size of each barge "container," it is apparent that the ship hold space must be essentially rectangular in order to maximize the cargo carrying capacity. Thus, in the preliminary design stage, the square-cornered cargo space is first defined in terms of the desired number of units, and the hull is wrapped around it in the best possible manner to provide a good speed form, consistent with the need for strength, stability, and space requirements for fuel and machinery. Optimization is therefore a piecemeal process, wherein the addition of one or more barges can override the nicety of an optimum hull form.

Essentially then, the cargo holds are box-like in shape, of dimensions carefully determined to just accommodate the maximum number of standard barge modules. The ship envelope then incorporates double bottoms, wing walls, bow and stern forms, etc. as required for strength and operational purposes. Arrangement details of these vessels are discussed more extensively in Chapter III.

b. Structural Considerations. For the SEABEE elevator type of barge ship and the float-on feeder ship, the absence of full transverse bulkheads through most of the vessel's length requires careful structural investigation of racking stresses, particularly under roll dynamic conditions. For the LASH type, which has full transverse bulkheads, a major structural consideration is the torsional effect at the corners of the large hatch openings, under quartering sea conditions. A further consideration is that, with the full cargo loading being concentrated at the barge corners, high shear loads develop downward at the barge posts in way of the ship's double bottom which is also subject to evenly distributed hydrostatic loading upward.

In addition, for the large ships in particular, the fact that concentrated loadings are significantly large for each barge module, attention must be given to disposition of loading, for strength and stability. Depending on the ports of call, and the desired transfer of empty or heavily laden barges, situations can occur where the loading pattern will induce high longitudinal bending stresses in the ship's hull girder, or possible critical stability with heavy loads mainly topside. Therefore, careful monitoring of loading must be established, and computerized loading schemes are a practical necessity in operations.

c. Size and Speed. The SEABEE and LASH vessels are similar in size to the containership, in dimensions, proportions and deadweight capacity, as can be seen from Table 4. However, their speeds are considerably less, being more in the order of general cargo liners. It is perhaps axiomatic that with a slower inland movement of predominantly bulk cargo, there is less demand for high ship speed.

d. Future Considerations. It is probable that, in the foreseeable future, the majority of barge carrying ships will be designed to accommodate one (or more) of the size barges that will have been designated as standard. This is a simple matter of practicality, for it should be anticipated that while barges are proprietary to the individual shipowner they may be traded (or at least swapped temporarily) between one operator and the other, similar to the practice developed in the container trade.

2.6 Tankers. The history of tanker design and construction through the years since World War II has been highlighted by continued and progressive increase in size and power. From the general standard of about 25,000 deadweight ton capacity in the '40s, to the *Jumbo* size 50,000 dwt in the mid-fifties, the size of tankers continued to increase rapidly in the following years through the 100,000 dwt range and continued to the point where the largest vessels in history are tankers of over one-half million deadweight tons. Installed power also increased from the range of 13,000 shp to over 50,000 shp in the largest vessels. (This comprises a decrease in shp/dwt from 0.52 to 0.10 which, although adequate for propulsive power, represents a significant loss in maneuvering capability.) Table 5 gives the major particulars of distinctive tankers, illustrating the range of sizes through the years, as also shown in Fig. 7 on page 63.

a. Structural Design. Structural design considerations

Table 4—Barge Carriers

	LASH Green Valley	SEABEE Doctor Lykes
Length O.A., m (ft)	272.3 (893.3)	267.0 (875.9)
Length B.P., m (ft)	243.0 (797.3)	225.6 (740.0)
Beam, m (ft)	30.5 (100.0)	32.2 (105.8)
Depth, m (ft)	18.3 (60.0)	22.8 (74.8)
Draft, m (ft)	9.5 (31.0)	11.9 (39.0)
VS, Knots	20.0	18.6
C_B	0.631	0.657
Froude Number	0.211	0.203
SHP	32,000	36,000
Propulsion	Steam	Steam
Number of Barges	89	38
Light Ship, tons	16,161	18,880
Deadweight, tons	46,153	38,410
Displacement, tons	62,314	57,290
Length/Depth	13.289	8.893
Length/Beam	7.973	6.992
BARGES	**LASH**	**SEABEE**
Length O.A., m (ft)	18.75 (61.50)	29.72 (97.50)
Beam, m (ft)	9.50 (31.17)	10.67 (35.00)
Depth, m (ft)	3.96 (13.00)	3.81 (12.50)
Draft, m (ft)	2.66 (8.73)	2.59 (8.50)
Deadweight, tons	369.6	833.9
Light Ship, tons	85.7	166.1
Displacement, tons	455.4	1,000.0
Bale Capacity, m^3 (ft^3)	546.5 (19,300)	1,108.3 (39,140)

have been a dominant factor thru these years of tanker growth. Reaching out beyond the 213.4 m (700 ft) length (50,000-dwt range) required extrapolation beyond the existing state of the art and beyond classification society rule application at that time for hull girder properties. With the realization that the classic wave bending criterion was too severe for ships of extraordinary length (that the ship should be poised on a wave of its own length, and with a height of 1/20 length) the hull girder requirements were gradually relaxed, until a number of bottom shell buckling failures occurred.

By analyzing the hull structural requirements on a more rational basis than had been done previously, using advanced technology and computer methods, the classification societies were able to permit reduced scantlings but with sufficient strength to withstand buckling loads.

While the present tanker rules of the classification societies delineate the major requirements and methods for establishing scantlings, the structure of a large tanker is so complex that the owner/designer is encouraged to perform his own detail and independent investigations on structural requirements, taking into account the wave induced bending stresses (in both longitudinal and transverse directions), torsional stresses in quartering seas, loading patterns representative of the tank arrangement and intended usage, including ballast, and the selected disposition of internal structure.

In effect, the trend of modern tanker design (particularly where proportions, arrangements, or environment deviate significantly from established norms) is for the designer to make his own rigorous investigations and detail structural determinations, using the classification rules as a guide and control. From an environmental protection standpoint, it is essential that structural integrity be doubly assured.

On the other hand, from an economic point of view, it is illogical to increase structural scantlings arbitrarily (as a substitute for design analysis to determine correct requirements). Hull steel weight represents 12 to 15 percent of the total tanker displacement, a highly significant cost factor in the large modern tanker. Intensive investigations are required therefore to optimize structural weights, structural arrangements, and structural details to minimize the cost of material and of construction.

An example of the types and intensity of structural design investigations required for the modern tanker is given by Laredo et al (1977).

b. Ship Proportions and Weight. Aside from the fact that the large tankers of today are simply bigger in all dimensions than their predecessors, there have been significant changes in ship proportions. Length to beam ratios are now typically 5.5 to 6.5, where previously they were in the 7.5 range. Similarly, length to draft ratios have reduced to under 16.0 (except for the specially designed shallow draft tankers) where the previous range was 18.0 and over. In addition, block coefficients have increased from the former 0.75 to the range of 0.8–0.85.

With design speed requirements remaining essentially constant through the years, less proportionate increases in length could be adopted along with higher block coefficients, without penalty in operating economy. The net effect has been a reduction in steel weight ratio and a resulting increase in deadweight capacity, from the earlier range of 0.75 of total displacement to 0.80–0.85.

c. Shallow-Draft Tankers. Beyond the range of 75,000 to 100,000 deadweight tons, the draft of normally proportionated tankers increases beyond 13.7 to 15.2 m (45 to 50 ft) draft, which then exceeds the limit of most American and foreign ports. Such tankers can be operated at reduced draft to enter these ports or tankers of special proportions may be considered for such restrictions. Being beyond the range of normal application of classification rules for structure, and model test series for speed and power, these designs require some early investigations to define roughly the range of particulars before more detailed design can begin. An excellent and fairly comprehensive treatment of determining preliminary ship characteristics is given by Roseman et al (1974).

d. Shape. Due to the lower speed-length ratios of the present long tankers, and the consequent reduction in wave-making resistance (per ton displacement), there has been a definite change in bow form. Instead of the classic fine entry type, bows are now essentially vertical semi-cylinders of fairly large radius (see Fig. 7, Chapter I). In addition to simplifying construction to some extent, the large

Table 5—Tankers

	Amanda Miller	Sea Spirit	Jade	Nisseki Maru	Esso Atlantic
Length B.P., m (ft)	228.0 (748.0)	251.0 (843.2)	329.2 (1080.0)	330.0 (1082.7)	406.6 (1334.0)
Beam, m (ft)	32.2 (105.7)	40.8 (134.0)	51.8 (170.0)	54.5 (178.8)	71.0 (232.9)
Depth, m (ft)	17.5 (57.5)	21.3 (70.0)	25.6 (84.0)	35.0 (114.8)	31.2 (102.4)
Draft, m (ft)	13.2 (43.2)	16.0 (52.4)	20.1 (65.8)	27.0 (88.7)	25.0 (82.0)
VS, Knots	15.0	16.95	15.70	15.0	16.0
C_B	0.828	0.803	0.836	0.862	—
Froude Number	0.163	0.174	0.142	0.136	0.130
SHP	20,000	28,000	32,000 (M)	40,000	45,000
Propulsion	Diesel	Steam	Steam	Steam	Steam
Light Ship, tons	15,060	19,700	33,150	52,150	—
Deadweight, tons	65,740	116,250	255,374	370,812	508,731
Displacement, tons	80,800	135,950	288,524	422,962	—
Length/Depth	13.009	12.046	12.858	9.428	13.032
Length/Beam	7.079	6.293	6.352	6.055	5.727

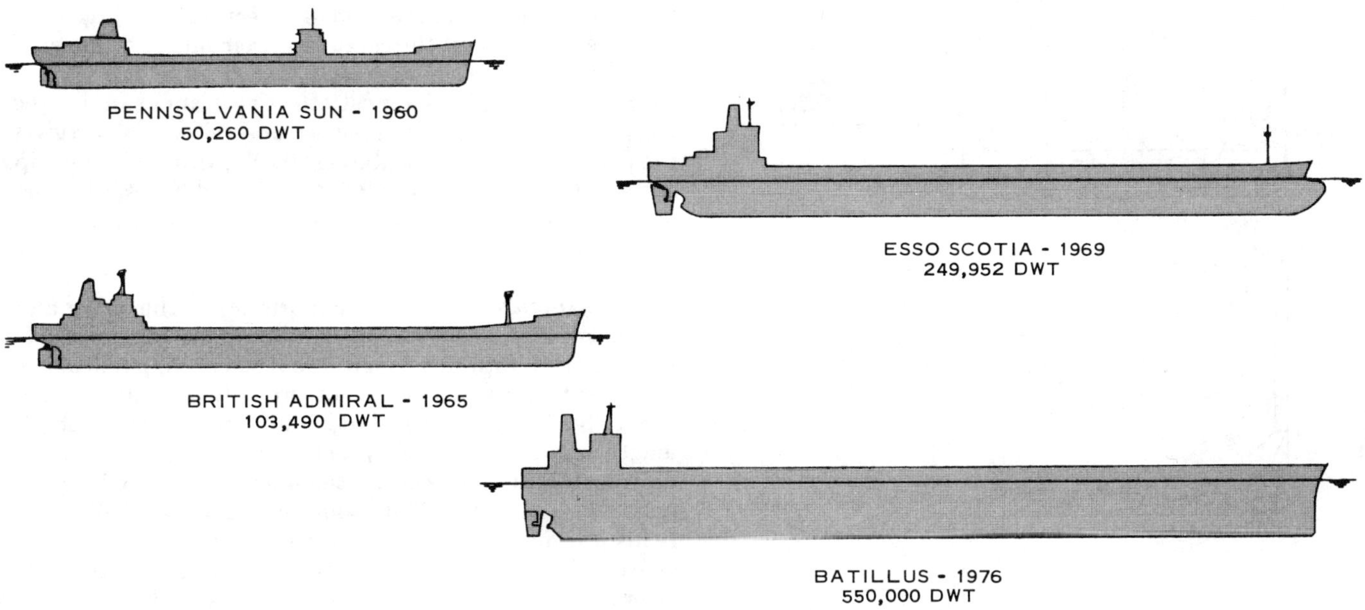

Fig. 7 Changing profiles of tankers

radiused bow provides easy lines leading aft, reducing the sharp *shoulder* that is normally a problem to fair, particularly in ships with high block coefficients; and which shoulder in itself tends to cause high resistance. In fact, it has been indicated that for these large, high block vessels, the cylindrical bow (some with bulbs, some without), provides the least resistance form.

There is a further advantage, in that the additional displacement at the bow brings the ship's center of buoyancy further forward; this provides a more natural balance with the cargo loading which is also predominantly forward. More than that, it allows for a finer stern, helping to provide better flow to the propeller(s) for greater efficiency and control, factors which are essential to the proper operation of the vessel.

e. Compartmentation and Regulatory Constraints. In determining the optimum subdivision of a tanker as to the number and size of cargo and ballast tanks, the major factors previously considered were the optimum tank size consistent with the cargo handling capabilities, the desired cargo mix, and the most effective bulkhead arrangements for strength under various loading conditions and environments.

Today, these factors must be weighed, to satisfy additional regulatory constraints that have been applied to tanker service as detailed in Chapter XI.

f. Design Data. Fairly comprehensive details of size, proportions, weights, arrangements, etc. for tankers up to 100,000 deadweight tons may be found in:

- Robinson et al (1948)
- Benford (1951)
- Nichols et al (1960)
- Long et al (1960)
- Johnson and Rumble (1965)

While these represent the practice of several decades ago, the information is valuable towards development of today's larger vessels.

2.7 Dry Bulk Carriers and OBOs. *Dry Bulk Carrier* is the general connotation for those vessels primarily intended to carry dry cargo, loaded into the vessel with no containment other than that of the ship's hold boundaries, as distinguished from the liquid bulk carrier or tanker. Dry bulk cargos may vary in nature and specific gravity, from iron ore at 3.49 (10 ft^3/ton) to grains at 0.36 (100 ft^3/ton), as shown in Fig. 8, and vessel proportions, internal arrangements, structure, etc. are strongly influenced by the specified cargo type, in addition to the usual logistic and economic constraints.

Bulk carriers engaged in long international trade are most frequently intended to be combination carriers, with one principal commodity carried on the outbound leg of the voyage and a different one inbound. Such cargos may be dry bulk both ways (ore and grain, for example) or dry bulk to liquid (coal and oil). In the latter case, which is prevalent in today's worldwide trade, the bulk carrier is generally designated as an *OBO* (*Ore/Bulk/Oil Carrier*).

a. Size and Configuration. Bulk carriers have shown a growth pattern through the years parallel to that of tankers, and are in fact practically identical in hull form and proportions, powering, etc. for a given cargo deadweight (with perhaps a slightly larger block coefficient and a fraction less speed).

The principal differences in configuration are in the incorporation of large hatches in the main deck leading into the dry bulk compartments, and in the shape and arrangement of the hold compartments themselves. Where the compartments of a pure tanker are relatively simple in pattern, with vertical transverse and longitudinal bulkheads from deck to bottom shell resulting in a rectangular eggcrate arrangement, those of bulker may be slope sided, oc-

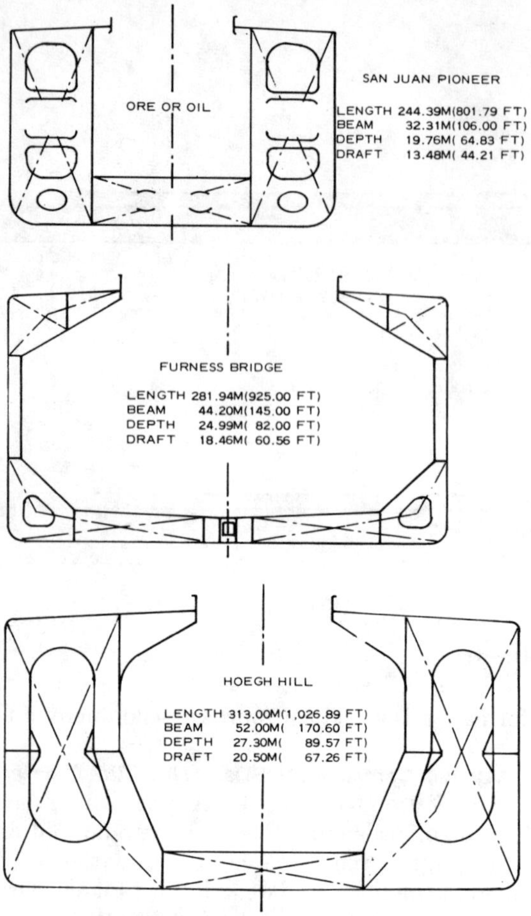

Fig. 9 Typical OBO midsections

tagonal shaped, extending to the hull sides as one transverse compartment, or relatively small in section with large wing tanks, etc. all to suit the cargo requirements as to density, handling methods, comparative quantities of the various cargos intended, in the most effective manner under the constraints imposed for the particular service. Table 6 gives the particulars of representative OBO's, with their midship sections shown in Fig. 9. Chapters III and X describe various arrangements, as well as valuable data on sizes, weights and arrangements. Additional data are given by Henry (1955) and Dorman (1966).

b. Structure. Of major importance in the sizing and arrangement of compartments for the various bulk products is the consideration of longitudinal strength, particularly where extremely heavy cargos are intended and which cannot be effectively distributed uniformly along the ship's length (due in part to the practical advantage of utilizing only such individual compartments as necessary for each particular product). Thus, some arrangement of alternate holds appears most prudent for dense cargo to raise the center of gravity and to minimize hull girder bending stresses in this loading condition. However, this results in the need to provide hull girder shear area and local strength for load transfer between buoyant and weighted compartments over a greater length of hull than required under more equally distributed loadings. The ultimate selection of hold arrangements to satisfy the load carrying requirements for the various cargos to be accommodated, consistent with the dictates of strength, trim, and stability, and all with an eye toward economy in the use of steel and construction details, is the most demanding task in the development of the design of a multi-purpose bulk carrier.

2.8 Liquefied Natural Gas (LNG) Tankers. Liquefied

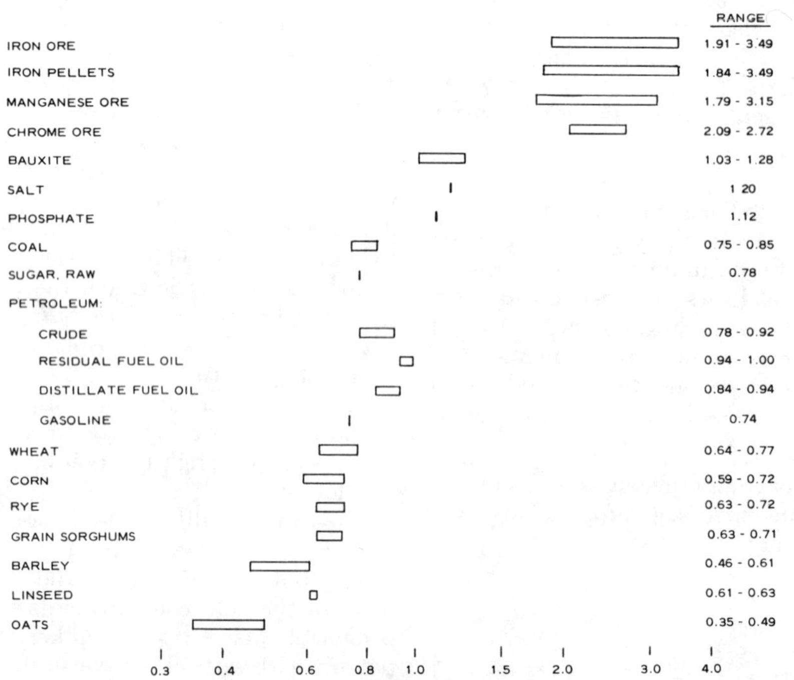

Fig. 8 Specific gravity of various bulk cargos

Table 6—OBO Ships

	San Juan Pioneer	Furness Bridge	Hoegh Hill
Length B.P., m (ft)	244.4 (801.8)	281.9 (925.0)	313.0 (1026.9)
Beam, m (ft)	32.4 (106.2)	44.2 (145.0)	52.0 (170.6)
Depth, m (ft)	19.8 (64.8)	25.0 (82.0)	27.3 (89.6)
Draft, m (ft)	13.5 (44.2)	18.5 (60.6)	20.5 (67.3)
VS, Knots	16.4	15.5	15.3
C_B	0.819	0.846	0.841
Froude Number	0.172	0.152	0.142
SHP	20,000	30,400	33,000
Propulsion	Steam	Diesel	Steam
Cargo Capacity Ore, m^3 (ft^3)	35,963 (1,270,037)	95,175 (3,361,060)	—
Cargo Capacity Oil, m^3 (ft^3)	91,172 (3,219,718)	167,888 (5,928,910)	—
Light Ship, tons	17,898	30,110	37,917
Deadweight, tons	70,254	166,064	245,323
Displacement, tons	88,152	196,174	283,240
Length/Depth	12.364	11.280	11.465
Length/Beam	7.546	6.379	6.019

natural gas was first transported at sea in 1959, and such transport became commercially practical in 1964. Complex technical considerations are involved in carrying flammable liquid cargo at a temperature of about −162°C (−260°F).

However, the rate of growth in ship size has been phenomenal beginning with the 5,000 m^3 *Methane Pioneer* in 1959, leaping to the 27,400 m^3 *Methane Princess* in 1964 and to the 125,000 m^3 *El Paso Paul Kayser* in 1975. Designs of up to 300,000 m^3 have been prepared and given basic acceptance by classification societies.

Due to the extremely low density of the cargo (specific gravity of 0.45 to 0.50) such large ships have relatively light full load drafts and can be accommodated in channels at numerous ports. Perhaps one of their greatest problems is the extremely high freeboard which results in high wind loading.

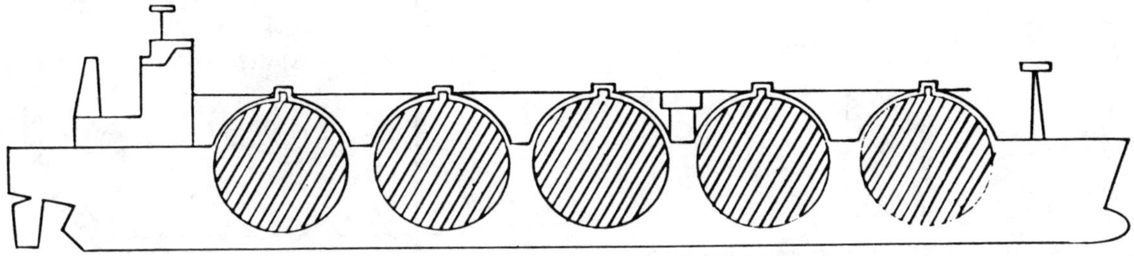

Fig. 10 125,000 m^3 LNG tanker—Kvaerner-Moss spherical tanks

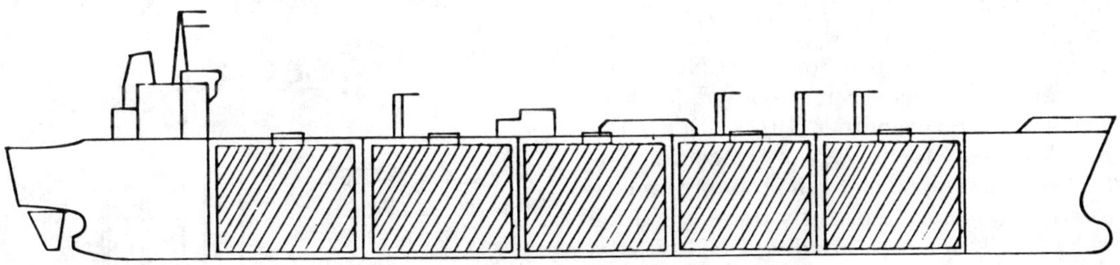

Fig. 11 125,000 m^3 LNG tanker—Conch free standing prismatic tanks

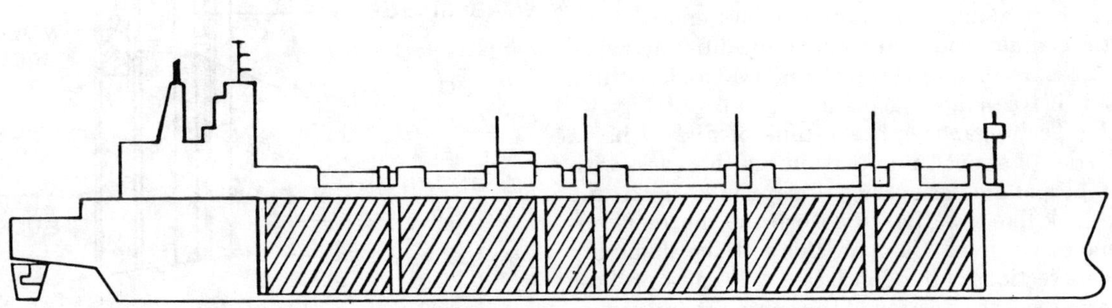

Fig. 12 125,000 m^3 LNG tanker—Technigaz membrane tanks

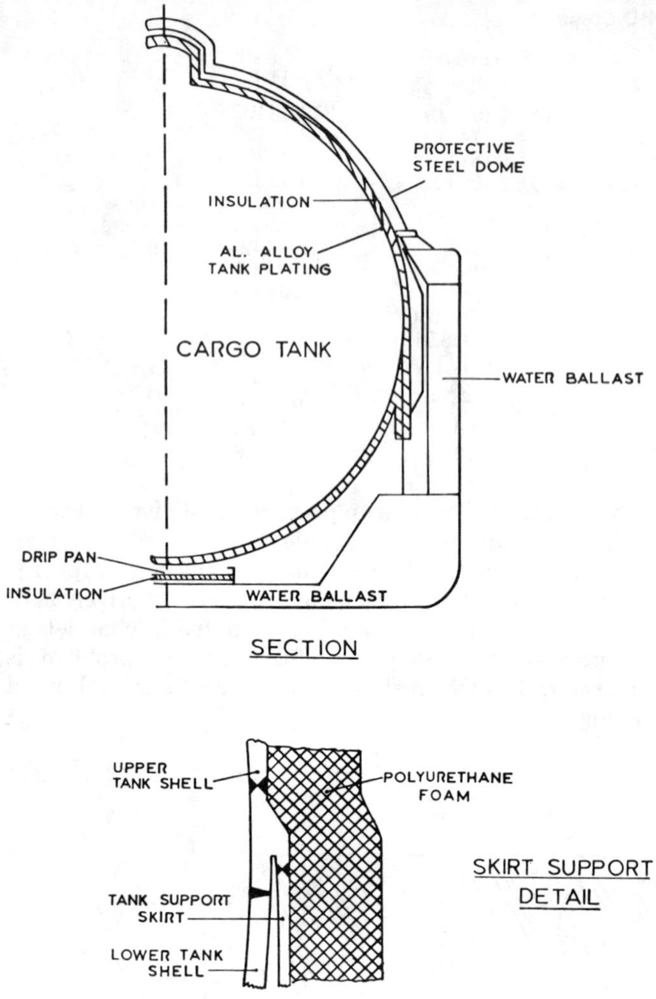

Fig. 13 Typical section Kvaerner-Moss spherical tank design

a. Containment Systems. In addition to rapid growth in size, inventors have devised a plethora of containment systems to serve as the tanks which hold the LNG. The systems used to date, as well as others proposed, are well described by Thomas and Schwendtner (1971). Due to the applicability of NASA space technology in cryogenics, new systems are still being developed and will probably replace some of the existing systems. Presently there are three systems in use by U. S. shipyards building LNG tankers (Connors, 1978). Figs. 10, 11, and 12 show the inboard profile of the ships and Figs. 13, 14, and 15 show the general nature of the corresponding containment systems.

Clearly the containment systems are quite different. The Kvaerner-Moss free standing tanks in Fig. 13 look nothing like the Conch free standing prismatic tanks of Fig. 15. Further, the Technigaz membrane tank of Fig. 14 has a primary barrier of about 1 mm (0.04 in.) in thickness with a series of perpendicular corrugations, which make the membrane look like a waffle, to absorb the thermal contractions and expansions. Compare this with the aluminum stiffened plate sections of the Conch tank. Because of these major differences, the selection of the containment system will have large impact on the final ship design. Here the naval architect must work closely with the overall system planners to choose the best overall system.

Among the factors which should be considered are the efficiency of cubic utilization in the ships, the vertical center of gravity of the tank and cargo, weight, material cost, need for secondary barriers, type of insulation, boil-off rate, visibility from the bridge, tank cool down time, susceptibility to mechanical damage, and ease of repair. Some of these factors affect construction, others operations, but they must be considered in total for an acceptable selection.

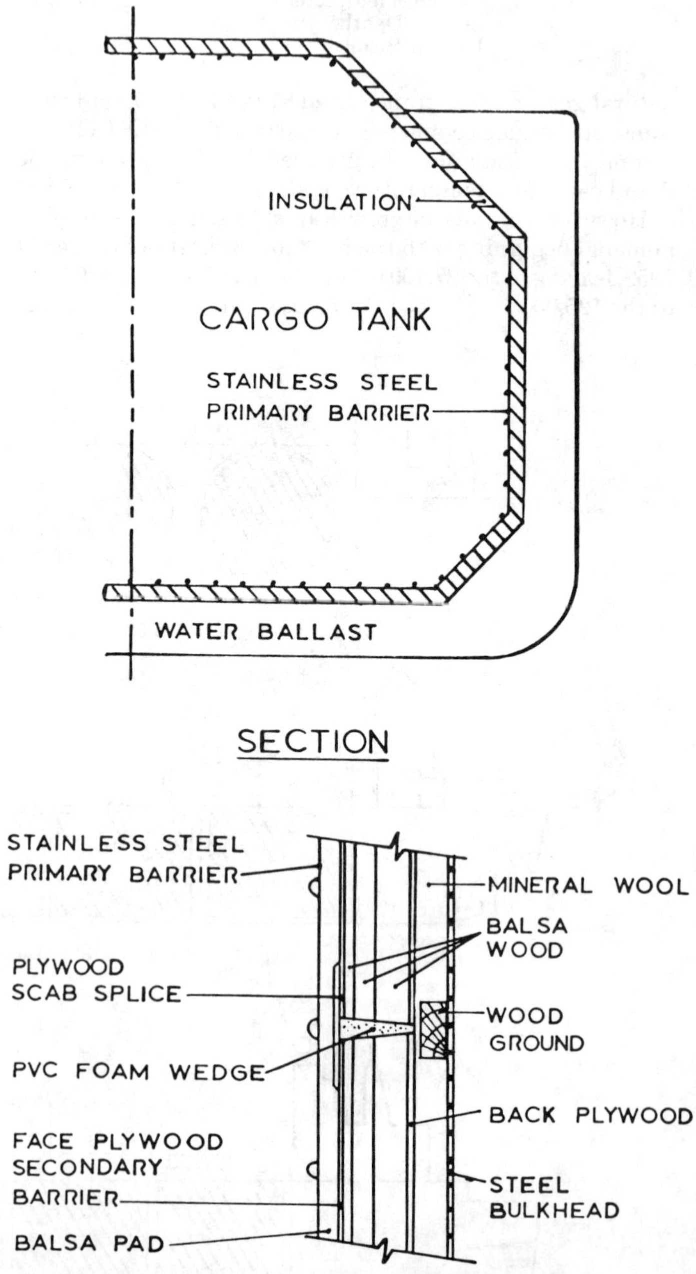

Fig. 14 Typical section—Technigaz membrane tank design

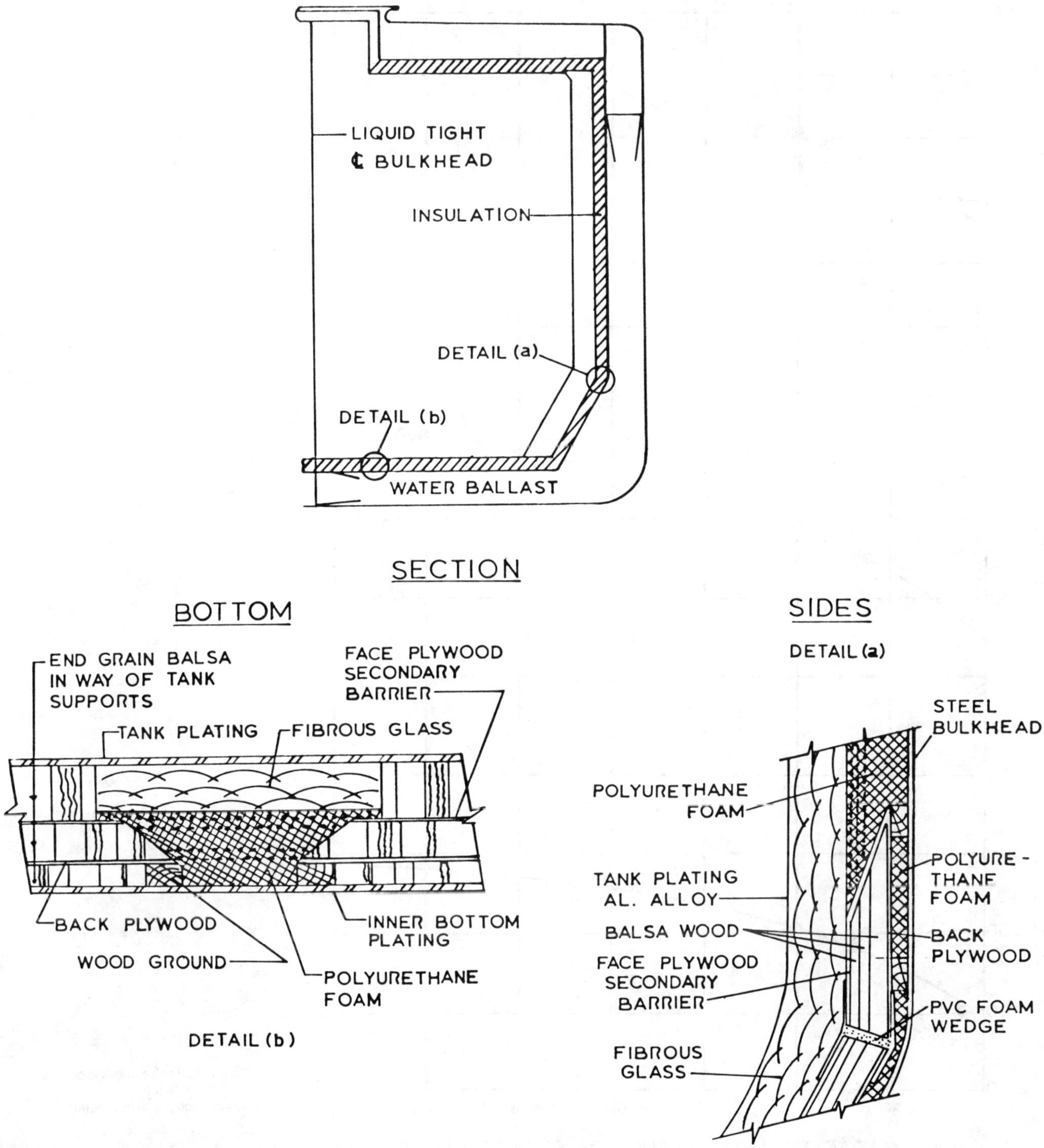

Fig. 15 Typical section—Conch free-standing prismatic tank design

b. Major Particulars. Fig. 16 shows the trend of principal dimensions (L, B, T) for existing and contracted for LNG tankers, and Fig. 17 provides preliminary light ship data. Designers must recognize that the containment system will exert a large influence on the final light ship weight. Depth has been omitted, since it should be made as large as possible subject to the limits of draft and stability. If port conditions or physical limitations of the building yard require significant changes in length and beam, they can be made without causing a large change in transportation cost. In contrast, a reduction in draft of only one ft was found to increase transportation cost by one percent.

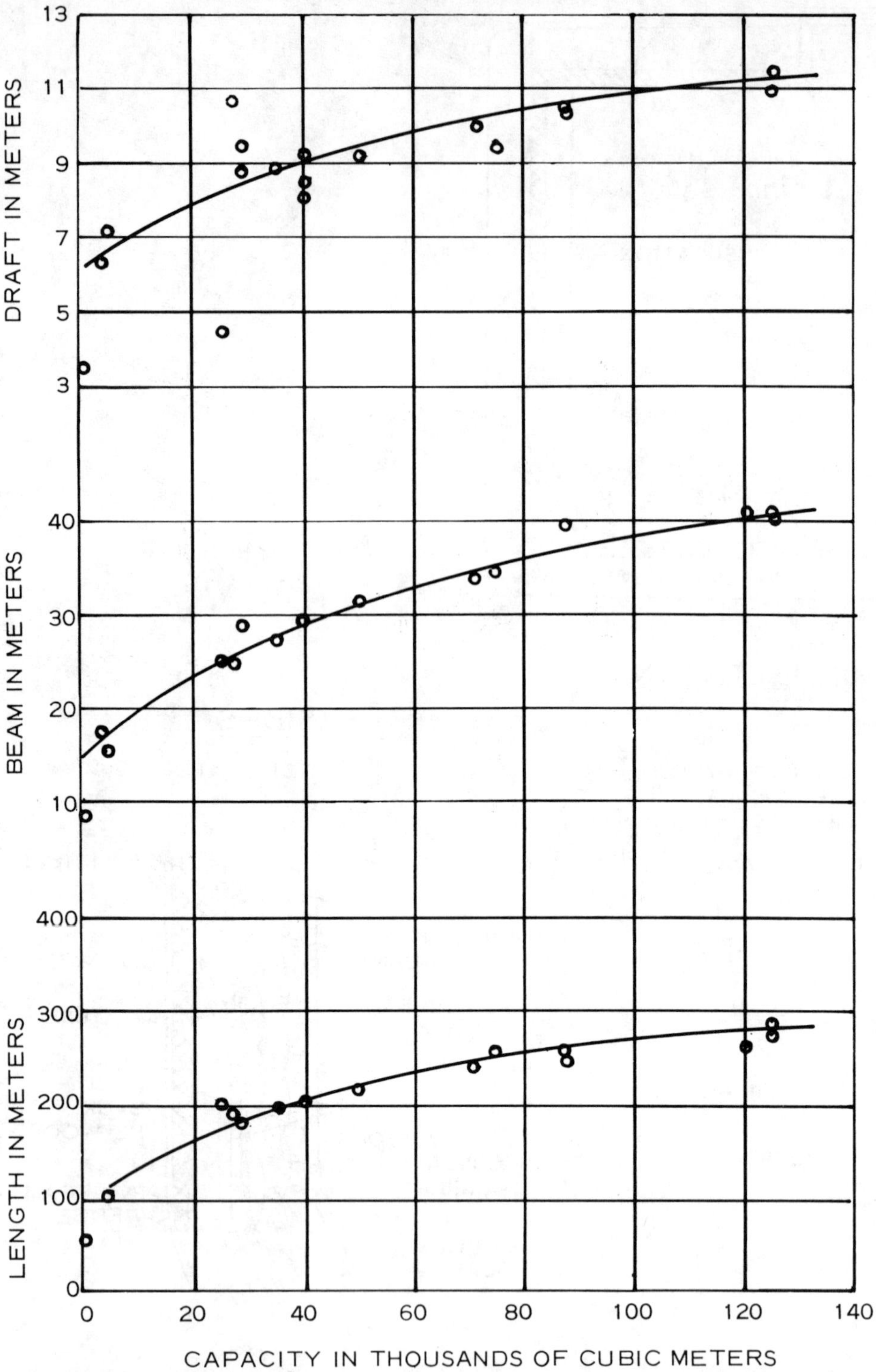

Fig. 16 LNG Tanker dimensions as a function of capacity in cubic meters

The choice of block coefficient, speed, and horsepower are closely related. Speed is normally determined in the systems analysis, and C_B and power first estimated in the basic design. Variations in C_B from 0.70 to 0.76 have not resulted in major changes in overall cost.

Due to relatively shallow drafts involved, the maximum acceptable power on a single screw is about 45,000 hp. Even this requires careful design and analysis of the lines and propeller. This power level imposes a speed limitation of about 20 knots on larger ships (21 knots trial speed). At least one design utilizes twin screw propulsion to achieve a sea speed of 23 knots.

The final major influence on LNG tanker design is the rules of the regulatory bodies, the classification societies, U. S. Coast Guard, and IMCO. These requirements are more fully discussed in Chapter XI.

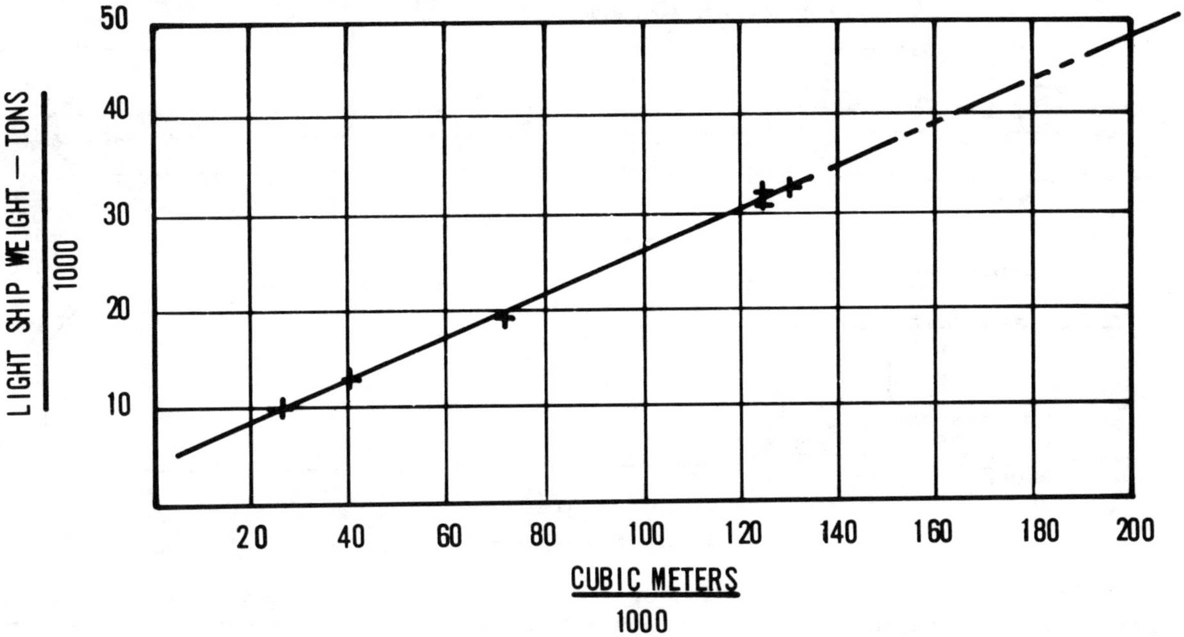

Fig. 17 LNG Tankers–Light ship weight

2.9 Integrated Tug/Barge. The term *tug/barge* has generally been adopted to describe push towing at sea, whereby the tug is positioned at the stern of the barge and provides positive control of thrust and steering through rigid or semi-rigid structural attachment. Its inland waterways counterpart is the typical push tow employed in relatively calm water, where the square bow of the towboat is simply snubbed up to the barge stern and secured by a relatively elastic wire rope system. This latter system will be described subsequently. The sea-going tug/barge must be configured so that there are no damaging effects due to bumping from surge differences, or grinding due to relative transverse, vertical, or rotational motions of the two units under wave action.

Present tug/barge configurations include the linkage arrangement such as *Sea-Link,* (Waller and Filson, 1972), (Glosten, 1967) which attaches the tug to the barge stern through a structural framework, but with the tug clear of contact with the barge. The linkage is such that the tug and barge act together in the three horizontal modes of surge, sway, and yaw; differential roll is partially restrained, but there is freedom to allow independent motions of pitch, and heave, Fig. 18.

Other arrangements such as *Artubar* (Waller and Filson, 1972) include trunnion mountings, which are extended from both sides of the tug's hull and socketed into wing wall extensions of the barge, affording full fixity of all motions between tug and barge with the exception of pitch. In this case, the tug is reasonably nested within the barge sides but with sufficient clearance to allow free pitching, Fig. 19.

The ultimate arrangement comprises the tug being completely nested into the barge recess, such as the *Breit/Ingram* configuration (Waller and Filson, 1972), (Rynn, 1975), (Giblon and Tapscott, 1973). Fig. 8, Chapter I shows a typical tug-barge system. Conversely, a twin-hull tug is used that straddles a tongue extension of the barge, such as the *Catug,* Fig. 20. The tug is fully restrained in all directions, so that it effectively forms a "power-navigation-accommodation" package of what is then a complete ship in all major respects. To this end, great attention is given to the interface between the two units to provide smooth, uninterrupted, and well-proportioned stern lines to take maximum powering advantage of the arrangement.

One method of mating the units consists of hydraulic rams mounted on the bow of the tug, which by engagement of pad-eyes on the barge deck physically pull the tug into forward position. During this movement, structural wedges built into the sides of the tug become pressed into wood and rubber lined sockets built into the surfaces of the barge extension, resulting in a firmly secured system against all relative motion between units. Retraction of the tug is accomplished by reversing the procedure.

a. Features and Comparisons. From a hydrodynamic standpoint the tug/barge has a distinct powering advantage over the conventional configuration of a barge being pulled by a tugboat, insofar as the push tug operates in the barge wake with reduced resistance and higher propulsion efficiency. In pull towing the tugboat has its own high resistance to overcome as an individual vessel as well as that of the barge it is moving, in addition to any apprended barge skegs needed for course-keeping stability in this configuration. Fig. 21 illustrates the performance characteristics. Resistance reductions for the combination of the tug and barge over the resistance of the barge alone plus the tug alone on the order of 40 percent are considered realistic. In particular, the locked tug/barge in a well-nested design has such low powering requirements as to be within 2 percent of an equivalent ship (Waller, 1972). Directional stability

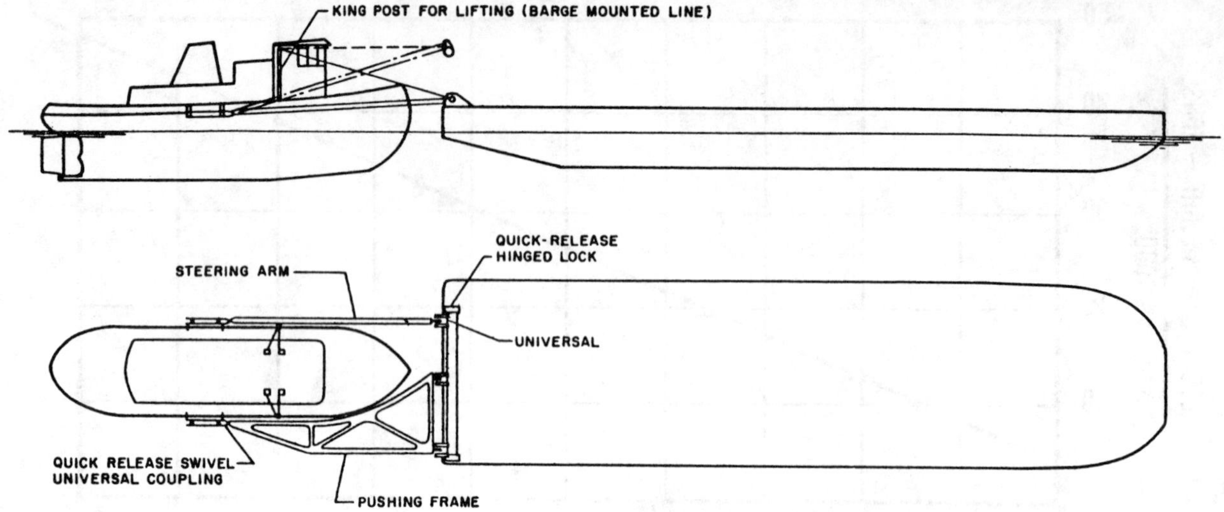

Fig. 18 Extended linkage SEA-LINK

and control underway is of course far superior to that of a towed barge and may be the most important consideration for operation in restricted waters.

At present, tug-barges under U. S. flag may operate with the normal tug complement instead of roughly twice that number required for a ship of the same size and power. This is a distinct advantage in manning. However, because of the integral nature of the configuration, no waiver of licensing (or of safety features) is allowed such as may normally be granted for a tugboat alone on its gross tonnage.

For the linkage type of tug/barge, no particular changes in size or shape of either the tug or the barge from the conventional form are mandated except that a higher pilot house than normally installed may be necessary for good visibility over the barge, particularly when the latter is in light condition. Sufficient local strength is required in the sides of the tug and the stern corners of the barge to take the forces imposed through the structural linkage.

The nested tug/barge however, has unique configurations in both the barge and tug units. The barge will have a ship-shape bow and its stern is deeply recessed (to about ⅔ the tug's length) with ship form stern lines at the outer shell and with molded inner shell except for the attachment devices for the tug. The tug is foreshortened to minimize the barge recess as far as practicable, must be deep enough to match the barge freeboard and still maintain draft for propulsion, and with its necessarily high deck house presents a fairly stubby and unorthodox appearance.

From the standpoint of logistics the linkage type of tug-barge is most suitable for the typical shuttle service of relatively short voyages, where loaded barges are swapped for empties, which service parallels inland river practice. The locked-in, nested tug-barge is specifically designed for longer ocean voyages, essentially remaining married in service, competing directly with sea-going ships on the basis of smaller crews and less initial construction cost.

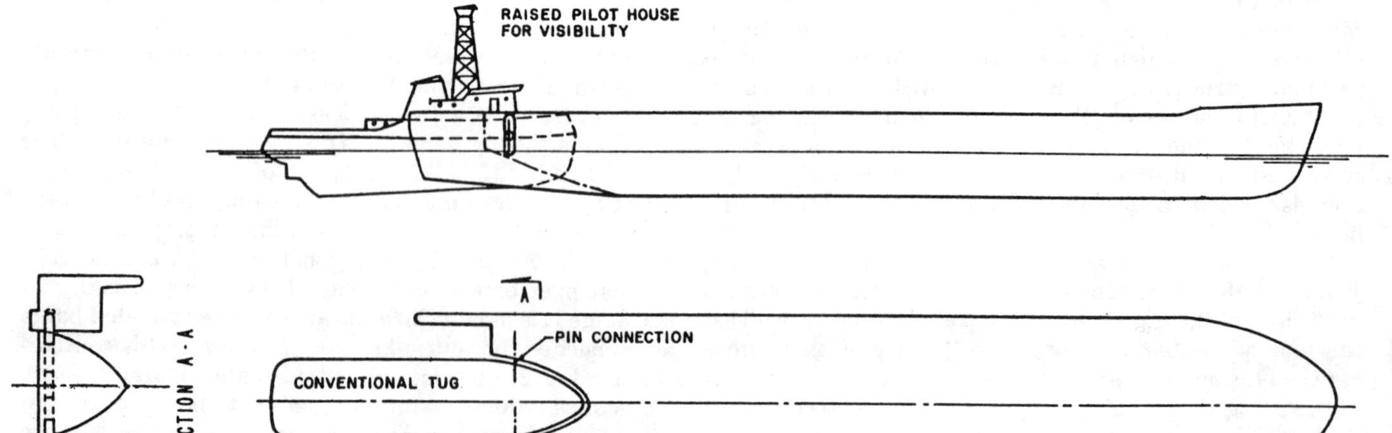

Fig. 19 Deep notch-pinned ARTUBAR

It has been well attested that the tug/barge costs less to build than a comparable ship for the same service, despite the extra structure, connecting mechanism, and other handling equipment. The main reason advanced for this anomoly is that there exists a large number of small shipyards that specialize in either tug or barge construction and can be selected on price for either unit. Shipyards that can (and frequently do) undertake the construction of both, must do so on a very competitive basis with the normally cheaper specialists. What is often overlooked, however, is the consideration that tug and barge specifications call for lesser quality than those of fully-formed ships (in the United States, at least)—a psychological factor that is difficult to assess.

c. Design Requirements. In tug/barge structural design it is necessary to determine the probable forces that are engendered on the barge and/or tug due to seaway, or due to normal propulsion and steering. This is needed in order to establish the structural requirements for the connecting devices and the subsequent reinforcements in the hulls, and to establish any additional hydrodynamic loading occasioned due to the restraint between units. While some generalized tests and analyses have been made (Eda, 1972) toward reasonable prediction, it is still necessary at this writing to conduct model tests of the specific configuration in waves of different direction to produce reliable force measurements (and in the case of the linkage system, relative motion maxima) to the satisfaction of the cognizant regulatory bodies.

Further, in the case of the fully locked-in tug/barge, the scantlings of the hull girder must be sized on the basis of the total combined length of the tug-barge configuration.

Propeller sizing and revolutions should be selected on the basis of the wake experienced in position behind the barge, and not on the typical independent tug requirement. Similarly, rudder area should be sized to the demands of the tug/barge as a single unit.

There are specific requirements for the tug itself, whether or not it is intended for the tug ever to leave its attached position at the barge stern. In order to qualify as a tug with reduced crew, it is required under present regulations to be able to disengage quickly in any dangerous situation, and to come around with attached towline to tow the barge in the conventional manner. In this respect, the tug must be a fully-found vessel, with necessary fuel for at least towing the barge to a port of haven within its routing, with all required life-saving equipment, and with the requisite stability—in effect, all of the necessary features for an independent vessel of its size and purpose. In addition, for the nested, locked-in, configuration sufficient ballast capacity must be included to control draft and trim differences between the tug and barge, during mating and, more importantly, as may be necessary for ready release.

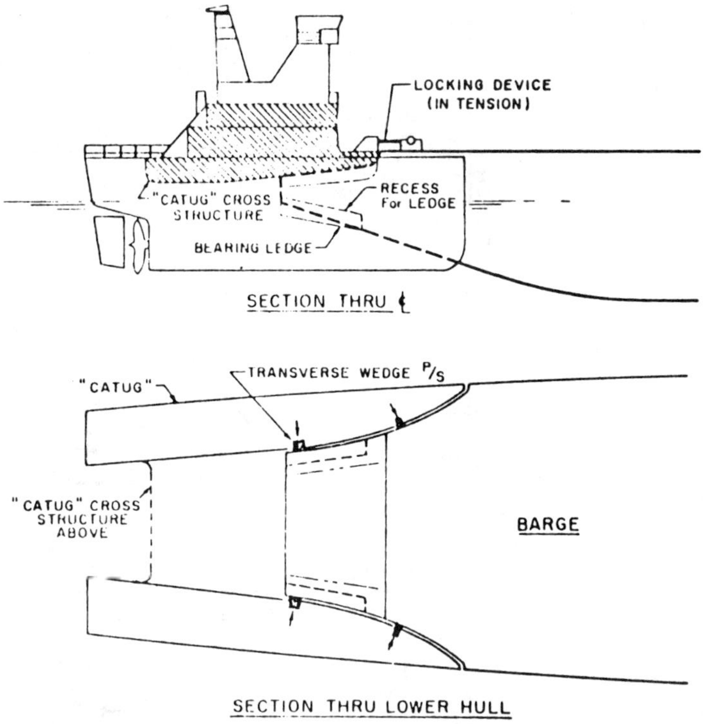

Fig. 20 Sectional views of close-connected-rigid seabulk CATUG system

2.10 Towboats. Inland river *towboats* (functionally they are pushboats), have been used since the early 1900s as the propulsive power for fleets of barges moving many types of cargo up and down the rivers of the United States. The early units were stern paddlewheelers of which the *Sprague* was perhaps the most famous, Fig. 22, and which were descended from the glamorous *packet boats* of an earlier era. These first towboats were powered by single cylinder steam engines which converted their reciprocating action into rotary motion at the paddlewheel. Steering was awkward but *flanking* was excellent, which maneuver involved backing on the paddlewheel, (now the propeller) and steering with flanking rudders forward of the paddlewheel. With the vessel moving ahead with the current of the river, control was maintained by swinging the stern either way to keep the tow in the center of the stream. This was done by controlling the flanking rudders forward of the paddle wheel and turning the paddle wheel astern. The same basic maneuver is used on modern towboats.

Each area of the country developed its own type of towboat to suit its particular environment. The rivers above Pittsburgh spawned boats that are small yet extremely maneuverable for use with small tows (six or less barges) in the narrow, twisting and swift rivers of that area. Similar small vessels predominate on the upper Mississippi River and its tributaries. Ohio River craft are larger and more powerful because the river and locks can accommodate larger tows (fifteen jumbo barges), while the largest boats are found on the middle and lower Mississippi River where tows of 40–50 barges are handled easily by boats of 8,000 to 10,000 hp, such as shown in Fig. 13 of Chapter I, and as described by van Mook and Courtsal (1975).

While push type towboats were developed and have their largest use in the Mississippi River system, they are also used extensively on the Gulf Intracoastal Waterway from Florida to Texas, principally between the New Orleans and

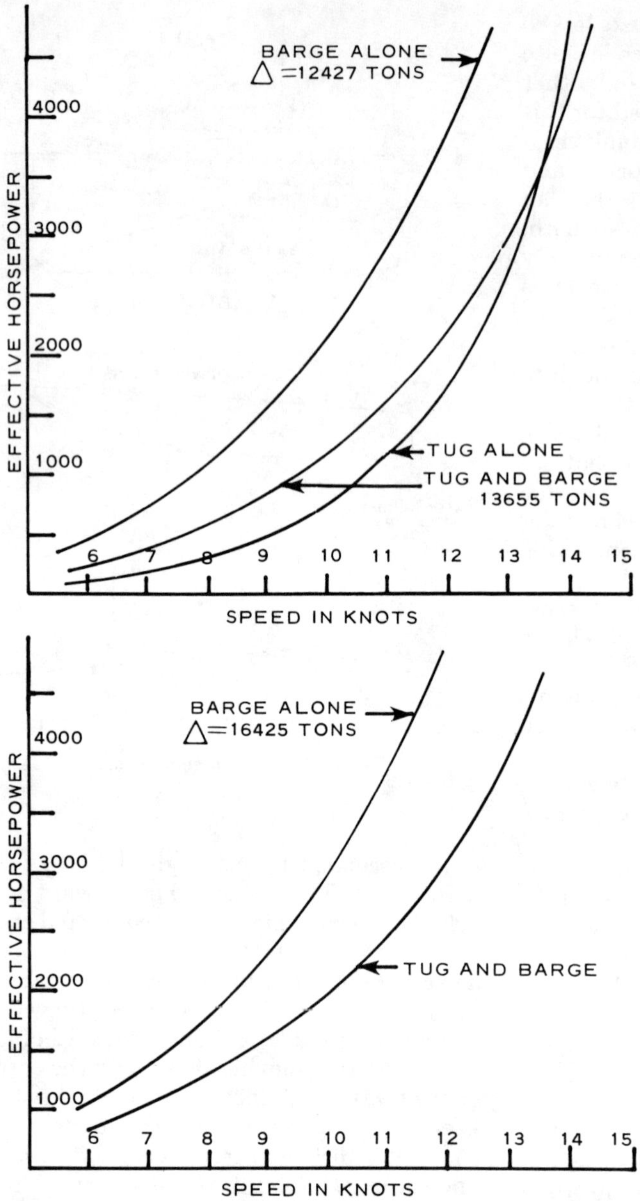

Fig. 21 Speed/Power relationship 12,000 deadweight ton timber carrier

Houston harbors. The Atlantic Intracoastal Waterway from Jacksonville, Florida, to the northeast also has push type towboats in use, and similarly from New York City up the Hudson River, but the boats are fairly small and the traffic relatively light. Reference should be made to Courtsal (1971) for a comprehensive treatment of the towboat/barge trade and design requirements. Push towing has recently been introduced in European rivers such as the Rhine.

A form of towboat which is about halfway between a normal harbor tugboat and an inland river towboat was developed for use in the West Coast area, and is similarly used on the lower Mississippi and in other bays and harbors where alternate uses of pushing and pulling may be required.

a. Design Considerations. Before the design of a towboat can be initiated, the parameters of its intended use must be established, and the owner must delineate his basic requirements. The following information is essential:

1. Rivers to be travelled,
2. draft limitations on those rivers,
3. speed anticipated up and down river to make the movement economically feasible,
4. size, make-up and type of tow (size of barges, extent of integration, draft of barges, expected fleet make-up),
5. crew anticipated,
6. engine manufacture preference, if any, and
7. Special equipment to be carried.

The size of the boat will be determined by the rivers to be run, the tow size anticipated, and the locks (if any) to be traversed. Horsepower can be determined from the speed and tow size.

b. Size and Speed. The size and speed of towboats have

Fig. 22 Barges were attached in "duck pond" fashion on early river boats. Main tow was tied to bows of barges lashed to boats' sides, leaving open water, or a pond, in front of bow

had unusual variations over the years. The early boats were large (the *Sprague* was 96.9 m (318 ft) long by 18.6 m (61 ft) beam) and had considerable horsepower for towing a typical fleet of 60 barges on the Mississippi River. Thereafter, the trend turned to smaller boats as they became more efficient and as more smaller rivers became available for navigation. However, as scientific methods were applied to towboat and barge fleet testing, coupled with the introduction of the diesel engine, boats began to increase in size and power, along with the size of the tows.

For up-river service, towboats are about 27.4 to 36.6 m (90 to 120 ft) in length and utilize 800 to 1,500 hp. The Ohio and upper Mississippi River boats range up to 48.8 m (160 ft) in length and 3,000 to 4,000 hp. The lower Mississippi River vessels are now up to 61 m (200 ft) in length and in the 10,000 to 11,000 hp range. However, towboats for use on locking rivers may have their size dictated by the anticipated conditions of traversing the locks.

As an example, in a river with 182.9 by 33.5 m (600 by 110 ft) locks, a tow of 15 barges of the 59.4 by 10.7 m (195 by 35 ft) standards can put nine barges (3 wide by 3 long) into the lock, back out, wait for that fleet to be warped out on the other side and when the lock is filled again, re-enter with the remaining six barges, plus the towboat. This limits the boat size to below the 59.4 m (195 ft) length of a barge in order to have room behind the boat for the propeller wash to escape the confines of the stern.

Towboat speeds vary from zero when the boat is pushing a fleet of barges into a lock, up to 19 to 21 km/hr (12 to 13 mph) when running free. The speed under tow is fairly well related to the river characteristics, ranging from about 10 km/hr (6 mph) for narrow rivers to 14 km/hr (9 mph) on the Ohio River and up to 16 to 19 km/hr (10 to 12 mph) for low resistance, integrated tows on the lower Mississippi River or the Gulf Intracoastal Waterway. Table 7 shows typical towboat particulars for a range of sizes and powers. Many special type boats are also in use which have sizes to suit their peculiar needs. There are boats for making up tows, feeding barges to barge carrying vessels, handling dredges, and many similar jobs. These are usually in the 15 to 21 m (50 to 70 ft) class with 400 to 600 hp, Fig. 23.

Boats used in push towing in rivers and pulling on a hawser in open waters are usually sized and designed for the open water use. They are 27.4 to 36.6 m (90 to 120 ft) in length with 800 to 2,000 hp. All the usual stability, watertightness and seaworthiness requirements are applied to this type of vessel and the only concession to push towing needs is the addition of towing knees built out from the pointed bow.

b. Configuration. A towboat is wide for its length and very shallow, due to river depth limitations. The draft is

Table 7—Towboat Dimensions

	Point Clear	Thurston B. Morton	Jason
Length O.A., m (ft)	21.3 (70.0)	45.7 (150.0)	57.9 (190.0)
Beam, m (ft)	7.9 (26.0)	10.7 (35.0)	16.5 (54.0)
Depth, m (ft)	3.2 (10.5)	3.8 (12.5)	3.8 (12.5)
Draft operating, m (ft)	2.4 (8.0)	2.6 (8.5)	2.7 (9.0)
Displacement—operating	215 t (237.0 s.t.)	768 t (846.0 s.t.)	—
SHP	2 @ 565	2 @ 2,200	3 @ 3,500
Type	Diesel	Diesel	Diesel
Speed—free running, mph	10 mph	12 mph	—
Crew	3	10	13
Eye level above water, m (ft)	9.8 (32.0)	—	13.2 (43.25)
Tow size	3,630 t (4,000 s.t.)	20,417 t (22,500 s.t.)	15,240 t (50,000 s.t.)

Fig. 23 Service type of river towboat

most important because it influences the propeller size, and the efficiency of propulsion is affected by the amount of water under the hull. Yet displacement must be achieved to carry the machinery and consumables required. If predetermined dimensions of length, width, and draft are forced on a designer he must achieve the required displacement by increasing block coefficient. This makes it very difficult to form acceptable lines to feed water to the wheels. Compromise here is essential but it must be on the basis of achieving the best performance.

Bow lines should be determined for the best flow conditions between the stern of the last barge in the fleet and the bow/midbody area of the boat, while keeping in mind that the towboat will frequently have to operate by itself. Spoon or modified spoon bows seem to be the best compromise to date.

Midbody shape is generally dictated by the need for engine room space and tankage requirements along with displacement needs. Square midship sections with a radius or double chine at the bilge knuckle are quite common.

The stern lines are the most important and, therefore, the most controversial area of design. While it is evident that smooth, uninterrupted water flow to the propellers is essential, there are many different ideas as to how the lines should be formed. Flanking rudders, stern tube, shafting, and struts all tend to block or interrupt the flow of water to the propellers, and any study or testing of stern lines must take these items into account.

Water to the wheels will not flow only from the bottom of the hull but in shallow water will flow mostly from the sides. Therefore, long flow lines from the midship section are necessary with easy and rounded shapes in both directions, longitudinally and transversely. Propeller size and the amount of water through it greatly influence the length of run from the midship section to the wheel.

Since steering and flanking are so very important to river towboats, the configuration of rudders must be carefully studied. It is well established that one steering rudder and two flanking rudders are needed for each propeller.

An essential for good towboat performance is the requirement for a seal at the aft end. That is, the transom should be immersed about 230 to 300 mm (9 to 12 in.) into the water when resting in still water. When going ahead the propeller race will flow directly off the bottom without creating eddying effects. However, when backing down, the seal will prevent the entry of air under the hull which could cause disastrous cavitation in the propellers and almost complete loss of reverse thrust.

In keeping with this necessity the placement of consumables should be made to allow the stern to remain at a constant draft no matter what condition of fuel and fresh water is aboard. Close control of the addition of ballast to assist in this problem can also be useful.

The bow deckline on a towboat is as broad as at the midbody except for a generous radius at each outboard corner. Each side of the bow will have one or two towing knees, which contact the stern of the barge directly ahead. The knees are usually lined on the forward side with hard rubber to reduce impact damage when boat comes against barge. When the boat turns, say to starboard, the port knees must take the ahead thrust while the starboard side wires from the winches are in tension to take the reverse pull caused by moving the stern to port. Therefore, the greater the spread of the towing knees the less the reaction forces.

Side winches as well as bow winches are used to give as great a spread of the couple forces as possible. The bow winches generally tie onto the barge just ahead of the boat to prevent sideways movement and the side winches lash to the outboard barges in the tow for the couple effect.

The superstructure for a towboat is used to house the engine room, (generally amidship providing the largest space in the hull); galley and mess, and the sleeping quarters extending to one or more levels above the main deck. The pilot house should be placed several levels above the main deck to give the pilot the best view possible of the river ahead of his tow and behind him. A typical arrangement of major towboat components for a medium sized towboat is shown in Fig. 24. When bridge clearances dictates, a retractable pilot house is used whereby the house can be raised above the main deck for good visibility between bridges and lowered when going under the bridge. A feature always found on river towboats is sloping forward and aft pilot house windows, to eliminate unwanted reflections of lights from the opposite direction.

c. Machinery. Despite an occasional steam or gas turbine plant, diesel engines are the established prime movers on river towboats. Medium speed and occasionally high speed diesels with reverse reduction gears are the rule, since slow speed diesels would have excess weight to cause unneeded problems in maintaining minimum draft. Control of the engines is always directed from the pilot house so that the pilot can react immediately to an emergency. Standby controls, of course, are installed in the engine room.

Dirty river water cannot be used for cooling systems in the diesels and even causes problems in heat exchangers. Therefore, most engines are cooled by a self-contained fresh water system that in turn is cooled by the river water in *keel coolers*. These can be channels on the inside or outside of the hull. The major requirement is to insure some flow over the cooler when the boat is pushing against a fleet or dock but not moving through the water, yet locating it so that it will not be damaged by grounding or by barges bumping alongside the towboat hull.

Another area affected by the condition of the river water is the stern tube and strut bearings. In some instances rubber bearings work very well, but in others the silt or acid in the water causes premature wear on the shaft or shaft sleeve. This has been solved in many instances by the use of oil lubricated bearings, properly sealed.

Stainless steel has become the universal material for propellers for river towboats. Its added strength is very useful in withstanding damage from river debris, and the

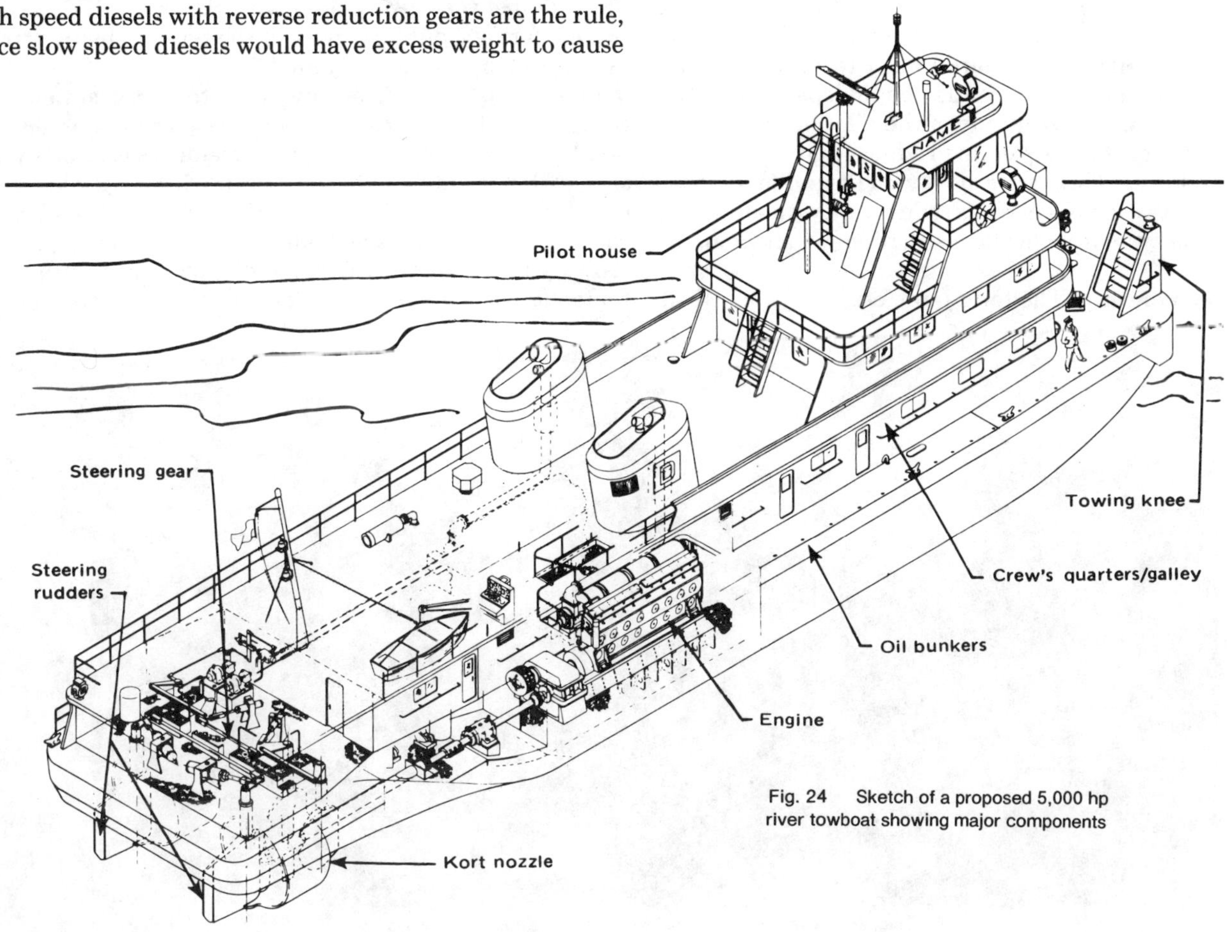

Fig. 24 Sketch of a proposed 5,000 hp river towboat showing major components

Fig. 25 Double-ended ferry for short runs

material stands up well to the river water conditions, where no salt water is encountered.

Most of the other machinery aboard the river towboats is commonplace for diesel propelled vessels except the steering system. Since there are two systems of rudders there are two systems of control and operation for the rudders. The steering rudders are linked together and the flanking rudders are linked together but independent of the steering rudders. On each side of the pilot house console is one tiller arm for the steering rudders and one for the flanking rudders. These arms move with the rudders so that the pilot knows exactly where his rudders are without taking his attention away from his fleet situation to look at an indicator.

2.11 Ferryboats. The term *ferry* originally described a vessel carrying passengers or cargo on short trips across rivers or harbors or coastwise to inland. The USCG defines a ferry as a vessel, "having provisions only for deck passengers and/or vehicles operating on a short run on a frequent schedule between two points with the most direct water route offering a public service of the type normally attributed through a bridge or tunnel."

Ferries now are primarily used to carry automotive equipment of all types and their passengers between two points on a highway, although the distances are not always short. For example, the Seattle to Sitka, Alaska route covers a distance of 1,407 km (874 miles). There are, of course, still some passenger only ferries operating, but these are very few in number compared to the combination automobile and passenger ferries. With the advent of the

Fig. 26 River ferry boat built for Louisiana Dept. of Highways

Fig. 27 Alaskan ferry COLUMBIA

energy crisis and the traffic congestion in major cities, it may be expected that their use will increase in the future. Typical characteristics of ferryboats for different services are shown in Table 8.

a. General Arrangements. Ferries can be single ended with a propeller or propellers at one end only, or double ended with propellers at both ends. Double-ended ferries (Fig. 25) are commonly used on short commuter runs where the time lost in turning around a single-ended ferry at each end of the crossing represents a considerable percentage of the time required for the trip, and, thus, the double-ended ferry can transport more units per hour than a single-ended ferry of the same size. Single-ended ferries are cheaper to build, are better sea boats in rough weather, and generally have higher propulsive efficiencies. They are therefore used on runs more exposed to the elements and those which are longer where the time lost in turning at the terminals is a relatively small percentage of the trip time.

Some ferries load through the sides of the vessel. These are generally small ferries on short runs across rivers where it is desirable to land head on to the current, and are of barge-like proportions providing adequate space to turn the cars around on the car deck, Fig. 26.

The length and beam of ferries is determined from the number of vehicles which must be carried. The vessel's freeboard is also determined mainly by the car loading and unloading facilities at the terminals. Thus, one of the common truisms in ferry design, which is, "first lay out a car

Table 8—Typical Ferryboat Particulars

SERVICE & TYPE	Alaska Single Ender	Puget Sound Double Ender	Staten Island Double Ender	Mississippi River Side Loader
Principal Characteristics				
Length O.A., m (ft)	127.4 (418.0)	134.1 (440.0)	89.6 (294.0)	46.0 (150.8)
Length B.P., m (ft)	114.3 (375.0)	127.4 (418.0)	—	45.7 (150.0)
Length Design Load Waterline, m (ft)	117.4 (385.0)	132.9 (436.0)	86.9 (285.0)	—
Breadth Extreme Overguards, m (ft)	25.9 (85.1)	26.5 (87.0)	18.5 (60.8)	—
Breadth Molded at DLWL, m (ft)	22.0 (72.0)	25.1 (82.5)	15.5 (50.9)	18.3 (60.0)
Depth Molded to Main Deck, m (ft)	7.3 (24.0)	7.7 (25.2)	6.3 (20.6)	3.0 (10.0)
Draft—Designed, m (ft)	4.9 (16.0)	5.2 (17.0)	4.1 (13.5)	1.8 (6.0)
Displacement, tons	6,700	4,336	2,260	800
Service Speed, knots	20	20	17.5	10
Capacities				
Vehicles	226	206	175	35
Passenger Berthing Capacity	324	—	—	—
Passengers Maximum	1,000	2,000	3,000	50
Fuel Oil, tons	745	504	90	81
Fresh Water, tons	165	60	16	22
Lub Oil, tons	36	20	—	—
Machinery				
Type	Geared diesel W/C-P Prop	Diesel-electric	Diesel-electric	Geared diesel
Total HP	18,000	9,350	6,000	900

deck, then design a hull and machinery deck under it, and the required passenger and navigation spaces above it."

All early ferries used single casings extending through the car deck for access to the space below the car deck and to the passenger spaces above the car decks, and to lead the exhaust stacks, piping, electrical, and ventilation services from the machinery space up above the upper decks. It is still the most efficient utilization of space since the combined width of two casings will always of necessity be greater than the width of a single casing and therefore will require a wider beam.

Single casings are therefore generally used in single-ended ferries designed for long voyages or for voyages in exposed waters. In recent years, however, as ferries have increased in size, and with the speed of loading and unloading playing an important part in the delivery capacity of the vessel and its operating costs, double casings have been more common. This permits a wide open area at the centerline of the vessel, with double lane loading at the approach ramps, allowing two cars or trucks to come aboard and move quickly and easily into position at the other end of the ferry without the necessity to turn or to dodge around casings. The lanes outboard of the casings are generally reserved for small passenger cars only.

With the need to provide ferries which can take any equipment that can operate on the highways, car deck heights have increased greatly in the past years, with 4.7 to 4.9 m (15.5 to 16 ft) clearance generally being the minimum provided. This requirement for trucks in the center lanes permits the installation of two level car decks outboard of the casings, with the upper levels served by fixed ramps at the ends of the vessel. With the increasing use of campers and recreational vehicles, ferries within the past 10 years have been designed with 2.3 m (7.5 ft) clearance at the lower outboard level and 3.2 m (10.5 ft) clearance in the upper outboard level in order to accept this type of vehicle.

The present practice is to provide a 2.7 m (9 ft) clear width for each car lane, and 5.5 m (18 ft) length of car lane for each unit of its rated capacity. The actual load carried may exceed the rated load by as much as 5 percent with certain mixes of cars, but the 5.5 m (18 ft) rated capacity will insure that the average number of vehicles carried will usually equal the rated capacity.

b. Propulsion. The frequent stops and starts of ferry vessels, including the common practice of shutdown for a good portion of the day, militates against the use of steam plants, resulting in the almost universal adoption of the diesel engine for propulsion power. Gas turbines, because of their very light weight, have made an impact on high-speed, light-weight passenger ferries such as the all aluminum 50.2 m (165 ft)—750 passenger ferry vessels operating in San Francisco Bay, and in the propulsion of passenger hydrofoils. Single-ended ferries generally use conventional drives, either geared diesel or controllable pitch propellers with single or twin screws. Because of the restricted depth of water in many harbors and areas where ferries operate, twin screw configuration is the predominant choice on single-ended ferries.

In early double-ended ferry installations, a steam engine or a diesel engine was directly connected to shafts leading to the propellers at the forward and after ends of the vessel. Directional change was accomplished simply by reversing the engine. This system was inefficient, however, and required from 35–50 percent more power to drive a vessel with the bow propeller operating at full rpm in reverse at the bow of the vessel than what was required for the same hull without the bow propeller. A number of ferries with diesel electric drive have since been built wherein Ward Leonard type controls were used, with only sufficient power delivered to the bow propeller to turn it at a reduced rpm approximating as close as possible a no thrust condition. With this type of installation, the loss due to the bow propeller was only some 7 to 10 percent.

More recently, diesel-electric systems on ferries have used alternating current generators with rectifiers to provide the direct current for the propulsion motors. On the most recent double-ended large ferries, feathering type controllable pitch propellers have been installed, and tests show that with the bow propeller in a feathered position located with two of the four blades vertically directly in line with the stern frame of the ferry, the loss due to the bow propeller is only 5 percent. Since this type of installation also eliminates the electrical losses, it is obviously the more efficient of the two systems and in these days of high fuel costs, the savings in operating costs are considerable.

In Europe, Japan, and in Canada, a number of ferry installations have been made using cycloidal propellers. They have excellent maneuvering characteristics but cannot always match the propulsive efficiency of conventional fixed pitch or controllable-pitch propellers.

c. Hull Form and Stability. Because of the broad beam required to carry a maximum number of cars and the clearance height required for trucks between the vehicle deck and the first passenger deck, ferries generally must have a relatively broad beam at the waterline. Draft restrictions generally require a shallow draft which calls for a high beam-to-draft ratio. In the past 25 years, straight line hull forms have generally been used either with a straight line from the guard leading down to the keel or straight lines from the guard to a chine line and from the chine to the keel. Subdivision and damaged stability requirements are generally not onorous although the USCG requires a one compartment standard for vessels below 45.7 m (150 ft) in length and a two compartment standard above. Quite often, the controlling factor from a stability standpoint is the USCG wind heel criteria at light drafts because of the high deck houses required for the carriage of trucks. Since the automobile and passenger load is relatively small the distance between light and load drafts is also usually small. However, there have been cases where the need to carry a large amount of fuel sufficient for two to three weeks operation, have increased this distance and in the light ship condition, the stability becomes marginal. This is particularly true in the straight line, vee form hull. Either permanent liquid or fixed ballast must be installed in these cases.

Ferryboats have one unusual characteristic of their hull forms that is peculiar to the class. They must be designed for minimum wave making. Most ferry vessels operate in areas close to shore, and waves from the vessels can cause erosion of the beaches. The waves must therefore be as small as possible, and, although ferries may have a total resistance higher than conventional hull forms, their wave making is small. Quite often such ferries can operate at higher maintained speeds in restricted waters than similar vessels which have a lower total resistance per ton of displacement.

d. Accommodation Requirements. Passenger accommodations on ferryboats will vary widely. Many small ferries have no accommodations whatsoever, and the driver and passengers are expected to stay in their car during the voyage. Toilet accommodations on ferries carrying under 100 people, or on ferries on short runs, are not stipulated by the USCG. Short runs have been defined as runs of 15 min or less. At the opposite end of the scale are the New York Staten Island ferries (Karppi, 1966) which have accommodations for 4,000 people to go along with a car capacity of only about 40 cars or some of the Alaskan ferries which have sleeping accommodations in staterooms for over 300 persons, and which by their size, decor, and service are more akin to cruise ships, Fig. 27.

e. Structural Considerations. Since most ferries are not designed for operating in the open sea, longitudinal bending in general, except on the very longest and shallowest vessels and on vessels with long and light superstructure, is not a governing criterion. The car deck structure presents the most serious problem confronting the designer in order to prevent buckling of the plating between beams due to heavy wheel loads. Present trends are toward the use of wide spaced beams together with higher tensile deck plating using I-beams or channels with their upper flanges left intact as deck longitudinals instead of inverted tees which, although ideal from a structural weight standpoint, leave too much width of unsupported plating between beams.

Racking needs careful attention on an automobile ferry. The only structural elements able to resist racking are the connections between transverse framing members and the decks at the side and the transverse bulkheads in the casings. While no specific criteria for racking forces exist, common practice has been to design the structure to withstand the loads imposed by a 15 degree roll in a period of seven seconds, which has generally been accepted by classification societies and the USCG as adequate for the service as a rule of thumb.

Finally, vibration needs special attention on ferries primarily because of the long unsupported spans of structure under the passenger decks. Natural frequencies of structural panels, particularly in the casings and upper decks should be checked and structural members and spacings selected to keep the panels well out of the frequency ranges of the propellers.

Section 3
Industrial Vessels

3.1 Introduction. Under the heading of "Industrial Vessels," Table 2 of Chapter I lists suction dredges, pipe laying vessels, drilling vessels, semi-submersibles, incinerator vessels, hopper dredges, fish processing vessels, fish catching vessels, fisheries research vessels, oceanographic vessels, hydrographic survey vessels, ocean mining vessels, seismic exploration vessels, and ocean construction vessels. This is obviously an abbreviated listing; for example, the Corps of Engineers Floating Plant List covers some twelve types of barges, six types of dredges, multiple launches, patrol boats, and survey boats, plus fourteen other categories of floating plants that can qualify as industrial vessels.

Obviously the characteristics of any of these types of industrial vessels are directly dictated by the mission that they are called upon to perform and there is not adequate space to deal directly with this wide variety of possible vessel configurations. Instead, only three generic types of industrial vessels are dealt with herein. These are fish catching vessels, offshore drilling vessels, and an ocean construction vessel.

Fish catching vessels of four general types are treated in an overall manner to indicate how the specific mission of catching a certain kind of fish dictates the characteristics that must be incorporated in each kind of vessel. Drilling rigs, of four different basic types, are discussed individually because of the markedly different characteristics that result from the task to be performed, the working environment, and the owners' preference; these include fixed platforms, jack-up rigs, column stabilized platforms, and drill ships. Finally, an ocean construction platform, designed and constructed to perform a specific series of missions in the ocean environment is described in some detail.

This total coverage shows the wide range of industrial vessels that the naval architect may be called upon to design and gives a rough idea of the range of missions and vessel characteristics with which he must have a reasonable familiarity.

3.2 Types of Fish Catching Vessels. Fishing vessels are the most numerous of all commercial vessels on the sea. Used in practically every body of water in all parts of the world, they come in all sizes from oar or pole-propelled dories or punts, through 76.2 m (250 ft) tuna catchers, to giant mother ships and processing ships with an infinite variety of forms, arrangements, propulsion machinery and deck gear. They can, however, be broken down into four generic groups with each genus classified by the type of fishing method employed:

1. By far the most numerous of these vessel types, is that employing a net towed behind the vessel. Fish enter the towed net which is then brought aboard and the fish released into the fish holds. This type includes trawlers and draggers of both side or stern trawling type and the nets either are

Fig. 28 North Sea stern trawler

towed along the bottom or at a predetermined depth below the surface. This type of vessel is used primarily for the production of bottom fish, or in the case of mid-water trawls, for trash fish, e.g., fish caught for the fishmeal and fish oil market. A subspecies of this class is the dredger used primarily for gathering oysters from the bottom, using a sled type dredge towed behind the vessel. Fig. 28 shows a typical North Sea stern trawler.

2. This type includes those vessels which carry a net to be set out in such a manner as to encircle a school of fish. After encirclement the net is closed at the bottom and brought back to the ship, hoisted aboard and the fish released into the holds. Generally called a seiner, or purse seiner this type of vessel is extensively used to catch tuna, salmon, herring, anchovies and any type or species of fish which have the characteristic of travelling together in *schools*. A sub-species of this class is the gillnetter. Nets are strung so that the fish swim into them, getting entangled by their gills.

3. Vessels that use hooks and lines to catch the fish. This type includes large number of sport fishing vessels which catch fish by trailing lines astern to catch salmon, swordfish, marlin, etc. Commercially, there are still some bait boats operating in the tuna trade which carry large bait tanks in the stern and throw over the bait to bring schools of tuna close to the stern of the boat where they are caught by hook and line on poles manned by the fishermen who surround the stern. The Japanese operate a large number of long line vessels which pay out long lengths of heavy line, buoyed at each end, and with separate lines secured at intervals which have baited hooks on the end of them. This method of catching fish is used not only in the tuna trade, but in catching halibut. Many years ago this was the common method of catching cod. Other commercial vessels called trollers will string up to 8 lines at the stern of the boat while moving ahead at slow speed. These are generally used for species such as salmon. Fig. 29 shows a modern tuna clipper.

4. The type of vessel that includes those which operate by setting out traps on the ocean bottom, each secured to a buoy on the surface. After an interval, the vessel returns, picks up the buoy and the pot. This type of fishing is used primarily for the harvesting of crabs, lobster and other crawling shell fish from the bottom. Fig. 30 shows a typical large crab boat. An excellent reference to the numerous types and sizes of fishing vessels, their methods and their handling gear is given by Blair and Ansel (1968).

a. Deck Arrangements. Initially, practically all vessels which towed nets behind them were set up to tow from the side. The side trawler was the most widely used vessel of this type in the North Atlantic, and the North European and New England steam trawlers were the predominant type of self-propelled fishing vessels in operation 75 years ago. Today, there are practically no side trawlers being built and they are being replaced by the stern trawler. The smaller stern trawlers have open decks aft and generally do no

Fig. 29 Modern tuna clipper by J. M. Martinac Shipbuilding Corp.

processing on board. The larger vessels are generally of the shelter deck type. Trawl nets are brought aboard through a chute at the stern and up to the top of the shelter deck. From here the fish are fed down to the interior of the ship where processing can take place out of the weather. Raised forecastles are common on the smaller vessels, usually providing space for crew accommodations. All trawlers require large winches to pay out and bring in the trawl net. In addition, they require extra drums or separate winches for bringing the net aboard. Mechanization of all parts of the operation has been developed with considerable success in reducing the number of crew required.

Fig. 30 Big crab boat by Marine Construction & Design Co.

Seine vessels, which use encircling nets, conventionally have large flat sterns, although herring vessels with poops and forecastles are used successfully in the North Atlantic. Modern tuna vessels are generally of the shelter deck type with the stern down to the main deck sloped to permit easy paying out of the net. The use of power blocks, which have hydraulically actuated sheaves and are suspended from the ship's boom, permit the net to be brought aboard quickly and to be flaked easily along the deck ready for quick launching after the net is all aboard. This allows considerable flexibility in the actual location of the seine winch which hauls in the cork and float lines.

Line fishing vessels are of a great variety of types. The tuna bait fishing boats all have raised forecastles with large bait tanks at the stern and with minimum freeboard at the stern to get the fishermen who are bringing the tuna aboard with hook line and pole, as close to the water as possible. Japanese long liners generally are poop and forecastle type with a well amidships from which the line paying and hauling gear is located, through complete shelter deckers of this service are not uncommon. Trollers generally have open decks aft from which the lines are streamed.

On the trap fishing type, lobster boats are generally small with a great variety of arrangements. They operate primarily in inshore waters. Crab boats operating in the North Pacific and the Bering Sea are generally of the poop and forecastle type with a midship well though they have been built either as full shelter deck vessels or with large flat sterns. Their deck gear is not as large or powerful as trawlers, but they do have pot haulers, line coilers, pit launchers, picking booms and cranes for handling the pots on deck.

b. Hull Form and Propulsion Machinery. Fish boats were originally practically all constructed of wood and molded hull forms were the rule, rather than the exception. Modern vessels are practically entirely constructed of steel, though there is still some use of wood and of glass reinforced plastic in the smaller sizes, and as a result, for reasons of economy of construction, chine and double chine hulls are now generally the rule. Because of the need for the payload to be at a maximum, hull capacity of fishing vessels is generally large and their displacement ratios are high.

Early self-propelled fishing vessels used steam reciprocating machinery if they were of the ocean going type and heavy duty slow speed gasoline engines if they were the inshore type. These have been entirely replaced with the advent of the diesel engine. Early diesel installations were heavy slow speed, direct connected engines, but now they are practically entirely of the geared diesel type with medium or high speed diesels and reverse reduction gears. With the development of synthetics for net lines, trawl nets have become bigger and the power required to pull the net under water has increased so that more and more attention is being paid in trawler design to getting the maximum thrust available at towing speeds. This has increased the use of propellers shrouded in nozzles and in the use of controllable pitch propellers to obtain good free running speed as well as optimized thrust in the towing condition.

The fishing vessel designer is always confronted with the problem of satisfying the desires of the owner for maximum speed using short waterline lengths and a correspondingly high displacement/length ratio. Since fishing vessels are generally individually owned, there is intense competition among various owners as to who has the fastest vessel. Quite often the power put in a fishing vessel is determined by the owner's pride, not by a cost effectiveness study by a naval architect. Owners will often consider that gaining a 2 percent advantage or a 0.2 knot gain in speed over a competitor is worth buying an engine that has 50 percent more power and which will consume 50 percent more fuel.

c. Fish Holds. The types of holds installed in fishing vessels to receive the fish and to carry them back to port vary widely. Some holds, notably those for bulk fish such as salmon, herring, or anchovies are bare with no insulation. These are obviously vessels which have short runs from the fishing grounds to port and which catch fish in the bulk and can thus load up in a short time. The standard method of holding fish for long periods in the early days was the use of ice, with the fish being gutted and headed if necessary, and packed in boxes in ice, or on shelves in the fish holds in ice. This method is still used for the fresh fish market in certain areas, but is gradually being replaced by methods of mechanical refrigeration. Many vessels, particularly in the bottom fish trade, will freeze the fish in quick freezers of the air blast or plate type and then store them in dry holds which are fully refrigerated.

Brine freezing is common in those vessels which operate in trades where the fish can be frozen in the round and do not need to be processed prior to being frozen. In this type of vessel the holds are actually tanks. The fish are placed in the tanks and refrigerated sea water is circulated through them. As the temperature lowers, salt is added to permit a lower circulating water temperature and, after the fish has been brought down to the desired temperature, the brine is then pumped out and the fish are maintained in the tanks in a dry frozen condition. The tanks must be heavily insulated and have a watertight inner lining.

Crab fishing vessels must keep the crabs alive until they can be processed on board or transported to port for processing. Holds in these vessels are also tanks and are generally large. Stability would not be satisfactory if such holds were permitted to go slack at one time. The water level is kept up in the hatch coamings at a level above the main deck. Water is circulated continually through them at a rate 3 to 4 changes per hour, and the discharge water is taken from the side of the hatch coamings to ensure the level is maintained.

d. Design Problems. There are two main problems confronting the naval architect in the design of a fishing vessel. The first has to do with his knowledge of the method of fishing and the type of gear to be used on the vessel. This is all important. He needs to know how the fisherman is going to fish, the type of equipment and gear he is going to use and it must be placed correctly on the vessel if the fisherman is to be successful. The location of the gear in the various areas required for maximum efficiency must be decided first and then the vessel developed around it.

The second major problem for the naval architect is to

design the vessel in a manner to ensure that it cannot be overloaded. There is no way to control a fisherman when he runs into a large supply of fish. If he has any space on board, he will fill every available spot with fish as long as he can catch them. Thus, the naval architect must know accurately the weight/volume relationship of the type of fish the vessel will be catching and must ensure the hold volume is such that if it is completely filled, the vessel will not be loaded beyond a safe draft and safe freeboard. Particular care must be taken in converting the vessels designed for use in one service, but being shifted into another type of fishery. Hold volumes suitable for iced fish are entirely too great if that same hold space is to be converted for use as a brine freezer.

3.3 Offshore Drilling Units—General. Offshore drilling is well into its second quarter century, having progressed from the simple deck barge grounded in the Louisiana marshes on which a typical land drilling rig was installed, to the highly complex mobile units of today that operate in hostile environments of the continental shelves of the world, with drilling equipment and methodology developed to meet the more exacting requirements of relatively deep water, platform motions, and wave and current effects.

An understanding of basic drilling practice, equipment, and systems is important to the naval architect engaged in the design of offshore drilling units. Of similar importance is an understanding of oil-field terminology (for instance, a *substructure* is the foundation under the drill floor and derrick, but it is located above deck, which to the naval architect makes it a *superstructure*). Reference should be made to Macy (1969) which gives a good description of drilling practices and equipment, in addition to a fairly comprehensive catalog of early types of offshore drilling units, many of which remain in operation today and whose basic principles are still being incorporated in new construction. Additional references on drilling fundamentals, as well as the adaptations made to suit the marine environment, are (The Petroleum Publishing Co., 1976a) and (Seymour and McCardell, 1967).

Further, for the serious student of offshore technology, whether he be directly concerned with drilling units or other associated aspects of ocean exploration, the continuing explosion of information found in the yearly proceedings of the Offshore Technology Conference will be of benefit.

Table 9—Jack-Up Drill Rigs

	Diamond M Gem	Western Polaris I	Key Gibraltar	Texas Star
Hull				
Length O.A., m (ft)	63.4 (208.0)	50.6 (166.0)	75.6 (248.0)	45.1 (148.0)
Beam, m (ft)	54.2 (178.0)	40.2 (132.0)	61.0 (200.0)	48.8 (160.0)
Depth, m (ft)	6.7 (22.0)	4.9 (16.0)	7.9 (26.0)	4.9 (16.0)
Legs				
Number	3	3	3	3
Type	Lattice Tubular	Caisson	Lattice Tubular	Lattice Tubular
Length, m (ft)	127.4 (418.0)	95.1 (312.0)	125.0 (410.0)	45.1 (148.0)
Chord Section	Square	Tubular	Square	Triangular
Aux. Footing m (ft)	14.6 (48) dia	64.0 × 51.8 × 3.7 (210 × 170 × 12)	14.0 × 10.1 (46 × 33)	11.0 (36) dia
Jacking System				
Type	Rack & Pinion	Hydraulic	Rack & Pinion	Rack & Pinion
Jacks Per Leg	2	4	4	3
Operating Conditions				
Hull Load Draft, m (ft)	3.8 (12.5)	7.0 (23.0)	4.6 (15.0)	3.7 (12.0) including footings
Drilling Water Depth, m (ft)	91.4 (300)	7.6 to 76.2 (25 to 250)	54.9 (180)	4.6 to 45.7
Maximum Drilling Depth m (ft)	9144 (30,000)	6096 (20,000)	7620 (25,000	4572 (15,000)
Number of Wells/Location	9	N.A.	9	12
Capacities				
Personnel	54	103	88	40
Bulk Mud/Cement, m^3 (ft^3)	170 (6,000)	212 (7,500)	269 (9,500)	113 (4,000)
Sack Mat'l. Sacks	3,500	3,000	3,000	2,000
Liquid Mud, m^3 (bbl)	223 (1,400)	249 (1,565)	223 (1,400)	12 (800)
Diesel Oil, m^3 (bbl)	651 (4,130)	277 (1,742)	318 (2,000)	127 (800)
Drill Water, m^3 (bbl)	879 (5,530)	748 (4,705)	795 (5,000)	604 (3,800)
Potable Water, m^3 (bbl)	143 (900)	142 (893)	159 (1,000)	114 (720)
Propulsion/Positioning				
Number	2	None	None	None
HP Total	1,500	None	None	None
Mooring				
Type	Wire	N.A.	Wire	Wire
No. of Lines	1	N.A.	4	2
Size, mm (in.)	41 (5/8)	N.A.	N.A.	N.A.
Length of Lines, m (ft)	411 (1,350)	N.A.	N.A.	N.A.
Anchor Weight, kg (lb)	3630 (8,000)	N.A.	N.A.	4536 (10,000)

a. Fixed Offshore Drilling Platforms. Fixed platforms are basically lattice type structural towers that are individually designed for specific locations, taking into account the sea bottom conditions, mean water depth, and anticipated maximum sea and wind. They are secured by piling driven deep into the ocean floor, and rise above the water surface to a height required where the drilling platform and all associated equipment and outfit are above the expected maximum wave crest elevation. Generally, fixed platforms are used for production drilling of multiple wells at locations previously explored by more adaptable mobile units, and are prevalent in water depths up to 90 m (300 ft), although a number have been built for depths to 300 m (1000 ft) or more. While the design of these units requires hydrodynamic analysis of the wave forces, and in some cases rather ingenious methods of floating and positioning by selective ballast control, (Metcalf, et al, 1979), they are basically civil engineering structures and generally beyond the scope of naval architecture. They are mentioned here for the sake of completeness, but reference should be made to other publications (The Petroleum Publishing Co., 1976b) for more detailed information.

b. Self-Elevating Drilling Platforms. More commonly known as *jack-up barges,* these units are described by the American Bureau of Shipping (1973) as:

> "the type of unit having a barge type hull with sufficient buoyancy and reserve buoyancy to safely transport the drilling equipment and supplies to desired location after which the entire unit is raised to a predetermined elevation above the sea surface. Units of this type may have legs which penetrate the sea bed or have legs with enlarged sections at their lower ends or be attached to individual bottom pads or a mat to minimize penetration."

Jack-up units have been the most popular and numerous of the various mobile types in existence. Drilling from a basically motionless platform and without the need for special equipment and outfit to cope with the drilling problems due to motions, is both more efficient and more economical, resulting in an overall least cost to the proprietor.

Jack-ups range in maximum water depth capability from about 15 m (50 ft) to over 90 m (300 ft) with installed drilling equipment capability for maximum well depths from about 2,300 m (7,500 ft) to 7,600 m (25,000 ft). This covers an extremely wide range of utility for meeting the needs of the exploration areas through the world's coastal waters, associated with a large variation in dimensional parameters, capacities and operating characteristics for the various units. Table 9 gives the particulars of some representative units, with additional data given by Danforth (1977).

1. Mat Supported Type. For use in very soft bottoms, such as prevail off the Mississippi Delta and in Indonesia waters, the bottom mat type of leg support is required. The mat is essentially a flat box structure having a platform approximately the same as the basic barge hull above, usually with open areas through the center to resemble a broad "A." Mat area is usually provided to give about 23.9 kPa (500 lb/ft^2) bearing pressure under normal working load, with a limiting storm criterion established so that the maximum pressure does not exceed twice that value at any point. Skirt plates of about 610 mm (24 in.) deep are sometimes provided all around the periphery of the mat to prevent loss of bearing due to possible scour and to resist any tendency for sliding off position under storm conditions. With these provisions, pre-loading (temporary overloading by adding ballast in the raised upper hull) is not normally required for mat types to prevent soil bearing failures under storm conditions.

Mats are normally compartmented to provide selected ballasting so that the mat is close to neutral buoyancy in all modes of raising and lowering, as well as on bottom position. This mitigates any instability problems that may otherwise occur. The remaining buoyant chambers are located in the vicinity of the leg attachment, thus helping to make the mat loading more uniform in bearing, as well as utilizing the heavier scantlings (required for pressure loading) in the attachment of the legs. Provision is made to de-ballast compartments when the mat is fully raised against the upper hull, for greater freeboard on long distance moves.

The structural advantage of a bottom mat is that it provides almost complete fixity of the legs, and thus affords legs of lighter weight and lesser dimensions (which in turn, reduces the wave force and moment). Its disadvantage is that the configuration cannot accommodate large trim angles (beyond one or two degrees) that would rack the structure, and thus the mat type is generally restricted to flat and even sea bottoms.

2. Independent Leg Type. For medium to hard bottoms, the independent leg type of jack-up is most suitable, and where such bottoms are uneven or sloped, it is essential. The size and shape of the footing (or spud can) under each leg depends on bottom conditions and considerations of impact loadings in the structure when touching down on location. The *Le Tourneau* type of footing is cylindrical with a shallow conical bottom except at the center which has a sharp conical point for providing initial penetration with minimum shock. Other types may simply be cylindrical with flat bottoms, where the leg jacking system is rubber mounted to accommodate any shock loading.

In general, footings are sized to about 95.8 kPa (2,000 lb/ft^2) normal bearing pressure, although a few installations have high bearing pressures of 478.8 kPa (10,000 lb/ft^2) or more with *cookie-cutter* bottom shape to ensure penetration into hard soils. In the latter case, provision may be made for attaching supplemental footing structure for operation in softer bottoms.

Penetration of the sea bottom is essential in order to ensure proper bearing, prevent scour, and provide against sliding under storm conditions. On the other hand, it must be established that under storm loading excessive penetration does not occur (such as may happen in breaking through a hard soil structure into a soft one). For both of these reasons, pre-loading of legs to the maximum design load plus storm overturning moment is required. On three-legged units, pre-load is accomplished by providing hull ballast

Fig. 31 Jack-up drill rig WESTERN POLARIS I

tankage at each leg to the full extent required. On units having four or more legs, pre-loading can be effected with minimum ballast, by selectively unloading opposite pairs of legs (lifting up on the jacking system) to transmit the normal weight through the hull structure unto the remaining legs. All pre-loading operations are, of course, undertaken with the hull jacked up clear of the water surface.

Because of the deep penetrations, high pressure water jetting systems are provided down to each footing to help free the legs from the surrounding soil at the time of leg-raising for moving off location. The footings are also normally ballasted to a neutral buoyancy condition for the same reasons as the mat.

The particular advantage of the independent leg type of unit is that it can locate on uneven bottoms, remaining level due to individual leg height control. It can also thereby adjust to changes in bottom soil conditions that may develop due to storms, tides, etc., without the need for relocating. Its main disadvantage is that the legs, having to be considered pin-jointed at the bottom rather than fixed, are significantly heavier than those with a bottom mat. This can cause excessive stressing due to dynamics when the unit is in long transit with the legs fully raised, and it is not unusual that top leg sections have to be removed (burned off, normally stowed on deck and rewelded later) for this condition, whereas this may not be necessary for the legs of an equivalent mat supported unit.

3. *Leg Structure and Jacking Systems.* There are two basic types of leg structure. Typically, the relatively slender tubular column is used predominantly with the bottom mat, where due to end fixity the bending stresses in the leg are less severe and the slenderness ratio of the member is not as critical. Further, local reinforcement can be added to the leg if required, at the precise storm position of the main hull, which is not subject to change as with independent legs. Tubular legs may range in diameter from about 0.9 to 1.2 m (3 to 4 ft) for the smaller units to about 3.7 to 4.6 m (12 to 15 ft) for the deeper, larger rigs. Fig. 31 shows a typical unit of this type.

The truss leg, having three or four chords each, depending on jacking requirements, is generally installed on independent leg units. The leg requires greater structural stiffness, as pin ended members, and high strength along a great part of its upper length (due to variations in penetration, etc.) and along its lower length (to take dymamic loads in transit). Center to center distances of chords may range 4.6 to 10.7 m (15 to 34 ft) or more.

While there have been a number of types of jacking systems used, the two most prevalent are the continuous rack and pinion and the stepwise hydraulic mechanical ratchet. The rack and pinion system is most commonly used with the truss type leg, with continuous racks built into the leg chords and with the pinions in a drive case attached to the hull. Drive through a series of spur gears is usually by individual electric motors, although hydraulic motors are sometimes used. Two basic installations are most common:

1. The single rack per chord, with the pinion case fixed to the hull, as employed in the *Le Tourneau* type system. With a rack on each chord, as normally installed, the pinions not only support the vertical load but absorb most of the leg moments due to wave and wind directly from the chords in a vertical couple. The bracings are thus relieved from transmitting leg moments in a horizontal couple between upper and lower guides fixed in the hull, but do have to resist high compressive loading due to the horizontal force component of the pinion reaction on the sloping rack teeth.

2. The opposed pinion, double rack jacking system, such as developed by National Supply, which eliminates the compressive effect on the bracing. Typically, however, the pinion cases are rubber mounted, top and bottom, to minimize shock loading and to provide easy alignment of teeth during leg raising and lowering. The pinions thus carry little more than the basic vertical load, requiring the bracing to transmit most of the leg moment in a horizontal couple between upper and lower guides, which further requires a substantial *jack-house* structure to carry the upper guide loads back down into the hull.

Variations on the above, such as opposed pinion cases fixed to the hull, and with racks used with cylindrical legs of moderate diameter, etc. have been used, but the above are the most common.

The stepwise, ratchet-type of jack usually employs structural pins top and bottom that alternately engage and disengage through pin holes (or occasionally lugs) spaced evenly along the leg length to climb or descend in steps. Hydraulic cylinders of sufficient capacity are used for the powering means and to provide the necessary stroke whereby with one pin engaged, the other pin is disengaged and pre-positioned in the next succeeding pin hole, and the movement proceeds in the desired direction, up or down. This system is usually employed with cylindrical legs associated with the mat type unit, although it has also been used on independent legs and individual chords. In position, the hydraulic cylinders may serve to maintain vertical loading or a structural lock-off system may be employed. In any case, the design usually considers that the leg moments are taken out principally as a horizontal couple (which for the mat type is of much less magnitude than for the independent leg).

4. *Hull Design Considerations.* The hull form and structure for a typical jack-up unit is almost exclusively that of a wall-sided barge, as being most suitable for its major purposes. Good depth of hull at the wide-spread leg locations provides the best structural arrangement for absorbing the leg-generated moment loadings, in jack-up position under wind and wave forces and in transit under motion dynamic loadings. Further, maintaining deep, simple girder construction between legs for effective support of the weights of equipment, supplies and drilling loads when jacked-up, along with providing the optimum usage of space and the optimum buoyant support when afloat, makes the barge-type hull the most reasonable selection, as well as the most economical.

Transit speed has not been a major consideration. Units basically remain in one general area of operation indefinitely, and long ocean transits may occur very few times in a lifetime. There have been some refinements in shape such as raked ends, and less extreme triangular plan forms for three-legged units, with transit speeds improving from about three knots for early rigs to about six to seven knots today (partially because of the more powerful tugs presently available). Some units have propulsion-assist drives to aid in towing speed, as well as to enable short, unattended moves, and several are fully propelled ships with higher speed forms, but the basic barge shape and construction remains prevalent.

5. *Wave Force Analysis.* The determination of wave forces and resulting overturning moments on the legs of a unit, when in jack-up position, has universally been based on the theory of regular waves of finite height in shallow water. The method of Bretschneider (which is reproduced in ABS Rules) provides a simple and fairly accurate evaluation of the forces and moments produced by a given wave, but it has largely been replaced by the more analytical Stokes theoretical wave (generally the 5th order approximation) which by modern computer methods is faster, and more accurately determines the maximum effect on the complete array of legs being passed by the design wave.

Because the leg forces are primarily due to viscous drag, which varies as the square of the water particle velocities, it is difficult to assess the effects by means of spectral analyses that are now utilized for other types of vessels. Further, shallow water spectra are not developed as yet to any great degree of confidence, or universality, for the various coasts around the world, and the regular wave approach remains the accepted standard for jack-up units.

6. *Seaworthiness Considerations.* The jack-up unit is subject to many modes of casualty that do not normally affect a surface vessel. These include:

- loss of bottom support due to soil phenomena, or due to *blowouts* during drilling that displace the sea bottom under the unit,
- increase in storm intensity beyond the design criteria established for the unit, while in a jacked-up position and unable to escape, resulting in leg failure,
- leg impact during touchdown, resulting in leg and/or upper hull damage, or
- rising weather while legs are being implanted in the soil, but with the main hull still waterborne, causing extraordinary loadings in leg/jack/hull.

These situations are in general beyond the designer's control, but they are not the major causes of jack-up casualty. Actually, it is with the unit in transit, particularly on a fairly long move with legs fully raised and all tow preparations made and approved, that major casualties and total losses occur. In strong seas, the dynamic loadings on the up-raised legs can be severe and repetitive, resulting in possible collapse or in fracturing of the supporting hull structure. Further, the large inertia of the gross barge structure augmented by the very high and heavy legs can result in minimum response to the sea and large shipping of green water, with possibly disastrous consequences.

This is the province of the naval architect and requires diligent attention. The only rule requirement extant to date is that the legs be structurally adequate to withstand a 15 deg roll (single amplitude) in a 10-sec period, which is not broad enough in scope to cover all eventualities for all configurations, and which does not address the seaworthiness of the entire unit.

3.4 Column Stabilized Drilling Platforms (Semi-Submersibles and On-Bottom Units). The ABS Rules describe this type of unit as one . . .

> "that depends on the buoyancy of widely spaced vertical columns, of similar attitude and shape, for flotation and stability. This may apply to the normal drilling position afloat or in the raising or lowering of the unit to other operating positions, where applicable. At the top, the columns are connected to a working platform or deck that contains the drilling equipment. At the bottom, hulls or footings may be provided for additional buoyancy or to provide sufficient area to support the unit when bearing on bottom. Bracing members of tubular or structural sections may be used to connect the columns or lower hull and provide a truss to support the working platform."

The first column-stabilized units were developed solely for on-bottom operation (designated as *submersible* in contrast to the later *semi-submersible*). The configuration essentially consisted of a rectangular barge hull, with a large cylindrical column at each corner extending upward to the working platform decks. Intermediate, small diameter columns helped to support the upper platforms and provided access to some of the pump control spaces in the hull. Major moves were made afloat on the barge hull, and short moves afloat on the columns (semi-submerged). The unit was lowered to the bottom by controlled ballasting of barge hull compartments, with the required stability being afforded by the columns during the operation. On bottom, the columns were then ballasted as required for good bottom bearing. Essentially, the unit provided the same function as today's mat-supported jack-up, except that the allowable water depths were expressly limited by the fixed elevation of the upper platform above the barge hull. For the early submersibles, design water depths ranged from about 12.1 to 12.3 m (40 to 70 ft).

Fig. 32 Model of proposed Mohole semi-submersible drilling platform

The next step in the development was the replacement of the lower barge hull with individual pontoons under each of the columns, with horizontal and diagonal bracings to effect full structural tie-in and support of the upper platforms. These units were designed for water depths of over 30 m (100 ft) but their service was extended by the early realization that they could operate in deeper waters in semi-submerged position with little change in concept. (The *Sedco 135* series of three-column units were so named for their on-bottom capability in 41 m (135 ft) water depth, but their usage through the years has almost exclusively been in semi-submerged operations in deeper waters.)

Through the 1960s, the development of the semi-submersible paralleled the growth of offshore drilling on a world-wide basis, and the need for greater transit speeds for long moves and far-flung operations resulted in the twin-hulled configuration for semi-submersibles. Pioneered by the Mohole design (McClure, 1965), as shown in Fig. 32, the twin-hull type has proven to be superior in transit speed, justifying the incorporation of self-propulsion to minimize or eliminate the need for tugs on long moves, and without sacrificing any of the minimum motion characteristics in semi-submerged drilling mode. This configuration remains the most popular today, and the design features and requirements that follow emphasize this arrangement (although the basic principles are essentially the same for all types). Arrangements of typical units built for world-wide operations are shown in Figs. 33 and 34. Table 10 shows characteristics of a number of units.

a. Structural Loading and Analysis. There are a large

Table 10—Semi-submersible Drill Rigs

	Pacesetter III	Sedco 700	Zapata Ugland	Aker H-3	Sedco 135
Lower Hulls					
Length O.A., m (ft)	79.2 (260)	89.9 (295)	111.9 (367)	108.2 (355)	30.5 (100)
Beam, m (ft)	15.2 (50)	15.2 (50)	12.2 (40	11.0 (36)	13.3 (60
Depth, m (ft)	6.1 (20)	6.4 (21)	7.6 (25)	6.7 (22)	7.6 (25)
Width O.A., m (ft)	61.0 (200)	74.7 (245)	64.0 (210)	67.4 (221)	103.6 (340)
Section Shape	Rectangular w/ Radius Corners	Oblong	Rectangular w/ Shape Corners	Rectangular w/ Shape Corners	3 Individual Footings
Stability Columns					
Number	6	8	6	8	3
Diameter, m (ft)	9.8 (32)	4 @ 9.1 (30) 4 @ 5.5 (18)	9.1 (30)	4 @ 7.9 (26) 4 @ 5.8 (19)	10.7 (35)
Weather Deck					
Elevation Above Keel, m (ft)	33.8 (111)	39.6 (130)	42.7 (140)	39.6 (130)	50.6 (166)
Length O.A., m (ft)	65.2 (214)	70.4 (231)	73.1 (240)	69.2 (227)	73.9 (242)
Width O.A., m (ft)	48.5 (159)	60.0 (197)	56.7 (186)	61.0 (200)	65.3 (280)
Operating Conditions					
Transit Draft, m (ft)	5.9 (19)	5.8 (19)	6.4 (21)	6.1 (20)	6.1 (20)
Operating Draft, m (ft)	18.3 (60)	24.4 (80)	24.4 (30)	21.3 (70)	24.4 (80)
Operating Displacement, tons	19,152	24,980	26,786	19,000	15,000
Survival Draft, m (ft)	13.7 (45)	18.3 (60)	19.8 (65)	18.3 (60)	24.4 (80)
Maximum Wave					
Drilling, m (ft)	15.2 (50)	12.2 (40)	22.9 (75)	15.2 (50)	—
Survival, m (ft)	30.5 (100)	30.5 (100)	30.5 (100)	30.5 (100)	30.5 (100)
Air Gap—Drilling, m (ft)	13.2 (43)	N.A.	13.7 (45)	15.2 (50)	22.9 (75)
Working Depths					
Water Depth, m	366	457	305	183	183
Water Depth, (ft)	(1,200)	(1,500)	(1,000)	(600)	(600)
Drilling Depth, m	7,620	7,620	7,620	7,620	7,620
Drilling Depth, (ft)	(25,000)	(25,000)	(25,000)	(25,000)	(25,000)
Capacities					
Personnel	84	90	90	75	48
Bulk Mud/Cement, m^3 (ft^3)	268 (9,480)	272 (9,625)	566 (20,000)	510 (18,000)	230 (8,105)
Sack Matl. Sacks	6,000	N.A.	4,500	6,000	3,250
Liquid Mud, m^3 (bbl)	273 (1,710)	318 (2,000)	270 (1,700)	231 (1,450)	262 (1,650)
Diesel Oil, m^3 (bbl)	1,057 (6,650)	1,065 (6,700)	1,590 (10,000)	2,687 (16,900)	808 (5,080)
Drill Water, m^3 (bbl)	1,908 (12,000)	1,892 (11,900)	1,590 (10,000)	2,290 (14,400)	1,615 (10,160)
Potable Water, m^3 (bbl)	207 (1,300)	191 (1,200)	111 (700)	N.A.	74 (467)
Propulsion					
Number of Screws	2	—	2	2	—
Number of Thrusters	—	4	—	—	—
H.P. (Total)	6,000	6,400	6,000	5,000	—
Mooring					
Type	Chain	Chain	Chain	Chain	Wire Rope
Number of Lines	8	8	8	8	9
Size, mm (in.)	76 (3)	76 mm (3)	82 (3 1/4)	76 mm (3)	76 (3)
Length of Lines, m (ft)	1,067 (3,500)	1,219 (4,000)	1,219 (4,000)	914 (3,000)	1,463 (4,800)
Anchor Weight kg	13,600	13,600	18,140	13,600	13,600
(lbs)	(30,000)	(30,000)	(40,000)	(30,000)	(30,000)

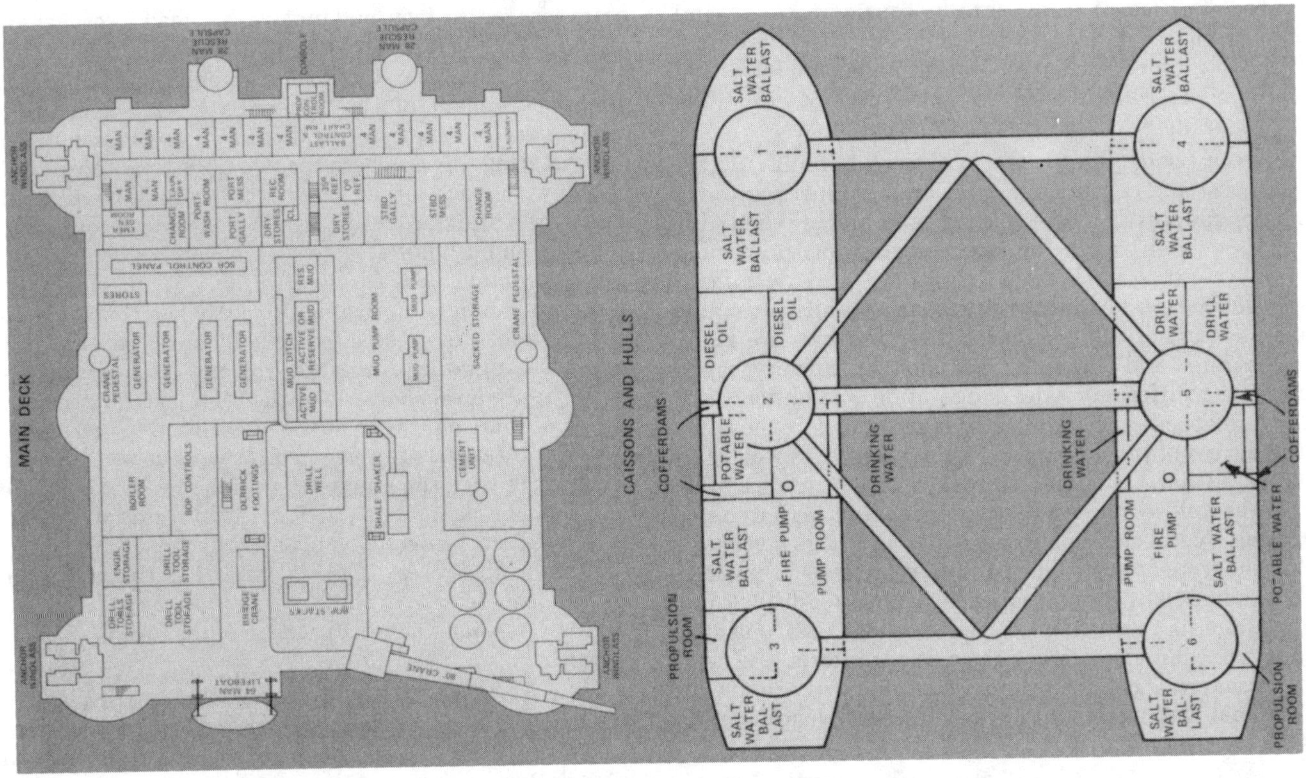

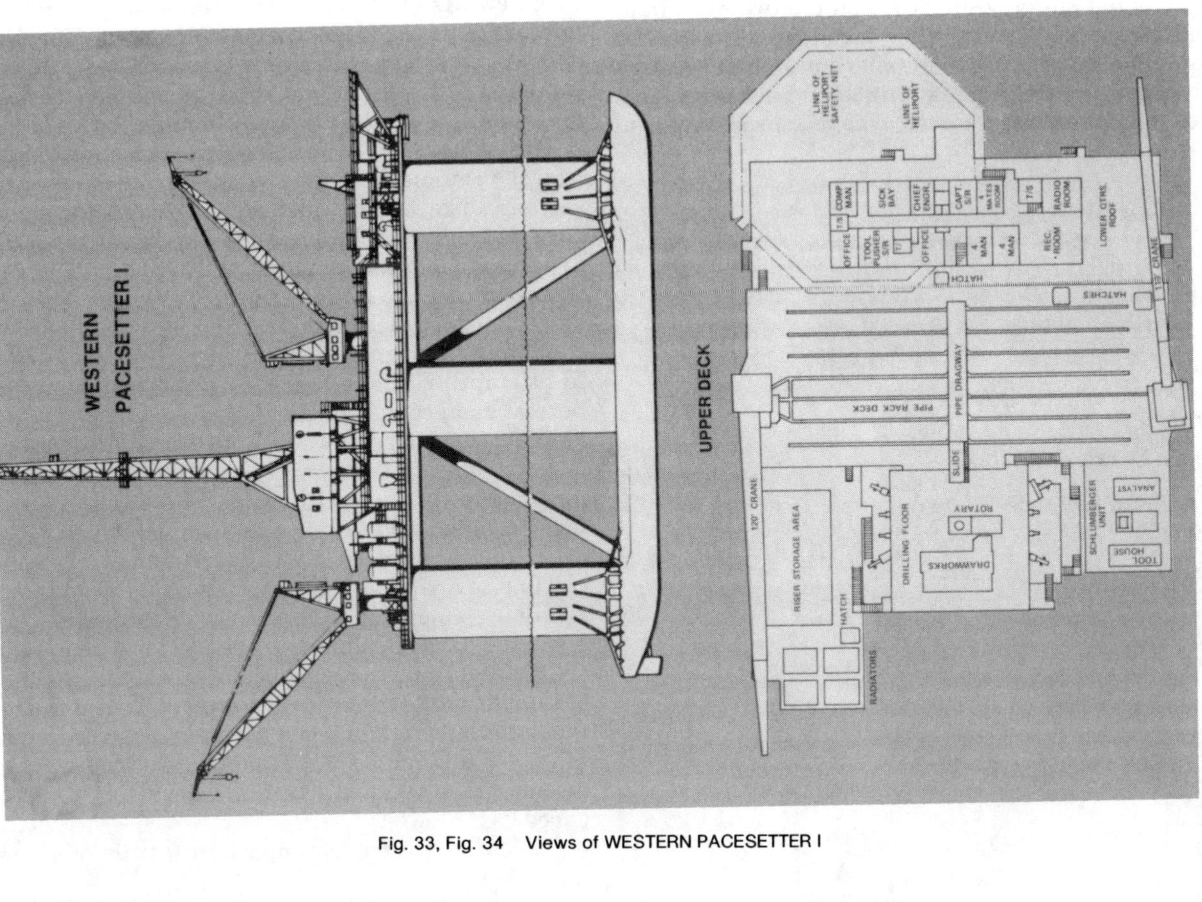

Fig. 33, Fig. 34 Views of WESTERN PACESETTER I

number of different loadings to be considered on a semi-submersible, that need to be analyzed individually and collectively, for the various operating modes and for various environmental conditions of wind, wave, current, etc. These may generally be categorized as follows:

• *Deck Load.* This is the total weight of all structure, equipment and consumable loads on the upper platforms, that are supported by the columns and bracing (including weights of and within the upper portion of these members).

• *Derrick and Substructure Load.* While ultimately taken as part of the deck load, these loadings are highly concentrated and of large magnitude that can vary significantly with different modes of operation, and under different wind speeds and rig motion dynamics. For the structural design of the derrick and substructure, as well as for the overall unit, the criteria normally include: transit, afloat on lower hulls—no load in the derrick, but high dynamic loadings due to short period motions and large lever arms from the waterline; drilling conditions at design semi-submerged draft, with various assigned loadings for drilling, drill pipe standing in derrick (setback loading) riser tensioner loads, etc., under different rig motion criteria and wind speeds; survival condition under maximum wave and wind speed, not drilling but with possible setback loading in the derrick.

A typical set of derrick/substructure criteria is shown in Table 11.

• *Wave Loads.* These are loadings of a quasi-static nature, of equal but opposite effect on the two hulls (and opposing columns). As such, they do not contribute to the motion dynamics of the overall unit, but do impose large forces in the supporting bracing members. Such wave loads include: spread-squeeze forces, which are maximum in beam seas for the twin-hull unit, with spread forces generally higher than squeeze forces, if in large waves, due to the effects of mooring lines; torsional moment, due to differential pitching moment of the two hulls when waves are approaching from a quartering direction; yaw moment, again in quartering waves, where the two hulls tend to yaw out of phase with each other; and longitudinal bending moment—the standard ship bending consideration, equally imposed on both hulls.

• *Dynamic Loads Due to Motion.* While the motions of a semi-submersible are significantly less than those of a surface vessel (in those seas normally anticipated), the dynamic loadings due primarily to roll and pitch can be severe at the upper levels because of the high lever arm from the center of rotation (taken at the waterline).

• *Wind Loads.* Determined for various wind speeds in accordance with the standard approach (see ABS Rules). These can be significant, particularly in regard to the forces developed on the derrick and substructure.

• *Mooring Forces.* This is defined as the normal pretension load on each of the mooring lines, as augmented by increased tension due to sway/surge of the unit in waves and current.

The wave forces, motion-dynamics, and mooring loads can be evaluated from hydrodynamic principles, for preliminary design purposes. (See Appendix to ABS Rules for wave force determination, and the following section herewith for motion estimates). However, it is a preferred and usual practice to conduct model tests of a fairly extensive nature on a new design, to determine the principal wave forces, motions, mooring loads, current effects, etc., for the above purposes as well as establishing maximum wave heights that can be passed (without impacting the upper structure) and the usual speed/power testing afloat on the hulls.

The loadings thus determined are then applied in various logical combinations to a computer-modeled structure, wherein all key structural members comprising main upper girders, bracings, columns, and lower hulls are represented as individual beam-columns with fixed-end joints, in a three-dimensional space frame. The stress values then are determined for each member for each of the applied loading conditions, which may number 10 to 20 or more. It is futile to attempt an analysis of this scope and complexity by normal hand computations.

Finite element analyses may also be required in such areas as the column/hull/bracing connection, or the bracing connections to the upper structure, where the joints may be fairly complex and the stresses nominally high.

b. Semi-Submerged Motions in Waves. The main attraction of a semi-submersible unit is its low motion characteristics, compared to those of a surface vessel. With the greatest part of its submerged mass and buoyancy well below the water surface at design drilling draft, the exciting forces are small, and with the low waterplane area of the columns the natural periods of heave, roll, and pitch are beyond the range of anticipated wave periods, so that synchronous motions of any consequence are not likely to be experienced.

Since this is the most important feature of a semi-sub-

Table 11—Derrick Criteria

With Pipe in Derrick (Drilling)	
Wind	75 knots
Roll & Pitch Amplitude	7 deg, 10 sec
Heavy Amplitude	3.7 m (12 ft), 10 sec
Center of Oscillation	27.4 m (90 ft) below drill floor
Hook Load	223 tons
Setback	150 tons
Tensioners	243 tons
No Pipe in Derrick (Towing)	
Wind	75 knots
Roll & Pitch Amplitude	14 deg, 10 sec
Heave Amplitude	7.3 m (24 ft), 10 sec
Center of Oscillation	39.6 m (130 ft) below drill floor
No Pipe in Derrick (Maximum Storm)	
Wind	100 knots
Roll & Pitch Amplitude	15 deg, 12.5 sec
Heave Amplitude	9.1 m (30 ft), 12.5 sec
Center of Oscillation	32 m (105 ft) below drill floor
Static (Max. Hook Load—No Motions)	
Hook Load	446 tons
Setback	351 tons
Tensioners	243 tons
Damaged (List Condition)	
Wind	50 knots
Setback	150 tons
Static List	25 degrees

mersible, it is proper to explore it in some detail and show how it affects the selection of configuration particulars.

Evaluation of hydrodynamic forces on the various submerged elements, and the equations that define vessel motions can be found in various texts (Korvin-Kroukovsky, 1961), (Comstock, 1967). However, a simplified equation can be presented here to show how major proportions and dimensions of a semi-submersible influence its resulting motion characteristics.

Thus, the heave (total) per wave height ($2Z/H$) is composed of three factors, as graphically shown in Fig. 35, and indicated as follows:

$$(2Z/H) = f_1 \times f_2 \times f_3$$

where

f_1 The depth factor, which indicates that the wave exciting forces in heave act essentially at the level of the lower hulls:

$$f_1 = e^{-[4\pi^2/T^2\ d/g]} \tag{1}$$

f_2 The spread factor, which averages out the force on the two hulls (or along the hulls for head seas). For waves abeam, as in the present example, this is:

$$f_2 = \cos\left[\frac{4\pi^2}{T^2} \times \frac{b}{2g}\right] \tag{2}$$

f_3 The usual solution for amplitude of motion, with sinusoidal excitation. Disregarding the effect of frictional damping for the moment, this factor can simply be shown to be:

$$f_3 = 1 - \frac{(4\pi^2 d/g)}{T_N{}^2 - T^2} \tag{3}$$

where T is the wave period
d is the mean depth of lower hulls below water surface
b is the center to center distance of the two hulls
T_N is the natural heave period of the unit in still water

From Fig. 35, it can be readily seen that:

- The decay factor is of primary importance at the usually anticipated wave periods, and deep submergence is a desirable feature;
- the spread factor is also significant, and the greater the distance between hulls, the greater reduction in motions over a longer range of wave periods;
- deep submergence is also a factor in reducing the basic amplitude of motion through the expected range of periods (although increasing it when synchronism is approached).

Increasing the natural period on the other hand will tend to increase the amplitude in the lower period ranges to some degree, but extends the range of generally low motions.

Now consider the effect of damping forces. These are produced by the vertical fluid flow past the hulls (as a function of the hull plane area, its shape coefficient, and the square of the relative velocity between the body motion and the fluid motion at the hull depth). It can be seen from the figure that in the low period range, there is little relative motion between the fluid ($f_1 \cdot f_2$) and the body ($f_1 \cdot f_2 \cdot f_3$) and

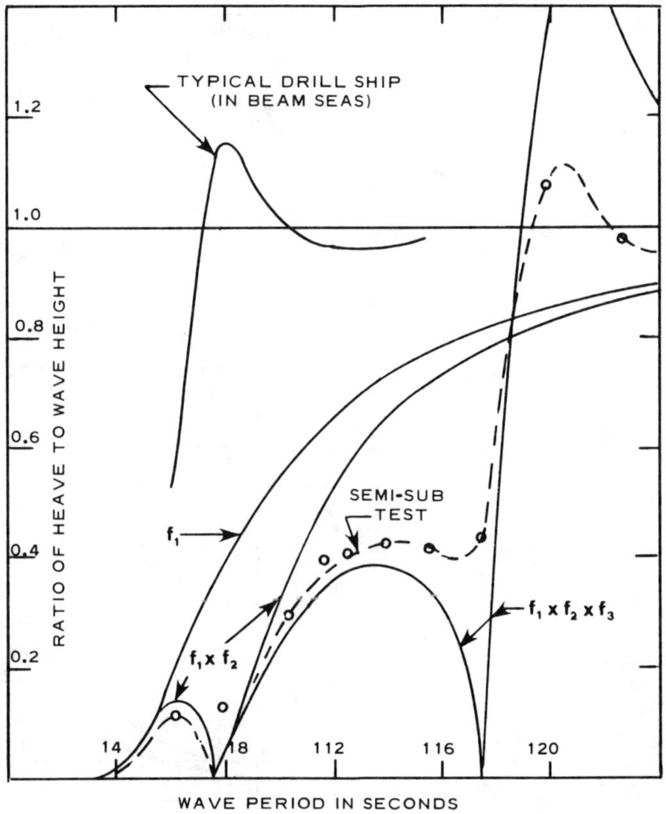

Fig. 35 Motion curves

thus little damping effect. However, with increasing period, the wave motions increase and the body motions tend to decrease, producing a high relative motion and a significant excitation force that increases the body motions (as indicated by the test results).

As synchronism is approached ($T_N = T$), the reverse situation occurs where the body tends to greatly exceed the wave motion, and the resulting damping force in this case tends to restrict the body motions.

The hull plane areas and its shape coefficient dictate the character of the body motion in this range of high period waves. Cylindrical hulls have a minimum area and coefficient, and thus will maintain lower motions through the range of anticipated waves, but will have extremely high motions in long swells. On the other hand, shallow and broad rectangular hulls will have high areas and high shape coefficients, which will tend to increase the body motions generally through the range of periods, except to greatly modulate the motions at synchronism.

Philosophies differ as to which characteristic motion curve near the synchronism range is most desirable. Some wish to maintain lowest possible motions through the wide range of lower periods, accepting the high motion characteristics at synchronism (considering that since these long waves are low swells in general, the synchronous motion will not necessarily be severe). Others consider that the very abrupt change from very low to very high motions that could occur in a mixed swell of slightly different periods may be dangerous during operations, and prefer to accept the higher but

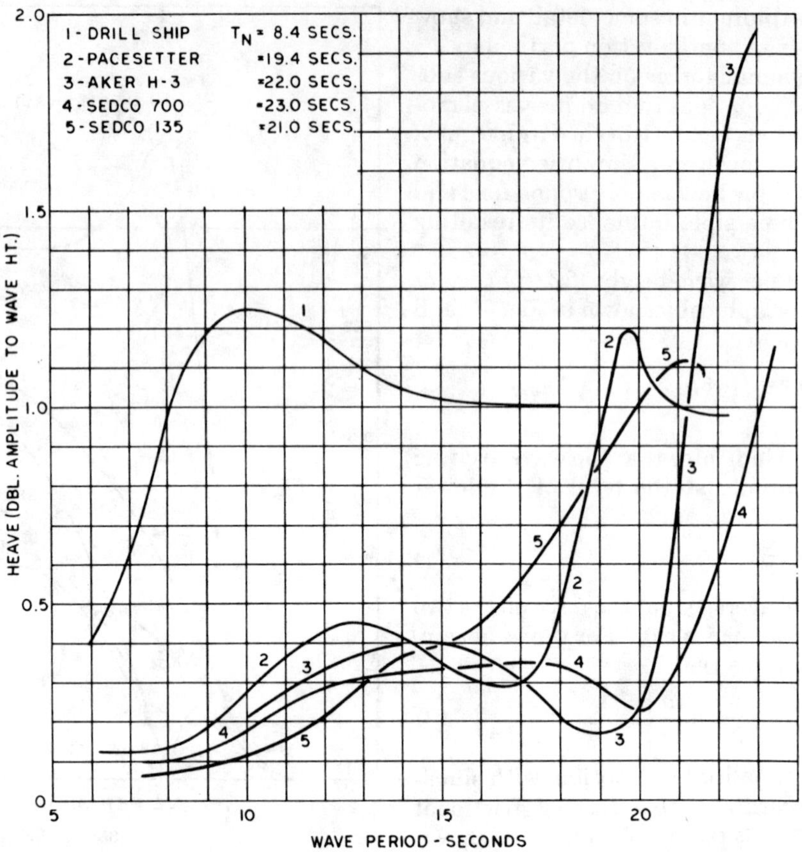

Fig. 36 RAO curves—heave

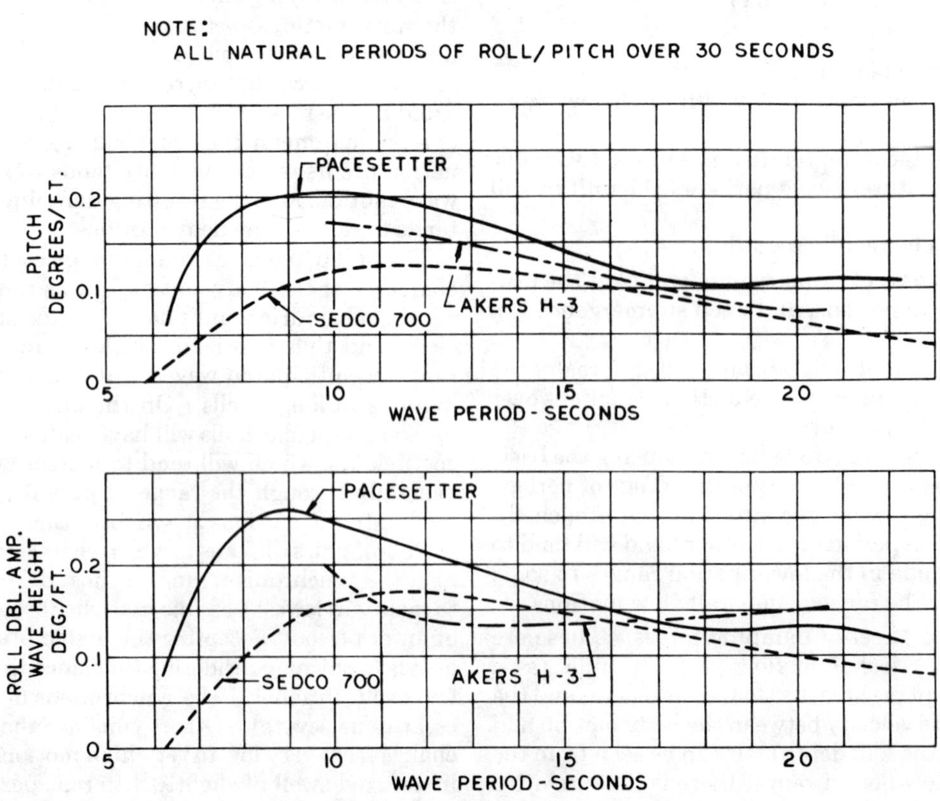

Fig. 37 RAO curves—pitch/roll

more gradual level of damping motions through the medium to high period ranges.

Fig. 36 shows the characteristic heave curves for different units. The influence of draft, hull shape, and natural period is evident in the comparison.

Roll and pitch characteristics are not as simple to establish as heave, due to the coupling effect of sway or surge, mooring restraints, and the variations that occur in metacentric height. In general, however, they are strongly influenced by the same major factors as in heave, and exhibit the same basic motion trend, as indicated in Fig. 37.

3.5 Drill Ships. The most prominent features that distinguish the surface drilling vessel from the normal seagoing ship are the large drilling derrick with massive supporting structure, extending up above the main deck amidships, and the large *moon pool* opening extending down through the hull to accommodate the major drilling operations. Fig. 38 is a typical example.

The large concentrated weight and drilling load, in association with the extensive loss of deck and bottom structure through the well opening, requires careful design attention to midship scantlings and reinforcement, for longitudinal strength and for proper local support of the loadings augmented by dynamics of the oscillating vessel.

Otherwise, structural considerations follow standard ship requirements, and the design process is more concerned with proportions and arrangements (to address stability and load-carrying capacity), and the maintaining of position with a minimum of motions.

Table 12 shows the principal dimensions, powering and other characteristics of a number of existing drill ships, with Fig. 10 of Chapter I showing the recognizable configuration. It is noteworthy that length to beam ratios are fairly low (in the order of $L/B = 6$). This is essential to provide sufficient beam to meet stability requirements, due to the high windage and high vertical center of gravity, as well as to accommodate the structural opening through the hull.

a. Ship Motions and Positioning. Typically, the drill ship has higher heave and angular motions than a semisubmersible due to its being excited by the more severe sea surface conditions. Similarly, its surge/sway motions are greater, which can severely limit its operating efficiency, particularly in relatively shallow waters where maintaining a position over the well is essential. To cope with these various requirements, a number of mooring arrangements and positioning control features have been used.

1. Moorings. Generally, an eight-point mooring system is used, sometimes utilizing a wire-rope, winch system, but more frequently a chain, windlass system. The ship (or barge) is positioned to be head on (hopefully) to the prevailing sea and current, to experience pitch (rather than roll) and reduced heave and sway than would otherwise result from quartering or beam seas. Adjustment of position to accommodate changes in sea direction are limited to about 15 degrees, before it is necessary to relocate anchors and lines. A good mooring system is nevertheless essential to maintain reasonable position over the well, in shallow waters under 91 m (300 ft) in any kind of rough seas.

Fig. 38 Model of Typical drilling ship

Table 12—Ship Type Drill Rigs

	Discoverer 534	Sedco 445	Glomar Conception	Saipem Due
Dimensions				
Hull Length O.A., m (ft)	163 (534)	136 (445)	122 (400)	131 (431)
Beam, m (ft)	24 (80)	21 (70)	20 (65)	22 (72)
Depth, m (ft)	10 (32)	10 (32)	8 (27)	12 (40)
Operating Conditions				
Draft—Load, m (ft)	7 (24)	8 (25)	6 (21)	7 (23)
Displacement, tons	20,800	15,100	11,000	13,519
Capacities				
Personnel	163	93	76	98
Bulk Mud/Cement, m^3 (ft^3)	476 (16,800)	745 (26,300)	373 (13,180)	566 (20,000)
Sack Matl., Sacks	13,000	20,000	11,000	3,600
Liquid Mud, m^3 (bbl)	668 (4,200)	471 (2,960)	482 (3,030)	572 (3,600)
Diesel Oil, m^3 (bbl)	2131 (13,400)	2409 (15,152)	1247 (8,158)	
Drill Water, m^3 (bbl)	1972 (12,400)	2624 (16,500)	2417 (15,200)	1988 (12,500)
Potable Water, m^3 (bbl)	223 (1,400)	159 (1,000)	81 (512)	572 (3,600)
Propulsion				
Number of Screws	2	2	2	4—
				—
Number of Thrusters	6 × 2,500 HP	11 × 800 HP	—	9,000
H.P. (Total)	16,000	9,000	4,500	9,000
Mooring				
Type	Chain & Wire Rope	Dynamically Stationed With Operational Wire Rope	Chain & Wire Rope	Chain
Drilling Characteristics				
Drilling Water Depth, m	914	1,067	183	610
Drilling Water Depth, (ft)	(3,000)	(3,500)	(600)	(2,000)
Maximum Drilling Depth, m	7,620	6,096	7,620	7,620
Maximum Drilling Depth, (ft)	(25,000)	(20,000)	(25,000)	(25,000)
Number of Lines	8	8	10	8
Size, mm (in.)	64 (2½) Chain 76 (3) Wire	64 (2½)	70 (2¾)	64 (2½)
Length of Lines, m (ft)	427 (1,400) Chain 914 (3,000) Wire	1158 (3800)	8 @ 762 (2500) Chain 2 @ 457 (1500) Chain + Wire	1006 (3,300)
Anchor Weight, tons	13.4	10.0	13.4	8.9

2. *Self-Positioning Power.* A system of propulsion for fore and aft control, combined with a number of thrusters at both bow and stern, has been utilized on several drill ships for maintaining complete control of well position and for orientation of the vessel for minimum motions, during all operations associated with drilling. In some cases, no mooring assembly has been installed (which may limit drilling operations to deep and/or relatively quiet waters). In others, such as *Sedco 445* (Williford, 1973) both mooring and self-positioning systems are installed, to allow more versatility, and in certain environments, to utilize both systems concurrently (with either one overriding the other, where the particular advantages of either is more demanding for changing conditions). In considering self-positioning systems in the design stage, it is essential to evaluate all of the motion and force characteristics of the vessel, in association with anticipated environmental requirements in order to assess the powering needs (the amount of power necessary and its distribution between main props and thrusters, whether thrusters should be rotating type or fixed, etc.). In this respect, comprehensive model tests that determine the hydrodynamic constraints for all six degrees of freedom are invaluable (determining the body coefficients by oscillating the model in still water, and the exciting forces by fixing the model in waves).

3. *Turret Mooring.* A unique method of holding position with a positive mooring system, but allowing orientation to the prevailing environment for minimum motions was developed by the Offshore Company, and is used in all of their *Discoverer Class* drill ships. It provides for all mooring gear to be located at the well center, with the lines led down through the well and dispersed from there. The ship is free to rotate around this center, by virtue of a roller bearing trunion ring, and is directed and maintained in position toward the oncoming sea (and wind) disturbances for minimum motions, by thruster power. While the mooring system is somewhat more difficult to handle and service than the normal *over-the-side* arrangement, and encumbers some of the drilling operations in the same moon pool area, it has proven to be very effective and in the larger size of drill ship, the resulting heave and pitch motions are competitive with semi-submersibles.

The major advantage of the drill ship is its high mobility, with least cost of moving and least elapsed time enroute. Further, it has much greater capacity for fuel, water and drilling supplies, particularly when on a long move, and is thus highly self-sufficient over a longer period. All of this is particularly beneficial when exploration drilling is required in remote areas, or far-flung locations in a large area such as Indonesia.

The major disadvantage is in its higher motions, which lead to more time waiting on weather in moderate to rough environments. In order to cope with such motion problems, more elaborate and expensive motion-compensating equipment is required, which in turn has nullified the cost advantage that drill ships had originally enjoyed.

For more details of offshore drilling rigs see Macy (1969).

3.6 Ocean Construction Platforms. The Naval Facilities Engineering Command's Ocean Engineering and Construction Project Office is charged with the responsibility for the design, acquisition, installation, maintenance, and recovery of any fixed Navy facility on or in the ocean. As ocean construction tasks multiplied over the last decade it rapidly became evident that there was a growing need for a vessel with the capability of performing such tasks at sea.

Early ocean construction jobs were completed successfully using combinations of crane barges, tugs, small craft such as Whalers, and Zodiacs, jury-rigged knock-down barges, and World War II landing craft. Although many such missions were accomplished, they demanded excessive use of both floating equipment and manpower with a concommitant lack of efficiency and often with borderline personnel safety. It became obvious that the ocean construction workload was building up to a point where a mission oriented ocean construction platform was becoming an essential asset.

Available time and funds dictated the conversion of an existing hull rather than the development of a new design. The hull selection criteria were tailored to the mission that the vessel was to perform. These criteria included:

- sufficient space to house 25 crew and 25 construction operations personnel,
- a large open deck space for roll-on equipment, equipment handling, and construction operations,
- space and structural integrity to permit installation of a center well at the longitudinal center of flotation large enough to handle equipment and appropriate submersibles,
- an arrangement adaptable to both a ship control bridge looking forward and an operations control room looking aft over the working deck,
- a shape conducive to the installation and operation of a dynamic positioning system, and
- a materially sound hull.

These criteria were delineated by Agdern and Sherwood (1979).

The hull selected for conversion was a Navy YFNB seagoing barge hull, last used by NASA for transporting Saturn rocket components to Cape Canaveral. It is interesting to note that this hull was of the same class as the first dynamically positioned vessel, with a similar centerwell, that was used in the Mohole experimental drilling program in 1961 (Taggart, 1961). The basic barge hull has a length overall of 79.2 m (260 ft) with a beam of 14.6 m (48 ft). It has a spoon shaped bow and a typical rake-ended barge stern, and is fitted with a pair of skegs for stability in towing.

Fig. 39 Ocean construction platform SEACON

Depth to the main deck, which runs the full length of the ship, is 4.6 m (15 ft) above the baseline. Ramps, port and starboard, run from the main deck up to the forecastle deck, which extends 7.9 m (26 ft) abaft the bow at a height of 7.2 m (23.8 ft) above the baseline.

a. Arrangements. The converted YFNB hull was renamed *Seacon* and was rearranged to adapt its basic hull characteristics to its ocean construction mission as illustrated in Fig. 39. The main deckhouse runs from the after end of the forecastle deck to the midship section. The remaining main deck aft is available for construction equipment and activities. A 9.75 by 4.87 m (32 by 16 ft) centerwell was cut from the main deck through the bottom amidships and was fitted with bottom doors and a closure at the main deck level that allows partial or full use of the deck space when the well is not in use.

Two passive antiroll tanks are fitted forward and abaft of the centerwell for roll control during the construction operations. Sixty percent of the original skeg area was removed to permit the installation of two cycloidal-type propulsion units aft; a third unit is installed forward with its blades extending below the baseline. All three propulsors are individually powered.

An enclosed work area is provided just forward of amidships covered by a roller curtain; this space can be used as a weather-protected staging area for deck operations and has easy access to a bosun's locker, scuba locker, and electronics shop. Work can also be done through the forward half of the centerwell from within this protected space.

The open deck space aft is 39.6 m (130 ft) long by 14.6 m (48 ft) wide. It has a 1.5 m (5 ft) freeboard for ease in working over the side and over the stern. Below the main deck at the stern are ballast tanks, forward of which is the after propulsion machinery room. Next forward is a space that can serve either as a storage compartment or a cable tank. There is sufficient main deck space between this compartment and the stern roller for the installation of cable handling machinery when required by a specific construction operation.

b. Machinery. The three vertical axis rotating blade propellers are each driven by a diesel engine through hydraulic power transmission and clutches. The engines turn at a constant speed of 1800 rpm which is reduced to 600 rpm input to the propellers. The propeller blade rotors turn at 140 rpm and all thrust and maneuvering control is exercised by varying propeller blade pitch. The forward diesel engine is rated at 360 horsepower at 1800 rpm. Each of the two after propellers is driven by twin diesel engines, also rated at 360 horsepower at 1800 rpm. Auxiliary power is furnished by two diesels each driving a 250 kW-450 volt A.C. generator. This power is transformed down to 115 volts for all shipboard uses.

c. Outfit. Seacon is outfitted for maximum effectiveness as an ocean construction platform. A stern roller is provided for work over the transom and foundations are fitted to the main deck to support a cross deck winch, which can work over the stern or through the centerwell, and a Pengo constant tension winch for cable handling. Rails are installed to support a traveling gantry crane to cover the entire portion of the main deck. The hold is rigged for stowage of cable on reels or to serve as a cable tank. Twin capstans are provided for anchor handling forward and a stern anchor can be handled over the stern roller. The ship is fitted with all modern aids for position finding by LORAN, satellite, or sonar navigation plus a gyrocompass and speed measuring equipment.

d. Maneuvering and Control. The ship is capable of dynamic positioning with controls located both in the pilot house and in the control room. Position can be established either by an acoustic location system, an acoustic transponder system, or a mini-ranger system. Heading control is automatically maintained by an autopilot working in combination with the steering system and translation is manually controlled with a joystick in response to lateral position indicators on each control console. This combination actuates the pitch changing mechanism on all three propellers to produce the required translational forces and rotational moments. The vessel has a maximum ahead speed of 7.50 knots but normally transits to an ocean construction station under tow.

Section 4
Service Vessels

4.1 General. Table 2 of Chapter I lists under "Service Vessels" the following: tugboats, offshore supply boats, offshore supply vessels, crane barges, diving support boats, fire boats, pilot boats, and buoy tenders. There are, of course, many additional types of vessels that perform an essential service to maritime operations. To illustrate the

Sea Swift Class
41.5 (136.2)
11.2 (36.7)
5.8 (19.2)
—
14
1,063
934,126 (210,000)
Diesel
7,000
8
x
x
x

design considerations involved in developing some of these service vessels to perform specific missions, three different types will be selected for discussion herein. These three are tugboats, crew boats, and a unique buoy tender.

4.2 Tugboats. Tugboats vary in size and performance, from the 15 m (50 ft), 300 horsepower barge jockeys, through the range of large harbor and offshore tugs that maneuver large cargo vessels and drill rigs, and up to the 76 m (250 ft), 20,000 horsepower mammoths that engage in world-wide deep ocean tows and salvage. Their basic function remains the same, that of towing or directing other floating equipment from one location to another, and while other uses and services may be incorporated, tugboat design is directed to that primary purpose.

a. Harbor Tugs. Tugs that are principally intended for harbor work are not required to have high free-running speed, high freeboards, or lavish accommodations. Typically, such vessels will be of relatively short length, low to the water (except possibly in pilot house height for visibility over barge tows) and of minimum quarters. Rubber *snouts* are usually fitted on the bow for working ships or barges into position (towboat type push knees are sometimes used). Powering is strictly geared to the requirement of providing sufficient bollard thrust in moving the larger vessels. Fig. 40 shows a typical tug, with characteristics as given in Table 13. The use of cycloidal, vertical axis propellers on tugs for harbor service is becoming widespread in Europe and many other countries. The reduction in bollard pull ahead when compared to conventional, ducted propellers is accepted because of the very high maneuverability obtained with the vertical axis propeller due to its ability to exert maximum thrust through a full 360 degrees. The underwater configuration generally employs twin propellers mounted forward of midship, with a large stabilizing fin at the stern, such as shown in Fig. 41 for a 4,000 HP tug of this type built for use in Saudi Arabian ports. Baer (1973) gives a good description of the operation of these vessels, which are commonly called *water tractors* in Europe.

While the size of harbor tugs used for docking ships has increased considerably with the increased numbers of large vessels, such as tankers, they now appear to have approached their limit in horsepower and consequent size. The largest docking tugs are now beweeen 3,000 and 4,000 horsepower, which appears to be the maximum power that can be safely used when pushing against a ship's hull. (Even now, the tugs must push with their bow fenders against the ship's frame, not against the unsupported plating between frames, to avoid damage to the ship's structure.)

Fig. 40 Typical harbor tug

Table 13—Tugboats

	Tappan Zee II	Jason Smith	Andrew Martin	Smit Rotterdam	Water Tractor
Length, m (ft)	14.6 (48.0)	30.5 (100.0)	457 (150.0)	74.8 (245.24)	35.8 (117.5)
Beam, m (ft)	4.9 (16.0)	8.8 (29.0)	12.2 (40.0)	15.3 (50.20)	12.8 (42.0)
Depth, m (ft)	—	4.3 (14.0)	6.7 (22.0)	7.6 (24.93)	5.5 (18.2)
Draft, m (ft)	1.6 (5.33)	3.5 (11.5)	6.6 (21.5)	6.4 (21.00)	—
Speed, knots	9.6	13.0	—	—	11
Displacement, tons	40.7	—	1,856	—	935
Bollard Pull, N (lb)	38,254 (8,600)	271,340 (61,000)	—	—	391,444 (88,000)
Propulsion	Diesel	Diesel	Diesel	Diesel	Diesel
SHP	2 @ 348	2 @ 850	2 @ 3,800	2 @ 11,000	4,000
Crew	2	8	—	—	14
Quarters		x	x	x	x
Towing Winch			x	x	x
Fire Fighting	x			x	x
Kort Nozzle			x	x	
Bow Thruster			x	x	
Oil Dispersant Equipment				x	x
Salvage Equipment	x			x	

x = have

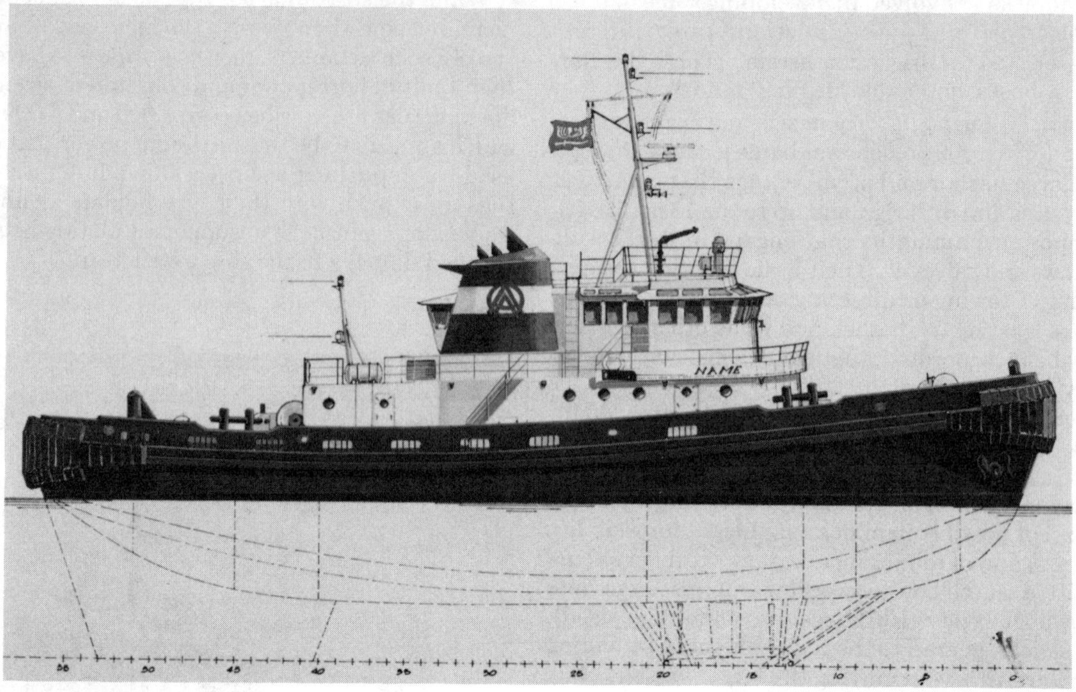

Fig. 41 European "Water tractor"

b. Seagoing Tugs. As the service requirements extend towards offshore uses, size and power increase as do most of the major particulars. Higher freeboards are evident, particularly at the bow (which ultimately leads to a high forecastle, for seakeeping in rough seas). Free-running speeds start to become important, for time-saving in moving from one basic mission to the next. Larger quarters are necessary for full-time crews on extended service. The snub nose is minimized, with emphasis directed toward the stern for heavy towing, anchor handling, and other rough water services. There is in fact a marked similarity in configuration and characteristics between the offshore tug and the offshore supply vessel (discussed in the next section). While the latter tends to be longer and broader for its major purpose of cargo supply and the tugboat more compact for its purpose of powering, each adopts some measure of the other's capabilities and usefulness, particularly in the servicing of offshore drilling units (Guarino, 1975).

Fig. 42 Oceangoing tug SMIT ROTTERDAM

The design of ocean-going tug boats under American ownership is governed to a large extent by regulations, which require licensed personnel on vessels of 200 gross tons and over; and by the law requiring USCG inspection and certification of motor vessels over 300 gross tons. Since operators want the greatest freedom in crew selection and manning and the minimum of regulation and inspection, the naval architect has to employ all the tricks of the tonnage trade to obtain the maximum exemption of spaces, and yet provide the owners with a tug of the maximum proportions possible under a limited tonnage consistent with horsepower, fuel, and accommodation requirements. The competent naval architect, through astute design, can construct a modern, high powered, seaworthy, maneuverable, ocean-going tug boat with more than adequate stability and cruising range, and still maintain tonnage under 200 gross tons. Fig. 11 of Chapter I illustrates such a vessel. Maximum lengths of such tugs are in the range of 46 m (140 ft), with maximum power of about 8000 shp, with twin screws.

The ultimate ocean-going tugs are in the range of 76.2 m

(250 ft) long with installed powers of 15,000 to 20,000 horsepower, Fig. 42. Their primary function involves towing of drill rigs, platform structures, heavy laden ocean barges, and other large vessels on long ocean voyages, with the further provision for conducting salvage operations on ships in distress. High bollard pull when under tow is of great importance, but also is high free-running speeds to arrive at the right place at the right time to pick up a tow or to assist in distress situations. Typically then, these large tugs are equipped with twin-screw Kort nozzles with controllable pitch propellers, to achieve the optimum dual performance.

c. Design Data. Table 13 shows typical particulars of tugs for various operations, including a listing of services aside from the basic towing function. Design information on tug proportions, resistance, and propeller characteristics are given by Roach (1954). Additional data on comparative particulars for a number of tugs in the medium size range, along with information on scantlings, arrangements and various machinery requirements can be found in the paper by Argyriadis (1957). Good information on the design aspects of ocean tug boats, proportions, tonnage, propulsion, and other major considerations is given by Spaulding (1973), while the practical aspects of offshore tug design, covering the shipyard designer's approach to lines, arrangements, scantlings, etc. may be found in the paper by Rook (1978). On the very important consideration of tug stability, reference should be made to Storch (1972), and to McGowan and Meyer (1979).

On water tractors, design parameters and performance information may be found in the literature of the J. M. Voith Company and by Sarchin and Tryner (1970).

4.3 Offshore Supply Vessels. The offshore supply vessel was developed in the early 1950s to meet the service requirements of drilling rigs that were at that time reaching out from the Louisiana Coast. The early designs basically consisted of a shallow draft, wide beam, river type deck barge structure, extended into a shaped bow section with a relatively high forecastle that contained quarters and pilot house. The hull space was devoted to tanks carrying fuel and fresh water *ballast* for rig use, in addition to the necessary but moderate propulsion and steering spaces, while the main deck was used for cargos of drill pipe, drilling mud, and other supplies. Typically, these early vessels were in the range of 30.5 to 36.6 m (100 to 120 ft) in length, with about 600 tons displacement at an eight-ft draft. Speeds were up to 10 knots with power in the vicinity of 700 shp.

Through the intervening years, with drilling operations moving further offshore and into rougher waters throughout the ocean areas of the world, supply vessels have steadily increased in size and sophistication. Vessel lengths have about doubled, with finer lines and deeper hulls, and with greatly increased seaworthiness. Powering has increased to tenfold over the early vessels, with significant increases in speed.

The modern supply vessel now serves additional functions principally related to the offshore drilling industry, such as towing and anchor and chain handling, although supply remains a foremost purpose. For augmenting tow performance, flow accelerating nozzles and controllable pitch propellers are being utilized and towing winches installed. Deck rollers and gantry cranes are frequently incorporated at the stern for lifting and handling the large mooring anchors of the drilling unit, along with windlasses and special

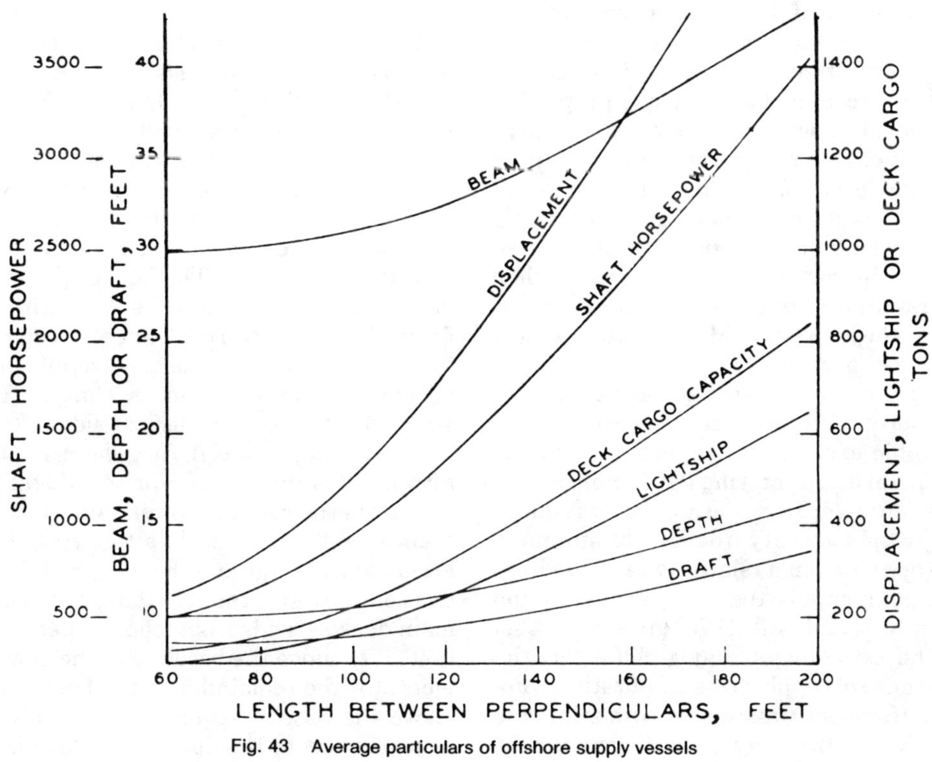

Fig. 43 Average particulars of offshore supply vessels

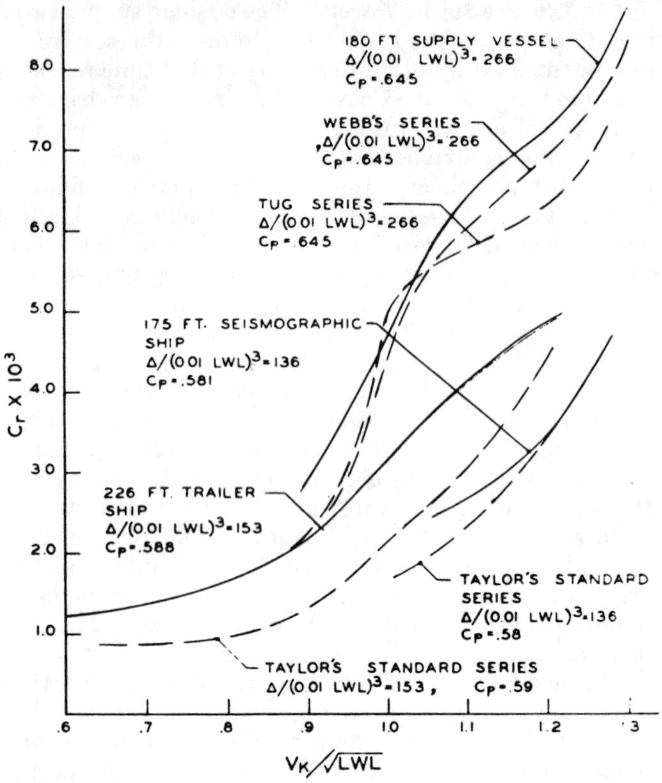

Fig. 44 Residuary resistance of supply and related vessels

chain lockers for carrying the mooring equipment to another drill site.

However, the general layout of the vessel remains virtually the same. The hull spaces are still largely devoted to liquids for supplying the drill rig, although much of the increase in hull volume has been to accommodate the larger propulsion system and for additional rig services, such as independent pressure tank installations for ready transfer of bulk mud and cement. The large clear after main deck is still retained for deck loads of drill pipe and other supplies and rig equipment. The quarters and pilot house remain located at the bow, although somewhat larger and less spartan than on the early vessels, resulting in an overall configuration that remains recognizably the same, as evidence that it is most suitable for the intended purpose.

The configuration has also been found attractive for other usages. It has been successfully adapted for oceanographic research, seismographic exploration, submarine tending, cable laying, and general cargo carrying (Guarino, 1975).

a. Selection of Characteristics. An owner's requirements for a supply vessel may vary from one of minimum size for short range operation in a relatively calm environment to a full ocean-going supply/tug of large capacity and power. Fig. 43 from Mok and Hill (1970) gives a good set of guidelines for the selection of major characteristics through the general range of supply vessel application. For the minimum vessel, (frequently associated with minimum cost and speedy delivery), this may be sufficient for the purpose, whereas for the more demanding maximum vessel, it will serve as the first set of particulars, in association with Fig. 44 for estimating speed/power requirements, from which the final characteristics can be developed to meet the specific design goals.

Deck cargo capacity remains a major consideration, both as to deck area and weight. Accordingly, the feature of the hull plan extending essentially square to the stern remains desirable for effective use of space for cargo as well as for the other requirements of handling equipment over the stern. However, achieving the optimum weight of deck cargo requires intensive study of ship proportions and loading conditions to insure satisfactory stability through the range of intended operations, which is a critical consideration for these relatively shallow, low freeboard vessels.

A second major consideration is the selection of powering. Good free-running speed is desirable for quick supply to offshore installations, while high bollard pull is necessary for towing and for running anchor lines. Thus, propulsion may range from the simple fixed wheel installation for the low-budget, restricted vessel to ducted, controllable pitch propellers for the high performance ship (along with bow thrusters for maneuvering control around structures). In any event, twin-screw propulsion is a practical necessity.

The paper by Mok and Hill (1970) provides much valuable information on proportions, arrangements, stability, structure, and powering of supply vessels, along with a comprehensive set of characteristics of some twenty such vessels covering the range of sizes and powers through the 1960s. The advances made since then may be typified by the more sophisticated, full ocean-going tug/supply vessels, such as the *Mammoth Tide* class, Fig. 45 whose characteristics are delineated in Table 14. Also see Fig. 12 of Chapter I.

4.4 Buoy Tenders. The maintenance of aids to navigation has been a basic mission of the USCG for many years and the familiar stubby buoy tender has performed this function quite satisfactorily. However, a recent development by the USCG was the design of a new vessel with characteristics tailored specifically to the inland buoy tending mission. It is thus an example of the effect of mission requirements on vessel configuration which is illustrated in Fig. 46. The USCG *Tern* is a prototype for a new class of inland buoy tenders with a house structure forward, a large open deck aft, and a cutaway stern serviced by a traveling gantry crane. Overall length is 24.6 m (80.8 ft) and the molded beam is 7 m (23 ft). Molded depth amidships to the main deck at side is 3.6 m (11.8 ft).

a. Geometry and Arrangements. The vessel sections forward of amidships comprise straight line elements running from the keel to a lower chine, to an upper chine, and thence to the main deck; aft of amidships the upper and lower chines merge and the bottom structure is formed into two concave semi-tunnels above the main propellers. The main deck extends from stem to stern. There is a flat 1.9 m (6.2 ft) above the baseline in the bow thruster compartment; for the remainder of the length a flat 1.4 m (4.5 ft) above the baseline comprises the only other deck. The displacement to the design waterline is 134 tons but with

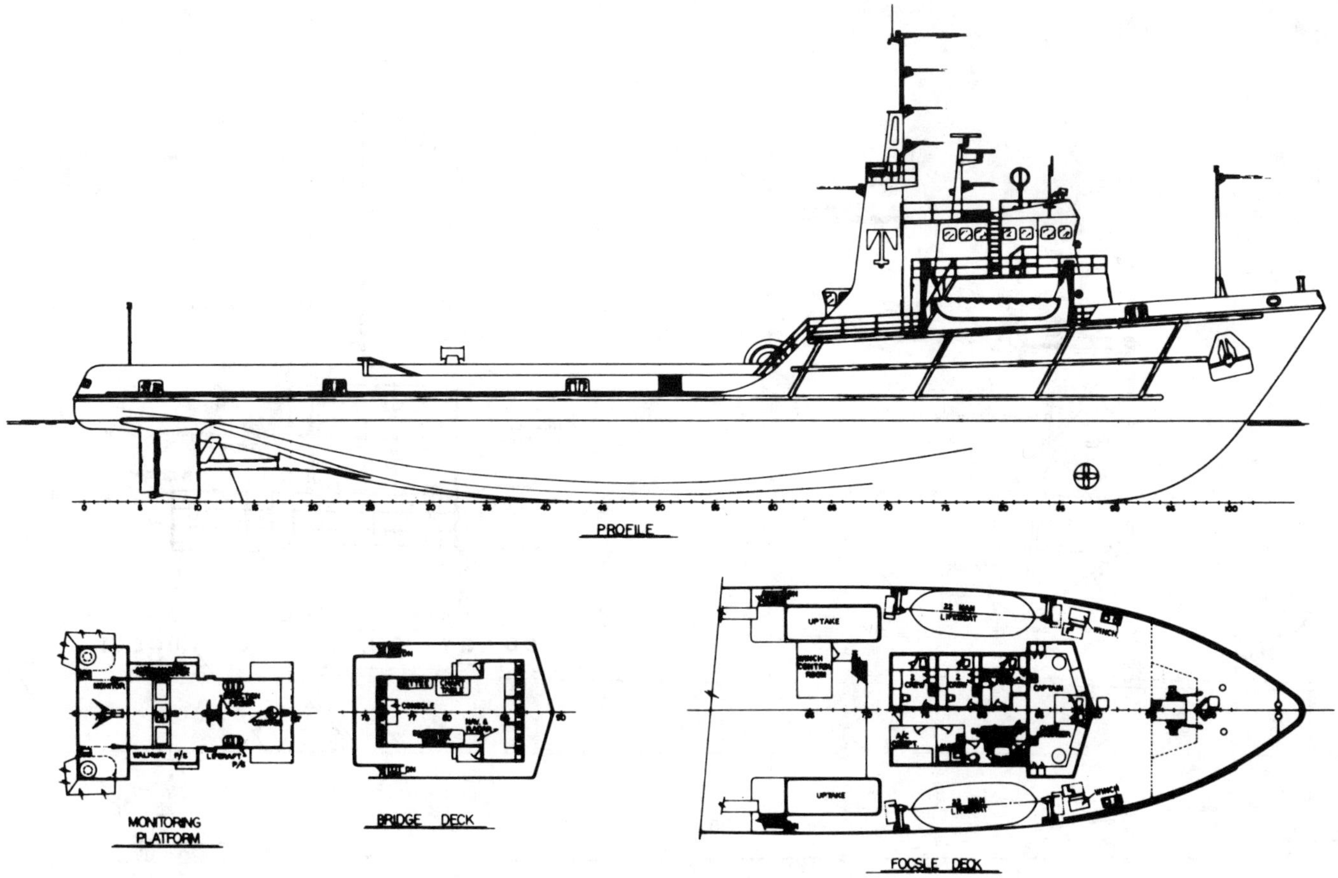

Fig. 45 Offshore supply boats SUN TIDE, MOON TIDE

Table 14—Offshore Supply Boats

	Boat "A"	Boat "B"	Tidewater 218′ Class	Acadian Mariner
Length O.A., m (ft)	50.3 (165.0)	53.6 (176.0)	65.4 (218.0)	66.0 (216.5)
Length B.P., m (ft)	47.9 (157.2)	51.0 (167.3)	61.0 (200.1)	61.0 (200.0)
Beam, m (ft)	11.6 (38.0)	11.6 (38.0)	13.1 (43.0)	13.4 (44.0)
Depth, m (ft)	3.8 (12.5)	4.0 (13.0)	5.8 (19.0)	4.9 (16.0)
Draft (Load), m (ft)	3.1 (10.2)	3.4 (11.0)	4.4 (14.5)	4.1 (13.5)
VS, knots	12.3	11.8	13.5	13.0
C_B	0.691	0.716	0.716	0.722
Froude Number	0.292	0.272	0.284	0.274
SHP	2 @ 850	2 @ 1,125	2 × 2 @ 2,100	2 @ 3,600
Propulsion	Diesel	Diesel	Diesel	Diesel electric—SCR
Lightship, tons	381.8	466.0	1268.7	1138.0
Displacement, tons	1248.0	1466.0	2554.0	2451.0
Beam/Draft	3.040	2.923	2.263	2.75
Length/Beam	4.34	4.63	5.070	4.92
Length of Afterdeck	123.0	123.83	116.0	146.0
Capacities				
Deck Cargo, tons	575.0	590.0	550.0	825.0
Fuel Oil, tons	106.8	152.4	719.0	710.0
Potable Water, tons	4.5	4.5	162.0	167.0
Ballast, tons	1052.7	1004.2	753.0	1155.0
Bulk Mat'l. m^3 (ft^3)	None	56.6 (2000)	169.9 (6000)	185.2 (6540)
Anchor Handling	No	No	Yes	No
Towing	No	No	Yes	Yes
Bow Thruster	No	No	Yes	Yes

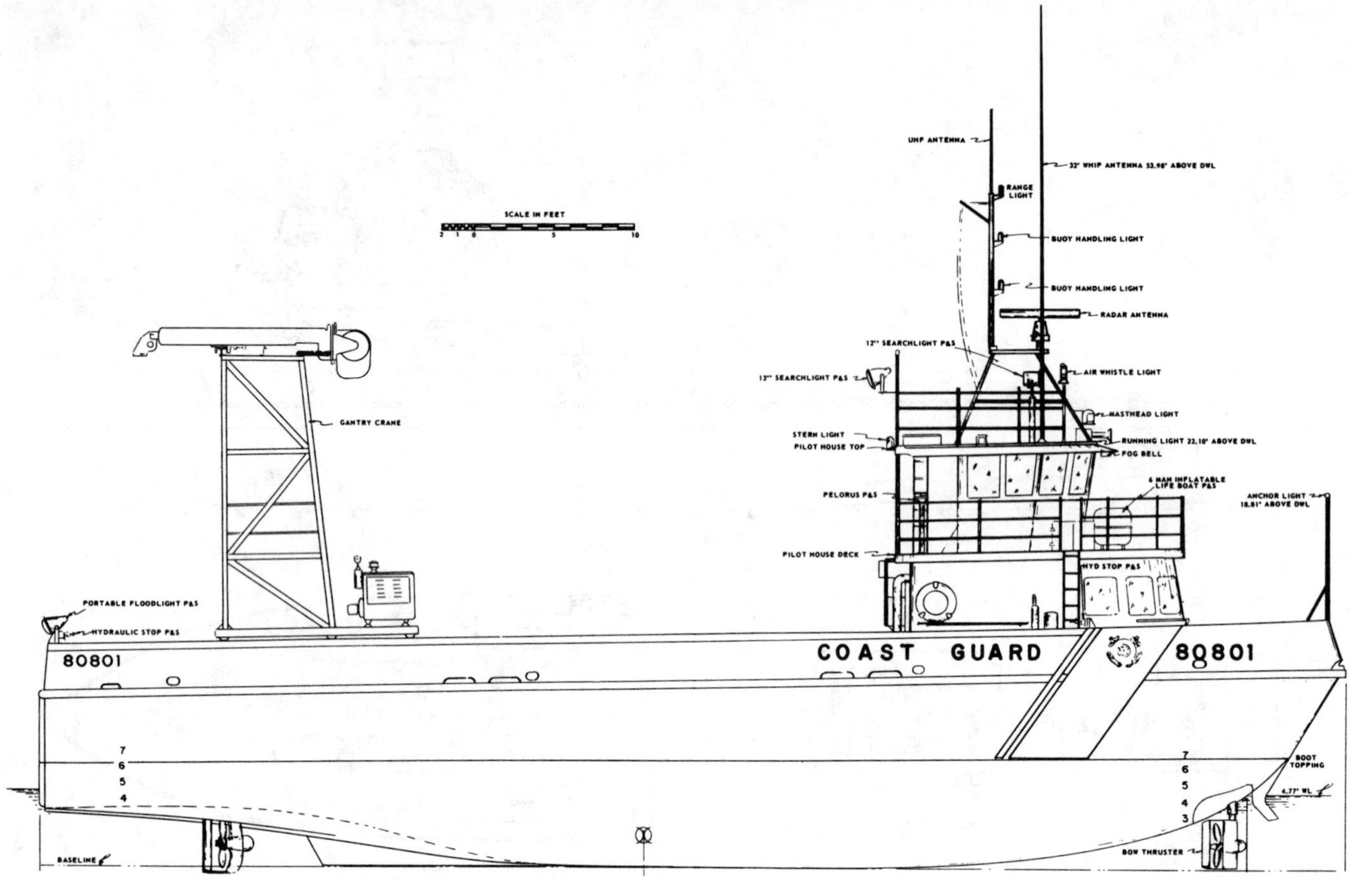

Fig. 46 Coast Guard buoy tender TERN

about one-third total tankage and 35 tons of buoy load, stores, and equipment the displacement is estimated at 168 tons. This same maximum displacement is reached with total tankage and no buoy load. Level trim conditions can be maintained from the design waterline to the maximum waterline with correct distribution of the buoy load and liquid in tanks. This degree of trim control is essential when handling buoys on deck and into the water.

b. Maneuvering and Propulsion. Main propulsion machinery consists of two diesel engines driving hydraulic pumps. These hydraulic pumps supply power to three hydraulic motor-driven right-angle drive propulsion units. The two after propulsors are the main propulsion units and are rated at 235 horsepower each. They extend down through wells in the after end of the machinery space. These units are fixed vertically but can rotate through 360 degrees. The bow thruster is also driven by a hydraulic motor and is rated at 125 horsepower. It can be retracted up to the hull below the flat with the maximum extended depth of its lower guard reaching the baseline. It too can be steered through 360 degrees. Maximum speed, with all three propellers in operation, is 10 knots in the fully loaded condition. Individual control is provided for the direction and magnitude of thrust from each propulsor. This results in excessive maneuverability which requires either an experienced operator or a new maneuvering control.

c. Outfit and Mission Support. The anchor is handled using a portable anchor davit with the anchor line stowed on a reel in the bow thruster compartment. Immediately aft of the house structure on the main deck is a sinker and chain winch used for hauling the buoy anchor chain over the stern roller and depositing it into the sinker and chain hold through the flush hatch amidships. A gantry crane provides the primary means of handling buoys over the stern. This crane rides on rails that run fore and aft above the main deck. The bottom of the cross truss-work clears the main deck by 5.5 m (18.0 ft). On top of the cross truss structure are mounted two booms running fore and aft off centerline. Each boom is mounted on a pivot at the forward end of the cross truss and is supported by wheels riding on a curved rail at the after end of the cross truss. Each boom can swing ±45 degrees from its fore and aft position driven by a hydraulic piston. This permits the after end of each boom, and its 3-ton lifting cable to swing from the centerline of the ship to the outboard edge of the main deck. With the fore and aft travel of the gantry this gives complete coverage of the working area. Tie down sockets are installed over the working area; wells are provided for buoy stowage.

References

Agdern, F. A., and W. G. Sherwood, (1979), "The Navy's

First Ocean Construction Platform, SEACON: Design and Operation," *Marine Technology,* April.

American Bureau of Shipping (1973), "Rules for Building and Classing Offshore Mobile Drilling Units."

Argyriadis, D. A. (1957), "Modern Tug Design," *Trans.* SNAME.

Baer, W. (1973), "Safety in Towing," *Proceedings* of the First North American Tug Conference, published by Ship & Boat International.

Benford, H. (1957), "Engineering Economy in Tanker Design," *Trans.* SNAME.

Blair, X., and X. Ansel, (1968), "*A Guide to Fishing Boats and Their Gear,*" Cornell Maritime Press.

Boylston, J. W., D. J. De Koff, and J. J. Muntjewerf, (1974), "SL-7 Containerships: Design, Construction, and Operatings Expenses," *Trans.* SNAME.

Comstock, J. P. (1967), Editor, "*Principles of Naval Architecture,*" SNAME, New York.

Connors, T. G. (1978), "Update: Domestic LNG Vessel Construction," *Marine Technology,* January.

Courtsal, D. P. (1971), "The Marine Business in the Central United States," *Trans.* SNAME.

Danforth, L. J. (1977), "Environmental Constraints on Drill Rig Configurations," *Marine Technology,* July.

Dillon, E. S., L. C. Hoffmann, and D. P. Roseman, (1976), "Forty Years of Ship Designs Under the Merchant Marine Act, 1936–1976," *Trans.* SNAME.

Dorman, W. J. (1966), "Combination Bulk Carriers," *Marine Technology,* October.

Eda, I. T. (1972), "Course Stability, Turning Performance, and Connection Force of Barge Systems in Coastal Seaways," *Trans.* SNAME.

Elbatouti, H. Y. Jan, and S. G. Stiansen, (1976), "Structural Analysis of a Containership Steel Model and the Comparison with the Test Results," *Trans.* SNAME.

Giblon, R. P., and R. J. Tapscott, (1973), "Design and Economics of Integrated Tug/Barge Systems," *Marine Technology,* July.

Glosten, L. R. (1967), "Sea-Link," Pacific Northwest Section, SNAME, December.

Guarino, S. J. (1975), "The Offshore Supply Vessel as a Naval Auxiliary," *Marine Technology,* October.

Henry, J. J. (1955), "Modern Ore Carriers," *Trans.* SNAME.

Johnson, R. P., and H. P. Rumble, (1965), "Weight, Cost and Design Characteristics of Tankers and Dry-Cargo Ships," *Marine Technology,* April.

Karppi, F. O. (1966), "New Ferryboats for Staten Island-Manhattan Service," *Marine Technology,* July.

Korvin-Kroukovsky, B. V. (1961), *Theory of Seakeeping,* SNAME, New York.

Laredo, D. Behin, and Garguet, M. (1977), "Design of the First Generation of 550,000 Tankers," *Trans.* SNAME.

Long, C. L., J. L. Stevens, Jr., and J. T. Tompkins, Jr. (1960), "Modern High Speed Tankers," *Trans.* SNAME.

Macy, R. H. (1969), "Drilling Rigs," *Ship Design and Construction,* SNAME, New York.

McClure, A. C. (1965), "Development of the Project Mohole Drilling Platform," *Trans.* SNAME.

McGowan, J. F., and R. B. Meyer, (1979), "Has Stability Delayed the Delivery of Your Tugboat?," Gulf Section, SNAME, February.

Metcalf, M. F., M. W. Praught, and W. O. MacDonnell, (1979), "On the Installation Planning of a Self-Floating Offshore Platform," *Marine Technology,* July.

Mok, Y., and Hill, R. C. (1970), "On the Design of Offshore Supply Vessels," *Marine Technology,* July.

National Academy of Sciences (1956), "Roll-On, Roll-Off Sea Transportation," Symposium, November 19.

National Academy of Sciences (1958), "Recent Research in Marine Transportation," Symposium, February 13, 1958.

Nichols, W. O., M. L. Rubin, and R. V. Danielson, (1960), "Some Aspects of Large Tanker Design," *Trans.* SNAME.

Novak, S. M., P. W. C. M. Van Oirschot, H. P. Young, and D. A. O'Neil, (1977), "Increased Profits for Gas Turbine Containerships by Unique Applications of Combustion Technology and Hydrodynamics," *Trans.* SNAME.

The Petroleum Publishing Co. (1976a), *The Technology of Offshore Drilling, Completion, and Production.*

The Petroleum Publishing Co. (1976b), *Offshore Platforms and Pipelining.*

Roach, C. D. (1954), "Tugboat Design," *Trans.* SNAME.

Robinson, H. F., J. F. Roesel, and A. S. Thaeler, (1948), "Modern Tankers," *Trans.* SNAME.

Rook, G. (1978), "Bayou Boat Building as Related to Modern Tug Design," *Marine Technology,* January.

Roseman, D. P., M. Gertler, and R. E. Kohl, (1974), "Characteristics of Bulk Products Carriers for Restricted-Draft Service," *Trans.* SNAME.

Russo, V. L., and E. K. Sullivan, (1953), "Design of the Mariner Type Ships," *Trans.* SNAME.

Rynn, H. S. (1975), "Review of Rigid Integrated Tug-Barge Combination," NE Coast Institution of Engineers and Shipbuilders, October 27.

Sarchin, T. H., and Tryner, H. G. (1970), "Cycloidal Propulsion and Control for Ship Handling Tugs," *ASNE Journal,* June.

Seymour, D. J., and McCardell, C. R. (1967), "Floating-Type Offshore Drilling Vessels," *Marine Technology,* October.

Spaulding, P. F. (1973), "Ocean Tug Boat Design," First American Tug Convention, published by Ship & Boat International.

Storch, R. L. (1972), "Stability of Offshore Tugboats," *Marine Technology,* October.

Taggart, R. (1961), "Positioning the Drilling Ship Over the Mohole," SNAME Chesapeake Section, February 2.

Thomas, W. D., and Schwendtner, A. H. (1971), "LNG Carriers: The Current State of the Art," *Trans.* SNAME, 1971.

van Mook, C., and D. P. Courtsal, (1975), "A New Dimension in River Towboats: 10,500 shp," *Marine Technology,* April.

Waller, D. B. (1972), "Integrated Tug/Boat Units for Ocean Transportation," Gulf Section, SNAME, February.

Waller, D. B., and J. J. Filson, (1972). "Advanced Ocean Tug/Barge Systems," Mar Ad Report MA-6-562, May.

Williford, F. B., and A. Anderson, (1973), "Dynamic Stationed—Drilling SEDCO 445," OTC, Volume II.

CHAPTER III

Robert J. Tapscott

General Arrangement

Section I
General

1.1 Definition. The general arrangement of a ship can be defined as the assignment of spaces for all the required functions and equipment, properly coordinated for location and access. Four consecutive steps characterize general arrangement; i.e., allocation of main spaces, setting individual space boundaries, choosing and locating equipment and furnishings within boundaries, and providing interrelated access. These steps progress from overall to detail considerations, although there is some overlapping. Generally, arrangement plans are prepared for the conceptual, preliminary, contract, and working plan stages. The data for early stages come from past experience, and the degree of detail increases as the design progresses.

It has often been said that ship design is inevitably a compromise among various conflicting requirements, and it is in the formulation of the general arrangement that most of the compromises are made. Ship design requires a melding of many arts and sciences, and most of this melding occurs in the general arrangement. The designer considers the demands for all the functions and subfunctions of the ship, balances the relative types and importance of the demands, and attempts to arrive at an optimum relationship of the space assignments within the ship hull. The general arrangement represents a summary and integration of information from other divisions and specialties in ship design, intended to provide for all the necessary functions of the ship in the most efficient and economical way.

The efficient operation of a ship depends upon the proper arrangement of each separate space and the most effective interrelationships among all spaces. It is important that the general arrangement be functionally and economically developed with respect to factors that affect both the construction and operation cost, especially the manpower required to operate the ship. Many other divisions of ship design provide the feed-in for the general arrangement, such as structure, hull engineering (hatch covers, cargo handling, etc), weights, stability, lines, engineering (machinery, uptakes), and specifications.

Problems of general arrangement are associated with the function of the ship and generally differ according to ship type. The arrangements of all types, however, have certain things in common. For example, the problems of accommodation and propulsion machinery arrangements are generally similar, although the different ship types impose different limitations.

1.2 The Problem and the Approach. The first step in solving the general arrangement problems of a cargo ship is locating the main spaces and their boundaries within the ship hull and superstructure. These spaces are:

1. cargo spaces,
2. machinery spaces,
3. crew, passenger, and associated spaces,
4. tanks,
5. miscellaneous.

At the same time, certain requirements must be met, mainly:

1. watertight subdivision and integrity,
2. adequate stability,
3. structural integrity,
4. adequate provision for access.

The general arrangement is evolved by a gradual process of trial, check and improvement. As for any other problem, the first approach to a solution to the general arrangement must be based on a minimum amount of information including:

1. Required volume of cargo spaces, based on type and amount of cargo,
2. method of stowing cargo and cargo handling system,
3. required volume of machinery spaces, based on type of machinery and shp,
4. required volume of accommodation spaces, based on number of crew and passengers and standard of accommodations,
5. required volume of tankage, mainly fuel and clean ballast, based on type of machinery, type of fuel and cruising range,
6. required standard of subdivision and limitation of main transverse bulkhead spacing,

7. approximate principal dimensions (length, beam, depth, and draft), and
8. preliminary lines plan.

The approximate dimensions and lines plan development are described in Chapter I.

The first general arrangement layout to allocate the main spaces is based on the above information. Peak bulkheads and inner bottom are established in accordance with regulatory body requirements. Other main transverse bulkheads are located to satisfy subdivision requirements, based on preliminary floodable length curves. Decks are located to suit the requirements of the spaces, cargo, machinery, accommodations, etc., and to satisfy strength requirements. Allowance for space occupied by structure must be deducted in arriving at the resulting net usable volumes and the clear deck heights.

Usually, in the first approach, several preliminary general arrangements are laid out in the form of main space allocations, boundaries, and subdivisions. These are checked for adequacy of volumes, weights and stability, and the changes to be made in the preliminary lines to make these features satisfactory. At this point, certain arrangements may be dropped, either because they are not feasible or are less efficient than other arrangements. The general arrangement process continues into more refined stages, simultaneously with the development of structure, machinery layout, and calculations of weights, volumes, floodable length, and stability, intact and damaged. The selection of one basic arrangement may come early in the process, or may have to be delayed and based on a detailed comparison of *trade-offs*. In any case, the selection is usually made in consultation with the owner so that consideration may be given to his more detailed knowledge of operating problems.

The approach and the steps for developing the general arrangement for ship types other than cargo ships are similar, although the types of space requirements will vary with the function and the mission.

Section 2
Cargo Spaces

2.1 Types of Cargo. In the maritime industry, the transportation of cargo across our oceans is of far greater importance than the carriage of passengers since airlines have taken over the transoceanic passenger traffic. In recent years the types of cargo and the method of handling and stowage have dictated the designs of new ship types and generally are the prime constraint of the general arrangement.

The types of cargo to be transported are myriad and may be classified in various ways. For purposes of this chapter, classification is discussed mainly from the point of view of how cargo may be carried. The two main classifications are *bulk cargo,* consisting of homogeneous materials in liquid or solid form with fairly small particle size, and *general cargo,* consisting of a mixture of numerous types of products in boxes, crates, bales and bundles, or not packaged at all.

2.2 General Cargo. The typical general cargo ship picks up and delivers a wide variety of commodities at a number of ports, but at none in such quantitites as to warrant a port-to-port service for one commodity. As previously stated, the primary objective of a ship's system and the arrangement of that system is least cost for its function. Costs may be reduced by improving the method of handling and stowing cargo and result mainly from savings in labor costs and ship time. The following methods of reducing cost of cargo handling for general cargo ships should be provided for by a good general arrangement.

a. Design of Cargo Spaces for Efficient Stowage. This involves making the cargo spaces as square as possible with the least number of interferences, such as pillars, protruding brackets, protruding vent ducts and piping. It is advantageous for deck beams and side frames to be of uniform depth, since interferences, such as web beams or frames extending into the cargo space, cause less efficient stowage, and in some cases increase the *broken stow* significantly.

b. Provide for Rapid Handling of Cargo in Port. As previously stated, most modern general cargo services involve handling many types of cargo in a number of ports. One of the most important aspects in rapid handling of cargo in port is the provision for access to a particular lot of cargo to be discharged at the port without restowing cargo destined for some other port. Therefore, one of the most important means of providing rapid cargo handling is to avoid the so-called *overstow problem,* that is, to provide access to a large number of lots of cargo without disturbing others.

c. Provide for Reduced Labor for Handling Cargo in Port. The objective is to handle more tons of cargo for less man-hours, and this is accomplished generally by providing a systematic means for handling cargo, utilizing mechanical equipment.

d. Provide for a Reduced Number of Handlings. The objective is to discharge cargo from the ship directly into the next carrier, which will carry it to a point inland, in such a way that it can be transferred without restowing. This principle, of course, also applies in reverse, with cargo arriving at the terminal in a form not requiring restowage for transferral to the ship.

The foregoing considerations lead to an overall saving, either directly in cargo handling labor costs, or in reducing ship's time in port and thereby saving ship costs. These considerations have led to improvements in breakbulk, general cargo ships and have resulted ultimately in the development of unitized cargo ships, such as containerships and ships to handle cargo stowed on standard-sized pallets.

2.3 Breakbulk Ships. Examples of ships of this type and

their general arrangements are discussed in Section 8 of this chapter. The following discussion concerns aspects of the general arrangement of the general cargo spaces of this ship type.

a. Cargo Stowage—General. Successful stowage should satisfy the following requirements:

1. Safety of ship and crew. The stability in various stages of loading and unloading must be considered. All precautions must be taken with respect to dangerous cargos and against cargo shifting, which could be injurious to the safety of the vessel.

2. Safety of cargo from damage. Stowage should be such that danger of fire, crushing, damage by moisture or tainting is minimized.

3. Orderly and rapid loading and unloading. A significant percent of the total operating cost is in cargo handling; the shorter the run, the higher the percentage. When a ship is tied up for cargo handling, practically all expenses continue except for propulsion fuel.

4. Ship's earning capacity should be used to best advantage. One of the principal factors conducive to profitable ship operation is the operator's ability to have his ships sail at full capacity, either on a cubic or weight basis, with the most profitable cargo available. The ship should be designed with a cubic/deadweight relationship most suitable for its average trade, but for any specific voyage the capacity will be controlled by one of these limits. For general cargo ships, it may be advisable to allow extra cubic beyond the theoretical full capacity of the ship to provide for more rapid cargo handling and some extra space allowance to reach a lot of cargo without having to restow. For all ship types, however it is usually also desirable to provide for extra deadweight capacity, since this can usually be done for little additional cost; i.e., scantling draft deeper than intended load draft.

b. Stowage Factor, Deadweight, and Measurement. The stowage factor is the volume per unit weight. In a given trade, a ship should be designed so that the ratio (cargo bale cubic/cargo deadweight) will exceed by 10 or 15 percent the overall stowage factor for the goods carried in the trade. This 10–15 percent margin is to compensate for *broken stowage,* which includes the spaces between and around the cargo packages, including *dunnage,* and those spaces which are not usable because of interferences, such as beam knees and hold brackets. Leeming (1942)[1] gives size of dunnage and ranges of broken stowage values for various types of cargo. He also gives a comprehensive list of stowage factors for both U.S. and non-U.S. goods.

c. Number of height of 'tween decks. The proper division of the height above the double bottom between hold and 'tween decks depends on the trade. At one extreme the bulk carrier may be said to be all hold, while at the other, ships such as the refrigerated banana carriers are all 'tween decks. The general *breakbulk* cargo carrier lies between these extremes. For a normal breakbulk general cargo ship ranging between 107 and 183 m (350 and 600 ft) in length, the depth is subdivided into an upper 'tween deck, lower 'tween deck, and a hold. The upper 'tween deck may range between 2.4 and 3.0 m (8 and 10 ft) in height, the lower 'tween deck between 2.7 and 4.6 m (9 and 15 ft), and the hold between 3.0 and 6.1 m (10 and 20 ft). Generally the hold is limited to about 5.5 m (18 ft) or less to minimize damage to cargo through crushing.

It should be pointed out that the more important hull characteristics, including depth and number of 'tween decks, are interrelated in some manner—indirectly through various regulations or more directly by considerations of economy and strength—so that the designer is not entirely free to determine deck heights from the viewpoint of utilization and cargo stowage alone.

d. Other design considerations for general cargo spaces for a breakbulk ship, all of which aim at providing efficient cargo stowage and quick handling, are outlined as follows:

1. Cubic Per Hold. Generally the amount of cargo cubic per set of cargo handling gear should not be greater than 1,700 m^3 (60,000 ft^3). A hatch size for two sets of cargo gear intended to be used simultaneously should not be smaller than 6.1 m by 9.1 m (20 ft by 30 ft) and preferably should be larger. The fact that the deck cargo gear and hatches require a certain length of ship sometimes presents a problem in the midship portion of the ship where it is desirable to make holds shorter, to limit the cubic per longshoreman gang to a reasonable figure. For this reason, the general arrangement sometimes assigns spaces amidships, particularly in lower holds, for purposes other than stowing general cargo; for example, such spaces may be assigned to fuel oil deep tanks or cargo oil deep tanks, or in some cases, clean ballast tanks.

2. Hatch size should be based on a study of the particular service and throughout the ship should be as standardized as possible. Large hatches mean quicker cargo handling, since a greater portion of the cargo is directly under the cargo hook and does not have to be moved laterally from the wing spaces (abreast the hatch) or the trunk spaces (fore and aft of the hatch). However, a certain amount of wing and trunk space are valuable for a multiport cargo service to avoid the overstow problem. The conflicting requirements of a large hatch opening for efficient handling and the avoidance of overstow can be resolved to some extent by multiple-opening hatch covers; that is, hatch covers which may open by thirds or halves at the choice of the operator. In this way, in effect, the size of the hatch and the amount of wing and trunk space can be varied at different points of the voyage. Another means of resolving these conflicting requirements is to provide multiple hatches abreast.

3. Different types of cargo handling rigs have been developed for more efficient and more rapid handling of general cargo, either in slings or on pallets or other types of lifts. These types of cargo gear include the improved conventional rig, the Ebel type rig, and the Farrell type rig. Cargo handling rigs of these types and others are discussed in detail in Chapter X of this book.

4. Other Features. In modern cargo liners the handling of cargo in the 'tween decks by forklift trucks has become

---

[1] Complete references are listed at end of chapter.

standard practice. For this operation, all 'tween decks should be made flush and as smooth as possible. Coamings should be eliminated at the boundaries of the hatch covers to permit the lift truck to operate between the hatch square and any part of the 'tween deck. The widespread use of forklift trucks has led to more attention to cargo hold ventilation, especially if the trucks are of the internal combustion type; however, it is still common practice to depend mainly upon portable blowers. Some general cargo ships carry their own electric-powered trucks and have assigned truck stowage spaces fitted with battery charging equipment. The trucks are transferred to the different holds and 'tween decks by means of the ship's cargo handling equipment. Electric-powered trucks, of course, do not present a ventilation problem, and it is believed that their use will become more prevalent for this reason.

Other aids for cargo handling, particularly for distribution of the cargo in the 'tween decks are commonly carried on general cargo ships as part of the cargo-handling system. These aids include conveyors of various types, some electrically powered, which are generally stowed at assigned locations in the 'tween decks or holds when not in use.

2.4 Containerships. Although most modern general cargo ships have some capability in carrying containers, the term *containership* usually is applied to a ship designed to carry almost all its cargo in containers. Some are non self-sustaining, meaning they must depend upon shore facilities to load and unload them; others have lifting equipment aboard and, with respect to loading and unloading containers, are self-sustaining. The two main types of containerships are the vertical cell type and the horizontal loading type.

a. Vertical Cell Containerships. This is the most common type of containership, and is arranged to have the containers stacked in vertical cells formed by angle corner guides. The advantages of cell type containerships are very rapid loading and discharge, and integration with inland transportation systems to eliminate multiple handling and restowing cargo, with its accompanying damage and pilferage.

Disadvantages of this type are that all the containers must be of a uniform length and width with uniform fittings for lifting, stacking, and locking; and that it cannot be used for any other type of cargo or even another size or type of container without extensive conversion.

The principal objective in laying out the container spaces is to fit the maximum number of containers within a given hull volume. The container cells are arranged so that the long dimensions of the containers are fore and aft; principally because this stowage is better suited to handling with a gantry crane over the side, particularly to trailer chassis. Fore and aft stowage also is easier to integrate with ship structure. The development of the final ship dimensions, hull form, and structure must be very carefully coordinated to attain the maximum count. The beam of the ship is finally selected to provide a certain number of cells athwartship, plus clearance, plus structural and stability requirements. The number of bays of containers in the length of the ship is governed by requirements for structure and clearances, as well as the practical limit of travel of the gantry crane on deck (if the ship carries cranes). As the hull form fines for and aft of amidships, containers must be omitted from the bottoms of the outboard cells. Often a few inches adjustment of the lines or special design variations of the structure at those critical points will add a significant number of containers which will pay for this design variation many times over in the life of the ship. It is obvious that the hull form and type of sections should be chosen from the point of view of enveloping container cells as well as for speed and power considerations.

The cells themselves are made up of corner guide angles attached to the ship structure. Normally the cell guides are installed after the main ship structure is complete, so they will have the proper clearances and tolerances for proper container stowage, without binding or allowing too much container movement due to ship motions. Hatch covers are normally of the lift-off pontoon type, dogged watertight by manual or hydraulic means. Since the covers may be handled normally by the gantry crane, they are usually large and span the length of one cell and the width of several cells. When opened, the covers are stowed on the adjacent cover forward and aft, so that the gantry crane has simultaneous access to a row of transverse cells or a complete bay of containers.

Most cell type containerships carry a maximum number of containers on deck, which is considered more economical since these can be containers which require no protection from the weather and the additional expense of the enclosing structural envelope is eliminated. However, up to certain limits, the depth of the ship can be increased to envelop a higher stack of containers below deck without significantly increasing the steel weight and the cost of the ship. The total height of the stack of containers, including those below and above deck, is ultimately limited by stability within acceptable ship proportions. Deck containers are carried up to 4-high generally with added complexity and manpower requirements for high stowages. Stacking of deck containers up to 4-high is simplified by the use of on-deck guide systems which form open type container cells similar to those below deck. (For deck container securing systems, see Chapter X.)

The hatch covers and deck structure must be designed for the extra weight of the deck containers, and the gantry cranes, either shipboard mounted or shore based, must have clearances for handling them. Reefer containers are normally carried on deck where the required ventilation for the electric-powered units can be supplied by natural means, and there is access to the units. Suitable electric connection boxes must be provided at the predetermined reefer container locations.

As stated above, the container space layout is intimately associated with structure. Careful consideration must be given to structural requirements, since it is obviously desirable from a cargo capacity viewpoint to have a ship bottom and shell, with no decks or internal structure. However, a certain amount of longitudinal structure is required for longitudinal strength, transverse structure for transverse wracking and torsional strength, vertical structure to support deck and deck loads, and tie-in structure to stabilize all major structural elements to keep them in working position and prevent buckling. The vertical loads of the stacked containers are normally transmitted directly to the double bottom structure, with the vertical cell guide structure taking only the transverse forces due to ship motion, list, and trim. However, if containers are stacked more than six high, which is the limit for the standard container, then movable supports off the vertical structure must be provided to support the upper containers. The strong incentive for minimum dimensional allowance for structure in container ships leads naturally to the application of higher strength steels, particularly in the upper deck structure which forms the flange of the ship girder. (For additional discussion of ship structure, see Chapter VI.)

Proper clearances must be provided between the bosoms of the container guide corner angles and the container corners to allow for easy loading and discharge, but at the same time to restrain the container stack against excessive motion. Normally, 13 mm (1/2 in.) clearance is allowed at each corner. In developing detail plans, plus and minus tolerances must be carefully specified, taking into consideration tolerances on container dimensions.

For cellular-type containerships, watertight subdivision of the ship presents no problem, since all cargo movement in the ship is vertical, and watertight bulkheads can be placed between container cells with no space sacrifice. These bulkheads extend to the weather deck, which becomes the bulkhead deck, with generally adequate freeboard for any subdivision or damaged stability requirement.

Examples of arrangements of cell-type containerships and a discussion of container handling and securing gear are included in Chapter X.

b. Horizontal Loading Containerships. In this type of ship, the containers are loaded on one or two levels above the bulkhead deck by fork truck or straddle truck, through stern ports or side ports. It is neither as common as the vertical cell type nor as distinct a ship, since it also can be used as a pallet ship or a roll-on/roll-off ship, and is classed by type on the basis of the prime purpose of the original design.

The ship is usually designed for certain size containers stowed in certain locations. However, the design is usually such that other sizes of containers can be stowed. An advantage is that the ship is not restricted to one specific size and type of container with specific fittings, but can take a variety of containers within certain limitations. As such, it is also readily adaptable to other types of cargo which can be handled by forklift trucks or rolling equipment; e.g. palletized cargo, automobiles, trucks and trailers, or military vehicles. Expensive gantry cranes, permanently mounted on the ship or on the pier, are not required, although these savings may be offset by the necessity of fitting folding, extensible ramps, usually on the stern quarter. Standard forklift or straddle trucks are used, and these can be carried on the ship or transferred from pier to pier, according to the demands of the service. Thus, this type of containership is much more flexible than the cellular type.

The horizontal loading containership is generally slower handling than the cellular type, and cannot be loaded and discharged simultaneously to the same degree. It is not as adaptable to delivering and receiving directly from trailer chassis. Accessibility to a large number of lots of containers can be attained by leaving aisles and truck maneuvering space empty, but the horizontal loading containership does not have this feature to nearly the same extent as a cellular containership, where there is access to each cell (after the deck containers are removed).

The space below the cargo deck or the bulkhead deck must be utilized for something other than containers. After allowance for machinery space and fuel oil deep tanks, the remaining space can be assigned to cargo oil tanks (an ideal cargo companion with containers for this type of ship), special cargo handled from above by rapid methods (such as reefer cargo handled by conveyor), ballast tanks, or void spaces. In this type of containership, the aisle space left unoccupied on the container deck and the void spaces left below the container deck may be compared to the void spaces outboard and below the container cells of the cellular type ship. In each case, excess space is built into the ship for more rapid cargo handling.

The layout of the cargo deck of a horizontal loading containership is again a matter principally of coordinating the dimensions, required clearances and structure locations to give the maximum container stowage. The containers are stowed with the long dimension athwartship to suit the fore-and-aft travel of the lift truck. The beam is determined by the number of containers athwartship, the requirements for side framing and pillars, and all working clearances. At the end of the deck, containers are dropped as required by the hull shape, but the deck line is adjusted to attain the maximum stowage.

The main structural feature of this type of ship is the strength of the container deck. Although the total load of a container, if distributed, would be moderate, the deck structure must be designed for the high concentrated loading of the lift truck wheels at any point where these trucks might travel. Also, the decks must be designed for the concentrated corner loads of the containers, which may be in predetermined locations if the containers are uniform and standardized, or may be anywhere if flexibility for random-sized containers is desired. Pillars and other interferences on the container deck must be minimized and coordinated with the container stowage.

This type of ship is designed with minimum freeboard, since the bulkhead deck is the cargo deck. Transverse watertight bulkhead spacing is chosen to suit the requirements of subdivision, stability, and the space assignments for machinery, deep tanks, etc.

Normally the weather deck is not designed for the carriage of containers by lift truck, because of the massive structure required for lift truck wheel loading and the difficulty of the lift trucks carrying containers up ramps. The deck may be designed for automobiles loaded by driving up a ramp, or for light containers, or other deck cargo loaded by ship's overhead lift gear or shore lift gear.

One problem with this type of ship, which requires attention during the arrangement design, is provision for CO_2

fire protection of the cargo space. Since all the cargo is, in effect, in one hold, the volume would require an inordinate amount of CO_2 for simultaneous flooding. Therefore, some screen bulkhead or bulkheads, reasonably gastight, must be provided to subdivide the spaces. These may be in the form of bulkheads with large sliding doors, accordion-type doors, overhead garage doors, etc, since it is desirable to have the cargo space as completely open as possible during the loading and discharging operation. These bulkheads must be coordinated with the container layout and must be capable of being closed when the ship is loaded and ready to go to sea.

c. Containers. Since the growth of the widespread use of containers and their incorporation in ship cargo handling systems was a gradual process, there are many types and sizes in present use. These range from fairly small plywood boxes to larger steel or aluminum *van* size containers; however, there are several standards of sizes and construction which are presently in use in large quantities.

The American National Standards Institute (ANSI) and the International Organization for Standardization (ISO) have developed standards for dimensions, strength, and fittings for a series of containers. ANSI (1971) container standards are illustrated in Chapter X. The most prevalent container size of this standard series is the 20-ft length although the present trend is toward more containers of the 40-ft length and containers of greater depth, up to 9 ft-6 in.

2.5 Pallet Ships. In this ship type, the cargo spaces are arranged for the rapid loading and discharge of cargo units, which consist of cargo items prestowed and secured to standard size pallets. The pallet itself and its sling (if overhead lifting is employed) form part of the cargo unit. Like containerships, there are two main general types of pallet ship, one designed for vertical loading and the other for horizontal loading.

a. Vertical Loading Pallet Ships. In this design, the ship is opened up as much as possible by the use of wide hatches, with a minimum fore-and-aft spacing between. In some cases, the hatches are also subdivided athwartship for the purpose of simplifying the covers or providing additional longitudinal structure. It is desirable that the design provide for a minimum horizontal movement of the cargo into the wings, with a maximum amount of cargo stowable in the hatch squares directly under the cargo hooks. It is also desirable for the cargo spaces to be squared off and to make the 'tween-deck heights uniform for stacking palletized cargo, usually to a total height of 2.4 to 3.0 m (8 to 10 ft).

Dimensions of the cargo spaces should be based on multiples of the pallet dimensions, with a suitable allowance for clearances between pallets and for some projection of cargo beyond the pallet edge. The cargo gear is selected to provide rapid handling of the relatively light pallet loads. Rotating cranes, either fixed or movable, or winch and boom burtoning gear are commonly used. Means of securing the stacked pallets in partially loaded cargo spaces must be provided.

b. Horizontal Loading Pallet Ships. This ship is similar to the horizontal loading containership in that the pallet cargo enters the ship through side ports or stern ports by one of various means, including forklift trucks, tractors or trailers, or a combination of sideport gear and lift trucks. The pallets can be transferred to lower decks by means of elevators or conveyors. Objectives of the general arrangement of the cargo spaces include large, unobstructed, warehouse-type spaces, uniform deck heights, and smooth decks, with the strength to support the concentrated wheel loads of lift trucks or tractors and trailers.

2.6 Roll-On/Roll-Off Ships. The term *roll-on/roll-off* covers a broad category of ships designed to load and discharge cargo which rolls on wheels. Broadly interpreted, this may include train ships, trailer ships, auto truck and trailer ferries, and ships designed to carry military vehicles. (Containerships and pallet ships, which are loaded by lift trucks or tractors and trailers, are sometimes considered in the roll-on/roll-off category since the cargo does move on and off the ship on wheels.) To meet the demands of the growth in import of foreign automobiles, principally from Japan and West Germany, special designs for autos featuring ships with low 'tween-deck heights and portable decks have been developed. Some of these vessels are loaded by driving the autos aboard, but most are loaded by lift-on/lift-off methods to avoid the hazards and problems of gasoline.

The roll-on/roll-off ship requires a high proportion of cubic compared to the cargo carried and is particularly suited to services with short runs and frequent loadings and discharges, although recent roll-on/roll-off ships have been designed for longer runs including trans-Pacific routes. Certain features in the arrangement of the cargo spaces are common to all roll-on/roll-off ships:

1. Clear decks without interruption by transverse bulkheads, and deck heights to accommodate vehicles.
2. Side ports, stern ports, and ramps of suitable dimensions and angles for the cargo transfer operation between pier and ship. Ramps are sometimes part of the shore installation, such as for train ships.
3. Decks designed to withstand the wheel loads of the vehicles.
4. Proper clearances for stowing and turning the vehicles.
5. Internal ramps or elevators to allow vertical distribution of the cargo over the various decks of the ship.

The requirement for clear decks (without interruption by transverse bulkheads) and 'tween-deck heights to accommodate the specific vehicles calls for a structure in the roll-on/roll-off ship which differs significantly from that of the standard, transversely framed cargo ship.

A more specific discussion of the general arrangements of the cargo spaces must be by specific ship types.

a. Train Ships. Special design considerations for the car spaces for train ships include:

1. Vertical clearance for the various types of freight cars and, in some cases, locomotives,
2. layout of tracks with proper limitations of curvature and the required clearances on each side,
3. clearances between tracks for access to and operation of the jacks and hold-down lashings.

The structure of the train ship is usually based on a longitudinal framing system. Main considerations in the design of the transverse beams and brackets are the required clearances for cars and the support to the weather deck, which generally does not carry cargo. The structure of the car deck itself is designed usually to supply the required structural support directly under each track, since all cargo loading is in line with the tracks. Machinery casings, access ladders, etc are arranged for minimum width and located in the way of pillar lines or structure at the side of the ship. Generally, selection of the ship beam is based on the number of tracks, the required clear width in way of each track, the requirements for access along the side of each track, plus the requirements for structure in way of the pillar lines and at the shell, port and starboard.

The most common type of train ship is the single-deck stern-loading ship. A variation of this type utilizes a lower deck with the cars transferred singly by an elevator. In another type of train ship the cars are lifted on and off by a shore crane through a large hatch amidships. When lifted on, the cars are rolled on tracks to the ends of the ship. Although not strictly of the roll-on/roll-off type, the same design principles are generally applicable to these ships.

In laying out the cargo spaces, it is important to have all the data on the dimensions and characteristics of the various cars to be carried. The widest rolling stock being about 3.0 m (10 ft), the tracks are generally not less than 3.5 m (11.5 ft) apart. 'Tween-deck heights are such as to accommodate the highest cars carried, usually about 5.5 m (18 ft) clear height. If flat cars with trailers are to be carried, the clear height must be greater. The radii of curvature of the tracks on train ships should generally be limited to about 49 m (160 ft) minimum. Maximum grades should be limited to about 5 percent, including the effect of trim of the ship during loading operations. Car securing fittings normally consist of jacks and holddown lashings applied to each corner of the car.

b. Trailerships. Most of the trailerships in service are for the carriage of trucks and trailers of assorted sizes. For standard size trailers, it is more economical to lift the trailer off the chassis and carry it as a container in a containership rather than to carry the complete trailer and chassis. The factors in the arrangement of the cargo space for a trailership are similar to those for a train ship, except for differences in clearances, allowable gradings, turning radii, etc. The trailers are loaded in rows, following wheel tracks similar to a rail track layout, utilizing a guide fitting on the deck which projects into the space between the dual tires on one side of the trailer. Trailers are loaded and discharged singly by means of special tractors, permitting a smaller radius of curvature of the trailer paths than for a car track. Also, greater allowable grade angles for internal ramps for loading trailers on lower decks permit multiple-deck trailer ship designs. The securing system and fittings for the trailers are similar to those used for railway cars.

2.7 Barge Carrying Ships. As discussed in Chapter II, there are two principal types of modern barge-carrying ships: the SEABEE type and the LASH type.

The arrangement problems and principles of the SEABEE type ship are comparable to those of a ro/ro ship, while those of a LASH type ship are comparable to a cell type containership. In each case the barge is comparable to the container. For discussion of components in barge handling systems, see Chapter X.

2.8 Bulk Cargo Ships—General. Bulk cargo ships exist in almost as many varieties as general cargo ships. These include tankers and other liquid bulk carriers, grain and other dry bulk carriers, and combination carriers for numerous types of specific dry and liquid bulk cargos. The philosophy of the general arrangement of the cargo spaces stems from the fact that the cargo is homogeneous and in particle or liquid form, and can be transferred by pumps, blowers, conveyors, or grab buckets. The cargo spaces are divided into tanks or holds of suitable size, with due regard to:

- Structural requirements,
- subdivision requirements,
- restraining cargo to avoid moments due to shift or free surface,
- number of different types of lots of cargo to be carried simultaneously,
- ballasting requirements when ship is light.

Special considerations in design and arrangement of the tanks or holds include:

- Minimum interferences or obstructions inside the tanks to catch and hold cargo, for rapid discharge of cargo and minimum cleaning. The ideal arrangement would be completely flush interiors with rounded corners.
- Shape for self-trimming or self-draining to the point of operation of (entrance to) the discharge equipment,
- self-loading with minimum hand trimming from the point of discharge of the loading equipment,
- hatches of size and location to suit type of cargo,
- distribution of cargo to limit the longitudinal bending moment on the ship girder,
- assignment of ballast spaces for proper distribution when ship is light.

The most outstanding recent trends in bulk carriers are:

- Larger size tankers for carriage of crude oil,
- design features to reduce pollution of the ocean by the bulk cargo, particularly petroleum products,
- large scale development of specialized carriers for liquefied natural gas (LNG ships),
- large size combination carriers which offer low-cost transportation and the flexibility to carry many types of cargo over a wide variety of routes, and
- very large self-unloading Great Lakes ships.

General arrangements aspects of the cargo spaces of the main types of bulk carriers are discussed below.

2.9 Tankers. Tankers are normally designed to carry petroleum products ranging from crude oil to gasoline, with specific gravities in the range of 0.73 to 0.97. In current designs, the cargo spaces are divided into tanks of regular length, usually with one centerline tank and wing tanks port and starboard. This division is dictated by considerations of a practical limit of separate lots of cargo, structural requirements, restrictions of free surface of liquid, and safety of ship in the event of collision. However, it is anticipated that the 1978 protocols will change significantly this traditional cargo tank arrangement. All tank boundaries must be oiltight and access is by raised oiltight hatches. Ease of cleaning and corrosion control are important considerations in the interior design of the tank.

The last decade has seen the development of very large crude oil carriers and ultra large crude oil carriers designated *VLCCs* and *ULCCs* up to more than 500,000 dwt. This development, together with a growing concern for the protection of the ocean environment from pollution by oil spills or discharge of oily ballast, has introduced requirements which affect the arrangement of the cargo spaces. The main requirements as set forth in the US Coast Guard Regulations are applicable to US-flag tankers over certain sizes and partially to foreign-flag tankers entering US waters. The international regulations as developed by Conventions sponsored by IMCO have also developed similar regulations to limit the probability of pollution. For detailed discussion of the design requirements, see Chapter XI.

Specialized tankers for products other than crude oil present special problems in arrangement of the cargo tanks dictated by the nature of the cargos and the service. Petroleum products carriers may be designed to carry a number of different products in various combinations which dictate the tank sizes, locations and cargo piping systems. Chemical carriers may be designed for the carriage of a wide variety of chemicals in various lot sizes. Characteristics of various chemicals require such tank features as special tank materials such as stainless steel, special interior tank coatings, cofferdamming to separate from adjacent tanks, heating systems, and cleaning systems. The optimum arrangement of tanks for such a carrier can only be arrived at by comprehensive studies based on a clear definition of the service and product requirements.

2.10 LNG Ships. The growth in demand for the carriage of liquefied gas in large quantities has led to the development of special designs for this purpose, including cargo space arrangement dictated by the type of LNG tank containment system selected, as discussed in Chapter II.

2.11 Dry Bulk Carriers. This term usually refers to ships designed to carry grain or similar cargos with specific gravities of 0.72 to 0.80. The cargo spaces are generally divided into regular lengths, with the transverse section of the hold shaped for self-trimming, restriction against transverse shifting of the cargo, and to facilitate loading and discharging the cargo. Large hatches are provided for handling cargo by grab bucket or possibly by blower in the case of grain. Hatches need not be oiltight (Murray, 1965).

a. Ore Carriers. This term usually refers to ships designed for bulk cargos with specific gravities of 0.69 and greater. Such cargos do not fill the full cubic of the ship when it is down to its marks; hence, the cargo spaces are normally arranged to provide rectangular ore holds on the centerline and above a deep double bottom, so that the metacentric height of the loaded ship is not excessive. Ore carriers require structure designed for the dense cargo. Large hatches must be provided as for bulk carriers.

b. Other Bulk Carriers. Numerous types of cargo, other than grains and ores, are carried in bulk. The arrangement of the cargo spaces is governed by the characteristics of the cargo and the loading and discharging system. The most important cargo characteristics to be considered are the density and whether it will *flow* beyond a certain angle of repose. Some examples of miscellaneous bulk cargos are phosphate rock, limestone, coal, sulfur, gypsum, kaolin, and wood chips.

c. Self-unloaders. Bulk carrier and ore carrier type ships can be designed as self-unloaders, which have cargo spaces trimmed to two- or three-belt conveyors which transfer the cargo to a bucket conveyor, then to another belt conveyor on a long unloading boom. Such an arrangement, common on the Great Lakes, provides for rapid discharge on the cargo in locations where no shore gear is available.

d. Combination Carriers. It is evident why this type of ship—latest in the general field of bulk carriers—was developed. They have great flexibility in a changing economic environment and can provide for a relatively high percent utilization by carrying different cargos on various legs of a complex service. (A pure bulk carrier or tanker designed for one commodity normally is loaded in one direction and in ballast on the return trip, giving a maximum utilization of 50 percent.) The requirements for certain types of cargo merge well with each other in a design; for example, an ore carrier, when down to its marks, has excess cubic which can be utilized for oil tankage on a portion of the return leg, and the complete ship can be used for grain on another leg or route. The various possible combinations have given rise to many arrangements of combination carriers, which are thoroughly discussed by Dorman (1966).

Section 3
Crew and Passenger Spaces

3.1 Crew Spaces. In discussing crew spaces, a distinction must be made between cargo ships (which can carry up to 12 passengers by US Coast Guard Regulations) and passenger ships. The discussion in this chapter is mainly applicable to cargo ships, although the principles generally apply to passenger ships as well. A passenger ship must carry a larger stewards' department crew, and there is less emphasis on minimum crews in other departments, since a high standard of passenger service depends upon people to supply the service. As a consequence of larger crews and limited space on passenger ships, the crew habitability standards are somewhat lower than on cargo ships, particularly in the stewards' department. For detailed discussion of access, see Section 6.

a. Size of Crew. The factors which determine the size of a crew for a ship of American registry are:

- US Coast Guard Regulations and rulings,
- owner's requirements for maintenance and stewards' duties,
- union requirements resulting from negotiations with the owner.

The Coast Guard is responsible for specifying the minimum manning of US merchant ships, on the basis of numerous statutes and amendments which clarify the application of the statutes. In developing the minimum manning of a ship as shown on the Certificate of Inspection, the Coast Guard is primarily concerned with the safety of life at sea; i.e., that the ship has sufficient qualified personnel to be capable of safely coping with the normal hazards of the sea.

The actual crew list is determined by the owner and the maritime unions with which he has contracts, considering mainly the maintenance and service to be provided beyond the safe operation of the ship. Many factors affect the final negotiations, including the size and arrangement of the ship; the type of propulsion system and the arrangement of the machinery spaces; the degree of mechanization and centralized controls of the machinery; the amount of reefer cargo or other special cargo facilities; the total number of crew to be served by the stewards' department; and, in the case of a passenger ship, the number of passengers and arrangement of passenger quarters. When the ship has about 1,130 m^3 (40,000 ft^3) or over of refrigerated cargo capacity, 1 to 3 additional men to service the machinery may be carried. On older ships, these men stand reefer watches while on newer ships they are day workers.

When passengers are carried (limited to 12 on a cargo ship), additional stewards may be added to provide for their needs. In lieu of adding stewards to the crew, some operators provide additional compensation to the regular stewards on voyages when passengers are carried. On some cargo ships, the passenger facilities may be quite elaborate, requiring an even greater number of stewards. The total number of men in the stewards' department is a function of the total crew in the engineering and deck departments plus the number of passengers.

The final crew list is generally the result of negotiations and compromise and is influenced largely by precedent. Table 1 shows examples of crew lists for vessels of different types. Crew lists for vessels other than larger cargo vessels and bulk carriers are based on the manning necessary for the function and safety of the vessel as proven by operating experience and economics over the years. A harbor tug normally has a working crew of about six while an ocean tug may have a crew of ten. Fishing vessel crews can range from 12 to 30, depending upon the size and type. The sizes and standards of accommodations must be reduced in proportion to the size of the vessel, but generally every effort is made to provide a high degree of habitability for vessels which are at sea for long periods.

b. Regulations and Standards. The Coast Guard is quite specific in its regulations for crew accommodations. Among other things, the regulations include the following:

1. Location. Crew quarters must not be located farther forward in a vessel than a vertical plane located at 5 percent of the vessel's length abaft the forward side of the stem at the designed summer load waterline, nor on a deck below the deepest load line, with certain exceptions. Space for crew is to be positively separated from other types of spaces, such as cargo and machinery spaces and, where practicable, is to be separate and independent from spaces for passengers and licensed officers. On tankers, the rules require that the crew be located aft of the cargo spaces.

2. Construction. Accommodation spaces must be constructed of fireproof materials and in accordance with specified methods of fire protection. Stairways and corridors, as a means of escape, have special fire protection requirements. The crew spaces are to be insulated from heat, cold, and condensation, and bulkheads are to be odorproof in appropriate locations.

3. Sleeping Accommodations. These should fulfill the following requirements: Separate for deck, engine and stewards' departments. Berthing by watches. Maximum of 4 persons per room. Each person to have 2.8 m^2 (30 ft^2) or 6.0 m^3 (210 ft^3). Clear headroom of 191 cm (6.25 ft). Beds not more than two-high 76 by 193 cm (30 by 76 in.). Each person to have locker 1935 cm^2 (300 $in.^2$) in area 152 cm (60 in.) high.

4. Washrooms and Toilets. Each 8 crew to have 1 toilet, 1 basin, 1 tub or shower. Separate toilets and showers for deck, engine, and steward departments when they exceed eight in number. Each toilet space should have at least one wash basin. Urinals, when fitted, cannot reduce number of toilets.

5. Messrooms. These are to be located near galley or suitably equipped serving pantry as is practicable, (except where messroom is equipped with a steam table), and of a size to seat number of persons scheduled to eat at one time.

6. Hospital. There must be a hospital when the crew exceeds 12 and voyages are longer than 3 days. This space cannot be used for any other purpose. It should have a berth for every 12 crew or fraction thereof, up to a maximum of 6 berths. It must be equipped with toilet, washbasin, bathtub or shower. The crew count does not include those members occupying single-berth rooms. A hospital is not required when all rooms are for single occupancy, but an emergency treatment room must be provided.

7. Miscellaneous. The crew must have a laundry, clothes drying facilities or drying space, and recreation accommodations. Heating of crew spaces is to maintain a minimum temperature of 21°C (70°F) under normal conditions. Crew is to be provided with berthlights. Except in such areas as are considered insect-free, air ports and doors are to be fitted with insect screens, except air-conditioned spaces. Other regulations concern adequacy of stairway size and number and location of escapes from spaces.

8. Officers' Quarters. The Coast Guard Regulations contain no specific instructions for officers' quarters other than a requirement that they be separate from crew accommodations and at least equivalent to them.

In addition to the Coast Guard regulations, various union agreements contain specific requirements for crew accommodations, and these vary somewhat among the unions. Modern practice is to provide single-occupancy rooms with adjoining toilet space for all unlicensed crew. Toilet spaces are usually between rooms and shared by two individuals, except higher ratings such as chief cook, bosun, and chief electricians, who may have private toilet spaces. Area provided is about 10 m^2 (110 ft^2) clear inside the room, not including the toilet space.

Licensed officer rooms are for private occupancy with a private toilet space and are generally sized at about 14 m^2 (150 ft^2). Quarters for captain and chief engineer are relatively sumptuous and usually include an office or day room in addition to a stateroom, and may be as large as 37 m^2 (400 ft^2), or more.

Officers should be quartered together by departments to provide ease of access to their place of duty. Deck officers and radio operators are housed abaft or below the navigating spaces; the purser, steward, and surgeon are located usually at passenger stateroom level. Engineering officers are located in the superstructure with the deck officers, with ready access to the machinery casing; by tradition, the deck officers on the starboard side and engineer officers on the port side.

c. *Habitability.* The term habitability is frequently used, particularly in connection with naval ship design, to refer to the collective features of crew accommodation and work spaces which make them habitable and physically suitable for their purpose. Habitability is a measure of the adequacy of work and off-duty facilities in terms of satisfying those personnel needs which are dependent on the physical environment. Habitability can affect health, motivation and performance, which in turn can affect ship

Table 1—Examples of Crew Lists for Ships of Various Types

TYPE OF SHIP	Cargo Ship (C4-5-69a) (12 passengers)	Container-ship (C6-5-85b) (12 passengers)	Ro/ro "Iron Monarch"	Trainship (Not built)	Bulk Carrier Cla. Sud. Amer. de Vapores	Products Carrier (Sun Ship)	Tkr. Crude Carrier (Not built)	LNG Carrier "El Paso Southern"
Principal Dims.								
L.B.P.	166.0 m	190.5 m	168.6 m	188.5 m	176.0 m	182.9 m	365.8 m	276.1 m
L.O.A.	174.3 m	209.2 m	179.3 m	194.8 m	185.9 m	192.0 m	385.6 m	289.0 m
Breadth (Mld)	25.0 m	27.4 m	25.0 m	29.0 m	27.8 m	27.4 m	59.1 m	41.1 m
Depth (Mld)	13.9 m	19.2 m		16.2 m	16.1 m	15.1 m	30.8 m	25.9 m
Draft (Design)	8.6 m	8.8 m	9.1 m	7.2 m	10.8 m	10.4 m	23.8 m	
DWT (Design)	10,830 T	18,215 T	15,453 T	11,220 T	35,120 T	31,000 T	383,600 T	63,460
Type of Engine	Geared Turb	Geared Turb	Gas Turb	Geared Turb	Diesel	Diesel	Geared Turb	Steam Turb
Horsepower	21,000	28,500	19,000	15,000	12,000	14,200	45,000	36,000
Number of Screws	1	1	1	2	1	1	1	1
DECK DEPT.								
Master	1	1	1	1	1	1	1	1
Chief Mate (Officer)	1	1	1	1	1	1	1	1
Second Mate	1	1	1	1	1	1	1	1
Third Mate	2	2	1	1	1	1	2	1
Boatswain	1	1	1	1	1	1	1	—
Seaman, A.B.	6	6	6	15	6	3	6	6
Seaman, O.S.	—	3	—	—	—	3	3	3
Carpenter	—	—	1	—	—	—	—	—
Deck Maintenance	—	1	—	—	—	—	—	—
Deck Storekeeper	—	—	—	—	1	—	—	—
TOTAL DECK DEPT.	12	16	12	20	12	11	15	13
STAFF								
Radio Operator	1	1	1	—	1	1	1	1
Purser	1	—	—	—	—	—	—	—
Radio Oper/Purser	—	—	—	1	—	—	—	—
TOTAL STAFF	2	1	1	1	1	1	1	1

Table 1—Examples of Crew Lists for Ships of Various Types (Continued)

TYPE OF SHIP	**Cargo Ship** (C4-5-69a) (12 passengers)	**Container-ship** (C6-5-85b) (12 passengers)	**Ro/ro** "Iron Monarch"	**Trainship** (Not built)	**Bulk Carrier** Cla. Sud. Amer. de Vapores	**Products Carrier** (Sun Ship)	**Tkr. Crude Carrier** (Not built)	**LNG Carrier** "El Paso Southern"
ENGRG. DEPT.								
Chief Engineer	1	1	1	1	1	1	1	1
First Asst. Engr.	1	1	1	—	1	1	1	1
Second Asst. Engr.	1	1	1	1	1	1	1	1
Third Asst. Engr.	1	3	—	1	1	1	2	2
Fourth Engineer	—	—	—	1	—	—	—	—
Ch. Electrician	1	—	—	—	—	1	—	—
Electrician	—	1	1	1	1	—	1	—
Second Electrician	1	—	—	—	—	—	—	—
Deck Engine Mech.	—	3	—	—	—	—	—	—
Deck Eng. Mech. (Day	—	1	—	—	—	—	—	—
Engine Maint.	—	—	—	2	—	4	—	—
Oiler	—	—	—	—	2	—	—	—
Wiper	—	2	—	—	2	—	—	2
Pumpman	—	—	—	—	—	1	—	—
Ch. Reefer Engr.	—	—	—	—	—	—	1	—
Reefer Maintenance	1	—	—	—	—	—	—	—
Eng. Storekeeper	—	—	—	—	1	—	—	—
Jr. Engineer	—	—	—	3	—	—	—	—
QMED	—	—	—	—	—	—	3	5
Oiler/Maint.	—	—	—	—	—	—	3	—
Jr. 3rd Engineer	1	—	—	—	—	—	—	—
Eng. Utility	3	—	—	—	—	—	—	—
TOTAL ENG. DEPT.	11	13	4	10	10	10	13	12
STEWARDS DEPT.								
Chief Steward	1	1	1	—	—	1	1	—
Second Steward	—	—	—	1	—	—	—	—
Chief Cook	1	1	1	—	1	1	1	—
Cook/Steward	—	—	—	1	—	—	1	1
Sec. Cook or Asst. Ck	1	—	1	1	1	—	—	—
Cook/Baker	—	—	—	—	—	—	—	1
Second Cook/Baker	1	1	—	—	—	—	—	—
Messman	1	3	3	2	4	—	—	—
Utility Man	2	—	—	—	—	2	—	3
Galley Man	—	3	—	—	—	—	—	—
Utility/Cook	—	—	—	1	—	—	—	—
Third Steward	—	—	—	1	—	—	—	—
Pantryman	1	—	—	—	—	—	1	—
Room Stewards	—	—	—	—	—	—	3	—
Passengers B.R.	1	—	—	—	—	—	—	—
Officers B.R.	1	—	—	—	—	—	—	—
Waiters	2	—	—	—	—	—	—	—
TOTAL STEWARDS DEPT.	12	9	6	7	6	4	7	5
GRAND TOTAL	**37**	**39**	**23**	**38**	**29**	**26**	**36**	**31**

mission effectiveness. The objective of habitability design is to provide physical working conditions and personnel support facilities which best contribute to total ship effectiveness for both work and off-duty environments. With respect to the work environment, the goal is to provide one which allows personnel to perform tasks effectively and insures health and safety. This requires adequate control of temperature, humidity, illumination, noise and vibration. With respect to the off-duty environment, the goal is to satisfy off-duty related needs (e.g., sleep, food, personal hygiene and recreation) to allow personnel to return to work physically and mentally prepared to carry out their tasks. This also requires adequate environmental control, plus appropriate facilities in terms of furniture, equipment, materials and internal arrangements. Appropriate habitability is dependent on a variety of factors including the ship's mission, where it operates, length of deployment, and living and work conditions ashore. See Weiler and Castle (1972) for naval ship habitability design.

For commercial ships, a recent trend has been the enhancement of habitability features for ships that spend a relatively small amount of time in port, such as long haul bulk carriers and larger size fishing vessels and factory ships. Such features as special recreation rooms, gymnasiums, and hobby shops are provided to compensate for lack of opportunity for shoreside recreation.

Control of the ship's environment to provide habitable conditions is covered in Chapter XIII. Arrangement of spaces to serve this purpose is evidenced in the typical layout of crew quarters in a cargo ship, Fig. 1, and typical layouts

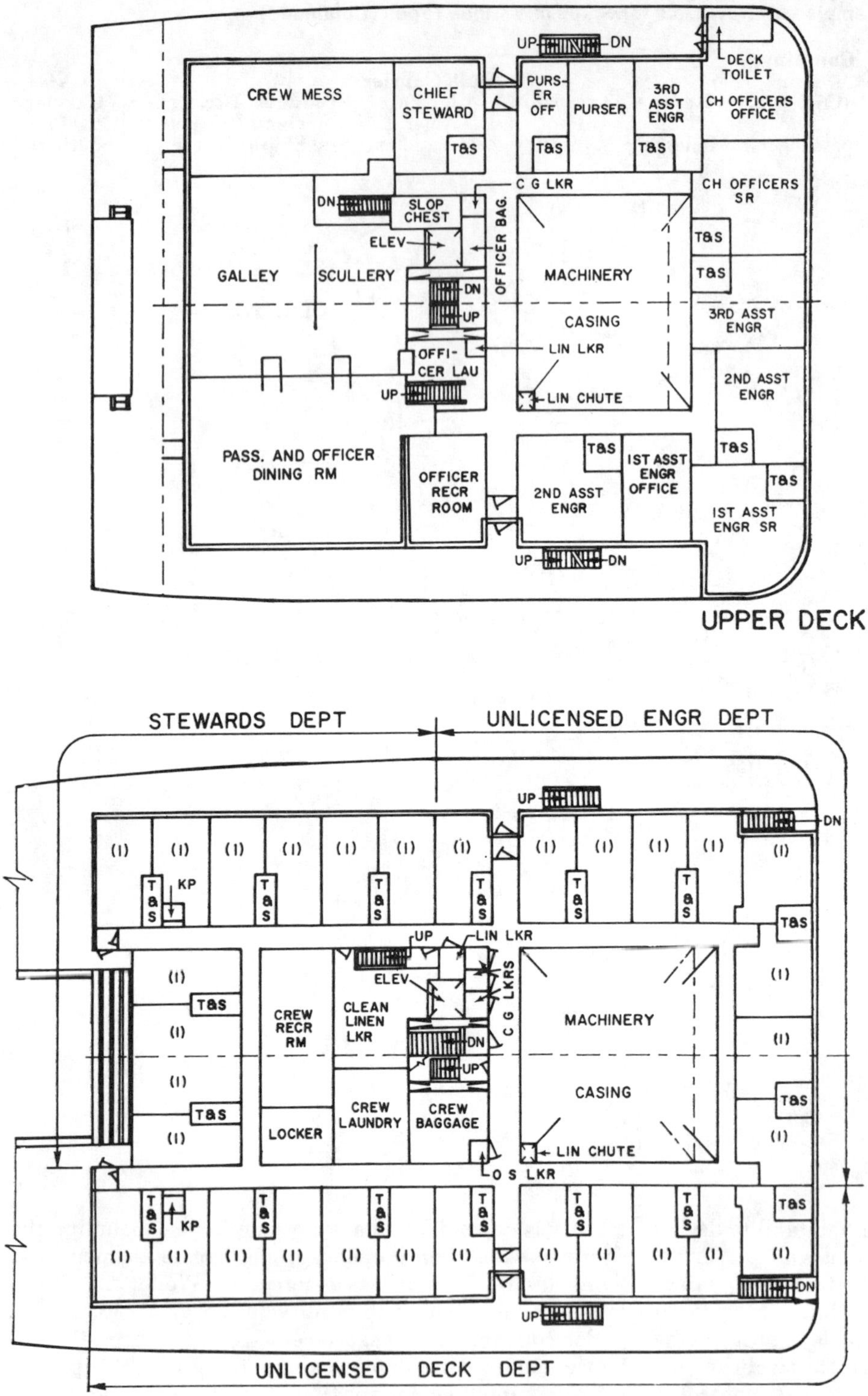

Fig. 1 Typical layout of crew quarters in a cargo ship

for crew and officer rooms, Fig. 2.

3.2 Passenger Spaces. The following discussion applies mainly to passenger liners and cruise ships but is also generally applicable to other types of passenger accommodations, such as those on passenger ferries and cargo ships. For detailed discussion of access, see Section 6.

a. Design Process and Integration. As a basis for the layout and arrangement of passenger quarters, the owner usually stipulates his requirements for the number of passengers, number of classes of passengers, numbers of various

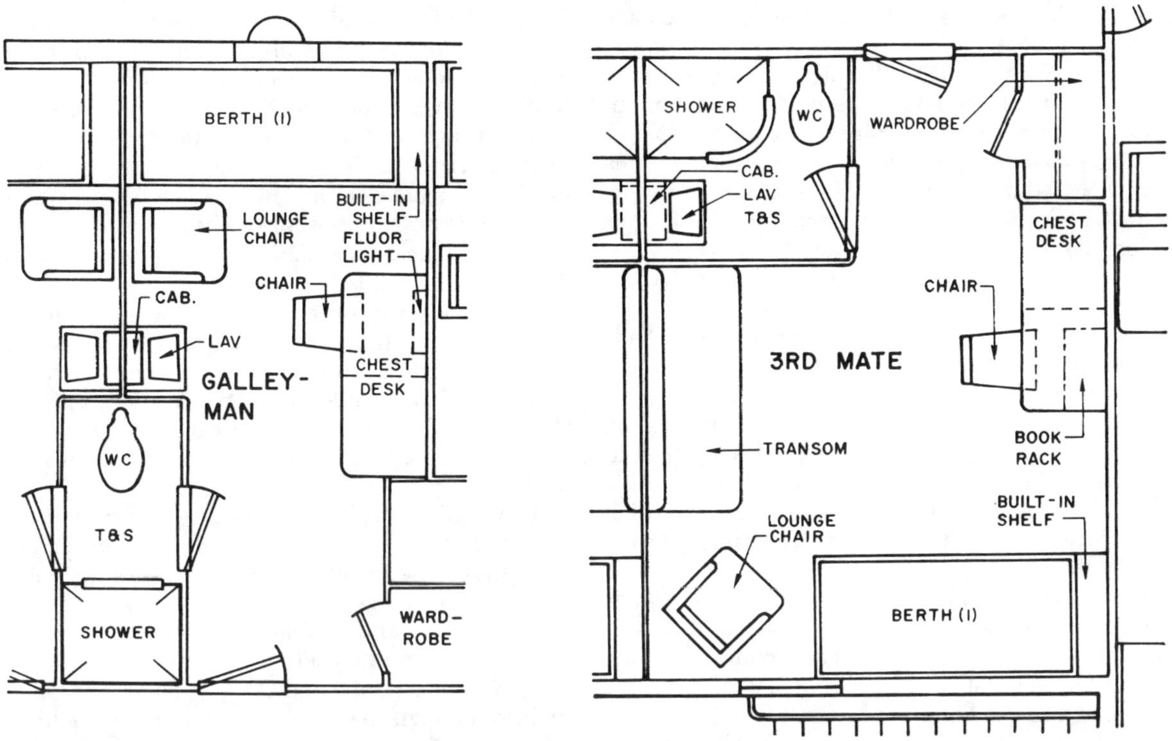

Fig. 2 Typical layouts for crew and officer rooms

types of staterooms, and general requirements for public rooms. He also stipulates the standards of space and qualities of furnishing, generally by comparison with an existing ship. Using these requirements, the naval architect first makes general assignments of blocks of staterooms and public rooms on the basis of overall areas per person or per stateroom, the areas based on data from previous ship designs. Since there is just so much space on a given ship which can be made available to passengers, it is important to proportion it to best advantage between staterooms, inside public rooms, and outside public areas. The assignments of spaces are in effect based on a three-dimensional grid which consists of the decks, main structural bulkheads, and main fire screen bulkheads which generally coincide with structural bulkheads. It is also important in the early stages of the design, to locate main vertical stairways, which form the fire control stair towers.

Normally, the process of design goes from the general to the particular; the successive steps in the design then fall in place in order of importance. There are, however, instances, such as in the design of a block of passenger staterooms, where structure, accommodation, access, and service trunks should be combined into a closely knit entity. To achieve this, it is necessary to further the design simultaneously along several fronts.

For example, from the standpoint of structure, a number of stateroom bulkheads in the superstructure should line up vertically with main transverse bulkheads and web frames in the hull proper to stiffen the superstructure against racking and torsional stresses. This is best accomplished when stateroom width bears a simple relationship to frame and web-frame spacing. Similarly, fore-and-aft bulkheads, such as passageway, machinery casing or toilet enclosure bulkheads should be carried in line with the hatch or main deck girder line where possible. This improves longitudinal strength, gives better headroom, and minimizes unattractive soffits to conceal girders.

The hotel services consist of ventilation, air conditioning, electric wiring, fresh water supply, soil and drain piping, telephone system, call bell systems, etc. Hotel services and the space required therefor have gradually increased with technology and generally higher standards of habitability and comfort. Horizontal and vertical service trunks for the hotel services must be allowed for in the early stages of the arrangement to assure economy in wiring, piping, duct work, etc., concealment of the service lines above ceilings and behind bulkheads and linings, access to the service lines without too many unsightly removable panels, minimum bulkheads, and minimum irregularities in stateroom arrangement. Worthen and Muller (1949) discuss the integration of hotel service zones and distribution of service lines. The end result of careful planning with an eye to detail in the initial stages is a *clean* vibration-free design and lower construction costs.

Although the passenger quarters are designed primarily for the particular service as defined by the owner, some thought also must be given to the trade of the ship from a long-range viewpoint. Since some trades are seasonal, features may be included to make the ship suitable for the cruise trade during the off-season. Because the useful life of a passenger ship is about 25 years, during which public taste can undergo considerable change, the appointments and decorations should be unpretentious, simple, and easy to maintain.

b. Passenger Staterooms. The design of stateroom layouts consists of fitting all the requirements of a stateroom into the smallest space commensurate with comfort, ease of movement, and pleasing appearance. Passenger staterooms are generally located on one or two decks for a moderately sized cargo-passenger ship, and may be spread over four or five decks on a large passenger ship and will vary in standards due to location, size and furnishing, as reflected in the range of room rates.

Requirements for higher class rooms include floor beds or convertible sofa beds and a private bathroom consisting of toilet, washbasin, and shower or tub. Pullman upper berths are sometimes added, more for the sake of gaining room flexibility than for actually increasing passenger count. One end of the room, outboard in the case of an outside room, is often furnished to serve as a sitting room, in which case a sofa is fitted rather than a pullman berth to provide an extra bed.

In lower class accommodations on excursion vessels or other vessels where space is at a premium, toilet and shower spaces are centralized to serve a number of rooms. In some ships, a combination toilet-washbowl unit, such as used on trains, is provided instead of a bathroom.

c. Dining Room. It is logical to place the gally between the crew messes and the dining room. Since space in the superstructure is at a premium on ships carrying passengers, it is common to place galley, messrooms, and dining room below decks, normally directly above the margin line, so that watertight doors on transverse bulkheads will not be required.

In some ships, the dining room is located in the superstructure, with the galley either adjacent or on the deck below, and with service to the dining room via escalators. The topside location allows a view from the dining room, and is particularly applicable to ships in the cruise trade where passengers spend a greater amount of time dining. The use of stabilizers has reduced ship motions so there is no longer a need to restrict the dining room to a location near the rolling axis.

As a general rule, every effort should be made by the designer to provide a dining room large enough so that all of the passengers can eat at one sitting. Since the dining room is often placed below decks and is one of the larger public rooms in a ship, it is frequently fitted with a dome over its central portion, or part of its deck within the girder line may be recessed. The object in either case is to avoid the shut-in feeling produced by a space where the height is low in proportion to the horizontal dimensions. The ceiling should be covered with a sound-deadening material.

Certain principles should be followed in laying out dining rooms: Passengers should not be seated so as to face bulkheads or shell at close range. Most tables should be for 4 persons; next in popularity are 2- and 6-person tables; tables for odd numbers are best made round. Waiter traffic in and out of pantries must be planned, and the layout of tables should allow for avenues wide enough to cope with this traffic. There should be one serving table per waiter, or for every 8–10 seats, and one sideboard for every 30–40 seats.

d. Public Rooms. Public spaces must provide suitable locales for social activities and entertainment. Public rooms tend to be noisy and should be kept apart from staterooms, or properly sound insulated from staterooms. Common practice is to locate most public rooms, the dining room excepted, on one deck and not to have any staterooms on this level. The possibilities for public rooms at the forward end of a superstructure are limited because of wind and weather. This location lends itself best to some kind of glassed-in observation room overlooking the weather deck forward. The choice location for public rooms with a sweeping view of the sea and communication with the open deck is at the after end of the several superstructure decks, protected from the wind. This is the best location for such features as swimming pools, sunbathing beaches, cafes and game decks.

The utilization of the after portion of the superstructure can be increased by stepping the decks in terraces. This allows increased deck heights for public rooms where required. The following subdivisions explain briefly the uses, requirements, and problems connected with the various public spaces found aboard seagoing passenger ships.

1. Entrance Lobby and Offices. Entrance lobbies should be large enough to handle crowds at arrival and departure. The lobby should contain the main passenger stairway as well as the purser's and chief steward's offices and should be in a central position with respect to the staterooms. The area of an entrance lobby, exclusive of offices, should be 0.37–0.46 m^2 (4–5 ft^2) per passenger. From the standpoint of passengers, the purser's office is the nerve center of the ship. It should have a counter and office furniture comparable to that found in the desk area of a hotel. The chief steward's office is smaller in size and importance than the purser's. It handles problems connected with service and steward personnel.

2. Lounges, Hall, Salon. There should be at least two such rooms, so that if one is being used for some function such as a ball, concert or show, the other is still available for conversation and lounging. This *quiet* lounge can be equipped with shelves for reading matter in one corner and writing desks in an alcove. If the lounge serves as a theater, it is preferable to locate it at one end of the superstructure to avoid creating a traffic problem when it is closed off at show time. In addition to these two main lounges, smaller lounging spaces, such as foyers, lobbies and verandas, give the arrangement more flexibility. Total lounge area should be about 50–60 percent of public rooms, other than the dining room.

3. Smoking Rooms, Cocktail Lounge, etc. These are spaces where drinks are served. A smoking room or club room is primarily a room for conversation and for playing cards or group games. A bar equipped with stools is normally adjacent to this space. A cocktail lounge or deck cafe usually includes a dance floor and sometimes a platform for musicians. In addition to a service bar, spaces can be equipped with a deck pantry for serving snacks and canapes. A cafe adjacent to the pool on some ships often has facilities for serving a light lunch. Rest rooms, opening into a foyer

or passageway, should be located nearby.

4. Children's Playroom. Some ships are fitted with a children's playroom outfitted with toys and supervised by a qualified crew member. The area of a children's playroom varies from 0.10 to 0.35 m^2 (1.0 to 3.5 ft^2) per passenger, depending on the percentage of children among the passengers carried.

5. Swimming Pool. On ships for cold weather runs, the pool is placed below decks, fairly low in the hull, with the gymnasium, baths, and dressing rooms placed close by. On warm weather runs, the pool is usually located on an open deck aft in the lee of the main superstructure.

6. Promenade and Open Deck. A promenade gives passengers a place to walk for exercise. For this purpose a complete loop is to be preferred over a horseshoe-shaped promenade. It is customary to enclose at least half of the loop at the forward end to protect walkers against the breeze caused by the ship's motion, or weather conditions. Full-height windows should be fitted at the sides and half-height around the forward end of the enclosure. In order to allow people to pass four abreast in way of deck chairs, the promenade should be not less than 4.3 m (14 ft) wide, if practicable. Where there are no deck chairs, a width of 2 or 2.5 m (7 or 8 ft) is sufficient. A promenade is usually the best boat embarkation area because of its width and length and its proximity to public rooms. Open-deck area for passengers, other than promenade, is provided with games such as shuffleboard, deck tennis, and quoits, as well as deck chairs. Provision must be made for stowage of gear for games in suitably located deck lockers, as well as for stowage space for deck chairs.

7. Theater. When the number of passengers approaches 250, the designer should consider including in the design a space with permanent seats, a stage, and a built-in projection room which may serve as theater, chapel, or concert hall. Lounges do not make very satisfactory theaters, since they are not designed for this specific function. If located in the superstructure, the theater should be at one end to avoid creating an access problem. To provide for good vision, a sloping deck should be fitted if possible. It may be possible to arrange access to the theater so that the crew can be entertained there during certain allotted hours. The seating capacity of theaters should be between 16 and 35 percent of passenger capacity.

8. Shops. The various shops generally found on the larger seagoing passenger ships and cruise ships include barber shop, hairdresser, gift and novelty shop, tailor shop and dry cleaning, tobacconist, flower shop, photo shop, and dark room. Of these the barber shop, hairdresser, and novelty shop are the most common. The others may be replaced by services provided by the purser's and steward's offices. For guidance a figure of one barber chair for every 100 potential customers may be used.

e. Interior Decoration. In addition to the architectural, engineering, and arrangement features of passenger accommodations, the interior design and decoration also receive considerable attention. In the U.S. it has been customary for the owner or the naval architect to employ specialists in this field. The various interior design firms show little conformity in their work on procedures. Some firms are concerned with the architectural features and the general arrangement, and must become involved in the early stages of the design in order to avoid changes at a later date. Other firms restrict themselves to furniture, furnishings, fixtures and finishes, with emphasis on selection of materials and color schemes.

Although some of the work of the interior designer is done during the design stage, most of his effort is during the construction stage when he makes the actual selections of furniture, furnishings, and materials, and reviews the shipyard's plans and purchase orders relating to interior design and decoration features. In all cases, the interior designer must work closely with the naval architect, who must coordinate all design activities.

f. Regulations. The Regulations which have the most effect on the arrangement of passenger quarters are those of the U.S. Coast Guard. The purpose of these Regulations is the protection and safety of passengers and crew, particularly in the event of a collision or fire at sea. The basic intent of the Regulations is to restrict the flooding of the ship in the event the hull is breached, to prevent the start or spread of fire, and to provide for the safe escape of passengers and crew, if necessary. Many of these Regulations have to do with materials and details of construction and do not affect the arrangements per se. (The fire control regulations largely dictate the materials and methods of joiner construction as described in Chapter IX.)

The main U.S. Coast Guard Regulations which affect arrangements of passenger accommodations (also applicable to crew quarters on a passenger ship) concern items listed below (since the complete regulations affect more than accommodation spaces, they may be discussed in other aspects elsewhere in this book):

1. Location of Passenger Quarters. Deckhead not to be below deepest subdivision load line.

2. Watertight Transverse Bulkheads. Spaced in accordance with subdivision requirements and extending up to the bulkhead deck.

3. Main Transverse Fire Control Bulkheads. Spaced not farther than 40 m (131 ft) apart to divide ship into main fire zones. Same types of bulkheads to separate accommodation spaces from other types of spaces. All openings in such bulkheads to have closures.

4. Decks. To be continuous to form fire barrier between one level and the next.

5. Main Stair Towers. At least one tower to be accessible from each main fire zone. Stair tower to be surrounded by fire control bulkheads with fire doors and to extend to weather deck above, so that a person can climb to the weather without coming out of the protective tower.

6. Stairways. Minimum widths and maximum angles of inclination are specified in relation to number of people who would be using stairway in emergency.

7. Corridors. Minimum widths specified in relation to number of people served. No dead-end corridor to be longer than 12 m (40 ft).

8. Doors. Minimum width and type of door specified, depending upon location and class of fire bulkhead.

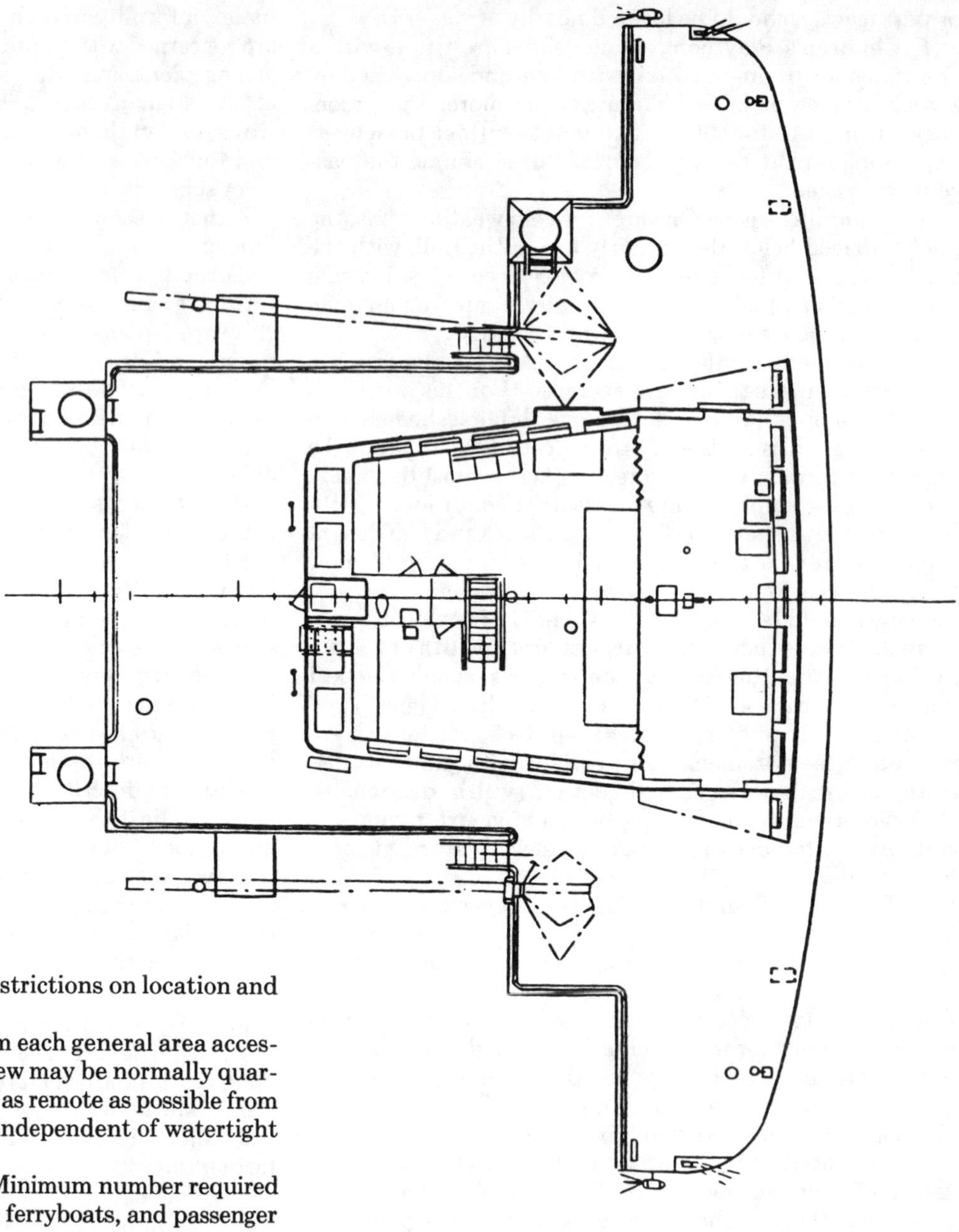

Fig. 3 Example of modern wheelhouse

9. Windows and Airports. Restrictions on location and size.

10. Escapes. At least two from each general area accessible to passengers or in which crew may be normally quartered or employed. Escapes to be as remote as possible from each other, and one escape to be independent of watertight doors.

11. Toilets and Washbasins. Minimum number required for passengers on excursion boats, ferryboats, and passenger barges.

For passenger ships, the Coast Guard requires the submittal of a *fire control diagram* showing the location of fire screen insulation, including main fire zones and subdivisions, stairways and elevator enclosures, control space enclosures, etc, and type of all doors in such subdivisions and enclosures.

The U.S. Public Health Service has regulations which apply to both passenger and crew spaces but have little effect on arrangements since they mainly concern sanitation and have to do with rat-proof construction, galley equipment, dishwashing facilities, and freshwater tanks and system.

3.3 Navigating Spaces. The pilothouse and chart room are at the forward end of the uppermost tier of superstructure. The radio office is normally in the same tier of superstructure as the chart room and pilothouse or nearby on the deck below. In some designs it has been incorporated into the forward stack. The fire-detecting and extinguishing control panels should be located in the pilothouse or in a room adjacent thereto. There should be an unobstructed view of the bow from the pilothouse, and the line of sight over the bow should intersect the level waterline not more than 1-1/2 ship lengths ahead. The flying bridge should overhang the sides of the ship by about 1 m (3 ft) if possible, and the helmsman should be able to see and hear the docking pilot standing at the end of the flying bridge.

A walkway from side to side in front of the pilothouse is a distinct advantage. Pilothouse doors are either hinged, with hinges forward, or sliding. Sliding doors are easier to open against the wind, but have a tendency to jam.

The chart room is placed directly abaft the pilothouse and

should have a minimum size of 2.4 by 2.4 m (8 by 8 ft) with the athwartship dimension being governed by the length of the chart table.

It is desirable to provide a door or sliding window between chart room and pilothouse for direct communication. The size of gyro room depends upon the size of equipment which may require 5 to 9 m^2 (50 to 100 ft^2). The radio room should not be less than 11 m^2 (120 ft^2) in area.

Fig. 3 is an example of a wheelhouse of a modern cargo ship (C8-S-85d Class). The owner considers that this wheelhouse incorporates nine features which are considered distinct advantages as follows:

1. 360 deg visibility,
2. a *head* on the deck so that watchstanders do not have to leave the deck during a watch,
3. a sofa which the Master can use should he wish to remain in the bridge area for extended periods,
4. two independent radars to assure availability of one at all times,
5. the master gyro to be on the bridge to assure ready access for prompt restarting in case of a momentary power failure,
6. to the greatest extent possible, all equipment and telephones are to be located so that their normal operation will provide the operator with forward visibility.
7. As many windows as possible should be sloped from

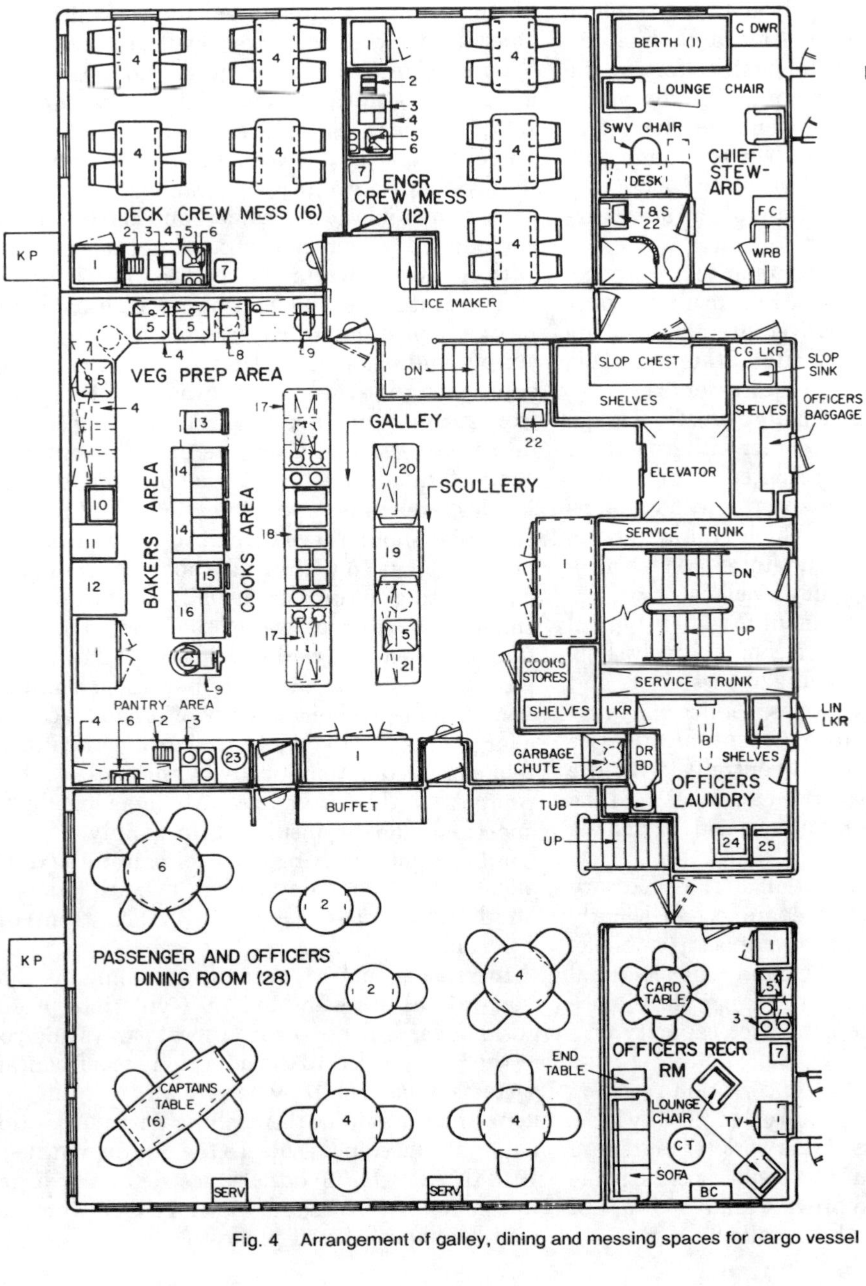

List of Commissary Equipment

No.	Description
1	Refrigerator
2	Toaster
3	Coffee maker
4	Dresser
5	Sink
6	Egg boiler
7	Deep fryer
8	Vegetable peeler
9	Mixer
10	Griddle
11	Pan rack
12	Oven
13	Vegetable steamer
14	Range
15	Fryer
16	Broiler
17	Cooks' table
18	Steam table
19	Dishwasher
20	Clean dish dresser
21	Soiled dish dresser
22	Lavatory
23	Hot water urn
24	Washer
25	Dryer

Fig. 4 Arrangement of galley, dining and messing spaces for cargo vessel

the vertical (preferably out at the top and in at the bottom) to prevent reflections of lights (from the opposite side or from within the bridge deckhouse) from appearing in the windows.

8. There should be space outside the wheelhouse doors having a deck over and forward protection with a window but with the outboard and aft sides clear for use in inclement weather, particularly for observing audible signals.

9. There should be bridge wing areas (other than the space noted in 8) without deck overheads to permit sight taking.

3.4 Galley, Messrooms, Dining Rooms, and Stores Spaces. The galley, messrooms, and stores must be arranged to provide all the hotel services necessary in a passenger or cargo vessel. The stores should allow convenient loading and should be readily accessible to the galley. The galley should be designed to serve conveniently all the messrooms directly or via pantries. Normal practice for cargo ships and tankers is to have one galley and several pantries serving the different ratings in separate messrooms. On the larger passenger ships, it is at times more convenient to have separate galleys for passengers and crew. On very large passenger ships with various classes widely separated, a number of galleys must be fitted and the locations of the stores arranged to suit.

Since each operator has problems peculiar to his service, it is usually necessary for the designer to consult the port steward. Labor unions vary in their requirements on the East, West, and Gulf coasts of the United States and it is important to ascertain their current requirements.

Figs. 4 and 5 show typical arrangement plans for a cargo ship and for a cargo-passenger ship, together with lists of the equipment. On the cargo ship represented in Fig. 5, the passenger and officer dining room and the crew messrooms are served directly from the galley, which eliminates the need for pantries and pantrymen. The stores are on a deck below and are transferred to the galley by elevator.

The cargo-passenger ship arrangement (Fig. 5) is typical of the requirements for a vessel having a comparatively short run, with a total complement of about 220 passengers and a crew of 130, and equipped for a one-class service of a fairly high standard. The economical arrangement for this type of ship is for the galley to be located centrally, with the passenger dining room forward and crew messes aft, and the stores in the immediate vicinity, on the crew end, loaded via side ports.

a. Galley. The galley should contain all the necessary mechanical and fabricated items to prepare food efficiently for all persons on board. It must be as compact as possible, light, airy, and easy to maintain and clean. The electrical load involved must be kept in mind in the design stage, since it is an appreciable factor in the generator capacity of a modern passenger ship.

The sheet metal work is, in general, of corrosion-resisting steel for all working surfaces, and galvanized steel elsewhere. Attention should be paid to details of this work to insure ease of cleaning, eliminate pockets and inaccessible spaces, make the installation verminproof, and provide ready stowage of the mass of portable gear required in a galley.

The subbases, deck covering, and drainage of the galley also should be designed for ease of cleaning. Sufficient drains and gutterways must be provided to rid the deck of water. The foundations for equipment may be either of the open or closed type. With the open type, the galley fixtures are set up on legs off the deck. In order to make the lower space accessible for cleaning, the height of the opening underneath should be at least 254 mm (10 in.). Where equipment is enclosed, it is desirable to make the foundations of the enclosed type, with toe room 100 mm (4 in.) high and 100 mm (4 in.) deep. With this type of foundation, the resulting void space is either filled with lightweight concrete or thoroughly covered with bituminous solution and enamel.

The deck covering in the galley is usually some form of nonslip tile. Practice has been to use an underlayment of about 32 mm (1 1/4 in.) of cement over which 19 mm (3/4 in.) thick quarry tile, red-grooved with ribbed back, is applied. Where weight is an important consideration, underlayment of latex or equivalent material may be used with greatly reduced thickness. The grooved type of tile has been favored because of its excellent nonslip qualities, in spite of its obvious disadvantages in cleaning. Special arrangements are required in way of steamers and kettles which are usually surrounded by cement curbs, suitably covered with tile, and with the deck covering pitched to drains.

The ventilation and lighting of galley and pantries should be of a high standard. It is customary to have mechanical supply and exhaust ventilation with exhaust in excess of supply, so that there is a movement of air into the galley from surrounding spaces, thus preventing emission of odors into those spaces. Canopies of corrosion-resistant steel, connected to the exhaust system, are fitted over ranges, kettles, steamers, and other heat-producing equipment. Canopies should have drip gutters and drain pipe, and be of heavy gage to withstand knocking from pots and pans. The lighting fixtures are usually built into the canopy and should be dustproof, waterproof and greaseproof. Grease filters should be fitted to all canopies and be of easily removable design to facilitate periodic cleaning. On large installations, where a battery of ranges is combined under a large canopy, steam jets or a carbon dioxide flooding system, actuated automatically by thermostat, are installed to protect this duct from fire. All ducts in galley and pantry spaces should be insulated; in supply ducts, to prevent condensation resulting from cold air in the duct and warm moist air in the galley and in exhaust ducts, to prevent reradiation of heat. Generally 2.54 cm (1 in.) of insulation is sufficient.

Increasing concern over the environment and the adoption of local, state and federal regulations prohibiting disposal of garbage overboard in waterways and coastal areas of the U.S. has led to the adoption of self-contained refuse systems aboard ships, whereby waste of this type must be kept aboard ship until reaching port. Adoption of international conferences such as the one on marine pollution in 1973 (MARPOL 73) has also restricted the disposal of such wastes to certain areas at sea and the conventional garbage chute previously fitted on most vessels is now only of limited use. These chutes, of sturdy construction with watertight covers and non-return valves at the shell, are generally fitted at the sides of the ship in the galley area. A flushing connection is provided for cleanout purposes. Garbage grinders are now normally installed in pantries of vessels and arrangements made for the retention of garbage onboard or the processing of food wastes, after grinding or pulverizing, through an approved marine sanitary device. Trash compactors and incinerators also are used to minimize storage requirements for garbage being held aboard for later disposal ashore and adequately sized garbage rooms located near the galley are provided. The garbage handling and stowage space should be completely closed off from food handling spaces in accordance with U.S. Public Health Service requirements.

The U.S. Public Health Service requirements cover in detail the construction of bulkheads, insulation, drainage, piping, lighting, ventilation, ratproofing equipment details, etc. in regard to galley and food preparation spaces. The rules are, in general, of a high standard.

b. Messrooms and Dining Rooms. The single-class cargo-passenger ship requires a main dining room of sufficient capacity to accommodate all of the passengers in one seating. Provision may be made to expand the facilities, if necessary, by means of fillers or tighter seating. In addition, the ship's officers may be seated with passengers, with officers' dining rooms omitted. In this arrangement, a small dining room is recommended to provide quick dining for officers in work clothes, with the added advantage that the main dining room may be closed off in port during turnaround periods.

The crew messing arrangement is varied to suit the individual operator. On modern cargo ships, the officers either have their own messroom, or, if passengers are carried, may be served in the passengers' dining room. In the U.S., a single messroom for all of the unlicensed crew is permitted on ships manned by East Coast labor unions, while West Coast unions require two messrooms, one for deck crew and one for engine crew. Some extra seating is fitted in the messroom to accommodate the steward's department. If messrooms are located adjacent to the galley, a pantry is not needed; if remote from the galley, a pantry is fitted for direct service to the messroom.

The crew messing also may be varied to suit individual operators. On most U.S. cargo ships, crew messing is broken down into crew, petty officers, and officers, with about two-thirds seating capacity for all ratings. On some ships, the stewards, deck, and engine crews mess separately, with common petty officer messing for all departments. Sometimes petty officers mess separately by departments. Still others combine all unlicensed personnel by departments.

c. Stores Spaces. The arrangement for stores on a cargo ship is more or less standard, the consumables being divided into dry and refigerated stores. In addition there are the usual bosuns' stores, deck gear lockers, and paint and lamp rooms. The dry and refrigerated food stores, as mentioned previously, are usually adjacent to or directly beneath the galley. Small linen lockers are adjacent to quarters. Also, every U.S. cargo ship, in accordance with laws governing marine inspection (Title 46 United States Code, Section 670), must carry a certain amount of clothing, tobacco products, toiletries, etc., for sale at fixed prices to the crew. The room carrying this material is known as the *slop chest*.

The capacity of each storeroom for consumables is dependent upon the number of days between provisionings and the number of passengers and crew to be accommodated. Operators seldom provision in foreign ports, except in emergencies.

The various storerooms are outfitted with bins, shelves, hangers, portable shifting tubes, etc., as required for each class of material and to suit the stewards' department of the particular operator. Stowage must be arranged to meet U.S. Public Health Service approval in regard to ratproofing and food handling.

The following is a brief description of the various types of stores and their fittings in common use on passenger ships:

1. Dry Stores. Part bulk and part broken stowage. Bulk material retained in place by shifting tubes. Broken material in shelves, bins, and large cans for loose sugar, flour, rice, and the like. The space is generally sparred all around, including the deck, so that stores do not come in contact with steel structure.

2. Flour. On ships with long runs, bulk flour stowage is provided, usually in a room completely lined on all interiors with zinc, monel, or other metal acceptable to U.S. Public Health Service. Shifting tubes are fitted for retaining bags of flour.

3. Crockery Stores. Shelving, with retaining battens between shelves. Part bulk stowage of size that the space warrants.

4. Sundry Stores. Part bulk and part shelving. Retaining battens between shelves.

5. Potatoes. No fittings required other than sparred platform about 230 mm (9 in.) above deck. Battens all around space.

6. Bonded Stores. These are of the type that must be sealed by custom agents when entering a port.

7. Refrigerated Stores. These are subdivided to segregate various types of foods, depending on the temperatures required and the operator's usage. In general, separate rooms are provided for meat, fish, butter and eggs, milk, ice and ice cream, poultry, fruit and vegetables, frozen vegetables, beverages, butcher shop, and thaw room.

8. Linen Rooms. These are arranged for stowage in such a manner that the linen will not come in contact with any steel surfaces. Shelving is usually constructed of 22 mm (7/8 in.) incombustible board faced with a plain marine veneer. Clean and soiled linen facilities should be provided.

9. Baggage rooms. These are important on passenger ships because of limited space available in staterooms, which are usually restricted to hand luggage. On vessels with a long itinerary, it is generally desirable to have a passenger baggage room which is available to passengers at regularly scheduled periods. This space must be accessible from the passenger accommodations. Sometimes crew baggage rooms are also fitted.

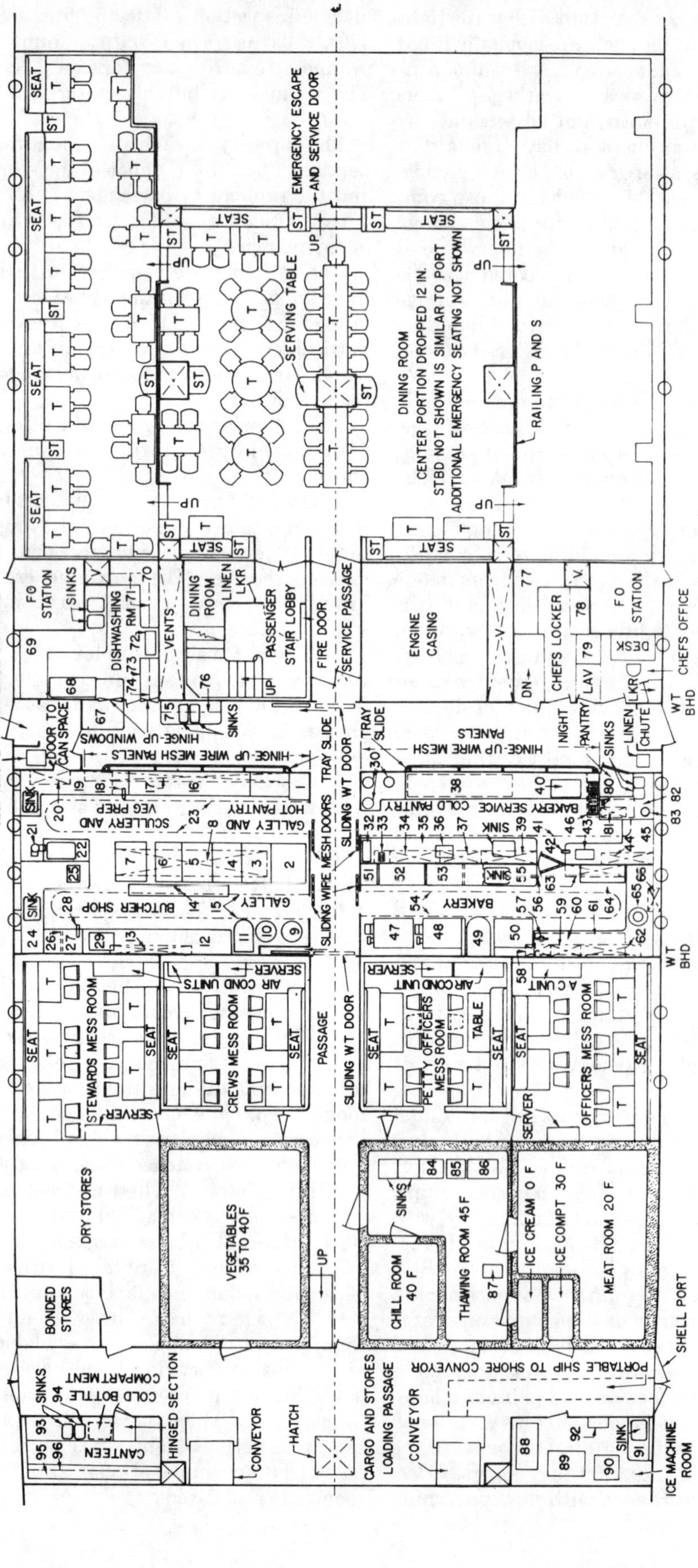

Fig. 5 Typical arrangement of commissary and dining spaces for a cargo-passenger ship

List of Commissary Equipment for Cargo-Passenger Ship Shown in Fig. 5

No.	Description	Remarks	Load, kw
		Galley and Hot Pantry	
1	Fry kettle	Electric	18
2	Broiler	With cabinet base—electric	10.5
3	Cooking top	3 hot plates—electric	15.9
4	Range	Electric	21.9
5	Range	Electric	21.9
6	Range	Electric	16.2
7	Cooking top	Same as item 3—electric	15.9
8	Pan rack	Overhead	
9	Steam kettle	30 gal	
10	Steam kettle	30 gal	
11	Mixer	60 qt—electric	1[a]
12	Dresser	Refrigerators under	
13	Shelves	Overhead	
14	Cooking table		
15	Canopy	Built-in lights and grease filters	
16	Steam table		
17	Shelves	Overhead	
		Scullery and Vegetable Preparation	
18	Can opener		
19	Dresser	With sink	
20	Vegetable locker	Overhead	
21	Potato peeler	Electric	¾[a]
22	Dresser	With sink	
23	Canopy	Built-in lights and grease filters	
		Butcher Shop	
24	Dresser	With sink	
25	Meat slicer	Electric	¼[a]
26	Cutting board	Under counter stowage	
27	Dresser		
28	Can opener		
29	Food cutter	Electric	⅓[a]
		Cold Pantry	
30	Coffee urns—set	Two 10-gal coffee; one 20-gal HW	
31	Urn table		
32	Egg boiler	Steam	
33	Toaster and shelf	2-slice—electric	1.2
34	Locker	Overhead	
35	Dresser	Lockers under	
36	Meat slicer	Electric	¼[a]
37	Shelves	Overhead	
38	Dresser	Refrigerators under	
39	Locker	Overhead	
		Bakery Service	
40	Griddle	Electric	4
41	Canopy	Built-in lights and grease filters	
42	Can opener		
43	Bread slicer	Electric	¼[a]
44	Locker	Overhead	
45	Dresser	Shelves under	
46	3 Toasters and shelf	4-slice each—electric	7.35
		Bakery	
47	Roast oven	Electric	15
48	Bake oven	Electric	12.6
49	Mixer	80-qt electric	2[a]
50	Dough trough		
51	Pan rack		
52	Proofing oven	Steam	
53	Dresser	Lockers under	
54	Canopy	Built-in lights and grease filters	
55	Dresser	With sink—drawers under	
56	Can opener		
57	Bakers' scale		
58	Shelves	Overhead	
59	Locker	Overhead	
60	Dresser	Refrigerators under	
61	Marble slab	Over dresser top	
62	Locker	Overhead	
63	Locker	Overhead	
64	Ice cream cabinet	15-gal electric	¼[a]
65	Dough divider	Hand type	
66	Shelves	Overhead	
		Dishwashing	
67	Soiled dish dresser		
68	Dishwasher	Electric	3[a]
69	Clean dish dresser	With twin sinks	
70	Dresser		
71	Shelves	Overhead	
72	Silver washer	Electric	7
73	Dresser		
74	Shelves	Overhead	
75	Dresser	With twin sinks	
76	Shelves	Overhead	
		Chef's Lockers	
77	Pots and pans		
78	Silver locker	Extra security	
79	Shelves		
		Night Pantry	
80	Toaster and shelf	4-slice—electric	1.2
81	Locker	Overhead	
82	Dresser	Twin sinks—refr under	
83	Coffee urn	5 gal	
		Thaw Room	
84	Double sink	With side board	
85	Meat block		
86	Cutting table		
87	Scale	300-lb spring	
		Ice Machine Room	
88	Ice maker	Six 15-lb ice cakes	
89	Ice maker	Nine 15-lb ice cakes	
90	Ice cuber	Electric	¾[a]
91	Dresser	With sink	
92	Thaw tanks	Two	
		Canteen	
93	Dresser	Twin sinks—cold bottle compt under	
94	Compressor unit	Under dresser—electric	¼[a]
95	Dresser	Shelves under	
96	Shelves	Overhead	

[a] Horsepower

NOTE: In addition, bar, bar pantry, and deck pantries are located elsewhere and not included in list.

Additional space must be assigned to various small stores such as rug lockers, steamer chairs, oilskins, games lockers, motion picture gear, cleaning gear, and others as necessary, depending upon the size and run of the ship.

With the advent of mechanization and reduced manning, there has been a strong trend toward increased use of precut meats and precooked and quick frozen foods and bakery items. In effect, this eliminates the need for a butcher or baker and the related equipment. The ultimate goal is almost complete elimination of galley operations, with all cooking on shore and provisioning based on the present aircraft practice. There have been isolated examples of this system in some non-U.S. ships, with shore catering staffs preparing and quick-freezing the full menus of an entire trip. Universal acceptance of this system is not yet in the offing and is further complicated by labor relations.

Section 4
Machinery Spaces

4.1 General. As stated in Subsection 1.4, the first step in formulating a *general arrangement* is locating the main spaces and their boundaries, and the *machinery space* is one of the main spaces. Since the objective is economic optimization for the function of the ship, the following principles should govern the selection of the machinery space location and shape:

- minimum volume consistent with adequate access for installing, operating, and servicing machinery components,
- minimum interference or conflict with main function of the ship, which, for a commercial ship, is to carry cargo or passengers,
- compatibility of machinery weights and watertight boundaries with stability and subdivision standards,
- avoidance of extreme trims in various loading conditions of the ship due to weight of machinery plant,
- suitability of machinery layout for minimum manning requirements,
- reasonable length of shafting between drive unit and propeller.

The proper balance of the foregoing principles depends upon many factors, principally the type and size of ship, and the type and power of the machinery plant. The fuel tankage location and arrangement should be considered in association with the machinery, and should depend upon the type of fuel and the rate of consumption. It is particularly important to consider all of these elements together when making the selection of machinery type.

Commercial ships are generally single- or twin-screw and there has been a definite trend toward extending usage of a single screw and a single engine room to higher powers (50,000 shp and above) unless the greater power, maneuverability or redundancy attainable by twin screws is required. The single screw arrangement results in the smallest space requirement, greater overall propulsive efficiency, and is the least expensive to install and operate. It also provides the operating personnel with an installation easy to supervise, and results in a smaller engine crew than is possible if the main machinery is distributed to drive more than one shaft. Two or more main machinery spaces are provided on large passenger ships, the number and location being dependent upon the power requirements, number of screws, and compartmentation desired.

Military features often include wide separation of machinery spaces. Furthermore, the length of machinery spaces is important, in addition to location and space requirements, since in many ship designs the machinery space length is a controlling factor in survival insofar as floodable length considerations are concerned.

4.2 Space Requirements. Since there are so many variables involved, there is no general rule to determine the space requirements for propelling machinery. For determining space requirements during the desing stage, small scale machinery arrangement plans must be developed indicating the principal machinery components that are involved.

Curves of approximate volumes required by various types of machinery plants are shown in Fig. 6.

4.3 Location. The optimum location for the machinery space is a function of the type of ship.

a. Passenger Ships. On passenger ships it has been generally desirable to locate the propelling machinery about amidships. A rectangular compartment is the most suitable for arranging machinery and results in the least waste of space. Also, it is desirable for passenger comfort to have accommodations amidships, which permit the machinery to be located under the largest part of the superstructure. Since overhead cargo handling gear cannot be installed economically in way of the superstructure, this is not a desirable space for carrying cargo. Also amidships machinery permits large areas aft for passenger utilization.

b. General Cargo Ships. Machinery is usually located amidships or aft of amidships. An amidship location generally results in the least length for machinery space, avoidance of trim problems, separation of the forward and after portions of the ship into two approximately equal cargo units, and a good location for the deckhouse and navigation spaces. The disadvantages are the length of shaft alley and the use of the space best shaped for efficient cargo stowage. With the machinery placed aft, the space is usually longer and of greater volume, and there may be a trim problem. Providing a suitable location for the navigation space may require another deckhouse forward. Machinery space location approximately one-quarter length abaft amidships has proved an efficient arrangement for modern, large, fast cargo ships, since with the fine form associated with high speed there are problems with placing the machinery all aft.

c. Bulk Cargo Ships. On most bulk cargo carriers, such as tankers, ore carriers, and colliers, the machinery is located aft. This location is mandatory for the tanker, including LVCC's and ULCC's, so as to avoid the complication of cofferdams aft of the machinery space and in way of the shaft tunnel as would be required with the machinery space amidships. On ore carriers and colliers, cargo handling is improved by having one continuous uninterrupted cargo space, and the obstruction offered by a shaft alley is avoided. For bulk carriers, it is important that the length of the machinery space be kept at a minimum to permit the cargo space to be as long as possible to reduce the sagging moment. Minimum machinery space volume is especially important for tankers with segregated ballast.

d. Containerships (cell type). The machinery space is usually located well aft with generally not more than one cargo hold between the machinery space and the stern. With the aft location there is no interruption of crane movement in way of the container stowage or interference of a deckhouse with a shore crane. Also, there is no shaft alley to interfere with container stowage; however, there may be some trim problem in light load conditions. If a deckhouse is required farther forward for navigation purposes, the machinery space may be placed under the deckhouse but should be kept as short as possible fore-and-aft, though it may have greater depth if necessary for the least loss in container stowage.

e. Roll-on/Roll-off Ships. On these, and any type on which the cargo is stowed on large, open 'tween decks through side or end ports, the machinery space should have minimum depth to allow location under the cargo decks with minimum casing interference in way of the cargo spaces above. For such ships there is little disadvantage in making the machinery space longer to make up for the loss in height.

f. LNG Ships. On these vessels the machinery space is located at the stern aft of all the cargo tanks, since the stern of the ship is not suitable for cargo tankage and so that the cargo tanks and systems can be continuous without interruption.

g. Barge-carrying Ships. On most barge-carrying ships the machinery is abaft amidships but not at the extreme stern.

4.4 Nuclear Power. The main features of a nuclear power plant which differ from others are:

a. The principal source of power is a reactor which must be heavily shielded, be accessible for refueling, be protected from collision, and not be located directly adjacent to accommodations.

b. The plant uses no fuel oil, so this adjustable weight is not available for trim and stability control.

Although there are many types of reactors, and more are being developed, it appears that future nuclear ships will probably incorporate pressurized water reactors (PWRs), as did the *N.S. Savannah*. However, it is possible that future commercial marine reactors will be of the integral type wherein the heat exchangers and primary coolant pumps are contained in the same pressure vessel as the reactor itself. This arrangement offers considerable savings in volume and

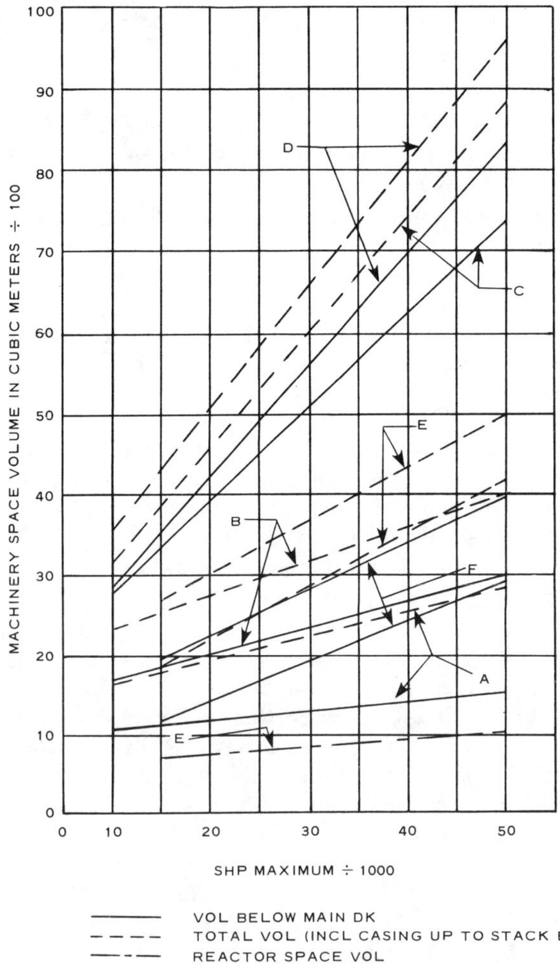

SHP MAX = MAX CONTINUOUS RATING
A = GAS TURBINE, AIRCRAFT TYPE, GEARED DRIVE
B = GAS TURBINE, AIRCRAFT TYPE, ELECTRIC DRIVE
C = DIESEL ENGINE, HIGH SPEED, GEARED TYPE
D = GEARED STEAM TURBINE, FOSSIL FUEL, COMMERCIAL DESIGN
E = GEARED STEAM TURBINE, NUCLEAR FUEL
F = COMBINED DIESEL AND GAS TURBINE (CODAG), GEARED DRIVE

Fig. 6 Approximate volume of machinery space

weight over the dispersed (or loop) PWR system utilized in the *Savannah* wherein the heat exchangers and primary pumps are separately mounted in the containment vessel, along with the reactor pressure vessel.

The reactor with its heavy shielding should be located amidships with structural collision protection at the sides and a clear deck overhead for refueling operations. In addition, American flag nuclear vessels must meet special provisions set forth by the U.S. Coast Guard in CRF Title 46, Part 99 which include requirements of a double bottom for the full length of the ship, two-compartment subdivision, increased requirements for emergency electric power, and increased reliability of the bilge pumping system. Also, it is necessary to consider the fore-and-aft location of the reactor installation on the basis of collision statistics analysis.

The propulsion machinery space has been normally located directly adjacent to and abaft the reactor containment compartment. Accommodations are normally above.

Section 5
Tanks

5.1 General. One of the main types of spaces to be allocated on a ship is for liquid-carrying tanks. The main considerations governing the location of tanks are:

- No access is required except manholes for cleaning and maintenance.
- Since the liquid contents are relatively dense, location low in the ship tends to improve stability.
- Free liquid surface causes a virtual loss in stability; therefore, the dimensions of the liquid surface should be restricted, particularly in the transverse dimension.
- In event of damage to the hull of the ship, flooding of the tanks or run-off of liquid from the tanks may result in an unsymmetrical moment which must be accounted for in the initial stability of the ship. Thus, it is desirable to locate tanks symmetrically on the centerline or to provide for cross-connecting port and starboard tanks.

5.2 Fuel Oil Tanks. Fuel oil tanks are normally located in the double bottoms, where the ship is fitted with double bottoms, since this space is relatively inaccessible, undesirable for cargo, and places the fuel weight low in the ship to aid stability. The double bottom tanks are subdivided as required to control free surface; however, excessive subdivision increases the piping requirements and the ship cost. The double bottom tanks may not have sufficient capacity for the required fuel oil, and it may be desirable to reserve some of these tanks for clean ballast to maintain required stability when fuel is consumed. For these and other reasons, deep tanks may be provided for fuel oil. It is generally desirable to locate such tanks symmetrically about the centerline of the ship for damaged stability considerations.

For certain types of ships such as roll-on/roll-off ships or warehouse ships where the cargo spaces are primarily above the bulkhead deck, there is normally excess space below the cargo deck. In such cases it is more economical to install deep tanks for fuel oil and reserve the double bottoms, if any, for clean ballast tanks or voids. This arrangement results in fewer tanks, less piping, and hence less cost.

Normally, if a tank is used only for fuel oil, there is little problem with corrosion. Such tanks are required to be cleaned infrequently to remove sludge and to be gas-freed when necessary for repair work. However, if a fuel oil tank must be used alternatively for saltwater ballast to preserve the required stability during the voyage, corrosion can be a serious problem. This is particularly true in the case of double bottom tanks which have a large amount of protruding structure in proportion to volume. For this reason, combination ballast and fuel spaces should be assigned to deep tanks if possible, since the internal protruding structure subject to corrosion can be reduced.

As further discussed in the following under ballast tanks, the ballasting of fuel oil tanks is still common practice for older ships and cannot always be avoided on newer ships. In such cases, the design must comply with U.S. Coast Guard Regulations to prevent the discharge of oily ballast water into the sea. In most modern ships, the practice is to provide separate tanks for clean ballast to avoid these problems.

5.3 Fresh Water Tanks. Fresh water is carried on a ship for three main purposes—potable water for cooking and drinking, washing water, and water for the machinery plant. The latter may be in the form of reserve feedwater for steam plants, cooling water for diesel or gas turbine plants, or shielding water for nuclear plants. In older ships, it was customary to carry the water as required for the complete voyage and to provide separate tanks and systems for each type of water since the required degree of purity and restrictions on location are different for the various types. Modern ships are fitted with evaporators to manufacture water of the required purity during the voyage so that the total quantity of tankage required is greatly reduced. The tendency is to combine the types of fresh water into a single system. with the resultant cost savings in tankage, piping, and pumps. This system, of course, must meet the maximum requirements for each type of water. The U.S. Public Health Service requires that potable water be in tanks separated from the shell and from fuel and ballast tanks. Reserve feedwater must have a high degree of chemical purity.

Just as for fuel oil tanks, the water tanks should be located with regard to both intact and damaged stability aspects, and normally are placed near the machinery space and accommodations to minimize piping.

5.4 Cargo Tanks. Petroleum products are carried in tankers, and special types of ships are built to carry other liquid cargos in large quantities; for example, wine tankers. Many modern cargo ships are fitted with special tanks to carry liquid cargos in limited quantities. Owing to the rapid development of new products by the chemical industry, the demand for tankage for new types of liquids is constantly increasing.

There are many types of tanks provided for liquid cargos on cargo ships. The type selected depends upon the requirements of the cargo with respect to such features as temperature control and protection from contamination. The simplest type is a *skin* tank, which is like an ordinary deep tank in that it extends to the hull of the ship, with frame stiffeners, etc. protruding into the tank. The structure may be designed to some extent to facilitate cleaning and may be provided with heating coils, possibly removable for ease in cleaning. The most complex and expensive type of tank is the *flush* tank, which is built entirely separate from the hull with a completely flush interior surface with rounded corners. Flush tanks are usually clad with stainless steel as a permanent coating for ease in cleaning and protection of the cargo from contamination. Cofferdams are usually provided all around the tank to facilitate inspection of its exterior for leakage, which is sometimes an underwriter's requirement for valuable cargos. Flush tanks usually have provision for temperature control of the cargo in the form of hot water in wells under the bottom of the tank or circulating in piping systems or channels on some of the exterior surfaces. The heating system must provide against overheating the cargo locally, as well as providing temperature control of the entire cargo within a certain range.

For liquid bulk ships, as discussed in Section 2 of this Chapter, the tanks comprise the complete cargo space and the tank design must be coordinated with requirements for structure, subdivision, cargo handling system, and other design aspects. For LNG ships, as discussed in Chapter II and Section 2 of this Chapter, the complete ship design in way of the cargo space is dictated by the tank system selected and the tanks must be constructed, installed, and tested in accordance with the requirements of the specific system selected. Chemical carriers may be fitted with numerous varieties of tanks of sizes, materials, coatings, and locations dictated by the chemicals to be carried, as discussed in Section 2.

For a complete discussion of liquid cargo handling and containment, see Chapter XI.

5.5 Ballast Tanks. The loading of a ship varies appreciably in the course of a voyage as the result of consuming fuel and stores, and also from one leg of a voyage to another as a result of loading and unloading cargo. Ballast, either liquid or solid, is carried to maintain stability or sea-kindliness.

a. Location of Ballast Tanks. For cargo ships, the use of fuel oil tanks for ballast can no longer be considered a satisfactory practice. Mixed water and fuel oil form a thick sludge difficult to remove. The presence of salt water in fuel oil, even in small amounts, interferes to some extent with combustion, even though most of the water is removed from the fuel oil by means of settling tanks. Most ports and countries have strict regulations against pumping out oily ballast in the port or near the shore, and there are also international regulations either in effect or imminent. The U.S. Coast Guard Rules (CFR, Title 33) contain a number of regulations aimed at requiring systems and equipment which will prevent any expulsion of petroleum to the environment. The principal points of these regulations are:

1. provision for containers at fill connections, overflows and fill pipes,
2. tanks for retention of oily waste and oily bilge slops,
3. provisions to isolate any oily-water system from oil and bilge systems,
4. provisions for exceptions to the above due to the installation of oily waste processing equipment.

In the case of colliers and ore ships, where the cargo is fairly dense, there is an excess of available cubic which is used up in ballast wing tanks. Some colliers tend to have excess stability in light condition, so topside tanks are assigned to ballast to reduce the metacentric height. The ballast must be distributed properly along the length of the ship to reduce the probability of excessive bending moments on the ship girder.

In general cargo ships, the ballast tanks should be distributed over the length of the ship for flexibility in controlling trim. To maintain proper stability, the combined center of gravity of the ballast tanks ideally should be near or below the combined center of the fuel oil tanks. Special tanks for ballast alone utilize cubic which could produce revenue; however, they are economically justified considering stability, trim, sea-kindliness and oily ballast.

Cargo oil tanks are usually piped for saltwater ballast and used for this purpose when not used for cargo. For flexibility and emergencies, it is still customary to pipe fuel oil tanks for saltwater ballast as well as fuel oil.

In the case of tankers, the U.S. Coast Guard rules require segregated ballast and other non-cargo carrying spaces to be so located and distributed as to limit the outflow of the cargo in the case of a hypothetical collision, in accordance with calculations carried out in a prescribed manner. This requirement for *defensive location* of ballast tanks has also been developed by IMCO in a slightly different form for international applications.

Section 6
Relationship Between Spaces and Access

6.1 Relationship Between Spaces. In Section 1, general arrangement was defined as the assignment of spaces for all the required functions and equipment, properly coordinated for location and access. Therefore, functional relationship between spaces is the essence of general arrangement and the prime factor to be considered in arriving at the final selected location of spaces; e.g., the reefer cargo spaces should not be remote from the reefer machinery. Aside from functional relationships among spaces, the general characteristics of a space must be considered in selecting its location on the ship; these include volume, shape, weight of contents, and requirements for ventilation. Requirements for access by personnel is another consideration discussed in detail below. After the main spaces are located, the principal problems of space relationships are in the accommodations.

a. Cargo ships and Cargo-passenger Ships. The problems of relationships between accommodation spaces are generally similar for these two types of ships, but more complex for a cargo-passenger ship. There are numerous factors such as number of passengers, weather on the route, degree of luxury, number of classes, length of voyage, which influence the design. In the following, a ship carrying 200 passengers with intermediate class accommodations will be considered. The design for a ship of this type stems from a few basic principles which are generally accepted.

1. Space in the superstructure is more suitable for accommodations, particularly for public rooms, than below.

2. The best passenger traffic conditions result when passenger accommodations are located on decks between dining room deck and public room deck.

3. The crew, other than deck officers, are usually located above the load waterline on the second and third decks on the same side of the galley as their messrooms.

4. The galley is centrally located, frequently just forward of the engine room.

The relationship between two spaces can only be satisfactory if the access between them is carefully thought out. In the course of a design the various spaces and groups of spaces cannot be disassociated from the horizontal and vertical access between them. Some further considerations on the relationship between spaces can be found in Subsection 3.2.

b. Tankers. The U.S. Coast Guard rules, aimed at maximum safety of personnel in event of collision and fire, largely dictate the arrangement and relationship of spaces of a modern tanker. The rules require that all accommodation, service, and control spaces be located aft of the cargo spaces so that the deckhouse is normally located over the main machinery space in the stern portion of the ship. The forward bulkheads of the deckhouse facing the cargo space are to be insulated to meet A-60 classification. The bulkheads, except for the wheelhouse, are to have no openings and portlights are to be of the fixed type with easily operable steel covers on the inside.

The deckhouse aft must have sufficient height to provide acceptable visibility from the wheelhouse, which for larger size tankers requires a deckhouse with six or seven levels. In a typical arrangement, the main deck level is generally assigned to engine crew and stores and service spaces; the second level to galley, messes and steward's crew; the third level to deck crew; the fourth level to officers; the fifth to the master, chief officers and radio room; and the sixth to the wheelhouse. The arrangement results in a compact deckhouse, somewhat like an apartment building, which can be served by one central stair tower, and sometimes an elevator, supplemented by outside ladders for secondary escapes.

The pump room is normally located low in the ship, just forward of the machinery space, with the pump control room on the main deck just forward of the deckhouse. Larger tankers generally do not have a forecastle deck since the freeboard to the main deck is adequate.

6.2 Access—General. Together with the relationship between spaces, access constitutes what is perhaps the heart of general arrangement design. A ship on which access problems have not been solved is lacking in comfort and efficiency. In an emergency, an awkward solution of access can constitute a serious hazard. The following discussion on access applies to a cargo-passenger ship. Crew access is similar for a cargo ship.

Access here means the ability of crew and passengers to move about the ship; more specifically, in the case of the crew, it means the possibility of getting in a simple and direct way from their quarters to their respective places of duty, from their quarters or place of duty to their messing and recreation spaces, from any of the foregoing spaces to the boat embarkation deck or to a weather deck, for emergency or drill, and to any space on the ship for inspection, maintenance, or emergency action.

The crew should be able to travel from their quarters to their place of duty at any time and under any weather conditions without traversing passenger or officer areas. The stewards' crew, of course, must traverse passenger areas as part of their normal duties. For access to be used under emergency conditions, the principle of segregation of crew and passengers may be relaxed.

In the case of passengers, access is satisfactory when they can go from their staterooms to the dining room and to the several public rooms, promenades, and boat-embarkation stations without confusion, without excessive walking, without crowding on stairs and in corridors, and without intruding upon crew areas or ship working areas.

The U.S. Coast Guard has definite requirements for access for safety and means of escape in case of fire or flooding. There is little technical literature dealing with access outside the Coast Guard Regulations; some phases are discussed by Sharp (1947), and by Kari (1948). Unfortunately, while it is possible to formulate a few basic rules and suggestions stemming from common sense, this is a subject where there are few substitutes for experience. The following general rules used for guidance should prove helpful.

1. Officers and crew should be quartered so as to provide simple access to their place of duty—seamen near the weather deck, engine crew near the machinery casings, etc.

2. It is desirable to provide fore-and-aft crew access below decks when possible.

3. The designer should be able to take advantage of the fact that purser and steward personnel may use passenger access to move about the ship.

4. Since messes are used three times daily by all crew, best access is assumed when these spaces are located near the intersection of the main crew stair with the main crew fore-and-aft passageway.

5. Main passenger stair towers serve main vertical zones. They should be spaced far enough apart to avoid confusion. For best efficiency they should be arranged so they are used by an approximately equal number of people in normal operation.

6. The fore-and-aft weather deck access by means of inclined ladders is used by longshoremen when in port, and should, if possible, bypass passenger deck-chair areas.

In connection with access, it should be mentioned that passengers are apt to be easily confused as to their whereabouts and this uncertainty concerning the interior arrangement of a ship can be dangerous in an emergency.

The U.S. Coast Guard regulates location and size of exit and boat embarkation direction signs, but in an emergency, the first concern of passengers may be to save their children, for example, rather than to find the lifeboat embarkation area. In any case, every possible means of helping passengers to become familiar with the arrangement should be used. Some of the devices employed in connection with this are the following:

1. Arrows indicating *Forward* and *Aft* are set into corridor deck covering under a light.

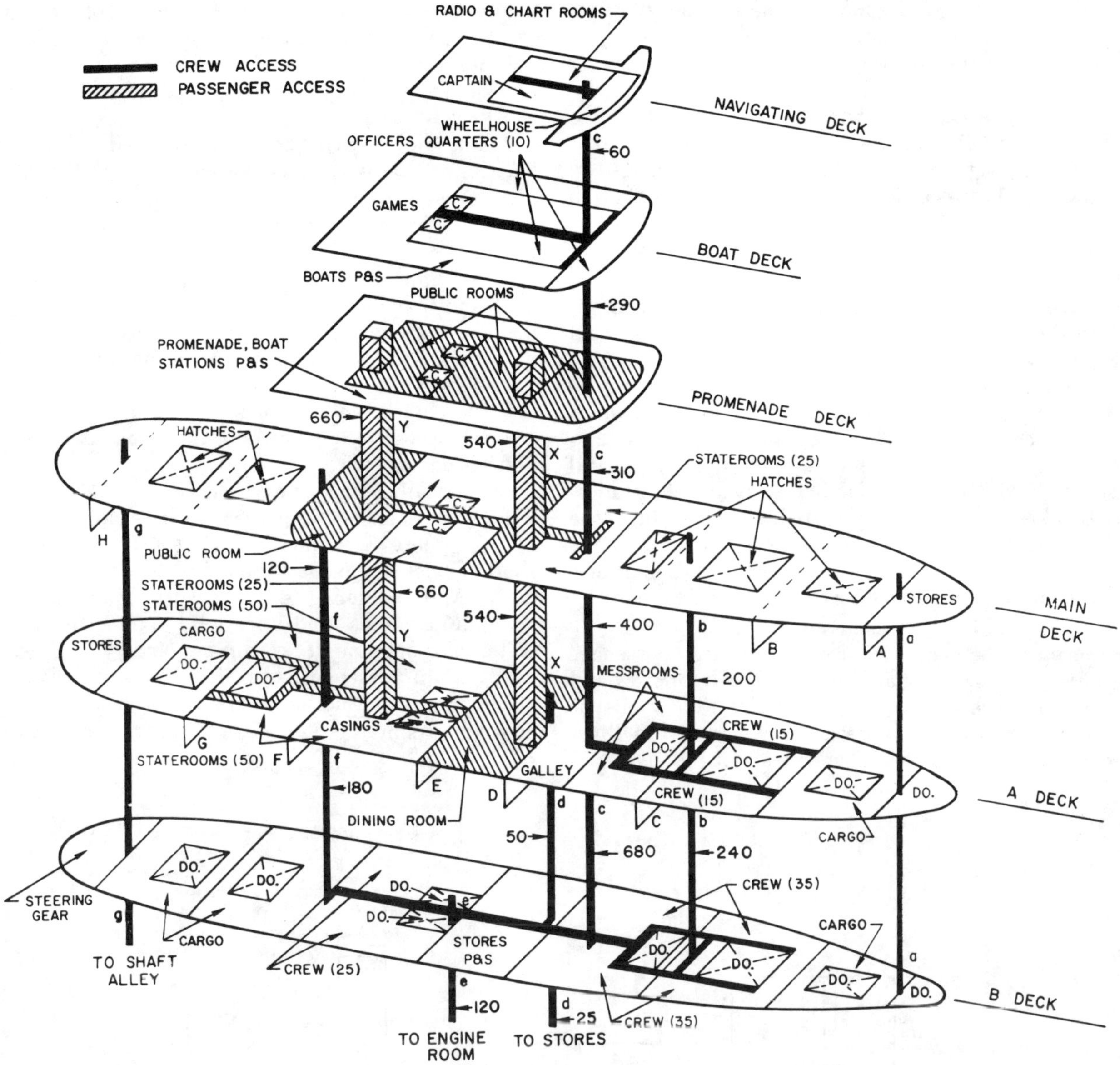

Fig. 7 Cargo-passenger ship access diagram

2. At every passenger stair landing there is a *You Are Here* type of plan for the deck in question, which also should show the next best alternate stairway if the immediate one is not usable.

3. At every stair landing, in addition to lifeboat direction signs, there are signs telling people to what public space the *Up* and *Down* stairs lead, respectively, as well as those to be found on the deck where they are now standing.

6.3 Existing Regulations Governing Access. The U.S. Coast Guard regulations are quite specific on access questions pertaining to escape in "Means of Escape." These are briefly summarized as follows:

1. There must be two independent means of escape from any general area where passengers or crew normally may be found, one of these escapes not being through a watertight door.

2. These two means of escape should be as far apart as possible.

3. Vertical ladders are not usually acceptable as a means of escape.

4. Doors leading to these escapes must not be locked, or if locked, devices such as crash panels, indicated by notices on either side, should be provided. Detail requirements for nonwatertight doors and also watertight doors are described in the U.S.C.G. regulations.

5. Stairs must be wide enough to cope with emergency traffic. Within each main vertical zone, a stair tower should give access to a weather deck and must have bulkheads and doors at each level. Stairs and elevators serving only two decks must have bulkheads and a door at one level. A stair fitted wholly within a public space cannot be considered to be a means of escape. Stairs cannot give direct access to

enclosed spaces where a fire may originate. Elevators are not considered a means of escape.

6. Dead-end corridors more than 12 m (40 ft) long are not permitted.

7. All public spaces over 28 m^2 (300 ft^2) shall have two exits giving egress to different spaces or corridors.

8. Stairs, passageways, and doors must be arranged to give ready and direct access to lifeboat embarkation areas.

9. The various weather decks shall communicate by means of inclined ladders.

In regard to cargo holds, the rules for access are complicated by the number of agencies involved. International and local organizations of longshoremen as well as the U.S. Department of Labor have regulations on this subject. In addition, the applicable regulations for all countries at which the ship will call should be checked to avoid possible delays when calling at a particular port. Generally, two vertical ladders are provided for each main hatchway and, in addition, a trunked vertical ladder is provided at one end of the hold. The trunked access is usually entered through one of the masthouses on the weather deck and is sometimes used at sea for examination of certain types of cargo.

6.4 Main Traffic Lanes. For a typical cargo-passenger vessel, Fig. 7 shows the main lanes of traffic for crew and passengers, both vertically and horizontally. The isometric diagram, which does not represent any actual ship, illustrates access on a cargo-passenger ship about 150 m (500 ft) long and carrying some 200 passengers, 30 officers and 150 crew. The figures in parentheses in accommodation blocks indicate assumed capacity. The figures on stairs are a rough index of frequency of daily use, which permits comparison on a relative basis.

From bow to stern, the successive stairs provide a link between various spaces as follows:

(a) Forecastle—forward boatswain's stores,

(b) messroom and weather deck access for crew in zone B-C,

(c) main crew stair. Access to mess for crew in zone C-D, and to embarkation area; access to fore-and-aft crew passageway on B deck; access to officers' quarters and navigating spaces; emergency passenger access to embarkation level,

(d) galley—galley stores,

(e) access to engine room in casing from crew fore-and-aft passageway,

(f) weather deck access for crew in area E-F; emergency weather deck access for passengers on A deck aft,

(g) poop—after boatswain's stores and steering; also, shaft alley escape (separate).

(X) Main passenger stair forward. Boat embarkation and public rooms—entrance lobby and offices, forward staterooms on main deck—dining room.

(Y) Main passenger stair aft. Boat embarkation and public room—public rooms and staterooms on main deck aft—staterooms on A deck, secondary access to dining room.

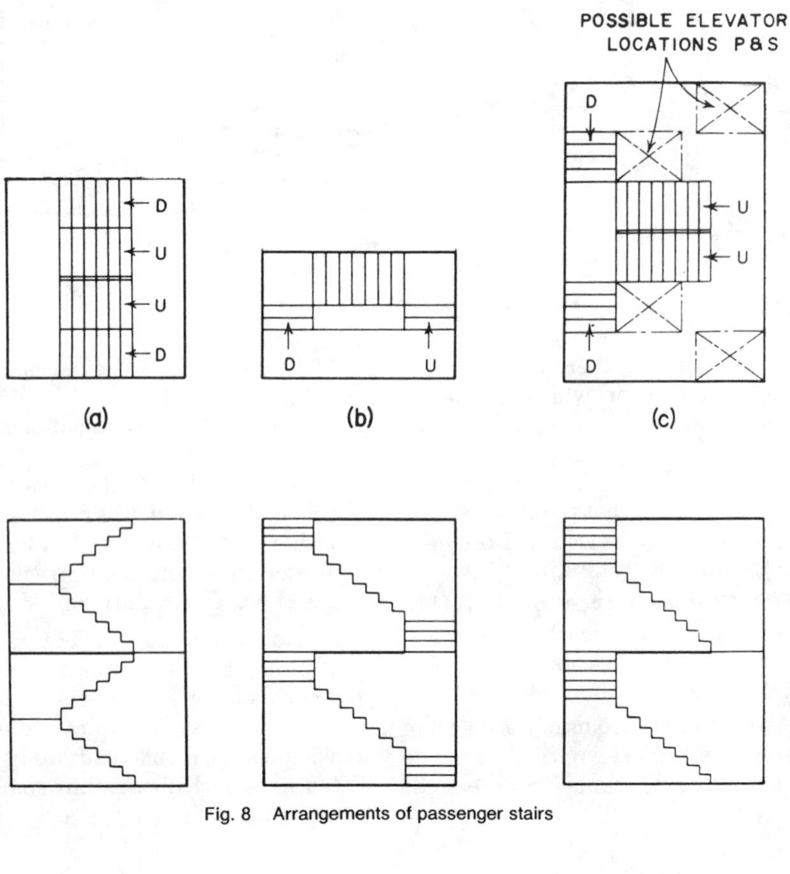

Fig. 8 Arrangements of passenger stairs

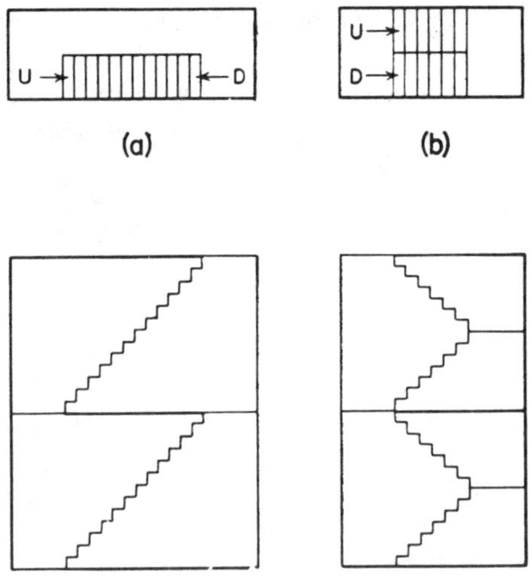

Fig. 9 Arrangements of crew stairs

6.5 Details. Design considerations and details of certain elements of access are discussed below:

a. Stairs. For location, the latest U.S. Coast Guard Regulations should be studied. For structural reasons, it is preferable to keep stairs inboard of the girder line. Athwartship stairs should be avoided, if possible, particularly for passengers. A high angle of roll added to the angle of stairs can produce an angle of 75 deg and make it difficult for a person to keep his footing.

The designer should avoid accommodations and stairs too far removed from amidships because the vertical forces due to pitching become sizable at the ends of the ship. On a stairway, this can result in serious accidents.

Marine practice allows crew stairs steeper than on land; it is also axiomatic that at sea every person on a stair should be able to hold onto a rail. This limits the width of each separate flight to about 107 cm (42 in.). The Coast Guard limits the angle of inclined ladders to 50 deg, except that steeper angles are permitted for isolated ladders at the ends of the ship. If traffic is such that more than two persons abreast will use the stairs, these should be made double. Passenger stairs should have at least one intermediate landing per deck height. Recommended proportions are shown in Table 2 with 25 mm (1 in.) nosings assumed on treads.

Fig. 8 offers a comparison on an equal basis between three common types of passenger stairs: example (a) is good, most compact, but unattractive; (b) is an attractive stair, but takes up more room for same effectiveness; (c) is usually designed with fore-and-aft flight of same width as athwartship flights, which is not logical. Although this stair layout seems uneconomical, not all the area is wasted. Frequently it houses the passenger elevators.

Fig. 9 compares crew stairs on an equal basis, and is self-explanatory. Circular stairs are not generally acceptable to the U.S. Coast Guard. All stairs are to be built of steel and incombustible materials and should have nonskid features in the treads and at landings.

b. Ladders. Outside inclined ladders should have the same nonslip features as stairs, with proportions as in Table 2, maximum width 107 mm (42 in.). Ladders should be fore-and-aft and protected from seas, where possible, by a structure or by being well inboard.

c. Elevators. They should duplicate the functions of main passenger stairs. In locations where double stairs are necessary, it may be advisable to fit two elevators. A design criterion is 930 cm^2 (1 ft^2) of net elevator area for every 5 passengers carried in the class.

d. Passageways. Passageway width should be in proportion to traffic carried. Widths given in Table 3 are for guidance and do not take special cases into account. The first figure is a minimum; the second figure is an average value.

For large passenger ships with a wide beam, there is generally a choice between layout with a centerline main passage or two main passages, one port and one starboard. The general layout of the staterooms is completely dependent on this choice.

e. Doors. Details of doors are given in Chapter IX. From the standpoint of access, doors to crew and passenger rooms are 66 cm (26 in.) clear width; to bathrooms, about 61 cm (24 in.); to hospitals, 91 cm (36 in.) for stretcher cases; fire doors, 91 cm and 107 cm (36 in. and 42 in.). Where doors have to close an opening more than 107 cm (42 in.) wide, such as the entrance to the main lobby, they are generally double.

Doors between pantries and messes, between galley and dining room, are frequently double-acting and controlled by a photoelectric cell as a help to waiters carrying trays. To

Table 2—Stair Proportions

	Riser		Tread		Angle, deg	
	Min mm (in.)	Max mm (in.)	Min mm (in.)	Max mm (in.)	Min	Max
Passenger stairs	152 (6)	197 (7 3/4)	279 (11)	381 (15)	30	40
Crew stairs	152 (6)	216 (8 1/2)	203 (8)	279 (11)	45	50
Inclined ladders	—	216 (8 1/2)	165 (6 1/2)	Open Open	50	50

Table 3—Widths of Passageways

	Minimum cm (in.)	Average cm (in.)
Crew, transverse, secondary	69 (27)	76 (30)
Crew, longitudinal, twin	76 (30)	107 (42)
Crew, transverse main, or centerline ..	91 (36)	122 (48)
Crew, working corridor*	137 (54)	183 (72)
Corridors, hospital area	107 (42)	
Passengers, transverse, secondary	76 (30)	
Passenger, longitudinal, twin	107 (42)	122 (48)
Passenger, transverse main, or centerline	137 (54)	152 (60)

* Where crew may be carrying loads as in way of galley and stores.

prevent collisions, this type of door should have a pane of glass set in at eye height.

Where there are glass doors between public rooms, the glass should be at least 6 mm (1/4 in.) thick and should not be plain; otherwise passengers will walk into them. This mishap can be prevented by etching or decorating the panes in some manner.

6.6 Trends. Fewer passenger ships are being built, and it appears that very few, if any, U.S. passenger liners will be constructed in the future.

Section 7
Miscellaneous Factors

7.1 Structural and Strength Requirements. The art of ship design requires that the structure be coordinated with the general arrangement to result in the least mutual interference possible; e.g., supporting bulkheads should be located where divisions between spaces are desired, pillars should fall on bulkhead lines or where clear space is not required, large structural webs and frames should not project into a space to detract from its utility. At the same time, the structural design should not be so compromised by general arrangement requirements as to be inefficient and should not be overdesigned. A ship with structure built around a preconceived general arrangement would suffer stress concentrations and would be subject to vibration. The structural design and the general arrangement must proceed simultaneously and be closely integrated. Structure is discussed in detail in Chapters VI, VII and XVI.

7.2 Watertight Subdivision, Floodable Length, and Damaged Stability. For ships carrying general cargo, the amount of cargo carried between transverse bulkheads must be balanced to some extent with the equipment to handle the cargo so that one slow hold will not delay the ship in port. Floodable length considerations alone would allow a greater length in the midship portion of a ship than at the ends or quarters. Usually however, the length of a midship hold must be kept below the allowable to keep the hold cubics in this full section portion of the ship from overbalancing the hatches and winches which serve the hold. This restriction would not apply to a lift-on/lift-off containership served by rolling ship cranes or by shore cranes.

The U.S. Coast Guard and SOLAS Rules require passenger ships to meet certain subdivision requirements for floodable length and damaged stability. MarAd requires a one compartment standard of subdivision for all ships receiving government support. The general effect of these requirements on arrangements is to require special attention to the location of watertight bulkheads, both transverse and longitudinal, and assignment of spaces with respect to their permeability.

There are no special requirements for longitudinal bulkheads for strength and watertightness, although some bulkheads frequently result from arrangement requirements and recesses in transverse bulkheads. These bulkheads must be taken into account in any calculations of the effect of flooding and damaged stability and, if necessary, provision must be made for symmetrical flooding. By U.S.Coast Guard regulations, a longitudinal bulkhead is considered intact after damage to the shell if the bulkhead lies entirely inboard of a vertical surface which is one-fifth of the beam inboard of the shell at the deepest subdivision load line. See Chapter I for a more detailed discussion of floodable length and damaged stability.

Tankers are generally more finely subdivided than cargo or passenger-cargo ships because the nature of the cargo and cargo handling generally requires segregation and also cuts down the forces due to shifting of liquids in partially filled tanks. The recent revision in the U.S. Coast Guard rules to reduce the probability of oil pollution requires tankers over a certain size to comply with certain subdivision standards. The longitudinal oil-tight bulkheads incorporated in the arrangement of a tanker present special damaged stability problems since they cause unsymmetrical flooding resulting in transverse list. See Chapter XI for a more detailed discussion.

Cell-type containerships are particularly adaptable to subdivision requirements. Since bulkhead spacing has no effect on cargo handling operations, the bulkheads can be spaced any multiple of container spacing down to the regulatory minimum of 10 ft plus 3 percent of length.

For a more detailed discussion of subdivision requirements and calculations, see Chapter I.

7.3 Fire Control or Protection Requirements. The fire protection requirements of both SOLAS (1974), and the U.S. Coast Guard Rules stipulate for passenger ships, tankers, and cargo ships that the hull, decks, and deckhouses shall be constructed of steel or other equivalent metal construction of appropriate scantlings. In addition, there are requirements that certain bulkheads throughout the ship be of steel.

Steel fire protection bulkheads are termed "A" class bulkheads and are defined as composed of steel or equivalent metal construction, suitably stiffened and made integral with the main structure of the ship, such as shell, structural bulkheads, and decks, and so constructed that, if subjected to the standard fire test, they would be capable of preventing the passage of flame and smoke for one hour. "A" class bulkheads are further subdivided according to insulation requirements, which depend on the types of spaces separated by the bulkhead. The locations where the Coast Guard requires "A" class bulkheads for passenger ships can be given most conveniently in the form of a list:

1. Transverse bulkheads subdividing the hull and superstructure into main vertical zones the mean length of which shall not, in general, exceed 40 m (131 ft). (These bulkheads are always made to coincide with the transverse

watertight bulkheads below the bulkhead deck, and insofar as practicable are kept in line with these bulkheads in the superstructure.)

2. Enclosure bulkheads for stairways and elevators. (A stairway tower extending to the weather deck is required to serve each main vertical zone.)

3. Bulkheads surrounding control stations or any space where a continuous watch is maintained.

4. Bulkheads adjacent to:

- lifeboat embarkation and lowering stations,
- public spaces of over 46 m^2 (500 ft^2) with combustible furnishings,
- galleys and main pantries,
- storerooms and workshops,
- motion picture-projection and film-stowage rooms,
- machinery spaces,
- dry cargo spaces,
- fuel and water tanks and voids,
- open decks and enclosed promenades.

SOLAS (1974) also requires "A" class transverse bulkheads spaced in general not over 40 m (131 ft). Requirements for "A" class bulkheads in other locations, while less specific, are in close agreement with Coast Guard requirements.

In both cases, the requirements are based on the objective of confining a fire within the space of origin such as a galley, cargo space, large public space, or containment within one main vertical zone at the most, while at the same time protecting stairs and corridors for escape of personnel or as a base for fire fighting operations.

Although "A" class bulkheads do not necessarily have to be structural in the sense of contributing to the strength of the ship, a good design will integrate structural and fire protection requirements; thus most steel bulkheads required for fire protection will have a structural function as well, and in the case of transverse bulkheads below the bulkhead deck, will also serve for watertight subdivision.

A few general design principles may be stated for saving steel and for making class "A" bulkheads coincide with structural requirements as nearly as possible:

1. Place galley and main stores spaces adjacent to one class "A" main transverse bulkhead.

2. Place large public rooms adjacent to one class "A" main transverse bulkhead.

3. Place main stair tower enclosure adjacent to one class "A" main transverse bulkhead, and inboard of and adjacent to main deck girder lines.

4. Place shaftways, wireways and elevator trunks where needed for structural supports.

7.4 Effect of Tonnage Regulations on Arrangements. The registered tonnage of a ship is proportional to its volume and is expressed in units of register tons of 100 ft^3. The *gross register* tonnage is, in general, the volume of all permanently closed-in spaces above the inner bottom. The *net register* tonnage is the gross register tonnage minus certain deductible spaces which do not produce revenue, principally the spaces occupied by crew quarters and propelling machinery.

Tonnage is important to the shipowner because it is the basis for certain items of operating expense. Harbor dues, dockage fees, and canal tolls are usually based on net tonnage. Gross tonnage is used as a basis for drydocking charges and is a factor for the determination of officers' wage scales and applicability of rules and regulatory bodies. Since it is always to the advantage of the shipowner to keep the tonnage figures as low as possible, the designer must have a thorough knowledge of the tonnage rules.

For a detailed discussion see Chapter V.

7.5 Ventilation and Air Conditioning. All spaces on a ship must be provided with ventilation with a few exceptions such as void spaces and, for most spaces on modern ships, mechanical ventilation is provided. Thus in the early stages of planning the arrangement, the runs of supply and exhaust vents and space for fan rooms must be allowed for. In most modern ships accommodation spaces are air conditioned. Air conditioning machinery is usually on a flat in the main machinery space or in an auxiliary machinery space.

The runs of ducts must be planned and coordinated with the structure, wireways, and piping runs so as to minimize interferences and preserve adequate headroom. Fore and aft runs of ventilation ducts are usually located overhead in corridors or alongside girders at sides of hatches.

The Coast Guard regulations cover the general requirements for ventilation and certain specific requirements aimed at maintaining the integrity of watertight structure, firescreen bulkheads, and preventing the transmittal of noxious fumes. For a detailed discussion of ventilation systems, see Chapter XIII.

7.6 Hull Piping. Although hull piping is not a major consideration in developing a general arrangement, since it occupies a relatively small amount of space, it will be treated here since some attention must be given to layout of tanks and spaces to permit an economical and practical layout of the piping systems. Also, development of recent regulations for pollution control requires tankage and equipment associated with hull piping systems and these must be provided for in the arrangement.

Hull piping systems are generally understood to be all piping systems which are not directly connected with operation of the main propulsion machinery. This division of piping systems into hull and machinery is not clear cut, since some portions of combined systems serve more than one function; for example, the fuel oil system in which portions of the piping are used for both transfer (a hull system) and boiler service (a machinery system).

Basically, the major hull systems are considered to be: bilge and ballast, fuel oil transfer, potable water, sanitary, fire, cargo, and cargo ballast. The basic requirements of most of these systems are covered by the rules of the various classification societies and by regulatory bodies, but these rules cover only the essentials. For machinery piping systems, piping details, and further discussion of hull piping systems, see Harrington (1971).

a. Fuel Oil Transfer Systems. For turbine powered ships, the Coast Guard requires the diesel oil for the emergency diesel generator and dead ship starting to be in a tank outside the machinery space. This tank is filled manually from drums or directly from lighters, so that no transfer system is required. Additional storage for auxiliary diesel generators is located in a double bottom or deep tank, with tank fittings, filling and pumping arrangements similar to the fuel oil system.

b. Freshwater Systems. On U.S.-flag ships, potable water can be carried above the deep load line in free-standing tanks, or in tanks below the waterline which can be either free-standing or provided with cofferdams. Nonpotable water may be carried in tanks against the shell. In countries other than the U.S., local law may permit potable water to be carried in structural tanks below the waterline. Where this is permitted, chlorination systems are required.

Cold drinking water is supplied by electric water coolers, except in older ships where this may be supplied through heat exchangers. The U.S. Coast Guard Rules and Regulations provide the approximate minimum rates of water consumption to be allowed for. Distilling plants must be sized to provide for the total consumption, with margin, and tank storage is customarily provided for a minimum of 2 days.

c. Fire Systems. Seawater is the basic fire-fighting medium, backed up by CO_2, foam and portable extinguisher systems in particular areas. The requirements for water systems are laid down by the various regulatory bodies and consist of fixed water lines entirely separate from other services, with hydrants, valving, and pumping requirements clearly defined for each type and size of ship. In addition to the fire pumps required by the Rules, bilge, ballast, and sanitary pumps can be arranged so that they can be used for fire service. The fire line is often used for services such as deck washing, tank cleaning, and anchor chain washing, which do not interfere with its basic function. Where the rules require separate emergency fire pumps, as on tankers, these are often fitted forward and are diesel driven. Where fitted aft, they must be in compartments separate from the other fire pumps. To permit fitting of the pump prime mover on deck, the forward fire pump often employs a two-pump arrangement, one for lifting the water to the deck and the other for providing the necessary pressure for fire service. Water spray systems are provided on deck and for protection of the after house on special purpose ships.

Additional protection for cargo holds, cargo tanks, and similar spaces is provided by CO_2, or foam systems. Such systems consist of main lines with individual branches to the spaces being protected. The branch lines are fitted with individual stop valves to permit isolation of nonaffected areas. CO_2 systems are normally used for protection of cargo holds, and inert gas systems and foam for cargo tanks. In-tank fixed foam systems have been superseded largely by foam monitor systems where the possibility exists that fixed systems may have the distribution branches destroyed by an explosion, such as on tankers. Where monitor systems are used, it is advantageous to provide a loop system arranged so that foam and seawater can be used at the same time from their respective lines, while also permitting supply of the monitors from the seawater line in the event of damage to the foam line.

Additional protection for machinery and working spaces outside the quarters area is provided by CO_2, foam, and water fog systems, selection being dependent primarily on the nature of the space involved. Because of the possibility of injury to personnel, steam smothering is limited to services such as boiler air casing protection. Either CO_2 or fixed foam systems are used for machinery spaces and similar areas, the selection being dependent on the type of equipment in the space, escape arrangements, and CO_2 bottle storage space limitation. Where foam systems are used, the foam tank should be outside the main machinery space area to prevent deterioration of the foam due to the high temperature in these spaces. Adequate coamings for retention of foam are also essential. Fixed water fog systems are used in paint lockers, lamp rooms, pump rooms and similar spaces which are not adjacent to areas protected by foam or CO_2 systems and which do not contain essential electrical equipment. Fixed or portable CO_2 cylinders are provided also for fighting localized fires in machinery spaces.

Within recent years total flooding halon fire extinguishing systems have been developed and one such system, Halon 1301, is now acceptable by the Coast Guard as a substitute for CO_2 for fires involving electrical equipment or gaseous and liquid flammable materials, in such spaces as pump rooms, paint and lamp lockers, machinery spaces, and vehicle stowage spaces. Although Halon 1301 is similar to CO_2 in application and storage, its extinguishing action is by direct interruption of the chemical combustion process, while CO_2 acts by displacing the atmosphere and reducing the oxygen below that required for combustion. Since Halon 1301 is somewhat toxic, essentially the same precautions are required as for CO_2 systems; including evacuation of the space before manual release of the Halon 1301, although automatic release is permitted for spaces less than 170 m^3 (6,000 ft^3). Diesel engines must be shut down before Halon 1301 is introduced since the engine would not stop automatically as in the case of CO_2, and the products of cylinder combustion would be toxic. At present, Halon 1301 is not permitted for general cargo spaces pending more research on the toxicity of the products of combustion. These halon type extinguishing agents show great promise for the future because of their rapid extinguishing action and their relatively low hazard to personnel as compared to CO_2.

Additional protection for quarters areas is limited usually to portable extinguishers. The type of extinguishers allowed is governed by local laws but they are usually of the foam or dry chemical type. On some passenger ships, cargo vessels, and tankers, a fixed spray system automatically operated by fusible plug is installed.

d. Cargo Systems. For a detailed discussion of cargo systems, see Chapter XI.

e. Sanitary Systems. The requirements of the various regulatory bodies for sanitary system connections, drains and vents are similar since they are basically concerned with prevention of pollution by intentional or accidental cross-connection of the various potable and nonpotable systems which intercommunicate at fixtures and drains, as well as the prevention of pollution of harbors or lakes.

Some ships have sanitary systems supplied by salt water (or raw water on lakes and rivers); however, freshwater systems have become more common because of the advantages in reduced installation and maintenance cost. Actual sanitary (water closet and urinal flushing) requirements on most vessels, except for passenger ships, are quite small, and where salt water is used it is normal to provide other nonsanitary services as part of this system. These include such services as air conditioning, boiler water test coolers, and garbage grinders. The air-conditioning requirement is normally the largest demand. The salt water supply may be from the salt water service main through a branch fitted with a nonreturn valve, by independent pumps, or by a hydropneumatic system. The hydropneumatic system is used only for small systems.

Where independent pumps are used, it is necessary to provide recirculation to ensure adequate flow when air conditioning and similar demands are low or when air conditioning is not in use. Pumps are normally of the centrifugal type and are fitted in duplicate, with each pump being capable of handling the entire system demand. Although of low head in comparison to fire pump rule requirements, sanitary pumps can be of fairly high capacity and they are often used as standby fire pumps.

Where sanitary services are provided by water from the potable water system, generally it is not feasible to use the system to supply the miscellaneous, nonsanitary services, and such services must be supplied by the saltwater service system or by independent pumps. Most of the precautions required by the rules, such as air-gaps for saltwater systems, should also be provided for potable systems because of the possibility of using saltwater for this service in the event of an emergency. Separate emergency systems can also be provided but are seldom warranted.

Normally pumps are run continuously and provision for recirculation of water to the storage tanks and use of flow-limiting sanitary valves is essential to prevent wastage of fresh water.

U.S. Coast Guard pollution control regulations require that all vessels fitted with toilet facilities be equipped with an approved marine sanitation device, which will process sewage products to an acceptable condition before discharge from the ship into the environment. Many such devices and systems are currently on the market.

Section 8
Ship Types

8.1 General. The discussion in this chapter on the principles of general arrangement is confined to general terms because of the large number of special ship types, some of which are discussed in Chapter II. The tendency in the last few years has been to develop ship types specialized for certain specific cargos, although the multipurpose general cargo ship will probably be in use for some time to come.

Ship types in regard to cargo spaces are discussed in Section 2, where it is apparent that certain types overlap and a cogent system of classification is difficult. In the following, examples of the arrangements and characteristics of certain ship types are presented for general guidance. The examples are intended to represent workable designs, most of which are constructed and are in operation. A number of the ship types included in this section are discussed and illustrated in Chapter II.

Instead of including total ship arrangement plans, the illustrations herein have been limited to those sections of the selected ships that are of primary interest from an arrangement viewpoint, leaving the overall characteristics to be derived from the tabular data supplied.

Large General Cargo Ship

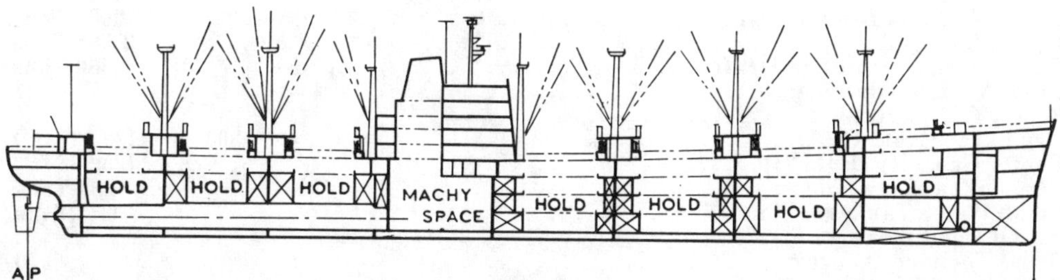

Fig. 10

Length Overall	**184.4 m (605.0 ft)**
Length Between Perpendiculars	**177.5 m (582.5 ft)**
Beam, Molded	**25.0 m (82.0 ft)**
Depth To Main Deck At Side, Molded	**14.0 m (46.0 ft)**
Draft, Full Load (Scantling), Molded	**10.7 m (35.0 ft)**
Light Ship	**9,787 Tons**
Passengers, Crew Effects & Stores	**60 Tons**
Fuel Oil	**3,596 Tons**

Fresh Water	**608 Tons**
Refrigerated Cargo	**218 Tons**
Liquid Cargo	**2,377 Tons**
General Cargo	**15,349 Tons**
Total Deadweight	**22,208 Tons**
Displacement, Full Load (Scantling) Draft	**31,995 Tons**
Cargo Volume, Bale	**30,645 m³ (1,082,207 ft³)**
Refrigerated Volume, Net	**618 m³ (21,839 ft³)**
Containers in Hold	**325**
Containers on Deck	**84**
Passenger Accomodations	**12**
Crew Accomodations	**41**
Shaft Horsepower, ABS	**24,000**
Speed, Knots	**20.8**
Propeller 4 Blades, diam	**6.7 m (22.0 ft)**
Propelling Machinery, Cross Compound, Double Reduction Geared Turbine	

8.2 General Cargo Ship, Large. The ship illustrated in Fig. 10 represents a large, high speed cargo vessel for multiport liner service with a variety of cargo demands both in terms of types of cargo and variations from voyage to voyage. The cargo holds and 'tween decks are suitable for break-bulk or palletized cargo. Containers can be stowed on deck and below deck in way of hatches. Deep tanks for cargo oils are fitted in holds 1, 5 and 6. Five holds (Nos. 2, 3, 4, 5 and 6) are specially fitted for carrying grains or other dry bulk cargos. Cargo gear of the boom and winch type, is designed for speed and flexibility for handling break-bulk, palletized, or container cargo. A heavy lift boom of 60 ton capacity is arranged to serve cargo hold 6. General arrangement details are on pages 142 and 143.

Small General Cargo Ship

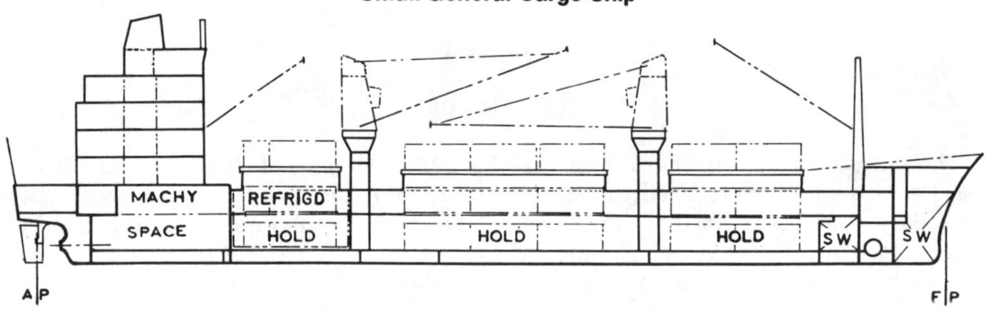

Fig. 11

Length Overall	**89.9 m (295.0 ft)**
Length Between Perpendiculars	**83.5 m (274.0 ft)**
Beam, Maximum Molded	**13.7 m (45.0 ft)**
Draft, Full Load	**4.5 m (14.8 ft)**
Depth	**6.7 m (22.0 ft)**
Designed Sea Speed, Knots	**13.75**
Shaft Horsepower, ABS Max.	**2800 Tons**
Dead weight At Design Draft	**2,062 Tons**
Light Ship	**1,588 Tons**
Displacement At Design Draft	**3,650 Tons**
Crew Accommodations	**11**
Factor Of Subdivision	**1.0**
Dry Cargo (Hold 1&2)	**3631 m³ (128,237 ft³)**
Reefer Capacity (Hold 3)	**678 m³ (23,938 ft³)**
Fuel Oil Tankage, Tons	**286 Tons**
Fresh Water Tankage, Tons	**11.5 Tons**
No. Of Containers (20 Ft)	**74**
Gross Tonnage, Approx.	**940**
Propeller, 4 Blades	**—**
Propelling Machinery, Single Diesel Engine	

8.3 General Cargo Ship, Small. This ship, illustrated in Fig. 11, is designed for carrying various types of cargo on short runs between a number of ports, and is especially suitable for feeder service. The hatch dimensions are compatible with the stowage of 20 ft or 40 ft standard containers, and the center hold is fitted with an extra large hatch for oversize cargo. Cargo handling gear consists of two sets of boom and winch rigs and two rotating cranes. General arrangement details will be found on pages 144 and 145.

Containership A

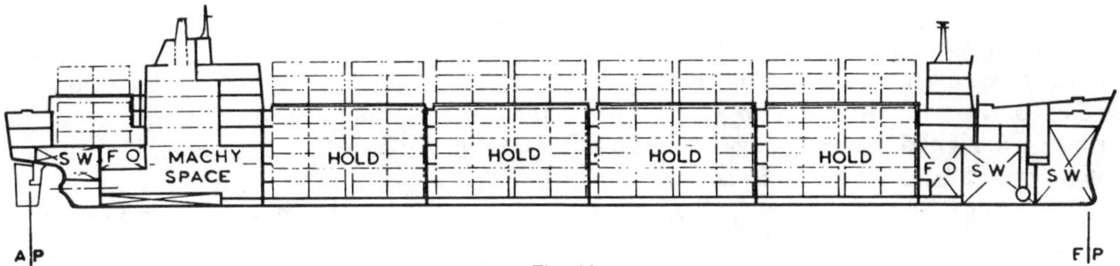

Fig. 12

Length Overall	**186 m (610 ft)**
Length Between Perpendiculars	**177 m (580 ft)**
Beam, Molded	**23.8 m (78 ft)**
Depth To Upper Deck At Side, Molded	**16.6 m (54.5 ft)**
Draft, Full Load (Scantling), Molded	**9.64 m (31.5 ft)**
Light Ship	**7,480 Tons**
Crew Effects & Stores	**50 Tons**
Fuel Oil	**3,380 Tons**
Fresh Water	**230 Tons**
General & Refrigerated Cargo	**10,940 Tons**
Total Deadweight	**14,600 Tons**
Displacement, Full Load (Scantling) Draft	**22,080 Tons**
Cargo Volume, Bales (20-ft Containers)	**27,800 m^3 975,840 ft^3**
Refrigerated Volume (20-ft Containers)	**1,795 m^3 63,360 ft^3**
Containers In Hold	**612**
Containers On Deck	**316**
Crew Accommodations	**42**
Shaft Horsepower, A.B.S.	**17,500**
Speed, Knots	**200**
Propeller 5 Blades, diam	**6.56 m (21.5 ft)**

Propelling Machinery, Cross Compound Double Reduction Geared Turbine

Containership B

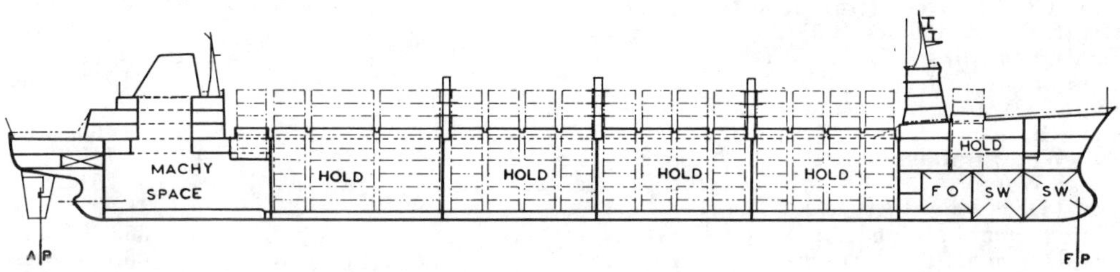

Fig. 13

Length Overall	**220 m (719 ft)**
Length Between Perpendiculars	**206 m (677 ft)**
Beam, Molded	**29 m (95 ft)**
Depth To Main Deck At Side, Molded	**16.5 m (54 ft)**
Draft, Design Molded	**10.4 m (34 ft)**
Lightship	**14,574 Tons**
Crew Effects & Stores (Bale)	**1100 m^3 (38,977 ft^3)**
Fuel Oil	**6,943.3 Tons**
Anti-Roll Tank	**1,790 Tons**
Fresh Water	**588.7 Tons**
Total Deadweight	**24,126 Tons**
Displacement At Design Draft	**38,700 Tons**
Cargo Volume, Bale (35-ft Containers)	**23,100 m^3 816,340 ft^3**
Cargo Volume, Bale (40-ft Containers)	**14,000 m^3 484,160 ft^3**
Containers In Hold (35-ft Containers)	**394**
Containers In Hold (40-ft Containers)	**178**
Containers On Deck (35-ft Containers)	**165**
Basic Manning	**37**
Crew Accommodations	**39**
Shaft Horsepower, ABS (Maximum)	**32,000**
Speed, Knots	**22.8**
Propeller, 6 Blades, diam	**6.87 m (22.5 ft)**

Propelling Machinery, Cross Compound, Double Reduction Geared Steam Turbine

8.4 Containerships. Two examples of vertical cell type containerships are illustrated in Figs. 12 and 13 and are designated Containerships A and B. These are the same ships that were designated A and B in Chapter I, Tables 6, 7, 8, 9, and 10. Both ships are essentially complete containerships of large capacity, with single screw geared turbine propulsion plants, designed for trans-ocean service. Containership A is designed for 20-ft containers including one hold abaft the deckhouse. A wheelhouse is located in a separate superstructure at the bow of the ship, which also houses the deck officers.

Containership B is designed for 35-ft and 40-ft containers and has two deckhouses, one at the stern and one well forward, utilized for the wheelhouse and the deck officer accommodations.

Neither ship is provided with cargo gear and depend on shore cranes for container handling. A limited amount of space for cargos other than containers is provided. Another large containership specially designed for frozen cargo carried in containers in insulated and refrigerated cargo holds is Containership C, illustrated in Chapter I. General arrangement details are on pages 146–149.

Roll-On/Roll-Off Ship

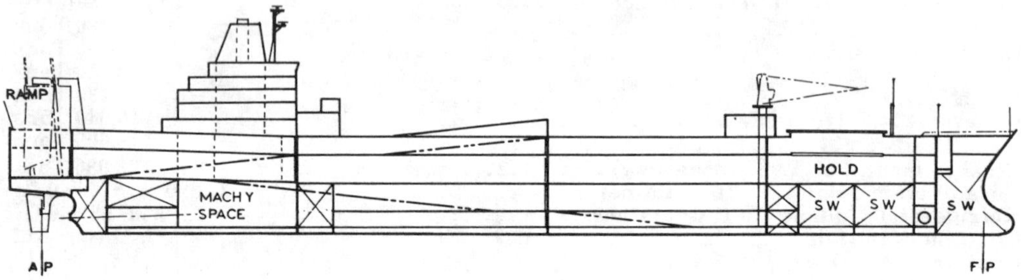

Fig. 14

Length Overall	**208.5 m (684.0 ft)**
Length Between Perpendiculars	**195.1 m (640.0 ft)**
Beam, Molded	**31.1 m (102.0 ft)**
Depth To "A" Deck, Molded	**21.2 m (69.5 ft)**
Depth To "B" Deck (Bulkhead Deck) MLD	**17.4 m (57.2 ft)**
Draft, Design Molded	**9.8 m (32.0 ft)**
Draft, Scantling	**10.4 m (34.0 ft)**
Lightship	**14,776 Tons**
Liquid Cargo	**802 m³ (28,330 ft³)**
Fuel Oil	**3,465 Tons**
Salt Water Ballast	**6,749 Tons**
Fresh Water	**251 Tons**
Total Deadweight At Design Draft	**18,989 Tons**
Displacement Molded, At Design Draft	**33,640 Tons**
Displacement, Total At Design Draft	**33,765 Tons**
Cargo Volume, Bale Holds	**55,416 m³ (1,956,982 ft³)**
Area Sq Ft (Including A Deck Ramps)	**14,248 m³ (153,360 ft²)**
Basic Manning	**41**
Crew Accommodations (Incl. Spares)	**42**
Shaft Horsepower Max. Cont.	**37,000**
Speed, Knots	**23.0**
Propeller 6 (Skewed) Blades, diam	**6.7 m (22.0 ft)**
Propelling Machinery, Cross Compound. Double Reduction, Geared Turbine	

8.5 Roll-On/Roll-Off Ship. The arrangement of a ship in which the cargo is loaded by roll-on/roll-off methods is shown in Fig. 14. The cargo is loaded by mobile equipment that delivers it to any location on one of four deck levels by internal ramps. Mobile equipment such as trucks, cars and tractors can be driven aboard under its own power and general cargo of all types can be unitized and taken aboard in large units. The ship is especially suitable for handling lumber in large quantities. The ship can carry containers of lengths from 10 to 40 ft.

Cargo handling equipment includes a 24 ft stern ramp and two 15-ton deck cranes to handle cargo too large for ramp loading. A hatch for break-bulk cargo loading is provided. An internal ramp system interconnects all five cargo decks. The ship-based cargo handling equipment includes two 20-ton straddle carriers, two 20-ton and six 10-ton forklift trucks. The straddle carriers are used for long length cargo and the lift trucks for containers, palletized and miscellaneous cargo. General arrangement details are on pages 150 and 151.

LASH Barge Carrier

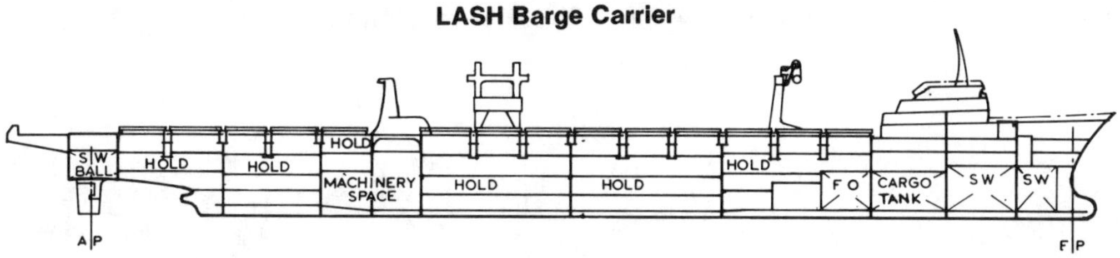

Fig. 15

Length Overall, Incl. 14.6m (48 ft.) Ext. 250.0m (820ft)
Length Between Perpendiculars 220.7m (724ft)
Beam, Molded 30.5m (100ft)
Depth to Main Deck at Side, Molded 18.3m (60ft)
Draft, Design, Molded 8.5m (28ft)
Draft, Scantling, Molded 10.7m (35ft)
Light Ship .... 14,230 Tons
Fresh Water .... 669 Tons
Fuel Oil .... 4,928 Tons
Anti-Roll Tank .... 300 Tons
Liquid Cargo .... 1,149 Tons
General Cargo (61 Lighters @ 180 tons each) .... 11,324 Tons
Total Deadweight @ 28 ft. .... 18,420 Tons
Total Displacement .... 32,650 Tons
General Cargo (61 Lighters @ 375 tons each) .... 22,924 Tons
Total Deadweight @ 35 ft. .... 30,020 Tons
Total Displacement .... 44,250 Tons
Cargo Volume, Bale (61 Lighters) 34,348 m^3 .... (1,213,000 ft.^{3})
(49 Lighters and 344,20-ft. containers 38,716m^3 .... (1,367,260 ft^3)
Containers in Hold 180
Containers on Deck 164
Crew Accommodations 46
SHP, ABS Max 32,000
Speed, Knots 22.5
Propeller diam 7.0m (23ft)
Propelling Machinery, Cross Compound, Double Reduction Geared Turbine

SEABEE Barge Carrier

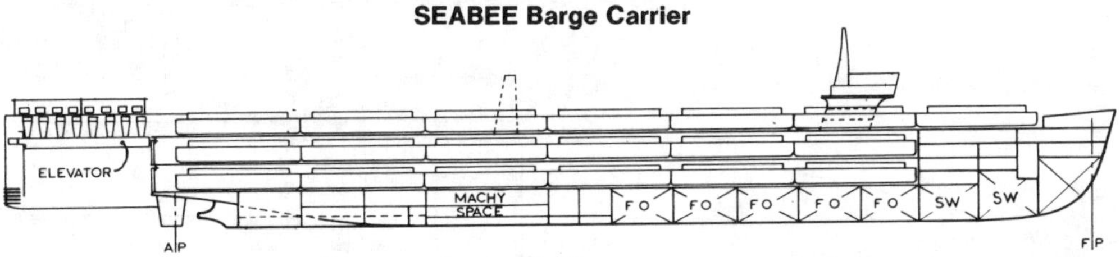

Fig. 16

Length Overall, Incl. 34.7m (114 ft.) Ext. 266.4m (873.8ft)
Length Between Perpendiculars 220.3m (722.7ft)
Beam, Molded 32.3m (105.9ft)
Depth to Upper Deck at Side, Molded 22.8m (74.8ft)
Draft, Design, Molded 9.5m (31.0ft)
Draft, Max. Full Load 11.9m (39.1ft)
Light Ship .... 18,880 Tons
Fresh Water .... 797 Tons
S.W. Ballast .... 1,016 Tons
Fuel Oil .... 5,997 Tons
Anti-Roll Tank .... 779 Tons
General Cargo (38 Barges @ 784 tons each) .... 29,791 Tons
Total Deadweight .... 38,410 Tons
Total Displacement .... 57,290 Tons
Cargo Volume, Bale .... 41,476 m^3 (1,464,710 ft.^{3})
Passenger Accommodations 4
Crew Accommodations 43
SHP, ABS Max 36,000
Speed, Knots 20
Propeller, 5 Blades, diam .... 7.0m (23 ft)
Propelling Machinery, Cross Compound, Double Reduction Geared Turbine

8.6 Barge Carrying Ships. Although generally referred to as a barge carrying ship, the LASH type is designed to carry lighters and containers which are carried on and below the large mechanical hatch covers which open up essentially the entire ship's hold. Heavy crane rails, running the entire

(Continued on page 168)

Large General Cargo Ship Afterbody Arrangements

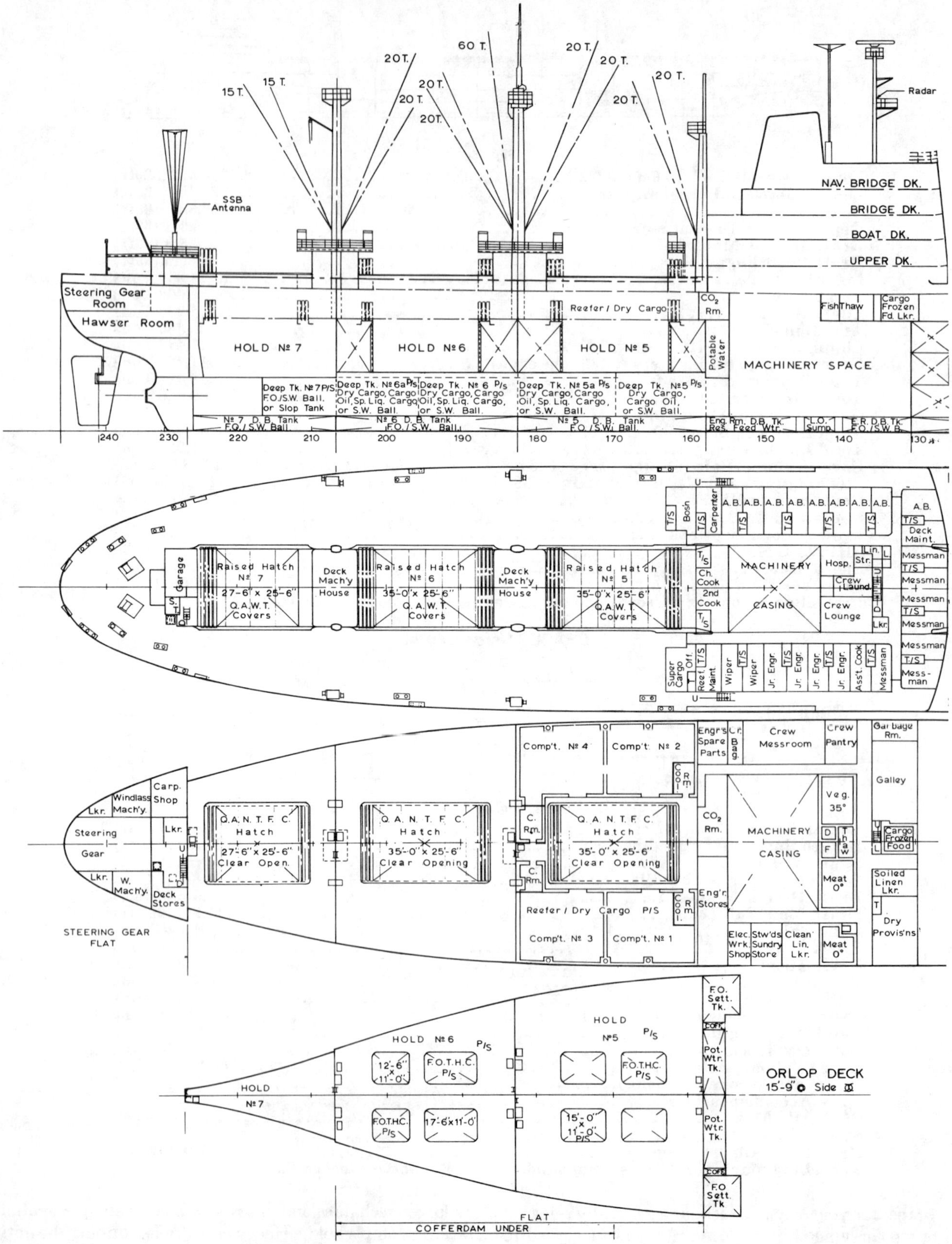

Large General Cargo Ship Forebody Arrangements

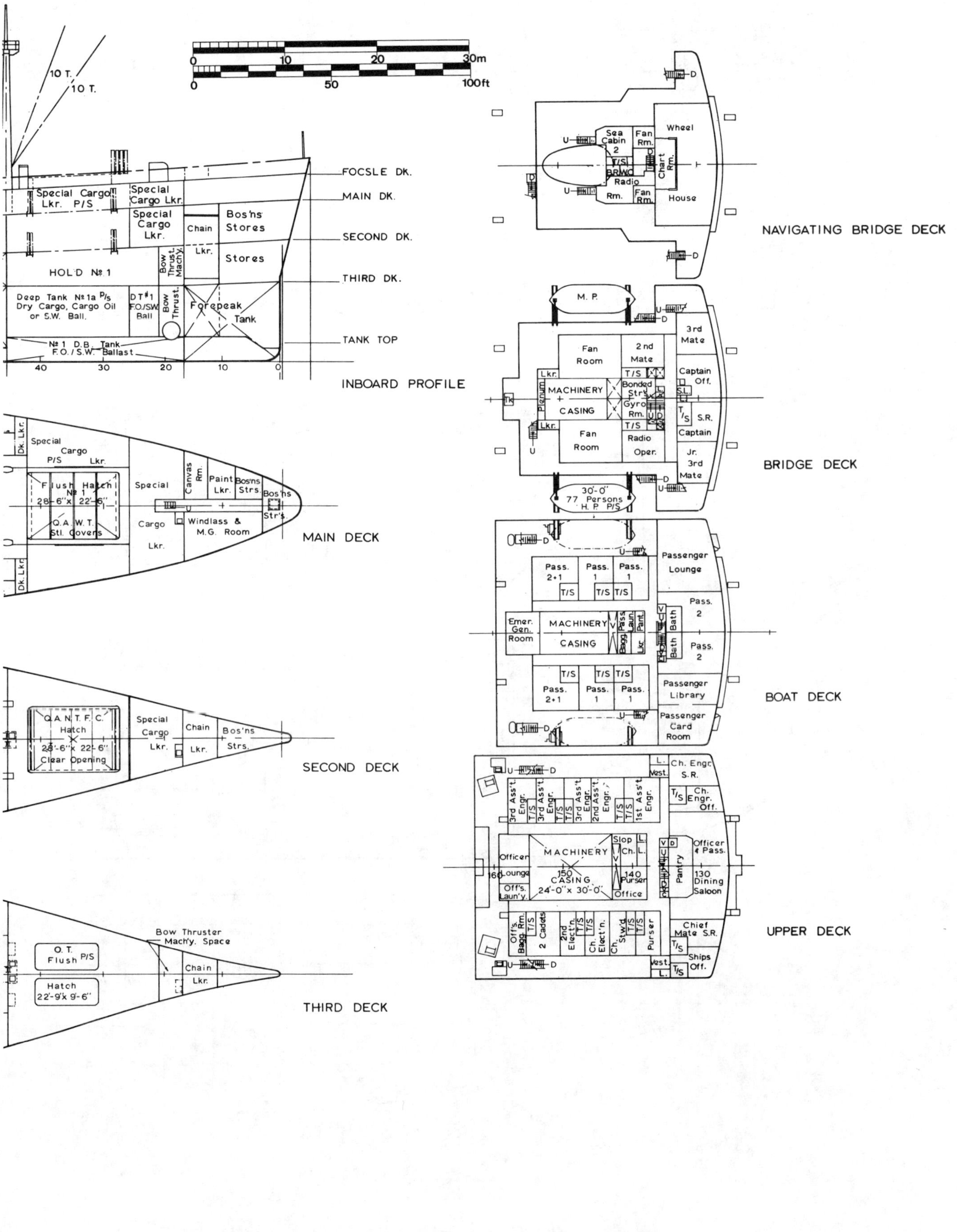

Small General Cargo Ship Afterbody Arrangements

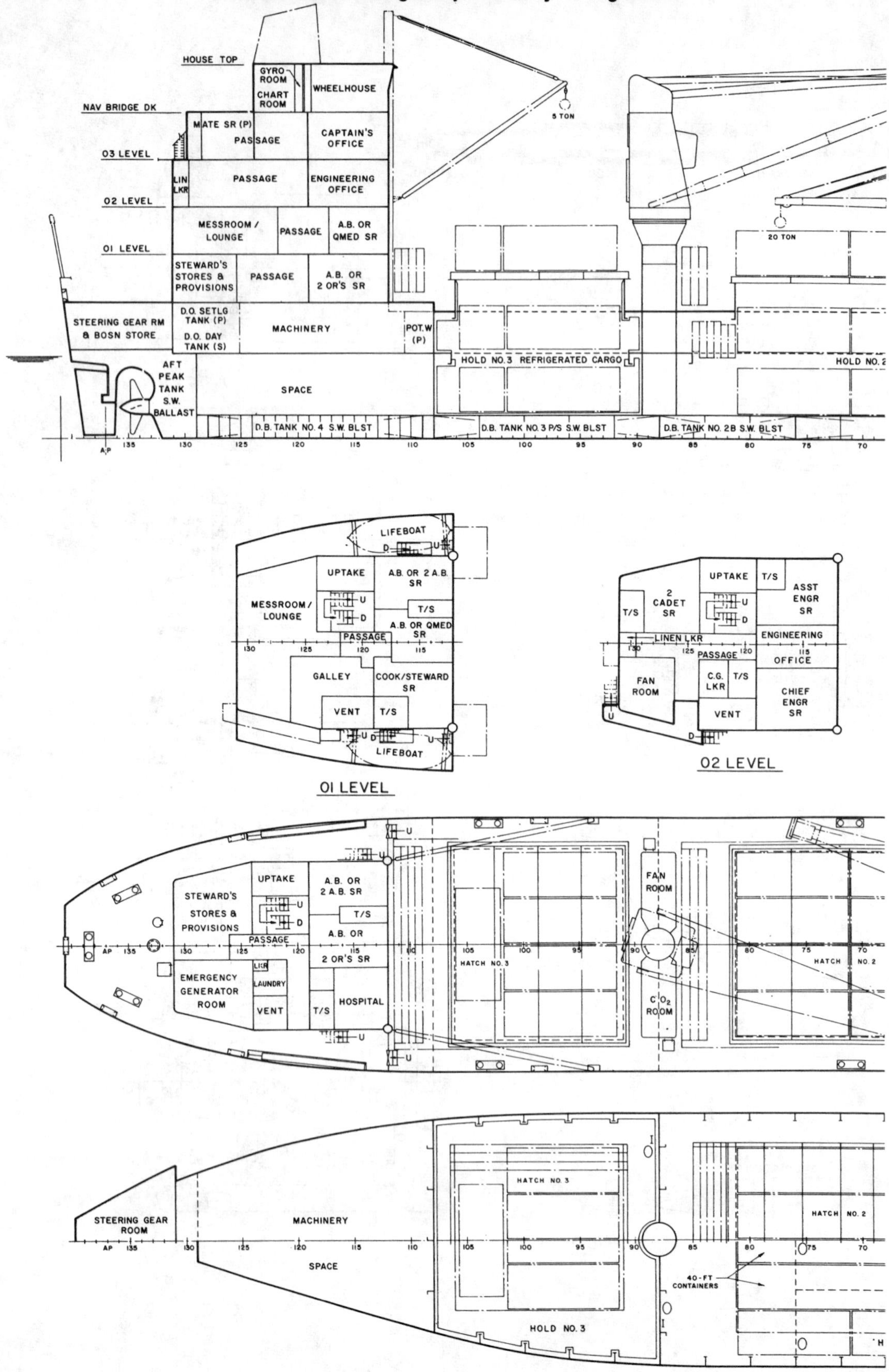

Small General Cargo Ship Forebody Arrangements

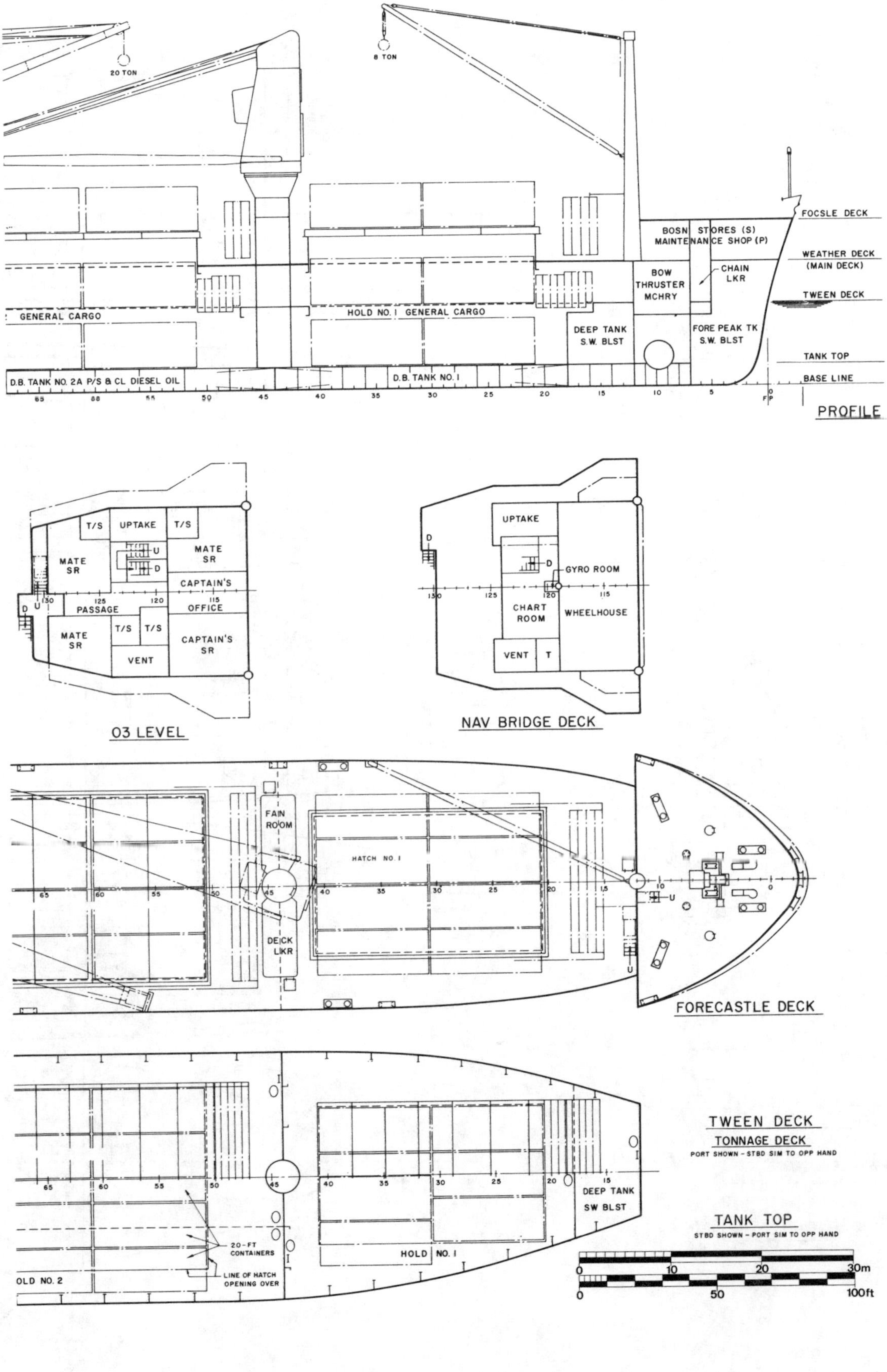

Containership A Afterbody Arrangements

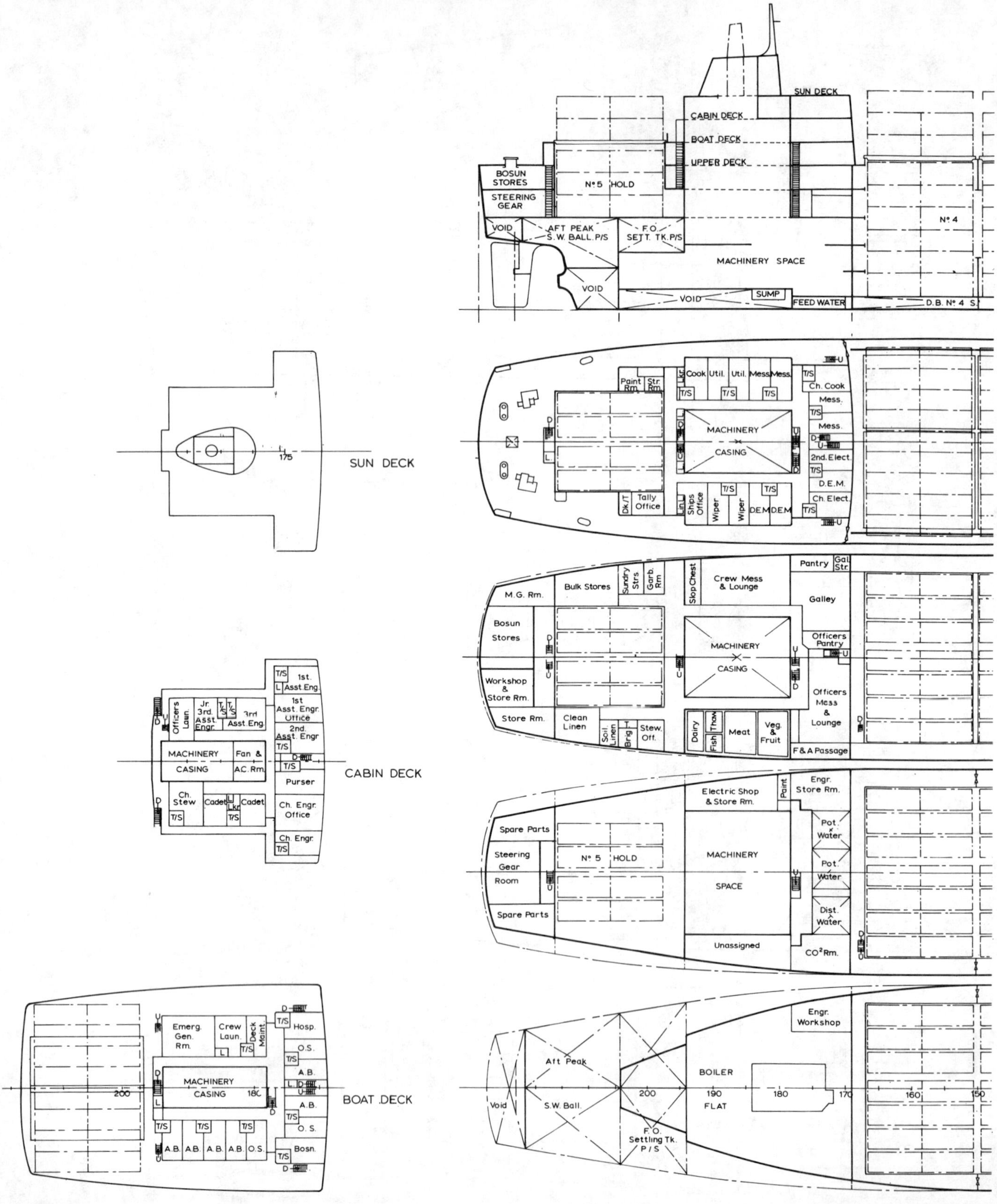

Containership A Forebody Arrangements

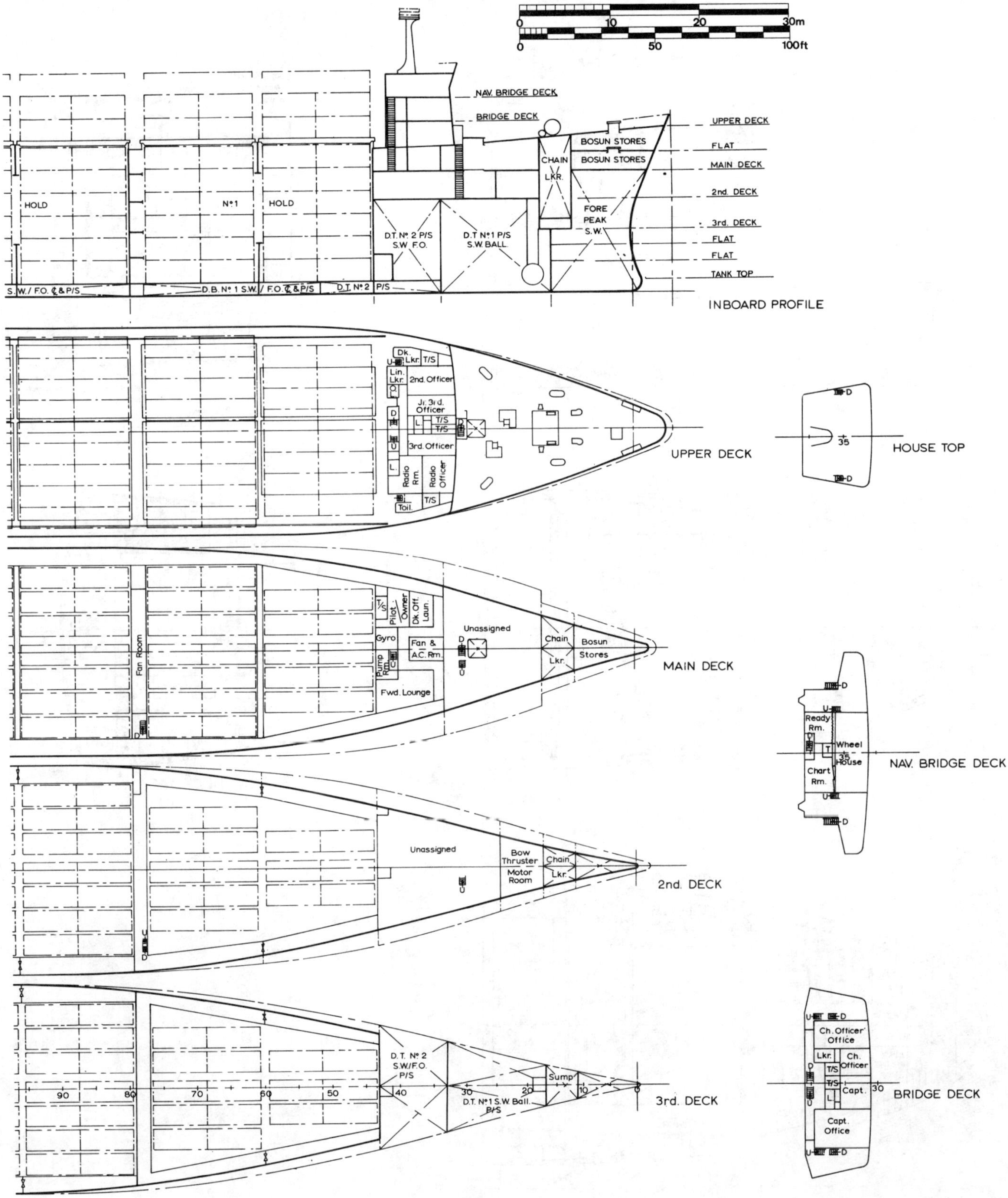

Containership B Afterbody Arrangements

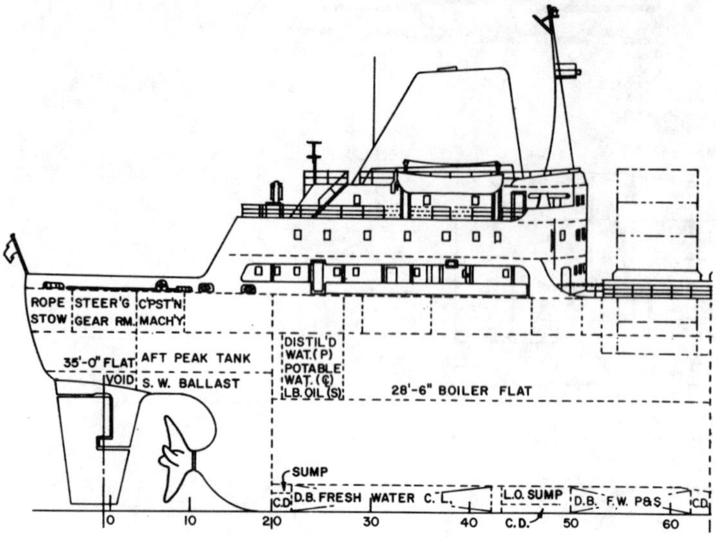

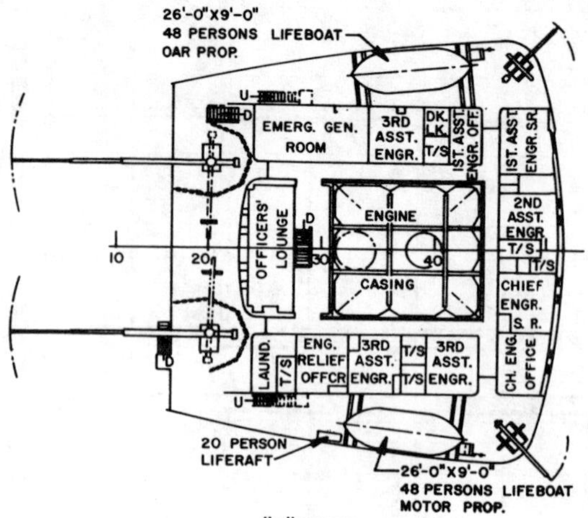

"B" DECK AFT

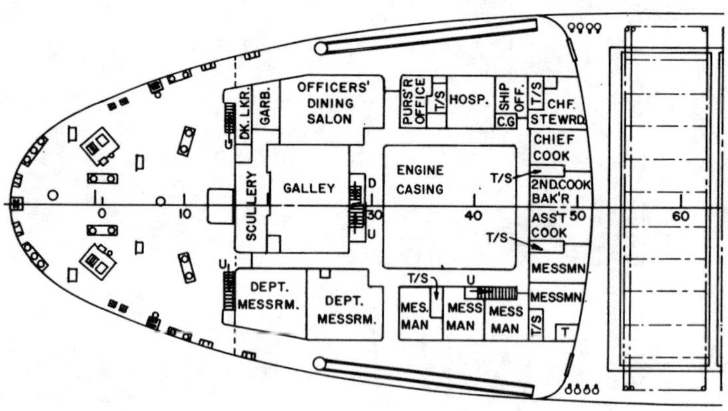

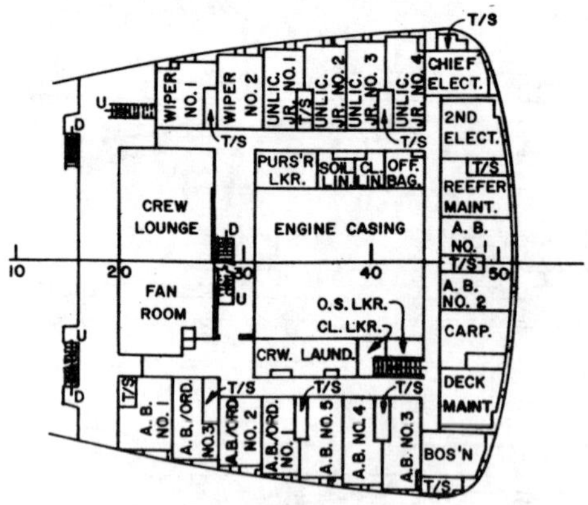

"A" DECK AFT

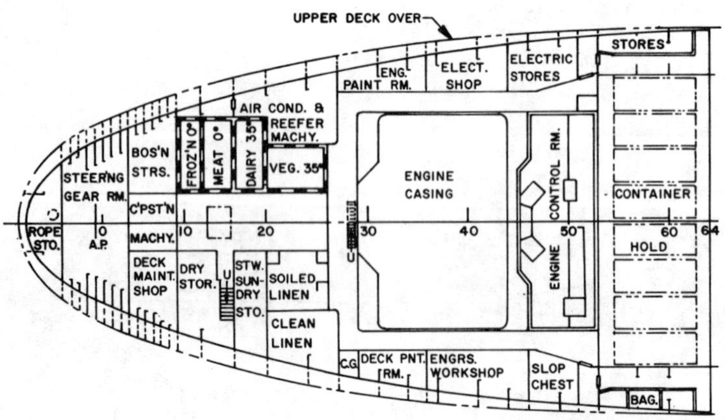

2ND DECK

Containership B Forebody Arrangements

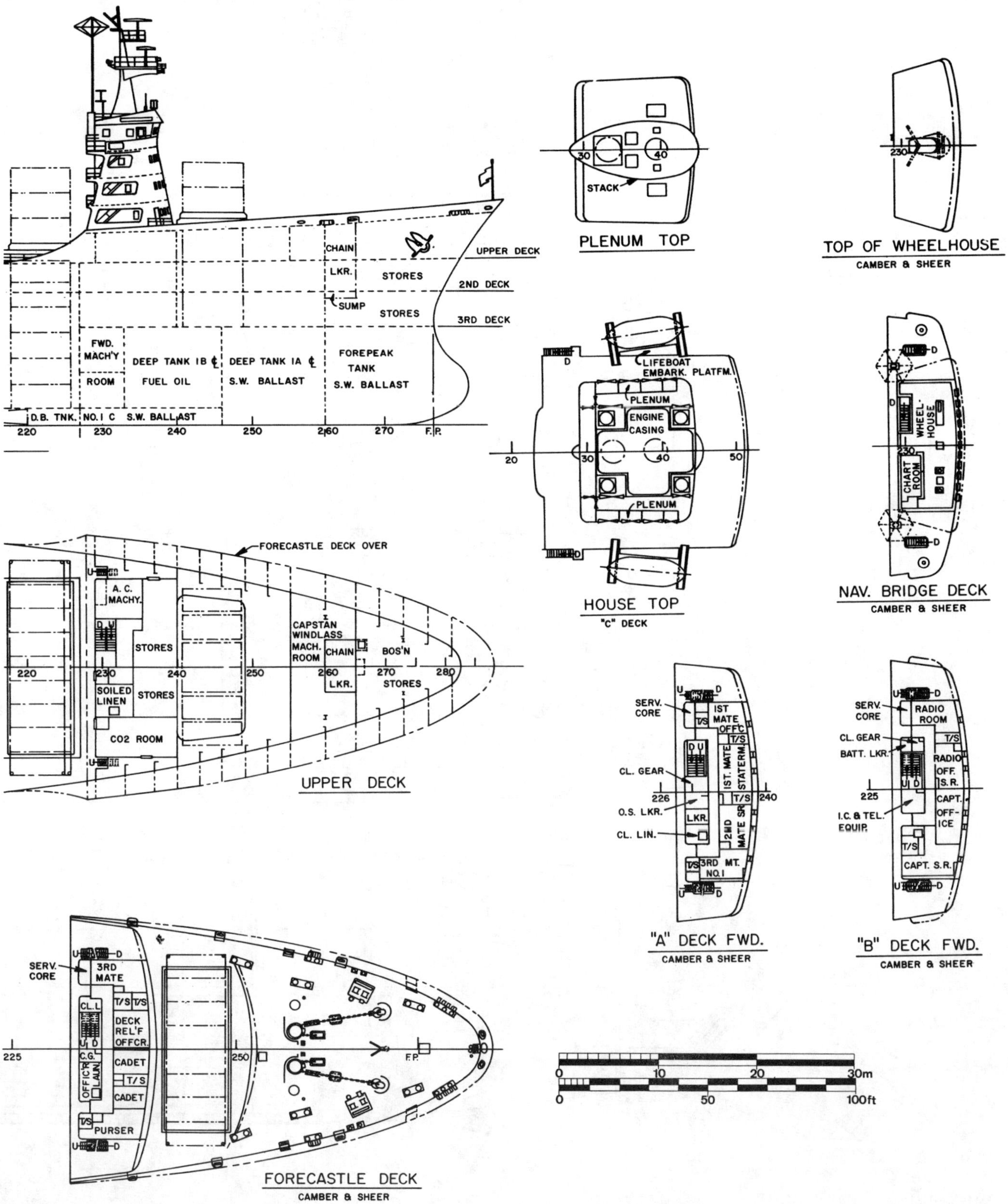

Roll-On/Roll-Off Ship Afterbody Arrangements

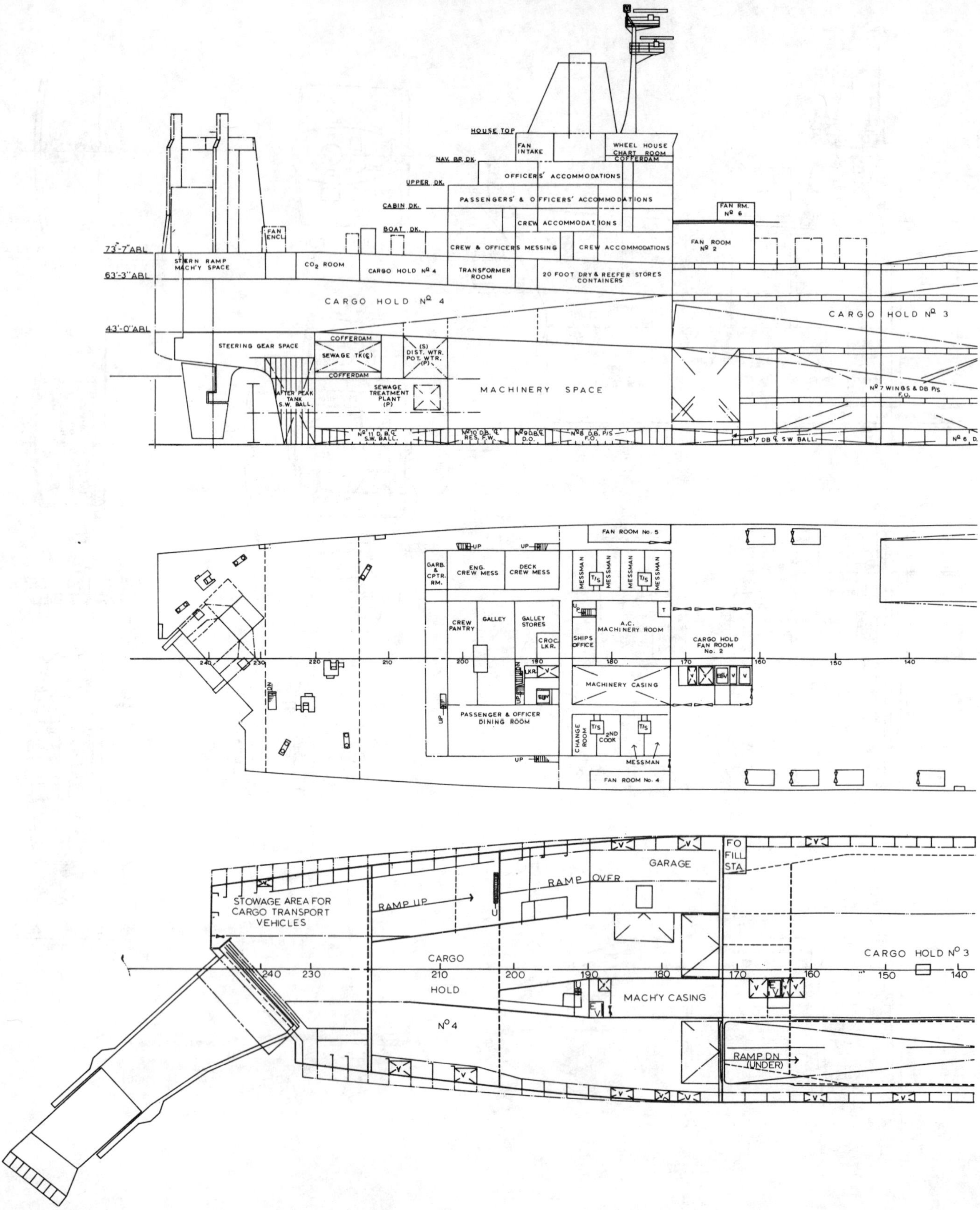

Roll-On/Roll-Off Ship Forebody Arrangements

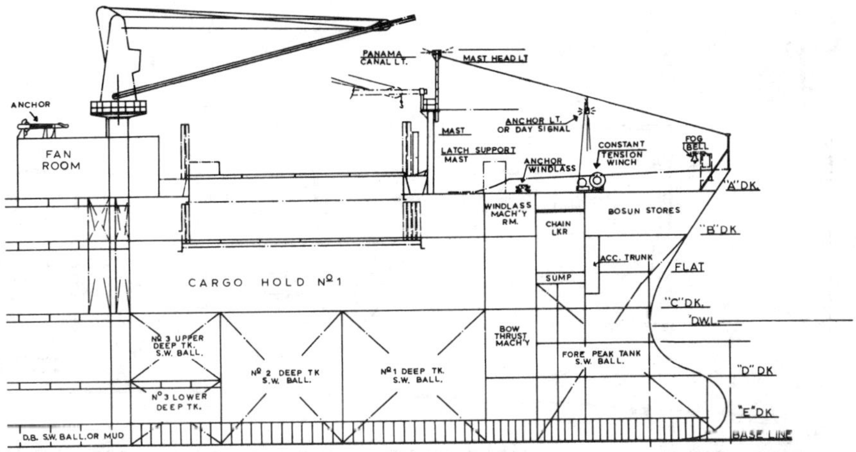

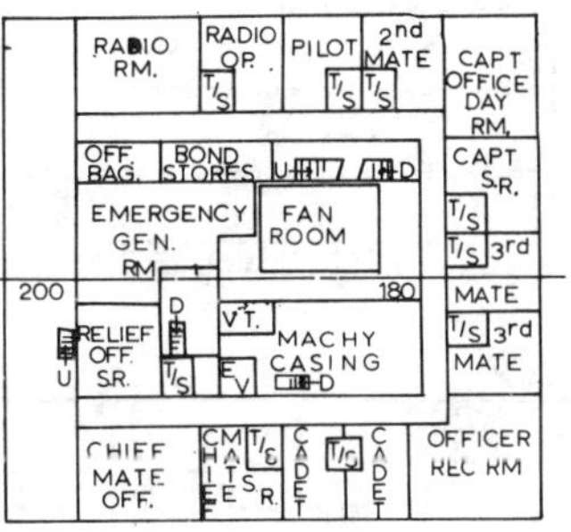

UPPER DECK

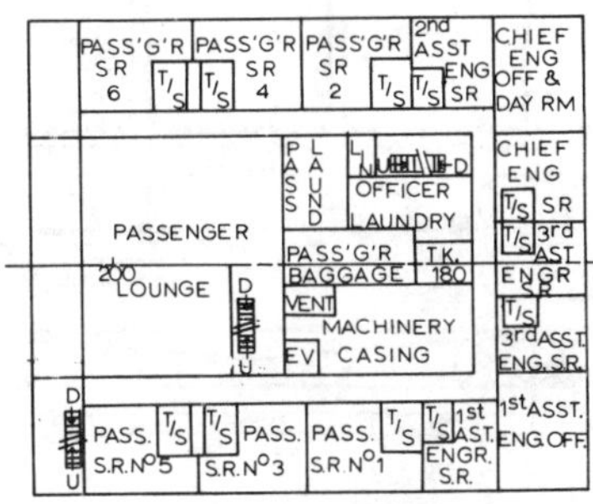

CABIN DECK

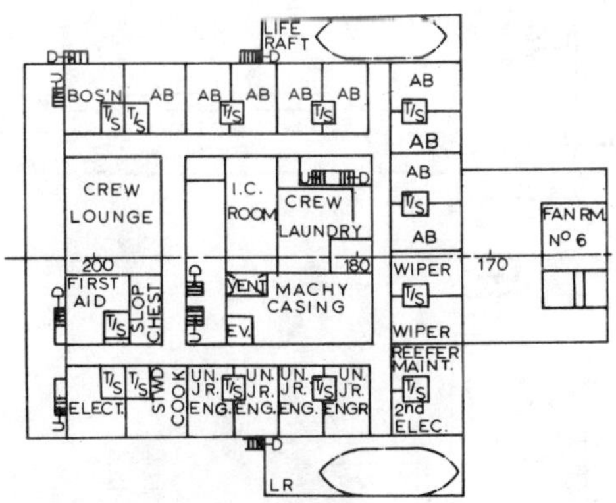

BOAT DECK

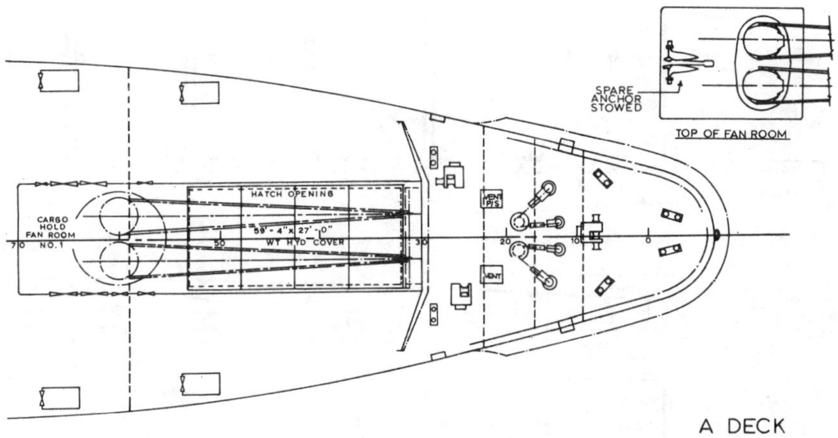

A DECK

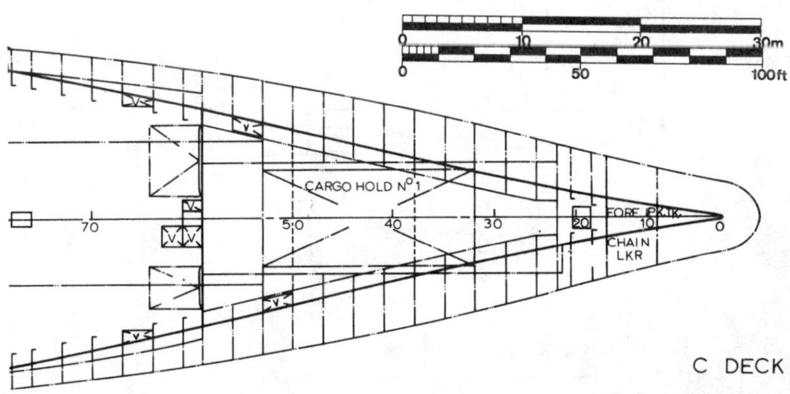

C DECK

LASH Barge Carrier Afterbody Arrangements

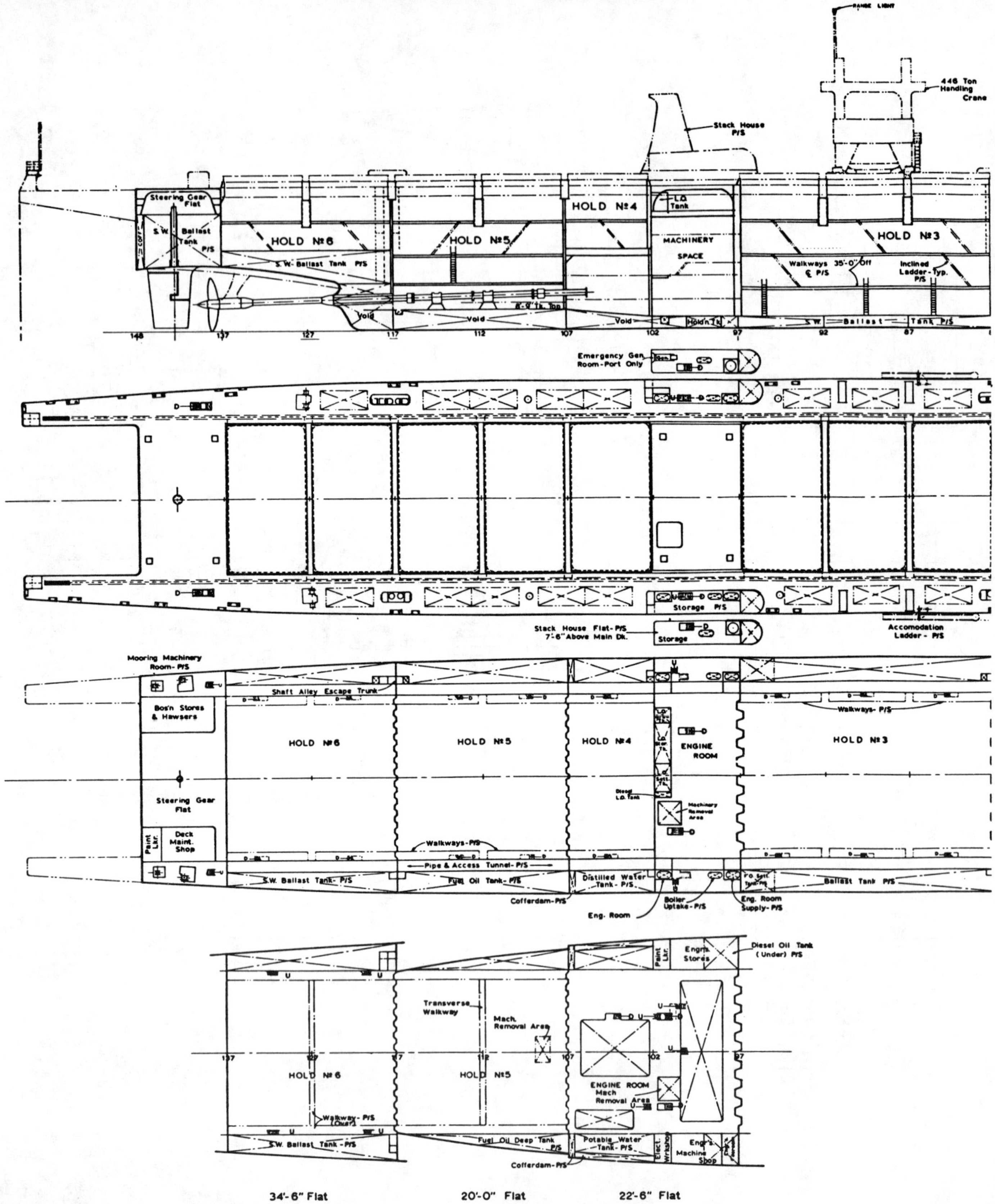

LASH Barge Carrier Forebody Arrangements

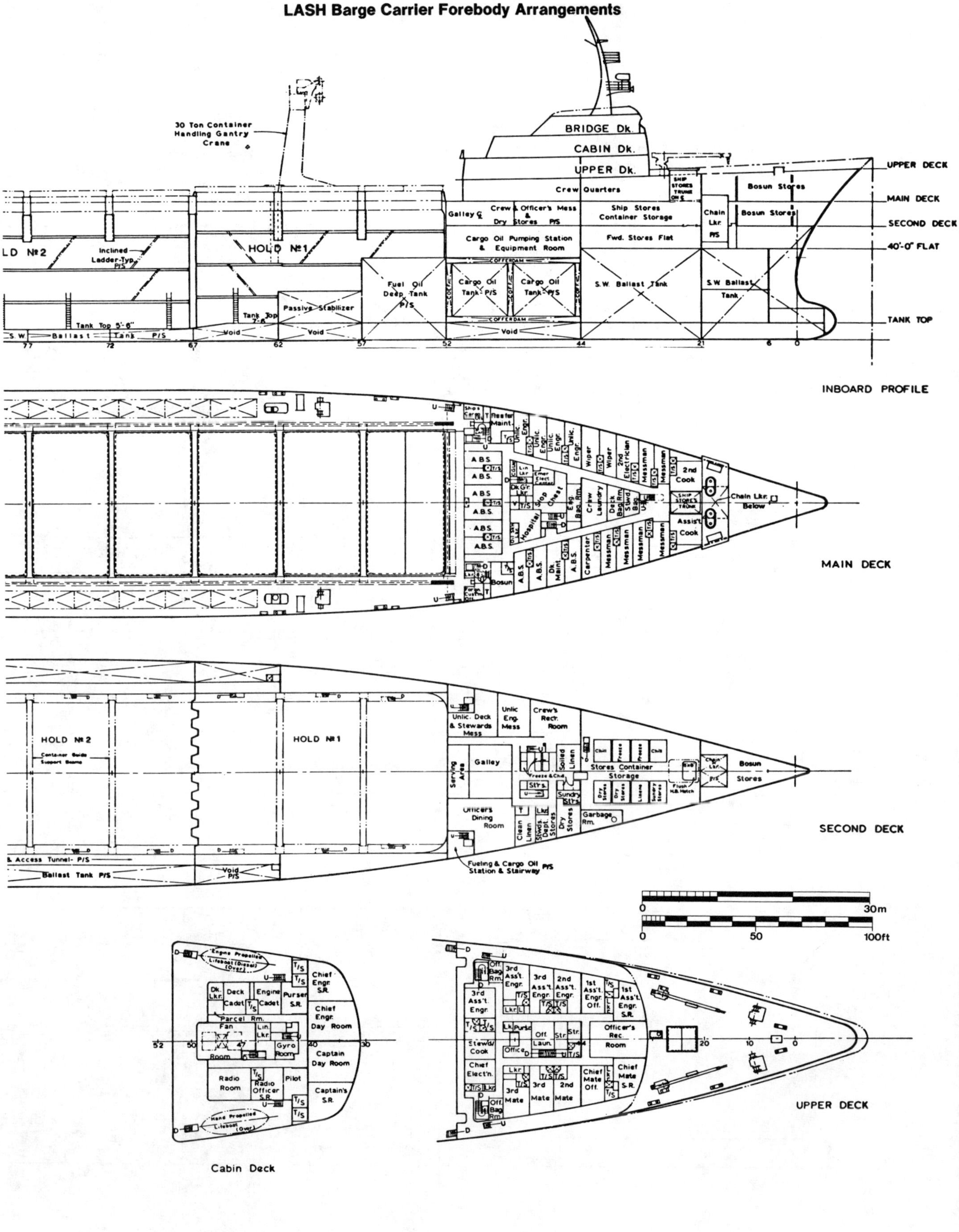

SEABEE Barge Carrier Afterbody Arrangements

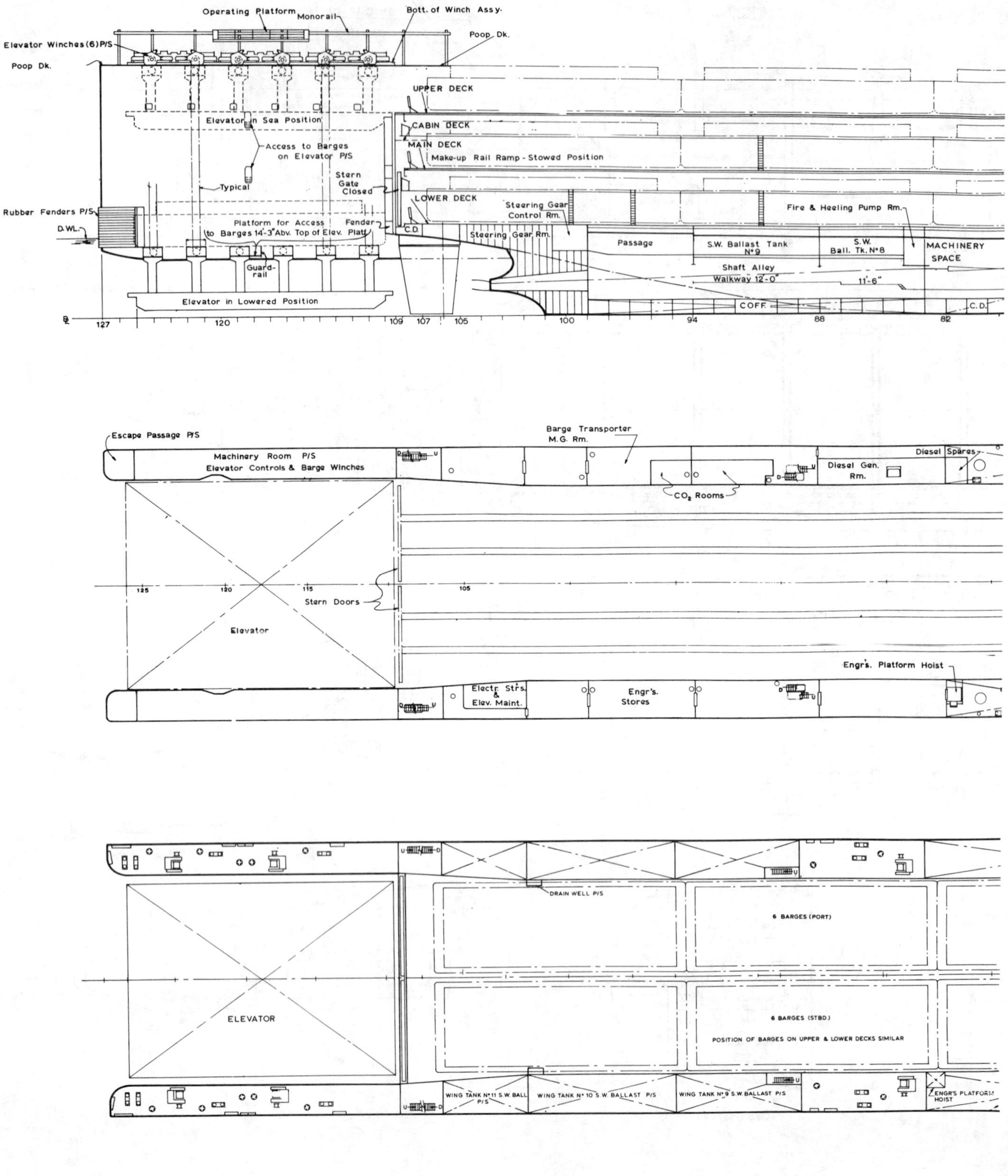

SEABEE Barge Carrier Forebody Arrangements

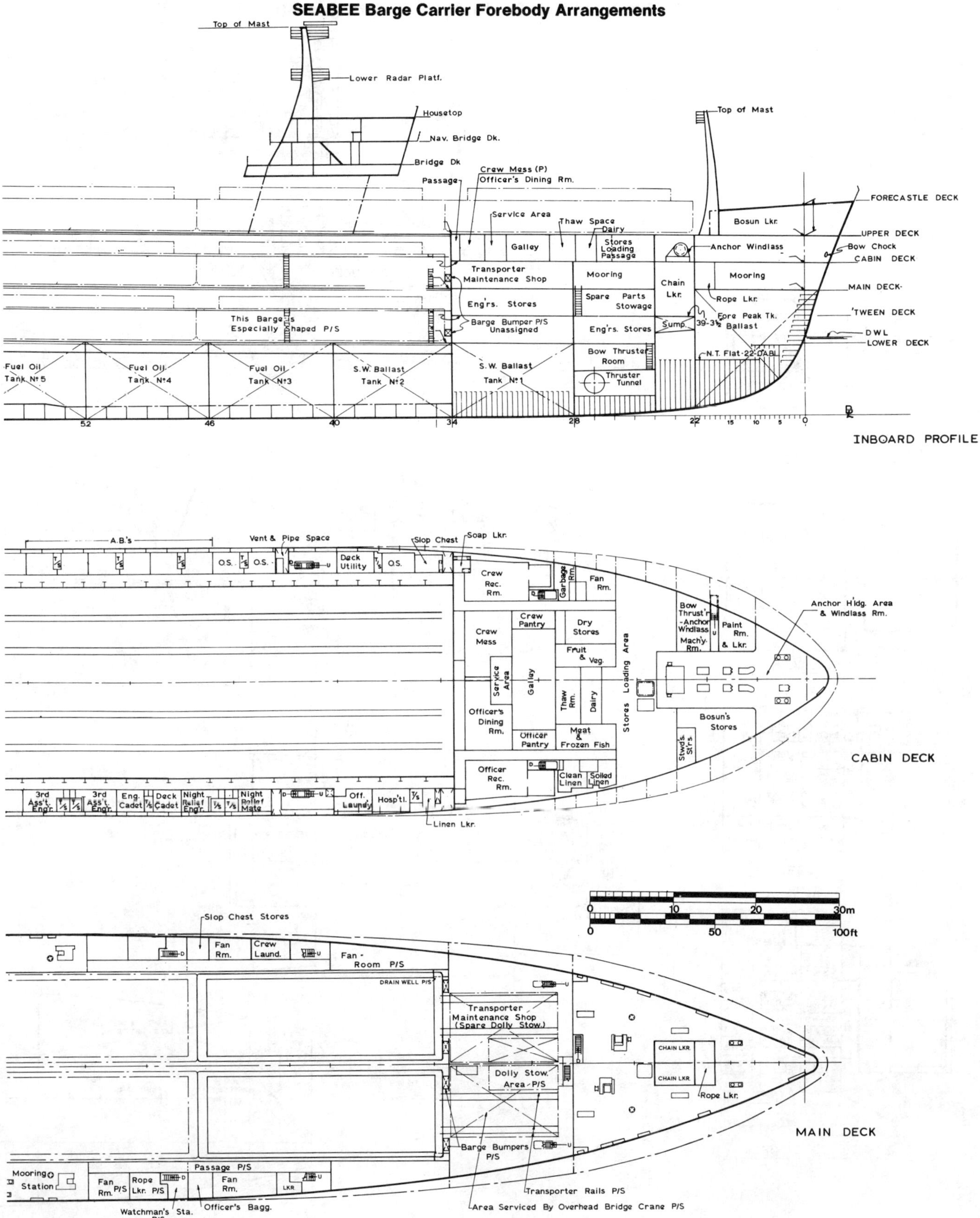

Tanker A, Products Carrier Afterbody Arrangements

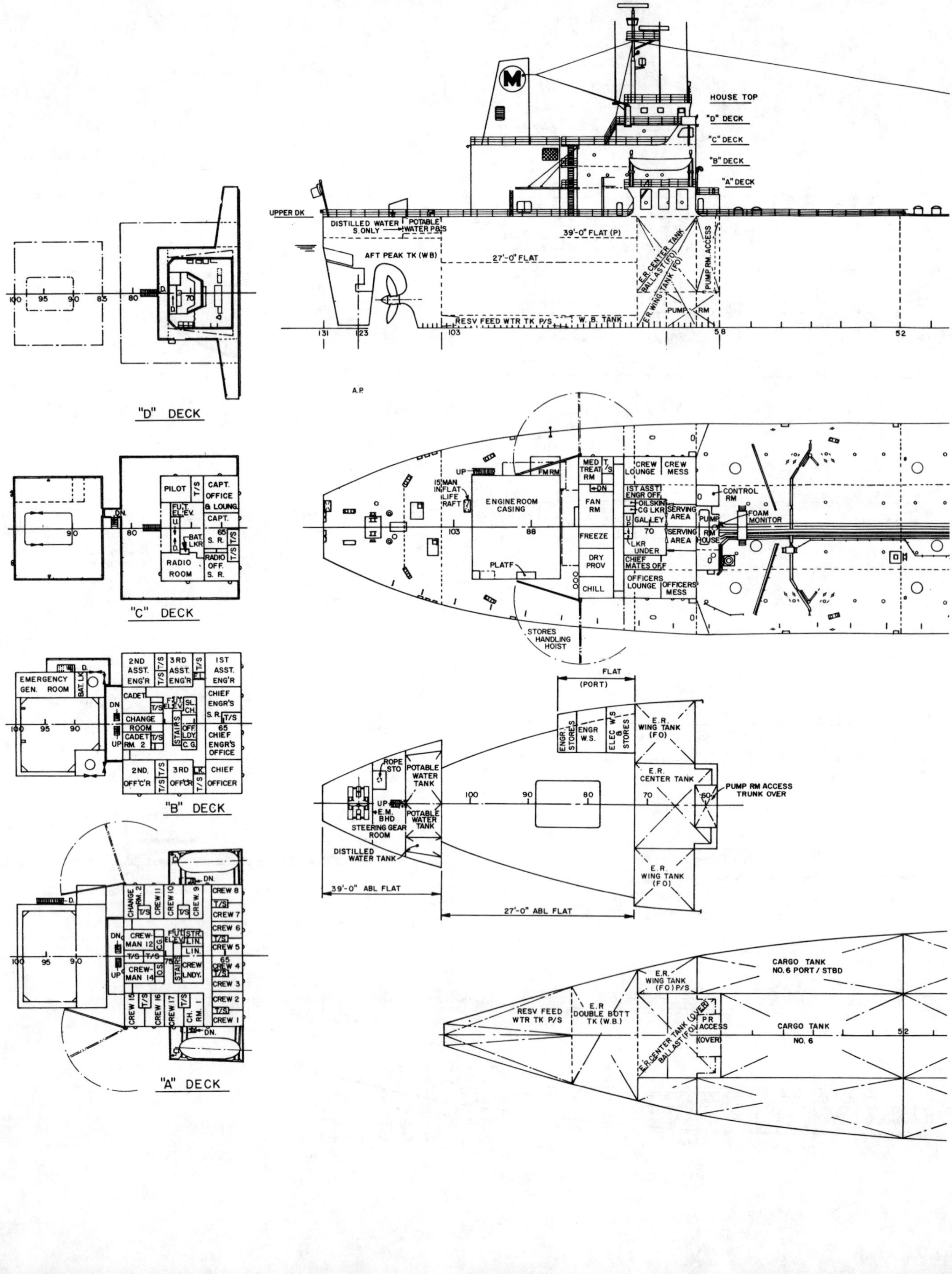

Tanker A, Products Carrier Forebody Arrangements

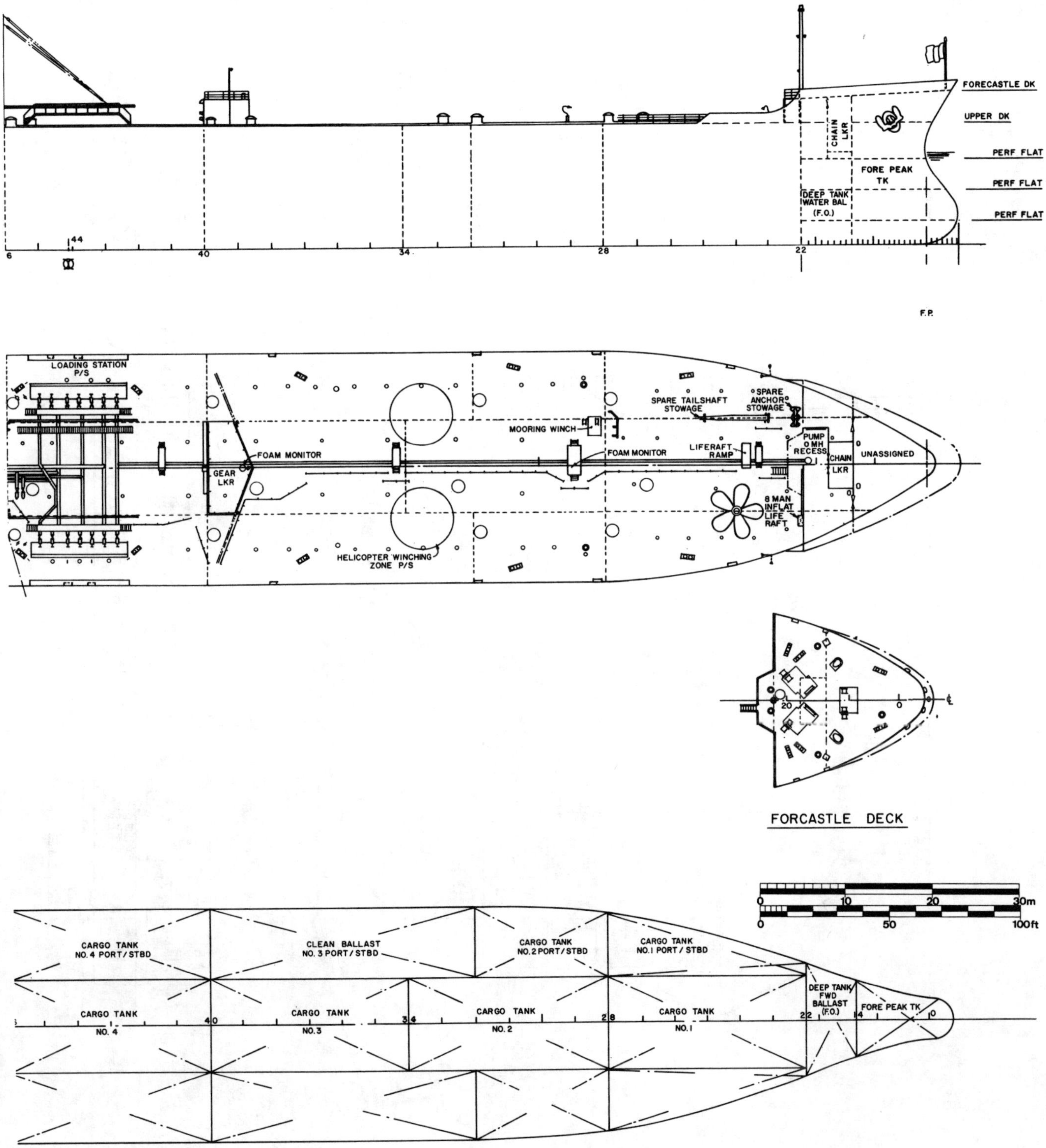

Tanker B, Crude Carrier Afterbody Arrangements

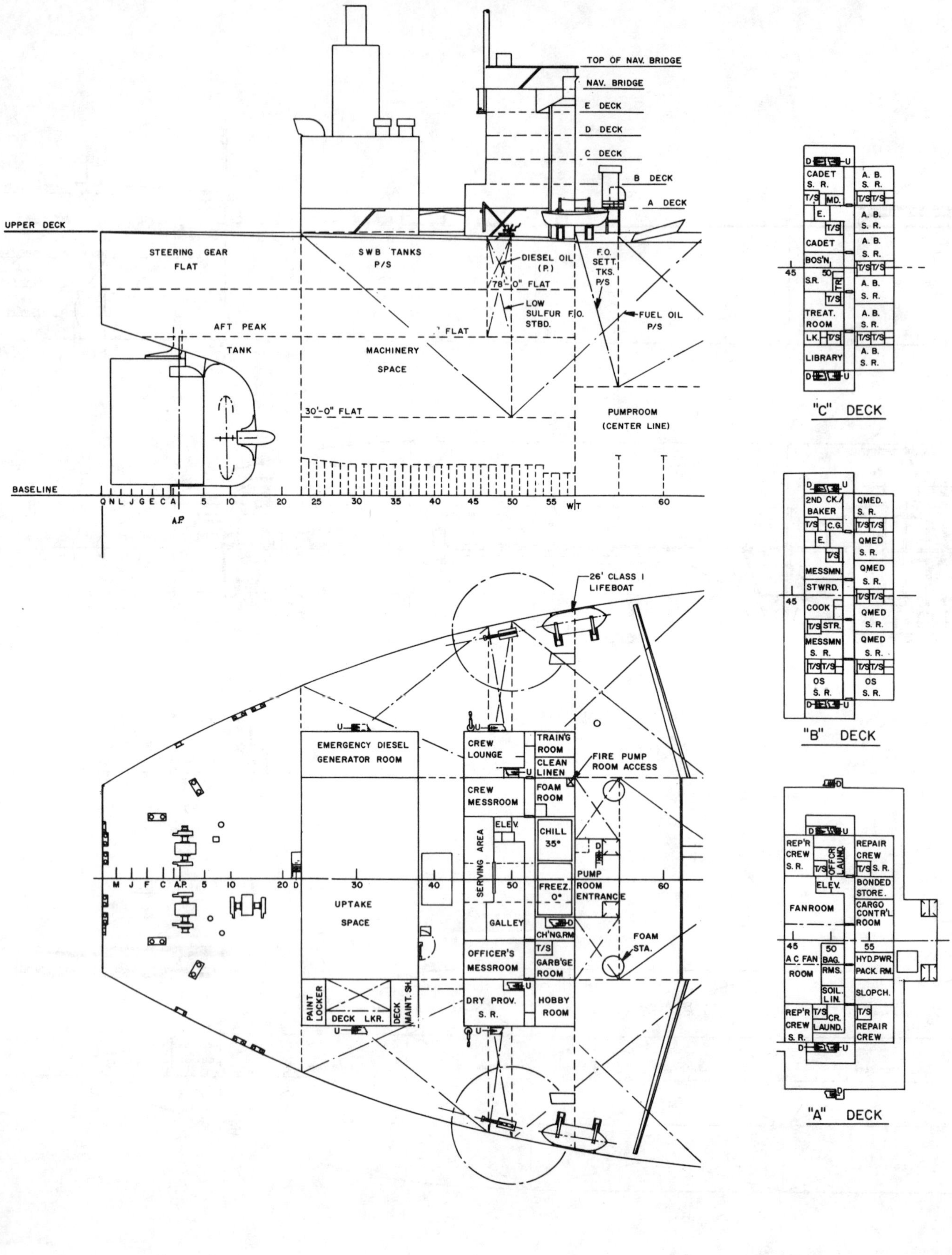

Tanker B, Crude Carrier Forebody Arrangements

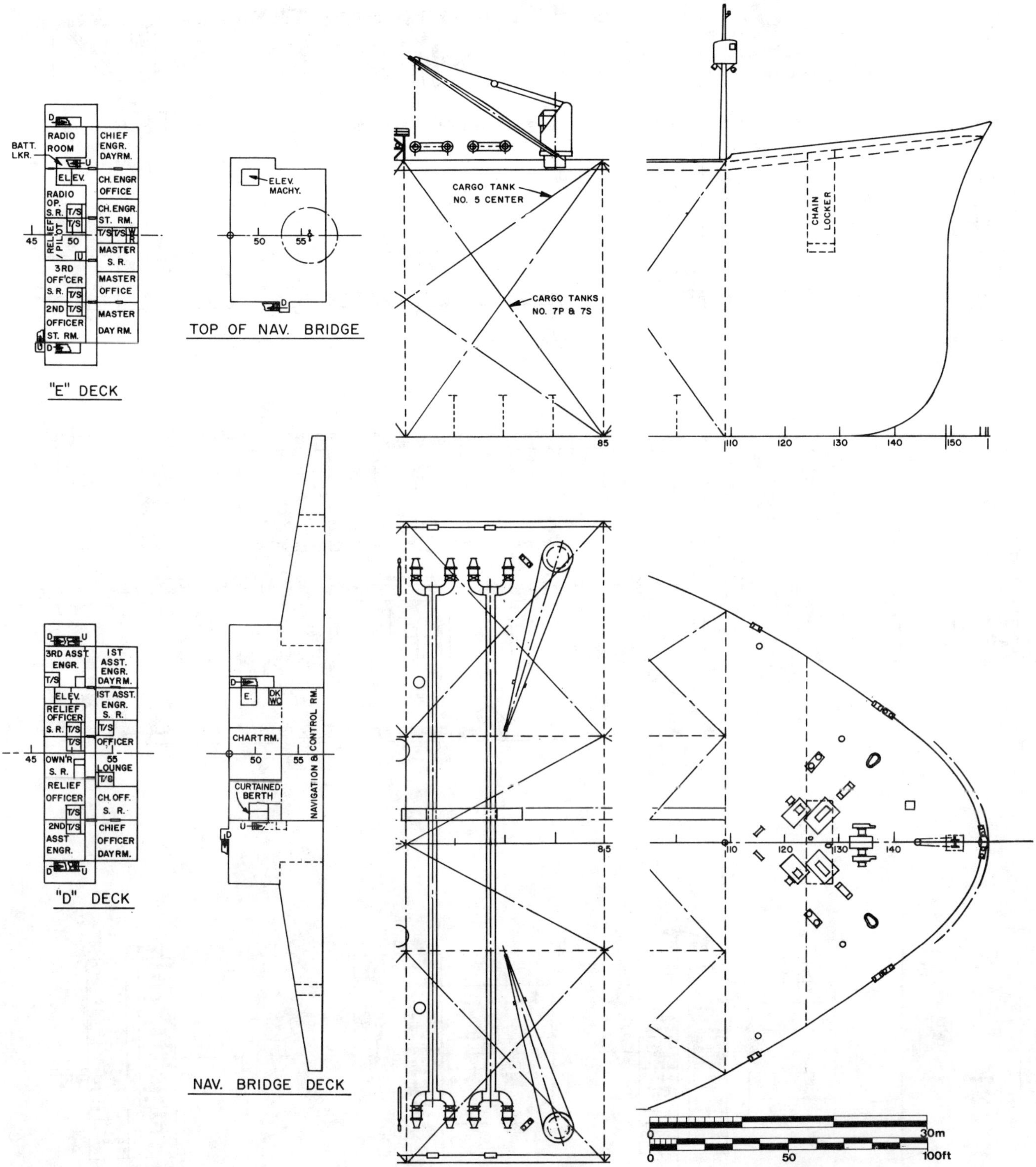

LNG Tanker Afterbody Arrangements

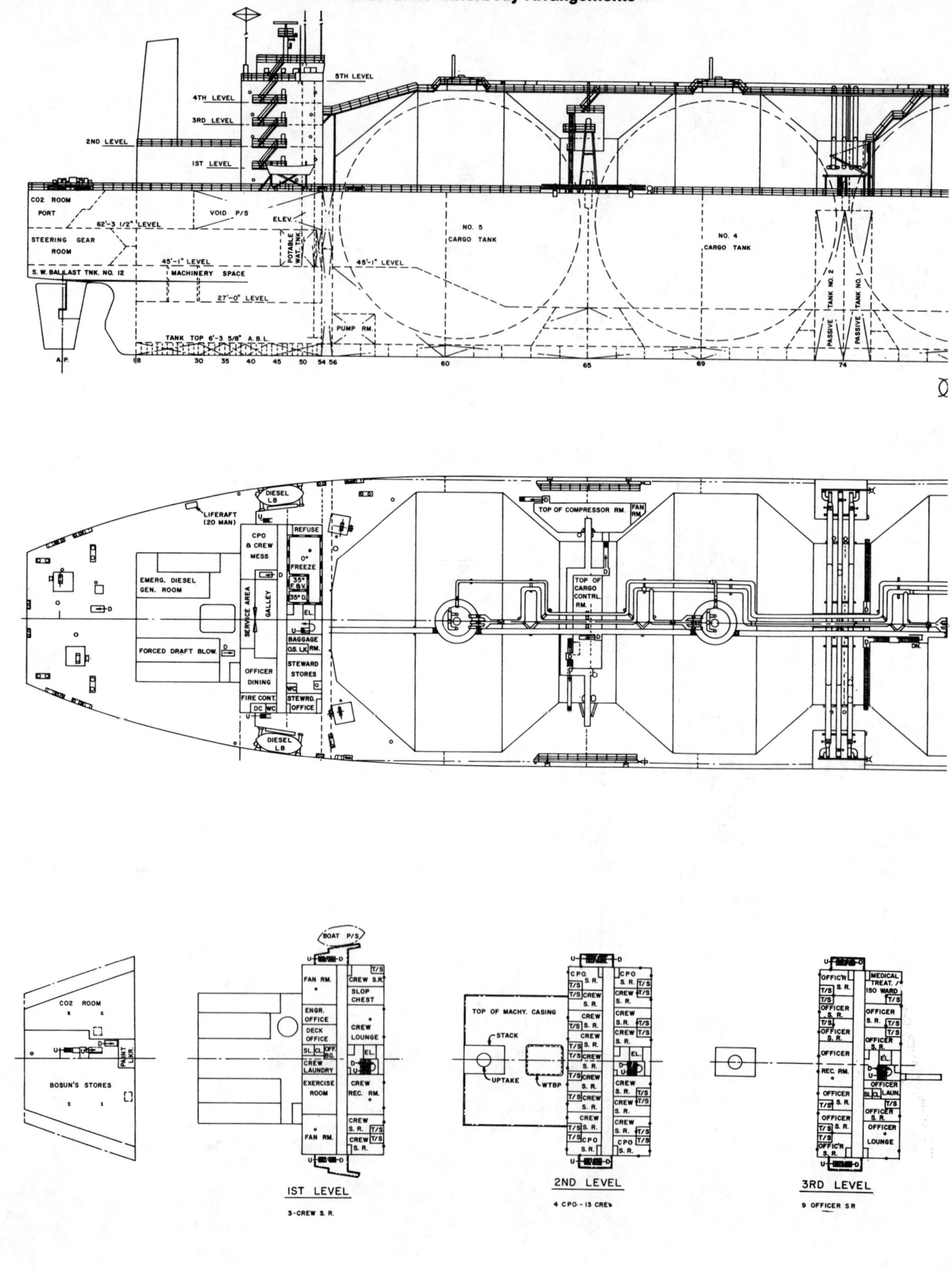

LNG Tanker Forebody Arrangements

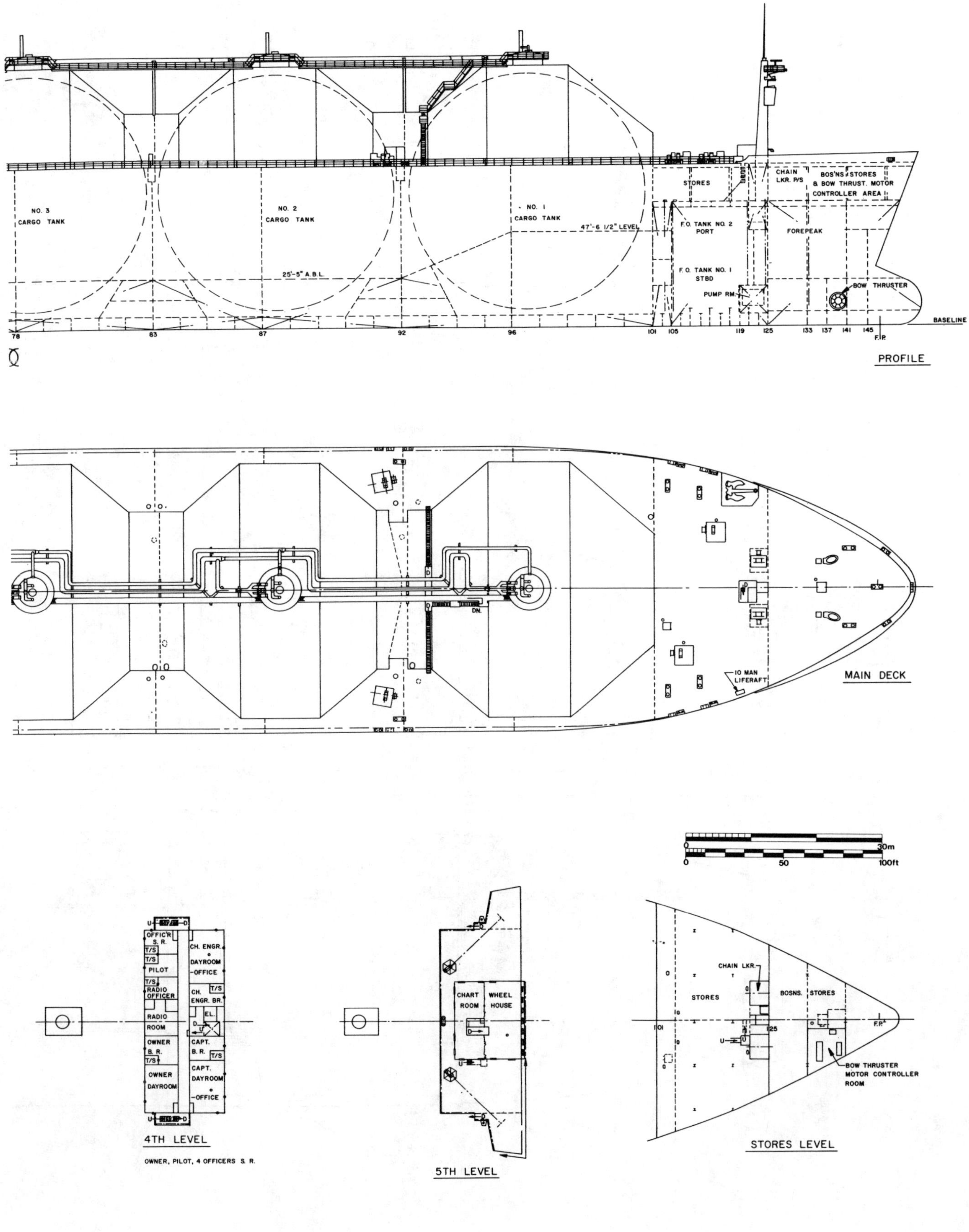

Ore/Bulk/Oil Ship Afterbody Arrangements

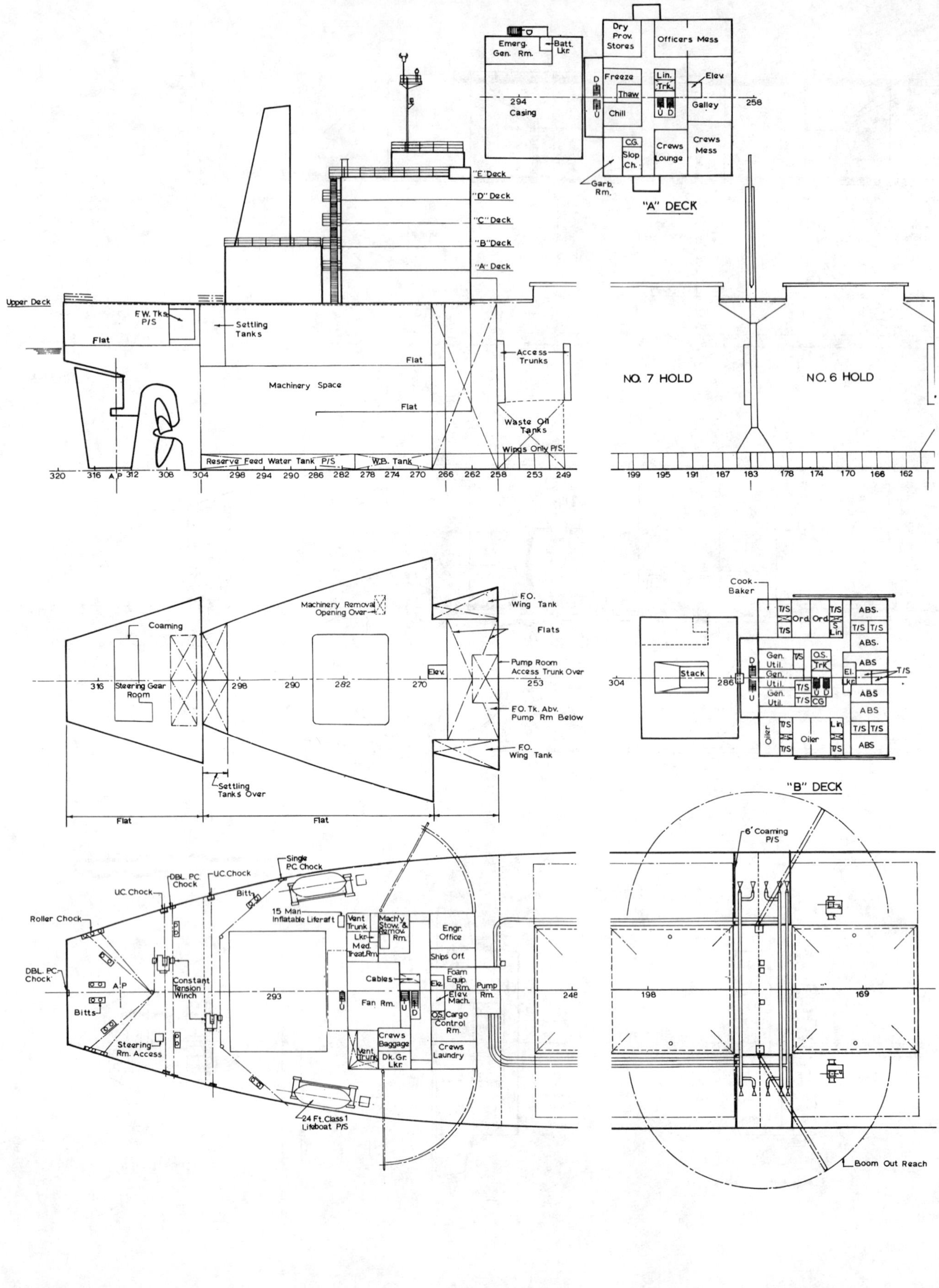

Ore/Bulk/Oil Ship Forebody Arrangements

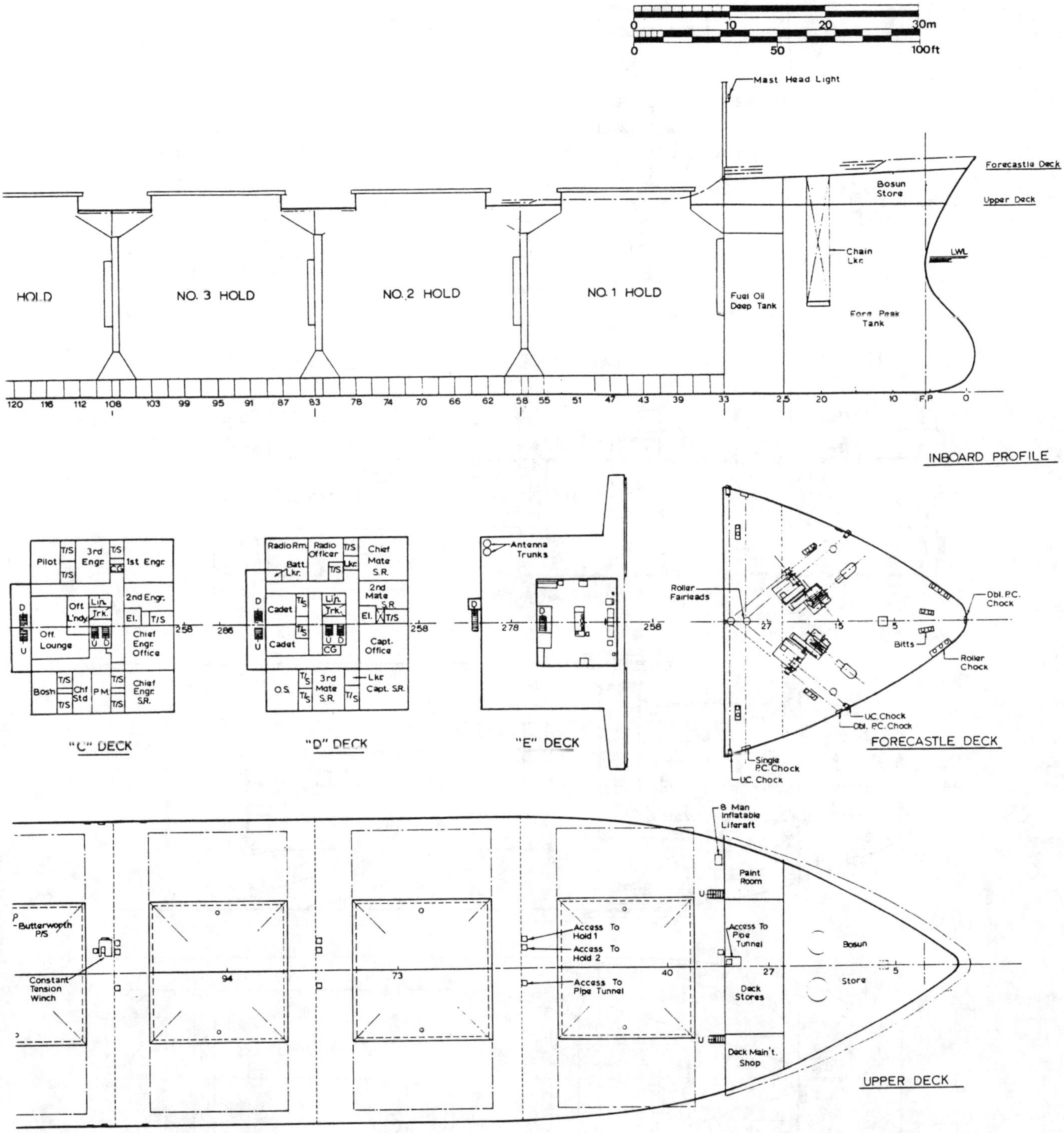

Bulk Carrier Afterbody Arrangements

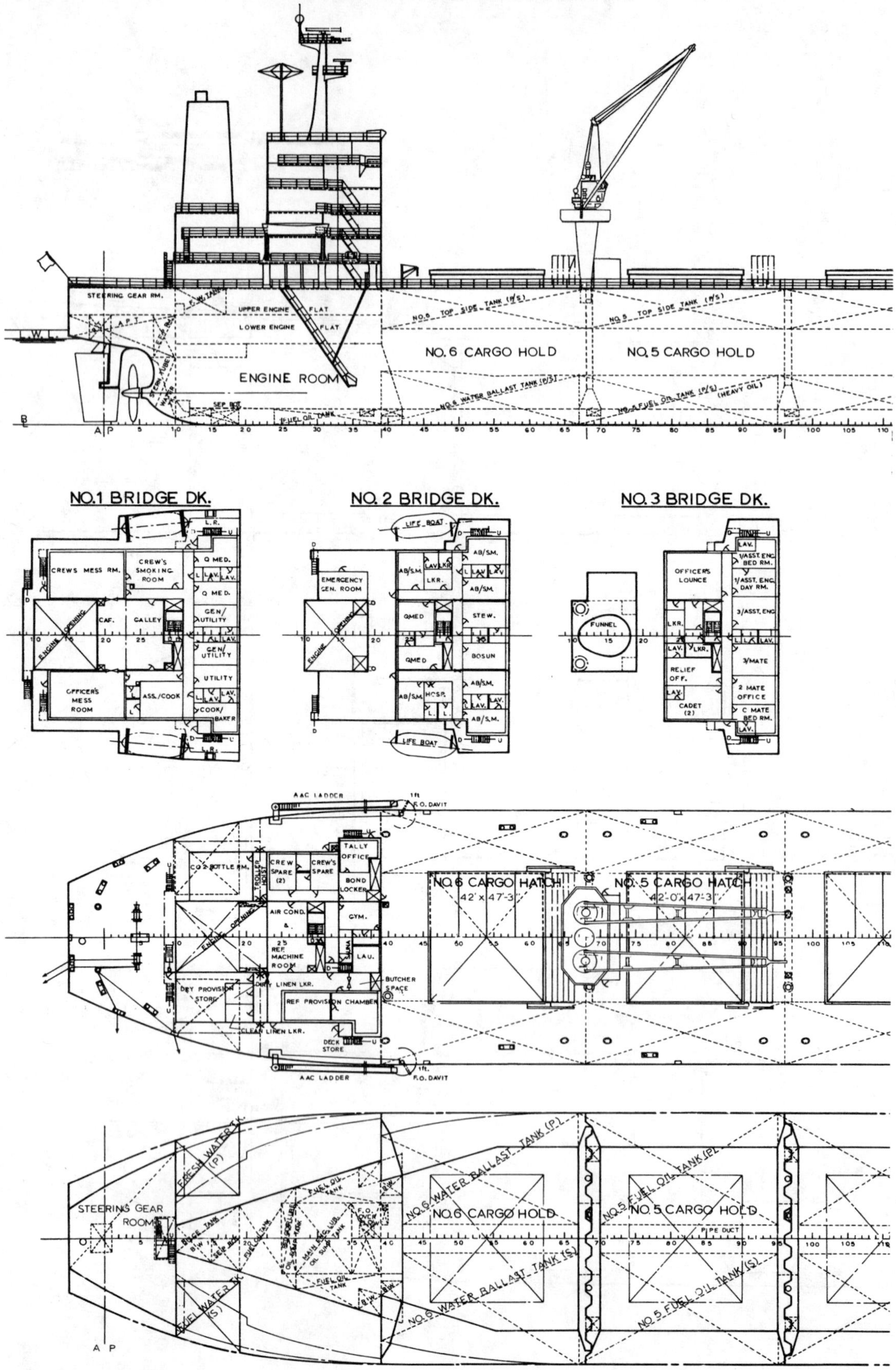

Bulk Carrier Forebody Arrangements

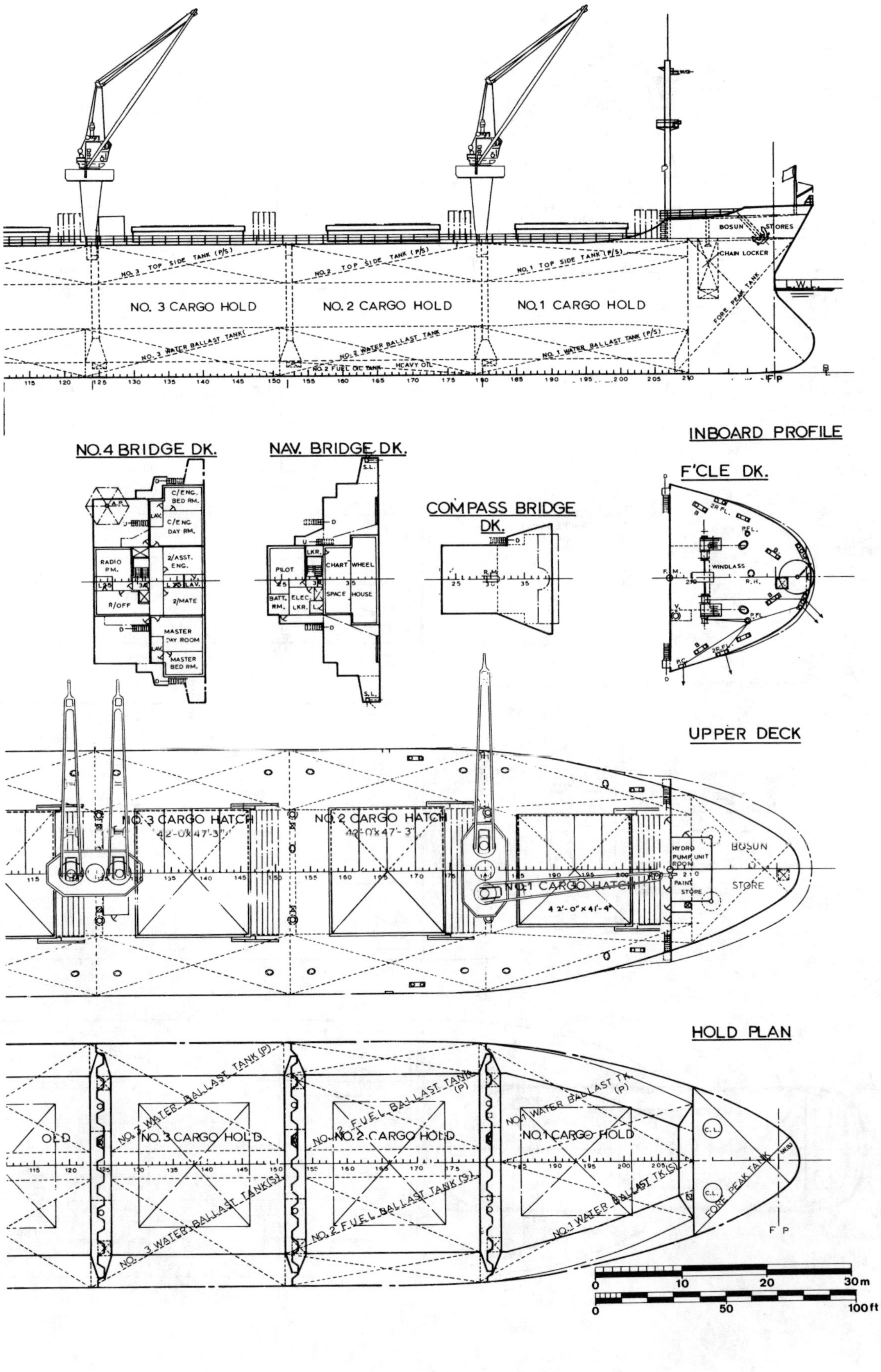

Integrated Tug-Barge; Afterbody Arrangements

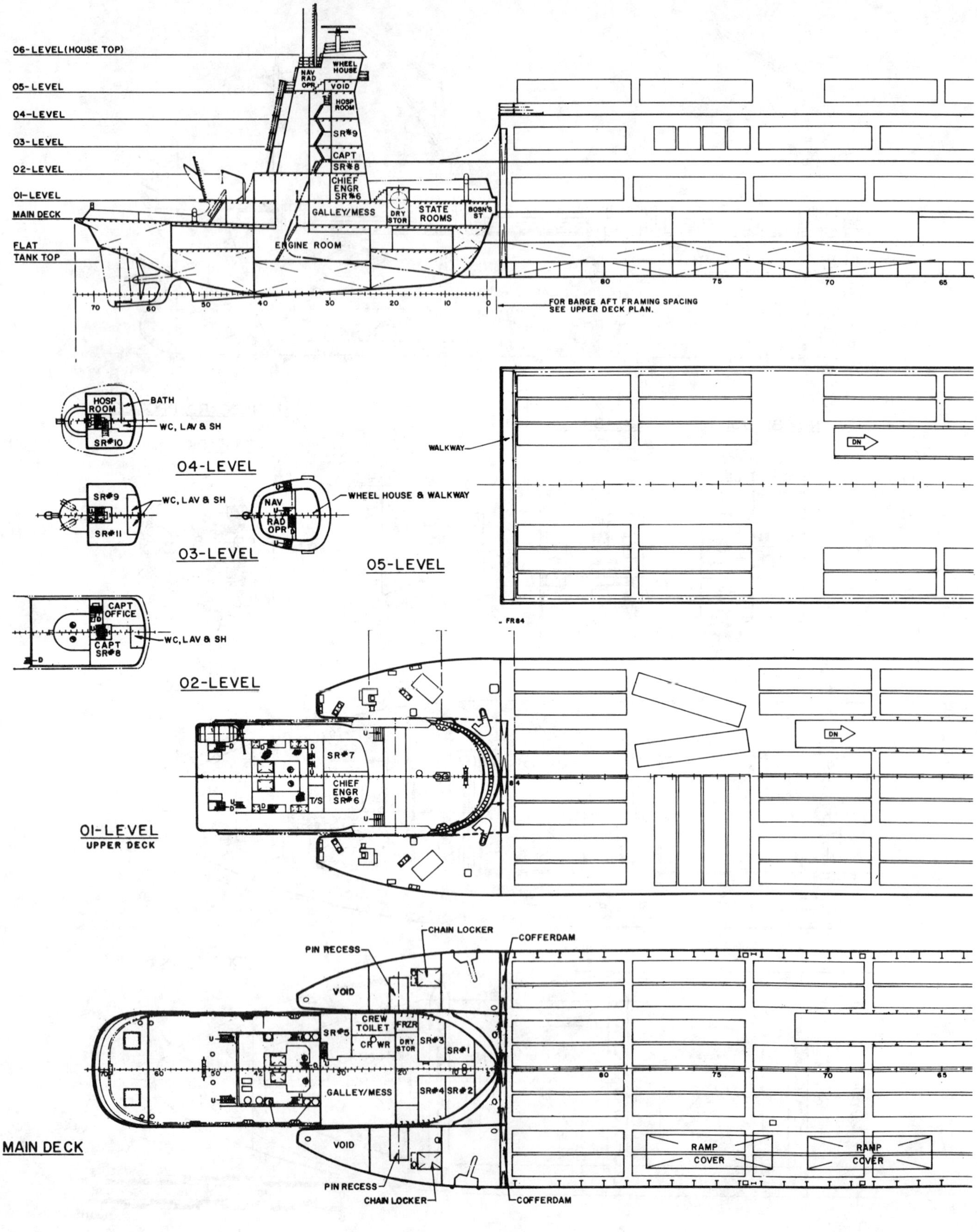

Integrated Tug-Barge, Barge Forebody Arrangements

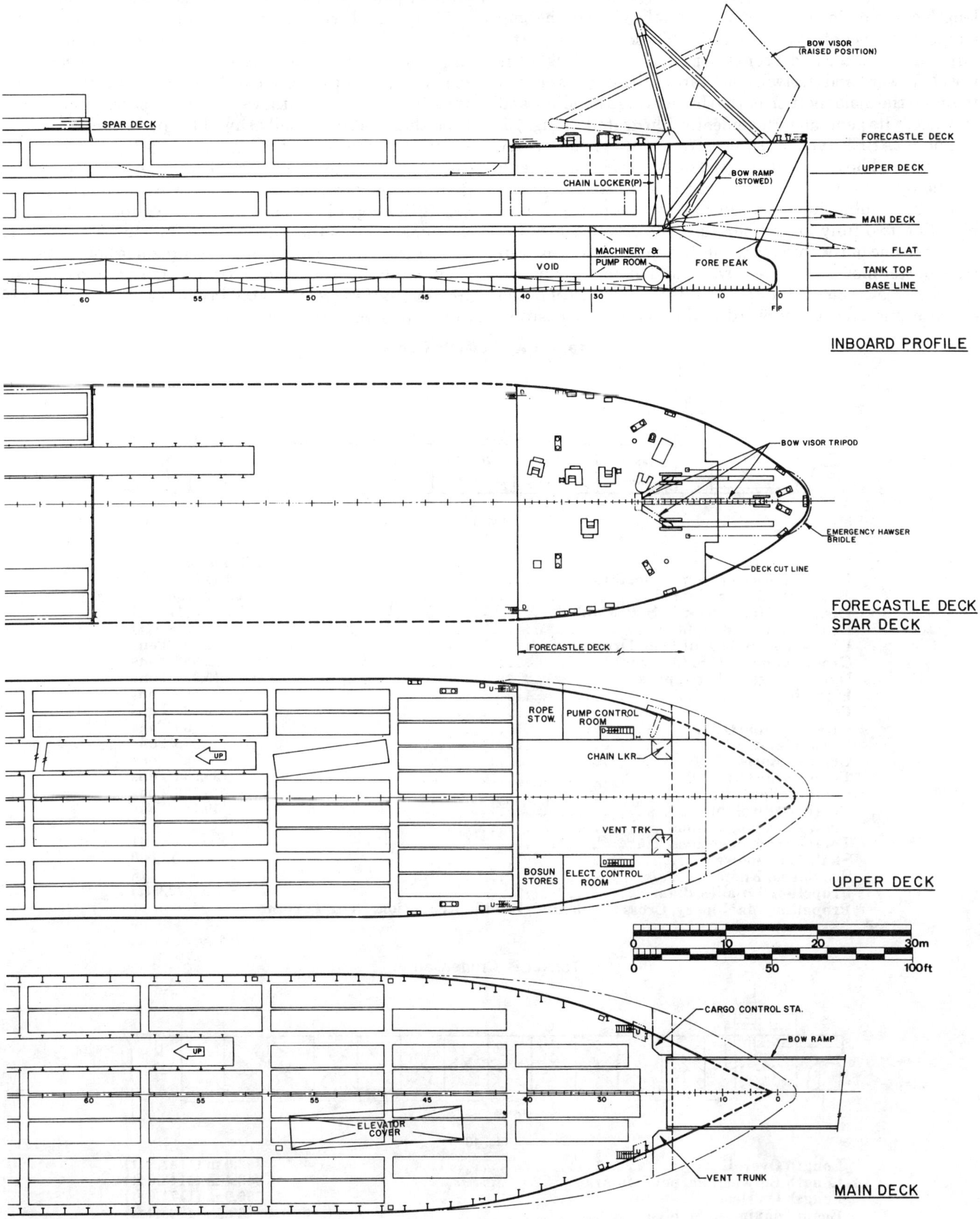

(*Continued from page 141*)

length of the main deck extend beyond the stern of the ship supported by cantilever structure. These rails support a large lighter crane and a container crane. The deckhouse is well forward and the twin stacks are at the side of the ship to retain the major portion of the ship for cargo handling and stowage. Its general arrangement is illustrated in Fig. 15.

The SEABEE type of barge carrier is designed to carry 38 barges. The barges are stowed on two lower decks, and on the open upper deck. The barges are lifted from the water by a submersible elevator at the stern, capable of handling two fully loaded barges. The elevator lifts the barges to one of the three cargo decks where a motor-driven transporter moves onto the elevator and under each barge. A series of jacks on the transporter lifts the barge clear of the elevator and moves it forward to the first empty position where the barge is lowered onto a series of pedestals.

The general arrangement and configuration of the SEABEE vessel, Fig. 16, is similar to the LASH in that the deckhouse and accommodations are forward to permit the major length of the ship to be kept clear for cargo, and each type incorporates cantilever stern structure. A SEABEE type ship is also specially suited for ports connecting with inland waterways, but the larger, more seaworthy barges are capable of traveling a greater distance in more open water than the smaller lighters of the LASH type vessel. The SEABEE barge dimensions were selected to coordinate with the dimensions of the standard U.S. inland waterway barge so that they can be integrated with regular barge tows.

Other features of these two types of barge carrying ships are described in greater detail in Chapter II. See pages 152 to 155 for arrangement details.

Tanker A, Products Carrier

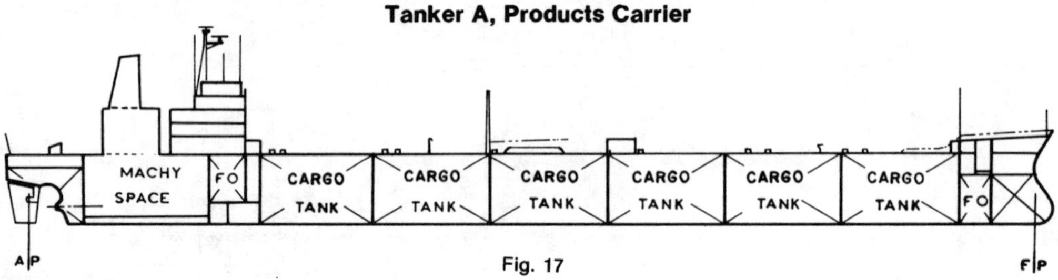

Fig. 17

Length Overall	**209.9 m (688.5 ft)**
Length Between Perpendiculars	**201.1 m (660.0 ft)**
Beam, Maximum Molded	**27.4 m (90.0 ft)**
Depth To Upperdeck At Side Molded	**14.3 m (47.0 ft)**
Draft, Full Load Molded	**10.7 m (35.0 ft)**
Displacement At Full Load Draft	**47,281 Tons**
Gross Tonnage, U.S. (Approx)	**22,358 Tons**
Net Tonnage, U.S. (Approx)	**15,951 Tons**
Lightship	**7,569 Tons**
Fuel Oil	**3,624 Tons**
Diesel & Lube Oil	**50 Tons**
Fresh Water	**275 Tons**
Crew & Stores	**50 Tons**
Total Cargo Oil	**39,934 Tons**
Clean Ballast	**13,500 Tons**
Total Deadweight	**40,760 Tons**
Crew Accommodations	**29**
Total Accommodations	**31**
Shaft Horsepower	**15,000**
Sea Speed, Knots	**16**
Propeller, 5 Blades, diam	**6.7 m (22.0 ft)**

Propelling Machinery, Cross Compound, Double Reduction, Gear Turbine

Tanker B, Crude Carrier

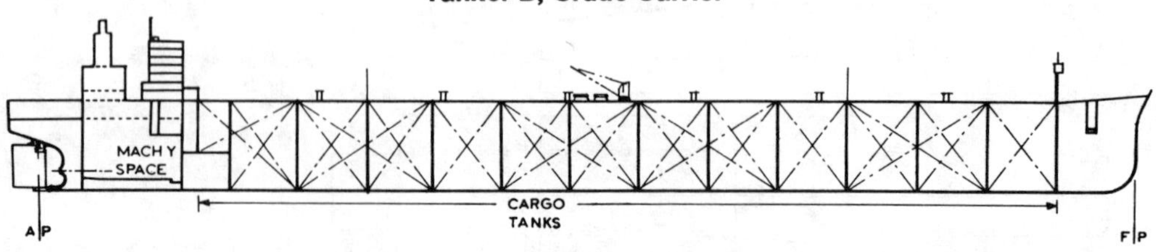

Fig. 18

Length Overall	**362.0 m (1187.5 ft)**
Length Between Perpendiculars	**348.4 m (1143.0 ft)**
Length On Design Waterline	**356.9 m (1171.0 ft)**
Beam, Maximum Molded	**69.5 m (228.0 ft)**
Depth To Upper Deck At Side Molded	**29.0 m (95.0 ft)**

Draft, Full Load Molded (Approx)	22.6 m (74.0 ft)
Displacement At Full Load Draft	450,910 Tons
Gross Tonnage U.S. (Approx)	—
Net Tonnage U.S. (Approx)	—
Lightship	60,140 Tons
Fuel Oil	17,857 Tons
Diesel & Lube Oil	315 Tons
Fresh Water	580 Tons
Crew & Stores	50 Tons
Total Cargo Oil	372,000 Tons
Clean Ballast	64,890 Tons
Total Deadweight	390,770 Tons
Crew Accommodations	27
Total Accommodations	38
Shaft Horsepower	45,000
Sea Speed, Knots	15.9
Propeller, 6 Blades, diam	9.6 m (31.5 ft)
Propeller Machinery, Cross Compound, Double Reduction Geared Turbine	

8.7 Tankers. The arrangement of two types of tankers are illustrated in Figs. 17 and 18. Fig. 17 is a product tanker, designated Tanker A, designed to carry a number of petroleum products, such as heating oil, diesel oil, and gasoline at the same time. The piping systems for loading and discharging must be designed to keep the cargos separated and to permit simultaneous handling. Clean ballast tanks are provided in accordance with regulatory requirements.

Tanker B, illustrated in Fig. 18 is a very large crude oil carrier (VLCC) which must off-load in deep water ports or at offshore terminals. The tanks are laid out to limited size, and clean ballast and slop tanks are provided in accordance with anti-pollution regulations. General arrangements are on pages 156 to 159.

Spherical Tank LNG Ship

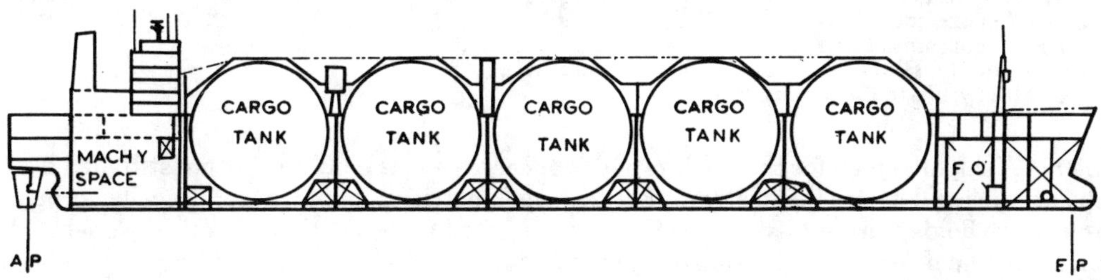

Fig. 19

Length Overall	285.3 m (936.0 ft)
Length Between Perpendiculars	273.4 m (897.0 ft)
Length On Designed Water Line	273.4 m (897.0 ft)
Beam Max. Molded	43.7 m (143.5 ft)
Depth Side Molded	25.0 m (82.0 ft)
Draft Full Load Molded (Approx)	11.0 m (36.0 ft)
Displacement At Full Load Draft	94,600 Tons
Lightship	31,000 Tons
Cargo Tank Volume At Minus 265°F	125,000 m^3 (4,414,333 ft^3)
Fuel Oil	44,000 Tons
Diesel & Lube Oil	80 Tons
Fresh Water	450 Tons
Crew & Stores, Lube Oil & Diesel Oil	250 Tons
Total Cargo	57,600 Tons
Clean Ballast	57,500 Tons
Total Deadweight	63,600 Tons
Crew Accommodations (Incl. Pilot)	34
Shaft Horsepower	43,000
Sea Speed, Knots	20.4
Propeller, Blades	—
Propelling Machinery, Cross Compound, Double Reduction Geared Turbine	

8.8 LNG Tanker. As discussed and illustrated in Chapter II, LNG tankers are constructed in accordance with a number of patented systems. Fig. 19 illustrates a tanker utilizing the Kvaerner-Moss spherical tank system, with the LNG carried in five identical tanks. The ship is designed to envelop, support, and protect the insulated aluminum tanks which are constructed and tested like pressure vessels. The main considerations in the arrangement design are structural, subdivision, and cargo handling system requirements. See page 160 for afterbody arrangement details and page 161 for forebody arrangement details. Principal characteristics are tabulated above.

Bulk Carrier

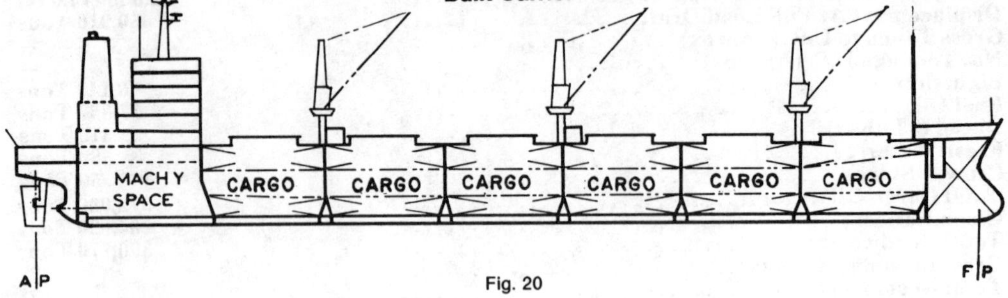

Fig. 20

Length Overall	186.5 m (611.8 ft)
Length Between Perpendiculars	178.0 m (584.0 ft)
Beam, Molded	28.4 m (93.2 ft)
Depth, Molded	15.3 m (50.2 ft)
Draft, Molded Designed	9.8 m (32.0 ft)
Draft, Molded Scantling	10.7 m (35.2 ft)
Sea Speed, Knots	16.9
Deadweight At Design Draft	32,100 Tons
Gross Tonnage U.S. (Approx)	23,500 Tons
Net Tonnage Panama Canal (Approx)	19,000 Tons
Net Tonnage Suez Canal (Approx)	21,000 Tons
Cargo Hold Capacity (Grain)	45,417 m^3 (1,603,880 ft^3)
Water Ballast Tank (Full)	19,763 m^3 (697,920 ft^3)
Fuel Oil Tank	2,010 m^3 (70,990 ft^3)
Fresh Water Tank	230 m^2 (8,120 ft^3)
Diesel Oil Tank	190 m^3 (6,710 ft^3)
SHP, ABS Max	15,288
Crew Accommodations	26
Total Accommodations	34
Propeller (1), Blades	5
Machinery, Twin diesel engines	

8.9 Bulk Carrier. A ship designed for the carriage of dry bulk cargos such as grains and coal is illustrated in Fig. 20. The ship has large, open holds, without 'tween decks, with the inner bottom sloped up at the shell and tanks fitted at the upper corners of the hold to provide for self-trimming and to prevent shifting of the cargo. The holds are designed with a minimum of structural obstructions to retain the bulk cargos e.g. the bulkheads are stiffened by their corrugated geometrical shape rather than by protruding stiffeners. Large hatches with mechanically operated covers are provided. Although the bulk cargos are normally loaded and discharged by shore-based gear, each hold is served by two 10 ton derrick booms which can be used for loading and discharging of bulks or other types of cargo (Pages 164-5).

Ore/Bulk/Oil Ship

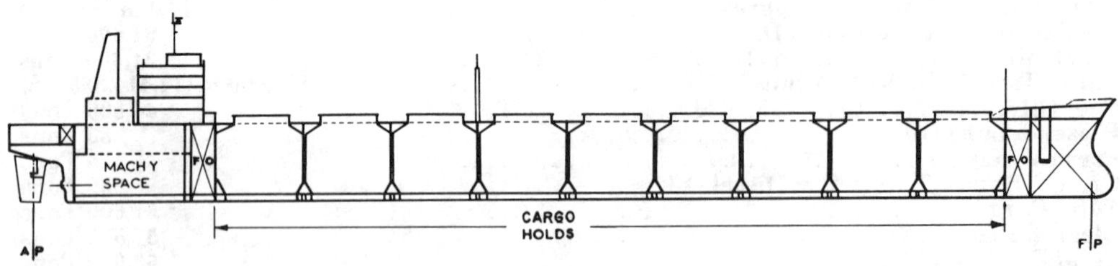

Fig. 21

Length Overall	272.0 m (892.5 ft)
Length Between Perpendiculars	260.6 m (855.0 ft)
Length On Designated Water Line	266.7 m (875.0 ft)
Beam, Max. Molded	32.2 m (105.8 ft)
Depth To Main Deck Aside Molded	19.0 m (62.5 ft)
Draft Full Load Molded	14.0 m (45.8 ft)
Displacement At Full Load Draft	99,210 Tons
Gross Tonnage, U.S. (Approx)	43,000 Tons
Net Tonnage, U.S. (Approx)	37,000 Tons
Lightship	18,710 Tons
Fuel Oil	4,845 Tons
Diesel & Lube Oil	50 Tons

Fresh Water	305 Tons
Crew & Stores	50 Tons
Cargo—Ore, Bulk or Oil	75,250 Tons
Clean Ballast	30,250 Tons
Total Cargo (Deadweight)	75,250 Tons
Total Deadweight (45′-11″ Mean, S.W.)	80,500 Tons
Total Deadweight (39′-0″ Mean, F.W.)	62,240 Tons
Crew Accommodations	27
Total Accommodations	31
Shaft Horsepower	24,000
Sea Speed In Knots	16.5
Propeller, 5 Blades	7.9 m (26.0 ft)
Propelling Machinery, Cross Compound, Double Reproduction, Geared Turbine	

8.10 Ore/Bulk/Oil Ship. Fig. 21 illustrates the arrangement of a ship designed to carry ore, oil, or bulk commonly known as an OBO ship. The flexibility for these three types of cargo allows the ship to be placed in more complex services than a single cargo carrier, where one of the cargo types is carried on each major leg of the voyage and the ship is travelling in ballast for a minimum percentage of the total voyage.

This ship combines the features of a bulk carrier (open self-trimming holds with large hatches) and is provided with the required features of a tanker including oil-tight structure and cargo piping and pumping system. (Pages 162-3).

ARTUBAR Tug-Barge

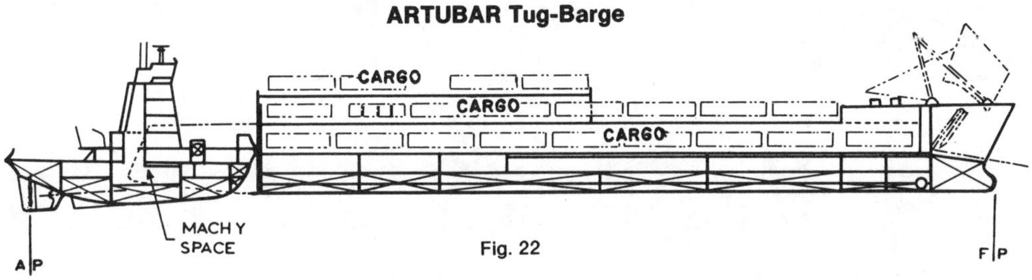

Fig. 22

	TUG	BARGE
Length Overall	45.7 m (150.0 ft)	173.1 m (568.0 ft)
Length Between Perpendiculars	41.0 m (134.5 ft)	— —
Length for Scantling	— —	141.1 m (463.0 ft)
Beam, Overall	12.6 m (41.2 ft)	— —
Beam, Molded	12.2 m (40.0 ft)	25.9 m (85.0 ft)
Depth, Molded	7.5 m (24.5 ft) above baseline	25.9 m (85.0 ft) — —
Depth, Molded to Upper Deck	—	12.6 m (41.5 ft)
Depth, Molded to Main Deck	—	7.2 m (23.5 ft)
Draft, Design	—	3.7 m (12.0 ft)
Draft, Scantling	7.2 m (23.6 ft)	— —
Draft, Max. of Assigned Freeboard (Scantling)	— —	4.1 m (13.4 ft)
Draft, Design, Molded Amidships	5.8 m (19.0 ft)	— —
Drag Aft, below Baseline	1.5 m (5.0 ft)	— —
Lightship	1,129 Tons	5,680 Tons
Gross Tonnage (Approx)	190 Tons	14,000 Tons
Net Tonnage (Approx)	135 Tons	13,250 Tons
Salt Water Ballast Tanks (Total Capacity)	—	5,051 Tons
Fresh Water Tonnage (Total Capacity)	12 Tons	1,936 Tons
Fuel Oil Capacity (Total Capacity)	430 Tons	—
Deadweight, Excluding Ballast	250 Tons	—
Cargo Deadweight and Water Ballast	—	6,450 Tons
Displacement at Scantling Draft	1,450 Tons	12,130 Tons
Permanent Ballast (Estimated Requirement for Stability)	71 Tons	—
Machinery	Twin Diesels	—
Crew Accommodations	14	—
Factor of Subdivision	1.0	—
Propeller (2) 4 Blades, One left-One right Hand		

8.11 Integrated Tug-Barge. There are a number of integrated tug-barge systems which are discussed and illustrated in Chapter II. The ARTUBAR system utilizes trunion mountings which extend from both sides of the tug hull and fit into sockets into the wing wall extensions at the sterm of the barge. The stern of the barge between the wing walls has a recess to fit the tug with clearance provided for the required relative motions. The ARTUBAR system does not permit fore and aft motion between the two hulls but does permit relative heave, pitch, and a certain amount of roll.

An example of an arrangement of the ARTUBAR tug-barge system, with the barge designed to carry trailers, is shown in Fig. 22. (See pages 166-7).

References

ANSI, American National Standards Institute (1971), "Basic Requirements for Cargo Containers," MH 5.1.

Dorman, W. J. (1966), "Combination Bulk Carriers," SNAME New York Metropolitan Section.

Harrington, R. L. (1971), Editor, *Marine Engineering,* SNAME.

Kari, A. (1948), "The Design of Merchant Ships & Cost Estimating," Crosby, Lockwood & Sons, London, England, 5th Edition.

Leeming, J. (1942). *Modern Ship Stowage,* U.S. Government Printing Office.

Murray, J. M. (1965), "Large Bulk Carriers," The Institution of Engineers and Shipbuilders in Scotland.

Roseman, Donald P., Alan K. Anderson, and James Lisnyk (1978), "Marine Transport of Bulk Commodities in Slurry Form—An Assessment," SNAME *Proceedings* of Third Ship Technology and Research Symposium.

Sharp, G. G. (1947), "Some Considerations in the Design of Modern Ships," SNAME.

SOLAS (1974), International Conference on Safety of Life at Sea 1974, Her Majesty's Stationary Office, London.

Weiler, D., and J. Castle (1972), "The Need for an Open Systems Approach to Naval Ship Habitability Design," *Trans.* SNAME, Volume 80.

Worthen, E. P., and W. H. Muller (1949), "Notes on the Design of Hotel Services for Passenger Ships," SNAME.

CHAPTER IV

William A. Cleary, Jr.
Angelo P. Ritola

Load Line Assignment

Section 1
General

1.1 Introduction. The *Load Line* is a formal term given to a mark located amidships on both sides of a ship to show the limiting draft to which the vessel may be loaded. This limiting draft is obtained by measuring from the uppermost continuous weathertight deck (normally the freeboard deck) down to the load line mark amidships. This distance is called the *Freeboard* of the ship.

The load line mark itself is required by the International Convention on Load Lines 1966 (ICLL, 1966) and, in the United States, by national law (Public Law 93-115, 1973). It is marked on the side of the ship in accordance with regulations issued by the government of the country whose flag the ship flies; in the United States, these regulations are issued by the U. S. Coast Guard and routine assignment is delegated to the American Bureau of Shipping. The ICLL, 1966 requires the Administration of each country accepting this convention to provide all facets of load line examination and control. Since it involves continuous knowledge of the ship throughout its life, countries with no inspection staff often delegate portions or possibly all load line activities to classification societies.

Since the load line regulations apply to almost all ships and embody a complete review of the general seaworthiness of the ship, it is important that the designer consider not just the desired freeboard but all facets of safety governed by load line regulations early in the preliminary design.

This chapter is concerned with the calculation, legal assignment, and marking of the minimum allowable freeboard in conjunction with an overall seaworthiness evaluation to ascertain that the vessel:

- is structurally adequate for its intended voyages;
- has adequate stability for its intended service;
- has a hull that is essentially watertight from keel to freeboard deck and weathertight above this deck;
- has a working platform (i.e., working deck for the crew) that is high enough from the water surface to allow safe movement on the exposed deck in heavy seas;
- has enough volume of ship, reserve buoyancy, above the waterline so that the vessel will not be in danger of *foundering* or *plunging* when in a very heavy seaway.

The above five basic rules have been the guiding principles for many decisions made during the past century by classification societies and national administrations relative to the proper minimum freeboard for ships of all kinds. The history of official load lines is well documented by the discussions and papers noted in Board of Trade (1906),[1] Norton (1942), and Ryan (1967) for the further guidance of those who wish to review the development of load lines.

The rules for determining the correct freeboard on any particular vessel are not scientifically exact. The freeboards determined under the international rules for vessels on international voyages evolved within the last 100 years. They are completely empirical and were initially based upon experience on vessels up to 91 m (300 ft) in length projected to 137 m (450 ft) in length. After the initial freeboards were set down as national legal requirements in Europe in 1890 they were amended from time to time and the tables extended gradually to lengths up to 229 m (750 ft) in 1915 when vessels still were generally well below that length.

Internationally, there have been only two load line conventions. The first was hosted by the United Kingdom in 1930 and the second was held in 1966 under the auspices of the United Nations' agency for marine safety, which is the Inter-Governmental Maritime Consultative Organization (IMCO) located in London.

The five basic rules of load line philosophy mentioned earlier are embodied in an agreed set of regulations published in the International Convention on Load Lines (ICLL, 1966) currently in force. As of December, 1979, ninety-three nations had ratified the Convention.

In the United States, load lines were originally established by Congress in 1929 for foreign voyages and in 1935 for coastwise and Great Lakes voyages (Public Law, 74-354, 1935). Currently, the International Convention on Load Lines, 1966 is implemented by Public Law 93-115 dated October 1, 1973 which is reproduced in the United States

[1] Complete references are listed at end of chapter.

Code 46 USC 86. The load line regulations implementing this law are detailed in the Code of Federal Regulations (46 CFR 42) which constitutes an almost verbatim copy of the technical Annex of the International Convention. All new ships 24 m (79 ft) or more in length which make an international voyage are required to be assigned load lines under this law. Warships, fishing vessels, existing ships of less than 150 gross tons and pleasure yachts are exempted.

1.2 Application. The principles of the Load Line Convention should be considered in the early design stages since they affect the choice of the major dimensions of the ship. Based on the geometric properties, a calculation is made and a freeboard determined. Upon completion of the vessel's construction, a Load Line Certificate is issued. The delivery of the Certificate to the ship means that the administration whose flag the vessel carries is fully satisfied that all the requirements of the Convention, i.e., hatches, superstructures, and doors topside are weathertight and in good condition; that stability information, fully adequate for all operating conditions, is available on board and that the basic structural scantlings used to construct the hull are adequate for full seaway conditions.

1.3 Certificates. There are two types of Load Line Certificates currently in use: the basic Certificate confirming that the vessel meets the requirements of the applicable regulations and the *Exemption Certificate* which exempts certain portions of the Convention.

In the U. S., the basic Load Line Certificate is issued in several forms, such as international voyage, coastwise voyage and Great Lakes.

The International Convention Exemption Certificate is specifically authorized for only three conditions:

- a single voyage from port to port, upon completion of which it automatically becomes invalid;
- research developments in ship design which would otherwise be prohibited by the load line regulations;
- voyages between adjacent countries which have completed a protected waters agreement or treaty.

1.4 Surveys. *Periodical surveys* are conducted every five years. A new centificate must be issued upon satisfactory completion of a detailed survey of the vessel which includes drydocking to ensure inspection of all the hull.

A complete periodical load line survey involves:

- physical survey of hull in drydock for structural maintenance and material adequacy;
- physical survey of topside openings for weathertightness;
- ensuring information is provided to enable proper loading, ballasting and assessing stability under varying conditions of loading and service.

Between periodical surveys, validation inspections, including endorsement on the back of the Load Line Certificate, are required annually. If this inspection is omitted, the Load Line Certificate itself becomes invalid three months after the yearly anniversary date. These inspections involve a review of the hull, fittings on all topside openings, reassurance that no new openings have been cut into the hull, or that any other changes have been made which would require a reevaluation of the freeboard, and finally that the load line mark itself is still properly positioned in accordance with the certificate.

1.5 Standard Ship. In order to assign a load line properly, it is necessary to compare the design to a geometric ship of Standard form. The concept of the Standard Ship with definite geometric proportions was evolved early in the discussions for standard freeboard.

a. Board of Trade Standard Ship. From a historical standpoint, the 1906 Board of Trade Rules (Board of Trade, 1906) in England used a Required Reserve Buoyancy to establish desired *winter freeboard* for both steamers and sailing vessels. This Reserve Buoyancy referred only to the intact weathertight ship and was deemed necessary for safe seakeeping. The freeboard to be assigned was such that the percentage of the total volume of the hull above the load line was equal to that required in the table. The required buoyancy was least for the shortest vessel, 20.4 percent at 22 m (72 ft), and increased to 35.8 percent becoming maximum at a length of 183m (600 ft). The required extra buoyancy for sailing vessels was 1 percent to 2 percent higher than for steamers. However, in lieu of making a complete volumetric calculation up to the freeboard deck, the designer was permitted to use certain tables of winter freeboard provided by the Board of Trade based upon a standard length to depth ratio (L/D) of 12.

Freeboard reductions of a very small order were allowed for summer weather. On the other hand, an arbitrary addition of 50 mm (2 in.) in winter time for the Mid North Atlantic area was required.

In addition to the regular reserve buoyancy due to the basic freeboard amidships, the regulations also prescribed a standard sheer curve adding buoyancy at the bow and stern. This buoyancy was considered effective in promoting the seakeeping properties of ships in heavy weather.

The freeboard for a given length and depth also varied slightly according to the "coefficient of fineness" which was defined (Board of Trade, 1906) as the ratio of all under freeboard deck volume to the product of L × B × D.

b. ICLL 1930 Standard Ship. The Standard Ship of the 1930 Convention had:

- an L/D of 15;
- a fineness coefficient of 0.68;
- a table of freeboards increasing with length of ship;
- a standard sheer;
- a standard camber of the main deck;
- a minimum percentage length of superstructure;
- a required forecastle for tankers.

In the International Convention on Load Lines 1930, the *coefficient of fineness* was specially defined only in English units as follows:

$$C = \frac{35 \times \text{Displacement}}{L \times B \times d_1} \tag{1}$$

where d_1 was the mean molded draft at 85 percent of the

molded depth. In the ICLL, 1966 the title *Coefficient of Fineness* was dropped and the correction is now called the *Block Coefficient* correction.

c. ICLL 1966 Standard Ship. The Standard Ship of the 1966 Convention (ICLL, 1966) is similar to the 1930 standard ship except for the camber requirement which was dropped and the forecastle requirement which was removed in favor of a minimum bow height for all manned vessels. The fineness coefficient was redefined as the block coefficient, as previously mentioned.

Some types of ships less than 100 m (328 ft) in length are expected to have a weathertight superstructure on at least 35 percent of their length which will add buoyancy and form a righting moment to resist extreme rolling. These ship types with superstructures covering less than 35 percent of the length must accept added freeboard.

Section 2
Considerations Affecting Freeboard

2.1 General. The minimum freeboard that can be assigned to a vessel is that derived from the regulations depending upon the dimensions and characteristics of the vessel.

The actual maximum *operating draft* permitted may coincide with minimum freeboard or it may be set by other federal regulations developed as a result of U. S. law or by international agreements such as the subdivision regulations in the 1973 Oil Pollution Convention, the IMCO Chemical Code or Gas Code.

Within the load line regulations the minimum freeboard may be affected by ship geometry, hull structure or stability. In no case can a freeboard less than the minimum geometric freeboard be assigned even though the scantlings of the vessel are heavier than required for the draft and the stability in excess of that required by an Administration for the intended freeboard.

2.2 Freeboard Tables. The tables issued for freeboard are based upon a comparison with a rule vessel having a standard sheer, length-to-depth ratio, block coefficient and reserve buoyancy. Adjustments are made for variations from the standard ship and there are different tables for different types of vessels.

2.3 Strength of Hull. The regulations assume that the strength of the vessel is satisfactory for the draft corresponding to the freeboard assigned. Ships which comply with the highest standards of a classification society recognized by the Administration (in the United States the American Bureau of Shipping and the U. S. Coast Guard respectively) are regarded as having sufficient strength for the minimum freeboards allowed under the regulations. Ships which do not comply with the highest standards of a classification society are to be assigned such increased freeboards as are determined by the assigning authority. The corresponding draft in such cases is often referred to as a *scantling draft.*

2.4 Protection of Crew. It should be noted that while the freeboard assigned is based primarily upon reserve buoyancy, the question of a suitable height of platform for the safe working of the vessel by the crew is automatically dealt with at the same time. Protection for the crew, in the strength of houses, gangways, guard rails, life lines, and the height of working platform itself, is a very important concern of the load line regulations and specific regulations are provided for each.

2.5 Stability. The original International Convention on Load Lines, 1930, presumed that specific stability approval was not a concern of the regulations. At that time it was assumed that those responsible had seen to it that the "nature and stowage of the cargo, ballast, and so on, are such as to secure sufficient stability for the ship."

The present Convention (ICLL, 1966) has reversed the position of the earlier Convention by including a specific regulation worded such that stability information must be provided the master of every new vessel, "in an approved form to give him guidance as the stability of the vessel under varying conditions of service." This requirement has been interpreted quite firmly by the U. S. Coast Guard to include an *inclining test* for almost all U. S. commercial ships, a full stability evaluation based on the inclining, and an official stability letter issued by them as a condition necessary to issuance of the official load line certificate. Many other administrations follow a similar procedure.

2.6 Passenger Ship Subdivision. A vessel engaging in international voyages and carrying more than twelve passengers is governed by a separate regulation. Internationally, a load line is assigned and marked depending upon a subdivision and damage stability analysis of the ship under the applicable regulations of the International Conference of Safety of Life at Sea, 1974 (SOLAS, 1974). Under U. S. regulations for certain ships, depending upon size or other limitations, a subdivision and damage stability examination is required if six or more passengers are carried. In no case may this subdivision load line be placed higher on a ship's side than the load line permitted under the load line regulations.

2.7 Geometry of Vessel (Reserve Buoyancy). The minimum freeboard is designed to provide a standard of *reserve buoyancy* (the volume of the watertight hull above the load waterline) that has been found by experience to be satisfactory in service. This minimum freeboard is based upon the geometry of the vessel. A comparison is made of the block coefficient, the length-to-depth ratio, bow height, and the sheer of the vessel with those of a standard vessel of the same length. Corrections are made to the basic freeboard, predicated on the length of the vessel, depending upon how these particulars vary from those of the standard vessel. Deductions are made from the freeboard depending upon the length of superstructures and the character of the closures in their end bulkheads. The resulting freeboard gives a height of working platform and a proportion of reserve buoyancy equivalent to that on vessels which have proven satisfactory in service.

a. Camber and Sheer. Both camber and sheer play a part in clearing water rapidly from the decks, but only sheer corrections to freeboard are made depending upon the differences in sheer from the standard. Since there is no standard for camber in the 1966 Convention, no adjustment need be made.

b. Superstructures. Superstructures can contribute to reserve buoyancy and offer protection to openings in the hull at the level of the freeboard deck under certain conditions. Deductions are made from the freeboard for these special superstructures depending upon the efficiency of the protection provided for access openings in the end bulkheads. Detached superstructures are also a consideration because there are differences in the deductions for superstructures depending upon their length and location. While the regulations do not require that a forecastle be fitted, a minimum height of the bow above the summer load water line is specified. In lieu of a forecastle the required bow height can be obtained by increasing the sheer curve of the main deck.

2.8 Openings in the Hull and Superstructure. A most important consideration in the assignment of freeboard is the protection of openings in the hull and superstructures, such as hatches, ventilators, air pipes, scuppers, overboard discharges, and the access openings in the end bulkheads of superstructures. Standards are laid down in the regulations for these, and the assigning authority must be satisfied with their efficiency before a minimum freeboard is assigned. The safety of the vessel depends far more upon their satisfactory maintenance than in any small differences in the freeboard assigned.

Section 3
Load Line Calculation

3.1 General. The formulas, definitions and tables are taken directly from the current Convention and are presented for the information of the reader.

The minimum geometric freeboard for any vessel is obtained by comparing the geometric particulars of the vessel under consideration with those of a standard vessel of the same length.

In this section no attempt is made to cover all the complications that arise in the application of the regulations. Only the basic concepts and the general approach are considered and many minor points in the regulations are not discussed at all. In any application to a particular vessel, the regulations (46 CFR 42) themselves must be referred to, but it is hoped that the presentation here will assist the reader to find his way through the regulations. Sufficient information is given to make possible a fair approximation where the particulars of a specific vessel are available.

3.2 Types of Ships. For the purposes of freeboard calculation, ships are divided into two types:

a. Type A Ships. A type A ship is one which is designed to carry only liquid cargoes in bulk, and in which cargo tanks have only small access openings closed by watertight gasketed covers of steel or equivalent material. Such a ship necessarily has the following inherent features:

- high integrity of the exposed deck;
- high degree of safety against flooding, resulting from the low permeability of loaded cargo spaces and the degree of subdivision usually provided;
- if over 150 m (492 ft) in length and designed to have empty compartments when loaded to its summer load waterline, shall be able to withstand the flooding of any one of these empty compartments at an assumed permeability of 0.95 and remain afloat in a satisfactory condition of equilibrium (Robertson, 1967);
- if over 225 m (738 ft) in length, the machinery space shall be treated as a floodable compartment with an assumed permeability of 0.85.

All ships which do not come within the provisions regarding type A ships are to be considered as type B ships.

A detailed discussion of the derivation of freeboard for a type A ship is given in Section 3.19 following that for a type B ship.

b. Type B Ships. A list of basic minimum tabular freeboards which increase with the length is given in the regulations (see Table 1). From this basic minimum freeboard, the final freeboard is obtained by applying corrections, which are explained later, depending upon the departures from the standard vessel. It should be noted that ships having hatchways closed by portable covers and secured weathertight by tarpaulins and battening devices located on the exposed freeboard deck or exposed superstructure deck situated forward of a point located one quarter of the length from the forward perpendicular will be assigned freeboards in excess of those found in Table 1. For the increment of increase see Table 2. From this basic minimum freeboard the summer freeboard is obtained by applying corrections depending upon the departures from the standard vessel as noted in Section 1.5 c. Deductions from the tabular freeboard are made for superstructures, taken as a percentage of a scale of allowances compared to a vessel 100 percent covered by superstructures on the freeboard deck. If a ship is less than 100 m (328 ft) in length and has superstructures which cover less than 35 percent of

(Continued on page 180)

Form L.L. 6-C Rev. 1-77

American Bureau of Shipping

1966 LOAD LINE CALCULATION

TYPE "______" FREEBOARD

DONE BY:__________ DATE:__________

CHECKED BY:__________ DATE:__________

NAME OF VESSEL: BUILDER: HULL NUMBER: YEAR BUILT:	OWNER: FLAG:	CLASSIFICATION: TYPE OF VESSEL:	PORT OF SURVEY: DATE OF SURVEY: SURVEYOR'S NAME:

L = ____ B = ____ D = ____ D_f = ____	FREEBOARD DEPTH: MOLDED D = STRINGER = ____ $\frac{T(L-S)}{L}$ = D_f =	BLOCK COEFFICIENT AT .85D: Metric Tons Use .976 × Long Tons Use 35. ____ = × ×	COEFFICIENT CORRECTION (REG 30) TAB. FBD. C_B = + .68 ____ × ____ = 1.36 =

SUPERSTRUCTURE CORRECTION: (REG 33-37)

	LENGTH(S)	LEAST HEIGHT	STD. HT.	EFFECTIVE LENGTH (E)
FOCSLE	=	=	=	=
BRIDGE	=	=	=	=
POOP	=	=	=	=
TRUNK	=	=	=	=
RAISED QTR. DK.	=	=	=	=

100 PERCENT ALLOWANCE (REG 37) =

PERCENTAGE ALLOWED = $\frac{\Sigma E}{L}$ = ____ =

SUPERSTRUCTURE DEDUCTION = 100 PER CENT ALLOWANCE × PERCENTAGE ALLOWED = ____

TABULAR FBD = ____

CORRECTED FOR REG 29 = ____

CORRECTED FOR C_B = ____ | TIMBER

DEPTH CORRECTION ± = ____

SUPERSTRUCTURE (−) = ____ ____

SHEER CORRECTION (±) = ____ ____

FREEBOARD = ____ ____

TIMBER SUPERSTRUCTURE CORRECTION: (REG 45)

PERCENTAGE ALLOWED =

SUPERSTRUCTURE DEDUCTION = 100 PER CENT ALLOWANCE × PERCENTAGE ALLOWED = ____

BOW HT: (REG 39)

RULE = RULE =

OFFERED = OFFERED = ____

PENALTY = ____

DEPTH CORRECTION: (REG 31)

D_f =

L = ____ = ____

15 × (R =)

=

REGULATION 29: (TYPE "B" SHIPS)

7.5 (100 − L =) OR .09 (328 − L =) × $\left(0.35 - \frac{E}{L} = ____\right)$ =

OTHER CORRECTIONS:

25% REDUCTION ____ ____

BOW HT. PENALTY ____ ____

STABILITY PENALTY ____ ____

SCANTLING PENALTY ____ ____

OTHER ____ ____

FREEBOARD ____ ____

SHEER CORRECTION: (REG 38)

SHEER	STANDARD * See Note below	No.	ORDINATE	MUL.	F(A)	Excess Ht REG 38 (12)	SUM	RULE REG 38 (10) + (11)
AFT	METRIC: $\frac{66.7}{8}\left(\frac{L = ____}{3} + 10\right)$ = ____	A.P.		1		POOP Y = L′ =	Σ F(A)/8 + S	
		1/6		3		$S = \frac{Y}{3} \times \frac{L'}{L/2}$		
	ENGLISH: $\frac{.2665 \times L = ____}{8} + \frac{26.65}{8}$ = ____	1/3		3		S =		
			Σ =	/8 =			=	
FWD.	METRIC: $\frac{133.4}{8}\left(\frac{L = ____}{3} + 10\right)$ = ____	1/3		3		FOCSLE Y = L′ =	Σ F(A)/8 + S	
		1/6		3		$S = \frac{Y}{3} \times \frac{L'}{L/2}$		
	ENGLISH: $\frac{.533 \times L = ____}{8} + \frac{53.3}{8}$ = ____	F.P.		1		S =		
			Σ =	/8 =			=	

Σ STD AFT + STD FWD =

= Σ RULE = ____

Σ RULE =

$\frac{____}{2}$ = $\left(.75 - \frac{S}{2L} = ____\right)$ = ____

* NOTE: Metric and English standard sheer formulas given represent a combination of standard ordinates and appropriate factors.

MOLDED DEPTH =

STRINGER PLATE = (+) ____ ____

(OR WOOD DECK)

FREEBOARD = (−) ____ ____

MOLDED DRAFT =

KEEL THICKNESS = (+) ____ ____

EXTREME DRAFT =

ALLOWANCES: (INCH OR M/M) (REG 40)

TROPICAL & WINTER = $\frac{____}{48}$ =

WINTER = $\frac{____}{36}$ =

TROPICAL = $\frac{____}{48}$ =

WINTER NORTH ATLANTIC =

WNA =

FRESH WATER ALLOWANCE: Δ = T = $\frac{\Delta =}{40 \times T =}$ =

FRESH WATER ALLOWANCE: Δ = T = $\frac{\Delta =}{40 \times T =}$ =

FREEBOARD RECOMMENDED AMIDSHIPS FROM CENTER OF RINGS TO TOP OF ____________ DECK: ____________

TROPICAL FRESH WATER LINE ________ ABOVE CENTER OF RING. CORRESPONDING FREEBOARD: ____________

FRESH WATER LINE ________ " " " " " " : ____________

TROPICAL LINE ________ " " " " " " : ____________

WINTER LINE ________ BELOW " " " " " : ____________

WINTER NORTH ATLANTIC LINE ________ " " " " " " : ____________

SUMMER TIMBER FREEBOARD AMIDSHIPS BELOW TOP OF ____________ DECK: ____________

TROPICAL FRESH TIMBER WATER LINE ________ ABOVE TIMBER SUMMER MARK. CORRESPONDING FREEBOARD: ____________

FRESH " ________ " " " " " " : ____________

TROPICAL " ________ " " " " " " : ____________

WINTER " " " ________ BELOW " " " " " : ____________

WINTER NORTH ATLANTIC TIMBER WATER LINE " " " " " " : ____________

Fig. 1 American Bureau of Shipping load line calculation form

Table 1—Freeboard Table for Type B Ships

Length of ship (meters)	Freeboard (millimeters)
24	200
25	208
26	217
27	225
28	233
29	242
30	250
31	258
32	267
33	275
34	283
35	292
36	300
37	308
38	316
39	325
40	334
41	344
42	354
43	364
44	374
45	385
46	396
47	408
48	420
49	432
50	443
51	455
52	467
53	478
54	490
55	503
56	516
57	530
58	544
59	559
60	573
61	587
62	601
63	615
64	629
65	644
66	659
67	674
68	689
69	705
70	721

Length of ship (meters)	Freeboard (millimeters)
71	738
72	754
73	769
74	784
75	800
76	816
77	833
78	850
79	868
80	887
81	905
82	923
83	942
84	960
85	978
86	996
87	1015
88	1034
89	1054
90	1075
91	1096
92	1116
93	1135
94	1154
95	1172
96	1190
97	1209
98	1229
99	1250
100	1271
101	1293
102	1315
103	1337
104	1359
105	1380
106	1401
107	1421
108	1440
109	1459
110	1479
111	1500
112	1521
113	1543
114	1565
115	1587
116	1609
117	1630
118	1651
119	1671
120	1690
121	1709

Length of ship (meters)	Freeboard (millimeters)
122	1729
123	1750
124	1771
125	1793
126	1815
127	1837
128	1859
129	1880
130	1901
131	1921
132	1940
133	1959
134	1979
135	2000
136	2021
137	2043
138	2065
139	2087
140	2109
141	2130
142	2151
143	2171
144	2190
145	2209
146	2229
147	2250
148	2271
149	2293
150	2315
151	2334
152	2354
153	2375
154	2396
155	2418
156	2440
157	2460
158	2480
159	2500
160	2520
161	2540
162	2560
163	2580
164	2600
165	2620
166	2640
167	2660
168	2680
169	2698
170	2716
171	2735
172	2754
173	2774

Length of ship (meters)	Freeboard (millimeters)
174	2795
175	2815
176	2835
177	2855
178	2875
179	2895
180	2915
181	2933
182	2952
183	2970
184	2988
185	3007
186	3025
187	3044
188	3062
189	3080
190	3098
191	3116
192	3134
193	3151
194	3167
195	3185
196	3202
197	3219
198	3235
199	3249
200	3264
201	3280
202	3296
203	3313
204	3330
205	3347
206	3363
207	3380
208	3397
209	3413
210	3430
211	3445
212	3460
213	3475
214	3490
215	3505
216	3520
217	3537
218	3554
219	3570
220	3586
221	3601
222	3615
223	3630
224	3645
225	3660
226	3675
227	3690
228	3705
229	3720
230	3735
231	3750
232	3765
233	3780
234	3795
235	3808
236	3821
237	3835
238	3849
239	3864
240	3880
241	3893
242	3906
243	3920
244	3934
245	3949
246	3965
247	3978
248	3992
249	4005

Freeboards at intermediate lengths of ship shall be obtained by linear interpolation.

Ships above 365 meters or 1200 feet in length shall be dealt with by the Administrations.

Length of ship (meters)	Freeboard (milli-meters)	Length of ship (meters)	Freeboard (milli-meters)	Length of ship (meters)	Freeboard (milli-meters)	Length of ship (meters)	Freeboard (milli-meters)
250	4018	279	4385	308	4714	337	5025
251	4032	280	4397	309	4725	338	5035
252	4045	281	4408	310	4736	339	5045
253	4058	282	4420	311	4748	340	5055
254	4072	283	4432	312	4757	341	5065
255	4085	284	4443	313	4768	342	5075
256	4098	285	4455	314	4779	343	5086
257	4112	286	4467	315	4790	344	5097
258	4125	287	4478	316	4801	345	5108
259	4139	288	4490	317	4812	346	5119
260	4152	289	4502	318	4823	347	5130
261	4165	290	4513	319	4834	348	5140
262	4177	291	4525	320	4844	349	5150
263	4189	292	4537	321	4855	350	5160
264	4201	293	4548	322	4866	351	5170
265	4214	294	4560	323	4878	352	5180
266	4227	295	4572	324	4890	353	5190
267	4240	296	4583	325	4899	354	5200
268	4252	297	4595	326	4909	355	5210
269	4264	298	4607	327	4920	356	5220
270	4276	299	4618	328	4931	357	5230
271	4289	300	4630	329	4943	358	5240
272	4302	301	4642	330	4955	359	5250
273	4315	302	4654	331	4965	360	5260
274	4327	303	4665	332	4975	361	5268
275	4339	304	4676	333	4985	362	5276
276	4350	305	4686	334	4995	363	5285
277	4362	306	4695	335	5005	364	5294
278	4373	307	4704	336	5015	365	5303

Table 1 (*Continued*)—Freeboard Table for Type B Ships

Length of ship (feet)	Freeboard (inches)	Length of ship (feet)	Freeboard (inches)	Length of ship (feet)	Freeboard (inches)
80	8.0	460	83.1	840	161.2
90	8.9	470	85.6	850	162.8
100	9.8	480	88.1	860	164.3
110	10.8	490	90.6	870	165.9
120	11.9	500	93.1	880	167.4
130	13.0	510	95.6	890	168.9
140	14.2	520	98.1	900	170.4
150	15.5	530	100.6	910	171.8
160	16.9	540	103.0	920	173.3
170	18.3	550	105.4	930	174.7
180	19.8	560	107.7	940	176.1
190	21.3	570	110.0	950	177.5
200	22.9	580	112.3	960	178.9
210	24.7	590	114.6	970	180.3
220	26.6	600	116.8	980	181.7
230	28.5	610	119.0	990	183.1
240	30.4	620	121.1	1000	184.4
250	32.4	630	123.2	1010	185.8
260	34.4	640	125.3	1020	187.2
270	36.5	650	127.3	1030	188.5
280	38.7	660	129.3	1040	189.8
290	41.0	670	131.3	1050	191.0
300	43.3	680	133.3	1060	192.3
310	45.7	690	135.3	1070	193.5
320	48.2	700	137.1	1080	194.8
330	50.7	710	139.0	1090	196.1
340	53.2	720	140.9	1100	197.3
350	55.7	730	142.7	1110	198.6
360	58.2	740	144.5	1120	199.9
370	60.7	750	146.3	1130	201.2
380	63.2	760	148.1	1140	202.3
390	65.7	770	149.8	1150	203.5
400	68.2	780	151.5	1160	204.6
410	70.7	790	153.2	1170	205.8
420	73.2	800	154.8	1180	206.9
430	75.7	810	156.4	1190	208.1
440	78.2	820	158.0	1200	209.3
450	80.7	830	159.6		

(*Continued from page 176*)

its length, an increase in freeboard is required as indicated in Section 3.7.

3.3 Calculation Form. Fig. 1 shows the form presently used by the American Bureau of Shipping for making load line calculations. The format for sheer correction has been changed slightly from previous forms; in particular, the sheer ordinate constants have been multiplied by the appropriate factors and summed for the forward and after portions of the ship. This expedites the process without changing the numerical value.

3.4 Freeboard Deck. The depth of the vessel used for the calculation of freeboard depends upon which deck is considered to be the freeboard deck. The freeboard deck is normally the uppermost continuous deck having permanent means of closing all exposed openings which would allow water to enter the hull or superstructures. A lower deck may be designated as the freeboard deck, provided it is a complete and permanent deck continuous in a fore and aft direction, at least between the machinery space and peak bulkheads, and continuous athwartships.

3.5 Dimensions for Freeboard. Freeboard dimensions in the computation of freeboard include length, breadth, and depth.

a. Length. The length L is taken as 96 percent of the total length on a waterline at 85 percent of the least molded depth measured vertically from the top of the keel, or as the length from the fore side of the stem to the axis of the rudder stock on the waterline, if that be greater. On barges the length is 96 percent of the overall length measured at 0.85 of the molded depth.

b. Breadth. The breadth B is the maximum breadth measured amidships to the molded line of the frame.

c. Depth of Freeboard. The depth for freeboard D_f is the molded depth amidships, plus the thickness of the freeboard stringer plate or, where the deck is sheathed, the molded depth plus $T(L\text{-}S)/L$, if that be greater, in which T is the mean thickness of the exposed sheathing clear of deck openings and S is the mean length of the part of the superstructure which lies within the length L.

3.6 Penalty Correction. If the minimum geometrical freeboard as finally calculated is less than the freeboard corresponding to the design draft and the strength of the vessel is only sufficient for the design draft, unless the scantlings are suitably increased, the freeboard assigned and marked on the vessel must correspond to the design draft. This is done by adding to the calculated minimum geometric freeboard a *penalty correction* to make the assigned freeboard correspond to the draft permitted by the scantlings of the vessel. Similarly, a penalty is sometimes necessary in order that a ship may meet the required stability standard.

3.7 Basic Minimum Freeboard. After the dimensions for freeboard have been determined, the first step in the calculation is to obtain from the applicable table for the ship type, A or B, the basic minimum tabular freeboard for the length of the vessel. The freeboard increases with the length of the vessel, and is given in the applicable table for each meter up to 365 m and for 10-ft increments of length for vessels up to 1200 ft in length. If the length falls between two increments in the table, the basic freeboard is obtained by interpolation. Vessels greater than 365 m (1200 ft) in length are to be dealt with as special cases by the Administrations.

The tabular freeboard for a type B ship between 24 m (79 ft) and 100 m (328 ft) in length having enclosed superstructures with an effective length of up to 35 percent of the length of the ship must be increased by:

$$7.5\,(100\text{-}L)\,(0.35\text{-}E/L)\text{ mm}$$

or

$$0.09\,(328\text{-}L)\,(0.35\text{-}E/L)\text{ in.}$$

where L is the freeboard length and E is the effective length of superstructure. *Effective length* is discussed in the corrections for superstructures in Section 3.11 b.

To this basic freeboard the various corrections are applied to arrive at the minimum freeboard to be assigned.

3.8 Correction for Block Coefficient. The block coefficient for freeboard purposes is taken by international agreement as the block coefficient at 0.85 of the molded depth to the freeboard deck:

$$C_B = \frac{\text{Molded volume at } d_1}{LBd_1} \tag{2}$$

where

d_1 = molded draft at 0.85 of the least molded depth
L = the length used throughout the freeboard calculation

When C_B is greater than 0.68, the basic freeboard is multiplied by:

$$\frac{C_B + 0.68}{1.36}$$

If C_B is less than 0.68, then 0.68 should be used.

3.9 Correction for Length-to-Depth Ratio. The standard vessel has a depth 1/15 of the length. Where D exceeds $L/15$ the freeboard is increased by $(D\text{-}L/15)R$, where R in millimeters is $L/0.48$ at lengths less than 120 m and R = 250 mm at 120 m and above, or expressed in English units, R in inches is $L/131.2$ at lengths less than 363.6 ft and R = 3 in. at lengths of 393.6 ft and above.

Where D is less than $L/15$ a reduction in freeboard is allowed only if there is an enclosed superstructure covering at least 0.6 L amidships, or a combination of intact superstructures and a trunk (a tight structure which does not extend to the sides of the vessel) extending the full length of the vessel. In such cases the reduction is at the same rate as for an excess of D.

3.10 Correction for Superstructure Height. When the height of superstructure, including a raised quarter deck (which is a stepped deck extending from the after perpendicular with no opening in the vertical bulkhead forming the forward boundary), or trunk is less than the standard height, the mean length of the superstructure is reduced in the ratio of the actual to the standard heights as given in Table 3.

The correction for variation of sheer from the standard is dealt with later in the calculation after the discussion of superstructures, since it depends upon the mean length of superstructures on the vessel.

Table 2—Freeboard Increase Over Tabular Freeboard for Type B Ships, For Ships with Hatch Covers Not of Steel or Equivalent Material

Length of ship (meters)	Freeboard increase (milli-meters)	Length of ship (meters)	Freeboard increase (milli-meters)	Length of ship (meters)	Freeboard increase (milli-meters)
108 and below	50	139	175	170	290
109	52	140	181	171	292
110	55	141	186	172	294
111	57	142	191	173	297
112	59	143	196	174	299
113	62	144	201	175	301
114	64	145	206	176	304
115	68	146	210	177	306
116	70	147	215	178	308
117	73	148	219	179	311
118	76	149	224	180	313
119	80	150	228	181	315
120	84	151	232	182	318
121	87	152	236	183	320
122	91	153	240	184	322
123	95	154	244	185	325
124	99	155	247	186	327
125	103	156	251	187	329
126	108	157	254	188	332
127	112	158	258	189	334
128	116	159	261	190	336
129	121	160	264	191	339
130	126	161	267	192	341
131	131	162	270	193	343
132	136	163	273	194	346
133	142	164	275	195	348
134	147	165	278	196	350
135	153	166	280	197	353
136	159	167	283	198	355
137	164	168	285	199	357
138	170	169	287	200	358

Freeboards at intermediate lengths of ship shall be obtained by linear interpolation.

Ships above 200 m (660 ft) in length shall be dealt with by the Administrations.

Length of ship (feet)	Freeboard increase (inches)	Length of ship (feet)	Freeboard increase (inches)
350 and below	2.0	510	9.6
360	2.3	520	10.0
370	2.6	530	10.4
380	2.9	540	10.7
390	3.3	550	11.0
400	3.7	560	11.4
410	4.2	570	11.8
420	4.7	580	12.1
430	5.2	590	12.5
440	5.8	600	12.8
450	6.4	610	13.1
460	7.0	620	13.4
470	7.6	630	13.6
480	8.2	640	13.9
490	8.7	650	14.1
500	9.2	660	14.3

3.11 Corrections for Superstructures. Deductions from the freeboard are made for superstructures on the freeboard deck only. A superstructure is a decked weathertight structure on the freeboard deck, extending from side to side or with superstructure side plating not inboard of the shell plating more than 4 percent of the breadth, B. A raised quarter deck is regarded as a superstructure. Structures that do not meet the requirements are considered deck houses and do not qualify.

a. Length of Superstructure. The length of superstructure, S, is the mean length of the parts of the superstructure which lie within the vessel's length, L. Where the end bulkhead of an enclosed superstructure extends in a fair convex curve beyond its intersection with the superstructure sides, the equivalent length of the superstructure may be increased on the basis of an equivalent plane bulkhead. This increase should be two-thirds of the fore and aft extent of the curvature. The maximum curvature which may be taken into account in determining this increase is one-half the breadth of the superstructure at the point of intersection of the curved end of the superstructure with its side, i.e., a semi-circle.

b. Height of Superstructure. The height of a superstructure is the least vertical height measured at side from the top of the superstructure deck beams to the top of the freeboard deck beams. Where the least actual height, h_a, of an enclosed superstructure is less than the standard height, h_s, determined from Table 3, the effective length is its length reduced in the ratio of the actual height to the standard height. Where the actual height exceeds the standard, no increase is made in the effective length of the superstructure. Therefore, if the superstructure extends from side to side,

$$E = S \times \frac{h_a}{h_s} \tag{3}$$

The effective length, E, of an enclosed superstructure of standard height is its length S.

c. Breadth of Superstructure. In all cases where an enclosed superstructure of standard height is set in from the sides of the ship not more than 4 percent of the breadth, the effective length is the length modified by the ratio of b/B_s, where

b is the breadth of the superstructure at the middle of its length; and

B_s is the breadth of the ship at the middle of the length of the superstructure.

Then, the effective length $E = S \times b/B_s$.

Where a superstructure is set in for a part of its length, this modification is applied only to the set in part.

If an enclosed superstructure of length S is set in from the side not more than 4 percent of the breadth *and* if its least actual height is less than standard height the effective length for superstructure would be

$$E = S \times \frac{b}{B_s} \times \frac{h_a}{h_s} \tag{4}$$

d. Bulkhead Effectiveness. Superstructures which are not enclosed do not have an effective length. For a superstructure to be credited as enclosed it must have bulkheads of efficient construction. If access openings are placed in these bulkheads the openings are to be fitted with doors of steel or other equivalent material, permanently and strongly attached and provided with gaskets and clamping devices so that the whole structure is of equivalent strength to the unpierced bulkhead and weathertight when closed. The doors should be arranged so that they can be operated from both sides of the bulkhead.

All other openings in the sides or ends of the superstructure are to be fitted with efficient weathertight means of closing.

A bridge or poop is not regarded as enclosed unless access is provided for the crew to reach machinery and other working spaces inside these superstructures by alternative means which are available at all times when bulkhead openings are closed.

e. Trunks. A trunk or similar structure which does not extend to the sides of the ship is given credit as for superstructures and is regarded as efficient provided:

1. it is as strong as a superstructure;
2. hatchways are in the trunk deck and comply with the requirements for coamings and closures;
3. the width of the trunk deck stringer provides a satisfactory gangway and sufficient lateral stiffness;
4. the trunk deck provides a working platform, with guard rails, by itself or in association with other superstructures;
5. ventilators are protected by the trunk, by watertight covers, or by other equivalent means;
6. open rails are fitted on weather portions of the freeboard deck for at least one half the length of the trunk;
7. the machinery casings are protected by the trunk, by a superstructure of at least standard height, or by a deckhouse of the same height and of equivalent strength;
8. the breadth of the trunk is at least 60 percent of the breadth of the ship;
9. where there is no superstructure, the length of the trunk is at least 0.6 L.

The full length of an efficient trunk reduced in the ratio of its mean breadth, b, to the breadth of the ship, B, is its effective length.

The standard height of a trunk is the standard height of a superstructure other than a raised quarter deck.

Table 3—Standard Height of Superstructures

Length		Raised Quarter Deck		All Other Superstructures and Trunks	
m	ft	m	ft	m	ft
30 or less	98.5 or less	0.90	3.0	1.80	5.9
75	246	1.20	3.9	1.80	5.9
125 or more	410 or more	1.80	5.9	2.30	7.5

While trunks are not superstructures by definition, the standard heights for superstructures are applied. The standard heights at intermediate lengths of the ship are obtained by linear interpolation.

Where the height of a trunk is less than the standard height its effective length is to be reduced in the ratio of the actual to the standard height. Where the height of hatch coamings on the trunk deck is less than height required based on its location, the actual height of the trunk, for the purpose of the calculation, is reduced in the ratio of the actual height of the coaming to the standard height.

f. Raised Quarter Deck. The effective length of a raised quarter deck if fitted with an intact front bulkhead is its length up to a maximum of 0.6 L. Where the bulkhead is not intact, the raised quarter deck shall be treated as a poop of less than standard height.

g. Deduction for Superstructures and Trunks. Where the total effective length of superstructures and trunks is less than 1.0L, the deduction is a percentage obtained from Table 4. The total effective length is determined by summing the individual effective lengths of all superstructures and trunks ΣE. The sum divided by the length, L, establishes the total effective length of superstructures and trunks as a percentage of the vessel's length, $\Sigma E/L$.

Applying the instructions, enter Table 4 at the appropriate Column I or II to determine the percentage of deduction.

This percentage of deduction is then multiplied by the deduction allowed for a vessel with 100 percent superstructure which is obtained as follows:

Where the effective length of superstructure and trunks is 1.0 L, the deduction from the freeboard is 350 mm (14 in.) at 24 m (79 ft), 860 mm (34 in.) at L equal to 85 m (279 ft), and 1070 mm (42 in.) at L equal to 122 m (400 ft) and above. Deductions at intermediate lengths are obtained by linear interpolation.

The resultant value becomes the superstructure correction, which is deducted from the corrected basic freeboard.

Where the effective length of a bridge is less than 0.2 L, the percentages are obtained by linear interpolation between Columns I and II.

Where the effective length of a forecastle is less than or equal to 0.4 L, the percentages are obtained from Column I.

Where effective length of a forecastle is more than 0.4 L, the percentages are obtained from Column II.

Where effective length of a forecastle is less than 0.07 L, the above percentages are reduced by:

$$5 \times \frac{(0.07\,L - E_f)}{0.07\,L} \tag{5}$$

where E_f is the effective length of the forecastle.

Table 4—Percentage of Deduction for Type B Ships Total Effective Length of Superstructures and Trunks

$\Sigma E/L$	I Ships and forecastle and without detached bridge percent	II Ships with forecastle and detached bridge percent
0.1	5	6.3
0.2	10	12.7
0.3	15	19
0.4	23.5	27.5
0.5	32	36
0.6	46	46
0.7	63	63
0.8	75.3	75.3
0.9	87.7	87.7
1.0	100	100

Percentage at intermediate lengths of superstructure is to be obtained by linear interpolation.

Table 5—Standard Sheer Profile

Station (L in m)		Ordinate (mm)	Factor
After half	A.P.	25 (L/3 + 10)	1
	⅙ L from A.P.	11.1 (L/3 + 10)	3
	⅓ L from A.P.	2.8 (L/3 + 10)	3
	Amidships	0	1
Forward half	Amidships	0	1
	⅓ L from F.P.	5.6 (L/3 + 10)	3
	⅙ L from F.P.	22.2 (L/3 + 10)	3
	F.P.	50 (L/3 + 10)	1

Station (L in ft)		Ordinate (in.)	Factor
After half	A.P.	0.1 L + 10	1
	⅙ L from A.P.	0.0444 L + 4.44	3
	⅓ L from A.P.	0.0111 L + 1.11	3
	Amidships	0	1
Forward half	Amidships	0	1
	⅓ L from F.P.	0.0222 L + 2.22	3
	⅙ L from F.P.	0.0888 L + 8.88	3
	F.P.	0.2 L + 20	1

3.12 Correction for Sheer. The standard vessel (ICLL, 1966) has a standard sheer profile consisting of two parabolas, one forward and one aft of amidships. If the actual sheer of the vessel is greater than the standard sheer the basic freeboard may be reduced; if it is less, the freeboard must be increased.

a. Sheer Line. The sheer is measured vertically from the deck at side to a line of reference drawn parallel to the base line through the sheer line at side amidships. The sheer, at amidships of the standard sheer profile, is zero. On

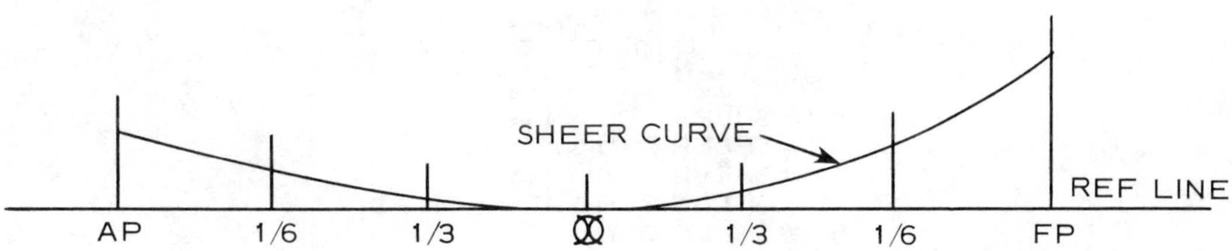

Fig. 2 Sheer curve

the actual vessel it is considered zero, even though the lowest point of the actual sheer may not be amidships. The sheer at any point forward or aft of amidships is always with reference to zero sheer amidships.

The length of the vessel is divided into six equal parts, see Fig. 2, and the ordinates for the standard sheer curve at each point of the division of length L and the end points are given in Table 5, together with Simpson's multipliers for determining the area under the sheer profile.

b. Sheer Comparison with Standard. The actual sheer profile on the vessel is compared with the standard sheer profile. The ordinates for each point of division of the length used in the table are determined from the sheer at those points compared with the zero sheer amidships. If the actual sheer curve at one of these specified ordinates falls below the sheer reference line a negative sheer ordinate value should be used.

c. Superstructure Equivalent for Sheer. In ships with a superstructure of standard height which extends over the entire length of the freeboard deck, the sheer is measured at the superstructure deck. Where the superstructure height exceeds the standard height, the least difference, Z, between the actual and standard height of superstructure is added to each end ordinate.

Similarly, the intermediate ordinates at distances of 1/6 L and 1/3 L from each perpendicular are increased by 0.444 Z and 0.111 Z respectively. See Fig. 3.

The values obtained are added to the standard height of superstructure to develop the equivalent sheer curve.

If the superstructure extends over the entire length of the freeboard deck, and its height is less than standard, the sheer is to be measured at the freeboard deck.

d. Poop and Forecastle Sheer Credit. Sheer credit is given for a poop or forecastle according to:

$$s = \frac{y}{3} \times \frac{L'}{L/2} \tag{6}$$

where

s = sheer credit, to be deducted from the deficiency or added to the excess of sheer
y = difference between actual and standard height of superstructure at the aft or forward perpendicular
L' = mean enclosed length of poop or forecastle up to a maximum length of 0.5 L
L = length of ship

The above formula provides the end ordinate of a parabolic curve which is tangent to the actual deck line at the point of intersection with the end bulkhead of the forecastle and poop. It intersects the end ordinate at a point below the superstructure deck a distance equal to the standard superstructure height. The forecastle or poop deck cannot be less than standard height above this curve at any point.

Fig. 4 is an example of this procedure with the forecastle deck sheer greater than the freeboard deck sheer and the forecastle height greater than standard.

e. Sheer Formula Computation and Comparison. Where the sheer profile differs from the standard the four ordinates of each profile in the forward or after half of the vessel are multiplied by the appropriate factors given in Table 5. The difference between the sums of the respective products of the actual profile and those of the standard is divided by 8 and if applicable the excess height adjustment, s, is added to the vessel's sheer profile.

Where the after half of the sheer profile is greater than standard and the forward half is less than the standard, no

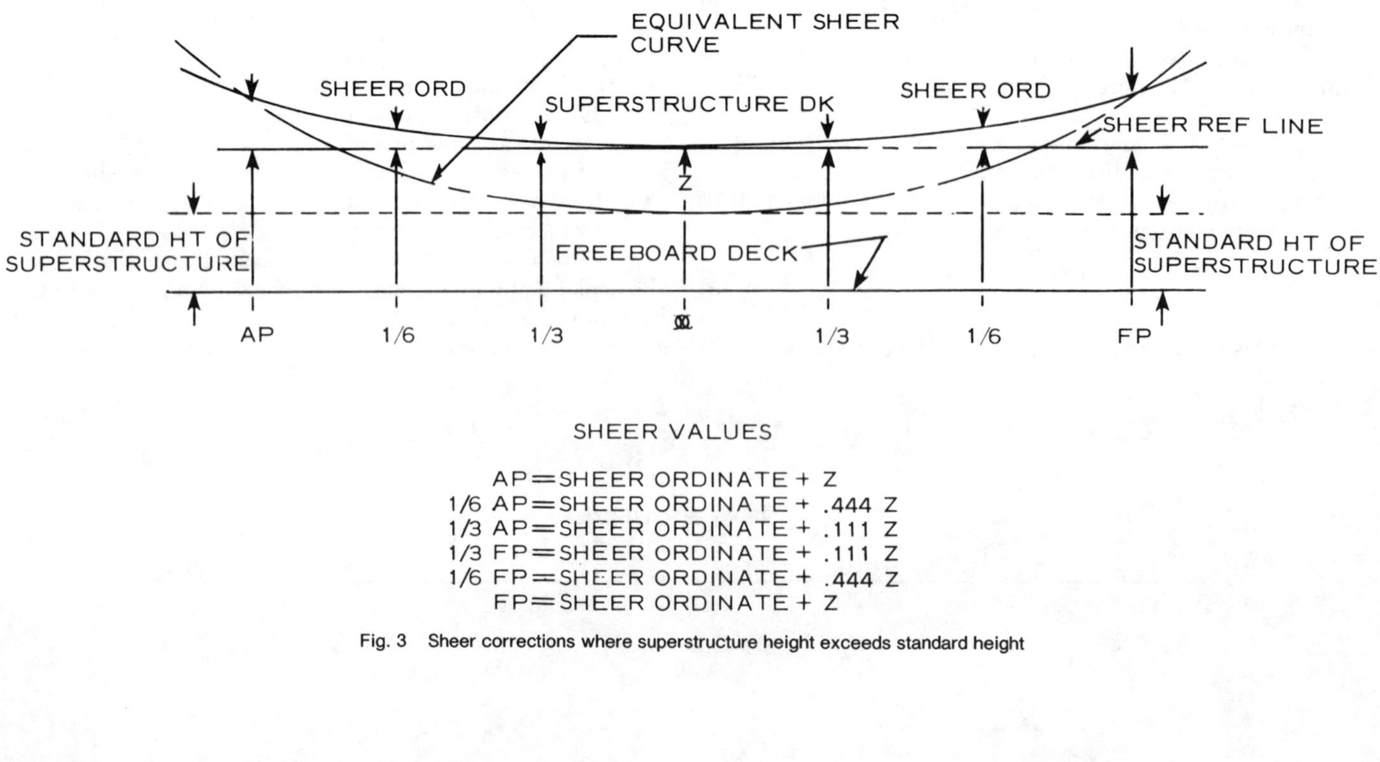

Fig. 3 Sheer corrections where superstructure height exceeds standard height

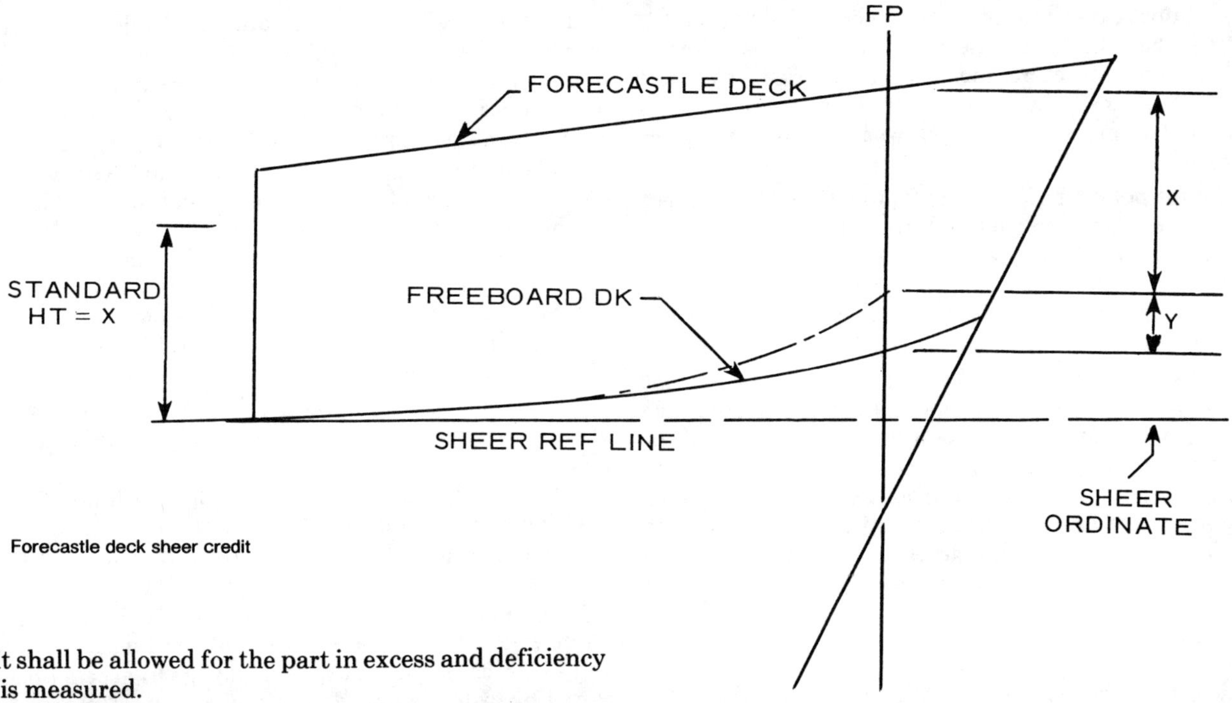

Fig. 4 Forecastle deck sheer credit

credit shall be allowed for the part in excess and deficiency only is measured.

If the forward half of the sheer profile exceeds the standard and the after portion is not less than 75 percent of the standard, credit is allowed for the part in excess. Where the after part of the sheer profile is less than 50 percent of the standard no credit is given for the excess sheer forward. Intermediate allowances may be granted for excess sheer forward when the after sheer is between 50 percent and 75 percent of the standard. These sheer applications are depicted in Table 6. The difference of these values measures the sheer for the forward and aft sections, the sum of which is the *rule sheer*.

The rule sheer subtracted from the sum of the standard forward and after sections divided by 2 results in the excess or deficiency of sheer.

The creditable sheer correction to be added or deducted from the freeboard is the excess or deficiency multiplied by $0.75 - S/2\,L$ where S is the total length of enclosed superstructure.

f. Midship Superstructure Correction. The creditable sheer correction is deducted from the freeboard in ships where an enclosed superstructure covers 0.1 L before and 0.1 L abaft amidships and the sheer correction is negative (excess of sheer). Where an enclosed superstructure covers less than 0.1 L before and 0.1 L abaft amidships, the deductions shall be obtained by linear interpolation. The maximum deduction for excess sheer shall be at the rate of 125 mm per 100 m of length (1.5 in. per 100 ft of length, L).

3.13 Minimum Bow Height. The bow height is defined as the vertical distance at the forward perpendicular between the waterline (corresponding to the assigned summer freeboard including the designed trim) and the top of the exposed deck at side.

For ships below 250 m (820 ft) in length this distance is not to be less than

$$56\,L\left(1 - \frac{L}{500}\right) \times \frac{1.36}{C_B + 0.68}\text{ mm}$$

$$\text{or, } 0.672\,L\left(1 - \frac{L}{1640}\right) \times \frac{1.36}{C_B + 0.68}\text{ in.}$$

and for ships of 250 m (820 ft) and above in length, not less than

$$7000 \times \frac{1.36}{C_B + 0.68}\text{ mm}$$

$$\text{or, } 275.6 \times \frac{1.36}{C_B + 0.68}\text{ in.}$$

where L is the length of the ship and C_B is the block coefficient which is to be taken as not less than 0.68.

Where the bow height requirement is obtained by sheer,

Table 6—Sheer Corrections for Various Deviations from Standard

Vessel Compared to Standard: *Aft*	*Forward*	Sheer to be applied: *Aft*	*Forward*
>Standard	<Standard	Standard	Actual
Deficient but ≥75% of Standard	>Standard	Actual	Actual
Deficient and <50% of Standard	>Standard	Actual	Standard
Deficient but Between 50 & 75% of Standard	>Standard	Actual	Standard Plus a Percentage of Excess
>Standard	>Standard	Actual	Actual
<Standard	<Standard	Actual	Actual

the sheer must extend for at least 15 percent of the length of the ship abaft the forward perpendicular. If it is obtained by fitting a superstructure, such superstructure is to extend from the stem to a point at least 0.07 L abaft the forward perpendicular and must comply with the following requirements:

- for ships not over 100 m (328 ft) in length it is to meet the requirements for an enclosed superstructure;
- for ships over 100 m (328) ft in length it need not meet the requirements for an enclosed superstructure but is to be fitted with closing appliances satisfactory to the administration of registry.

3.14 Barges. A barge or other ship without independent means of propulsion is assigned a freeboard in accordance with the provisions of the regulations. If a barge is unmanned certain requirements, such as those pertaining to the fitting of gangways, access from gangways to accommodation spaces and deckhouses and guardrails for the protection of the crew, are not necessary. Additionally, the requirement for minimum bow height above the waterline is not in effect.

Unmanned barges which have virtually an intact deck except for small access openings with weathertight covers may be assigned freeboards 25 percent less than those outlined in the regulations.

3.15 Reduced Freeboards for Type B Ships. A Type B ship over 100 m (328 ft) in length may be assigned a freeboard less than that usually assigned for a ship of the same geometric particulars provided:

- The ship is fitted with appropriate steel, weathertight hatch covers;
- The freeing port arrangements are adequate;
- the ship, when loaded to its summer load waterline, will remain afloat in satisfactory condition of equilibrium after flooding of any single damaged compartment at an assumed permeability of 0.95 excluding the machinery space;
- and, in ships over 225 m (738 ft) in length, in addition to the above, the machinery space shall be treated as a floodable compartment but with a permeability of 0.85.

The relevant calculation is based upon the assumptions that:

1. the vertical extent of damage is equal to the depth of the ship plus any superstructures credited as being buoyant in way of the damage;
2. the penetration of damage is not more than $B/5$;
3. no main tranverse bulkhead is damaged;
4. the height of the center of gravity above the base line is assessed allowing for homogeneous loading of cargo holds, and for 50 percent of the designed capacity of consumable fluids and stores.

3.16 Type B Ship 60 Percent Freeboard Reduction. In the calculating of freeboards for Type B ships which comply with the criteria stated above, the values from Table 1 can be reduced by not more than 60 percent of the difference between the B and A tabular values for the appropriate ship lengths (see Table 7).

3.17 Type B Ship 100 Percent Freeboard Reduction. This reduction in tabular freeboard may be increased up to the total difference between the values in Table 7 and Table 1 provided the ship complies with the machinery casing, freeing port arrangements, and gangway and access requirements for Type A ships. Further, the conditions set forth for a ship receiving not more than 60 percent reduction must be met except that the damage can occur to any two adjacent fore and aft compartments, neither of which is the machinery space. For ships over 225 m (738 ft) in length the machinery space is taken separately for the flooding calculation as previously described.

3.18 Damage Stability Analysis. To comply satisfactorily with the damage stability analysis the ship in its final condition of equilibrium must demonstrate that:

- the final waterline after flooding is below the lower edge of any opening through which progressive flooding may take place;
- the maximum angle of heel due to unsymmetrical flooding is in the order of 15 degrees;
- the metacentric height in the flooded condition is positive.

3.19 Type A Ships. As previously defined, a Type A ship is one which is designed to carry only liquid cargos in bulk, and in which cargo tanks have only small access openings closed by watertight gasketed covers of steel or equivalent material.

In addition, the inherent features of the ship should include:

- a high integrity of the exposed deck;
- a high degree of safety against flooding resulting from the low permeability of loaded cargo spaces and the degree of subdivision usually provided.

All Type A ships which are over 150 m (492 ft) in length and are designed to have empty compartments when loaded to their summer load waterline must be able to withstand the flooding of any one of their empty conpartments at an assumed permeability of 0.95 and remain afloat in a satisfactory condition of equilibrium as defined for Type B ships. In ships over 225 m (738 ft) in length, the machinery space is treated as a floodable compartment but with a permeability of 0.85.

Type A ships are assigned freeboards in accordance with Table 7 and must, in addition to the requirements for a Type B vessel, comply with the following specific requirements:

- a gangway at the level of the superstructure deck, with safe access for the crew to quarters and working spaces from that level, extending between the poop and midship house, poop and forecastle, and where the crew is quartered forward from the midship house to the forecastle;
- special protection for machinery casings;
- steel hatch covers for hatchways on exposed freeboard of superstructure decks;
- open rails for at least one half the exposed weather deck, or the whole length where a trunk connecting superstructures is fitted.

The deductions for superstructures are calculated by the same method as for Type B vessels except that the percentages of the deductions permitted for each 10 percent of effective length of superstructure are as given in Table 8.

Table 7—Freeboard Table for Type A Ships

Length of ship (m)	Freeboard (mm)	Length of ship (m)	Freeboard (mm)	Length of ship (m)	Freeboard (mm)	Length of ship (m)	Freeboard (mm)
24	200	80	841	133	1684	187	2474
25	208	81	855	134	1702	188	2486
26	217	82	869	135	1719	189	2497
27	225	83	883	136	1736	190	2508
28	233	84	897	137	1753	191	2519
29	242	85	911	138	1770	192	2530
30	250	86	926	139	1787	193	2541
31	258	87	940	140	1803	194	2552
32	267	88	955	141	1820	195	2562
33	275	89	969	142	1837	196	2572
34	283	90	984	143	1853	197	2582
35	292	91	999	144	1870	198	2592
36	300	92	1014	145	1886	199	2602
37	308	93	1029	146	1903	200	2612
38	316	94	1044	147	1919	201	2622
39	325	95	1059	148	1935	202	2632
40	334	96	1074	149	1952	203	2641
41	344	97	1089	150	1968	204	2650
42	354	98	1105	151	1984	205	2659
43	364	99	1120	152	2000	206	2669
44	374	100	1135	153	2016	207	2678
45	385	101	1151	154	2032	208	2687
46	396	102	1166	155	2048	209	2696
47	408	103	1181	156	2064	210	2705
48	420	104	1196	157	2080	211	2714
49	432	105	1212	158	2096	212	2723
50	443	106	1228	159	2111	213	2732
51	455	107	1244	160	2126	214	2741
52	467	108	1260	161	2141	215	2749
53	478	109	1276	162	2155	216	2758
54	490	110	1293	163	2169	217	2767
55	503	111	1309	164	2184	218	2775
56	516	112	1326	165	2198	219	2784
57	530	113	1342	166	2212	220	2792
58	544	114	1359	167	2226	221	2801
59	559	115	1376	168	2240	222	2809
60	573	116	1392	169	2254	223	2817
61	587	117	1409	170	2268	224	2825
62	600	118	1426	171	2281	225	2833
63	613	119	1442	172	2294	226	2841
64	626	120	1459	173	2307	227	2849
65	639	121	1476	174	2320	228	2857
66	653	122	1494	175	2332	229	2865
67	666	123	1511	176	2345	230	2872
68	680	124	1528	177	2357	231	2880
69	693	125	1546	178	2369	232	2888
70	706	126	1563	179	2381	233	2895
71	720	127	1580	180	2393	234	2903
72	733	128	1598	181	2405	235	2910
73	746	129	1615	182	2416	236	2918
74	760	130	1632	183	2428	237	2925
75	773	131	1650	184	2440	238	2932
76	786	132	1667	185	2451	239	2939
77	800			186	2463	240	2946
78	814					241	2953
79	828					242	2959
						243	2966
						244	2973
						245	2979
						246	2986
						247	2993
						248	3000
						249	3006
						250	3012
						251	3018
						252	3024
						253	3030
						254	3036
						255	3042
						256	3048
						257	3054
						258	3060
						259	3066
						260	3072
						261	3078

Freeboards at intermediate lengths of ship shall be obtained by linear Interpolation.

Ships above 365 m or 1200 ft in length shall be dealt with by the Administrations.

Length of ship (m)	Freeboard (mm)	Length of ship (m)	Freeboard (mm)	Length of ship (m)	Freeboard (mm)	Length of ship (m)	Freeboard (mm)
262	3084	288	3211	314	3312	340	3382
263	3089	289	3215	315	3315	341	3385
264	3095	290	3220	316	3318	342	3387
265	3101	291	3224	317	3322	343	3389
266	3106	292	3228	318	3325	344	3392
267	3112	293	3233	319	3328	345	3394
268	3117	294	3237	320	3331	346	3396
269	3123	295	3241	321	3334	347	3399
270	3128	296	3246	322	3337	348	3401
271	3133	297	3250	323	3339	349	3403
272	3138	298	3254	324	3342	350	3406
273	3143	299	3258	325	3345	351	3408
274	3148	300	3262	326	3347	352	3410
275	3153	301	3266	327	3350	353	3412
276	3158	302	3270	328	3353	354	3414
277	3163	303	3274	329	3355	355	3416
278	3167	304	3278	330	3358	356	3418
279	3172	305	3281	331	3361	357	3420
280	3176	306	3285	332	3363	358	3422
281	3181	307	3288	333	3366	359	3423
282	3185	308	3292	334	3368	360	3425
283	3189	309	3295	335	3371	361	3427
284	3194	310	3298	336	3373	362	3428
285	3198	311	3302	337	3375	363	3430
286	3202	312	3305	338	3378	364	3432
287	3207	313	3308	339	3380	365	3433

Table 7 (*Continued*)—Freeboard Table for Type A Ships

Length of ship (ft)	Freeboard (in.)	Length of ship (ft)	Freeboard (in.)	Length of ship (ft)	Freeboard (in.)
80	8.0	460	71.1	840	120.1
90	8.9	470	73.1	850	120.7
100	9.8	480	75.1	860	121.4
110	10.8	490	77.1	870	122.1
120	11.9	500	79.0	880	122.7
130	13.0	510	80.9	890	123.4
140	14.2	520	82.7	900	124.0
150	15.5	530	84.5	910	124.6
160	16.9	540	86.3	920	125.2
170	18.3	550	88.0	930	125.7
180	19.8	560	89.6	940	126.2
190	21.3	570	91.1	950	126.7
200	22.9	580	92.6	960	127.2
210	24.5	590	94.1	970	127.7
220	26.2	600	95.5	980	128.1
230	27.8	610	96.9	990	128.6
240	29.5	620	98.3	1000	129.0
250	31.1	630	99.6	1010	129.4
260	32.8	640	100.9	1020	129.9
270	34.6	650	102.1	1030	130.3
280	36.3	660	103.3	1040	130.7
290	38.0	670	104.4	1050	131.0
300	39.7	680	105.5	1060	131.4
310	41.4	690	106.6	1070	131.7
320	43.2	700	107.7	1080	132.0
330	45.0	710	108.7	1090	132.3
340	46.9	720	109.7	1100	132.6
350	48.8	730	110.7	1110	132.9
360	50.7	740	111.7	1120	133.2
370	52.7	750	112.6	1130	133.5
380	54.7	760	113.5	1140	133.8
390	56.8	770	114.4	1150	134.0
400	58.8	780	115.3	1160	134.3
410	60.9	790	116.1	1170	134.5
420	62.9	800	117.0	1180	134.7
430	65.0	810	117.8	1190	135.0
440	67.0	820	118.6	1200	135.2
450	69.1	830	119.3		

Section 4
Conditions of Assignment

4.1 General. It is of vital importance that there be provided efficient means of protection for all openings to the hull and superstructures, for the protection of the crew in heavy weather and for the rapid freeing of water from the weather decks. The regulations to ensure compliance with these considerations are grouped together as *Conditions of Assignment*.

By the terms of the International Load Line Convention (ICLL, 1966), freeboard may not be officially assigned until the ship has been inspected and a qualified surveyor representing the Administration of the flag country is willing to certify that the Conditions of Assignment have been met.

These Conditions of Assignment must not only be complied with initially but they must be at all times maintained in satisfactory condition. Their vital importance is recognized in the regulations which call for annual inspections to be made by the assigning authorities' Surveyors to ensure that, in fact, they have been maintained in satisfactory condition for the continued validity of the Load Line Certificate.

In the Convention (ICLL, 1966) repeated in the Regulations (46CFR42), are the specific formulas for minimum height of openings, strength, deflection, etc. for each of the items covered by the following subsections. They are described here in general form so that one realizes that a significant portion of load line review and freeboard assignment is dependent upon the watertight integrity of the hull, and the weathertight integrity of the ship's topside area.

4.2 Hatchways. Most important, because of their size, are the cargo hatchways. Standards are set forth in the regulations for the construction, heights of coamings, the covers, and the fittings of all exposed hatchways on the freeboard and superstructure decks. Hatchways inside superstructures must meet standards which depend upon the type of closing appliances fitted on the access openings in the end bulkheads. The requirements for hatch coaming heights and hatchway covers, and their supports, comprise a standard of strength and protection. Coamings may be reduced in height, or eliminated altogether, in association with gasketed metal covers, subject to the approval of the flag administration. While the regulations specify strength criteria, the rules of most of the classification societies contain formulas for hatchcover and beam design which provide equivalent strength.

4.3 Machinery Casings. Machinery-space openings on the exposed portions of the freeboard deck or superstructure decks, or within open structures, must be provided with steel casings, with any opening fitted with steel weathertight doors. Openings in required machinery casings must have specified minimum sill heights. Machinery access hatch openings are to have permanently attached steel weathertight covers. Exposed machinery casings of Type A ships cannot have direct access from the freeboard deck to the machinery space.

4.4 Other Openings in Deck and Shell. Other openings include ventilators, air pipes, hull piping, and air ports.

a. Ventilators. Ventilators on exposed positions on the freeboard and superstructure decks leading to spaces below the freeboard deck, or to enclosed superstructures, are to be fitted with coamings of minimum height, depending upon the location, and with provision for temporary means of closing.

b. Air Pipes. Air pipes from ballast tanks or other tanks below the freeboard deck, which extend above the freeboard deck or superstructure deck, are to be provided with a permanently attached means of closing.

c. Hull Piping. In the sides of the vessel, the numerous small openings required present a problem in maintaining the intactness of the vessel. Each overboard discharge pipe leading from spaces below the freeboard deck must have an automatic non-return valve with positive means for closing from an accessible position above the freeboard deck, or in some instances, two automatic non-return valves without positive means of closing may be allowed if the inboard valve is always accessible in service. Scuppers or sanitary dis-

Table 8—Percentage of Deduction for Type A Ships

Total Effective Length of Superstructures and Trunks

$\Sigma E/L$	*% of deduction superstructures and trunks*
0.1	7
0.2	14
0.3	21
0.4	31
0.5	41
0.6	52
0.7	63
0.8	75.5
0.9	87.7
1.0	100

Percentages at intermediate lengths of superstructures is obtained by linear interpolation.

Table 9—Total Effective Length of Superstructures

$\Sigma E/L$	*% of deduction superstructure*
0	20
0.1	31
0.2	42
0.3	53
0.4	64
0.5	70
0.6	76
0.7	82
0.8	88
0.9	94
1.0	100

Percentages at intermediate lengths of superstructures are obtained by linear interpolation.

charges from superstructures or deckhouses may also be required to have similar protection.

d. Air Ports. Portholes in superstructures on the freeboard deck or in the hull are required to be fitted with hinged deadlights. Any other openings in the shell below the freeboard deck such as gangway or cargo ports must have closures designed to ensure watertightness and structural integrity.

4.5 Miscellaneous Conditions of Assignment. Guard rails or bulwarks, gangways, lifelines, or other means must be provided for the protection of the crew in its operation of the vessel and for getting to and from their quarters. Deckhouses used for the accommodation of the crew are to be of adequate strength. Where bulwarks on the weather portions of freeboard or superstructure decks form wells, ample provision is made for freeing the decks rapidly of water and for draining the wells. The draining area requirement is based on the length and height of the bulwark. In ships with no sheer or in those fitted with trunks restricting free flow of water across the deck, adjustments are made by increasing the freeing port area requirements to a degree.

4.6 Information to be Supplied to the Master. The master is to be furnished with sufficient information in an approved form satisfactory to the administration of registry. This is to enable the master to arrange for the loading and ballasting of his ship in such a way as to avoid the creation of any unacceptable stresses in the ship's structure. In some cases, based on the length, design, or class of ship, the administration may consider this requirement unnecessary.

In addition, the master is to be supplied with sufficient information in an approved form to give him guidance as to the stability of the ship under varying conditions of service.

4.7 Timber Deck Cargos. Timber deck cargo refers to a cargo of timber carried on an uncovered part of a freeboard or superstructure deck. Such cargo may be regarded as providing the ship with certain additional buoyancy and a greater degree of protection against the sea. For that reason, ships carrying a timber deck cargo may be granted a reduction of freeboard calculated basically in accordance with the requirements for Type B ships with additional conditions stipulated relating to construction, stowage of the timber cargo, stability, protection of crew, and access to machinery and other such spaces necessary for the safe operation of the ship.

The percentage of deduction of superstructures is modified in accordance with Table 9.

Section 5
Seasonal, Fresh-Water, and Timber Freeboard Marks

5.1 Freeboard Marks. After the application of all the corrections to the basic Minimum Summer Freeboard are made, the result is the Minimum Summer Freeboard in salt water which will bc assigned to the vessel. This freeboard may in no case be less than 50 mm (2 in.). For ships having hatchways with covers other than steel or equivalent material on the freeboard deck or on superstructure decks situated forward of a point located a quarter of the ship's length from the forward perpendicular, the freeboard is not less than 150 mm (6 in.).

5.2 Load Line Format. The freeboard is measured from the top of the deck amidships to the top of the line through the center of the load-line ring. Forward of the ring is a grid composed of lines indicating the maximum loadings in fresh water and for the different seasons, including the summer line which will be at the same level as the center of the ring, Fig. 5.

5.3 Zones and Seasons. The regulations require a vessel to be so loaded when departing upon a voyage that at no time during any portion of the voyage will the applicable seasonal mark be submerged. The oceans of the world are divided under the regulations into various zones and seasonal areas according to the probable severity of the weather. In certain areas and times of the year where during the winter more severe weather may be expected the vessel is required not to load as deeply as is permitted in the summer. These areas are the seasonal winter zones, and the times of the year when the winter mark is applicable are shown on a map attached to the regulations. During other times of the year, the summer mark at the center of the ring is applicable. Other zones, roughly corresponding to the temperate zone in the northern and southern hemispheres, are permanent summer zones under the regulations where the summer mark is applicable the year around. Another zone, on either side of the equator, is a permanent tropical zone where the anticipated weather is generally less severe than might be expected in the summer in the temperate zone. In these zones vessels are permitted to load somewhat deeper any time during the year. Other areas between the permanent summer and tropical zones are considered one or the other at specified different times of the year.

5.4 Calculation of Seasonal Marks. The various seasonal and fresh water marks are obtained as follows:

a. Tropical Freeboard Mark (T). The line to mark the maximum loading in the tropical zone is obtained as a deduction from the Summer Freeboard of one forty-eighth ($^1/_{48}$) of the summer draft measured from the top of the keel to the center of the ring. Again, the freeboard must not be less than 50 mm (2 in.) or 150 mm (6 in.).

b. Winter Freeboard Mark (W). The line to mark the maximum loading in winter zones is obtained by an addition to the Summer Freeboard of one forty-eighth ($^1/_{48}$) of the molded summer draft.

c. Winter North Atlantic Freeboard (WNA). The minimum freeboard for ships of not more than 100 m (328 ft) in length which enter any part of the North Atlantic de-

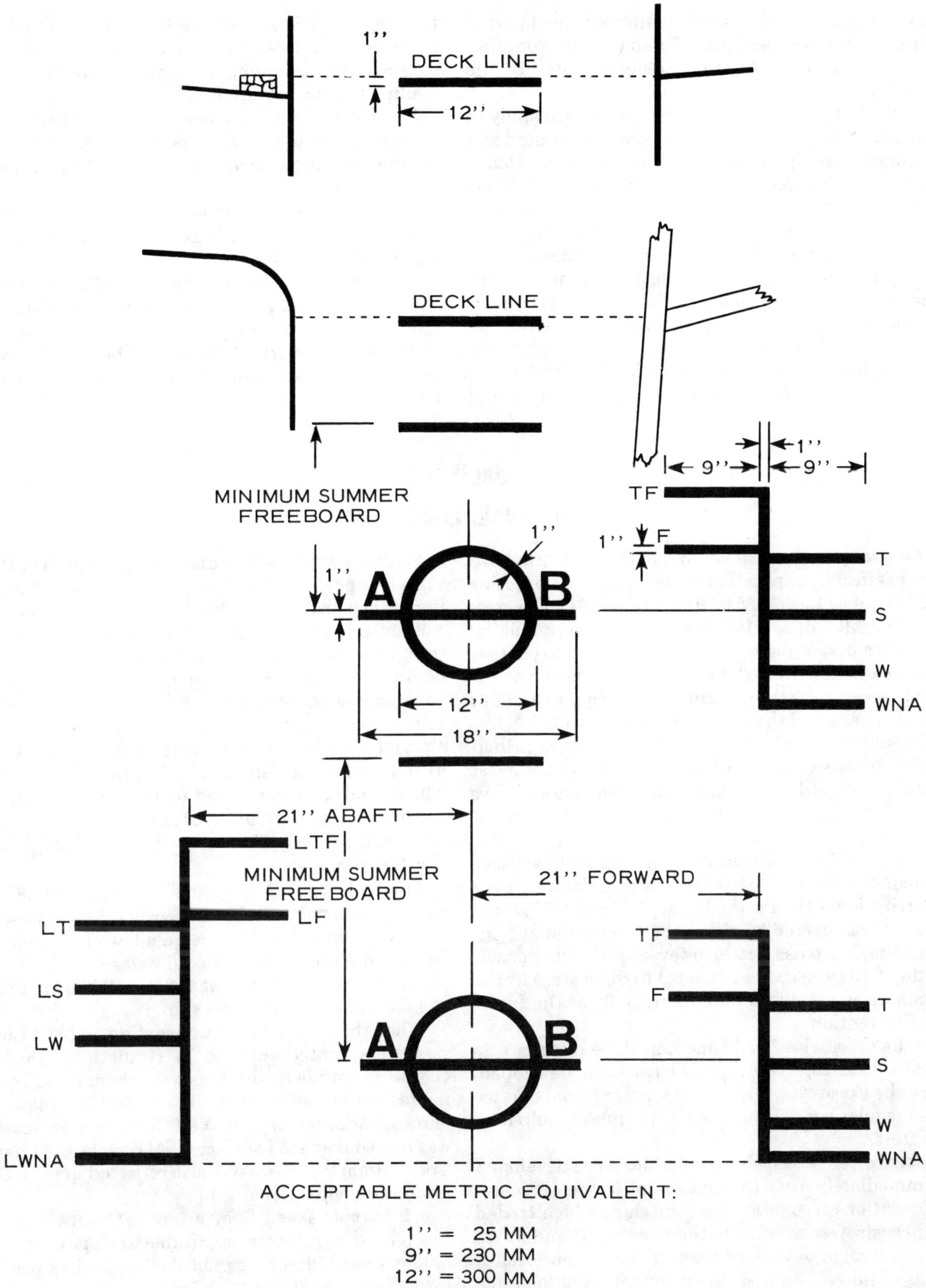

Fig. 5 Load line markings on a ship's side

fined in the regulations during the winter seasonal period shall be the winter freeboard plus 50 mm (2 in.). For other ships, the Winter North Atlantic Freeboard shall be the winter freeboard.

d. Fresh-Water (*F and TF*). The regulations apply to vessels at sea in salt water. An allowance is computed as a guide in compensating for deeper draft when a vessel loads in fresh water. The fresh water allowance in cm is determined by dividing the displacement at the summer load waterline by forty times the tons per cm immersion at the draft. The fresh-water allowance in inches is determined by dividing the summer load line displacement by forty times the tons per inch immersion at that draft. If the basic information is not available, the allowance is taken at one forty-eighth ($1/48$) of the summer draft. The fresh-water allowance applies to all seasonal freeboards, but fresh-water lines are marked on the vessel only for the summer and tropical conditions. Where the water is brackish, a proportion of the fresh-water allowance is used, and a measurement of the actual density of the water at the place of loading may be required.

e. Steaming Allowance. In addition to the fresh-water allowance, deeper loading is also permitted to allow for consumables used between the point of loading and the open sea.

5.5 Timber Freeboards. If a Timber Freeboard is assigned it is marked in addition to the ordinary load lines with some slight variation.

The Winter Timber Freeboard is obtained by adding to the Summer Timber Freeboard one thirty-sixth ($1/36$) of the molded summer timber draft.

The Winter North Atlantic Timber Freeboard is not to be less than the ordinary Winter North Atlantic Freeboard.

Section 6
Domestic Load Lines

6.1 General. Domestic load lines in the U. S. are those which are in effect by application of the Coastwise Load Line Act of 1935 (Public Law 74-354, 1935). In this Act, Congress specified that all ships of 150 gross tons or more making voyages from a port or place in the United States to another port or place in the United States were to meet the requirements enacted by the government taking into account the type of voyage and the conditions of operation.

The Coastwise Load Line Act originally applied principally to the contiguous states along the East, Gulf, and West Coasts and the Great Lakes. Since Alaska and Hawaii have become states, voyages to these two states are now included.

The Coastwise Load Line Act does not apply to voyages to the Commonwealth of Puerto Rico, Guam, Samoa, or the Trust Territories of the Pacific Islands. Until 1975, these islands were not covered by either the International Convention or the Coastwise Act. However, in 1975 the provisions of the ICLL, 1966, were extended to these areas by the United States as permitted under Article 32 of the International Convention.

Under the Coastwise Load Line Act, three distinct sets of domestic load line regulations have been developed. These are the Coastwise, Great Lakes, and Special Service Coastwise regulations. There are no load lines required on inland waters.

6.2 Coastwise. Coastwise regulations were developed almost immediately after the enactment of the 1935 law. These regulations recognized that the ships which traded along our coasts were exposed to the same winds and waves as ships on foreign voyages; therefore, the same level of safety was required. Indeed, the identical set of load line regulations was used for coastwise voyages as for foreign voyages. The freeboard tables and all the corrections originally used were precisely the same as those used for foreign voyages.

With the experience gained in application of the regulations for approximately thirty years and with developments in new construction, in 1962 the Coastwise Load Line Act was amended to allow domestic shipping on the East Coast to sail on summer marks year round. In 1965, a new freeboard table for tankers on U. S. coastwise/intercoastal voyages was adopted which allowed U. S. tankers on non-international voyages to sail at lesser freeboards. The amount varied from a zero difference at 91.4 m (300 ft) to a 61 mm (2.4 in.) difference at a length of 304.8 m (1,000 ft). These freeboards were used by tankers operating between the oil ports of the Gulf of Mexico and the East Coast and from Southern California north no farther than Seattle, Washington.

Also, in 1965, U. S.-flag ships longer than 112.8 m (370 ft), having watertight steel hatch covers, forecastles, and one-compartment subdivision were allowed a special reduced freeboard for non-international voyages. The deeper draft allowed varied from zero at 112.8 m (370 ft) to 685 mm (27 in.) for a 304.8 m (1,000 ft) ship.

When the International Convention on Load Lines, 1966, was incorporated into U. S. regulations, the domestic coastwise voyage regulations were changed and essentially made identical with the newly-enacted International regulations. Additionally, the West Coast summer seasonal zone was moved north to Dall Island, Alaska, by the International Convention; this move was also reflected in U. S. Coastwise Load Line regulations.

6.3 Great Lakes. Soon after the Coastwise Act of 1935 was passed, regulations applicable to ships operating solely on the Great Lakes were published and subsequently placed into effect in 1936.

These regulations differed from the coastwise regulations for several reasons. First, it was recognized that Great Lakes ships were shallow-draft vessels and, as a result, did not have the same geometry as the majority of ocean-going vessels. Secondly, they were an existing fleet of ships unique in form and operation, which were built between 1904 and 1936. From the start of federal regulation in 1936, these were looked upon as ships especially bred for the Great Lakes and not suitable for ocean service.

Soon after the U. S. adopted load line standards for the Great Lakes, Canada also developed Canadian national regulations for the Great Lakes. Since the initial U. S. and Canadian regulations differed somewhat, the two nations conducted an official *Exchange of Notes* between 1938 and 1940 whereby both nations formally recognized each other's Great Lakes load line regulations as being equivalent.

In May 1973, in continuation of the above longstanding agreement between the U. S. and Canada, a new set of Great Lakes regulations was developed and jointly put into effect (46 CFR 45). The load line concept contained in these new regulations is similar to the new International requirements. The significant differences are in the basic freeboard formula and in the formula for sheer on ships greater than 152.4 m (500 ft) in length. The basic freeboard is corrected, following the principles of the International Convention, for superstructure and sheer, predicated on their extent and configuration.

a. Great Lakes Standard Ship. Great Lakes regulations originally recognized that the standard ship had an L/D of 15. However, the Great Lakes rules recognize that ships less than 121.9 m (400 ft) long often have a lesser L/D ratio. Therefore, a separate correction is included assuming that a standard ship 24.1 m (80 ft) long has an L/D of 6.5 and that the L/D ratio increases to 15 for a ship 121.9 m (400 ft) in length.

Additionally, the rules recognize that ships longer than 121.9 m (400 ft) have an increasing L/D ratio. An L/D of 21 is the maximum allowed for strength purposes for any ship 213.4 m (700 ft) long or more.

b. Great Lakes Freeboard Formula. The formula for basic dry cargo freeboards in inches is $F = 12 \times P_1 \times D$, where D is the depth of the ship and P_1 is determined from coefficients given in tabular form. Tankers use the formula $F = 10.2 \times P_1 \times D$, where F is expressed in inches. Neither of these formulas has yet been validated in metric units.

The freeboard formula is adjusted by coefficients given in tables which were so constructed as to eliminate the block coefficient correction and to modify the L/D correction to fit the *standard* Great Lakes ship. It should be noted that a Block Coefficient of 0.864 is standard on the Great Lakes compared to 0.68 which is the universally accepted standard for ocean-going ships. Thus, by utilizing the formula all adjustments for hull form and strength due to hull geometry are automatically included. The remaining corrections are all concerned with the topside geometry of the ship.

c. Penalty for Short Superstructure. The freeboard is corrected taking into consideration the effective length of the superstructure which adds reserve buoyancy compared to the length of the ship. The freeboard is increased slightly for ships less than 152.4 m (500 ft) long which have a superstructure less than 25 percent of the length L.

d. Correction for Superstructure. The superstructure correction is a reduction in freeboard to give credit for the extra buoyancy of the additional topside structure. If the ship has 100 percent superstructure on the freeboard deck, then the maximum credit for one-half the height of a standard superstructure, for that length of ship, is allowed. The standard superstructure height H_s is from the following:

$$H_s(\text{ft}) = \left(6.0 + \frac{L}{300}\right) \text{(no metric equivalent is given in the regulations)}$$

e. Correction for Sheer. The correction for sheer may be an addition to or a subtraction from the freeboard summation. If the ship has sheer that is less than the standard, a mandatory addition to freeboard is made. If its sheer is greater than standard, the freeboard may be reduced.

The sheer calculation follows Simpson's second rule, as in the 1966 International Convention, to determine the area of the sheer profile.

An important distinction between the new Great Lakes and International regulations is that the standard sheer value for ships 152.4 m (500 ft) or more in length remains the same. Thus, the required sheer value for a 304.8 m (1,000 ft) ship is the same as that for a 152.4 m (500 ft) ship. This is a direct result of studies of probable seaway occurrences in relation to bow height.

f. Deck Line Correction. Where the depth to the upper edge of the reference deck line is greater or less than D, the difference between the depths must be added to or deducted from the freeboard.

g. Conditions of Assignment. Great Lakes load line regulations, just like their international counterpart, require much more than a simple assignment of freeboard. Freeboard cannot be assigned until the ship is found satisfactory for sea in respect to all load line considerations, namely strength, stability, watertight integrity of hull, and weathertight integrity of topsides and crew protection in the form of railings or bulwarks.

Load Line Regulations may impose a greater freeboard than minimum if it is necessitated by stability or if the ship is not structurally adequate to operate at the draft allowed by the calculated minimum geometric freeboard. These two considerations have been part of load line review since the first discussions in the past century.

In addition, superstructures and deck houses must be structurally sound; all openings into these structures as well as openings in the deck must be weathertight.

Adequate freeing area is required to allow for rapid removal of boarding seas.

The new Great Lakes regulations adopted in 1973 have followed the International Convention in requiring that both strength and stability information be provided to the master.

h. Seasonal Load Line Marks. Each vessel must have four seasonal marks calculated and marked. These are Summer (latter half April to September), mid-Summer (May to September), Intermediate (October and first half April), Winter (November to April).

i. Special Freeboard Allowance. Since the Great Lakes are fresh water, no fresh water allowance is made and none is allowed for salt water ships while voyaging on the Great Lakes. Neither is the steaming allowance for river travel allowed on the Great Lakes since severe storms and wave systems do occur on the Great Lakes.

6.4 Special Service. Special Service Load Lines were the results of a petition to Congress in 1937 which amended the Coastwise Load Line Act to remove the language requiring all coastwise vessels to be treated identically with vessels making foreign voyages. It was determined that the judgment of the administration was to be exercised as to the nature and conditions of the voyage, being particularly conscious of the distance the vessel would operate offshore.

The petitioners were acting on behalf of the very large coastal collier trade which existed at that time. Also included in Special Service were manned and unmanned barges; however, tugs were not included.

It was decided to allow such vessels which traveled not more than 20 mi offshore to have load lines with the same freeboards as were used on the Great Lakes. This allowed the U. S. operator in purely domestic service to load to a deeper draft, reasoning that by remaining close to shore a ship would have the opportunity to seek a safe harbor from the more severe coastal storms and, with proper discretion exercised by the master, would not encounter full ocean storms.

The Special Service Load Line Regulations (46 CFR 44) require the applicant to state in his application special items of design and operation such as sea speed, voyage limits, cargo to be carried, weather conditions expected, and weather the vessel will be manned or unmanned.

Special Service Load Line Certificates are allowed in only four specific coastal zones not more than 20 miles offshore.

1. Eastport, Maine to Norfolk, Virginia;
2. Jacksonville, Florida to Key West;
3. Key West, Florida to Rio Grande (River), Texas (Zones 2. and 3. are not open during hurricane season, July 1 to November 15),
4. San Diego to San Francisco, California.

Special Service freeboards for steam colliers, barges, and self-propelled barge-shaped vessels are calculated using the Great Lakes freeboard tables in effect prior to May 10, 1973. Therefore, the Great Lakes freeboard tables which were made obsolete by the 1973 Great Lakes regulations remain in use for the Special Service assignment.

Section 7
Subdivision Load Lines

7.1 General. Earlier it has been noted that freeboard can be limited in several ways, such as:

- The numerical calculation of ship's geometry;
- the structurally limited design draft;
- stability limitations.

Another method of fixing draft, hence limiting freeboard, is by requiring a particular level of subdivision safety for the ship.

Although subdivision is not actually required by the International Convention on Load Lines, the Convention does recognize subdivision as one of the several reasons for allowing tankers a reduced freeboard and permits other ships properly subdivided to approach tanker freeboards (see Sections 3.16 and 3.17).

As a load line matter, subdivision is somewhat controversial from two points of view. The first is the fact that some national administrations require the subdivision calculation only at the load line draft. The second difficulty sometimes expressed is that the freeboard assigned to the ship is automatically spoken of as *reserve buoyancy* without an examination of the internal bulkheading in the ship. Since bulkheading is not actually required for load line purposes, there are ships which do not have adequate internal compartmentation to qualify for any degree of subdivision.

The bulkheading that exists in many ships provides an unknown level of subdivision safety which can only be assessed by calculation.

When any loss of buoyancy occurs, the ship must depend for its survival on the volume of ship above the normal waterline, which is not a part of the flooded portion of the ship. Thus, part of the same volume that is used for reserve buoyancy of an intact ship in a seaway earlier in this chapter is now used for protection against sinking from loss of internal buoyancy. This could be termed residual buoyancy.

There is an important distinction between the two types of buoyancy. Reserve buoyancy for seakindliness and safety in a storm is considered fully intact at all times and the crew of the ship is responsible for properly maintaining all closures in order that this buoyancy remains effective against the sea.

Residual buoyancy for resistance to flooding (loss of buoyancy) is, by definition, a matter of internal compartmentation of the hull.

The hull may lose some or all of its watertight integrity at any time due to stranding, collision, shell fracture, firefighting action, ruptured seawater piping, loss of hatch covers, etc.

The rationale for subdivision is *not* to provide an unsinkable ship. The folly of that idea should have been permanently established with the loss of the *Titanic*, some seventy years ago. Instead, subdivision is used to provide a level of safety which enhances the probability of survival from flooding or prolongs the time for damage control action to be taken by the crew or by ship salvage crews and finally provides extra time for the people on board the ship to be safely removed if this becomes necessary.

For ships required to have load lines which also require a subdivision analysis, a check is made to ascertain that the permissible geometric load line draft does not exceed the subdivision draft.

7.2 Passenger Ship Subdivision Load Lines. The International Convention for the Safety of Life at Sea (SOLAS, 1974) requires that every ship carrying more than 12 passengers on an international voyage must be examined and must satisfy a particular level of subdivision. The level required for each ship is set by formula and is dependent upon the number of passengers which the ship intends to carry and the length of ship.

There are two ways to approach this subdivision draft. The first is via Part B of Chapter II of the 1974 SOLAS Convention. The second is through the use of an alternative method explained and contained in IMCO Resolution A.265(VIII). It utilizes the same basic parameters of persons, that is, passengers and crew, and length of ship. However, additional factors have been considered based on probable statistical occurrence of damage. Factors from previous casualties such as damage position along the length of the ship, extent of damage, existence of longitudinal bulkheads, loading and permeabilities, sea states, and finally a corroboration of formulas by model test were all used to decide the final format and coefficients for the formula replacing the SOLAS 1974 method of determining compliance. The advantages of the new method are principally that established shipbuilding techniques, such as longitudinal bulkheading, which were not creditable under SOLAS 1974 are now officially recognized in the formulas as a subdivision asset. Also, greater latitude in internal bulkhead arrangements can be accounted for.

In addition to the international rules for subdivision, which apply to all passengers ships making international voyages, United States regulations require subdivision calculations for all passenger ships on the navigable waters of the United States, whether on foreign, coastwise or Great Lakes voyages, or for ferryboats on inland waters.

7.3 Environmental Subdivision Protection. Until recently, subdivision has been used primarily to provide assurance that passenger ships in collision would not sink rapidly, thereby allowing time for a salvage attempt and, if necessary, abandonment in a safe and orderly fashion.

In the last few years, subdivision and damage stability have been increasingly used to provide some assurance that hazardous pollutants such as chemicals and petroleum products carried in bulk would not be immediately lost to the sea in the event of an accident. The recommendations are contained in several Inter-Governmental Maritime Consultative Organization documents:

- 1971—Code for the Construction and Equipment of Ships Carrying Dangerous Chemicals in Bulk (IMCO, 1971).
- 1973—International Conference on Marine Pollution (MARPOL, 1973).
- 1975—Code for the Construction and Equipment of Ships Carrying Liquefied Gases in Bulk (IMCO, 1975).

In general, these ships must be capable of surviving damage conditions and, in the case of minor injury, successfully contain their cargoes. This includes bulk chemical carriers, tankers, and liquefied gas carriers. For details, see Chapter XI.

Section 8
Non-Standard Ships

8.1 General. The most important area for load line improvement is that concerning the *non-standard* ship. Such ships might be treated casually by signatory nations but for two factors of great impact. The first factor is the legal realization that the International Load Line Convention, 1966 is the most widely applicable marine convention, since it requires a load line on every floating object with a horizontal dimension greater than 24 m (79 ft) unless it is a warship, fishing vessel or yacht. The second factor is the realization that the number and variety of ships which do not carry cargo in a weathertight interior have been increasing. Therefore, it is of some importance to outline these differences and explore those parts where new international agreement is needed most.

Let us examine those areas in which some of the new non-standard ships must be treated differently than the standard ship. First consider offshore drilling units. These can be subdivided into three general categories: *Drillship, jack-up,* and *semi-submersible.* The ship type fits the L/D, C_B, and sheer of a standard ship but may not be fully maneuverable in a storm; therefore, sheer and bow height requirements are not utilized as contemplated by the Convention. The jack-up type is a barge with limited stability while under tow and is also usually lacking in sheer and bow height. A semi-submersible has an L/D of no use in the strength evaluation, a very small C_B so as to be transparent to waves, and no need for sheer nor bow height since normally the weather deck may be between 9 m (30 ft) and 30 m (100 ft) above the water whether under tow or drilling. Its loading capacity is often stability limited and it is essential for survival that no wave ever impact fully on the upper structure. While all of these types can be prepared for storms of various degrees of intensity, they are all dependent on advance weather warning for survival.

Surface-effect ships and hydrofoils need two separate evaluations as to their ability to weather a storm at sea. The first is in the waterborne mode and the second is in the out-of-water mode which involves high-speed interaction with waves and wind. In the latter mode, L/D, C_B, sheer, etc., are essentially meaningless in the sense of the standard ship seakeeping evaluation assumed by Load Line Regulation.

High-speed planing hulls also defy the theory of seakeeping in the sense of the standard ship, having a varying relation of C_B and sheer to the seaway since their speed changes their displacement.

Catamarans and other multi-hull ships have no direct relation to C_B, while ocean mining ships and pipe-laying barges are moored at a fixed heading allowing little opportunity to utilize bow height sheer effectively by turning into a storm sea.

Semi-submersible ships are now being built for several different services, some of which envision submerging at sea; this, incidentally, being prohibited by the Load Line Convention. Their purpose is primarily for the carriage of continental shelf equipment, heavy industrial items, and other barges. If these ever engage in submerged operations while at sea, they must be designed to certain limits in obvious contradiction of the Load Line Convention and the government of registry must report the design limits to IMCO.

8.2 Solution. Thus far, the solution to finding a proper load line for these non-standard ships has been a separate evaluation of each design in the proposed maximum seaway conditions by the designer himself. Since a pure textbook solution to many of these problems is not feasible, it often means a series of intricate and expensive model tests to determine within generalized limits the adequacy of the design and geometric parameters for safety at sea.

In order to find the best approach to overall safety at sea for each of these non-standard ships and at the same time retain the value of the Load Line Convention as a legally acceptable certificate of evaluation for seaworthiness around the world, there is a need for further international agreements. This can be accomplished as a series of special codes or as additional annexes to the Convention; however, it remains to be seen if the many different types of non-standard ships can be successfully generalized.

Generalizing the large numbers of non-standard ships may be done in a variety of ways. Legally, in order to retain for all vessels the general safety foundation established for ship-shaped hulls by the Convention, it may be necessary to drop the definition of international voyage now used (e.g., that of a point to point movement with more than one flag state involved) and to substitute a definition which includes operation on and exposure to the sea anywhere in the world as both standard and non-standard ships have a need for the governmental protection offered by the Load Line Convention. As a function of design, they may be categorized by differences in form, geometry, stability, and seakeeping limits. Regarding service, they may be categorized by type of cargo, length of voyage (ferry vs. ocean voyage) or area of operation (tropics vs. high latitude voyages). Finally, it is evident to the authors of this chapter that non-standard ships will be proliferating and will most certainly be operationally limited by the Administrations until research and development has progressed to a point where administrations can predict the effect of seaway operation on these ships and will therefore permit wider latitude in their use.

References

Board of Trade 1906, "Rules and Instructions for Applying the Tables of Freeboard to the Various Types of Steam and Sailing Vessels," Instructions to Surveyors-Load Line, 1909 Edition.

CHEMCODE 71, Code for the Construction and Equipment of Ships Carrying Dangerous Chemicals in Bulk, Resolution A.212 (XII), IMCO, London, 1971.

Code of Federal Regulations, 46 CFR 42; Technical and administrative load line regulations reflecting the present International Convention on Load Lines and the full ocean requirements of the Coastwise Act.

Code of Federal Regulations, 46.CFR 44, 45; Special Service Coastwise and the Great Lakes Administrative and Technical Regulations.

GASCODE 75, Code for the Construction and Equipment of Ship Carrying Liquified Gases in Bulk, IMCO, London, 1975.

ICLL 66, International Convention on Load Lines, IMCO, London, 1966.

MARPOL 1973, International Conference on Marine Pollution, IMCO, London, 1973.

Norton, Harold F., (1942), "Freeboard, Tonnage, and Capacity," *Principles of Naval Architecture,* SNAME, New York.

Public Law 74-359, 46USC88, "The Coastwise Load Line Act of 1935."

Public Law 93-115, 46USC86, "Foreign Voyage Load Line Act of 1973."

Robertson, James B., (1967), "Subdivision and Stability in Damaged Condition." *Principles of Naval Architecture,* SNAME, New York.

Ryan, Robert T., (1967), "Load Line Assignment," *Principles of Naval Architecture,* SNAME, New York.

SOLAS (1960), International Conference on Safety of Life at Sea—IMCO, London.

SOLAS (1974), International Conference on Safety of Life at Sea—IMCO, London.

CHAPTER V

R. T. Cunningham
Phillips Stitt

Tonnage Measurement

Section 1 Introduction

1.1 General. Practically all seagoing merchant vessels and many vessels operating on the bays, rivers, lakes, and sounds are measured for assignment of *national gross* and *net register* tonnages. Vessels intending to transit the Panama Canal and the Suez Canal are measured according to the rules of the respective Canal Authorities. This chapter is intended to define the tonnages, outline their histories briefly and, in a general way, point out some of the ways that tonnage factors influence ship design and what is being done to develop tonnage parameters that will not adversely affect design.

Most English language dictionaries, in defining the word *ton* clearly distinguish between a register ton as a measure of volume equivalent to 100 ft^3 or 2.83 m^3 and those tons (long ton, short ton, and metric ton) which are measures of weight. A resumé of the bases used to tax shipping in England since the 15th century and in the United States since the acts of the first Congress in 1789 reveals the etymology of the term *register ton* and of new meanings for gross and net tonnages.

Historians tell us that as early as 1423 British law required that imported wine be carried in casks, then called *tuns,* of a specified size. Although measures then were not very precise, the tuns held about 252 gal of wine, a weight of about 2,240 lb. It was the *custom* of the time for the Crown to fix the price it would pay for a part of the cargo. It became the *duty* of the importer to submit to a *tonnage tax* (Blocksidge, 1942) (Van Driel, 1925).

a. Displacement Tonnage. Because it was not possible simply to count the tuns on a ship with cargo other than wine, commencing in the latter part of the seventeenth century, as a revenue raising measure, dues were assessed on the approximate deadweight of the ship. Various methods had been used to determine the deadweight, including loading the vessels with a known weight of iron or lead and applying a freeboard mark. By an Act of the British Parliament of 1720 a formula known as the Builders Old Measurement Rule was adopted and remained in use until 1835. The formula which had been used to determine carrying capacities of vessels for owners was:

$$DWT = \frac{L \times \frac{B^2}{2}}{94} \qquad (1)$$

That formula was derived from the basic ship displacement formula:

$$\Delta = \frac{L \times B \times T \times C_B}{35} \qquad (2)$$

in which C_B = block coefficient. The average block coefficient for a ship of that time was taken as 0.62.
L = length of keel
B = maximum breadth
T = draft. The average draft was estimated to be $B/2$ for a two-deck ship. 35 = number of cubic feet of sea water in a long ton. Therefore,

$$\Delta = \frac{L \times B \times \frac{B}{2} \times 0.62}{35} \qquad (3)$$

The light ship was taken to be 40 percent of the displacement and the deadweight 60 percent, hence

$$DWT = \frac{L \times \frac{B^2}{2} \times 0.62 \times 0.60}{35} = \frac{L \times \frac{B^2}{2}}{94} \qquad (4)$$

A very similar formula for computing tonnages of United States ships was adopted by the first Congress in 1789; the formula remained in use until 1864 when the United States adopted the Moorsom System which had been developed and adopted by the British in 1854 after experimenting with other simple formulas (Blocksidge, 1942).

b. Moorsom System. Since the prescribed formula for computing the displacement tonnage of a two-deck vessel substituted one half the breadth for the depth factor, many owners had their vessels built full, long, very narrow and very deep in order to have large vessels with low tonnages. Of course, the carrying capacities came to exceed the assigned tonnages of the vessels and the vessels became clumsy and unstable. In order to eliminate the adverse effect of the tonnage formula on ship design and at the same time to stabilize the relationship of a vessel's carrying capacity to tonnage, the British turned to measurement of the internal capacity of the ship and the closed-in superstructures, a

[1] Complete references are listed at end of chapter.

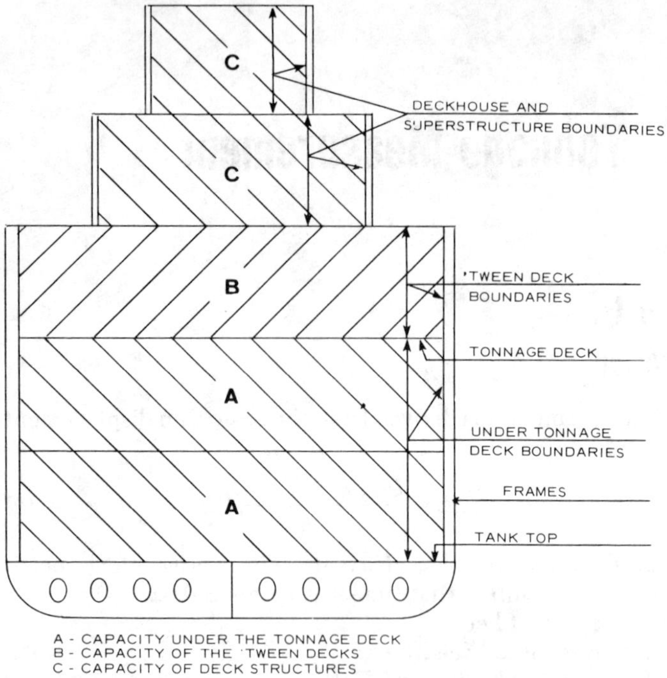

Fig. 1 Tonnage boundary definitions

system developed by Mr. George Moorsom. In developing the Moorsom System, the total volume of the British merchant fleet in cubic feet was determined. That volume was divided by the total assigned tonnage of the fleet and the resulting quotient, 98.22, provided a coefficient that would convert the volume of a vessel fairly close to the tonnage computed under the previous system. For convenience, however, and to reduce the number of vessels which would receive higher tonnages under the new system, 100 was selected as the divisor.

c. League of Nations Study and Oslo Convention. Studies on the unification of tonnage measurement systems had been initiated by the League of Nations as early as 1925 and a draft convention with regulations had been drawn up in 1939. A conference on this subject was prevented by the outbreak of World War II. Following the war, however, a convention based on those regulations was signed June 10, 1947 at Oslo, Norway. The Convention came into force December 30, 1954. By September 1966, sixteen governments had accepted or acceded to the Oslo Convention. While this afforded a degree of uniformity in tonnage measurement among the adherents to the Convention, a provision requiring unanimous acceptance of any amendments to the Convention or regulations made it necessary, as a practical matter in many cases, for the adherents to be guided by a series of recommendations lacking the force of agreed regulations.

d. Current Rules. The tonnage measurement rules currently in use evolved from the Moorsom System. Rules used by the major maritime nations such as British rules and rules derived from the Oslo Convention and United States rules (Code of Federal Regulations, 1978) are referred to as national rules. Rules used by the Panama Canal and Suez Canal Authorities are referred to as canal rules. For both national rules and canal rules the unit ton is 100 ft^3 or the metric equivalent, 2.83 m^3. The basic gross tonnage comprises the space under the *tonnage deck* bounded by the under surface of the tonnage deck, the tops of the floors or the tank top fitted to the floors and the line of the side frames or lining fitted to the frames plus the space in the 'tween decks if any, and finally the closed-in space above the tonnage deck available for cargo and stores or for the berthing or accommodation of passengers or crew, Fig. 1. The sum of the tonnages of non-revenue spaces is subtracted from the gross tonnage and the remainder is the net tonnage.

Although the various national and canal rules are based on the Moorsom System, there are significant differences in interpretations of the rules resulting in very different tonnage assignments for similar vessels of similar sizes and missions. The differences stem from the varying interests of both the measuring authorities and of businesses vital to the well-being of the nations. For example, national authorities have to balance interests of port authorities who need stable bases for assessing charges, on the one hand, against the interests of shipowners who, on the other hand, must compete with owners from other nations that may allow ships registered there to enjoy competitive advantages stemming from subsidies, liberal interpretation of tonnage rules, less stringent regulation, or lower labor costs.

Canal authorities, however, find it relatively easy to accommodate their interests and for that reason find it easier to maintain rational tonnage measurement rules. Canal authorities, for example, do not have ships in competition with other ships and each time the pertinent canal tonnage is being used as a base for assessing transit tolls the vessel is available for verification that the tonnages are correct. There are, therefore, comparatively few ways to reduce canal tonnages and comparatively few options to be considered by the designer.

1.2 History of Some Changes. Tonnage measurement rules generally are spelled out in great detail in national laws. More detailed criteria for application of the laws are spelled out in regulations. The laws and regulations are interpreted by administrations established for the purpose.

Since tonnages are used to determine the applicability of provisions of treaties, laws, and regulations and as bases for assessing charges, fees, and duties, any change in rules that would result in substantially different tonnage assignments for many vessels would disrupt the shipping industry. Historically, administrative decisions and rule changes have favored the shipowner, probably, because other segments of the shipping industry can protect their individual interests merely by changing rates for charges or by adopting parameters other than tonnages.

a. Transition from Deadweight to Volumetric Tonnage. As pointed out in section 1.1*b* of this chapter, the use of one half the breadth for the draft in the formula for approximating deadweight in the system preceding the Moorsom system led owners to acquire vessels that were poorly designed to obtain official tonnage assignments that were much less than the actual deadweight tonnages. When making the register tonnage a simple function of the volume of a vessel, the British opted for a coefficient (1/100) that would, on the average, slightly reduce the existing tonnages instead of opting for a coefficient that would more precisely approximate the deadweight. The reasoning apparently was that if a system yielding higher, more precise deadweight tonnages were adopted, shipowners would find themselves burdened with higher bases for being assessed charges with no assurance that charging authorities would correspondingly reduce their rates. Other governments in amending their laws followed the example set by England. As shown in Section 3 of this chapter, while the philosophy of avoiding radical changes in the tonnages assigned merchant vessels continues, governments can be persuaded to make their rules more logical. In seeking a universal tonnage measurement system, the International Conference on Tonnage Measurement held in London during May and June 1969 decided to do away with the system of exemptions and deductions from gross tonnage. Moreover, the conference adopted a formula that would yield gross tonnages closely approximating those of vessels measured under present national rules without exemptions for shelter 'tween decks, deck spaces opened by tonnage openings, passenger spaces, and water-ballast spaces. On the other hand, the conference decided to maintain the net tonnage advantage enjoyed by shelter deck types and to extend that advantage to other types of vessels having low draft to depth ratios. That decision has already caused some charging authorities to shift their charge bases from net tonnage to gross tonnage.

b. Resistance to Illogical Changes. With varying degrees of success, governments have, at times, resisted illogical rule changes forced upon or willingly adopted by administrations of other governments.

National rules usually provide that a space outside the double bottom adapted only to carry water ballast shall be included in the gross tonnage and *deducted* to arrive at net tonnage. They also provide that a space above deck shall be included in the gross tonnage if it is closed in and is available for the carriage of cargo or stores or for the berthing or accommodation of passengers or crew.

The United States adopted the Moorsom system by an Act of Congress dated May 6, 1864. That act specifically required passenger spaces to be included in the register tonnage. By an act dated February 28, 1865, however, the U.S. provided for the exemption of passenger spaces on or above the first deck which is not a deck to the hull. Only Liberia and Panama have followed that example. Pursuant to a law enacted February 6, 1909, until 1915 the U.S. included in the gross tonnage then *deducted* to arrive at net tonnage space adapted only for carrying water ballast outside the double bottom. In 1915, however, although the law had not been changed, the Commissioner of Navigation published regulations *exempting* water-ballast space from gross tonnage. Again, of the other countries now having large merchant fleets, only Liberia and Panama adopted the practice. In order to facilitate movement in and out of foreign ports, vessels with passenger space and water-ballast space exemptions must, at times, carry special appendices to their registers or British style tonnage certificates showing those spaces included in the gross tonnage.

The history of the open shelter deck exemption and IMCO Resolution A.48(III) dated October 18, 1963, titled *Recommendations on the Treatment of Shelter-Deck and other 'Open' Spaces* illustrates how governments can be persuaded in the interests of their shipowners, so to amend their rules that they become exactly contrary to a basic provision of their laws.

As mentioned above, basic tonnage measurement laws require that the volume of a closed-in structure available for the carriage of cargo or stores or for the berthing or accommodation of passengers or crew shall be measured and added to the gross tonnage. Under a British House of Lords decision of 1875 relating to the *SS Bear*, however, it was held in effect that a cargo space could be deemed to be under cover but open to the weather, that is not enclosed and not included in the gross tonnage. Thereupon, to protect the interests of charging authorities, the British enacted a deck cargo law in 1876 providing for measuring the volume of any cargo found in an open exempted space and including that volume in the taxable tonnage. More important for tonnage measurement considerations, the British Board of Trade adopted regulations establishing dimensions of tonnage openings and the degree to which the openings could be protected by coamings and temporary closures without making the deck space closed. Similar regulations were quickly adopted by other governments but not by the U.S.

The U.S. Treasury steadfastly held that deck structures having bulkhead tonnage openings with or without temporary closures and so-called open shelter deck spaces were, in fact, closed spaces suitable for carrying cargo within the meaning of U.S. law. Except in cases where bilateral treaties provided for mutual acceptance of national tonnages, Collectors of Customs were ordered to add the volumes of *open* deck structures and *open* shelter decks to the register tonnages when computing tonnage duties.

In 1915, however, after responsibility for the administration of the tonnage measurement rules had been transferred to the Department of Commerce, the Commissioner of Navigation published regulations for exemption of open spaces similar to but more liberal than the British regulations. The principal difference was that the U.S. regulations did not, and to this day do not, require an otherwise open, exemptible space to be included in the gross tonnage if it is accessible through a hinged or watertight door. Such easy access without loss of exemption makes it possible to exempt crew space as open under the U.S. rules but not under the British and Oslo rules. According to the 1915 Annual Report of the Commissioner of Navigation, the rationale for adoption of those criteria was that it removed a handicap for U.S. vessels and facilitated registration under U.S. laws of shelter deck vessels acquired by Americans from European owners at the outset of World War I.

Until shortly after World War II, classification societies permitted a shelter-deck vessel to be of substantially lighter scantlings than those required for a vessel of similar dimensions having its freeboard measured from the weather deck. After World War, II, classification societies in Europe required shelter-deck vessels to have increased strength. That led to the development of the *open-closed* shelter-deck vessel. Such a vessel has sufficient strength to qualify for assignment of a minimum geometric freeboard measured from the weather deck when its tonnage openings are sealed and from the second deck when the openings meet the criteria of the tonnage measurement rules. When in the closed condition, a convertible vessel is permitted to carry more weight than in the open condition but it does so at the cost of higher gross and net register tonnages than it would have in the open condition.

Open-closed shelter-deck vessels did not develop in the United States. In fact a number of vessels built as open shelter-deck vessels after World War II had their tonnage openings closed because MarAd adopted a policy to discourage vessels that could not meet a one-compartment subdivision standard. Without the ability to open their upper 'tween-deck bulkheads, U.S. vessels were assigned tonnages as much as 50 percent higher than their European counterparts in the open condition.

U.S. shipowners operated for years under that handicap until after the international maritime community through IMCO Resolution A.48(III) of October 18, 1963, recommended an interim solution to the problem of the reduced safety of open shelter-deck vessels pending development and adoption of a simplified, universal system of tonnage measurement to replace the several national systems.

The Soviet Union was the first country to adopt the recommendation, and the U.S., by a law enacted September 29, 1965, was the second. Most countries with merchant marines adopted the recommendation in one form or another as quickly as they could. In its most widely accepted form it is known as the *Tonnage Mark Scheme.*

The Tonnage Mark Scheme provides for assignment and recognition of open shelter-deck tonnages for a full scantling vessel without tonnage openings while operating at a freeboard approximating that it would be assigned if it were fitted with tonnage openings.

In practice under the scheme, a vessel may either be issued a certificate reciting two sets of gross and net tonnages, one set corresponding to the tonnages of an open shelter-deck vessel and the other set corresponding to the tonnages of a closed full scantling vessel, or the vessel may be issued a certificate reciting only the lower tonnages.

In the case of the dual tonnage assignment, the freeboard is calculated and the load line assigned with the upper deck as the freeboard deck and a tonnage mark, comprising a horizontal line 380 mm (15 in.) long surmounted by a 300 mm (11.8 in.) equilateral triangle resting on its apex, is placed on each side a prescribed distance below the second deck. When the horizontal line is submerged, the higher tonnages apply for all purposes for which register tonnages are used. When the tonnage mark is not submerged, the lower set of tonnages may be used as bases for assessing duties and other charges. Since a dual tonnage vessel measured under the Tonnage Mark Scheme needs only to submerge its tonnage mark to be of the higher tonnages, governments use the higher tonnages to determine the applicability of treaties, laws, and regulations relating to safety.

In the case of the single low tonnage assignment, the second deck is used as the freeboard deck and the tonnage mark is fixed at the upper level of the load line grid. Since, in such a case, the tonnage mark cannot be submerged without submerging the load line, the assigned low tonnages are used to determine the applicability of treaties, laws and regulations relating to safety, as well as, to assess charges.

Closed spaces for dry cargo and stores in detached superstructures and deck houses are exempted under the resolution whether or not the 'tween-deck spaces are exempted. Although the concept that a closed space available for dry cargo and stores may be exempted from gross tonnage cuts directly across the basic tonnage measurement tenet that such a space should be included in both gross and net tonnage, the resolution was enthusiastically adopted. The U.S. Government supported the resolution because it enabled full scantling U.S. vessels to be assigned tonnages approximating those of open shelter-deck vessel of other countries without the reduction in safety attending tonnage openings. Other governments supported the resolution because it made it possible for their vessels to be made safer without an increase in tonnage.

Section 3 of this chapter shows how the 1969 Tonnage Measurement Conference attempted to do away with the concept of tonnage openings and the concept of exempting any closed space on a vessel.

Section 2
History Leading to the 1969 Convention

2.1 Need for a Standardized System. The need for a standardized international system of tonnage measurement of ships is evidenced by the fact that small ships of identical size and form may measure less than 200 gross tons or more than 1,000 gross tons and the fact that exemptible and deductible spaces are treated differently under various national rules. The variations in tonnages cause inequities in the assessment of charges and in the application of provisions of treaties and laws.

This need for a standardized system was recognized in the initiation of the League of Nations study and in the Oslo Convention. However, there were many differences in national systems and in those systems evolving from the foregoing international activities that were yet to be resolved.

2.2 Work by the Intergovernmental Maritime Consultative Organization. In the meantime, the question of tonnage measurement had often been discussed by the Transport and Communications Commission of the United Nations. After the Intergovernmental Maritime Consultative Organization (IMCO) came into being in 1958, the task of developing a universal system of tonnage measurement of ships was taken over by the Organization as the United Nations had intended. Against this background, IMCO formed a subcommittee of its Maritime Safety Committee in 1959 to study the problem and to draw up recommendations for a system of tonnage measurement suitable for worldwide application, which would be just and equitable between the individual ships and groups of ships, and would not hamper good design or mitigate seaworthiness, and which would take account of the economics of the shipping industry generally.

Over a period of years, the Subcommittee and its working group considered a number of proposals for a universal system of tonnage measurement. Finally the International Conference on Tonnage Measurement of Ships, 1969, was held in London during a four-week period beginning May 27, 1969.

The Conference adopted the International Convention on Tonnage Measurement of Ships (ICTM, 1969), which the delegations felt largely met the above-listed criteria for a satisfactory system.

Section 3
International Convention on Tonnage Measurement of Ships, 1969

3.1 Gross Tonnage. Gross tonnage as defined in the 1969 Convention is a function of the total volume of all enclosed spaces of the ship. No exemption of enclosed spaces is permitted although there are certain open spaces that are carefully defined in the Regulations contained in Annex 1 to the Convention that are permitted to be excluded (ICTM, 1969). The gross tonnage of a ship is determined by the formula $GT = K_1 V$ where V is the total volume of all enclosed spaces in cubic meters and $K_1 = 0.2 + 0.02 \log_{10} V$, Table 1. All volumes included are to be measured, irrespective of the fitting of insulation or the like, to the inner side of the shell or structural boundary plating in ships constructed of metal, and to the outer surface of the shell or to the inner side of structural boundary surfaces in ships constructed of any other material. The volumes of appendages are to be included but the volumes of spaces open to the sea may be excluded.

The numerical value of the volume V exceeds the numerical value of the gross tonnage found by applying any of the national systems because those systems express gross tonnage in units of 100 ft^3 or 2.83 m^3 and take into account only the volume of space above the floors and inboard of the hull frames and inboard of deck structure frames less the volume of certain exempt spaces. The K_1 coefficient was developed, therefore, to convert the molded volume in cubic meters to a numerical value approximating the gross tonnage computed according to the national systems.

The fact that convention gross tonnage is a function of the molded volume of the entire vessel permits reasonably good approximation of the gross tonnage at an early stage of design of vessels of more than 10,000 dwt. In the preliminary design stage the total volume of spaces below the deck can be approximated as soon as the principal dimensions of length, breadth, depth, and draft are determined. Even though these dimensions may only be reasonably close to those finally selected, by assuming a block coefficient in the range of that to be finally determined, a displacement in metric tons Δ may be calculated. Using the calculated displacement, the volume under the weather deck Vu may be estimated by the formula:

$$Vu = \Delta\left(1.25\frac{D}{d} - 0.115\right) \tag{5}$$

in which:

D = Molded depth as defined at Regulation 2 (2) and used at Regulation 4 (1) of the 1969 Convention,

d = Molded draft amidship as defined at Regulation 4 (2) of the 1969 Convention.

The gross tonnage could then be estimated by substituting $Vu + V_H$ for V in the formula $GT = K_1 V$; where V_H represents the volume of the houses or enclosed spaces above the weather deck and Δ is the displacement in metric tons. The term, V_H, can be approximated as the product of the length, breadth and height of deck structures or by using the volume from a vessel with a similar size crew. V_H as a percentage of underdeck volume varies in a range of about five to 15 percent with the smaller percentage applicable to larger vessels such as tankers over 100,000 dwt.

When the design has progressed to the point that the lines have been reasonably well settled, the underdeck volume can be computed. The displacement curve can be extended to the upper deck amidships and this volume will be much more accurate, lacking only the sheer and camber, for use in the GT formula given in the Convention.

3.2 Net Tonnage. Net tonnage as defined in the 1969 Convention is primarily a function of the volume of cargo spaces and the number of passengers.

The formula for net tonnage NT is

$$NT = K_2 Vc\left(\frac{4d}{3D}\right)^2 + K_3\left(\frac{N_1}{1} + \frac{N_2}{10}\right) \tag{6}$$

in which:

V_c = total volume of cargo spaces in cubic meters

K_2 = $0.2 + 0.02 \log_{10} V_c$ (See Table 1)

K_3 = $1.25\,(GT + 10{,}000)/10{,}000$

D = molded depth amidships in meters as defined in Regulations

d = molded draft amidships in meters as defined in Regulations

N_1 = number of passengers in cabins with not more than eight passengers
N_2 = number of other passengers
$N_1 + N_2$ = total number of passengers the ship is permitted to carry as indicated in the ship's passenger certificate.

In applying the formula:

- the factor $(4d/3D)^2$ shall not be taken as greater than unity;
- The term $K_2V_c(4d/3D)^2$ shall not be taken as less than 0.25 GT, and
- NT shall not be taken as less than 0.30 GT.

Volumes of cargo spaces are to be measured as discussed under Section 3.1 for gross tonnage.

The draft to depth ratio permits a reduction of NT for those vessels with high freeboards and in effect maintains to varying degrees the shelter deck or the tonnage mark concept in previous regulations without any structural requirements. In some vessels with high freeboards the effect of squaring this ratio is excessive, therefore, the NT shall not be taken as less than 0.30 of the GT.

Another concept that is indirectly maintained is the water-ballast exclusion from net tonnage in that only cargo spaces are to be marked and included in V_c. (See Regulations).

In the preliminary design stage the cargo space volume V_c may be approximated as 0.66 times the underdeck volume approximated for the gross tonnage. This approximation may be as much as 10 percent low or 10 percent high and should be checked as soon as an arrangement plan is developed. The designer may have his own approximation to cargo space volume depending on the purpose of the volume and his own knowledge of required stowage factors. The approximation given applies to cargo vessels, bulk carriers, tankers and container vessels. Special types such as roll-on-roll-off require other consideration.

Provision is made in the Convention for occasional changes in the net tonnage if the characteristics of the ship such as V, V_c, d, N_1 or N_2 are altered (ICTM, 1969). The passenger part of the net tonnage formula requires no approximation since it is not a volume function but simply a function of the number of passengers.

A sample of the International Tonnage Certificate (1969) is given in Fig. 2.

3.3 Tonnage Coefficient. It should be noted that the Tonnage Conference adopted the logarithmic coefficients K_1 and K_2 in order to produce curves reasonably representing plots of molded volumes against gross tonnages and cargo cubics against net tonnages as contained in data furnished by IMCO members during studies preceding the Tonnage Conference. These coefficients are listed in Table 1.

Section 4
Precautions to Minimize Adverse Economic Impact of the Tonnage Convention

4.1 Precautions Taken by the 1969 Tonnage Conference. The 1969 Tonnage Conference, in order to minimize the adverse economic impact of adopting a truly international uniform system of tonnage measurement, sought a system in which the new tonnages would closely match those found under the various national systems. Since the Conference decided to do away with the shelter deck concept and other exemptions from gross tonnage, however, it realized that certain types of vessels would have substantially higher gross tonnages and, in some cases, higher net tonnages when measured under the new system.

Accordingly, the Conference decided that to be effective the Convention should be widely accepted and that so to minimize the impact there should be a relatively long period of transition from the national systems to the new system.

a. Coming into Force. In Article 17 (1) of the Convention it is provided that the Convention shall come into force twenty-four months after the date on which not less than twenty-five governments of states the combined merchant fleets of which constitute not less than sixty-five percent of the gross tonnage of the world's merchant shipping have deposited instruments of acceptance or accession. The Secretary General of IMCO in a document (TM. 2/Circ. 45), July 22 1980, announced that Japan had deposited an instrument of acceptance on July 17. With that action 44 states, with combined merchant fleets which exceed 65 percent of the gross tonnage of the world's merchant shipping, have become parties to the Convention. The Convention shall, therefore, come into force July 18 1982.

b. Transitional Period. Resistance to the Convention in countries which have not deposited instruments of acceptance is largely from owners and operators of *paragraph ships* who are concerned about the tonnage boundaries listed at Section 1.2 of this chapter. The Tonnage Conference anticipated some of the problems with paragraph ships and provided in Article 3(2):

> "The present Convention shall apply to:
> (a) New ships;
> (b) Existing ships which undergo alterations or modifications which the Administration deems to be a substantial variation in their existing gross tonnage;
> (c) Existing ships if the owner so requests; and
> (d) All existing ships, twelve years after the date on which the Convention comes into force, except that such ships, apart from those mentioned in (b) and (c) of this paragraph, shall retain their then existing tonnages for the purpose of the application to them of relevant requirements under other existing international conventions."

INTERNATIONAL TONNAGE CERTIFICATE (1969)

(Official seal)

Issued under the provisions of the International Convention on Tonnage Measurement of Ships, 1969, under the authority of the Government of

....

(full official designation of country)

for which the Convention came into force on...19....

by ....

(full official designation of the competent person or organization recognized under the provisions of the International Convention on Tonnage Measurement of Ships, 1969)

Name of Ship	Distinctive Number or Letters	Port of Registry	*Date

* Date on which the keel was laid or the ship was at a similar stage of construction (Article 2(6)), or date on which the ship underwent alterations or modifications of a major character (Article 3(2)(b)), as appropriate.

MAIN DIMENSIONS

Length (Article 2(8))	Breadth (Regulation 2(3))	Moulded Depth amidships to Upper Deck (Regulation 2(2))

THE TONNAGES OF THE SHIP ARE:

GROSS TONNAGE...

NET TONNAGE ....

This is to certify that the tonnages of this ship have been determined in accordance with the provisions of the International Convention of Tonnage Measurement of Ships, 1969.

Issued at ....19....

(place of issue of certificate) (date of issue)

....

(signature of official issuing the certificate)
and/or
(seal of issuing authority)

If signed, the following paragraph is to be added:

The undersigned declares that he is duly authorized by the said Government to issue this certificate.

....

(Signature)

Fig. 2a First page of International Tonnage Certificate

SPACES INCLUDED IN TONNAGE

GROSS TONNAGE			NET TONNAGE		
Name of Space	Location	Length	Name of Space	Location	Length
Underdeck	—	—			

NUMBER OF PASSENGERS
(Regulation 4(1))

Number of passengers in cabins with not more than 8 berths ....

Number of other passengers ....

EXCLUDED SPACES
(Regulation 2(5))

An asterisk (*) should be added to those spaces listed above which comprise both enclosed and excluded spaces.

MOULDED DRAUGHT
(Regulation 4(2))

Date and place of original measurement ....

Date and place of last previous remeasurement ....

REMARKS:

Fig. 2b Second page of International Tonnage Certificate

4.2 Unilateral Actions. Governments appear to be free to deal unilaterally with problems arising from tonnage boundaries relating to domestic laws and standards. Governments will probably, as a long range solution, raise some of the tonnage boundaries and replace some with other more relevant parameters. In the meantime, for a vessel required to be measured under the new system a government may, at the owners request, measure the vessel also under the previous national rules and use the tonnages so found to determine the applicability of the various domestic standards.

4.3 Limiting Parameters. Although the Tonnage Convention does not specify that its tonnages should be used as limiting parameters for determining the applicability of provisions of existing international conventions, there is little doubt that was one of the ultimate aims of the 1969 Conference. For that reason administrations appear understandably reluctant to recommend that their governments take unilateral action with respect to use of national tonnages as limiting parameters in connection with other international conventions after the 1969 Tonnage Convention comes into force. Accordingly the IMCO Assembly during its tenth session in November 1977 adopted the following Resolution A.389(X):

Table 1—Appendix 2 of the Convention Regulations

Coefficients K_1 and K_2 Referred to in Regulations 3 and 4(1)

V or V_c = Volume in cubic meters

V or V_c	K_1 or K_2	V or V_c	K_1 or K_2	V or V_c	K_1 or K_2	V or V_c	K_1 or K_2
10	0·2200	45 000	0·2931	330 000	0·3104	670 000	0 3165
20	0·2260	50 000	0·2940	340 000	0·3106	680 000	0·3166
30	0·2295	55 000	0·2948	350 000	0·3109	690 000	0·3168
40	0·2320	60 000	0·2956	360 000	0·3111	700 000	0·3169
50	0·2340	65 000	0·2963	370 000	0·3114	710 000	0·3170
60	0·2356	70 000	0·2969	380 000	0·3116	720 000	0·3171
70	0·2369	75 000	0·2975	390 000	0·3118	730 000	0·3173
80	0·2381	80 000	0·2981	400 000	0·3120	740 000	0·3174
90	0·2391	85 000	0·2986	410 000	0·3123	750 000	0·3175
100	0·2400	90 000	0·2991	420 000	0·3125	760 000	0·3176
200	0·2460	95 000	0·2996	430 000	0·3127	770 000	0·3177
300	0·2495	100 000	0·3000	440 000	0·3129	780 000	0·3178
400	0·2520	110 000	0·3008	450 000	0·3131	790 000	0·3180
500	0·2540	120 000	0·3016	460 000	0·3133	800 000	0·3181
600	0·2556	130 000	0·3023	470 000	0·3134	810 000	0·3182
700	0·2569	140 000	0·3029	480 000	0·3136	820 000	0·3183
800	0·2581	150 000	0·3035	490 000	0·3138	830 000	0·3184
900	0·2591	160 000	0·3041	500 000	0·3140	840 000	0·3185
1 000	0·2600	170 000	0·3046	510 000	0·3142	850 000	0·3186
2 000	0·2660	180 000	0·3051	520 000	0·3143	860 000	0·3187
3 000	0·2695	190 000	0·3056	530 000	0·3145	870 000	0·3188
4 000	0·2720	200 000	0·3060	540 000	0·3146	880 000	0·3189
5 000	0·2740	210 000	0·3064	550 000	0·3148	890 000	0·3190
6 000	0·2756	220 000	0·3068	560 000	0·3150	900 000	0·3191
7 000	0·2769	230 000	0·3072	570 000	0·3151	910 000	0·3192
8 000	0·2781	240 000	0·3076	580 000	0·3153	920 000	0·3193
9 000	0·2791	250 000	0·3080	590 000	0·3154	930 000	0·3194
10 000	0·2800	260 000	0·3083	600 000	0·3156	940 000	0·3195
15 000	0·2835	270 000	0·3086	610 000	0·3157	950 000	0·3196
20 000	0·2860	280 000	0·3089	620 000	0·3158	960 000	0·3196
25 000	0·2880	290 000	0·3092	630 000	0·3160	970 000	0·3197
30 000	0·2895	300 000	0·3095	640 000	0·3161	980 000	0·3198
35 000	0·2909	310 000	0·3098	650 000	0·3163	990 000	0·3199
40 000	0·2920	320 000	0·3101	660 000	0·3164	1 000 000	0·3200

Coefficients K_1 or K_2 at intermediate values of V or V_c shall be obtained by linear interpolation.

THE ASSEMBLY,

"NOTING Article 16 (i) of the IMCO Convention concerning the functions of the Assembly,

NOTING FURTHER that the International Convention on Tonnage Measurement of Ships, 1969, may come into force in the near future,

REALIZING that tonnages determined under the 1969 Tonnage Convention can be sufficiently different from those determined under tonnage regulations presently in force as to create difficulties in connection with the application of the International Convention for the Safety of Life at Sea, in force,

RECALLING Recommendation 2 of the International Conference on Tonnage Measurement, 1969, which, inter alia, recognized that the transition from existing tonnage measurement systems to the new system provided in the 1969 Tonnage Convention should cause the least possible impact on the economics of merchant ships.

DESIRING to overcome possible difficulties which might arise with regard to the application of the safety requirements in force for certain ships when measured in accordance with the 1969 Tonnage Convention in comparison with the national tonnage rules in effect prior to the coming into force of that Convention.

NOTING that the International Convention for the Safety of Life at Sea does not specifically define the gross tonnage of ships which should be measured for the purpose of application of the provisions of that Convention,

HAVING CONSIDERED the recommendations by the Maritime Safety Committee at its thirty-sixth session,

ADOPTS the following interim scheme:

(a) At the request of a shipowner, the Administration may allow a ship required to be measured under the provisions of the International Convention on Tonnage Measurement of Ships, 1969, to use the gross tonnage measured under the national tonnage rules which were in effect prior to the coming into force of the 1969 Tonnage Convention, for the purpose of application of the International Convention for the Safety of Life at Sea, such tonnage however shall not be shown on the 1969 Tonnage Certificates;

(b) For such a ship, the appropriate box in the pertinent Ship Safety Certificate of the International Convention for the Safety of Life at Sea, in force, may show only the gross tonnage measured under the national tonnage rules which were in effect prior to the coming into force of the 1969 Tonnage Convention, with the following footnote:

'The above gross tonnage has been measured by the tonnage authorities of the Administration, in accordance with the national tonnage rules which

were in force prior to the coming into force of the International Convention on Tonnage Measurement, 1969,
AGREES that the above interim scheme will expire on 31 December 1985,
INVITES Member Governments and governments of States Parties to the aforementioned Convention to take cognizance and to accept the use of this interim scheme, for the purpose of application of the provisions of the International Convention for the Safety of Life at Sea,
REQUESTS the Maritime Safety Committee to continue its work on this subject with a view to developing amendments to the International Convention for the Safety of Life at Sea, 1974, to obviate problems which may be encountered upon the entry into force of the 1969 Tonnage Convention."

The 1969 Tonnage Convention has the potential, ultimately, to eliminate gross tonnage as an important design influence. In the meantime, at least through 1985, designers may find it necessary to be familiar with the national tonnage measurement rules where the vessels they design are to be registered.

References

Blocksidge, E. W. (1942), "Hints on the Register Tonnage of Merchant Ships," *The Journal of Commerce,* Liverpool.

Code of Federal Regulations (1978), Title 46, Part 69, U.S. Government Printing Office.

IMCO (1969), International Conference on Tonnage Measurement (ICTM), London.

Ryan, Robert T. (1969), "Tonnage Admeasurement," *Principles of Naval Architecture,* edited by J. P. Comstock, SNAME.

Van Driel, A. (1925), "Tonnage Measurement Historical and Critical Essay," The Hague Government Printing Office.

CHAPTER VI

David B. Bannerman, Jr.
Hsien Y. Jan

Analysis and Design of Principal Hull Structure

Section 1
Ship Types

1.1 Wooden Ships. Up to the end of the 18th Century, wood was the only material used in shipbuilding since no other material was so readily available and easily worked with simple hand tools. Although the study of wooden ships is mainly of historical interest, it is included since wood practice had an effect on the development of *scantlings,* distribution of material, and method of construction of iron ships.

Apart from the inherently lower strength of wood, one main weakness of wooden ships was that the fastenings (i.e., spikes, bolts, wooden treenails, etc.) could not develop the full strength of the material. But more lately, the development of techniques to make use of strong and durable adhesives has overcome this limitation in the few wooden craft that are left, such as minesweepers.

The old wooden ships were relatively small, the clipper ships of a century ago averaging around 61 m (200 ft) in length. The size of later ships, structurally reinforced by iron, was limited to around 91 m (300 ft) by the difficulty of obtaining heavy scantlings and long lengths of timber, even with a length-to-depth ratio around 9.

The midship section of a wooden sailing ship in Fig. 1, reproduced from the early American Bureau of Shipping Rules, shows the massive keel construction and preponderance of material in the bottom as compared with the deck.

1.2 Iron Ships. It was not until the 1830's that the construction of iron seagoing vessels began. The advent of steam propulsion, necessitating more inboard space for the machinery and bunkers, accelerated the change from wood to iron.

The first builders of iron ships routinely continued to use the transverse system of framing. The iron scantlings reflected their idea on the relative strength of iron compared to wood. The midship section of an early iron ship, shown in Fig. 2, resembles its wooden counterpart, with considerable material in the bottom as compared to the relatively light top flange of hull girder. The introduction of steamers, with their greater length and length-to-depth ratio, intensified and brought out the deck weakness.

One of the first applications of beam theory to iron ship structural design was to the *Great Eastern,* built in 1858, and although very much larger than any ship with which there had been any experience, the scientific approach resulted in a structure that was satisfactory in service. Arnott (1955)[1] describes this remarkable ship in more detail.

1.3 Steel Ships. Steel was introduced to shipbuilding around 1870. The change from iron to steel was, like that from wood to iron, very gradual, but less revolutionary. Steel, then made by the Bessemer process, besides costing about 50 percent more than iron, was more difficult to fabricate and tended to be brittle. Developments in steel manufacture, especially the open-hearth process, improved the quality and gradually reduced the cost. By 1890, steel had replaced iron in British shipyards. Classification requirements for certification specified rigid testing at the mill which insured uniformly high quality ship steel. Such tests had never been required for iron.

Except for the much larger rolled plates and wider variety of shapes available, substitution of steel for iron did not affect structural arrangements or methods of construction. The greater strength of steel allowed a 20 percent reduction in most scantlings.

1.4 Introduction of Welded Construction. Until World War II, classification society rules described structures primarily joined by riveting. Although welding had been in limited use for many years, it became universally accepted during the World War II emergency shipbuilding program as a strategic and economic necessity. The change brought its own problems (Bannerman and Young, 1946) but the advantages were great. Numerous parts could be dispensed with, particularly boundary bars, clips connecting webs to

[1] Complete references are listed at end of chapter.

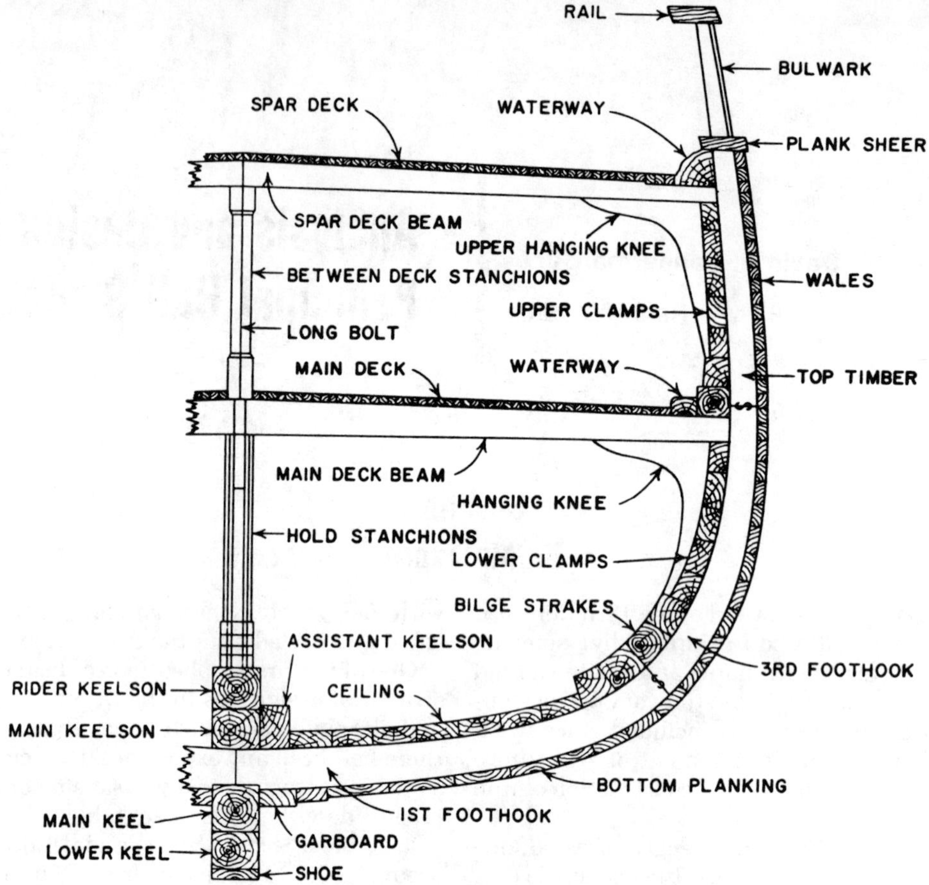

Fig. 1 Wooden sailing ship midship section

plating, and many others which served only to take the rivets. The laying out of rivet holes, punching, reaming, and countersinking the holes, bolting up and finally driving the rivets were eliminated. Edges of plates and bars at boundaries of watertight and oiltight spaces which had to be caulked with riveted construction were now monolithic and required no caulking. Overlaps or straps needed for riveted connections of plates were eliminated in favor of flush-welded butt joints that minimized the number of pieces and reduced weight to about the minimum needed for a serviceable ship structure. A very important advantage is that welded construction can be quite severely damaged and still remain watertight, whereas rivets would be loosened and leak. The application of welding has been by far the most important change in ship construction since the advent of steel.

1.5 Unique Nature of Ship Structures. In addition to being the largest moving structures made by man, the design of ships differs considerably from that of other structures for several reasons. Awareness of these peculiarities will make the subsequent presentation of strength requirements for ships, in this chapter, more readily understood.

Unlike many structures, to which known loads from wind, snow, and dead load can be applied with reasonable accuracy, the most critical loading on the ship's structure is imposed by the sea. This is added to the known loads but its magnitude is uncertain; furthermore, the sea loading is made even more complex by the motions of the ship itself in response to the loading from the sea.

The structure of a ship is complex, its geometry is unique; it cannot be expressed in mathematical terms. Most of the structure consists of broad expanses of plating, stiffened by a variety of structural shapes, whereas most other large structures contain little plating.

An important additional requirement is that the structure must remain tight for the containment of liquids within and the exclusion of water from without besides having sufficient strength to withstand the pressures from these liquids. Both of these requirements are affected in an imprecise but significant way by the corrosive nature of the salt water environment in which the structure operates, for which allowance must be made after determining the necessary initial strength of the intact structure.

While strength to withstand all expected loadings is the primary criterion for design, the stiffness of the structure may also be a constraint on the geometry quite apart from strength. Reference should be made to Subsection 1.6c for a discussion of the ratio of length to depth and its influence on stiffness.

Since the mission of most ships is to carry cargo, the

weight of the ship itself is kept as low as possible consistent with necessary strength and stiffness. Optimum distribution of material is a matter of higher priority in ship design than in some other structures.

Notwithstanding the peculiarities of ship design, ships have been used for centuries, most of which were successful as structures. Design, therefore, has become a rationalization of existing systems extrapolated to each new design. Fortunately, in recent years, as a result of statistical methods and many wave observations taken in real seas, much more is known and a relatively accurate description of the short- and long-term behavior of the sea can be had, from which load estimates may be obtained.

1.6 Development of Classification Rules.

a. History. Classification of wooden ships involved periodic maintenance surveys rather than complying with rules for new construction. The quality and probable life of these small sailing ships depended more on the kind of wood, type of fastening, and standard of workmanship than on the scantlings. When Lloyd's Register issued the first wood rules in 1835, based on a tonnage numeral, most British sailing ships were similar and relatively short (about 100 ft long). These rules worked out satisfactorily.

Early iron ships, designed and built without the benefit of classification or other rules, varied considerably structurally. The individual designer's idea of the iron equivalent to a wooden ship of the same size determined the scantlings and structural arrangements. While most ships had transverse framing, some had longitudinal or diagonal frames. A few ships had no framing at all, which was practical only because they were small, with rounded midship sections and heavy shell plating.

Lloyd's Register classed the first iron ship in 1832 and in 1855 published its Rules for building iron ships. These were of a simple form and patterned after those for wood. Tabulated scantlings were in sixteenths of an inch, by tonnage,

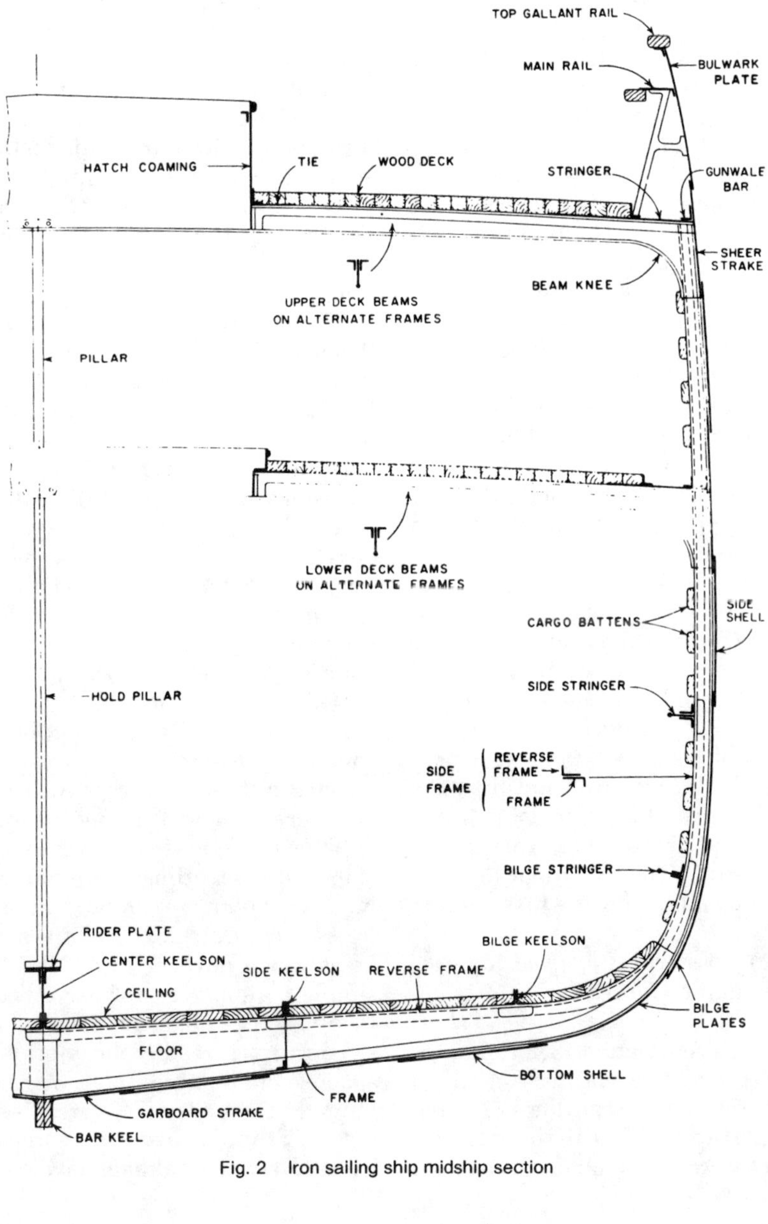

Fig. 2 Iron sailing ship midship section

for ships of 6-, 9-, and 12-year grades. Ships complying with these rules were assigned the character *A*. The retention of this character was dependent on the results of periodic surveys.

Lloyd's Register explained that because of the lack of experience with iron ships, the new Rules were based to some extent on experience with wooden ships. It insinuated also that these Rules would be revised when adequate information on hull corrosion and on unascertained points became available. This approach has been characteristic of the development of classification society rules. Classification societies claim a responsible balance between the acceptance of innovations, based on technological progress, and the need to insure the safety and reliability of ships's structures through the aid of service experience.

In 1863, Lloyd's Register Rules were amended. The designation of class by years was abolished and symbols substituted to designate probable durability and to indicate length of time between special surveys.

In 1870, Lloyd's Register issued new Rules for iron ships with the scantlings based on numerals determined by dimensions. New classification symbols were introduced, the familiar 100A1, 90A1, and 80A1 being preceded by a maltese cross (✠) where ships were built under the supervision of the surveyors.

The first iron Rules of the American Shipmasters' Association (renamed American Bureau of Shipping in 1898) were published in 1872.

It was remarked that, "these Rules are based upon the belief that properly constructed iron vessels of good materials, of sufficient strength, by the use of cement inside, and with necessary attention to the outside coating, accidents excepted, will last, in good condition, for a period of 20 years." The scantlings of keel plates and the diameter of rudder stocks were based on a tonnage numeral. The framing was based on depth of hold, the shell plating on the sum of the half-breadth and depth, and the keelsons and stringers of the various decks on ship's length. The longitudinal scantlings were augmented when the length-to-depth ratio exceeded 12.

Lloyd's Register 1885 Rules for Iron and Steel Vessels are important because for many years they constituted the legal standard of strength for the British Load Line Regulations. Two sets of numerals were used for scantlings. The first, or transverse number, was obtained by adding the measurements in feet of the half molded breadth amidships, the depth from the keel, and the girth of the half midship frame sections, measured from the centerline at the top of the keel to the upper deck *stringer plate*. The second, or longitudinal number, was obtained by multiplying the first number by the length of the vessel.

A general reduction of 20 percent was allowed for steel ships; e.g., the sixteenths of an inch for iron scantlings become twentieths for steel.

In the American Shipmaster's Association 1888 Rules, the tables were greatly extended and a special section added which outlined in detail the additions to scantlings required for ships of excessive proportions. This is the first appearance of steel rules which were drawn up on the basis of a reduction of 20 percent from the scantlings for iron ships. The scantlings given for steel ships were tabulated in pounds per square foot.

The advent of the British Corporation Register of Shipping in 1890, when steel shipbuilding was well advanced, signaled the birth of a new spirit in classification work. It started with a clean slate, unhampered by practices and administrative methods dating back to the wooden ship era. The whole classification structure was thoroughly investigated and simplified. The principle was laid down that, provided a ship met the required strength standard for her maximum draft, she was fit to be classed and to be retained in class as long as her hull and machinery were maintained in good condition. Only one class was specified in association with a service or draft limitation where necessary. A more specific approach to ship structural problems than had been evident in the past was reflected in the first issue of the Rules for Steel Ships, published in 1893. The system of numerals was abandoned and scantlings were determined in relation to the work that the various structural components were expected to do. The table scantlings were graded in $1/40$ in. in lieu of $1/20$ in. and the rules were applicable to ships whose length-to-depth ratio did not exceed 14. A formula was given for the diameter of rudder stocks. Detail structural plans were required to be submitted. Plan approval was centralized in the head office and not left to the surveyors in the field, as had been classification practice in the past.

In the American Bureau of Shipping Rules of 1900, steel ships were given first place. The old tonnage numeral was dropped as a basis for certain scantlings and scantling numerals similar to Lloyd's, including the half-girth, appeared for the first time in the ABS Rules.

In 1909, the Lloyd's Register Rules were completely revised and in all respects very much improved. Lloyd's Register went one step further than the British Corporation by specifying gradations of 0.02 in. While table scantlings were still determined by dimension numerals, the latter were very much simplified, the half-girth measurement being eliminated. Rules were laid down for two basic types, full-scantling ships and ships with complete superstructures. Scantlings for ships with intermediate drafts were determined by interpolation.

In 1916, the British Corporation Rules were revised completely. Extended use was made of simple formulas for arriving at the scantlings of such component parts of the ship structure as hold framing, beams, girders, and bulkhead stiffeners. Following the precedent set by Lloyd's Register 1909 Rules, scantlings were tabulated in increments of 0.02 in. The design maximum draft entered directly as a factor in assessing scantlings, and the specifying of minimum deck areas was a novel feature. While the Rules were not as simple to apply as former rules, they were much more flexible. They had the decided advantage of enabling designers to gage readily the effect of variations in draft, frame spacing, etc. These Rules, with some modifications to suit United States practice, were adopted by the American Bureau of Shipping after its reorganization in 1916, and also by the Registro Italiano Navale and NKK.

Classification society rules are promulgated only after consideration and approval by technical committees, thoroughly representative of the shipbuilding and allied industries. This insures their practicability and suitability for up-to-date standards. Classification societies have continued to grow in influence and importance because of their usefulness to the shipping fraternity (Murray, 1955); (American Bureau of Shipping, Annual).

b. Purpose of Classification. The reason for the existence of classification societies is not always understood, probably because there is nothing quite like classification service in other fields of activity. From the beginning of maritime commerce it has been in the interest of the shipowner and the shipper of goods, and later the marine underwriter, to assure the soundness and seaworthiness of ships. The forces to which the ship is subjected by the sea are not wholly understood; therefore, the only criterion by which a ship can be appraised reliably is by comparison with similar ships known to have been successful in service.

Forward strides are being made through research, both analytically and by instrumentation of ships in service, in learning the nature and magnitude of the forces of the sea. Also, continual review has made the rules of the classification societies, referred to throughout this book, more precise in comparing one ship with another, and in comparing individual components on the basis of recognized engineering principles.

Through application of the record of successful experience of ships in service and of the theory of structures, standards for the construction of ships and their machinery have evolved which are acceptable to all parties interested in ships. These standards are referred to as Rules. It may be noted in Subsection 1.6a that the standards have changed greatly and continually over the years, as experience was gained with new types of ships, new materials, and different services.

When a ship is built to the requirements of the rules of a classification society, under the survey of the society's surveyors, the hull material and other components are tested to specifications given in the rules. If all tests and trials prove satisfactory, the society grants "classification" to the ship by formal action of its committee in accepting the recommendations and reports of the surveyors. This fact is then published in the society's register book, where anyone may see that the ship in question conforms to recognized standards of sound construction.

The classification of the ship helps the owner, in the event of a casualty, to establish that he has used "due diligence" required of him; it informs the shipper that he is not taking a disproportionate risk by sending his goods aboard that particular ship; and it helps the underwriter decide the nature of the risk involved when he is asked to insure the ship, especially if it is a wholly new type of vessel or an unusual ship.

While the relation of each society to governments varies, in general they are independent, a fact which makes their services valuable. The major societies have offices throughout the world and representatives are available at almost any port. For this reason, many governments have authorized classification societies to carry out on their behalf the technical survey and certification required by the various international conventions with respect to ships under their registry (IMCO, 1966) (IMCO, 1974). This delegation of duties is an indication of the recognition given to the competence and integrity of the societies.

The societies are nonprofit in nature, and are supported by fees charged to the owners or to the builders of a new ship for their services.

The rules also provide for periodical surveys of ships in service to ascertain that they are being maintained "in class." Not only does this tend to insure proper maintenance of the ships, but furnishes valuable information on the adequacy of the rule requirements and points out areas in which revision of the rules should be considered. In such cases the staff prepares studies to support a change shown to be needed, and then submits them to the cognizant committee of the society for its approval. This is the process of evolution of the rules.

c. Length-to-Depth Ratios. As mentioned earlier, the ratio of length to depth (L/D) of the comparatively small ships of iron and steel of a century ago was quite small, and therefore, required no upper limit. As lengths increased, however, the application of compulsory freeboards resulted in wasted cubic contents, unless the depth were kept as low as permissible. Rules of the classification societies indicate the proportions of ships to which they apply and, if a ship is designed outside of those limits, class may not be granted or specially determined increases in scantlings may be required.

For oceangoing ships, the length-to-depth ratio has been held to a maximum of around 14, later extended to 15.

Smaller ships in limited, short, coastwise services such as interisland and the Gulf of Mexico have been permitted greater ratios of length-to-depth (about 16 being general in the United States). Some of the barges in this trade are so large, and the experience of the smaller barges has been so successful, that certain of them have been approved for more extensive services even with higher ratios. This is particularly true of the coastwise trade on the Pacific coast of the United States, where one class of railroad car barges having longitudinal strength much above the minimum has operated to Alaska with an L/D ratio of 20.

The L/D ratio is a rough measure of the stiffness of the vessel and, of course, for a given set of scantlings, a deeper hull will provide equal longitudinal strength with less weight. It is known from experience that the extremely shallow hulls of river barges will not survive large ocean waves. However, concern for the need for stiffness has been examined critically in the light of the long, successful operation of Great Lakes bulk carriers having L/D ratios of 18 and over, up to an extreme of 21. These ships exhibit visible deflections but suffer no apparent damage. The difference has always been justified on the basis of shorter waves on the Great Lakes, but this too is being carefully studied.

d. Influence of Draft. Even before the enactment of load line laws, draft entered into the determination of scantlings. The British Corporation Rules of 1916 introduced draft as a direct factor in determining strength requirements. This concept continued for many years until it became recognized that bending moments induced by waves were little affected by draft, and the formulas for required section modulus began to give progressively less effect to draft. More effect was given to stillwater bending moments which are a function of weight distribution rather than total weight to which draft is related, until draft has all but disappeared as a factor in arriving at scantlings.

e. New Design Concepts. With rapid advances in technology, the traditional concept of classification as a compendium of satisfactory experience is not sufficient to cope with the totally new concepts which make their appearance quite frequently. Some of the more important of these are: nuclear energy as a power source, with its attendant hazards; liquids at extremely low temperatures as bulk cargo; completely novel structures for use in drilling for oil and gas in deep water; and submersible vehicles capable of descending to the greatest depths of the ocean. The lack of experience is not likely to deter the pioneers in these new fields, but they will seek some standards and be glad to refer to whatever previous experience that bears any relation to their problems.

When new concepts are presented, the classification societies are not in a position to grant class in the usual manner. Instead of evolving requirements for satisfactory service wholly on previous experience with similar ships, the societies participate in the development of the new concepts. They use their experience wherever applicable to materials, components, etc. When the industry is prepared to adopt standards, the mechanism of the technical committees of the societies is brought into play to formulate them.

A somewhat similar procedure is used when new developments take the conventional type of ship design far beyond the size with which present experience has been accumulated. Typical examples are those of oceangoing tankers and Great Lakes bulk carriers.

Section 2
Framing Systems

2.1 Transverse Framing. Essentially, this system consists of a series of closely spaced ribs girding the ship. These ribs, comprising floors, side frames and beams (Fig. 2), stiffen the shell and deck plating upon which longitudinal strength primarily depends, and help in supporting hydrostatic and cargo loadings, besides keeping the structure in shape. The side frame (Fig. 2) consisted of an angle iron extending the full depth of the ship, riveted back-to-back to a reverse angle, which in the larger ships extended alternately to the upper and second decks. In early iron ships, great emphasis was laid on the number of decks or tiers of beams, regulated by hold depth, following wooden ship

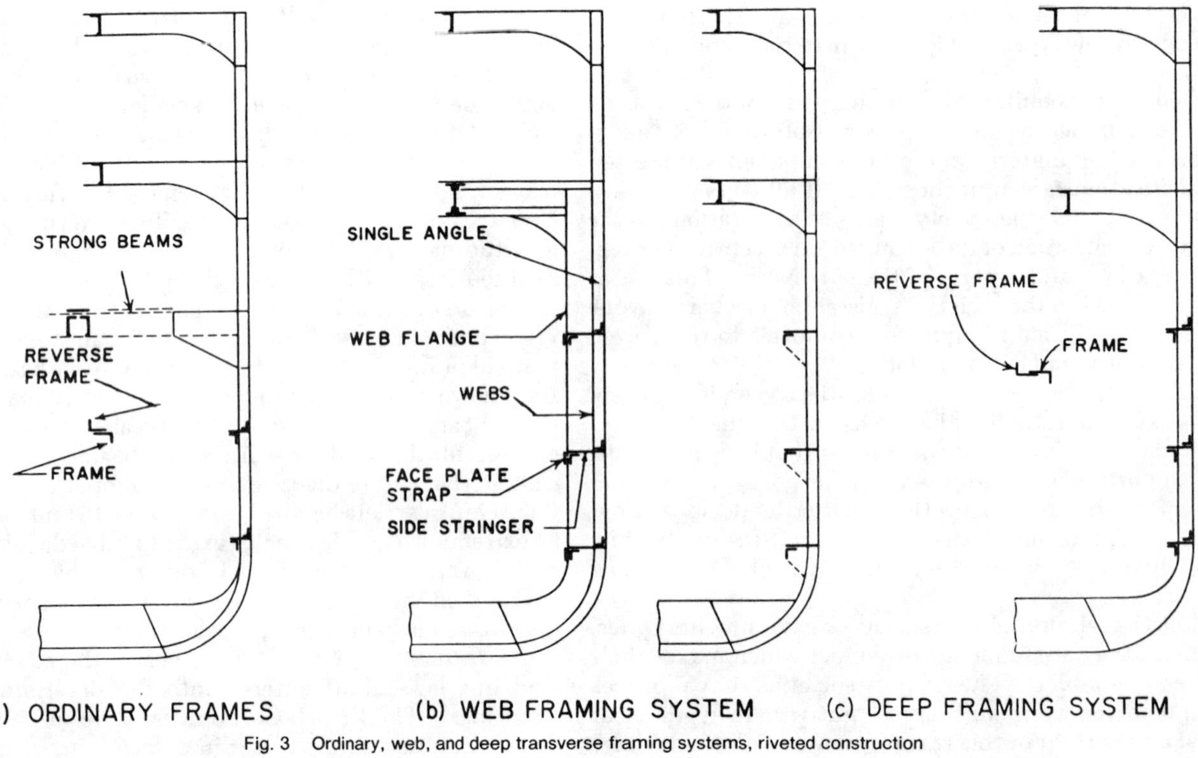

Fig. 3 Ordinary, web, and deep transverse framing systems, riveted construction

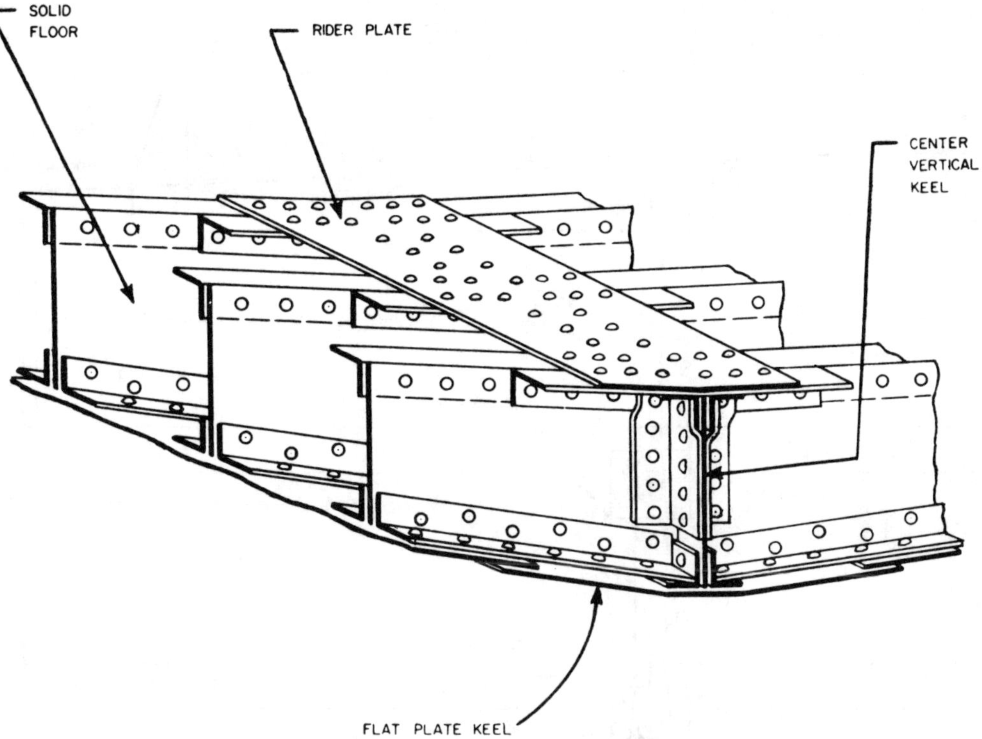

Fig. 4 Single bottom of riveted construction

practice. The lower tier, built of strong beams 8 to 12 frame spaces apart, facilitated cargo handling, Fig. 3a. Later, this was replaced by the *web-frame* or a *deep-frame* construction system. Fig. 3 illustrates these three systems of transverse framing. Web frames (Fig. 3b) are continuous, with horizontal girders (side stringers) fitted *intercostally* between them. A narrow plate strap, instead of the former large *diamond plates* at the intersections, provides continuity to the girder face bars. The intercostal side stringer shown in Fig. 3b is simpler and less massive than those originally associated with deep framing. The web framing system, though structurally efficient, interfered with cargo stowage and, therefore, was replaced by the deep framing system of Fig. 3c.

The deep frames of Fig. 3c, built up from two angles, were superseded by single rolled sections such as large bulb angles or channels at every frame, usually without side stringers. The *faying flanges* of these sections are removed for welded construction. In several countries other than the U. S., bulb-plate sections are frequently used.

a. Single Bottoms. Single bottoms in early iron ships, patterned after those of wooden construction, were inefficient for strength and cargo stowage, see Fig. 2 and Arnott (1955). Figure 4 shows an improved single bottom of riveted construction.

b. Double Bottoms. Liquid ballast was carried originally in separate tanks. As a logical development, these were integrated with the ship structure, thus resulting in the adoption of the double bottom.

Early riveted double bottoms had *solid floors* at every

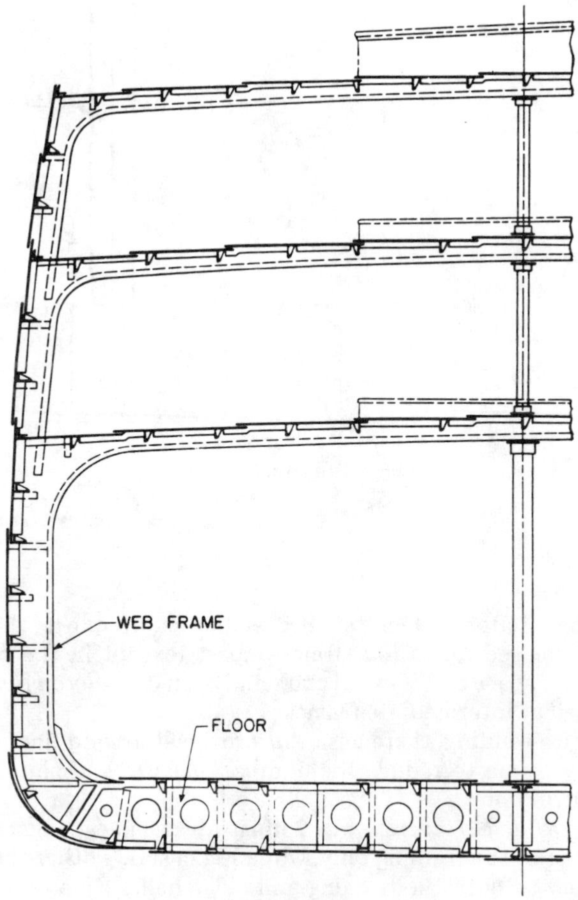

Fig. 5 Cargo ship, Isherwood system

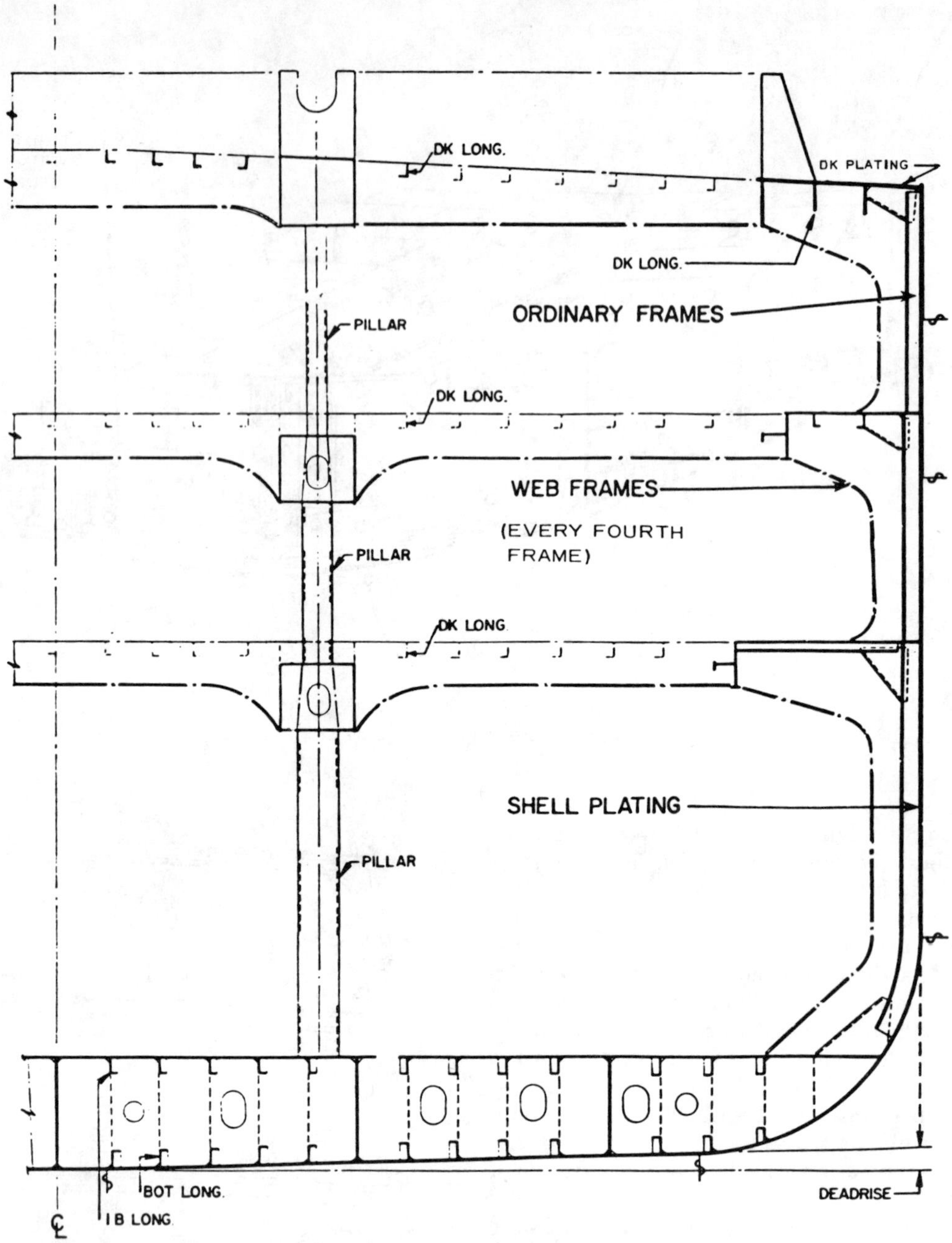

Fig. 6 Midship section of an all-hatch cargo ship

frame. Later, to facilitate access and maintenance, these were spaced up to four frames apart (except in the machinery space and forward end), and open or *skeleton floors* placed at intermediate frames.

With welding, there has been a reversal to solid floors at every frame to reduce the number of parts and simplify construction.

c. Side Stringers. Fig. 2 illustrates a side stringer attached to the framing only, while Fig. 3 shows others connected to both the framing and side shell. Riveted side stringers were inefficient structural members. However, those attached to the side shell helped prevent panting and tripping of the hold frames at the ship's ends.

d. Beams. Early iron ships had wooden decks laid on iron stringers and tie plates. Beams were fitted on alternate frames and consisted usually of a bulb plate with double angles riveted to it, as shown in Fig. 2. The beam knees (Fig. 2) were formed originally by splitting and spreading the bulb plate, filling the gap by forge-welding a piece of plate. Simpler and cheaper innovations were the adoption of tee bulb beams and plate brackets (either plain or edge flanged), the latter to replace the costly forged knees. With the advent of unsheathed iron decks and wide frame spacing, deck beams were fitted at every frame.

e. Pillaring. In early iron ships, each tier of beams was supported by a row of centerline, small, solid pillars with forged heads and heels, spaced every second frame, Fig. 2. As ships became broader, two or three rows of these closely spaced pillars were used, interfering with cargo handling.

Most cargo ships today have widely spaced pillars capped by deck girders. Exceptions are ships, such as the Great Lakes bulk carriers, with arched beams (Fig. 19) or where deck beams are cantilevered from the side frames. Pipe stanchions and simple, flanged plate deck girders are widely used and are adequate for most purposes. Pillars should be fitted vertically in line with one another and proper support should be included in the double bottom.

2.2 Longitudinal Framing. In this system, the ribs mentioned in Subsection 2.1 are replaced by a series of closely spaced, longitudinal frames supported transversely by bulkheads and widely spaced, deep web frames. The latter also provide transverse strength (see Fig. 5).

While some early iron ships, the *Great Eastern* among them, had longitudinal framing, this system was never widely used because of the difficulties involved in erection.

With the introduction in 1906 of the Isherwood system (Fig. 5), attention was focussed again on longitudinal framing. In this system, the deep webs or transverses were spaced 3.7 m (12 ft) apart and the longitudinal frames 760 mm (30 in.) apart. The longitudinals were attached to the transverses and kept continuous throughout, but cut short of the bulkheads and bracketed to these. This framing reinforced the shell and deck plating more efficiently than the transverse framing to resist longitudinal compressive stresses.

Despite the saving in weight involved, the Isherwood system was seldom used for cargo ships because the deep hold transverses interfered with cargo stowage. However, it was widely adopted for tankers and wherever structural interference was unimportant.

Subsection 3.6 has a more detailed discussion of longitudinal framing in tankers.

2.3 Combination Framing. In this system of framing, Fig. 6, the bottom plating, and often the deck plating are longitudinally framed, while the sides are transversely framed. Thus, the two superior features of the longitudinal framing are capitalized upon; this is, better stiffening of deck and bottom plating for compressive loads and effectiveness of longitudinals as a part of the hull girder. In addition, objectionable cargo interference of deep side webs is eliminated.

Section 3
Development of Ship Types

3.1 Introduction. This section is intended to show some of the wide variety of structures brought about by changes in shipbuilding practice in the past and more recently, as influenced by economics and experience with earlier types. Great Lakes vessels are discussed herein because, although they are highly specialized for a unique environment, they illustrate well how the structure has been determined by their mission, discussed in Chapter I.

3.2 General Dry-Cargo Ships. Up to the end of World War II, general dry-cargo ships were usually nonspecialized. They accommodated general cargo in almost any form, and loaded and discharged it with their own cargo handling gear. Thenceforth many ships were built with this same service in view, but with an increasing tendency toward specialization, as indicated in Chapters I and II. This has brought about a need for changes in structural design and arrangement.

As pointed out by Arnott (1955), inner bottoms were not always fitted for the entire length of earlier dry-cargo ships. But for many years up to this time however, double-bottom structure has been fitted throughout, between the forepeak and afterpeak bulkheads. This provides a safeguard against flooding of the holds from bottom damage, and in addition the double-bottom tanks may be used for fuel oil, water ballast or fresh water. The inner-bottom plating is included in the longitudinal material contributing to the strength of the ship.

Most general dry-cargo ships have been fitted with one or more deep tanks used for fuel oil, water ballast, or liquid cargos such as latex, coconut oil, or palm oil. The tops of the tanks are often fitted with hatches and watertight covers of sufficient size to permit the carriage of dry cargo alternatively. Some general dry-cargo ships have been fitted with sophisticated tanks surrounded by cofferdams, coated internally with special inert materials, or made of steel clad with noncorrosive materials. They may be provided with heating arrangements and independent means of handling the cargos of potable liquids and other products susceptible to contamination. Some of these cargos were not carried previously in bulk.

As indicated in Subdivision 2.1e, to reduce obstruction to the stowage of cargo, general dry-cargo ships have deep girders, supported by widely spaced pillars. Going one step further, sometimes girders in line with the hatch side are made the only support of the deck beams inboard of the ship's side, and these are in turn supported by massive transverse members at the hatch ends, called hatch-end beams. They may be supported by a single large pillar at the centerline, of rolled H-section, built up of bars or most commonly of cylindrical tubing. This results in a hold having only two pillars in its entire area.

The trend has been toward wider hatches or multiple hatches, even in general cargo ships, since the bottleneck in cargo handling is known to be the moving of units of cargo

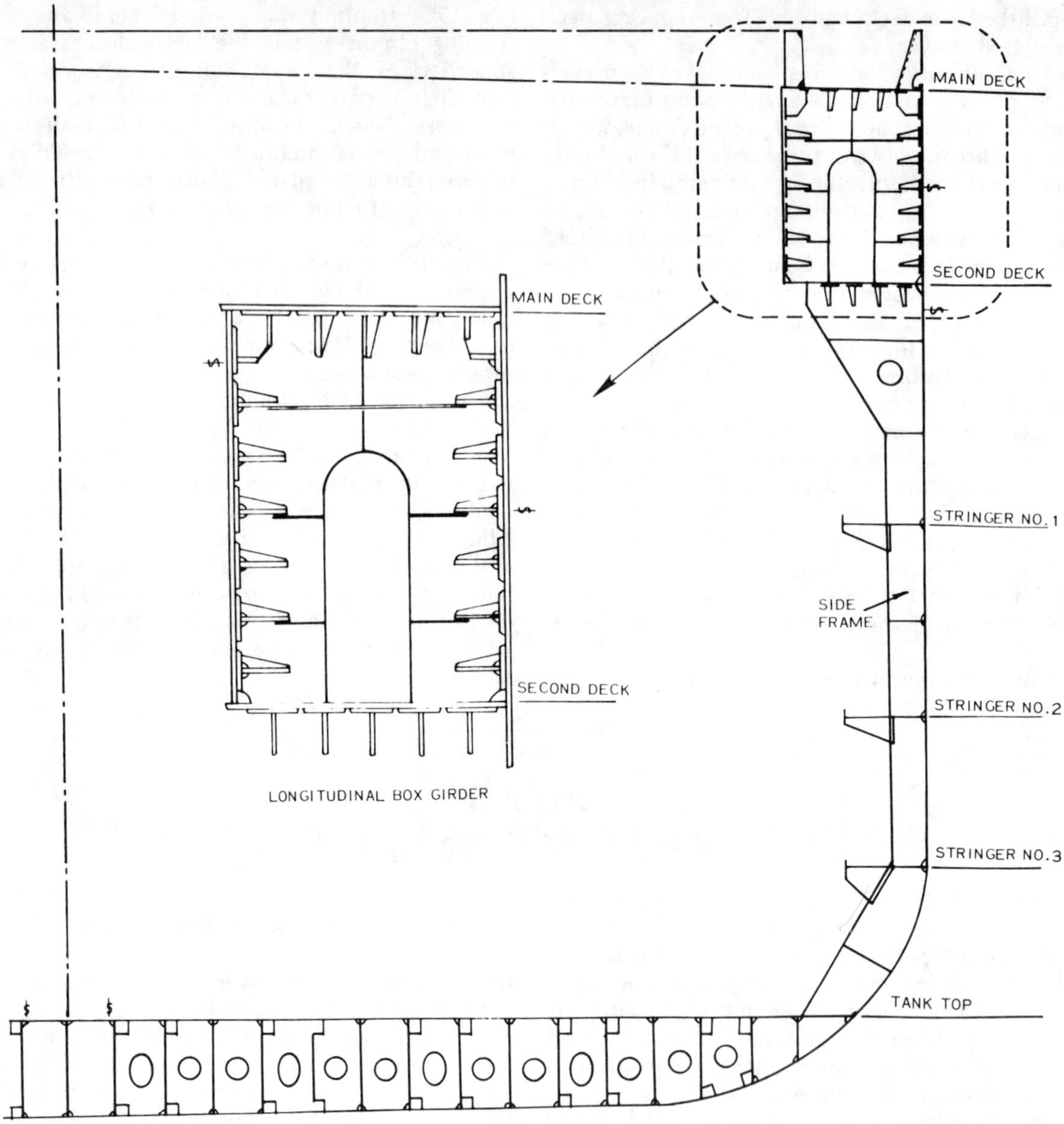

Fig. 7 Midship section of SL-7 containership

in the holds and 'tween decks from the square of the hatch to the wings. This makes for shorter spans of deck beams but confines the effective sectional area of deck plating needed for longitudinal strength to a width which necessitated fairly thick plating.

The tendency toward wider hatches has gone to the extreme, resulting in the "all-hatch" ship. The arrangement usually consists of two or even three hatches abreast in all decks so that all cargo can be lowered directly into place. The problems of providing adequate longitudinal strength within a limited width of deck and sufficient resistance to torsion and *racking* have been overcome.

Several designs provide for the optional stowage of containers in some of the holds, and some of these allow conversion to full containerships. A midship section of an all-hatch ship is shown in Fig. 6.

The improved efficiency of machinery and the premium on speed brought about by the increasing cost of ships has led to considerably greater speeds. For example, the highly successful Victory class of World War II made about 16 knots, but some later liner ships are often capable of operating speeds of 24 knots or more. However, the price of fuel in 1980 has largely reversed the trend to higher speeds.

The possibility of slamming is increased greatly, since even at half speed the newer ships are moving as fast as some of the prewar types at full power. A problem area exists in

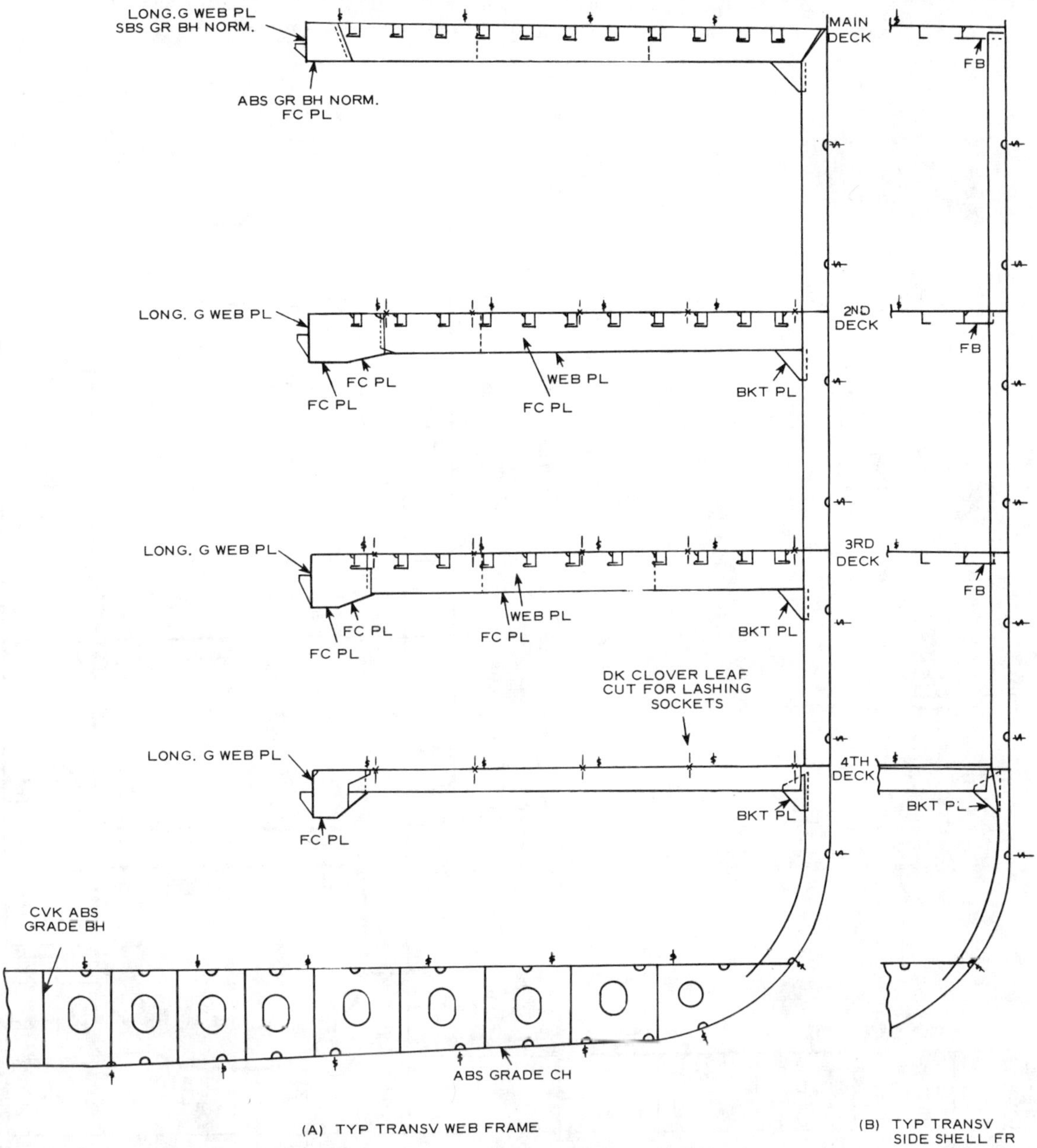

Fig. 8 Midship section of roll-on/roll-off ship

providing adequate strengthening to resist the forces of slamming, and it is being given close attention in research.

Riveted construction included faying flanges of angles connecting floors and girders to bottom shell, but the change to all-welded construction had the effect of enlarging the unsupported panels of plating through omission of the faying flanges. Earlier rules called for intermediate frames as one method of resisting slamming damage, but this arrangement has not been used of late owing to the physical limitations on space which prevent the welding operator from making welds of acceptable quality. It has recently been shown that the form of the bottom has an important influence on susceptibility to damage; even a slight curvature or slope to the bottom plating seems to reduce the maximum slamming pressures considerably. Another approach which has been effective in a limited number of applications is the use of higher strength steel shell plates. Unless backed by adequate internal structure, however, the damage will merely be transferred from the plating to the internals. Classification rules specify increased thickness of shell plating in the slamming area, but there seems to be little benefit from further increasing thickness unless the size of panels is also reduced. Damage to bottom plating seems to occur most frequently on the so-called "flat of bottom forward" even when the truly flat area extends no further for-

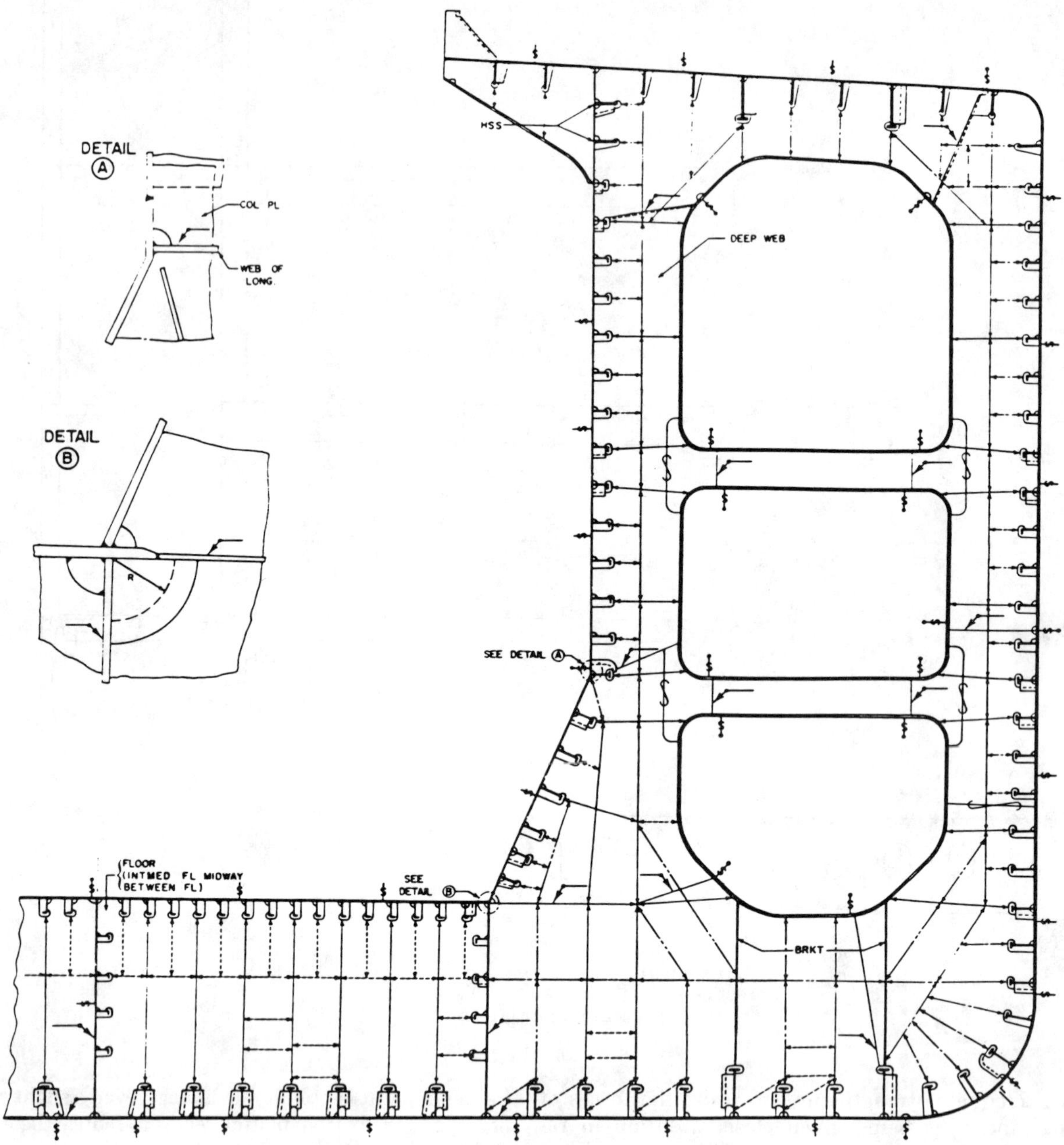

Fig. 9 Oceangoing dry bulk carrier with deep double bottom

ward than the midship half-length. Bottom buckling of welded ships led to almost universal use of longitudinal framing throughout the double bottom areas.

3.3 Containerships. The advantages of carrying cargo in large containers of uniform size have led to the design of ships to carry containers exclusively (Henry and Karsch, 1966). A midship section of a high speed containership is shown in Fig. 7 (Boylston et al, 1974). Almost the entire deck space is occupied by hatches, usually only one in the breadth of the ship, leaving only a narrow strip of deck plating outboard. This necessitates a box girder construction, often of very heavy plating, at the deck and sheer strake junction to provide longitudinal material in tension, stiffness against lateral and torsional loads from the sea, and resistance to buckling in compression when the ship is in sagging condition. The containers are stacked as many as six high and weigh around 30 tons each, with resulting loads on the inner bottom applied entirely through the four corner posts of the containers. Extra stiffening within the double bottom has to be provided at the points where these high, concentrated loads are taken. The size of containers to be carried determines largely the arrangement of structure.

Vertical cell guides formed of angles to hold the stacked containers against ship motions are the only other structures in the holds. Bulkheads are spaced according to multiples of the length of containers, plus the bulkhead structure itself with a strip of deck above it. Hatch covers are quite massive, sometimes acting as support for containers stowed on deck.

A specialized form of containership is the barge-carrying ship, in which the containers are barges and are removed to and from the ship by floating them. One such type, the LASH, carries its own rolling gantry crane which lifts the barges from the water at its stern; the operation is otherwise similar to handling ordinary containers, which can be carried on the ship at the same time. Another type, the SEABEE floats the barge into the ship aft, where it is lifted by an elevator and then rolled to position (see Chapter II). Both of these arrangements involve special structural problems.

3.4 Roll-on/Roll-off Ships. Vehicle carriers are described in Chapter II. A midship section of a roll-on/roll-off ship is shown in Fig. 8. The requirement for clear decks, without interruption by transverse bulkheads, and 'tween-deck heights to accommodate the specific vehicles, calls for a structure for the roll-on/roll-off ship which differs significantly from that of the standard, transversely framed cargo ship. The transverse bulkheads, which in a conventional design are carried up beyond the freeboard deck to the uppermost continuous deck (in order to provide transverse strength), in the case of a roll-on/roll-off ship must be replaced by closely spaced deep web frames in conjunction with deep beams. Spacing of such webs varies between 2.4 and 3.7 m (8 and 12 ft) depending upon the size of the ship. With the web frames, the plating between the webs at decks and side is normally reinforced by longitudinal framing, although intermediate transverse side frames may be used. The deck thicknesses required to take the panel load of concentrated wheel loads of heavy vehicles, such as trains or trailer trucks, are considerably more than in the standard cargo ship. To keep the increased deck thicknesses to a minimum, the spacing of the deck longitudinals is reduced

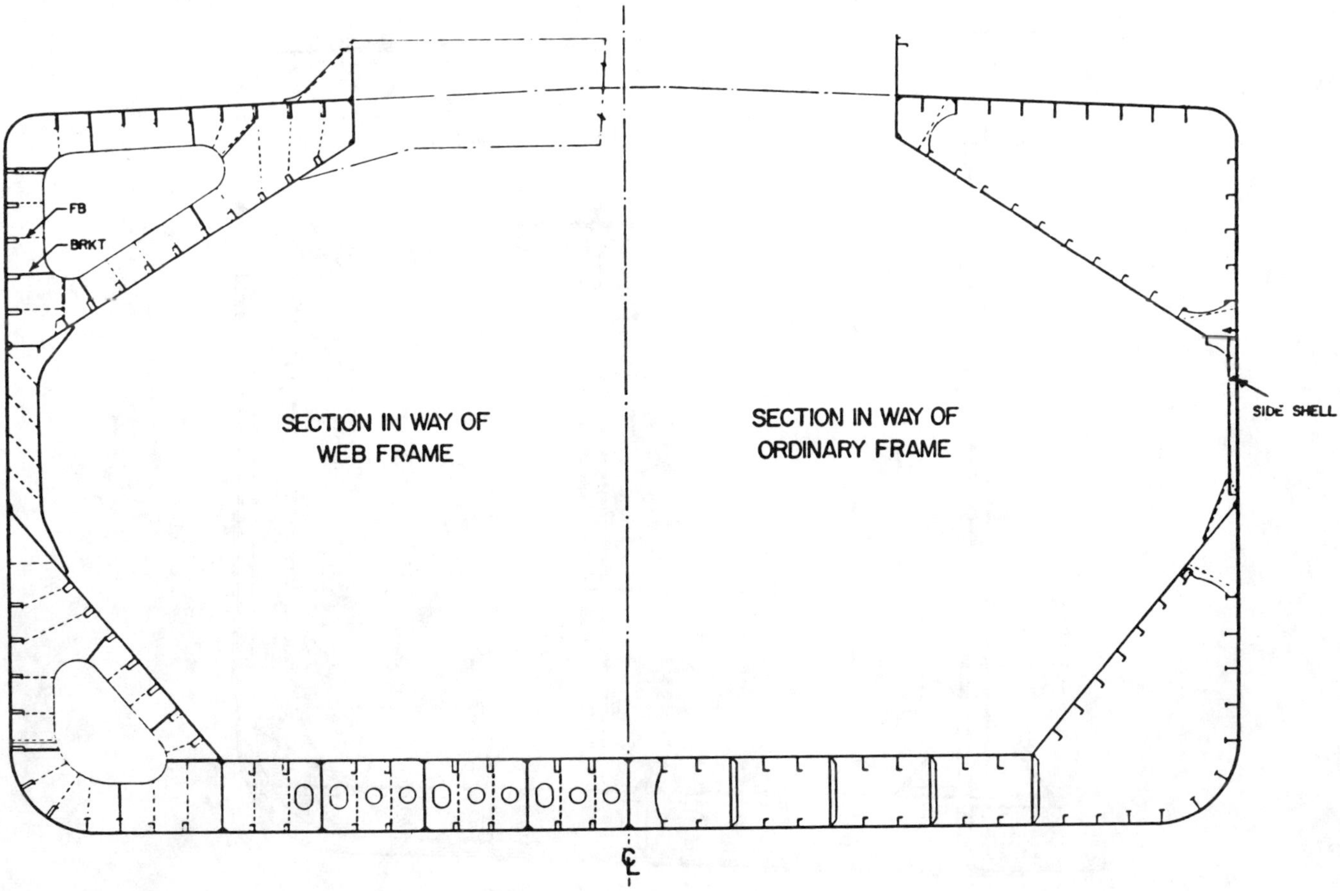

Fig. 10 Oceangoing dry bulk carrier with shallow double bottom

somewhat from the spacing commonly used on a standard break-bulk ship of the same size. The deck longitudinals are also larger than in general cargo ships, to take wheel loads at the middle part of their spans.

Because of the heavy deck thicknesses, longitudinal framing, and the great depth of the ship caused by the deep 'tween decks, the roll-on/roll-off ship has longitudinal strength characteristics well in excess of the requirements.

In order to make use of roll-on/roll-off capability to best advantage, particularly at ports without well developed facilities, the ramps for loading and unloading have become rather massive structures themselves. They are so long as to require being folded when stowed, often in a vertical position at the stern, and are provided with gear by which they may be positioned fore-and-aft or swung around to rest on the quay where the ship is lying alongside.

3.5 Dry Bulk Carriers. Dry bulk carriers are arranged to suit the density of the anticipated cargo which controls the location of the inner bottom. For dense cargo such as some types of ore, the volume required to carry a given weight is relatively small. In order to prevent problems of cargo shifting, it is desirable that the hold be narrow at the top. In order to prevent violent motions which would result from excessive metacentric height, it is desirable that the center of gravity of the cargo be relatively high. Both of these considerations lead to a configuration such as shown in Fig. 9, incorporating a high inner bottom and large wing tanks.

The low-density bulk carrier, on the other hand, needs much more volume to carry her cargo, which results in a much lower inner bottom, Fig. 10. This is typical of many bulk carriers, characterized by the high sloping inner bottom at the bilge and by the topside tanks. This configuration has been adopted for bulk carriers over a wide range of sizes. The engine room is invariably aft, and often there is no midship bridge. Variations include the omission of the topside tanks in some; in the larger ships an inner side shell may be fitted, making for easier cleanup of cargo and providing added space for water ballast (a problem in this type of ship).

When shallow double-bottom bulk carriers are designed

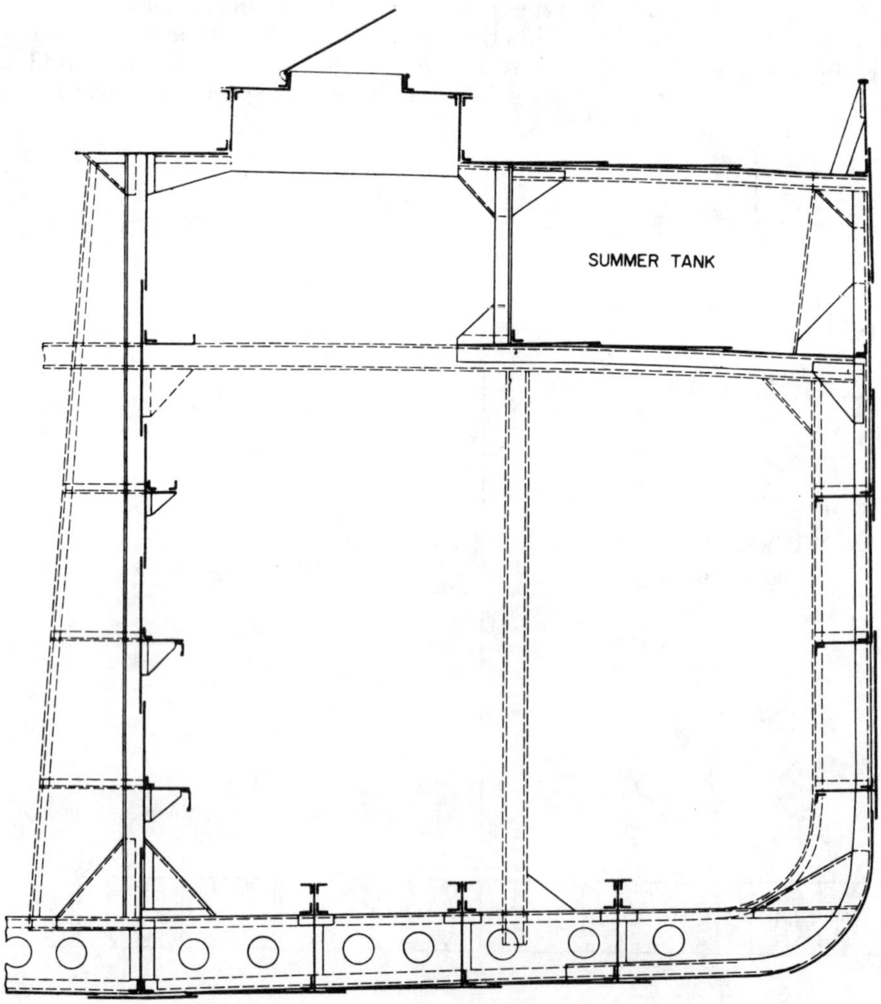

Fig. 11 Midship section of early transversely framed tanker

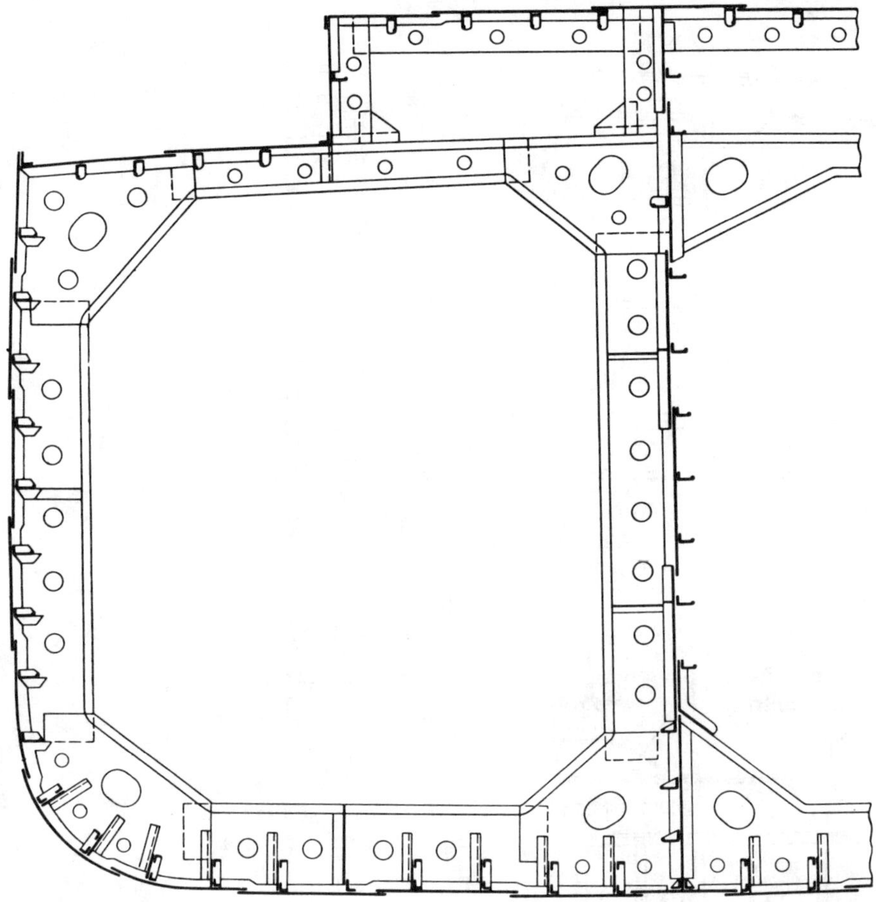

Fig. 12 Midship section of the first Isherwood tanker, *PAUL PAIX*

to carry high density cargo, they are arranged with alternately long and short holds. High density cargo is loaded only in alternate holds in order to bring the center of gravity high enough to result in an acceptable metacentric height. This distribution creates very high vertical shear forces near the bulkheads, which may call for increases in shell plate thickness. In these designs, the double bottom structure in the holds intended for heavy cargos is augmented.

Another variation on the arrangement illustrated in Fig. 10 is to fit bolted plates over openings in the sloping plating of the topside tanks. When light cargos are carried, the bolted plates are removed and the topside tanks are also filled with cargo, which feeds by gravity to the hold when discharging.

One of the most efficient of ships, economically, is a variation on the dry bulk carrier, called the OBO (ore/bulk/oil) ship. While tankers and ore carriers usually carry cargo one way and return in ballast, the OBO often carries cargo both ways or on two or even all three legs of a three leg route. The structure is arranged much like that of the dry bulk carrier, but all bulkheads are designed to tank requirements and the hatch covers are oiltight. Side tanks and topside tanks are usual in order to make the center cargo space suitable for liquid cargo with proper stability characteristics.

3.6 Oil Tankers. Oil was at first transported in separate tanks aboard ordinary cargo ships. In 1886 the first specialized bulk oil carriers were designed. These early tankers were built on the web-frame system where the side webs supported horizontal plate girders which, in turn, carried the ordinary transverse frames, Fig. 11.

This type of tanker was costly to build because of the high degree of workmanship required for oiltightness. Maintenance of sound tight riveting was expensive, especially the attachments of keelsons and stringers to the bulkheads and, as ships became longer, deck and bottom plating butts.

Striving for sounder and more leakproof riveting led to the Isherwood framing system in 1906. The first application of this was in the tanker *Paul Paix,* 108.2 m by 15.0 m by 8.5 m (355 ft by 49 ft 3 in. by 28 ft), completed in 1908 (midship section shown in Fig. 12).

The transverse bulkheads were spaced 9.9 m (30 ft) apart, with two transverses in each tank. To avoid expensive collaring and to facilitate drainage, the longitudinals were cut and bracketed to the bulkheads.

In the Isherwood "bracketless" system, patented in 1925, the transverses inside the tanks were unequally spaced with the short spans adjacent to the bulkheads, as indicated in Fig. 13. This was used only on relatively small vessels.

The characteristic tanker deep transverses were generally

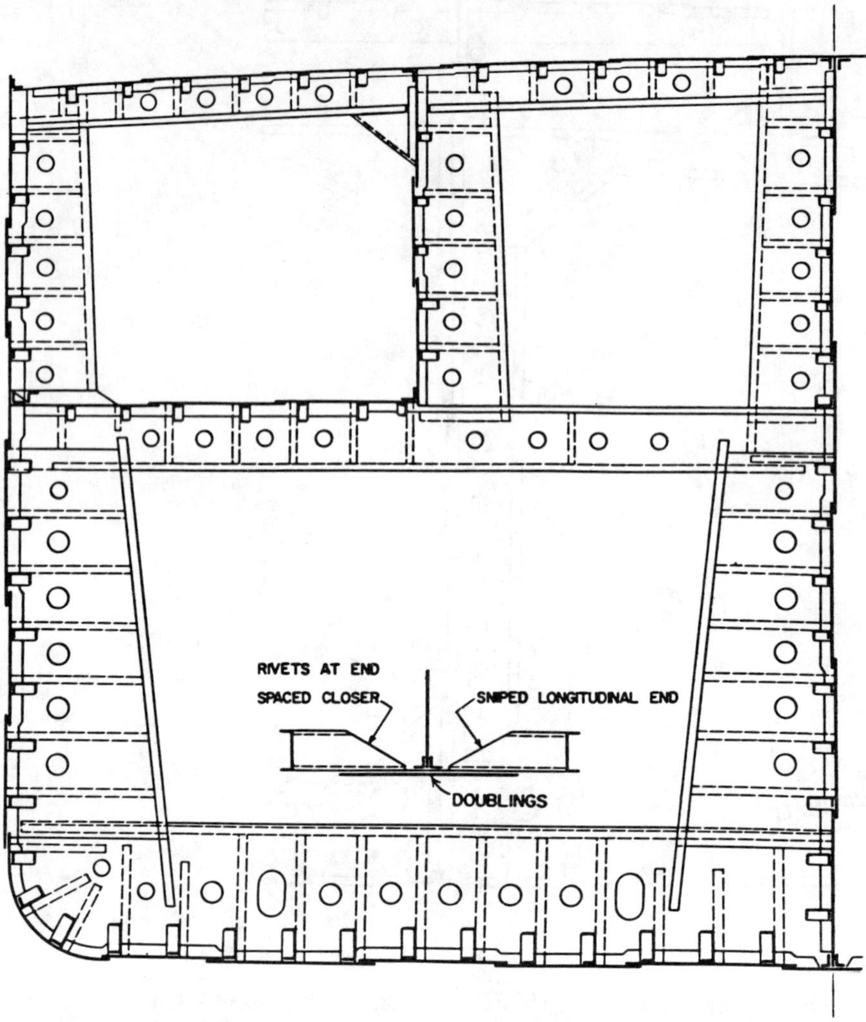

Fig. 13 Midship section of a tanker on the Isherwood "bracketless" system

fitted in preference to shallower girders with large brackets, as shown for the *Paul Paix*. The longitudinals were not connected to the bulkheads. To compensate for the omission of brackets, the shell plating was reinforced by fitting doublers in way of the bulkheads, whereas riveting at the sniped ends of the longitudinals was spaced more closely and (or) a short back bar fitted.

Other essential considerations in tanker design, besides structural soundness and oiltightness in service, are the provision of adequate margins against the serious corrosion of bare steel which takes place with certain oil cargos and the provision, insofar as is practicable, of that type of structure which can most readily be drained, cleaned, and gas-freed in order to keep maintenance costs to a minimum. Vertical stiffening of vertical plating obviously makes for better drainage, and this is probably the reason why the combination system was often applied to tankers built in Europe. A number of effective systems of coating cargo tank interiors have helped reduce corrosion appreciably in recent years. See Chapter XIV for a discussion of these coatings.

As tankers increased in size, it became advantageous to rearrange the structure to provide two full-depth, longitudinal bulkheads throughout the tank space instead of a single, centerline bulkhead and expansion-trunk bulkheads. Omission of the second deck provided a center tank and wing tanks extending the full depth of the ship. This same arrangement has been used in tankers of the largest sizes.

With moderate beam, the bottom transverse webs could be designed to span the distance between the longitudinal bulkheads, but, since a bottom centerline girder was needed for docking purposes, general usage was to adopt a high, substantial, center girder supporting relatively short transverses between the longitudinal bulkheads and the centerline. A similar structure was adopted for the deck. Both of these centerline girders contribute to longitudinal strength.

The girders were, in turn, supported by the main transverse oiltight bulkheads, which were limited in their spacing up to about 13.2 m (40 ft), since increasing the bulkhead spacing appreciably creates problems in developing the centerline girders. A few designs resorted to a third longitudinal bulkhead on the centerline in lieu of the girders, and then this practice was dropped.

With the aim of facilitating drainage and tank cleaning

and, at the same time, reducing welding and the number of pieces, often bulkheads were made of corrugated plates without stiffeners. Several configurations were used, the corrugations being disposed horizontally in longitudinal bulkheads and either vertically or horizontally in transverse bulkheads. While any welded structure such as this is dependent upon the most careful design of the details and execution of the fabrication, these corrugated bulkheads were so prone to cracking that there has been a general return to flat plate bulkheads with welded stiffeners.

As mentioned earlier, vertical stiffening of vertical plating is advantageous for drainage, but as ships increased in size, it became necessary to dispose all possible material in such a way as to contribute to longitudinal strength, and as a result the shell and longitudinal bulkheads are stiffened horizontally. Transverse bulkheads are often stiffened horizontally for matching, although the deep, vertical, supporting webs are sometimes longer than if the arrangement were reversed. The latter would result in a better distribution of material. These very long deep webs require support at their ends, and it has been usual to fit substantial coplanar bottom and deck girders.

Toward the end of the 1950s, the economic advantage of tankers of very large size became progressively more pronounced as the demand for petroleum soared and the cost of ships and their operation continued to rise. Since then, the dimensions of a tanker in the "large" category changed upward at frequent intervals and by very large increments. It became apparent that projections of the longitudinal strength requirements established by experience with ships 183–213 m (600–700 ft) in length would result in structures that were unduly heavy for ships of 274 m (900 ft) or more.

This increase in size initiated studies of longitudinal strength more extensive and detailed than any in the past, and led to the methods of design discussed fully in Section 6.

The increase in size of ships brought a natural tendency to longer tanks. The limitations imposed by the classification societies were liberalized, and tanks as long as 0.15 to 0.20 L were approved, but with nontight or swash bulkheads at intermediate locations. Likewise, the spacing of transverse webs tended to increase in accord with the practice for many years with respect to most transverse members; that is, the spacing increased with ship length. This brought about new problems of shearing forces and local loads at bulkheads and bottom transverses imposed by the large areas of bottom supported at great drafts.

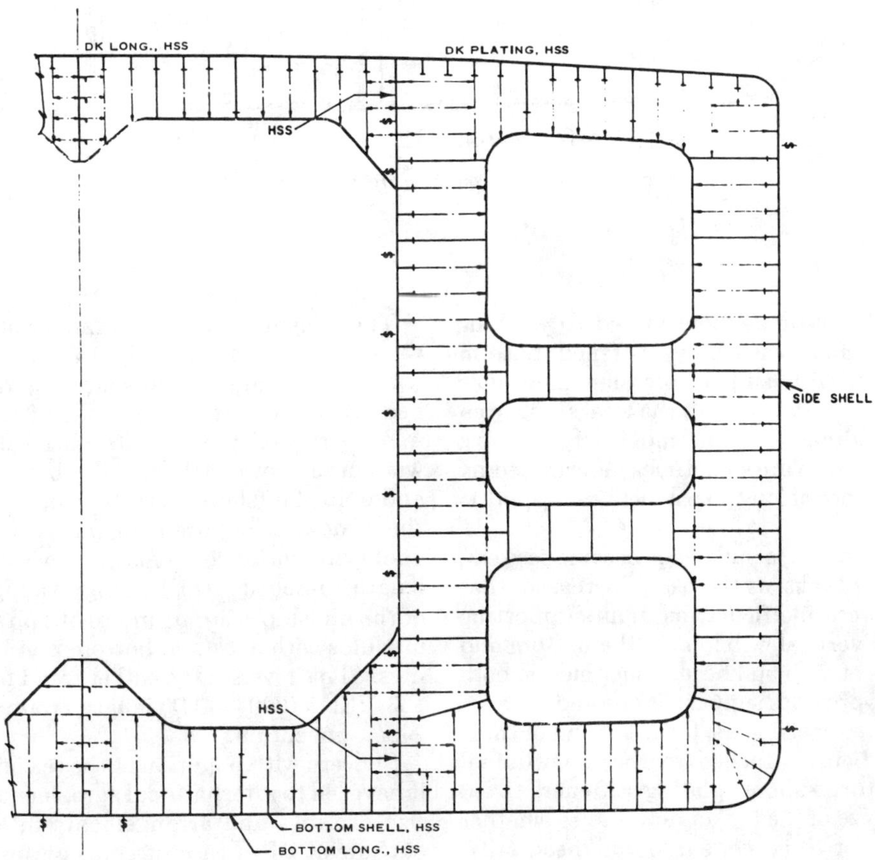

Fig. 14 Midship section of a tanker of about 175,000 dwt

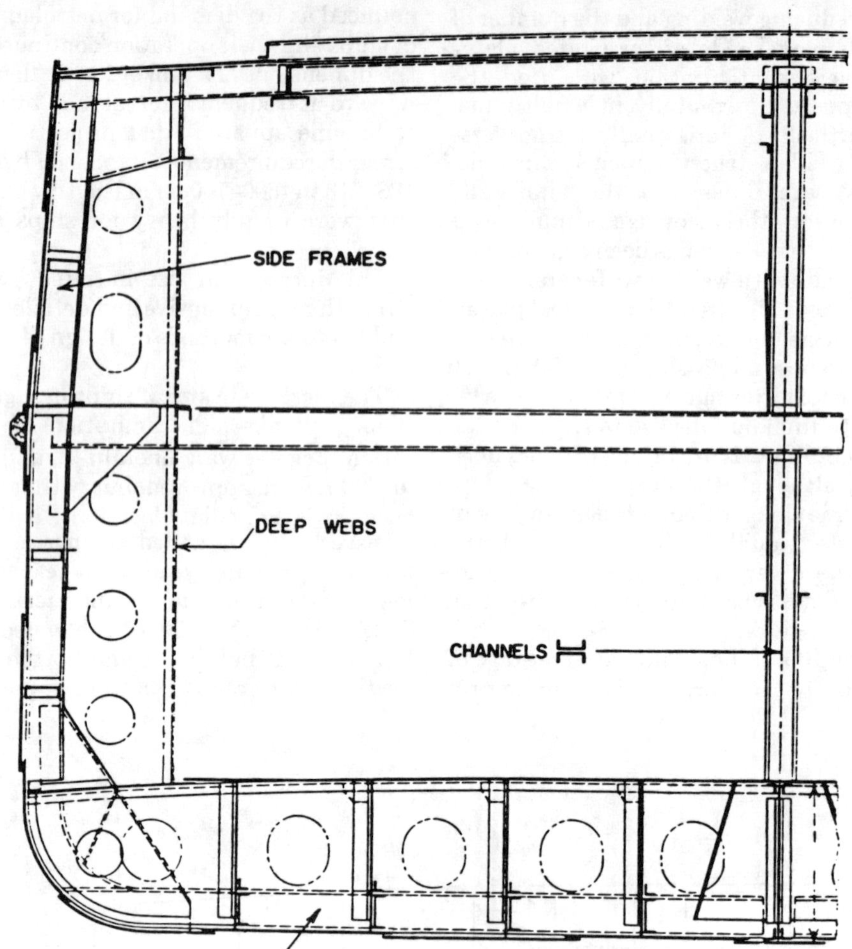

Fig. 15 Midship section of the *FRANK H. GOODYEAR*

Previously, bulkhead scantlings, determined to working hydrostatic pressure, and reasonably stiffened bottom transverse web plates, with added corrosion allowance thickness, had ample strength to absorb the local shear from the individual longitudinals. In the much larger ships, scantlings of these members are sometimes determined by the local condition because of longer span between supports of longitudinals.

The dimensions of transverse bulkheads became so great, especially in the center tanks, as the beam increased, that vertical webs alone were inefficient as main supporting members. A massive vertical web joining the bottom and deck centerline girders at each bulkhead, sometimes on both sides of the bulkhead, provides support at one end for a series of horizontal shelves or stringers, the other end of these horizontal members being attached to the longitudinal bulkhead. These in turn support plating stiffeners fitted vertically. Thus the size of the tanker determines whether it is to have one, two, or three systems of bulkhead stiffening.

The midship section of a tanker of about 175,000 deadweight tons is shown in Fig. 14.

Some very large tankers are constrained by major European ports to drafts on the order of 21 m (70 ft). However, many very large tankers introduced after 1965 are designed without reference to depth of harbors, being intended to load and unload offshore. Depth is no longer a limitation, and the structural requirement for freeboard becomes of secondary importance. It is thus possible to utilize a small length-to-depth ratio, leading to a high moment of inertia of the midship section, and to obtain the necessary section modulus with deck and bottom plating of moderate thickness. This reverses the earlier trend found in tankers of the 183–213 m (600–700 ft) length group under the 1930 Load Line Convention.

Concern with pollution of the sea and efforts to minimize it have led to international agreement regarding maximum size of tanks and arrangements intended to restrict the outflow of oil in case of collision or grounding, as far as practicable. These measures will have an important effect on the structure of future oil tankers when they become fully effective. See Chapter XI for details of these requirements, and IMCO (1978).

3.7 Liquefied Gas Carriers. Owing to the hazardous nature of most liquefied gasses, the design of ship structures and details of containment of the liquids require the greatest care. Not only are the more commonly carried cargos, liquefied natural gas (LNG) and liquefied petroleum gas (LPG), explosive in mixtures with air, but the low temperatures at which these and other such cargos are carried are hazardous to the structure of the ship itself. All grades of shipbuilding steel become brittle at reduced temperatures, and even a small spill of liquefied gas cargo can cause brittle fracture in important structural members. For this reason, rules for liquefied gas carriers have required a second barrier to contain temporarily any leakage from the primary container.

In order to provide the massive insulation needed to maintain the low temperature of the cargo, primary containers must necessarily be independent of the ship's structure.

There are a number of systems of containment: free standing tanks with sufficient internal framing to make them self-supporting; spherical tanks requiring no framing; thin membrane tanks supported by the hull structure (but separated from it by insulation).

There are various forms of secondary barrier, but all are most readily applied to an unobstructed inner hull surface. Most gas carriers are constructed with double skin hulls; this arrangement also provides space for ballast water in large quantities. Since the specific gravity of most gas cargos is very low, a large volume is needed but relatively little draft, and ballast capacity is vital. The double skin also provides a degree of protection to the hazardous cargo against moderate damage to the shell.

Longitudinal strength is not a problem because of the light loading and large depth of hull needed to obtain the necessary internal volume.

3.8 Great Lakes Ships. Bulk freighters carrying ore, coal, stone, and grain constitute the major part of the Great Lakes fleet. Perhaps nowhere else are bulk cargos carried as efficiently or economically. This type of ship has been influenced largely by improvements in locks and waterways, allowing ever increasing dimensions, and by cargo handling

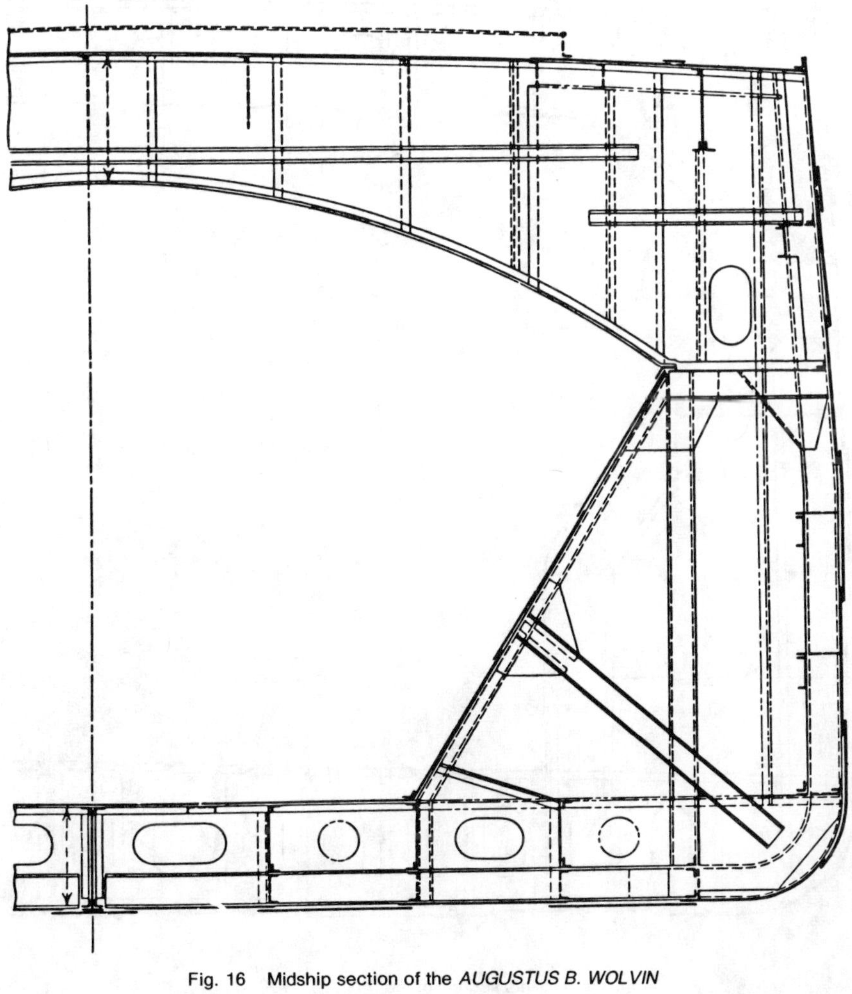

Fig. 16 Midship section of the *AUGUSTUS B. WOLVIN*

installation at cargo terminals determining the hatch arrangements and internal structure.

Some of the outstanding developments on Great Lakes ships date back to the introduction of the Huelett bulk unloader around 1900. Fig. 15 shows a midship section of the *Frank H. Goodyear* of dimensions 127 by 15.2 by 8.5 m (416 by 50 by 28 ft), built in 1902 when it was still considered essential for structural reasons to fit beams and stanchions in holds. This ship represented the highest development of the bulk carrier at that time. The extensive use of channels in her construction is worthy of note, as is also the unusually wide spacing of the side frames at 1.2 m (4 ft).

A radical advance in the structure of bulk carriers was made in 1904 when the *Augustus B. Wolvin* was built of the then unprecedented dimensions of 170 by 17 by 9.8 m (650 by 56 by 32 ft) with 33 hatches spaced on 3.7 m (12 ft) centers. As shown in Fig. 16, very substantial arched plate webs were fitted between the hatches to compensate for the omission of the hold beams and stanchions in previous designs. The inner bottom was carried up to the main deck stringer, and the space between the sloping sides and the side shell was utilized for carrying additional water ballast. The unobstructed holds, to facilitate cargo handling and the provision of large water ballast capacity in side tanks, are desirable features which have been incorporated in all later bulk carriers, along with improvements in structural arrangements and details.

Another interesting development took place in 1913 when a number of bulk freighters were built on the Isherwood system, including the steamship *James A. Farrell* of dimensions 177 by 18 by 9.8 m (580 by 58 by 32 ft), the midship section of which is shown in Fig. 17. Experience in service showed that the longitudinally framed sides were more vulnerable to lock damage than transversely framed sides. As a corrective measure, additional webs had to be fitted at the ends where such damage was prevalent. Apart from this local condition, the structure proved satisfactory in service and influenced the design of bulk freighters where the inner bottom and decks are longitudinally framed with transverse framing on the sides. Fig. 18 shows a typical web frame and

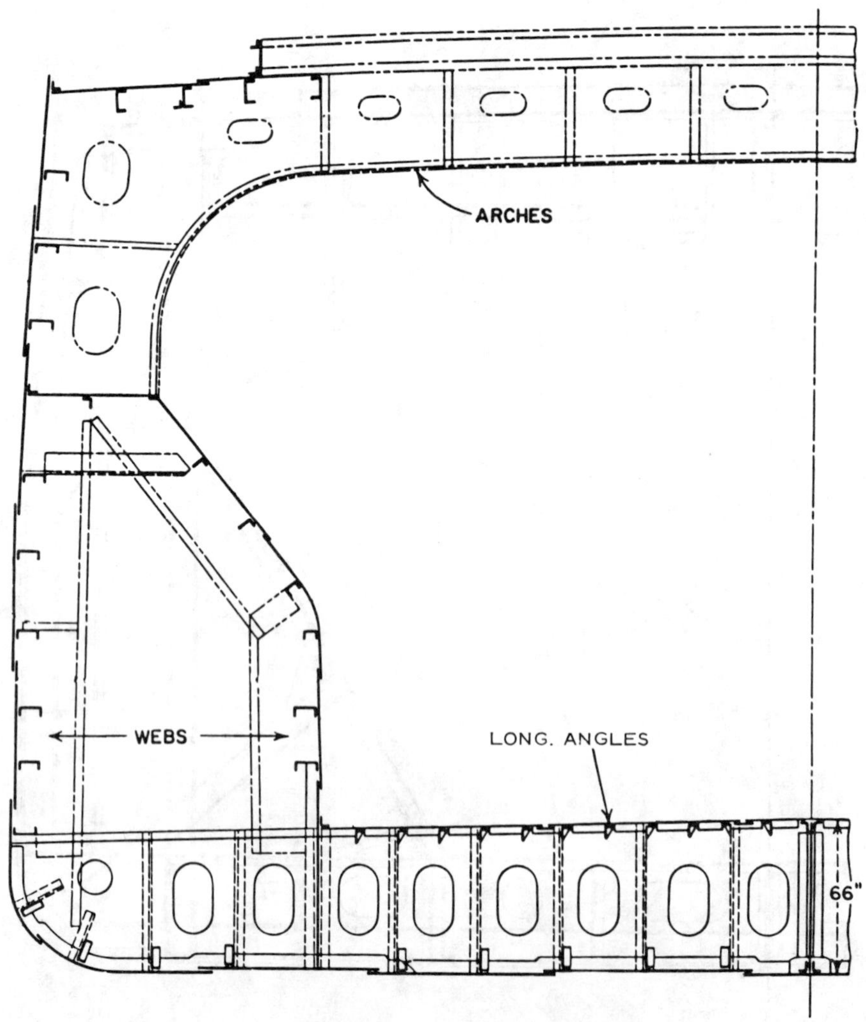

Fig. 17 Midship section of the *JAMES A. FARRELL*

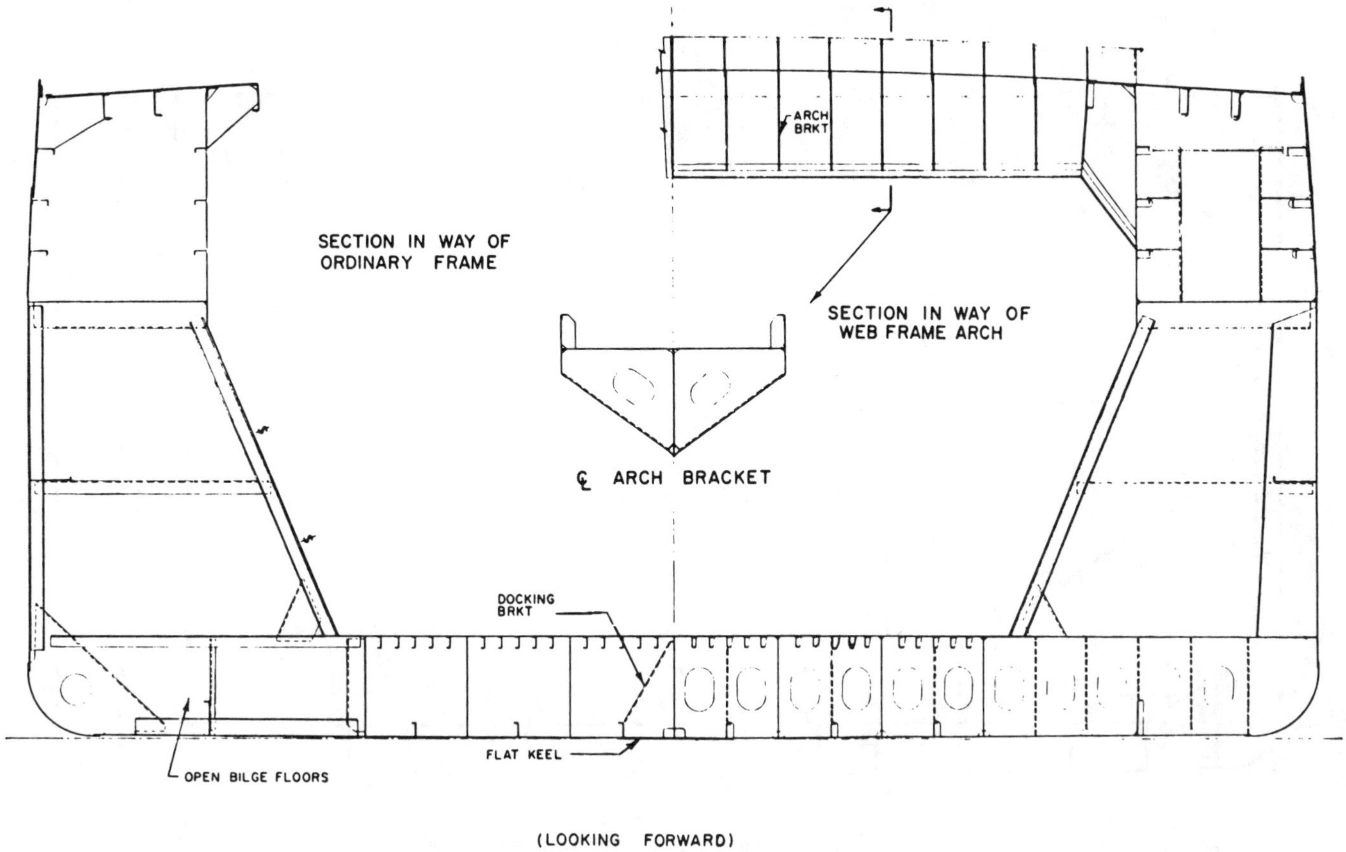

Fig. 18 Typical Great Lakes bulk carrier

ordinary frame of such a vessel.

In later bulk carriers, much more attention was paid to the means provided for the protection of openings. In earlier ships, the only protection to the machinery openings was that afforded by a wooden house usually with a door entering directly into the engine room. In addition, the numerous deck hatches had long wood covers in association with low coamings. The administration of the Great Lakes Load Line regulations has been beneficial in remedying conditions which became of primary importance with increased operating drafts.

Great Lakes ships generally have a useful life of fifty or sixty years, since they are not subject to the corrosive atmosphere which limits the life of oceangoing ships. This situation does not provide incentive for drastic changes in design concept. Two events in the 1950s, the opening of the St. Lawrence Seaway and the enlargement of the Poe Lock at Saulte Ste. Marie (the Soo), have furnished incentive for change.

Prior to the Seaway, Great Lakes ships were built and remained on the Lakes. After the Seaway opened in 1959, the existing Lake vessels could navigate as far east as Seven Islands, Quebec, on the St. Lawrence River (their proportions and strength standards precluded their going to sea). Great Lakes ships could also be built elsewhere and brought to the Lakes via the Seaway. (Some converted oceangoing ships were brought to the Lakes in a partially completed condition, by way of the Mississippi and Chicago Rivers, in the 1950s, as a stopgap measure.) A limiting length of 222.5 m (730 ft) overall and a beam of 22.9 m (75 ft) were established for ships transiting the Seaway, and some ships were built taking advantage of these maxima for trade between the Lakes and St. Lawrence River ports. A draft of only 8.2 m (27 ft) below low water datum was allowed for in the Seaway, the same as in the channels connecting the Great Lakes. The characteristics of these new ships were not greatly different from their predecessors except for more modern and efficient machinery.

The enlarged Soo lock can accommodate ships 305 m (1000 ft) long and 32 m (105 ft) wide. The depth of water over the sill is 9.8 m (32 ft). Designs for new ships and lengthening of existing ones were a natural result of the opening of the new lock. Prior to the opening date, a number of older vessels were retired as uneconomical in the new situation.

A type of bulk carrier unique to the Great Lakes is the self-unloader, in which the cargo is carried in continuous hoppers from which it can be fed to conveyor belts running the full length of the cargo space. It is then carried forward or aft to a bucket-chain elevator, and deposited on the belt of a boom—as much as 81.7 m (268 ft) long—which delivers the cargo ashore. Long used for limestone for the steel and

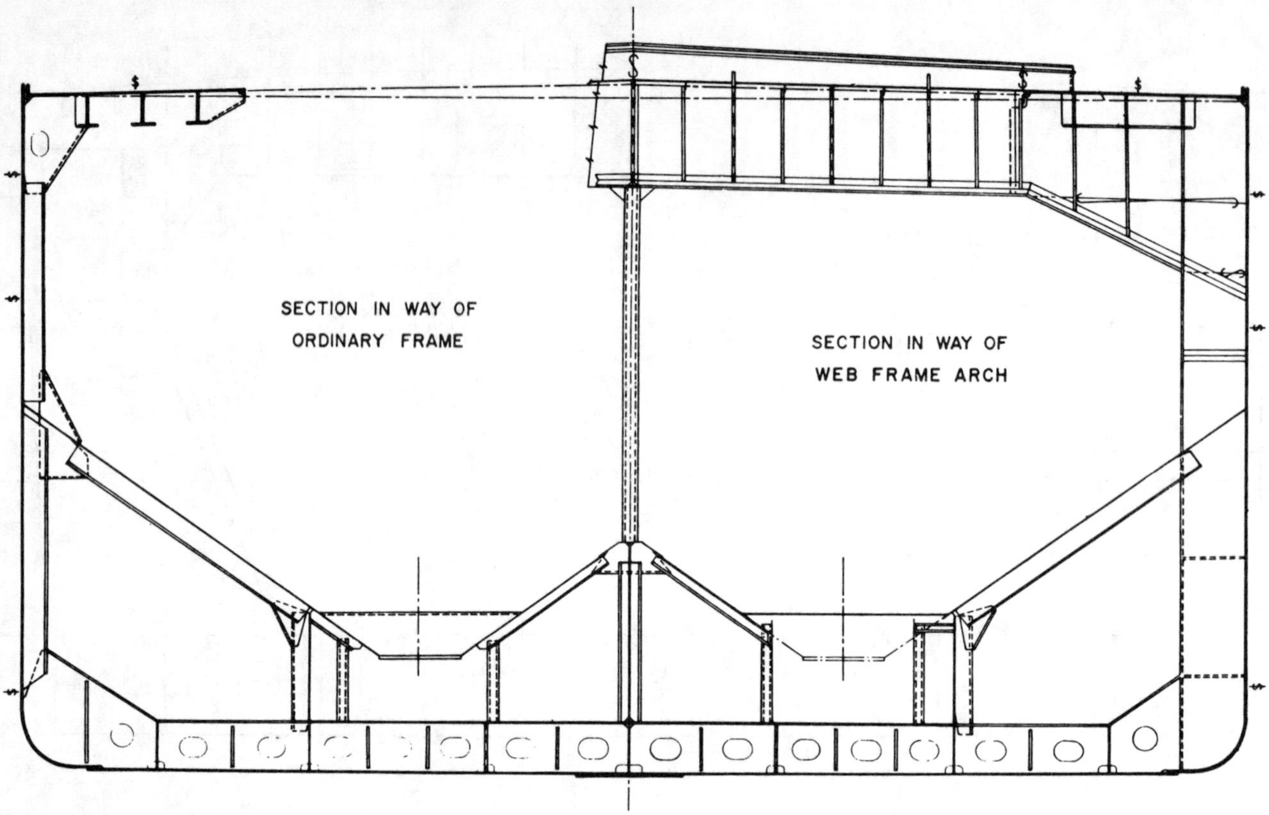

Fig. 19 Midship section of typical Great Lakes self-unloader

chemical industries and for coal, this type is finding increasing use in the transportation of pelletized ore, which lends itself to conveyor belt handling. The structural arrangements of self-unloaders are highly specialized, there being usually no tight transverse bulkheads and the double bottom being unusually shallow. Large ballast tanks are formed by the sloping hopper plating and the side shell. Fig. 19 shows the midship section of a typical Great Lakes self-unloader.

3.9 Ocean Barges. There was considerable trade on the Atlantic coast of the United States for many years, ending with World War II, using wooden barges of ship shaped form, manned and fitted with steering arrangements, ground tackle, and even sails. They carried coal largely, and were towed in tandem on long hawsers, several barges to a tow. There were also several large ship shaped barges on the Great Lakes, which were towed behind cargo steamers in many cases, some of them of steel. (At least two have been converted to self-propelled vessels after more than 50 years of service as barges.)

There has been an extension of service, with increases in scantlings, of the rectangular barges with rake ends, commonly used in harbors and on the rivers, to short coastwise runs on the Atlantic and Pacific coasts of the United States, and particularly in the Gulf of Mexico, where much of the coastal route is in protected waters. A great deal of experience has been built up in these services over the last several years, and the sizes of the barges and the length of the voyages have gradually increased to the point where barges make voyages from U. S. West Coast ports to Alaska and even to Hawaii. These barges may be over 91 m (300 ft) long and have a number of advantages over self-propelled ships. They can enter some of the smaller harbors where lumber is loaded, have lower crew costs, etc. (Foley, 1965).

A little-noted provision of the 1966 Load Line Convention permits certain unmanned barges to obtain greatly reduced freeboards, and this may further stimulate the use of tug and barge operation in place of self-propelled ships in some services.

There are tank barges for delivering petroleum, dry cargo barges with hatch covers carrying bulk cargos like coal and phosphate rock, coaming deck barges used for nonperishable cargo, such as scrap metal, etc. Some of the largest barges have a cargo deckhouse. The most versatile example is a large tank barge which carries oil in the hull, has a deckhouse used for grain or other dry cargo, and carries lumber around and on top of the house.

Although it has been found necessary to increase some of the scantlings over those acceptable in more limited services, especially where damage from slamming may occur, the length-to-depth ratio has been increased with satisfactory results in barges having intact decks and ample longitudinal bulkheads and trusses.

The economies of transporting some bulk cargos, such as cement and sugar, favor tug and barge operation over self-propelled ships, and, with the development of very powerful twin-screw tugs, there have been put into service on the Atlantic coast several ship shaped barges of unprecedented sizes as well as the tug-barge combinations described in Chapters II and III.

Practically all of these barges are unmanned, and have no steering arrangements. Anchors and chain or wire rope are provided however, and in some cases can be released by radio-operated remote control.

Sufficient experience has been accumulated to make it possible to promulgate classification rules for these ocean barges (American Bureau of Shipping, 1973). Since they have no machinery space, the distribution of cargo within the hull can be more favorable. Being towed at relatively low speeds, the slamming and other forces due to motions will be less than in powered ships of similar sizes. Scantlings, therefore, are a little lighter than in their self-propelled counterparts.

3.10 Inland Waterways Craft.

a. The Waterway. For the most part, this section refers to the long rivers and their tributaries which connect the central part of the United States with both the Great Lakes and the Gulf of Mexico, but, of course, much of it applies to other inland waters such as the New York State Barge Canal or the Atlantic Intracoastal Waterway. Although largely consisting of fresh water with its relative freedom from corrosion and fouling, much of the waterway is salt water and this has an influence on scantlings and on maintenance systems.

The craft involved are intended for operation in smooth water only, no provision being made for encountering bending moments created by waves, nor for excluding water from boarding seas. There are none of the "perils of the sea" which oceangoing ships are designed to withstand. A port, or at least the shore, is always within a short distance. There are other hazards, however, for which provision must be made. Most important is the contact between barges handled together as a tow, and contact with the numerous locks by which craft pass the dams constructed to maintain the level of rivers at different seasons.

Physical conditions in river and canal travel limit barge size and draft. Nine-foot channel depths are maintained in many major waterways by the U. S. Army Corps of Engineers, permitting at least 2.6 m (8.5 ft) draft towing throughout the year. Actually, most rivers have considerably deeper average depths, requiring dredging to maintain the 2.7 m (9 ft) limit only at shoal areas. Some tributaries such as the Missouri River have only 2.1 m (7 ft) minimum depths at shoals, limiting tows to 2.0 m (6.5 ft) drafts. Dredged canals such as those in the Intracoastal Waterway are, on the other hand, of relatively uniform depth and width. Usually these are at least 3.7 m (12 ft) deep.

b. Barge Service and Characteristics. There are practically no self-propelled cargo ships on U. S. inland waterways. Unlike river towing in many other parts of the world, the cargo-carrying barges are not towed by a hawser from the stern of the towboat but pushed by the towboat, which is in direct contact with the nearest barges. Groups of barges are lashed tightly together to make up tows consisting of, perhaps, four barges wide and six or eight long, all pushed as a unit by a single towboat.

There has been considerable standardization of dimensions to facilitate making up tows of barges of equal breadths and lengths, but this does not preclude some barges being of other dimensions. A common size is 53.5 m (175 ft) in length, 10.7 m (35 ft) beam, with a depth of 3.4 m (11 ft). Barges intended to be operated together in a regular service are sometimes built as units of an "integrated tow;" that is, there is a lead barge having a forward rake, any number of square-ended barges for the main part of the tow, and a shorter after-end barge to which the towboat is secured. Most other barges are built with a rake, or slope, of the bottom at one end, or both ends, to minimize resistance when being propelled and, incidentally, to provide buoyant space. The form of the rake has been developed from model tests of single barges and entire tows.

The most common type is the open hopper barge, a rectangular structure having an inner bottom and longitudinal bulkheads bounding the cargo space, with narrow wing void spaces outboard. They are usually fitted with a *coaming* around the cargo space. Similar barges intended for dry cargo are fitted with covers. Another common type is the deck cargo barge, having an intact deck and carrying its cargo of bulk materials, such as sand, within a coaming on the deck. Petroleum and other liquids are carried in tank barges, also rectangular, but having a centerline oiltight bulkhead and several transverse bulkheads. Pumping arrangements may be provided on board the barge or left to shore equipment.

A common sight in major coastal harbors are carfloats, a type of deck barge equipped with tracks to move 10, 20, or more railroad cars from one rail to another or to shipside for transferring loads.

There are barges fitted for particular cargos to be carried in independent tanks; that is, those which are not part of the hull structure. They are usually cylindrical but may take other forms. Some of these independent tanks are pressure vessels, and some carry refrigerated cargo or liquids at high temperatures. Precautions must be taken to protect these special cargos, some of which are dangerous, against the results of possible damage during handling of the barges in making up a tow and in transiting the numerous locks. Not only the barge and its cargo are involved, but, in the service under consideration, personnel and property on shore in the vicinity are subject to the dangers of releasing such a cargo to the air or water (Steinman and Carman, 1965).

Double skin tank barges have, as the name implies, an inner and outer shell. The inner shell forms cargo tanks free of appendages and they are thus easy to clean and to line. Poisons and other hazardous liquids such as liquid chlorine require the protection of the void compartments between the outer and inner shells.

c. Structures. The structure of all but the most highly specialized barges is extremely simple, and owing to the uniform sections, it is largely repetitive. Competition in this field is intense and has resulted in an arrangement making use of a minimum of steel and welding. Production of these simple structures is rapid, as they lend themselves to practices which would be uneconomical in ordinary shipbuilding.

They are almost entirely of light plate and standard rolled sections, except that channels are serrated by automatic methods to produce lightweight members and a highly efficient use of steel. There is very little plate forming. Maximum use is made of prefabrication of subassemblies, production line techniques, and efficient material handling. Welding is used exclusively; the advantages over earlier riveted construction are pointed out by Arnott (1955). This construction is able to withstand without leaking the minor damages to which the service necessarily exposes it.

d. Towboats. The towboats for this service are derived from the steam-driven stern paddle wheel type. The hull is rectangular in plan and has little freeboard. Quarters are located in one or more tiers of deckhouses surmounted by the pilothouse. The structure is generally similar to that of barges; there are normally some tanks for fuel oil abreast the machinery space, which contains foundations for the main propulsion and auxiliary units. The unusual feature of construction in these craft is the "tunnels" in which the propellers are fitted to minimize draft. These semicircular inverted troughs reduce the depth of the hull appreciably in way of the propellers, where stiffness is needed to resist vibration resulting from the large horsepowers being used. The comparatively shallow structure over the propellers is heavily reinforced by plate members, both longitudinal and transverse. In addition, the same area contains steering gear for the several large rudders required by the service of push towing. A pair of knees of ample strength and which are high enough to engage barges of various depths extend upward from the deck at the bow.

Section 4
Design Loads

4.1 General. The development of a satisfactory ship structure involves generally the following steps:

• Establish the sizes or scantlings of and combining effectively the various component parts of the structure so that it can resist the hull-girder loads, in terms of longitudinal and transverse bending, torsion and shear in still water and among waves.

• Design each component part so as to withstand the loads imposed upon it from the weight of cargo or passengers, hydrodynamic pressures, impact forces and other superimposed local loads such as deck house, heavy machinery and masts.

To establish the required sizes or scantlings of the structure, it is necessary to determine first the maximum loads imposed on it. The major loads imposed on a ship structure may be conveniently divided into the following groups:

• Stillwater bending moments and shearing forces;
• wave-induced moments and shearing forces;
• springing and impulsive loads;
• thermal loads;
• dynamic loads, i.e., hydrodynamic pressures, sloshing;
• other loads, i.e., launching, docking, grounding, mooring, and collision.

A combination of the most severe loads is usually selected as the nominal design load for a given structure. The other important loads may be either accounted for in the assigned safety margin or treated statistically in conjunction with all the load components in a refined analysis of the final design.

Some background data and calculation methods for the essential load components will be discussed in Subsections 4.2–4.8.

4.2 Standard Bending Moments (Background Data). In the consideration of longitudinal strength, the ship's hull may be treated as a nonuniform girder having a varying load throughout its length corresponding to the weight of the structure and its contents. Unlike most beam problems which assume finite supports at fixed positions, the supporting forces are those resulting from the buoyancy distribution throughout the length. When there is a balance between the total downward forces of weight and mass acceleration, and the total upward forces of buoyancy and hydrodynamic reaction, and the lines of action of these two forces coincide, the ship is in equilibrium; even so, normally an excess of buoyancy over weight will result throughout certain portions of the length, counterbalanced by an excess of weight over buoyancy throughout the remaining portions. These excesses of upward or downward forces create conditions of shear and bending moments in the same manner as in ordinary beams.

It is quite possible, by means of simple approximations, to establish the general characteristics of the weight curve with a reasonable degree of accuracy, and, in the case of ships operating in smooth water exclusively, the buoyancy curve can be determined easily so that the shearing forces and bending moments under any condition of load can be estimated with reasonable accuracy. Until recent years there was a dearth of information concerning the most severe wave conditions that a ship might reasonably be expected to encounter. Therefore, at that time, to assess longitudinal strength requirements on a comparative basis between structures of the same general type, a so-called "standard wave" was adopted in association with which buoyancy curves, or curves of supporting forces, could be developed.

This standard wave was assumed to be quasi-static, to have a length from crest to crest equal to the length of the ship, and the contour of a trochoid with a depth from crest to trough equal to $1/20$ of the length. Depending upon the type of ship, as mentioned hereafter, it was customary to consider the vessel either poised on a wave, with a crest at the middle of the ship length and a trough at each end, or as bridging two crests, with one at each end and a trough at amidships. The former condition induced what is termed a *hogging moment* (one which tends to depress the ends of the ship, inducing tension in the deck and compression in the bottom), and the latter condition induced what was designated as a *sagging moment* (one which tends to depress the ship amidships, inducing compression in the deck and tension in the bottom).

In more refined investigations, calculations were made with the crest of a standard wave located at regular intervals along the length of the ship, and the peaks of the bending moment curves so obtained for both hogging and sagging were then connected into two curves called the "envelope curves" for the maximum hogging or sagging conditions. However, it was found that the standard calculations, with either a crest or trough at the middle of the ship length, were sufficient to indicate any abnormal conditions of form or weight distribution which would require margins in the longitudinal strength over and above the established standards.

A complete description of the underlying theories of bending moment calculations as well as the standard wave assumptions for oceangoing vessels is given in Chapter IV of *Principles of Naval Architecture* (McNaught, 1967).

It should be recognized that what are termed *standard bending moment calculations* are actually only calculations to give numerical values which reflect the distribution of the weights in the ship itself in association with the distribution of the supporting forces as affected by the hull form. Thus, the structural strength requirements for ships of unusual form, or having unusual distribution of weight, may be evaluated properly by comparison with those of ships of normal weight and buoyancy characteristics.

The foregoing methods of calculation indicate that for the normal type of oceangoing cargo vessel with machinery amidships, the bending moment with a well distributed load will be a maximum in the hogging condition, with the crest of the standard trochoidal wave amidships. Under the same condition of loading, the sagging moment with the trough of the wave amidships is generally less than one-half the hogging moment and is usually disregarded. When loaded to less than the full load draft, unless the load distribution is very unfavorable, the moments, whether they be hogging or sagging, are less than the hogging moment in the fully loaded condition, so that ordinarily this latter bending moment is taken as the criterion for the longitudinal structural strength standard. This standard was found to give a reasonable approximation of the calculated bending moment for those ships having a normal length of machinery space with the cargo weights nearly uniformly distributed in proportion to the volumes of the holds.

In the types of vessels having machinery amidships, the major bending moment is a hogging moment. Barring extraordinary conditions of loading or arrangement, the maximum moment obtained by detail calculations can be approximated reasonably closely by a simple, standard bending moment formula. In the case of seagoing ships having machinery aft, such as tankers, bulk carriers, and ore carriers, however, the conditions when fully loaded are such as to provide an excess of weight over buoyancy throughout the midship portion with the necessary excess of buoyancy over weight extending for some distance at either end. This creates a sagging moment. In investigations of longitudinal bending moments in such ships among waves, it is usual to assume the trough of the standard wave at the middle of the length in order to obtain the sagging moment for the *full and down* condition.

Unlike vessels having machinery amidships, where the hogging moments only need be considered and the sagging moments are of secondary importance, under normal conditions, tankers, bulk carriers, and ore carriers have a multiplicity of spaces available for partial cargos or for ballast. As a result, sometimes it becomes necessary to investigate the hogging moments with the wave crest amidships in association with certain conditions of partial loads, or of ballasting. Leaving a group of tanks or ore holds entirely empty at the middle of the length, with the cargo or ballast concentrated near the ends, even when the vessel is not loaded to the full load draft, might easily induce hogging moments on the crest of a wave of such serious magnitude as to warrant special consideration, even though normally the longitudinal strength of such ships is assessed in terms of the sagging moment in the fully loaded condition.

The bending moments for these types as obtained from detail calculations are influenced to a much greater degree by variations in the form and arrangement than is the case in the machinery-amidships vessels, and these influencing variables occur much more frequently in tankers and ore carriers than in the other ship types. The introduction near amidships of a pump room in a tanker, or a deep ballast tank in an ore ship, which normally would be practically empty in the fully loaded condition, will provide an excess of buoyancy at a section where it is of greatest advantage in reducing the sagging moment. The relation that the length of the ship over which the cargo is distributed bears to the displacement length also has a decided effect on the bending moment; the shorter the cargo length, the greater becomes the concentration of weight producing the sagging moment. For the sagging moment, the block coefficient may have a decided effect opposite to that for the hogging moment.

In very full ships, by comparison with fine ships, the buoyancy curve with wave crests at the bow and stern will show much higher peaks, with a corresponding lowering at the middle of the length. The weight curve at the ends may be affected in a similar manner, but to a lesser extent, and the overall result in most cases will be that the calculated sagging bending moment, when expressed in terms of the principal dimensions of L, B, and T, will vary directly with the block coefficient instead of in the opposite direction as in the case of a hogging moment. As a result, the inclusion of an arbitrary average value for the block coefficient in a standard bending moment formula may lead to further inaccuracies.

In some cases, these effects may tend to offset each other. For instance, in some of the high speed tankers with fine lines, the tendency is to reduce the length of the portion of the ship devoted to cargo tanks in proportion to the length of the ship, with the result that the effect of the fine lines in creating more favorable moment is offset by the comparatively shorter tank space.

All of these commonly occurring variables have such a pronounced effect on the longitudinal bending moment, as indicated by the standard methods of calculation, that it is not possible to express in simple terms of either length (L) times the displacement, or of the length squared (L^2) times the breadth (B) times the draft (T), a value for a standard bending moment which will have anywhere near the accuracy of that obtained with the simple form of expression of a standard bending moment, as is commonly used for the types of break-bulk cargo or mixed cargo-and-passenger vessels having machinery amidships.

4.3 Stillwater Bending Moments and Shearing Forces. In Subsection 4.2 the standard bending moment was explained at length, and its use as a practical and convenient means for comparing strength requirements for various designs was indicated. As more knowledge of wave bending moments and their relation to the total bending moments was gained, there gradually developed a procedure wherein the classification societies adopted an approach in which both stillwater and wave bending moments are explicitly allowed for and then combined.

Any treatise on assessing longitudinal strength should touch on the subject of loading manuals and loading instruments. The loading manual may have had its beginning with the famous T2 tankers, at which time it was deemed advisable to provide seagoing personnel with some form of loading guidance. Hence a manual was prepared, the purpose of which was to indicate, by means of a simple calculation, whether a particular loading or ballasting arrangement would impose a bending moment in excess of that for which the ship originally was designed. This "T2 Loading Guidance Manual" contained several diagrams indicating various loading and ballasting conditions, together with examples as to how to calculate the *loading numeral* for any arrangement.

The manual served as an immediate and useful tool for mates who otherwise might not be aware of the extent to which certain arrangements of cargo or ballast could be critical from a strength standpoint.

Some loading manuals are much more comprehensive than others. However, the majority present only a nominal amount of information and provide no assurance as to absolute maximum values of the bending moment, since the maximum values are frequently located at some distance from amidships.

Loading manuals were introduced primarily to serve tankers, so many of which were built during World War II. This development naturally spread to other general bulk types, such as ore carriers and bulk carriers. It would be difficult to provide comprehensive loading manuals for general cargo ships, owing to the many combinations of loading and wide variations in densities of cargos.

Recognition of the need for an improved method of providing loading guidance information has given rise to the development of loading instruments. These comprise generally a mini-computer and are capable of indicating both the shearing forces and bending moments, as well as their locations, for any given distribution pattern of cargo or ballast. The specified limit values of both the bending moments and shearing forces can be also incorporated in the display screen.

Other instrument systems which are recently in the development stage consist of a mini-computer and strain gages and are capable of calculating and monitoring stillwater and wave-induced stresses at critical locations. These systems are intended to avoid undue stresses in conjunction with both stillwater and wave loads. The day may not be far distant when these instrument systems will be mandatory for safety measures and also be required as a condition of classification.

At the present time, computer programs are employed as routine tools in ship designs. The bending moments and shearing forces can be accurately calculated for a given weight and buoyancy distribution in still water. A loading manual with detailed information is usually provided by designers for a new ship. A number of empirical formulas with various degrees of accuracy is now available for approximating the maximum stillwater bending moments and shearing forces in the early design stages, among them the following equations:

$$M_{sw} = C_s L^{2.5} B (C_B + 0.5) \qquad (1)$$

and

$$F_{sw} = 5.0\, M_{sw}/L \qquad (2)$$

where M_{sw} = stillwater bending moment, in ton-m (ton-ft)

F_{sw} = maximum hull-girder shearing force in still water, in tons

$$C_s = \left[0.618 + \frac{110 - L}{462}\right] 10^{-2} \quad 61 \le L \le 110 \text{ m}$$

$$\left\{\left[0.312 + \frac{360 - L}{2990}\right] 10^{-3} \quad 200 \le L \le 360 \text{ ft}\right\}$$

$$= \left[0.564 + \frac{160 - L}{925}\right] 10^{-2} \quad 110 < L \le 160 \text{ m}$$

$$\left\{\left[0.285 + \frac{525 - L}{6100}\right] 10^{-3} \quad 360 < L \le 525 \text{ ft}\right\}$$

$$= \left[0.544 + \frac{210 - L}{2500}\right] 10^{-2} \quad 160 < L \le 210 \text{ m}$$

$$\left\{\left[0.275 + \frac{690 - L}{16400}\right] 10^{-3} \quad 525 < L \le 690 \text{ ft}\right\}$$

$= [0.544]10^{-2} \qquad 210 < L \leq 250$ m

$\{[0.275]10^{-3} \qquad 690 < L \leq 820$ ft$\}$

$$= \left[0.544 - \frac{L - 250}{1786}\right] 10^{-2}$$

$250 < L \leq 427$ m

$$\left\{\left[0.275 - \frac{L - 820}{11600}\right] 10^{-3}\right.$$

$\left. 820 < L \leq 1400 \text{ ft}\right\}$

L = scantling (Rules) length of vessel, in m (ft) measured as *LBP*, is not to be less than 96 percent and need not be greater than 97 percent of the length on the summer load waterline.

B = breadth of vessel, in m (ft).

C_B = block coefficient at summer load waterline. For this equation, C_B is not to be taken as less than 0.64.

It should be emphasized that M_{sw} in Equation (1) is a *standard* value and can be used only in the early design stages. A detailed calculation of stillwater bending moments is usually required for the final design.

In general practice, the cargo spaces and ballast tanks are so arranged that the maximum stillwater bending moments for all anticipated loading cases would not exceed a limit value derived from successful past experience. Alternatively, a design may be based on a permissible stillwater bending moment, so that the loading manual can be provided accordingly at later dates. In any event, the maximum stillwater bending moment is customarily treated as a constant in the determination of the maximum total bending moment. In a refined analysis the stillwater bending moment should, however be treated as a variable due to the following reasons:

- Ships are seldom loaded exactly in accordance with loading manuals.
- The magnitudes of stillwater bending moments also vary slowly during voyage due to changes of ballast and consumption of stores and fuel.

To perform a statistical analysis, a long-term distribution of the stillwater bending moments is preferable.

4.4 Wave-Induced Moments and Shearing Forces. In Subsection 4.2, considerable emphasis was given to the observation that for many years naval architects were faced with an unusual set of circumstances. Unlike other branches of engineering, information was lacking as to the exact magnitude of the loads imposed on the structure when in service, and this was true of the loads on the hull girder as well as of the local loads on various component parts of the structure. It is readily understandable, therefore, why the experience factor was so important, and why so great a degree of reliance was placed on classification requirements which had been predicated largely on experience.

Experience always will be a sound and dependable source of guidance, even in projecting to ships of much greater size. However, the general picture is changing; new developments in water transportation have reached the point where experience either has limited application, or is nonexistent. The recent trend toward much larger ships naturally led to considerable extrapolation of existing requirements. This, in turn, necessitated careful consideration of possible hidden size effects for which past experience offered little help. Since extrapolation of existing requirements was not feasible, a basic approach had to be adopted. Fortunately, this transition period was associated with additional knowledge, and any discussion of modern trends in assessing required longitudinal strength would not be complete without special mention of research developments.

a. Research Developments. In contrast to the low level of research in the marine industry in earlier years, there seems to have been a worldwide upsurge in this type of activity commencing in the 1950s. This development has grown steadily and the advent of the mammoth ship has given the effort added impetus. Research efforts concerned with longitudinal strength have been directed largely along three principal channels; i.e., sea conditions, response, and model testing.

1. Valuable oceanographic data obtained from weather ships and stations in scattered locations have greatly facilitated the study of wave conditions. These studies have led to the development of trends in long-term predictions of wave bending moments and, owing to the mysteries associated with the wave portion of the total bending moment (stillwater moment plus wave moment), considerable importance has been attached to this particular field of research. It is fortunate that the phenomenal growth in the size of ships, particularly tankers, has coincided with an increase in knowledge of ocean waves.

2. Closely related to research on actual sea conditions is the matter of ship response to those conditions, and this represents a second branch in which there has been a marked effort in the past few years. A number of ships, including larger bulk carriers have been fitted with instrumentation systems to measure, among other features, the actual degree of bending in the hull when subjected to known (simultaneously observed) wave conditions. These instrumentation systems vary considerably in respect to their degree of sophistication. In at least one instance, provisions have been made to register simultaneously the actual wave pattern during severe conditions at sea.

3. There is also the model testing phase of research, which constitutes a third and related phase of longitudinal strength investigations. The purpose here is to be able to establish the correlation pattern between model response to simulated wave conditions and the measured degree of response in the ship. Once this relationship is established between ship and model, particularly in the case of the larger ships, it can be utilized with considerable confidence to establish longitudinal strength requirements for the much larger ships contemplated.

The foregoing explanation is confined to longitudinal strength considerations. There are, of course, research efforts along numerous other lines of endeavor within the marine industry and these have followed a similar growth pattern.

A gradual change has taken place in classification rules, with particular reference to the numerous tables of scantlings formerly present. Many of these tables have disappeared as a result of substituting formulas to facilitate computer solutions to the problems. In addition, it should be noted that in many cases previously it was possible, from the rather extensive tables, not only to determine the required thicknesses of plating for shell, deck, inner bottom, etc., but to be reasonably certain that, if the ships were of normal proportions, they also would meet longitudinal strength requirements. This situation has changed. Increased knowledge of longitudinal strength requirements to suit the type, size, and service of a particular design, together with certain modifications and refinements to those requirements as they previously existed, has resulted in the required hull girder modulus being dealt with separately. Therefore, from the standpoint of classification requirements, it is necessary to check most new designs against two separate sets of requirements: one for longitudinal strength, and one for local considerations. Consequently, tables of scantlings where existing, or the formulas which may have been substituted, are in many cases an indication of minimum scantlings predicated on considerations independent of longitudinal strength.

For many years it was customary to require that ships arranged with machinery aft have their deck areas (cross-sectional area of longitudinal effective deck plating) increased by 20 percent over the normal rule. This was an outgrowth of the accepted theory that with machinery aft this type of ship (mostly tankers and colliers), when fully loaded and in the *sagging* condition, would be subject to bending moments somewhat greater than would result in ships with their machinery located in the vicinity of amidships. The additional deck area not only resulted in a greater section modulus, but provided increased material in the deck as a precautionary measure against the buckling tendency associated with the large sagging bending moment. The thinking behind this increase prevailed for many years until the more modern and larger designs gave rise to more thorough investigations of the actual calculated bending moments under various loading and ballasting arrangements. Such investigations showed that not only was too much emphasis being placed on the aforementioned approach, but that in some instances, especially where it was necessary to carry appreciable amounts of ballast near the ends of the vessel, the reverse situation could occur. This, then, was part of a gradual development toward a more comprehensive calculation of bending moments that might result from all loaded and ballasted conditions in actual service. It also led to abandonment of the requirement for additional deck area in the case of ships with machinery aft.

b. Quasi-Static Approximations. As discussed previously, the nominal wave-induced vertical bending moments were customarily determined by poising the ship on a $L/20$ trochoidal wave. Since this standard wave had been considered as too conservative for larger vessels, the standard wave height was then expressed by either $0.6L^{0.6}$ or $1.1\sqrt{L}$, where L is the length of the vessel. Starting in the early

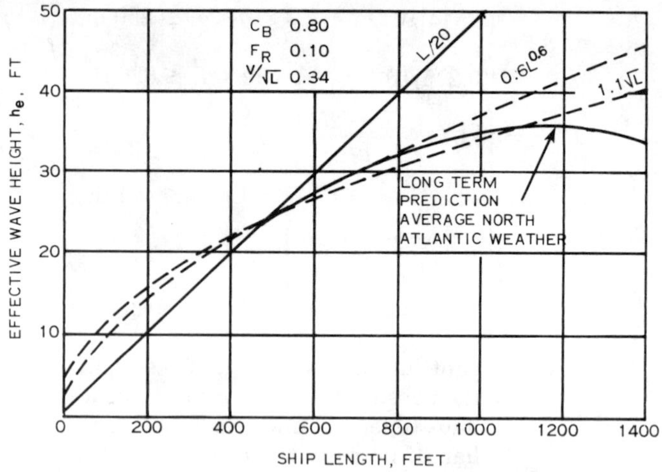

Fig. 20 Variation of effective wave height with length for tankers predicted from long-term probability curves and compared with various formulations

1960s, ship motion computer programs and probability approaches based on measured wave data were developed and were capable of predicting the long-term trend of the wave-induced bending moments. In analogy to the customary quasi-static calculation, the predicted wave-induced bending moments were expressed in terms of *effective* wave heights, as shown in Fig. 20 (Lewis 1967). Here, the long-term prediction is compared with the above mentioned criteria related only to ship length. The *effective* wave is defined as a static wave with length equal to the ship's length and the specified height, which in turn gives the same amplitude of the predicted maximum wave-induced bending moment in a quasi-static calculation. In this calculation, either a trochoidal or a sinusoidal wave can be used, without *Smith effect* corrections. Fig. 21 shows a similar comparison with calculated values for specific ships, based on full-scale stress data (Little and Lewis, 1971).

It is worth noting that a sinusoidal wave is generally utilized instead of the trochoidal wave nowadays for simplifying the computation procedure. A recent study indicates that the effect on the wave moments of the difference between these two wave profiles is negligible.

Due to the availability of ship motion computer programs and realistic wave spectra, the quasi-static wave moment calculation which had been widely employed by naval architects for many years becomes gradually obsolete. At the present time, some of the major classification societies would no longer accept a quasi-static wave moment calculation for plan approval. This is also partly due to frequent misinterpretations of the calculated results by many designers. The effective wave height is, however, still used as a coefficient in the calculation of the wave-induced bending moments in classification rules and in some textbooks.

c. Ship Motion Calculations. One of the most fruitful results of the intensified research efforts devoted to wave loads in the past 20 years is the establishment of computation procedures for predicting ship motions and wave loads. These procedures comprise essentially an establishment of wave spectra, ship motion calculations and a probability analysis. The computation sequence can be divided into the following three steps:

- calculation of transfer functions,
- calculation of short-term responses,
- calculation of long-term responses.

A brief description of the general procedure is given below.

The ocean wave data have been collected by three different methods: observation, measurement, and hindcasting. A vast body of visually observed data is documented in tabular form, such as those presented by Hogben and Lumb (1967) and Yamanouchi et al (1965). Measured wave data are recorded by ocean weather ships, wave buoys, and offshore platforms. Hindcast wave data are generated by a technique whereby the wind fields are used as a basis of updating synoptic data for predicting wave characteristics, using theoretical methods. Of the available methods, the Spectral Ocean Wave Model (SOWM) developed by the U. S. Navy Fleet Numerical Oceanography Center (FNOC) is currently operational, and is used to hindcast wave spectra for the Northern Hemisphere over the past twenty years. Some of the FNOC wave data were correlated with measured wave spectra measured at Station India. As discussed by Hoffman and Chen (1978) and also Chen et al (1979) the FNOC wave spectra compare very well with the measured data at Station India, and can be applied to investigate the response of ocean-going ships. More recently, Cummins and Bales (1980) presented the analysis of 10 years' hindcast wave data for the North Atlantic and Pacific Oceans. Similar hindcast wave data are also available for the Great Lakes region. It is anticipated that the vast volume of wave data generated by hindcasting techniques will provide more realistic wave environments for assessing the longitudinal strength of vessels intended for either restricted or unrestricted services.

The major classification societies have established their own wave spectra in conjunction with the determination of wave loads. For example, ABS is currently using a set of wave spectra derived from the measured data in the North Atlantic region for ocean-going vessels and a different set of wave spectra for Great Lakes ships.

Most of the ship motion computer programs developed to date are based on the two-dimensional strip theory, where the hydrodynamic interaction between two adjacent ship sections is neglected. Computer programs available in the United States for this kind are, for example, SCORES, MIT/MarAd Seakeeping and NSRDC Shipmotion and Seaload. All these programs calculate ship response including motions and wave loads with 5 or 6 degrees of freedom. For a given ship with selected loading cases, these computer programs are all capable of calculating the transfer functions (ship's responses in sinusoidal waves with unit amplitude and various frequencies) and short-term responses with respect to each individual wave spectrum specified.

Some ship motion computer program systems, such as ABS/SHIPMOTION, contain also a probability approach for calculating long-term responses.

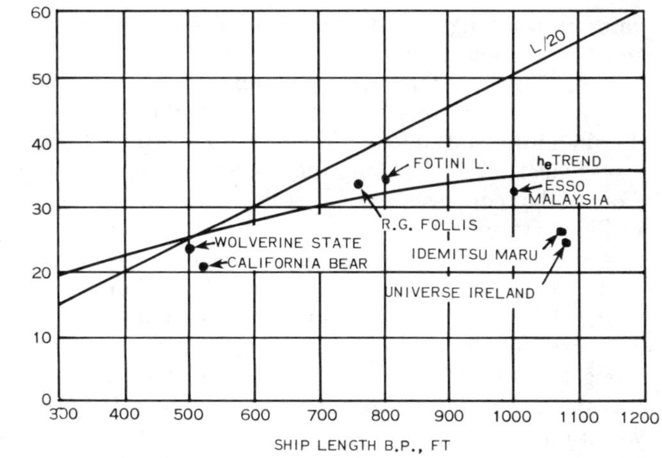

Fig. 21 Variation of effective wave height with length for specific ships, based on full-scale stress data

The basic assumptions involved in the direct computation method are: 1 the sea conditions can be represented by a wave spectrum for a short period of time, 2 the ship response is a linear function of wave height, and 3 the short-term probability characteristics of ship response follow a certain pattern of distribution in a longer time duration. The first two assumptions are necessary for calculating the short-term ship responses corresponding to specified wave spectra. The third assumption is intended to facilitate the long-term predictions.

For design purposes, the maximum wave loads can be determined either by the short-term or by the long-term prediction techniques. The application of the short-term approach was suggested by Ochi (1978). In general, the short-term approach can only be applied to a specific ocean location where the exact characteristics of extreme sea conditions are known. The long-term prediction method, however, has a number of different forms, depending on what kind of short-term statistical characteristics are being used. Those where the root-mean square value is used can be found from Band (1966) and Nordenstrom (1963). Also, Hoffman and Lewis (1969) and Ochi (1978) applied extreme

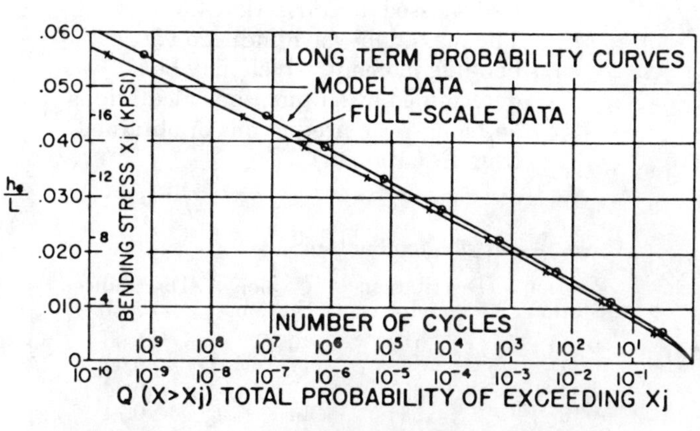

Fig. 22 Long-term possibility of exceeding he/L or peak-to-mean stress prediction for "Wolverine State" compared with full-scale trends for average North Atlantic weather

value in the long-term trend calculation. A summary of these approaches was presented by Stiansen and Chen (1979). The long-term approach can take the variety of sea conditions along the ship route into consideration, and the detailed information of characteristics of sea conditions is less critical. A typical long-term probability curve derived from both model tests and full-scale measured data is shown in Fig. 22 (Lewis, 1967).

Although the ship motion computer programs are presently capable of calculating all wave load components in terms of vertical, horizontal and torsional moments together with vertical and horizontal shearing forces at any stations along the length of vessel, only the vertical components (moments and shears) are generally selected as the nominal wave loads in designs of conventional-type vessels. This approach is partly due to the dominating stress amplitudes induced by the vertical bending moments and shearing forces, and partly due to the successful development of analytical methods for predicting the vertical motions and vertical wave load components. The other hydrodynamic force components resulting from roll, yaw, and sway can not yet be predicted with the same degree of accuracy as those due to pitch and heave.

It should be noted that the nominal design loads are determined in conjunction with the specified permissible stresses as discussed in Section 5 and cannot be interpreted as the *extreme* loads. Customarily, a certain safety margin is incorporated in the allowable stress level, and the structure is expected to be capable of withstanding loads greater than the maximum design load. Therefore, the limit values of both loads and stresses are established based on successful service experience for many ship-years, and should be regarded only on a comparative basis.

d. Classification Rules. In the current ABS Rules, the longitudinal strength requirements are specified in terms of the maximum total bending moment (M_t) and permissible bending stress (σ_p) (See Subsection 5.4). The total bending moment is determined by the following equation for oceangoing vessels.

$$M_t = M_{sw} + M_w \tag{3}$$

where M_{sw} = The maximum stillwater bending moment as discussed in Subsection 4.3

M_w = The maximum wave-induced vertical bending moment, which may be determined by shipmotion calculations and long-term predictions or obtained from Equation (4).

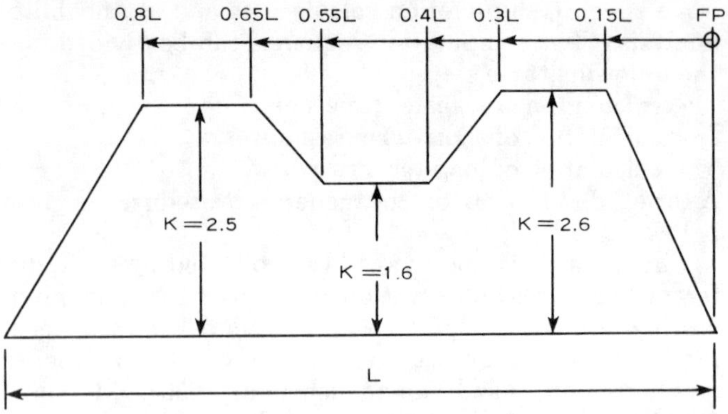

Fig. 23 Envelope of wave-induced shearing forces for oceangoing vessels

The maximum wave-induced vertical bending moment amidships may be determined by

$$M_w = C_2 K_b H_e B L^2 \text{ in ton-m (ton-ft)} \tag{4}$$

where K_b = 1.0 for $C_B \geq 0.80$
= $1.4 - 0.5C_B$ for $0.64 \leq C_B < 0.80$

C_2 = $[2.34C_B + 0.2]10^{-2}$ metric units
($[6.53C_B + 0.57]10^{-4}$ inch/pound units)

L = Rules length of vessel, in m (ft)

B = breadth of vessel in m (ft)

C_B = block coefficient at summer load waterline, for this equation, C_B is not to be taken as less than 0.64.

H_e = effective wave height in m (ft)
= $0.0172L + 3.653$ $\quad 61 \leq L \leq 150$ m
($0.0172L + 11.98$ $\quad 200 < L \leq 490$ ft)
= $0.0181L + 3.516$ $\quad 150 < L \leq 220$ m
($0.0181L + 11.535$ $\quad 490 < L \leq 720$ ft)
= $[4.50L - 0.0071L^2 + 103]10^{-2}$ $\quad 220 < L \leq 305$ m
($[4.50L - 0.00216L^2 + 335]10^{-2}$ $\quad 720 < L \leq 100$
= 8.151 $\quad 305 < L \leq 427$ m
(26.750 $\quad 1000 < L \leq 1400$ ft)

Equation (4) was derived based on a long-term trend predicted by means of analytical calculations and also analysis of full-scale measured data as shown in Fig. 21. The term H_e, effective wave height, in Equation (4) is intended only as a necessary coefficient, and cannot be regarded as an

Table 1—Distribution Factors

Position Station	Distribution Factor	Postion Station	Distribution Factor
0 (AP)	0	10 ⊗	1.00
2	0.10	11	0.99
4	0.35	12	0.94
6	0.68	14	0.74
8	0.95	16	0.43
9	1.00	18	0.13
		20 (FP)	0

Table 2—Distribution Factors

Position Station	Distribution Factor	Position Station	Distribution Factor
0 AP	0	11	1.00
2	0.17	12	0.96
4	0.57	14	0.76
6	0.76	16	0.44
8	0.95	18	0.12
9	1.00	20 FP	0
10 ⊗	1.00		

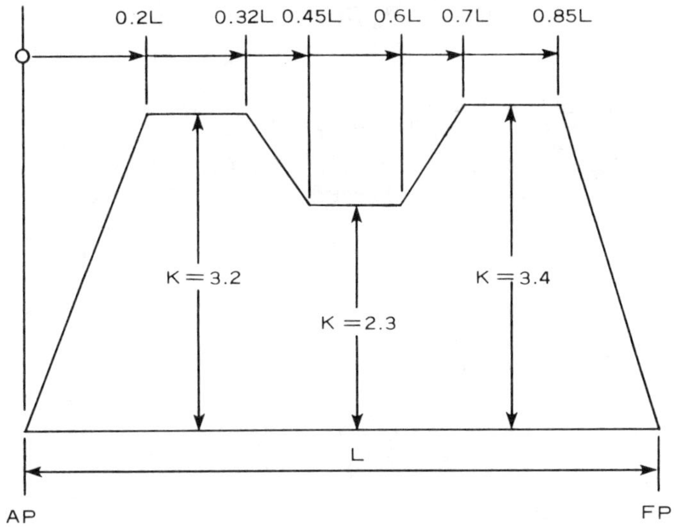

Fig. 24 Envelope of Dynamic shearing forces for Great Lakes bulk carrier

implication for the acceptance of quasi-static wave moment calculations.

The maximum wave-induced bending moments at stations other than amidships may be determined by the distribution factor given in Table 1 for vessels having C_B not less than 0.64.

The total vertical shearing force (F_t) is expressed by

$$F_t = F_{sw} + F_w \tag{5}$$

where F_{sw} = maximum stillwater shearing force

F_w = maximum wave-induced shearing force which may be either determined by long-term predictions or obtained from the following equation:

$F_w = KM_w/L$

M_w and L are as defined previously

K is given in Fig. 23 for vessels having C_B not less than 0.64.

For Great Lakes bulk carriers, the maximum total bending moment (M_t) is specified in ABS Great Lakes Rules, 1978, in the following form:

$$M_t = M_{sw} + M_c$$

$$M_c = C_s\sqrt{M_w{}^2 + M_{sp}{}^2} \tag{6}$$

where M_{sw} = maximum stillwater bending moment,

M_c = maximum combined dynamic bending moment,

M_w = maximum wave-induced vertical bending moment amidships,

M_{sp} = maximum springing bending moment amidships

C_s = correlation coefficient

= $0.995 - 0.172[(L/305) - 0.4]^2$ metric units ($0.995 - 0.172[(L/1000) - 0.4]^2$ inch/pound units)

L = Rules length of vessel, in m (ft)

The maximum wave-induced vertical bending moment

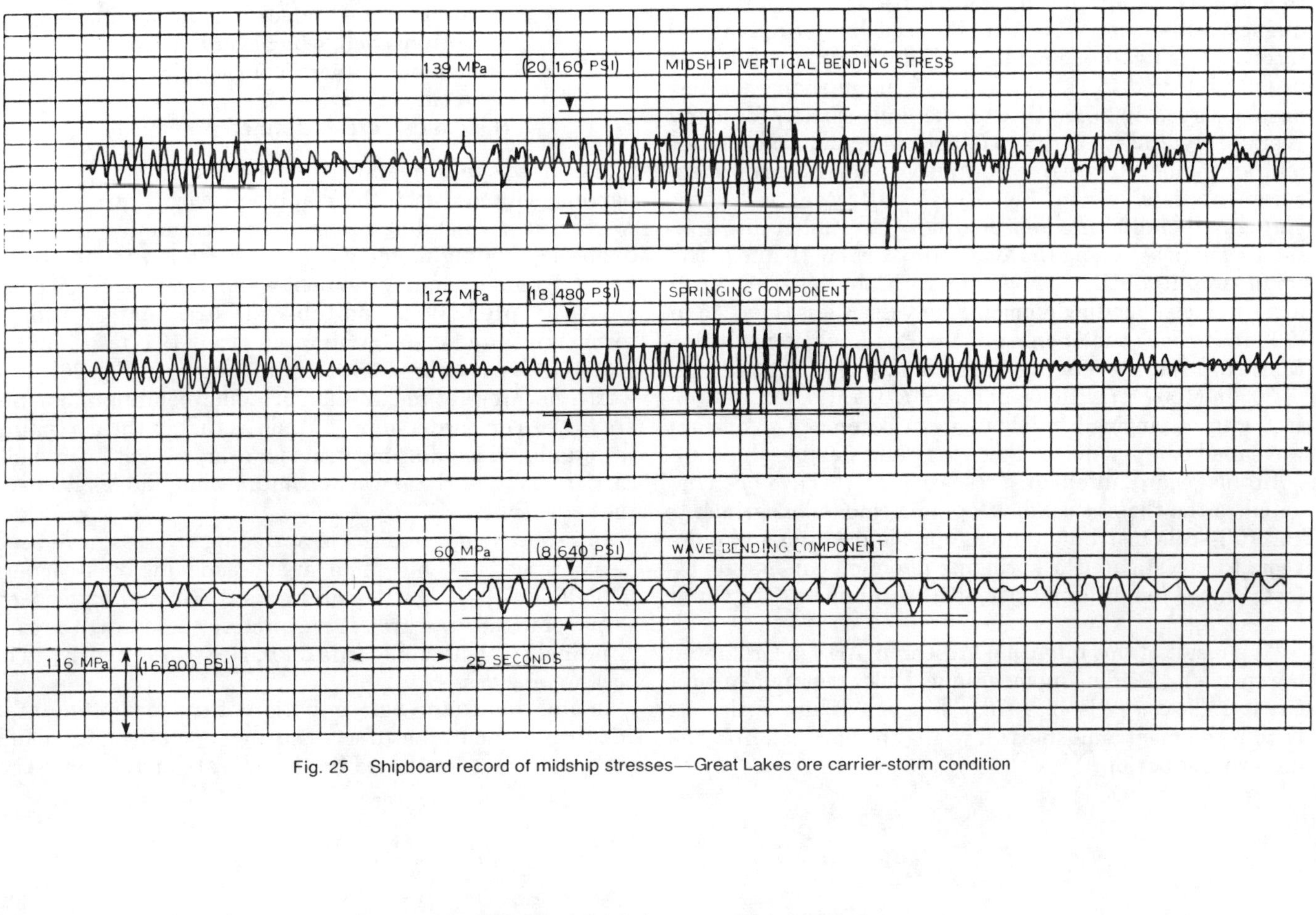

Fig. 25 Shipboard record of midship stresses—Great Lakes ore carrier-storm condition

amidships may be determined by the following equation:

$$M_w = C_w B(L/305)^2 \text{ ton-m}$$

$$[C_w B(L/1000)^2 \text{ ton-ft}] \quad (7)$$

where B and L are as previously defined and

C_w = 9261 − 4.700L $122 \le L \le 183$ m
(9113 − 1.410L $400 \le L \le 600$ ft)
= 8993 − 3.240L $183 < L \le 244$ m
(8850 − 0.972L $600 < L \le 800$ ft)
= 8803 − 2.460L $244 < L \le 305$ m
(8663 − 0.738L $800 < L \le 1000$ ft)
= 8656 − 1.977L $305 < L \le 366$ m
(8518 − 0.593L $1000 < L \le 1200$ ft)

The springing bending moment, M_{sp}, will be discussed in Subsection 4.5.

The distribution of the combined dynamic bending moments along the length of vessel may be determined by the distribution factors given in Table 2.

The maximum dynamic shearing forces (F_d) may be obtained from

$$F_d = KM_c/L \quad (8)$$

where M_c and L are as defined previously
K is a coefficient as given in Fig. 24.

4.5 Springing and Impulsive Loads. The wave-induced bending moments and shearing forces discussed in the previous subsection are in the same frequency range as the ships' motions, and may be regarded as wave loads of low frequencies. Another type of wave load is due to high-frequency hull-girder vibrations induced by hydrodynamic forces. The two-node mode vertical vibration of a hull is usually referred to as *springing*. This phenomenon has been observed on board Great Lakes bulk carriers for many years. A typical stress record of both wave-induced bending and springing stresses as obtained from full-scale measurements is shown in Fig. 25. Until recent years, the springing bending moment had not been taken into consideration in assessing the longitudinal strength due to the small amplitude in comparison with the low-frequency wave-induced bending moments for vessels less than 215 m (700 ft). Since the opening of the Poe Lock at Sault Ste. Marie in 1969, the length of Great Lakes vessels has grown to 305 m (1000 ft) with a maximum L/D ratio of 21. Consequently, springing bending moment becomes a significant wave load component for those long and flexible ships.

In this regard, intensified research efforts have been devoted to the theoretical development, model experiments and full-scale instrumentations for the past ten and more years to investigate the springing phenomenon and its effects on the longitudinal strength requirements for Great Lakes ships.

As a result of this intensified research, ABS incorporated the springing bending moment in its Rules requirements for Great Lakes vessels in 1978. The maximum springing bending moment amidships (M_{sp}) may be obtained from the following equation:

$$M_{sp} = CC_{sp}B(L/305)^3 \quad \text{ton-m}$$

$$(CC_{sp}B(L/1000)^3 \quad \text{ton-ft}) \quad (9)$$

where C = 2333 − 1.2798L $122 \le L \le 183$ m
(2296 − 0.3839L $400 \le L$ 600 ft)
= 2260 − 0.8800L $183 < L \le 244$ m
(2224 − 0.2640L $600 \le L \le 800$ ft)
= 2208 − 0.6670L $244 < L \le 305$ m
(2173 − 0.2001L $800 < L \le 1000$ ft)
= 2168 − 0.5353L $305 < L \le 366$ m
(2134 − 0.1606L $1000 < L \le 1200$ ft)
C_{sp} = 5.58 − ω $1.0 \le \omega \le 2.0$
= $5.06/\sqrt{\omega}$ $2.0 < \omega$

and

$$\omega = \frac{C_f}{27.33}\sqrt{\frac{ID}{2Y}\Big/\sqrt{B^3T}}\Big/(0.1L)^2$$

metric units

$$\left(C_f\sqrt{\frac{ID}{2Y}\Big/\sqrt{B^3T}}\Big/(0.1L)^2\right.$$

inch/pound units)

C_f = 1645 − 2.4768L $122 \le L \le 183$ m
(1645 − 0.7549L $400 \le L \le 600$ ft)
= 1483 − 1.5867L $183 < L \le 244$ m
(1483 − 0.4836L $600 < L \le 800$ ft)
= 1374 − 1.1411L $244 < L \le 305$ m
(1374 − 0.3479L $800 < L \le 1000$ ft)
= 1294 − 0,8784L $305 < L \le 366$ m
(1294 − 0.2678L $1000 < L \le 1200$ ft)
Y = distance, in m (ft), from the neutral axis to the strength deck at side or to the bottom shell, whichever is greater
I = moment of inertia of the midship section in cm^2-m^2 (in.^{2}-ft^2)
L, B, D and T are as defined previously

As shown in Equation (6), the maximum springing bending moment is to be combined with the maximum wave-induced bending moment in determining the total dynamic bending moment. The expression for the combined dynamic bending moment was derived based on both analytical predictions and full-scale stress data. A long-term trend predicted by Stiansen et al (1977) for a 305m (1000 ft) Great Lakes bulk carrier is shown in Fig. 26. It is worth noting that maximum springing bending moments are of the same order of magnitude as that of the maximum wave-induced bending moments for a large Great Lakes bulk carrier, and that these two maximum values do not occur at the same time.

In recent years, significant springing stresses have been also observed onboard some large oceangoing bulk-carriers (Goodman, 1971) (Little and Lewis, 1971) and high speed ships. Further research is required for assessing the longitudinal strength standards for various sizes and types of ocean-going vessels.

In addition to the determination of the maximum total bending moment, the effect of springing on the fatigue life of a ship structure should be also investigated. Since the frequency of springing is several times higher than that of the wave-induced bending moments, the structure will experience many more stress cycles during its life span than it would due to normal wave bending alone.

Other wave loads which may induce additional hull-girder bending and shears are impulsive wave loads imposed on the forebody, such as bottom slamming, bow flare impact and shipment of *green water*. All these loads are of short duration and are generally referred to as transient loads. In addition to the direct loads imposed on local structures, these impacts may also excite high frequency hull-girder vibrations, generally referred to as *whipping*. A typical full-scale record of vertical bending stresses with slamming is shown in Fig. 27 (Lewis et al, 1973). Several computer programs have been developed recently for calculating hull-girder responses to these impact forces (Ochi and Motter, 1973) (Mansour and d'Oliveira, 1975) (Stavovy and Chuang, 1976). The occurrence of deck wetness and slamming can be predicted using ship motion theory. The magnitudes and distributions of these loads, however, remain uncertain.

Current studies on bow flare impacts have indicated that the nonlinear hydrodynamic pressures imposed on a large bow flare above the water line may induce significant hull-girder shearing forces and bending moments, in addition to the possible vibratory responses. The nonlinear effect may significantly change the distribution of the longitudinal bending moments and shearing forces. This explains why the distributions given in Subsection 4.4.d are restricted to vessels having a value of C_B not less than 0.64. Major classification societies have either approximated this nonlinear effect in the rules or require additional calculations

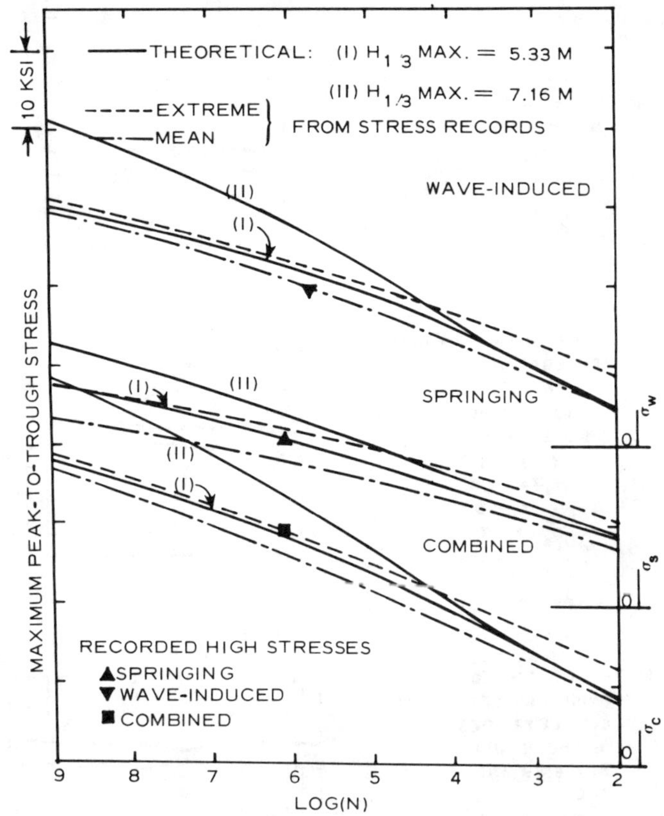

Fig. 26 Comparison between long-term recorded and theoretical stress of the *STEWART J. CORT*

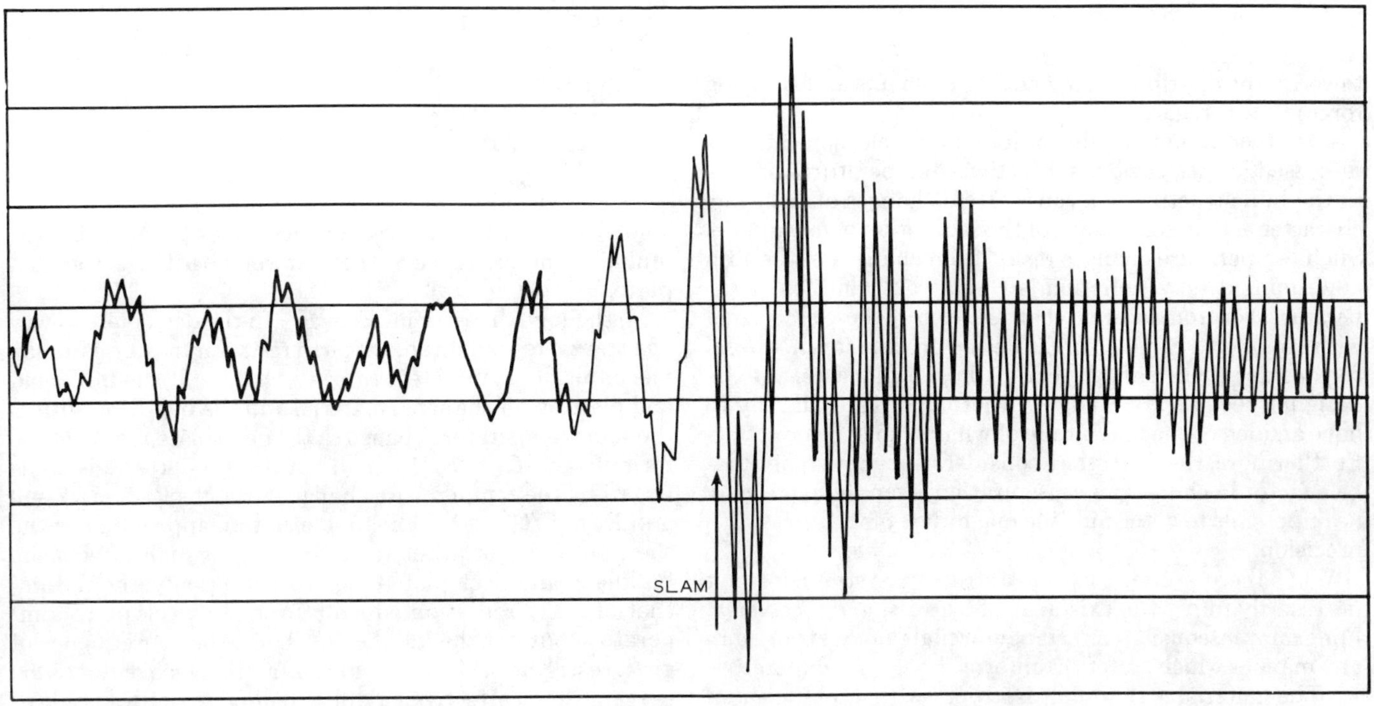

Fig. 27 Typical record of midship vertical bending stress with slamming, *M. V. Fontini L.*

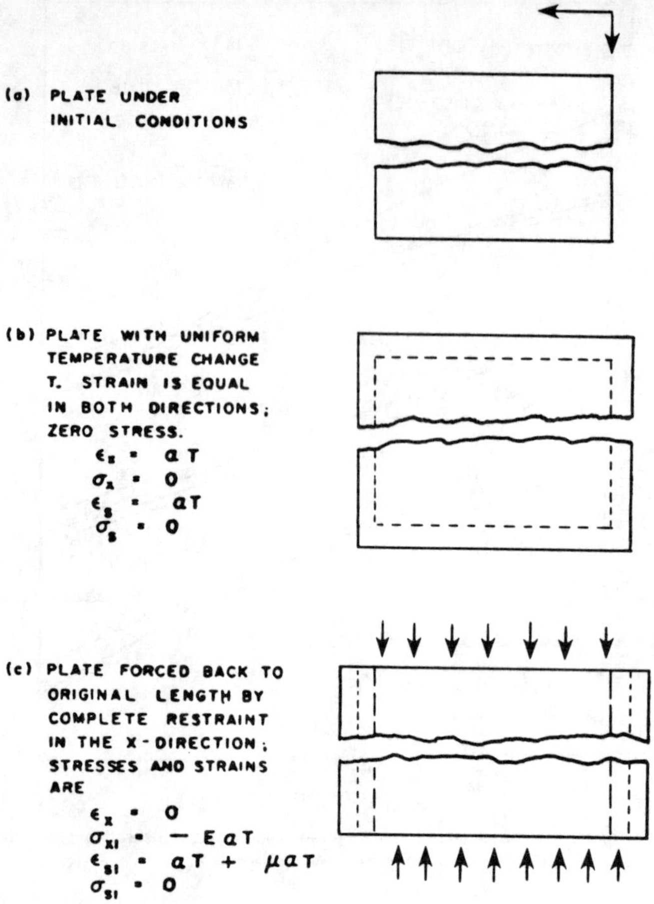

Fig. 28 Plate with complete thrust fixity in the X-direction under thermal change

to verify the distribution of bending moments and shearing forces in this regard.

4.6 Thermal Loads. In addition to the mechanical loads discussed in the previous subsections, temperature changes in the hull structure can cause stress changes of the same character and, in some cases, of the same order of magnitude. Such temperature changes result from changes in the ship environment (sky radiation, air and water ambient temperatures) or from internal changes, such as those associated with hot or cold cargos. In the design process, it is not usual to consider these changes except in special cases, yet the designer needs a knowledge of thermal stresses, if only to have an idea of the magnitude of what he may be omitting. Furthermore, thermal stress consideration may sometimes be a factor in choosing a configuration, even though it may be impossible to determine the magnitude of the stress with precision.

When thermal effects exist in a structure, a strain does not necessarily imply the existance of a stress, and vice versa. For example, consider a rectangular plate not restrained in any manner which suffers a uniform change of temperature t. The material is thermally isotropic, with a coefficient of thermal expansion α. According to the definition of α, the temperature change will produce a strain αt in all directions, but no stress. But if the plate were completely restrained in one direction, the temperature change can produce no strain in that direction, and instead a stress will be generated. The stress can be computed by imagining that the plate is first unrestrained when the temperature change t takes place, creating uniform strain αt then applying boundary forces in one direction (say the x-direction) to bring the plate back to the condition of zero strain in that direction, mechanically, creating the stress $\sigma_{x1} = -E\alpha t$ if there is no restraint in the girth (s) direction. See Fig. 28.

When such a plate is a component of ship structure, it is usually under some restraint in the direction normal to x; in a deck plate, for example, such restraint may be offered by transverse deck beams, or transverse bulkheads. To approximate the effect of such girth restraint, an effective elastic modulus E_x, may be used. This elastic modulus (effective) E_x may also be variable across the section. However, in this case the thermal coefficient α in the plate and restraint may be unequal if the materials are different, and the temperature change in the plate and restraint may be unequal, with the result that the effects of girth restraint on plates subject to thermal loading may be much greater than on plates subject to mechanical loading only. The effective elastic modulus which reflects these possibilities is:

$$E_x \equiv \frac{1 + r^* + \mu r^* \left(1 - \dfrac{\bar{t}^*}{t}\tau\right)}{1 + (1 - \mu^2) r^*} E \qquad (10)$$

where $r^* = \dfrac{\bar{E}\bar{A}}{EA}$

μ = Poisson's ratio.
A = section area

$\bar{t}^* = \dfrac{\bar{\alpha}}{\alpha}\bar{t}$

The symbols with the bar over them refer to the girth stiffening member; those without it refer to the associated plate.

Without girth restraint, $E_x = E$. In the deck plate in way of a transverse bulkhead, the girth restraint may be assumed mechanically complete; i.e., $r = r^* \rightarrow \infty$. If the bulkhead and plate are of the same material and the bulkhead suffers the same temperature change as the deck plate, $t = \bar{t}^*$ in the formula and $E_x = E/(1 - \mu^2)$, but if the bulkhead is unaffected by the temperature change in the deck plate, $\bar{t}^* = 0$ and $E_x = E/(1 - \mu)$. The first situation approximates the thermal effect of a hot (or cold) cargo which affects the bulkhead and deck plate temperature equally; the second, that of a sky radiation which affects the deck plate temperature but not the bulkhead. Thus, the consequence of girth restraint of the latter kind is an increase of nearly 50 percent in the effective modulus, compared with zero girth restraint, and a corresponding increase in the stresses.

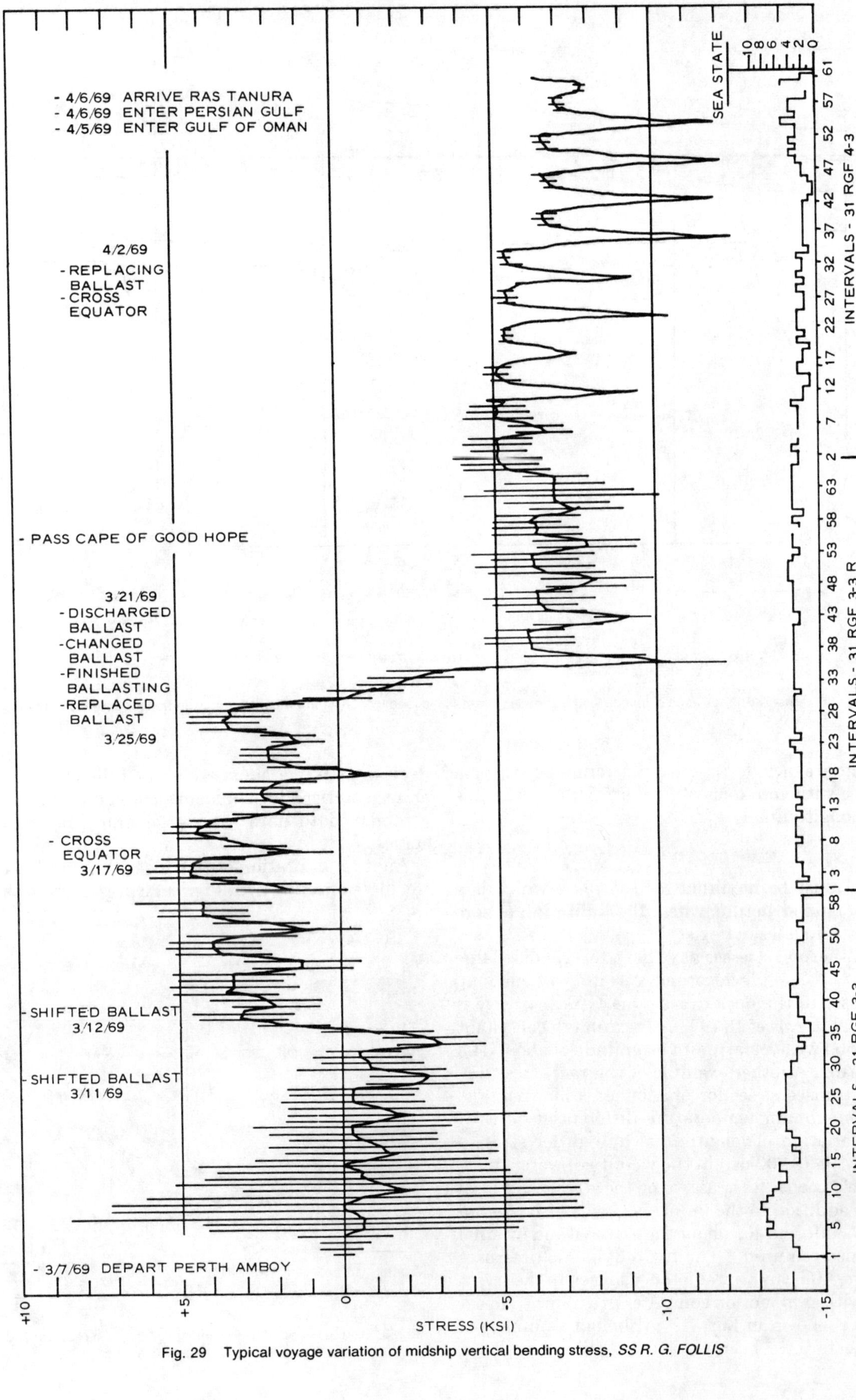

Fig. 29 Typical voyage variation of midship vertical bending stress, *SS R. G. FOLLIS*

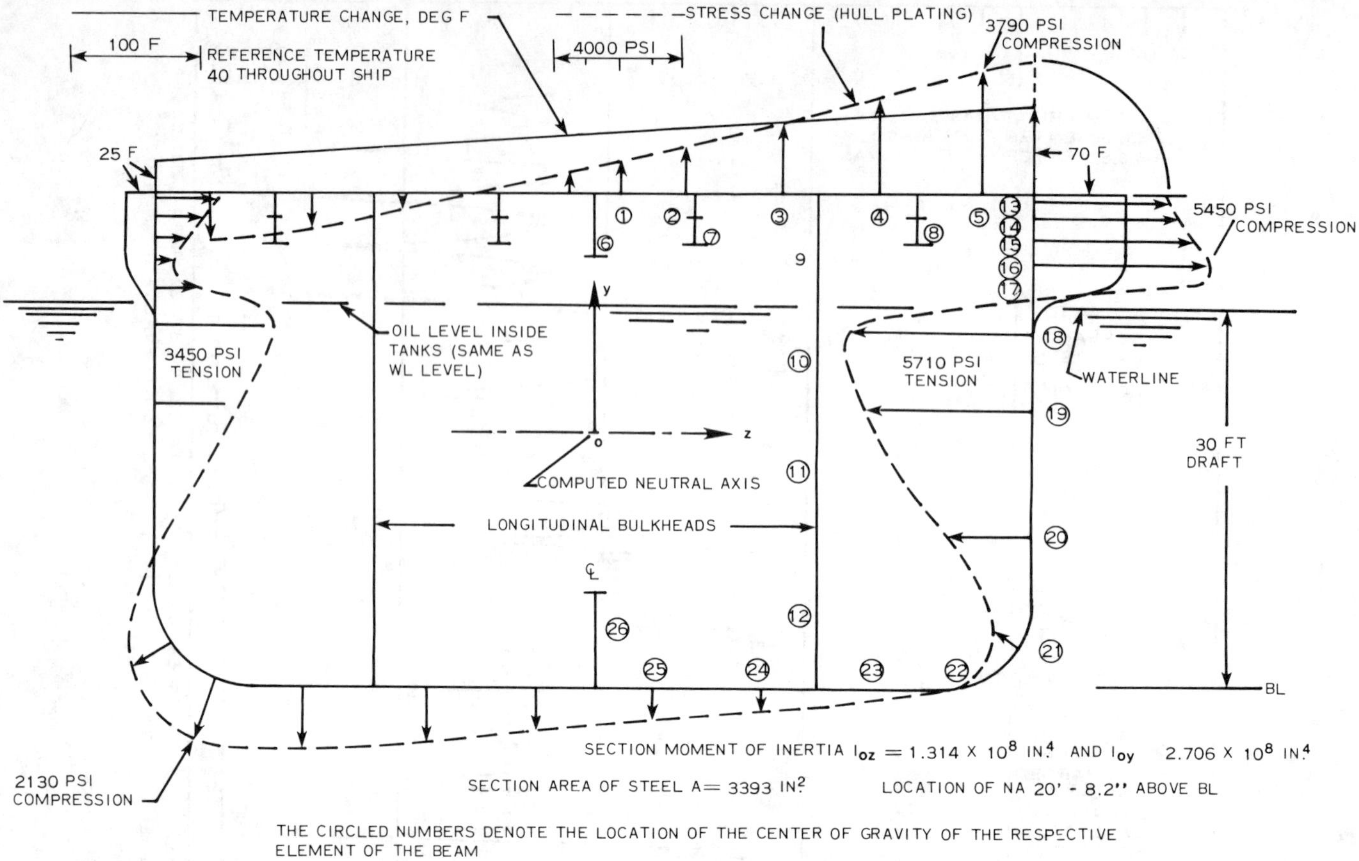

Fig. 30 Theoretical variation in longitudinal hull-girder stresses induced by nonsymmetrical variations in temperature distribution

These effects are included in the stress formula for a girth restraint plate with zero longitudinal strain; i.e., which is thrust-fixed longitudinally.

$$\sigma_{x1} = -E_x \alpha t \tag{11}$$

This formula must be modified for the plate which is a component of ship structure, where the limitation of zero longitudinal strain does not exist.

Significant thermal stresses have been recorded on large bulk carriers. A sample recorded variation of midship bending stresses in the deck during one typical voyage is shown in Fig. 29 (Lewis et al, 1973). It exhibits a consistant diurnal variation with a maximum magnitude of about 34.5 MPa (5,000 psi). Another example of thermal stress distribution in a transverse section of a tanker is shown in Fig. 30. With a maximum temperature differential of 21°C (70°F), the variation of longitudinal hull-girder stresses reaches 39.3 MPa (5,700 psi) in the vicinity of water line.

4.7 Dynamic Loads. For assessing the strength of local structures, in addition to the impulsive loads discussed in Subsection 4.5, information about the external and internal pressures is also required. The hydrodynamic pressures imposed on the hull can be predicted, using strip theory, in conjunction with ship motion computer programs. An example of this is shown in Fig. 31. Although a number of research projects have been carried out in recent years, dealing with model experiments, full-scale measurements and theoretical developments, the computation procedures established to date remain generally in the developing stage.

Another limitation of the linear ship motion theory is to calculate only the hydrodynamic pressures on the envelope

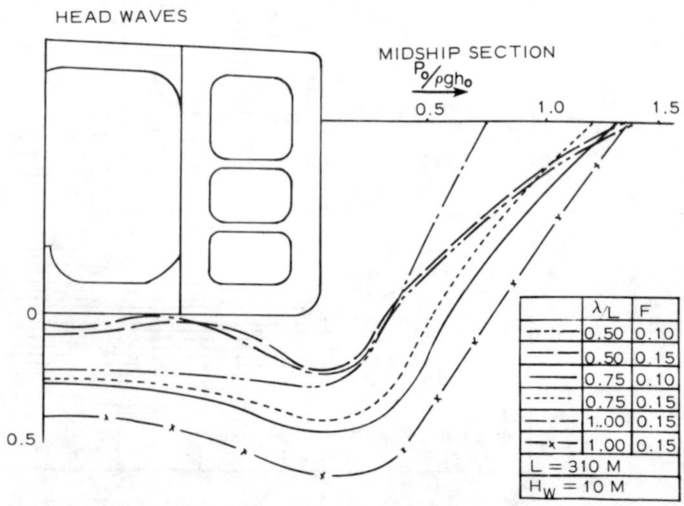

Fig. 31 Amplitudes of hydrodynamic pressure in regular head waves (ϕ = 180°), midship section

of the hull under the stillwater line. In performing structural analysis, the calculated pressures should be extrapolated for the portion above the stillwater line. An example in this regard is shown in Fig. 32 (Stiansen et al, 1979). Regarding the internal pressure on the boundary of a tank, it can be determined based on the accelerations and motions of the ship as obtained from ship motion calculations for 100 percent filling. For a partially filled tank, the motion of the liquid cargo induces additional dynamic pressure and possibly sloshing. For a given tank geometry, the internal tank pressures at any point may be determined by:

$$p = p_0 + \rho h\sqrt{(a_{xs} + a_x)^2 + (a_{ys} + a_y)^2 + (a_{zs} + a_z)^2} \quad (12)$$

where

p_0 = vapor pressure or the pressure relief valve setting,

ρ = mass density of the fluid cargo,

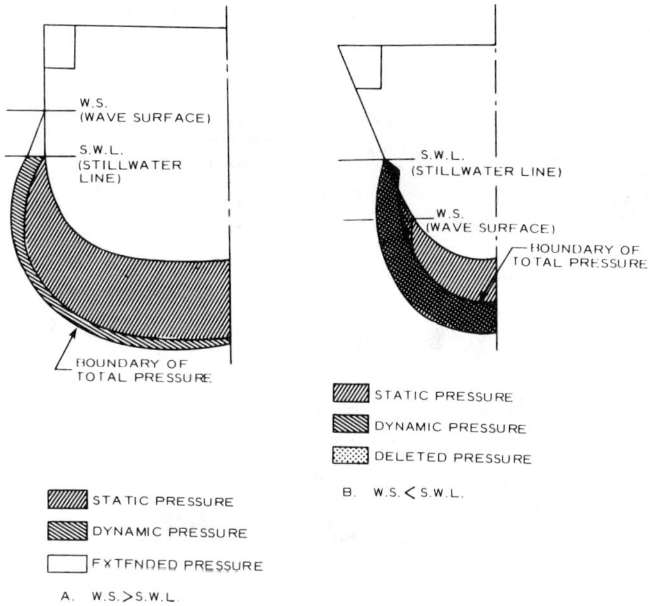

Fig. 32 Distribution of static and wave-induced dynamic pressures in head seas

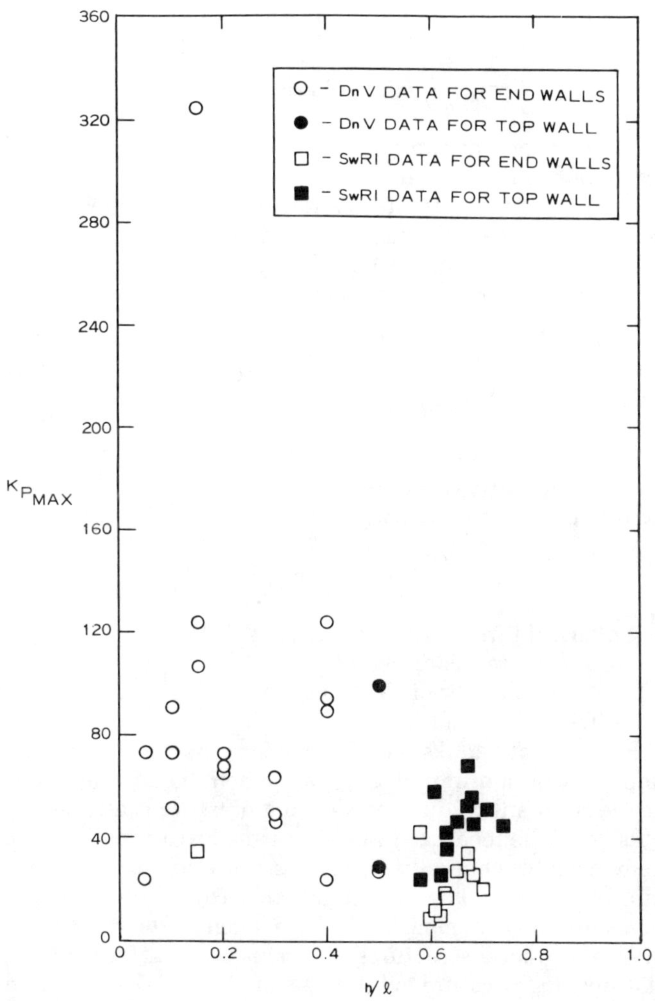

Fig. 33 Maximum pressure coefficient vs tank filling level for ϕ or $(x/l) \leq 0.30$

h = height of the column of the fluid above the tank boundary point in the direction of the resultant acceleration, including position changes resulting from pitch and roll,

a_{xs}, a_{ys}, a_{zs} = static (gravitational) acceleration components in x, y, z direction,

a_x, a_y, a_z = dynamic acceleration components, due to the ship's motion in x, y, z direction.

In current designs of large tankers and liquefied natural gas (LNG) carriers, the internal loads induced by motions of liquid cargo on the boundary structure of tanks are given due consideration, especially when a resonant sloshing mode may be present. To prevent possible sloshing in transporting liquid cargo, it is advisable to restrict the filling levels in tanks to either empty or full. However, the necessary operational requirements make partial filling unavoidable. The designer of all types of liquid cargo carriers must be aware of the possible occurrences of sloshing and its consequences. Presently, adequate analytical methods for predicting sloshing loads are not available. Design guidances can only be derived based on model experiments.

Several model experiments have been conducted by researchers in recent years to investigate the sloshing phenomenon in LNG cargo tanks. In most of the model experiments, the critical ship motion components, such as pitch, roll, and yaw are simulated individually with various amplitudes and frequencies. The test fluid has usually been fresh water. Consequently, the scaling factor in converting model data to full-scale applications is solely based on Froude numbers. The effects on the scaling factor of the density, viscosity, and vapor pressure are subject to further investigations.

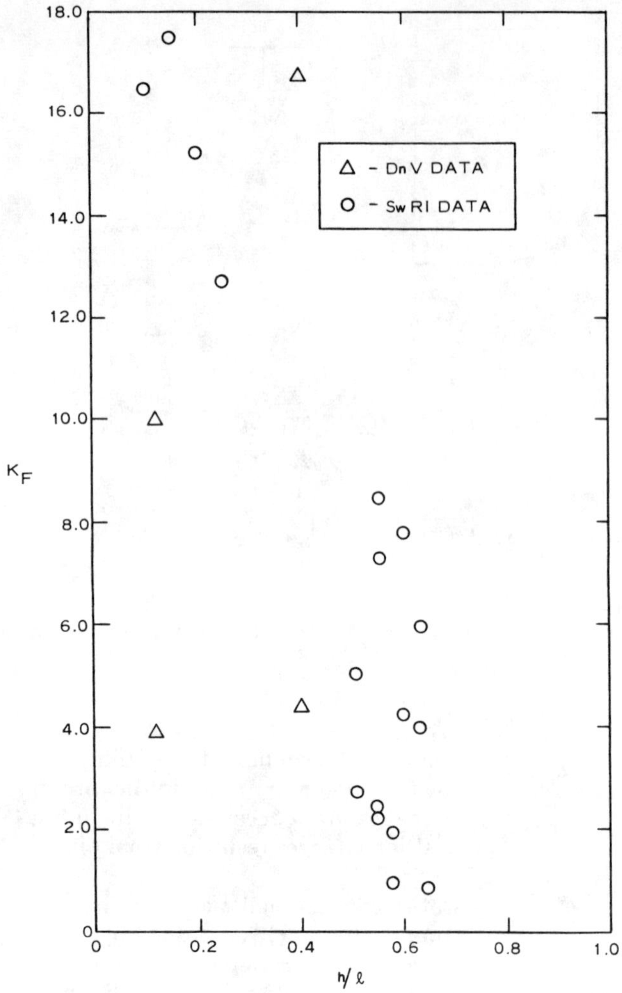

Fig. 34 Average force coefficient vs tank filling level for ϕ or $(x/l) \leq 0.30$

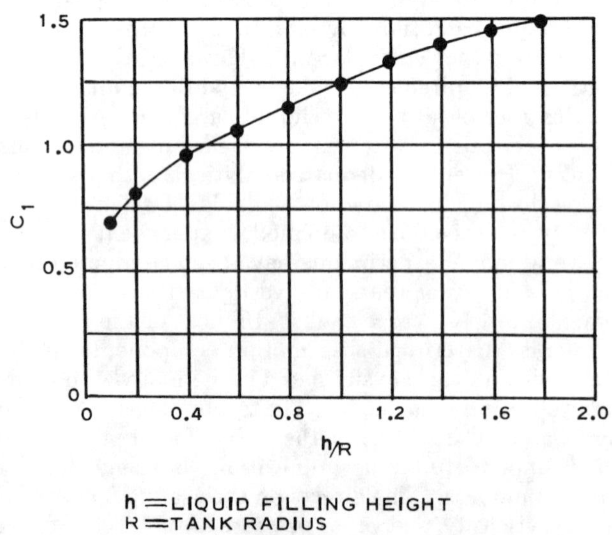

Fig. 35 Values of C_1 in equation (15)

In general, the measured pressure data are presented in a non-dimensional form or as a pressure coefficient defined by the following equation:

$$K_p = \frac{p}{\rho g l \phi} \tag{13}$$

where p = sloshing pressure,

ρ = mass density of the liquid cargo,
g = gravitational acceleration,
l = length of the tank,
ϕ = angle of pitch, roll or yaw.

For rigid body translations, such as heave, surge and sway, the variable $(l\phi)$ can be replaced by the translational amplitude. A sample plot of the maximum K_p for a prismatic tank is shown in Fig. 33 (Cox et al, 1979).

Another important parameter for the design of a tank structure is the total dynamic force induced by sloshing on the tank boundary. This is generally expressed by a non-dimensional force coefficient K_F which is defined as:

$$K_F = \frac{F}{\rho g l h b \phi} \quad \text{for prismatic tanks} \tag{14}$$

$$= \frac{F}{\rho g V} \quad \text{for spherical tanks}$$

where F = total dynamic force,
h = filling depth,
b = tank width, in the transverse direction with respect to the liquid motion,
V = volume of the liquid cargo.
ρ, g, l and ϕ are as defined above,

A sample plot of K_F for a prismatic tank is shown in Fig. 34. The natural period of the liquid cargo (T_R) may be approximated by Equation (15).

$$T_R = 2\pi \left[\frac{\pi g}{l} \tanh\left(\frac{\pi h}{l}\right)\right]^{-1/2}, \tag{15}$$

for prismatic tanks with $h/l \geq 0.1$
or

$$T_R = \frac{2\pi}{C_1}\sqrt{\frac{D}{2g}}\sqrt[4]{4\left(\frac{h}{b}\right) - \left(\frac{2h}{D}\right)^2}$$

for spherical tanks with $0.05 < h/D < 1.0$
where g, l, h are as defined above,
D = diameter of the tank,
C_1 is given in Fig. 35.

For prismatic tanks, sloshing pressure can be of either the impulsive or non-impulsive type. When the tank oscillates at the resonant frequency, sloshing pressure is always impulsive. The measured pressure-time history in this case demonstrates generally a sharp pressure rise at the beginning of sloshing followed by a smooth decay. This impulsive pressure with a remarkably high magnitude may cause damages to local structures. At non-resonant frequencies, the sloshing pressure exhibits a smooth and slow rise and decay.

For spherical tanks, the pressure induced by sloshing on the shell is generally of the non-impulsive type. In the design of tank structure and its supports, only the total dynamic force induced by sloshing should be considered.

It should be noted that the above equations and test data shown in Figs. 33, 34 and 35 are applicable to tanks having smooth internal walls. For tanks with internal structures, such as those in tankers, the determination of sloshing loads is subject to future research.

4.8 Other Loads. Other types of hull-girder loads, such as those indicated by launching, grounding, mooring, docking, and collision, which vary essentially case by case, must be considered individually. In general, the characteristics of both the load and the corresponding structural responses are nonlinear. The situation can be investigated either by means of model or full-scale experiments or by performing a nonlinear analysis, generally using a linear computer program with small increments of load or changes of position. A brief summary of the current research carried out in this regard is given in ISSC *Proceedings,* 1979.

Section 5
Stresses and Deflections

5.1 Stress and Deflection Categories. The structural response of the hull girder and the associated members can be subdivided into three components:

1. *Primary response* is the response of the entire hull, when bending as a beam under the longitudinal distribution of load. Taking as an example Fig. 36, with the plate in an x-y plane with the z direction normal to it, displacements are limited to those parallel to the unloaded midsurface of the plate (x and y displacements only).

2. *Secondary response* comprises the stress and deflection of a panel of stiffened plating, e.g., the panel of bottom structure contained between two adjacent transverse bulkheads. The loading of the panel is normal to its plane and the boundaries of the secondary panel are usually formed by other secondary panels. In Fig. 36 the configuration displaces out of the x-y plane (z-displacements), but the plate and stiffener displacements match.

3. *Tertiary response* describes the out-of-plane deflection and associated stress of an individual panel of plating between stiffeners. The loading is normal to the panel, and its boundaries are formed by the stiffeners of the secondary panel of which it is a part. Individual plate panels displace out of the mean surface created by the stiffeners.

Primary and secondary stresses in plate members are membrane stresses, uniform (or nearly uniform) through the thickness. Tertiary stresses, which result from the bending of the plate member itself vary through the thickness, but may contain a membrane component if the out-of plane deflections are large compared to the plate thickness.

The above discussion is intended to demonstrate that the resultant stress at a given point in the ship structure is composed of several parts, each of which may arise from a different cause. In many instances, there is little or no interaction between the three (primary, secondary, tertiary) component stresses or deflections, and each component may be computed by methods and considerations entirely independent of the other two. The resultant stress, in such a case, is then obtained by a simple superposition of the three component stresses. An exception is the case of plate (tertiary) deflections which are large compared to the thickness of plate. In this case, the primary and secondary stresses will have an effect on this tertiary deflection and resulting stress, thus simple superposition may no longer be employed to obtain the resultant stress. Fortunately for the ship structural analyst, such cases rarely occur with the load magnitudes and member scantlings used in ships, and superposition of the three components may usually be performed. For example, when the girder bends under the action of hydraulic envelope pressures, the bottom plating direct stresses may be divided in the following way:

• Primary stresses (σ_1): The plate itself develops plane stresses only. (The ship axis curvature is neglected).

• Secondary stresses (σ_2 and σ_2*): The stiffener system bends under the applied lateral loading and the plate develops additional plane stresses since it acts as flange plating with the stiffeners. In longitudinally framed ships there are

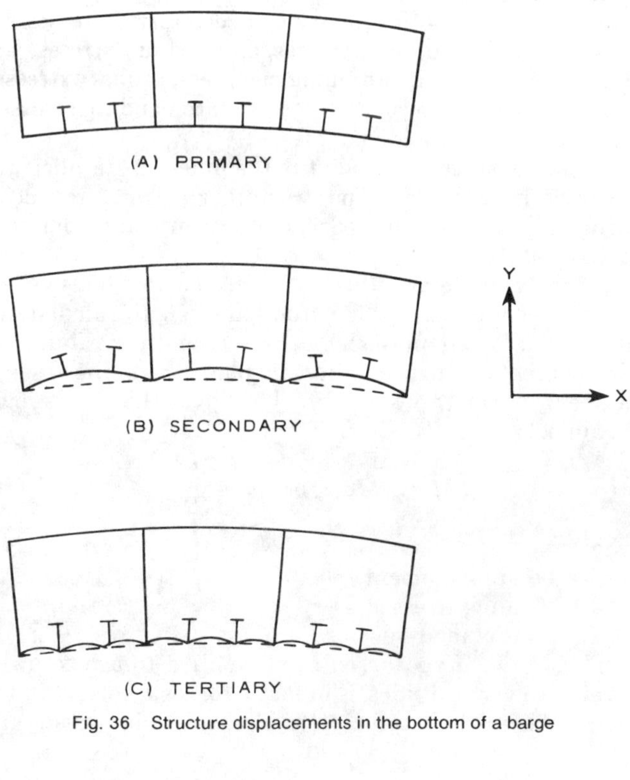

Fig. 36 Structure displacements in the bottom of a barge

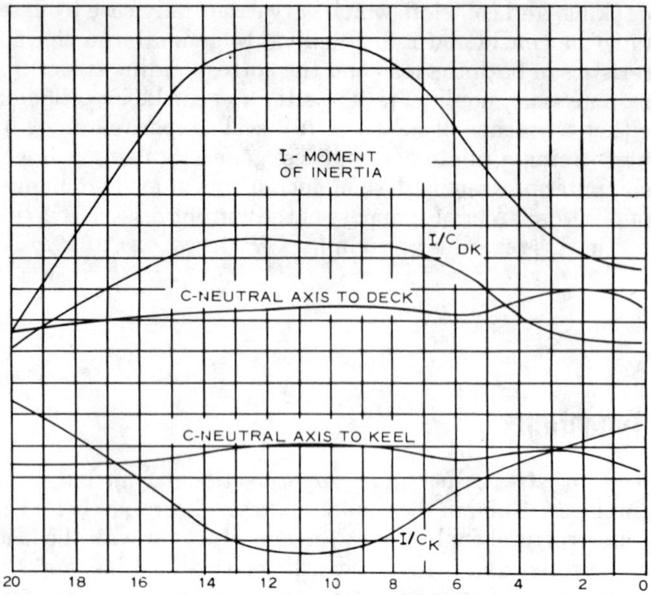

Fig. 37 Moment of inertia and section modulus

two types of secondary stresses: σ_2 corresponds to the bending of the plate-stiffener field between transverse bulkheads; $\sigma_2{}^*$ corresponds to the superposed bending of the plate-stiffener system under the hydrostatic pressure.

In Fig. 36 the three types of displacement corresponding to these three types of strain are shown on the inboard profile of a transversely framed three-compartment barge. In this application, the primary stresses are those which are usually called the longitudinal bending stresses, but the general category of primary does not imply a direction. Thus, in plates, both primary and secondary stresses are midplane or membrane stresses, and tertiary stresses are bending stresses. In stiffening members, primary stresses are axial stresses, secondary stresses are bending stresses, and tertiary stresses do not exist.

These considerations point to the inherent simplicity of the underlying theory. The structural naval architect deals principally with beam theory, plate theory, and combinations of both.

5.2 Simple Beam Theory Calculation Methods. The simple beam theory for longitudinal strength calculations of a ship is based on the hypothesis (usually attributed to Navier) that plane sections remain plane and in the absence of shear, normal to the axis. This gives the well-known formula:

$$M = SM\sigma, \quad \text{or} \quad \sigma = \frac{M}{SM} \tag{16}$$

where:
M = bending moment
σ = bending stress
SM = section modulus

The *Navier hypothesis,* when applied to a box-girder form with vertical sides, is in fact rigorously correct in the sides, provided that complete girth restraint exists and vertical shear stresses are constant in the vertical direction.

In applying Equation (16) to ships, the necessity for comparison of results as between ship and ship must be borne in mind. The methods used for calculating the bending moments and section moduli for any two vessels must be the same in all respects, if the resulting calculated stresses are to be compared.

For a given bending moment at a given cross section of a ship, at any part of the cross section a stress may be obtained ($\sigma = M/SM = Mc/I$) which is proportional to the distance c of that part from the neutral axis (a line parallel to the baseline drawn through the centroid of all the effective longitudinal strength members comprising the section). It is customary, however, to use the terms σ and c as referring to their values at the extreme fibers, so that, unless otherwise denoted,

σ = bending stress at the extreme top or bottom member, corresponding to
c = distance from the neutral axis to the extreme member.

It will be convenient in many cases to use I/c instead of SM, where I is the sectional moment of inertia about the neutral axis. The neutral axis will seldom be located exactly at half-depth of the section; hence two values of c and σ will be obtained for each section for any given bending moment, one for the top fiber (deck) and one for the bottom fiber (keel).

Theoretically, a thorough analysis of longitudinal strength would include the construction of a curve of section moduli throughout the length of the ship as shown in Fig. 37. Dividing the ordinates of the maximum bending-moments curve (the envelope curve of maxima) by the corresponding ordinates of the section-moduli curve yields stress values, and by using both the hogging and sagging moment curves four curves of stress can be obtained; viz., tension and compression values for both top and bottom extreme fibers.

It is customary, however, to assume the maximum bending moment to extend over the midship portion of the ship. Accordingly, the classification societies ordinarily require the maintenance of the midship scantlings throughout the midship four-tenths length. However, consideration is given to accepting section modulus variations within this portion of the ship based on the bending moment envelope curve (see Section 6).

This practice maintains the midship section area of structure practically at full value in the vicinity of maximum shear as well as providing for possible variation in the precise location of the maximum bending moment.

It usually is necessary, therefore, to ascertain only the least I/c value for any section in the midship four-tenths length. This most often occurs in the way of a hatch or deck opening, and the section modulus clear of hatches or deck opening in the midship length is usually larger. Often several calculations are required to determine the minimum I/c value, especially when the uppermost strength deck changes from one level to another, as with a bridge partially

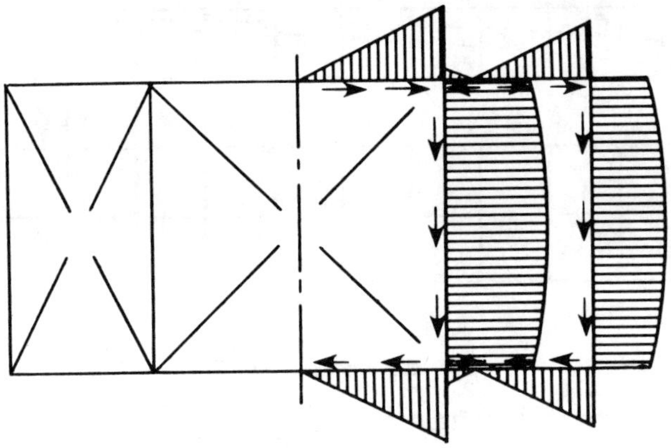

Fig. 38 Shear flow on a simple transverse section (phantom tanker)

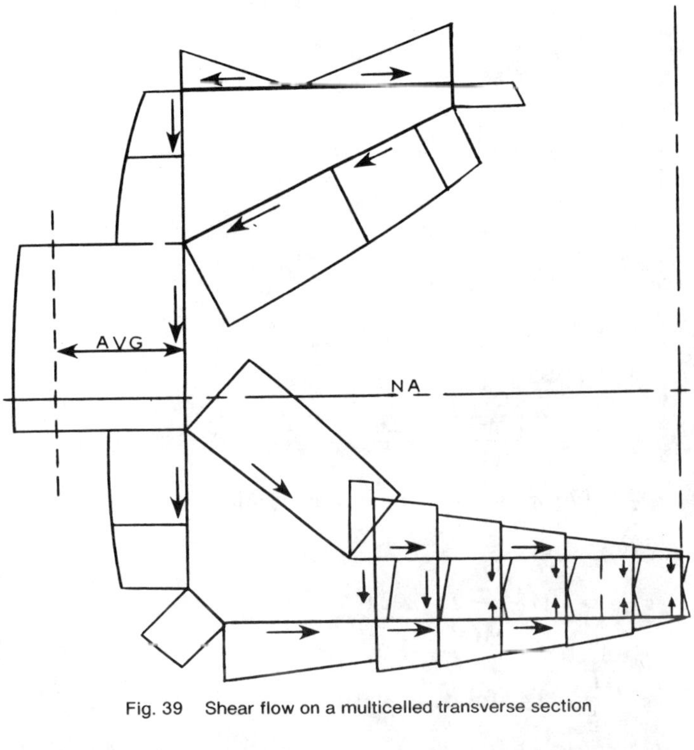

Fig. 39 Shear flow on a multicelled transverse section

covering the midship four-tenths length. The total vertical shearing force at any point in the ship's length may be obtained by the integration of the load curve up to that point. Ordinarily the maximum value of the shearing force occurs at about one quarter of the vessel's length from either end.

Since only the vertical, or nearly vertical, members of the hull girder are capable of resisting vertical shear, this shear is taken almost entirely by the side shell, any continuous longitudinal bulkheads, and by the webs of any deep longitudinal girders.

For vessels without continuous longitudinal bulkheads, the expression for the shear stress τ at any point in the cross section is

$$\tau = \frac{Vac}{tI} \tag{17}$$

where

V = total shearing force (from shear curve),

ac = moment of area above shear plane under consideration taken about neutral axis,

t = thickness of material at the shear plane,

I = moment of inertia of the entire section.

The maximum value of τ occurs in the vicinity of the neutral axis, where the value of t is usually twice the thickness of the side plating. For vessels with continuous longitudinal bulkheads, the expression for shear stress is more complex and requires the assumption of a point of zero shear stress in the multi-cell cross-section. A calculation process for the maximum shear is given in ABS Rules, Section 6.3.3d and Appendix C. Some shear flows in transverse sections are illustrated in Figs. 38 and 39.

a. Effective Breadth of Stiffened Plating. The effective breadth of a rectangular plate that is shear-loaded in its own plane along one side and shear-free along the opposite side is the breadth of a uniformly stressed phantom plate of the same thickness stressed to the same maximum stress, and sustaining the same total force as in the real plate. In the typical real plate loaded in this way, the longitudinal stress attenuates, or falls off, with increasing distance from the shear-loaded edge, and this is called shear lag, Fig. 40. The ratio of the effective breadth to the real breadth is useful to the designer in determining the longitudinal stress along the shear-loaded edge. It is a function of the kind of external loading applied and of the boundary conditions along the plate edges, but not of its thickness. Figure 40 gives the effective breadth ratio at mid-length for column loading and harmonic-shaped beam loading, together with a common approximation for both cases:

$$\frac{b_e}{b} \cong \frac{1}{6} k \frac{L}{b} k \frac{L}{b} \leq 6 \tag{18}$$

The category *column loading* means the situation shown in Fig. 40 where the shear is transmitted to the plate from a member which at its end is either axially loaded at its centroid or loaded by a pure moment, or both. The category *beam loading* means that the loaded member is laterally loaded with a sinsusoidal distribution (approximately).

For a box girder configuration, with two webs and flange plate between, b_e/b is equal to or slightly greater than shown, and for a wide plate stiffened by a repetitive system of stiffeners identically loaded, b_e/b is about 10 percent greater, because of girth restraint exerted by the plate. Therefore the data are a conservative composite for general use. They imply, for example, that for bulkhead stiffeners the plating at mid-length is 100 percent effective if spacing equals or is less than $k(L/3)$.

The beam loading curve in Fig. 40 is constructed for a harmonic or sinusoidal distribution of lateral loading, but is sufficiently accurate for any lateral load distribution that corresponds to a bending moment without points of inflection (that does not reverse its curvature).

The factor k is obtained from the bending moment curve. For example, $k = 0.58$ for a beam with fixed ends, with uniform lateral loading, since at $0.21L$ from each end the

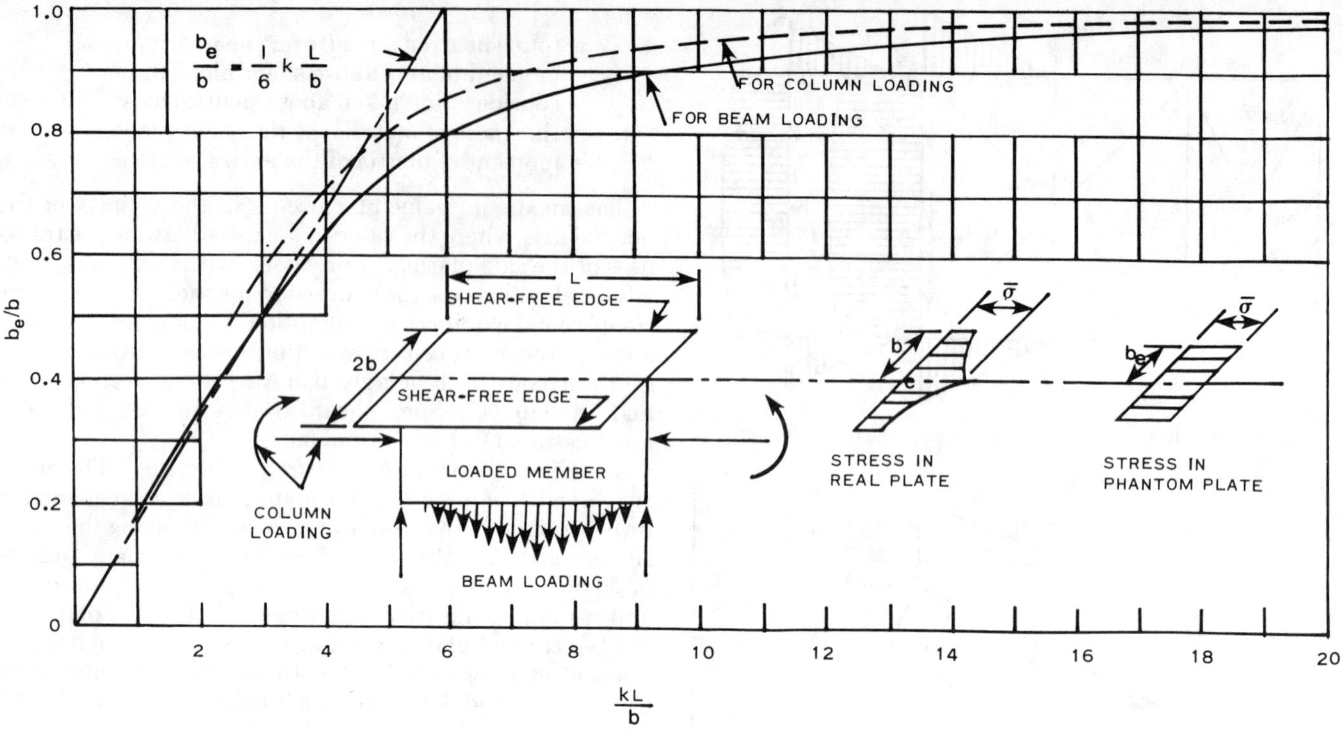

NOTE: CL IS THE DISTANCE BETWEEN POINTS OF ZERO BENDING MOMENT UNDER BEAM LOADING.

Fig. 40 Effective breadth ratios at mid-length

bending moment is zero. However, the end support forces represent concentrated loads, which reduce flange effectiveness near their points of application. If the beam is a bulkhead stiffener, the bulkhead plating in the vicinity of fixed ends should therefore be assigned 100 percent effectiveness only if spacing equals or is less than $k(L/4)$, where k is 0.42 for uniform loading. The factor 1/4 is entirely arbitrary. If spacing is greater than this, the effective breadth should be determined according to a suitable method.

b. Shear Lag in Deck and Bottom Plating. In most ships of normal form the deck and bottom plating, treated as flanges of a box girder in bending can be assigned 100 percent effectiveness for design purposes. However, some extreme cases of loading may make this questionable. Fig. 41 shows a stillwater bending moment curve, M_b for a 202.6 × 29.6 × 16.7m (664.7 × 97.1 × 54.8 ft) ore carrier. To approximate the length constant c for an effective breadth computation, the bending moment is divided into two components:

1. A single-lobed component curve, M_a, between stations 5 and 17.5, constructed as shown.
2. The remainder, stations 11 to 13.5, which surmounts the M_a curve.

For (1.), the effectiveness of bottom or deck plating is 0.88 or 0.97 respectively, according to Fig. 40, since

$$k\frac{L}{b} = \frac{(17.5 - 5)}{20}\frac{202.6}{14.80} = 8.5 \text{ for bottom plating}$$

and

$$k\frac{L}{b} = \frac{(17.5 - 5)}{20}\frac{202.6}{7.42}$$

$$= 17 \text{ for deck plating in way of openings}$$

For (2.),

$$k\frac{L}{b} = \frac{(13.5 - 11)}{20}\frac{202.6}{14.80} = 1.71 \text{ for bottom plating}$$

$$k\frac{L}{b} = 3.42 \text{ for deck plating}$$

and from Fig. 40, b_e/b of bottom and deck plating are 0.30 and 0.59 respectively.

The section moduli at station 12 with 100 percent effectiveness of bottom and deck plating are as follows:

$$SM_N = 27.92 \text{ m}^3 \ (986.1 \text{ ft}^3) \text{ at bottom}$$

$$SM_N = 21.63 \text{ m}^3 \ (763.8 \text{ ft}^3) \text{ at deck}$$

The corresponding section moduli at station 12 for (1.) and (2.) with deck and bottom effectiveness diminished in accordance with the foregoing can be combined with the corresponding moment components to give stresses:

$$\sigma_a = \frac{M_a}{SM_a} = \frac{39.38}{27.92} \times \frac{10^6}{0.88} = 15.97 \text{ MPa}$$

(2,316 psi) at bottom

$$\sigma_a = \frac{-M_a}{SM_a} = \frac{-39.38}{21.63} \times \frac{10^6}{0.97} = -18.70 \text{ MPa}$$

(−2,712 psi) at deck

$$\sigma_b = \frac{M_b}{SM_b} = \frac{147}{27.92} \times \frac{10^6}{0.30} = 17.55 \text{ MPa}$$

(2,545 psi) at bottom

$$\sigma_b = \frac{-M_b}{SM_b} = \frac{-147}{21.63} \times \frac{10^6}{0.59} = -11.52 \text{ MPa}$$

(−1,671 psi) at deck

$\sigma = \sigma_a + \sigma_b = 33.52$ MPa (4,862 psi) at bottom

$\sigma = -30.22$ MPa (−4,383 psi) at deck

The stresses corresponding to the same total bending moment and moduli with 100 percent effectiveness of bottom and deck plating are:

$$\sigma_N = \frac{M_a + M_b}{SM_N} = 19.32 \text{ MPa (2,802 psi) at bottom,}$$

$\sigma_N = -24.94$ MPa (−3,617 psi) at deck

Thus shear lag increases these stillwater stresses 73 percent at bottom and 21 percent at deck over stresses computed without accounting for it.

c. Hull-Deckhouse Interaction. The prediction of the structural behavior of a deckhouse or superstructure constructed above the strength deck of the hull has facets involving both the general bending response and important localized effects. Two opposing schools of thought exist concerning the philosophy of design of such erections. One either attempts to make the deckhouse effective in contributing to the overall bending strength of the hull, or it is purposely isolated from the hull so that it carries only localized loads and does not experience stresses and deflections associated with bending of the main hull. This may be accomplished by cutting the deckhouse into short segments by means of expansion joints. Aluminum deckhouse construction is another alternative.

The behavior of a deckhouse, acting in conjunction with the ship hull can be described as follows: essentially, as the ship hull experiences a bending deflection in response to the wave bending moment, the deckhouse is forced to bend also. However, the curvature of the deckhouse may not necessarily be equal to that of the hull but depends upon the length of deckhouse in relation to the hull and the nature of the connection between the two, especially upon the vertical stiffness or *foundation modulus* of the deck upon which the

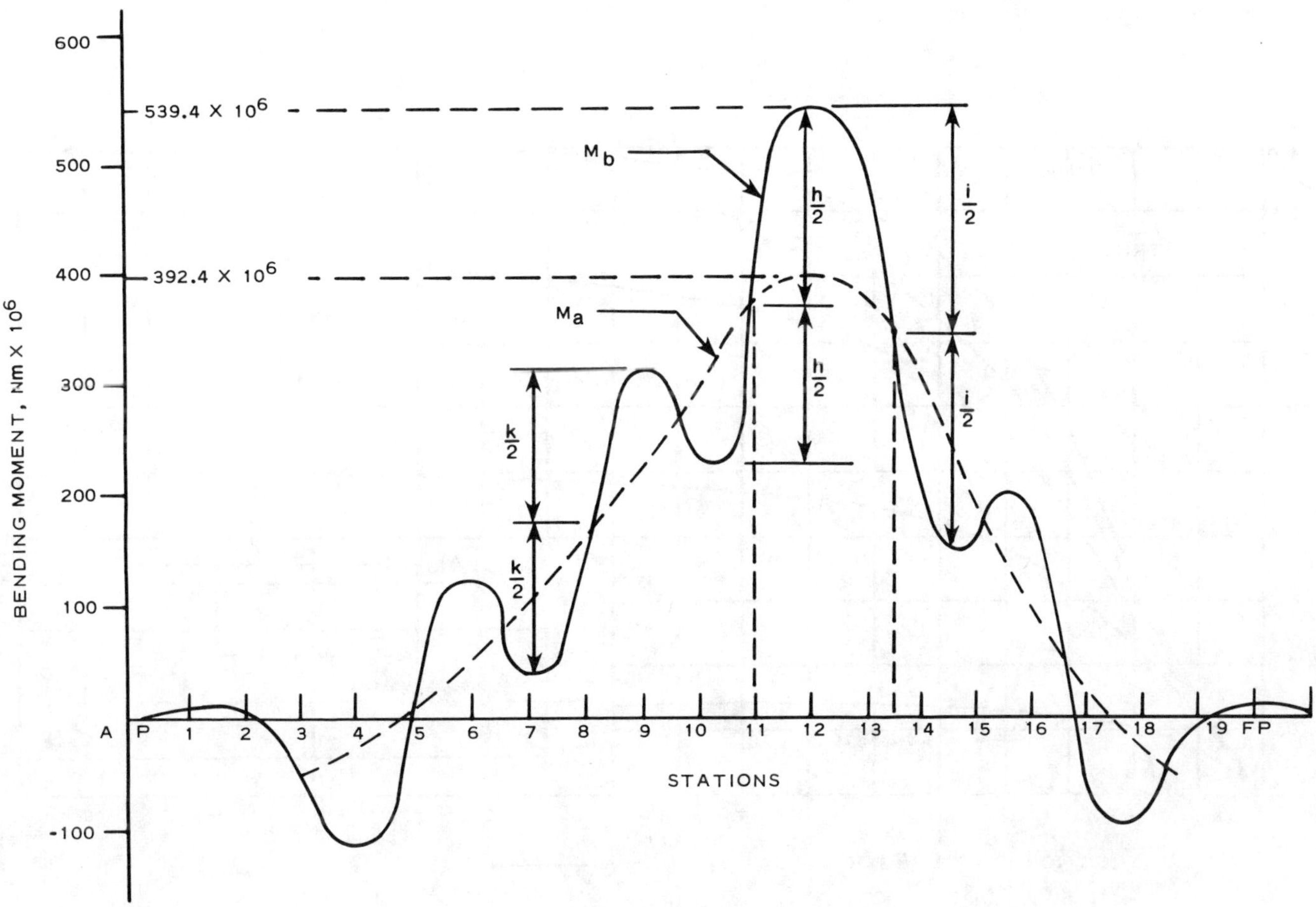

Fig. 41 Stillwater bending moment curve for an ore carrier

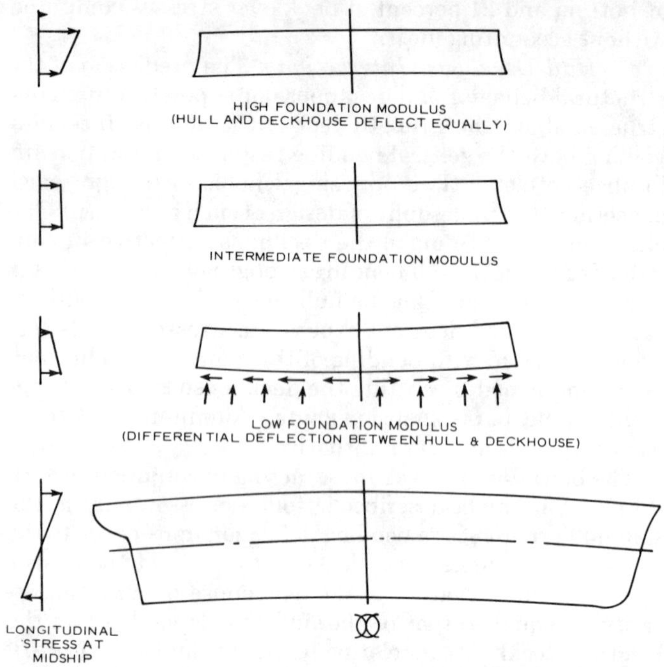

Fig. 42 Bending behavior of deckhouse and hull

house is constructed. The behavior of the deckhouse is similar to that of a beam on an elastic foundation loaded by a system of normal forces and shear forces at the bond to the hull. The nature of this loading and the differential deflection between deckhouse and hull are shown in Figure 42. From this, it is apparent that the following factors will be of importance:

- bending stiffness of hull and deckhouse,
- foundation modulus of the deck,
- shear stiffness of deck and deckhouse sides.

Most of the design analysis methods which have been published, have been concerned with the prediction of stress distribution at the midlength of the deckhouse. The stress distributions at the midlength of the deckhouse for three different degrees of deckhouse effectiveness are sketched in the left portion of Fig. 42 and show:

1. deckhouse fully incorporated with the hull so that the two bend together as a beam;
2. deckhouse partially effective, i.e., connection between hull and deckhouse such that some differential deflection is permitted;
3. vertical foundation stiffness of deck so low that deckhouse responds primarily to shears at the joint to hull resulting in bending in opposition to hull.

In all three cases, high localized loads are developed at the

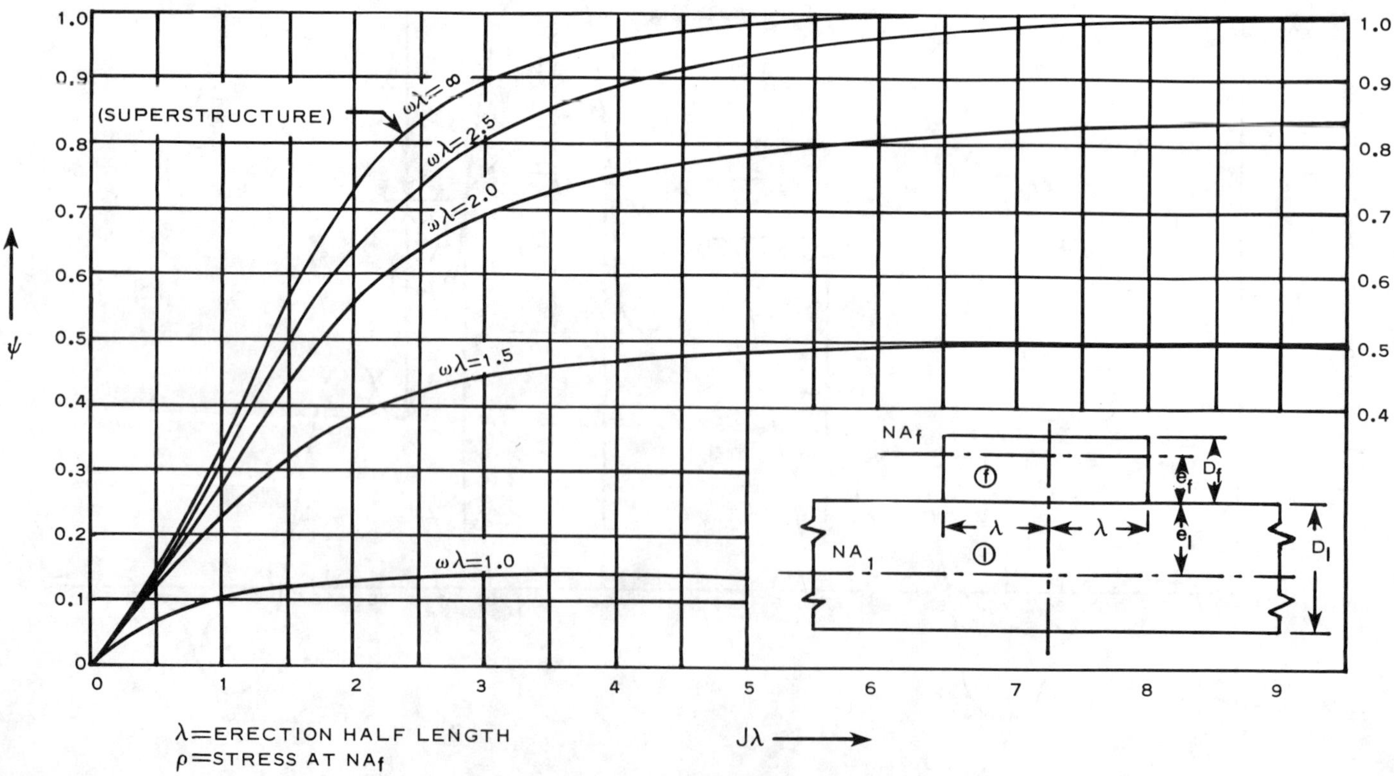

Fig. 43 Erection efficiency ψ

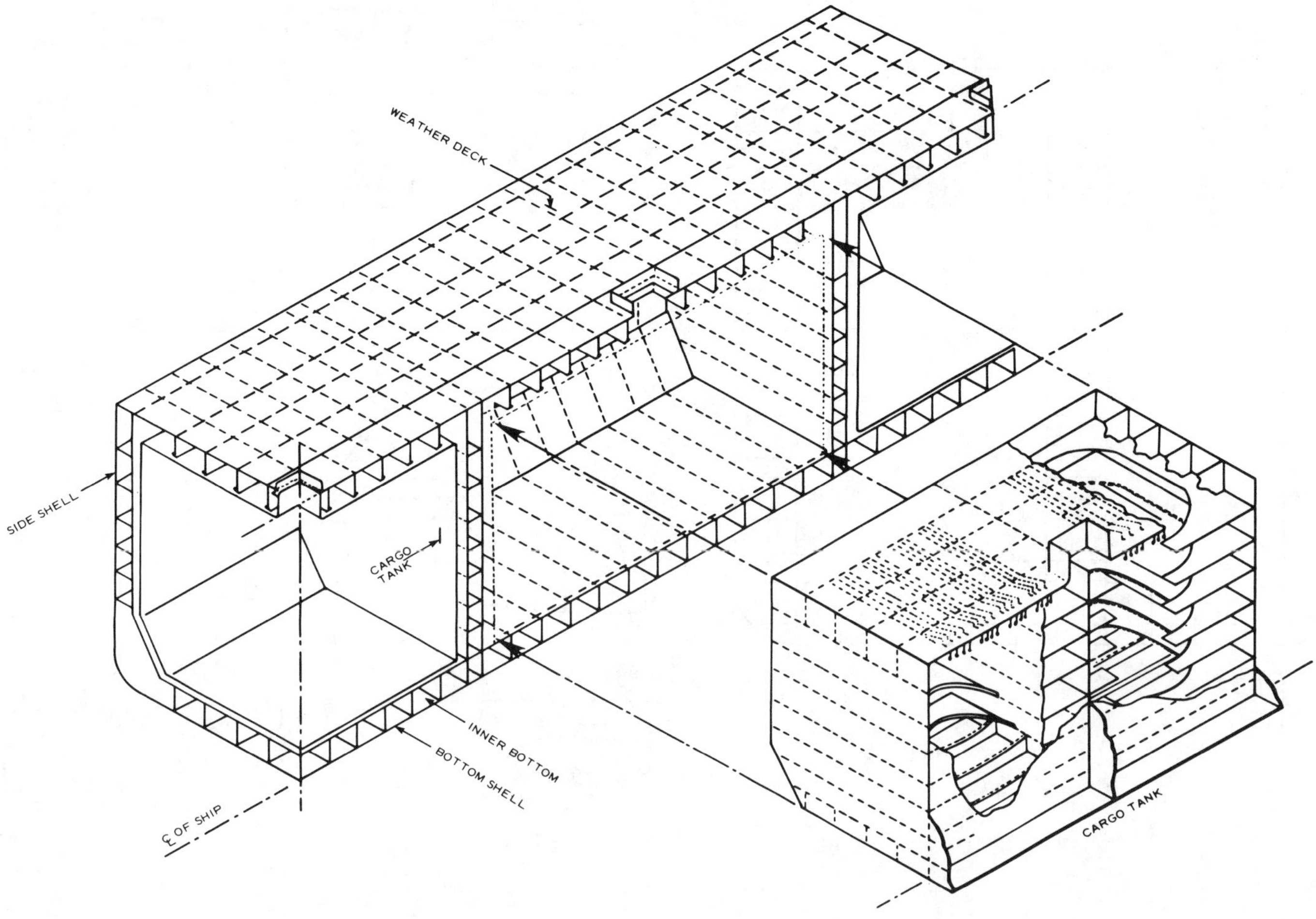

Fig. 44 Extent of three-dimensional model for independent tank LNG carrier

ends of the deckhouse, which most theories of hull-deckhouse interaction fail to predict. While this is usually dealt with as a local problem using empirical information for its solution, it should be considered as contributing to the secondary and tertiary stress magnitude in deck panels in the vicinity of the ends. Further details on the design considerations for deckhouses may be found in Chapter VII.

The treatment of the influence of deckhouses and superstructures upon the bending strength and stiffness of the hull is far from standardized, and most of the rules and design requirements are intended to provide safeguards against the effects of the discontinuities at the ends. There are really two problems here: 1. stress diffusion into the erection (deckhouse or superstructure) away from its ends, with consequent modification of the stresses in the hull; and 2. the local end effects.

The erection responds mainly to longitudinal shears and vertical forces along the connection between its sides and the hull, and is therefore a special kind of shear lag problem.

The method used in this section for estimating the efficiency of the deckhouse to the primary hull was discussed by Schade, 1966, and focuses on the stress diffusion problem. The theory, like all theories used for this situation, does not properly describe what happens near the erection ends, and there the designer must usually depend on empirical data. It rests on the assumption (seemingly well confirmed by full-scale and model observations) that a deckhouse bends similarly to and simultaneously with the hull, but with a different curvature. Shear deflection is accounted for, and may affect diffusion strongly. The superstructure responds by bending with the same curvature, but shear deflection affects stresses there also.

Fig. 43 gives the erection efficiency in terms of the principal factors that affect it; in this context, efficiency is defined as the ratio of the longitudinal resultant force of the stresses across the midlength section of the erection to the corresponding force, computed as if the hull and erection behaved as a single Navier beam. Efficiency is also the ratio of actual stress at the neutral axis of the erection to the single-beam computed stress at the same point. Thus, stress at the erection neutral axis is computed on the single-beam basis, and efficiency taken from Fig. 43, by entering the figure with the parameters discussed in the following. Single-beam stress at the neutral axis of the erec-

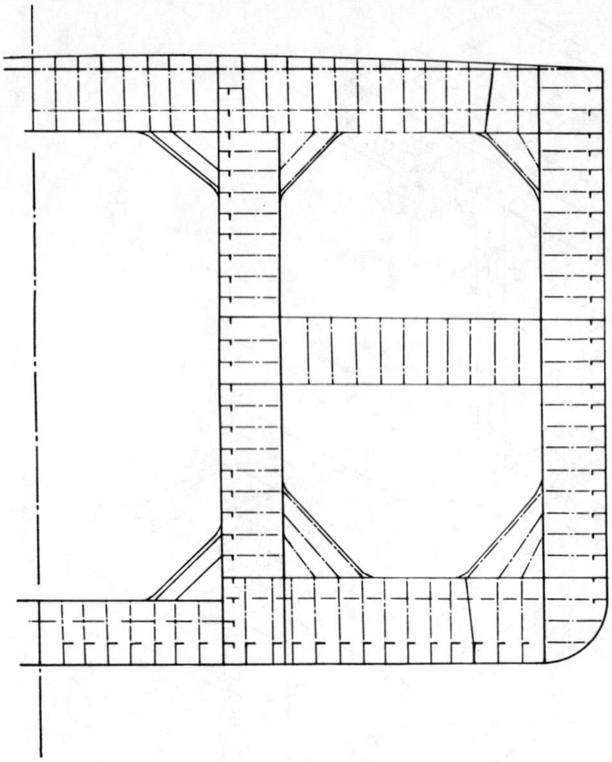

(A) SHIP'S STRUCTURE

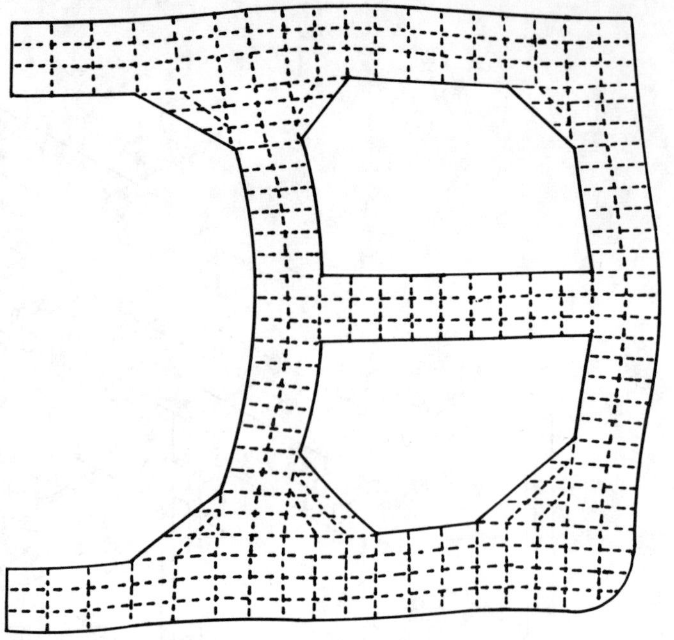

(B) DISPLACEMENT OF MODEL

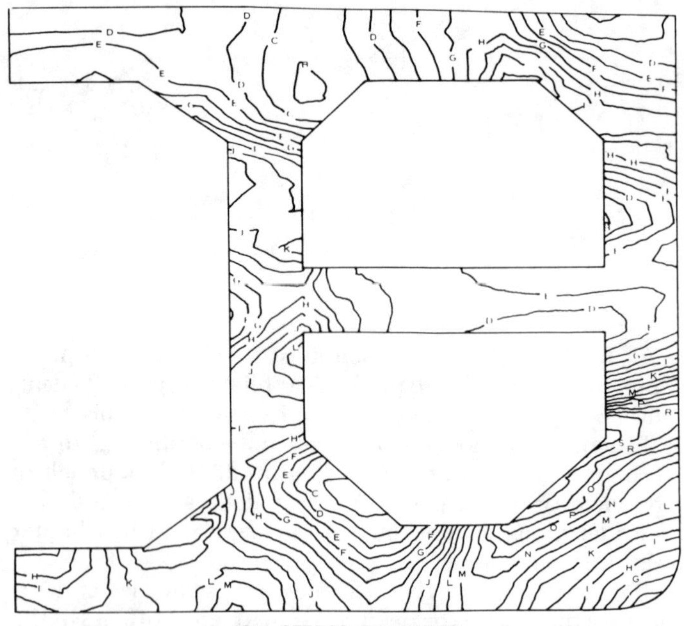

(C) STRESS CONTOURS

Fig. 45 Two-dimensional web frame analysis

tion for bending moment M is simply calculated by usual naval architecture methods or by the formula for p/ψ as given in Fig. 43.

The relevant parameters are defined as:

$$\omega^4 = \frac{k}{E_1} \cdot \frac{\Omega}{4}, \text{ in.}^{-4}, \quad J^2 = \frac{1}{\dfrac{1}{a_1} + \dfrac{1}{a_f}} \cdot \frac{\Omega}{2.6}, \text{ in.}^{-2}$$

$$\Omega = \frac{(A_1 + A_f)(I_1 + I_f) + A_1 A_f (e_1 + e_f)^2}{(A_1 + A_f) I_1 I_f + A_1 A_f (I_1 e_f^2 + I_f e_1^2)}, \text{ in.}^{-4} \quad (19)$$

$$p = \psi M \frac{A_1(e_1 + e_f)}{(A_1 + A_f)(I_1 + I_f) + A_1 A_f (e_1 + e_f)^2} \cdot \frac{E_{(\text{erection})}}{E_{(\text{hull})}}$$

where A_1 and A_f are the independent sectional areas of hull and erection, respectively; a_1 and a_f are the corresponding vertical shear areas; I_1 and I_f are the corresponding section moments of inertia; and e_1 and e_f are the vertical distances from the main (upper) deck down to the neutral axis of the hull and up to the neutral axis of the erection respectively, Fig. 43.

The areas and inertias can be computed to account for shear lag in decks and bottoms by using the "beam loading" curve in Fig. 40. If the erection material differs from that of the hull (aluminum on steel, for example) the geometric erection area A_f and inertia I_f must be reduced according to the ratio of the respective material moduli; i.e., by multiplying by E (aluminum)/E (steel) (approximately one-third).

5.3 Finite Element Calculation Methods. Finite element theory and applications have been increasingly used in all phases of ship analysis and design. The following is excerpted from Liu and Wojnarowski (1976).

The finite element technique is a relatively new and very useful method for stress analysis of structural continua. The method relies strongly on the matrix formulation of structural analysis introduced mainly as a result of the increasing use of computers. There has been a concurrent and rapid development of electronic computers, matrix methods and finite element techniques.

In the finite element method, a solid continuum is subdivided into an assemblage of discrete elements of finite dimensions. In effect, the real system or structure is modeled by a simplified, idealized system for which a solution is available. The idealized model is then analyzed by one or more of the available methods of analysis.

The finite element technique has a sound theoretical basis within the framework of the classical theory. It may be intepreted as a close relative to the well-known *Ritz method,* in which the displacement field in a continuum is usually described by means of a sum of preselected functions, each multiplied by a constant. The constants are determined by means of the condition of minimum potential energy.

While in the classical Ritz procedure one set of functions describes the displacement field in the entire continuum, the finite element method assumes individual displacement fields for each of the elements. The internal displacements in the elements are uniquely defined in terms of the nodal points values, and the entire displacement field is assumed to consist of a large number of piecewise continuous fields, each covering the extent of one element. The conditions of equilibrium of the nodes may be shown to yield the displacement field corresponding to minimum potential energy for the selected displacement pattern. As in the Ritz procedure, the solution will generally be approximate, but the method converges toward the correct solution.

Many new developments and refinements are constantly being added to the existing finite element programs. One of the most promising of these extensions, extremely useful in the analysis of complex structures, is the substructuring capability. Substructures are assemblages of basic elements (beams, plates, etc.) which serve as building blocks for the representation of larger, more complex structures. Substructuring decreases the amount of input data required to generate the structure, particularly for repetitive structural arrangements. It also decreases the number of unknowns encountered in the solution of the problem, with a corresponding reduction in computer requirements.

Discretization is an essential part of the structural analysis problem. In general, discretization introduces approximations into the analysis; for example, a finite element idealization generally involves approximation both of the geometric form of the structure and of the displacements which it develops. The degree of refinement which is employed in the finite element mesh should be based on the analysis requirements, i.e., a fine mesh is needed in regions where the structural behavior is most complex. Moreover, the mesh must be particularly fine if the primary objective of the analysis is stress distribution rather than deflection, because the derivatives of the displacements are represented less accurately than are the displacements themselves when

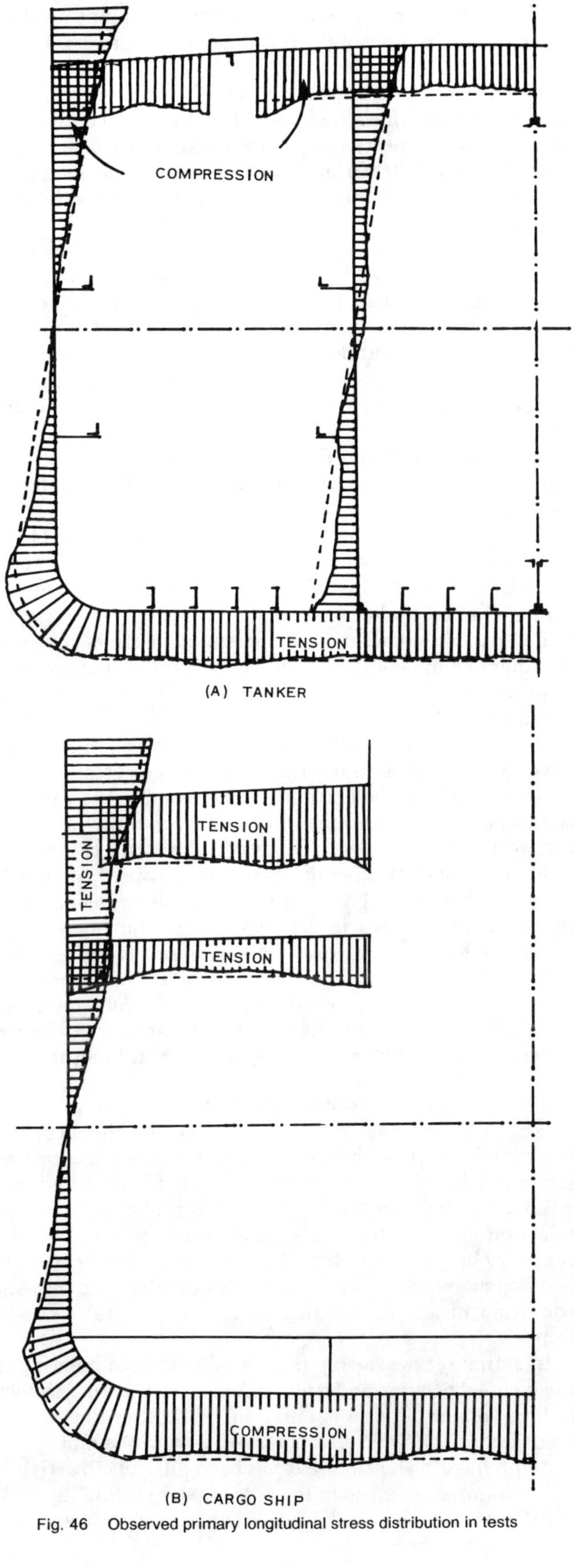

Fig. 46 Observed primary longitudinal stress distribution in tests

applying the more commonly used displacement method. A simplified example of the finite element approach is given in Section 7.2.

The finite element method has been extensively applied to all types of marine structures. The first applications were for analysis of supertankers (large tankers with deadweight tonnage above 200,000), as a result of their rapidly increasing size and associated changes in ship configuration. Classification society rules based on previous design and experience were not adequate for these new ships, and considerable structural damage was sustained by some of the earlier supertankers. Computerized structural analyses have provided the means to overcome rule limitations and to provide the required structural integrity and design efficiency.

Other marine structures such as containerships, ore or bulk carriers, liquefied natural gas (LNG) carriers, submarine structures, surface effect ships, hydrofoils, ice breakers, drilling rigs (fixed and floating), offshore platforms, etc., have been extensively analyzed by computerized methods.

Some examples of these analyses are shown in Figs. 44 and 45. Fig. 44 shows the extent of a three-dimensional model of an LNG carrier. Fig. 45 shows a typical transverse web frame of an oil tanker, the calculated displacements of its finite-element model, and the corresponding stress contours. Another example of the application of finite element techniques to the design of a tanker midship section is given in Section 6.

The need to make certain simplifying assumptions in developing theories and formulas implies a certain degree of uncertainty in the accuracy of the results, the range of their applicability and the limitations beyond which they cannot be extrapolated. One way to quantify this uncertainty is to conduct experiments to prove, disprove or amend theoretical results. For example, full scale experiments to measure the stresses in actual structure on board large tankers subjected to static and dynamic loadings were conducted to correlate the measured values with those obtained from analytical calculations, and the correlation, as reported by Little and Stiansen (1971), appears to be excellent. Fig. 46 illustrates the type of general picture that can be derived from test data.

5.4 Permissible Bending Stresses. Once a probable longitudinal bending moment has been established, the designer is faced with the problem of assigning a proper stress as a basis for the determination of the required longitudinal strength in the design under consideration. In the selection of a suitable stress value, it has to be recognized that very broad assumptions had to be made in the past for bending moment calculations and that they covered only the condition of simple bending in an upright static condition.

In actual service, a ship may be subjected to bending in the inclined position and to other forces, such as those which induce torsion or side bending in the hull girder, not to mention the dynamic effects resulting from the motions of the ship itself. Heretofore it has been difficult to arrive at the minimum scantlings for a large ship's hull by first principles alone, since the forces that the structure might be required to withstand in service conditions are uncertain.

Accordingly, it must be assumed that the stress includes an adequate factor of safety, or margin for these uncertain loading factors, so that what is commonly termed the "stress" in the structure must be considered as a relative value only. The stress to be assigned to a particular design should be determined only from ships of similar type and service which have proved satisfactory. Today's designers are fortunate in having available a wealth of information covering the longitudinal strength characteristics of ships of many types and services. Occasionally, however, they may be faced with designs of vessels of unusual types where it is essential to make complex and sophisticated analyses, complemented by sound judgment, for the selection of a proper stress to be related to the bending moments determined by *standard* type calculations.

As a result of theoretical, experimental, and service history considerations, classification societies have recently established longitudinal strength standards in terms of the section modulus, total bending moment, and permissible bending stress of the hull girder. In the formulas following, permissible stresses, constants, and range of applicability have been arrived at to give results which correspond to these standards.

The required section modulus, SM, is the quotient of the total bending moment M_t and the nominal permissible bending stresses. In equation form,

$$SM = M_t/\sigma_p \tag{20}$$

The permissible bending stress σ_p is a function of the Rules length of the vessel. For ships between 61m and 240m (200 ft and 790 ft) in length this stress is:

$$1.663 - \frac{240 - L}{1620}\ \text{tons/cm}^2 \left(10.56 - \frac{790 - L}{845}\ \text{tons/in.}^2\right)$$

For ships between 240m and 427m (790 ft and 1400 ft) in length, the permissible bending stress is:

$$1.663 + \frac{L - 240}{4000}\ \text{tons/cm}^2 \left(10.56 + \frac{L - 790}{2045}\ \text{tons/in.}^2\right)$$

There is also a requirement for a minimum section modulus as a function of the ship's length, breadth and block coefficient (see Section 6).

Another consideration must be taken into account in the case of Great Lakes vessels which may be subject to springing stresses. For these Great Lakes vessels, the permissible bending stress is a function of the length of the ship, and is defined as follows in American Bureau of Shipping (1978).

For ships between 112m and 213m (400 ft and 700 ft) in length,

$$\sigma_p = 1.948 + 0.221\frac{L}{305} - 0.42\left(\frac{L}{305}\right)^2 \text{tons/cm}^2$$

$$\left[12.367 + 1.4\frac{L}{1000} - 2.667\left(\frac{L}{1000}\right)^2 \text{tons/in.}^2\right] \tag{21-1}$$

for ships between 213m and 259m (700 ft and 850 ft) in length,

$$\sigma_p = 2.001 + 0.045\frac{L}{305} - 0.28\left(\frac{L}{305}\right)^2 \text{ tons/cm}^2$$
$$\left[12.709 + 0.287\frac{L}{1000} - 1.778\left(\frac{L}{1000}\right)^2 \text{ tons/in.}^2\right] \quad (21\text{-}2)$$

for ships between 259m and 320m (850 ft and 1050 ft) in length,

$$\sigma_p = 2.478 - 0.714\left(\frac{L}{305}\right) \text{ tons/cm}^2$$
$$\left[15.74 - 4.533\left(\frac{L}{1000}\right) \text{ tons/in.}^2\right] \quad (21\text{-}3)$$

for ships between 320 m and 366 m (1050 ft and 1200 ft) in length,

$$\sigma_p = 2.478 - 0.714\left(\frac{L}{305}\right) \text{ tons/cm}^2$$
$$\left[15.74 - 4.533\left(\frac{L}{1000}\right) \text{ tons/in.}^2\right] \quad (21\text{-}4)$$

5.5 Permissible Shear Stresses. Permissible shear stresses due to stillwater and wave-induced loads in the side shell and longitudinal bulkhead plating are 1.065 tons/cm^2 (6.75 tons/in.^{2}) according to ABS Rules. For longitudinal bulkhead plating within the middle eight-tenths depth of the bulkhead, the total shear stresses may be increased to 1.225 tons/cm^2 (7.765 tons/in.^{2}) if the critical shear buckling stress of the bulkhead plate panel between stiffeners is satisfactory.

For oil carriers, permissible shear stresses q in the webs of supporting members are as follows:

s/t	MPa	(tons/in.^{2})
80 and less	85.3	(5.5)
160 maximum	53.9	(3.5)

where s = spacing of stiffeners or depth of web plate, whichever is the lesser,
t = thickness of web plate.

For intermediate values, q may be obtained by interpolation.

For longitudinal supporting members, the permissible shear stresses are 80 percent of the above indicated values.

5.6 Permissible Local Stresses. Where a complete structural analysis is performed, the stress intensity (σ_i) and the stability of the local panel must be checked against the permissible stress (σ_p) and the buckling strength. σ_i and σ_p are calculated by the following formulas:

$$\sigma_i = \sqrt{f_a{}^2 + f_b{}^2 - f_a f_b + 3f_{sab}{}^2} \quad (22\text{-}1)$$

$$\sigma_p = \alpha f y m/Q, \quad (22\text{-}2)$$

$$\sigma_i \leq \sigma_p \quad (22\text{-}3)$$

where

f_a = computed total direct stress along the a-axis,
f_b = computed total direct stress along the b-axis (a is perpendicular to b)
f_{sab} = computed shear stress along the a or b-axis,
f_{ym} = specified minimum yield point of mild steel,
Q = conversion factor for higher strength material.

α is a coefficient which depends on the loading conditions and method of analysis. Typically, for a detailed three-dimensional analysis of a tanker with loading conditions as specified in ABS Rules, α can be determined as follows:

1. For longitudinal supporting member $\alpha = 0.75$. When additional loading conditions which account for the wave-induced bending moments are imposed on a three-dimensional analysis the α-value may be taken as 0.85.
2. For transverse, vertical and horizontal supporting members, $\alpha = 0.85$.
3. Where effective methods of corrosion control are provided, the α-values specified by 1. and 2. may be increased by 10 percent or 300 percent/t whichever is less, where t is the actual thickness of the plate in mm.

In way of end brackets and at the connection of the structural members, the α-values specified above may be increased provided that the detail design is able to transmit the loads smoothly and that the highly stressed region is localized.

It is worth noting that the a-b coordinates used in the expression for the stress intensity can be any set of perpendicular axes in the plane. For instance, if a and b axes coincide with the directions of the principal stresses f_1 and f_2, the stress intensity can be expressed by

$$\sigma_i = \sqrt{f_1{}^2 + f_2{}^2 - f_1 f_2} \quad (23)$$

5.7 Buckling Stresses. Where the computed shear stresses in the web plating of a main supporting member are high or where stability of a local panel is in doubt, the buckling strength of the individual plate panel may be checked by an interaction formula. As a result of numerous investigations, many such formulas are available. Typically, the formula given below can be used, and it should be noted that this formula is equally applicable to either vertically or horizontally stiffened main supporting members, or members utilizing a combination of both systems.

$$\left(\frac{f_c}{f_{crc}}\right)^2 + \left(\frac{f_s}{f_{crs}}\right)^2 \leqq 1.0 \quad (24)$$

where

f_c = calculated maximum compressive stress due to axial compression and bending,
f_s = calculated average shear stress,
f_{crc} = critical buckling stress corresponding to axial compression and bending loads,
f_{crs} = critical buckling stress corresponding to pure shear loading.

Table 3—K-Value

(A) For plate panels between stiffeners;

Type of Loading	Description	K

1. For evaluating f_{crc}: corresponding to axial compression and bending

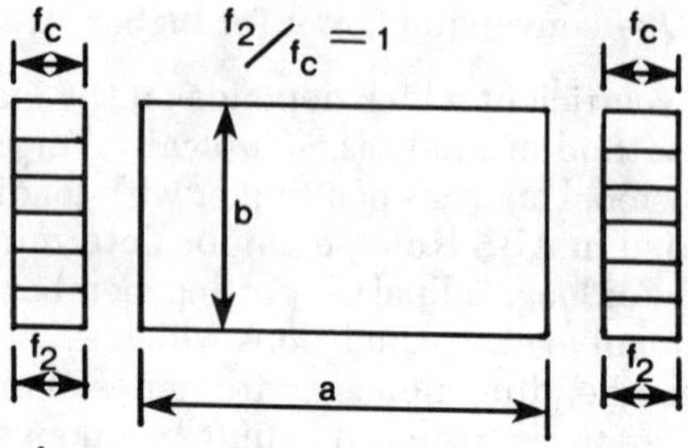

where $a/b \geqq 1.0$ - - - - - - - 4

where $a/b < 1.0$ - - - - - - - $\left(\frac{a}{b} + \frac{b}{a}\right)^2$

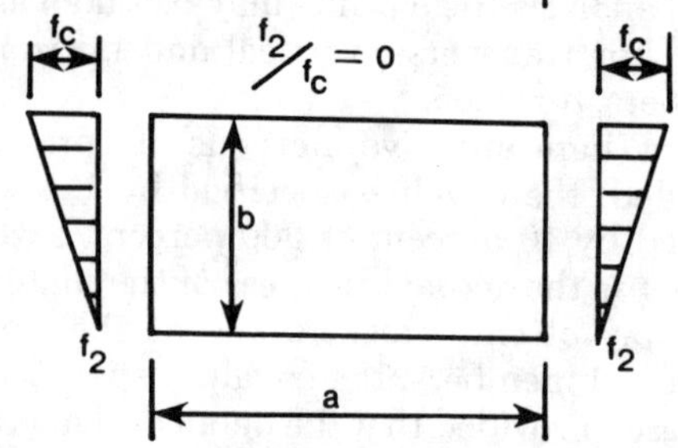

where $a/b \geqq 1.0$ - - - - - - - 7.7

where $a/b < 1.0$ - - - - - - - $7.7 + 33\left(1 - \frac{a}{b}\right)^3$

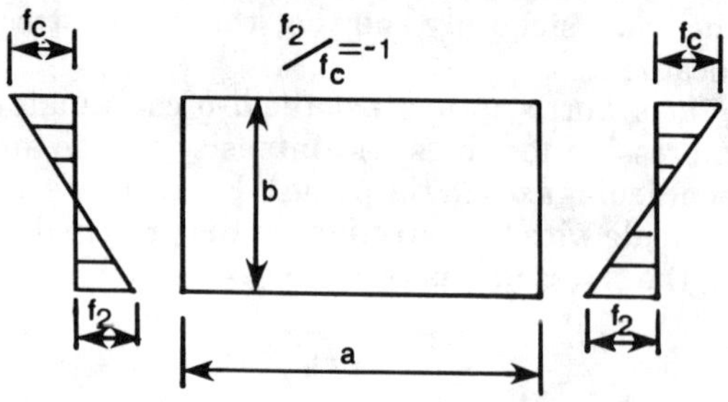

where $a/b \geqq 2/3$ - - - - - - - 24

where $a/b < 2/3$ - - - - - - - $24 + 73\left(\frac{2}{3} - \frac{a}{b}\right)^2$

For any linear distribution of in-plane stress other than shown above, K may be computed by interpolation.

2. For evaluating f_{crs}:

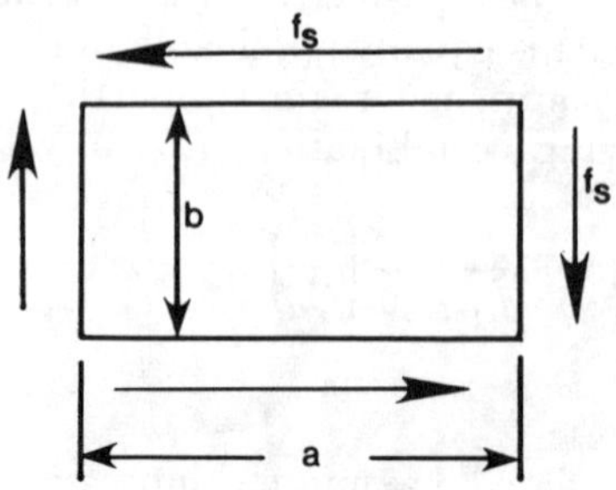

K
corresponding to shear loading

where $a/b \geqq 1.0$ - - - - - - - $\sqrt{3}\left[5.34 + 4\left(\frac{b}{a}\right)^2\right]$

where $a/b < 1.0$ - - - - - - - $\sqrt{3}\left[4 + 5.34\left(\frac{b}{a}\right)^2\right]$

where a = width of the plate panel
b = as defined above.

(B) For plate panels between heavy supporting members: Where the plate panel is supported by members heavier than ordinary stiffeners, the listed K values, whether for axial compression and bending or for shear, may be increased by the following percentages:

Where one edge is supported by plating or by *heavy* face plate	10 percent
Where two parallel edges are supported by plating or by *heavy* face plates	20 percent

The critical buckling stresses, f_{crc} and f_{crs} may be approximated as follows:

$$f_{crc} = f_t \quad (25\text{-}1)$$

$$f_{crs} = f_t/\sqrt{3} \quad (25\text{-}2)$$

where

$f_t = f_e$ when $f_e/f_y \leqq 0.75$ and

$f_t = f_y \left(1 - \dfrac{3f_y}{16f_e}\right)$ when $f_e/f_y > 0.75$,

$f_e = 1.88 \times 10^6 \left(\dfrac{t}{b}\right)^2 K$ kg/cm²

$\left[26.75 \times 10^6 \left(\dfrac{t}{b}\right)^2 K, \text{psi}\right]$

f_y = specified minimum yield point of the material or 72 percent of the specified minimum ultimate strength whichever is less,

t = thickness of web plate reduced by 10 percent as an allowance for corrosion

b = depth of the plate panel.

K is a function of types of loading, aspect ratio and boundary conditions. It is necessary to determine a value for K for each of the two basic types of loading, namely, one for combined axial compression and bending, and one for shear loading. Values of K may be evaluated in accordance with Table 3.

When a/b is equal to or greater than 2, no credit should be given to the actual degree of constraint along the short (or "b") edges. When a/b is less than 2 and *heavy* supporting members are provided at both the "b" edges, an additional 20 percent may be added to the K values as increased per Table 3.

5.8 Special Considerations. For special ship types, different permissible stresses may be specified for different parts of the hull structure. For example, for oil carriers, the maximum permissible local bending stresses are 139 MPa (9 tons/in.^{2}) for deep transverses and 92.7 MPa (6 tons/in.^{2}) for longitudinal girders. For LNG carriers, there are special strain requirements in way of the bonds for the containment system, which in turn can be expressed as equivalent stress requirements.

For local areas subjected to many cycles of load reversal, fatigue lives must be calculated and a reduced permissible stress may be imposed to prevent fatigue failure.

5.9 Longitudinal Deflections. The bending deflection of a ship girder is obtainable from the appropriate curvature equation by integrating twice. A semi-empirical approximation for bending deflection amidships is:

$$w_{\otimes} = k\frac{M_{\otimes}L^2}{EI} \quad (26)$$

where the dimensionless coefficient k may be taken as 0.09. In terms of the stresses in the top and bottom flanges of the ship girder and its depth D:

$$w_{\otimes} = \frac{kL^2}{ED}(\sigma_B - \sigma_T) \quad (27)$$

so that deflection (or flexibility) for a given stress level is proportional to the L/D ratio and the length. For a given bending moment, obviously, it is proportional to L/I and the length. Positive stresses are tensile; negative, compressive. The deflection w is positive when deflection is downward (sagging) and negative when upward (hogging).

A still more approximate deflection formula derives from the last equation for stress values of 144 MPa (9.3 tons/in.^{2}) and 154 MPa (10 tons/in.^{2}) for the bottom and top flanges, respectively.

Thus:

$$w_{\otimes} = \frac{L}{535} \quad (28)$$

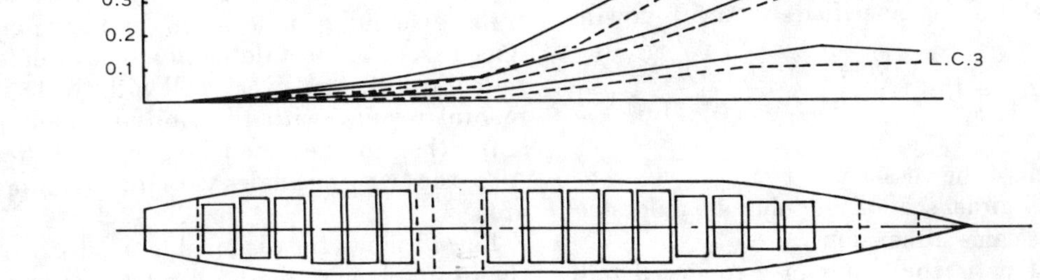

Fig. 47 Comparison of twisting, angles along the model

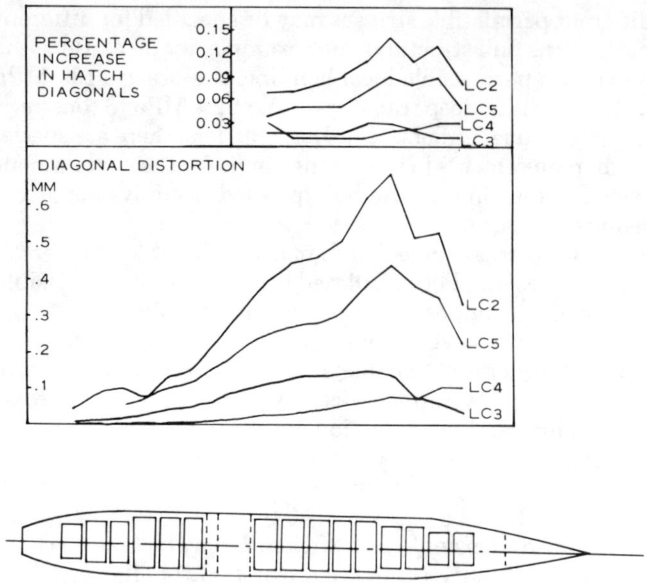

Fig. 48 Hatch diagonal distortion

These values correspond to a reasonable design total bending moment (stillwater plus wave), and an L/D value of 14.

Actual deflection in service is affected also by thermal influences, rigidity of structural components, and workmanship; furthermore, deflection due to shear is additive to the bending deflection, though its amount is usually relatively small.

The same influences which are gradually raising nominal design stress levels are also raising flexibility. Additionally, draft limitations and similar factors may force the L/D ratio up as ships get larger. In general, therefore, modern design requires that more attention be focused on flexibility than formerly.

Presently, no specific limits on hull girder deflections are given in the classification rules. The required minimum scantlings however, as well as general design practices, are based on a limitation that the L/D ratio is not greater than 15 for oceangoing vessels. For Great Lakes bulk carriers this limitation has been extended to 21.

For vessels constructed either of higher strength materials or aluminum alloys, the major classification societies generally impose a minimum requirement on the hull-girder moment of inertia I. For example in the ABS Rules the minimum value of I amidships for vessels constructed of higher strength materials is specified by the following equation:

$$I_{hts} = L(SM)/34.1 \tag{29}$$

where

L = Rules length of the vessel
SM = required hull-girder section modulus of a mild-steel vessel of the same dimensions

5.10 Racking. This is the result of a torsional hull distortion and is caused by either dynamic loads due to rolling of the ship or by the transverse impact of seas against the topsides. Bulkheads prevent racking if the bulkhead spacing is close enough to prevent deflection of the shell or deck plating in its own plane. Racking introduces primarily compressive and shearing forces in the plane of bulkhead plating.

With the usual spacing of bulkheads the effectiveness of side frames in resisting racking is negligible. However, when bulkheads are widely spaced or where the deck width is small in way of very large hatch openings, side frames, in association with their top and bottom brackets, contribute significant resistance to racking.

Racking stresses due to rolling reach a maximum in a beam sea each time the vessel completes an oscillation in one direction and is about to return. The angle between a deck beam and side frame tends to open on one side and to close on the other side at the top and reverses its action at the bottom.

5.11 Twist and Warping. Twist and warping may occur when the slenderness of the hull girder, which is desirable for higher sea speeds, is accompanied by wide hatch openings in containerships, which are desirable for fast loading and unloading operations.

In order to increase the torsional rigidity of the containership cross sections, longitudinal and transverse closed box girders are introduced in the deck structure. Hatch opening flexibility and the abrupt change in torsional rigidity due to ship end closed sections and to the engine housing may cause undesirably high stresses at certain locations on the deck.

In the classical types of hatches, where the hatch opening was less than half of the deck beam, the main deck structure was designed to withstand longitudinal bending stresses and generated local stresses in the vicinity of the hatch corner. Torsional stresses were small, and the theoretical investigation of the deck torsional strength based on approximating the hull girder as a closed thin-wall box was adequate. When the size of hatch openings was increased, with deeper hatch coamings and heavier, side deck stringers, analytical methods had to include consideration of cutting the structure into modules, thus ending the traditional assumption of treating the hull girder as a closed box.

In order to correlate the analytical and measured response of the open deck structure under various loading conditions, a 1:50 steel model of a containership was instrumented and studied (Elbatouti et al., 1976).

One of the items studied was the twisting angle, and Fig. 47 shows its distribution along the ship for various loading cases. The twisting angle was calculated at various transverse bulkhead locations and compared with that deduced from the test deflection measurements. The theoretical results agree very well with the trend of the experimental results, with the finite-element model generally stiffer than the steel model regarding the torsional response. Also, the twisting angles vary linearly along the open deck area.

Fig. 48 shows the diagonal hatch distortion calculated by the finite-element method and also the percentage change in the diagonal length of the hatch. In general, the forward hatches exhibit more deformation than the aft ones. This is an indication of the torsional flexibility of the containership sections at forward hatches.

From the above-mentioned and other studies, it has been established that more attention should be paid to the torsional rigidity distribution along the hull. Usually, toward the ship's ends, the section moduli are justifiably reduced. On the contrary the torsional rigidity, especially in the forward hatches, should be gradually increased to keep the warping stress as small as possible.

5.12 Local Deflections. Local deflections must be kept at reasonable levels in order for the overall structure to have the proper strength and rigidity. Towards this end, the classification society rules may contain requirements to ensure that local deflections are not excessive.

For example, a requirement of the ABS Rules, applicable to oil carriers, is that webs, girders and transverses have a minimum depth to span ratio of 0.125 for side and deck transverses, for webs and horizontal girders of longitudinal bulkheads, and for stringers, and 0.20 for deck and bottom centerline girders, for bottom transverses, and for webs and horizontal girders of transverse bulkheads. Special requirements also apply to stiffeners. Tripping brackets are provided to support the flanges, and they should be in line with or as near as practible to the flanges of struts. Special attention must be given to rigidity under compressive loads in order to avoid buckling. This is done by providing a minimum moment of inertia of the stiffener and associated plating which is:

$$I = 0.19lt^3(l/s)^3 \qquad l/s \le 2.0$$
$$= 0.38lt^3(l/s)^2 \qquad l/s > 2.0 \qquad (30)$$

where l = length of stiffener between effective supports,
t = thickness of web plating,
s = stiffener spacing.

Section 6
Application of Classification Rules

6.1 General. As indicated in Section 1, the selection of sizes and scantlings of hull structural members is generally based on classification rules. In this section, an example is given for the design of a midship section of a tanker in accordance with ABS Rules (1980). The procedures for applying rules of other major classification societies may slightly differ from those described in this section.

The prototype ship of this example is a typical tanker having the following principal dimensions:

Length (LBP)	310.00 m (1017.0 ft)
Length (Rules)	306.16 m (1004.5 ft)
Breadth (Mld)	48.71 m (159.8 ft)
Depth (Mld)	24.50 m (80.6 ft)
Draft (Design)	18.87 m (61.9 ft)
Block Coefficient (C_B)	0.835
Deadweight	230,000 tons

Design variables considered in this section are limited to hull-girder section modulus and stiffness, shell and deck plating, longitudinal bulkheads, longitudinals, and transverses. The structural configuration, such as locations of longitudinal bulkheads, arrangements of struts and brackets, spacings of transverses and longitudinals, are selected directly from the prototype design.

6.2 Hull-Girder Section Modulus. *a. Minimum Requirements.* The longitudinal strength of a ship is specified in terms of minimum hull-girder section modulus requirements. The hull-girder section modulus amidships is not to be less than that obtained from either Equation (20) or the following equation:

$$SM = 0.01C_1L^2B(C_B + 0.70) \text{ cm}^2\text{-m (in.}^2\text{-ft.)} \qquad (31)$$

where

L = Rules length of the vessel, as defined in 4.3, in m (ft),
B = breadth of the vessel, in m (ft),
C_B = block coefficient at summer load waterline, which is not to be taken less than 0.60.

$$C_1 = 10.75 - \left(\frac{300 - L}{100}\right)^{1.5} \qquad 90 \le L \le 300 \text{ m}$$

$$\left\{0.01441\left[10.75 - \left(\frac{984 - L}{328}\right)^{1.5}\right] \qquad 295 \le L \le 984 \text{ ft}\right\}$$

$$= 10.75 \qquad 300 < L \le 350 \text{ m}$$
$$\{0.01441\,[10.75] \qquad 984 < L \le 1148 \text{ ft}\}$$

$$= 10.75 - \left(\frac{L - 350}{150}\right)^{1.5} \qquad 350 < L \le 427 \text{ m}$$

$$\left\{0.01441\left[10.75\left(\frac{L - 1148}{492}\right)^{1.5}\right] \qquad 1148 < L \le 1400 \text{ ft}\right\}$$

Assume that the maximum still-water bending moment for all the anticipated loading conditions for this vessel is determined by Equation (1) which gives a value of 546,750 ton-m (1,760,535 ton-ft) amidships. The maximum wave-induced bending moment as obtained from Equation (4) is 801,790 ton-m (2,581,760 ton-ft). Therefore, the maximum total bending moment is 546,750 + 801,790 = 1,348,540 ton-m (4,342,295 ton-ft). The permissible bending stress, σ_p, as specified in Equation (20) is 1.68 tons/cm² (10.65 tons/in.²) for the ordinary mild steel. Consequently, the required section modulus by Equation (20) is 802,920 cm²-m (408,330 in.²-ft).

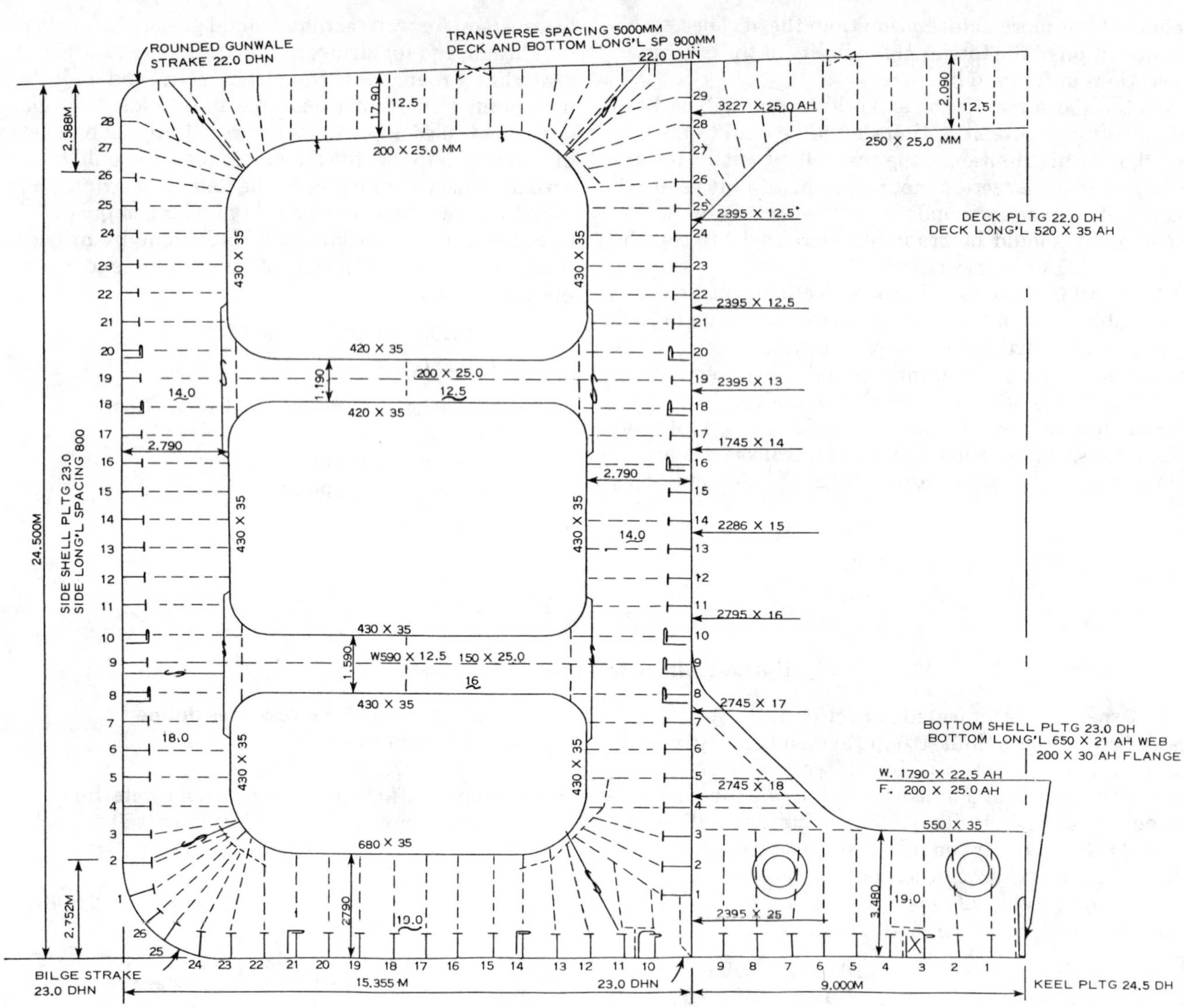

Fig. 49 Midship section designed based on ABS Rules 1980

On the other hand, Equation (31) yields a value of 753,530 cm^2-m (383,210 in.^{2}-ft) for this vessel, which is less than the value obtained from Equation (20). Therefore, the required section modulus is 802,920 cm^2-m (408,330 in.^{2}-ft) provided that ordinary mild steel is used.

For higher strength materials, the required hull-girder section modulus can be reduced based on the following equation.

$$SM_{hts} = Q(SM) \tag{32}$$

where $Q = 49.92/(Y + 2U/3)$ in metric units
[$70900/(Y + 2U/3)$ in inch/pound units]
Y = specified minimum yield strength, in kg/mm^2 (psi) of the higher-strength material, or 72 percent of the specified minimum tensile strength, whichever is less,
U = specified minimum tensile strength of the higher-strength material, in kg/mm^2 (psi).

As shown in Fig. 49, the upper and lower flanges of the hull girder are of ABS H32 steel, for which the specified minimum yield strength and tensile strength are 32 and 48 kg/mm^2 (45,500 and 68,000 psi) respectively. Consequently,

$$Q = 49.92/(30 + 2 \times 48/3) = 0.78,$$

and $SM_{hts} = 0.78 \times 802{,}920 = 626{,}280$ cm^2-m (318,500 in.^{2}-ft)

The corresponding permissible bending stress, $(\sigma_p)_{hts}$ is, as determined by the following equation, 2.15 tons/cm^2, (13.62 tons/in.^{2}).

$$(\sigma_p)_{hts} = \sigma_p/Q \tag{33}$$

The higher-strength material is to be extended from the upper and lower flanges to a level where the bending stress is less than that allowed for mild steel, which is 1.68 tons/cm^2 (10.65 tons/in.^{2}) in this case. The section modulus obtained above is generally required to extend throughout the 0.4 L

amidships and gradually tapered beyond. For fine form ships, an extension of deck area and longitudinal strength members throughout the 0.4 L amidships may be considered as acceptable.

In the case that the envelope curve of stillwater bending moments is provided by the designer, an envelope curve of the total bending moments can be constructed by superimposing it on the wave-induced bending moment envelope specified in the rules of major classification societies. In that case, the required section modulus can be taken in such a manner that the bending stresses remain within the permissible limits. When higher-strength materials are used, their extent in the longitudinal direction is also to be governed by the permissible stress limits.

b. Section Modulus Calculation. In the calculation of hull-girder section modulus, the following items may be included provided they are continuous and effectively developed;

- deck plating (strength deck and other effective decks),
- shell and inner bottom plating,
- deck and bottom girders,
- plating and longitudinal stiffeners of longitudinal bulkheads,
- all longitudinals of deck, sides, bottom and inner bottom.

In general the net sectional areas of the above items are to be used in the hull-girder section modulus calculations, except that small isolated openings need not be deducted provided the openings and the shadow area breadths of other openings in any one transverse section do not reduce the section modulus by more than 3 percent. The breadth or depth of such openings is not to be greater than 1200 mm (47 in.) or 25 percent of the breadth or depth of the member in which it is located, whichever is less. The length of small isolated openings not required to be deducted is generally not greater than 2.5 m (100 in.). Scallops with a maximum of 75 mm (3 in.) may not be required to be deducted.

The shadow area of an opening is the area forward and aft of the opening enclosed by the lines tangential to the corners of the opening intersecting each other to form an angle of 30 degrees.

For performing section modulus calculations, the minimum sizes and scantlings as specified by the classification rules may be first selected for all the longitudinal strength members. If the calculated section modulus based on the minimum scantling requirements is insufficient, an increase of the plating thicknesses, or more economically, the sizes of longitudinals in the regions of the upper and/or lower flanges of the hull girder is required. On the other hand, if the first calculated section modulus is greater than that required, a reduction in plating thicknesses and sizes of longitudinals should be examined by decreasing the spacings of longitudinals.

In determining the minimum thicknesses of the deck and shell plating, it is also necessary to meet the buckling strength and shear strength of the plate panels as required by classification rules. The calculated section moduli of the midship section shown in Fig. 49, which is designed in accordance with the Rules requirements are 626,720 cm^2-m (318,720 in.^{2}-ft) and 632,110 cm^2-m (321,460 in.^{2}-ft) to the deck and to the bottom respectively. It can be seen that the value of the section modulus to the deck is only slightly greater than the minimum requirement. For actual cases, designers generally incorporate in their designs an appropriate safety margin.

c. Engineering Analysis. Alternatively, instead of the specific section modulus requirements as discussed above, the major classification societies also consider and accept designs derived from sound engineering principles, provided that they meet the overall safety and strength standards predicated by the classification rules. Various levels of engineering analyses have been employed by different designers to verify their designs. For longitudinal strength requirements, the *acceptable* engineering analyses can be generally grouped into two categories. In the first category, the analyses and calculations are limited to wave load predictions. In the second category, both wave load predictions and structural analyses are required.

For wave load predictions, the analysis involves ship motion calculations utilizing computer programs developed based on the two-dimensional strip theory, and probabilistic approaches based on the available statistical wave data, as discussed earlier. At the present time, the major classification societies have all developed their own ship motions computer programs and wave spectra for wave load predictions. Because of the different features and characteristics of different computer programs and wave spectra, each individual classification society has established an acceptable probability level based on its past experience. However, it is generally expected that the predicted maximum wave load in conjunction with plan approval reflects a maximum value which may be probably exceeded only once during the entire life span of the vessel considered.

Direct ship motion calculations and statistical analyses have been performed for the prototype vessel, using a set of wave spectra derived from the measured wave data in the North Atlantic region and a six-degree of freedom ship-motion computer program. The calculated maximum wave-induced vertical bending moments are 791,490 ton-m (2,548,600 ton-ft) and 861,590 ton-m (2,774,320 ton-ft) for the full load condition and a heavy ballast condition respectively at a probability level of 10^{-8} which corresponds to a ship life of 20–25 years. This represents a 1.1 percent decrease in the maximum wave-induced vertical bending moments in one case and an increase of 7.5 percent in the other case in comparison with that obtained from the Rules formula (Equation 4).

In this case the required section modulus depends on the magnitudes and distributions of the stillwater bending moments. If the assumed maximum stillwater bending moment occurs amidships in the full load condition and the maximum total bending moment for the heavy ballast condition is less than that for the full load condition, the designer may apply for a section modulus reduction in accordance with the specified permissible bending stress limits. On the other hand, if the assumed maximum stillwater bending moment occurs amidships in the heavy ballast condition, the hull girder section modulus may be required to be increased, unless a structural analysis has been also carried out (2nd category) and the total stresses, including primary, secondary, and tertiary stress components in the longitudinal strength members are within the permissible limits specified by the classification society.

In the case considered, the maximum stillwater bending moment for the full load condition is 76,730 ton-m (247,070 ton-ft) greater than that for the heavy ballast condition. Assume that the maximum stillwater bending moment for the full load condition is 546,750 ton-m (1,760,535 ton-ft) and occurs amidships. The maximum total bending moment will be 1,338,240 ton-m (4,309,135 ton-ft) which is about one percent less than that obtained in 6.2*a*. Consequently, the required hull-girder section modulus may be reduced by one percent.

Where the maximum stillwater bending moment does not occur amidships and an envelope curve of the wave-induced bending moments is provided, the maximum total bending moment may be determined by superimposing the envelope curve on the stillwater bending moment curve for the loading case condition. In calculating the section modulus requirement, it is, however, necessary to examine all other anticipated loading cases to ensure that a maximum value is used.

It is important to note that a classification society may require a direct wave load calculation and/or structural analysis for special types of vessel or for vessels intended for special load conditions which may not be covered by the rule formulas.

6.3 Hull Girder Stiffness. For vessels constructed of higher-strength materials, the hull-girder moment of inertia amidships is to be not less than that obtained from Equation (29).

For L = 306.16 m (1,004.5 ft) and SM = 802,920 cm^2-m (408,330 in.^{2}-ft), the required I_{hts} is 7,208,856 cm^2-m^2 (12,021,160 in.^{2}-ft^2). The midship section shown in Fig. 49, which meets all the requirements specified in ABS Rules gives a hull-girder moment of inertia of 7,704,219 cm^2-m^2 (12,847,210 in.^{2}-ft^2) which is about 7 percent greater than the required values.

As discussed previously, there is no specific hull-girder stiffness requirement in classification rules for vessels constructed of mild steel, provided the L/D ratio is not greater than 15.

6.4 Shell Plating. The shell plating which is intended to provide watertightness and also to contribute a major portion of longitudinal and transverse strength of a hull girder constitutes one of the principal structural elements of a vessel. The determination of the required shell plate thickness is generally a compromise of several governing aspects, namely, the longitudinal strength requirements with respect to both hull-girder bendings and shears, local bending (secondary) and pressures (tertiary), buckling strength, steel weight, and assembling cost.

Although the rationale in determining the thickness requirements is not clearly stated in the classification rules, all the influential parameters are generally incorporated in the rule formulas, and the constants selected to give results corresponding to known acceptable values.

Shell plating within the midship 0.4L is to be not less in thickness than is required for longitudinal hull-girder strength, or than the thickness obtained from this subsection.

a. Bottom Shell Thickness. The thickness t of the bottom shell plating of a bulk oil carrier is to be not less than obtained from Equation (33) or Equation (34) for mild steel:

$$t = s(L + 8.54)/(42L + 2318) \text{ mm} = [s(L + 28)/(42L + 7602) \text{ in.}] \quad (33)$$

$$\text{or } t = 0.006s\sqrt{0.7T + 0.02(L - 50)} + 2.5 \text{ mm} = [0.00331s\sqrt{0.7T + 0.02(L - 164)} + 0.1 \text{ in.}] \quad (34)$$

where s = spacing of bottom longitudinals in mm (in.),
L = Rules length of the vessel, in m (ft),
T = design draft in m (ft).

In applying Equation (33) which is intended to cover buckling strength of plate panels, a minimum value of s is specified in ABS Rules. For this example, the value of s is to be taken not less than 864 mm (34 in.).

For s = 900 mm (35.4 in.), L = 306.16m (1,004.5 ft), T = 18.87 m (61.9 ft), the required bottom shell thickness is 18.7 mm (0.74 in.) by Equation (33) or 25.62 mm (1.01 in.) by Equation (34) for mild steel. Therefore, Equation (34) governs in this case. For higher strength materials, the required plate thickness may be converted by Equation (35).

$$t_{hts} = (t - 4.3)Q + 4.3 \text{ mm} \quad (35)$$
$$[(t - 0.17)Q + 0.17 \text{ in.}]$$

for Q = 0.78, the required minimum thickness is
t_{hts} = (25.62 − 4.3) × 0.78 + 4.3 = 21.0 mm (0.83 in.).

However, in order to meet the hull-girder section modulus requirement shown in 6.2*a* either the sizes of bottom longitudinals or the thickness of the bottom shell, or both are to be increased. For simplicity, only the thickness of bottom shell is increased from 21.0 to 23.0 mm (0.83 to 0.91 in.) in the sample design (Fig. 49).

b. Side Shell Thickness. The thickness t of the side shell plating is to be not less than obtained from Equation (36) or Equation (37) for ordinary-strength mild steel.

$$t = 0.01L\,(6.5 + 21/D) \text{ mm} = [0.0003937L\,(2.0 + 21/D) \text{ in.}] \quad (36)$$

$$\text{or } t = 0.0052s\sqrt{0.7T + 0.02L} + 2.5 \text{ mm} = [0.00287S\sqrt{0.7T + 0.02L} + 0.1 \text{ in.}] \quad (37)$$

where L, s and T are as defined in 6.4*a*,

D = molded depth of the vessel in m (ft).

For s = 800 mm (31.5 in.), D = 24.50 m (80.4 ft), L and T as above, the required side shell thickness is 22.5 mm (0.89 in.) by Equation (36) or 20.8 mm (0.82 in.) by Equation (37). Apparently Equation (36) governs in this case. In the final design shown in Fig. 49, the side shell thickness of 23.0 mm (0.91 in.) is selected. This slight increase is due to insufficient longitudinal strength based on minimum shell thickness.

For ships other than tankers, the thickness requirements for bottom and side shell are determined by somewhat different equations. The side shell thickness obtained from the above equation is to be checked with the hull-girder shear strength requirements. The total shear stress, including both stillwater and wave-induced components, in the side shell and longitudinal bulkhead is to be generally not greater than 1.065 tons/cm^2 (6.75 tons/in.^{2}) except that for longitudinal bulkhead plating within the middle eight-tenths depth of the bulkhead, the total stress may be increased to 1.225 tons/cm^2 (7.765 tons/in.^{2}) if the critical shear buckling stress of the bulkhead plate panel between stiffeners is satisfactory.

c. *Sheer Strake.* The thickness of the sheer strake is to be generally not less than the thickness of the deck stringer plate or the side shell plating, whichever is greater. To compensate for stress increases at structural discontinuities, the sheer strake thickness is to be increased 25 percent or 6.5 mm (0.25 in.) in way of breaks of superstructures, whichever is less.

The width of the sheer strake for the midship 0.4 L is to be not less than obtained from the following equation for all types of vessels:

$$\begin{aligned} b &= 5L + 916 \text{ mm}, L < 120 \text{ m}; \quad (38)\\ &[0.06L + 36 \text{ in.}, L < 395 \text{ ft}]\\ &= 1525 \text{ mm}, 120 < L \le 427 \text{ m}\\ &[60 \text{ in.}, 395 \le L \le 1400 \text{ ft}] \end{aligned}$$

where b = width of sheerstrake in mm (in.)
L = Rules length in m (ft).

The sheerstrake is, as shown in Fig. 49, selected to be of the same thickness as the deck stringer plate, and is extended 2.588 m (101.9 in.) from the deck edge at side to cover the high stress region.

d. *Keel Plate.* The thickness of the flat-plate keel is to be maintained throughout and is to be generally not less than 1.5 mm (0.06 in.) greater than the bottom-shell thickness amidships except that where the bottom shell thickness exceeds 37 mm (1.44 in.), this requirement may be modified. Where this strake is increased over the minimum required value for longitudinal strength, the flat-plate keel may be gradually reduced, forward and abaft the midship 0.4 L, to the requirement amidships. Therefore, the required thickness for the keel plate is 23.0 + 1.5 = 24.5 mm (0.97 in.) for H 32 steel. Since the thickness of the bottom shell has been increased 2.0 mm (0.08 in.) for longitudinal strength, the keel plate may be reduced to 22.5 mm (0.89 in.) forward and abaft the midship 0.4 L.

The width of the keel plate is generally determined by the docking practice of each individual shipyard. There is no specific requirement imposed by classification rules.

For ships other than tankers, the above requirements for sheerstrake and keel plate are also applicable.

6.5 Deck Plating. Similar to the requirements for shell plating, the strength deck plating within the midship 0.4 L is to be of not less thickness than is required to provide the deck sectional area necessary for longitudinal strength, nor is the thickness to be less than determined by Equation (39) or (40) for mild steel.

$$t = 0.0016s\sqrt{L-53} + 0.32(L/D) - 2.5 \text{ mm}; \quad (39)$$
$$[0.000883s\sqrt{L-174} + 0.0126(L/D) - 0.1 \text{ in.}]$$

$$\begin{aligned} t &= \frac{s(30.48+L)}{4981+40L} \text{ mm}, L \le 152.4 \text{ m} \quad (40)\\ &\left[\frac{s(100+L)}{16339+40L} \text{ in.}, L \le 500 \text{ ft}\right]\\ &= \frac{s(259.1+L)}{22641+15L} \text{ mm}, 152.4 < L \le 305 \text{ m}\\ &\left[\frac{s(850+L)}{74264+15L} \text{ in.}, 500 < L \le 1000 \text{ ft}\right]\\ &= \frac{s(L+8.537)}{42L+2318} \text{ mm}, 305 < L \le 427 \text{ m}\\ &\left[\frac{s(L+28)}{42L+7602} \text{ in.}, 1000 < L \le 1400 \text{ ft}\right] \end{aligned}$$

where t = plate thickness in mm (in.),
s = spacing of deck longitudinals in mm (in.),
L = Rules length, in m (ft),
D = molded depth in m (ft).

For s = 900 mm (35.4 in.), L = 306.16 (1,004.5 ft), and D = 24.50 m (80.4 ft), Equations (39) and (40) give thicknesses of 24.5 mm (0.95 in.) and 18.7 mm (0.74 in.) respectively. Therefore, the minimum deck thickness amidships is 24.5 mm (0.95 in.) for mild steel. For H32 steel, Equation (35) is applied. In this case:

$$t_{hts} = (24.5 - 4.3) \times 0.78 + 4.3 = 20.0 \text{ mm } (0.79 \text{ in.})$$

To meet the longitudinal strength requirement, this minimum thickness is to be increased to 22 mm (0.87 in.), as shown in Fig. 49. Alternatively, to meet the required deck area it is optional to increase the sizes of deck longitudinals provided that the buckling strength of the deck plating is satisfactory.

For vessels other than tankers, the required minimum plate thickness for the strength deck is generally much less than required for the longitudinal strength consideration.

6.6 Longitudinal Bulkheads. Both the longitudinal and transverse bulkheads of a tanker are generally designed based on the hydrostatic pressure in accordance with the requirements for deep-tank bulkheads which are applicable to all types of vessels. The required thickness is expressed by the following equation:

$$t = (s\sqrt{h}/254) + 2.54 \text{ mm}$$
$$[(s\sqrt{h}/460) + 0.1 \text{ in.}] \quad (41)$$

where t = thickness of bulkhead plating in mm (in.),
s = stiffener spacing in mm (in.),
h = the distance, in meters (ft) from the lower edge of the plate to the top of the hatch or to a point

Table 4—Sizes of Side and Bulkhead Longitudinals (mm)

(see Fig. 49)

	SIDE LONGITUDINAL		LONGITUDINAL OF LONG. BULKHEAD
SL-28	300 × 25 AH	LS-29	300 × 25 AH
SL-27	300 × 25 AH	SL-28	300 × 25 AH
26	290 × 11.5 + 100 × 16	27	300 × 25 AH
25	290 × 11.5 + 150 × 16	26	290 × 11.5 + 125 × 16
24	290 × 11.5 + 100 × 25	25	290 × 11.5 + 150 × 16
23	390 × 11.5 + 100 × 16	24	390 × 11.5 + 100 × 16
22	390 × 11.5 + 125 × 16	23	390 × 11.5 + 100 × 16
21	440 × 11.5 + 100 × 16	22	390 × 11.5 + 125 × 16
20	440 × 11.5 + 150 × 16	21	390 × 11.5 + 150 × 16
19	440 × 11.5 + 150 × 16	20	440 × 11.5 + 125 × 16
18	440 × 11.5 + 100 × 16	19	440 × 11.5 + 150 × 16
17	540 × 11.5 + 125 × 16	18	440 × 11.5 + 100 × 25
16	540 × 11.5 + 125 × 16	17	540 × 11.5 + 125 × 16
15	540 × 11.5 + 150 × 16	16	540 × 11.5 + 125 × 16
14	540 × 11.5 + 150 × 16	15	540 × 11.5 + 150 × 16
13	540 × 11.5 + 125 × 25	14	540 × 11.5 + 150 × 16
12	540 × 11.5 + 125 × 25	13	540 × 11.5 + 125 × 25
11	540 × 11.5 + 125 × 25	12	540 × 11.5 + 125 × 25
10	540 × 11.5 + 150 × 25	11	540 × 11.5 + 125 × 25
9	540 × 11.5 + 150 × 25	10	540 × 11.5 + 150 × 25
8	540 × 11.5 + 150 × 25	9	540 × 11.5 + 150 × 25
7	590 × 12.5 + 125 × 25	8	540 × 11.5 + 150 × 25
6	690 × 12.5 + 150 × 16	7	590 × 12.5 + 125 × 25
5	690 × 12.5 + 150 × 16	6	690 × 12.5 + 150 × 16
4	740 × 12.5 + 150 × 16	5	690 × 12.5 + 150 × 16
3	740 × 12.5 + 150 × 16	4	790 × 12.5 + 100 × 16
SL-2	740 × 12.5 + 150 × 16	3	790 × 12.5 + 100 × 16
SL-1	650 × 21AH + 200 × 30 AH	LS-2	790 × 12.5 + 150 × 25
		LS-1	790 × 12.5 + 150 × 25 AH

located 1.22 m (4.0 ft) above the deck at side amidships, whichever is greater, for tankers; (for other types of vessels, see ABS Rules).

The thickness of the top strake of a complete longitudinal bulkhead is generally required to provide a proper margin for corrosion. Furthermore, a large thickness margin of the top strake and the bottom strake of a longitudinal bulkhead is also desirable for the longitudinal strength consideration. The selected thicknesses of the longitudinal bulkhead plating are shown in Fig. 49.

6.7 Longitudinals. Deck, bottom and side longitudinals and bulkhead stiffeners are generally designed based on the maximum hydrostatic pressure. The structural section of a longitudinal or a stiffener in cargo spaces, in association with the effective plating to which it is attached, is to have a section modulus (SM) not less than obtained from the following equation:

$$SM = 7.90chsl^2 \text{ cm}^3$$
$$[0.0041chsl^2 \text{ in.}^3] \quad (42)$$

where c = 1.40 for bottom longitudinals,
= 0.95 for side longitudinals,
= 1.25 for deck longitudinals,
= 1.00 for vertical frames,
= 1.00 for horizontal or vertical stiffeners on transverse bulkheads and vertical stiffeners on longitudinal bulkheads.
= 0.90 for horizontal stiffeners on longitudinal bulkheads.

h = distance in m (ft) from the longitudinals, or from the middle of l for vertical stiffeners, to a point located 1.22 m (4.0 ft) above the deck at side amidships in vessels of 61 m (200 ft) in length, and at a point located 2.44 m (8.0 ft) above the deck at side amidships in vessels of 122 m (400 ft) in length and above; at intermediate lengths, h is to be measured to intermediate heights above the side of the vessel. The value of h for bulkhead stiffeners and deck longitudinals is not to be less than the distance in m (ft) from the longitudinal, or stiffener to the top of the hatch.

s = spacing of longitudinals or stiffeners in m (ft).

l = length between supporting points in m (ft).

The selected sizes of longitudinals and bulkhead stiffeners for the sample case are shown in Fig. 49 and Table 4. It should be noted that the deck longitudinals are generally oversized to provide the required deck area for longitudinal strength considerations.

6.8 Transverses. The deck, side, and bottom transverses together with vertical webs on longitudinal bulkheads are intended to provide sufficient transverse strength of a tanker and to provide support to the longitudinal strength members. In general, these transverses and webs are designed in accordance with the permissible bending and shear stresses specified by classification rules. As discussed previously, the maximum bending and shear stresses in the main portion of the transverses are limited to a value of 135 MPa (9.0 tons/in.^{2}) and 89 MPa (5.5 tons/in.^{2}) respectively for mild steel. The stiffness of a transverse is specified in terms of the depth/span ratio. For side and deck transverses, and vertical webs on longitudinal bulkheads, the depth of the member is to be generally not less than 12.5 percent of the span defined in the Rules. For the bottom transverses, the required depth of the member is 20 percent of the span.

Equations for calculating the required section moduli and shearing forces are given in the classification rules. However, for determining stress distributions in the end connection regions and for verifying the specified bending moments and shearing forces, the major classification societies also require a direct calculation employing sound engineering analysis. A brief discussion of these two approaches is given below.

a. Classification Rules. The section modulus of a transverse in association with the effective plating is to be not less than obtained from the following equations. The effective breadth of the plating may be taken as the transverse spacing or 33 percent of the span, whichever is less.

$$SM = M/f \tag{43}$$

or

$$SM = 4.74chsl_b{}^2 \text{ cm}^3, \quad [0.0025chsl_b{}^2 \text{ in.}^3] \tag{44}$$

where M = maximum bending moment along the member between the toes of end brackets as computed by acceptable methods of engineering analysis, in kg-cm (ton-in.),

f = permissible maximum bending stress, 139 MPa (9.0 tons/in.^{2}) for transverse members,

c for bottom and deck transverses as shown in Fig. 50.

c for side transverses and vertical webs on longitudinal bulkheads
= 1.50 without struts
= 0.85 with one horizontal strut
= 0.65 with two horizontal struts
= 0.55 with three horizontal struts

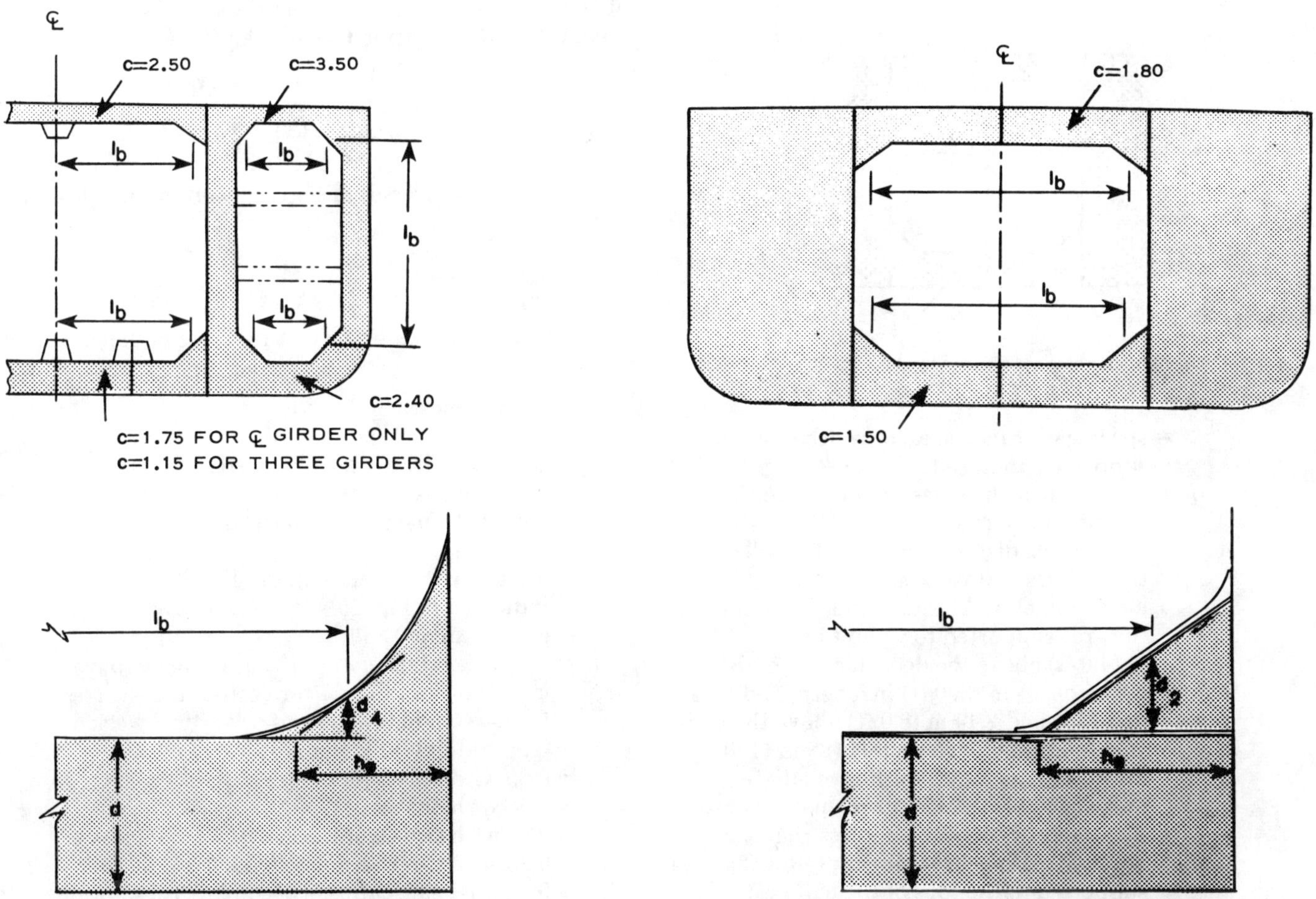

Where face plate area on the member is carried along the face of the bracket

Where face plate area on the member is not carried along the face of the bracket, and where face plate area on the bracket is at least one-half the face plate area on the member

Fig. 50 Coefficients and lengths for transverses

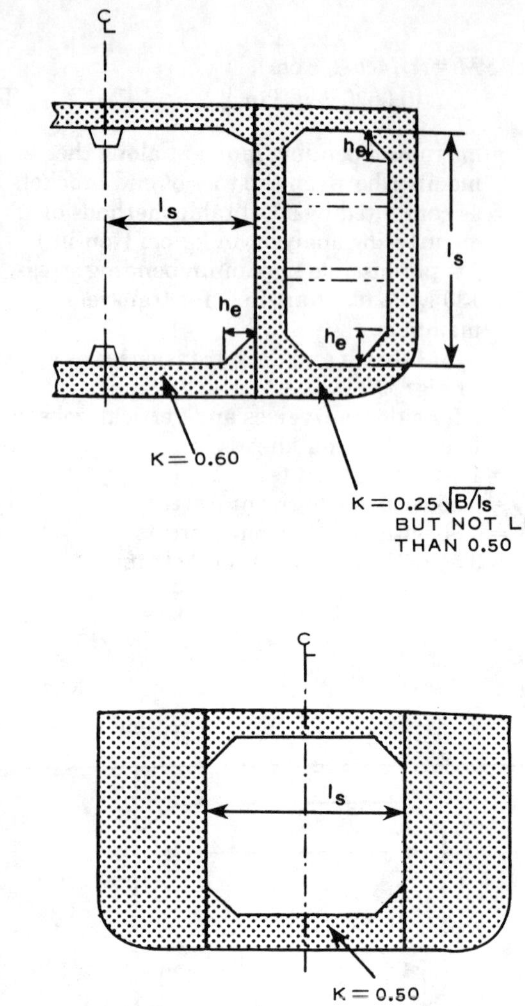

Fig. 51 Spans of members and effective lengths or heights of brackets

s = spacing of transverses, or width of area supported, in m (ft),

h = the depth of the vessel, in m (ft), for bottom transverses,

= the vertical distance in m (ft) for side transverses and vertical webs on longitudinal bulkheads, from the center of the area supported to a point located 1.22 m (4.0 ft) above the deck at side amidships in vessels 61 m (200 ft) in length, and to a point located 2.44 m (8.0 ft) above the deck at side amidships in vessels 122 m (400 ft) in length and above; for intermediate lengths, intermediate points may be used. The value of h is to be not less than the vertical distance from the center of the area supported to the tops of the hatches,

= the vertical distance in m (ft), for deck transverses measured as indicated above for side transverses, except that in no case is it to be less than 15 percent of the depth of the vessel.

l_b = span of the member, in m (ft) measured between the points of support as indicated in Fig. 50. The span of the deck and bottom transverses in wing tanks is to be not less than $0.125B$ or one-half the breadth of the wing tank, whichever is greater.

The net sectional area of the web portion of the member, including effective brackets where applicable, is not to be less than that obtained from the following equation.

$$A = F/q \tag{45}$$

F = shearing force at the point under consideration,
q = allowable average shearing stress in the web of the supporting member, as indicated above.

The thicknesses of the web portion of the members are not to be less than 12.5 mm (0.5 in.) for the vessel considered. The magnitude and distribution of the imposed shearing forces shall be determined by means of an acceptable method of engineering analysis. Where this is not practicable, the following equations may be used as guides in approximating the shearing forces in kg (tons).

$$F = C_2 s D(Kl_s - h_e) \text{ for bottom transverses,} \tag{46}$$

$$F = C_2 s\left[K_L l_s h - h_e\left(h + \frac{l_s}{2} - \frac{h_e}{2}\right)\right]$$

for lower side transverses or vertical webs on longitudinal bulkheads,

$$F = C_2 s\left[K_u l_s h - h_e\left(h - \frac{l_s}{2} + \frac{h_e}{2}\right)\right]$$

for upper side transverses or vertical webs on longitudinal bulkheads

C_2 = 1025 in metric units, (0.0285 in inch/pound units)
s = spacing of transverses in m (ft)
D = depth of vessel, in m (ft)
l_s = span of transverse, in m (ft) as indicated in Fig. 51
h_e = effective length or height of bracket, in m (ft) as indicated in Fig. 50. In no case is h_e to be greater than $0.33l_s$
h = vertical distance, in m (ft) as defined above
K is as shown in Fig. 51 for bottom transverses.
K_l for lower side transverses or vertical webs on longitudinal bulkheads
= 0.65 without struts
= 0.55 with one strut
= 0.43 with two struts
= 0.38 with three or more struts
K_u for upper side transverses or vertical webs on longitudinal bulkheads
= 0.35 without struts
= 0.25 with one strut
= 0.20 with two struts
= 0.17 with three or more struts

To illustrate the application of the above equation, a sample calculation for the bottom transverse in wing tanks as shown in Fig. 49 is given below:

for $c = 2.4$, $h = 24.5$ m (80.4 ft), $s = 5.0$ m (16.4 ft), and $l_b = 8.39$ m (27.5 ft), the required section modulus obtained from Equation (44) is 98,096 cm^3 (5,986 in.^{3}).

The effective breadth of bottom plating is $0.33 l_b = 2.77$ m (9.1 ft). With a web of 2,790 × 19 mm (109.8 × 0.75 in.) and a flange plate of 680 × 35 mm (26.8 × 1.38 in.), the actual section modulus of the bottom transverse in the wing tank is 98,140 cm^3 (5,989 in.^{3}) for this design.

Similarly, the shearing force in the transverse is 495,290 kg (486.5 tons) by Equation (46) with $s = 5.0$ m (16.4 ft), $D = 24.5$ m (80.4 ft), $K = 0.558$, $l_s = 9.775$ m (32.1 ft), and $h_e = 1.51$ m (5.0 ft). By Equation (45) the required net sectional area of the web portion is 569 cm^2 (88.2 in.^{2}) for $q = 870$ kg/cm^2 (5.5 tons/in.^{2}). For this design the sectional area at the effective bracket toe of the web plate is 570 cm^2 (88.4 in.^{2}) without any deduction for openings. Consequently, collar plates are required for the longitudinal cut-outs in the high shear regions.

The sizes of other transverses and webs on longitudinal bulkheads, obtained from similar calculations, are shown in Fig. 49. It should be noted that some of the sizes and scantlings selected in this design are in excess of the minimum rule requirements, for example, the bottom transverse in the center tank, to account for possible concentrated loads not being accounted for in the rule formulas.

As shown in Fig. 49, the thicknesses of the web plates in the upper portion of the side transverses and of the vertical webs on the longitudinal bulkheads are 14 mm (0.55 in.). This thickness is determined based on buckling strength considerations. If an effective vertical stiffener is added on the side transverses and the vertical webs, the plate thickness can be reduced to 12.5 mm (0.5 in.).

b. Finite Element Analysis: In addition to the scantling requirements derived directly from the rule equations and tables, the major classification societies require also an engineering analysis to verify the imposed loads and the corresponding stress distributions. This generally consists of a three-dimensional analysis of a portion of the hull girder and subsequent two-dimensional analyses of typical web frame and longitudinal sections (see examples shown in Figs. 44 and 45). A general-purpose finite element computer program or a space frame computer program may be utilized for these analyses. In the analysis, the imposed loads may be determined either as specified by the classification rules or based on an independent engineering analysis.

A simplified three-dimensional finite element analysis of a two-tank portion of the hull girder had been performed at the American Bureau of Shipping for the prototype vessel. A subsequent two-dimensional analysis of the web frame shown in Fig. 49 was then carried out. The mathematical representation of the web frame is similar to that shown in Fig. 45.

The calculated stresses in the web frame are all within the permissible stress limits indicated earlier. Since the maximum calculated shear stresses in the bottom transverses in wing tanks and the lower portion of the side transverse are below the permissible limits, the thicknesses of web plates in these regions can be reduced by 1 to 2 mm (0.04 to 0.08 in.). This indicates that rule equations in calculating shearing forces are on the conservative side for this particular case. This difference should, however, not be interpreted as discrepancy between classification rules and engineering principles, since the classification societies actually require designers to provide independent engineering analysis in their routine procedure.

6.9 Other Considerations. *a. Special Materials.* In selecting material grades for the deck and shell plating, it is necessary to consider the toughness of the material, especially, in high stress regions, with respect to crack propagation characteristics. The major classification societies require special applications of material grades with superior toughness in certain critical regions of the hull girder, such as sheer strakes, deck stringer plates, bilge shell plates, etc. For large tankers, special material is also required for the deck and bottom shell strakes in way of each continuous longitudinal bulkhead.

For general application, certain material grades, such as grades A & B, are being restricted based on the required thicknesses for longitudinal strength members. The additional requirements are discussed in Chapter VIII.

b. Corrosion Control. Where special protective coatings are adopted for corrosion control in a tanker, the thicknesses of the deck and shell plating may be reduced by 10 percent or 3 mm (0.12 in.), whichever is the lesser. The allowable reduction for longitudinal bulkhead plating is 3 mm (0.12 in.), except where the required thickness is less than 12.5 mm (0.5 in.), for which the allowable maximum reduction is 20 percent.

The section modulus of longitudinals or stiffeners on longitudinal bulkheads, as obtained from Section 6.6, as well as the thicknesses of web plates in transverses are also allowed to be reduced by 10 percent if an effective corrosion control method is adopted.

Section 7
Other Design Criteria and Procedures

7.1 Fatigue Strength. Ship fatigue involves a series of stress reversals of varying frequency. *Low-cycle fatigue* relates to long-term reversals extending from those stress changes that take place with temperature changes and ship loading up to those associated with wave induced bending moments. *High-cycle fatigue* relates to those stress reversals that are associated with dynamic loads such as whipping, and springing.

Fatigue characteristics of marine structures should be examined in the design stages. Two key characteristics are fatigue strength, the value of stress beyond which the material will fail at a specified number of stress cycles; and fatigue or endurance limit, the fatigue strength for an infinite number of stress cycles.

Fatigue strength relates stress ranges which are the algebraic differences between the minimum and maximum stress values, and the number of cycles to failure for each stress range. This relationship is usually represented as a series of straight lines on log-log paper, Fig. 52. The S-N relationship is a function of the material and the type of connection, including edge preparation, weld penetration, notches, etc. A typical group of S-N curves for steel is shown in Fig. 52, adapted from the American Welding Society Code D1.1, Section 10, for different stress categories.

Steel has a definite fatigue limit and theoretically can be subjected to an infinite number of stress cycles without failing so long as this limit is not exceeded. The higher strength steels have a higher fatigue limit than do lower strength steels. The fatigue limit of polished medium-carbon steel is about 50 percent of its ultimate strength. Higher strength steels have a yield strength of approximately 80–95 percent of the ultimate strength while medium-carbon steels have a yield strength of about 50 percent of the ultimate strength.

Where working in the high stress levels of ultra-high strength steels, over 1,241 MPa (180,000 psi) yield strength, the fatigue limit is much lower than 50 percent of the ultimate tensile strength. In the application of ultra-high strength steels, fatigue becomes an even more important factor and the designer must use an even lesser percentage of the yield strength than in the case of high yield strength steels.

The endurance limits of actual ship steels, with their inherent toughness, are less than those obtained by laboratory tests with polished specimens, and they are further reduced by any corrosion, as may occur in sea water.

The fatigue strength of a structure subject to various cyclic stresses is generally quantified by means of a *usage factor* or *damage ratio,* which provides a direct measure of how much of the structure's available strength has been *used up* along the way to possible fatigue failure.

The usage factor can be calculated by means of the Palmgren-Miner linear cumulative damage hypothesis, which assumes that each stress cycle results in a small irreversible increase of the usage factor and that the total usage factor can be calculated by linear addition of the usage factor increments for the various cyclic stress levels.

In equation form:

$$U = \sum_i \frac{n_i}{N_i} \qquad (47)$$

where

U = usage factor,

n_i = number of cycles of occurrence at stress level i

N_i = number of cycles to failure at stress level i.

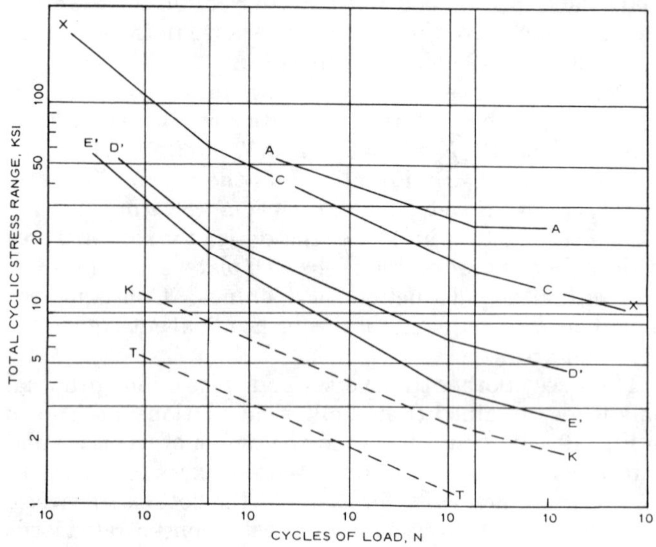

Stress categories for type and location of material for circular sections

Stress category	Situation
A	Plain unwelded pipe, and butt splices, no change in section, complete joint penetration groove welds, ground flush, and inspected by RT or UT (see Chapter VIII for inspection methods).
D′	Connections designed as simple T-, Y-, or K-connections with complete joint penetration groove welds.
E′	Simple T-, Y-, and K-connections with partial joint penetration groove welds or fillet welds; also, complex tubular connections in which the punching shear capacity of the main member cannot carry the entire load and load transfer is accomplished by overlap (negative eccentricity), gusset plates, ring stiffeners, etc.
K	Simple K-connections in which gamma ratio R/t of main member does not exceed 24.
T	Simple T- and Y-connections in which gamma ratio R/t of main member does not exceed 24.
X	Intersecting members at simple T-, Y-, and K-connections; any connection whose adequacy is determined by testing an accurately scaled model or by theoretical analysis (e.g., finite element), unreinforced cone-cylinder intersection.

Fig. 52 S-N curves for steel

A usage factor of 1.0 would correspond to fatigue failure, and in practical applications a safety factor is incorporated in designing for fatigue. For example, IMCO recommends a safety factor of 2.0 on fatigue for LNG structural components, so that their fatigue design criterion is

$$\sum_i \frac{n_i}{N_i} \leq 0.5 \tag{48}$$

All ships' hulls are subjected to fatigue-producing loadings during their service life. However, both experience and experiment involving ships in normal service indicate that fatigue is not an important consideration with respect to the primary hull-girder structure.

For example, data obtained on the *Ocean Vulcan* (MacNaught, 1967) show that for a twelve month period of North Atlantic service, for only the time recorded did the waves exceed 1.6 m (5 ft) in height, and that during the twelve months the sheerstrake stress experienced a stress of:

15.4 MPa (1 tons/in.^{2})	267,884 times
30.9 MPa (2 tons/in.^{2})	7,105 times
46.3 MPa (3 tons/in.^{2})	1,329 times
61.8 MPa (4 tons/in.^{2})	102 times
77.2 MPa (5 tons/in.^{2})	5 times
92.7 MPa (6 tons/in.^{2})	2 times

From the foregoing it can be seen that the storm conditions producing the stress peaks for which the ship must be designed occur for only an exceedingly small percentage of the ship's life. Therefore, the endurance limit of the hull-girder structure will not be approached in the normal service life under normal service conditions. Also, an analysis of the incidence of failures against the length of service made during the study of the epidemic of brittle fractures of World War II ships showed no increase in the incidence of fractures with length of service. If fatigue were a factor, incidence of failure would be expected to increase with length of service.

On the other hand, for ship members and details subject to repeated reversal of high stress, fatigue may be of paramount importance. Fatigue failures involve little, if any, plastic flow of the material prior to failure. Ductile fractures, on the contrary, are preceded by plastic flow and *necking down* of the material prior to failure.

Lacking a precise knowledge of the nature of fatigue failure in the steel of ship structures, there is always some question as to the merit of any proposal that the design stress level be raised. This is a serious problem in attempting to reduce costs of ships by reducing the scantlings of ship structures. Economic considerations include not only the higher cost of fatigue resistant materials but the additional costs associated with appropriate fabrication and welding techniques and possibly the provision of special crack arresters of various types.

The prevention of fatigue failure is of paramount importance in submarine design, where the designer must deal with the fatigue life of highly stressed details in the 20,000-cycle range. This relatively low fatigue life is commonly termed low-cycle fatigue. Since any fatigue crack will destroy the watertight integrity of the pressure hull or of any other structures, such as tank bulkheads which experience submergence pressure, and in some cases, may impair the capacity of the structure to resist applied loadings, special attention must be given to the fatigue resistance of the many critically stressed structural members and details of submarines.

The mechanisms involved with fatigue-crack initiation and propagation are not yet fully understood. Although the mechanisms may be different for low- and high-cycle fatigue phenomena, the engineering considerations are essentially the same. Fatigue failures are predominantly stress dependent and usually occur at regions of geometric discontinuity and where the nominal applied stresses are augmented by the stress concentration associated with the discontinuity.

A geometric discontinuity frequently is associated with a welded joint which introduces pyramiding adverse effects such as surface irregularities (especially at the toe of the weld), residual welding stresses, weld porosities and cracks, undercutting, slag inclusions and incomplete fusion of the material joined by the weld. These metallurgical faults and geometric discontinuities at the welded joint are an inherent source of fatigue cracks.

Tests have shown that fatigue failures occur in longitudinally oriented welds at approximately one-third the cycle endurance limit of the basic plating. The fatigue life of a butt-welded joint is much greater than that of a lapped joint and butt-welded joints are to be preferred at areas of structure subjected to a substantial number of stress reversals.

It is important to recognize that, in general, the higher the intensity of local stress at a discontinuity or stress concentrator of any kind, the lower the number of cycles that will be required to initiate and propagate a fatigue crack.

Pressure vessels which experience many cycles of change in temperature are subject to fatigue cracks which generally originate at a stress concentrator.

Prudence in design dictates that the total augmented stress at any geometrical discontinuity of structure should be limited to the elastic limit of medium-carbon steel. The many variables associated with the direction of applied loadings, residual stresses, other stress concentrations, and so on, preclude any guarantee that the conditions for which the structure at the discontinuity was designed will not be exceeded during its service life. A conservative approach and thorough investigation are therefore warranted in all cases involving design for fatigue loadings.

Stress concentrations can be reduced by judicious placement of adequate materials, easing of the geometrical discontinuities by providing a gradual transition fillet and, in the case of a penetration, by providing suitable local reinforcement at the opening. Although the provision of reinforcements at penetrations tends to reduce local stress concentration, it often creates a steep gradient adjacent to the reinforcement which may adversely affect the fatigue resistance of the structure.

There is always a question as to what constitutes a bal-

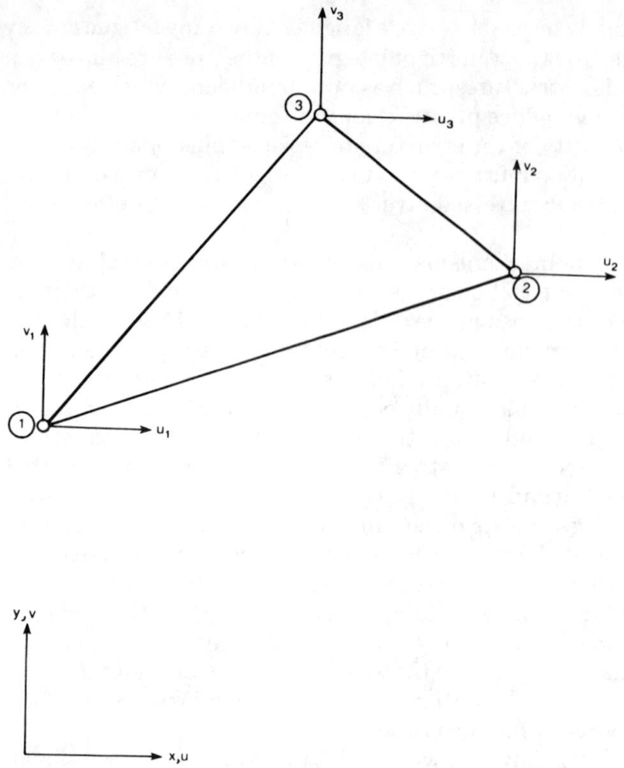

Fig. 53 Nodal displacements for the constant strain triangle

anced reinforcement. Theory provides partial answers for simple geometries in a simple loading system, but there are practical limitations. First, the theory is not valid in cases involving high stress when local plasticity occurs with a consequent hysteresis loop. Secondly, the theory does not consider the effect of interaction between the load-induced stress and the existing residual welding stress. When this residual stress is tensile (as may be the case at the surface of a welded joint), the interaction between the residual stress and the load-induced stress may have a marked effect on the fatigue resistance of the structural detail. In such a case, the metal could experience an alternating range in stress from tension to compression when the applied stress alone would indicate a stress range from zero to compression or to tension.

Another problem which may be encountered in some intersections of structural members is lamellar tearing or separation, which involves local material failure in the form of lamellae (layers) due to loadings applied normal to the material surface (see Subsection 4.4 of Chapter VIII). Even though this is not specifically a fatigue problem, it is a factor in the long-range resistance of the material at critical locations.

Well known techniques relating to post-treatment of welds at geometric discontinuities will do much to improve the fatigue resistance of highly stressed structural details involving welding. Two techniques worthy of mention are grinding the weld to a smooth contour to eliminate localized surface notches, especially at the toe of the weld, and peening the weld surface to produce a state of local residual surface compression. These techniques are known to be beneficial but to what extent they should be used depends on the importance of the structural detail and upon economic considerations. The designer must use his judgment, based on the literature and results of recognized experiments, to decide on the proper course of action for the problem at hand. There is no easy solution applicable to all conditions involving fatigue as applied to structural details.

7.2 Finite Element Example. The mathematical foundations of the finite element method are well-established and have been the subject of numerous textbooks such as Zienkiewicz (1971), Gallagher (1975), and Bathe and Wilson (1976), to mention a few. The finite element displacement method can be summarized by the following six basic steps:

1. Discretization of the physical structure into elements connected at nodal points,
2. generation of element stiffness matrices,
3. formulation of the complete structure stiffness matrix and applied forces,
4. application of boundary conditions,
5. nodal displacement solution of the resultant system of equilibrium equations,
6. calculation of strains and stresses based on the nodal displacements.

The first step of the method consists of reducing a structure to a mathematical model by discretizing the physical continuum into an assemblage of finite elements interconnected at discrete node points. The behavior of the finite element model can only approximate that of the real, physical model. A finite element model generally involves approximations both of the geometric form of the structure and of the displacements which it develops. The process and procedures used to discretize a structure into individual finite elements and select the mesh of element node points is termed "modeling technique."

The fact that the structural behavior of the finite element model is an approximation can best be illustrated in the development of the element stiffness matrix. The stiffness matrix represents the nodal load-displacement relationships of the element. It is formulated by defining the element displacement field in terms of the unknown nodal displacements and applying the principle of conservation of energy. The mechanics of the formulation can best be illustrated by an example.

The simplest and perhaps most versatile two-dimensional element is the plane stress three-node triangle, Fig. 53. To formulate the element stiffness matrix, an appropriate assumed displacement field is expressed in terms of polynomial equations which interpolate nodal displacements within the domain of the element. The simplest expression that can be used is a first order polynomial equation describing a linear displacement field. The behavior of the element, Fig. 54, is described by the six nodal displacements:

$$u_1, v_1, u_2, v_2, u_3, v_3 \tag{48}$$

The interior displacements u and v within the element may be described as a linear displacement field:

$$u = a_1 + a_2x + a_3y$$
$$v = a_4 + a_5x + a_6y \tag{49}$$

The six coefficients $a_1, \ldots, a_6$ in Equation (49) can be found by evaluating the u and v displacements at the three vertices of the triangle. Imposing the conditions that

$$u = u_1 \text{ and } v = v_1 \text{ at } (x_1, y_1)$$
$$u = u_2 \text{ and } v = v_2 \text{ at } (x_2, y_2) \tag{50}$$
$$u = u_3 \text{ and } v = v_3 \text{ at } (x_3, y_3)$$

in Equation (49), the unknown coefficients can be determined in terms of the coordinates and displacements at the nodes. Substituting these coefficient values back into Equation (49) gives

$$u = \frac{1}{2A}\{[y_{32}(x - x_2) - x_{32}(y - y_2)]u_1 + [-y_{31}(x - x_3) + x_{31}(y - y_3)]u_2 + [y_{21}(x - x_1) - x_{21}(y - y_1)]u_3\}$$

$$v = \frac{1}{2A}\{[y_{32}(x - x_2) - x_{32}(y - y_2)]v_1 + [-y_{31}(x - x_3) + x_{31}(y - y_3)]v_2 + [y_{21}(x - x_1) - x_{21}(y - y_1)]v_3\} \tag{51}$$

$$\text{where } x_{ij} = x_i - x_j,\ y_{ij} = y_i - y_j \tag{52}$$

$$2A = x_{32}y_{21} - x_{21}y_{32} = 2 \times \text{area of triangular element} \tag{53}$$

It follows from Equation (51) that the assumed displacements along any edge vary linearly, and they depend only on the displacements of the two vertices on that particular edge. Thus when two adjacent triangle elements are joined along a common edge, continuity of displacements is ensured.

Equation (51) can be used to find the relationship between the strains in the element and the six nodal displacements. Applying the strain-displacement equations of plane elasticity given by:

$$\epsilon_x = \frac{\partial u}{\partial y}, \quad \epsilon_y = \frac{\partial v}{\partial y}, \quad \epsilon_{xy} = \frac{\partial u}{\partial y} + \frac{\partial v}{\partial x} \tag{54}$$

Evaluating Equation 54, the strains can be expressed in matrix form as

$$\begin{bmatrix} \epsilon_x \\ \epsilon_y \\ \epsilon_{xy} \end{bmatrix} = \frac{1}{2A}\begin{bmatrix} y_{32} & 0 & -y_{31} & 0 & y_{21} & 0 \\ 0 & -x_{32} & 0 & x_{31} & 0 & -x_{21} \\ -x_{32} & y_{32} & x_{32} & -y_{31} & -x_{21} & y_{21} \end{bmatrix}\begin{bmatrix} u_1 \\ v_1 \\ u_2 \\ v_2 \\ u_3 \\ v_3 \end{bmatrix} \tag{55}$$

or symbolically

$$[\epsilon] = [B][\delta] \tag{56}$$

where $[\epsilon]$, $[B]$, $[\delta]$ are the strain, strain-displacement and nodal displacement matrices, respectively. Equation (56)

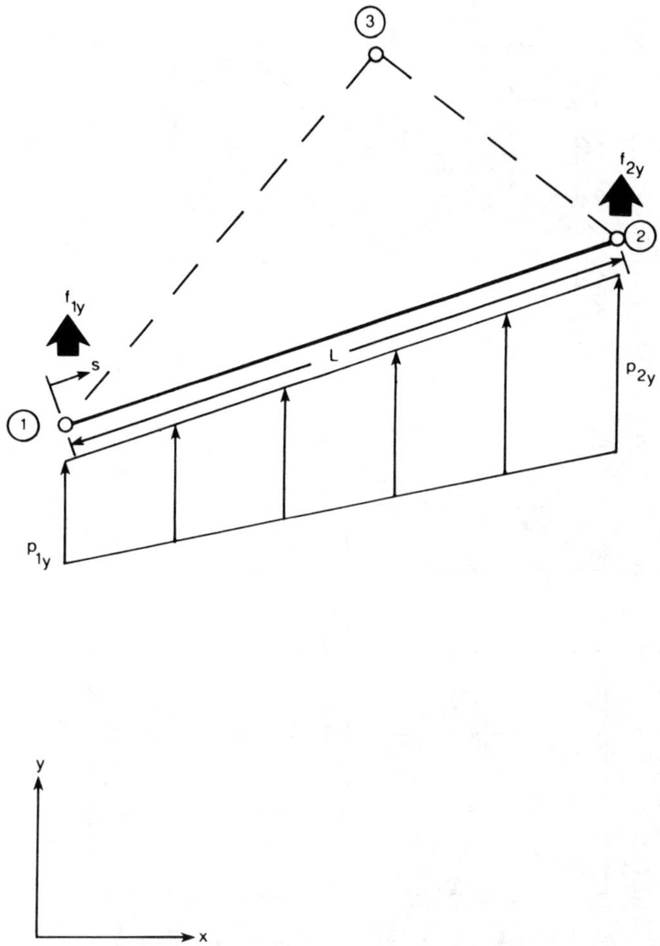

Fig. 54 Distributed vertical load acting on Side 1–2 represented by equivalent nodal loads

symbolizes the transformation from nodal displacements to strains. Equation (55) shows that the assumption of a linear displacement field in the triangular element leads to a constant strain, as well as constant stress since the $[B]$ matrix is constant. This element is therefore commonly referred to as the constant strain triangle.

The next step in the element formulation is to establish the relationship between the stress field and the nodal displacements. For a two dimensional isotropic material, the mechanical properties can be expressed by *Hooke's Law* relating stress to strain. For the case of plane stress:

$$\begin{bmatrix} \sigma_x \\ \sigma_y \\ \sigma_{xy} \end{bmatrix} = \frac{E}{1 - \mu^2}\begin{bmatrix} 1 & \mu & 0 \\ \mu & 1 & 0 \\ 0 & 0 & \dfrac{1 - \mu}{2} \end{bmatrix}\begin{bmatrix} \epsilon_x \\ \epsilon_y \\ \epsilon_{xy} \end{bmatrix} \tag{57}$$

or symbolically

$$[\sigma] = [D][\epsilon]$$

where

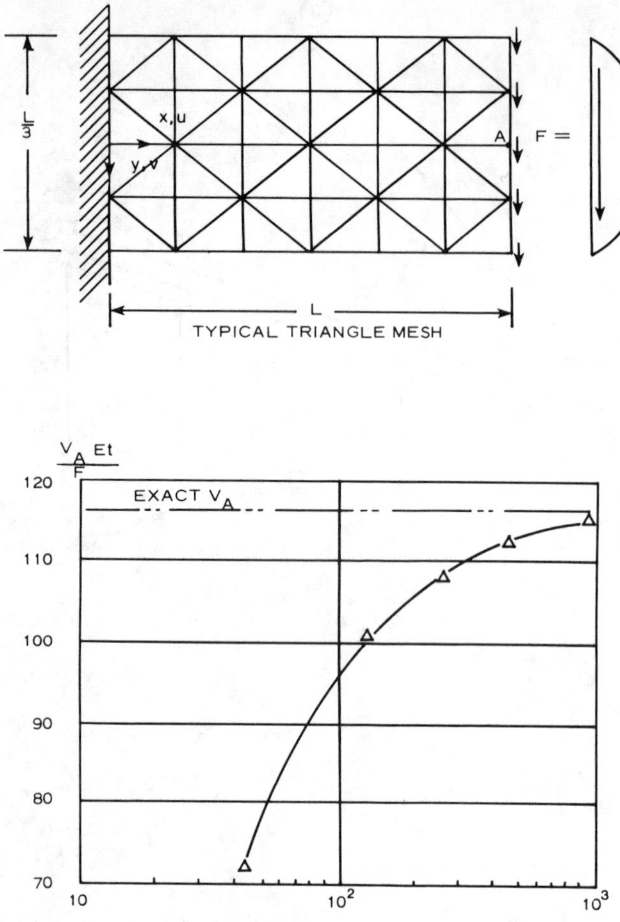

Fig. 55 Cantilever beam analysis using constant strain triangle

μ = Poisson's ratio,
σ = stress,
$[D]$ = material stiffness matrix.

The strains from Equation (55) can be substituted into Equation (57) to give the stress-displacement relationship:

$$[\sigma] = [D][B][\delta] \tag{58}$$

The strain-displacement Equation (55) and stress-strain Equation (57) from the theory of elasticity are fundamental in developing the element stiffness matrix. Formulation of the element stiffness is best explained using work and energy considerations. When loads are applied to the element, strains and stresses develop in the element. As it deflects, the loads do external work, W, and strain energy, U, is stored in the element due to deformation. Applying the principle of conservation of energy:

$$U = W \tag{59}$$

Expressing U and W in terms of the element stiffness, nodal displacement and external nodal load matrices $[k]$, $[\delta]$ and $[f]$, Equation (59) becomes the well-known equilibrium condition

$$[f] = [k][\delta] \tag{60}$$

For the constant strain triangle, the above equation represents a system of six simultaneous equations relating the six component nodal displacements (also called the element degrees of freedom) to six corresponding nodal forces. The matrices $[k]$ and $[f]$ are known whereas $[\delta]$ is unknown and remains to be solved.

In applying external loads to the element, the loads must act at the nodal points. Where the external loads are distributed loads, equivalent nodal loads must be calculated to be used in Equation (60). For the constant strain triangle, the displacement variation along a side is linear and is dependent only on the displacement of the two end points. Therefore, for a linearly distributed loading acting in the y-direction along side 1–2 shown in Fig. 54, this loading can be expressed as

$$p_y = p_{1y}\left(1 - \frac{s}{L}\right) + p_{2y}\left(\frac{s}{L}\right) \tag{61}$$

where s = distance along side 1–2.

Since the displacement variation is linear,

$$v = v_1\left(1 - \frac{s}{L}\right) + v_2\left(\frac{s}{L}\right) \tag{62}$$

The work done by the nodal loads at points 1 and 2, f_{1y} and f_{2y}, must be equivalent to the work done by the distributed loads. Hence

$$f_{1y}v_1 + f_{2y}v_2 = \int_0^L v p_y ds \tag{63}$$

Substituting Equations (61) and (62) into (63) leads to the following work-equivalent or consistent nodal loads,

$$f_{1y} = \frac{L}{6}(2p_{1y} + p_{2y}) \tag{64}$$

$$f_{2y} = \frac{L}{6}(p_{1y} + 2p_{2y}) \tag{65}$$

The consistent nodal loads given by Equations (64) and (65) are the static resultants of the linear load. This is true only because the displacement variation is linear.

The next step in the finite element method is the formulation of the stiffness matrix for the entire structure. This requires the assemblage of the individual element stiffness matrices $[k]$ into the structure stiffness matrix $[K]$. The assemblage process can be symbolically written

$$[K] = \sum_i [k_i] \tag{66}$$

where the matrix $[k_i]$ is the stiffness matrix of the ith element expressed in a common (global) system of coordinates and the summation extends over all elements in the assemblage. The mechanics of carrying out the assembly of the element stiffness matrices into the complete structure stiffness matrix involves summing for each node of the structure the corresponding stiffness contribution of the elements having that node in common.

In an analogous manner the structure nodal loads $[F]$ are

assembled from the element nodal loads

$$[F] = \sum_i [f_i] \qquad (67)$$

For analysis of the complete structure, the element nodal displacements $[\delta]$ are assembled into a set of overall nodal displacements which shall be defined as $[\Delta]$. Equilibrium of the complete structure requires that

$$[F] = [K][\Delta] \qquad (68)$$

The next step in the analysis is to impose the displacement boundary conditions on the physical structure. We can rewrite Equation (68) in a partitioned form

$$\begin{bmatrix} F_u \\ F_b \end{bmatrix} = \begin{bmatrix} K_{uu} & K_{ub} \\ K_{bu} & K_{bb} \end{bmatrix} \begin{bmatrix} \Delta_u \\ \Delta_b \end{bmatrix} \qquad (69)$$

where Δ_u are the unknown displacements and Δ_b are the known boundary displacements. From Equation (69) we have

$$[K_{uu}][\Delta_u] = [F_u] - [K_{ub}][\Delta_b] \qquad (70)$$

The unknown nodal displacements $[\Delta_u]$ are calculated from Equation (70), symbolically written

$$[\Delta_u] = [K_{uu}]^{-1}[F_u] - [K_{uu}]^{-1}[K_{ub}][\Delta_b] \qquad (71)$$

In practice the solution to Equation (70) is commonly obtained by a Gaussian elimination procedure. The boundary reaction forces corresponding to $[\Delta_b]$ are then

$$[F_b] = [K_{bu}][\Delta_u] + [K_{bb}][\Delta_b] \qquad (72)$$

Once the nodal displacements have been obtained, element stresses are calculated using Equation (58).

The foregoing briefly presents the formulation and fundamentals of the finite element method using one of the simpler finite elements. The primary advantage of the finite element method over other analysis techniques is its generality and applicability to digital computer implementation. Using the constant strain triangle as an example, a structure of any plane geometric form consisting entirely of these triangles requires only a repetitive application of the element stiffness routine with only the nodal coordinate values being changed. The reliance on the use of matrix methods is also highly adaptive to digital computer applications.

In general, the accuracy of a finite-element analysis can be improved if a more refined mesh is used. This is particularly true for areas of high stress gradients. This can be realized when we consider the constant strain triangle in which the stress field is constant. More nodes and hence elements would obviously better represent the high stress gradients. The convergence to the exact solution with increasing number of elements is shown in Fig. 55 for the cantilever beam modeled by constant strain triangles. The beam is subjected to an end load applied as a parabolically varying shear stress, and loads are assigned to the nodes in a work-equivalent manner employing Equation (63). Results for the vertical displacement at mid-depth at the end of the beam are shown as a function of the number of degrees of freedom for the beam model.

With a more exact representation of the element displacements by means of a higher order polynomial (using a linear strain element, for example), the same accuracy can be obtained by using a smaller number of elements for the same structure.

References

American Bureau of Shipping (1973), *Rules for Building and Classing Barges for Offshore Service.*

American Bureau of Shipping (1978), *Rules for Building and Classing Bulk Carriers for Service on the Great Lakes.*

American Bureau of Shipping (Annual), *Rules for Building and Classing Steel Vessels.*

American Welding Society (1980), "Structural Welding Code," Publication AWS D1.1-80.

Arnott, D. (1955), "Development of Ship's Structures" in *Design and Construction of Steel Merchant Ships,* SNAME.

Band, E. G. U. (1966), "Analysis of Ship Data to Predict Long-Term Trend of Hull Bending Moments," American Bureau of Shipping Report, November.

Bannerman, D. B., Jr. and R. T. Young (1946), "Some Improvements Resulting from Studies of Welded Ship Failures," Welding Journal, American Welding Society, Vol. 25, No. 3, March.

Bathe, K. J., and E. L. Wilson (1976), "*Numerical Methods in Finite Element Analysis,*" Prentice-Hall, Inc., Englewood Cliffs, New Jersey.

Boylston, J. W., D. J. DeKoff, and J. J. Mintzewerf (1974), "SL-7 Containerships: Design, Construction and Operational Experience," *Trans.* SNAME.

Caldwell, J. B. (1965), "Ultimate Longitudinal Strength," *Trans.* RINA.

Chen, Henry, H. H. Chen, and D. Hoffman (1979), "The Implementation of the 20-Year Hindcast Wave Data in the Design and Operation of Marine Structures," Offshore Technology Conference, Houston.

Cox, P. A., E. B. Bowles, and R. L. Bass (1979), "Evaluation of Liquid Dynamic Loads in Slack LNG Cargo Tanks," SSC Report, SR 1251.

Cummins, W. E., and W. L. Bales (1980), "Extreme Value and Rare Occurrence Wave Statistics for Northern Hemispheric Shipping Lanes," SNAME Spring Meeting/STAR Symposium, Coronado, Calif., June.

Elbatouti, A. M. T., H. Y. Jan and S. G. Stiansen (1976), "Structural Analysis of a Containership Steel Model and Comparison with the Test Results," *Trans.* SNAME.

Foley, J. L. (1965), "Barges in Ocean Service," *Trans.* SNAME.

Fukuda, J., et al (1971), "Theoretical Calculation on the Motions, Hull Surface Pressures and Transverse Strength of a Ship in Waves," *Journal of the Society of Naval Architects of Japan,* Vol. 129, June.

Gallagher, R. H. (1975), *Finite Element Analysis Fundamentals,* Prentice-Hall, Inc., Englewood Cliffs, New Jersey.

Goodman, R. A. (1971), "Wave Excited Main Hull Vibration in Large Tankers and Bulk Carriers," *Trans.* RINA.

Hannan, W. M., H. Y. Jan, and A. Bakker (1972), "The Application of Computer Techniques in Hull Structural Analysis," Symposium on the Computer in Finite Element Analysis of Ship Structures, University of Arizona.

Henry, J. J. and H. J. Karsch (1966), "Container Ships, *Trans.* SNAME.

Hogben H., and F. E. Lumb (1967), "Ocean Wave Statistics," Her Majesty's Stationery Office, London.

Hoffman, D., and H. H. Chen (1978), "The Use of Directional Hindcast Spectra as Design Wave Data," Offshore Technology Conference, Houston,

Hoffman, D., and E. V. Lewis (1969), "Analysis and Interpretation of Full-Scale Data on Midship Bending Stresses of Dry Cargo Ships," Ship Structure Committee Report, SSC-196, June.

IMCO (1973), "International Convention for the Prevention of Pollution from Ships," Sales No. 74.01.E, and the Protocol of 1978 Relating to that Convention.

IMCO (1974), "International Convention for the Safety of Life at Sea," Publication No. 75.01.E.

IMCO (1966), "International Convention on Load Lines," London.

ISSC Report of Committee I-3 (1979), *ISSC Proceedings*.

Jasper, N. H. (1955), "Temperature-Induced Stresses in Beams and Ships," DTMB Report 937, June.

Lewis, E. V. (1967), "Predicting Long-Term Distributions of Wave-Induced Bending Moment on Ship Hulls," SNAME Spring Meeting *Proceedings*.

Lewis, E. V. et al (1973), "Load Criteria for Ship Structure Design," Ship Structure Committee Report, SSC-240.

Little, R. S., and E. V. Lewis (1971), "A Statistical Study of Wave-Induced Bending Moments on Large Oceangoing Tankers and Bulk Carriers," *Trans.*, SNAME.

Little, R. S., and S. G. Stiansen (1971), "Correlation of Full-Scale Experimental and Computed Stresses of Main Supporting Structure on Large Tankers," 16th Annual Tanker Conference, Transportation Division, American Petroleum Institute, Ponte Vedra Beach, Florida.

Liu, D., and M. E. Wojnarowski (1976), "Computer Analysis and Evaluation of Marine Structures," National Computer Conference, New York, AFIPS Conference Proceedings, Vol. 45.

MacNaught, D. F. (1967), Strength of Ships in *Principles of Naval Architecture,* edited by J. P. Comstock, SNAME.

Mansour, A., and J. M. d'Oliveria (1975), "Hull Bending Moment Due to Ship Bottom Slamming in Regular Waves," *Journal of Ship Research,* June.

Murray, J. M. (1955), "A Hundred Years of Lloyd's Register Ship Rules," *Trans.* RINA.

Nordenstrom, N. (1963), "On Estimation of Long-Term Distributions of Wave-Induced Midship Bending Moments in Ships," Chalmers Tekniska Hogskola.

Ochi, M. K. (1978), "Wave Statistics for the Design of Ships and Ocean Structures," *Trans.*, SNAME.

Ochi, M. K., and L. E. Motter (1973), "Prediction of Slamming Characteristics and Hull Responses for Ship Design," *Trans.*, SNAME.

Schade, H. A. (1966), "Two-Beam Deckhouse Theory with Shear Effects," *Schiff and Hafen,* Vol. 18, Heft 5, May.

Stavovy, A., and S. L. Chuang (1976), "Analytical Determination of Slamming Pressures for High Speed Vehicles in Waves," *Journal of Ship Research,* December.

Steinman, G. C., and T. E. Carman (1965), "Design Considerations for Barges Transporting Hazardous Commodities," *Marine Technology,* October.

Stiansen, S. G., and H. H. Chen (1979), "Application of Probabilistic Design Methods to Wave Loads Prediction for Ships Structures Analysis," Report prepared for SNAME Panel HS-4 (Design Procedure and Philosophy), November.

Stiansen, S. G., H. Y. Jan, and D. Liu (1979), "Dynamic Stress Correlation for the SL-7 Containership," *Trans.*, SNAME.

Stiansen, S. G., A. Mansour, and Y. N. Chen (1977), "Dynamic Response of Large Great Lakes Bulk Carriers to Wave-Excited Loads," *Trans.*, SNAME.

Yamanouchi, Y. et al (1965), "Statistical Investigations on Wind and Waves in North Pacific Ocean and Coastal Seas of Japan," Shipbuilding Research Vol. 7, No. 2, November.

Zienkiewicz, O. C. (1971), *The Finite Element Method in Engineering Science,* McGraw-Hill Book Company, New York.

CHAPTER VII

Stanley G. Stiansen

Structural Components

Section 1
The Function of Ship Structural Components

1.1 Definitions. The strength deck, bottom, and side shell of a ship act as a box girder in resisting bending and other loads imposed on the structure. The weather deck, bottom, and side shell also form a tight envelope to withstand the sea locally, and to provide the buoyancy which keeps the ship afloat. The remaining structure contributes either directly to these functions, or indirectly by maintaining the main members in position so that they can act efficiently.

The bottom plating is a principal longitudinal member constituting the lower flange of the hull girder. It is also part of the watertight envelope, and subject to the local water head. At the forward end, it must withstand the additional dynamic pressure associated with slamming, and there the plating thickness is usually increased to provide the necessary strength.

When fitted, the inner bottom also makes a significant contribution to the strength of the lower flange. The inner bottom and bottom shell, together with the bottom floors and girders, work as a double-plate panel to distribute the secondary bending effects caused by hydrostatic loads and cargo loads to main supporting boundaries, i.e., bulkheads or the side shell. The inner bottom provides local support when it forms a tank boundary for the double-bottom tanks and is subject to the local pressures of the liquid contained therein. In addition, it is subject to the local loads from above, usually from cargo placed in the holds.

One or more strength decks form the principal members of the upper flange, usually provide the upper watertight boundary, and are subject locally to water, cargo, and equipment loadings. The remaining decks, depending upon their extent in the longitudinal direction, their distance from the neutral axis of the hull, and their effective attachment to the main hull, contribute to a greater or lesser extent in resisting the longitudinal bending loads. Locally, internal decks are subject to the loads of cargo, equipment, stores, living spaces, and, where they form a tank boundary or barrier against progressive flooding, liquid pressure.

The side shell provides the webs for the main hull girder and is an important part of the watertight envelope. It is subject to static water pressures, as well as the dynamic effects of pitching, rolling, and wave action. Particularly forward, the plating must be able to withstand the impact of the seas. Aft, extra plate thickness is beneficial in way of rudders, shaft struts, and stern tubes for strength, panel stiffness, and reduction of vibration. Additional thickness is necessary between the maximum winter and minimum service waterlines for navigation in ice and, more locally, for resisting the loads imposed by striking quays, piers, locks, and vessels alongside.

Bulkheads are one of the major components of internal structure. Their function in the hull girder depends on their orientation and extent. Main transverse bulkheads act as internal stiffening diaphragms for the girder and resist in-plane torsion loads, or *racking* loads, but do not contribute directly to longitudinal strength. Longitudinal bulkheads, on the other hand, if extending more than about one-tenth of the length of the ship, do contribute to longitudinal strength, and in some ships are nearly as effective as the side shell itself. Bulkheads generally serve other structural functions, such as forming tank boundaries, supporting decks and load-producing equipment such as king-posts, and adding rigidity to reduce vibration. In addition, transverse bulkheads provide subdivision to prevent progressive flooding. All applicable loads must be considered during design.

The foregoing structural elements of a ship are basically large sheets of plate whose thicknesses are very small compared with their other dimensions, and which, in general, carry loads both in and normal to their plane. These sheets of plate may be flat or curved, but in either case they must be stiffened in order to perform their required function efficiently. Corrugated bulkheads, stiffened by the corrugations, may also be used.

The various stiffening members have several functions: the beams stiffen the deck plating; the girders, in turn, support the beams, transferring the load to the pillars or bulkheads; for transverse framing, the transverse beams stiffen the side shell and support the ends of transverse deck beams and are, in turn, supported by the decks and stringers; for longitudinal framing, the frames supporting the plating run fore-and-aft and are in turn supported by transverse members. As discussed in detail in Section 5, the stiffening members are generally rolled, extruded, flanged, flat, or built-up plate sections with one edge attached to the plate they reinforce.

Vertical plates connect the bottom shell and inner bottom. Those fitted transversely are called *floors,* and those fitted longitudinally are called *center girders* or *side girders*, as appropriate.

1.2 Interaction of Structural Components. Stiffening members do not, of course, act independently of the plating to which they are attached. A portion of the plate serves as one flange of the stiffener, and properties such as section modulus and moment of inertia used in the strength analysis of the stiffener must reflect this.

Stiffening members serve two functions, depending on how they are loaded. In the case of loads normal to the plate, such as fluid loading on a transverse bulkhead, the stiffeners provide edge restraint for the plate. In the case of in-plane loads, such as those induced in the deck by longitudinal bending of the hull girder, the beams serve to maintain the deck plating in its designed shape. If the deck beams are longitudinally fitted, they will, of course, carry the same hull bending stress as the plating, and may contribute substantially to the hull girder strength.

The decks, side shell, inner bottom, bottom, and bulkheads interact to provide overall edge restraint for each other. For example, a transverse bulkhead's ultimate support is provided by the side shell, decks, and bottom. At the same time, the bulkhead provides edge restraint for the large stiffened plate panels of the decks, side shell, and bottom which span between major transverse structural elements such as bulkheads. This interaction causes a complex stress pattern at stiffened plate intersections.

Pillars are used to support deck girders or deck transverses. These supports, in addition to carrying local loads from cargo, equipment, etc., serve to keep the deck and bottom from moving toward each other as a result of longitudinal bending of the hull girder.

In general, the concept of which structural component supports which other structural component is a simplified description of the actual structural interaction. On a ship or any other structure, all the elements tend to act together to provide the proper support and to carry the loads for which they are designed. This structural interaction, which in general can be very complex, can be very well represented by comprehensive finite-element three-dimensional mathematical models analyzed with the help of structural computer programs.

Section 2
Design Philosophy and Procedures

2.1 Design Based Upon Engineering Calculations. Since the classification rules do not cover all design aspects in specifics, and in order to encourage innovative designs, most of the classification societies will review under special consideration any design supported by rational calculations. Such design may deviate from the published classification rules, and yet be accepted if the supporting engineering analysis proves it to be structurally sound. For example, according to the ABS Classification Rules (American Bureau of Shipping, Annual),[1] alternative arrangements and scantlings will be considered if "they can be shown through . . . a systematic analysis based on sound engineering principles, to meet the overall safety and strength standards of the Rules." The design procedure, in this case, combines both intuition based on past design experience and structural analysis aimed at determining satisfactory structural response.

Nowadays, many computer programs are available for rational engineering analyses. The number and capabilities of these programs are constantly increasing. Typical capabilities cover various types of analyses, such as small displacement, large displacement, incremental plasticity, creep, thermal effects, temperature-dependent materials, natural frequencies, mode shapes, transient response and structural instability.

The complexity of engineering problems encountered in the marine industry has led to extensive and ever-expanding computer usage. In addition to the conventional static and dynamic problems of design, the problems encountered in assessing marine structural response are compounded by the unpredictable nature of sea, and sometimes cargo loads. Special phenomena associated with the dynamic interaction of waves and ships at sea must be taken into account. Springing, for example, is a vibration of the complete vessel induced by the wave frequency in conjunction with the ship's elastic properties. Other areas of concern are local vibrations, which may be induced by waves or action of the propeller and drive shafts. Other loading conditions include those due to thermal effects, sloshing of liquid in cargo tanks, bottom slamming, and sea ice.

Mathematical techniques, such as matrix methods, finite-element methods, and statistics, have been available for a long time. The advent of electronic digital computers makes possible the full utilization and implementation of these techniques in an efficient engineering approach to the solution of the numerous problems associated with the design, construction, and analysis of ships and other marine structures.

The computer allows a more rigorous determination of structural response to the specified loads than has been previously possible, even though the load specification remains somewhat indeterminate. In effect, the use of computers lessens the need to make simplifying assumptions in one area of the total problem, and hence the accuracy of the final solution is improved, even though a certain level of possible inaccuracy relating to the uncertainty of the load input must still be acknowledged.

2.2 Optimum Design Using Numerical Methods. Rational design of ship structures has forced the designer to determine quantitatively as many as possible of the factors affecting the safety and performance of the structure

[1] Complete references are listed at end of chapter.

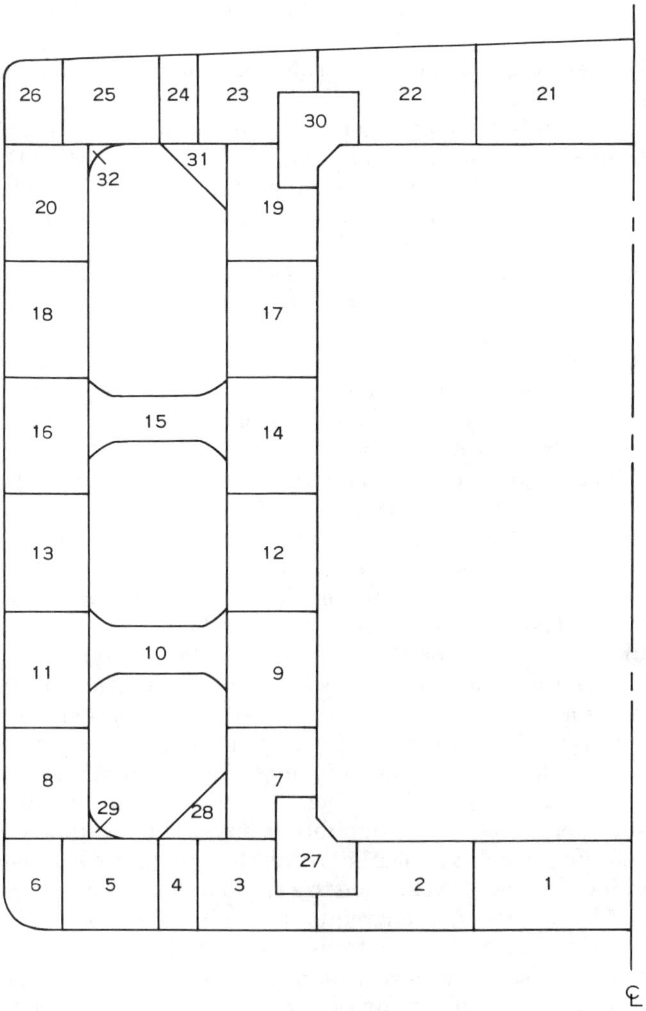

Fig. 1 Zones of constant thickness for structural optimization

throughout its life, and to use this information to determine that particular design which optimizes performance and provides adequate safety. This process involves many calculations, but the use of a computer can simplify the task, providing an automated optimum rational design.

Ship structural optimization is a complex task involving structural analysis to determine the response to multiple loads, and application of optimization techniques for resizing of the individual structural members to achieve an optimum design. Different optimization techniques combined with various structural analysis methods for ship design have been used.

Structural optimization determines the design variables which will minimize (or maximize) a specified objective function, while satisfying a constraint condition. Typically, the objective function is the weight or the cost of the structure, and the constraints are the stresses, displacements, or other response characteristics. Various optimization methods and their theoretical backgrounds are described in Chapter VI.

An example of optimization design for the web frame of a tanker is shown in Figs. 1 and 2. The web frame was divided into zones of constant thickness, as indicated by each numbered panel in Fig. 1. The extent of this decomposition is arbitrary, and it depends on the size and availability of steel plates, convenience, ease of construction, etc.

An optimality criterion based on fully-stressed design was used to minimize the weight of the web frame with a double-iteration procedure developed for the efficient use of the finite-element analysis in the optimization program. The shell, deck, and bottom plating which are attached to the web frame were not allowed to vary in the optimization procedure, since their thicknesses are determined from longitudinal strength requirements and other considerations.

The main conclusion is that it is possible to minimize the weight of the web frame, with the reduction in weight being dependent on the minimum allowable plate thickness, not on the allowable stress in the web frame. Three different thickness requirements were used, and the corresponding reductions in weight are shown in Fig. 2.

The optimization procedure is general and is applicable to any web frame or similar structure. Possible extensions of the procedure could relate the number of stiffeners required to prevent shear buckling and vibration to the minimum plate thickness of the webs.

2.3 Classification Society Rules. Although direct engineering design might be preferable, the design of structural members of merchant ships is greatly influenced by the rules

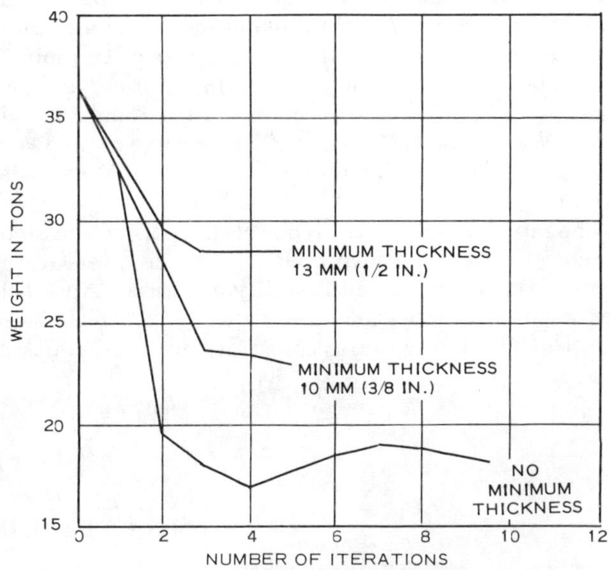

Fig. 2 Reductions in weight using structural optimization

of classification societies; in fact, the principal scantlings of most merchant ships are based directly on these rules. Classification societies were created to serve the marine industry by establishing certain standards which provide assurance that a vessel possesses the structural and mechanical fitness for its intended service. Classification societies were recognized by governments under a provision of the 1930 Load Line Convention and were urged to confer from time to time "... with a view to securing as much uniformity as possible in the application of the standards of strength on which freeboard is based."

Since the 1930, Load Lines Convention was signed, great changes have occurred in ship design and construction, shipbuilding technology, and ship operation. New types of closing appliances, in particular metal hatch covers, have improved the watertight integrity of ships. Other technical developments (the extensive use of welding, rounded gunwales, etc.) have also become widespread. The vast increase in the size of ships, particularly tankers and bulk carriers, has made it necessary to extend the existing freeboard tables to cover ships up to a length of 366 m (1200 ft). All these considerations, together with the experience gained from the use of the 1930 Convention, merited a thorough examination with a view to the adoption of an up-to-date Convention on Load Lines.

Another international conference on load lines was therefore convened in 1966, in order to draft a new convention and bring the load line regulations into accord with the latest developments and techniques in ship construction. The regulations developed by this convention are now in force. Since the societies were not only recognized as the source of strength requirements, but were also designated as Load Line Assigning Authorities by their respective governments and the governments of many other countries, application of the Load Line Convention was the major subject of discussion, with a view to achieving the intended uniform treatment of all ships under its provisions. It was only natural, however, that other subjects of common interest should be discussed and, as the shipping and shipbuilding industries became more international following World War II, the desirability of some degree of uniformity among the classification requirements became readily apparent.

The rapid increase in the scope of the cooperation among the societies, and the creation in 1959 of the Inter-Governmental Maritime Consultative Organization (IMCO), led classification societies to join as a group to establish liaison with IMCO. This group is called the International Association of Classification Societies (IACS) and was formed in 1968. As of mid-1979, it had nine members and three associate members. Its work falls into two general categories: development of uniform classification rules called Unified Requirements, and collaboration with outside organizations.

The unification of classification requirements is a long-term effort. To date, more than one hundred Unified Requirements have been adopted by all the members, ranging from uniform specifications for hull steel to the maximum steam temperature in tanker pump rooms. Unified Requirements developed by the various working parties or correspondence groups are submitted to the IACS Council for approval. Following this, they are subject to the normal rule-making procedure of each classification society before they may be incorporated into the Rules, since the governing bodies of the individual societies still retain control over their own rules.

The majority of merchant vessels are classed under the Rules of the American Bureau of Shipping (ABS) or Lloyd's Register of Shipping (LR). These rules are listed as ABS and Lloyds Rules, and most of the specific rule requirements in the sections that follow are taken from these two sources. Other classification societies have similar rules, and sometimes several standards are applied to the same ship.

Classification society rules contain a great deal of useful information relating to the design and construction of the ship's various structural components, so that determinations can be made of some rule scantlings, i.e., the dimensions of a ship's frames, girders, plating, etc., directly from equations and tables given in these publications. The possibility of choosing structural members directly from tables has been reduced in recent years due to a trend for classification societies to present their requirements in terms of section properties rather than in terms of actual sizes required. In many cases, for various structural components, the classification rules indicate by sketches and descriptive matter *good-practice* construction methods for the designer.

In recent years, modifications to Rule thicknesses have been permitted when high-strength steel is used or where special protective preservative coatings are used to reduce the need for corrosion allowances. In such cases, careful consideration must be given to plate instability, since standard Rule panel sizes are based on full Rule plate thicknesses. Other modifications and special requirements have been incorporated for vessels intended to carry oil, ore, bulk cargos, liquefied gases, etc. The study and use of these rules is essential in merchant ship structural design.

Section 3
Relation of Structure to Molded Lines

3.1 Nature of Molded Lines. The *molded lines* are a delineation of the ship's form. They are first drawn to relatively small scale in the early stages of a ship's design. In order to construct the ship, these small-scale lines must be redrawn at a large enough scale to define the shape accurately so that structure may be cut and formed, systems located and designed, etc. Traditionally, this larger scale was full size, the lines being laid down on a mold loft floor and templates made from the full-scale lines for use in guiding cutting and forming operations. This system is still used, but in many large shipyards the full-size mold loft has been replaced by other systems, as described in Chapter XVI.

3.2 Need for Molded Lines. Lines, however produced, define mathematical *surfaces* of no thickness rather than the actual shell, deck and bulkhead plating, frames, longitudinals, and so on. For these structures to fit together when assembled, the surfaces represented by the molded lines must be clearly defined and due allowance made for the thickness of the structure. In order to do this, when the loftsmen and ship fitters undertake to transform detailed working drawings of the ship's structure into actual structural members, they must have very definite understandings regarding the relation of the structure to the molded lines. The molded line represents the junction between the shell, deck, and bottom plating and the supporting structure; hence, one line defines the inner line of the shell, deck, and bottom as well as the outer line of the supporting structure. Conventional relationships in this respect have developed which are fairly standard throughout the industry. Figs. 3 and 4 illustrate some of these practices.

The spatial disposition of the molded lines of the hull and internal structure is described by their vertical distances from the horizontal reference plane indicated on drawings as the base line, and by their horizontal distances from the vertical reference planes, one located at the longitudinal centerline of the vessel and one located transversely at the mid-length of the vessel. The horizontal reference plane generally coincides with the molded line of the horizontal bottom-shell plating.

3.3 Molded Line. *a. Shell Plating.* The inner surface of welded shell plating is usually flush and is on the molded line. This arrangement eliminates the necessity of joggling shell frames crossing welded seams where the plates vary in thickness.

b. Double Bottoms. The underside of the inner-bottom plating is usually flush, and is placed on the molded line.

The vertical keel plate is located on the centerline of the hull, with half its thickness on each side. Side longitudinals and a sloping margin plate are normally set with the inboard side of the plate to the molded dimension.

c. Deck Plating. The underside of deck plating is normally set to the molded line of the deck with differences in thickness showing above the deck. Where the stringer strake is thicker than the remainder, this results in a ledge at the inner edge of the stringer strake which, as seen in the midship section, would seem to interfere with drainage, but actually the sheer and camber of the deck are such that this condition is not objectionable.

When unusual thicknesses of deck plating are used, it is not uncommon to make special definitions of the molded line of the deck plating to suit conditions. For instance, in naval vessels where 6 mm (0.25 in.) plating may be used at the ends in association with heavy plating amidships, the molded line sometimes has been run 6 mm below the top of the plating throughout the deck, regardless of the thickness of the plating. Similarly, if a thin deck covering is used, such as rubber tile, with a very minimum of leveler beneath it on a deck with substantial differences in plate thickness, the plating may be arranged to be flush on top with the differences in thickness showing on the underside of the plating. Where such departures from usual practice of a shipyard are followed, they will be defined on the structural drawings.

d. Bulkhead Plating. Transverse bulkhead plating

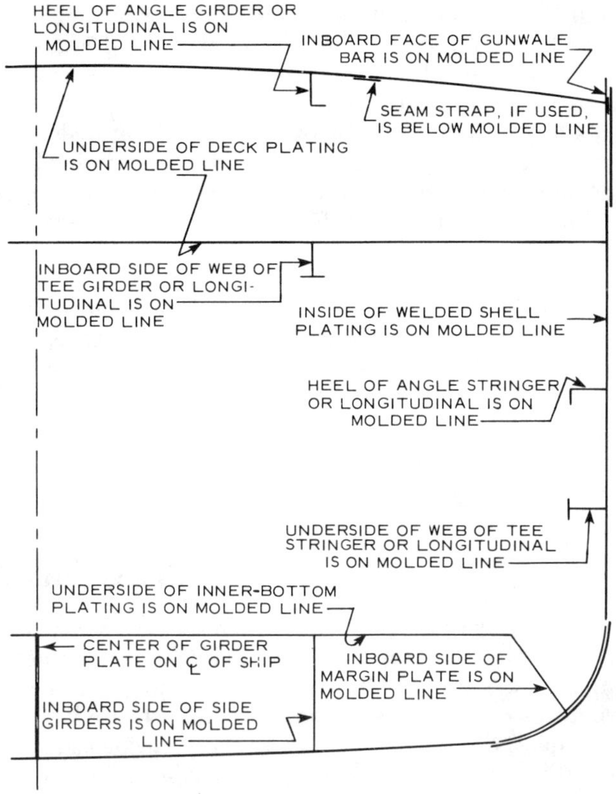

Fig. 3 Relation of longitudinal structure to molded lines

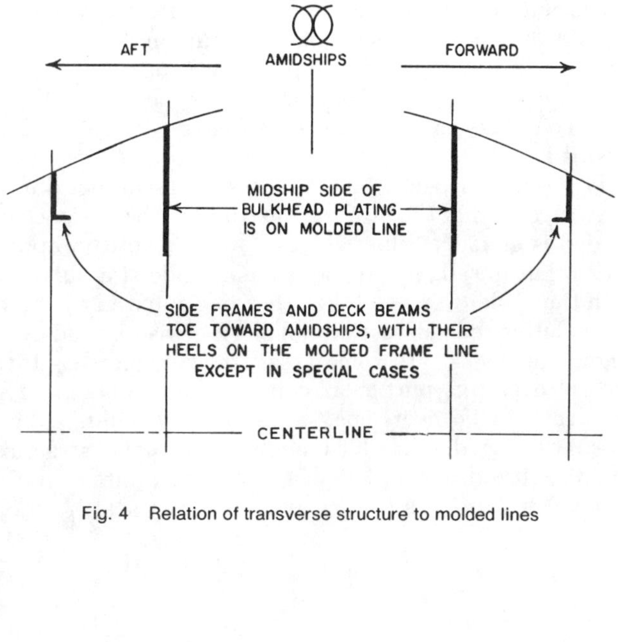

Fig. 4 Relation of transverse structure to molded lines

usually has its after surface on the molded frame line in the forebody and its forward surface on the molded frame line in the afterbody. Longitudinal bulkhead plating generally has its inboard surface on the molded line.

If a bulkhead has varying thicknesses of plating and the stiffeners are located on the side away from the molded line, the bulkhead can be made flush on the stiffener side to avoid joggling or notching the stiffeners, and the thinner plates can be moved from the molded line.

e. Frames and Beams. Shell frames and deck beams are normally toed toward amidship with their heels on the molded frame line, as in Fig. 4. This facilitates access for welding and inspection of side frames at the forward and aft ends of the vessel where there is more shape to the shell. Where the side shell is longitudinally framed using angles or bulb plates, the angles or bulb plates toe down and the heel is on the molded line; when tees are used, the lower side of the web is on the molded line. Where decks are longitudinally framed using angles or bulb plates, the angles or bulb plates toe outboard, and the heel is on the molded line; when tees are used, the inboard side of the web is on the molded line.

Good coordination must exist among draftsmen, loftsmen, and shipfitters, based on a clear understanding of the importance of the relationship of structural members to the molded lines. This relationship must be standardized, as further emphasized in Chapter XVI, in order to avoid assembly problems and discontinuity of structure.

Section 4
Structural Alignment and Continuity

4.1 Introduction. The classification society rules and the methods of structural analysis provide means for determining the size and thickness of various parts of ship structures. When these sources are properly used, the designer may be reasonably sure that he has provided adequate strength. Of equal importance, however, are structural alignment and continuity.

Structure is aligned when the loads in structural members have a direct path to the supporting structure. Alignment usually concerns two connected members in the same plane. Their connection may be direct by a butt weld to each other, or intercostal by a fillet weld connection of each member to the opposite sides of a continuous member in a plane perpendicular to the members. This is fundamental, but many ships have suffered severe problems when other design requirements have been assigned undue priority over structural alignment. The problem is twofold, i.e., alignment during design, and alignment during construction.

During design, support for vertical loads at one level must be aligned with support below. Thus, a bulkhead designed to prevent racking of a deckhouse is less effective when not aligned vertically with adequate structure in the main hull. Similarly, pillars must, if at all possible, be placed one under the other. If a line of pillars must be stepped, specially reinforced girders or other means must be provided to transfer the load.

The second aspect of alignment is the assurance during construction that the position of structure the designer intended is actually achieved. Not every shipfitter appreciates that misaligning a chock on one side of a bulkhead with the girder flange on the other side subjects the bulkhead plating to loads that would be otherwise carried by an alignment chock. In some areas, the designer has little control over this particular construction problem. The problem can be overcome, however, by avoiding where possible design details which require alignment of structure on opposite sides of a plate where the fitter putting in the backing structure has no convenient means, visual or simple direct measurement, for precise location of the back-up member.

The alignment which the designer controls directly must be provided early in the development of plans. It is more expedient to develop arrangements within the framework of given bulkhead, pillar, and girder locations which provide the necessary support, than to develop a set of arrangements and then try to find consistent locations for structure within them. Close collaboration between structural and arrangements design people in the early stages is essential for a balanced design. If one group or the other goes too far in development without consultation, unsatisfactory results are inevitable.

4.2 General Description of Problem. Structure has continuity when it is capable of transferring the loads in the structure without creating abrupt changes in stress levels. It is axiomatic that poorly aligned structure has little continuity. Good alignment, however, does not insure continuity. A 150-by-10-mm flat bar may be perfectly aligned with a 50-by-10-mm flat bar, but continuity would be lacking unless the deeper member is tapered to 50 mm. Generally, a designer compensating for a cut in a bulkhead is preserving continuity, while the designer assuring that this bulkhead is supported by structure on the far side of an intervening deck is said to be providing alignment (and, hopefully, continuity). The distinction between alignment and continuity is not important, but providing both is essential to good design and construction. The subject of alignment and continuity in construction is covered in Chapter XVI, whereas this chapter relates more specifically to design requirements.

The superstructure provides the classic case of an unavoidable discontinuity where one or more decks are added to the ship. These decks participate in resisting longitudinal bending of the ship as a whole, but by their nature end abruptly and thus form a discontinuity. This discontinuity requires local increases in the sheerstrake, the strength deck stringer plate, and increases in the superstructure deck

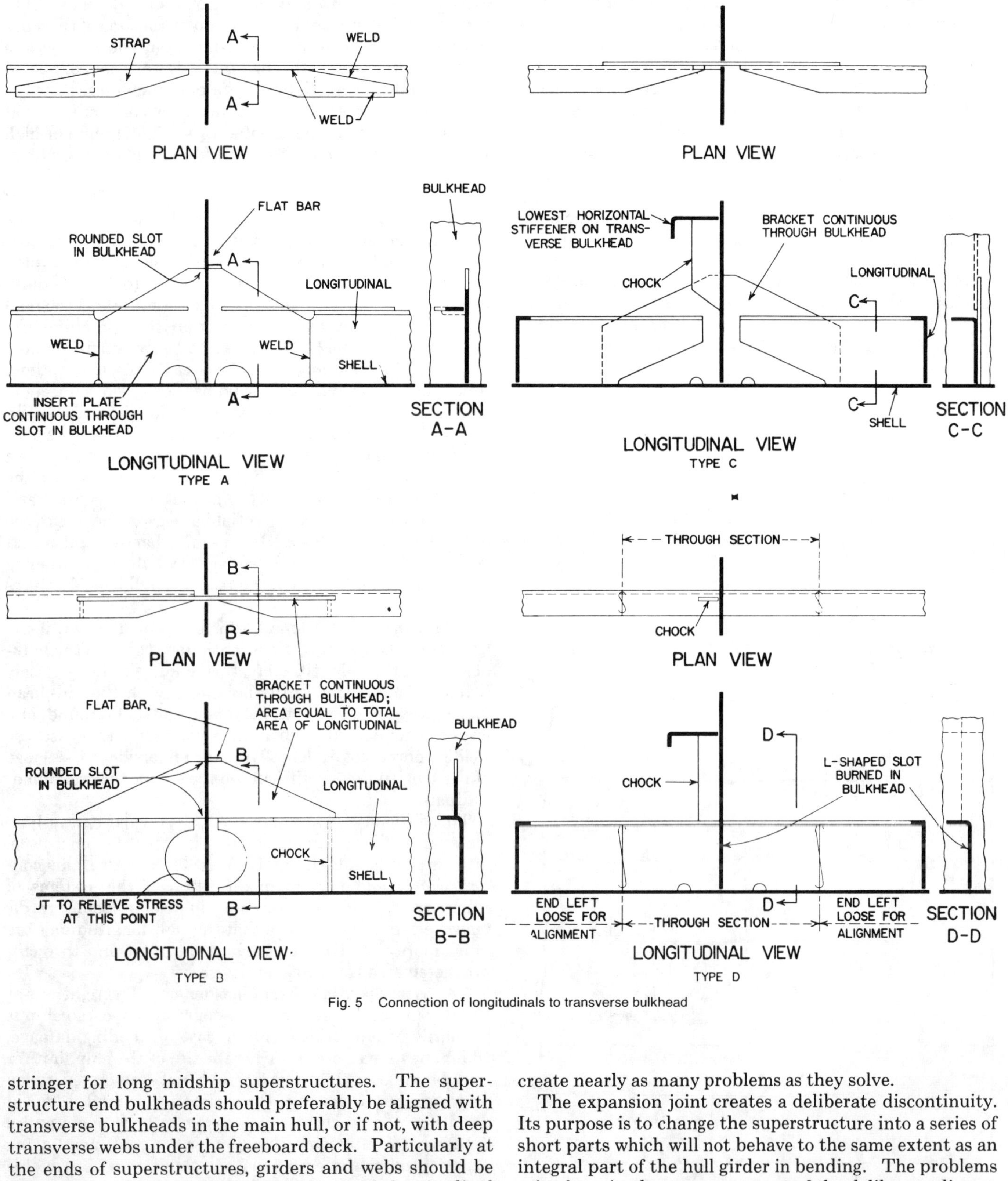

Fig. 5 Connection of longitudinals to transverse bulkhead

stringer for long midship superstructures. The superstructure end bulkheads should preferably be aligned with transverse bulkheads in the main hull, or if not, with deep transverse webs under the freeboard deck. Particularly at the ends of superstructures, girders and webs should be fitted under the freeboard deck in line with longitudinal bulkheads in the superstructure and with webs on superstructure end bulkheads. Attempts to avoid the problem have led to the introduction of *expansion joints*, which create nearly as many problems as they solve.

The expansion joint creates a deliberate discontinuity. Its purpose is to change the superstructure into a series of short parts which will not behave to the same extent as an integral part of the hull girder in bending. The problems arise from inadequate treatment of the deliberate discontinuities. Aluminum superstructures offer another solution (see Section 17 of this chapter, which treats deck houses and superstructures).

The structural designer must always be alert to the fact that structure must allow for the flow of force from one member to another. If a member is present and under stress, that stress must be transferred where the member ends. This problem must be given due attention at all levels of design, from the development of preliminary arrangements to the development of end connections during the detail working plan stage and the follow-up construction, to ensure that the location of material envisioned actually occurs in the ship.

The following two subsections discuss two specific problem areas which fall under the general category of alignment and continuity.

4.3 Intersection of Longitudinals and Bulkheads. The construction at the intersection of the bottom and side longitudinals and the transverse bulkheads has sometimes been troublesome because of minor fatigue cracking due to working of the ship and flexing of the bulkheads (Ship Structure Committee, 1978). Many variations of this construction have been tried, ranging from the *bracketless* construction in which the longitudinals were not connected to the bulkheads at all, to the present much more satisfactory concept of obtaining a high degree of continuity and at the same time avoiding, insofar as possible, points of high stress concentration in the longitudinal and *hard spots* on the bulkhead.

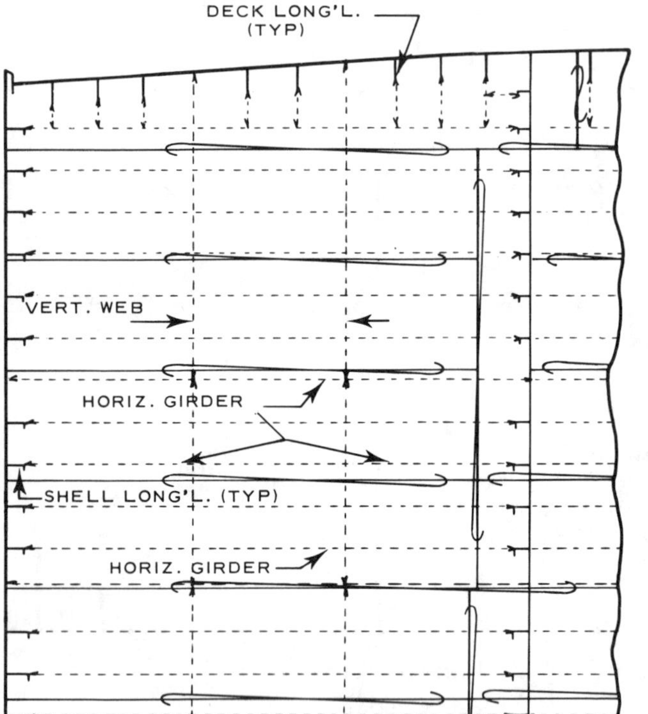

Fig. 6 Arrangement of horizontal bulkhead stiffners

Fig. 5 shows several types of connections. The type B connection presents difficulties in developing fully the strength and area of the longitudinal. It is not therefore recommended, except where it is not necessary to develop fully the strength and area of the longitudinal at the bulkhead connections. In general, this limits type B to small vessels. Types A and C connections are potential improvements of type B. However, to be so, attention must be given to developing fully the area of the longitudinal throughout the connection, including the welded connection. Also, the section modulus of the bottom longitudinal should be developed throughout the connection. Of all the connections, type D is preferable as it ensures full development of the strength and area of the bottom longitudinal with the least structural discontinuity. In connections types A and C, it is necessary to fit substantial flat bars at the toes of the brackets on the bulkhead. These flat bars should be extended and attached to adjacent vertical bulkhead stiffening members. Their omission can result in bulkhead fractures at the bracket toes.

Ideal continuity of longitudinals can be attained if the longitudinals themselves are continuous through the bulkheads. With angle stiffeners, this requires either a closely fitted L-shaped slot in the bulkhead, with the bulkhead plating welded directly to the longitudinal all around, or a large rectangular cut with a collar plate. In either case, the welding of the slot or the collar to the underside of the flange of the longitudinal is difficult on the smaller sizes of longitudinals.

This difficulty is largely avoided if heavy flat bars or bulb plates are used instead of angles. This construction is usual in many yards outside the United States. Also, this construction would require modification of the systems of subassembly and erection now in use at some yards. However, tests have shown that through-longitudinals are much more effective than the best of the cut-and-through-bracketed arrangements yet devised.

4.4 Hard Spots in Tanker Construction. The tightness of subdivision bulkheads in a passenger or dry-cargo ship is primarily of importance only in case of flooding damage, which may never occur during the life of the ship, but for tank boundaries tightness is vital to satisfactory everyday performance. Consequently, every effort is made to avoid details of construction which result in hard spots and a tendency for fatigue cracking under the continual flexing experienced by these bulkheads in service. A hard spot is defined as a point at which the deflection curve of a plate is abruptly changed in a step-like form by the effect of a very rigid member landing on the plate. Such an abrupt change of curvature induces high local stresses.

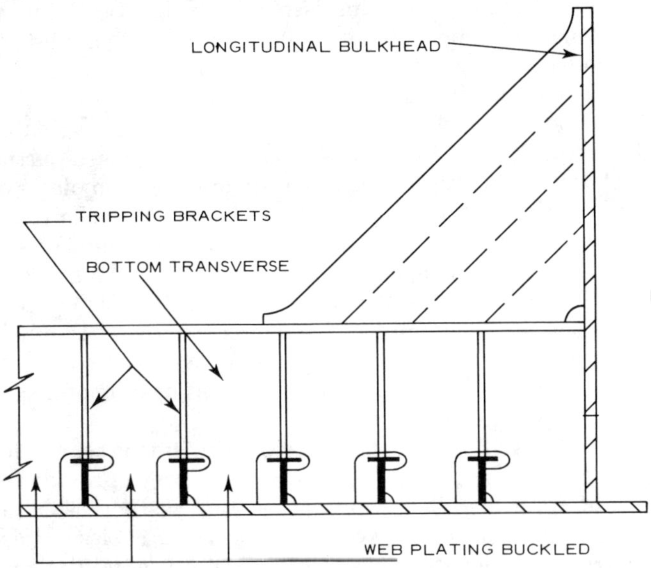

Fig. 7 Buckling of bottom transverse web frames

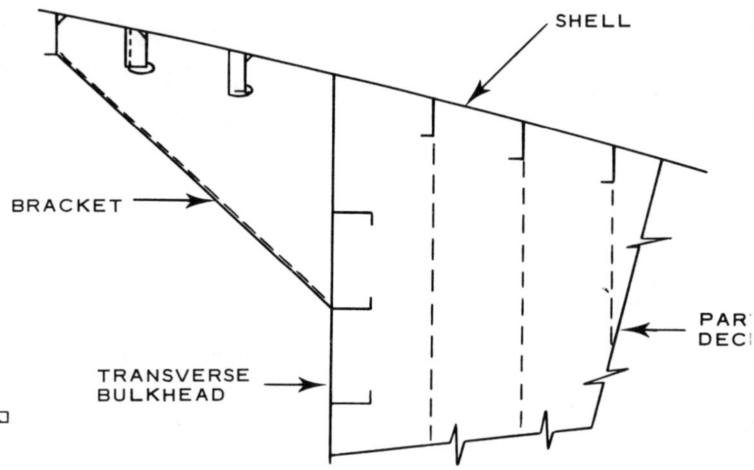

Fig. 9 Shell bracket at partial deck ending

One advantage of the use of vertical stiffeners on the transverse bulkheads is that each stiffener is in line with and attached to a bottom longitudinal, and can effectively prevent working of the bulkhead plating. However, with this arrangement, there may be cases in which the bulkhead flexes excessively in way of the rigid side longitudinals and, unless flat bar stiffeners are introduced to connect the side longitudinals to the adjacent vertical bulkhead stiffeners, fractures may occur in the excessively flexed plate. Similarly, when horizontal bulkhead stiffeners are used, it has been found desirable to fit the horizontal stiffeners in line with the side shell longitudinals, as shown in Fig. 6, and efficiently connect the two. While this arrangement can be worked out easily in the parallel midbody, it becomes difficult near the ends where the construction is no longer parallel.

4.5 Designing Components to Minimize Repairs During Life of Vessel. The definition of damage adopted by the 1967

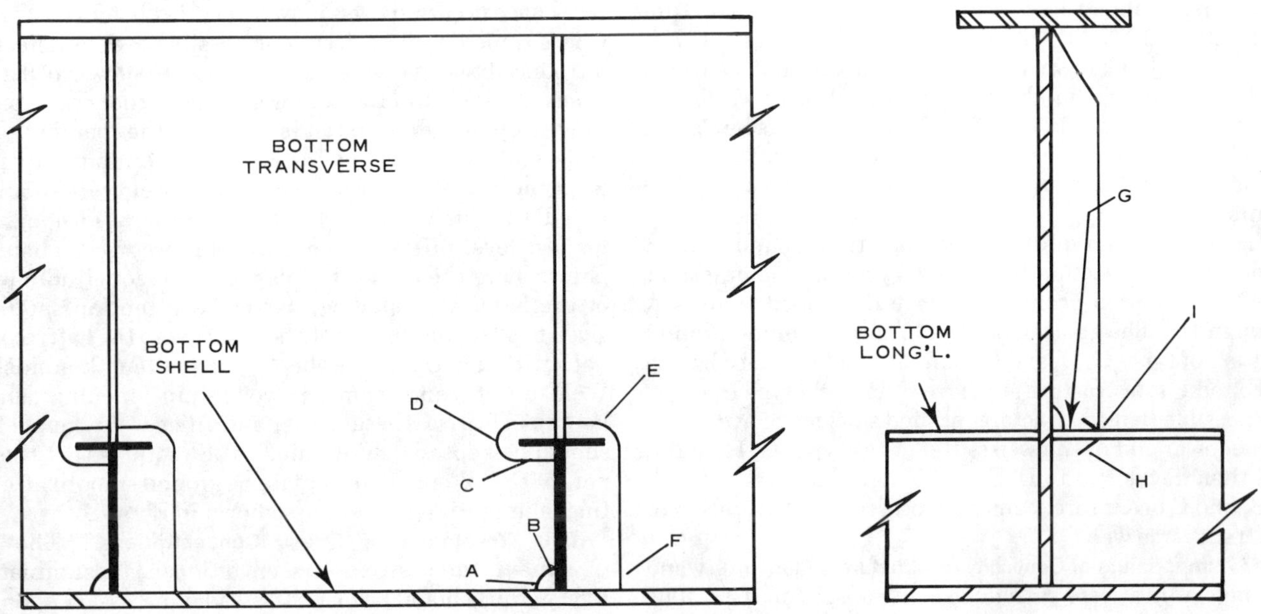

Fig. 8 Cracks occurring at the junction of bottom longitudinals and bottom transverses

International Ship Structures Congress is as follows:

> "*Damage*. A structure is damaged if its original form has changed in a way which is detrimental to its future performance, even though there may be no immediate loss of function. Examples of damage include excessive permanent deformations resulting from local yielding or buckling, or the appearance of cracks due to fatigue or local brittleness. In such cases the structure may still be able to sustain its design loads, but because of the possible adverse effects on performance or appearance, and hence on the confidence of operators and users, repairs should be effected as soon as convenient."

Collapse, a more critical form of damage, is defined as follows:

> "*Collapse*. This occurs when a structure is damaged so badly that it can no longer fulfill its function. This loss of function may be gradual, as in the case of a lengthening fatigue crack or spreading plasticity; or sudden, as when the failure occurs through plastic instability or through propagation of a brittle crack. In all cases the collapse load may be defined as the minimum load which will cause this loss of function."

Proper design can minimize the chances of damage or collapse; proper choice of proportions of members used as stiffeners, particularly built-up sections and flat bars, and the addition of stabilizing devices, such as chocks, will contribute to avoidance of local buckling, for example.

4.6 Cutouts in Transverse Frames for Longitudinal Stiffeners. A common location of damage is the intersection of transverse primary members and longitudinal frames or stiffeners, as illustrated in Figs. 7 and 8 (Ship Structure Committee, 1977-b), (Ship Structure Committee, 1978). Fig. 7 shows buckling in the web plate of a transverse frame near a longitudinal bulkhead due to high shear in way of the cutouts. To avoid this, it is advisable that all slots be cleanly cut with rounded corners and be collared in way of the areas of high shear. This is a part of the ABS Rule requirements.

Fig. 8 depicts a typical bottom longitudinal and bottom transverse intersection where low-cycle fatigue cracks at points of stress concentration have developed (points A through I). The great majority of fractures found around slots are of types G, H, and I. Since most of the webs having D, E, and F type fracture also have G, H, and I type fracture, it is possible that the fractures around slots may have begun at the lower end of the web stiffeners as types G, H, and I, and then developed to D, E, and F type fractures. Types A, B, and C occur rarely and may be a result of vibration of the transverse webs.

4.7 Importance of Continuity of Structure. Continuity and its importance were introduced in Subsection 4.1. The proper order of importance in considering aspects of continuity is: First, can discontinuity be avoided? From a structural standpoint, this is obviously the best way of handling the problem, but it cannot always be accomplished. Second, if the discontinuity is unavoidable, what can be done to alleviate its effects?

Various typical cases of alleviation of unavoidable discontinuities include brackets to the shell when a partial deck must end, Fig. 9, and provision of transition plates of intermediate thickness in deck plating where a change in deck cross-sectional area is required. Such a requirement may be due to a change in hatch width or in shell plating, or for reinforcement in way of a rudder horn or other concentrated loads. The common characteristic in these seemingly different cases is the provision of an area in which average stress levels are kept reasonably constant in spite of necessary changes in geometry.

4.8 Stress Concentrations and Cuts in Structures. Holes must be cut in the steel structure of a ship to provide for access, cargo handling, and engineering systems, such as ventilation ducts, piping systems, and electric cables. Holes in structure are potential sources of structural failure, and it is necessary to guard against unduly weakening the structure by cutting these holes.

4.9 Stress Concentrations Including Application to Hatch Corner Design. Openings in structure increase stresses in two distinctly different ways: first, by reducing the amount of material available to support the load, and second, by causing concentrations of stress. These stress concentrations will, in many cases, be more serious than the effect of removing material. In the classic case of a circular hole in an infinitely wide plate under tension, the change in stress due to lost area is infinitesimal, but the stress at the edge of the opening is three times the nominal value. Poor design of intersection of plates and stiffeners is also a cause of stress concentration areas, an example of which are the hatchway corners. Badly designed or constructed hatchway corners are a frequent source of low-cycle fatigue fracture. Fractures have occurred even in well-rounded hatchway corners, where some detail of design or possibly a weld defect has introduced a stress concentration. Fig. 10 shows such a case, where the source of fracture was at the corner of the part of the deck plate which extended through the coaming. This is not only a point of stress concentration, but a difficult position in which to produce a sound weld. It should be noted that in rounded hatchway corners, although the highest level of stress concentration occurs at about the point where the corner radius meets the longitudinal side of the hatchway opening, particular attention should be given to all details in way of and adjacent to the hatch corner, both within and outside the 40-percent length amidships. Welded butts and seams, as well as pipe openings, should be kept clear of these areas, and fillet weld connections should be kept to a minimum. ABS requires that the corners of the extended deck plate be ground smooth after fitting, eliminating the notches shown in Fig. 10.

4.10 Treatment of Stress Concentrations. The best *treatment* for a stress concentration is to eliminate it. Every effort should be made to avoid unnecessary openings, particularly in critical structural members such as the keel, sheerstrake, bilge strake, strength deck in way of superstructure breaks, house ends, or hatch corners. The opening in the sheerstrake for the accommodation ladder, Fig. 11, is an example of poor design, because:

- it could have been eliminated;
- it is in a highly stressed area;
- it has square corners;
- it has no compensation.

When openings are unavoidable, they must be designed and oriented to minimize the resulting stress concentrations. For the design of hatchway corners, certain guidelines are suggested:

- detail design should be simple, avoiding multiple connections resulting in an accumulation of difficult welds;
- attachments should not be welded to a free deck edge;
- elliptical corners in the deck are preferred to radius corners with respect to hull-girder bending; however, recent analysis has shown that radius corners are preferable with respect to torsional moments.

4.11 Effect of Radii. The degree of stress concentration is primarily a function of the abruptness of the discontinuity. Thus, the stress concentration factor at the end of a sharp notch such as a crack may mathematically approach infinity, and a very low nominal stress will cause local stresses in the plastic range. In some cases, the local stress level will cause failure and the crack will advance across the plate. This action can sometimes be stopped by drilling a circular hole at the end of the crack. The effect is a radical increase in the radius at the end of the crack, which markedly reduces the stress concentration.

4.12 Avoiding Square-Cornered Cuts. From the foregoing, it is obvious that square-cornered cuts constitute an invitation to failure and should never be permitted. All holes in strength members and in stressed plating should have generous corner radii. A radius equal to one-eighth of the dimension of the opening perpendicular to the stress is good practice, but for large openings 610 mm (2 ft) is usually acceptable. Recent studies of clearance openings for longitudinals passing through main transverse webs have resulted in a variety of details recommended for lessening the notch effect (Ship Structure Committee, 1977-b).

4.13 Opening Orientation. Openings which must be made in stressed areas should always be oriented to minimize the stress concentration. According to test results (Ship Structure Committee, 1977-a), an opening with a length-to-width ratio of 2 to 1, placed so that its long dimension is parallel with the direction of stress, creates a stress concentration only 50 to 60 percent of that caused by the same opening placed with the long dimension perpendicular to the stress.

4.14 Compensation for Openings. Compensation for openings generally takes one of three forms: doubling, heavy insert plates in the plane of the plate, or flat bar rings welded around the periphery of the opening and normal to the plane of the plate. The flat bar rings are generally used in naval construction. The flat bar reinforcement is convenient from a construction standpoint, since it is much

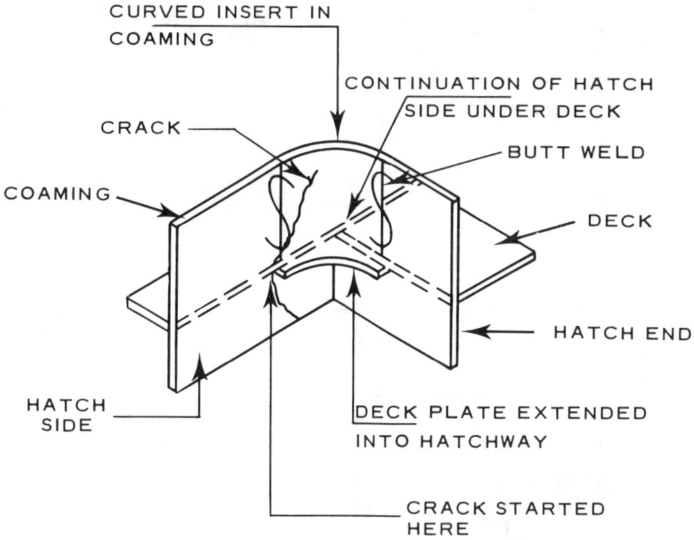

Fig. 10 Fractures occurring in hatchway

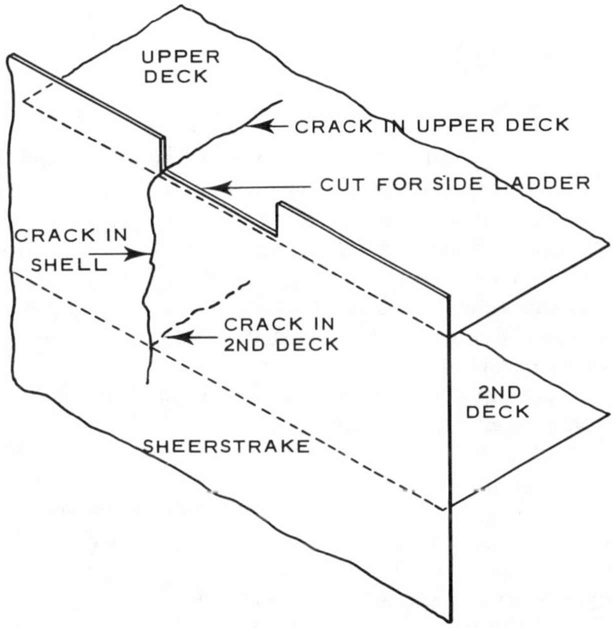

Fig. 11 Cut for side ladder

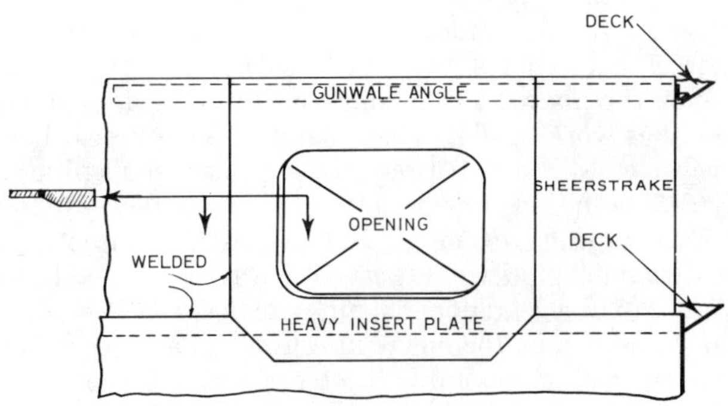

Fig. 12 Compensation at side ports

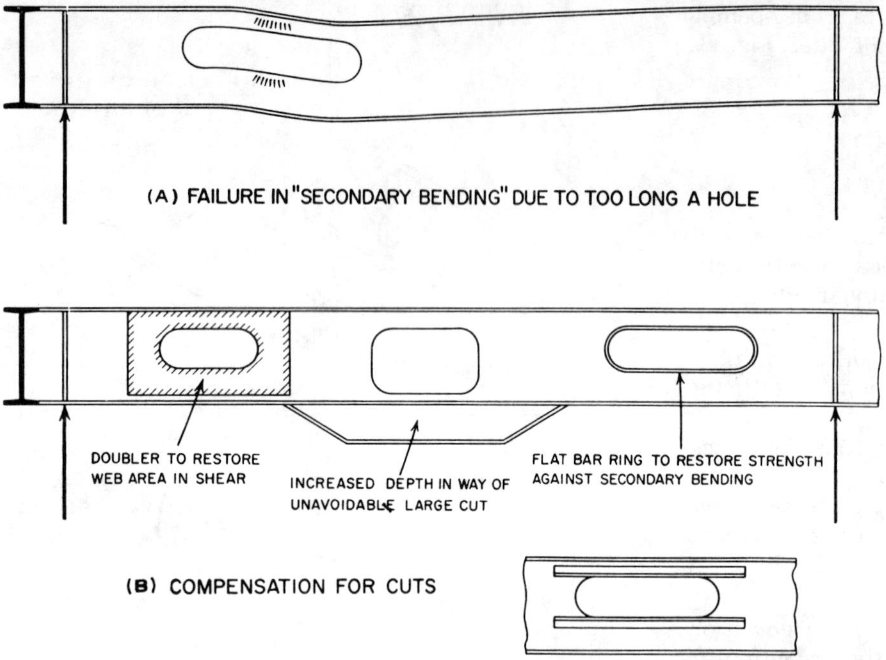

Fig. 13 Cuts in strength structure

easier to add a flat bar than an insert plate when a hole is made in a completed area of the deck or shell. The flat bar is, however, a tripping hazard in some areas, and may also be awkward from the standpoint of adjacent structure. Insert plates are the standard means of reinforcing holes in current merchant ship construction. The doubling plate is not as efficient as an insert plate, and often presents a tripping hazard when used on decks. However, it is a useful means of reinforcing where openings are made as a modification to existing structure.

Typical compensation for shell openings is shown in Fig. 12. The examples in Fig. 12 show the increased thickness carried around the full length of the ends of the opening. The extra material is not needed at the middle of the forward and aft edges of an opening from a strength point of view, since these areas are largely relieved of fore-and-aft stress by the presence of the opening. However, the advantages of a flush surface for fitting the sideport frame make it desirable to fit the doubler or insert plate all around the opening, as shown.

4.15 Pipe Penetration. Tests have demonstrated that pipes which penetrate and are welded to plating do not weaken the plating as a diaphragm, and may be spaced as closely as working clearances permit. Such pipes should have a clear distance between pipes, or between a pipe and adjacent bulkhead or deck plating, approximately as follows: nominal pipe size 76 mm (3 in.) and less: 102 mm (4 in.) clear; nominal pipe size between 76 and 127 mm (3 and 5 in.): 127 mm (5 in.) clear; nominal pipe size between 127 and 203 mm (5 and 8 in.): 152 mm (6 in.) clear. These spacings are for working clearance only. Insofar as strength is concerned, pipes or sleeves welded to plating may be spaced as closely as desired, provided that the aggregate width of cuts in any one line of plating does not exceed about 15 percent of the total width of plating. Depending upon the structural members the pipes penetrate, insert or double-plate-area compensation may be required.

Pipes passing through bulkhead or deck plating should be kept close to boundaries to minimize the effect of deflection of the plating. However, the practical necessity of keeping pipes out of the way takes care of this condition automatically in most cases. Regulations for damage stability of the vessel may require certain pipes to be located at specific distances from the side shell.

4.16 Openings in Stiffening Members. Openings such as lightening holes, scallops, and snipes are often made in all non-tight members (deck beams, girders, floors, and stringers) in order to provide ventilation, free liquid drainage, and access to piping. These openings may be sources of weakness because of their abundance, and should therefore be located with adequate control by those responsible for design.

Holes in side girders and floors should be made at mid-depth, their length should not exceed 50 percent of the web depth, and their height should be so limited as to provide a plate width of at least one-quarter of the double-bottom depth. These holes are to be kept well clear of end brackets and areas of high shear. The holes are to be well-rounded with a minimum radius of about one-third of the depth of the hole.

Air holes are to be made in the double-bottom side girders and floors at their intersections with the inner bottom. The arrangement and area of these air holes must be sufficient to prevent pressure build-up on the tank boundaries and

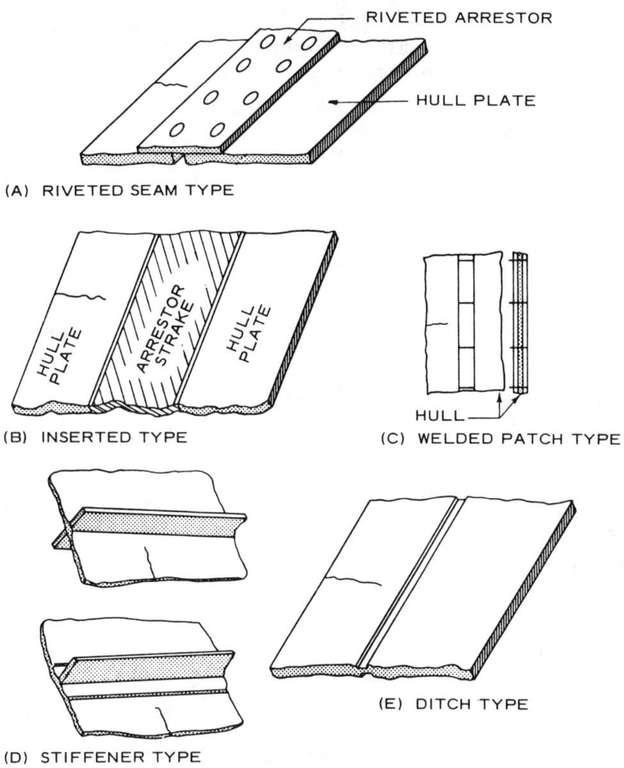

Fig. 14 Crack arrestors

filling lines. Drain holes and welding scallops in longitudinal girders are to be arranged as far away from the center of lightening holes as possible. Cut-outs for frames, longitudinals, or stiffeners deeper than 508 mm (20 in.) should be stiffened at their free edge.

Openings in beams, girders, longitudinals, and stiffeners should be located in the web near the neutral axis of the member. These and more detailed guidelines are given by the Ship Structure Committee (1977-b). Figure 13 (A) illustrates the effect of *secondary bending* on shallow beams where the length of the opening exceeds twice its depth. Situations like this should clearly be avoided. It will be noted that these general rules pertain to the usual case of members with distributed loads. Large concentrated loads may change the shear distribution significantly, and openings in beams loaded in this manner require special consideration.

4.17 Compensation for Openings in Stiffening Members. Openings exceeding the recommended depth will generally require a doubler plate on the web, Fig. 13 (B), to make up the deficiency in shear strength of the web in way of the opening; a face bar around the edge of the opening would not accomplish this. Openings exceeding the recommended length, unless known to be in areas of low shear stress, will require a flat bar ring around the edge of the opening, Fig. 13 (B), to strengthen the remaining portion of the beam against secondary bending. To permit effective reinforcement, depth of the openings should not exceed 50 percent of the depth of the beam, and length of the openings should not exceed 150 percent of the depth of the beam. When these dimensions must be exceeded, the depth of the beam should be increased locally, Fig. 13 (B). No openings in the flanges of beams or stiffeners should be made without compensation, unless it is determined that the member has excess strength in way of the opening. A simple means of compensation for such openings is the addition of a welded flat bar to the remaining portion of the flange. To accommodate wiring which may be run on the ship without detail plans, a single row of round holes in the web of beams or stiffeners is usually acceptable anywhere in the beam without compensation, provided that the diameter of the holes does not exceed one-eighth of the depth of the beam, and that the holes are not less than three diameters apart or away from other openings.

4.18 Causes of Cracking. The occurrence of cracks is usually associated with stress concentrations, and poor workmanship and materials. Since modern shipbuilding steels are designed to be crack-resistant, this problem is not as serious as it was during World War II. Currently, brittle fracture and fatigue are the major causes of cracking. Cracks are generally initiated at points of stress concentration which can be due to:

a. *Design deficiencies*
- cuts in highly stressed areas
- abrupt changes in continuity

b. *Poor workmanship*
- faulty welding
- rough plate edges
- misalignment of structure

For example, the welding of chock foundations to the highly-stressed top edge of the sheerstrake is a possible means of starting a crack.

4.19 Need for Crack Arrestors. Since a crack, once started, becomes a traveling stress concentration which can cause the loss of a ship, all classification societies require that means be incorporated in the design to prevent loss of an unacceptable portion of the effective longitudinal material making up the hull girder.

Although preventing initiation of cracks may be difficult in some areas of the hull, attention to structural details can alleviate the problem. Great effort has also been put into the investigation of the crack propagation mechanism, in order to find remedies (crack arrestors) of more effective nature, and consequently to prevent ship hull fracture. The basic principle behind the use of a crack arrestor is to reduce the crack-driving force (strain, kinetic, external) to a magnitude lower than the resisting force that must be overcome to extend a crack (Ship Structure Committee, 1977-a). The resisting force represents fracture energy which is closely related to the fracture toughness of the material.

A few of the existing crack arrestor design practices are illustrated in Fig. 14. The location and number of these required crack arrestors vary with ship size and type, i.e., whether tanker or dry cargo ship. The type B crack arrestor is the most commonly used; types A and B are known to have been accepted by classification societies.

4.20 Notch-Tough Steel Strake Crack Arrestors. Classification societies permit the substitution of a welded strake of special steel for a riveted crack arrestor. The special steel is designed to be much more resistant to failure by fracture, and therefore is much more able than the adjacent plating to prevent further crack propagation. The classification societies specify the location of the strakes, as well as the composition and physical properties of steels to be used for this purpose.

Section 5
Sections Used for Frames, Beams and Stiffeners

5.1 Commonly Used Sections. Fig. 16 illustrates some of the sections commonly used in present-day construction. The sections range from a simple flat bar, edge-welded to the supported plating, to a flanged plate girder with additional face-bar reinforcement. Girders are also built-up by welding a flange to a lighter web, producing a more efficient section than the simple flanged plate. *Inverted* unequal leg angles, with the edge of the long leg welded to the plating, are relatively cheap stiffeners through the available range of sizes. Bulb-plate stiffeners are frequently used in ships built outside the United States in this size range. Larger angle sizes may be obtained by cutting one flange from a channel section. The most efficient section is a T-section, but its use has practical disadvantages. A horizontal T-section in a tank space will have poor drainage; T-sections in smaller sizes are more difficult to weld than other shapes due to restricted access; T-sections require larger slots than other shapes when passed through supporting structure and require more work in their attachment to supporting members. If tested to failure, T-sections may be expected to resist bending better than inverted angles, channels, or flanged plates because of their symmetrical shape.

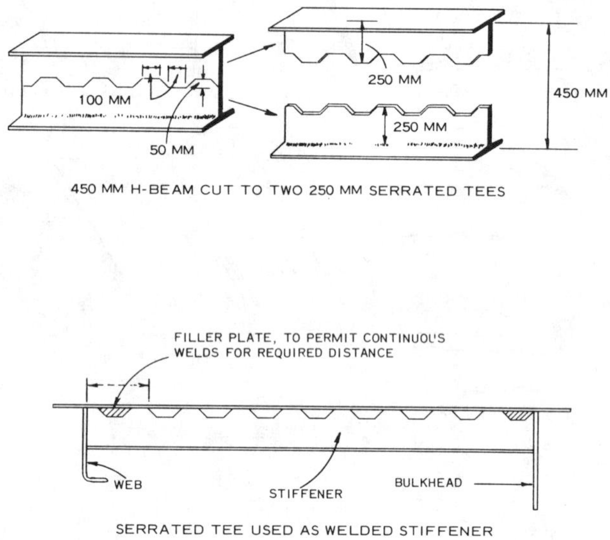

Fig. 15 Serrated tees

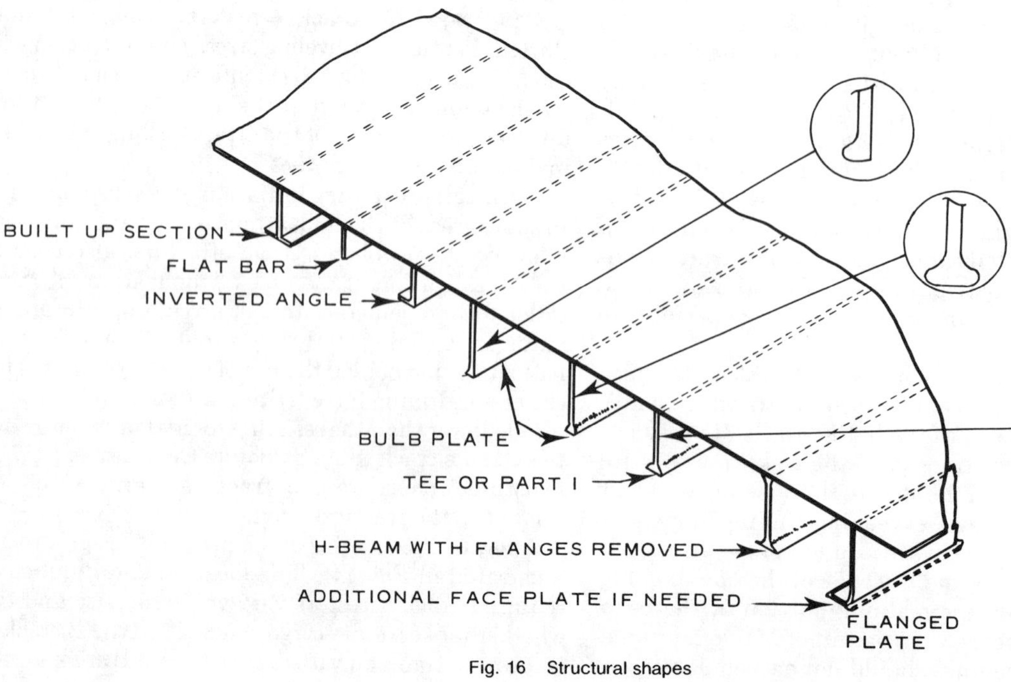

Fig. 16 Structural shapes

The greater tendency of the unsymmetrical rolled shape to *trip* is not necessarily detrimental to its use. The present size and proportion limits of these shapes ensure that if the section modulus and support spacing requirements of the classification societies are met, there will be no tripping of the unsymmetrical rolled sections. The classification societies therefore accept symmetrical and unsymmetrical rolled sections of equal section modulus as equivalent. For built-up sections that replace rolled sections, however, particular attention needs to be given to the proportions to ensure adequate stability. For all secondary members, where the primary supporting member is on the opposite side of the plating to the secondary member, the secondary member is to be supported by a tripping bracket, fitted in alignment with the web of the primary supporting members.

5.2 Serrated T-Sections. A modification of the T-section formed by cutting H-(wide flange) beams is the *serrated* T-section, in which the H-beam is cut as shown in Fig. 15, with two 250 mm (10 in.) T-sections obtained from one 450 mm (18 in.) H-beam. Since continuous welding is usually desirable for a short distance at the end of a beam or stiffener, the serrations must be omitted at the ends, if this construction is used. The only practicable way of doing this is to let the zig-zag cut continue right to the end, and weld in one or more small filling pieces, as shown in the figure. Even with this inconvenience, the resulting construction is efficient and economical.

Serrated members are particularly advantageous in oil tanks, where they offer free flow of liquid cargo when pumping, free venting of vapors and a minimum of obstruction to gas-freeing. Experience has indicated that serrated members should not be used as framing for decks or inner bottoms subject to forklift-truck loads or the impact of unloading buckets, or as bottom frames of river barges subject to grounding damages, because the local loads will cause stress concentrations at the serrations.

ABS Rules for River Vessels accept serrations with cutouts not exceeding one half of the depth of the finished member. In seagoing practice, the serrations are usually from one-fifth to one-third of the depth of the member. The serrations of principal framing members are spaced from 300 to 380 mm (12 to 15 in.) on centers, and fillet-rounding at the corners of the serrations is essential.

Section 6
Transverse Frame Spacing

6.1 Frame Spacing on Oceangoing Vessels. Considerable latitude is allowed by the classification societies regarding frame spacing. The ABS Rules give a standard frame spacing amidships for transversely framed vessels varying linearly from 540 mm (21.3 in.) for a 50 m (164 ft) ship to 1000 mm (39.4 in.) for ships 270 m (886 ft) or more in length. Spacing in the peaks must not exceed 610 mm (24 in.) or the standard frame spacing, whichever is less. Lloyds Rules specify that frame spacing should not, in general, exceed 1000 mm amidships or 610 mm in the peaks. Whenever the actual spacing differs from the Rule spacing, corrections are applied to the various scantlings affected by frame spacing.

Investigations in several particular cases have indicated that the ABS Rule spacing is not far from the spacing associated with minimum weight, which, however, may not be the most economical. A slight increase in spacing will reduce the number of framing members to be fabricated and handled, as well as the amount of welding. Therefore, a somewhat greater spacing than that associated with minimum weight may be more economical. In practice, it is common to adopt a somewhat greater spacing than that given in the Rules and to accept the resulting increases in scantlings.

6.2 Frame Spacing on Great Lakes Ships. On the Great Lakes, a side-frame spacing of 915 mm (36 in.) has been adopted for the cargo holds of bulk cargo ships regardless of length. Transverse frames are spaced 610 mm (24 in.) in the afterpeak and forepeak, but in the latter, spacing is measured along the shell rather than parallel to the centerline. Great Lakes ships are designed to operate between the same loading and unloading docks throughout their life, and to drydock in the long-established docks on the lakes. At all of the loading docks, the chutes are spaced 3.66 m (12 ft) on centers. The keel and bilge blocks in all Great Lakes dry docks are 1.83 m (6 ft) on centers. Since the cargo hatches must be located at 3.66 m or 7.32 m (12 ft or 24 ft) on centers, and the floors at 1.83 m (6 ft) on centers, it is convenient to adopt a side-frame spacing of 915 mm (3 ft). In the case of ships around 200 m (650 ft) in length, this spacing is close to both the Rule spacing and the most economical spacing. The strength deck and bottom shell are usually longitudinally framed, and this practice has been extended to the side framing within the length of the cargo hold spaces.

6.3 Frame Spacing on River Vessels. In river vessels, transverse framing is, in general, spaced 610 mm (24 in.). Longitudinal framing normally is based on 610 to 686 mm (24 to 27 in.) spacing of bottom shell longitudinals as dictated by the service intended. River towboats are transversely framed at about 610 mm (24 in.) spacing, except in way of special stiffening at the bow to withstand grounding and ice. Reduction in frame spacing at the bow generally is accomplished in two steps of 51 to 76 mm (2 to 3 in.) per step, with an accompanying increase in plating thickness of approximately 1.5 mm (1/16 in.) at each reduction in frame spacing.

6.4 Frame Spacing on Offshore Barges. For offshore barges, the typical frame spacing used is 534 to 610 mm (21 to 24 in.). The frame spacing is based upon the deck load, and for heavy deck loads the larger 610 mm with appropriate scantlings is generally chosen. One important consideration in the determination of frame spacing is the tendency to use identical stiffeners throughout the barge, based on material availability, to simplify assembly of components during construction. The strength deck and bottom shell of offshore barges are generally longitudinally framed in modern construction.

Section 7
Longitudinal Framing

7.1 Advantages of Longitudinal Framing. When the frames which stiffen and support the shell and inner-bottom plating, and the beams which stiffen and support the decks, are run longitudinally instead of transversely, and are made effectively continuous through transverse bulkheads, they contribute in large part to the section modulus of the hull girder and thus assist in resisting the longitudinal bending of the ship's hull. Where the primary plating is subject to high in-plane compressive stress, longitudinal frames not only aid in resisting the longitudinal bending of the hull girder directly, they also greatly increase the critical compressive buckling strength of the plating to which they are attached. Lloyds Rules permit a reduction in the thickness of longitudinally framed deck and bottom shell plating. ABS Rules provide for reduction of the required thickness of deck and bottom plating when longitudinally framed. For oil tankers, bulk carriers, general cargo vessels, containerships, and Great Lakes ore carriers, longitudinal framing is now generally adopted, at least for the strength-deck plating and bottom shell. An exception is made at the ends of the vessel, where the advantages of longitudinal framing disappear and where transverse framing is simpler to build.

7.2 Disadvantages of Longitudinal Framing. Longitudinal framing is so efficient that one might question why longitudinal framing of side and deck plating is not standard practice. For most merchant ships, the deep side transverses required to support the longitudinal side framing have serious disadvantages. In cargo ships and refrigerated cargo vessels, they interfere with cargo stowage. As a rule, in passenger vessels, longitudinal framing does not lend itself to satisfactory architectural treatment of the joiner work in passenger spaces. Also, in passenger ships, the longitudinals interfere with running engineering service systems, such as wiring, ventilation, and piping. These supply systems generally run fore-and-aft over the passageways, and longitudinal framing interferes with the transverse branches to the various rooms and spaces being serviced. A practical solution for modern cargo ships is to frame the bottom shell, inner bottom and strength deck longitudinally utilizing transverse floors every third frame. Transverse framing is then applied to the side shell and 'tween decks. Longitudinal framing is generally used on warships, where strength and weight saving are of paramount importance, where cargo stowage is not a problem, and where difficulties with system runs are secondary to weight saving.

7.3 Spacing of Longitudinal Framing. With longitudinal framing, the bottom, inner bottom, and deck longitudinals are run fore-and-aft at a spacing which may vary from 686 to 1000 mm (27 to 39.4 in.), and they are supported by deep *transverses* usually spaced from 3.05 to 4.57 m (10 to 15 ft) apart in merchant vessels and from 4.57 to 5.25 m (15 to 17.2 ft) in large oil carriers.

7.4 Great Lakes Practice Regarding Longitudinal Framing. On the Great Lakes, cargo ships have been framed using a combination system of framing. The vessel has longitudinal framing in the bottom, inner bottom, upper side plating, upper side-tank bulkhead, and upper deck. Transverse framing is used in the lower side shell and side tank bulkheads for ease of construction and, in the case of the side shell, to better resist docking damage, and damage from swinging ore grab buckets.

7.5 River Practice Regarding Longitudinal Framing. On river vessels, the selection of transverse framing versus longitudinal framing is in most instances dictated by the type of ship. Longitudinal frames are supported by regularly-spaced transverse frames formed either by channels extending across the inner faces of the longitudinal frames or by flanged plates notched over the frames and attached to the shell or deck. In many designs, a combination of transverse and longitudinal framing is used. For instance, in the case of barges having longitudinally continuous wing bulkheads, it may be found desirable to use transverse framing in the wings and longitudinal framing in the center compartments. In the case of longitudinal framing, the longitudinals are usually spaced from 609 to 762 mm (24 to 30 in.) on centers, and the transverses are spaced from 1.83 to 2.13 m (6 to 7 ft) apart.

Section 8
Double-Bottom Construction

8.1 Advantages of Double-Bottom Construction. Double-bottom construction, such as shown in Figs. 17 to 20, offers several advantages over single-bottom construction. It results in a strong bottom that is well adapted to withstand the upward pressure from the sea as well as the longitudinal hull girder bending stresses, especially the compression resulting from *hogging* stresses. It provides tankage for liquids such as fuel oil, fresh water (other than potable water), and salt-water ballast, thus using space that is ill-suited for other purposes. It results in a structure which can withstand a considerable amount of bottom damage caused by grounding without flooding of the holds or machinery spaces, provided the inner bottom remains intact. And finally, a smooth inner hull free of stiffening structure is produced which provides easier cleaning accessibility.

The International Convention for the Safety of Life at Sea, 1974, includes conditions for the fitting of double bottoms extending from the forepeak bulkhead to the afterpeak bulkhead, insofar as this is practicable and compatible with the design of the vessel. Classification societies require, in general, double bottoms in way of dry cargo and machinery spaces.

8.2 Design Considerations in Double-Bottom Construction. The section on double bottoms in the ABS Rules contains information on double-bottom design. The bottom shell and inner bottom are subject to both longitudinal and transverse loads. The longitudinal loads result from the hogging and sagging bending moments of the ship. The transverse loads are caused by water pressure or docking loads on the bottom of the vessel, and distributed or concentrated internal cargo or ballast weights on the inner bottom.

The ABS requirements for double bottoms illustrate the combination of theoretical and practical considerations incorporated in classification society rules.

For example, the depth of the double bottom, in millimeters, is given by the formula

$$d_{DB} = 32\,B + 190\sqrt{d}\ \text{mm} \qquad (1)$$

where:

B is the breadth of the vessel in meters, representing the transverse span of the double bottom;

d is the molded draft in meters, which determines the bottom hydrostatic pressures on the shell and the internal

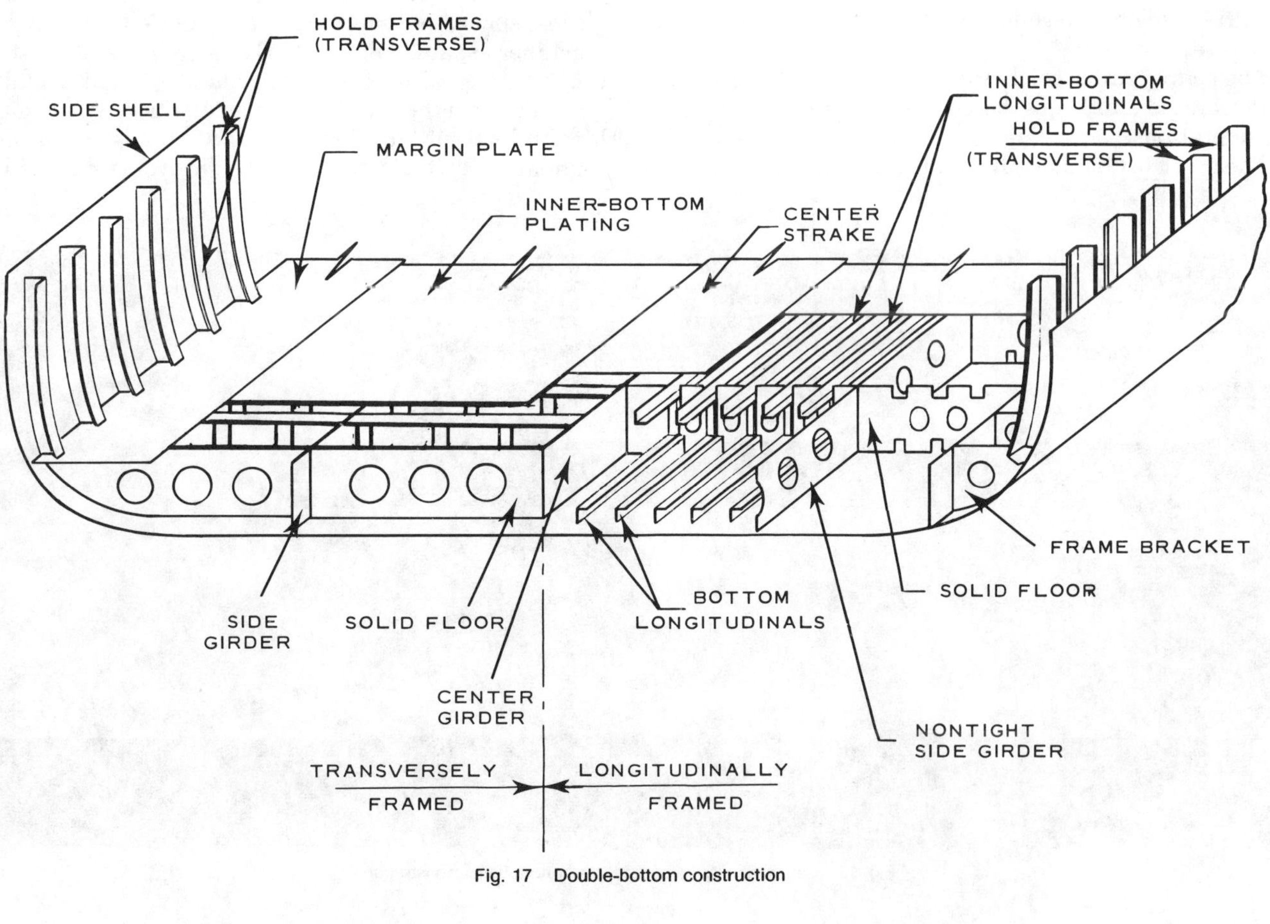

Fig. 17 Double-bottom construction

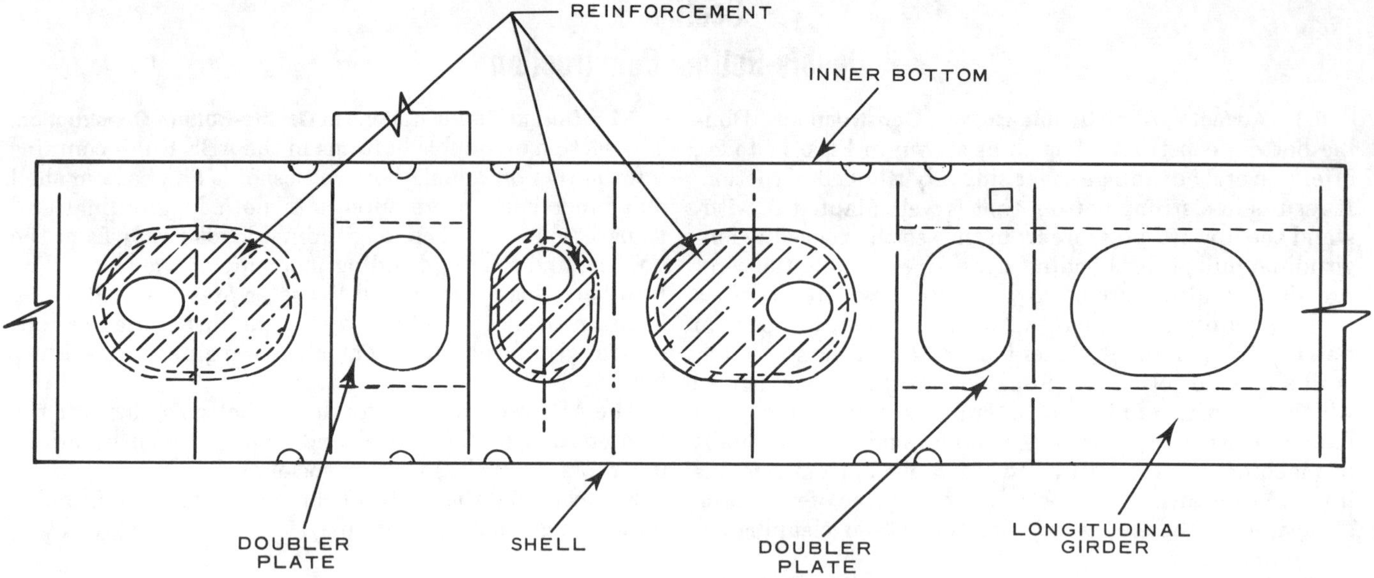

Fig. 18 Reinforcement of longitudinal girders

cargo loads on the inner bottom. The center girder thickness amidships, in millimeters, is given by the formula

$$t = 0.056\,L + 5.5 \text{ mm} \qquad (2)$$

where: L is the length of the vessel in meters, related to the draft which determines the net load on the bottom structure.

The center girder thickness at the ends is 85 percent of that required amidships, based on empirical and practical considerations.

The construction illustrated in Fig. 17 consists of an inner skin extending across the bottom of the ship from bilge to bilge (and, in some cases, part way up the sides). With transverse framing clear of the machinery space, there are transverse bottom frames and reverse frames in line with every side frame, solid floors at every 3.05 to 3.65 m (10 to 12 ft), and a center girder and one or more longitudinal girders spaced not more than 4.57 m (15 ft) apart. In the machinery space, solid plate floors are generally fitted on every frame, while the arrangement of center and side girders used in the cargo holds is maintained, with additional longitudinal girders in way of the propulsion machinery foundations. The longitudinal side girders may be inter-

Fig. 19 Longitudinally-framed double bottom under construction

costal between solid floors for this type of construction, whereas the center vertical keel is continuous.

With longitudinal framing, also shown in Figs. 18 and 19, solid floors are spaced from about 3.05 to 3.9 m (10 to 12.8 ft), except where it is intended to carry heavy cargo or have alternate hold loading, in which case the floors are generally spaced less than 3.05 m (10 ft). Longitudinal side girders are spaced about 4.6 m (15 ft); where the vessel is intended to carry heavy cargo or have alternate hold loading, their spacing should not exceed about 3.05 m (10 ft). Longitudinal framing of the double bottom is considerably more efficient in resisting plate buckling under compressive stress than a standard transverse floor system. The classification societies recognize this by permitting a reduction in bottom shell thickness when longitudinals are used. For construction reasons, the longitudinals are usually stopped near the ends of the ship and transverse solid floors are installed on every frame. Fig. 19 shows a longitudinally-framed double bottom under construction.

The thickness of bottom plating is determined not solely by the water pressure, but also by the length of the ship. This is because its paramount duty is forming part of the bottom flange of the hull girder. It must, of course, be checked for special local loads where these exist. The outboard strake of the inner-bottom plating, called the margin plate, is generally thicker than the remainder of the plating where it is sloped down outboard to form a bilge space for drainage; this provides an extra allowance for corrosion. Where approximately horizontal, the margin plate may be of the same thickness as the rest of the inner-bottom plate. Fig. 20 shows the two different arrangements of margin plates.

8.3 Design Based Upon Direct Engineering Calculations. One way to evaluate the effect of large lightening holes on the stress levels of a double-bottom structure is to conduct computer-aided engineering calculations. After a large-scale analysis indicates the general level of stresses in the transverse floors and longitudinal girders of the double bottom, a portion of the more highly stressed structure is usually analyzed for the determination of local stresses. Fig. 18 shows a portion of the bottom longitudinal girder of a bulk carrier indicating the areas of highest stress as obtained from a fine-mesh finite-element analysis. Reinforcements in these areas were added to reduce the stress levels to acceptable limits.

8.4 Solid Floors in Double Bottoms. Solid floors consist of vertical plates extending from the bottom shell to the inner bottom, welded directly to the shell and inner bottom, Fig. 20 (B). Such floors, unless they form watertight or oiltight divisions between compartments, usually have large holes for access and weight saving. Floors forming the ends of water or oil tanks within the double bottom are, of course, watertight or oiltight. Their scantlings are specified by the ABS Rules to be the same as those of floors under the boiler space. A similar requirement covers solid floors; they must be fitted on every frame under boilers and engines, under the toes of bulkhead stiffener brackets, and throughout the forward one-fifth length where the bottom is transversely framed. This may be extended to the forward one-quarter length if the machinery is aft, or if the ship has relatively high speed.

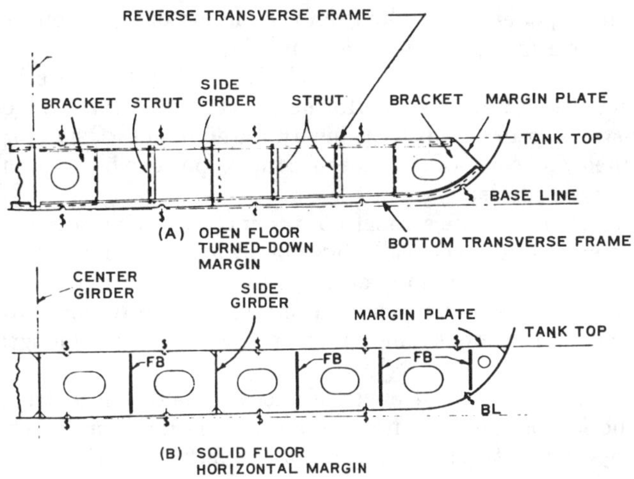

Fig. 20 Double-bottom floors

8.5 Duct Keel Construction. A *duct keel,* Fig. 21, forming a centerline pipe tunnel within the double bottom, is sometimes fitted to provide a convenient place to run fore-and-aft piping. The side plating of such a tunnel should be of the same thickness as that of the watertight floors, and, for continuity, it should overlap the normal center girder at the ends of the duct keel construction for about three frame spaces. The Rules also require an increase in flat keel thickness in way of the duct keel.

8.6 Strengthening of Forward End. Damage to ship forebody structures has caused concern, and the increasing demand for vessels of higher speed and larger size has magnified the inadequacy of some forebody arrangements to meet the service requirements under heavy weather conditions. The most common forms of damage to the forebody are the setdown of deck beams and longitudinal plating, the buckling of pillars between decks, and various damages to the bow flare region. On the whole, forebody

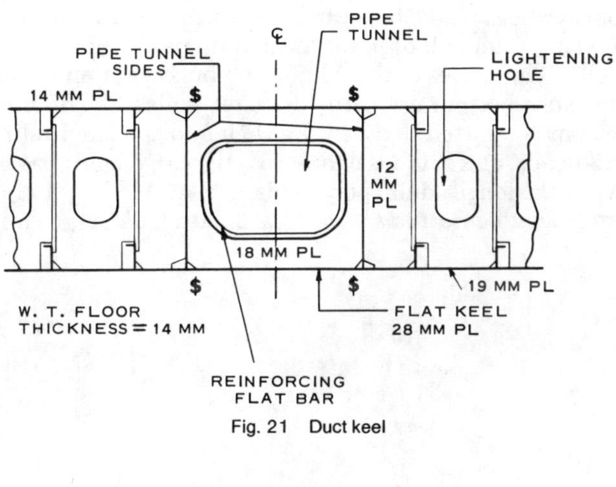

Fig. 21 Duct keel

damage generally results from the impact of waves on the vessel during severe weather conditions.

a. Direct Engineering Calculations. Direct engineering calculations have been performed by ABS to determine realistic pressure loads to be incorporated in new design formulas, and some of the general conclusions can be summarized as follows:

1. The ultimate strength of the forebody structure of a damaged ship is generally, but not always, less than that of the forebody of an undamaged vessel of similar length.
2. The maximum expected impulsive load tends to increase as the ship's length increases, within typical length ranges.
3. The ultimate strength of a local structure is a strong function of the stiffening systems. The influence of the supporting structure should be properly taken into account.
4. For determining design loads for deck beams and side frames in the forebody region, effects of the ship's freeboard, speed, and bow flare shape must be taken into consideration.

b. Rule Requirements. The frequency of repairs required to the forward end, of which the forefoot is one of the most vulnerable areas, has led to the requirements for increased strength in the ABS Rules under "Fore-end Strengthening," and in the Lloyds Rules under "Strengthening of Bottom Forward."

8.7 Bilge Plating. On river vessels, additional thickness for the bilge plating is advisable for grounding reasons. The preferred stiffening for large-radius bilge plates is an inverted fore-and-aft angle, welded at about the 45-deg line. Bilge plates bent to a radius of 305 mm (12 in.) or less are normally left unstiffened, forming what is known as a *soft bilge,* which, on heavy contact, permits some deflection and deformation but is seldom damaged to the point of rupture.

If a river barge is expected to traverse open water for a portion of the voyage, it is advisable to carry the grounding frames the full length of the rakes up to the headlog, as defense against pounding.

8.8 Great Lakes Bulk Carriers. Requirements for double-bottom construction of Great Lakes bulk carriers are given in the ABS Rules for Great Lakes Bulk Carriers. They differ appreciably from the requirements for oceangoing vessels. Lake bulk carriers are usually provided with two longitudinal bulkheads located inboard from the side shell. The spaces outboard of these bulkheads and below the inner bottom are used for water ballast. The inner bottom is omitted between the shell and the longitudinal bulkheads, and extends only across the cargo hold area between the longitudinal bulkheads.

a. Double Bottoms. In Great Lakes bulk cargo ships, the floors are considered to span between the longitudinal side tank bulkheads rather than from side shell to side shell. Since there are no cargo 'tween decks, the entire cargo deadweight in vessels not having self-unloading cargo gear is carried on the inner bottom, which must be designed to carry cargo loads plus the cargo-unloading rig shock loads. These loads are greater than the hydrostatic pressure on the bottom shell. In vessels with self-unloading cargo gear, the cargo load is carried by the side or centerline hoppers or both. The double bottom is subject to the external hydrostatic pressure only and is supported by the bulkheads under the hoppers. To provide a maximum of ballast space, the depth of the double bottom often exceeds the Rule requirements for normal oceangoing ships, and is usually between 1.52 and 1.83 m (5 and 6 ft), except for self-unloading bulk carriers, in which the depth of the double bottom is generally about 0.915 m (3 ft).

b. Inner-Bottom Plating. With respect to the scantlings of the inner bottom, Great Lakes ships again depart from the Rule requirments for oceangoing ships. In addition to its contribution to the longitudinal strength of the vessel, the inner bottom is subject to local damage resulting from the impact of the unloading equipment in use on the lakes. Therefore, the ABS Rules for these ships have included in the inner-bottom plating requirement an increase for such loads. For self-unloading bulk carriers, where the inner-bottom and sloped-wing tank boundaries in cargo spaces are not subject to such severe loading, the required thicknesses do not include this increase. The inner bottom is sometimes constructed of abrasion-resistant or corrosion-resistant steel where it is exposed to cargo loads. In many rebuilt Great Lakes bulk carriers, the inner bottom is constructed of channels oriented transversely, or, in some cases, longitudinally, with the backs of the channels forming the upper surface of the inner bottom.

8.9 Oceangoing Bulk and Ore Carriers. The ABS Rules give requirements for the thickness of the inner-bottom plating for these vessels. Where there is no ceiling and where cargo is handled by mechanical means, it is recommended that the inner-bottom thickness be increased throughout the cargo holds. For ore carriers, the Rules also recommend an increased minimum thickness for the inner-bottom plating because of the impact loads.

8.10 Inner-Bottom Plating in Barges. In the case of coastal or harbor barges where *high tower* rigs are used for unloading coal, it has been found that 19 mm (0.75 in.) inner-bottom plate is the minimum acceptable in association with 610 nm (24 in.) frame spacing; heavier plating up to 25 mm (1 in.) is sometimes used. ABS Rules recommend that the inner-bottom thickness in river and offshore barges be increased by specified amounts where the cargo is handled by grabs.

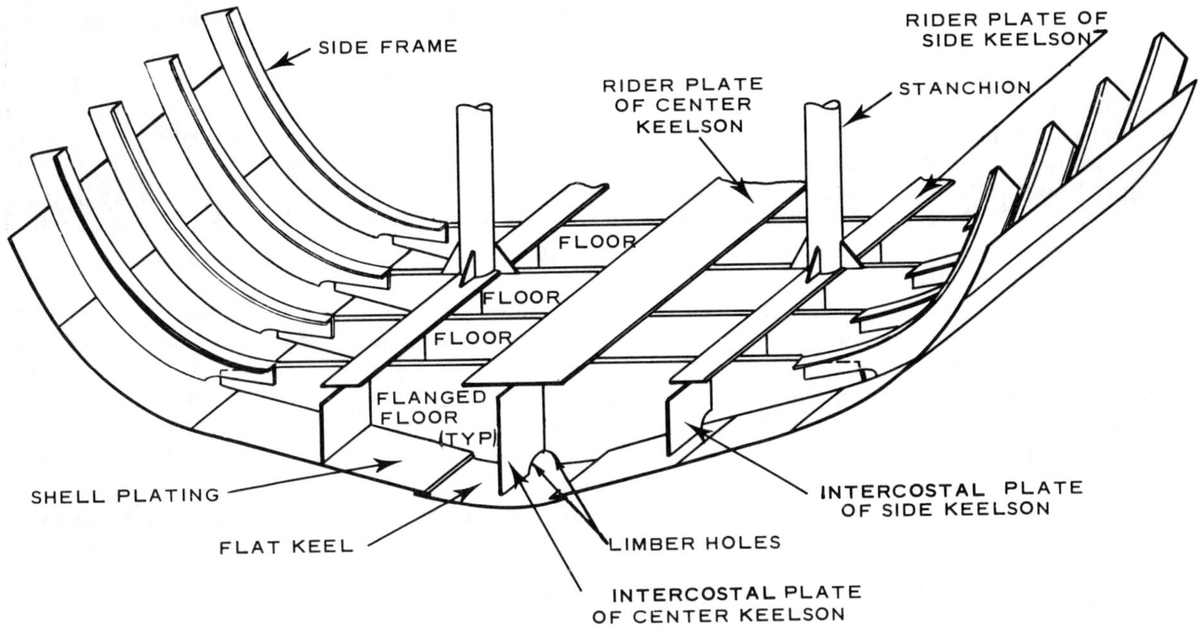

Fig. 22 Small vessel single-bottom construction

Section 9
Single-Bottom Construction

9.1 Use of Single-Bottom Construction. The advantages of double-bottom construction discussed in Section 8 are so great that single-bottom construction is found principally in relatively small ships, tankers, and other special vessel types as illustrated in Fig. 22.

9.2 Design Considerations in Small-Vessel Single-Bottom Construction. The single bottoms of small vessels are either transversely or longitudinally framed. They are designed for the external water pressure, for liquids in tanks, and tacitly for keel block loads while docking. This basis of

Fig. 23 Typical tanker bottom construction

Fig. 24 Typical tanker midship construction

design is evident in the ABS "Rules for Building and Classing Steel Vessels Under 61 Meters (200 Feet) in Length." Where transversely framed, floors or bottom frames are spaced from about 457 to 610 mm (18 to 24 in.). The design length of floors is taken as the distance between their intersections with side frames, or frame brackets, and they are required by the above Rules to be provided with center and side keelsons of specified scantlings. Bottom transverse frames are not required to be provided with side keelsons, but a center keelson the same depth as the bottom frames is required for docking. The design length of bottom frames is either the same as for floors or is often reduced by providing effective supporting girders. Longitudinal frames are generally spaced from about 305 to 610 mm (12 to 24 in.) and are supported by transverse webs spaced from about 1.2 to 2.4 m (4 to 8 ft.). A center girder or center keelson is required for docking. The length of the ship is not a factor in the scantlings of these floors.

The center keelson and side keelsons are intended primarily to keep the floors from buckling, or *tripping,* and to assist them in acting together. When the ship is in dry dock, the center keelson has the additional duty of transferring the pressure from the keel blocks into the floors and bulkheads, and thus into the hull as a whole. The scantlings of these members increase with the length of the ship because of their contribution to the longitudinal strength of the hull.

9.3 Design Considerations in Tanker Single-Bottom Construction. Traditionally, tankers have carried their ballast in their cargo tanks; hence, there was no need for segregated ballast areas and single-bottom construction was utilized to provide maximum cargo capacity. Recent regulations will require segregated ballast in tankers, and in future tanker designs a partial double bottom may be included to provide bottom ballast tanks. Figs. 23 and 24 show typical tanker structure. The bottom longitudinals are supported by bottom transverse webs, which are in turn supported by the side shell, longitudinal bulkheads, and in many cases by the centerline or side girder. The design load results primarily from the difference between the external water pressure and the cargo load. The bottom transverses are assumed to have equal fixity at the side shell and longitudinal bulkhead, but the flexibility of a centerline or side girder reduces the support provided by these members to the bottom transverse webs. A two-dimensional web frame computer analysis is often used for transverse structure design.

Careful consideration must be given to cargo drainage and complete air circulation for gas freeing when designing a tanker structure. To facilitate this, many cutouts will be necessary. Care must be taken in designing the cutouts to insure proper structural continuity and adequate local strength of the cut member.

Section 10
Shell Plating

10.1 Classification Society Rules for Shell Plating. The thickness of shell plating as specified in classification society rules is based upon engineering calculations and research, and is heavily influenced by operational experience as well. In general, the plating thickness in large ships is based on hull girder requirements, but hydrostatic and hydrodynamic water-head criteria may take precedence around the ship mid-depth. For small ships, these criteria are the determining factor. For all types and sizes of ships, a margin for corrosion is generally provided, so that after a reasonable period of service sufficient plating will remain for structural purposes. This allowance is not greatly affected by the size of the ship.

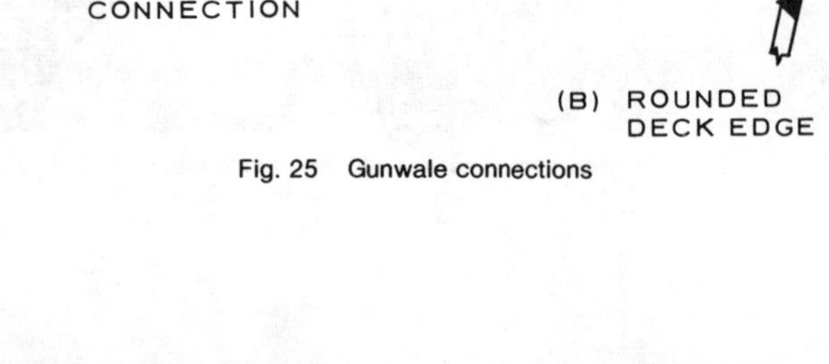

Fig. 25 Gunwale connections

The classification society rules give thicknesses for the midship portion of the bottom and side shell; the width and thickness of the sheer strake for various lengths and depths of hull; the width and thickness of flat plate keel, and the thickness of various special areas, such as shell plating at ends, bottom plating forward, immersed bow plating, bossing plating, and plating attached to the spectacle frame and stern frame. The midship thicknesses are normally maintained throughout the midship 40-percent length. Forward and aft of the 40-percent length, the plating thickness is gradually reduced to the *end* thickness, which extends for 10 percent of the length by the ABS Rules and 7.5 percent of the length by Lloyds Rules.

In the case of Great Lakes ships, which are designed for different wave and loading conditions, the midship thicknesses are maintained for about two-thirds to three-quarters of the ship's length.

At critical locations such as the lower turn of the bilge and sheer strakes, notch-tough steel is required for vessels 137 m (450 ft) or more in length. For these strakes, according to ABS Rules, the special materials are to be extended throughout the midship 40-percent length.

10.2 Strengthening for Navigation in Ice. If a ship is to navigate in ice and to receive the special classification for such service, the shell plating thickness is increased between the lightest and the deepest winter service waterlines to the degree specified in classification society rules, with additional intermediate frames to the extent required for the particular class of strengthening against ice.

a. Requirements Based Upon Finnish-Swedish Ice Class Rules. These requirements were developed for ships operating in the northern Baltic in winter. The increased thickness of shell plating and its extent are functions of frame spacing, material strength, and ice belt pressure, as indicated in the ABS Rules.

b. Navigation in Ice-General Service. For navigation in ice, both ABS and Lloyds Rules specify increased shell plating thickness. For the highest class of ice-strengthening, these thicknesses exceed the midship Rule thickness by 80 percent forward, 40 percent amidships, and 25 percent aft; specified plate thicknesses for this highest class are between 14 and 32 mm (0.55 and 1.25 in.) according to ABS Rules and between 12.5 and 25.5 mm (0.5 to 1.0 in.) according to Lloyds Rules.

10.3 Breaks in Shell Plating. The shell plating, as the principal member of the hull girder, is subject to especially high concentrated stresses wherever abrupt discontinuities exist. Such discontinuities occur principally at the ends of partial superstructures and in way of large openings, such as loading ports.

The need to provide a gradual, easy transition from one topmost strength deck to another at the ends of a bridge or a long poop or forecastle has long been recognized. At best, these are areas of high concentration of stress when the hull girder is bending in a seaway, and sudden discontinuities are potential sources of plate fractures. Classification society rules clearly emphasize this.

10.4 Shell Plating in Way of Hawsepipe. The shell plating surrounding and directly below the hawsepipe is subject to severe blows from the anchor as it swings against the ship while being hoisted, and later as it is snugged hard into the hawsepipe. Doublers or heavy insert plates are customary around and below the hawsepipe. The plating below the hawsepipe is usually flush at the seams or plated vertically. Short headers may be required behind the shell at points where the anchor contacts the shell when it is fully stowed, although the hawsepipe bolster should prevent anchor contact with the shell.

10.5 Gunwale Connections. The gunwale connection joining the side shell to the weather deck is one of the most important in the ship. A typical welded joint, Fig. 25 (A), shows notch-tough steel for the sheerstrake and stringer plate.

A variation of this connection using a rounded plate welded to shell and deck, Fig. 25 (B), is sometimes used. It has the advantage of eliminating the exposed plate edge of the sheer strake. Disadvantages include a reduction of deck area, difficulty of making a transition to a tee joint at the ends of the parallel midbody, and additional construction cost. The radius of this plate is governed by a strain criterion, which is expressed as:

$$E = \frac{0.65t}{R} \tag{3}$$

where:

t = plate thickness

R = bend radius

$E \leq 0.03$; otherwise, stress-relieve.

Section 11
Deck Plating

11.1 Classification Society Rules for Deck Plating. Classification society rules specify minimum thicknesses for decks, depending on their function, and also give the required section modulus of the hull girder as a whole. This section modulus requirement usually controls the thickness of the strength deck, and in some cases the thicknesses of the lower decks which are effective for longitudinal strength because of continuity.

In the case of *platform* decks, e.g., those which do not contribute to the longitudinal strength of the ship, the scantlings are based on local loading and vary with beam spacing and 'tween-deck height. Special consideration must be given to plating of any deck on which cargo-handling forklift trucks will operate during cargo-handling operations. ABS gives a specific guidance for determining deck scantlings for the highly concentrated loads of forklift trucks where they are used.

Decks forming watertight or oiltight boundaries are also designed to withstand the appropriate design head, with a special corrosion allowance required for decks forming the tops of tanks. Fig. 27 shows weather-deck plating under construction.

11.2 Application of Direct Engineering Calculations for Unusual Configurations. The case of a containership, with hatch openings which can approach 85 percent of a ship's beam, is an example of an unusual configuration which lends itself to direct engineering calculations for the determination of deck plating thicknesses. Abrupt changes in deck stiffness also accentuate longitudinal stresses due to warping restraint present at the closed ends of the vessel and at the engine room housing. The torsional rigidity of the ship's hull girder must be thoroughly investigated in conjunction with an oblique wave loading on the ship.

11.3 Deck Plating of River Vessels. In river vessels, there are certain commercial minimums which have been established by experience; for instance, in the case of barges carrying sand, gravel, or similar cargos on deck where grab buckets are used in handling the cargo, it has been deter-

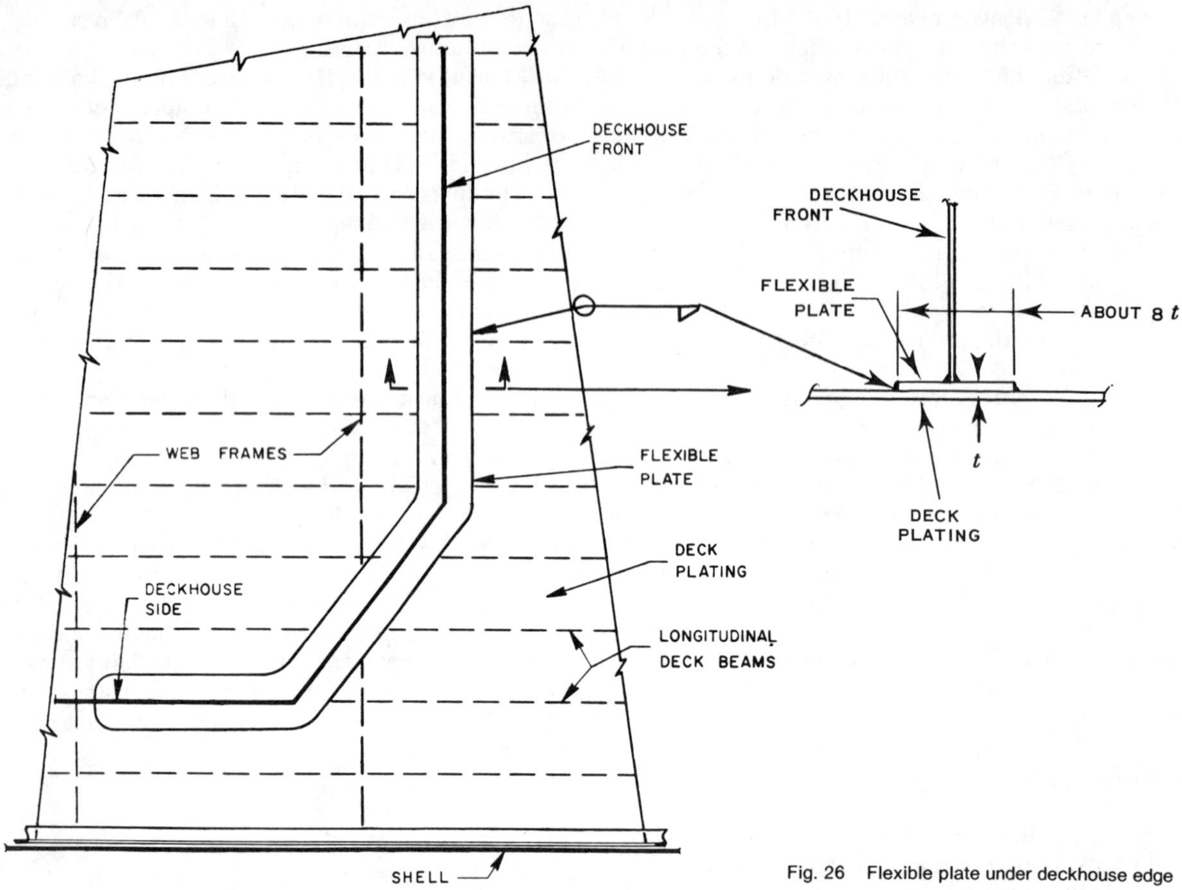

Fig. 26 Flexible plate under deckhouse edge

Fig. 27 Weather-deck plating

mined that the minimum acceptable thickness of deck plating is about 9.5 mm (0.37 in.) supported by framing at not over 381 mm (15 in.) on centers. The 9.5 mm minimum plating will show some denting and deformation under the blows imposed by the buckets normally used at river terminals for such bulk items as sand, gravel, coal, and sulphur, and the use of 11 or 13 mm (0.44 or 0.50 in.) plating is recommended.

11.4 Deck Plating of Ocean Barges. The decks of ocean barges may be subject to uniform loading such as cut lumber or to concentrated loads such as those imposed by wheeled or tracked vehicles.

a. Uniformly-Loaded Decks. The deck plating thickness within the midship 40-percent length must be sufficient to satisfy the longitudinal hull girder strength requirements. In addition, the ABS Rules for Steel Barges for Offshore Service specify minimum thicknesses based on the kind of framing (longitudinal or transverse) and the frame spacing, and a smaller minimum thickness within the line of openings. Specific requirements for the thickness of deck plating are also given in the above-mentioned Rules.

b. Concentrated Loads on Decks. The deck plating in way of deck concentrated loads must also satisfy minimum thickness requirements given in the ABS Rules for Steel Barges for Offshore Service. If the height of cargo exceeds 2.44 m (8 ft), the minimum thickness of plating in cargo spaces is increased by 0.83 mm for each meter (0.01 in. for each foot) of excess height.

11.5 Openings in Deck Plating. The large openings in strength decks necessary for cargo hatches, machinery casings, and stairway and elevator wells in large passenger ships are potential sources of weakness. The stress concentrations at the corners of such large openings demand particular attention to workmanship and detail in these areas. As in way of openings in the shell plating, generous corner radii are required.

For the hatch corner structure to be relatively free from fracture, it has been the practice to ensure generous corner radii, with the plate edge in way of the radii smooth and without welded attachments to it. In addition, deck butt welds and pipe or other openings should be kept well clear of the hatch corners. Also, hatch side girders, deck longitudinals, and longitudinal bulkheads should always be carried well past the corners, and the hatch coaming should preferably be extended beyond the hatch corners in the form of vertical tapered brackets.

Where extra-large hatches are provided, as in containerships, special steels resistant to cracking may be used for the deck plating at the hatch corners. Large hatches also interrupt a large percentage of the deck available for longitudinal strength. Providing sufficient area for the section modulus may result in very thick deck plating or doublers. For this reason, longitudinal box girders or continuous longitudinal bulkheads are sometimes fitted between the strength deck and second deck, replacing the outboard hatch girders and providing the cross-sectional area necessary to meet the section modulus requirements.

11.6 Deck Plating at Ends of Superstructures and Deckhouses. Section 10 mentioned the necessity of strengthening the shell in way of the ends of superstructures. Precautions must be taken in these areas to ensure that sufficient strength exists in the deck plating.

The construction at the ends of deckhouses, i.e., erections which do not extend to the side of the ship, is not so well covered in the Rules, and there has been some difficulty in these areas. As discussed in Section 4, provision must be made for transfer of a considerable but indeterminate amount of load from the ends of the deckhouse to the main hull girder. The situation can sometimes arise in which it is very difficult to support the corner of a deckhouse with an adequate amount of structure. In this case, use is sometimes made of a flexible plate, Fig. 26. The end of the deckhouse is brought down in the center of a deck doubler plate which is attached to the deck only at the edges. The doubler plate then relieves the hard spot formed by the deckhouse end landing on relatively flexible deck plating by flexing upward when a tensile load is applied. It also provides extra thickness to reduce the abrupt change in curvature of the deck plate under compression. As explained in Section 4, this abrupt change in curvature produces a high stress concentration. Deck girders and transverses in line with deckhouse sides and ends help prevent deck fractures at the ends of deckhouses.

11.7 Decks in Refrigerated Ships. Some fractures have occurred in refrigerated ships because the plating of uninsulated lower decks in refrigerated holds is subject to extreme cold while the shell plating to which it is attached is much warmer. The resulting unequal contraction has sometimes resulted in brittle fracture of the deck plating. In some refrigerated vessels, the plating of the lower platform decks has been isolated completely from the shell plating by stopping it at the inboard side of the frames, with apparent success. In other ships, a riveted seam placed well outboard to serve as a crack arrestor has been used.

11.8 Deck Plating Between Openings. Deck plating between and within the outboard edge of main openings, such as cargo hatches and machinery casings, usually is made lighter than the continuous plating outboard of the openings, and this arragement is undoubtedly an efficient distribution of material. However, it should not be assumed that this lighter plating always experiences a lower stress than the heavier outboard plating. As the ship bends in a seaway, this plating must necessarily experience approximately the same elongation and, therefore, approximately the same stress as the heavier outboard plating. The exception is a small tapering area just forward and aft of the openings. It follows that, if higher-strength steel is adopted for an upper deck to permit higher working stress, this plating should extend a suitable distance inboard of the line of hatch openings.

On Great Lakes ships, the hatches are very closely spaced throughout about two-thirds of the ship's length. In this case, the earlier approach was that the light plating between the hatches was not long enough to develop the full hull bending stress through shear from the heavier outboard plating. Therefore, the assumption was that effectively there was only one opening. Major treatment was applied at the ends, where stress is low. Trouble at the intervening hatch corners was largely eliminated by lapping the light plating between openings onto the heavy stringer plates and riveting the seam. With the adoption of all-welded construction (elimination of riveting even for crack arrestors), the practice is to provide rounded corners and extend the heavy plates inboard of the hatch sides. The light plating between hatches is welded to the heavier outboard plating.

Fig. 28 Framing decks and bulkheads near stern

Section 12
Transverse Side Framing

12.1 Functions of Transverse Side Framing. The side frames of a ship have two principal functions. First, they help to resist water loads applied to the outside of the shell. This function is recognized in the ABS requirement that "Frames are not to have less strength than is required for bulkhead stiffeners in the same location in association with heads to the bulkhead deck." In addition, they serve to stiffen the side plating against buckling. Second, they furnish vertical support to the outboard ends of the beams supporting the several decks. Thus, the Rule formula for frame properties takes into consideration the load supported by the beams. In addition, side frames resist loads imposed by contact with piers, docks, and lock sides. This loading is particularly important for Great Lakes ships and vessels engaged in inland navigation.

12.2 Application of Direct Engineering Calculations in Design. The connection of the side framing to the deck and horizontal stringers involves problems of continuity in transmitting loads. The brackets used for these connections must be carefully designed to avoid discontinuities and provide for the proper transition of loads. Since shipyard practices relative to brackets vary widely, engineering calculations are necessary to evaluate the resulting local stresses and stress concentrations for the various loading conditions. If the calculated stresses are excessive, minor local modifications, such as the extension of bracket face plates, can usually alleviate the problem. A simple method of insuring that there are no areas of local overstress is to plot the bending moment, section modulus, and resulting stress along the member in question. From this, the extent of bracketing necessary will be obvious.

12.3 Racking Stresses in Side Framing. A third function commonly attributed to side frames in association with their top and bottom connections and the shell plating is that of contributing resistance to racking stresses caused by rolling of the ship and by impact of seas against the topsides. Actually, the rigidity of the main transverse bulkheads against racking is so much greater than that of the frames that the bulkheads do practically all of the work of holding the hull against racking distortion; the contribution of the frames in this respect is almost negligible, except when the bulkheads are widely spaced or omitted entirely.

12.4 Panting Stresses in Side Framing. The sides of the hull near the bow, and, to a lesser extent, near the stern, are alternately deeply immersed and under heavy pressure and, moments later, completely out of water and under no pressure at all when the ship is pitching heavily. This action causes what are called *panting stresses*. The conditions are similar to those requiring special strengthening of the forward bottom structure mentioned in Subsection 8.6. The effect is aggravated in the bow side framing by the heavy impact of the seas as they surge against the ship at the combined speeds of the ship and of the oncoming waves.

The ABS Rules require panting webs and stringers to be fitted aft of the forepeak bulkhead and forward of the afterpeak bulkhead "As may be required to meet the effects of sheer and flatness of form." Two kinds of arrangements are recommended: either *panting beams* in association with stringer plates fitted between the beams and supporting the intermediate frames, or web frames at a gradually increasing spacing aft of the forepeak bulkhead in association with narrow stringers in line with the stringers (breasthooks) within the forepeak. Framing in the peaks is also specially considered in the Rules, with maximum span limited and special supporting structure required.

The Lloyds Rules contain a special section, titled "Panting," requiring strengthening for this purpose for 15 percent of the ship's length aft of the bow and aft of the afterpeak bulkhead.

12.5 Strengthening for Navigation in Ice. Ships intended to navigate in ice require special strengthening of the shell and framing from a nominal distance above the winter loadline to a nominal distance below the lightest service waterline to resist the action of the ice. The ABS and Lloyds Rules have special requirements for this purpose. For side framing, the strengthening required by both so-

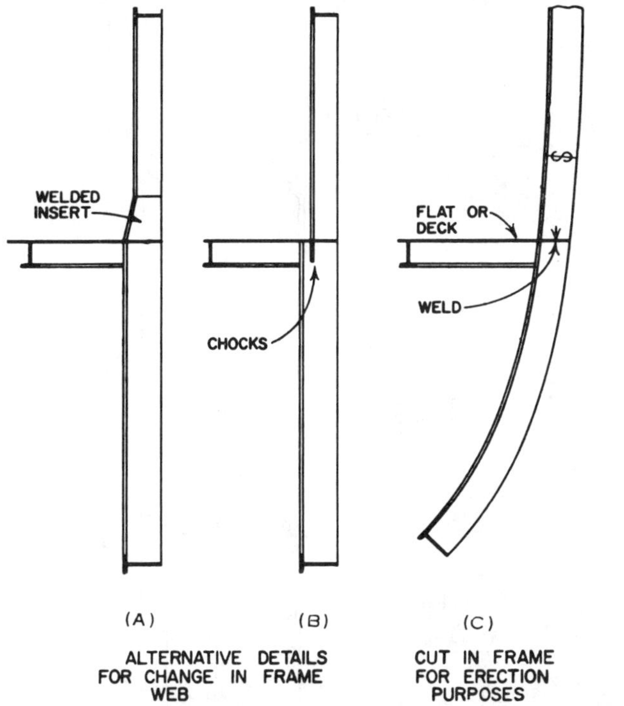

Fig. 30 Joints in side frames

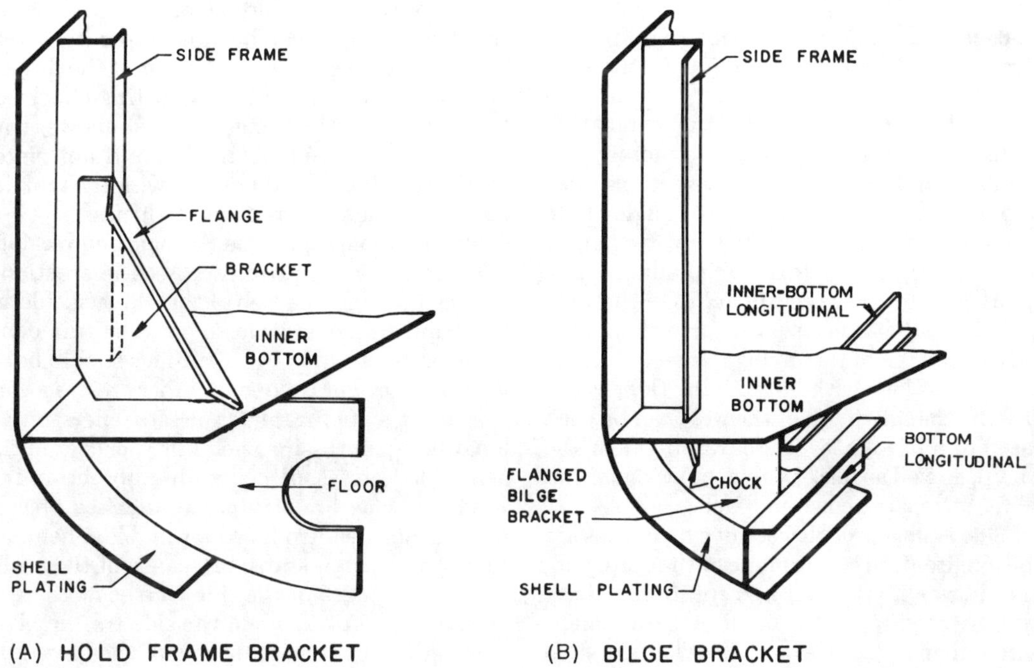

Fig. 29 Lower connections of hold frames

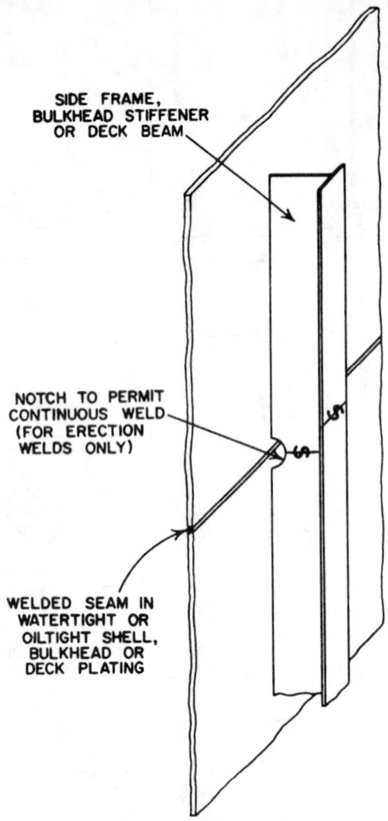

Fig. 31 Cutout in way of weld

cieties consists principally of intermediate transverse frames extending over the area from above the winter loadline to near the inner bottom.

12.6 'Tween-deck Frames. The section modulus of 'tween-deck frames plus plating, as given in both the ABS and Lloyds Rules, depends on the height and vertical location of the 'tween decks and the spacing of the frames. In the ABS Rules, the length of the ship is also a factor.

12.7 Sections Used for Side Framing. The sections most commonly used for frames are flat bars, inverted angles, bulb plates, flanged plates, and T-sections. The inverted angle may be either a rolled section or cut from a channel. The latter is usually cut so as to leave the full depth of the web. In some cases, channels have been used for welded side frames with the heel and toe of the faying flange welded to the shell plating. H-sections with both faying flanges cut off, if desired, leaving the full depth of the web, can be used as T-sections, as shown in Fig. 15. Side framing near the stern of a large ship, as well as deck beams, bulkheads, and other structural features, are shown in Fig. 28.

12.8 Bevel of Side Framing. The webs of side frames are set normal to the longitudinal centerplane of the ship, and, as a result, are not normal to the shell, except in way of the parallel midbody of the ship. This facilitates alignment with transverse members, while the angularity at the shell does not cause a severe construction problem.

12.9 Lower Connections of Hold Frames. The lower ends of hold frames are usually attached to the margin plate of the inner bottom by large *hold frame brackets,* usually flanged plates as shown in Fig. 29 (A). The web of the bracket is welded to the inner bottom and the flange cut clear of the inner bottom.

Where the hold frame brackets interfere with cargo stowage, they are eliminated by running the frame down to the inner bottom and welding it. In this case, the flange must be welded down and a chock provided. Fig. 29 (B) shows the bilge bracket used as a connection for the lower end of the hold frame in cargo ships with a longitudinally-framed double bottom. The bilge bracket is fitted at every frame not having a solid floor. The scantlings of the frame must be increased when the hold bracket is eliminated due to the longer span considered, since, for rule purposes, the length is measured to the top of the bracket.

12.10 Joints in Side Frames. The hold frames are generally longer, and therefore heavier, than is necessary for the upper 'tween-deck frames, and, even if the hold size could be carried through in one long length, considerations of subassembly and erection usually make it desirable to fit the frames in more than one piece. At the resulting joints, effective continuity should be maintained.

In welded construction, it might seem that a 100-percent butt-welded joint could be accepted at any desired place in the frame, but a backing strip should be used to ensure full penetration of the weld in the back side of the flange, where welder's access and visibility are poor. ABS will accept a frame butt made in way of a deck or flat, as in Fig. 30 (C), when there is assurance that the frames above and below the decks are lined up accurately. Careful workmanship, checked by inspection, is necessary in this connection. Where the frame is not considered as a continuous beam and the depth of the frame changes by more than about 25 mm (1 in.), provision should be made to maintain continuity of the flanges. This can be done either by widening the shallower section, as in Fig. 30 (A), or, if the difference in depth of section is considerable, by fitting chocks on the deeper frame opposite the flange of the shallower frame, as shown in Fig. 30 (B). Side frames should not pierce the tops of deep tanks; they should be cut, with the tank top continuous out to the shell to ensure its tightness.

12.11 Crossing of Side Framing and Welded Shell Seams. Wherever a welded frame crosses an erection seam weld in the shell plating, the web of the frame should be notched out to permit the seam weld to pass the frame intact and continuous, as shown in Fig. 31. This should be done generally where a nontight welded member crosses a continuous weld in a tight structure, such as stiffeners on a bulkhead or longitudinals crossing shell butts, except in the case of subassembled plates, discussed in Subsection 14.5.

12.12 Side Framing of Great Lakes Ships. Special requirements are introduced for the side shell framing of Great Lakes bulk cargo ships. Because of the absence of 'tween decks, the length of the side frames might appear to be excessive. However, when the side framing is considered in association with the stiffening of the longitudinal side tank bulkheads, it becomes apparent that the framing can be considered as a unit, and by the introduction of horizontal struts and longitudinal stringers the whole can be considered as a truss. Since Great Lakes ships are in almost daily contact with either docks or lock walls, it is customary to increase the scantlings of the side frames to resist local damage.

Section 13
Transverse Deck Beams

13.1 Functions of Transverse Deck Beams. Transverse deck beams have three principal functions: working with the plating to resist in-plane compressive loads, working with it for local loads normal to the deck, and acting as struts holding the sides of the ship apart against inward pressure of the sea at deep draft and in heavy weather. The latter function is only indirectly reflected in classification society rules (through a minimum span to be used for calculations), presumably because the combination of beams and deck plating, sufficiently strong to carry the design deck loads is, in general, more than sufficient as struts giving horizontal support to the side frames. This is particularly true where several decks extend throughout most of the vessel's length. Unusual cases, such as a large ship with few decks, should receive special consideration. Of course, deck beams also act as supports for the side framing.

13.2 Transverse Deck Webs—Longitudinal Framing. Longitudinal deck stiffeners, although they satisfy the bending stress criteria, need support to withstand a variety of local loads. Hence, the established practice has been to use transverse deck webs at intervals not exceeding the specified maximum as obtained from the ABS Rules. It is also recommended that deep girders be arranged in line with webs and stringers to provide complete planes of stiffness. Although these deck webs are very common on tankers and bulk carriers, they present special problems associated with the use of space in cargo vessels, as described in Subsection 7.2.

13.3 Strengthening of Weather Deck Forward. The weather deck at the forward end is subject to extremely heavy loads in heavy weather, because of the impact of seas on the deck at the end of a downward plunge of the bow into an oncoming sea. Rule requirements take this condition into account by specifying additional strengthening in the weather deck forward.

13.4 Effect of Suspended Cargo. If cargo is suspended from a deck, as in the case of meat hanging from the deck overhead, the deck beams must be strengthened to carry the suspended load in addition to the normal load on the deck. This is done, when using the ABS Rules, by increasing suitably the height of cargo in the formula for obtaining the Rule scantlings for the beams. The deck beams under the hanging load cannot be correspondingly reduced, however, since during its life the ship may carry a normal deck load.

13.5 Hatch End Beams. At the forward and after ends of main hatches, heavy hatch end beams are sometimes fitted to support the ends of the hatch side girders. If hold pillars are fitted at the hatch corners to support the hatch side girders, or if the distance to the supporting transverse bulkheads is short, special hatch end beams may not be necessary. Typically, hatch end beams are supported either at the centerline or at the hatchway corners by pillars, and at the sides of the ship by web frames. They carry heavy concentrated loads from the hatch side girders and must be designed accordingly.

13.6 Transverse Deck Webs on Great Lakes Ships. Great Lakes bulk carriers require completely open cargo holds with a minimum of transverse bulkheads and no pillars or stanchions. These requirements lead to special problems in the support of the main deck (sometimes called the *spar* deck on the Great Lakes). The solution is to provide heavy transverse arches between cargo hatches. These are spaced either on 3.66 or 7.31 m (12 or 24 ft) centers, depending upon the hatch spacing, and are designed to carry the deck loads and hatch coaming loads. These loads are transmitted to the arches through large fore-and-aft brackets extending from the arch to the hatch end coamings. These arches also

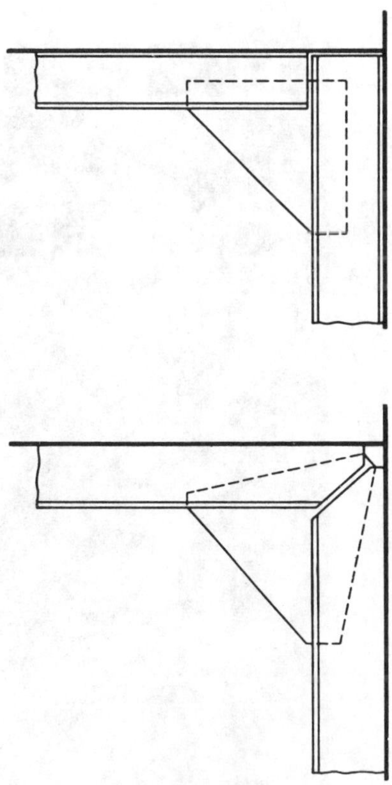

Fig. 32 Beam knees

assist materially in resisting the racking stresses in the ship, and compensate for the lack of transverse bulkheads.

13.7 Beam Knees. Beam knees or brackets facilitate the attachment of the beams to the frames, and serve to support the outboard end of the beams. While they contribute in some degree to the ship's resistance to racking, as discussed in Subsection 12.3, analysis will show that, compared to the great resistance to racking provided by the transverse bulkheads, the contribution of the small beam knees is negligible. Their main function is to support the beams and increase their strength somewhat by reducing the effective span of the beams and the frames. The size of beam knees is that required to obtain the necessary amount of weld (Fig. 32). The thickness, as specified in the Rules, depends on the size of the bracket, and the larger brackets are flanged on the unsupported edge. When special circumstances have made it desirable, such as improved construction of insulation in refrigerated ships, ABS has approved omitting the brackets entirely, since ample shear strength can be obtained by welding the beams directly to the frames. In this case, due allowance must be made for the increase in effective span of the beams and frames.

Section 14
Bulkhead Stiffeners and Plating

14.1 Subdivision and Compartmentation. The subdivision of passenger ships is subject to the requirements of the International Convention for the Safety of Life at Sea (SOLAS), and is rigidly controlled under the administration of government authorities of the various maritime countries (in the United States, the U. S. Coast Guard). Subdivision bulkheads, according to Robertson (1967) and the 1974 SOLAS Convention (premised on a compartmentation philosophy going back at least to the 1929 SOLAS Convention), are to be so spaced as to enclose a definite portion of the *floodable length* of the vessel. This approach now has been discredited as not providing a continuous improvement in the survival ability of the ship with an increase in the pertinent requirements.

An entirely different concept, premised upon probability of enduring damage, has been developed by IMCO and is presently a desirable alternative to the foregoing procedure. In the International Convention on Load Lines (1966), there are provisions that certain dry cargo vessels may be permitted reductions in freeboard provided the vessel can survive the flooding of one compartment at an increased draft. Some of these provisions apply to oil carriers with

Fig. 33 Typical transverse strength and subdivision bulkhead

empty compartments when at full-load draft. In addition, the 1973 International Convention for the Prevention of Pollution from Ships includes limitations on the size of individual cargo tanks in tankers and on the floodability of their machinery spaces. Similar subdivision limitations are included in the IMCO Codes pertaining to ships carrying liquefied gases and dangerous chemicals (see Chaper XI).

14.2 Strength Bulkheads. Transverse *strength* bulkheads, extending from the inner bottom to the strength deck, are required by the ABS Rules, in general, for all ships. They are to be formed where practicable by fitting substantial 'tween-deck bulkheads immediately over the main transverse watertight subdivision bulkheads. If transverse strength bulkheads interfere seriously with the ship's function, equivalent strength is to be maintained by fitting deep webs or partial bulkheads at the sides of the ship, or a combination of these, so as to maintain effective transverse continuity of structure.

14.3 Subdivision Bulkheads. Fig. 33 shows a typical subdivision bulkhead which also serves as a strength bulkhead on a cargo ship. The number and approximate location of the transverse subdivision bulkheads are recommended in the Rules, with some latitude permitted for the spacing of cargo hold bulkheads. However, the *collision* bulkhead near the bow, *afterpeak* bulkhead near the stern, enclosing the shaft tubes, and watertight bulkheads at the forward and after ends of the machinery space, are mandatory. In this connection, the ABS Rules contain the following statement: "It is recognized, however, that for certain types of cargo vessels in special services it may be impracticable to adhere to the number and arrangement of hold bulkheads as recommended. In such cases, the Bureau is prepared to consider an alternative arrangement or even the omission of certain bulkheads in order to meet the requirements of special trades." Lloyds Rules contain a similar provision.

A case in point is the typical Great Lakes ore carrier, the main hold of which sometimes will extend throughout the middle 70 percent of the length with no watertight bulkheads whatever. In these particular ships, the subdivision of the outboard ballast spaces is relied upon to protect the ship in case of accident, and the transverse arches, referred to in Subsection 13.6, provide the required transverse strength in the absence of the bulkheads.

a. Collision Bulkheads. The most important subdivision bulkhead is the *forepeak* or collision bulkhead. In nearly every collision, the bow of one of the ships involved is damaged, and it is a common occurrence for ships with considerable bow damage to reach port safely, with the forepeak open to the sea and the collision bulkhead keeping the sea out of the remainder of the ship. Fig. 34 shows such a ship.

To provide for this increased possibility of damage, as well as for the impact effects of proceeding underway with the forepeak flooded, the Rule scantlings for both the plating and the stiffeners of the forepeak bulkhead are heavier than required for other subdivision bulkheads. For passenger vessels, this bulkhead is required to be located not less than 5 percent the ship length and not more than 5 percent of the

Fig. 34 Ship with bow damage

length plus 3.05 m (10 ft) aft of the stem. For all other vessels, for ship lengths under 200 m (61 ft), the collision bulkhead can be not less than 5 percent nor more than 8 percent of ship length abaft the forward perpendicular; for ship lengths greater than 200 m, the bulkhead can be not less than 10 m (32.8 ft) nor more than 5 percent of ship length

(*Continued on page 308*)

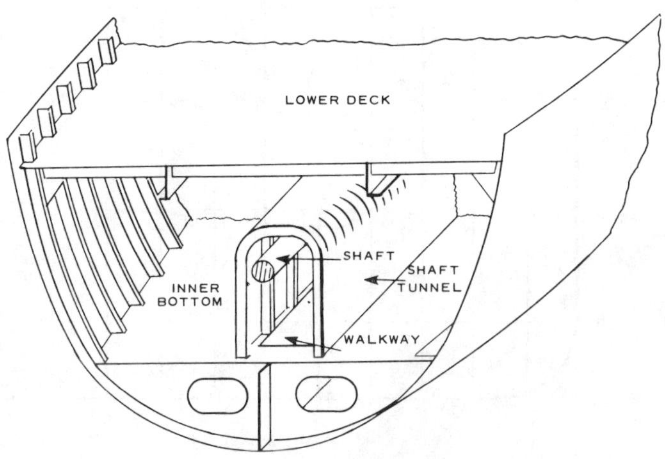

Fig. 35 Shaft tunnel

Fig. 36 Shaft tunnel under construction

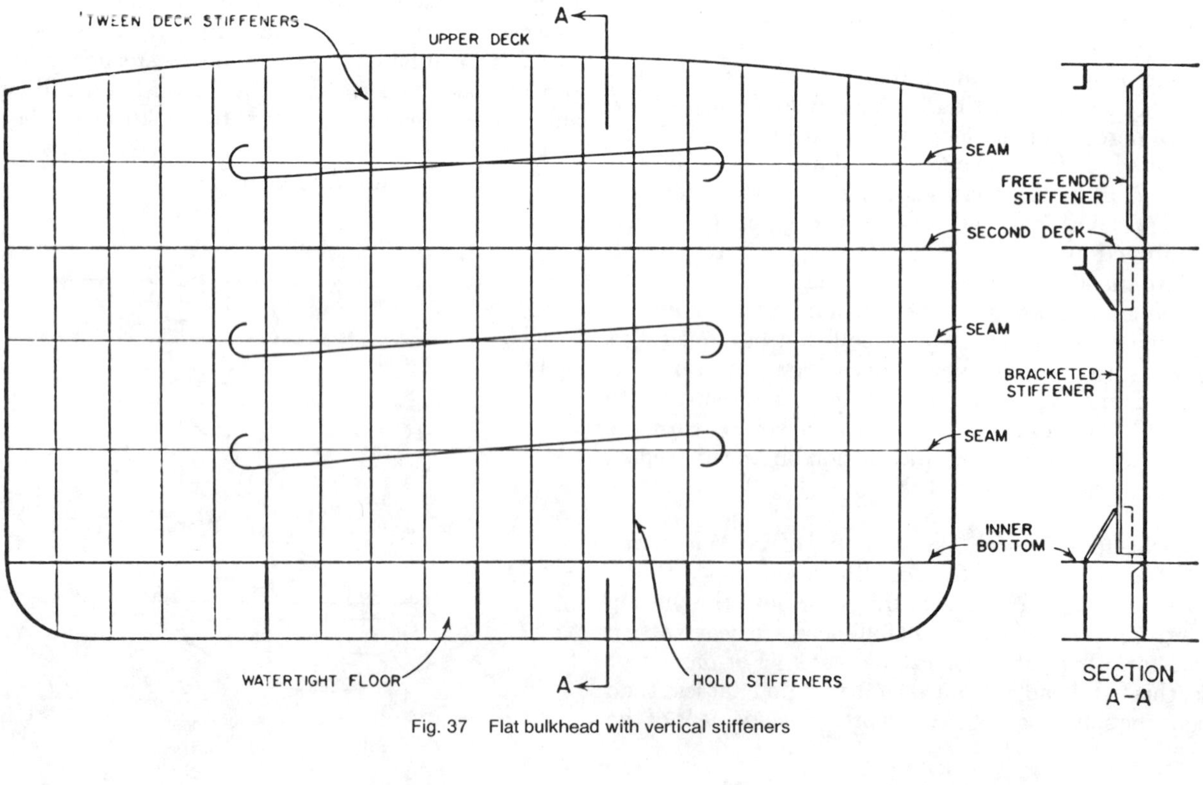

Fig. 37 Flat bulkhead with vertical stiffeners

Fig. 39 Corrugated bulkhead construction

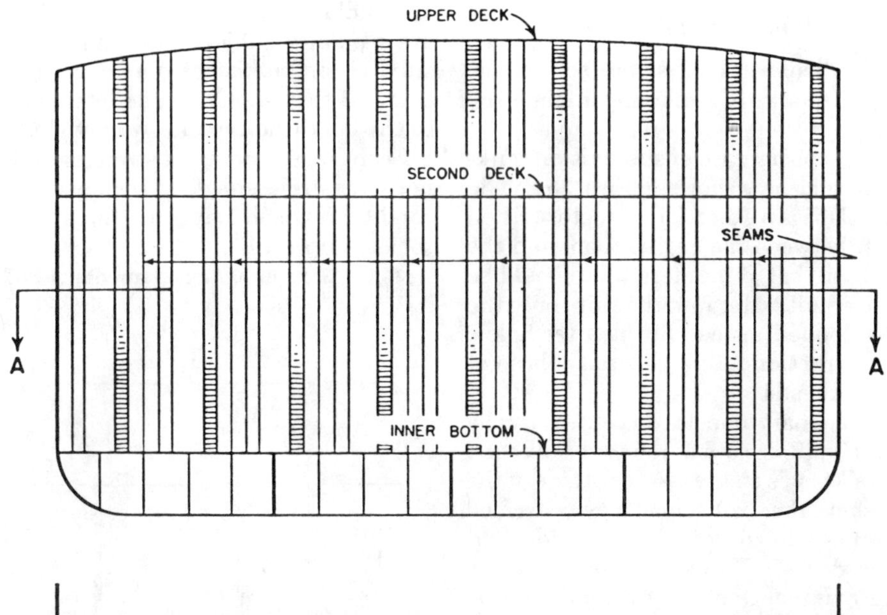

Fig. 38 Corrugated bulkhead

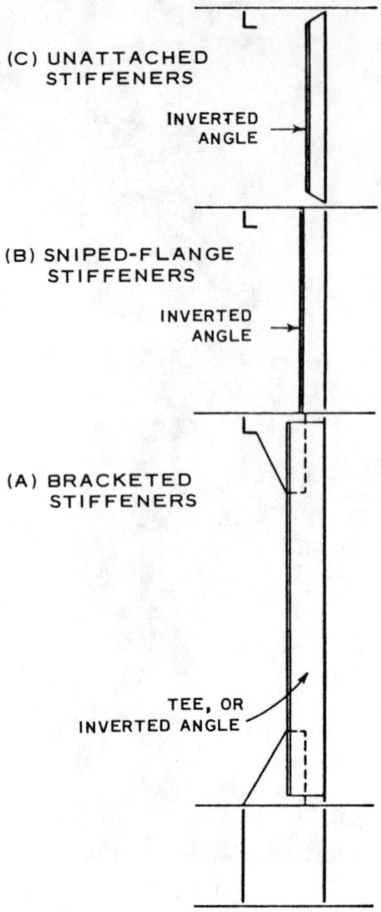

Fig. 40 Bulkhead stiffener connections

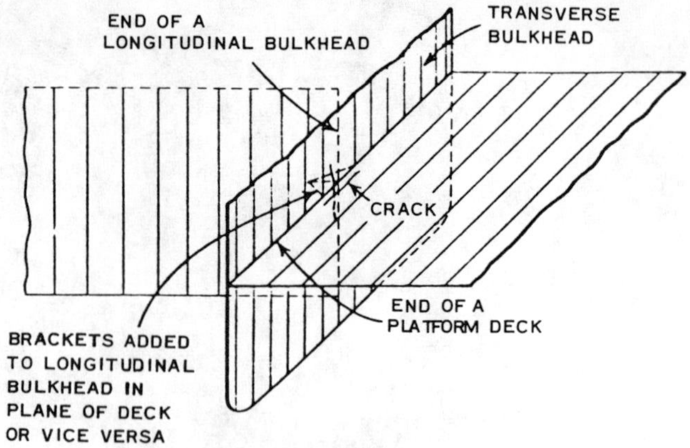

Fig. 41 "Point" meeting of two plated bulkheads

(*Continued from page 305*)

abaft the forward perpendicular. This is done so that the bulkhead will have a good chance of remaining intact if the bow is damaged.

b. Shaft Tunnels. A special case of subdivision bulkhead is the *shaft tunnel*, such as is shown in Figs. 35 and 36 for a single-screw ship. This is a watertight structure fitted around the entire shaft between the engine room and the stuffing box at the stern, so that, if the stern tube should be damaged so badly as to admit water into the ship, only the shaft tunnel would be flooded instead of the after holds. The scantlings and construction follow generally those of ordinary subdivision bulkheads.

In single-screw ships, the shaft tunnel is off center, usually to starboard, so that the shaft is close to the port bulkhead of the tunnel and the walkway is on the starboard side of the shaft. In multi-screw ships, the shaft tunnel may be formed in large part of ordinary longitudinal bulkheads and water-tight portions of decks. Two emergency exits are required from all shaft tunnels, one by a *tunnel escape trunk,* usually at the after end of the tunnel, extending up to above the bulkhead deck, the other by means of the sliding watertight door required between the tunnel and the machinery space. This sliding door is required to be operable from above the bulkhead deck and is fitted at the forward ends of shaft tunnels.

14.4 Bulkhead Scantlings. For the principles underlying the design of watertight bulkheads, the reader is referred to MacNaught (1967). In practice, the designer of merchant ships needs only to follow the rules of the classification societies, which have evolved from a combination of long experience, extensive special tests, and calculations. These rules give plating thicknesses and stiffener sizes for all practical design conditions and types of stiffener end attachments.

Both the ABS and Lloyds Rules have formulas for scantlings for subdivision (watertight) bulkheads and for the plating of tank (*deep tank*) bulkheads. The latter are of heavier construction and are required to be tested by a head of water instead of the hose test required for subdivision bulkheads. The reason for this is that tank bulkheads are regularly subject to liquid pressure in service, while subdivision bulkheads may never be subject to such pressure. If they should be, deflection is unimportant so long as the bulkhead remains watertight.

Scantlings of nontight strength bulkheads are, in general, not given in classification society rules because of the varied conditions which may arise, but are subject to classification society approval.

14.5 Structural Arrangements of Bulkheads. A variety of

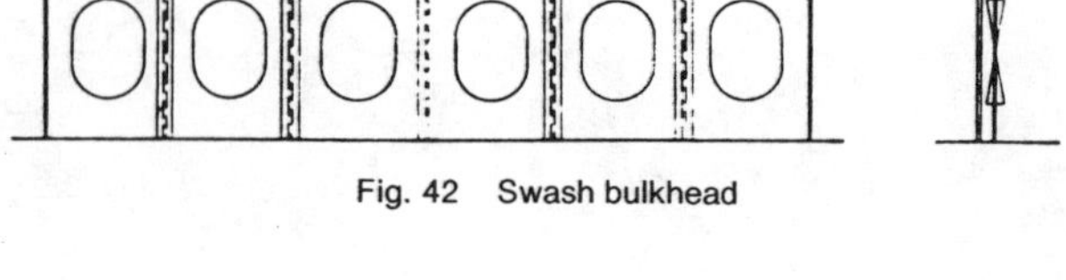

Fig. 42 Swash bulkhead

Fig. 43 Swash bulkhead construction

structural arrangements for bulkheads is found in practice. The most common, illustrated in Figs 37 and 38, consists of horizontal strakes of plating stiffened by a system of vertical stiffeners, the usual spacing being about 762 mm (30 in.). The discussion of optimum frame spacing in Section 6 of this chapter is equally applicable to bulkhead stiffener spacing. This construction permits making each strake of plating the thickness required by the head of water affecting that particular strake, the bottom strake being heaviest and the top lightest.

The long vertical stiffeners in lower holds may be supported by one or more horizontal girders, so spaced as to equalize the bending moment in the several segments of the stiffeners, resulting under certain conditions in a reduction in the weight of the framing system. Fabrication costs, however, are increased.

a. Corrugated Bulkheads. Several patented constructions have been devised, one of which is the *fluted* or corrugated bulkhead, in which stiffeners are omitted and their function served by deep corrugations in the plating, Fig. 38. If corrugated bulkheads are plated vertically, they are somewhat wasteful of plating as the thickness cannot be reduced gradually toward the top of the bulkhead because of the economics of the manufacturing process. However, a well-designed corrugated bulkhead is generally somewhat lighter than a stiffened bulkhead. It has fewer parts, and some owners prefer it for tankers because it is easier to clean. Corrugated bulkheads are extensively used in dry bulk cargo carriers because they do not interfere as much as stiffened bulkheads with shedding and freeing cargo. They also expand freely under thermal loads, which makes them desirable for holds containing high-temperature cargo.

Experience indicates that the many cases of minor but expensive fractures in corrugated bulkheads have for the most part occured in horizontally-corrugated transverse bulkheads. This is explainable, because one of the principal functions of transverse bulkheads is to support the bottom and top of the hull girder when the hull is subjected to bending. Horizontal corrugations obviously impair the effectiveness of the bulkheads in resisting this compressive load. Vertically-corrugated transverse bulkheads are reported to have given relatively little trouble. Longitudinal corrugated bulkheads should have horizontal corrugations if they are expected to share the longitudinal hull girder stresses resulting from longitudinal bending. Fig. 39 shows a corrugated bulkhead under construction.

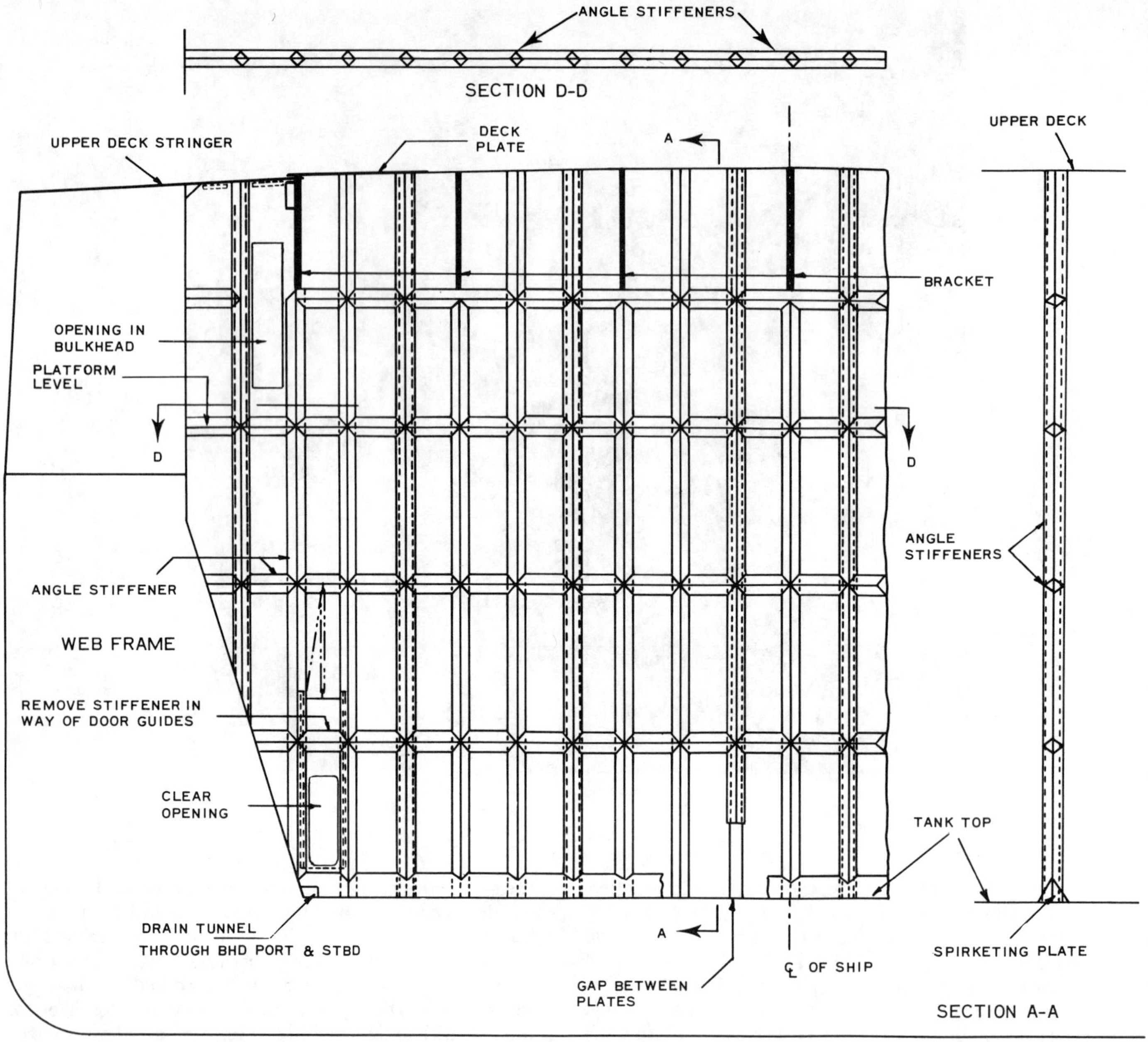

Fig. 44 Ore carrier screen bulkhead

b. Stiffening Details of Bulkheads. Bulkhead stiffeners usually are inverted angles, bulb plates, or tees. Welding may be intermittent, except that double continuous welding is required for unbracketed stiffeners for a distance from each end equal to one-tenth of the length of the stiffener. Classification societies will approve departures from Rule welding which provide equal strength, such as longer segments of smaller welds. However, the Rule weld for the end one-tenth length of unbracketed stiffeners obviously cannot be reduced without reducing strength.

In tanks that carry liquid cargo, fresh water or salt water ballast, continuous welding is recommended to prevent contamination and corrosion resulting from liquid penetration between the web of the stiffener and the bulkhead plates, an area which cannot be readily cleaned or painted.

In general, where welded stiffeners cross welded seams of plating, a semi-circular cutout of about 25 mm (1 in.) radius should be made in the stiffener in way of the welded seam to allow the latter to continue past the stiffener without interruption, as illustrated in Fig. 31. An exception to this is the case of seam welds of subassembled plates which are welded together, edge to edge, usually by automatic submerged-arc welding, prior to attaching the stiffeners. In this case, the completed seam weld may be chipped or ground reasonably flush in way of the stiffener, and the latter welded across the seam without the cutout. For *field* welds, that is, those joining subassemblies together on the ship, the cutout is essential to good workmanship.

c. End Connections of Bulkhead Stiffeners. The ends of bulkhead stiffeners may be either bracketed, Fig. 40 (A), *sniped-flange,* Fig. 40(B), or unattached, Fig. 40 (C). Long hold stiffeners are usually bracketed to reduce their span. The size of the stiffener depends primarily on the spacing and the design head to the middle of the stiffener. The type of end attachment is also a consideration, the stiffener size being smallest if efficient brackets are fitted and greatest if there are no end attachments. To be effective, brackets must land on structure strong enough to provide the necessary rigidity, or resistance to rotation, of the support. Tank bulkhead stiffeners are generally expected, on all but the shortest spans and smallest heads, to have their ends attached.

d. Smooth Side of Bulkheads. The choice of which side of a bulkhead the stiffeners should be welded to is primarily a matter of minimum interference with arrangements. Often stiffeners must be kept out of tanks which require frequent cleaning, i.e., those carrying edible oils or other liquids. For other tanks, it is preferable to put stiffeners on the tank side if the opposite side is part of a cargo hold and maximum stowage is required. If no particular reason to do otherwise exists, stiffeners are generally placed on the midship side of transverse bulkheads to take advantage of the shape of the shell beyond the parallel midbody, thus obtaining better welding clearances for the outboard stiffeners. This obviously becomes more important towards the ends of the vessel.

General practice for tankers is to place all stiffeners on the after side of transverse bulkheads. As tankers generally trim aft when running in ballast, this practice promotes drainage.

e. Knife-Edge Intersections. A condition extremely likely to give trouble is illustrated in Fig 41. This condition results whenever two plated structures, such as a longitudinal bulkhead and a partial deck, end at the same plane and meet at one point only. When the ship *works* in a seaway, either structure moves relative to the other except at the point where they meet. This crossing forms an unyielding point, and one or the other of the two structures is almost sure to tear or crack as indicated. Such fractures have even occurred on the building ways during the construction period. The problem of local high stress concentration is avoided by the provision of transition brackets in the plane of one or both of the plate structures.

14.6 Swash Bulkheads. Another type of bulkhead is the open or *swash* bulkhead, which is fitted within an oil or water tank to reduce the surge of the liquid when the ship rolls and pitches, and thus to reduce the dynamic impact of the liquid on the surrounding structure. Two types of swash bulkheads are shown in Figs. 42 and 43. Since these bulk-

Fig. 45 Longitudinal bulkhead construction

heads are open and not subject to a calculable static pressure, there may be a tendency to underestimate the required strength. Sometimes swash bulkheads have been fitted and have not stood up against the surge.

Swash bulkheads are particularly important in tanker design. Not only do they reduce the dynamic cargo pressures, but they also support the centerline longitudinal girder, as can be seen in Fig. 43. A practical rule for designing small swash bulkheads is to provide one half of the scantlings that would be required for a watertight bulkhead at the same location. For larger swash bulkheads, it may also be necessary to provide the equivalent column areas to carry the vertical concentrated girder loads. The deep transverse ring girders fitted in tankers are also subject to sloshing loads. Extra web stiffening of these members is considered good practice in tanker design.

14.7 Shifting Bulkheads. A less common type of bulkhead is the so-called shifting bulkhead. This is a longitudinal bulkhead, usually fitted at the centerline, whose function is to prevent the athwartship shifting of grain, coal, sand, or other bulk cargo, or ballast, when the ship rolls considerably in heavy weather. A most interesting and informative account of trouble resulting from shifting of 'tween-deck solid ballast on a *Liberty* ship in a hurricane and a discussion of the strength requirements of shifting bulkheads can be found in Beattie (1950). The latest Rules for Carriage of Grains applied today are the ones issued by the Intergovernmental Maritime Consultative Organization (IMCO, 1974), which are further discussed in Chapter XI.

14.8 Screen Bulkheads of Ore Carriers. It is sometimes desirable that a non-watertight bulkhead be designed with some flexibility. This is the case with the transverse screen bulkheads of Great Lakes ore carriers, such as that shown in Fig. 44. These bulkheads divide the cargo holds into separate compartments. They are non-watertight and are not required for transverse strength. Because of the nature of the unloading machinery and the frequency of unloadings, these bulkheads are subject to heavy damage and must be replaced frequently.

It has been found useless to endeavor to build enough strength into these bulkheads to render them invulnerable to damage. Therefore, they are built with considerable flexibility, a factor which increases their life materially. It is found that they give under impact and thereby receive a minimum of damage.

14.9 Longitudinal Bulkheads in Tankers. Typical tanker construction includes two longitudinal bulkheads extending the full length of the cargo tank spaces, as shown in Fig. 45. These bulkheads provide additional hull girder shear area, support the transverse deck girders and bottom girders, and reduce the transverse dimensions of the cargo tanks, thus reducing the sloshing loads. A shear flow calculation of the hull, a pillar calculation, of the section of the bulkhead supporting the transverse girders, as well as local plate and stiffener hydrostatic load calculations are required for the proper design of the longitudinal bulkhead.

The longitudinal bulkheads are stiffened by longitudinal framing which is in turn supported by vertical web frames. The large span of these web frames makes it necessary to fit intermediate struts along their length. Since the principal loading on these struts consists of axial tension and compression, they are designed using typical column considerations. The struts will also experience some bending action due to longitudinal sloshing of the cargo, requiring a safety margin for column eccentricity to be included in the design. Experience has shown that end connections of the struts require careful detailing to ensure suitable structural continuity. Consideration must therefore be given to the transfer of strut axial forces as well as end bending moments from the struts to the transverse webs.

Section 15
Pillars, Girders, and Hatch Coamings

15.1 Pillars. Pillars are relatively long columns subject to compressive loading and as such are subject to failure by buckling rather than failure in compression. Their strength is governed by their length as well as by their cross-sectional dimensions, a longer pillar having less load-carrying ability than a short pillar of the same cross-section.

a. Classification Society Rules. The classification society rule scantlings for pillars, while calculated on the basis of their primary function of supporting the decks, take into account other factors discussed in Section 1, and actually represent the size found necessary and desirable by long experience with ships in service.

The ABS Rules specify the permissible load that can be carried, as a function of the pillar area, its radius of gyration, and the unsupported length of the pillar. Lloyds Rules specify a required area as a function of the other three variables.

For pillars and girders under accommodations, a reduced height is used in the Rule formula in view of the lighter load carried in such spaces, and under weather decks an arbitrary height is used representing the load assumed to be imposed by the weight and impact of seawater on the deck. In the case of suspended cargo, such as meat, in a hold or 'tween-deck space, the deck above has to carry this suspended cargo as well as the load on the deck. This double load affects the pillars and girders supporting the deck, and may even require extra strength in the floors in the double bottom supporting the feet of these pillars because the pillars in this case are carrying concentrated loads which normally would be distributed over the inner bottom.

b. Pillars Under Weather Deck at Bow. One of the most common types of storm damage is the *setting down* of the forecastle deck by impact of seas on the deck when the bow encounters an oncoming wave, as pictured in Fig. 46.

Fig. 46 Impact of sea on bow

This may cause failure of the pillars or the girders below the forecastle. Special consideration with regard to strength should be given to large, fast ships and to ships with a flared bow (passenger, high-speed general cargo, and containerships) where the minimum rules requirements might be exceeded under severe weather conditions in head seas. This strengthening should be carried well down into the main hull structure, so as to distribute the loads into the main fore end structure formed by the shell and bulkheads.

It is interesting to note that tension failures in the pillars may also be induced by the action of the waves in head seas. If the ship has a pronounced flare, an impact load is more likely to occur on the shell and shell frames in the level below the weather deck than on the deck itself, deflecting them inboard, and, by rotation of the beam knees, deflecting the center of the weather deck upward. For this reason, the ABS Rules require connections capable of transmitting tensile loads for these pillars in ships whose form and speed are such that they may be subject to so-called *bow flare slamming*. Examples of such connections are discussed in Subsection 15.1*g*.

c. Systems of Pillaring. As pillars interfere seriously with the stowage and handling of cargo, large, widely spaced pillars in association with deck girders are now in general use instead of the closely spaced pillars found on earlier ships. The system most commonly found on moderate-size general cargo ships is a two-row system with cargo hatches about one-third of the beam in width, and with the pillars located

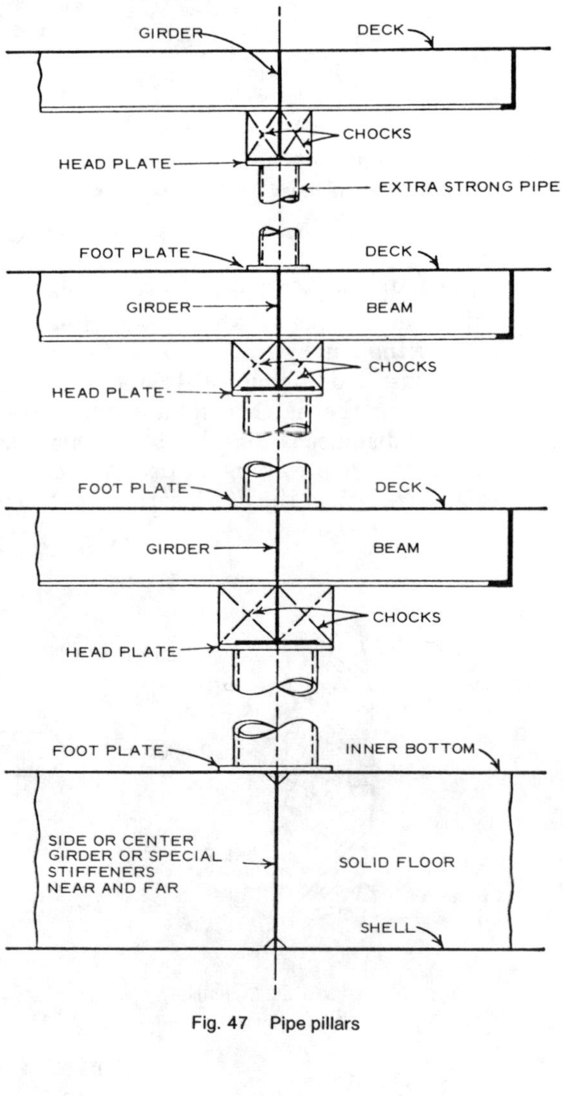

Fig. 47 Pipe pillars

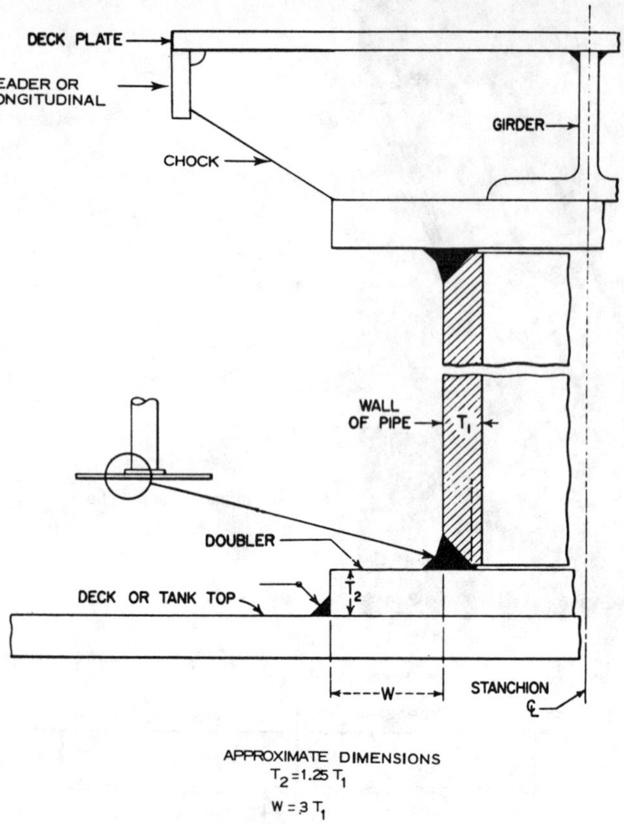

Fig. 48 Details, heads, and feet of pipe pillars

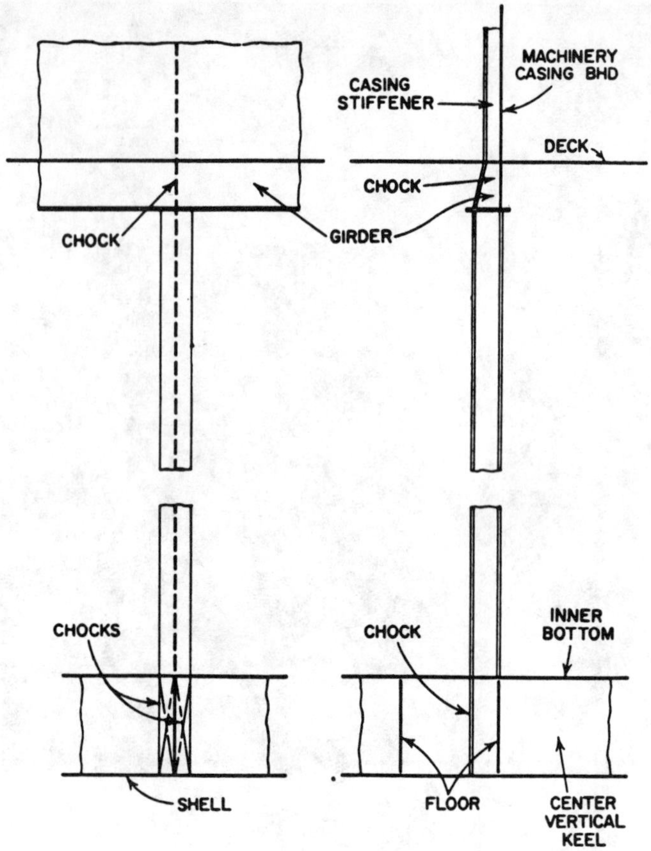

Fig. 50 Typical II-section pillar

at or near the hatch corners. The hatch side girders carried by these pillars should line up where practicable with the girders, clear of the hatchways, and with the side of the machinery casings. Additional widely spaced pillars may be needed between the hatch ends and the transverse bulkheads if this distance is considerable. This arrangement is well adapted to providing the desirable continuity of longitudinal strength at the hatch corners, as the hatch side beams and the girders form a continuous system. For large ships, three or even four rows of pillars and girders may be found desirable.

In some ships, a long hatch opening in relation to hold length has led to severe unbalance of girder spans within and beyond the hatch. When the girder is analyzed as a continuous member and the pillar reactions checked, it is sometimes found that the pillar load is substantially higher

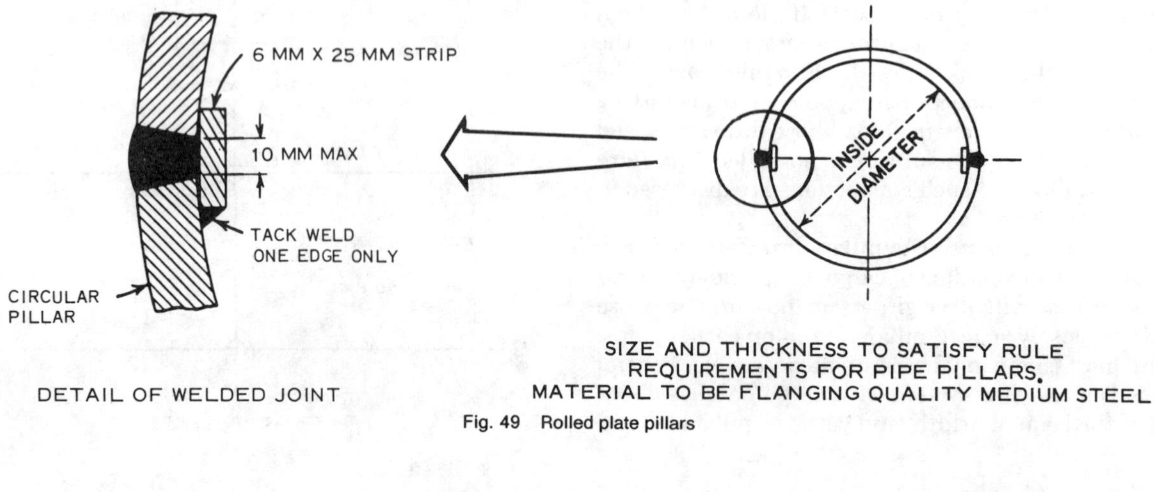

Fig. 49 Rolled plate pillars

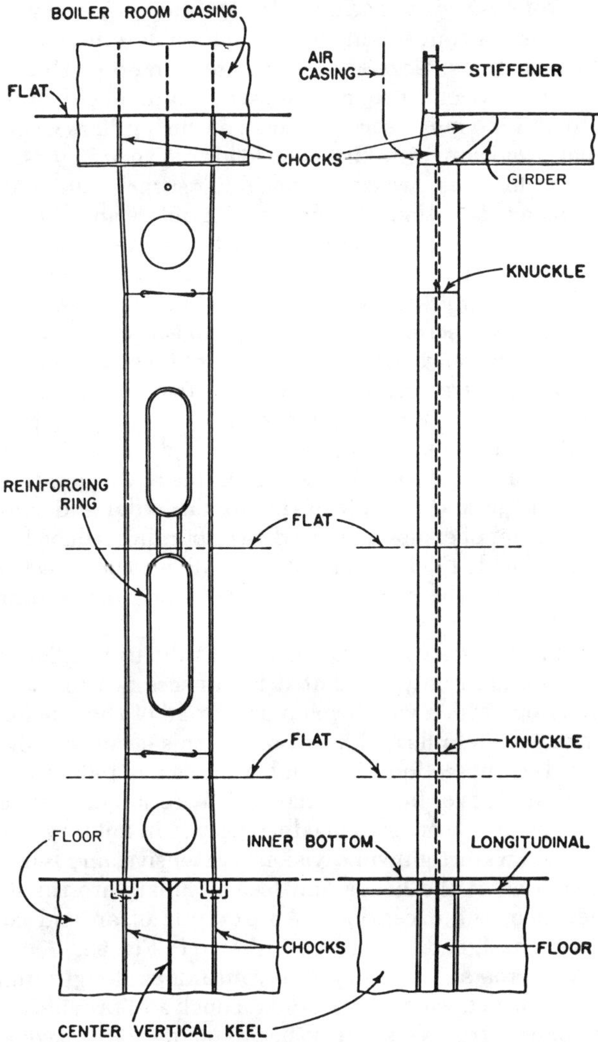

Fig. 51 Heavy built-up pillar

than indicated by the usual Rule assumption that it supports the load on one-half of each of the girder spans involved. Hatch end pillars, if sufficiently close to the bulkhead, may be replaced by deck-to-deck brackets welded to the transverse bulkhead in line with the girders. Such brackets cause little (if any) more interference than the pillars and eliminate the problem of heavy pillar loads at the inner bottom by transferring the load to the bulkhead.

d. Alignment of Pillars. Both the ABS and Lloyds Rules have provisions requiring that pillars in 'tween-deck spaces be directly over pillars below wherever practicable, a sound principle which is violated far too often in practice. Pillars landing on girders instead of over pillars below, even though the girders are suitably increased in strength, result in a more flexible construction which may lead to vibration.

e. Tubular Pillars. Tubular sections are especially suitable for columns, since stiffness is equal in all directions and is greater for a given cross-sectional area than the stiffness in the weakest direction of rolled sections. Pipe sections are, therefore, used extensively for pillars in holds and accommodation spaces. Fig. 47 shows a typical pipe pillar and Fig. 48 shows details of construction at the tip and bottom of such a pillar. These end connections are designed primarily for compression only; tension connections are discussed later in this section. If pillars larger than those available in tubular sizes are required, they may be fabricated at the shipyard with welded seams, as shown in Fig 49.

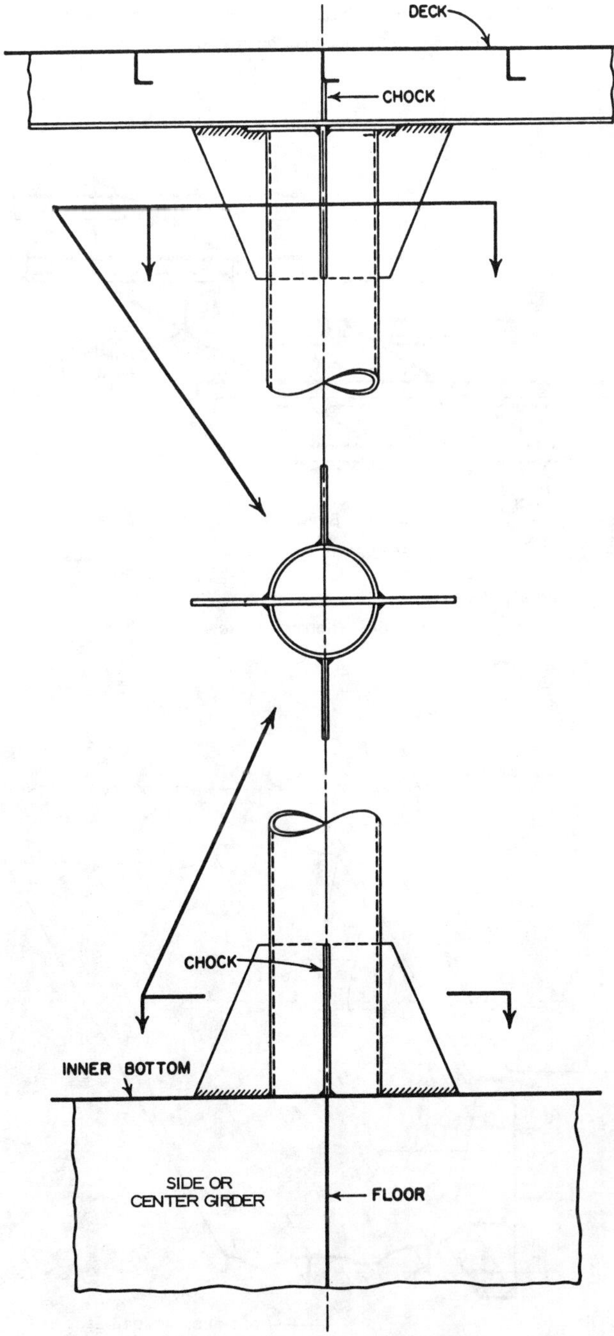

Fig. 52 Pillar with tension connections

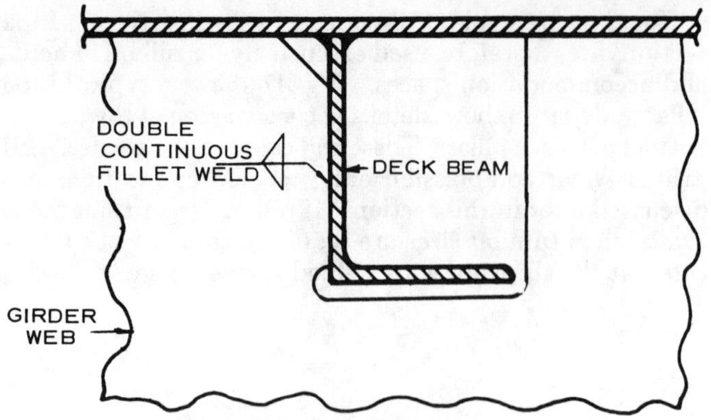

Fig. 53 Intersection of beam and girder

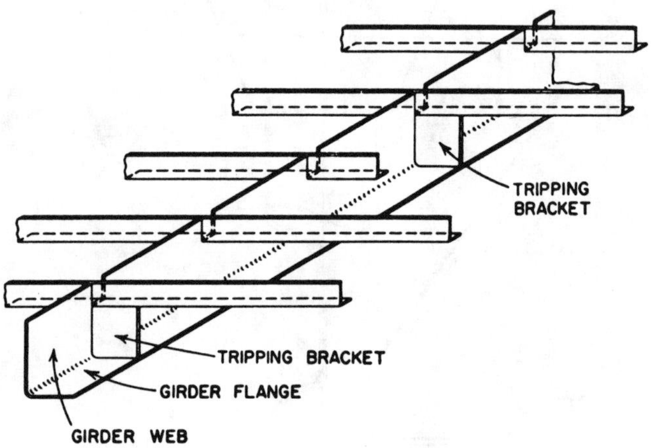

Fig. 54 Tripping brackets

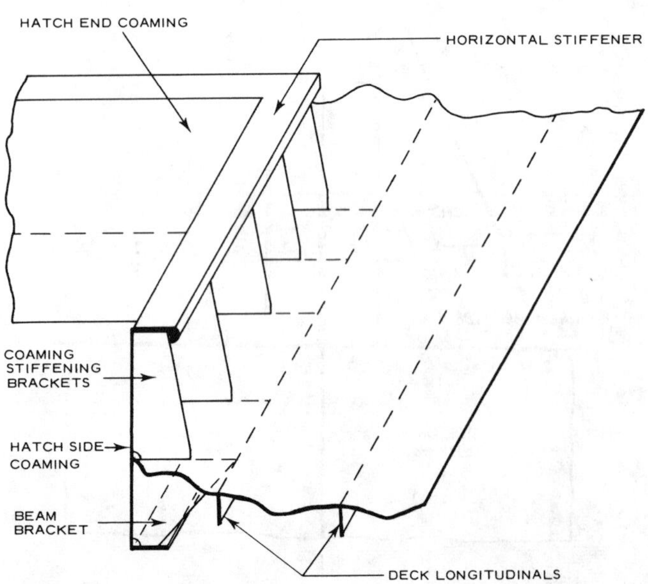

Fig. 55 Section through hatch coaming

f. Rolled-Section Pillars. In machinery spaces, where pillars form a convenient support for small pumps, tanks, and other items, large H-sections are common. Built-up structural sections also may be used. Care must be taken not to add too much eccentric load on such pillars without proper provision for taking such loads. A seemingly small eccentric load may weaken a long pillar surprisingly. Also, when using H-sections as combination pillars and supports for miscellaneous auxiliaries, it must be remembered that an H-section has relatively little resistance to torsion or twisting, and, unless suitable bracing is added, a heavy item hung on an H-section pillar may be subject to excessive vibration. Fig. 50 shows a typical H-section pillar, and Fig. 51 shows an extremely heavy, built-up structural pillar such as might be used in a very large ship.

Rolled-section pillars are also used in deep tanks where the Rules do not permit hollow tubes due to the possibility of oil leakage to the inside of the tubular pillar and subsequent danger of explosion if welding or burning is done later. Leakage inside the tubular pillars would also be objectionable in cases where liquid cargo could become contaminated.

g. Pillars Designed for Tension. Ordinarily, pillars act in compression only, so that it is unnecessary to fit end connections which will develop part or all of the tensional strength of the pillar. Exceptions to this occur, however, as noted in Subsection 15.1*b*, and in the case of pillars under watertight flats or in deep tanks which are subject to a test head or service head or to flooding below. In military ships, horizontal watertight subdivision is extensive, and tension connections at the heads and feet of pillars are the rule rather than the exception. An example of an end construction designed for tension is shown in Fig. 52.

15.2 Girders. Girders are continuous longitudinal members which usually run under a deck and provide support for the transverse deck beams or deck transverses. Girders are in turn supported by pillars or bulkheads. Cargo ship girders are frequently built-up sections. Girders may also be flanged plates or, for the smaller sizes, inverted tees or angles. Where transverse deck beams pass through girders, it is usual to cut a rectangular notch, radius-cornered, in the web of the girder, the beam being attached by welding, as shown in Fig. 53.

a. Hatch Side Girders. In way of long hatches with corner support only, the hatch side girders tend to become large. To facilitate cargo handling, these hatch side girders usually are not supported by pillars between hatch ends, and, if the hatch is long, an extremely heavy section becomes necessary. Any unnecessary depth of girder is undesirable in this area because it interferes with stowage of cargo. The conflicting needs of adequate strength and shallow depth are usually met by using a wide flange on the bottom of the girder, sometimes with a plate doubler for additional area. The ABS Rules, however, limit the ratio of span to depth to not more than about 17. Long, shallow girders, while they may be sufficiently strong, are inherently flexible, and may result in undesirable deflection. They should also be examined for possible excessive vibrations.

b. Tripping Brackets. Girders, especially if unsym-

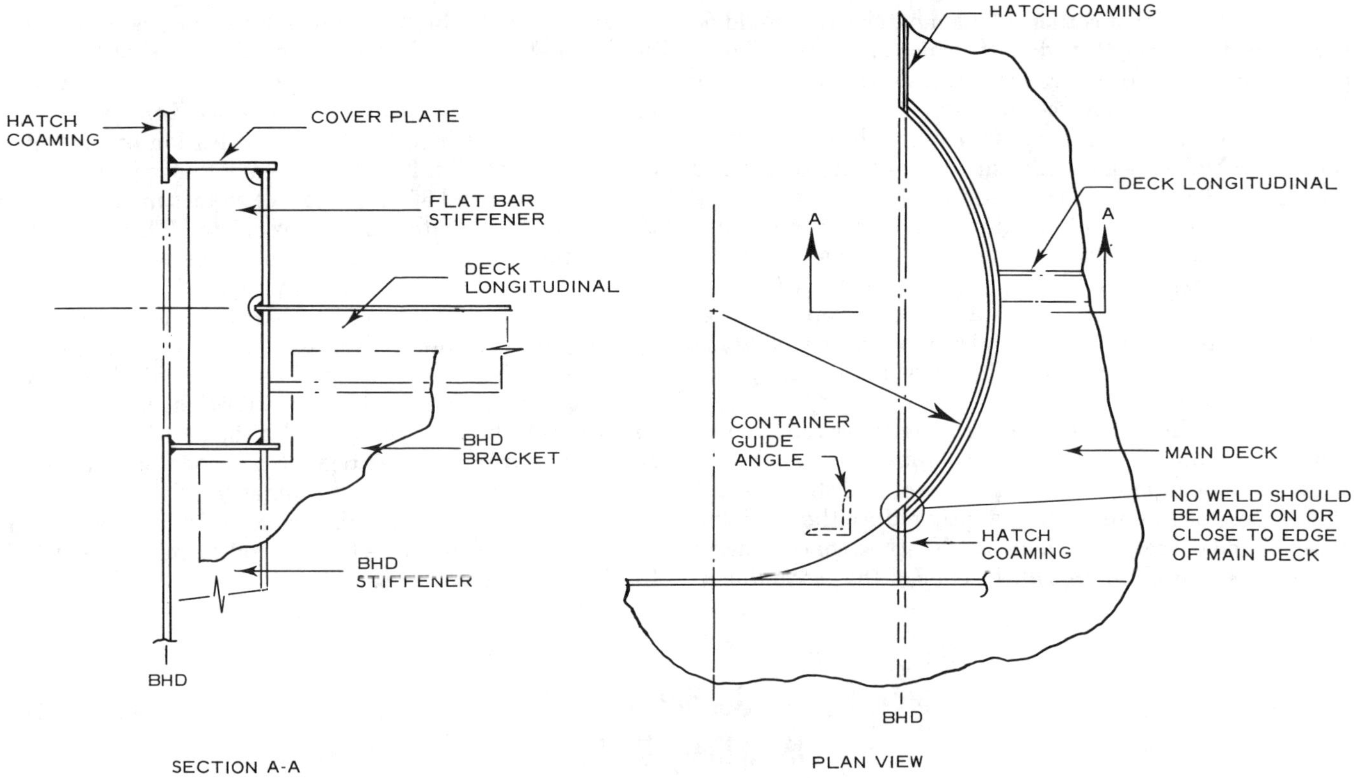

Fig. 56 Detail of hatch corner end

metrical, must be supported against possible *tripping* by brackets, as shown in Fig. 54, spaced usually not more than 3.05 m (10 ft) apart; where the width of the flange exceeds about 150 mm (6 in.), these brackets should be connected to the flange as well as to the web.

15.3 Hatch Coamings. Hatch coamings usually perform the combined functions of girders (supporting the adjacent areas of decks and hatch covers) and of keeping water out of the hatch. The latter function is particularly the duty of weather deck cargo hatch coamings, which must protect the hatch covers and the hatch opening from the sea in heavy weather. Classification society rules specify minimum heights, as established by the International Convention on Load Lines, and thicknesses of plating of hatch coamings required for seaworthiness, according to the location and degree of exposure of the hatch. The strength of the coaming as a girder must also conform to the Rule requirements for girders.

a. Weather Deck Hatch Coamings. A section through a typical fore-and-aft weather deck hatch coaming is shown in Fig. 55. On the weather portions of freeboard decks, the coaming must extend at least 610 mm (24 in.) above the deck if the hatchway is covered and battened. It is nevertheless usual to have higher coamings to avoid having to fit a temporary chain rail for personnel protection when the covers are open. In less exposed locations, a lesser height is permitted. Where efficient watertight steel covers are fitted and made tight by means of gaskets and clamping devices, the heights of the hatch coamings may be reduced.

The portion of the coaming plate below the deck is either flanged or fitted with a face plate. In such cases, brackets are fitted as required by the classification society rules. The brackets are sometimes extended to the longitudinal bulkhead of the torsion box to provide girder support, as has been the practice with the latest wide-hatch cargo vessels.

Near the top of the coaming, an inverted angle, bulb plate, or flat bar is fitted to stiffen the coaming, and fastened to this will be fittings for hatch covers, such as wedge cleats, dogs, or roller tracks.

b. Hatch Corner Construction. Particular care is needed at the hatch corners to avoid sudden discontinuity of longitudinal strength at the ends of the side coamings. Fractures in welded ships have originated at hatch corners, and considerable attention has been devoted to designing hatch corner construction which will minimize stress concentration.

The hatch side girders are to be arranged not to end abruptly at the ends of the hatches, even if they are not required to support beams beyond the hatch end. They are to be extended a suitable distance past the hatch end for structural continuity. It is also good practice to extend the hatch coaming forward and aft of the hatch corner by means of taper brackets.

Extensive structural analysis and data collection has led

to the conclusion that corners of main hatchways should be surrounded by strengthened plates which are to extend over at least one frame space fore and aft, as well as athwartships (Ship Structure Committee, 1977–b). Well-rounded corners on strength deck openings are required by classification societies. Make the plate edge smooth and keep deck butts and seams clear of the corner radius.

The most recently built multi-purpose cargo vessels have one or two long parallel hatch openings with a very short distance between longitudinally adjacent hatches. On such vessels, special hatch corners of elliptical shape are installed. Guidelines for such opening and compensating plate arrangements are given by the Ship Structure Committee (1977-b).

Speedy loading and unloading operations required of today's containerships are best accomplished through very wide hatch openings. This requirement and the inherent slenderness of containerships greatly reduce the torsional rigidity, and consequently increase the stress concentration factor at the hatch corners at the ends of the vessel due to the exceptionally high level of torsional stresses at these points of structural discontinuity (Elbatouti et al, 1976). Several studies have been implemented and experimental results collected, particularly on the *Sea Land SL-7* class containerships, which led to a modification of the hatch corner illustrated in Fig. 56. A heavy flange plate 48 mm (1.88 in.) thick and 305 mm (12 in.) wide is installed along the contour of the cutout in the deck plate. This 2.59 m-(8.5 ft) long plate follows the circular cut of the deck around the hatch corner and tapers down to a width of 51 mm (2 in.) at both ends, thus creating a smoother transition from a deck opening to a continuous deck section.

c. Lower Hatch Coamings. Lower deck hatch coamings were required in conjunction with watertight steel covers on shelter deck ships. However, with the elimination of this type of ship from tonnage regulations, coamings below the weather deck are obsolete. Elimination of these coamings is highly desirable for handling cargo which is stowed in the 'tween-deck spaces across the edge of the hatch and allows full use of forklift trucks.

Section 16
Machinery Casings

16.1 Function of Machinery Casings. The primary function of the machinery casings is to protect the openings which are fitted in the weather deck over the engines and boilers (for access, for light and air, and for the uptakes from the boilers) against the sea entering the ship through these openings in heavy weather. Below the weather deck, the machinery casing separates machinery spaces from accommodation and cargo spaces. Classification society scantlings for machinery casing depend primarily on the degree of exposure, as exemplified in the requirements regarding heights of door sills in the casings, which in the ABS Rules vary from 600 mm (23.5 in.) to 380 mm (15 in.).

The machinery casings also serve to support the various decks attached to them as well as the uptakes and stacks. This support is transmitted into the ship's hull either by pillars fitted under the casing sides and ends down through the machinery space to the inner bottom, or by extending the fore-and-aft bulkheads forming the casing sides to the adjacent main transverse bulkheads, and designing them to act as deep girders between the transverse bulkheads, Figs. 50 and 51.

16.2 Classification Society Rules for Casings. Classification society rules contain considerable descriptive material as to what constitutes good practice in construction of machinery casings, in addition to the required scantlings.

Rule scantlings, however, are established only to provide the necessary protection of the openings, and any special arrangements required by the support system are subject to approval by the classification society. The wide variation in possible arrangements of casings, even those considered solely as protection for the openings, makes it necessary for the rules to be somewhat flexible. This is exemplified by the provision in the ABS Rules that, "Hatchways within deckhouses are to have coamings and closing arrangements as required in relation to the protection afforded by the deckhouse from the standpoint of its construction and the means provided for the closing of all openings into the house."

Nevertheless, the rules are quite explicit in that all openings in the bulkheads of enclosed superstructures should be provided with efficient means of closing, so that water will not penetrate the vessel in any sea condition, and that opening and closing appliances are to be framed and stiffened so that the whole structure is equivalent to the unpierced bulkhead when closed.

16.3 Stiffness of Machinery Casing Support. If long beams are used to support the casings, or the uptakes and stacks within the casings, there is danger that even when such beams are sufficiently strong they may be so flexible as to result in excessive vibration. Such vibration may be transmitted to nearby structures in such a manner that the point of origin will be extremely obscure. In one case of a large cargo ship, the entire midship house and bridge vibrated badly at propeller blade frequency at full power. The affected structure was so extensive that several unsuccessful remedies were tried before it was found that the simple expedient of fitting a single sloping strut, port and starboard, under an unduly flexible uptake foundation stopped the vibration.

Any lack of rigidity of support of the casings is likely to result in damage to the joiner work and jamming of doors. The designer must insure the provision of a continuous, rigid, and substantial system of support all the way down to the double bottom structure or to the side shell plating for carrying the weight not only of the casings but of the decks, houses, uptakes, stacks, and all other weights supported by the casings, which may, in the aggregate, be greater than at first realized.

16.4 Size of Machinery Casings. The length and width of the casings should be only that needed for the space requirements of boilers, engines, and uptakes, and for such unshipping of machinery as the owners desire so as to minimize the weakening of the upper strength decks. The Rule deck area must be maintained past the casings, and, if the casings are relatively wide, the required thickness of deck plating is correspondingly increased.

16.5 Insulation of Machinery Casings. Machinery casings are almost invariably insulated, usually on the inside, to avoid objectionable transfer of heat and noise from within the casing to the surrounding spaces. On large ships, an air casing frequently is incorporated with the machinery casing to form a space through which the air supplied to the machinery spaces is drawn, in which case the insulation usually will be fitted on the main structural casing. This arrangement, with the cool air flowing down between the heated space within the casing and the spaces outside of the casing, provides an effective means of keeping the heat out of the surrounding spaces.

Section 17
Superstructures and Deckhouses

17.1 Classification Society Rules for Superstructures and Deckhouses. Unless a midship superstructure is comparatively short, it will act as part of the hull girder in absorbing the hull girder bending and should be designed accordingly. Classification society rules contain appropriate scantlings. "Comparatively short" is the phrase used in the ABS Rules; Lloyds Rules are more explicit, defining an "effective" superstructure as one which exceeds 15 percent of the ship's length. In practice, ABS Rules consider a midship superstructure exceeding about 10 percent of the ship's length to be effective for hull girder bending.

The ends of a midship superstructure or deckhouse, where the stress carried by the bridge deck and bridge sheer strake has to be transferred down to the upper deck and its sheer strake, are areas of high stress. Both ABS and Lloyds Rules contain detailed instructions regarding special provisions at these *breaks*. Extreme care is called for in designing the shell and deck structure in these areas to avoid abrupt discontinuities and potential proneness to fractures in service. The general type of construction illustrated in Fig. 57 has proved satisfactory in service.

17.2 Bridge or Deckhouse Fronts. Bridge or deckhouse fronts are particularly vulnerable to impact loads of *green water* in heavy weather. One has to see the damage on ships that have experienced exceptionally heavy weather to appreciate the force of these loads. Damage to the house front of a passenger liner is illustrated in Fig. 58. Rule scantlings provide extra strength for superstructure, deckhouse and unprotected poop front bulkheads.

17.3 Deckhouses. Deckhouses also are divided into short and long houses in classification society rules. *Long* houses are defined as those which exceed about 10 percent of the ship's length in the ABS Rules and 15 percent of the length in Lloyds Rules. The structure of all houses, short or long, must provide transverse racking strength against the effects of rolling and beam winds. This is provided by transverse bulkheads, either complete or partial, and by webs, shown in Fig. 59.

In addition, long houses participate in the ship's bending action, resulting in direct tension or compression of the house structure, and in large horizontal and vertical shear forces near the ends, which force the house to bend with the upper deck of the main hull. These forces at the ends of the house probably are more important than the direct tensile or compressive stresses in the deckhouse, and require adequate provisions, such as shown in Fig. 60, to transmit the stresses effectively into the main hull below.

The horizontal forces require fore-and-aft webs below the

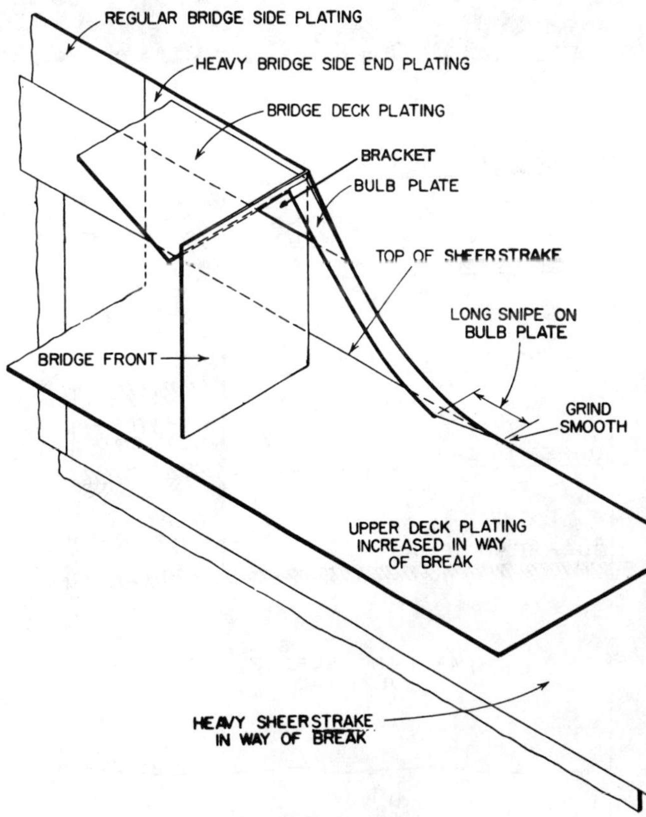

Fig. 57 Side plating at end of superstructure

Fig. 58 House front damage

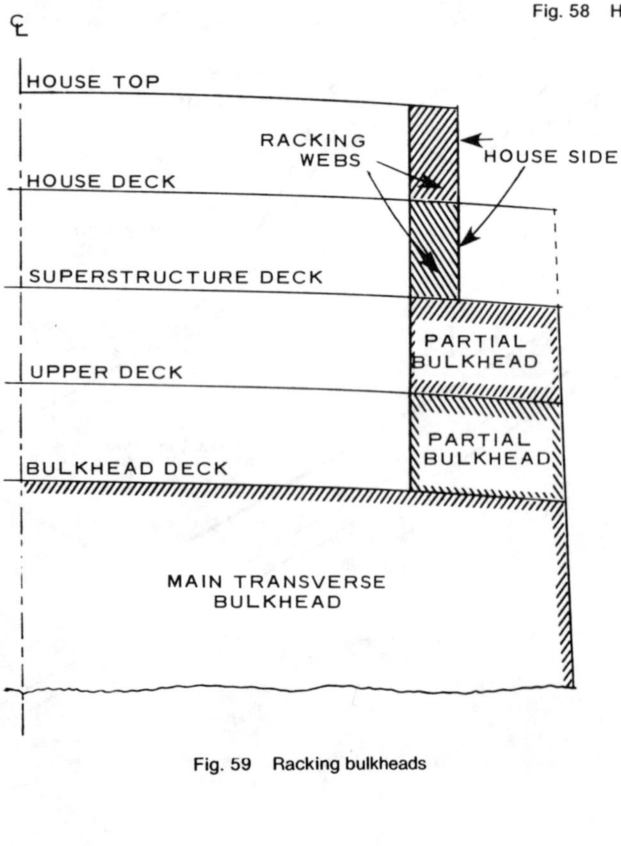

Fig. 59 Racking bulkheads

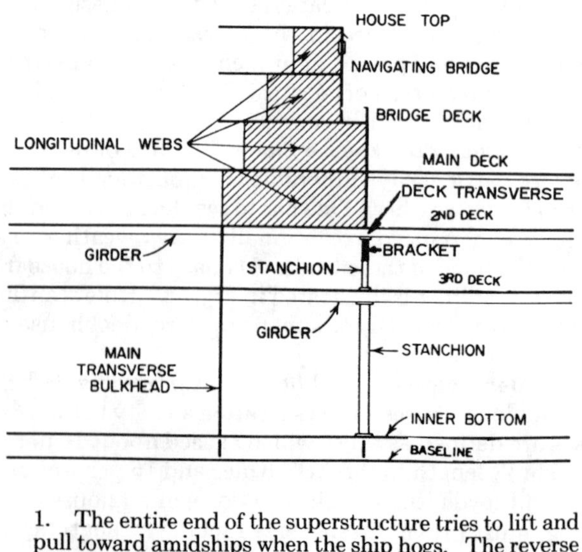

1. The entire end of the superstructure tries to lift and pull toward amidships when the ship hogs. The reverse happens in sagging.
2. A system of longitudinal webs, port and starboard, in line with the hatch side girder system is recommended.

Fig. 60 Web construction at ends of houses

house sides and, if practicable, also near the centerline so that all the stress in the deck will not have to be carried out and down through the house sides. The vertical forces can best be carried down by webs either in line with the house sides and house front or in line with fore-and-aft webs within the house. The magnitude of the forces to be withstood by such webs cannot be predicted accurately. However, if the houses are of considerable length, the scantlings, details, and disposition of such webs should be based on the assumption that they will be highly loaded.

17.4 Openings in Deckhouse Plating. A common difficulty in deckhouse structures is cracking of the plating in way of square-cornered door and window openings. The only effective remedy is to use rounded corners, the larger the radius the better. Fitting doublers to square-cornered openings will not suffice since the crack is likely to repeat itself in the doubler, as has been demonstrated many times. Although stated earlier, it bears repeating that square corners should be strictly avoided in openings in shipboard structures of any kind.

Deckhouses cannot be assumed to be uniformly supported by the deck structure on which they rest. Transverse bulkheads and webs below will be so much more rigid vertically than the intervening deck beams that the house will be supported almost entirely on these more rigid support points, and act as a girder between them. If this is recognized and the structure designed accordingly, with due regard for areas of high shear near supports, there will be less likelihood of cracking and buckling of house side plating with accompanying jamming of doors and windows and other related problems. All these considerations are more serious in large ships and long houses than in small ships or short houses.

17.5 Expansion Joints. Expansion joints have been used extensively on large vessels with long houses. Such a joint is simply a cut in all the structure above the upper deck of the main hull, in association with suitable slip joint arrangements to provide weather tightness, and slip joints or expansion bends in all interior work, such as joiner work and piping systems. The purpose of such joints is to relieve the houses of fore-and-aft bending stress so that they may be designed simply for vertical loads and racking stresses, resulting in considerable saving of topside weight and improvement in stability.

The principal considerations against the use of expansion joints are:

- they cannot be spaced close enough practicably to serve their intended purpose of relieving the houses of bending stresses;
- they introduce severe concentrations of stress at the bottom of the joints;
- the working of the joiner work and other details of the joints in a seaway is usually accompanied by objectionable creaking and leaking.

On the whole, it seems preferable to design the house structure so that it will be able to take hull bending stress and to accept the resulting increase in topside weight, and to provide the desired stability by other means.

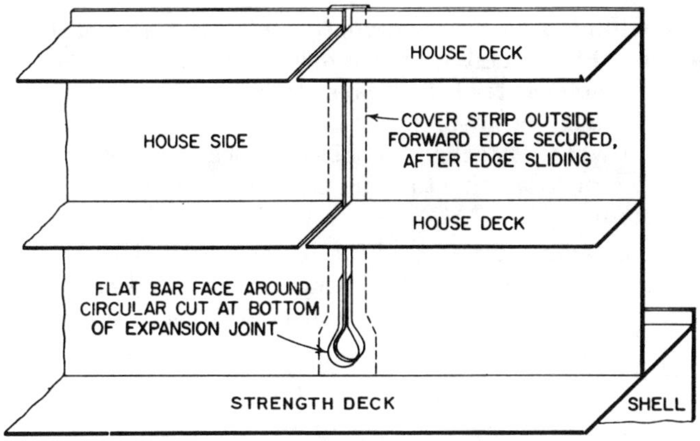

Fig. 61 Detail at bottom of expansion joints

If expansion joints are fitted, the construction at the bottom of the cut requires careful treatment because an expansion joint is, in effect, a severe large-scale notch. The continuous length of deckhouse between expansion joints should also be short enough to ensure that it does not bend appreciably with the main hull. A construction similar to that shown in Fig. 61 has been used successfully.

17.6 Design of Deckhouses. If and when sufficient knowledge is obtained to determine and utilize the contribution of the house structure to hull girder strength, there might be no increase in weight involved in designing the house structure to work as part of a single, deep hull girder with a corresponding reduction in the scantlings of the main hull, because the deepest girder is the lightest for a given strength.

The interaction of main hull and houses is a complex problem that has not been completely resolved. The ABS Rules require that deckhouses whose lengths are greater than 10 percent of the ship's length and are located amidships have effective longitudinal scantlings to give a hull-girder section modulus through the deckhouse equal to that of the main hull girder. In addition, the main hull is generally designed to be strong enough to take the bending moments without assistance from the houses. This represents sound but probably conservative design.

17.7 Use of Aluminum in Deckhouses. A successful method of aluminum deckhouse construction used on the *SS United States* was to make the fore-and-aft material of the deckhouses of an aluminum alloy having a yield strength comparable to that of steel, but with a modulus of elasticity only one-third as great. This results in a stress in the aluminum equal to only about one-third of the stress which steel would experience if subjected to the same strain or elongation. This effect should go far toward eliminating the need for expansion joints.

This advantage of using aluminum is accompanied by the disadvantages that:

- its coefficient of thermal expansion is almost twice that of steel, which causes assembly difficulties and enhances the possibility of distortion with temperature variations in service;

• aluminum loses its strength at elevated temperatures and must be insulated against heat to be considered equivalent to steel as regards fire resistance;

• it must be insulated from steel or other dissimilar metals where attached to them in locations exposed to weather or to wetting to avoid electrolytic corrosion of the aluminum.

When aluminum is used for deckhouses which are long enough to participate in hull bending, care must be used to avoid introducing rigidly connected steel fore-and-aft members into the aluminum structure. Such a member would experience the same strain as the aluminum structure to which it is attached, but three times the stress. Thus, steel transverse beams are permissible in an aluminum house, but longitudinal members must be aluminum. Steel transverses are sometimes desirable to reduce the depth of structure, but their use requires riveting and is therefore rare. The problems associated with joining and isolating steel and aluminum structural members are discussed in Chapter VIII.

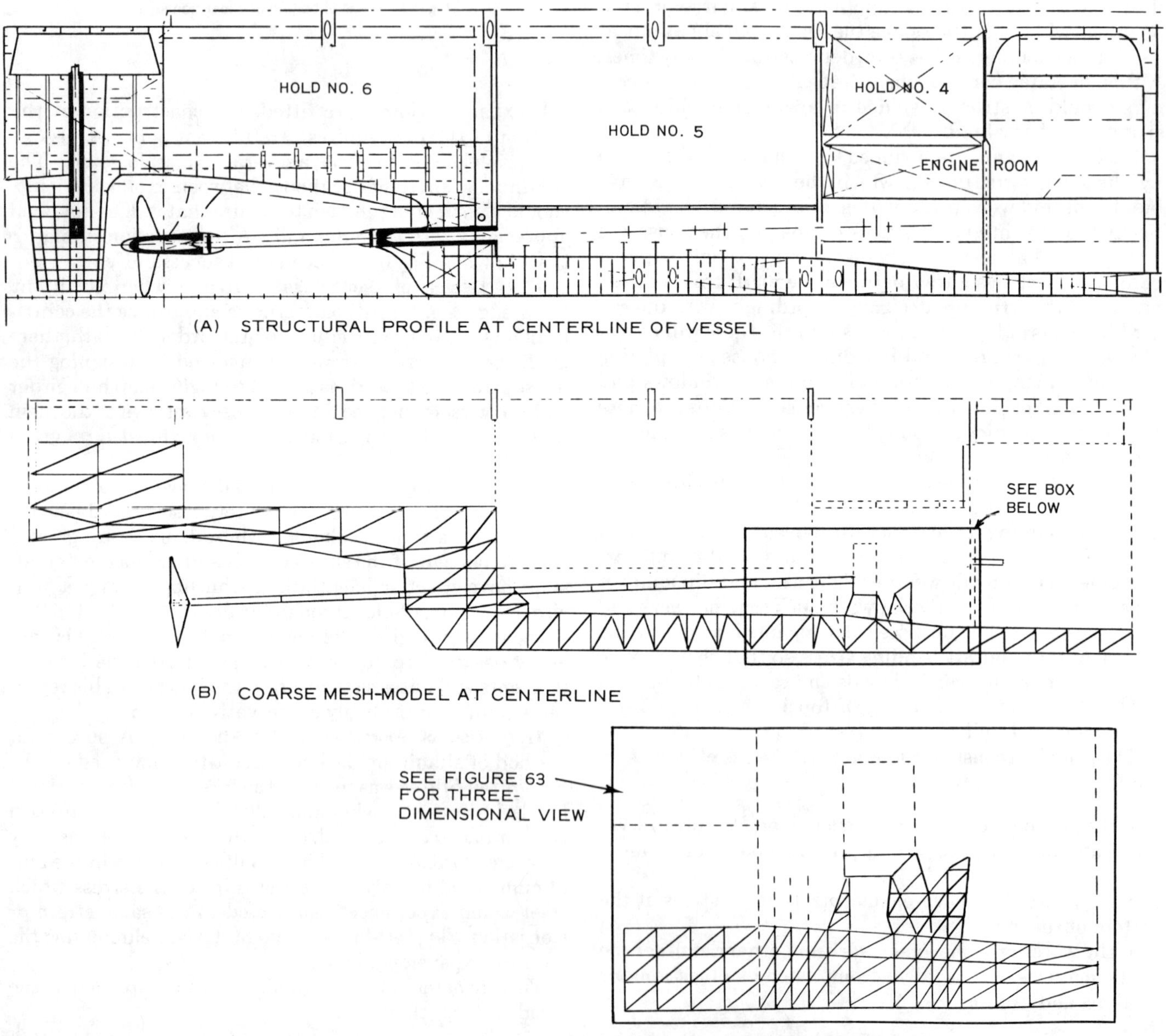

(A) STRUCTURAL PROFILE AT CENTERLINE OF VESSEL

(B) COARSE MESH-MODEL AT CENTERLINE

(C) FINE-MESH MODEL AT CENTERLINE

Fig. 62 Centerline section of ship and finite-element models

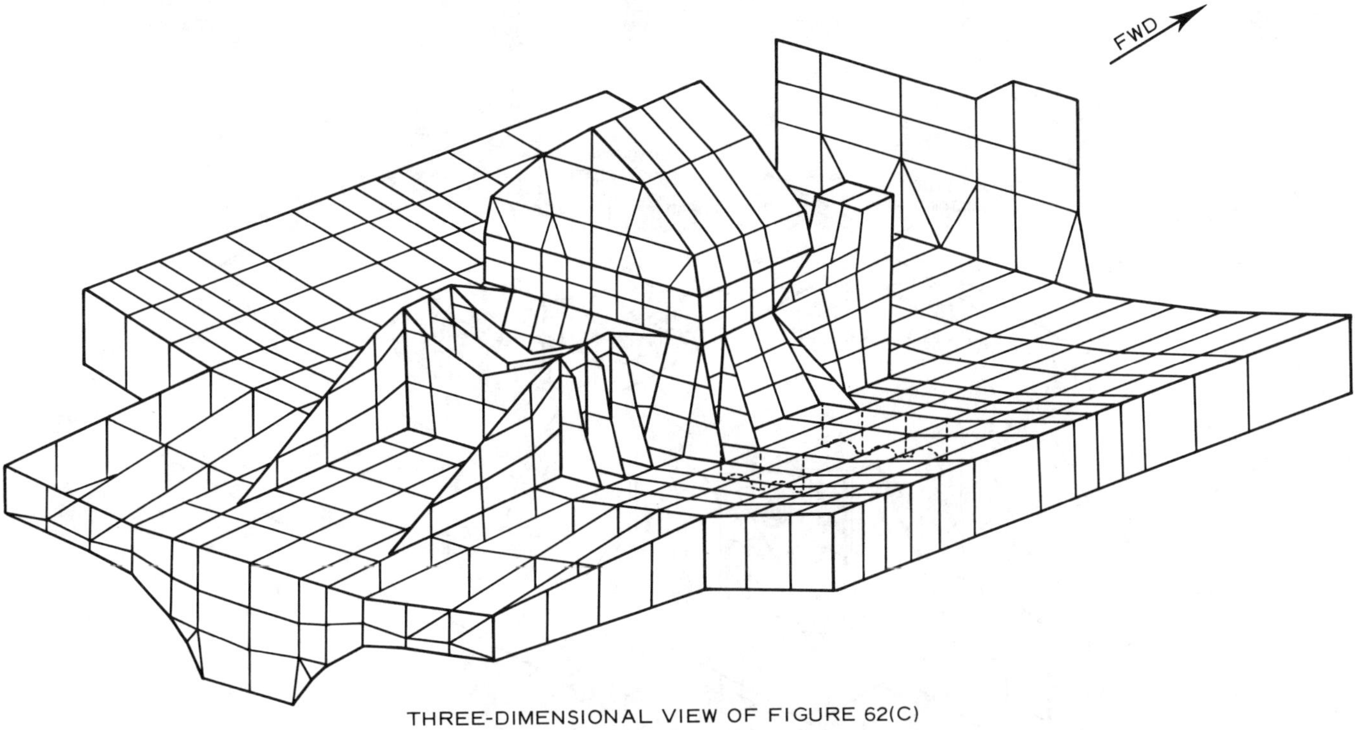

THREE-DIMENSIONAL VIEW OF FIGURE 62(C)

Fig. 63 Fine-mesh finite-element model of machinery foundation

Section 18
Foundations

18.1 Functions of Foundations. The term *foundations* refers to those structures fitted to support and to secure equipment to the main hull structure. The equipment ranges from the main machinery to ventilation fans. Shipboard foundations must be able to carry their loads when the ship is subjected to extreme sea conditions, that is, inclined to a large angle and subjected to high roll, pitch, and heave accelerations, as well as alternating thrust loads transmitted through the propeller shaft. The primary function of foundations is to restrict deflections to levels that can be tolerated by the supported machinery units, particularly main engines with gearing and shafting attached.

18.2 Design Considerations for Foundations. To fulfill its function, a foundation must be designed with sufficient stiffness to limit local deflections, and the supporting structure must be capable of limiting the overall deflections of the foundation. The foundation loads must be transmitted into adjacent beams or other structure and followed through until they are well distributed into the main hull. The beams or pillars to which foundations are attached may be already loaded to their design capacity and may need further strengthening to take the extra load (see Subsection 15.1*f*). If additional pillars below are required, they themselves must land on a sufficiently strong and rigid structure. This seems obvious but is sometimes overlooked.

In designing foundations for machinery with moving parts producing pulsating forces, consideration must also be given to the prevention of excessive vibration in the foundation and in adjacent structure. This problem arises with all kinds of reciprocating machinery, including diesel engines, pumps, and compressors.

18.3 Use of Numerical Methods to Design Foundations Compatible with Machinery Systems. Difficulties experienced with the main reduction gear assemblage of new vessels, with greater horsepower and finer lines than conventional cargo vessels, have given rise to a more sophisticated approach to foundation arrangement and design, namely a computer-aided numerical method. The use of a general-purpose finite-element computer program is a good example of this approach.

On one occasion, the existing shaft, thrust bearing, and bull gears with their foundation arrangement were analyzed by means of three-dimensional coarse-mesh and fine-mesh models, Figs. 62 and 63, under various loading conditions, resulting in the computation of the corresponding stresses and deflections. This analysis led to the conclusion that an alternative foundation arrangement was necessary in order to make it compatible with the vessel's machinery system.

Similar analyses and measured deflections of ship power plant foundations have exhibited remarkable agreement,

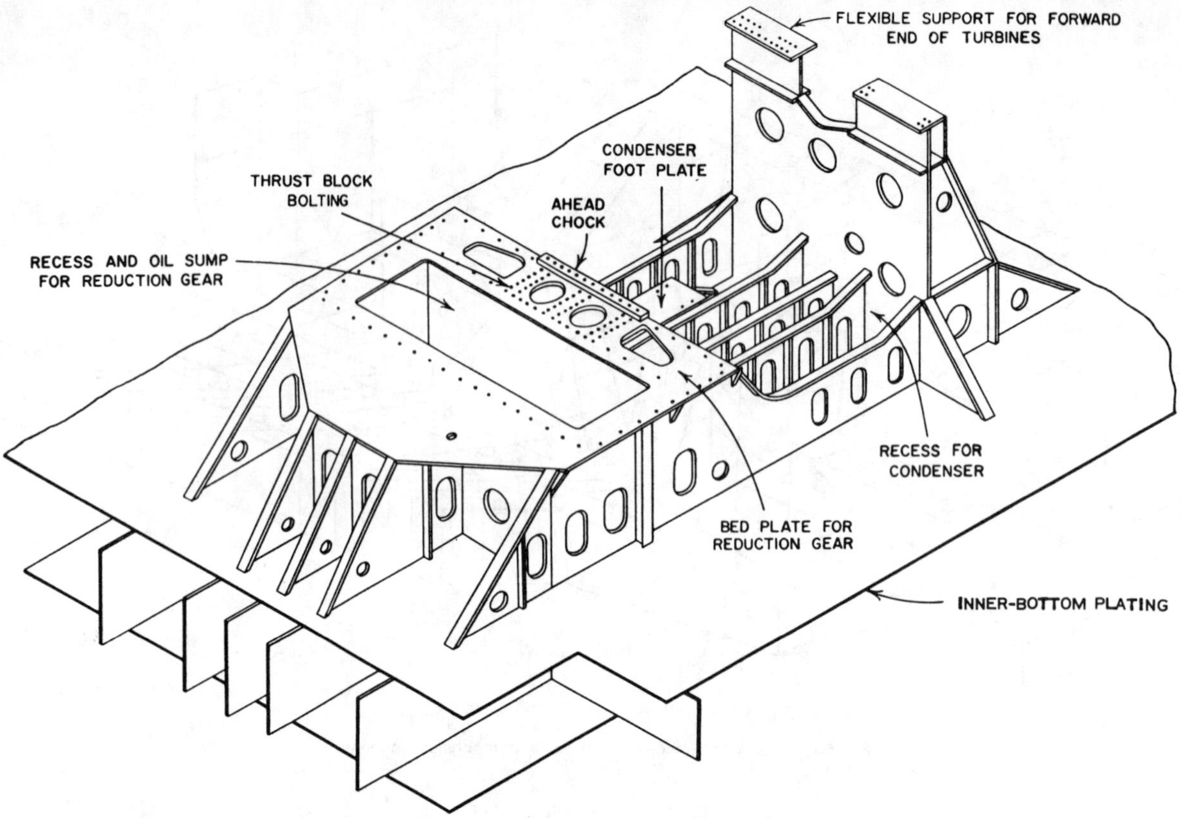

Fig. 64 Combined turbine, condenser, and gear foundation

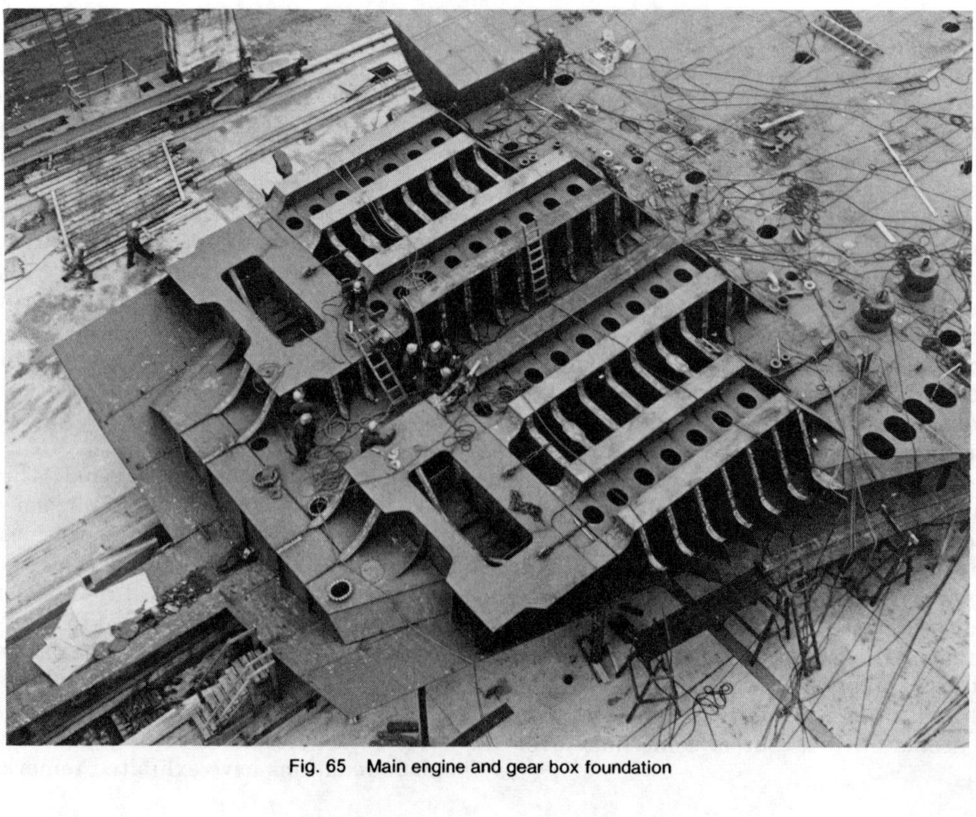

Fig. 65 Main engine and gear box foundation

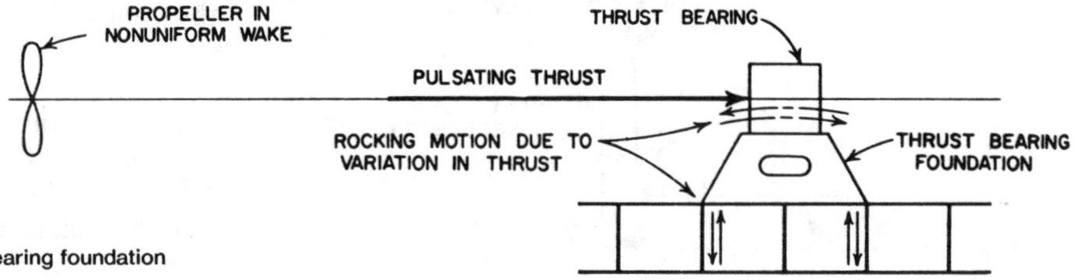

Fig. 66 Motion of thrust bearing foundation

which establishes this analytical procedure for determination of foundation stiffnesses and main propulsion shafting arrangements as an extremely useful design aid.

18.4 Main Engine Foundations. Construction of engine foundations may vary from a single pair of fore-and-aft girders supporting a small diesel engine, to the multiple foundation required for the high-, intermediate- and low-pressure turbines, the condenser, and the reduction gear shown in Fig. 64. The condenser is hung from the low-pressure turbine. The thrust bearing is located immediately forward of the main gear with its bolting designed to take the entire thrust. The *ahead chock* is fitted as indicated and is also capable of taking the entire ahead thrust. Fig. 65 shows the main engine and gear box foundation for a twin-screw diesel-powered vessel. The assembly is designed to accommodate one 12-cylinder and one 16-cylinder engine driving each shaft.

A foundation with fore-and-aft flexibility or a sliding foot is generally provided under the forward end of the turbine casing to permit thermal expansion of the machinery. An oiltight compartment is built under the main gear to serve as a sump for lubricating oil for the reduction gears. The interiors of the lubricating oil sump, as well as those of the storage tanks, are usually prepared by sandblasting down to bright metal to remove any mill scale, then wiping down with light oil. The application of paint on the interiors is not recommended because the lubricating oil is recirculated and must be free of any foreign matter.

The top plates of the foundations shown in Fig. 64, at least in the larger ships, were formerly made approximately 6 mm (0.25 in.) thicker than the design thickness to allow for machining the tops of the foundations to a true plane after fabrication was complete. It is now common practice to obtain alignment with welded chocks used in conjunction with fitted chocks in way of the bolts. This limits the machining to the bearing surfaces of the chocks.

18.5 Thrust Bearing Foundations. The foundation for the thrust bearing may be incorporated in the engine itself, as in the case of small diesel engines. It may be located in the main gear foundation aft of the reduction gear or forward of it, as in the turbine-driven reduction gear installation shown in Fig. 66. With steam reciprocating engines, it is often an independent unit, relieving the engine of the task of transmitting the thrust of the shaft into the hull.

The thrust in propeller-driven ships is of a pulsating nature, with the pulsating forces maximized at propeller blade frequency. If the thrust bearing foundation is not rigid enough, a vibration may result which is severe enough to prevent operating the machinery at its maximum speed. For practical reasons, these foundations are invariably below the line of the shaft. As a result, when the thrust pulsations move the shaft and foundation in the fore-and-aft direction, a *rocking* motion of the thrust bearing and its foundation occurs, Fig. 66. Thrust bearing foundations should be long in a fore-and-aft direction, rigid, and firmly anchored at their ends to minimize this rocking tendency.

18.6 Boiler Foundations. Boiler foundations are divided into two distinct groups, depending on whether the boiler is located close to the inner bottom or is a substantial distance above it.

- *Low boiler* foundations are relatively simple structures designed to distribute the weight into the bottom structure and prevent relative movement as the ship rolls and pitches. The boiler manufacturers furnish bolting plans, and the machinery arrangement determines the desired location. The ABS Rules require the boilers to be 460 mm (18 in.) clear of adjacent structure, if possible. Since the maximum roll and pitch of the ship is indeterminate, a reasonable basis of design is to assume a static inclination of 45 degrees with a working stress one-half of the yield stress.
- *High boiler* foundations usually consist of an open framework of heavy H-beams which transfer the loads from

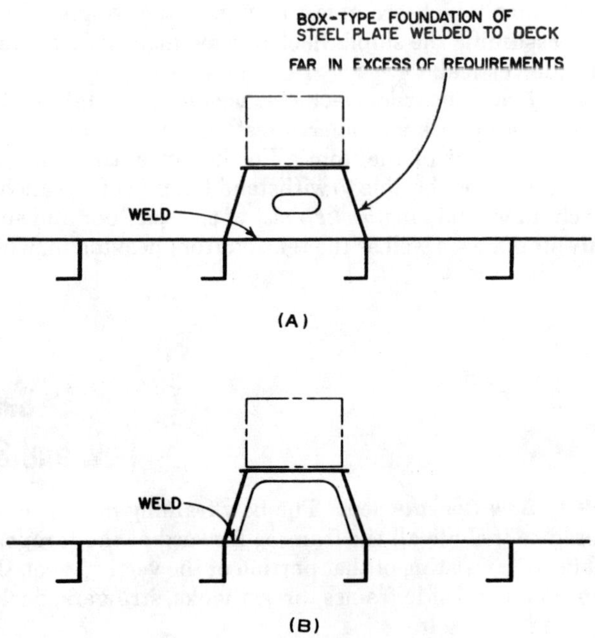

Fig. 67 Small equipment foundation

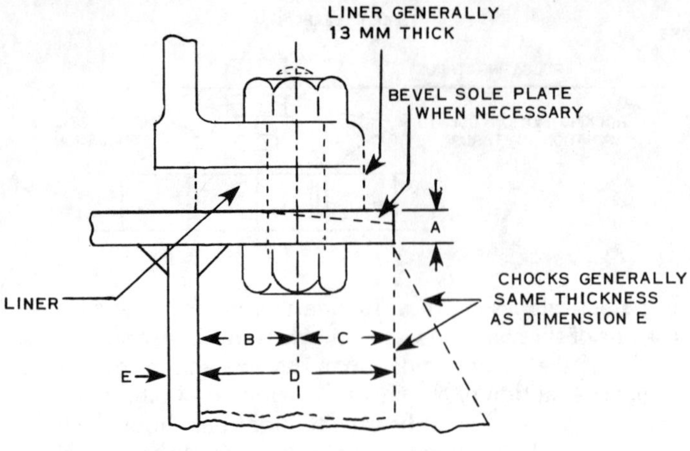

GENERAL WELDED FOUNDATION DATA
EACH FOUNDATION IS SUBJECT TO INDIVIDUAL STUDY

A = ONE HALF OF BOLT DIAMETER
B = 2 TO 3 TIMES BOLT DIAMETER
C = 2 TIMES BOLT DIAMETER
(MINIMUM IS 1 5/8 TIMES BOLT DIAMETER)
D = 4 TO 5 TIMES BOLT DIAMETER
E ≐ THREE-EIGHTHS TO ONE-HALF OF BOLT DIAMETER

Fig. 68 Standards for small foundations

the bolting pads to stanchions or other supporting structure. This type of foundation must be very carefully checked to insure against serious vibration, since such vibration can result in difficulties with the main steam piping.

18.7 Steady Bearing Foundations. Steady bearings, which support the line shafting at suitable intervals, have little to do except to carry the relatively small weight and to steady the shaft against any tendency to *whip* (the term used to describe a circular movement of the shaft centerline at midlength). As this latter tendency is unpredictable, a conservative basis for design is to provide steady bearing foundations which are able to support the weight of the shaft, assuming the ship is heeled to an angle of 90 degrees from the vertical.

18.8 Foundations for Deck Machinery. Foundations for many items of deck machinery must provide for forces other than the weight of the item. For instance, the windlass foundation must be able to withstand the pull of the anchor and chain when being *broken out* of the sea floor and suddenly stopped, as well as the tension from heaving in, while the chain stopper foundation must take the pull when the ship is anchored in a storm.

Steering-gear foundations must take any thrusts that are not self-contained by the steering gear itself. Winches and capstan foundations must take the breaking strength of the cables handled.

18.9 Foundations for Miscellaneous Small Items. While care must be taken to make large foundations for heavy items sufficiently strong and rigid, foundations for small items are sometimes too strong and unnecessarily heavy and expensive. Foundation for items weighing a few hundred pounds are frequently able to carry several tons. Also, excessive welded attachment is often provided.

As an example, the foundation shown in Fig. 67 (A) might be amply strong if cut away as shown in Fig. 67 (B), and the saving resulting from the reduced amount of welding to the deck pays several times over for the cost of cutting away the useless steel as indicated. On the other hand, foundations for small items must have adequate strength for loads which may be reasonably anticipated in addition to the design load. For example, a bracket to support an 11.3 kg (25 lb) item, but which a man might step on, should be sufficiently strong to take this possible load.

The dimensions in Fig. 68, in which the size of the bolt governs the scantlings, have proved useful as a means of obtaining some degree of consistency in the design of foundations for small miscellaneous items. Such data can only serve as a guide, and each foundation should be considered on its merits.

To reduce vibration, past practice has been to *bed* certain auxiliaries and deck machinery, such as winches and windlasses, on wood. It is increasingly common practice to eliminate all such wood beds and to bolt the machines directly to steel foundations, which in turn are supported and stiffened by suitable headers, or other framing, beneath the deck.

In some instances, aligning and positioning has been accomplished by the use of cast-in-place chocks of fiberglass-filled resins. These are noncorroding and noncorrosive, not vulnerable to attack by oil, and their use eliminates expensive machining for many foundations. In the case of foundations exposed to weather, the resins cover the entire top surface of the foundation, reducing maintenance costs. Details of the fitting of resin chocks are given in Chapter XVI.

Section 19
Bow and Stern Structures

19.1 Bow Construction. The bow framing may be considered to include all the framing forward of the forepeak bulkhead, consisting of that portion of the vertical keel, the deep floors and side frames, breasthooks, stringers, decks, and the stem itself.

The stem is usually a plate stem, a cast or forged stem, or, for some small ships, a rectangular bar. Classification society rules give sizes for such stem bars for small vessels based solely on the length of the vessel. For large ships, a cast-steel trough-shaped forefoot casting is sometimes used for the lower portion of the stem, with the upper portion

(*Continued on page 330*)

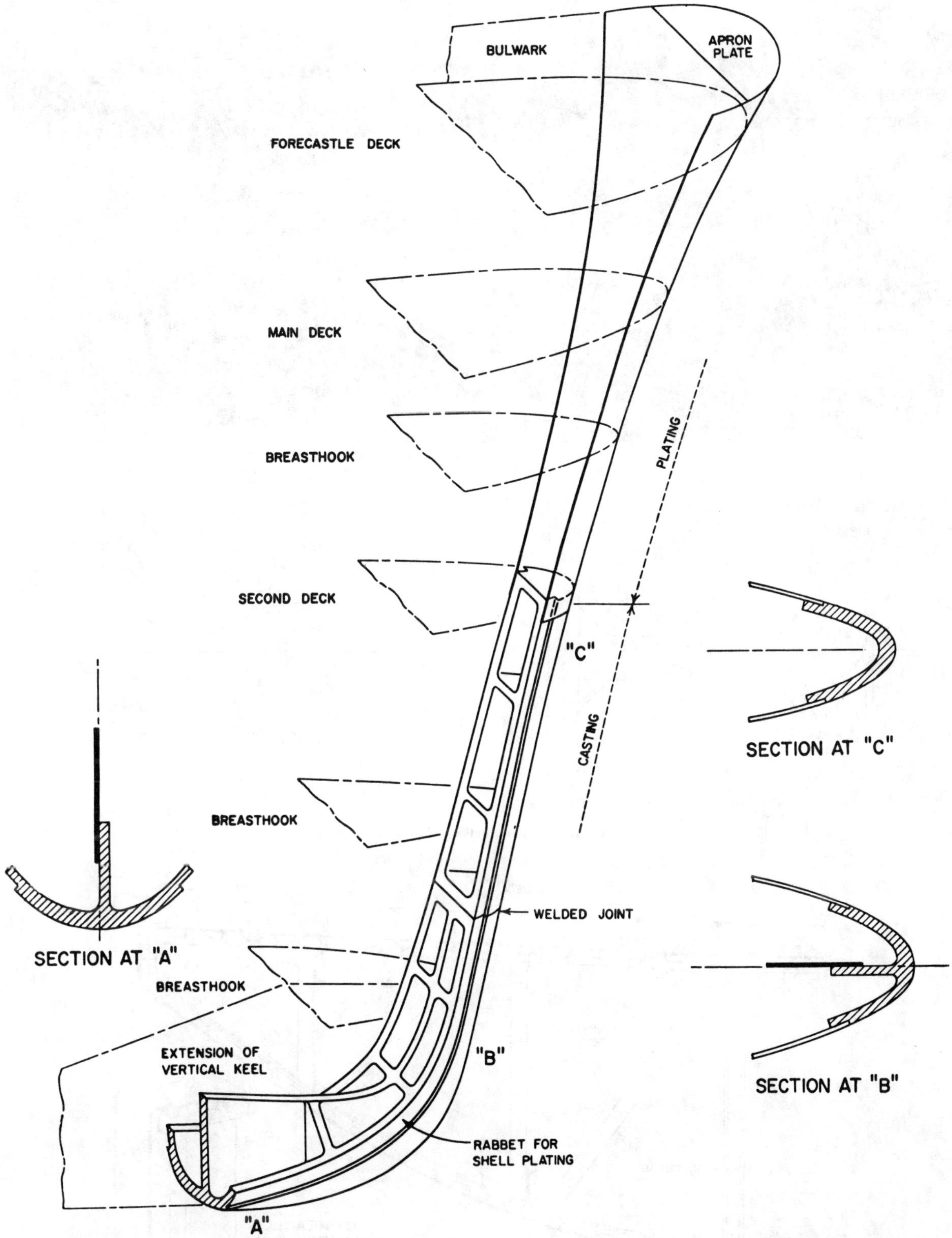

Fig. 69 Combination cast and rolled stem. To simplify construction and reduce cost the rabbets are being eliminated. Backing strips are used where necessary

Fig. 70 Bow framing subassemblies

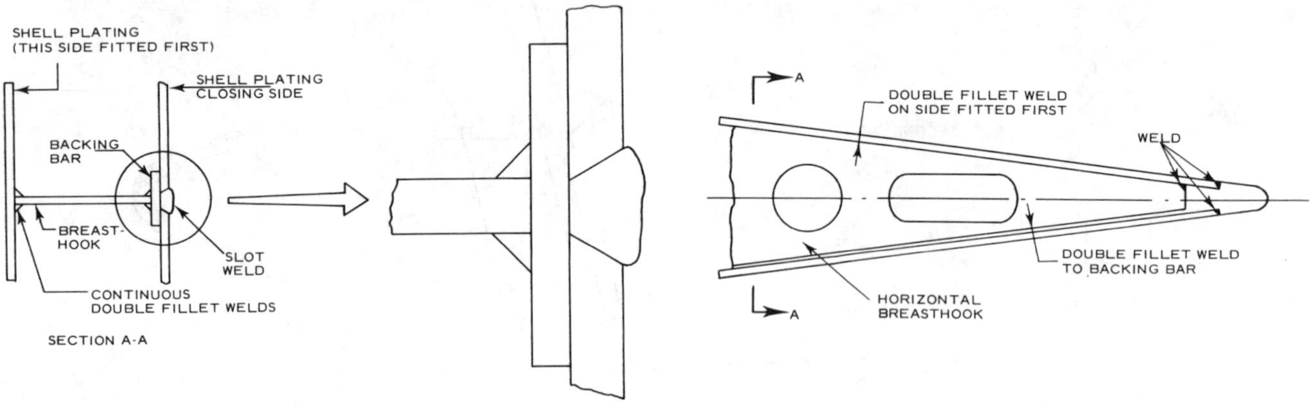

Fig. 72 Closing plate in narrow structure at stem

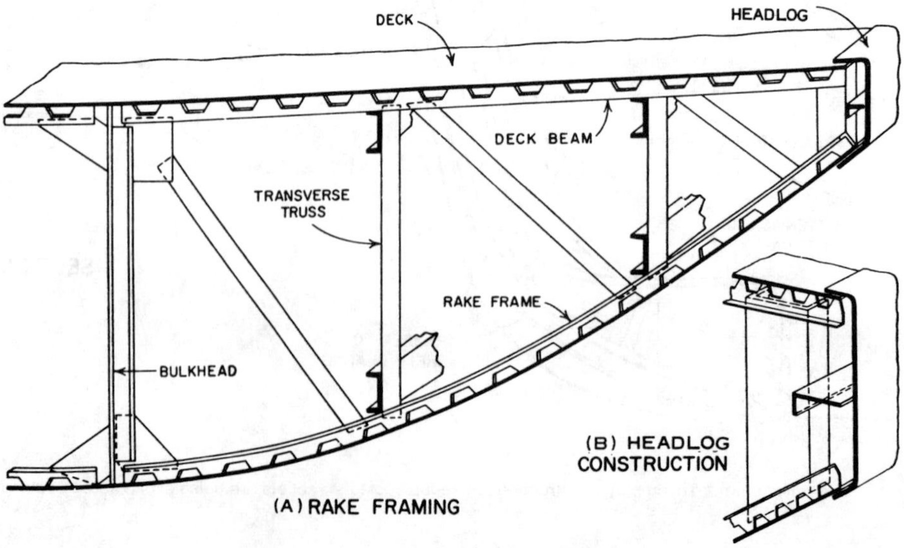

Fig. 73 Bow and stern construction for river barges

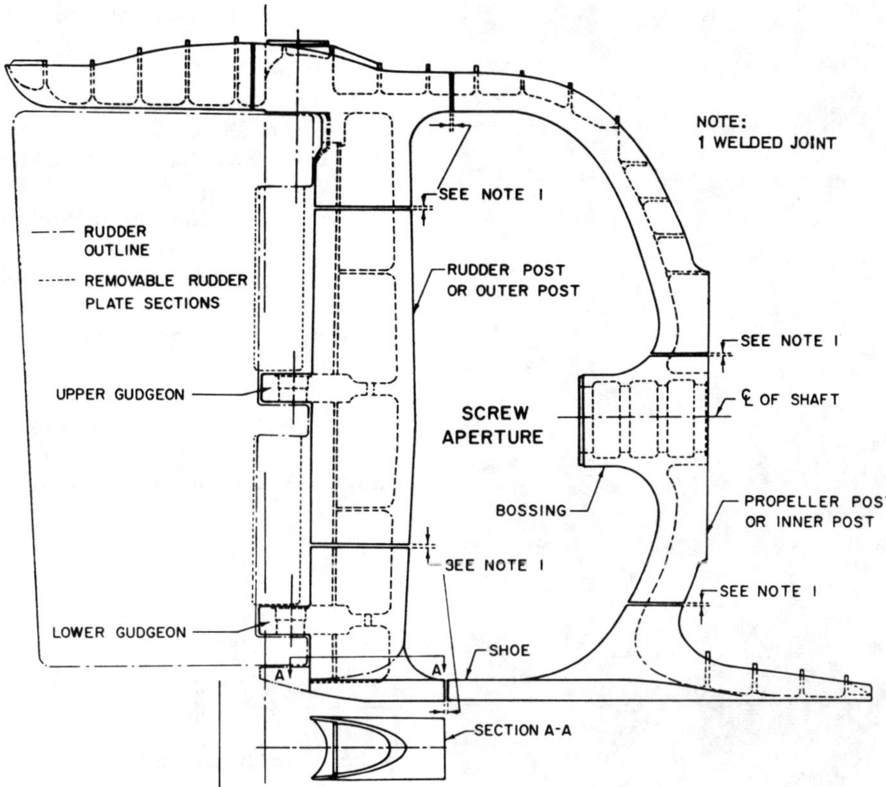

(A) STERN FRAME WITH PROPELLER APERTURE AND UNBALANCED RUDDER

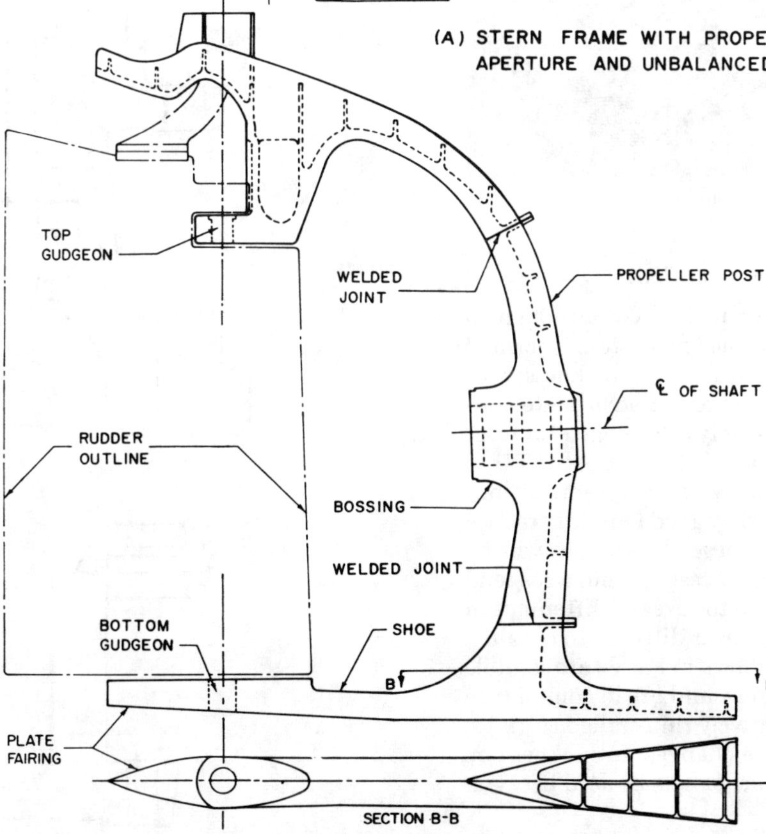

(B) STERN FRAME FOR BALANCED STREAMLINED TWO-BEARING RUDDER

Fig. 74 Stern frames of single-screw ships

Fig. 71 Tanker bow construction

(*Continued from page 326*)

formed of heavy plate, as shown in Fig. 69, although in general even for the largest ships the entire stem is formed of heavy furnaced plates. Lloyds Rules give thicknesses for a cast-steel forefoot casting as well as for steel plate construction, but ABS Rules require only a "strength" at least equivalent to that of a plate stem.

The thicknesses of walls and webs of steel stem castings must not be less than that required by good foundry practice for sound castings, and usually these thicknesses will be ample for strength. In the process of casting, thin webs tend to solidify first during cooling, and too great a difference in thicknesses of adjacent parts makes it difficult to obtain a sound casting. The thicknesses shown in Fig. 69 are suitable for a ship about 152 m (500 ft) long, and are in compliance with Lloyds Rules. As in all relatively thin-walled castings, it is important that sudden large changes in thickness be avoided, and that ample fillets and gussets be used between the flanges and the webs.

The thickness of the upper, plated portion of the stem in way of the hawsepipe is determined by the need to resist anchor or anchor-chain damage to the stem when the anchor is being raised or lowered with the ship rolling slightly and the anchor swinging against the ship. Another such load occurs when the ship is riding at anchor with the chain leading across the bow. On a 152 m (500 ft) ship, this thickness may be about 19 mm (0.75 in.). In a large ship, it may be as much as 25 mm (1 in.).

19.2 Bow Shape for High-Speed Ships. The shape used for the bow of a ship often has a direct effect on the type of structure used and the scantlings chosen. The construction of the stem, the choice of rolled frames or side-to-side solid floors, and the manner and sequence in which the structure is erected, are all directly related to bow shape.

Scantlings of bow framing and of supports for the weather deck forward are dependent to some extent on the degree of bow flare, as discussed in Section 15. A raked or *clipper* stem permits reasonably fine entrances on topside waterlines in combination with a reasonable degree of flare. Too large a radius at the upper end of the stem will result in a bluff

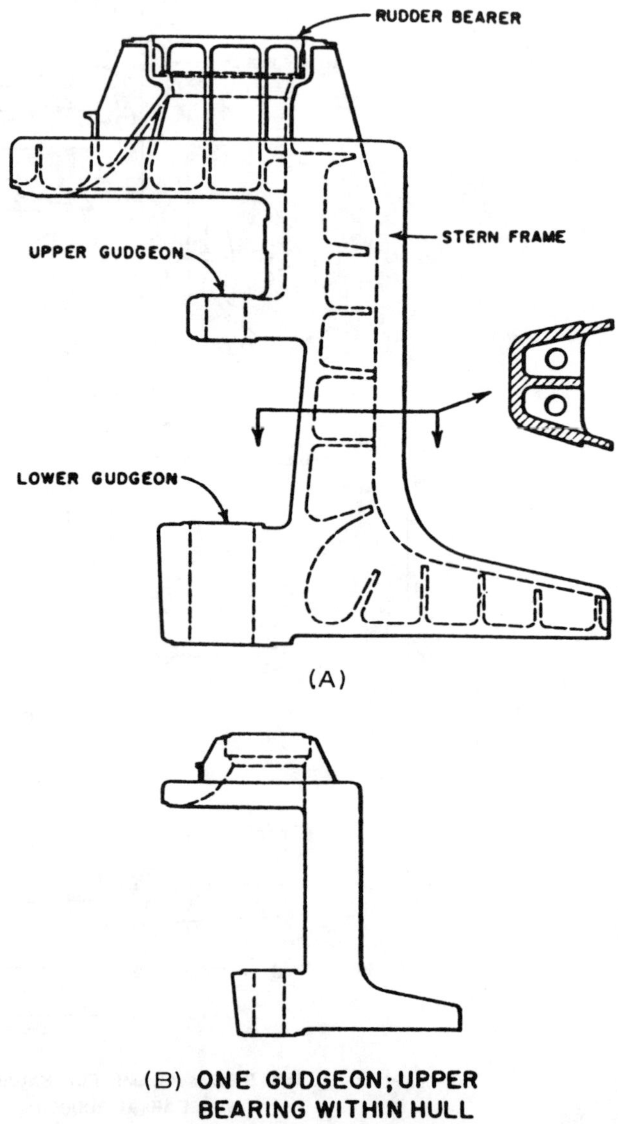

Fig. 76 Stern frame for multi-screw ship

Fig. 75 Single-screw fabricated rudder horn

Fig. 77 Stern frame casting for multi-screw ship

topside entrance and high dynamic loads during heavy weather.

19.3 Stem Construction on Great Lakes Ships. On Great Lakes bulk carriers, the stem is generally formed by a rolled semicircular plate or a heavy pipe, and it is usually very close to being vertical. There are practical reasons for this custom, the principal one being the custom of *winding* the ship to turn it around in a narrow channel. In this operation, the ship is put full ahead with the rudder hard over and a bow line secured to the dock. As the ship comes around, the stem is under heavy stress as it rides against the dock. Also, during this operation a heavily raked stem would tend to make the ship climb up on the dock, or overhang the dock far enough to foul the dock unloading equipment.

19.4 Bulbous Bow Framing. When a bulbous bow is used in conjunction with a fine, hollow waterline entrance, as may be desirable for certain combinations of speed and length, extreme care is needed to support the lower part of the bow adequately against the slamming effects of the seas. This can be done by providing ample horizontal fore-and-aft framing below the narrow load waterline in association with heavy vertical webs as far forward as is practicable. Fig. 70 shows three subassemblies of bow structure for a large fine-entrance ship, including the framing and the many breasthooks used in this particular vessel.

19.5 Full-Ended Ships. Recent construction trends for tankers have included very blunt bow entrances in conjunction with large bulbs, as shown in Fig. 71. These bows are often called cylindrical and are characterized by a 90-degree entrance angle. They are constructed entirely of formed plates stiffened by large breasthooks. Because of their large entrance angle, there is adequate room for supporting structure.

19.6 Closing Plate Construction. In fast, fine ships, the structure at the lower levels near the stem will become so narrow as to prevent ready access for securing the *closing* side of shell plating. The construction shown in Fig. 72, in which one side is plated with full access to both sides, and the other, or closing side, is so arranged as to require access from the outside only, is useful under these circumstances.

19.7 Rake Framing in River Vessels. In river vessels, the *rake* framing corresponds to the bow framing of an ocean-going ship. Studies of rake frame construction show that the use of a longitudinal truss at each frame without the addition of transverse trussing will not distribute local loads properly over a sufficiently large area, and that widely-spaced truss frames should be used, as illustrated in Fig. 73.

Headlog construction is commonly 19 mm (0.75 in.) thick plating, sometimes flanged-over at both deck and rake plating, as illustrated in Fig. 73. It requires substantial framing at not over 305 mm (12 in.) centers and heavy doubling at the corners. The connection of the headlog to the rake bilge plating requires special attention. It has been found that the best defense against damage at this critical area is the insertion of heavy plate, usually varying from 38 mm (1.5 in.) to 57 mm (2.25 in.) in thickness, at the intersection of the rake bilge plating and the corner plating.

19.8 Stern Frames of Single-Screw Ships. Scantlings for rectangular cross-section stern frames of single-screw ships are given in both the ABS and Lloyds Rules. In addition,

Fig. 78 Bossings, stern frame and rudder

Lloyds Rules give scantlings for streamlined stern frames, while ABS Rules indicate that the latter should be proportioned so as to be at least equal in strength to bar-type stern frames.

Fig. 74 (A) illustrates the stern frame of a single-screw ship with propeller aperture and unbalanced rudder. This stern frame consists of: a rudder post or outer post on which the rudder is hung, a propeller post or inner post which is swelled out into a bossing at the shaft line level to permit the tail-shaft to pass through it, and a shoe piece supporting the lower end of the outerpost and taking the severe athwartship loads imposed by the rudder.

Fig. 74 (B) illustrates the stern frame of a single-screw ship for a balanced, streamlined, two-bearing rudder. In this stern frame, the rudder post is eliminated and the rudder supported entirely by the top gudgeon and a bottom gudgeon at the after end of the shoe, which must then be considerably stronger than if an outer post were fitted. This arrangement has the advantage of permitting the rudder to extend forward of the center of the bearings for any practical degree of balancing.

There is a trend toward fabrication of the entire sternpost of plate. While the fairing forward of the propeller is not as good as with castings, the problems of casting are eliminated.

On many ships, the support at the bottom of the rudder is eliminated. In such designs the semi-balanced rudder is supported by a large cast or fabricated *horn*, as shown in Fig. 75, which is in turn attached to the ship. This horn-to-ship attachment is usually made by continuing the horn vertically to the underside of the steering gear flat, and welding heavy floors and longitudinal bulkheads to it. The shell thickness in the area of the horn must be increased to withstand the heavy rudder loads.

19.9 Stern Frames of Multi-Screw Ships. The stern frames of twin- or quadruple-screw ships do not have to support the tail-shaft and propeller; their only function is to support the rudder and contribute to the strength and ruggedness required for the stern of the ship. A common arrangement using two rudder bearings is shown in Fig. 76 (A). Fig. 77 shows a large stern frame of this type used on a large passenger liner, and Fig. 78 shows the same type of stern frame on a smaller ship. The upper bearing may be within the hull, as in Fig. 76 (B).

The stern frames of multi-screw ships, as shown in Figs. 76 and 77, are almost invariably steel castings of heavy construction. The ABS Rules give thicknesses of stern post castings for ships with no bossing and shoepiece in general, but Lloyds Rules specify minimum thicknesses based on the length of the vessel. As in the case of steel forefoot castings, these thicknesses usually are determined as much by foundry requirements for good sound castings as by strength considerations.

19.10 Use of Weldments. In welded construction, if the cast-steel stern frame were erected by itself and then the adjacent structure were erected piece-by-piece and welded to the casting, an indeterminate state of stress would exist within the casting, resulting from the intense local heat of the welds and the consequent shrinkage. This might have a tendency to cause cracks in the casting at or near the welds.

To minimize this tendency, it is good practice to weld as much of the adjacent structure as is practicable, including the aftermost shell plates, onto the casting to form a subassembly, and then to stress-relieve the entire weldment by heating it in an oven to about 620°C (1150°F) and slow-cooling it in the oven. This procedure is equally desirable, where practicable, for all large hull castings such as rudder castings and spectacle frames.

19.11 Framing Adjacent to Stern Frame. For low resistance and good propulsive efficiency, it is desirable that the underwater part of the ship immediately forward of the propeller be as fine and narrow as is practicable, with the result that the structure in this area is often not much more than a skeg, and is not well adapted to take the severe dynamic and vibratory loads imposed on it by the propeller and the rudder. This inherent weakness is overcome by making the floors heavy and deep, extending them well above the shaft to where the stern begins to widen out, and using flats tying the floors together, extending the flats forward to the aft-erpeak bulkhead.

The shell is made heavier in this local area than elsewhere. As a result, instead of two flat stiffened-plate sides of the ship, a single strong structural unit is formed. Similarly, solid floors are used in the after portion of bossings in way of the stern bearing, and special care is taken to tie them into the hull extending across the narrow width of the ship at that location and into a flat above the shaft.

Section 20
Bossings and Struts

20.1 Function of Bossings and Struts. Bossings and struts have the common function of supporting the outboard end of the propeller shaftings as well as the propeller itself. In this section, discussion of bossings will relate only to the bossings of multi-screw ships, since the bossing of single-screw vessels consists only of the swelling of the stern post and of the adjacent frames, as illustrated in Fig. 75.

The difference between a bossing and a strut is that a bossing is plated-over, while a strut is an exposed support which is located close to the propeller with open shafting forward of it. Large merchant ships generally are fitted with bossings, and smaller ships, particularly naval ships, with struts. The two may be found on the same ship, a short portion of the shaft being enclosed by a bossing, and the remainder of the shaft exposed and supported by a strut at the after end.

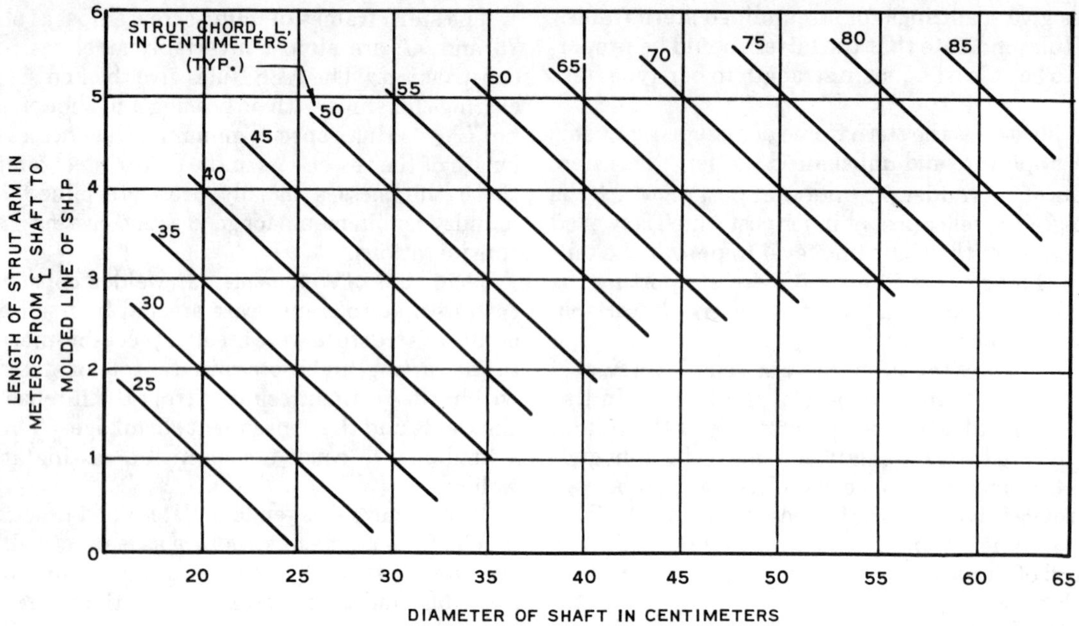

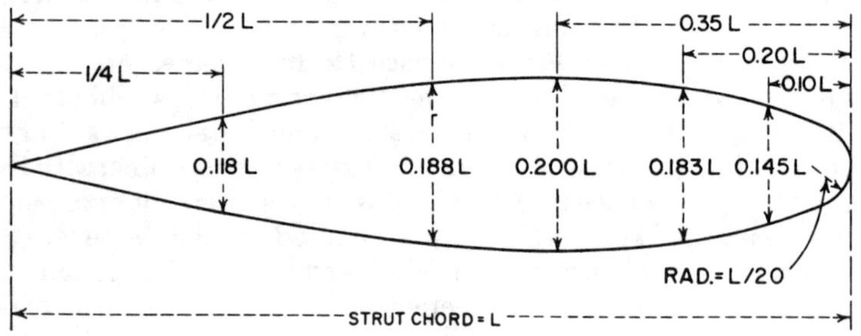

Fig. 80 Streamline contour and sizes for strut arms

20.2 Spectacle Frame. A bossing ends in a *spectacle frame,* so called because, when extended from port and starboard to the centerline of the ship, it may resemble a pair of spectacles. The spectacle frame need not extend to the centerline. In large ships, it usually extends only far enough into the hull for an adequate connection to the structure, the connection across the ship being formed by substantial floors and flats. This frame is often a steel casting, but it may be a weldment consisting of a combination of castings and structural plates, such as a stern bearing made of heavy rolled steel plate and a web made of a steel casting. If it is such a weldment, it is desirable to stress-relieve the entire weldment after fabrication to eliminate residual stress in the castings resulting from welding. This is similar to the usual practice for large hull castings, such as stern frames, referred to in Section 19.

20.3 Design of Bossings and Struts. A logical basis for the design of either struts or spectacle frames would be the unbalanced force that would result from the loss of one propeller blade, but, in practice, the size or strength of the shaft is used as a convenient basis. For bossings, the rigidity found by experience to be necessary to prevent excessive vibration will in itself provide more than sufficient strength by any reasonable design basis. In general, bossings and not struts should be used for vessels designed to operate in ice.

The two principal objectives in the design of a bossing are directed toward the prevention of the transmission of vibration from the propeller into the hull. The first objective is to provide the greatest practicable transverse rigidity of the bossing; the second is to provide ample clearance between the after edge of the bossing and the propeller, Fig. 79. These considerations are to some extent contradictory, because a boss frame kept well forward of the propeller, as in Fig. 79, cannot be quite as rigid a support as one carried out square to the end of the shaft close to the propeller. Nevertheless, to avoid objectionable vibration, it is desirable to keep the end of the bossing clear of the propeller and to obtain the necessary strength and rigidity by rugged construction.

a. Shape of Trailing Edge of Bossing. In model tests, a slightly higher propeller efficiency is obtained by carrying the bossing fairly close to the propeller and so shaping it as to deflect the water against the propeller in the opposite direction to the sweep of the blade. However, it has been found that if this is done there is risk of severe vibration, far outweighing the advantage of any slight gain in efficiency. It is better to locate the bossings as nearly as possible in the

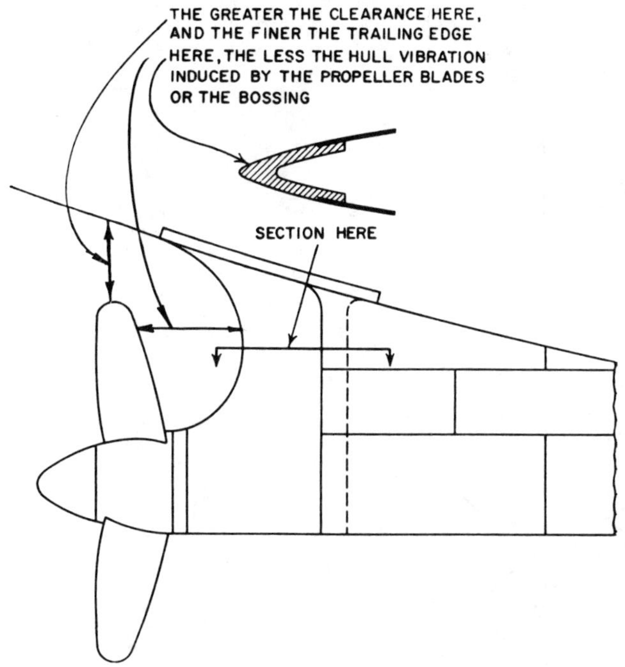

Fig. 79 Spectacle frame and propeller clearance

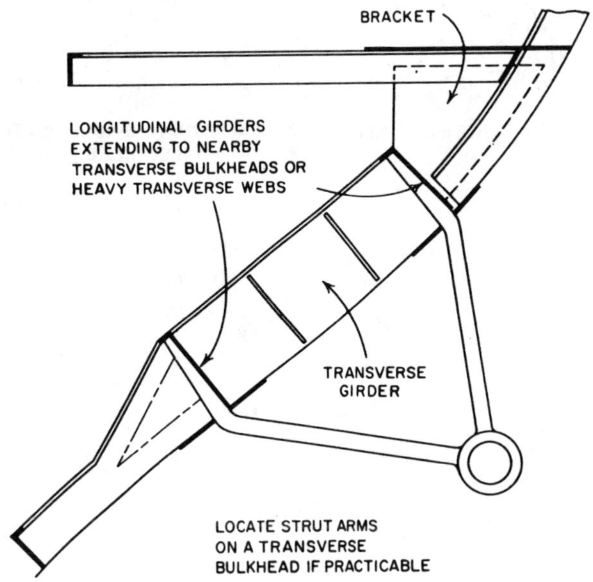

Fig. 81 Method of attaching strut arms to hull structure

line of the natural flow of the water, and to end them in a symmetrical taper as far ahead of the propeller, as fine and sharp as structural strength permits, as indicated in Fig. 79, thus allowing the flow of water to the propeller to be as undisturbed as possible. The resulting smooth running is worth more than a possible slight gain in efficiency.

b. Two-Armed Versus Single-Armed Struts. In general, struts should be two-armed. It is possible in some cases to design single-armed struts of sufficient strength for any reasonable design basis, and they are used for small high-speed vessels. However, for other than small vessels, their rigidity would be deficient and excessive vibration almost inevitable. Single-armed struts occasionally are used for very short intermediate struts.

c. Cross Section of Strut Arms. Strut arms invariably are streamlined to minimize disturbance of the water just ahead of the propeller. Strut requirements for small vessels are given in the ABS Rules for Vessels Under 61 m (200 ft) in Length. Fig. 80 gives suitable sizes and proportions based on the strut arms acting as columns in compression, assuming a solid shaft made of steel with an ultimate tensile strength of 414 MPa (60,000 psi) and an angle between the strut arms of not less than 65 degrees. If a stronger steel is used for the shaft, the strength of the strut should be increased correspondingly. The strut barrel is also designed to provide the shaft bearing casing, similar to that of stern tubes.

20.4 Construction at Inboard Ends of Struts. The inboard ends of shaft struts are usually carried through the shell and attached to special framing arranged to distribute the load from the arms into a sufficiently broad and rigid structure to eliminate any likelihood of local vibration, approximately as suggested in Fig. 81. The hole in the shell plating is closed by a heavy plate around the strut arm.

Section 21
Bilge Keels and Fenders

21.1 Description. *Bilge keels* are fin-like structures fitted along the outside of the shell plating near the turn of the bilge and extending throughout the midship portion of the ship to reduce rolling. *Fenders,* in the sense used here, are rubbing strips of various kinds, the purpose of which is to prevent local damage to the shell and framing as a result of contact with piers, quays, locks, or vessels alongside.

21.2 Bilge Keels. There are two principal types of bilge keels: single-plate and double-plate. These two types are illustrated in Fig. 82. They are designed to come well within the area enclosed by the vertical line of the ship's side and the deadrise line extended. Figs. 82 (A) and 82 (B) show common construction methods for single-plate bilge keels. The construction in Fig. 82 (B) is preferred because it prevents shell damage in the event that the bilge keel is torn off.

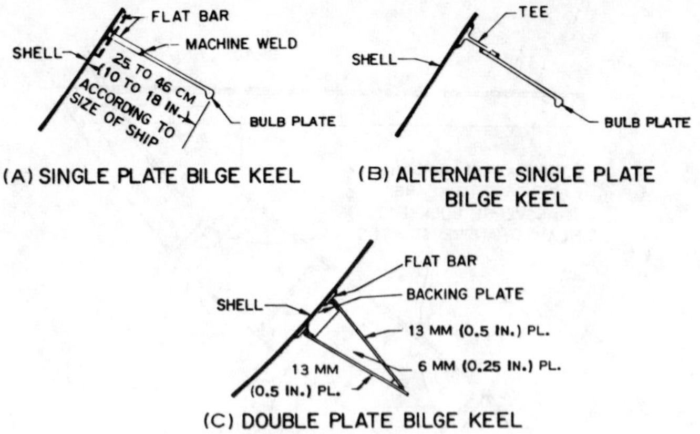

Fig. 82 Bilge keels

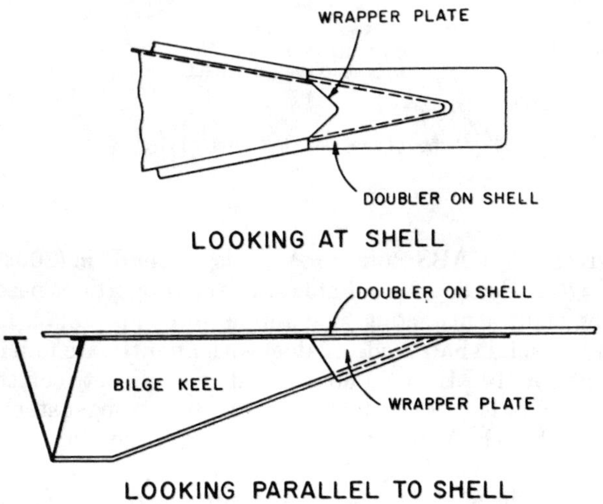

Fig. 83 End construction of double-plate bilge keel

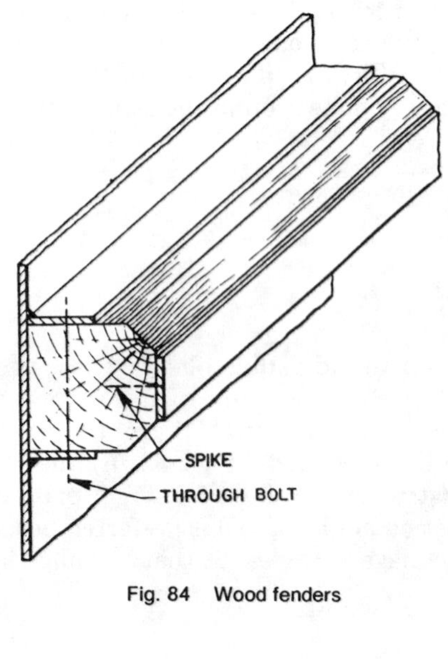

Fig. 84 Wood fenders

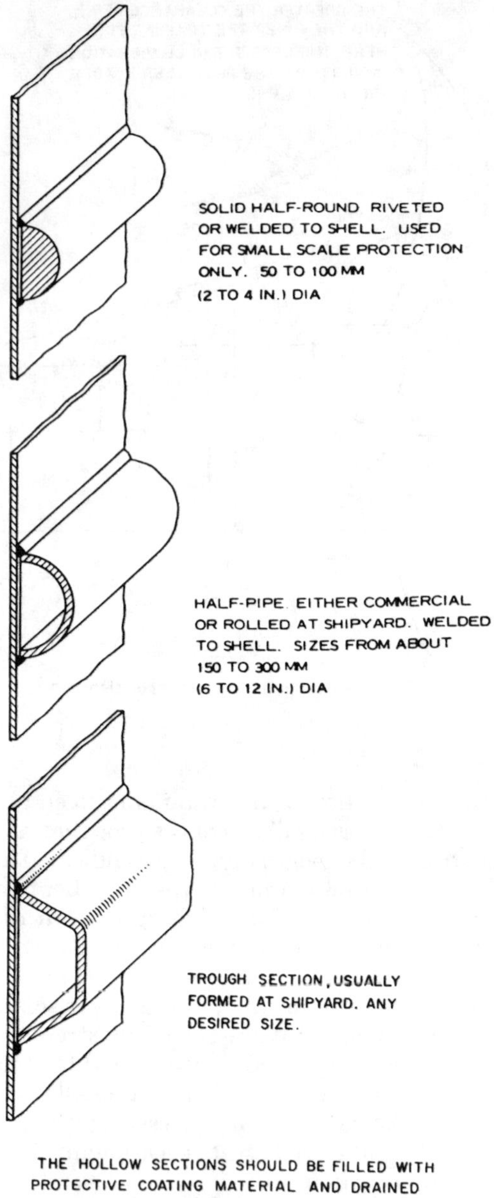

Fig. 85 Steel fenders

a. Depth of Bilge Keels. Experience indicates that about 460 mm (18 in.) is the maximum practicable depth for a single-plate bilge keel of the type shown in Figs. 82 (A) and 82 (B), and about 1.22 m (4 ft) for the double-plate type shown in Fig. 82 (C). Within these limits, the deeper the bilge keel, the more effective it will be. In general, a short, deep bilge keel is more effective than a long, narrow keel of the same area. In addition to limiting the depth, careful design of the tapered forward end is necessary, and the bilge keel should not extend into the forward one-third length. The end construction shown in Fig. 83 for the double-plate bilge keel has been found satisfactory.

b. Streamlining Bilge Keels. To minimize resistance, bilge keels should conform as nearly as is practicable to the streamlines along the bilge at the design speed, and the trace of the bilge keel should always be determined by model tests on large important ships. If located without benefit of model tests, the trace in the afterbody should be run approximately normal to the frame lines as seen in the body plan, or, what is practically the same thing, along the bilge diagonal, while in the forebody the trace should be considerably steeper than a bilge diagonal.

c. Filling Double-Plate Bilge Keels. Double-plate bilge keels were formerly filled with wood, balsa if available, and the interstices pumped full of bitumastic compound to keep water out of the keel in case of damage or leaks. Modern practice is to foam-in-place a unicellular foam such as polyurethane or to fill and drain the void spaces with a corrosion-resistant liquid.

d. Forward Ends of Bilge Keels. In heavy weather, with the ship pitching and rolling, the forward end of the bilge keel is liable to take a severe loading. At an annual dry-docking of a large transatlantic liner, the forward 12.2 m (40 ft) portion of bilge keel on both sides of the ship was found to have been torn off. It was replaced and a year later was again found to be gone. Then it was left off, and the forward end of the remainder was suitably finished off with an easy taper and a strong connection, similar to that shown in Fig. 83. No further difficulty has been reported. It is of interest to note that the behavior of the ship was not perceptibly affected, since no one on board suspected that considerable portions of both bilge keels were missing. Similar losses of the forward portions of single-plate keels have been found on several ships.

e. Bilge Keels on Great Lakes Ships. It is standard practice not to fit bilge keels on Great Lakes ore carriers because of their vulnerability to damage in the restricted channels in the rivers between the lakes. Fortunately, the percentage of time when these ships experience rough weather is small.

21.3 Fenders. There are two common types of fenders: one is a wood fender, secured between steel flat bars welded to the shell plating, Fig. 84; the other is a trough-shaped steel section, Fig. 85, which may or may not be filled with a bitumastic compound. Half-pipes, solid half-rounds, or flat bars, Fig. 85, are also used for steel fenders.

Fenders are used more on small ships than on large ones. Tugboats may have two or three fenders along their sides. However, fenders may be found on any ship where rubbing damage is anticipated. For instance, in a ship with a riveted bilge strake but otherwise all-welded side shell, the upper bilge seam may be the widest part of the vessel, with the result that the rivet points in this seam may be subject to rubbing against pier piling so that paint cannot be kept on them and excessive rivet point corrosion results. A fender fitted close to this seam, as shown in Fig. 86, would prevent this condition.

If wood fenders are used, it is undesirable to use oak directly against steel, because the tannic acid of oak is corrosive. If oak is desired, it should be used as a facing only, with a backing of pine or other suitable wood.

On river barges, deep-sectioned side fenders have proved undesirable for fleet towing. The added space between the barges caused by the projection of such side fenders increases the resistance of the fleet and also creates a drift-catching crevice where drift will pack and build into a large mass protruding below the bottom of the barges and greatly increasing the resistance. For this reason, side fendering on such vessels is confined to patch plates, varying from 19 to 25 mm (0.75 to 1 in.) in thickness, so located as to accept the normal wear from rubbing against the concrete lock walls and against adjoining barges, and so mounted that they may be replaced easily when worn.

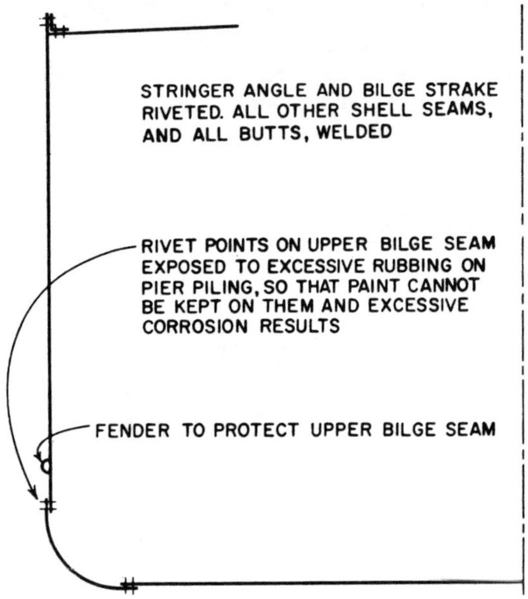

Fig. 86 Riveted seam-protection fenders

References

American Bureau of Shipping (Annual) *Rules for Building and Classing Steel Vessels.*

American Bureau of Shipping (1973) *Rules for Building and Classing Steel Barges for Offshore Service.*

American Bureau of Shipping (1973) *Rules for Building and Classing Steel Vessels Under 61 Meters (200 Feet) in Length.*

American Bureau of Shipping (1978) *Rules for Building and Classing Bulk Carriers for Service on the Great Lakes.*

Elbatouti, A. M. T., H-Y Jan and S. G. Stiansen (1976) "Structural Analysis of a Containership Steel Model and Comparison with the Test Results," *Trans.* SNAME.

International Maritime Consultative Organization (IMCO), "International Convention for the Safety of Life at Sea," Chapter VI, 1974.

Lloyd's Register of Shipping (Annual) *Rules and Regulations for Construction and Classification of Steel Ships.*

MacNaught, D. F. (1967) *Principles of Naval Architecture,* Edited by J. P. Comstock, SNAME.

Robertson, J. B., (1967) *Principles of Naval Architecture* Edited by J. P. Comstock, SNAME.

Ship Structure Committee (1977a) "A Study of Ship Hull Crack Arrester Systems," Report No. SSC-265.

Ship Structure Committee (1977b) "Review of Ship Structural Details," Report No. SSC-266.

Ship Structure Committee (1978) "In-Service Performance of Structural Details," Report No. SSC-272.

CHAPTER VIII

Irving L. Stern

Hull Materials and Welding

Section 1
Prefacing Remarks

1.1 General. A major change in ship construction occurred over 100 years ago when steel was introduced to replace iron and wood as a hull material. Subsequent important developments in materials and ship construction were the all-welded ship and the application of concepts of notch toughness for the prevention of the brittle hull fractures experienced with the all-welded steel ships of the 1940's. Over the past thirty years, many new designs have been introduced such as container ships, liquefied gas carriers, high speed surface-effect ships, and mobile and fixed offshore structures. To meet requirements for such designs, high strength-to-weight ratio alloys and alloys intended for low temperature service have been introduced into shipbuilding. The complex networks of intersecting members in offshore structures require that consideration be given to material properties when the applied tensile loads are perpendicular to the plate surfaces. The increasing size of ships, such as the VLCC tankers, and the concern for economy, stimulated automation of fabrication processes. The relatively simple concept of toughness developed to answer the brittle fracture problems in ordinary strength steel hulls required refinement and extensive development before it could be applied to the newer hull materials and structures. For some structures increased consideration had to be given to such factors as fatigue and corrosion. Accompanying all these changes, was an increased demand for the designer to provide for quality assurance by nondestructive methods. To meet the challenges presented by these new developments, the designer had to become familiar with metallurgy, welding engineering, nondestructive testing, and the materials sciences. An appreciation of the basic principles of these fields will provide more efficient and reliable hull designs through selection of appropriate materials, joining, and quality assurance requirements.

Section 2
Material Properties and Tests

2.1 Tensile Properties. The properties most commonly used for design calculations and material acceptance are determined from the tensile test. Specimens and procedures for tensile testing vary between different products, those described herein being applicable to hull steel plate. For other materials and for more complete requirements and details the reader's attention is directed to the American Society for Testing and Materials (ASTM) specifications E8 and A370 covering the test methods applicable to the specific material of concern. Typical rectangular 200 mm (8 in.) gage length and round 50 mm (2 in.) gage length tensile specimens are illustrated in Fig. 1; behavior of such specimens under an applied load is illustrated in Figs. 2A and 2B.

a. Proportional Limit and Elastic Limit. When a tensile load is applied to a specimen it produces a proportional amount of stretching or increased strain between the gage points. The maximum unit stress at which the strain remains directly proportioned to stress is known as the proportional limit. In the case of the commonly used ship steels, this stress also approximates the elastic limit, which is the maximum unit stress attainable without any permanent strain remaining upon complete release of the stress. Proportional and elastic limit tests are not usually required in production testing of structural materials.

b. Yield Point. As stress increases above the proportional limit, a given increase in stress produces a relatively greater amount of strain. In the case of the ordinary strength structural steels, a stress is reached where an increase in strain occurs without any increase in stress; in some cases a decrease in stress may occur as the material stretches. The stress at the point of increased strain without increased stress is designated as the yield point (Fig. 2A). Alternate methods for determining conformity to a yield point re-

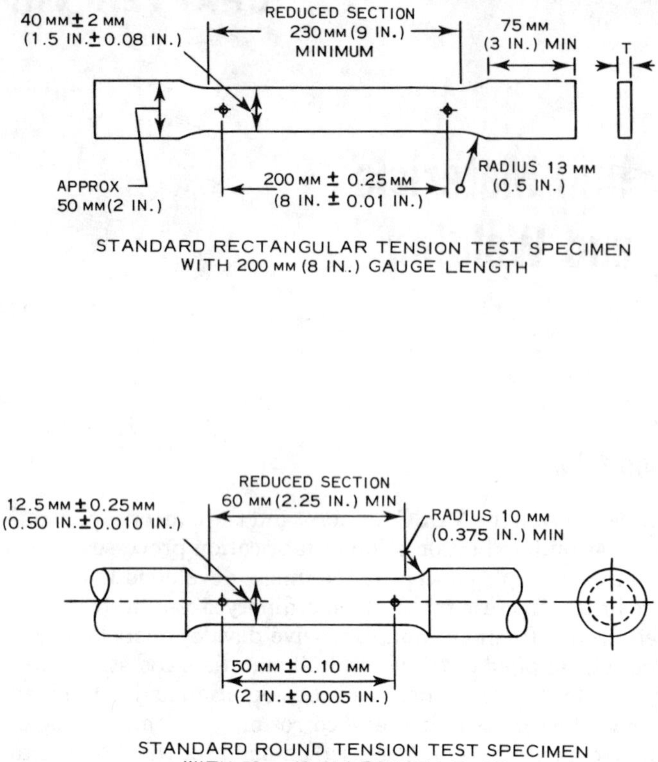

Fig. 1 Typical tension test specimens

quirement, which are generally accepted in testing normal strength hull steels are the divider, extension under load, and drop of the beam methods.

c. Yield Strength. Materials such as high strength steels and nonferrous alloys do not exhibit a definite (yield) point at which strain occurs without increased stress. For these materials a related value of yield strength is pertinent. Yield strength is the unit stress (stress/area) at which a material exhibits a specified limiting deviation from the proportionality of stress to strain; the strain is usually expressed in terms of 0.2 percent offset as illustrated in Fig. 2B or as a 0.5 percent extension (strain) under load.

d. Tensile Strength, Elongation, and Reduction in Area. Tensile strength refers to the maximum unit tensile stress which a material is capable of sustaining and it is calculated from the maximum load divided by the original cross sectional area of the specimen. Percent elongation and reduction in area are calculated on the percent difference in gage length or cross section area, respectively, of the specimen before and after test. Tensile strength is independent of specimen dimension. Percent elongation and percent reduction in area are highly dependent upon gage lengths and specimen dimensions and should be determined with specimens of appropriate standard dimensions.

2.2 Bend Test. The bend test is a qualitative method of measuring ductility in which a specimen is bent around a mandrel of a specified diameter. The bend test has been eliminated as a hull steel specification requirement, although it is widely used for evaluation of weld joints in procedure and operator qualification tests. In such tests a rectangular bar is bent around a mandrel which will vary in diameter depending on the strength level of the weldment; the higher the strength, the greater the diameter. For ordinary strength steels a 38 mm (1.5 in.) wide by 9.5 mm (0.375 in.) thick specimen and a 19 mm (0.75 in.) radius mandrel are used.

2.3 Hardness Tests. *Hardness* of steel is determined by indenting the surface with an indentor with a specific geometry under a specific load, and measuring the resultant impression. In the *Brinell Test* the diameter of the impression made by a steel ball is measured; in *Rockwell hardness testing,* the hardness is indicated directly from the depth of an impression from a diamond cone or steel ball indentor. In both Brinell and Rockwell tests, different loads and indentors are used for different hardness levels. Brinell and Rockwell Hardness Tests may be used to estimate tensile strength in steels, to check the uniformity of a material, to indicate the thermal effects of heat treating or welding on the base metal, as well as to determine hardness where abrasion is of concern. Table 1 indicates the general relationship between hardness values and tensile strength of steel.

2.4 Fatigue Tests. When a material is exposed to repeated cyclic loading at stresses below the tensile strength of the material, it may fracture from *fatigue.* Fatigue is not observed in a non-corrosive environment below the endurance limit; below this limit fatigue fracture will not occur regardless of the number of cycles of loading, Fig. 3. Fatigue failures initiate as small cracks which propagate until eventual failure under continued cyclic loading. Fatigue is a complex phenomenon which depends on such factors as material, size effects, stress concentration, environment, order of loading of varying stresses, stress range, material homogeneity, and overall design. When extrapolating fatigue test data to predict service performance, the interactions between the various significant factors should be taken into account. A wide variety of specimens ranging from the small rotating beam and flat cantilever (Kraus) specimens to full-scale models are used in fatigue testing. The occurence of fatigue cracks in ships is usually associated with points of stress concentration derived from an undesirable design detail or from poor workmanship (Jordan and Cochran, 1978).[1]

When failure occurs as a result of many millions of loading cycles, it is designated as a *high-cycle fatigue failure;* such conditions may occur under highly repetitive low stresses, such as vibratory stresses. When failure is associated with cyclic stresses on the order of hundreds of thousands of cycles or less, it is referred to as a *low-cycle fatigue failure.* The latter is of particular importance to the hull designer, since low-cycle fatigue may be a cause of cracking in areas of high localized stress, after a relatively few years of service life.

---

[1] Complete references are listed at end of chapter

As fatigue life does not increase in direct proportion to tensile strength, and since the tendency for stress concentrations to reduce fatigue life is increased as the strength of the steel increases, fatigue becomes of greater concern for high-strength steel structures of complex design, such as mobile offshore drilling units. In addition regulatory bodies may require appropriate fatigue analyses for cargo tanks such as those carrying liquefied natural gas (LNG).

2.5 Toughness Properties. The occurrence of brittle fractures in ship hulls during World War II stimulated an exhaustive investigation to determine the causes of the failures and to develop recommendations for their prevention (Government Printing Office, 1947). This investigation, which revealed the importance of the tendency of ship steels to change in fracture mode over a narrow temperature range, led to significant improvements in hull steels, and stimulated development of the science of fracture mechanics. It was found that a steel which fractures in a fibrous (ductile) mode with the absorption of a large amount of energy will at some lower temperature fracture in a crystalline (brittle) mode with the absorption of very little energy (Fig. 5A and 5B). The range in which the fracture mode changes from ductile to brittle is referred to as the *transition temperature* range. Within this range a specific transition temperature value is defined by an arbitrary level of performance in a selected toughness test, such as 27 J (20 ft-lb) CVN level. Numerous tests have been devised to measure transition temperature and relate transition temperature to service performance. Transition temperature, however, is not a material constant since it is influenced by factors such as rate of loading, notch acuity, flaw size, structural and local restraint, alloy microstructure, and nature of the loading. Prevailing practice is to use an empirically established toughness criterion which can be related to service performance. Extensive hull materials research effort is being directed toward establishing quantitative toughness criteria that can be incorporated into design calculations. The field of fracture mechanics is concerned with quantitative relationships between fracture propagation, stress, and flaw size, shape and acuity.

a. Charpy V-Notch Tests. The Charpy V-notch test, CVN, is the most widely used empirical toughness test and forms the basis for evaluation of many ship steels. An extensive background of CVN data is available which relates

Table I—Approximate Relationship Between Hardness and Tensile Strength of Steel

Brinell Hardness No. 10 mm Standard Ball	*Rockwell B 100 kg Load 1.6 mm (1/16 in.) Ball*	*Rockwell C 150 kg Load Brale Penetrator*	*Tensile Strength MPa*	*(ksi)*
293	—	31	1000	(145)
285	—	30	965	(140)
273	—	28	931	(135)
262	—	27	896	(130)
255	—	26	862	(125)
245	—	24	827	(120)
235	99	22	793	(115)
224	97	21	758	(110)
217	96	—	724	(105)
207	95	—	690	(100)
197	93	—	655	(95)
187	91	—	621	(90)
173	88	—	586	(85)
163	85	—	552	(80)
154	82	—	517	(75)
143	79	—	483	(70)
130	72	—	448	(65)
121	70	—	414	(60)

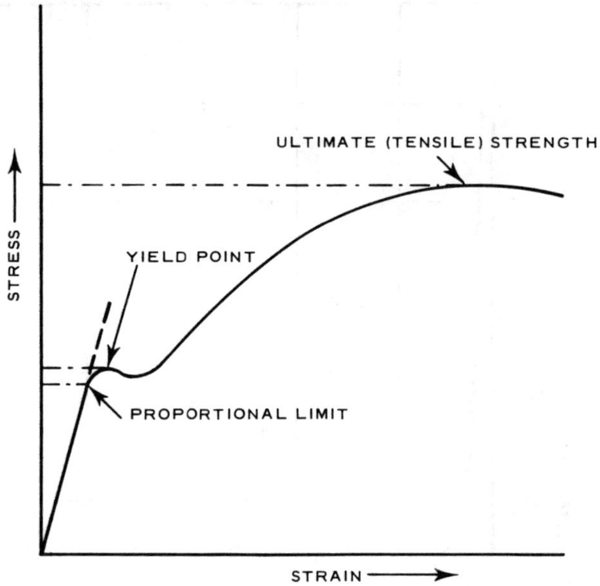

(A) STRESS-STRAIN DIAGRAM FOR ORDINARY STRENGTH

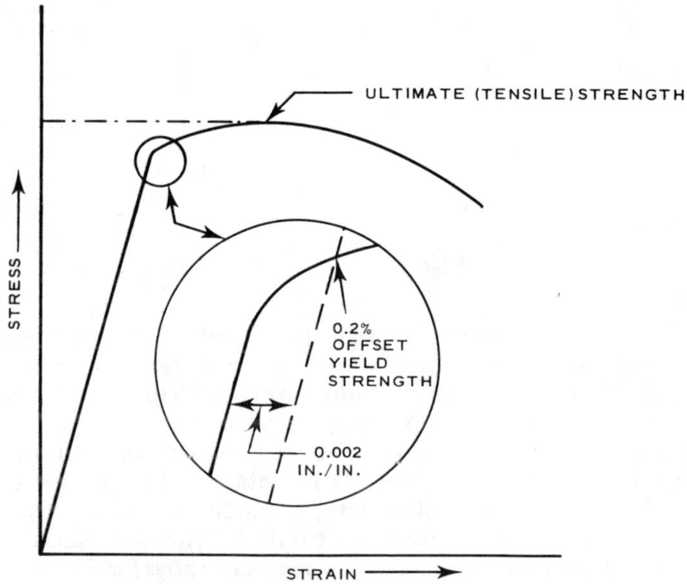

(B) STRESS-STRAIN DIAGRAM FOR HIGH STRENGTH STEEL (SHOWING AN ARBITRARILY SELECTED 0.2% OFFSET YIELD STRENGTH)

Fig. 2 Stress-strain diagram for steel, illustrating yield point and yield strength

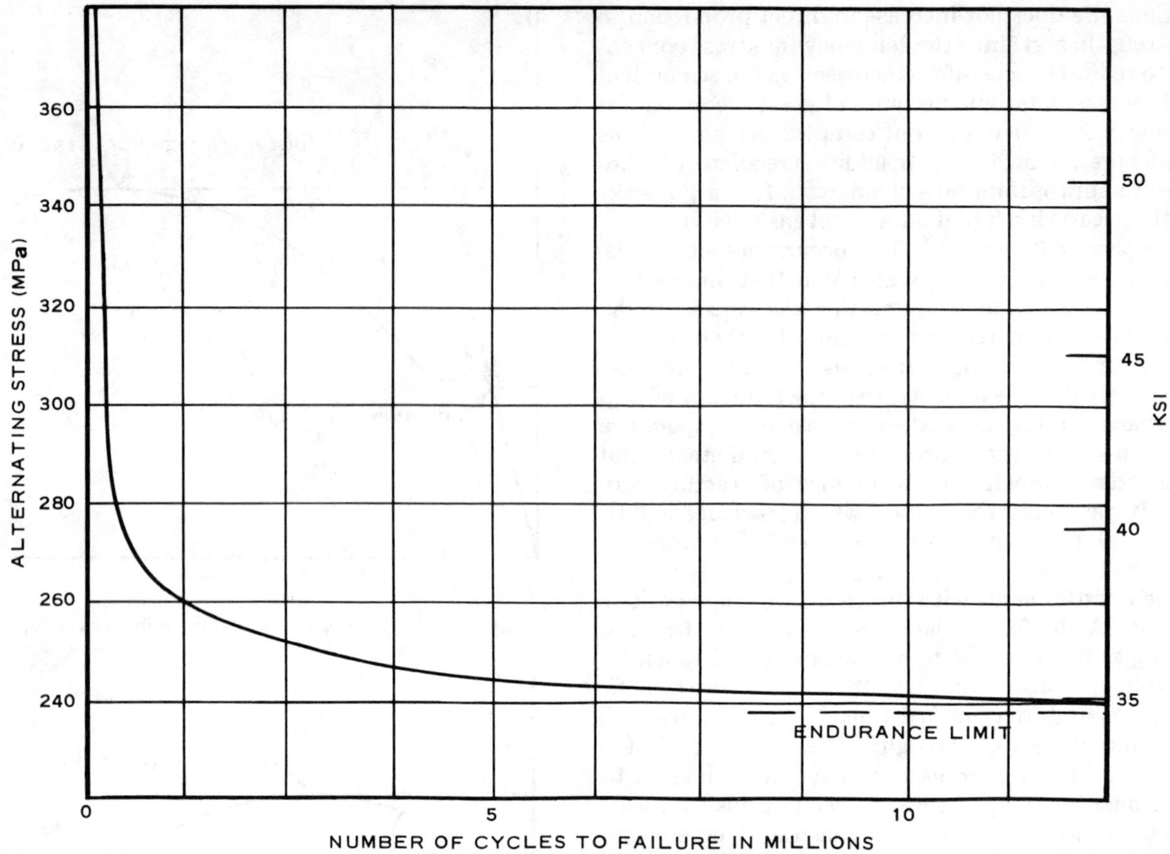

Fig. 3 Representative fatigue curve—structural steel

hull steel toughness to service performance. In addition, it is a rapid, simple and economical test that is accepted worldwide. Studies of World War II ships indicated that plates having low CVN energy values were associated with catastrophic brittle ship fractures (Government Printing Office, 1947). The primary disadvantage of the CVN test is that it is only indirectly related to fracture mechanics concepts and cannot be used quantitatively in design. Also the significance of a specific test energy value varies for different families of alloys and strength levels. However, because of its advantages, the CVN test is the principal toughness test

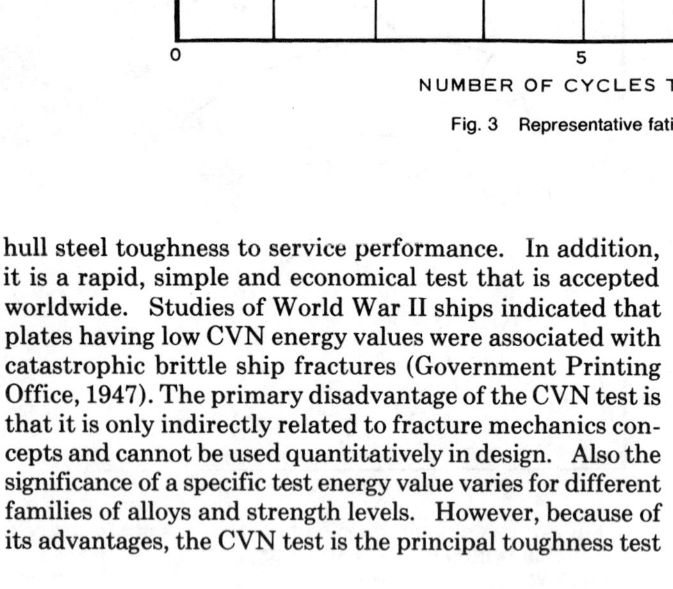

Fig. 4 Charpy V-notch specimen

specified for materials and welds in shipbuilding as well as in most structural and pressure vessel codes. The CVN specimen, Fig. 4, is supported as a simple beam and broken by a single blow of a swinging pendulum weight released from a fixed height. The difference between the initial height of the weight and the height to which it rises after breaking the specimen is a measure of the energy absorbed in breaking the specimen. In some instances the lateral expansion of the specimen in the area of the fracture may be used as the criterion. In general, for a given steel and strength level, lateral expansion will be proportional to energy absorbed. Percent crystallinity shown in the fracture may also be reported for information. A schematic CVN curve and typical CVN curves for several steels are shown in Figs. 5A and 5B. CVN values are sensitive to plate rolling direction. The values in Fig. 5B are for longitudinal specimens (specimen length parallel to plate rolling direction); transverse specimens (specimen length perpendicular to plate rolling direction) would exhibit lower values (Fig. 8).

b. Drop Weight Test. In the Drop Weight Test, DWT, the specimen of Fig. 6, with a notched brittle crack starter bead, is subjected at various test temperatures to an impact load from a falling weight. The highest temperature at which a crack forms and propagates to a specimen edge is

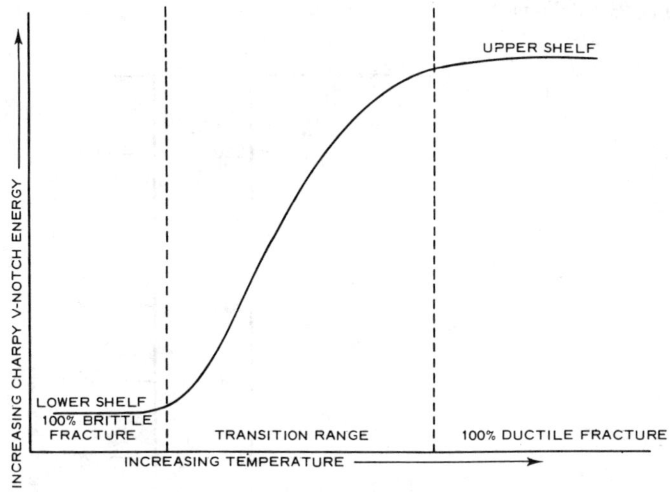

Fig. 5a Schematic Charpy V-notch energy curve

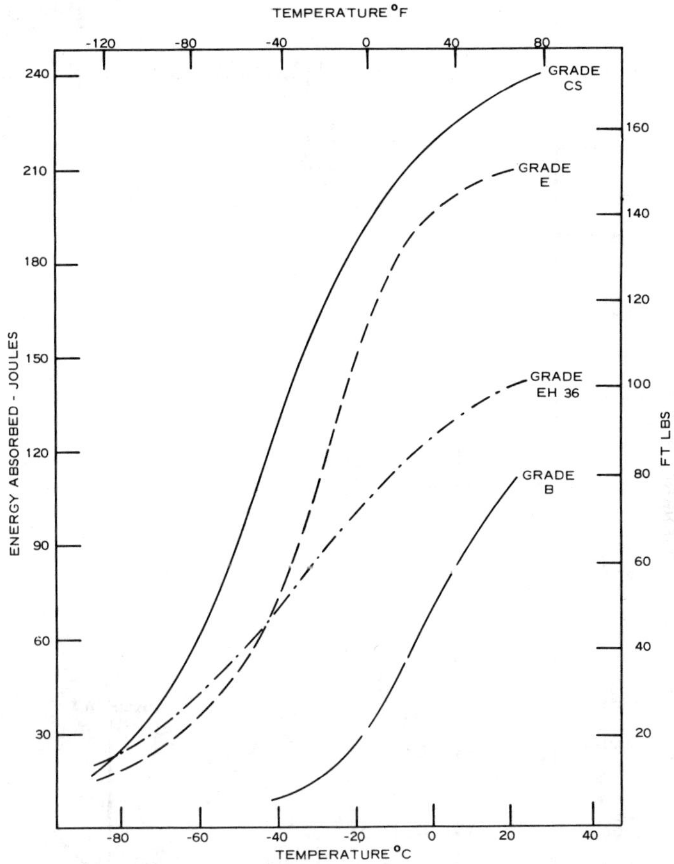

Fig. 5b Average Charpy V-notch (CVN) curves (ABS steels, longitudinal specimens)

defined as the *nil-ductility temperature, NDT*. The NDT represents the highest temperature at which a material will exhibit brittle performance in the presence of a small flaw at low levels of applied stress. For normal strength hull steels, at a temperature approximately 33°C (60°F) above NDT, applied stress must generally exceed yield strength for fracture propagation; at approximately 67°C (120°F) above NDT, fractures are fully ductile when tensile strength is exceeded. The Drop Weight Test which provides a direct measurement of the NDT is often accepted as an alternate to the CVN test. Some of the factors that limit its use are:

- Test facilities are not as available as for the CVN tests;
- it does not provide information as to energy absorption;
- the background of service related experience is not as extensive as that of the CVN.

In addition, anomolous behavior may occur in a material which develops a tough heat affected zone at the edge of the crack-starter weld bead used in the test. The principal advantage of the test is that it can accurately establish the NDT on a wide variety of ferritic steels; this NDT is more directly related to design analyses involving fracture mechanics concepts.

c. Special Fracture Mechanics Tests. The need to

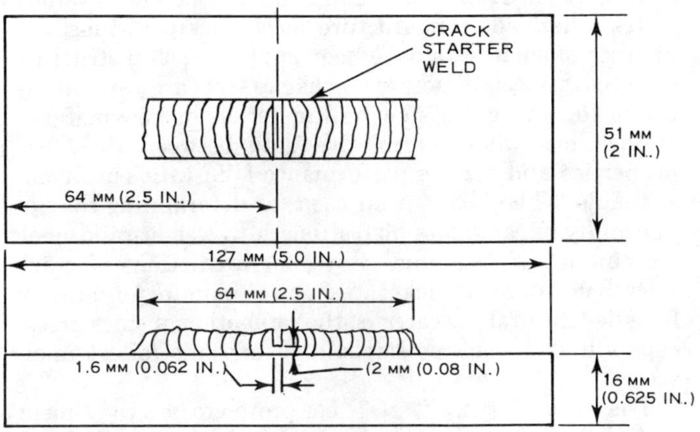

Fig. 6 Drop weight (DWT) specimen

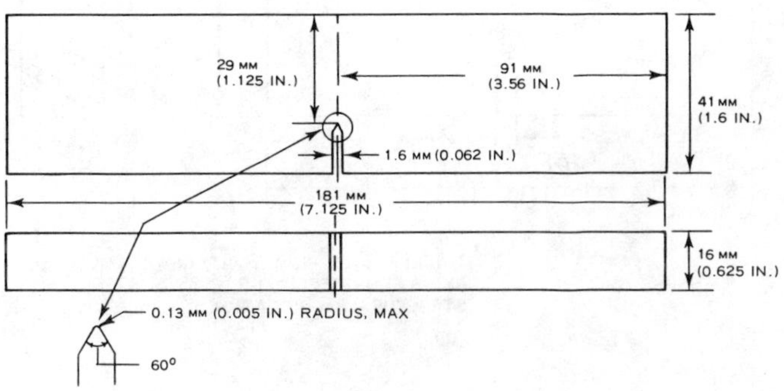

Fig. 7 Dynamic tear (DT) specimen

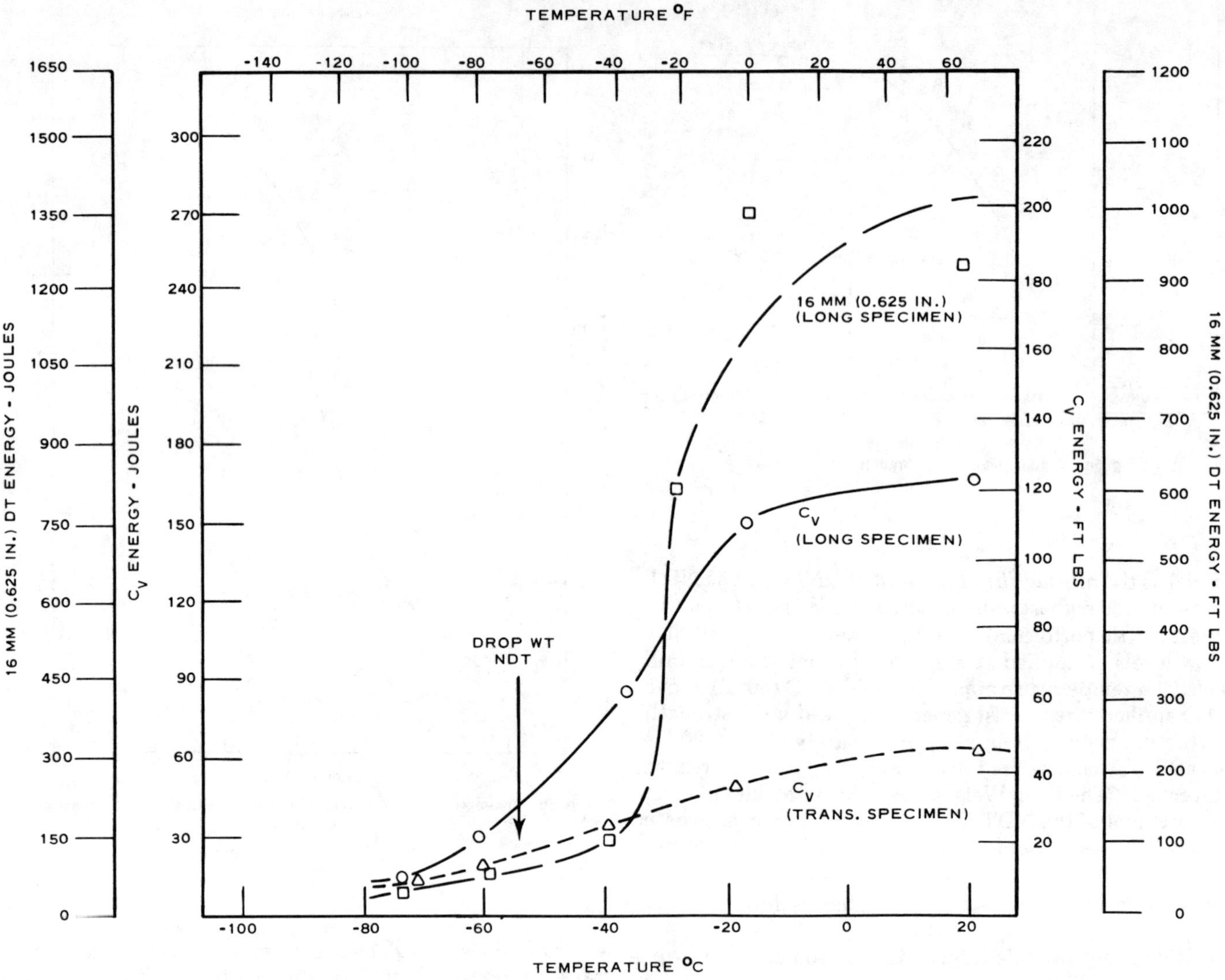

Fig. 8 DT and C_V curves (ABS grade CS specimen)

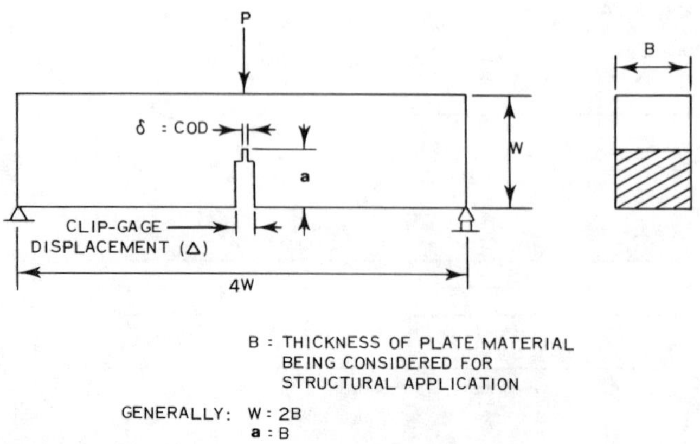

Fig. 9 COD test specimen

characterize fractures and fatigue crack propagation in terms of parameters which could be incorporated into design analyses, such as stress and flaw size, has generated a variety of tests derived from fracture mechanics principles. A number of such tests have been applied to ship structure research. Special fracture mechanics tests are particularly useful for evaluating new hull materials or new material applications where correlative data between the CVN properties and service performance is insufficient or not available. They have been used for determining the applicability of candidate high strength-to-weight ratio steel, aluminum and titanium alloys, for predictions of crack growth in 9 percent nickel steel and aluminum for tanks in liquefied natural gas carriers, the estimation of crack arrest capabilities of various steels, and for some failure analyses.

The *dynamic tear test, DT,* has proven to be a convenient and useful test to characterize fracture behavior. The test measures the energy absorbed in fracturing a specimen, Fig. 7, held at a specified temperature by a falling weight or swinging pendulum. A DT energy temperature curve for a hull steel and associated CVN, DT and DWT (NDT) data are shown in Fig. 8.

Relationships have been developed between DT fracture energy values and the *stress intensity factor, KI_d,* for dynamic or impact loading. KI_d can be related mathematically to applied dynamic stress, crack geometry, crack size, and the configuration in the immediate vicinity of the crack front. Using the DT test energy at a given temperature, such as the lowest expected service temperature, the designer can then estimate the tolerable flaw size for structural members at an assumed dynamic stress; conversely the design stress level appropriate for an assumed flaw size in the structure could also be calculated.

Another test used extensively in hull structural materials research is the *Crack Opening Displacement, COD,* test. This test in which a static load is applied to a type specimen shown in Fig. 9, can be used to establish a *critical stress intensity factor, KI_c;* this factor can be used in static loading relationships in the same manner as KI_d is used for dynamic loading. In addition to the foregoing, large-scale tests such as the explosion tear test, explosion bulge test and various notched wide-plate tests have been used to study fracture.

Steel transition temperature increases with loading rates, and a steel which exhibits a ductile performance and high fracture energy absorption at a given temperature at a slow loading rate, may fracture in a brittle manner with little or no energy absorption with a faster rate of imposition of the same load. Similar differences in fracture performance are associated with increases in notch acuity. These factors should be taken into account in comparing results of different fracture toughness tests, and in projecting results of such tests to service performance.

Section 3
Structural Steels

3.1 Types of Steels and Metallurgy. Ordinary strength hull steels such as the American Bureau of Shipping (ABS) Grades A, B, D, DS, CS and E are the most extensively used group of shipbuilding steels. The properties of these plain carbon steels depend on their chemical content and microstructure. In addition to carbon, these steels contain manganese, silicon, phosphorus, and sulfur; minor amounts of other elements may also be present. Higher strength steels with yield strengths up to 350 MPa (51,000 psi), such as ABS grades AH, DH and EH are also widely used. The higher working stresses permitted with these steels allow for reduction in section thickness and weight. A major difference between these steels and the ordinary strength steels is that the higher strength steels have special additions such as aluminum, columbium, and vanadium, which promote micro-structural improvements and strengthening.

High strength low alloy steels with yield strengths in the 415 MPa (60,000 psi) to 690 MPa (100,000 psi) yield strength range are occasionally used in marine applications. These steels utilize alloy additions and usually a quench and tempering heat treatment to achieve the specified strength level.

a. Microstructure. The microstructure of shipbuilding steels consists of iron-carbide (*cementite*) dispersed in a matrix of *ferrite* (the metallographic name for one form of iron in steel). As the temperature of a steel increases to a *transformation temperature,* the iron which is in the ferrite phase transforms to another form of iron (*austenite*) in which the cementite is highly soluble. Upon cooling below the transformation temperature, the austenite with dissolved cementite reverts back to ferrite and precipitated cementite. A laminated microstructure of cementite and ferrite, referred to as *pearlite,* is a major constituent of the common ship steels. In general, the carbon content and rate of cooling influence the microstructure which in turn determines the strength and hardness of the resulting steel. Most hull structural steels are cooled in air after hot rolling or heat treatment. However, some high strength hull steels above 350 MPa (51,000 psi) yield strength are water quenched from above their transformation temperature and then tempered by heating to a temperature well below the transformation temperature. This quenching and tempering treatment produces a microstructure called *tempered martensite* which is characterized by high strength and toughness.

b. Steelmaking. In low carbon steels, in the absence of deoxidizers, the reaction of carbon with oxygen produces carbon monoxide during ingot solidification. The resulting ingot has an outer rim free of voids, and an inner zone containing voids derived from shrinkage and occluded gases. Such steels which are identified as *rimmed steels* are generally not used as hull steels in thickness over 13 mm (0.5 in.), because of their relative unsoundness. *Semi-killed steels,* which are derived from ingots that are partially deoxidized are sounder than rimmed steels and are commonly used as hull structural steels. ABS Grades A, B, and AH and some forms of Grade D are examples of semi-killed steels. *Killed steels* which are completely freed of the gassing reaction by additions of strong deoxidizing agents such as silicon or aluminum, are the soundest of the three steel types. *Fine grain practice* is the addition of elements such as aluminum, niobium, or vanadium to limit grain size during the period of grain formation. Steel quality may be further enhanced by subjecting the steel to a *normalizing* heat treatment which homogenizes and refines the grain structure. Normalizing involves reheating steel to a temperature above its transformation range and cooling in air. Fine grain practice, fully killing and normalizing enhance

Table 2—ABS Ordinary Strength Hull Structural Steels (Based on 1978 Rules)

Grades	*A*	*B*	*D*	*E*	*CS*	*DS*
Approximate Relative Cost ABS Gr.A = 1.0	1.0	1.03	1.3	1.4	1.3	1.1
Process of Manufacture	For all grades: open hearth, basic oxygen, or electric furnace					
Deoxidation	Any method except rimmed		Semi-killed or killed (2)	Killed, fine grain practice	Killed, fine grain practice	Killed, fine grain practice
Chemical Composition, Percent (Ladle Analysis)	(1)					
Carbon	0.23 max.	0.21 max.	0.21 max.	0.18 max.	0.16 max.	0.16 max.
Manganese	— (6)	0.80–1.10	0.70–1.35	0.70–1.35	1.00–1.35	1.00–1.35
Phosphorus	0.04 max.	0.04 max.	0.04 max.	0.04 max.	0.04 max.	0.04 max.
Sulfur	0.04 max.	0.04 max.	0.04 max.	0.04 max.	0.04 max.	0.04 max.
Silicon	—	0.35 max.	0.10–0.35	0.10–0.35	0.10–0.35	0.10–0.35
Heat Treatment		—	Normalized over 35.0 mm (1.375 in.) (4)	Normalized	Normalized	Normalized over 35.0 mm (1.375 in.)
Tensile Test						
Tensile Strength	For all Grades: 400–490 MPa (58,000–71,000 psi) (5)					
Yield Point, Min.	For all Grades: 235 MPa (34,000 psi) (5)					
Elongation, Min.	For all Grades: 21 percent in 200 mm (8 in.); 24 percent in 50 mm (2 in.); 22 percent in 5.65 $\sqrt{A}$ (A equals area of test specimen)					
Impact Test Standard Charpy V-Notch						
Temperature		0°C (32°F)	−10°C (14°F) (3)	−40C (−40F)	—	—
Energy, Min. Avg.		27J (20 ft-lb)	27J (20 ft-lb)	27J (20 ft-lb)	—	—
No. of Specimens		(7)	3 from each 50 tons	3 from each plate	—	—

Notes

1. A maximum carbon content of 0.26 percent is acceptable for Grade A plates equal to or less than 12.5 mm (0.5 in.) and all thicknesses of Grade A shapes.
2. Grade D may be furnished semi-killed in thickness up to 35 mm (1.375 in.) provided steel above 25.0 mm (1.00 in.) in thickness is normalized. In this case the requirements relative to minimum Sl and Al contents and fine grain practice do not apply.
3. Impact tests are not required for normalized Grade D steel when furnished fully killed fine grain practice.
4. Control rolling of Grade D steel may be specially considered as a substitute for normalizing in which case impact tests are required for each 25 tons of material in the heat.
5. Upper limit of tensile strength is 552 MPa (80,000 psi) for Grade A shapes; for cold flanging quality the tensile strength range is 379–448 MPa (55,000–65,000 psi) and yield point minimum is 207 MPa (30,000 psi).
6. For Grade A plates over 12.5 mm (0.50 in.) the Mn should be 2.5 × C % (min).
7. Impact tests for Grade B only required in thicknesses over 25.4 mm 8).0 in.); 3 from each 50 tons.

steel quality. The relationship of the various treatments to grade and thicknesses of hull steels and impact characteristics are indicated in Tables 2 and 3.

3.2 American Bureau of Shipping Steels. The American Bureau of Shipping specifications for structural steels are intended to provide steels with adequate toughness without being excessively costly and which can be readily fabricated with shipyard welding techniques. The 1978 hull steel specifications are shown in Tables 2 and 3. The history of these steels can be traced back to 1947 when, after an extensive investigation of brittle fractures of World War II ships (Government Printing Office, 1947) (conducted by a government board of inquiry that included ABS, Navy and Coast Guard representatives) the first consideration of toughness was given to hull steels. The investigation revealed that the brittle ship fractures were associated with steels which were relatively brittle at the temperature of fracture as evidenced in Charpy V-notch tests. As a result, ABS steel specifications since 1948 have provided for steels with controlled levels of notch toughness.

A variety of toughness levels are provided by controlling the manganese to carbon ratio, requiring specific deoxidation, grain refining and heat treatments or, in some cases, by requiring impact testing of each plate or heat. With the exception of Grades DS and CS, which are only included in ABS Rules, similar steels are specified in the rules of all the world's major classification societies. ABS permits unrestricted substitution without impact testing of Grade DS and CS, for Grades D to 35 mm (1.375 in.) and E to 51 mm (2 in.) respectively (which require impact testing) because of the more favorable chemistry of the Grades DS and CS grades. In the development of the current ship steels, the European community did not rely on refinements of chemical composition to provide assurance of the desired steel toughness levels. The approximate relative material costs, included in Table 2, indicate that Grades DS and CS have a significant economic advantage over Grades D and E. The more favorable compositions to assure the desired toughness levels in Grades DS and CS are less costly than the confirmatory impact tests required for Grades D and E. Typical Charpy toughness values for various ABS steels are shown in Fig. 5B. Such toughness data have proven useful in projecting ABS steels for new applications and for comparing ABS steels with other steels. Since 1948, four ABS grades have been eliminated from the Rules; Grade R which was eliminated when its requirements became similar to those for Grade A; Grade C which approximated the current Grade DS and Grades BH and CH which were dropped entirely.

Structural shapes and bars are generally made to the same chemical composition and mechanical property requirements as the corresponding grade of plate steel. However, in the case of ABS Grade A shapes, which is by far the most frequently used shape grade in shipbuilding, a slightly higher maximum carbon content (0.26 vs. 0.23 percent) is permitted, the manganese requirement is waived, and the upper

Table 3—ABS Higher Strength Hull Structural Steels (1978 Rules)

Grades	AH32 or AH36	DH32 or DH36	EH32 or EH36
Approximate Relative Cost ABS Gr.A = 1.0	1.2	1.4	1.5
Process of Manufacture	For all grades: open heart, basic oxygen, or electric furnace		
Deoxidation	Semi-killed or killed	Killed, fine grain practice	Killed, fine grain practice
Chemical Composition, Percent (Ladle Analysis)	For all grades:		
Carbon	0.18 max.		
Manganese	0.90–1.60		
Phosphorus	0.04 max.		
Sulfur	0.04 max.		
Silicon	0.10–0.50	{ AH to 12.5 mm (0.50 in.) may be semi-killed in which case 0.10 percent min. Si does not apply	
Nickel	0.40 max.		
Chromium	0.25 max.		
Molybdenum	0.08 max.		
Copper	0.35 max.		
Aluminum (Acid Soluble)	0.06 max.		
Columbium (Niobium)	0.05 max.		
Vanadium	0.10 max.		
Heat Treatment	Normalizing req'd. over 12.5 mm (0.50 in.) if Ns treated	Normalizing req'd. over 25.5 mm (1.0 in.) if all treated over 12.5 mm (0.50 in) if Ns treated over 19.0 mm (0.75 in.) if V treated	Normalized
Tensile Test			
Tensile Strength	For 32 Grade: 471–588 MPa (68,000–85,000 psi) For 36 Grade: 490–618 MPa (71,000–90,000 psi)		
Yield Point, Min.	For 32 Grade: 314 MPa (45,500 psi) For 36 Grade: 353 MPa (51,000 psi)		
Elongation, Min.	For all Grades: 19 percent in 200 mm (8 in.); 22 percent in 50 mm (2 in.) 20 percent in $5.65\sqrt{A}$ (A equals area of test specimen)		
Impact Test Standard Charpy V-Notch			
Temperature	—	−20°C (−4°F)	−40°C (−40°F)
Energy, Min. Avg.	—	34J (25 ft-lb)	34J (25 ft-lb)
No. of Specimens	—	3 from each 40 tons	3 from each plate

Table 4—Examples of ASTM Substitutes for ABS Steels

A36	May be substituted for Grade A plate under 12.5 mm (.5 in.) and shapes of all thicknesses. Equivalent to Grade A plate over 12.5 mm (.5 in.) when Carbon is 0.23 percent Max., and Manganese at least 2.5X Carbon. May be substituted for Grade B up to 25.5 mm (1 in.) when Carbon is 0.21% Max. and Manganese 0.80 percent Min. (0.60 percent Min. if killed steel).
A131	Equivalent to ABS Grades (A = A, B = B, DH = DH etc.)
A242 Type 2	Similar to AH32 to 38.1 mm (1.5 in.)
A441	Similar to AH 32 to 38.1 mm (1.5 in.)
A537 Class I	Similar to DH 32 and DH 36 when Carbon is 0.18 percent Max. Similar to EH 32 and H 36 if Carbon is 0.18 percent Max. and supplementary CVN impact requirement is invoked.
A572 Gr.45	Similar to AH 32
A572 Gr.50	Similar to AH 36

limit of the tensile strength range is higher. With these modifications the requirements for ABS Grade A shapes are compatible with those of the ASTM structural steel Grade A36 which is the most widely used and available industrial structural steel shape. In the case of cold flanging steel, requirements as to tensile strength range and minimum yield point are reduced approximately 10 percent as compared to ordinary plates.

a. Castings. Heavy structural members of complicated shapes such as rudder parts are generally produced as steel castings. The grade of steel casting specified in the *ABS Rules for Building and Classing Steel Vessels* is substantially similar to the ASTM A27 Grade 60-30 which is readily weldable and has mechanical properties that approximate those of ordinary steel. Higher strength steel castings are usually purchased to the requirements of ASTM or other recognized commercial specifications. In designing large complex castings, it is often advisable to confer with foundry personnel to assure that the final design selected is compatible with the foundry techniques necessary to provide sound castings. It may be desirable to divide a large casting into simpler units to allow for optimum casting and then weld the units together.

In spite of the foregoing precautions, castings are likely to be nonhomogeneous and, upon fabrication at the shipyard, such conditions as cracks, sand inclusions, gas holes, and internal shrinkage may be revealed. The extent of repairs required in such cases must be determined by consideration of the service conditions and the location and extent of the nonhomogeneous areas in each individual case. Because of their potent stress-raising effect, cracks should be excavated completely and the area repaired by welding. Internal shrinkage, sand and gas holes are less objectionable in this respect, and complete excavation may be unnecessary when they occur in sections of low stress.

However, these defects are likely to interfere with sound welding, and castings which are to be incorporated into the hull structure by welding should be examined closely and conditioned in the welding areas. Prompt fabrication and inspection of castings upon receipt at the shipyard are necessary because of the time delay in procuring large castings. If the initial castings received prove to be unsuitable to the extent that repair welding is uneconomical, the time required for replacement may interfere seriously with building schedules.

Steel castings used for critical applications, such as stern frames and rudder horns may be required to be subjected to nondestructive test examination. The designer may find it desirable to specify supplementary nondestructive tests for other castings which are involved in critical welded assemblies in order to assure soundness in way of welded connections. To improve weldability and reduce residual stress, castings are required to be subjected to a homogenizing annealing or normalizing heat treatment before welding or delivery.

b. Forgings. These are used for applications where the shape is comparatively simple (such as anchors and rudder stocks), but not sufficiently so for adaptation to a rolling process, and where there is a desire for better homogeneity than can be obtained in castings. While forgings are made in a wide variety of alloy steels of different mechanical properties, those used for structural applications are usually of low carbon steel (0.35 max.), of welding quality, and with mechanical properties about the same as those of structural plates and shapes. Hull steel forgings are usually annealed or normalized and tempered to ABS or the comparable ASTM A668, Grade BH requirements.

Large forgings are made directly from a cast ingot and unless a sufficient amount of work is done in forging to close and weld the porosity of the ingot, evidence of this condition may appear in the forging. American Bureau of Shipping Rules require that the forging be not more than one-third the area of the ingot, except for large flanges, palms, and similar enlargements which may be not more than two-thirds the area of the ingot. If the interior of these enlargements is exposed, as by machining, some of the ingot porosity may be evident. When this occurs, the condition must be evaluated as to its extent and the service condition for the section involved.

Forgings are also likely to contain nonmetallic inclusions, which are generally elongated in the direction of the forging and of relatively small cross section. Their terminations are not as sharp as a crack; mechanical tests, including fatigue tests, have indicated that they are not particularly harmful if of moderate size and concentration. When encountered, such inclusions should be evaluated by size, concentration, and location.

3.3 ASTM Specifications. Certain ASTM grades of steel have been used as substitutes for ABS steels and to meet requirements for strength levels above those provided by the classification society steels. Table 4 indicates examples of specifications used and the modifications generally required when they have been used as substitutes for classification society steels.

Steels from over 353 MPa (51,000 psi) yield strength to 689 MPa (100,000 psi) yield strength have been found particularly advantageous for structures such as containerships where relatively small deck areas are available for the development of required strength and legs of jack-up drilling units where the strength to weight ratio of the leg structure may be particularly important. In considering use of the high-strength steels, fabricability and the proportion of increased strength which can effectively be utilized in the design should be taken into consideration. Typical steels used when strength levels over 353 MPa (51,000 psi) yield strength have been required are shown in Table 5.

3.4 Military Specifications. Military specifications MIL-S-22698 and MIL-S-16113 cover the steels analogous to those of ABS and ASTM grades. In addition MIL-S-16216 (HY80 and HY100) and MIL-S-24371 (HY130) are occasionally used. "HY" steels which provide steel yield strength levels of 550 MPa (80,000 psi) to 896 MPa (130,000 psi) provide superior fracture toughness. However, commercial applications have not been extensive because of the relatively high cost of materials and fabrication, and the absence of need for the extraordinarily high toughness provided by these materials.

3.5 Foreign Specifications. Major classification societies have specifications for steels which are essentially the same as ABS Grades A, B, D, E, AH, DH and EH. (See Table 1 Chapter XVI). In some of these specifications, the grade of higher strength steel may be a letter designation followed by a numerical designation indicating yield strength in kg/mm^2; for example the equivalent of AH36 steel in these systems may be A36. These grades should not be confused with ASTM steels, such as A36, which may bear the same grade designation but represent a completely different steel. In addition, there exists in most industrial nations, a series of standards corresponding to the ASTM Standards (ISO—International), (BSI—British), (CSA—Canadian), (DIN—German), (NF—French), (JIS—Japanese). Compilations are available which relate ASTM and foreign steel grades (Ross, 1972).

Table 5—ASTM Steels over 353 MPa (51 ksi) Yield Strength (Used in Shipbuilding)

Specification (1)	*MIL-S-16216 HY-80 (2)*	*A517 (3)*	*A678 Gr.C*	*A633 Gr.E*	*A537 Cl.2*
Description	Cr-Ni Mo	High Yield Q & T	C,Mn,Si	V, N	C,Mn,Si
Yield MPa (ksi) (min)	552 (80)	690 (100)	621 (90)	414 (60)	414 (60)
Tensile MPa (ksi) (min)	621 (90)	793 (115)	621 (90)	552 (80)	552 (80)
Elongation 5mm (2 in.) %	20	16	19	23	22
Heat Treatment	Q & T	Q & T	Q & T	Norm.	Q & T

Notes:
1. Specification are ASTM unless otherwise noted.
2. Also available as HY-100 @ 690 MPa (100 ksi) yield strength.
3. Similar steel A514 also available.

Section 4
Special Steels

4.1 General. The common structural steels are intended for the service normally encountered by most ships and marine structures. Special steels with enhanced properties are available where service conditions involve exposure to unusual temperatures, corrosion, or loading conditions. In some cases, the use of special steel may be mandated by requirements of a regulatory agency or in other cases it may be a design selection for the purpose of achieving improved serviceability.

4.2 Steels for Low Temperature Applications. In general, steels equivalent to the ABS grades, when used in applications appropriate for each grade, may be used for all applications where the lower limit of service temperature is primarily related to the lowest possible sea temperature. Where extraordinary cooling effects exist, consideration should be given to the use of steels with fracture transition temperatures and toughness characteristics appropriate to the service temperature involved. Such special requirements for low temperature service may be derived from the cooling effects of cargo, such as in refrigeration ships and liquefied gas carriers. They may also be derived from service where steel temperatures are not moderated by ocean

Table 6—Refrigerated Cargo Vessel Steels

Steel Service Temperature °C	°F	*ABS Hull Steel Grade*
5	40	B
−1	30	B
−7	20	D, DS and DH
−12	10	D, DS and DH
−18	0	DN and DHN
−23	−10	DN and DHN
−29	−20	CS, E and EH

Table 7—Requirements for Liquefied Gas Carriers

Plates, Sections and Forgings for Cargo Tanks, Secondary Barriers and Process Pressure Vessels for Design Temperatures Below 0°C and Down to −55°C

Maximum thickness 20 mm (0.75 in.)

Chemical Composition and Heat Treatment

Carbon-Manganese Steel — Fully killed. Aluminium treated fine grain steel.

Chemical composition, percent (ladle analysis)

C	Mn	Si	S	P
0.16 max.	0.70–1.60	0.10–0.50	0.035 max.	0.035 max.

Optional additions: Alloys and grain refining elements may be generally in accordance with the following:

Ni	Cr	Mo	Cu	Nb	V
0.80 max.	0.25 max.	0.06 max.	0.35 max.	0.05 max.	0.10 max.

Normalized or quenched and tempered.

Tensile and Toughness (Impact) Test Requirements

Plates	Each "piece" to be tested
Sections	Batch test
Charpy V-Notch Test	Test temperatures 5°C below the design temperature or −20°C whichever is lower
Plates	Transverse test pieces. Minimum average energy value (E) 2.8 kpm
Sections and Forgings	Longitudinal test pieces. Minimum average energy value (E) 4.2 kpm

Table 8—Requirements for Liquefied Gas Carriers

Plates, Sections and Forgings for Cargo Tanks, Secondary Barriers and Process Pressure Vessels for Design Temperatures Below −55°C and Down to −165°C

Maximum thickness 20 mm

Minimum design temperature (°C)	Chemical composition and heat treatment	Impact test temperature (°C)
−60	1½ percent Nickel steel—normalized	−65
−65	2¼ percent Nickel steel—normalized or normalized and tempered	−70
−90	3½ percent Nickel steel—normalized or normalized and tempered	−95
−105	5 percent Nickel steel—normalized or normalized and tempered	−110
−165	9 percent Nickel steel—double normalized and tempered or quenched and tempered	−196
−165	Austenitic steels, such as types 304, 304L, 316, 316L, 321, 327 and 347 Solution treated	−196
−165	Aluminium alloys; such as type 5083 annealed	Not required
−165	Alloy steel −36 percent nickel	Not required

Tensile and Toughness (Impact) Test Requirements

Plates	Each "piece" to be tested
Sections and Forgings	Batch test
Charpy V-Notch Test	
Plates	Transverse test pieces. Minimum average energy value (E) 2.8 kpm
Sections and Forgings	Longitudinal test pieces. Minimum average energy value (E) 4.2 kpm

Table 9—Requirements for Liquefied Gas Carriers

Plates and Sections for Hull Structures Cooled by Low Temperature Cargo

Minimum design temperature of hull structures C°	(F°)	Thickness, mm (in.)	Classification Society steel grades
0 and above			Normal practice
−10	(14)	t ≤ 12.5 (0.5)	B
		12.5 (0.5) < t ≤ 25.5 (1)	D
		t > 25.5 (1)	E
−25	(−13)	t ≤ 12.5 (0.5)	D
		t > 12.5 (0.5)	E
Below −25	(−13)	—	in accordance with table 7

Note:
1. The thickness limitation of Table 7 does not apply to plates and sections for hull structures.

temperatures, as in the case of upper structure of mobile offshore drilling units.

a. Ships. The guidelines shown in Table 6 are applicable to refrigeration ships for steel forming and adjacent to the refrigerated areas. ABS V-039, V-051, and V-060 steels are useful for service temperatures of −34°C (−30°F), −46°C (−50°F), and −55°C (−67°F) respectively. In liquefied gas carriers, low temperature service requirements may be encountered in the cargo tanks, secondary barriers and portions of the hull affected by the cargo. When the outer hull serves as the secondary barrier, such as in some liquefied petroleum gas (LPG) carriers, it has the same toughness requirement as the tank material.

With the adoption of international conventions such as IMCO, general requirements for liquefied gas carriers such as those in Tables 7 to 9 have become applicable worldwide and have been adopted by many regulatory bodies.

b. Mobile Offshore Drilling Units. In mobile offshore drilling units material selection is based on the anticipated stresses, criticality of the application, and lowest steel service temperature. The service temperature of the parts of the structure which are submerged or in the vicinity of splash zones is assumed to be that of the sea. Minimum service temperature for areas away from the moderating effects of the sea is based on minimum average air temperature. Moderating effects of any supplementary heating are taken

Table 10—Charpy V-Notch Joules (ft-lbs) Guidelines for Structural Steels used in Mobile Offshore Drilling Units

Application (1)	*Steel Type (2)*		
	ORDINARY STRENGTH	HIGHER STRENGTH	HIGH STRENGTH (3)
Secondary (Least Critical)	27.2 at Service Temp. (20) at Service Temp.	34.3 at Service Temp. (25) at Service Temp.	34.3 at −40 C (25) at −40 F
Primary (Intermediate)	27.2 at 10°C below Service Temp. (20) at 18°F below Service Temp.	34.3 at 10°C below Service Temp. (25) at 18°F below Service Temp.	34.3 at −40C (25) at −40F
Special (Most Critical)	27.2 at 30°C below Service Temp. (20) at 54°F below Service Temp.	34.3 at 30°C below Service Temp. (25) at 54°F below Service Temp.	34.3 at −50C (25) at −58F

1a. Secondary—Applications such as internal framing, upper platform decks, deck, side and bottom of upper hull.
1b. Primary—Applications such as external shell, diagonal and horizontal braces, plating of lattice type legs, jack house supporting structure.
1c. Special—Applications such as external shell in way of intersections of critical columns.
2. Ordinary strength = 235 MPa (34 ksi) yield strength
Higher Strength = 314–353 MPa (45.5–51 ksi) yield strength
High Strength = 412–690 MPa (60–100 ksi) yield strength
3. Guidelines shown for high strength steels only apply to service temperatures down to −20C (−4F).

Table 11—ABS Steel Selection Guidelines for Mobile Offshore Drilling Units

APPLICATION AREAS (See Note 1, Table 10)

SECONDARY

ABS STEEL GRADE	Service Temp. C°/(F°) 0 (32)	−10 (14)	−20 (−4)	−30 (−22)
A		19 (.75)	12.5 (.50)	shaded
B		25.5 (1.00)	19 (.75)	12.5 (.50)
D, DS		35 (1.40)	35 (1.40)	22.5 (.89)
DN	51 (2.00)			27.5 (1.08)
DSN, CS, E				
AH		25.5 (1.00)	19 (.75)	12.5 (.50)
DH				19 (.75)
DHN	51 (2.00)			27.5 (1.08)
EH				

PRIMARY

ABS STEEL GRADE	Service Temp. C°/(F°) 0 (32)	−10 (14)	−20 (−4)	−30 (−22)
A	19 (.75)	12.5 (.50)	shaded	shaded
B	25.5 (1.00)	19 (.75)	12.5 (.50)	shaded
D, DS	35 (1.40)	35 (1.40)	32.5 (.89)	12.5 (.50)
DN	51 (2.00)		27.5 (1.08)	22.5 (.89)
DSN, CS, E				
AH	19 (.75)	19 (.75)	12.4 (.50)	shaded
DH			19 (.75)	12.5 (.50)
DHN	51 (2.00)		27.5 (1.01)	22.5 (1 9)
EH				

SPECIAL

ABS STEEL GRADE	Service Temp. C°/(F°) 0 (32)	−10 (14)	−20 (−4)	−30 (−22)
A	shaded	shaded	shaded	shaded
B	16 (.63)	shaded	shaded	shaded
D, DS	22.5 (.89)	16 (.63)	shaded	shaded
DN	27.5 (1.08)	22.5 (.89)	16 (.63)	shaded
DSN, CS, E	51 (2.00)		27.5 (1.08)	16 (.63)
AH	19 (.75)	shaded	shaded	shaded
DH	19 (.75)	16 (.63)	shaded	shaded
DHN	27.5 (1.08)	22.5 (.89)	16 (.63)	shaded
EH	51 (2.00)		32.5 (1.08)	16 (.63)

Numbers in table are maximum thicknesses in mm (in.). Shaded areas indicate no application

Table 12—Corrosion Resistant Steels
AISI Types

TYPE	*304*	*304-L*	*316*	*316-L*
Mechanical Properties				
Yield Strength MPa (psi) (0:2 percent offset)	172–276 (25–40,000)	172–276 (25–40,000)	207–310 (20–45,000)	207–310 (30–45,000)
Tensile Strength MPa (psi)	483–621 (70–90,000)	483–621 (70–90,000)	552–655 (80–95,000)	517–655 (75–95,000)
Elongation in 2 in. percent	50–60	50–60	40–60	45–66
Reduction of Area percent	60–75	60–75	55–75	55–70
Chemical Comp., (Percent)				
Carbon max.	0.08	0.03	0.08	0.03
Chromium	18–20	18–20	16–18	16–18
Nickel	8–12	8–12	11–14	10–14
Manganese max.	2.00	2.00	2.00	2.00
Molybdenum	—	—	2.00–3.00	2.00–3.00
Silcon max.	1.00	1.00	0.75	1.00
Phosphorus max.	0.045	0.045	0.03	0.040
Sulphur max.	0.030	0.030	0.03	0.030

into account. Table 10 indicates general guidelines that have been considered satisfactory for a range of mobile off-shore drilling unit service temperatures and applications. Table 11 indicates the applicable ABS grades of steel suitable for the ranges of service temperatures usually encountered (American Bureau of Shipping, 1978-b), (Alia et al, 1976). Equivalent criteria are contained in publications by other regulatory bodies.

c. Submersibles. The special requirements associated with submersibles and related underwater systems have led to the promulgation of special rules for such service (American Bureau of Shipping, 1979).

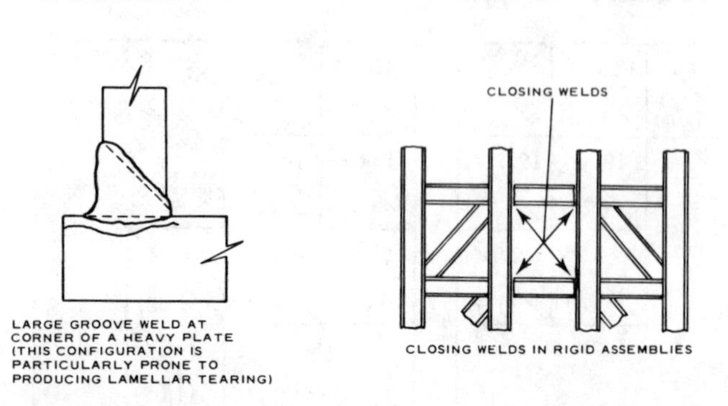

Fig. 10 Examples of welded joints that increase the probability of producing lamellar tears

4.3 Corrosion Resistant Steels. The major structural application of corrosion resistant steels in merchant ships is to provide a surface which is resistant to chemical action from a liquid cargo. It is commonly used in the form of a 1.3 mm (0.05 in.) to 2.5 mm (0.10 in.) protective cladding on ordinary steel plates. However, it may be used in solid form for relatively thin plates and for shapes where the clad product is not available. In cases where cargo tanks are also used as ballast tanks, consideration should be given to the corrosive effects of both media. Although there are a wide variety of corrosion resistant steels available, the alloys shown in Table 12 are those most commonly used. The 316 types are especially resistant to sulfurous acid and are superior to the 304 types in flowing sea water and for phosphoric or acetic acids. The extra low carbon (0.03 max.) grades, designated by the suffix L after the grade number, 3XXL, are generally recommended for welded construction. Welding of stainless steels can produce carbide precipitation in the heat affected zone which in turn reduces corrosion resistance. Providing for extra low carbon minimizes this adverse effect.

4.4 Steels with Improved Through Thickness (Short Transverse) Properties. In some marine applications, such as the intersections of principal members of mobile offshore drilling units, loads are imposed perpendicular to the plate surfaces either as service loads, or by residual welding stresses. Conventional steels may exhibit a weakness under such loads which is evidenced by lamellar separation, Fig. 10. Through thickness weakness of a plate material may be indicated by reduced elongation, reduction in area or a woody fracture appearance in tensile tests of a specimen whose axis is perpendicular to the plate surfaces. When a structural steel with superior resistance to through thickness loading is required, it can be provided commercially by means of a variety of special melting practices; the degree of resistance achieved is dependent on the particular practice used.

Attempts have been made to use ultrasonics as a nondestructive method for assurance of adequate through thickness properties. Such methods may be useful for those cases where weakness is due to gross plate laminations; they are not useful for cases where the weakness is related to metallurgical components in the microstructure. This topic has been discussed at length in the American Institute of Steel Constructors (AISC) Engineering Journal (1973), Australian Welding Research Association (1976), and Jubb (1971).

4.5 Abrasion Resistant Steels. The most common application for abrasion resistant materials is for components associated with the loading and unloading of bulk cargo. Two types of materials are available for abrasion resistance in such applications. The non-weldable type with high carbon, manganese, or chromium is not generally used for structural applications. The weldable type, similar to steels covered by ASTM A514, is available in the standard structural condition or quenched and tempered to high hardness levels for superior abrasion resistance. For special cases where localized wear is encountered, hard facing structural steel with local weld overlays may be considered (SNAME, 1976).

Section 5
Nonferrous Alloys

5.1 Aluminum Alloy Applications. Aluminum alloys find use where their special attributes such as low density and high strength to weight ratio, corrosion resistance in certain environments or retention of toughness at low temperatures (and in some cases their relatively low modulus of elasticity) are of value. Development of inert gas arc welding processes has facilitated the use of aluminum alloys for various ship structural applications. Aluminum alloys are frequently used in superstructures of large ships and for the entire hull structure of some ferries and small boats such as those serving the offshore industry. The low density of aluminum alloys makes them particularly attractive for applications where high strength to weight ratios are of particular concern such as in surface effect and hydrofoil craft. Since aluminum alloys increase in strength and maintain toughness as temperature decreases, they have proven particularly suitable for cryogenic services such as containment of liquefied natural gas. Aluminum alloys for plates, extrusions, forgings and castings are shown in Table 13. Details of compositions, properties and methods of inspection are contained in American National Standards Institute (ANSI/ASTM (1977) and SNAME (1974).

5.2 Non-Heat Treatable Aluminum Alloys. Aluminum–magnesium alloys which are the alloys most widely used for marine structures, 5083 (4.5 percent Mg), 5086 (4.0 percent Mg), and 5456 (5.0 percent Mg) alloys, acquire increased strength from cold work and not from heat treatment. The 5454 alloy is used for applications where service temperatures above 65°C (150°F) are anticipated. These alloys, which have good weldability characteristics, are usually used in the mildly cold worked (1/4 hard) temper to provide the desirable combination of strength and corrosion resistance. Higher strength forms of these alloys, attained either by additional cold work (up to fully hard) or by magnesium contents over 5 percent are not generally used, since they tend to exhibit an undesirable increased susceptibility to stress corrosion. Where special corrosion problems are anticipated, such as in stagnant bilge areas, the alloys may be provided in special tempers (5083-H116, 5086-H117, 5454-H116) which are particularly resistant to exfoliation, a special form of intergranular corrosion which produces delamination. In general, the base plate in the vicinity of welds in non-heat treatable alloys (such as the 5XXX series) are transformed to an annealed condition by the heat of welding. The effect is to reduce tensile properties in the vicinity of the weld to the annealed or non-work hardened values. The effect, which should be taken into account in design, is reflected by the fact that the minimum ultimate tensile strength properties expected for transverse butt welds (Table 14) are annealed plate properties.

5.3 Heat Treatable Aluminum Alloys. Heat treatable aluminum alloys such as 6061-T6 develop strength by heating to an annealing temperature, water quenching and then reheating to a lower temperature to achieve a controlled precipitation of intermetallic compounds. The 6061-T6 alloy is occasionally used in marine service, particularly for extrusions, since it extrudes more readily than the 5083 or 5086 non-heat treatable alloys. The strength of the 6061-T6 alloy is higher than that of the 5083 or 5086 alloys; however, in the 6061-T6 alloy, the strength, ductility and corrosion resistance of the area in the vicinity of welds are severely degraded by the heat of welding. The extent of degradation is indicated by the reduction from the specified minimum tensile strength of 289 MPa (42,000 psi) for 6061-T6 base plate to the 165 MPa (24,000 psi) indicated for the weld joint in Table 14. Such adverse effects limit the applicability of the 6061 alloy for welded applications.

5.4 Corrosion of Aluminum Alloys. Aluminum alloys generally do not experience excessive corrosion under normal operating conditions. However, aluminum alloys in contact with dissimilar metals, may corrode at an accelerated rate. Such conditions may occur between faying surfaces of aluminum and other metals, between aluminum hulls and non-aluminum piping or when non-aluminum piping passes through aluminum bulkheads, decks, etc. In such cases, aluminum should be isolated from the other metal by means of suitable non-water absorbing insulating tapes or coatings or gaskets or by use of special pipe hangers or fittings. The 1974 SOLAS Convention contains certain stipulations on the use of aluminum.

Table 13—Mechanical Property Limits of Sheet and Plate Aluminum

NON-HEAT-TREATABLE ALLOYS

Alloy and Temper	*Thickness millimeters*	*Thickness (inches)*	*Ultimate Tensile Strength MPa (ksi) minimum*	*Ultimate Tensile Strength MPa (ksi) maximum*	*Minimum Yield Strength MPa (ksi) minimum*	*Minimum Yield Strength MPa (ksi) maximum*	*Minimum Elongation in 50 mm (2 in.) percent*
5052-0	3.0–6.5	(0.114–0.249)	172 (25.0)	214 (31.0)	66 (9.5)		20
	6.6–75.0	(0.250–3.000)	172 (25.0)	214 (31.0)	66 (9.5)		18
5052-H32	3.0–6.5	(0.114–0.249)	214 (31.0)	262 (38.0)	159 (23.0)		9
	6.6–12.5	(0.250–0.499)	214 (31.0)	262 (38.0)	159 (23.0)		11
	12.6–51.0	(0.500–2.000)	214 (31.0)	262 (38.0)	159 (23.0)		12
5052-H34	3.0–6.5	(0.114–0.249)	234 (34.0)	283 (41.0)	179 (26.0)		7
	6.6–25.0	(0.250–1.000)	234 (34.0)	283 (41.0)	179 (26.0)		10
5052-H112	6.5–12.5	(0.250–0.499)	193 (28.0)		110 (16.0)		7
	12.6–51.0	(0.500–2.000)	172 (25.0)		66 (9.5)		12
	51.1–75.0	(2.001–3.000)	172 (25.0)		66 (9.5)		16
5083-0	1.5–38.0	(0.051–1.500)	276 (40.0)	352 (51.0)	124 (18.0)	200 (29.0)	16
	38.1–76.5	(1.501–3.000)	269 (39.0)	345 (50.0)	117 (17.0)	200 (29.0)	16
5083-H112	6.5–38.0	(0.250–1.500)	276 (40.0)		124 (18.0)		12
	38.1–76.5	(1.500–3.000)	269 (39.0)		117 (17.0)		12
5083-H116	4.5–38.0	(0.063–1.500)	303 (44.0)	386 (56.0)	214 (31.0)	296 (43.0)	12
5083-H117	38.1–76.5	(1.501–3.000)	283 (41.0)	386 (56.0)	200 (29.0)	296 (43.0)	12
5083-H323	1.5–3.0	(0.051–0.125)	310 (45.0)	372 (54.0)	234 (34.0)	303 (44.0)	8
	3.1–6.5	(0.126–0.249)	310 (45.0)	372 (54.0)	234 (34.0)	303 (44.0)	10
5083-H343	1.5–3.0	(0.051–0.125)	345 (50.0)	407 (59.0)	269 (39.0)	338 (49.0)	6
	3.1–6.5	(0.126–0.249)	345 (50.0)	407 (59.0)	269 (39.0)	338 (49.0)	8
5086-0	1.5–6.5	(0.051–0.249)	241 (35.0)	303 (44.0)	97 (14.0)		18
	6.6–51.0	(0.250–2.000)	241 (35.0)	303 (44.0)	97 (14.0)		16
5086-H112	4.5–12.5	(0.188–0.499)	248 (36.0)		124 (18.0)		8
	12.6–25.5	(0.500–1.000)	241 (35.0)		110 (16.0)		10
	25.6–51.0	(1.001–2.000)	241 (35.0)		9 (14.0)		14
	51.1–76.5	(2.001–3.000)	234 (34.0)		97 (14.0)		14
5086-H116 and H117	1.5–6.5	(0.063–0.249)	276 (40.0)	324 (47.0)	193 (28.0)		8
	6.6–51.0	(0.250–2.000)	276 (40.0)	324 (47.0)	193 (28.0)		12
5454-0	3.0–76.5	(0.114–3.000)	214 (31.0)	283 (41.0)	83 (12.0)		18
5454-H32	1.5–6.5	(0.051–0.249)	248 (36.0)	303 (44.0)	179 (26.0)		8
	6.6–51.0	(0.250–2.000)	248 (36.0)	303 (44.0)	179 (26.0)		12
5454-H34	4.0–6.5	(0.162–0.249)	269 (39.0)	324 (47.0)	200 (29.0)		7
	6.6–25.5	(0.250–1.000)	269 (39.0)	324 (47.0)	200 (29.0)		10
5454-H112	6.5–12.5	(0.250–0.499)	221 (32.0)		124 (18.0)		8
	12.6–51.0	(0.500–2.000)	214 (31.0)		83 (12.0)		11
	51.1–76.5	(2.001–3.000)	214 (31.0)		83 (12.0)		15
5456-0	1.5–38.0	(0.051–1.500)	290 (42.0)	365 (53.0)	131 (19.0)	207 (30.0)	16
	38.1–76.5	(1.501–3.000)	283 (41.0)	359 (52.0)	124 (18.0)	207 (30.0)	16
5456-H112	6.5–38.0	(0.250–1.500)	290 (42.0)		131 (19.0)		12
	38.1–76.5	(1.501–3.000)	283 (41.0)		124 (18.0)		12
5456-H116 and H117	4.5–15.5	(0.063–0.624)	317 (46.0)	407 (59.0)	228 (33.0)	317 (46.0)	12
	15.6–32.0	(0.625–1.250)	317 (46.0)	386 (56.0)	228 (33.0)	310 (45.0)	12
	32.1–38.0	(1.251–1.500)	303 (41.0)	386 (56.0)	214 (31.0)	296 (43.0)	12
	38.1–76.5	(1.501–3.000)	283 (41.0)	386 (56.0)	200 (29.0)	296 (43.0)	12
5456-H323	1.5–3.0	(0.051–0.125)	331 (48.0)	400 (56.0)	248 (36.0)	317 (46.0)	6
	3.1–6.5	(0.126–0.249)	331 (48.0)	400 (58.0)	248 (36.0)	317 (46.0)	8
5456-H343	1.5–1.0	(0.051–0.125)	365 (53.0)	434 (63.0)	283 (41.0)	352 (51.0)	6
	3.1–6.5	(0.126–0.249)	365 (53.0)	434 (63.0)	283 (41.0)	352 (51.0)	8

HEAT-TREATABLE ALLOYS

Alloy and Temper	*Type*	*Thickness millimeters*	*Thickness (inches)*	*Minimum Tensile Strength MPa (ksi)*	*Minimum Yield Strength 0.2% Offset MPa (ksi)*	*Minimum Elongation in 50 mm (2 in.) percent*
6061-T4	Sheet	0.5–6.3	(0.021–0.249)	207 (30.0)	110 (16.0)	16
6061-T451	Plate	6.4–25.4	(0.250–1.000)	207 (30.0)	110 (16.0)	18
		25.4–76.2	(1.001–3.000)	207 (30.0)	110 (16.0)	16
6061-T6	Sheet	0.5–6.3	(0.021–0.249)	290 (42.0)	241 (35.0)	10
6061-T62 and T65	Plate	6.4–12.7	(0.250–0.499)	290 (42.0)	241 (35.0)	10
		12.7–25.4	(0.500–1.000)	290 (42.0)	241 (35.0)	9
		25.4–50.8	(1.001–2.000)	290 (42.0)	241 (35.0)	8
		50.8–76.2	(2.001–3.000)	290 (42.0)	241 (35.0)	6

Aluminum in contact with wood, insulating materials or concrete should be protected against the corrosive effects of impurities in these materials by suitable coverings or coatings; concrete should be free of additives for cold weather pouring. Suitable precautions should be taken to avoid arrangements that could induce crevice corrosion in wet spaces. In certain stagnant water applications, such as bilge spaces or chain lockers where exfoliation corrosion may be of concern, use of the alloys specially heat treated to resist this form of corrosion should be considered (American Bureau of Shipping, 1975a). For generalized protection, sacrificial anodes or cathodic protection systems may be considered. To minimize adverse corrosive effects of stray currents, it is advisable, when possible, to keep such systems in operation while the vessel is in dock.

5.5 Fire Protection. Compared with steel, aluminum alloys have relatively low melting points and tend to lose strength rapidly upon exposure to elevated temperatures. In considering use of aluminum, due consideration should be given to applications where retention of structural integrity would be required in fire exposure. The use of appropriate insulation protection should be considered for such applications (ANSI/ASTM, 1977).

5.6 Copper-Nickel Alloys. Copper-nickel alloys have been used as solid plate to 10 mm (0.375 in.) thick, and as copper-nickel clad steel in thicknesses over 10 mm (0.375 in.) for small boats; copper-nickel clad steels have been suggested for large seagoing vessels. Significant economic advantages of copper-nickel hulls were reported for a fleet of small, 20-m (65-ft) shrimp boats operating in highly fouling Central American waters. Benefits were attributed to eliminating frequent drydocking for removal of barnacles and repainting, and to avoidance of reduction in speed and operating efficiency associated with increased hull frictional resistance derived from barnacle growth and accumulation (Manxolillo et al, 1976). Steel hulls have been reported to be inferior to copper-nickel hulls in the following respects:

Table 14—Minimum Mechanical Properties for Butt-Welded Aluminum Alloys

Values shown are for welds in plate thicknesses up to 38 mm (1.5 in.) unless otherwise noted.

	Ultimate Tensile Strength (U_{al})		*Yield Strength* (Y_{al})[3]	
Alloy	*MPa*	*(psi)*	*MPa*	*(psi)*
5083[1]	275	(39000)	145	(21000)
5086[1]	240	(35000)	95	(14000)
5454[1]	215	(31000)	85	(12000)
5456[1]	290	(41000)	130	(19000)
6061-T-6[2]	165	(24000)	105	(15000)

Notes:
1. All tempers.
2. Values when welded with 5183, 5356, or 5556 filler wire.
3. Based on 254 mm (10 in.) gage length.

- Barnacles grow and accumulate on steel,
- steel eventually corrodes in service,
- cleaning and repainting of a steel hull does not restore the hull to its original smoothness; this results in increasing hull resistance.

The retention of hull smoothness over the life of a ship, which may be provided by a copper-nickel hull, could result in cost benefits associated with increased operating speeds and decreased fuel consumption. An economic study has indicated that the potential annual cost benefit of a copper-nickel hull, compared with a steel hull, may possibly justify the extra cost of the copper-nickel material for some larger ships with high capital costs such as liquefied natural gas carriers and roll-on/roll-off ships (Copper Development Association, 1976).

To date the alloy used has been 90 percent copper- 10 percent nickel; new alloys with lower copper and higher nickel contents which offer promise of increased resistance to cavitation erosion have been proposed as possible candidates for the larger ships. For larger ships, where plate thickness exceeds 6.3 mm (0.25 in.), it is generally more economical to use copper-nickel clad steel instead of copper-nickel plate.

Section 6
Non-Metallic Materials

6.1 Glass Reinforced Plastics. Glass reinforced plastics, *GRP,* are a form of fiber reinforced plastics, *FRP,* which were introduced for marine structural applications in the 1940's in the form of Navy personnel boats. Since that time GRP have found widespread acceptance for yachts and small boats such as fishing trawlers up to 34 m (110 ft) in length. Details of the requirements are available in the form of Classification Society Rules (American Bureau of Shipping, 1978a). Although the future of GRP for large ship structures is promising, economic factors, and to some degree, questions of durability, limit their applicability. Reinforced plastics used for ship structures are composed of glass fibers embedded in unsaturated polyester resins. Typical properties are shown in Table 15 (Scott and Sommella, 1971). Other pertinent properties and variation of properties with sea water exposure time are discussed throughout the technical literature (SNAME, 1977), (Scott and Sommella, 1971). Properties of GRP that are particularly useful for marine service, and have led to their extensive use for small boats, are high strength-to-weight ratio combined with good resistance to deterioration upon prolonged exposure to sea water. Lower maintenance costs for GRP hulls compensate for their relatively high initial cost as compared with steel or wood.

Although GRP have been established as useful materials for small hulls, their application to large ships is limited by several factors. The modulus of GRP is of the order of one tenth that of steel. This tends to increase GRP hull de-

Table 15—Physical Properties of Typical Marine GRP Laminates[(a)]

AVERAGE VALUES FOR GUIDANCE ONLY

Physical Property[(b)(d)]	*Chopped Strand Mat Laminate Low Glass Content*	*Composite Laminate*[(c)] *Medium Glass Content*	*Woven Roving Laminate High Glass Content*
Percent Glass by Weight	25–30	30–40	40–55
Specific Gravity	1.40–1.50	1.50–1.65	1.65–1.80
Flexural Strength, MPa	124–172	172–207	207–241
(ksi)	(18–25)	(25–30)	(30–35)
Flexural Modulus, MPa	5516–8274	7584–10,342	10,342–15,168
(ksi × 10³)	(0.8–1.2)	(1.1–1.5)	(1.5–2.2)
Tensile Strength, MPa	76–103	124–172	193–221
(ksi)	(11–15)	(18–25)	(28–32)
Tensile Modulus, MPa	6205–8274	6894–9653	10,342–13,790
(ksi × 10³)	(0.9–1.2)	(1.0–1.4)	(1.5–2.0)
Compressive Strength, MPa	117–145	117–145	117–152
(ksi)	(17–21)	(17–21)	(17–22)
Compressive Modulus, MPa	6205–8963	6894–11,031	11,721–16,547
(ksi × 10³)	(0.9–1.3)	(1.0–1.6)	(1.7–2.4)
Shear Strength Perpendicular, MPa	69–90	76–97	90–103
(ksi)	(10–13)	(11–14)	(13–15)
Shear Strength Parallel, MPa	69–83	62–83	55–76
(ksi)	(10–12)	(9–12)	(8–11)
Shear Modulus Parallel, MPa	2758	3103	3447
(ksi × 10³)	(0.4)	(0.45)	(0.50)

(a) Properties from short term loading tests—wet condition. Composite and woven roving values for warp direction.
(b) Tested in accordance with ASTM Standard Specification or equivalent Federal Standard LP-406b.
(c) Based on typical alternate plies of 0.06 g/cm² mat and 0.08 g/cm² woven roving.
(d) Strength values are ultimate strengths.

flections to the extent that it could present problems in large ships in propeller shafting and piping arrangements as well as uncertain effects due to appreciable hull deflection. In addition, GRP have a tendency to creep upon prolonged exposure at high stress; this may be of special concern at points of stress concentration. Current methods of assembly are not oriented toward the production of large GRP hulls, and mechanized production techniques appropriate to such hulls would have to be developed. In addition, any consideration for larger hulls must recognize that GRP laminates in general use rapidly lose strength at elevated temperatures and will support combustion. GRP are not as yet cost effective for large ships (Scott and Sommella, 1971). A study of the feasibility of a 152m (470 ft) GRP cargo ship concluded that it would not be economically competitive with the equivalent steel ship and would not meet governmental regulations relative to fire resistance. However, the report did indicate economic potential of GRP for structural compenents such as deck houses, hatch covers, kingposts, and bow modules. The high strength-to-weight ratio of GRP could make them particularly attractive for some applications in weight critical hulls such as hydrofoils or hovercraft. Considerable research effort is directed toward increasing the modulus by substituting high modulus fibers such as boron or carbon for glass fibers. Although such substitutions increase the modulus significantly, their high costs preclude their use for general hull applications.

Table 16—Structural Performance of Reinforced Concrete

	Compressive		*Tensile*	
Age, Days	*MPa*	*(psi)*	*MPa*	*(psi)*
1	28–41	(4,000–6,000)	2.8–4.1	(400–600)
7	41–55	(6,000–8,000)	3.4–4.8	(500–700)
28	48–83	(7,000–12,000)	4.1–6.2	(600–900)
90	55–97	(8,000–14,000)	4.8–7.2	(700–1050)
180	62–103	(9,000–15,000)	5.5–7.6	(800–1100)

6.2 Concrete. Concrete consists of a mixture of stone aggregate bonded by a hardened cement (Portland cement is normally used for marine applications). The aggregate consists of sand, gravel, and crushed stone. Specific gravity of concrete normally varies between 2.2 and 2.5, primarily depending upon the sizes and density in the stone mixture. Lighter weight concrete with specific gravities in the 1.6 to 2.0 range are made by using clay and shale aggregate. The ratio of water to cement is one of the most significant factors in determining concrete quality and properties. Ordinary structural concrete with a water/cement/ratio approximating 0.40 by weight is usual for marine work. The long term durability of concrete in sea water has been well established on the basis of service experience and testing of samples from structures that have been submerged (Morgan, 1977). However, in certain situations when concrete is exposed to sulfate in soils or fresh water, it may react with the sulfate, and degrade. Sea water, however, minimizes or prevents such deterioration. Where sulfate deterioration is of concern, special sulfate resistant concretes are used.

a. Ferrocement. Ferrocement is a form of reinforced concrete wherein layers of steel mesh are used as the reinforcing medium. The material has been used for making small boats up to 50m (164 ft) with skin thickness of 10 mm (0.375 in.) to 40 mm (1.5 in.). The low cost and availability of the mesh and concrete ingredients make ferrocement particularly attractive where sophisticated industrial facilities are not available. Its most extensive commercial use has been for fishing vessels up to 50 m (164 ft) in length (Morgan, 1977), (United Nations, 1972).

b. Reinforced Concrete. Reinforced concrete consists of cement reinforced by structural grade steel bars. It is usually used in thicknesses of 90 mm (3.5 in.) or greater. For applications where loading is in compression the concrete as well as the steel is effective in providing the required compressive strength. However, for ships, where applied loading alternates between tension and compression, the cement provides no significant resistance to the tensile forces; its principal function under such loading condition being to provide a watertight structure. Despite these limitations, a variety of oceangoing ships, including oil tankers up to 7,500 dwt have been built to classification society rules and successfully operated (Tuthill, 1945). Because of the inability of concrete to sustain significant tensile loads, sufficient knowledge of anticipated service stresses coupled with good design are required to avoid exposure of the concrete to service tensile stresses, which induce cracking of the concrete. In addition to strength reduction, cracking of concrete enables sea water to reach and initiate corrosive attack on the reinforcing bars; this in turn weakens the bond between cement and reinforcing bar and induces separation of the concrete. With the advent of prestressed concrete, use of reinforced concrete is now only generally considered for such structures as pontoons, floating docks, small ships, in conjunction with prestressed concrete, or in applications where appreciable alternating loading is not anticipated (Anderson, 1975). Table 16 indicates some typical properties of marine types of reinforced concrete.

c. Prestressed Concrete. In prestressed concrete high strength, up to 2,068 MPa (300,000 psi) tensile, reinforcing wires which are prestressed well in excess of 862 MPa (125,000 psi) replace the structural grade bars used in reinforced concrete. When such prestressed wires are embedded in concrete, the resulting effect is to impose a high compressive load on the concrete. With the application of alternate cyclic loading equivalent to that experienced by a floating structure such as a ship, the force on the concrete will vary from the initial high compression, to a higher compression (with external compression loading) to a lower compression (as the external tensile loading is applied). Under such conditions the concrete remains in compression with externally applied tensile and bending loads and the threat of cracking of the concrete from tensile loading is eliminated. Concrete resists imposed shear and compression loads.

Prestressed concrete develops several attractive properties as a result of its heterogeneous prestressed wire/cement structure. Loads imposed locally are dispersed through the structure via the numerous supporting metallic wires, thereby preventing or minimizing damage from concentrated shock loads, i.e., the structure is highly resistant to fracture propagation.

The combination of the foregoing, and the relatively heavy thicknesses and large mass associated with prestressed concrete structures make the material an attractive candidate for applications where resistance to shock, collision damage or sudden failure is of concern (Shaw and McGarey, 1971). Prestressed concrete also has high damping properties which are beneficial in minimizing vibration.

The applicability of prestressed concrete to cryogenic applications is well documented by its successful application to liquefied natural gas storage at −160°C (−260°F). Brittle fracture of the reinforcing wires does not occur at cryogenic temperatures because of the thinness of the wire's cross section, and the supporting concrete matrix.

The relatively thick sections required of reinforced concrete and prestressed concrete structures as compared to steel, are obstacles to their use for many marine applications. In addition, while service experience has been generally satisfactory for the reinforced and prestressed concrete marine structures built to date, extensive service experience and supporting data will be required for some of the more ambitious designs proposed such as a 300 m (1000 ft) LNG tanker. Concrete has been successfully used in fixed offshore structures and guides for such use are available (American Concrete Institute, 1978).

An area of required research pertinent to reinforced and prestressed concrete is the possible corrosion of steel in splash zones. Concrete normally inhibits corrosion of imbedded steel by maintaining an alkaline environment that produces a passivated surface condition. However, at the splash zone, cyclic exposure to oxygen and chloride levels above those of subsurface sea water may lead to corrosion of the reinforcing wires or bars and loss of the synergistic effect of the concrete/steel bonding, this in turn may lead to premature spalling and a need for repair. In addition, concrete tends to exhibit greater permeability and could become more readily saturated with sea water at the splash zone. This in turn tends to make such areas more susceptible to frost damage. In view of the preceding, it appears, that for some applications, consideration may be required to provide suitable protective covering at the splash zone areas.

Section 7
Joining Metallic Materials

7.1 Weldability, General. In the course of the welding cycle three events occur:

1. Filler metal and contiguous base metal are melted and resolidify to form a fused connection.

2. The heat effect of welding subjects the adjacent base metal to a thermal gradient ranging from above the base metal melting point to ambient temperature. The heat affected zone, *HAZ*, and deposited weld metal then cool to ambient temperature. As indicated in Fig. 11, the variety of metallurgical structures produced in this heat affected zone include those exposed to the various temperatures which produce resolidification of melted base metal, grain growth, grain refinement, or modification of metallurgical microstructure.

3. The solidification of molten metal, as well as the metallurgical phase changes, induce plastic flow and develop residual stresses which may exceed the yield point in magnitude; thus resulting in structural distortion.

7.2 Hot Cracking. When cracking occurs at elevated temperatures, the crack is usually intergranular (between grain boundaries). Such cracking is associated with excessive solidification and cooling stresses acting on constituents present at the grain boundaries which are relatively weak at elevated temperatures. The weakened grain boundary may consist of specific low melting constituents such as sulfides in steel. In other cases the deposition of a weld bead of unfavorable geometry may impose excessive cooling stresses on the hot weld deposit which has relatively low strength at elevated temperature. For example in submerged arc welding, weld beads such as those shown in Fig. 12A, would tend to form a center section which solidifies last and remains at an elevated temperature after the surrounding metal has solidified and cooled. The low strength at the grain boundaries of the material at elevated temperature is inadequate to resist the thermal stresses, and hot cracking occurs. Such cracking, can usually be readily prevented by changing weld parameters to produce a bead of more favorable contour, Fig. 12B.

7.3 Cold Cracks (Hydrogen Cracks or Delayed Cracks). The role of hydrogen is an important consideration in the welding of ship steels. Hydrogen-bearing compounds such as water or organic compounds present on the filler metal surface, in electrode coverings, or on base metal surfaces may dissociate in the welding arc to form atomic hydrogen. The atomic hydrogen penetrates and is highly soluble in molten steel weld metal and the zone of adjacent heat affected steel which has been transformed to a phase known as austenite; the austenite forms when the heat affected zone of a steel is heated above a critical temperature, (approximately 900°C (1,640°F), for structural steels. As the solidified weld metal and austenitized heat affected zone cool to ambient temperatures, they are transformed into non-austenitic phases which release most of the dissolved hydrogen from solution, since hydrogen is practically insoluble in these phases. When hydrogen is released from solution in the presence of a hard zone in the microstructure and a high residual stress field, a condition known as hydrogen cracking may occur. Since the time of such cracking varies from

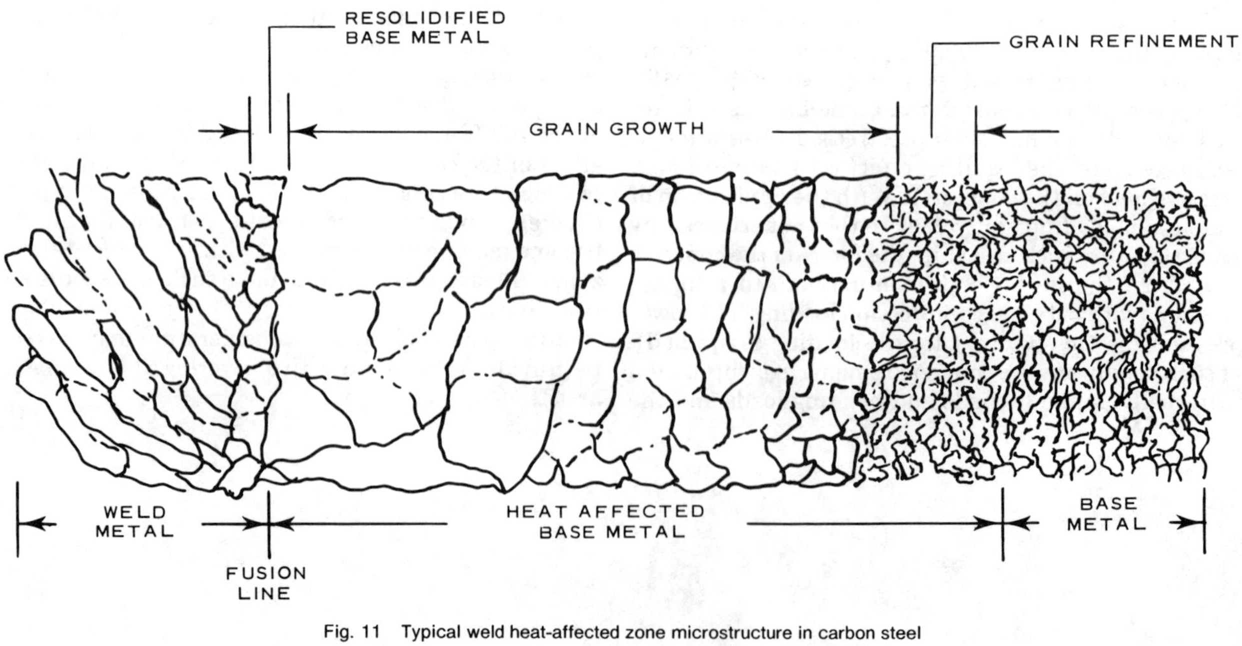

Fig. 11 Typical weld heat-affected zone microstructure in carbon steel

immediate to several days or weeks after the completion of welding, the phenomenon is also known as *delayed cracking.* The tendency for such cracking varies directly with the magnitude of:

- hydrogen concentration,
- local metal hardness,
- residual stress.

Hydrogen delayed cracking is the most important and troublesome form of cracking encountered in welding of the higher strength ship steels (Mishler, 1976).

7.4 Ordinary Strength Steels. Steels equivalent to ordinary strength shipbuilding steels, such as ABS Grades A, B, D, DS, CS, E are usually readily weldable with normal procedures. Because of the relatively low strength, and the absence of hardened areas in the heat affected zone, the tendency for hydrogen cracking under most conditions is minimal. However, where heavy thicknesses are involved, such as over 50 mm (2 in.), or when base metal temperatures are unusually low, the accelerated quench rate would tend to produce a harder heat affected zone with increased residual stress levels; high residual stresses also occur in welds in highly restrained structures. Under such conditions consideration should be given to precautionary measures such as the use of low hydrogen welding processes (use of low hydrogen electrodes in shielded metal arc welding) and preheat to minimize adverse quench effects and reduce residual stresses. Preheat is usually not required for processes such as submerged arc or electroslag welding where the higher heat input rates and relatively large area heated in the weld vicinity provide conditions analogous to some degree of superimposed preheat.

7.5 High Strength Steels. Steels with specified yield strengths above 234 MPa (34,000 psi) similar to the ABS Grades AH, DH and EH higher strength grades should generally be welded with low hydrogen electrodes. Preheat may be advisable where heavy base plate thicknesses or high restraint could result in high residual stress. Where there is a possibility of moist base plate surfaces, it is considered good practice to dry the surfaces by heating with an oxy-gas torch. This will prevent hydrogen cracking by reducing possible hydrogen contamination from condensed moisture. As steel strength increases, sensitivity toward hydrogen induced cracking increases and the need for preheat and reduced moisture content (hydrogen source) of electrode coverings increases. Preheat tends to reduce weld and HAZ hardness and residual stress. For example, for steels of 414 MPa (60,000 psi) yield strength, the moisture content of an electrode covering should be limited to 0.4 percent maximum, for steels of 689 MPa (100,000 psi) (such as ASTM A514 or 517) a 0.2 percent limit should be applied. Recommendations for selecting electrode types and preheat conditions for the various ship steels as well as other steels are available in the technical literature (Ott and Snyder, 1974).

7.6 Stainless Steel. The stainless steels most widely used in marine construction; i.e., American Iron and Steel Institute (AISI) 304 and 316 types, are readily weldable by the inert-gas arc and shielded metal arc-welding processes with standard techniques, using filler wires of compatible composition. As was noted in 4.2, the extra low carbon (0.03 max.) modifications of these grades are generally recommended for welded construction. When such low carbon grades of base plate are to be welded, filler metals used should also be of extra low carbon grade (i.e., 308L, 316L).

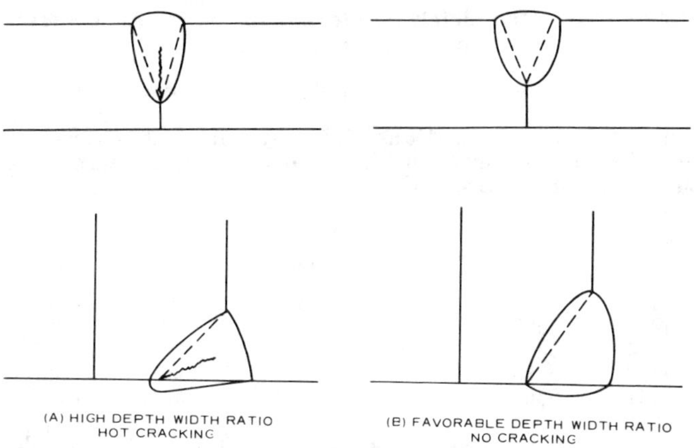

Fig. 12 Effects of weld geometry

7.7 Aluminum Alloys. Aluminum alloys used in marine construction are readily weldable with the inert-gas arc-welding processes; Gas Metal Arc, *GMAW* or *MIG,* and Gas Tungsten Arc, *GTAW* or *TIG.* The former process pre-

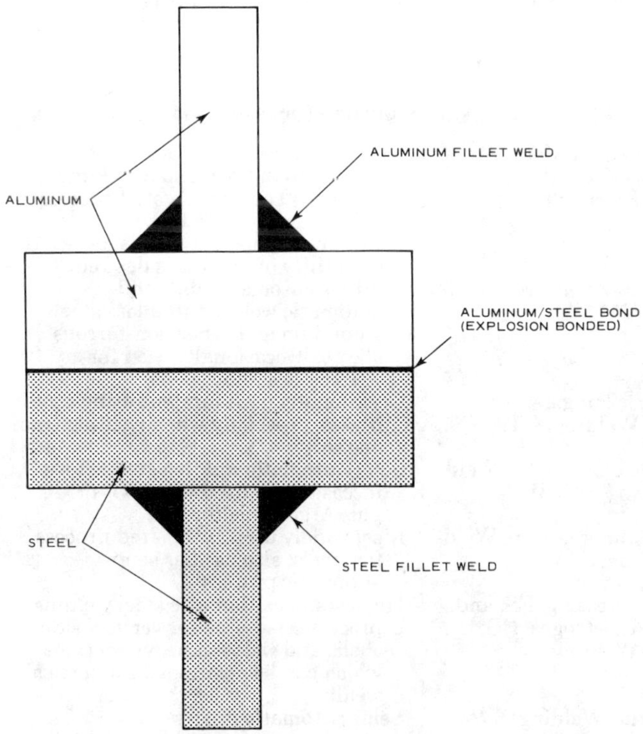

Fig. 13 Aluminum bonded steel

Table 17A—Filler Metals for Welding Aluminum Alloy Sheet, Plate, and Extrusions

Recommendations in this table apply to gas shielded-arc welding processes.
Filler metal alloys 5183, 5356 and 5556 may be used interchangeably provided that strength, ductility and corrosion resistance are suitable for the service conditions.

Base Metal Alloys	*5083*	*5086*	*5454*[1]	*5456*	*6061*
5083	5183	5356	5356	5183	5356
5086	5356	5356	5356	5356	5356
5454[1]	5356	5356	5554[1]	5356	5356
5456	5183	5356	5356	5556	5356
6061	5356	5356	5356[2]	5356	4043[2]

Notes
1. 5454 aluminum alloy welded with 5554 filler metal is generally recommended for service applications above 65°C (150°F) such as for smoke stacks and engine room enclosures.
2. 5183 or equivalents may be used.

Table 17B—Filler Metals for Welding Aluminum Alloy Castings to Castings and Plate

ASTM American Society for Testing and Materials
AA Aluminum Association

Castings *ASTM*	*AA*	*SG70A, SG70B, 357.0 (Note 1)*	*5154, 5454, 6061 (Note 2)*	*5456, 5083, 5086 (Note 3)*
SG70A	356.0	4043	4043	5356
SG70B	A356.0	4043	4043	5356
	357.0	4043	4043	5356

Notes:
1. Filler metal with same analysis as base metal is sometimes used.
2. 5183, 5356, 5554, 5556 and 5654 may be used. In some cases they may provide higher weld ductility and higher weld strength, 5554 is suitable for elevated temperature service.
3. 5183, 5356 or 5556 may be used. 4013 may be used for some applications where filler metal properties are not of primary concern.

dominates because of its higher production speeds and greater economy. In welding aluminum, particular care should be taken to see that all surfaces in the way of welding are clean and free of contaminants, such as water stains, oxide films, and anodized layers; when present, they may be removed as described in more detail in Chapter XVI. Preheat is not generally needed except when welding exceptionally thick sections, under conditions of high restraint, when humidity is very high, or when temperatures are below 0°C (32°F). For the 5000 series alloys, prolonged preheating or exposure in the 65°C to 200°C (150°F to 400°F) range should be avoided, since it could sensitize the alloys to corrosion. Filler metals appropriate for welding the various aluminum alloy sheets, plates, extrusions and castings encountered in marine construction are detailed in Table 17. Requirements and recommendations for welding in aluminum hull construction are contained in the literature (American Bureau of Shipping, 1975a).

a. Welding Effect on Base Plate. Welding of the 5000 series alloys, where strength is usually derived from work hardening, (see subsection 5.2) produces a zone within approximately 13 mm to 25 mm (0.5 in. to 1 in.) of the weld where yield and tensile strength of base metal are reduced to values approximating annealed base plate properties. This zone of reduced strength must be taken into account in design calculations. In the case of heat treatable alloys, such as 6061-T6, a greater degree of tensile and yield strength degradation occurs; in addition, ductility and corrosion resistance are also severely degraded. Minimum tensile properties for butt welds in aluminum alloys are indicated in Table 14.

Table 18—Shipyard Welding Processes and Applications

Shielded Metal Arc, SMAW (Covered Electrode)	Universally used in steel ship construction for manual welding whenever semi-automated processes are not used due to lack of equipment, operator skills, or where versatility of process is desired.
Gas Metal Arc Welding, GMAW	Used for automatic and semi-automatic welding stainless steels, aluminum and other non-ferrous alloys. Occasionally used for welding low alloy steels.
Gas Tungsten Arc Welding, GTAW	Occasionally used for depositing root passes, especially when welding from one side.
Flux Cored Arc Welding, FCAW	Semi-automatic welding steel alloys (occasionally used with CO_2 or CO_2 plus Argon shielding).
Submerged Arc Welding, SAW	Most widely used automated process for steels; also used as semi-automatic process.
Electroslag, ES, and Electrogas, EG Welding	Highest deposition rate steel welding processes suitable for vertical side shells and welds in heavy sections which can be positioned for vertical welding.
Stud Welding, SW	Semi-automatic process for attachment pins, studs, strips, etc. to structures.

7.8 Welding Dissimilar Metals. The possibility of adverse effects resulting from galvanic corrosion should be considered whenever dissimilar metals are joined. In some instances special precautions, such as the application of coatings in the vicinity of the dissimilar metal joint may be indicated. The most common dissimilar metal combinations that are used in shipbuilding are stainless steel to carbon steel, and aluminum to carbon steel.

a. Stainless Steel to Carbon Steel. In welding stainless steel to carbon steel appropriate precautions should be taken to minimize deleterious effects associated with dilution of the stainless steel by the carbon steel base metal. Excessive dilution can produce crack-sensitive weld metal near the carbon steel interface. When stainless steels similar to the 18 percent chromium–8 percent nickel type commonly used in shipbuilding are joined to carbon steel, nickel-rich stainless filler metals such as type 309 (22 percent chromium–12 percent nickel) or 310 (25 percent chromium–20 percent nickel) are generally recommended for any stainless steel weld layers which come in contact with the carbon steel. When butt welding stainless clad steels, the carbon steel side

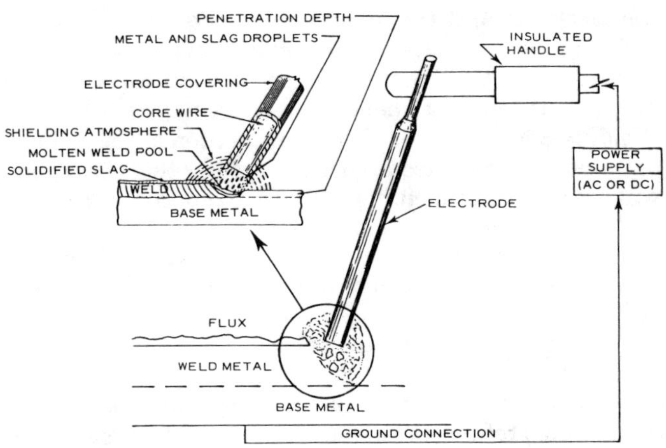

Fig. 14 A Shielded metal arc welding

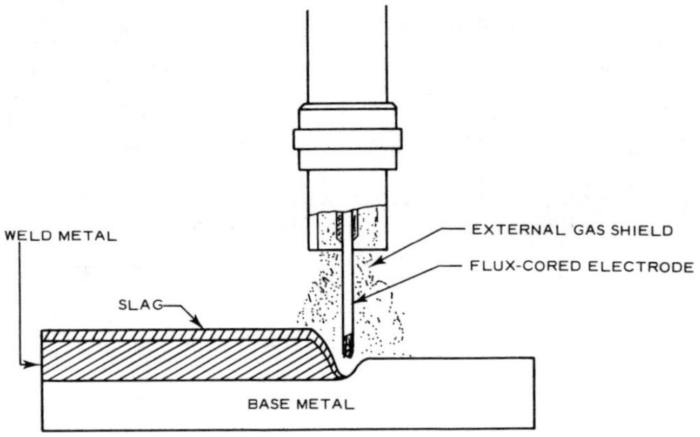

Fig. 14 D Flux-cored arc welding

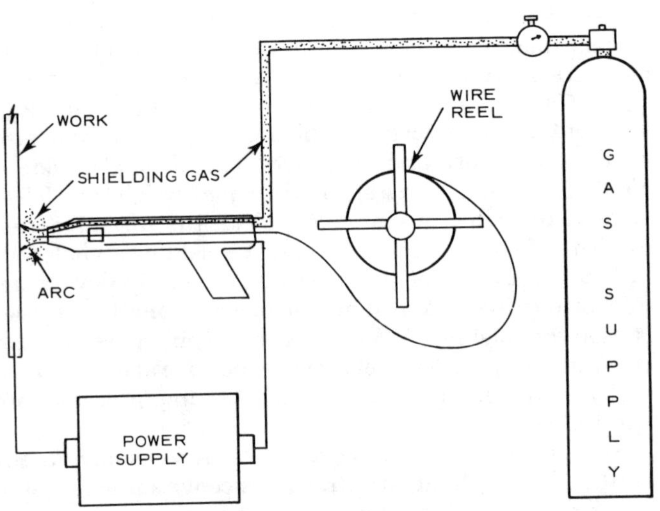

Fig. 14 B Gas metal arc welding

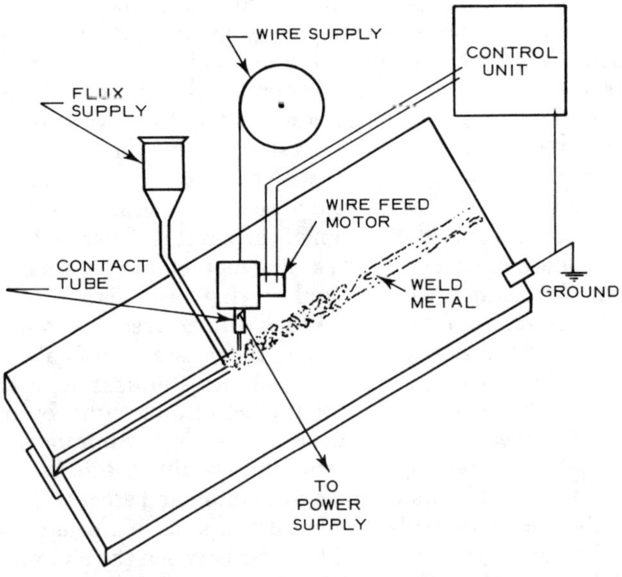

Fig. 14 E Submerged arc welding

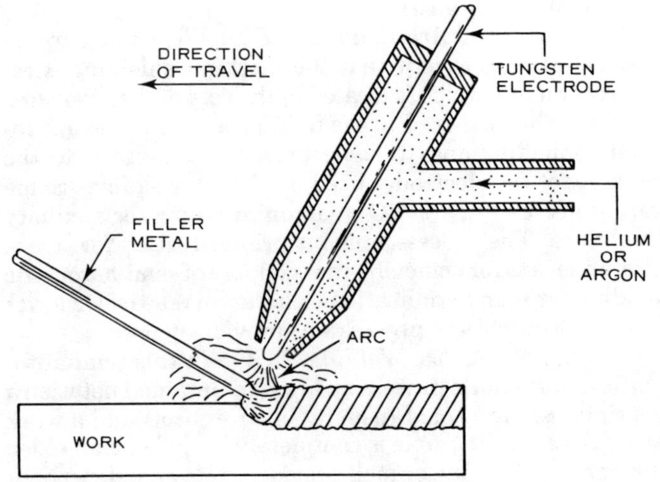

Fig. 14 C Gas tungsten arc welding

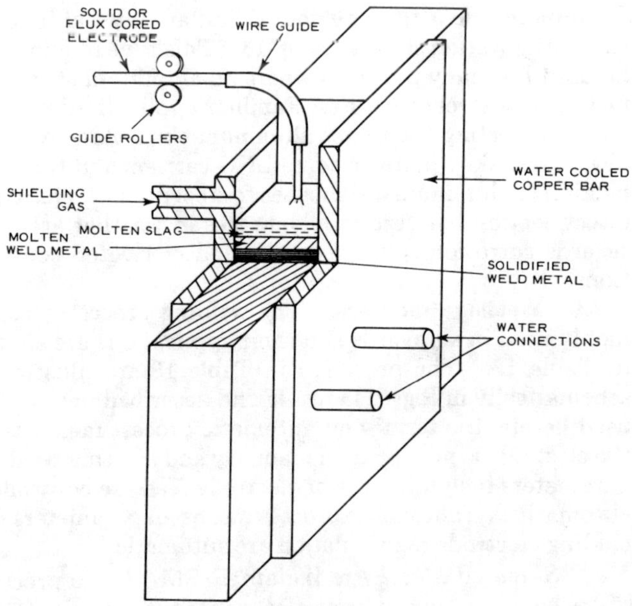

Fig. 14 F Electrogas welding

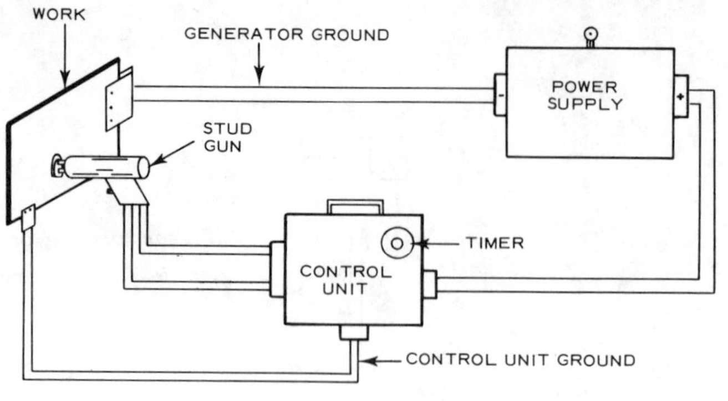

Fig. 14 G Stud welding

is usually welded first with the appropriate carbon steel filler metal; particular care must be exercised to prevent the carbon steel weld deposit from impinging on the stainless steel overlay. The second side (stainless steel side) is then welded with a nickel-rich stainless steel filler wire such as type 309 or 310. If the carbon steel layer is relatively thin, the entire weld may be made with the 309 or 310 filler metal. Similar procedures are usually used for welding other clad carbon steels; i.e., deposition of carbon steel filler metal on the cladding is avoided, and a filler metal that is compatible with the cladding and the underlaying base metal is used.

b. Steel to Aluminum. When joining steel to aluminum consideration should be given to the possibility of galvanic or crevice corrosion. Galvanized and stainless steel fasteners have been used to join steel to aluminum by mechanical means using rivets or fasteners. Aluminum is not weldable to steel by conventional welding methods. To effect welded joints between aluminum and steel an intermediate composite plate material consisting of an aluminum and a steel layer is used. The bond between the aluminum and steel in the composite aluminum-steel plate, is made by special manufacturing processes such as explosion bonding. Each plate side is then welded to similar material to form the configuration shown in Fig. 13. This type of joint can be used for many purposes where an aluminum structure fastens to a steel structure. Familiar applications are the joint connecting the upper aluminum skirt plate and the lower steel skirt plate on some LNG carriers and the connection of aluminum deckhouses to steel decks. This type of connection has certain advantages in weather areas as regards corrosion compared to bolted or riveted connections.

7.9 Welding Processes. The welding processes commonly used in shipyards and their applications are shown in Table 18. The processes in Table 18 are illustrated schematically in Figs. 14a to 14g and described below. As used herein, the term semi-automatic process means that the electrode is manipulated manually and all other welding parameters including rate of electrode feed are controlled automatically; automatic process means all parameters including electrode manipulation are automatic.

a. Shielded Metal Arc Welding. SMAW is a process where heat is produced by an electric arc between a covered metal electrode and the work. The arc melts the metal of the electrode and the spray or droplets formed transfer across the arc to coalesce as a molten pool before solidifying as weld deposit. The transfer mechanism involves a combination of complex phenomena of arc physics (American Welding Society, 1978). The formulation of the cellulosic or mineral base electrode covering assures that the covering will decompose or melt in the arc in an appropriate manner and rate, and accomplish the following:

- Provide a gas or slag environment which shields the metal from the atmosphere during metal transfer and solidification;
- establish a favorable electrical environment for arc stability;
- provide a slag covering for the deposited molten weld metal which refines the metal and may, in some cases, provide alloying additions; and,
- influence the fluidity of the molten weld metal which in turn influences the shape and contour of the deposited weld bead. Since the covering has a great influence on the transfer and nature of the resulting weld deposit, it is important that coverings be kept free of contaminants such as moisture or grease which could alter their characteristics.

b. Gas Metal Arc Welding. GMAW is an automatic or semi-automatic process in which a welding arc is formed between the work and bare electrode. The electrode is continuously fed from a spool which may weigh from 0.5 to 23 kg (1 to 50 lb). An inert gas shields the arc and molten weld area from the atmosphere; such shielding is analogous in function to that of the covering in the SMAW welding. Carbon dioxide, argon, or helium or a combination of gases is used for shielding. When argon or helium are used for shielding the process is relatively expensive and is not generally used when more economical welding processes are applicable.

c. Gas Tungsten Arc Welding. GTAW is similar to the aforesaid Gas Metal Arc Process except that a tungsten electrode is substituted for the continuously fed filler metal electrode of the GMAW process; in GTAW the filler metal is provided by a weld rod which is fed (usually manually) so that its end is melted by the welding arc maintained between the tungsten electrode and base metal. Argon or helium are used as shielding gases.

d. Flux Cored Arc Welding. FCAW is similar to the GMAW process except that the inert gas shielding is replaced by a flux which is located in the core of the filler wire; the flux, when exposed to the welding arc, provides appropriate shielding and to some extent is analogous to the covering of the electrodes used in SMAW welding. Some variations use CO_2 or CO_2+ argon mixtures for auxiliary shielding. The process which is primarily used for steels, offers a means for achieving the economy of semi-automatic welding for many applications where the relatively slower but versatile SMAW process was previously used.

e. Submerged Arc Welding SAW. In this semi-automatic or automatic process an arc is maintained between a continuously fed spool (usually in wire form) and a work area. The welding zone is completely buried and shielded under a granular flux or melt provided from an independent feed tube. The flux or melt, when molten, maintains an electrical path of high current density which generates a great quantity of heat. The insulating characteristics of the flux concentrate the heat in the weld area and induce significant melting of base metal as well as welding electrode. Under such conditions, high welding speeds, high deposition rates, significant melting of base metal, and deep weld penetration can be achieved. SAW with two- or three-wire electrodes instead of a single wire provides even higher welding speeds and deposition rates. Because of these features SAW is a frequently used automated welding process in steel ship construction. Common practice in making a subassembly of full penetration welds in a panel line is to submerge arc weld a subassembly of several plates from one side, turn the subassembly and complete each weld from the second side; welding from both sides is usually necessary to assure complete weld fusion. However, SAW welds with sound roots can be made from one side only, thereby eliminating the cost and time consumed in subassembly turning and rewelding from a second side. This form of the process designated as *one-side welding* requires close control of joint fit, plate waviness, and weld parameters. Additionally, a special backing or tape on the back side of the joint is usually necessary to contain the molten weld metal at the root so that it forms a sound weld deposit of satisfactory contour.

f. Electroslag, ES and Electrogas EG Welding. These are high deposition rate processes analogous to the SAW and GMAW processes respectively, except that the molten weld pool is contained within movable copper *shoes* at each side of the weld joint. A variation of the electroslag method, which uses a consumable guide tube instead of a permanent tube has been used for applications such as butt welding of underdeck longitudinals. Because of the exceptionally high deposition rates and large molten weld pools, arrangements are only available for vertical welding. In ES welding, a bar or strip is occasionally substituted for the one or more electrodes. Exceptionally thick materials may be welded in a single pass and in the case of ES welding, materials in excess of 400 mm (16 in.) thick have been welded in a single pass. Fig. 15 indicates the relatively high rates of welding speeds available. Appreciably higher speeds are attained as plate thickness decreases and as thicknesses approach 12 mm (0.5 in.) welding speeds of more than 10 times those shown are attained. Because of their relatively high heat input rates, ES and EG welding cause a greater degree of grain growth and other metallurgical changes in the weld heat affected zone, HAZ, than other processes. In some cases these may adversely affect HAZ properties, such as toughness, to the extent that use of the process must be restricted. Research work is underway to minimize these effects and broaden the applicability of the processes.

g. Stud Welding. SW as used in shipbuilding, is an arc welding process wherein an arc is maintained between a stud or similar piece and the work, for a predetermined time so that both are properly heated. The stud is then brought to the work by spring pressure. A ceramic ferrule is sometimes used to provide partial shielding and some contour. The process is accomplished with an automated welding gun, power source, and control panel; the control panel regulates electrical parameters, welding arc time, arc distance, and the imposition of pressure between stud and work at the end of the welding cycle. The process is widely used in shipbuilding for attaching a wide variety of items such as studs, clips, hangers, and insulation pins to structural members.

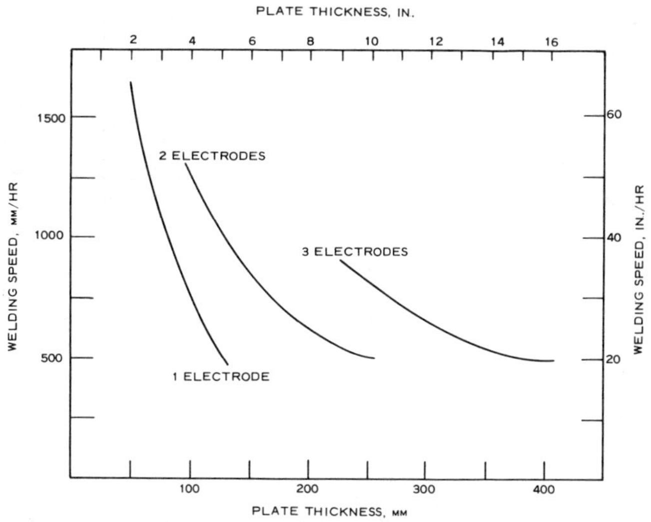

Fig. 15 Electroslag welding speeds

7.10 Welding Filler Metals. Specifications for welding filler metals used in shipbuilding are issued by ship classification societies such as ABS or by national technical societies such as the American Welding Society, (AWS) or by governmental agencies. ABS filler metal mechanical property requirements, which are similar to those of the foreign classification societies, are shown in Table 19. The lower Charpy test values specified in Table 19 for automated welds, as compared with manual or semi-automated welds, are based on the assumption that, in production, automated welds are more consistent in quality. Selection of filler metals for the various grades of steel used in shipbuilding is based on the principle that the weld deposit should be comparable in properties to the base metal being joined. The filler metals applicable to the different grades of ABS steels are shown in Table 20. Appropriate AWS or MIL specifications may be used when steels have strength or toughness requirements in excess of the ABS steels, as well as for stainless and non-ferrous alloys. The ABS annual publication "Approved Welding Electrodes, Wire-Flux and Wire-Gas Combinations" lists specific electrode brands meeting requirements for the various ABS filler metal grades and also contains ABS filler metal specifications. The publication also indicates the relationships between ABS and AWS filler metal grades.

7.11 Designing for Welding. Economy in ship construction and improvements in the serviceability and service life of ship structures can be enhanced if several principles basic to welded construction are observed in the design process. These principles are derived both from service experience and from studies of the causes and prevention of structural failures in ships.

a. Base Metals. The mechanical toughness and corrosion properties of the base metals selected should resist excessive degradation from welding and forming practices.

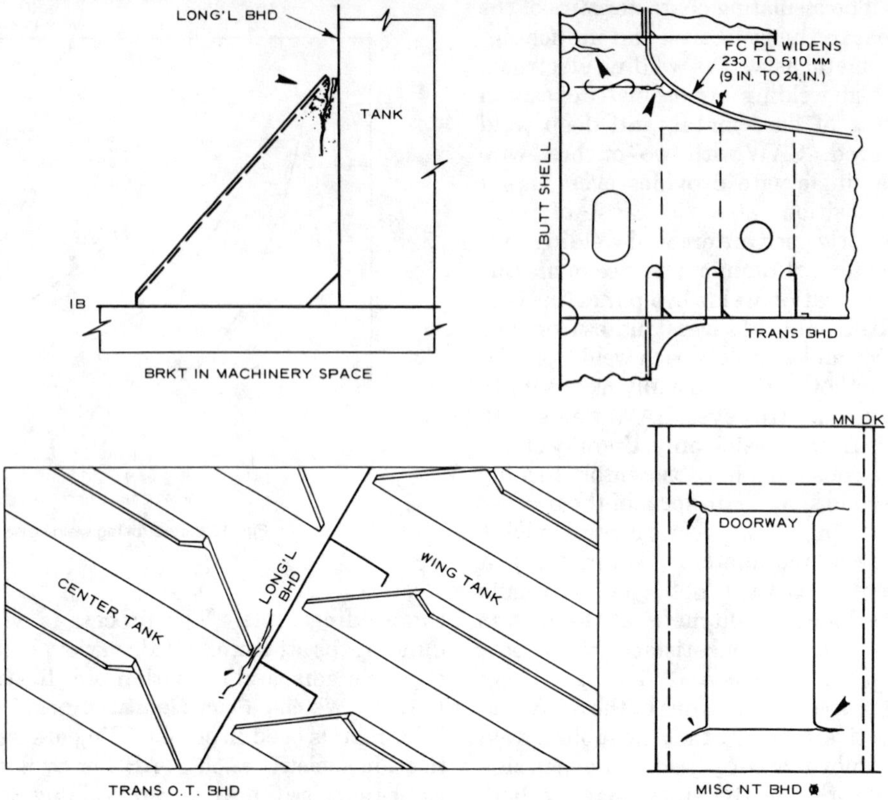

Fig. 16 Fatigue crack in areas of stress concentration

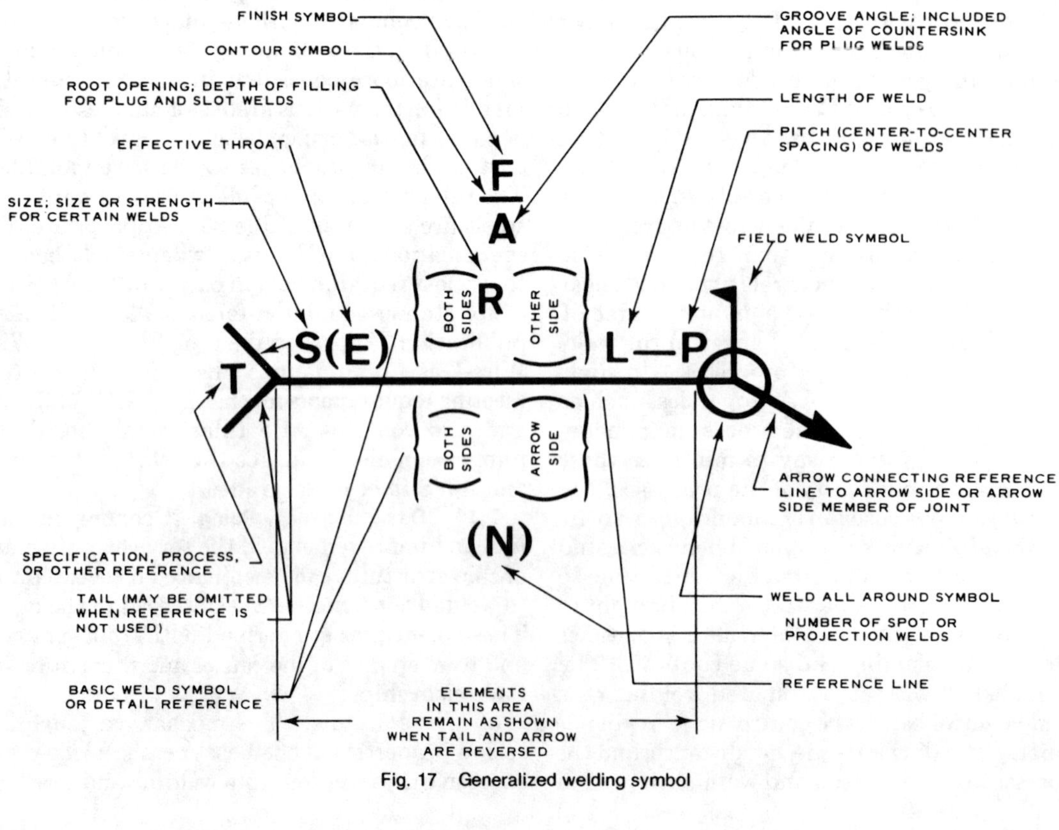

Fig. 17 Generalized welding symbol

Table 19—Mechanical Property Requirements for ABS Filler Metal Grades (1978)

ORDINARY STRENGTH STEELS

Tensile Requirements for All Grades

Tensile strength, MPa(psi)	400–560(58,300–81,100)
Yield point, min., MPa(psi)	305(44,100)
Elongation in 50mm sample min. percent	22

Impact Requirements: Grades	1	2	3	
Test temperature, C(F)	20(68)	0(32)	−20(−4)	−10(14)
Energy absorbed Manual and semi-atomatic, J (ft-lb)	47(35)	47(35)	47(35)	61(45)
Automatic, J (ft-lb)	34(25)	34(25)	34(25)	44(33)

HIGHER STRENGTH STEELS

Tensile Requirements for All Grades

Tensile strength, MPa(psi)	400–755(71,000–95,000)
Yield point, min.	370(54,000)
Elongation in 50mm sample min. percent	20

Impact Requirements: Grades	1y			2y			3y			
Test Temp. °C(°F)	20(68)	0(32)	10(50)	0(32)	−20(−4)	−10(14)	−20(−4)	−10(14)	−40(−40)	−30(−22)
Energy absorbed, manual and semiautomatic	54(40)	27(20)	—	54(40)	27(20)	—	54(40)	68(50)	27(20)	—
automatic, J (ft-lb)	40(30)	—	27(20)	40(30)	—	27(20)	40(30)	52(38)	—	27(20)

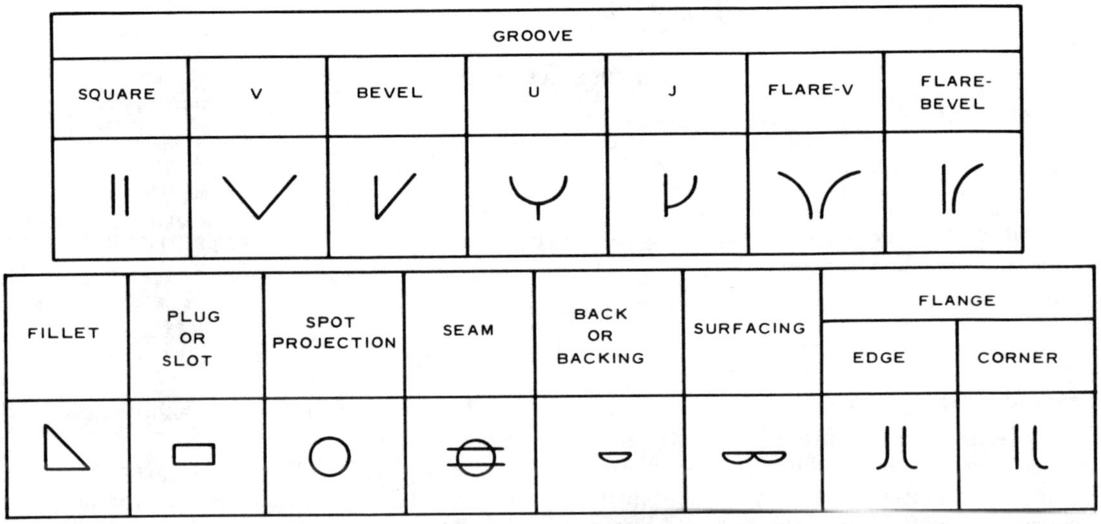

Fig. 18 Basic weld symbols

This precaution is particularly applicable to those materials whose properties have been enhanced by heat treatment or cold work. In addition, when materials of widely differing corrosion resistant characteristics are joined, possible adverse galvanic corrosion effects should be considered.

Examples of degradation which may be encountered are:

- Loss of toughness in the HAZ of some steels, particularly some higher strength steels, where weld procedures with excessively high heat input rates have been used.
- Loss of strength, ductility, and corrosion resistance in the HAZ of the heat treatable aluminum alloys such as 6061.
- Accelerated corrosion attack on a carbon steel located adjacent to an area overlayed with a stainless steel.
- Loss of ductility and toughness in materials subjected

Table 20—Application of Filler Metals to ABS Steels

ABS Hull Structural Steel	*Acceptable ABS Filler Metal Grade*
Ordinary Strength	
A to 12.5mm (1/2 in.) inclusive	1, 2, 3
A over 12.5mm (1/2 in.), B, D, DS	2, 3
DN, CS, E	3
Higher Strength	
AH to 12.5mm (1/2 in.) inclusive	1Y, 2Y, 3Y
AH over 12.5mm (1/2 in.), DH	2Y, 3Y
EH	3Y

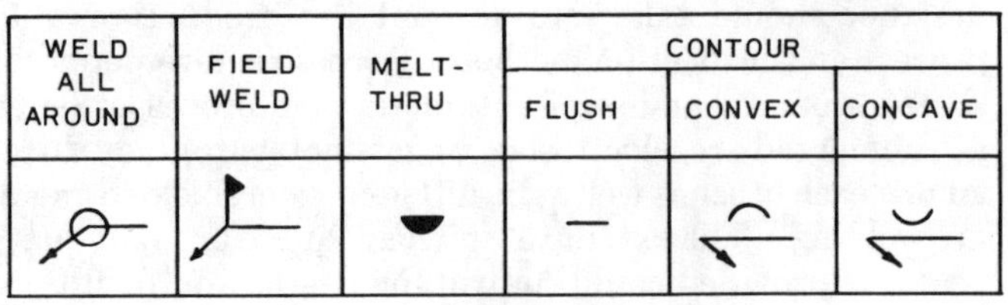

Fig. 19 Supplementary symbols

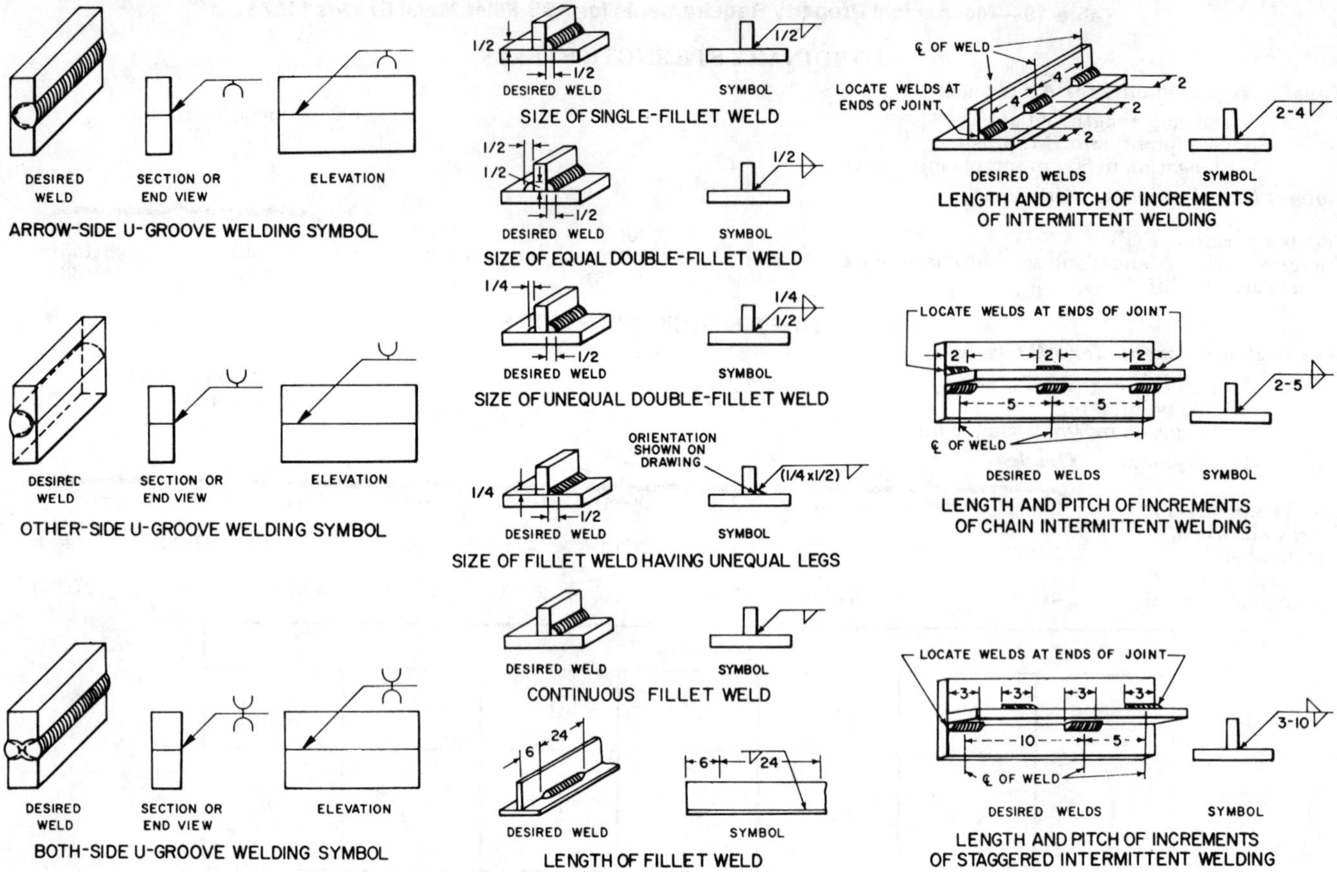

NOTE: ALL DIMENSIONS IN IN. (MULTIPLY BY 25 TO OBTAIN MM)

Fig. 20 Use of welding symbols

to excessive cold forming.

b. Stress Concentration. Points of high stress concentration such as may be introduced by a flaw or an abrupt change of geometry at an intersection have been identified as potential sources of brittle fracture initiation. Surveys have shown that they can be a primary source of fatigue cracking which produces many of the nuisance cracks (Jordan and Cochran, 1978). Such cracks can represent significant cost items in respect to their interruption of normal ship operations, and the time and effort required for their repair. Typical illustrations of the above are shown in Fig. 16. Areas of high restraint where high weld residual stresses can develop should also be minimized.

c. Joint Design. The attainment of a sound weld joint and its proper inspection can only be achieved if appropriate clearances are provided. In considering this aspect, the designer should take into account the production weld process and inspection method. Access requirements for some automatic and semi-automatic weld processes (such as submerged arc, electroslag, or gas-metal arc) may differ among each other as well as be different from shielded-metal arc welding. The extent to which adequate or inadequate access is provided could control the degree and facility to which an automated weld process could be considered which could in turn represent a significant production cost factor.

d. Overall Design. Direct overall design to minimize the probability of transverse (athwarthship) fractures. In some cases this may involve the use of material of superior toughness or the judicious incorporation of designs of special geometry or redundant structure which would interrupt a transversly running crack.

e. Avoid Excessive Welding. In some cases overwelding may result in the imposition of excessively high welding stresses.

f. Through Thickness Loading. Since most conventional hull steels are not provided with minimum specified through thickness properties, they may exhibit weakness under such a loading condition. Where through thickness loading in a structure cannot be avoided by a design modification, special materials with enhanced through thickness properties should be considered (SNAME, 1976).

g. Welding and Nondestructive Testing Symbols. Welding symbols are used to communicate a designer's and fabricator's requirements to those concerned with design, design review, and fabrication of a structure. While preliminary design may require few details of weld joints, the requirement for inclusion of more details increases as the development of plans progress through the preliminary design to contract design and working plan stages. Detail design plans, when used in the shipyard, should contain complete details of the welds and any nondestructive tests that may be required. When plans form the basis of a contract, omission of any special requirements in respect to extent of penetration, finish, post weld nondestructive test examination etc. could lead to disputes between the purchaser and fabricator. When such details are omitted in final fabrication plans of the shipyard, such omission may allow for inadequately penetrated, finished or inspected welds.

A system of symbols for welding and nondestructive testing has been developed which provides the designer with a means of communicating complete welding information on drawings (American Welding Society, 1976a). The generalized symbols shown in Fig. 17 which are universally applicable, may be used for this purpose; in many cases, only a few of the elements of the symbol are required for a particular application. Fig. 18 and 19 illustrate basic and supplementary weld symbols. Fig. 20 indicates their application to butt and fillet welds. A similar system is also available for specifying nondestructive testing requirements.

Section 8
Qualification Tests

8.1 General. Qualification tests in welding have a twofold purpose:

1. Weld procedure qualification tests to determine whether the welding process and procedure will produce welds of satisfactory soundness and properties;
2. Welder performance tests to determine whether an individual welder has the required skill to make satisfactory welds.

In the case of automatic welds, the ability of the machine operator may be determined. In the interests of economy and convenience many regulatory agencies have similar qualification requirements, and in most cases each may accept the qualifications of the other.

a. Procedure Qualification Tests. Procedures may be qualified on the basis of proven satisfactory use for similar work under similar conditions. In other cases, formal procedure qualification tests are required. When such tests are required, ABS minimum requirements usually call for two reduced section tension tests and four guided bend tests for each position involved, and in some cases, fillet weld tests. Where an approved filler metal is not used, additional all weld-metal tensile and Charpy V-notch tests may be required to establish the adequacy of weld deposit properties. Depending upon the application and the process, all weld-metal tension, Charpy V-notch impact, macro-etch, and hardness tests may be required for special high strength or low temperature steels or for certain processes such as for some electroslag welds. In the case of procedure qualification tests for welds for low temperature service, Charpy V-notch impact tests are usually required at the weld metal fusion line and at a distance of 1, 3, and 5 mm from the fusion line (American Bureau of Shipping, Annual), (IMCO, 1975), (U.S. Coast Guard, 1973).

b. Welder Qualification or Performance Tests. Welders are generally qualified on the basis of their ability to fabricate welds, with the procedures, welding positions and general type of base metals of concern, that will either satisfactorily pass guided bend tests or alternatively exhibit satisfactory soundness upon radiographic examination. Additional welder qualification tests may be required if there is a change of welding process, change of welding position, or doubt relative to the ability of the welder.

c. Production Weld Tests. In addition to the above, production welds may be subjected to nondestructive tests or sampled and tested for mechanical properties. Such a requirement may be imposed for certain applications such as welding for some low temperature service applications.

8.2 Production Controls. Supplementary to the mechanical weld tests, quality assurance is attained throughout production by means of appropriate checks and supervision to verify that the conditions, materials, and procedures of qualification tests are maintained during production. Visual examination is used to determine the quality of prefabrication fits, and final weld appearance and sizes. As required, weld soundness may be verified by nondestructive evaluation.

Section 9
Nondestructive Evaluation

9.1 General. Nondestructive evaluation is widely used in shipbuilding to assess the soundness of welds during construction and repair. A nondestructive testing schedule is frequently required as part of the plan approval submission for new designs or where new or unusual service requirements are anticipated. It may also be included in drawings where the designer desires to achieve increased structural reliability by specifying a greater extent of non-

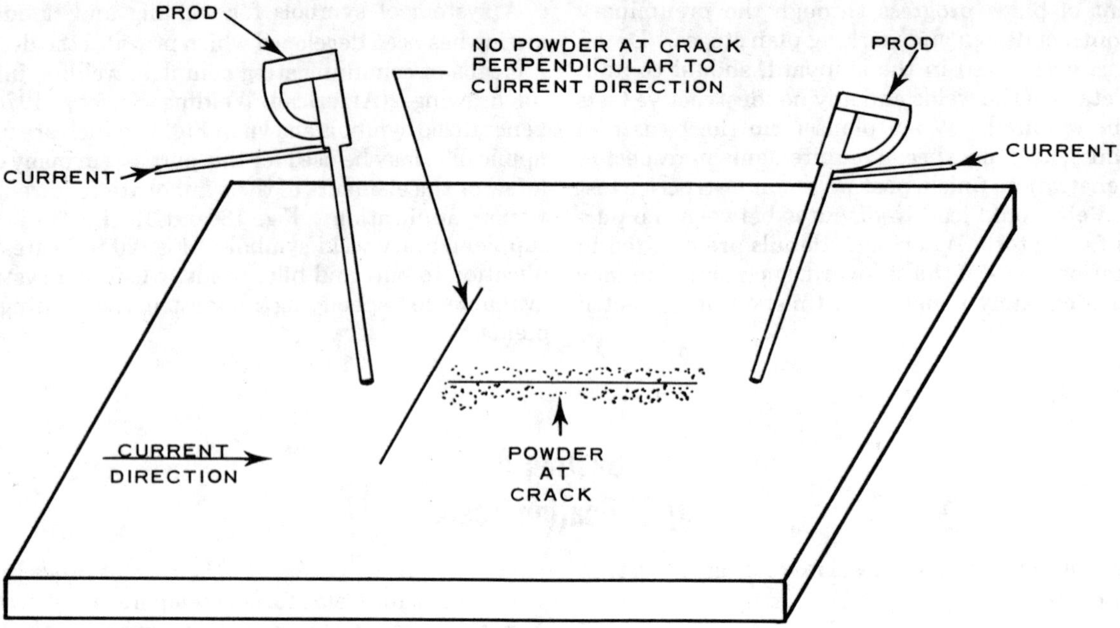

Fig. 21 Effect of current direction on magnetic particle sensitivity

destructive testing than would be achieved in normal inspection.

9.2 Visual Inspection. All welds are subjected to visual inspection, ranging from the casual inspection of the welder to a formal inspection by a qualified weld inspector or ship surveyor. Visual inspection, when properly accomplished, is considered by many to be one of the most important methods of quality assurance, since it provides important information not readily available from other methods. Visual inspection of the weld joint prior to welding will prevent welding in joints which have been improperly cleaned, prepared, or fitted. Completed welds are examined for surface soundness, regularity, geometry, and alignment. If deficiencies observed are reported promptly, timely corrective action of production operations can be effected.

9.3 Magnetic Particle Inspection. In this method, the base metal (steel plate) is magnetized as a superimposed electrical current is passed through, and finely divided magnetic particles are applied to the plate surface. A flaw at or near the surface will form a pair of magnetic poles which will act as a magnet and attract the applied magnetic particles. The technique is highly directional in sensitivity and, in accordance with the laws of magnetism, is most sensitive to linear flaws approximately parallel to the direction of the imposed current; it is practically insensitive to flaws perpendicular to current direction, see Fig. 21. Use of the technique requires a qualified technician, since many factors have to be considered such as surface conditions, selection of appropriate test conditions, interpretation of indications and the ability to differentiate between false and real indications. Requirements for qualification of operators (American Society for Nondestructive Testing, 1978), specifications relative to procedures (ASTM, 1976a), and examples of various types of indications (ASTM, 1976b) may be found in the publications of various technical societies. Magnetic particle inspection is used in shipbuilding to verify the soundness of root passes, intermediate weld passes, back gouged areas as well as completed welds and to inspect large steel castings and forgings. It is most widely used for the inspection of fillet welds, since such welds are not ordinarily subjected to ultrasonic or radiographic inspection.

9.4 Dye Penetrant Inspection. In this method a liquid penetrant of low surface tension is used to penetrate surface cracks. After excess penetrant is removed from the surface, a suitable developer is applied which draws the liquid penetrant from the crack and holds the wetted developer; the remainder of developer is freely released from the area. Depending on the system used, the indication will appear as an accumulation of developer around the crack or fissure and may be white, colored or, in some systems, fluorescent under an ultraviolet light. In shipbuilding, dye penetrant systems are used for surface inspection of non-magnetic materials such as non-ferrous alloys (aluminum) and corrosion resistant steels. They may also be used for inspection of steels in lieu of magnetic particle inspection. Requirements for qualifications of operator, technique for applications, and examples of typical indications are contained in the publication of various technical societies (ASNT, 1978), (ASTM, 1974a), (ASTM, 1976c).

9.5 Radiographic Inspection. Radiographic inspection is used for the examination for internal soundness of weldments, castings, and forgings. This method employs a source of electromagnetic radiation (such as x-rays from an x-ray machine, or gamma rays from a radioisotope source) capable of penetrating the thickness of material under investigation; a suitable film records the amount and pattern of radiation transmitted. Discontinuities in the material such as cracks, porosity, lack of fusion, as well as areas containing low density material (such as entrapped slag) will present less of a barrier to the radiation; the greater amount of radiation through such sites will be indicated by more dense (darker) areas on the film negative. Entrapped materials higher in density than base metal (such as tungsten in aluminum welds) will appear as less dense (lighter) areas.

a. Radiographic Examination Factors. Some of the more important factors to be taken into account in radiographic examinations include:

1. Safety—Both x-ray and gamma rays present potential hazards to the operator as well as to other personnel working in the area of exposure. In addition, special precautions and regulations are applicable to the storage, handling, and disposal of the radioisotope used as a gamma ray source.

2. Selection of Appropriate Source—The penetrating characteristics of the radiation source selected must be appropriate for the density and thickness of the material being examined. The sensitivity of the procedure is decreased substantially if the penetrating characteristics are excessive or insufficient.

3. Radiographic Technique—The penetrameter, which consists of a series of holes in strips of various thicknesses or a series of wires of graded diameters, is superimposed on the work during exposure and is indicated as a set of graded images on the final film. It provides a permanent indication of the sensitivity of the inspection.

4. Interpretation of Indications—The interpretation of the indications on the final radiograph in terms of type of internal imperfection, and decisions as to their conformity with a given specification, involve subjective judgements for which appropriate training and experience are required. This facet of the inspection is particularly difficult in borderline cases of acceptability where disagreements between recognized experts are not unusual.

b. Radiographic Requirements. Requirements for radiographic inspection of ship hulls are contained in the radiographic section of ABS publication "Nondestructive Inspection of Hull Welds." These requirements are only applicable to full penetration welds and are not intended for

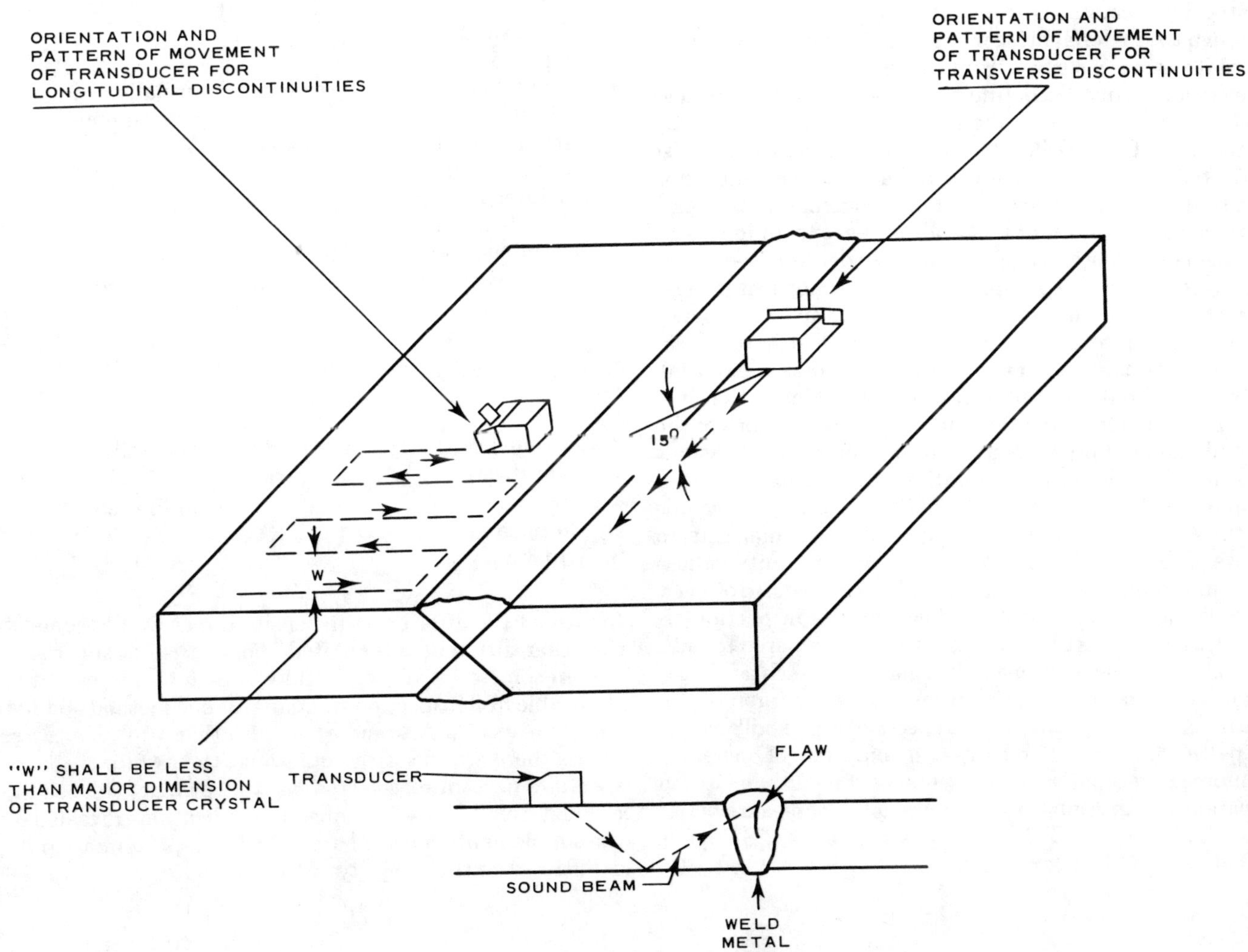

Fig. 22 Method of ultrasonic inspection of weld

corner or fillet welds. Other acceptance standards are also available such as those of the Navy and other classification societies. Information on qualification of personnel, details of testing techniques, and examples of various levels of radiographic indications of typical discontinuities are contained in References (ASNT, 1978), (ASTM, 1974b), and (ASTM, 1975). In European practice, it is common to use selected levels of indications as acceptance standards (International Institute of Welding, 1962). Radiographic inspection of ships is carried out mainly in important locations such as intersections of butts and seams in sheer strakes, bilge strakes, deck stringer and keel plates, and butts in and about hatch corners in main decks, and in the vicinity of breaks in superstructures. In other marine structures, it is mainly carried out in highly stressed areas, and at butt and seam intersections. Complete (100 percent) radiography is usually used only in specialized cases such as for liquefied gas containment or the shell of a submersible.

9.6 Ultrasonic Inspection. Ultrasonic inspection is used as an alternate to radiography for the examination of welds, castings and forgings; it is also used to measure thickness and detect laminations in plate. The principle of ultrasonic inspection is shown in Fig. 22. An ultrasonic impulse generated by a crystal is transmitted at a prescribed angle through the material being inspected; the impulse continues until it reaches a surface from which it is reflected back to the crystal. Any discontinuity in the path of the impulse will also act as a reflector; the size, orientation, and geometry of the discontinuity will determine the proportion of impulse reflected back to the crystal. For base metal examination the beam is usually transmitted perpendicular to the plate surface (compression or longitudinal wave technique) and for weld examination, angles of 45 to 70 deg are used (shear wave technique). Appropriate electronic apparatus measures the amplitude of the reflected impulse and the distance of the surface from which it was reflected. As in radiography, use of qualified personnel and procedures is essential. One of the limitations of ultrasonic inspection is that it is highly dependent upon the skill and interpretations of the technician, and errors relative to improper transmissions of sound impulse or interpretation of the signals received cannot usually be reviewed. In addition, permanent records of the sensitivity attained or of the indication, analogous to those of the penetrameter and actual discontinuity indication in a radiograph are not usually provided. However, because of the convenience and economy of inspection offered, as well as its greater sensitivity, as compared to radiography, to important linear discontinuities such as cracks, its use in shipyards as an alternative to radiography is increasing. Rules for ultrasonic inspection of ship hulls as well as pertinent references relative to qualification of operators, equipment, techniques, and acceptance standards are issued by appropriate regulatory agencies and technical societies (ASNT, 1968) (ASNT, 1978), American Bureau of Shipping, 1975b), (ASTM, 1974c).

9.7 Radiography vs Ultrasonic Inspection. Although radiography and ultrasonic methods are both accepted for inspection of hull welds, each has distinct advantages and limitations and their sensitivity to various types of discontinuities differ. Some of the principal differences are tabulated below:

Qualities	*Radiography*	*Ultrasonics*
• Sensitivity to Discontinuity—(cracks, linear slag, & imperfections)	less sensitive	more sensitive
• Generalized lack of fusion, slag inclusions	more sensitive	less sensitive
• Interpretation of indications	indicates shape & size	interpretations from indications may be more difficult
• Distance at indication from weld surface	not indicated	indicated
• Availability of review for sensitivity of measurement & interpretation of indications	permanent records available	usually not available
• Dependence on subjective skill & judgement of technician	less dependent	more dependent
• Safety	greater precautions required	
• Convenience, speed, cost	generally inferior	generally superior
• Range of applicable thickness	limited	unlimited

In view of the differences in sensitivity, as well as some of the other differences tabulated above, it is not surprising that in some cases a given weld will be found to meet the applicable hull inspection standard for one method and not meet the applicable standard of the other. In such cases, unless there are specific stipulations, the results of the inspection procedure selected as the primary inspection method usually governs, unless the indications revealed by the supplemental method are shown to represent a significant threat to the integrity of the structure.

Section 10
Miscellaneous Processes

10.1 Flame Cutting and Gouging. Flame cutting and gouging are used throughout shipbuilding to cut steel to required size, to prepare bevels for welded joints, and to remove metal as required for such operations as back-gouging to sound metal in preparation of the second welded side of the joint.

a. Metal Cutting or Removal. Metal cutting or removal is effected by the following sequence of events:

1. Using a cutting or gouging tip, the surface of steel in the vicinity of the cutting or gouging tip is heated to a *kindling* temperature by several relatively small oxy-gas flames (usually oxy-acetylene or oxy-propane).
2. A high pressure stream of pure oxygen is directed by the tip toward the heated surface. The effect produces ignition of the heated steel; the heat generated by combustion melts the adjacent iron which is swept away by the high pressure oxygen stream.
3. The oxidizing action generates sufficient heat to support combustion along the oxygen stream. Directing the oxygen stream perpendicular to the plate surface produces a square edge; appropriate bevels for joint preparations or fit are produced by positioning the cutting tip to direct the oxygen stream at the appropriate angle. Gouging is accomplished by directing the oxygen stream approximately parallel to the plate surface.

10.2 Effects of Flame Cutting and Gouging. Whenever flame cutting is used, slag produced in the process should be removed to prevent adverse effects on subsequent operations such as welding and painting. In the course of flame cutting or gouging, base metal in the immediate vicinity is heated above the transformation temperature and then cools as the cut progresses. As the strength and hardenability of the steel increases, there is a greater tendency for hardening to occur in the immediate vicinity of the cut surface. However, hardening of the heat affected zone of steels equivalent to the ordinary and higher strength ship steels is not critical and the affected metal need not be removed. The above is also applicable to straight and bevel cutting of quenched and tempered steels. However, flame gouging of quenched and tempered steels is not generally permitted, due to the adverse metallurgical effects associated with excessive exposure to heating.

10.3 Oxy-Gas-Iron Powder Cutting. While oxy-gas cutting is most commonly used for metals which will maintain oxidation (combustion) in the oxygen stream, modifications of the process have been utilized for other materials. For oxy-gas cutting of materials resistant to oxidation, such as stainless steels, an oxy-gas process is used wherein a stream of iron powder is fed around the oxygen stream; the combustion of iron powder providing the heat to melt the metal which is washed away by the high pressure oxygen stream.

10.4 Arc Cutting. In arc cutting, the melting or severing of metals is produced by the heat of an arc between an electrode and base metal. A variety of electrodes, including covered electrodes and carbon electrodes are used; in some cases a compressed gas such as air or oxygen is used either to wash away or to oxidize metal. A process commonly used in shipyards is an air-carbon arc method which is applicable to most metals. In this process the melted metal is blown away by high velocity jets of compressed air. When arc-air gouging is employed, techniques which minimize carbon build-up should be used. Weld joints prepared by arc-air gouging may require additional preparation by chipping or grinding prior to welding to minimize the possibility of excessive carbon on the scarfed surfaces; this precaution is particularly pertinent for high strength steels.

10.5 Forming and Fairing. Hot or cold forming methods are used to produce plating of desired contour. When plating is subjected to excessive strain during cold forming, degradation of toughness properties may occur. In such circumstances a stress relief heat treatment is used to eliminate degradation resulting from cold forming. Misalignment and excessive distortion may be corrected by the appropriate application of loads by hydraulic jacks, levers, etc. However, fairing by heating or flame shrinking may also be used (Patee et al, 1969), (Rothman, 1973). When heat fairing is used on higher strength carbon steels and high strength low alloy steels, particular care should be taken to avoid treatments which will degrade strength and/or toughness. The suitability of the heat straightening procedure should be verified by an appropriate procedure approval trial prior to its use in production. Applications by forming and fairing to ship construction are discussed at length in Chapter XVI.

References

Alia, Bruno L. et al (May 1976), "Material Selection, Welding, and Nondestructive Testing Guidelines for Offshore Mobile Drilling Units," *Proceedings* of Offshore Technology Conference, OTC 2647.

American Bureau of Shipping (Annual), *Rules for Building and Classing Steel Vessels,* Section 30.

American Bureau of Shipping (1975a), *Rules for Building and Classing Aluminum Vessels.*

American Bureau of Shipping (1975b), *Rules for Nondestructive Inspection of Hull Welds.*

American Bureau of Shipping (1978a), *Rules for Building and Classing Reinforced Plastic Vessels.*

American Bureau of Shipping (1978b), *Rules for Offshore Mobile Drilling Units.*

American Bureau of Shipping (1979), *Rules for Building and Classing Underwater Systems and Vehicles.*

American Concrete Institute (1978), Guide for the Design and Construction of Fixed Offshore Structures, *ACI Journal.*

American Institute for Steel Constructors (1973), "Commentary on Highly Restrained Welded Connections," *AISC Engineering Journal,* 3rd Quarter.

American National Standards Institute (1977), "Standard Specification for Aluminum Alloy Sheet and Plate," ANSI/ASTM B-209-77.

American Society for Nondestructive Testing (1978), "Guidelines for the Qualification and Certification of Nondestructive Testing Personnel," Recommended Practice No. SNT-TC-1A, Evanston, Illinois.

American Society for Testing and Materials (1968), "Evaluating Performance Characteristics of Pulse-Echo Ultrasonic Testing Systems," ASTM E317-68.

American Society for Testing and Materials (1974a), "Recommended Practice for Liquid Penetrant Inspection Method," ASTM E165-74.

American Society for Testing and Materials (1974b), "Recommended Practice for Radiographic Testing," ASTM #94-68.

American Society for Testing and Materials (1974c), "Ultrasonic Contact Examination of Weldment," ASTM E164-74.

American Society for Testing and Materials (1979), "Reference Radiographs for Steel Fusion Welds," ASTM E390-75.

American Society for Testing and Materials (1976a), "Methods for Magnetic Particle Inspection," Dry Powder, ASTM E109063.

American Society for Testing and Materials, (1976b), "Reference Photographs for Magnetic Particle Indications on Ferrous Castings," ASTM E-125-63.

American Society for Testing and Materials (1976c), "Reference Photographs for Liquid Penetrant Inspection," ASTM E433-71.

American Welding Society (1976a), "Guide for Steel Hull Welding," AWS D3.5-76.

American Welding Society (1976b), "Symbols for Welding and Nondestructive Testing," AWS A2.4-76.

American Welding Society Handbook (1978), Seventh Edition.

Anderson, A. (1975), "Prestressed Concrete Floating Structures," SNAME Spring Meeting.

Australian Welding Research Association (April 1976), "Control of Lamellar Tearing," AWRA Technical Note 6.

Copper Development Association (1976), "The Potential Market for Copper Nickel Ship Hulls through 1980."

Government Printing Office (1947), "Final Report of a Board of Investigation to Inquire into the Design and Methods of Construction of Welded Steel Merchant Vessels," 15 July 1946, Washington, D. C.

Inter-Governmental Maritime Consultative Organization (1975), "Code for the Construction and Equipment of Ships Carrying Liquified Gases in Bulk," Resolution A.328(IX), November 12, 1975, London.

International Institute of Welding (1962), "Collection of Reference Radiographs of Welds in Steel."

Jordan, C. R., and C. S. Cochran, (1978), "In-Service Performance of Structural Details," Ship Structure Committee Report SSC-272.

Jubb, J. E. M. (1971), "Lamellar Tearing," Welding Research Council Bulletin No. 168.

Manxolillo, J. L., E. A. Thiele, and A. H. Tuthill,(1976), "CA706 Copper-Nickel Alloy Hulls: The Copper Mariner's Experience and Economics," SNAME Annual Meeting, New York.

Mishler, H. W. (1976), "Preventing Delayed Cracks in Ships Welds, Part II," Ship Structure Committee Report SSC-262.

Morgan, R. J. (1977), "History and Technical Development of Concrete Ships," *The Naval Architect*, January.

Ott, C. W., and D. J. Snyder, (1974), "Suggested Arc Welding Procedures for Steels Meeting Standard Specifications," Welding Research Council Bulletin 191.

Patee, H. E., R. M. Evans, and R. E. Monroe, (1969), "Flame Straightening and its Effect on Base Metal Properties," Ship Structure Committee Report SSC-198.

Rolfe, S. T., D. M. Rhea, and B. O. Kuzmanovic, (1974), "Fracture-Control Guidelines for Welded Steel Ship Hulls," Ship Structure Committee Report SSC-244.

Ross, R. B. (1972), *Metallic Materials Specification Handbook*, John Wiley & Sons, New York.

Rothman, R. L. (1973), "Flame Straightening Quenched and Tempered Steels in Ship Construction," Ship Structure Committee Report SSC-247.

Scott, R. J., and J. H. Sommella, (1971), "Feasibility Study of Glass Reinforced Plastic Cargo Ship," Ship Structure Committee Report SSC-224.

Shaw, S. T., and F. J. McGarry, (1971), "Griffith Fracture Criteria and Concrete," J. Engr. Mechs. Division, *Proceedings, ASCE*, Vol. 97.

Society of Naval Architects and Marine Engineers (1974), "Aluminum Fire Protection Guidelines," SNAME, T&R Bulletin No. 2-21.

Society of Naval Architects and Marine Engineers (1976), "Guide for High Strength and Special Applications Steels for Marine Use," SNAME T&R Bulletin No. 2-20.

Society of Naval Architects and Marine Engineers (1977), "Guide for Quality Assurance Fiberglas Reinforced Plastic Structures," SNAME T&R Bulletin 2-23.

Tuthill, A. H. (1945), "Concrete Operations in the Concrete Ship Programme," *ACI Proceedings*, Vol. 16.

United Nations (1972), "Boats from Ferrocement," Publication E 72.II B.23.

U. S. Coast Guard (1973), "Marine Engineering Regulations," Subchapter F, CG-115.

CHAPTER IX

Daniel T. Mallett
Norman J. Thompson
Norman W. Lemley

Hull Outfit and Fittings

Section 1
Closures for Hull Openings

1.1 Introduction. Closures to openings in the main hull, deck houses, tanks, subdivision bulkheads, and elsewhere affecting the safety of the ship, or persons aboard, are subject in the United States to regulations of the U. S. Coast Guard, the American Bureau of Shipping, and in some instances the U. S. Public Health Service. The applicable standards of construction are adequately described in these regulations and thus will not be duplicated here. However, the sketches and descriptive material which follow are representative of standard practice approved by the regulatory bodies.

Requirements for closures may be divided into two general categories:

1. Closures which are normally opened and closed while at sea, such as access and escape doors in accommodation and work spaces;

2. Cargo space closures including hatch covers, sideports, sluice gates, conveyor belt gates, etc., which normally remain secured while at sea.

a. Closures Operated While at Sea. Such closures; i.e., doors, hatches, scuttles and manholes are designed to the requirements of ABS and USCG, of strength equivalent to the surrounding structure. Hinged doors are fitted in the upper part of a vessel, equivalent to USCG Class 1—Hinged doors. Sill heights, coaming heights, etc. are prescribed in the International Load Line Convention and in the ABS rules. Sliding doors, Class 2 or Class 3 are fitted in the lower portions of the vessel.

b. Closures Secured While at Sea. Cargo hatches, sideports, bulkhead doors, etc. must meet USCG subdivision requirements and must be classed by ABS as closures in a watertight bulkhead. Gasketing and dogging must thus be designed to provide a 100 percent watertight seal. The only instances where absolute watertight requirements are not strictly followed are in the case of tanker free-flow sluice gates which are Drop Tight, Class 2 and Class 3 sliding doors and watertight conveyor gates on self-unloading vessels, which are built to USCG leakage limitation requirements. Even though not strictly classified as watertight, owners and regulatory bodies, including Marad, now favor use of conveyor-belt doors, which are designed to reduce the maximum leakage when closed and secured around the rubber conveyor belts to approximately 25 percent of the capacity of the bilge system serving the flooded compartments.

1.2 Watertight Doors. Watertight doors include those required for personnel access on weather decks and through watertight bulkheads and those required to move cargo into or out of the ship.

a. Deck House Access. Hinged watertight metal doors are fitted at all exterior deck-house openings on the weather deck level as required by the regulatory bodies for the particular ship and service. This type of door is fitted with a rubber gasket and secured with suitably spaced hand-operated dogs, which may be manipulated from either side of the door, Fig. 1. It has a minimum sill height of 610 mm (24 in.) or less as approved, depending on the space protected, and the height above the assigned load waterline and distance from the bow (USCG Rules, Current),[1] (American Bureau of Shipping, Annual).

b. Watertight Bulkhead Access. Hinged watertight doors also are fitted in subdivision bulkheads below the ship's bulkhead deck. The number, location, and method of closing such doors are determined by the regulatory bodies in accordance with the characteristics, number of passengers, and service of the ship (USCG Rules, Current), American Bureau of Shipping, Annual).

Fig. 2 shows a mechanically operated, sliding watertight door which is used in place of a hinged door as required by USCG Rules, when the door sill is less than the prescribed distance above the subdivision load line, or when the size of the opening is too large to make a hinged door practicable. If a remote controlled door, located in the lower part of the ship, is to be opened while at sea, the sliding watertight door is mandatory. The most common case is a door from the machinery space to the shaft alley which must be operated from above the bulkhead deck.

c. Side Port Access, Stores, and Fueling Doors. Side ports are used to provide access to a cargo hold, fueling station, personnel or vehicle embarkation, and stores spaces

[1] Complete references are listed at end of chapter.

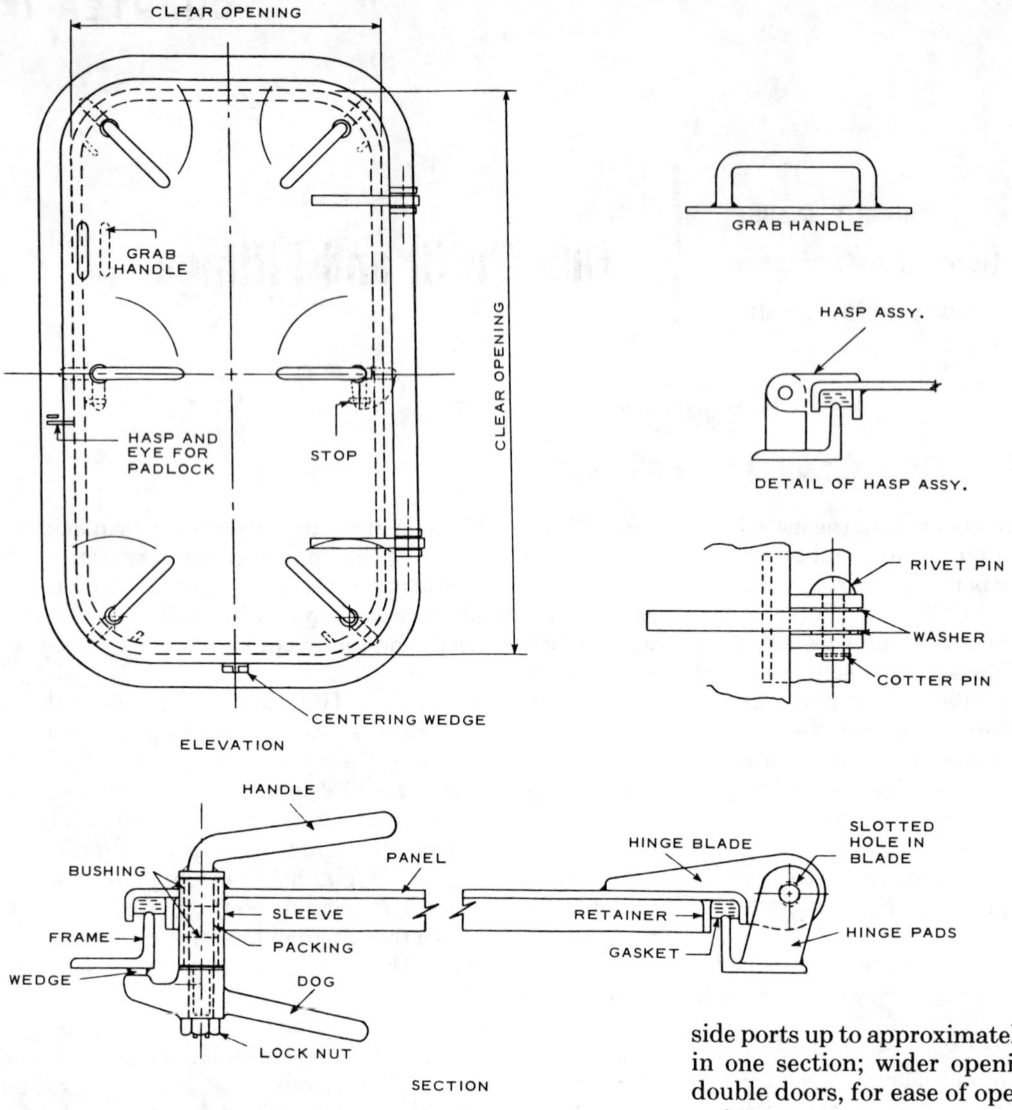

Fig. 1 Hinged watertight door

where overhead access cannot be fitted conveniently. The USCG requires side port sills to be located above the deepest subdivision load line.

d. Side Port Cargo Doors. Side ports ordinarily are fitted at the upper 'tween-deck level. Cargo is run through the port by truck, conveyor, or other means and then lowered to the decks below by elevator, vertical conveyor, or a hoist acting through a hatch opening. As a general rule, the type of cargo thus handled is limited to small packaged units such as canned goods and refrigerated cargo, easily moved on conveyors, etc. Construction in way of a side port opening must be the equivalent of the adjacent ship side structure. Thus the opening in the shell must be adequately compensated, by heavier structure, as required by the classification society. Hinged side ports can swing either inboard or outboard and are secured by closely spaced dogs which force the door against a gasket making a watertight seal, Fig. 3. Recessed side ports which swing inboard are less vulnerable to damage from docks, or barges tied alongside, but must be reinforced with *strongbacks* in addition to dogs. Hinged side ports up to approximately 1.2 m (4 ft) in width are built in one section; wider openings generally are fitted with double doors, for ease of operation.

Other variations in side port construction, particularly for large-size openings, include horizontal and vertical sliding power-operated doors. Such doors are most frequently used in Ro/Ro vehicle carriers where door size and deck space limitations often do not permit use of swinging hinged doors.

e. Stern Port Cargo Doors. On Ro/Ro trailer ships and military vehicle carriers, a stern loading port is usually fitted, off center or at the center line of ship through a flat, transom stern. The stern port may, if the available pier facilities permit, be used in combination with the side ports to minimize *in-port* time by simultaneous loading and discharge.

Stern port doors generally are hinged at the loading deck level and are secured against rubber or neoprene gaskets around the boundaries of the stern opening to form a watertight closure. The door also serves as a loading ramp, when lowered down onto the pier, and the inside or road surface thus is fitted with a non-skid surface such as diamond plate or closely spaced welded cleats. The raising and lowering of the stern port is normally accomplished by multi-part tackles actuated by special deck winches with

Fig. 2 Mechanically operated sliding watertight door

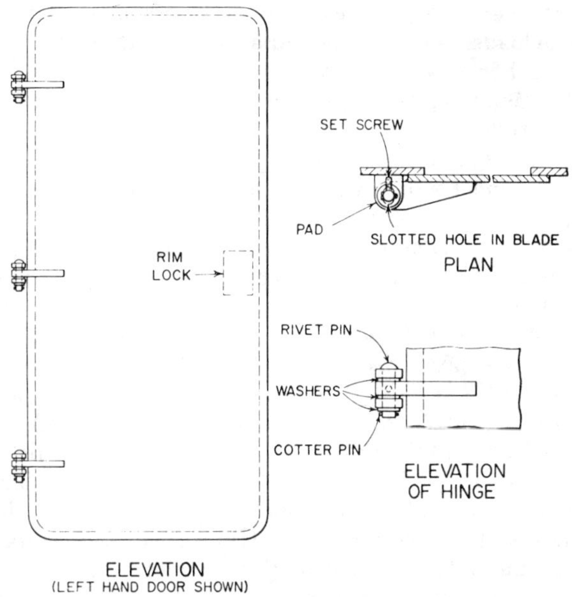

Fig. 4 Nonwatertight steel door

Fig. 3 Hinged cargo loading side port

controls suitably located at the stern rail. To reduce the slope of the ramp during vehicle loading, it is frequently necessary to make the stern port door somewhat longer than required to close the opening in the ships stern. Thus in the closed position the end of the door, or ramp, projects above the weather deck. Since stern port doors must withstand vehicle loads, as well as the impact of seas, they tend to become rather massive structures.

1.3 Miscellaneous Type Doors. These include gastight, weathertight, and non-watertight doors.

a. Gastight Doors. These are of lighter construction but must meet more exacting tests for tightness than watertight doors and are usually limited to the hinged type suitable for installation in the boundary bulkheads of a space containing poisonous or objectionable fumes, such as a battery room.

b. Weathertight Doors. Wood weathertight doors may be fitted at exposed deck-house openings where a watertight door is not required by regulations. Such a door usually consists of solid oak or teak planks, about 5 cm (2 in.) thick, edge-bolted, and so built and drained as to exclude driving rain and spray, and capable of withstanding the impact of a boarding sea. A fixed light should be provided in all doors opening to the weather deck. Hinged wooden doors are most commonly used but sliding wooden doors are also frequently used for convenience of operation. The openings to deckhouses are usually recesses inboard of the house side for protection from the weather. Hinged exterior doors always should be fitted with the hinges on the forward, outboard side so that wind or sea striking the door will tend to close it.

Steel or aluminum doors with painted finish are more frequently used in place of wood doors, especially where appearance is not a prime consideration. They perform the same function and must meet the same general requirements as outlined for wood doors. They are similar in construction to steel watertight doors, but are of somewhat lighter construction and have fewer securing dogs. Stainless steel doors are easier to maintain than ordinary steel doors, but are considerably more expensive.

c. Non-Watertight Steel Doors for Stores and Shops. Non-watertight doors of dished construction or made up of plate 4.8 to 6.4 mm ($^3/_{16}$ to $^1/_4$ in.) thick, suitably stiffened, are installed at entrances to stores spaces and working areas. Either type of door can be secured with a padlock rather than a rim lock, Fig. 4.

d. Non-Watertight Joiner Doors for Accommodations. Metal joiner doors are used extensively for staterooms, toilets, and living quarters where fire-protection regulations prohibit the use of joiner wood doors. Such doors are formed of light gage sheet steel and are specially insulated for soundproofing, to prevent *drumming,* and to meet fire resistance requirements where necessary. Magnetic hold-back devices are used on doors, remote controlled where required by the rules for fire control. Door hardware should be of substantial steel or bronze construction. Ajar hooks are usually fitted on stateroom doors to hold the doors partly open for ventilation.

1.4 Windows, Air Ports, and Port Lights. Rectangular windows with cast or extruded bronze frames or aluminum frames with steel retaining clips, are usually fitted in lieu of air ports, in the upper levels of a superstructure, and the pilot house. They are fitted with wire inserted heat treated plate glass, at least 6 mm ($^1/_4$ in.) thick. The glass may be tinted to exclude solar glare in all locations except the wheelhouse. The sliding windows descend into a metal pocket or drain pan below the window and the pan is drained to the exterior. Two or more of the front windows in the pilot house are usually fitted with wipers, or rotating disc inserts knows as *clear views,* for visibility in rain or snow.

Hinged air ports, complete with air scoops and screens, are fitted in living quarters. Air ports in the shell of the ship must be fitted with deadlight covers. Air ports with sills located below the bulkhead deck are under the master's control (USCG Rules, Current) and can be opened only by personnel having access to a special key. Above this level, air ports can be opened at will by the passengers or crew, except for air conditioned spaces where fixed lights or fixed windows are fitted.

Air ports and fixed lights generally have frames constructed of cast bronze or aluminum fitted with steel retaining clips. They are fitted with plate glass at least 6 mm ($^1/_4$ in.) thick, heat treated in exposed locations on seagoing vessels, and of regular glass on inland vessels in locations and on runs where there is no danger of water impact from the seas. Large view windows in passenger vessels in lake, bay, and sound service are commonly fitted with double pane windows with an air space to minimize heat loss. Screens are made of bronze and air scoops made of galvanized sheet iron. Deadlight covers are made of cast bronze, aluminum, or steel; they hinge upward and inboard, in which position they are secured with a keep chain. Pivoted type air ports are occasionally fitted in deck houses, to provide natural ventilation.

Eyebrows are fitted over air ports on the outside of the hull and deck houses to drain water away from the opening. If the air ports spigots project at least 19 mm ($^3/_4$ in.) past the shell, eyebrows may be omitted. Drip pans, with exterior drains, are provided under air ports.

Any portlight having a sill below a line 2.5 percent of the beam above the deepest subdivision load line must be fitted as a fixed light (USCG Rules, Current). Fixed lights and/or fixed windows are commonly fitted in air-conditioned accommodation spaces.

1.5 Access Hatches, Companionways, and Manholes. Companionways are used to provide a sheltered landing when the space below is reached by means of an inclined ladder or stair. A hinged watertight steel door with a 460 or 610-mm (18 or 24-in.) sill and a fixed light, as described in Subsection 1.2, is standard for a weather deck companionway. Full standing headroom is provided over the landing inside the companionway.

Hatches are fitted on weather decks over bosum's stores, work spaces, steering gear, and other spaces where access from the weather deck is essential and/or entrance cannot be made conveniently from below deck. For ease of operation the covers of access hatches, if frequently used, are generally fitted with counterbalance weights or springs with

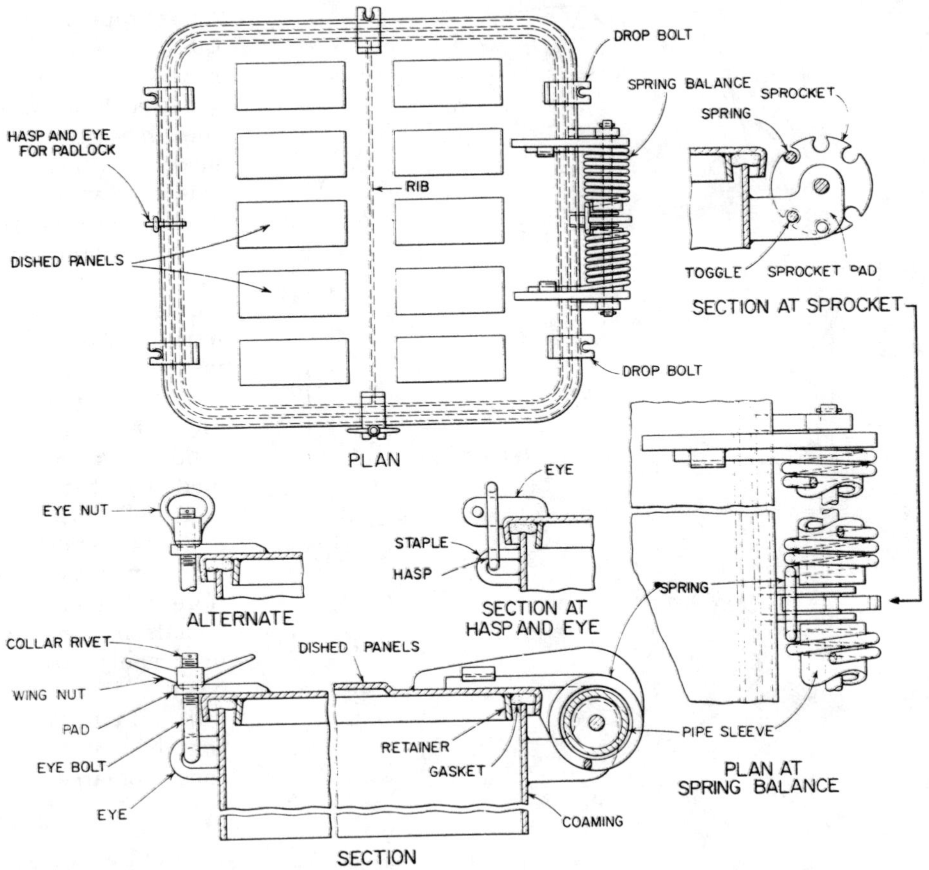

Fig. 5 Raised, hinged, watertight access hatch cover (stores) spring balanced.

hold-back hooks to prevent them from closing accidentally and injuring personnel. The covers, Fig. 5, are constructed of steel plate, fitted with dogs and flexible gaskets. Weather deck hatch coamings are 610 mm (24 in.) high or less, depending upon the location, as specified in the USCG regulations.

Manholes are fitted for access to tanks, cofferdams, and void spaces. To insure adequate means of escape and to facilitate ventilation, it is standard practice to provide two manholes to each tank or void space, located at diagonally opposite corners.

Manholes are either oval or round depending upon the configuration of the space available for the installation. The covers are bolted down on flexible gaskets for access to ballast tanks and other spaces, where required to be watertight. Neoprene gaskets are used for oil-tight applications. The minimum clear opening for a round hole should not be less than 460 mm (18 in.) in diameter, and for an oval hole, 380 by 580 mm (15 by 23 in.). Manholes can be flush, raised, hinged, or recessed.

The flush manhole cover, Fig. 6, is a flat steel plate secured with studs threaded or welded into a reinforcing ring welded around the edge of the opening in the tank top or bulkhead. Raised manhole covers are bolted to the flange of an inverted angle welded around the edge of the opening. The boundary angle prevents liquids or dirt spilling into the tank when the cover is removed. The bolts through the angle flange are easily replaced and are less liable to damage than studs.

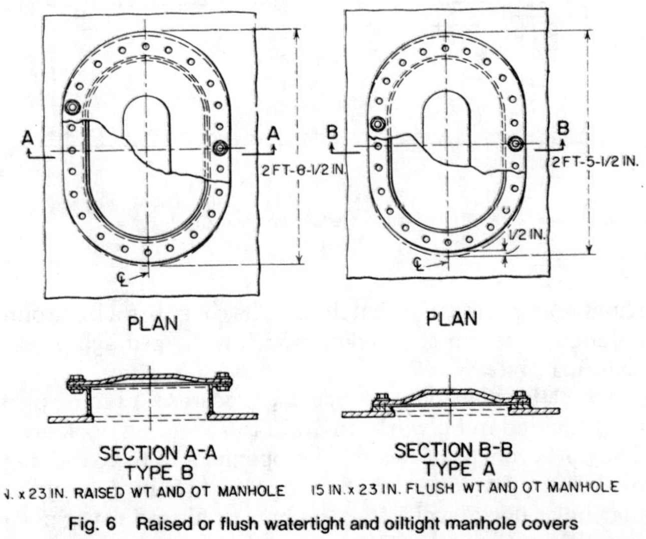

Fig. 6 Raised or flush watertight and oiltight manhole covers

Portable manhole guards, made of 6 to 10- mm (1/4 to 3/8-in.) plate, are fitted over manholes in cargo holds, work spaces, etc. where the cover might be damaged, and are secured by bolts or studs.

Hinged manholes are used for access to stores spaces where a door or larger hatch opening is not justified by frequent use. The height of coaming is governed by USCG

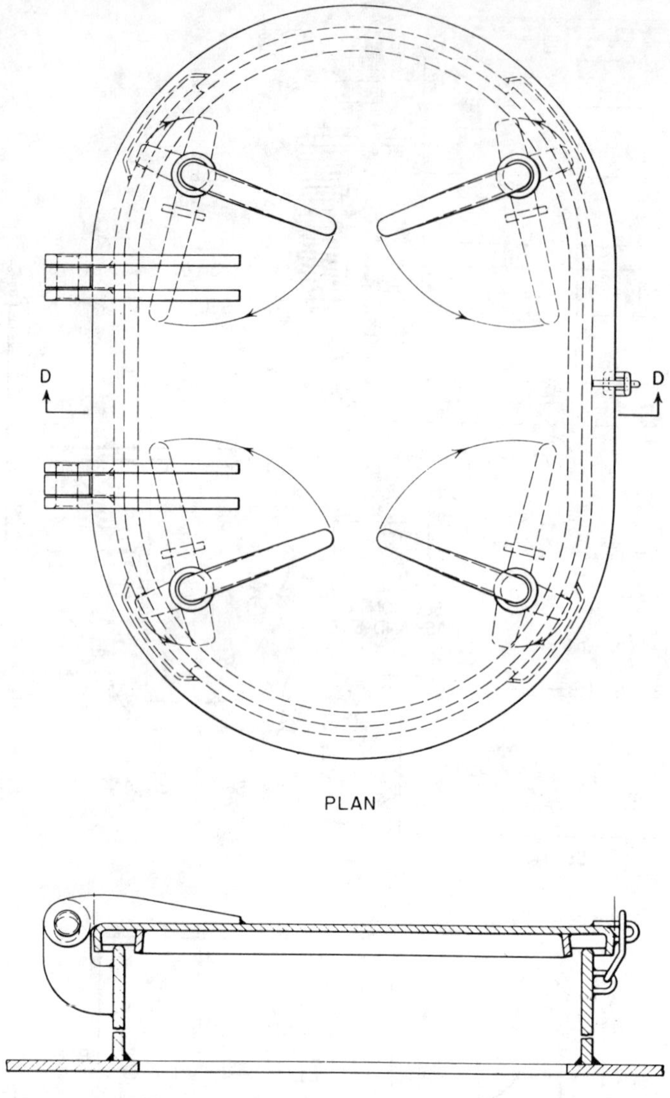

Fig. 7 Raised, hinged watertight manhole cover

Rules applying to deck hatches. The cover has a lip around its edge to retain the gasket, which is dogged against the coaming plate, Fig. 7.

1.6 Miscellaneous Hull Opening Closures. Freeing ports are provided in bulwarks to drain the enclosed deck areas. The ports consist of rectangular openings 15 to 20 mm high by 0.6 to 1.2 m long (6 to 8 in. by 2 to 4 ft) in the bulwark plating at deck level. Vertical bars or hinged cover plates usually are fitted.

Rope scuttles are fitted through the weather deck on ships where a stores hatch to the stowage space below is not available or is not suitably located for passage of heavy mooring lines. The construction of a rope scuttle is basically similar to a hinged raised manhole, as discussed above, except that it need only to be large enough to permit easy passage of a steel or fiber rope-end eye splice. Also, the inside surfaces of the coaming are rounded to ease the passage of ropes as well as to avoid damaging them. To facilitate passing heavy wire ropes, rollers are frequently fitted inside the scuttle coaming, similar to a roller chock.

Through-hull mooring ports are watertight closures in the shell plating, used in mooring arrangements where the mooring winches are located below the weather deck or where the weather deck would be too high, in relation to the mooring bitts on the pier, to afford a proper slope to mooring lines below the horizontal.

Through-hull mooring ports are of generally similar construction to the rope deck scuttles described above, but hinge inboard and are dogged tight against a gasket from the inside. Heavy roller fairleads are built integral with the port opening since the mooring lines are lead through the opening, directly to the mooring winches.

1.7 Cargo Hatch Covers. Wood and steel pontoon covers made weathertight by means of tarpaulin covers are now essentially obsolete. Hatch covers on modern vessels generally consist of single-piece lift-off pontoons, hinged pontoon sections which fold to the open position, rolling covers, (fore and aft or athwartship), and cover sections which stow on a drum. In all cases the covers are made weathertight by dogging with gasketing against a hatch coaming seal bar.

a. Single Piece Pontoon Covers. Pontoon covers for containerships and modern break-bulk cargo ships usually consist of large steel sections of sufficient size to cover the

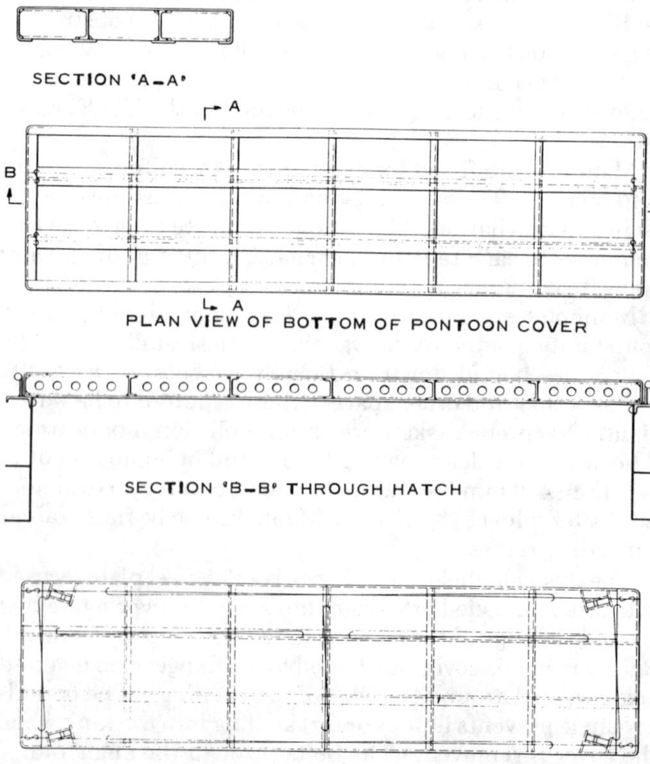

Fig. 8 Cargo hatch covers, steel pontoon

Fig. 9 Direct-pull weather deck hatch covers

entire hatch opening in one piece. The pontoon is open on the under side and is designed to take the span of the hatch width without the aid of auxiliary beam supports. Lifting fittings are attached to each pontoon which is then handled by ship or shore cranes.

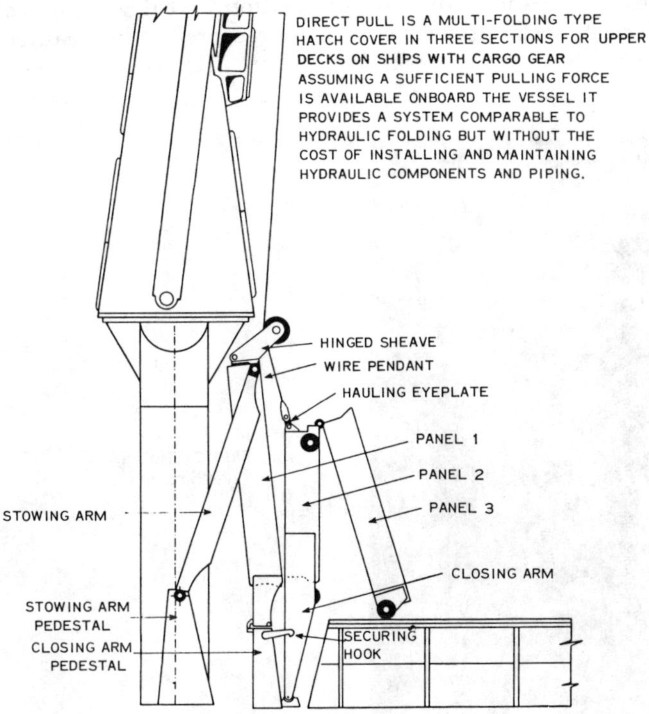

Fig. 10 Direct-pull weather deck hatch cover

Large, single-piece pontoon covers are preferred aboard container and heavy-lift ships where special crane facilities are available for handling large units. In both cases, opened cover sections are stowed on top of adjacent or nearby hatches which need not be worked simultaneously with cargo gear or alongside on the pier. Construction details of this type of cover are generally similar to that of the steel pontoon cover shown in Fig. 8.

b. Mechanical Covers. Mechanically operated hatch covers are of two basic types; those used on weather decks which are weathertight and either mounted on raised coamings, or flush with the weather deck, and those used on lower decks which are non-weathertight and flush with the surrounding deck area. Typical mechanical hatch covers are illustrated in Figs. 9 and 10.

Both of the above types of covers are fitted with natural or synthetic rubber gasketing, cross joint wedges at the panel joints and quick acting dogs around all sides to attain a weathertight seal. Because of the large deck areas occupied by the cover when the hatches are open, the side and end rolling covers are normally fitted only on dry bulk and ore-bulk-oil carriers (OBO). Fig. 11 illustrates a hatch crane used on Great Lakes bulk carriers to handle large, heavy hatch covers of this type. Automatic mechanical dogging mechanisms are frequently built so as to be integral with the hatch closing mechanism on large hatch covers, thus eliminating manual dogging.

c. Rolling Covers. End or side-rolling hatch covers, of similar construction, are usually arranged to split at the center, with half-rolling to each side or end of the hatch on permanently fitted wheels and rails. After undogging, but

Fig. 11
Hatch crane

before rolling, the covers must be raised by special jacking mechanisms built into the cover or coamings, or by portable hand jacks, in order to clear the gasketing from the seal bar on the hatch coaming. The cover is then rolled to its stowage location, as shown in Fig. 12.

d. Hinged and Folding Covers. Ships handling break-bulk general cargo or containerized cargo, frequently have hinged, power-actuated steel hatch covers in order to obtain fast economic operation. Weather deck hatch covers of this type are constructed with integral gasketing at all cross joints and the periphery which makes a weathertight seal when the covers are in position and dogged down. 'Tween deck hatch covers are not required to be weathertight, thus have no gasketing, and are fitted flush with sur-

Fig. 12 Rolling hatch covers

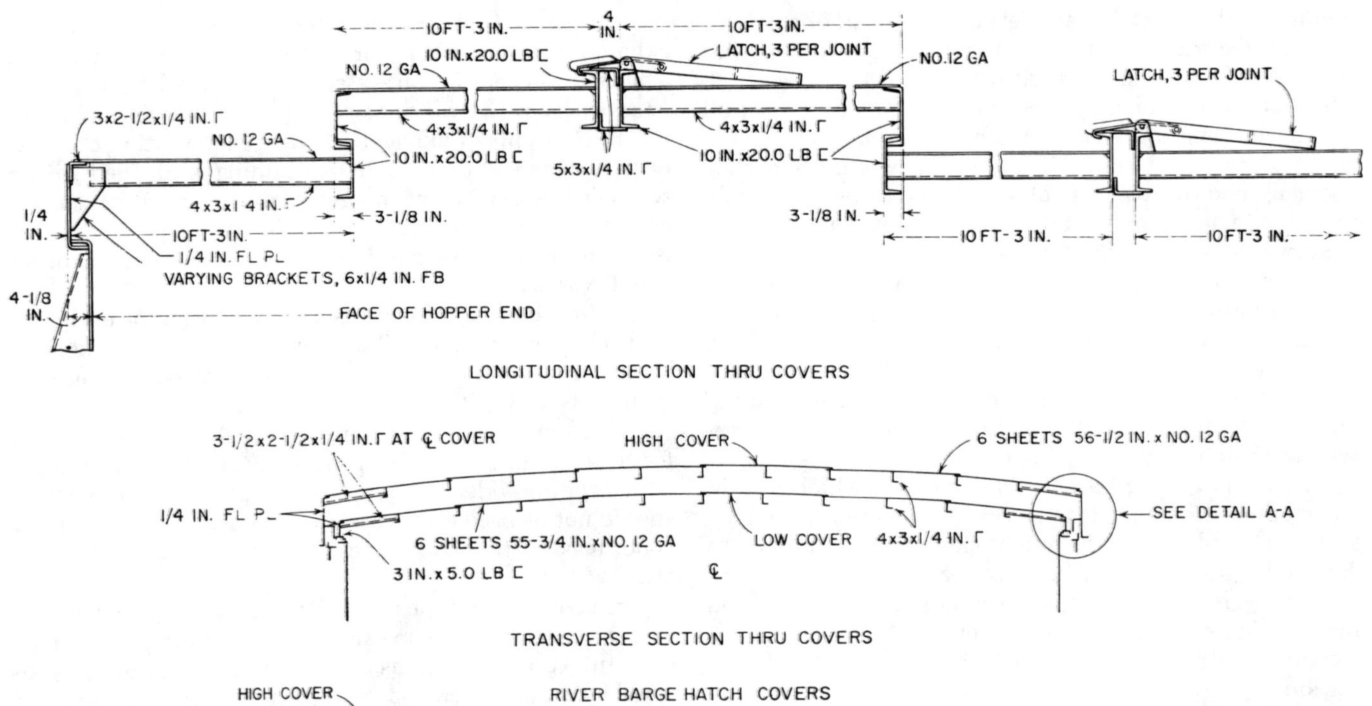

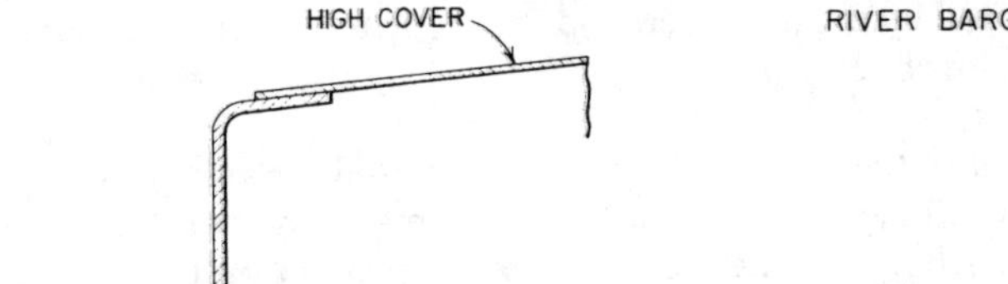

Fig. 13 River barge hatch covers (weathertight)

rounding deck area to facilitate vehicle handling and the use of forklift trucks for moving cargo to and from the hatch square and wings of the hold.

Hinged, folding hatch cover installations are made up in a multiple number of panels. The number of panels is dependent upon the length of hatch opening and the horizontal and vertical clearance at hatch ends in which to accommodate the cover panels when stowed in the open position. The panels normally operate in two groups, forming pairs. The panels at the fore-and-aft hatch extremities are hinged to the ship's hatch-end structure and the intermediate panels are connected to the end pairs by hinges. The parting joint can be arranged so that half or two thirds of the cover section fold to one end of the hatch independently of the remainder, which fold to the opposite end of the hatch. Selective, partial, opening of the covers can thus be achieved to permit access to cargo below without having to restow cargo on the remainder of the hatch cover, destined for another port.

e. Cover Actuating Mechanism. Hinged folding hatch covers externally mounted on the ship's decks or coamings, are actuated by hydraulically powered mechanisms accommodated within the hatch covers, or by wire ropes operating through fairleads to a cargo falls hook. The former are driven by one or more electro-hydraulic power stations centrally located near the respective hatchways.

In order to minimize the installation of hydraulic interconnecting piping aboard ship, compact built-in electro-hydraulic power units are sometimes fitted within the hatch covers themselves. Hydraulic control stations are usually located on winch platforms and frequently on each deck level to provide the operator with a clear view of each hatch cover during operation. As might be expected, the hinged folding hatch covers are generally more expensive than the single-piece types previously described. Hydraulically powered multileaf covers are more expensive than non-hydraulically powered types, but are preferred by ship operators for their ease, flexibility, and convenience of operation. Mechanical dogging devices, make the closing and opening of these hatches entirely automatic in operation.

f. Single Pull Cover. Single pull covers for weather deck hatchways consist of multiple, interlocking, gasketed, rolling sections, which by application of a continuous pull from a wire rope or driven electro/hydraulically by chain are readily drawn to one or both ends of the hatch, where they disengage and tilt to a vertical position and stow compactly on a rack. Closing action follows a reverse sequence. The primary advantages of this type of cover are speed of opening and closing, plus the unlimited length of hatch to which they can be applied. As in the case of end and side-rolling covers, they must be jacked off the coaming gasket before rolling. More sequenced operations and external mechanisms are therefore required than is the case with hinged folding covers, but this installation is, in most instances less expensive than hydraulic folding covers.

g. Drum Stowing Covers. Hatch covers which stow in the open position on a drum, at either the fore/aft end or/ port/starboard side of the hatch opening, can be fitted to ships or barges. They can be fitted for weathertight service. The undogging/uncleating from the closed to open position and dogging/cleating from the open to closed position is accomplished automatically with the opening/closing operation.

h. Watertight, Oiltight, and Special Type Covers. Small watertight or oiltight cargo hatch covers are fitted at hatch openings of holds which are used alternatively for dry and liquid cargo, or water ballast. To withstand the internal pressure, such hatch covers are fitted with synthetic rubber gaskets and the securing bolts are more closely spaced than typical for weather deck weathertight hatches. Tonnage regulations prescribe the maximum size of hatch openings in ballast tanks which are permitted to be exempted from tonnage measurement. On tankers and OBOs a larger oiltight hatch is generally permitted.

Various types of patented hatch covers, both oiltight and watertight, are in use. A special type of steel cover developed for use on river barges, is shown in Fig. 13. This type of cover is described as being raintight, being intended primarily to exclude driving rain, since boarding seas are not encountered in river service.

1.8 Ventilation System Terminals. Weather deck ventilation fittings, or *terminals,* are used to provide a ship with adequate air intakes and discharges, suitably located and protected from boarding seas. Ventilation systems are described in Chapter XIII.

All ventilators are provided with wire-mesh screens for protection against rats. Ventilation fittings are generally constructed of welded steel and are of sufficient strength to withstand normal shipboard wear and tear, plus corrosion due to exposure to the elements. Steel parts are usually galvanized after assembly. The height and support of coamings as well as the closing arrangements must be approved as a condition of the ships' freeboard assignment.

a. Cowls. Cowls are used for natural air supply to and exhaust from cargo holds or storage spaces, not fitted with a mechanical ventilation system. A cowl can be turned into or away from the wind, as required to provide a positive means of supply or exhaust. During bad weather the cowl head is removed and the deck coaming is plugged with a screw-in metal disc or a hinged plate cover. With the increased use of mechanical ventilation this type of terminal is becoming obsolete, and is seen only on fishing and other small vessels.

b. Goosenecks. Goosenecks can be used for both natural and mechanical ventilating systems, supply or exhaust. They can be provided with watertight covers, if necessary, secured by dogs.

c. Mushrooms. The standard mushroom can be used for either natural or mechanical supply or exhaust terminals. This type provides adequate protection against rain but should not be used where exposed to boarding seas, since it is not feasible to incorporate a watertight closure in its design.

The screw-down type of mushroom (bucket type) is used exclusively for ventilating small compartments and can be used in exposed locations. Water-tight features are provided by means of screw-down-type top, operated above and below deck.

d. Louvers. Louvers may be used for air supply intakes for all types of ventilation systems. They are installed in weather bulk-heads and provide protection from rain, but must be restricted to protected locations, not vulnerable to seas. They are generally used for large-volume systems, are provided with wire-mesh screens, and are of sturdy construction, consisting of frame and louver blades set at about 45 deg with 50 mm (2 in.) openings. It is advisable to construct louvers of noncorrosive materials and design them to be removable, since experience indicates that louvers must be replaced several times during the life span of the ship (USCG Rules, Current).

Ventilation ducts or openings penetrating subdivision bulkheads below the *Margin Line* must be fitted with watertight closure devices, operable from above the Margin Line, and must be provided in accordance with USCG requirements.

e. Induced Draft Exhausters. Inductive-type exhausters are of standard design, rotating or nonrotating vaned construction, thereby inducing air flow. They are mainly used for small deck compartments. The *Liverpool* head is a common example of the rotating head type.

f. Air Lifts. These are used in place of louvers at points on deck exposed to seas. They consist essentially of a steel box with a central baffle plate over which the stream of air must pass. Any water entering the box is excluded by the baffle plate and drained out of the box back onto the open deck.

Section 2
Deck Fittings

2.1 Introduction. The term *deck fittings* includes a broad assortment of items consisting of structure and hardware, attached to the hull on the weather deck to perform various ship functions, as noted below.

2.2 Bulkwarks, Rails, and Stanchions. Bulwarks are fitted on the weather deck as a protection from seas for personnel and deck cargo. They are of heavier construction than those on superstructure decks, owing to the greater possibility of damage from seas as well as to provide sufficient strength for attachment of rigging fittings, lashing deck cargo, etc. Adequate freeing ports must be provided for drainage in accordance with load line regulations.

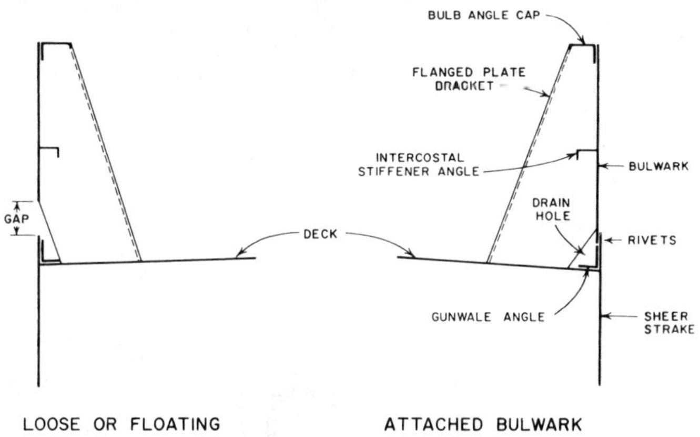

Fig. 14 Typical floating and attached bulwarks

Bulwarks, shown in Fig. 14, are usually constructed of steel plate not less than 6 mm (0.25 in.) thick supported by flanged plate brackets spaced 1.5 to 1.8 m (5 to 6 ft) apart. On vessels subject to load line regulations, the upper edge of the bulwark must be at least 1 m (3 ft 3.5 in.) above deck and suitably stiffened with a flat bar, angle, bulb angle, or channel. (An exception reducing this to 0.76 m (30 in.) for towboats exists.) Where extra strength is needed, such as on the forecastle, a longitudinal intercostal member is fitted at mid-height. On passenger decks a teak or a polished noncorrosive metal cap rail, is sometimes fitted directly over the steel top member, or mounted on short pipe stanchions 15 to 23 cm (6 to 9 in.) above the steel to form a *monkey rail.*

In free or floating bulwarks, the plating is not attached directly to the main structure of the ship. The advantage of this arrangement is that the bulwarks are not welded to the sheer strake and thus do not act as part of the ship girder, and can therefore be of somewhat lighter construction. This construction also prevents cracks, which may originate in the light bulwark plating, from progressing down into the deck stringer and sheer strake. The 15 cm (6 in.) of space between bulwark and sheer strake also provides freeing space in place of the customary freeing ports.

Bridge front bulwarks are usually fitted with a venturi-type rail which deflects the wind upward, thus minimizing the airflow striking personnel on the open bridge.

Breakwaters are of similar construction to weather deck bulwarks. However, they are of considerably heavier construction, to withstand the direct impact of seas and are of plow form to deflect the seas laterally. The height of the breakwater is made greater than that of the hatches or equipment it protects.

Open rails, Fig. 15, are fitted where bulwarks are not essential. When used on the weather deck of a cargo ship, they are usually made portable in way of the cargo hatches to facilitate loading, and to minimize damage. The USCG Rules specify three-course rails, of definite height, at the shell and prescribe the rail spacing. Two-course evenly spaced rails are prescribed at other than exposed peripheries.

Open rails generally are constructed of galvanized-steel pipe attached to stanchions spaced approximately 1.5 m (5 ft) apart. The upper rail is located 1.07 m (3 ft 6 in.) above the deck and is heavier than the intermediate rails. The stanchions can be made up of steel-pipe weldments, flat bars, structural tees, etc., see Fig. 15. On decks used by passengers or officers, a teak or polished noncorrosive metal cap rail is frequently fitted. The portable section of an open pipe railing is made up in sections about 1.8 to 3 m (6 to 10 ft) in length, convenient for handling manually. The stanchions are set into deck sockets and secured with brass toggle pins, attached with keep chains. The ends of a portable section are laterally secured by pipe braces bolted to deck lugs. Chain rails constitute a convenient type of portable railing, commonly used at a ship's side abreast of cargo hatches, around deck openings, etc. The stanchions are set into deck sockets and galvanized chains are rove through eyes in the stanchions. One end of a length of chain is shackled to a lug while the other end is set up with a turnbuckle secured to the deck. The chain size is usually 8 or 9 mm (5/16 or 3/8 in.), depending upon the nature of the duty.

Guard rails are fitted around openings in decks, at side ports, escape trunks, etc., and may be made of either pipe or chain as required. Portable guard rails are also fitted around exposed moving parts of deck machinery for protection of personnel. The guard rail around a magnetic compass platform must be entirely of brass or other nonmagnetic material to avoid deviation. Grab rails or storm rails are fitted around the outside of deckhouses as well as along one side of interior passageways, service spaces, etc., to provide a safe hand grip for personnel walking about the ship in heavy weather. Where passageways are 1.8 m (6 ft) wide or more,, a handrail must be fitted on each side. They are usually located about 0.9 m (3 ft) above the deck and secured with bulkhead-mounted brackets or sockets spaced 1.2 to 1.5 m (4 to 5 ft) apart. The rails are set away from the bulkhead not less than 50 mm (2 in.) to afford a convenient hand grip. Galvanized-steel pipe is commonly used for grab rails on weather decks and in crew quarters. In the wheel-

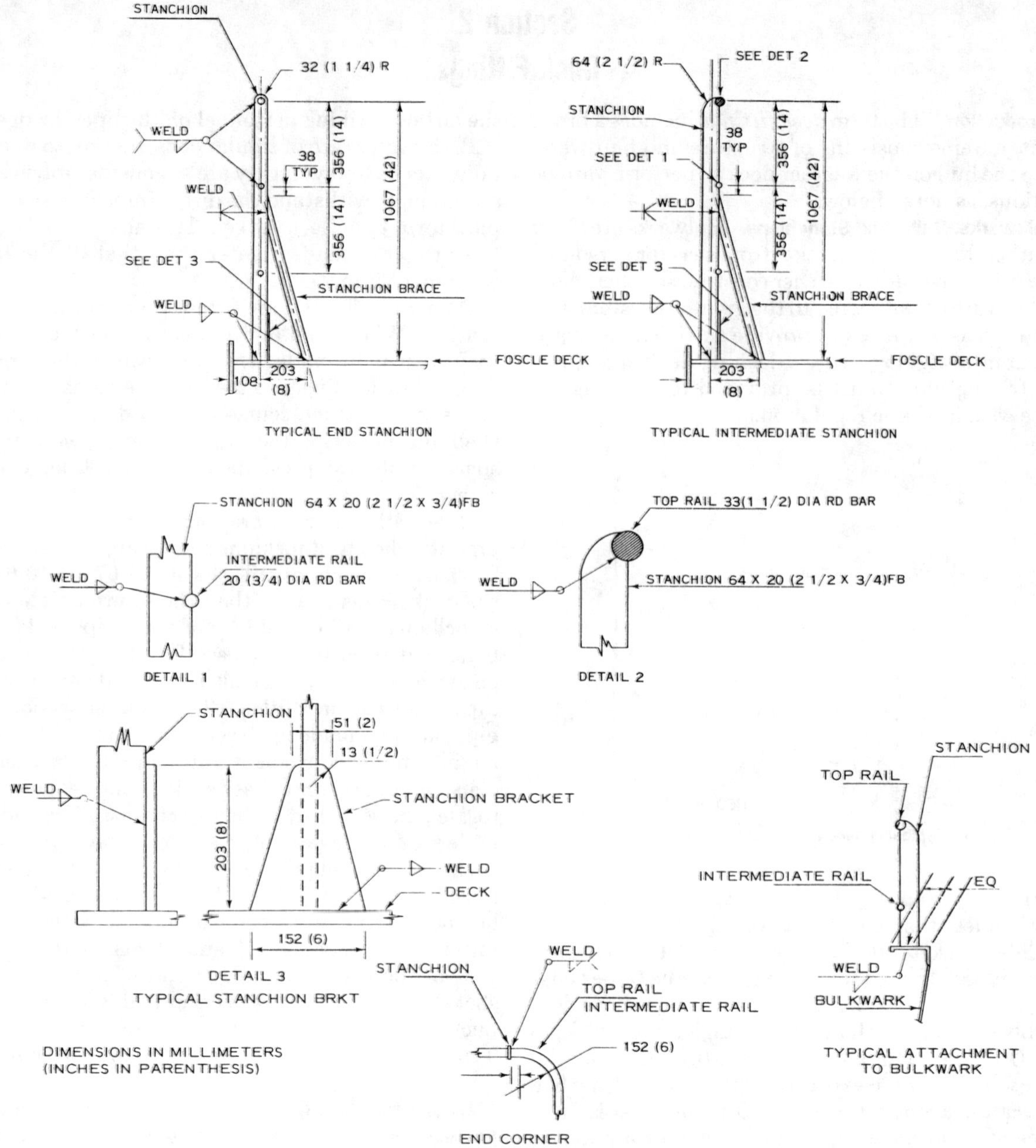

Fig. 15 Details of open rails

house and in officers and passenger areas, the grab rails are made of mahogany, brass, or aluminum.

Provision is usually made for fitting awnings of canvas, nylon, dacron, or other rigid noncombustible material over certain deck areas for the comfort of the ship's personnel. Common locations are the poop deck, the weather deck aft of the deckhouse, the boat deck inboard of life boats, and over the wings of the navigation-bridge deck. On passenger ships, awnings also are installed adjacent to outdoor swimming pools. Large portable awnings frequently are divided into several sections for convenience of handling, the section edges being overlapped to avoid leaks. The material is reinforced with doubling pieces in way of ridge poles to prevent wear from chafing. Nylon or dacron are now widely used in place of canvas, being lighter in weight and less prone to deterioration.

Most commercial ship operators prefer premanent, rigid awnings made of a variety of materials such as corrugated aluminum, translucent corrugated glass fiber or (GRP), flat GRP panels bolted to a suitable framework of galvanized structural sections. Aboard passenger vessels, however, specific fire protection regulations apply to such enclosed spaces, and the use of plastics in lieu of glass must be carefully evaluated. Such awning structures must be of a substantial nature capable of withstanding hurricane force winds. The increased cost of this construction is justified on a life cycle basis by elimination of the periodic cost of repairs and/or periodic replacement, and are in use.

Weather cloths, consisting of panels of heavy canvas, dacron, or nylon are frequently fitted around navigating-bridge rails, at the sides of the wheelhouse where bulwarks are not fitted, and along accommodation ladders.

2.3 Ladders and Stairs. Ladders are installed where required to provide suitable access or means of escape as required by U. S. Coast Guard Rules.

a. Fixed Vertical Ladders. Vertical ladders are fitted for access to all cargo holds, bosun's stores, tanks, etc., where horizontal access is not possible. As a general rule, not less than two ladders are provided to each space. In cargo holds, a vertical ladder is installed at each hatch end in way of the structural pillars, where the ladder is best protected from damage by cargo and at the same time imposes the minimum interference with cargo stowage. On some ships, additional hold ladders are fitted at the transverse bulkheads with a deck hatch or hinged manhole overhead. In containerships the ladders are located at the bulkheads.

Access to bosun's stores, and in some instances steering-gear compartments, is provided through a hinged deck hatch with ladder below, while access to tanks is made through a bolted manhole.

A common method of constructing a vertical ladder for access to a cargo hold consists of two 150 × 100 × 10 mm (6 × 4 × 3/8 in.) angle stringers 280 mm (15 in.) apart with 16 mm (5/8 in.) square rungs on 300 mm (12 in.) centers welded to the stringers. The rungs can be placed with either the flat or the corner up; the latter is generally considered to be safer since it affords a more certain footing, particularly in deep tanks where the ladders are frequently oily. Vertical ladders usually are made up in sections and bolted to lugs or clips which are welded to the ship's structure. In some instances it is possible to omit the stringers and merely weld the rungs directly to bulkhead stiffeners or the shell frames. In shallow tanks, such as double bottoms, the rungs generally are formed into a stirrup with the ends welded to the bulkhead or floor plate.

b. Spar Ladders. Ladders must be installed on masts, kingposts, stacks, etc., to provide access for adjustment and servicing of rigging fittings, blocks, lines, etc. On large masts the ladders are usually made up of 76 × 10 mm (3 × 3/8 in.) flat-bar stringers spaced 300 mm (12 in.) apart with 16 mm (5/8 in.) square rungs on 300 mm (12 in.) centers, and are bolted to lugs welded to the mast, similar to a hold ladder. Spars less than about 400 mm (18 in.) in diam generally are fitted with stirrups shaped in such fashion that a person's foot will not slide off sideways. In way of rigging fittings, etc., where work is to be performed, hoops are installed around the mast or kingpost for convenience and safety. Care must be taken to arrange ladders and hoops so that will not foul the blocks, lines, etc., in any operating position.

Spar ladders are provided for interior inspection and maintenance of large-diameter masts and kingposts. Usually this type of ladder consists of rungs or stirrups welded to the inner wall on one side of the spar. Access is obtained through a manhole and bolted cover, located near the point at which the bending stress is minimum.

c. Inclined Ladders. Inclined ladders are provided on all weather decks for access from one deck level to another. This type of ladder is also found in stores spaces. Generally, an inclined ladder resembles a stair, the principal differences being steeper pitch and the omission of risers. The ladder is made up as a unit, complete with stringers, treads, and handrails. The assembly is then welded at the top and bottom of the stringers to the decks. The stringers generally are light channels spaced 760 mm (30 in.) apart. Treads are steel with nonslip covering and are welded or riveted to the stringers. A nonslip deck pad is located at the top and bottom of all inclined ladders. Rails usually are made up of 32 mm (1.25 in.) galvanized pipe. For safety, weather deck inclined ladders are located to avoid tripping hazards on deck, as well as cargo gear and other deck machinery.

d. Accommodation Ladders. Accommodation ladders by which the ship is boarded from a boat or low pier are

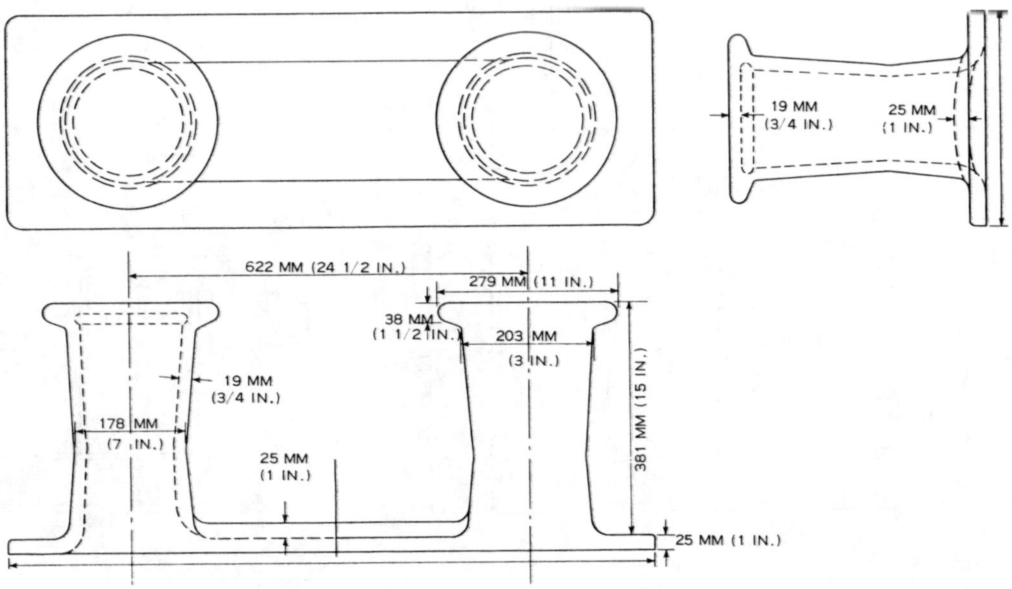

Fig. 16 Cast steel 203 mm (8 in.) barge type double bitt

designed to reach from the weather deck level to the light operating draft line at an angle of approximately 45 deg to the horizontal and along the hull. The length of the ladder is adjustable to variations between light and full-load draft by modifying its angle, by telescopic action, or by extending an adjustable lower section.

The side stringers are made of lightweight metal (chiefly aluminum) channel sections, or of wood, and are usually designed for an assumed load of 136 kg (300 lb) per tread. The treads are shod with nonskid material and are either positioned to be horizontal when the ladder is in working position of 45 deg or *feathering treads* are arranged to pivot automatically to the horizontal for any working angle.

The upper end of the ladder pivots from a portable platform grating attached to the ship's side at deck level. A similar platform is fitted at the lower end to facilitate boarding from small craft on the pier. The ladder and platforms are equipped with sockets to take portable guard-rail stanchions. The wire rope guard rails are rigged through eyes on the stanchions. Canvas, dacron, or nylon weather cloths along the rails are fitted and fastened with braided nylon lacing cord.

When in its working position, the lower portion of the ladder is supported by wire-rope slings attached to an adjustable tackle. The upper end of the block and tackle is shackled to a davit head or fixed point overhead. The upper ends of the slings are spread apart by a spacing bar or bridle to provide suitable head room.

An accommodation ladder is stowed by hauling it up to a horizontal position at deck level with the supporting tackle. It is then lifted inboard by means of davits and tackles, and usually is stowed on its side on deck adjacent to the rail.

Normally, the accommodation ladder is located on the upper deck with provision for rigging it port or starboard. Its working position is carefully selected to keep it well clear of overboard discharges and cargo loading side ports.

2.4 Mooring Fittings. Mooring arrangements fall into two broad categories, mainly those featuring constant tension winches which permit constant, automatic adjustment of mooring lines, and those mooring arrangements in which fiber and/or wire-rope mooring lines are manually adjusted periodically when necessary with the aid of capstans plus warping heads on the anchor windlass, and secured to mooring bitts. As might be anticipated the latter (manual) method is the less expensive to install and is thus widely used on break-bulk general cargo ships and on small to medium size passenger ships in which changes in draft at the loading pier due to loading or discharging cargo are relatively slow or small. Constant tension winches become desirable for large ships of most types and are essential for tankers, dry-bulk, and containerships which can load and/or discharge their entire cargo in a few hours, thus requiring continual adjustment of mooring lines to compensate for large, rapid changes in draft. Constant tension winches also enable a dry bulk carrier to be conveniently moved fore and aft to various loading stations along the pier, as is standard practice for Great Lakes ore ships.

Changes in ship draft are automatically accommodated

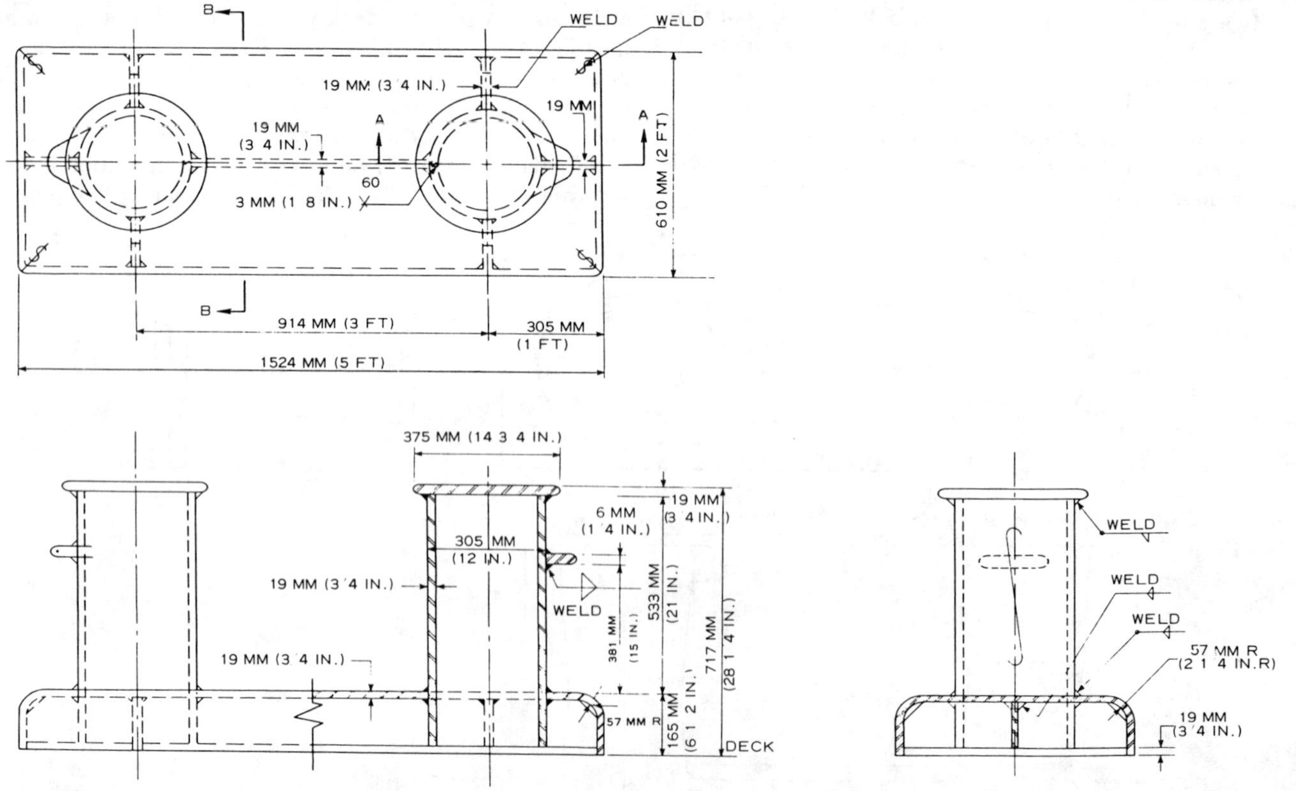

Fig. 17 Welded-12 mooring bitt (12 in. size shown)

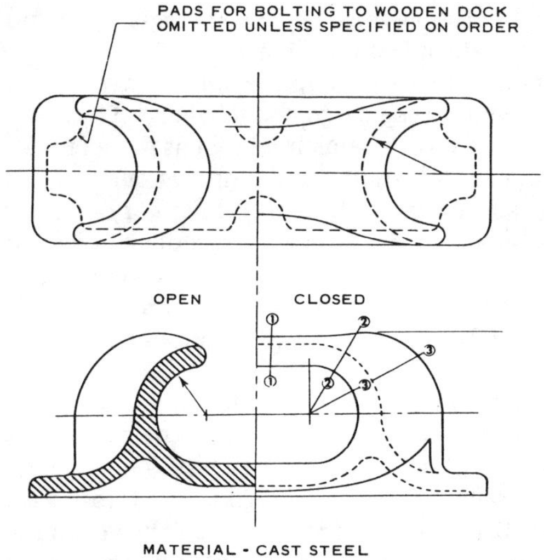

Fig. 18 American Marine Standard plain chocks

by setting the mooring winches to pay out and reel in the mooring cables at a pre-set load. The ship's position is changed by manually controlling certain winches to pay out in one direction while other winches take in line in the opposite direction, as conditions dictate. Once the ship has moved to the temporary new position, the winches are set to resume automatic constant tension operation.

a. Mooring Bitts, Cleats, and Rings. Mooring bitts are cast steel or fabricated of vertical heavy pipe barrels welded to a dished base plate which in turn is welded to the deck. Local reinforcement is usually added directly below deck to properly distribute the reactions into the adjacent deck framing members. The size of the bitt installed varies with the diameter and strength of the mooring lines to be used, and is designated by the diameter of the two pipe barrels, Fig. 16 is a typical cast steel bitt. A typical welded bitt is shown in Fig. 17.

Cleats perform a similar function but are used primarily for mooring barges or other small craft alongside a ship. The cleats are either cast steel or built up weldments and usually range between 36 and 61 cm (18 and 24 in.) in length, depending upon the size of barge lines. Cleats are mounted on the top of the bulwarks or on deck near the side of the ship as conditions indicate.

Mooring rings are installed in the side shell between the water line and the weather deck at a convenient height for barge or small craft crews to handle their mooring lines. They usually consist of a recessed steel casting with a horizontal bar across the opening, and are set between frames flush with the shell plating.

b. Chocks, Fixed and Roller. Chocks are installed at the sides of the ship to lead the mooring lines from their fixed point on shore to the hauling end or attachment aboard ship. The latter point may consist of a mooring bitt, warping head, or capstan in the case of fiber rope mooring lines or a constant tension winch in the case of wire rope. Fixed (or non-roller type) chocks, Fig. 18, are used for fiber rope hawsers which do not require adjustment under load, but a roller type chock is preferable to minimize friction on ships where the mooring lines are wire-rope and are normally hauled in and payed out frequently.

Fixed chocks consist of a steel casting or weldment, having an oval shaped opening with well rounded edges to reduce chafing action. If the chock is to be mounted on deck, it will have a flat base which is welded to the deck. If the chock is to be bulwark mounted its periphery will be oval shaped and will be welded directly to the bulwark plating with local stiffening between brackets.

Roller chock frames are steel castings or weldments in which rollers are mounted. Roller chocks used in conjunction with fiber rope generally have only vertical cast concave shaped rollers at the end of the opening whereas chocks intended for constant tension winches with wire lines are fitted with four pipe rollers (two vertical, two horizontal) so that a roller is brought into play for any direction of line pull to the pier. In Great Lakes ore carriers and other ships in which mobility during loading is essential, the patented type *Berger* fairlead which acts like a large snatch block, is generally used, Fig. 19.

Supplementing the fixed or roller chocks as discussed above, roller deck fairleads are installed where required for leading lines to a warping winch or capstan.

2.5 Stores Handling Gear. Davits, cranes, and tackle are commonly used to handle bosun's stores and on smaller ships, to hoist or lower single items or packages through stores hatches. The out-reach of a davit is relatively short, being intended to swing loads over a hatch opening from a position on deck alongside the hatch. Cranes, on the other hand, generally have sufficient outreach to hoist loads from the pier to points on deck near a stores hatch, or over the hatch opening.

Davits are of simple construction consisting of a rotating bracket arm pivoted from a deck-house bulkhead or other conveniently located structure. A block and tackle is suspended from an eye at the end of the arm. Davits are fabricated by the shipbuilder whereas cranes tend to be more elaborate, frequently having extendable, telescoping booms with hydraulic drive. Cranes usually are stepped on the weather deck about half way out from the center line to the side of the ship port and starboard, to minimize boom length. Cranes of this type are purchased by the shipyard from a machinery manufacturer.

Blocks and tackle are used occasionally to hoist spare

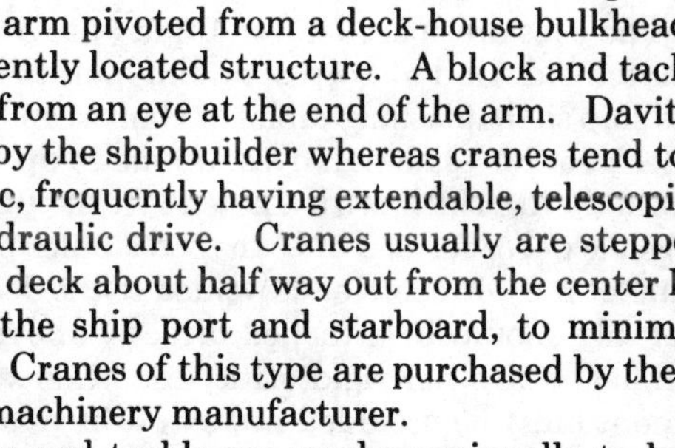

Fig. 19 Deck-mounted balanced head fairleader

parts or to skid heavy items horizontally along deck on rollers and planks during ship maintenance and repair.

Elevators are in wide use on ships built in recent years due to the trend towards large ships with many deck levels which make stair climbing a tiring, time consuming chore. Shipboard elevators ride on guide rails in an "A" Class steel trunk and are of similar size and operation to those used in small apartment houses. Ship elevators are frequently used to move stores as well as personnel.

Dumb-waiters are extensively used to move food from a galley or pantry on one deck level to dining or mess rooms located on other deck levels. The dumb-waiter is also used to move cartons of food products from the stores loading deck to the galley area. Controls are located on each deck served by the dumb-waiter.

Conveyors are used to load food stores and for package cargo handling. Horizontal-belt conveyors rigged through side ports provide a rapid means of loading packaged items, such as cartons of canned goods or crates of fruit, and are fitted in the upper 'tween decks. Vertical conveyors move similar unitized cargoes through hatch openings from one deck level to another. Horizontal and vertical conveyors are thus commonly used in combination on fruit ships and other types of unitized package cargo carriers. Conveyors are also used for food stores handling on large passenger or naval ships on which striking down stores for a large number of people must be accomplished in a relatively short time.

A simplified form of the conveyor consists of portable sections of rollers mounted in metal frames for manually moving packages on deck. Chain hoists rolling on fixed overhead tracks are another means by which ship stores may be handled.

2.6 Deck Stowages. On break-bulk general cargo ships, lashing pad eyes are fitted on weather decks abreast of the hatches. The eyes are spaced 1.2 to 1.5 m (4 to 5 ft) apart near the hatch side and at corresponding points near the sides of the ship so that lashing cables or chains can be rigged over the deck cargo. The pad eyes are made up of plate, about 19 mm (3/4 in.) thick, with a hole to fit the pin of a 25 mm (1 in.) shackle, or formed of U-shaped staples of 25 mm (1 in.) diameter rod with both ends welded to the deck. Pad eyes preferably should be attached directly above a deck beam, bulkhead, or other rigid structurc; otherwise local reinforcement must be provided in way of the pad eye.

On containerships, the deck containers are secured to interlocking, malleable cast-steel deck sockets. The container corner castings, built integrally with the container corner structure, have slots for locking. The lower tier of deck containers is set into place and secured with twist-locks to the deck sockets. Additional layers of containers are then secured to those below with twist-lock fittings which engage the slots in the container corner castings slots when turned with a lever. Wire-rope and turn-buckle lashings are frequently added as a precautionary measure but are not usually essential if containers are carried only one or two high on deck and are well secured at the corners with twist locks, as just described. A more detailed discussion of stowage of deck containers appears in Chapter X and in Henry and Karsch (1966), which shows also typical arrangements and details of container-deck lashing fittings.

On general break-bulk cargo ships, *boom stowage chocks* are provided for securing the cargo boom heads. The lower half of the chock consists of a semicircular bent piece of steel flat bar. The upper half is similar and hinges at one end, the other end being secured with a brass toggle pin or butterfly nut. When the boom head is stowed over a deck house, or alongside a mast, the chock is mounted on a bracket or a low foundation. If the boom head is stowed over an open deck, the chock is mounted on a crutch of sufficient height to give proper head room.

2.7 Spare Parts Stowage. In general, miscellaneous spare parts are properly labeled and stowed out of the weather at convenient locations. For example, spare pipe, boiler tubes, valves, etc., are stowed in racks in the engineers' stores room. Spare armatures, brushes, etc., are stowed on shelves in the electrical stores room. Steering-gear spare parts are stowed in chocks or on shelves in the steering-gear room. Other heavy spare parts are similarly stowed in chocks in convenient locations. Parts vulnerable to damage by salt air are suitably protected with grease, sealed packaging, or other appropriate means.

2.8 Miscellaneous Rigging Fittings. The following covers rigging fittings in general, (other than cargo gear), such as used for stores handling and other ship functions. Typical cargo boom rigs are illustrated and discussed in Chapter X.

a. Padeyes and Cleats. Padeyes are used for attachment of standing rigging, snatch blocks, etc. Padeyes are also installed on the shell plate above the rudder and propeller for use in installing and removing these units. They are made up of heavy pieces of steel plate welded to the deck, or spar. Those subjected to relatively heavy loads, such as for shrouds or stays, require the deck plating to be reinforced locally with angle or channel headers between deck beams. The padeye should be positioned to fall in line with the applied load.

Plate padeyes are drilled to take the pin of a shackle or clevis. The diameter of the hole should provide a clearance of approximately 1.6 mm (1/16 in.) for pins up to 25 mm (1 in.) diameter and about 3 mm (1/8 in.) for larger pins. Similarly, the thickness of the plate selected for the pad eye should provide 1.6 to 3 mm (1/16 to 1/8 in.) clearance in way of the jaws of the shackle or clevis to obtain an easy fit without excessive slack. Excessive slack, particularly between jaws, greatly increases the bending moment on the pin. Steel washers are welded to either side of the plate to increase the padeye thickness or width when the jaw spacing is large.

Cleats of various size and types are used to secure the hauling ends of halyards and tackle. Small cleats mounted at the corner base of the masts for flag and signal halyards are generally a stock type cast steel galvanized attached with machine screws or bolts. Larger cleats intended for heavy loads such as boom, vang tackle, etc., are usually fabricated of welded pieces by the shipyard and are welded to the deck or bulwarks with local reinforcing if necessary.

b. Turnbuckles. Turnbuckles used for setting up

standing rigging, etc., are galvanized-steel forgings purchased complete from the manufacturer. The ends of the turnbuckles may be either a clevis with pin or an elongated eye. When both ends are a clevis, the turn buckle is known as *jaw-and-jaw* if both ends are an eye, it is called *eye-and-eye;* if one end is a clevis and the other an eye, it is known as a *jaw-and-eye* turnbuckle. The choice of one of the three types depends upon the application. The jaw-and-jaw turnbuckle is perhaps the most widely used, since one end slips over a pad eye blade welded to the ship structure, while the other end connects to a closed-type socket attached to the end of the wire rope. However its use is limited to a fixed direction of pull. The eye-and-eye type with connecting shackles has the advantage of flexibility of alignment and is thus commonly used for deck-cargo lashings, *devils claw* attachments on anchor chains, and similar applications, requiring variation in alignment.

The size or thread diameter of the turnbuckle is determined by the load. A turnbuckle should be slightly stronger that the wire rope or chains to which it is attached to prevent deformation or damage of the turnbuckle. Lock nuts should be provided on each end of the threaded end pieces, and set up hard against the center body section after the turnbuckle has been adjusted to the desired tension.

d. Shackles. Shackles are U-shaped steel forgings with a pin through an eye on each leg of the U. They are usually galvanized and serve as connecting links for various rigging components. They are purchased complete from a manufacturer. Two principal types are in common use for ships' rigging: *anchor* and *chain* shackles. The loop or *bail* of an anchor shackle is bowed out to provide a loop wider than the spacing between jaws. The bail of a chain shackle is the same width as the jaw spacing. Owing to its straight sides, a chain shackle is somewhat stronger than the same size anchor shackle, the size being the diameter of the rod forming the body of the shackle.

The large bale opening of the anchor shackle makes it particularly suitable for reeving through chain slings, and for attachment to eye splice thimbles in ends of manila of fiber ropes. It also provides greater freedom of motion and thus is used for attaching blocks to pad eyes, etc.

2.9 Ro/Ro Vehicles Securing Fittings. Vehicles carried on ocean-going Ro/Ro ships must be secured in such manner as to prevent motion or shifting of the vehicles at sea. In general, the method of securing is designed to take the weight off the axle springs and to transfer the loading onto fixed supports. A simplified method consists of jacking-up the trailer or other vehicle off its wheels and transferring the weight to a wooden sawhorse type structure placed under the chassis frame. Hold-down turnbuckles are then attached to the chassis frame and to securing lugs on deck. Motion is eliminated by tightening down the vehicle chassis frame against the supporting wood structure.

Patented type systems have been developed which speed up the vehicle loading operation by eliminating construction of the wood supporting structure and by replacing same with portable steel mounts. The design of the portable steel mounts is such that the weight of the vehicle is automatically transferred from the trailer-tractor saddle to the portable steel mount by proper maneuvering of the tractor unit. Turnbuckle lashings are then applied, similar to those described earlier.

Section 3
Hold Sparring, Ceiling, and Gratings

3.1 Introduction. The items covered in this section serve primarily to restrain and protect the various types of cargo carried in break-bulk cargo ships.

Ceiling and sparring, Fig. 20, are required in cargo holds and storerooms to protect cargo and stores from damage due to condensation, contamination from previous cargo, or injury caused by abrasive action in way of stiffeners, brackets, beam knees, etc., and to insure proper ventilation. As a general rule, the cargo is stowed so as not to come in contact with steel, except for various types of dry bulk cargoes such as grain, coal, and ore plus some chemicals such as sulphur and phosphate.

Certain types of vessels, such as refrigerated ships, have all the necessary fittings for proper stowage installed during construction. In a ship designed for general cargo, the fittings necessary for cargo protection vary for each type of cargo to such an extent that only a limited installation is made during construction. During operation, it becomes necessary to provide supplementary protection in the form of dunnage, arranged to suit the type of cargo carried on each voyage.

3.2 Sparring or Battens. The term sparring includes wood protection of all vertical surfaces in way of shell frames in cargo holds, in way of all sides in storerooms used for bulk stowage, in way of fuel oil, lube oil, peak tanks, settling tanks, and distilled-water tanks where exposed in cargo holds or forming the boundary of storerooms. Sparring is fitted in refrigerated stores, even though completely wood lined, to allow for adequate ventilation.

Cargo batten is the term used for units of sparring on shell frames in cargo spaces. These battens are about 150 by 50 mm (6 by 2 in.) usually of Douglas fir, fitted horizontally with about a 230 mm (9 in.) space between them. Battens are bevelled on all edges to prevent tearing or chafing of bagged or other types of cargo, and are secured to frames by means of clips as shown in Fig. 20(b). The large number of clips required has led to the development of special types which can be attached quickly to frames by spot-welding. The life of cargo battens is relatively short because of rough handling by long shoremen and, because they are readily removable, they are misused by stevedores for dunnage. For this reason some operators insist on bolting them.

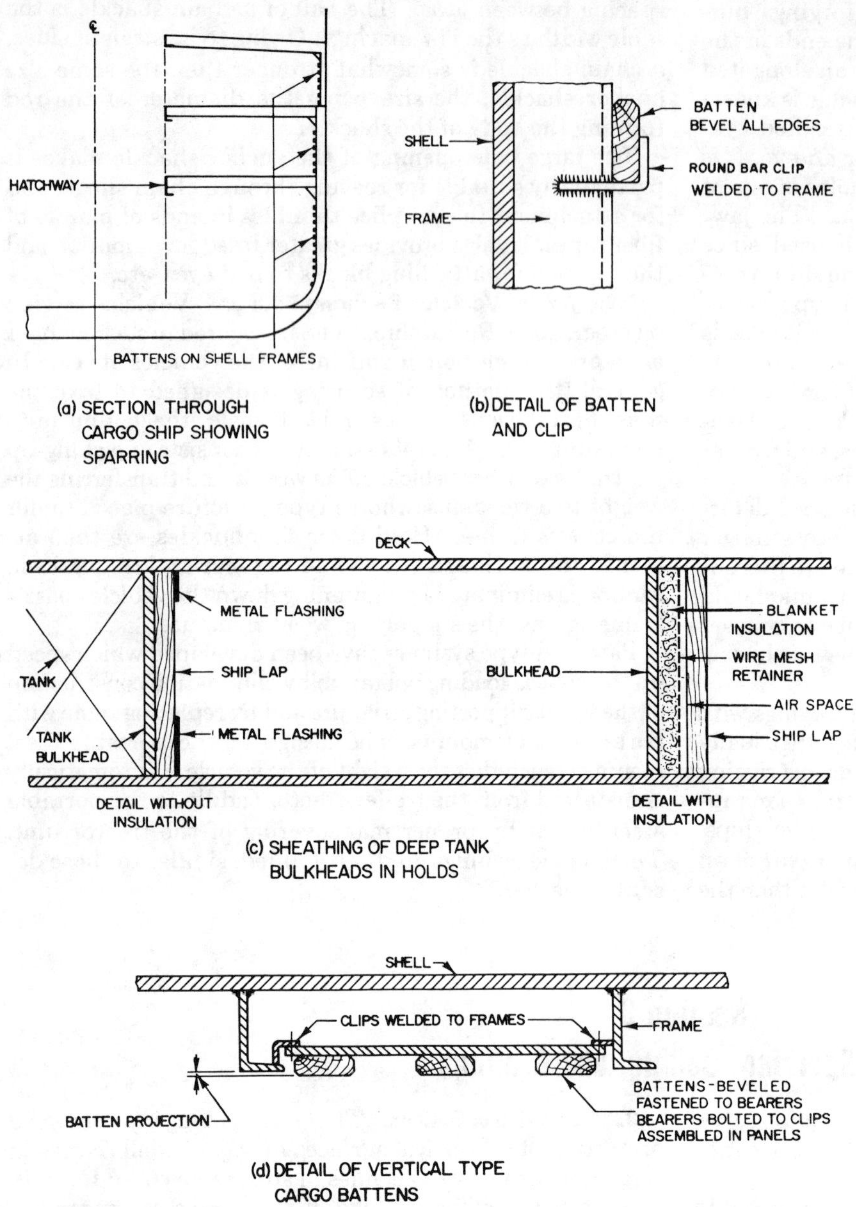

Fig. 20 Section showing cargo hold ceiling and method of attachment of sparring or cargo battens

At the ends of a general cargo ship with large hatches and fine form, the hold battens are very near the hatch-landing area and are readily smashed by swinging cargo or dislodged in retrieving the cargo hook. A vertical type of cargo batten, Fig. 20(d) has been developed which is unlikely to be caught by a cargo hook and resists a blow from cargo due to its better construction. The additional cost is offset in part due to the gain in cargo cubic because this type only extends 13 mm (½ in.), whereas the horizontal type extends 50 mm (2 in.) beyond the frame. On a large general cargo ship, this may amount to as much as 142 cm (5000 cu ft) of bale cubic. Some operators prefer this type for an entire ship, while others have compromised by fitting the conventional horizontal type in all spaces except in end lower holds.

In baggage rooms, all bulk storerooms and in refrigerated stores, vertical sparring is fitted on all walls. Usual practice is to use 50 × 50 mm (2 × 2 in.) wood spaced about 200 mm (8 in.), except that in refrigerated areas the width is increased to 76 mm (3 in.) or more and the spacing is increased to about 305 mm (12 in.) to suit the ventilation requirements. The sparring is fitted around all structural obstructions and over cooling coils and other fittings. When access to these fittings is required, the protective sparring is made portable.

3.3 Dunnage and Cribbing. Cargo protection in general cargo ships is usually provided for each loading by the stevedoring concern and consists of wood planks, temporary bracing, special shoring for deck cargos, temporary wood bulkheads, shifting boards, etc. In recent years, special patented systems have been developed to eliminate the excessive use of dunnage and wood sheathing which is very costly and causes some loss in cargo dead-weight and cubic.

Such systems consist of wire rope netting with quick-acting lashings on deck and deckhead. The fittings are closely spaced so that practically any type or combination of cargoes may be segregated and permit rapid partial filling and discharge of holds. Plastic inflatable dunnage is occasionally used as well.

There are numerous Coast Guard regulations governing the stowage of cargoes consisting of alcoholic liquors, coal and coke, cotton, grain, explosives, and other flammable or dangerous cargos. SOLAS 1974 has regulations for the carriage of grain and dangerous goods similar to USCG Rules. Chapter XI discusses these subjects in more detail. The National Cargo Bureau also issues rules for stowage of grain and cotton cargos to satisfy insurance underwriters. This discussion will be limited to those fittings usually provided by the shipbuilder for a general cargo ship during construction. Leeming (1942) provides further information in regard to stowage and handling of specific types of cargo and Henry and Karsch (1966) for container stowage on board.

Fig. 20(a) shows a section through a general cargo ship hold and typical locations of sparring and ceiling. The fitting of sparring in all ships of this type and of ceiling in single bottom ships, and under hatch openings in ships with double bottoms, in cargo holds, is a requirement of the American Bureau of Shipping. Dry bulk carriers especially designed for bulk cargos such as coal, ore, etc., do not require sparring and ceiling for protection of cargo. The Rules permit elimination of ceilings provided the thickness of inner bottom plating is suitably increased under the hatch openings.

3.4 Bulkhead Sheathing. Transverse bulkheads in holds are not sparred during construction, unless they form boundaries of tanks, in which case they are completely sheathed with 50 mm (2 in.) ship lap. Fuel oil, settling, lube oil, and distilled water tanks which contain heated liquids are sometimes provided with 50 mm (2 in.) of blanket-type insulation, sheet metal lagged under the 50 mm (2 in.) ship-lap covering. Although this may not appear necessary on certain routes, the installation is recommended on dry cargo ships designed for unlimited service and subject to extreme variations in temperatures. Spoilage of cargo due to condensation in holds is of primary consideration since, according to underwriters, sweat damage ruins more cargo each year than any other ocean-shipping hazard. The insulation and ship lap also protect certain types of cargos which cannot withstand heat.

On some ships, deep tanks also are fitted for carriage of dry cargo. Since these tanks are fitted with heating coils, on bottom and sides it is necessary to provide ceiling and sparring over the coils. To avoid repeated dismantling when changing from dry to liquid cargo, the sparring and ceiling are made of steel and further isolation of dry cargo from the steel is made with dunnage.

3.5 Gratings. Wood gratings are installed in bosun's stores, dry stores, refrigerated spaces, and on certain portions of the navigating deck, to provide a dry walking or working surface. Those in stores or refrigerated spaces usually are made up of ash battens secured to cross cleats with galvanized screws. The space between and under the battens should not be more than 13 mm ($\frac{1}{2}$ in.), to exclude rats, (USPHS, 1965). Gratings on the steering stand, compass platform, etc., generally are constructed of teak strips mortised into the bearers and fastened with plugged brass screws. Molded GRP gratings are some times used in place of wood, based upon specific location and other fire protection considerations.

Metal gratings consisting of galvanized expanded metal, perforated sheets, or subway-type construction are often used in place of wood gratings and frequently are used in preference to wood in refrigerated cargo spaces where loads are carried on hand trucks or forklift trucks. Metal gratings are used on tanker walkways and access platforms on the weather deck.

Machinery space gratings in engine and boiler rooms are made up of portable sections of steel flat-bar stringers with closely spaced square bars welded between. The stringers are about 6×76 mm ($\frac{1}{4} \times 3$ in.) and the rungs 16 to 19 mm ($\frac{5}{8}$ to $\frac{3}{4}$ in.) on 51 to 76 mm (2 to 3 in.) centers. Gratings are supported by an auxiliary framework, built up of angles.

Section 4
Deck Coverings

4.1 General. The decks of living and working spaces of merchant ships, with the exception of machinery spaces, are covered by suitable material for comfort, safety, appearance, and in some areas, fire resistance. The coverings usually have some insulating value, provide some sound deadening, and where required include a measure of fire protection. In locations under which cargo spaces are located, the fire protection requirements are met by a Coast Guard approved thickness of noncombustible deck covering material. The use of deck covering for fire protection is dealt with in more detail in Section 5. American Bureau of Shipping Rules permit composition deck covering to be laid over steel not exposed to weather, provided the composition is not corrosive to steel.

The requirements for safety, protection of steel, appearance, minimizing topside weights, plus the ability to provide long-wearing service characteristics for a wide variety of conditions, have made it necessary to develop exacting specifications for deck coverings. Standard government specifications are available for both commercial (Marad, 1979) and naval deck coverings and are invoked by the designer in the ship's construction specification. The leading marine decking manufacturers supply various products under their trade names but each product complies with the

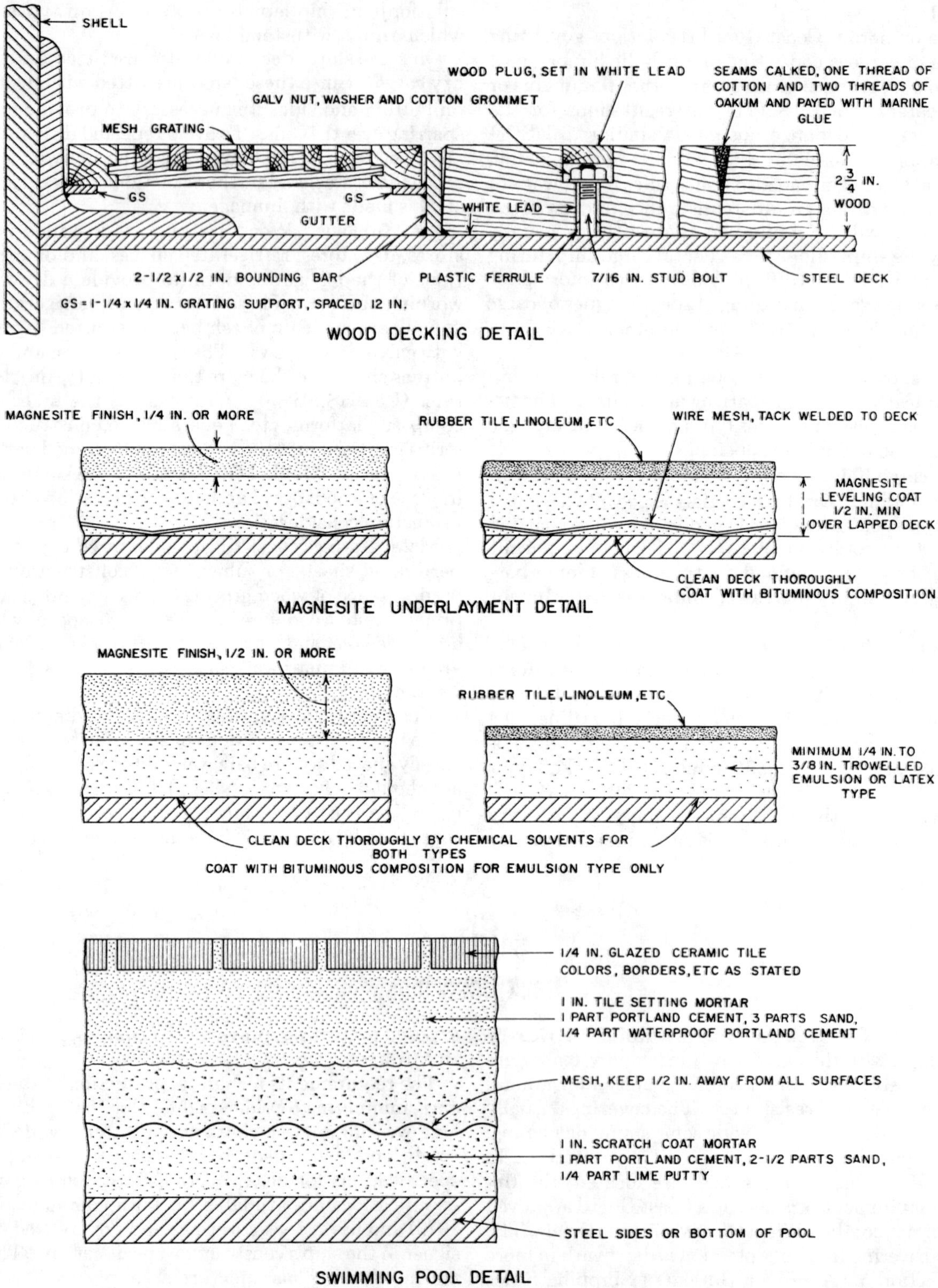

Fig. 21 Typical deck-covering details

standard specification and, in addition, is tested and approved by the U. S. Coast Guard before being used on ships certificated by them. Certain coverings also must be approved by the American Bureau of Shipping because of possible deleterious effects on the steel deck. Satisfactory decking is produced only when specific materials are properly applied. For this reason, it is generally the practice of American shipyards to assign the reasponsibility of the deck-covering installation to one sub-contractor who both supplies and installs all the material. The shipyard as a rule limits itself to the preparation of the steel surfaces.

In the following paragraphs, a brief description is given of the various deck coverings and their application, as well as a table of approximate installed weights for each type, plus a table of suggested locations. New developments in special non-skid lightweight materials, for deck coverings, are continually being made with resulting improvement in appearance, wearing characteristics and reduced weight. Typical details are shown in Fig. 21.

4.2 Wood. Wood decking is used primarily on passenger ships for enclosed promenades and in passenger weather deck areas and sometimes in weather portions of the superstructure of deluxe cargo ships. From the point of view of appearance and comfort, wood is considered the most desirable deck covering. The American Bureau of Shipping permits slight reductions in superstructure plating thicknesses where wood sheathing is applied. On superliners and special craft with large superstructures however, the weight and fire protection problem has made it necessary at times to prohibit wood in enclosed spaces in favor of lightweight mastic deckings, which are described in 4.9.

In spite of its limitations and high initial cost, wood decking is still considered superior and probably will continue as the preferred weather deck covering for passenger ships which cruise in the tropics, where tiling and plastic compositions in the vicinity of swimming pools and on sun-decks become unbearably hot.

Wood decking generally consists of top-quality Douglas fir, 67 mm (2⅝ in.) thick by 117 mm (4⅝ in.) wide, finished size, laid edge grain up, in lengths of not less than 4.88 m (16 ft). In a first-class installation the margin plank against houses should be of teak, 203 × 76 mm (8 × 3 in.), tapering to 67 mm (2⅝ in.) for proper drainage. The ABS permits slight reductions in superstructure plating thickness in cases where wood sheathing is applied.

The planking is attached to the steel deck by means of steel studs welded to the deck, replacing through bolts for this purpose. The studs are fitted with grommets and washers and the stud holes filled with wood plugs of the same material as the planking. Plugs are set in white lead with the grain in the same direction as that of the plank. Seams are caulked with one thread of cotton and two of oakum, then payed with special marine type seam composition. The planking is laid in a thick coat of rust preventive compound which also is used to fill uneven spots in the plating.

Just prior to delivery of a new ship, the planking should be cleaned, planed, scraped, and sanded to produce a smooth, clean, top surface. Ash strips are fitted as stops for deck chairs, and gutterways are covered with portable sections of teak gratings. Meticulous work by experienced craftsmen under proper supervision is vital in all extensive wood-deck installations.

4.3 Magnesite. Magnesite, once the most common decking in crew quarters on cargo ships, is no longer in favor, having been supplanted by tiling material, described in the following, which has been found more attractive in appearance and easier to clean. Magnesite is still used as an underlayment and as fire insulation, but a suitable tile or other covering, not exceeding 10 mm (⅜ in.) in thickness, is applied over the the underlayment for finish purposes.

4.4 Terrazzo. Terrazzo is a thin set deck covering consisting of inorganic powders and marble chips combined with a synthetic binder. It is trowel-applied, ground smooth, and polished after curing to make a finished deck. It has excellent decorative properties, as well as being of low fire hazard; it has excellent adhesion, and is economical to maintain. It is applied in thicknesses of 6 to 13 mm (¼ to ½ in.). Terrazzo is used extensively in all wet spaces, passages, swimming pool beaches, and, occasionally in crew staterooms. A cast variety of this material is often used for flooring in shower stall spaces. A special type of non-conductive synthetic terrazzo is used in hospital spaces.

4.5 Ceramic Tiles. Ceramic tiles of various kinds are widely used as finish deck coverings in the typical wet spaces listed at the end of this section. In spaces requiring frequent wash-down (such as galleys), quarry tile-red-grooved, ribbed back, 152 mm (6 in.) square and 19 mm (¾ in.) thick is used. Quarry tile has extraordinary wearing qualities, is non-slip, and has proved to be highly successful in service. Cove tiling is fitted at bulkheads and curbings. The underlayment is required to be somewhat higher than the minimum in order to obtain satisfactory slope for drainage to gutterways. For this reason, some designers prefer cement to a thickness of about 44 mm (1¾ in.) in preference to building up excessive thickness of the latex type underlayment.

Ceramic non-slip tile, 6 mm (¼ in.) thick, is used extensively and is very satisfactory in service. It may be hexagonal or square and of various colors which may be worked into pleasing patterns by the interior decorator. One piece covers are fitted around bulkheads and in way of shower stalls and a gentle slope is provided to the drains. In general, passenger and cargo ships utilize this type of treatment for toilet and shower spaces. On superliners an attempt was made to do away with this tiling because with its cement or latex underlay it substantially increased topside weight. Lightweight non-ferrous alloys of non-skid pattern have been used as shower-stall flooring, with the remainder of the bathroom covered with vinyl tile over minimum underlay. Generally speaking, however, this is done to minimize weight and to satisfy military considerations and should not be considered as normal practice for the intermediate type of passenger or cargo-passenger ships.

A special tiling problem is encountered in swimming pools, particularly the weather type used on semi-tropical runs. Not only is weight an important factor, but the adhesion of the tiling to the cement or latex underlay has repeatedly failed. None has proven too successful, requiring

Table 1—Typical Selection of Deck Covering Materials

Space	Material	Comment
PASSENGER AREAS		
Dining room, foyers	Vinyl tile, rubber tile	Some owners prefer carpeting
Theater, library, quiet lounges	Carpet	
Staterooms	Vinyl tile, carpet	
Bathrooms	Ceramic tile, terrazzo	On Luxury liners, tiling is provided on bhds. also
Stairs and landings	Vinyl tile, rubber tile	
Passageways	Vinyl tile, rubber tile	
Swimming pool tank	Ceramic tile	
Swimming pool beach	Ceramic tile, terrazzo	
Weather decks	Wood-latex	
Enclosed promenade decks	Wood, molded rubber tile	
CREW SPACES		
Staterooms, and passageways	Vinyl tile, terrazzo	Owners' preference
Recreation spaces	Vinyl tile	Owners' preference
Galleys, pantries	Quarry tile, ceramic tile, terrazzo	Owners' preference
Laundries	Terrazzo, ceramic tile	
Wet service spaces (lockers)	Terrazzo, ceramic tile	
Dry service spaces (linen lockers, baggage rooms, etc)	Gratings on painted steel deck	Generally wood

continuous and costly repairs. The tiling, with its coloring scheme, borders, etc., in combination with underwater lighting gives an extremely pleasing appearance, however, in a short time some of the tiling and underlay loosen up, fall off, and rust streaks begin to appear which are extremely difficult to correct without extensive repairs. One piece monel all-welded pools have been substituted in lieu of steel covered with tiling. While these are advantageous in reducing weight and cost of maintenance, they do not present as pleasing an appearance. Present day epoxy resin coatings have by now proven superior for pool applications.

4.6 Rubber Tile and Sheet. Rubber tile, available in three thicknesses, 6, 5, and 3 mm (1/4, 3/16 and 1/8 in.), forms a decorative covering, is very resilient, and has a smooth surface and dull gloss. A wide variety of designs is possible because the material is laid in individual blocks or tiles. It resists the action of water for cleaning but is not intended for prolonged immersion. It has excellent abrasive qualities and is long wearing. For large public spaces, sheet rubber with good service characteristics may be used to make a less expensive installation, lending itself to decoration by inlays. Rubber coves of dark color usually are fitted around the deck boundaries of all spaces where rubber tiling or sheeting is fitted. The cove can be eliminated in all spaces, except public rooms, by using stainless steel or nonferrous metal baseboards. Sheet rubber tile is often used in hospital spaces.

4.7 Vinyl Tile. Vinyl tile is the most widely used of all finished deck covering on all types of ships. There are various types, the best of which is a homogeneous vinyl, 3 mm (1/8 in.) thick available in plain, marbleized, or terrazzo-effect colors. Due to its dense surface, it is easy to maintain. A rubber or vinyl-set cove is used in the periphery of the space, available in 100 or 150 mm (4 or 6 in.) height. Laminated vinyl tile is a thin veneer of vinyl laminated to a backing. It is also available in 3 mm (1/8 in.) total *thickness,* has decorative qualities comparable to homogenous tile and is more economical but is not as long wearing. A light weight fire-retardant type of vinyl asbestos tile, 2 mm (0.8 in.) thick, used mainly on naval ships, also is available.

4.8 Carpeting. On both passenger ships and cargo ships, carpeting is used extensively in both staterooms and public rooms. Carpeting is usually fitted wall-to-wall with concealed nailing strips and customary carpet padding. Carpets are fitted over latex underlayment where no fire protection requirements exist or over the required thickness of Coast Guard approved deck covering where required for fire control. Carpeting is used extensively in public spaces and also in dining rooms, main stairways, and passageways. In some instances, carpets and paddings must be of wool, or

Table 2 — Underlays for Deck Covering

Latex mastic	from 8.5 to 12.2 kg/m^2 for 6 mm thick (from 1$^3/_4$ to 2$^1/_2$ psf (lb/ft^2) for 1/4 in. thick)
Magnesite scratch coat	13.7 kg/m^2 for 10 mm thick (2.8 psf for 3/8 in. thick)
Portland cement	61 kg/m^2 for 25 mm thick (approx. 12.5 psf for 1 in. thick)
Asphalt emulsion	12.2 kg/m^2 for 6 mm thick (2$^1/_2$ psf for 1/4 in. thick)

Table 3—Installed Weight of Deck Coverings

(The weight of underlay chosen must be added, per Table 2)

Magnesite	22.0 kg/m^2 for 10 mm thick (4.5 psf for 3/8 in. thick)
Latex mastic	12.2 kg/m^2 for 6 mm thick (2.5 psf for 1/4 in. thick)
Latex terrazzo	14.6 kg/m^2 for 6 mm thick (3 psf for 1/4 in. thick)
Sheet rubber or rubber tile	7.3 kg/m^2 for 3 mm thick (1.5 psf for 1/8 in. thick)
Ceramic tile	13.4 kg/m^2 for 6 mm thick (2.75 psf for 1/4 in. thick) (Cove 2.34 kg/m, 100 mm high) (Cove 1.57 pf 4 in., high; 2.34 pf 6 in. high)
Vinyl tile, homogeneous	9.3 kg/m^2 for 3 mm thick (1.9 psf for 1/8 in. thick)
Vinyl asbestos tile	8.5 kg/m^2 for 3 mm thick (1.75 psf for 1/8 in. thick)
Quarry tile	41.5 kg/m^2 for 19 mm thick (8.50 psf for 3/4 in. thick)
Mastic finish refrigerated spaces	Approx. 73 kg/13 mm thick (Approx. 15 psf for 1/2 in. thick)

other fire resistant materials, approved by the Coast Guard.

During foul weather at sea and during in port periods when the ship is crowded, carpet runners, cocoa mats, rubber matting, and the like are often rolled out in traffic areas. Outdoor carpeting of nylon or similar plastic material is occasionally used in passenger and/or officer decks in unenclosed areas.

4.9 Lightweight Outdoor Decking. As mentioned in the description of wood decking used aboard large passenger liners with extensive superstructures and weather decks, stability considerations have led to development of lightweight decking for closed promenades and weather decks. A corrugated rubber tile has been developed for closed promenades, which is constructed of units approximately 760 mm (30 in.) square, subdivided into smaller squares by grooves of uniform width and depth. The overall effect resembles a continuous sheet of rubber flooring because the grooves coincide in both directions. These installations have proved successful and have eliminated the need for constant planing and caulking of wood decks. In weather areas, a lightweight aggregate in rubber-latex, in common use on large naval craft, has been used as a substitute for wood decking. It is resilient, adheres well to steel, is waterproof and weatherproof throughout the weather temperature range, is very light in weight, will not burn, and is nonslip whether wet or dry. It is available in fast colors and lends itself to markings for deck games, directional signs, etc.

Other types of trowel-on finished deckings recently developed, offering light weight and good appearance, are vinyl plastics, magnesite with terrazzo mixture, latex with marble chips providing a terrazzo finish, and other combinations to produce various finishes. A special nonsparking aluminum abrasive tile tile is used in operating rooms.

4.10 Typical Deck Coverings. Table 1 indicates typical practice for deck coverings in various categories of passenger and crew spaces. Where more than one covering is indicated for any space, each type has been found satisfactory. Most ship operators have their own preferences so that a rigid standard finish for commercial ships is not possible.

4.11 Underlayments. For interior deck coverings, it is necessary to use underlayments under the finished covering for the purpose of smoothing out surface irregularities of the steel. Lapped steel decks seams always require an underlayment and flush welded decks, are seldom smooth enough to make a satisfactory surface for resilient coverings. When resilient deck coverings have been cemented directly to steel without underlayment, a drumming sound is caused by foot traffic. Humps and hollows are magnified, particularly when the finished deck has a polished surface, and changes in temperature tend to deteriorate the adhesive cement with which the deck coverings are bonded. The underlayments used are of the plastic, trowel-on type in the following categories:

- Those with magnesium oxychloride as a binder, commonly known as magnesite underlayments;
- asphalt emulsion as a binder, known as emulsion underlayments;
- liquid latex (rubber) binder, known as latex underlayments.

Of these three underlayments, the magnesite type requires welded clips or wire mesh for attachment with an average thickness or 19 mm (3/4 in.) and a minimum of 13 mm (1/2 in.). It is heavier than the others but cheaper than the latex type. The asphalt-emulsion type does not require anchoring clips if the steel surfaces are thoroughly cleaned of mill scale, oil, grease, etc., and may be applied to a minimum of 10 mm (3/8 in.) thickness. The latex type has come into general use in the United States because of its light weight, good bonding quality directly to steel and, relatively small thickness required. Generally, a 10 mm (3/8 in.) thick underlayment will suffice for most installations. It is highly resilient, and was originally developed and used extensively on naval ships. It is used as a backing for tile, carpeting, magnesite, and all of the deck covering material described herein. Whenever evaluating materials for deck coverings, consideration must be given to Coast Guard fire protection regulations, as they place specific limitations or thicknesses and location of combustible deck covering materials aboard some vessels. Only approved deck coverings evaluated under Coast Guard Rules 46 CFR 164.006 may be used without restriction.

Fig. 21 shows a typical deck covering detail with magnesite, emulsion, latex-type underlayment, and other deck coverings of the customary types. The underlayments for all finished deck coverings are listed in Table 2. Table 3 gives the installed weight of various types of deck coverings, including the underlay used, thicknesses, etc.

Section 5
Joiner Bulkheads, Linings, Ceiling, and Insulation

5.1 Introduction. *Joiner work* is the term generally applied to those materials used for the construction of the finished interiors of compartments which were formerly referred to as the woodwork in a steel ship. Bulkheads especially in accommodation areas generally consist of joiner work which provides livable, workable, decorative spaces and in addition, the joiner work can be used to subdivide larger into smaller areas. Such bulkhead panels, linings and ceilings with their connecting devices form the joiner work of a compartment. The current use of noncombustible materials in ship construction, in place of wood, involves a variety of materials such as inorganic composition panels, metallic core section materials clad with decorative board, light gage steel plates and shapes, and decorative hard and soft plastic laminates of specified thickness and fibrous insulation. Joiner work thus involves a complex collection of materials grouped together primarily because they have replaced wood, the design, construction and erection of these materials constitutes joiner work.

Joiner work, originally a means of subdividing a ship for reasons of utility or privacy, nowadays provides fire protection, thermal insulation and acoustical protection, comfort, sanitation and decor. The achievement of a satisfactory balance of all these functions is difficult for the ship designer bearing in mind that a thorough knowledge of regulatory body requirements is of prime importance also.

5.2 Regulatory Body Requirements. The most important regulations affecting joiner work deal with fire safety measures. Both the United States Coast Guard and the International Convention for the Safety of Life at Sea (SOLAS) exert a dominating influence in specifying the locations for and types of structural fire protection and in defining the materials which may be used.

There have been several revisions in the SOLAS requirements for fire safety starting with the 1929 Convention followed by those of 1948, 1960 and 1974 involving successive improvements in structural fire protection. Present day USCG Rules and SOLAS regulations embody the following principles, having regard for the type of ships and the potential fire hazard involved:

- Division of passenger ships into main vertical zones by thermal and structural boundaries;
- Separation of accommodation spaces from the remainder of the ship by thermal and structural boundaries;
- Restricted *use* of combustible materials;
- Detection of the fire in the zone of origin aboard passenger ships;
- Containment and extinction of any fire in the space of origin;
- Protection of means of escape or access for fire fighting;
- Ready availability of fire-extinguishing appliances;
- Minimization of possibility of ignition of flammable cargo vapor.

While the above principles apply to all types of vessels, specific regulations are graduated depending upon the age, type and size of ship, route of operation, cargo carried, and the number of passengers and crew. The maximum degree of fire and life safety is required by large passenger ships operating in ocean service; less stringent requirements apply to cargo ships.

a. Classes of Fire Integrity of Bulkheads and Decks. Definitions have been established for the integrity of bulkheads and decks which apply to all types of ships. Generally the definitions are common to both the 1974 SOLAS and USCG Rules except where stated.

"A" Class Divisions must be constructed of a minimum 11 USSG steel or equivalent metal construction, suitably stiffened and capable of preventing the passage of smoke and flame for one hour when exposed to a standard fire test. They should be made intact with the main structure of the vessel such as shell, structural bulkheads and decks. Additionally they may require insulation with approved non-combustible materials capable of meeting limited temperature-rise requirements intended to prevent the advance of fire from one space to another laterally or vertically. Higher temperature rise time limits are employed where the spaces affected are critical or have significant amounts of combustibles. The divisions are designated alpha-numerically. The numerical portion of the designation will indicate the amount of time that the insulation is capable of limiting the temperature rise as follows:

Class "A-60"	60 minutes
Class "A-30"	30 minutes
Class "A-15"	15 minutes
Class "A-0"	0 minutes (uninsulated)

"B" Class Divisions must be constructed with a minimum 16 USSG steel or approved non-combustible materials capable of preventing the passage of flame for one-half hour when exposed to a standard fire test. They should be made intact from deck to deck (or ceiling to ceiling) and to shell or other boundaries. In addition their insulation value shall be capable of limiting the temperature rise as follows:

Class "B-15"	15 minutes
Class "B-0"	0 minutes

The USCG Rules require the use of approved non-combustible materials for all "B" Class divisions.

"C" Class Divisions must be constructed of approved non-combustible materials and need meet no requirements as regards the passage of smoke and flame nor the limiting of temperature rise. Non-combustible materials are Coast Guard approved under 46 CFR 164.009, and neither flame nor give off flammable vapors in sufficient quantity for self-ignition when heated to approximately 750° C (1,382° F). Any material with more than a very small percentage of organic content cannot qualify as being non-combustible.

The definitions of "A" and "B-" Class Divisions necessitate careful attention to piping, electrical cable, and ventilation ducting during penetrations so that the soundness of the fire integrity is not impaired.

b. SOLAS 1974 Construction Methods. The minimum requirements for fire protection are established here for all countries which have signed the 1974 Convention (or acceded to it of any signature or deposited any instrument of ratification, acceptance, approval or accession). In many cases the requirements of the 1974 Convention are included as part of the national Rules and in certain cases the national Rules go beyond the requirements of the 1974 Convention.

Main vertical zones are those sections into which the hull, superstructure, and deck houses are divided by "A-" Class divisions, the mean length of which on any one deck does not in general exceed 40 meters (131 ft). High risk spaces such as machinery spaces, galleys, cargo holds, and spaces in which flammable liquids are stowed must be enclosed by "A-" Class divisions. Stairways also are enclosed in "A-" Class divisions to ensure a protected means of escape.

The most important improvement in fire safety requirements in the SOLAS 1974 regulations from a ship design standpoint is the change from three separate methods of construction prescribed by SOLAS 1960. For a better understanding these are summarized as follows:

• *Method I.* The construction of all internal bulkheads, other than those required to be "A" Class, of incombustible "B-" Class divisions, generally without the installation of detection and sprinkler systems in the accommodation and service spaces. All bulkheads, grounds, and insulation were required to be of non-combustible material except for very limited quantities of combustible trim and veneer. Low flame-spread properties were required of exposed surfaces in corridors, stairway enclosures and concealed spaces. This concept called for very stringent control of the combustible materials used in interior construction and insulation with the intention of reducing the scale of any fire, and preventing it from progressing beyond the space of origin. This method was the only one permitted for U. S. construction from the time of the investigation (about 1936) of the *Morro Castle* disaster.

• *Method II.* The fitting of an automatic sprinkler and fire alarm system for the detection and extinction of fire in all spaces in which a fire might be expected to start with no restriction on the type of internal bulkheading in the spaces so protected. Concealed surfaces of all bulkheads, linings, etc., were required to have low flame-spread characteristics. This method, preferred in the United Kingdom, relied on action by a mechanical system. While the sprinklers were conceded to be capable of extinguishing a fire, these systems are vulnerable to being overridden once a fire gains any scope. Furthermore, ships so fitted often lacked fire protection during critical in-port or shipyard periods.

• *Method III.* A system of subdivision within each main vertical zone using "A" and "B-" Class (non-combustible) divisions together with the fitting of an automatic detection system in spaces where a fire might be expected to start. The use of combustible materials for the construction of other bulkheads and furnishings was only marginally restricted. This method, used in France and other countries, relied on (1) early warning of a fire by a mechanical system and (2) positive follow-up action by the crew. The failure of one, the other or both was not unusual.

SOLAS 1974 adopts one method with the central requirement constraining the amount of combustible material employed in ship construction. Alternatives are permitted within this method so that it is possible to include sprinkler systems in certain specified spaces for which installation a reduction in boundary insulation will be permitted.

SOLAS 1974 effectively eliminates Methods II and III and adopts the long standing U. S. principle (Method I) that a minimum amount of combustible material should be employed in ship construction. However, sprinklers and fire alarm systems are still called for as an ancillary protection against catastrophic fires.

c. U. S. Coast Guard (USCG) Rules. These are based upon minimizing the probability of a fire assuming major proportions by arranging structural elements to confine any outbreaks of fire to as small an area as possible. This is done by specifying fire endurance capabilities of structural elements in accommodation and service spaces in recognition of the likelihood of a fire arising in or advancing through the spaces involved.

In the U. S. Coast Guard Rules, fire-resisting bulkheads and decks are defined in greater detail than the general classifications given in Section 5.2 a. Exact requirements are contained in Coast Guard Rules 46 CFR 164.007 (structural insulation), 46 CFR 164.008 (bulkhead panels) and 46 CFR 164.006 (deck coverings). Since steel has a relatively high heat conductivity, the required insulation value must be achieved by addition of bulkhead panels, insulation, or deck coverings arranged to limit heat transmission through the bulkhead or deck.

A bare steel bulkhead could, of course, qualify as an "A-" 0 division since there are no heat transmission limitations. Bare aluminum, on the other hand, could not qualify as an "A"-0 bulkhead, because aluminum melts at a relatively low temperature, 649° C (1200° F), and loses its structural strength at an even lower temperature, 232° C (450° F), and will not remain intact for the one-hour exposure to the standard fire test required of "A-" Class divisions. Thus, to qualify as being an "A-" Class division or as being "equivalent to steel," aluminum must be insulated.

"B-" 15 and "B-" 0 divisions are generally various types of panels which have been tested and found to meet all of the classification requirements. In the case of a "B-" 15 division, the required insulation value is generally achieved by the panel core and not by external insulation. Bare steel would qualify as a "B-" 0 division, but would require insulation to meet "B-" 15 requirements. Aluminum, again because of its low melting point, requires insulation to meet either "B-" or "B-" 15 classifications.

The construction classifications required for any bulkhead or deck aboard passenger vessels can be determined from the U. S. Coast Guard Rules. There are two tables in the Code of Federal Regulations, Title 46 one covering all bulkheads which form a portion of the main vertical zone bulkhead, and another for all other bulkheads. Similarly, one table classifies decks which form part of a stepped main vertical zone, and another classifies all other decks.

The tables are entered with the types of spaces to be separated by the division in question, all spaces being divided into 14 categories. Table classifications and resulting boundary requirements are based upon consideration of the following:

- Probability of fire originating in a particular space;
- combustibility and quantity of probable contents;
- size of space;
- combustibility and quantity of probable contents in the adjacent space;
- importance of the space in terms of the overall safety of the ship.

The highest degree of protection is required between a space which must be protected, such as a stairway enclosure or control station, and a space where a considerable fire hazard exists, such as a machinery space or galley. Since heat and smoke generally travel upward, the required deck insulation is greater when a more hazardous space is located below, rather than above, a less hazardous one.

A difference in boundary requirements of accommodation spaces exists, depending upon the furnishings and decor used in the space. If, for example, fire-resistant furnishings and surface finishes are employed, boundary requirements are less stringent than if combustible furnishings are used.

USCG Rules also affect joiner design in other respects. Combustible veneers on panels may not exceed 1.5 mm ($^1/_{16}$ in.) thickness on each side of the panel. The total amount of combustibles used in a space for trim, moldings, and decorations, including veneers, cannot exceed a volume equivalent to 2.5 mm ($^1/_{10}$ in.) thick veneer on the combined area of the walls of the space. Veneers used in passageways, stairway enclosures, and hidden spaces, in addition to being limited in thickness, must display low flame-spread and smoke development characteristics when tested in accordance with 46 CFR 164.012, a standard rate of flame-spread procedure. A number of materials have been tested and found to meet the required rating; a listing may be obtained from the U. S. Coast Guard.

"B-" Class bulkheads, with the exception of those forming passageway boundaries, may stop at the ceiling or lining and need not continue to the deck or shell, provided that close-fitting draft stops are fitted behind the ceiling or lining at intervals not exceeding 13.72 m (45 ft). This draft stop resembles a vertical bulkhead of "B-" 0 construction or alternatively 22 USSG steel.

All methods of construction "A-" and "B-" Class divisions and all non-combustible materials are subject to approval by the U. S. Coast Guard. Fire-resistant draperies and linings required to exhibit low flame-spread characteristics are generally tested in accordance with U. S. Coast Guard Rules or national consensus specifications by independent laboratories.

Wood hatch covers may be used between cargo spaces, but hatch covers in other locations must meet the required deck classification in that location. All insulation must be non-combustible except insulation of certain refrigerated cargo spaces.

The categorization of vessels into passenger, cargo, tank, and other ships under the USCG Rules have the same basic principles, except that the application of these principles varies. SOLAS 1974 embodies some different requirements than U. S. Coast Guard Rules.

d. U. S. Public Health Service. All ships built under the U. S. flag are required to follow the standards laid down by the U. S. Public Health Service (USPHS, 1965). Customary practice during the design and construction of a ship is to submit a plan schedule to the Public Health Service which will in turn indicate which particular plans will be required for review and approval. Compliance with USPHS regulations can be accomplished without excessive cost or delays. These regulations concern themselves mainly with structural or joiner work construction to prevent the harboring of dirt, pests (in the form of insects or rats and efficient sanitary maintenance of decks, bulkheads and ceilings) with considerable emphasis on the rat proofing.

5.3 Design Rules. The designer must not only assure that regulatory structural fire-protection requirements are met but must also minimize the amount of joiner work required, integrating fire protection with other functions required of the joiner work, considering maintenance required with various construction methods, and providing an attractive decor. The designer may reduce the required joiner work and insulation through basic arrangement and specifications of the ship. This is done by controlling relative positioning of spaces, size of public rooms, arrangement of corridors and stairways, and specifying furniture and furnishings of noncombustible or fire-resistant type.

A further control is exercised through use of the required insulation and joiner work for other purposes such as temperature insulation and sound reduction. If a ceiling is to be fitted for habitability reasons, it should be made to contribute to fire resistance. The same approach applies to deck coverings. If fire protection of the deck is required, use of a magnesite-type deck covering will contribute a portion of that protection. If insulation is to be fitted for heat or sound control, use of proper materials will also permit the insulation to form a portion of the required fire insulation.

Material handling and maintenance are very important considerations. Routine maintenance required by various types of joiner construction should be considered. Use of a melamine or vinyl finish may mean, for example, that little attention need be given to maintaining decor. On the other hand, painted surfaces have a lower initial cost but require more maintenance, which can be an important economic factor over the life of the ship. It is essential to have finish materials, particularly on bulkheads that can withstand the abuse of traffice in the spaces where the surface finish is employed. Ease of installation of various types of joiner panels, insulation, and surface finishes is also an important factor. Good joiner design also has a beneficial effect on the atmosphere of the ship, since the mood of the passengers and crew can be greatly influenced by their surroundings. Economy of fire-control joiner work, then means careful attention to all aspects of the functions which a bulkhead or deck is required to perform.

a. Joiner Bulkheads, Types and Details. The following discussion is limited to types of joiner-bulkhead construction which have been tested and approved by the U. S. Coast Guard and applies to "B-" and "C-" Class bulkheads which are primarily intended for use in accommodation spaces as non-loadbearing members and designated "B-"15, "B-"0, and "C". A "B-"15 bulkhead is required to have an insulating value which can prevent the transmission of heat for 15 minutes. Bulkheads to meet "B-"0 classification are similar, but in the case of composition panels, need not be as thick, and in the case of hollow metal panels, need not be insulation filled. Bulkheads of "C-" Class are used when it is desired to install bulkheads as room dividers. They need serve no structural fire protection purpose but must not contribute to the fuel load of a compartment in which they are installed and are required to be non-combustible. Such bulkheads fall into four types of construction.

1. Single composition-panel.
2. Double composition-panel.
3. Hollow metal-panel.
4. Double metal-panel.

Single composition-panel type of construction consists of panels which are manufactured to standard size, cut to size on the job, and erected on a framework of steel channels and shapes. The steel framework must receive its support from the basic structure of the ship. Joints may be covered finally with metal moldings and trim pieces. Panels are constructed of an inorganic composition of gypsum, and vary in thickness 19 to 22 mm (3/4 to 7/8 in.), excluding veneers. The gypsum panel normally has a steel, aluminum, or equivalent veneer.

1. Inorganic Material Marine Joiner Panels. Panels of reinforced calcium silicate and other inorganic materials are not required to have veneers, but veneers may of course be applied. All types of wood have been used. Sheet steel or marine veneer of a harder grade than the core material make suitable veneers for locations subject to knocks and scrapings, such as passages and stairways. Plastic veneers which are practically indestructible are now available in many colors and patterns. If a veneer is applied to one side, the panel will warp unless the other side is also veneered. The two veneers need not be the same, so that a concealed veneer may be a cheaper material than an exposed one. It will be recalled that veneers used in corridors, stairways, rooms with fire-resistant furnishings, and hidden spaces are required to exhibit low flame-spread characteristics.

In spite of the relatively higher first cost, current practice is to avoid painted surfaces in accommodation spaces by using permanently finished surfaces, generally of melamine plastic.

2. Composition-Panel Bulkheads. Composition panel bulkheads generally of gypsum and steel construction are erected by different methods, varying in detail but the same in principle. The bottom of the panel rests in a base channel secured to the steel deck. The channel may follow the camber or be cut to form a series of horizontal steps, so that bottom edges of panels need not be cut at an angle. The tops of the panels are retained by channels or clips secured to the steel structure above. The vertical edges of the panels are retained by vertical steel members called pilasters, which must overlap the entire edge of the panel front and back by 19 mm (3/4 in.). The pilaster is made of two pieces in such a way that, by taking up on screws, the two pieces are drawn together and squeeze the panel edge, holding it securely. For "B-"15 bulkheads, a joint must be selected which will limit temperature rise on the unexposed face to 225° C (405° F) above ambient when exposed to the standard fire test for 15 minutes. Metal trim or molding pieces, usually of the snap-on type, are applied along the base of the panel, along the intersection of bulkhead and ceiling, and on the faces of pilasters. Another type of joiner construction procedure once again becoming popular is the spline joint. With this method of joining panels, a groove is cut in the edge of each panel. The groove is fitted over a steel spline anchored at the deck and deck head.

The double composition-panel type of joiner bulkhead is constructed of panels similar to those just described but of less thickness. The panels are secured to both sides of a steel framework and an space of about 50 mm (2 in.) is left between. The total thickness of this type of bulkhead is approximately 76 mm (3 in.)

3. Hollow Metal Panels. Hollow metal panels are of either steel or aluminum sheet or extrusions suitably stiffened and insulated. In the case of hollow steel panels, insulation, if required, is generally applied internally. For aluminum panels, insulation may be applied either internally or externally. The thickness of this type of panel generally varies from 31 to 61 mm (1¼ to 2⅜ in.). The insulations used are mineral and glass wool and combinations of these. In some, the wool insulation fills the inside of the panels, and in others an air space is left unfilled. Hollow metal panels may be erected in a framework of channels and pilasters as used for the single composition panels, with one important difference; since most hollow metal panels are made in standard widths with the metal formed to make the panel edge, they cannot be cut to width on the job. Therefore the pilasters must be spaced to suit the ordered panel widths. Most of the panels are trimmed to length. Another and more common method of erecting hollow metal panels dispenses with the pilasters altogether. Edges of panels interlock in such a way that only a line appears at the intersection.

The double metal-panel type of joiner bulkhead is constructed of single thickness metal panels, edge stiffened and insulation backed. The panels are secured to both sides of a steel framework and an air space is left between. Thickness varies from 38 to 61 mm (1½ to 2⅜ in.). Fig. 22 shows the various types of panels described.

5.4 Practical Requirements. Aside from the required fire-resistance qualities, an ideal "B-"15 or "B-"0 bulkhead must combine ample strength with minimum weight, be immune to mold and vermin under moist conditions, and afford satisfactory resistance to heat and sound. Other desirable characteristics are ease of handling, working, and installing, a satisfactory screw-holding power and an absence of drumming. A number of these additional performance characteristics are required by Maritime Administration specifications for composition panels. Of the many joiner bulkheads available from the manufacturer, the designer must make a selection for a specific ship and specific locations on a ship. This selection must be based on cost, weight, space requirements, thickness, durability, maintenance, surface treatment desired, and appearance in general. Each type has its advantages and no one type would be the best selection for every situation. The single-composition panel has the advantages of simplicity, lightness, minimum thickness, and of being easily cut and fitted on the job. It has one important disadvantage compared to the other types; i.e. lack of space for concealed wiring and piping.

5.5 Electrical Systems Installation. Wiring is commonly

(*Continued on page 402*)

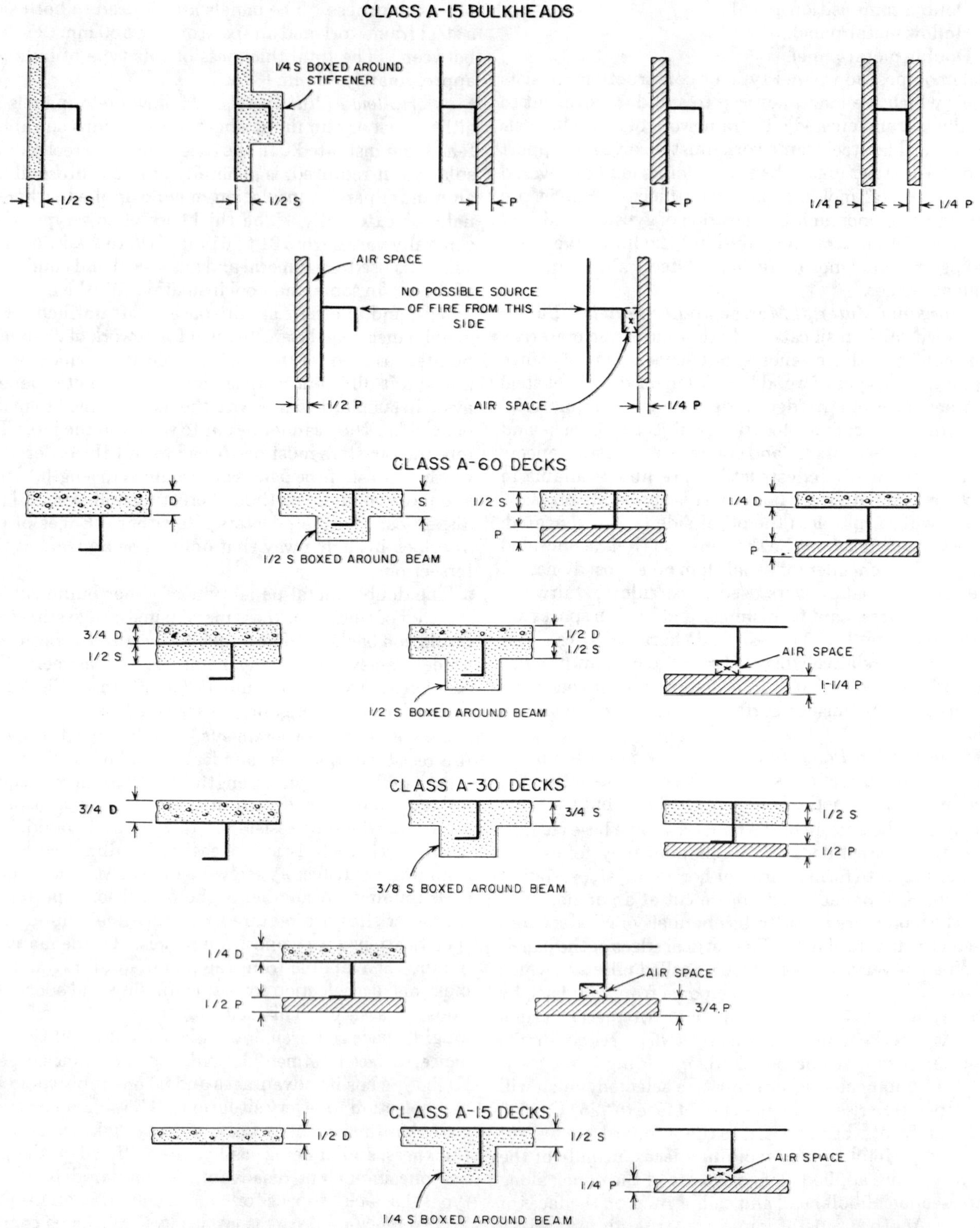

Fig. 22 Application of insulation, linings, ceilings, and deck coverings to Class A fire divisions

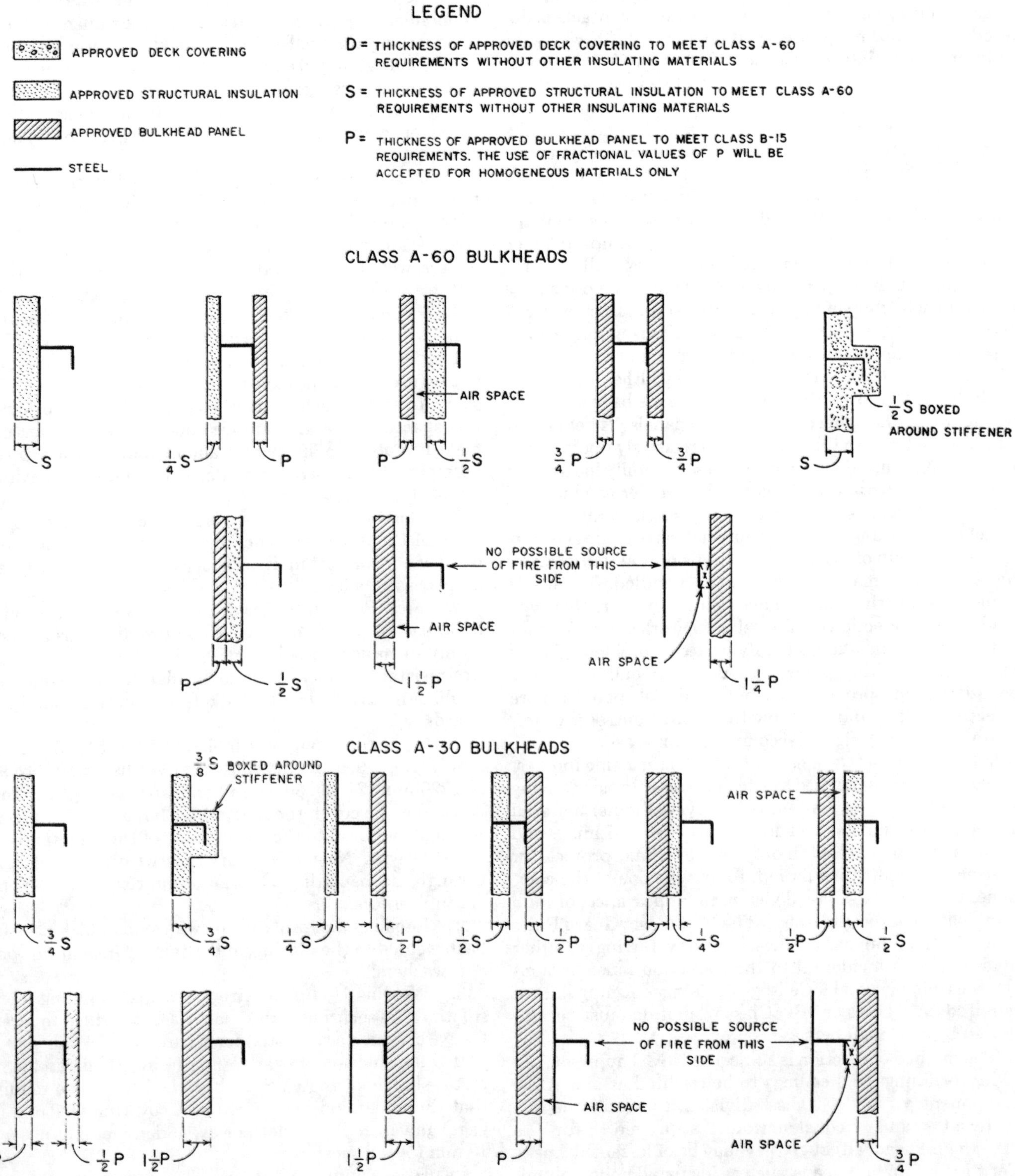

Fig. 22 Application of insulation, linings, ceilings, and deck coverings to Class A fire divisions (cont.)

(Continued from page 399)

concealed within double or hollow joiner bulkheads construction, but careful planning is required. All vertical wiring must run in the small space inside the pilasters. Switches and connection boxes must be of the minimum size for recessing in the pilasters, which must be located in advance at every position where a switch or receptacle is desired. Each normal pilaster will take care of the wiring for only one such fixture. Corner pilasters and door posts, where switches are usually placed, are a little roomier. Since the metal trim for pilasters projects about 2 mm ($^1/_{16}$ in.) from the face of the bulkhead panel and may not have the same surface finish as the panel, the pilasters are quite obvious. For this reason, in passenger quarters, it is customary that the pilasters be located to bear some visual relation to furniture along the bulkhead or at least to achieve an appearance of symmetry. Thus, for passenger ships, pilaster locations must be planned carefully to satisfy all requirements, although the planning would have to be done only once for each group of identical staterooms. In crew quarters and in cargo ships, where exposed wiring may be acceptable and appearance is less important, pilasters may be placed to conform with standard panel widths.

Hollow metal or double-panel bulkheads have space for wiring within the panel or between the panels. These types lend themselves to a flush effect with relatively unobtrusive panel joints which do not have to be so carefully located for appearance. However, the generally greater thickness of these bulkheads as compared to the single composition-panel type can amount to an appreciable percentags of lost space in a group of staterooms. In the double-panel bulkheads, twice as many panels must be handled, fitted, and secured as for the single-panel types; however, this type lends itself to erection of all-steel framework and deferment of the panel installation, simplifying access problems during installation of wiring, piping, etc. All "B-"15 bulkheads have adequate sound-proof factors, but the double panel is more effective for this purpose now that noise control is of importance in the design of accommodation areas.

5.6 Linings and Ceilings. A lining, as the name implies, is a joiner material applied as a facing over ship side framing and steel bulkheads. A ceiling is a layer of joiner material applied to the underside of an overhead deck. Linings and ceilings may be applied in order to cover and protect, for thermal or comfort insulation, to provide one of the components of an "A-" Class division, for appearance, or for a combination of these reasons. The U. S. Coast Guard Rules requires that linings, ceilings, and any furring or other holding pieces incidental to their erection must be noncombustible material. Ceilings or linings forming part of required "A-" Class or "B-" Class insulation must be supported by steel members.

The method of erection is subject to USCG approval. In order for linings and ceilings to be credited as one of the components of an "A-" Class division, they must be of a material and construction approved for a "B-"15 bulkhead, although they may be of lesser thickness. Certain "B-"15 bulkheads, such as insulated hollow aluminum, are not suitable as a component of "A-"60 divisions.

It is quite common for a designer to specify linings to be constructed of panels suitable for use as components of free standing "B-"15 bulkheads, even where regulations would permit thinner panels. Thinner panels would require more closely spaced steel furring at the shell or steel bulkhead so that the independent, free-standing construction is more economical. Such construction has the added advantage of matching other "B-"15 bulkheads bounding the same space, and giving greater insulating value. Much of the foregoing discussion of "B-"15 bulkheads also applies to linings except that no matter what lining is used, there is space for concealed wiring and piping between the lining and the steel.

Ceilings are usually applied to steel furring strips attached to the steel deck overhead. When ceilings form a portion of the required fire insulation, they must not depend upon screw attachments for their support, because panels which are screwed in place tend to crack under fire exposure and fall away. Where the joints between panels are not required to be covered with a steel strip overlapping the edge of the panel by 19 mm ($^3/_4$ in.), the panel edges may be beveled and left exposed. The heads of the screws fastening the panel to the furring may be exposed or countersunk and spackled over. In any case, joints between panels usually are obvious. For spaces where appearance counts, the designer must make layouts of ceiling panels, aiming toward symmetry and harmony of room arrangements, consistent with maximum use of standard panel widths.

5.7 Insulation. Insulation is installed to perform one or more of the following functions, i.e., fire protection, temperature or comfort in living spaces, sound deadening, etc., as discussed below.

a. Fire Protection Insulation. Insulations used on bulkheads or decks for "A-" Class fire divisions are commonly of mineral wool or glass fiber, held together by a fire-resistant binder. These insulations are commonly applied in the form of blankets or batts or semi-rigid boards.

Batts are commonly available in 25, 38, 51, 76, and 102 mm (1, 1.5, 2, 3, and 4 in.) thickness, widths of 610, 762, 915, or 1220 mm (24, 30, 36, 48 in.) and various lengths. Insulation may be cut in the field with a large, sharp knife and is usually secured to the steel by one of three methods:

1. Driving Nelson pins or studs with attached heads through the insulating material on approximately 305 mm (12 in.) centers;

2. Impaling the insulation on wire studs which have first been welded to the plating, and retaining insulation *speed clip* washers;

3. Welding U-clips of wire to plating, impaling the insulation and securing with square washers drilled to receive the wires which are twisted for securement.

If the insulation is over 25 mm (1 in.) thick, and is not metal-sheathed, the U. S. Public Health Service requires that the insulation have a ratproof covering of 1 mm (18 gage) galvanized wire cloth or expanded metal of not over 13 mm ($^1/_2$ in.) mesh.

Insulations approved for "A-" Class construction are also available as sheet or block glass fiber and in plaster form.

Although this latter generally requires less thickness, the installed weight is greater than the batt type because of the higher density of the materials. The sketches in Fig. 22 show how insulation is used in "A-" Class fire divisions alone and in combination with linings, ceilings, and deck coverings.

b. Thermal or Comfort. Thermal or comfort insulation is insulation used to protect living and working spaces from cold or heat in adjacent spaces or the outside environment. Some insulation is required between each heated or air-conditioned living and working spaces and the following:

- Outside air or water;
- Unheated spaces;
- Heat-producing spaces such as machinery spaces and galleys.

Further, since living spaces are now customarily air-conditioned, they must be insulated from any adjacent uncooled space. The design conditions and theory of heat transmission are dealt with in detail by Holm and Woodward (1971). This section deals only with the practical aspects and construction practices.

Comfort insulations are similar to fire insulations except that lighter densities may be used. Densities as low as 8 kg/m^3 ($\frac{1}{2}$ lb/ft^3) have been used but 32 to 48 kg/m^3 (2 to 3 lb/ft^3) is more common. Installation methods are the same as for fire insulation. Materials used for linings, ceilings, and deck covering contribute also to the insulating value of the bulkhead or deck. It is particularly important that construction be such that there is no short circuit for heat through beams and stiffeners. This all-steel short circuit can be avoided by the following means:

1. Applying insulation to flush side of steel. This is impractical in many places such as shell and exposed decks.
2. Carrying insulation around beams and stiffeners.
3. Using insulation of excess thickness to project beyond beams and stiffeners and finishing off flush.
4. Applying insulating board lining or ceiling across face of stiffeners or beams.
5. Applying steel lining or ceiling, separated from flanges of beams or stiffeners by insulating board blocks.

In general for any one type of insulation, the desirable thickness would depend upon the difference in temperature between the hot and cold side and the exact details of construction. In marine work it has become common practice to use a minimum of 51 mm (2 in.) of insulation at the shell and deckhouse sides, and 76 or 102 mm (3 or 4 in.) against machinery spaces and galleys. Under exposed decks, a minimum of 51 mm (2 in.) is used.

The designer seldom specifies insulation and lining with one purpose only in mind. Around machinery spaces and galleys, the fire control requirement would normally govern, so that fire insulation would also serve as comfort insulation. The problem is always to pick the minimum insulation which will satisfy all requirements, including that of appearance.

c. Fire Insulation for Aluminum Structures. The procedure for establishing adequate fire protection for the hull structure of aluminum vessels and/or aluminum deck houses mounted on steel hulls, is different from the standard procedure in the U. S. Coast Guard regulations, discussed above, which assumes that the fire loading in given spaces is the same for every vessel. The SNAME T&R Bulletin "Recommended Guidelines for Aluminum Construction" is followed by the U. S. Coast Guard, as well as ship designers and builders, in estimating the proper amount of insulation for adequate fire protection of aluminum under various service conditions. The SNAME Guidelines are based on a direct approach which evaluates the actual fire loading conditions for each particular space and circumstance, taking into account the quantity of combustible material in the compartment in relation to the surrounding deck and bulkhead areas. The amount of insulation used to protect the aluminum structure is then decided in accordance with the specific amount of combustibles per unit of deck or bulkhead area, enclosing the space in question. As an added reference, SOLAS 1974 also discusses fire protection in aluminum structures.

d. Sound or Acoustic Insulation. The development of all-steel ship structure, the introduction of relatively lightweight, high-speed propulsion machinery, and the widespread use of air-conditioning equipment with its associated fans and airducts, have produced a need for noise reduction by means of sound or acoustic insulation. Fortunately, the modern fireproof insulations of the glass fiber types are effective sound absorbers. Fire and comfort insulation, when applied to the inside of a machinery casing and lined with sheet steel, becomes a reflecting surface which aggravates the machinery noise. When lined with a material perforated to admit the sound waves, the insulation reduces the internal noise level considerably without any sacrifice in its other functions.

Lowering the noise level in a space wherein the sounds originate is commonly done by applying sound-absorption treatment to the overhead, bulkheads or decks and by insulating the sound generator from the structure. Public spaces are treated overhead with sound-absorption insulation to lower the noise level and permit normal conversation. Sound-absorption treatment of boundary bulkheads, ceiling, or deck, does not technically form a barrier to the passage of sound but has some effect in the space sound control by lowering the reverberant level within.

Lowering the noise level in a space, when the sound originates outside that space, requires some degree of sound isolation, either of the space to be quieted or the point of origin. Severe cases may require both sound-absorption and sound-*isolation* treatments. A case that occurs frequently on a ship is a noisy space bounded by steel bulkheading and adjacent to accommodations. Here it is desirable to put sound-absorption treatment in the noisy space and install joiner lining on the accommodation side, entirely independent of the steel bulkhead. This treatment lowers the noise level in the accommodation space far more effectively than merely applying insulation between the steel bulkhead and a joiner lining connected thereto.

The insulations used for the control of sound are also commonly of mineral or glass wool. For sound absorption of conversation, 25 mm (1 in.) of insulation on the overhead is commonly used. This insulation must be sheathed with a material perforated to admit the sound waves into the insulation. Perforated non-combustible inorganic board 4–5 mm (3/16 in.) thick is commonly used in accommodation spaces. The insulation and sheathing already combined into a unit are available commercially. These units are secured to metal furring with sheet metal screws through metal grommets which fit in holes in the sheathing.

For sound absorption on a larger scale, such as in machinery spaces or fan rooms, the insulation is kept separate from the sheathing and applied directly to the steel. Two 50 mm (2 in.) layers of insulation separated by felt are frequently used for very noisy conditions.

e. Sound Isolation. Sound isolation is achieved by sound-absorption treatment as described, plus the use of isolators. Isolators are nonrigid attachment fittings made of springs, rubber, or felt which do not transmit vibration and reduce the telegraphed, direct-contact noise. Isolators are applied to both the source of noise and to the partition between the noise-producing space and the space to be protected. For example, to sound-isolate an accommodation space from a noisy fan room, the fans are mounted on isolators, and joiner treatment is applied as follows depending on the relative positions of the rooms:

1. Accommodation space directly below fan room. Apply insulation and perforated sheathing to ceiling of fan room. Apply insulation overhead in accommodation space. Tight joiner ceiling, bulkheads, and lining in accommodation spaces to be acoustically separated from ship's structure by mounting on isolators.

2. Accommodation space directly above fan room. Apply insulation and tight joiner ceiling, suspended by isolators, overhead in fan room. Apply insulation and perforated ceiling overhead in accommodation space. Lay carpet on deck of accommodation space if possible.

3. Accommodation space next to and on same level as fan room. Apply insulation and perforated ceiling overhead in fan room. Apply insulation and solid, sound-isolated ceiling to overhead in accommodation space. Apply insulation and sound-isolated joiner lining on accommodation space side of steel bulkhead next to the fan room.

Noise from a fan room may be transmitted some distance through the duct structure unless flexible connections are installed between the fans and ducts. Insulation lined sound traps may be installed in the duct on each side of the fan. Fast-moving air inside of ducts and air entering or leaving ducts may be responsible for considerable noise. This noise may be minimized by proper design of ducts, interior directional baffles, inlets, and outlets, and by keeping air velocities reasonably low. Ducts and plenums in noisy locations frequently are lined with thin layers of noise-absorbing insulation or felt.

Machinery spaces and fan rooms are the most common sources of noise on a ship and, on a passenger ship, almost always require sound treatment. Whether treatment of other spaces is required depends on the individual arrangement. Spaces which should be checked in the design stage include the galley, pantries, radio room, and steering-gear room. Insulation used for sound control should be coordinated with comfort and fire control insulation.

5.6 Refrigeration Joiner Work. Apart from the refrigerated storerooms common to all classes of vessels many general cargo ships have holds or 'tween-deck spaces devoted to refrigerated cargo and a number of containerships are partially refrigerated. The design and specifications for refrigeration joiner work have become increasingly important as greater quantities and more diversified perishable products are carried. The growing practice of carrying products in quick-frozen form, (especially in containerships) with the associated low temperatures, tends to increase the emphasis on insulation efficiency.

Joiner work in refrigerated spaces may be divided into two: 1. The insulation itself, applied to the deck, bulkheads and overhead; 2. the interior finish of the space, the lining of the bulkheads and overhead, and the deck covering.

a. Types of Insulation. The quality which good insulations have in common is low heat conductivity, which is due to air trapped in the insulation, no matter of what material or in what form. Variations in materials, densities, binders, and size, color, and arrangement of fibers and cells, all affect conduction, convection, and radiation, which make up the insulation conductivity. Currently acceptable marine refrigeration insulations, the best of which have a conductivity of 0.14 to 0.28 at 21° C (70° F), fall into the following:

- Mineral and glass wool in batts and boards of various densities;
- cellulated glass blocks;
- polystyrene boards or foamed-in-place;
- polyurethane, sprayed, foamed-in-place or board type.

b. Properties to be considered. Conductivity, of course, is only one factor to be considered in the choice of an insulation. Insulations have other properties to be considered and a choice is seldom justified on the basis of a single property. Some of the factors to be considered are as follows:

- *Piece Size.* Piece size should be appropriate to spacing of frames, beams, and stiffeners to avoid waste.
- *Workability.* Must be easy to cut and handle on board ship. Must be resilient and flexible.
- *Combustibility.* A combustible insulation is a fire hazard, particularly during installation or repair work. Non-combustible linings, furring, etc., are preferred, but if polystyrene or foamed in place of polyurethane insulations are used, a steel or other non-combustible covering is recommended.
- *Sanitation.* Insulation should be odorless and not susceptible to mold, bacterial growth, or termites.
- *Specific Heat.* A high specific heat means a longer *pull down* time but greater temperature stability after pull down.
- *Compressive Strength.* Important for insulation on decks which must bear a load.
- *Density.* Weight on a ship is always important. Insulation weight on an all-refrigerated ship can vary by several hundred tons depending on type of insulation.

• *Physical Stability.* Insulation should not pack, crumble, or otherwise change its form over a reasonable period of time.

• *Cost.* Cost comparisons should be based on installed insulation costs, so that waste, workability, etc, are accounted for as well as material costs.

The problem of what thickness of insulation to use under given conditions is discussed at length in the Maritime Administration Standard Specifications. The plastic polystyrene and polyurethane foams are in great favor because of their light weight, multicellular construction, and, in general, favorable characteristics for all the factors mentioned previously. These plastic foams are combustible. Fire retardant additives do little to alter this characteristic and may lead to toxic vapors when exposed to fire. In addition, the polyurethane, foamed in place or sprayed on has the lowest K factor of all the available materials so that lesser thicknesses are required. On a ship with large refrigerated capacity requirements, there are great advantages in using this material, since the basic dimensions of the ship will be affected by 50 to 100 mm (2 to 4 in.) of insulation less than for insulations with the higher K factor.

c. Details of Construction. The variations in details of construction of refrigerated space are considerable but the principles followed are generally the same. Insulation is applied to the steel shell, bulkhead, or overhead to the required thickness. If insulation is in batt form, furring strips must be secured to stiffeners so that they project to total thickness of insulation. The finish lining secured to the furring strips retains the batt insulation. Furring is commonly commercial stock galvanized steel especially in stores spaces lined with sheet metal (CRES or galvanized). Spruce or fir suitably pressure treated for fire resistance and preservation is also used. If the insulation is board or block type, the lining may be secured directly to the insulation, thus eliminating furring, which has a higher conductivity than the insulation. It is of the greatest importance that the insulation be installed solid, leaving no air voids. This, of course, can be done more easily in the small space between and around the stiffeners with a flexible insulation. For these reasons, some installations have used two insulating materials; i.e., batt insulation between the stiffeners, then foamed polystyrene in two layers to the required total depth.

The foamed-in-place material requires that the lining and ceiling be erected in sections and shored as necessary to retain the foam as it expands and sets. Special care must be taken to insure that no voids are created and certain lining and ceiling portions are removable to permit a thorough examination of the insulation.

Linings form the finishing surface inside the refrigerated cargo space and must be fairly strong, tough, sanitary, easy to clean, and verminproof. Actually the lining gets considerable protection from contact with cargo by the battens or sparring installed on the face of the lining to allow an air circulation space around the cargo. The following materials have been used for linings:

• Spruce or pine—25 mm (1 in.) T&G—two layers at right angles;
• Spruce or pine—25 mm (1 in.) T&G—one layer covered 16 mm (5/8 in.) fir plywood;
• Bituminous emulsions on metal lath;
• Spruce or pine—25 mm (1 in.) T&G—one layer covered with bituminous emulsion on metal lath;
• Sandwich panel consisting of two layers of Fiberglas reinforced plastic with a core of expanded steel;
• Galvanized or corrosion-resisting steel (CRES);
• Reinforced plastic sheet material.

The wood linings with proper pressure treatment can be made sanitary but their weight and thickness are disadvantages. The emulsion on metal lath is equivalent to a plaster finish, which is subject to damage and is rarely used. Aluminum is used generally on banana carriers for its ease of maintenance and cleanliness. The reinforced plastic has the advantage of light weight, minimum thickness, and ease of cleaning. However, it is not as sturdy as the other finishes mentioned. Sandwich panel is light weight also good in strength and requires no painting or maintenance.

Insulation for the deck underfoot is usually in precut board form or may be poured-in-place foam. If of board form, successive layers are mopped with asphalt to prevent air pockets. The top of the deck insulation is finished for strength and ruggedness to withstand heavy, moving loads. The usual construction is about 38 mm (1½ in.) of concrete, reinforced with expanded metal, and a finish of 25 mm (1 in.) of mastic. The reinforced concrete gives strength, and the mastic gives a tough, durable finish. Asphalt paper has been used to form a watertight membrane with considerable saving in weight and cost over the lead pan once used. The multicellular insulation eliminated the need for a lead pan because it will not absorb water in quantity. A lead pan is fitted only in way of drains.

d. Liquefied Gas Carrier Cargo Tank Insulation. The entire subject of insulating the ship from the low temperatures associated with transporting liquefied gas and that of insulating the liquefied gas cargo from the environment is extremely complex. A number of solutions to these problems have evolved that differ as do the types of containment devised for the cargo. The design aspects of the insulation systems devised and the underlying principles involved are discussed in detail in Section 4 of Chapter XI.

5.7 Miscellaneous Joiner Work. Many other items of ship outfitting are broadly included in joiner work. Furniture, for example, which is a subject in itself, as well as doors, windows, and portlights are discussed elsewhere in this chapter. Door hardware (handles, coat hooks, holdbacks, stops, etc.,) and bathroom accessories (soap dishes, mirrors, cabinets, toilet paper holders, etc.,) would fall into this category. The coordination and selection of many of these items within an overall scheme of decoration and surface finishes, materials, etc., particularly in passenger quarters or owner suites, is usually in the domain of an interior designer.

5.8 Ratproofing. All ships engaged in foreign commerce are required to carry an acceptable certificate indicating freedom from rats or be subject to deratization if, on inspection by the U. S. Public Health Service, such measure

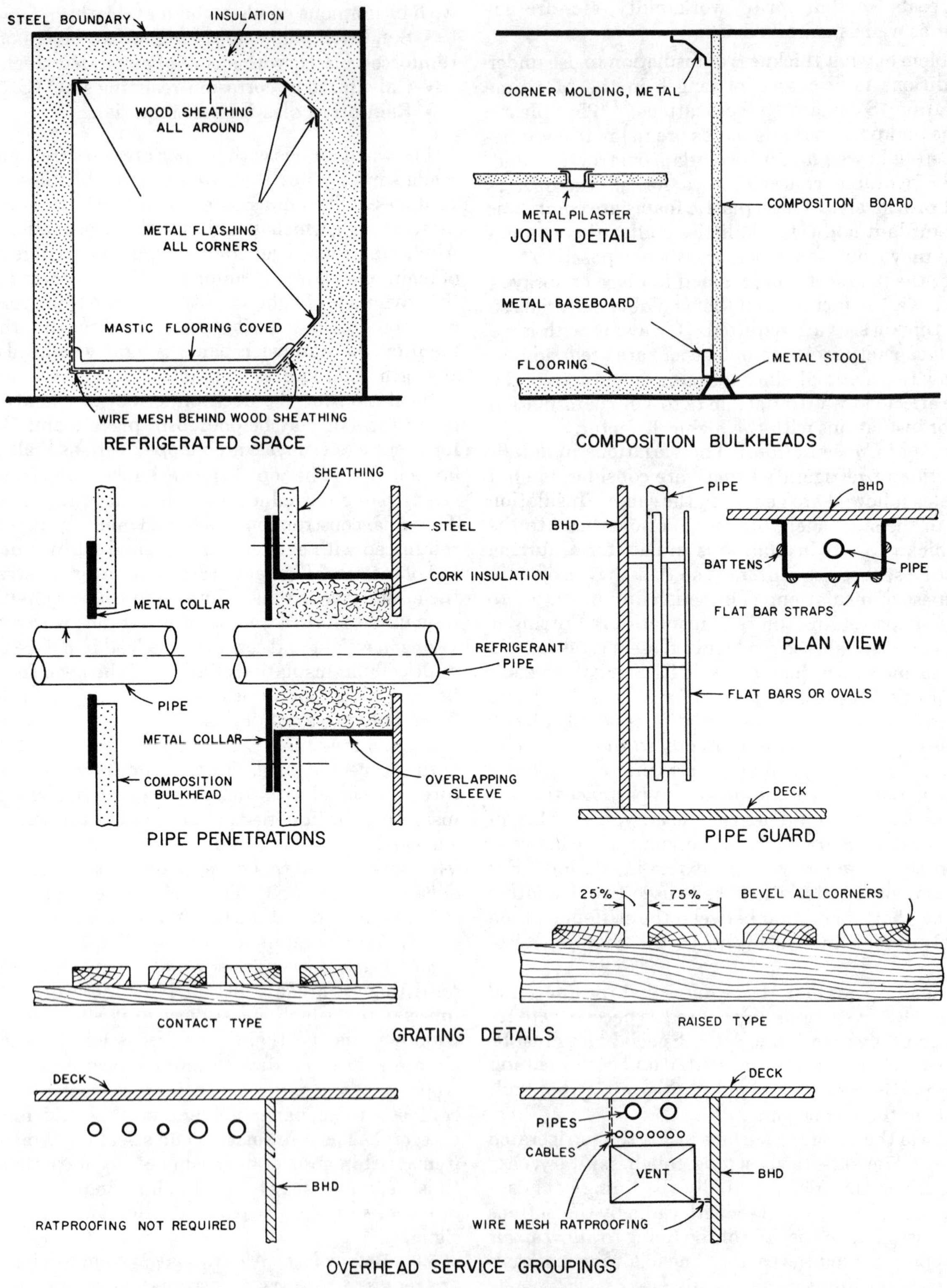

Fig. 23 Typical ratproofing details

is deemed necessary. The certificate either indicates that the ship has been recently fumigated (deratization certificate) or that inspection disclosed no sign of rat infestation (exemption certificate). Fumigation is usually accomplished by the use of hydrocyanic gas, which has been found highly effective for this purpose. Fumigation will have to be repeated periodically as the ship eventually will become reinfested with rats. An exemption certificate, on the other hand, attests to the fact that the ship is ratproof; e.g., it is so difficult for rats to live and multiply that they leave or ultimately suffer extinction.

All ships built under MarAd jurisdiction are ratproofed, the specifications requiring that an exemption certificate be obtained before delivery. In the long run, ratproofing has proved economical in that it eliminates the cost of repeated fumigations, and as the designers and builders have become more familiar with ratproofing principles, compliance with USPHS regulations usually is accomplished without excessive cost or delays. Customary practice is to submit a plan schedule to the USPHS which in turn will indicate to the builder which particular plans will be required for review and approval. With the increased emphasis on fireproofing, it is well to point out that the uses of ratproof materials will make for a fireproof ship.

Harborage of rats within the ship's structure can be entirely eliminated in design and construction. Maintenance free materials and well-lighted spaces materially assist in promoting cleanliness.

In principle, ratproofing consists of either protecting or eliminating partially or totally enclosed spaces, Fig. 23. Ratproofing by elimination of enclosed spaces is the most desirable method. Generally speaking, the ship may be made ratproof in the drafting room preparing the detail plans. However, many minor details do not show on the plans and small piping, ventilation trunks, wireways, etc., are not exactly located on plans so that outfitting personnel who are otherwise familiar with the principles of ratproofing may have difficulty in avoiding creating harborages, particularly during the late stages of outfitting. But experience has shown that ratproofing may be accomplished with few minor corrective measures required on final visual inspection per the USPHS (1965).

Section 6
Stewards Outfit

6.1 General. That part of the hull outfit which pertains to the outfitting of the living spaces in a vessel is often treated under the heading of "furniture & furnishings" which would include not only the berths, tables, dressers, desks, sofas, chairs, etc., but also bedsprings and mattresses, upholstery, cushions, mirrors, wastebaskets, lamps, curtains and miscellaneous fittings. Also pertaining to the living quarters are items such as bedding (sheets, blankets, pillowcases, towels etc.,) which are often referred to as items of "stewards outfit" which term would also include the commissary equipment contained in the galleys, pantries, sculleries, dining rooms, and mess rooms.

For the purpose of this section all these items will be reviewed under the heading of "stewards outfit" since this is the department within a vessel's organization which would be concerned with these features.

6.2 Furniture. Furniture for passenger, officer and crew accommodation is usually of steel, aluminum or hardwood with exposed hardware of stainless steel, bronze, brass or anodized aluminum. Metal case goods are often insulated with a mineral base material at least 1.5 mm ($^1/_{16}$ in) thick. All furniture except chairs should be secured to decks or bulkheads and portable furniture should have securing devices for lashing down. Drawers should have positive means to prevent opening in heavy seas. Resin laminate (such as melamine) tops on tables, dressers, and desks is often specified for scratch, burn, and liquid resistance.

a. Staterooms. Furniture in passenger and senior officer's (such as the master and chief engineer) staterooms are commonly of wood and often custom built by joiner constructors to designs of an interior decorator. Normal beds are sometimes provided in such spaces rather than built in berths of standardized dimensions. Furniture in other officer and in crew staterooms is usually of metal with dresser table and desk tops of resin laminates. In general upholstered stateroom furniture has self supporting spring construction. Upholstery coverings of synthetic leather of selected color are typical. Berths are commonly supplied with 190 mm (7.5-in.) high box springs and inner spring mattresses, 152 mm (6-in.) thick although the use of foam mattresses, also of 152 mm (6-in.) thickness are coming more into use especially in passenger and senior officer quarters. In this regard cushions of neoprene (latex foam) are in general use. Wardrobes for passengers, officers and crew should be of full deck height, often built in. Minimum size is 0.6 × 0.6 × 1.8 m (2 × 2 × 6 ft) provided with shelf, coat rod, hooks and shoe rack with a tie rack and a mirror on the door. Berth lights, desk and table lights are provided in all rooms, where appropriate.

The furniture in a typical single crew room would consist of: Berth, lounge chair, chest/desk, side chair, wastebasket, berth shelf and bookrack. The typical officer room would contain a similar list of items of better grade and increased size (such as the berth) plus an extra lounge chair. More senior officers such as the chief officer and the first assistant engineer would be provided with a separate dresser and desk as opposed to a combination desk/dresser plus an end table with lamp. The captain and chief engineer would be provided with the foregoing, often custom built to interior designers schemes, plus such items as a safe or a bookcase as may be necessary or as may be possible in the available space.

b. Public Rooms. This category includes dining rooms, lounges for passengers, officers or crew, recreation rooms, card rooms, libraries and offices.

1. Dining room. The room is often referred to as the saloon and is usually shared by passengers and officers and contains tables (sometimes a special captain's table as well as others of varying size or adaptability or designation such as for use solely by duty officers), sideboards, dining chairs, dressers or cupboards, mirrors, pictures, and possibly sofas, depending on the interior decoration scheme selected.

2. Lounges. While the quality or materials used might differ, passenger, officer or crew lounges contain basically the same items namely, sofas, lounge chairs, occasional chairs, card tables & chairs, cabinets, end tables, lamp tables and coffee tables.

3. Card rooms. Where such rooms might exist, as on passenger vessels special card tables, games cabinets and arm chairs would be provided.

4. Recreation rooms. Since these spaces act as social centers and are often provided in addition to officers and crews mess rooms or as the equivalent of a lounge, they should combine the facilities of lounge, reading and writing room and card room in one space.

5. Libraries. On a ship with passengers the library would contain sofas, lounge chairs, writing desks and chairs, book cabinets, magazine racks, coffee or occasional tables, lamps, and other decorator items such as planters, pictures etc.

6. Offices. The captain and chief engineer should be provided with offices where ship's business can be transacted and visitors or officials entertained. In addition to a desk, chairs, filing cabinets, key lockers, book racks, and book cabinets, a safe should often be provided in the captain's office.

A ship's office and an engineer's office are often provided adjacent to the first mate's and the first assistant engineer's offices respectively, generally outfitted as above, but with typewriter desks and often with plan desks (suitable for spreading out long plans). The steward's department office when provided should be similar to these.

6.3 Upholstery, Draperies, Carpeting. Decorative fabrics, similar to those found in corresponding spaces in an office, private home or hotel, are installed aboard ship, as follows:

a. Upholstery. Sofas, lounge, easy chairs and, if required, ordinary chairs are either upholstered with cloth or artificial leather. In general, cloth coverings are provided in passenger and senior officer's compartments and artificial leather in junior officer's and crews rooms. Upholstered furniture usually has self supporting spring construction with a horse-hair covering and on top of this a protective covering and the upholstery. Cushions are nowadays of neoprene or latex foam conforming to U. S. Government MIL specifications. Other top quality materials are used and loose cushions, where specified, are usually made reversible. Buttoning of upholstered furniture should be specified in such a way that the securing method will last for the life of the upholstery fabric.

b. Curtains. Curtains are used in way of doors, side-lights, windows, bunks, baths and showers. In some cases special blackout curtains or blinds are used in certain locations on the fronts of houses to prevent the radiation of light into the wheel house or from the chart room into the wheel house. Curtains in most locations are made of single fabric but are often lined in way of passengers' and senior officers' quarters. Door curtains are frequently omitted in current ship designs with air-conditioned accommodations. Bath and shower curtains are made of plastic. Curtains are supported on rods with fittings and rings. Chart rooms and other spaces which are fitted with special blinds of the roller type to prevent reflections are often provided with regular curtains in addition for a more pleasing appearance. Curtain materials are usually selected with an inherently fireproof quality or are treated to be fire resistant.
fireproof quality or are treated to be fire resistant.

c. Carpeting. Floors of passenger, senior officers, and certain public spaces are generally covered with carpet either fitted wall to wall or laid loose in individual areas but in both cases they are secured to the deck by special fasteners. Wool carpeting has proven most fire resistant but is now becoming relatively too expensive for marine applications so that synthetic materials equivalent to wool in fire resistance are now in general use.

d. Covers. In passenger and senior officer's quarters the upholstery of sofas, easy and occassional chairs is often provided with loose washable covers fastened by buttons or tapes. These are often used when the ship is in port for protection and removed when the ship is at sea. Artificial leather upholstery is not provided with covers since it can easily be cleaned.

6.4 Linen Supplies. The bedding, including sheets, pillows, pillow cases, blankets, towels etc., are normally provided by the shipowner and not the shipbuilder, as an "owner supplied item." Standards used by individual shipping companies can vary depending on quality desired. Normally these would be equal to first class shore hotel standards often with special requirements as to company insignia, etc. In any event U. S. Government MIL specifications could be utilized to specify adequate standards for these items of the ship's outfit.

6.5 Commissary Equipment. One of the most important functions covered by the steward's department includes spaces such as galley, bakery, pantries, mess rooms, dining room, and store rooms.

a. Galley. On most cargo vessels a single galley provides food for passengers (if carried), officers, and crew. Particular emphasis in the design and layout of all commissary spaces, particularly the galley, has to be made to USPHS requirements. Good commercial design of equipment, modified to suit shipboard conditions is needed. Most equipment and all surfaces coming into contact with food and drink should be of CRES material and special attention should be paid to the design of all equipment to prevent the lodgement of grease and food particles in corners, cracks and joints so as to contribute to the maintenance of sanitary conditions. Typical equipment would consist of the following:

- Sinks should be of CRES material and of sufficient

number to handle manual dishwashing procedures (including water heaters for sterilization purposes) in case of temporary breakdown of automatic dishwashing equipment. A separate lavatory should be provided for galley staff for hand washing purposes.

- Steam Tables, complete with hot water pans for keeping meats and vegetables warm or, alternatively, dry heat (electric) hot food tables.
- Racks for pots, dishes, cups and glasses.
- Refrigerators for ready-use food and food awaiting cooking after removal from reefer stores.
- Shelves for general storage in the open.
- Overhead Cabinets for general storage with hinged or sliding doors and sectional removable shelving.
- Tilting Bins for storage of flour, rice and sugar.
- Electric Range with hot plates (at least 3), handrails, searails, and overhead switches.
- Electric oven with 2 sections, on stand.
- Electric convection oven, single deck, on stand.
- Electric griddle, on counter.
- Fry kettle, counter type with full size basket.
- Combination steam cooker/kettle, one pressure cooker and one five gallon stainless steel tilting kettle with electric steam generator.
- Mixer, 36l (20 qt) bowl, flat beater wire whip, meat chopper, knife sharpener and dough hook.
- Meat slicer, single feed.
- Proofer, steam or electric built into baker's dresser.
- Garbage disposer, built in under dresser.
- Baker's scale, 3.6 kg (8 lb) capacity.

In the scullery area:

- Dishwasher, combination automatic and manual control, with door lock control and electric booster heater.
- Garbage disposer, built in under dresser.
- Ice cuber.

b. Pantries. Pantries are normally located adjacent to the dining room or mess rooms and provide a station where food can be received from the galley (often via a dumb waiter) and from there served to the dining or mess room. There are often facilities in the pantries for cooking snacks or serving simple meals to those on watch outside of normal galley hours. Pantries usually contain a refrigerator, hot plates, coffee maker, toaster and warmer units or a steam table with inserts for meat pans plus racks for dishes, cups and glasses and sinks.

c. Dining Rooms and/or Mess Rooms. Officers and passengers commonly share a dining room or saloon equipped with tables, chairs, sideboards and various small appliances such as a coffee maker, a 2 burner warming unit, a toaster and a small refrigerator. Mess rooms for petty officers and crew can be in separate rooms or with a section designated for the petty officers in a combined mess. They are normally provided with tables, chairs, sideboard, and a drinking fountain.

Section 7
Lifesaving Systems

7.1 General. The lifesaving system on a ship is a vital part of the ship and is an extension of the emergency escape route. A designer should view the fitting of lifesaving equipment in concert with the fire protection design and the emergency escape routes. For the lifesaving equipment to perform as intended it must be provided in a planned manner, specifically suited and arranged to assure catering to the particular features of the ship's operation, complement, route, and carriage.

7.2 Phases of the Lifesaving Problem. To develop a planned, rational lifesaving system design the basic phases of the problem must be examined, i.e., *pre-abandonment, abandonment, survival, detection,* and *retrieval.* The combination of life-saving equipment carried on the ship must be such that collectively all of the needs in these five areas are met.

Dividing the subject into these phases permits an evaluation of the requirements and lays down a technical approach for use by ship and equipment designers in developing new solutions. Each of the five phases should be examined. No one piece of equipment should be expected to answer all problems and the mission of the vessel will have an impact on the degree of importance of each of these phases.

The following defines each of the five phases of the vessel abandonment problem as it relates to lifesaving equipment:

a. Pre-abandonment concerns the training, maintenance, stowage, capacity, ship arrangement, protection, and provision for effective usage of lifesaving system components under operating conditions.

b. Abandonment comprises all operations required for breaking out of stowage and the safe disengagement and clearing away of the lifesaving equipment with a full complement from the stricken ship.

c. Survival is the preservation of groups of persons and of individual persons at sea until rescued.

d. Detection is the accurate determination of the location of survivors.

e. Retrieval is the safe and expedient transfer of survivors to a position of safety.

Each of these phases can be subdivided into elements which a ship designer must account for in the total system or which the designer of equipment must consider when designing one of the pieces of the system. These design elements are best viewed in relation to each of the phases of the problem and are given below in these categories:

a. Pre-abandonment. Survival craft should be provided with sufficient capacity and so distributed that all persons on board can be accommodated under adverse ship and

Fig. 24 Fire protected lifeboat on drill rig

weather conditions. They should be stowed in a state of readiness for launching with all components and equipment required for efficient use. Furthermore, these craft should be stowed at locations where they are secured and sheltered from damage but from which they can function as designed.

Ample space should be available for marshalling and supervision of personnel and suitable lighting should be provided in the boarding areas and descent areas. Active means should be available for alerting all personnel.

Spare parts and equipment should be available for reconditioning and reoutfitting after training operations as required, and for maintaining the system in a state of readiness. Necessary descriptions and instructions should be suitably located concerning inspection, maintenance, and special precautions. Additionally, equipment should be arranged to permit necessary inspection, maintenance, and functional tests; minimal maintenance should be necessary. Furthermore, equipment should be easily operable under emergency conditions and require minimum prior training and experience.

b. *Abandonment.* Survival craft should be launchable within the shortest period of time even if the ship machinery is not running and when adverse ship and environmental conditions exist. Means for safe and expedient transfer of personnel from ship to water level should be such that no one need remain aboard the ship to launch survival craft and the system should be designed to permit a single person to evacuate alone in safety.

Survival craft occupants should be protected when launching from great heights to minimize the danger of being thrown overboard. Rapid embarkation should be afforded for all personnel prior to launch and embarkation of injured persons should be possible.

Stowage of survival craft should be in well lit areas with ample space for supervision and mustering. These craft should be launchable from a location or in a manner that does not put the craft in danger of being damaged by the ship's propeller, by overhanging hull structure, or by launching equipment. Neither should its complement be exposed to harmful acceleration.

A reasonable percentage of the survival craft should be released automatically to an operative condition if the ship sinks with adequate mobility to be able to clear away from ship's side and attain a safe distance from the ship or hazardous areas under adverse environmental conditions.

c. *Survival.* Group survival equipment should accommodate all survivors from the abandoned vessel with sufficient room for operation of equipment components. Each unit should have adequate stability and buoyancy, under rough sea conditions, to support out of water, all survivors carried. The equipment also should afford sufficient mobility and manueverability to maintain position at sea under adverse weather conditions, as well as to have the capability of rescuing persons in man-overboard situations. Group survival equipment must provide proteotion against climate and environment, first-aid care of injured persons, and provide means for the survivors to assist each other in gaining access to survival craft after launching. It should permit the survivors to be self-sufficient for a period of time within which rescue is highly probable and therefore include survival equipment which may be used with a minimum of prior instruction. It must include instruction for survival with the proviso that all instruction material be easily readable under emergency conditions. This means providing adequate internal lighting for general purposes and for reading instructions.

Personnel survival equipment should include provisions for individuals to abandon ship without jumping or climbing from unsafe heights and also include buoyant articles of bright color, for all persons on board which provide adequate buoyancy for any person wearing them.

d. Detection. Group survival equipment should include active means of alerting ships, aircraft, and shore automatically with adequate power. There should be means for guidance of ships and aircraft to the location of the survivors, suitable for day, night, fog, and severe weather and climatic conditions. This includes active detection, by electronic, visual, and audible means, and passive detection by electronic and visual means. All detection systems should be easily operable and require a minimum training and experience.

Personnel survival equipment should include active means for visual and audible detection of persons in the water. Visual signals should be activated automatically and be possible to control manually. The equipment should include passive means of visual detection of persons in the water.

e. Retrieval. Survival craft should be arranged to be taken safely in tow by an assisting vessel in severe weather. They should be arranged for helicopter pickup of survivors without hazardous exposure to remaining survivors. Additionally, survival craft should be designed for the transfer of survivors to a rescue vessel under severe weather conditions.

Survival craft should be readily accessible for rescue use and be launchable and retrievable under severe environmental conditions; they should have rescue capacity for easy pickup of persons in distress and facilities for towing. These craft should afford sufficient mobility and maneuverability to provide efficient assistance under severe weather conditions and loading. In addition, rescue equipment should include means for persons in the water to be supported by the equipment.

Rescue equipment should be available for passing a line from one vessel to another and should provide means to transfer persons or equipment directly from ship to ship. This equipment should be easily reconditioned and restored to a state of readiness after use or drill.

7.3 Solutions to Savings of Life at Sea. There are several modern solutions to the lifesaving system problems which are depicted in Figs. 24, 25, and 26. Although there is a continuing need for innovation this does not mean that traditional solutions are not to be considered. The current minimum standards established by international treaty or those set by the flag the ship will fly should be taken as the prime reference. The following should be included in specific design considerations.

Ships with all living spaces aft are required to provide a small capacity inflatable life raft near the bow. It must be stowed for protection from boarding seas, yet be capable of being rolled across the deck and over the side by one man. An embarkation-debarkation ladder or knotted life line should also be stowed in this area.

a. Longitudinal location. In general, lifeboats should not be closer to the bow than one-quarter of the ship's length and should not be located so far aft as to be endangered by the propellers. The after end of a lifeboat must be at least one and one-half times its length forward of the ship's propeller position longitudinally. In some cases even this distance will be found to be insufficient, due to the overhang of the vessel in way of the stern sections.

b. Vertical Location. The vertical location of lifeboats and life rafts should be selected with care, inasmuch as boarding seas can and do cause considerable damage to lifesaving equipment. The uppermost deck extending to

Fig. 25 Free fall covered fire protected lifeboat

Fig. 26 Slide/raft for high density ferry

the sides of oceangoing passenger ships is generally sufficiently high to cause no concern. However, on cargo and tank ships, such a deck could be so close to the waterline that any lifeboat or life raft stowed thereon would be endangered during adverse weather.

c. *Clearances.* Once a suitable deck has been selected, there are a number of other problems affecting the design which must be resolved. The boat deck should be adequately supported to deal with the concentrated loads in way of the lifeboat davits and winches. Consideration must be given to minimizing the extent of openings on the ship's sides between decks so that lifeboats will not hang up in such openings when launching under an adverse list. The house sides in way of the lifeboats must be located so that none of the lifesaving equipment will extend beyond the sides of the ship, and so that there is sufficient distance between the inboard side of the lifeboat and the house side to provide a suitable passageway, keeping in mind that sea rails on the house sides and skates on the lifeboat can use up as much 0.3 m (1 ft) of this distance. Gravity davit installations having the inboard end of the trackways anchored to a housetop eliminate this problem.

Pad-eyes, cargo ports, overboard discharges, and other items on the sides of a vessel that interfere with the launching of lifeboats and life rafts must be kept to a minimum. When overboard discharges cannot be located except in way of the lifeboats and life rafts, means should be provided to deflect the flow down the side of the ship or other arrangements made to close the discharge openings during the launching operation.

d. *Launching Obstructions.* The rail should be demountable in way of the life raft stowage, regardless of whether the raft is located inside or outside of the rail. In any case, there should be an unobstructed passage across the deck. It should not be necessary to pursue a devious route in order to take advantage of the low side of a listing ship.

Since the inflatable life rafts are designed and stowed to float free in the event there is not time to launch them, it is essential that there be no canopies or other encumbrances above or below them. Normally, a hydrostatic release is provided, although other satisfactory arrangements can be made.

7.4 Installations on Passenger and Cargo Ships. On such ships it is generally necessary to string the lifeboats out in a line using almost the entire length of the superstructure. Since the carrying capacity of the largest size lifeboat (approximately 11.3 m (37 ft) long) is limited to 150 persons, it becomes relatively simple to determine the maximum number of persons which can be accommodated for a given stowage distance, keeping in mind that the leadoff boat on each side is likely to be a 7.9 m (26 ft) 41-person emergency boat and the lifeboats immediately aft of these on each side would be motor boats having a capacity of less than 150 persons. In this respect, Tables 4 and 5 give useful data on covered lifeboats and survival capsules.

a. *Inflatable Life Rafts Substituted for Lifeboats.*

Table 4—Typical Dimensions, Capacity, and Weight of Covered Class I Lifeboats

Persons Capacity		23	33	44	58	64
Length,	m	6.50	7.32	8.00	8.50	9.14
	(ft)	(21.33)	(24.00)	(26.25)	(27.88)	(30.00)
Beam,	m	2.36	2.67	2.69	2.97	3.05
	(ft)	(7.75)	(8.75)	(8.83)	(9.75)	(10.00)
Draft,	m	0.86	0.95	1.09	1.25	1.32
	(ft)	(2.83)	(3.13)	(3.58)	(4.10)	(4.33)
Height,	m	2.72	2.92	2.79	2.62	1.93
	(ft)	(8.92)	(9.58)	(9.17)	(8.58)	(6.33)
Lift Hook Ctrs,	m	6.10	6.71	6.78	7.26	7.92
	(ft)	(20.00)	(22.00)	(22.25)	(23.83)	(26.00)
Equipped Weight,	kg	2597	3567	3420	4753	4351
	(lb)	(5723)	(7861)	(7537)	(10475)	(9590)
Lowering Weight,	kg	4319	6037	6714	9095	9142
	(lb)	(9518)	(13306)	(14797)	(20045)	(20150)

Substitution of inflatable life rafts for up to 25 percent of the required life boatage is a way to relieve space limitation problems. Such rafts must be capable of being lowered to the water fully loaded by an approved launching device one of which must be installed on each side of the vessel. The entire launching operation, boats and rafts included, must not exceed 30 minutes. This normally dictates that no more than five rafts be served by each davit.

b. Embarkation into Lifeboats. It is important in the design stage of a large passenger ship to determine where the passengers will be assembled before they are embarked into the lifeboats. If an enclosed promenade deck just below the boat deck level is to be used for embarkation, as well as an assembly area, then the side of the vessel between the boat and embarkation decks must have two doors hinging inboard in way of each boat, and the windows in way of the ends of each boat must be capable of being opened so that the frapping lines can be readily handled. A large cleat in way of each of these window openings is necessary. The swing of the doors from the interior spaces to the enclosed promenade must also be considered in allowing sufficient room for the persons assembled in this area. If, on the other hand, it is intended to assemble the passengers in the enclosed promenade deck area and then lead them up to the boat deck, it is necessary to provide outside stairways for this purpose. In any case, the preferred arrangement is to have the crew prepare the boats for embarkation unencumbered by a milling crowd of excited and sometimes panic-stricken passengers. Most of the extra "5 percent life preservers" required by SOLAS should be stowed in the embarkation area. Considerations for cargo vessels are not significantly different than for passenger ships other than the amount of lifeboats and life rafts carried. The provision of 100 percent boat capacity for all the crew on each side is in recognition of the problems which can result from vessel damage (greater and more rapid heel and trim) and exposure to fires in the cargo area.

7.5 Lifeboats. Lifeboats are either enclosed or of open construction and are classified according to construction material, such as steel (galvanized), aluminum, and glass reinforced plastic (GRP). These in turn are further classified by the means of propulsion used, such as oar, hand crank, and motor propelled. The basic design constraints are contained in the SOLAS treaty and government marine safety regulations. As the capacity of a lifeboat increases, so does the capability of its propulsion system; oar-propelled lifeboats are permitted up to 60-person capacity, hand-crank propelled up to and including 100-person capacity, and finally diesel-propelled when more than 100 persons are carried. Fig. 27 is a typical motor-propelled enclosed life boat.

a. Sizes. Lifeboats range in size from 3.7 m (12 ft), 6-person boats used on tugs and other small vessels plying inland waters to 11.3 m (37 ft), 150-person boats used on large seagoing passenger ships.

b. Emergency Lifeboats. SOLAS passenger ships are required to have two emergency lifeboats suitable for performing emergency work at sea. These boats must be between 7.3 and 8.5 m (24 and 28 ft) in length and the ratio of length to beam must not be less than 3.3. They may be counted as a part of the required lifesaving capacity if fully equipped.

c. Releasing Gear. The releasing gear for lifeboats should be of a type which will release both ends simultaneously whether under load or not. All release gear on lifeboats of a ship should be of the same type.

7.6 Life Rafts. Life rafts are either inflatable or rigid with the basic design constraints for both types established in the SOLAS treaty and in government marine regulations. Life rafts serve several purposes:

- They provide backup survival craft capacity for per-

Table 5—Dimensions, Capacity, and Weight of Survival Capsules

Persons Capacity, Seated		14	28	50
Diameter,	m	3.35	4.27	4.27
	(ft)	(11.00)	(14.00)	(14.00)
Length,	m	3.35	4.27	6.02
	(ft)	(11.00)	(14.00)	(19.75)
Door Sill Height,	m	1.17	1.50	1.55
	(ft)	(3.83)	(4.92)	(5.08)
Overall Height,	m	2.41	3.00	3.15
	(ft)	(7.92)	(9.83)	(10.33)
Equipped Weight,	kg	1440	2242	4015
	(lb)	(3173)	(4942)	(8850)
Lowering Weight,	kg	2488	4338	7759
	(lb)	(5483)	(9562)	(17100)

Fig. 27 Modern fire-protected covered lifeboat

sonnel working or sleeping remote from lifeboat locations or where access to lifeboats is cut off.

• They serve as survival craft ready for boarding if a ship sinks suddenly without time for orderly abandonment.

• Either in conjunction with lifeboats, or as the only available means, they provide for ship abandonment; in the latter cases the rafts are off-loaded by a davit which permits personnel to board at deck level and to lower themselves into the water. Fig. 28 depicts a typical life raft designed for handling by means of the davit shown in Fig. 29.

Where life rafts are substituted for all of the lifeboats the following should be assured:

• Davit launch capability should be provided if personnel would have to descend more than 15 m (4.5 ft) or enter the water before entering the raft.

• Where davits are provided, there should be at least one on each side of the ship and it should not be necessary for anyone to remain on board following launch.

• Davit rafts should not replace the float-free rafts.

• A suitable rescue boat should be provided to replace the man-overboard and raft marshalling function of the lifeboat.

7.7 Rescue Boats. Rescue boats are provided when lifeboats are not fitted to perform man-overboard retrieval and raft marshalling duties. The characteristics of the boat selected of necessity are dictated by the trade and route of the ship on which fitted. Ships operating in protected waters may only need a small lightweight boat with built-in buoyancy, capable of being manhandled over the side without davits. Ships operating in more hazardous waters should carry a motor-propelled rescue boat suitable for ocean service with a davit or other suitable launching gear capable of launch by no more than three persons.

7.8 Davits. Davits are designed to move lifeboats from their stowed position to an embarkation position, and subsequently to lower them to the water. More recent developments permit boarding in the stowed position and then lowering, Fig. 30. Details related to design constraints of the davits and their attachments to the ship may be found in the SOLAS treaty and Government marine safety regulations. Davits are classified as gravity, mechanical, or radial type.

a. Gravity Davits. Gravity davits, Fig. 31, must be capable of being swung out without the use of manual, electric, steam, or other power supplied by the ship. They consist generally of arms rolling on trackways or arms with one or more pivoting links. They must be used whenever the weight of the lifeboat and its equipment exceeds 227 kg (500 lb), and they are required for tankers on international voyages regardless of the weight of the lifeboat. Gravity davits are the most economical of space since they are located within the length of the lifeboat, and the winch associated with such davits can be centered under one of the trackways. Fixed, outrigged davits which lower the boat by gravity are commonly fitted on drill rigs or ships which do not need to have a boat stowed inboard.

b. Mechanical Davits. Mechanical davits are swung out manually by cranks, operating screws, gears, or other mechanical devices; however, one type of pivot davit only requires the lifting of the winch brake handle to put it in operation after the gripes are cast off.

7.9 Winches. Lifeboats are supported under davits by wire falls led through a series of sheaves to the drums of a

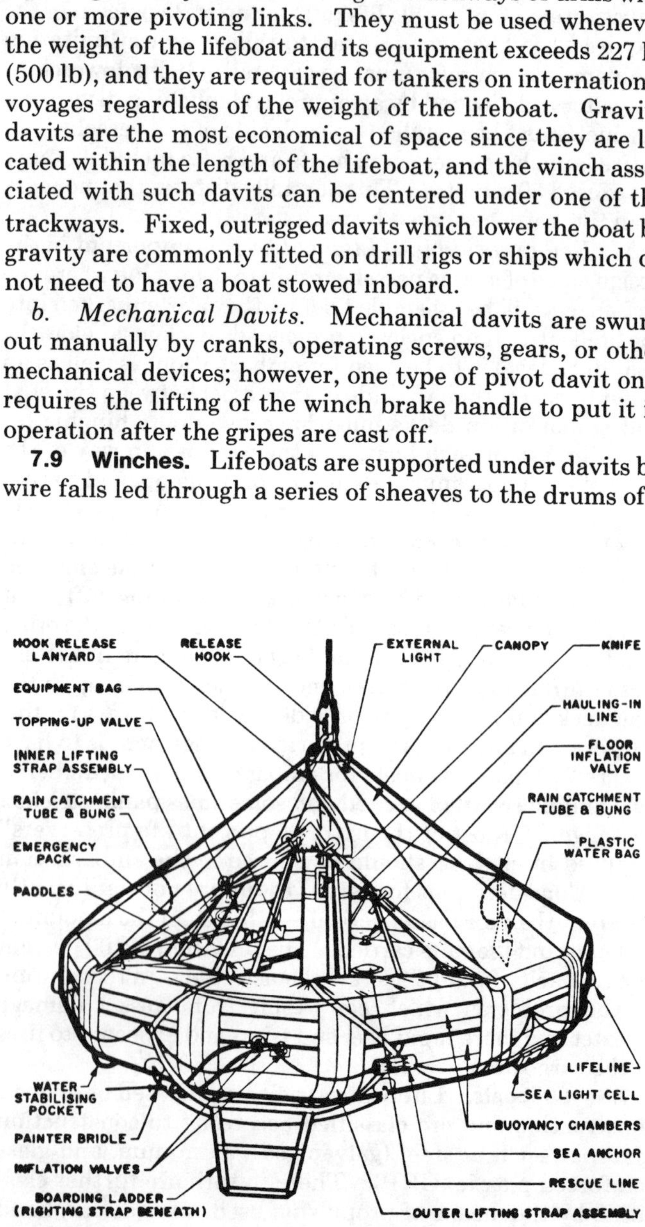

Fig. 28 Typical davit raft

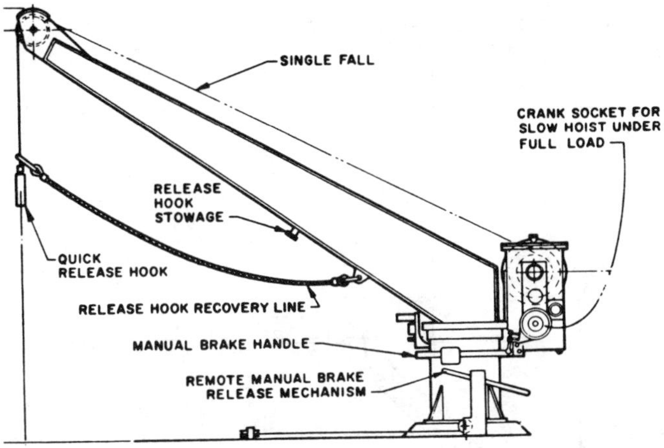

Fig. 29 Raft davit

lifeboat winch. Winches are required whenever gravity davits are used, and whenever the height of the deck on which lifeboats are carried exceeds 6 m (20 ft) from the lightest seagoing draft.

a. Types of Winches. Winches for use with gravity davits have grooved drums and only one wrap of the falls on the drums. Winches used in conjunction with mechanical davits generally have smooth drums and many wraps of the falls on each drum. Both types of winches are normally powered by electric motors, even though this is not required by regulation. A lifting speed of 0.1 m/sec (20 fpm) is the norm and is required for emergency lifeboats on passenger ships. In some cases, particularly on tankers, shipowners have specified manual winches, capable of being operated by a portable air or electric motor which can be moved from winch to winch when necessary to retrieve the boats. In all cases, limit switches should be fitted on davits which require hand return for the final few inches to prevent overstressing of falls.

b. Location. All winches must be located so that the operator can view the entire launching operation.

Gravity winches are generally centered under one of the gravity davit trackways near the rail, with controls clear of the area between the trackways. Such a location will generally provide for the necessary maximum 8-deg fleet angle for the lifeboat falls. The falls are led through a narrow slot

Fig. 31 Typical installation of gravity davits

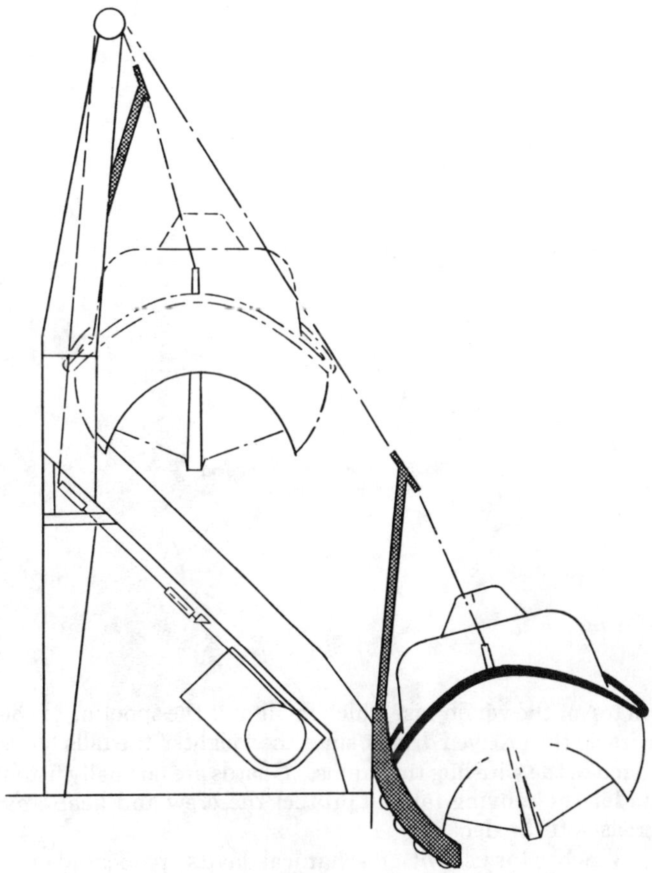

Fig. 30 Miranda gravity davit

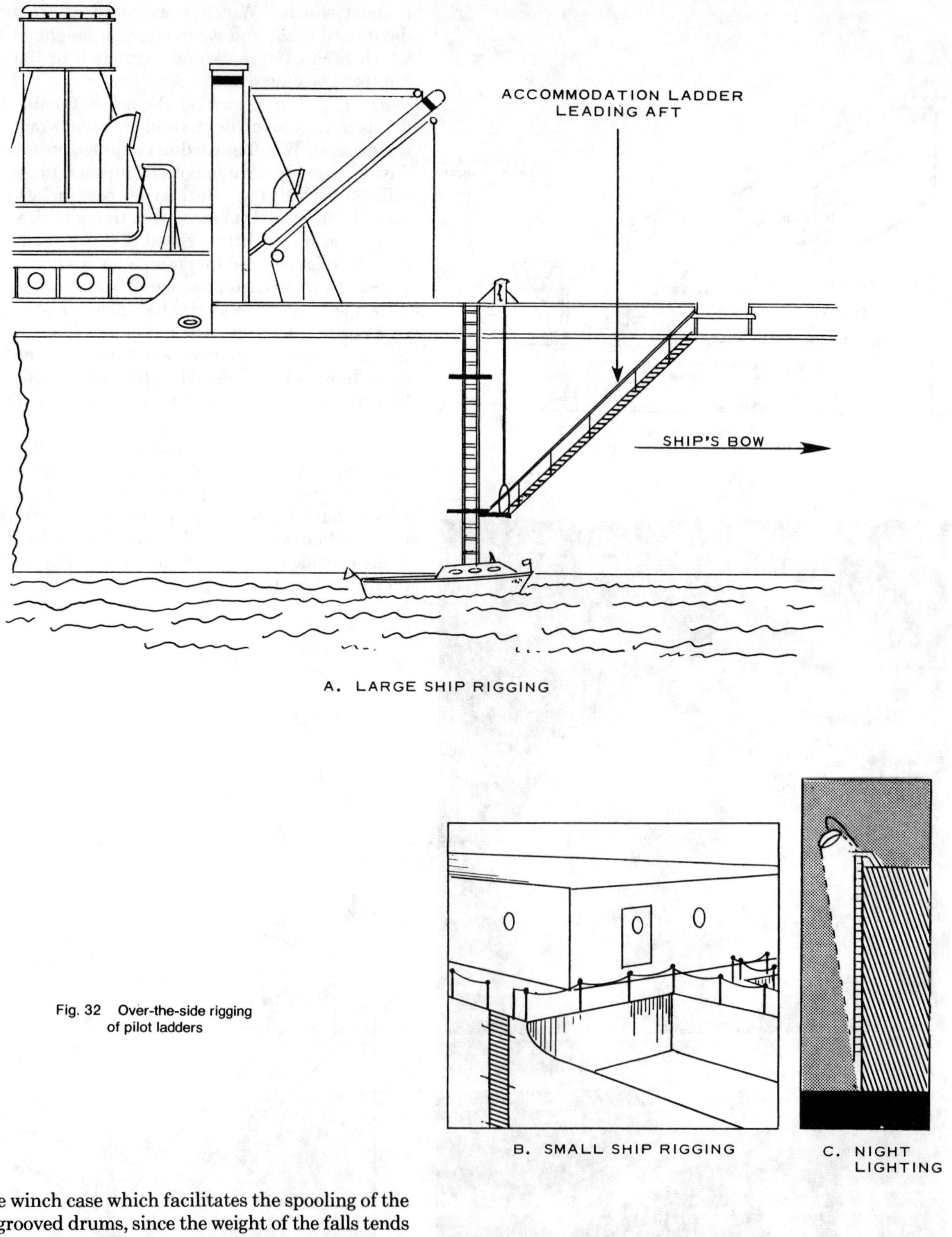

Fig. 32 Over-the-side rigging of pilot ladders

on top of the winch case which facilitates the spooling of the wire on the grooved drums, since the weight of the falls tends to make the wire hug the drums. Guards are normally fitted under the moving falls to protect the crew and keep any grease off the deck.

Winches for use with mechanical davits are located outside the area between the davits. The falls are led along the deck and the sheaves are so located as to insure a fleet angle of nor more than 4 deg. In order to minimize hazard to the crew, the falls are kept close to the deck and are covered. This causes some difficulties, since the winch must be bolted to a foundation which is high enough to make proper use of the winch hand cranks. Since the falls must lead to the bottom face of the drums and not exceed a 4-deg fleet angle, it is generally necessary to introduce a guide roller between the last sheave and the winch drum.

7.10 Life Preservers and Miscellaneous Equipment. The several items of secondary lifesaving equipment such as life preservers, ring buoys, and water lights must be considered in design primarily as they relate to stowage. Stowage must be such that rapid access to each item is possible under emergency conditions.

a. Life Preservers. Life preservers should be stowed under the bunks or in lockers in each passenger or crew cabin. On ferries and excursion boats, where large numbers of day passengers are carried, they are stowed in boxes and/or under the overhead in passenger areas. In the latter case, the stowage must be such that they are immediately available, by resorting to ladders or similar arrangements, if it is deemed necessary in order to reach them.

b. Ring Life Buoys. Ring life buoys are provided to facilitate the recovery of a man overboard. The number needed will vary with the length of the ship, and a certain percentage should be fitted with self-igniting water lights for nighttime operation. Additionally, two of the ring buoys fitted with water lights must also be fitted with a smoke signal and stowed so as to be capable of quick release from the navigating bridge. The stowage of ring buoys should be such that they can be cast loose quickly. Location must be such that there is a clear direct fall to the water for those units which are automatically released.

c. Emergency Position Indicating Radio Beacons provide a capability for automatic, float-free, distress-alerting in the event of a ship sinking. The EPIRB will continue sending a signal following the distress. Stowage must permit float-free release without fouling in the ship structure during sinking or capsizing. Ordinarily this is not a design problem.

Section 8
Pilot Boarding

8.1 General. Pilot boarding, while not an emergency situation, involves similar personnel risks to those dealt with in lifesaving system design; it is included here since the design considerations are similar. Provision for boarding at sea of persons other than pilots should be given the same consideration.

Pilot boarding is a frequently repeated occurrence during the life of a ship; it often involves considerable risk to human life. Since it is routine, it is often neglected in basic ship design and thus can pose a number of problems to owners and operators–not to mention pilots.

8.2 Design Features. The Safety of Life at Sea (SOLAS) Convention and government marine safety regulations specify the design details for pilot ladders and powered pilot hoists and their arrangements. The basic ship design principle is to provide a straight, clear side to permit the ladder to lie flat against the vessel from the bottom of the ladder to the point of access with a straight forward direct transfer over the gunwale. Provision for boarding at either side should be made.

Pilots should not climb ladders for more than 9 m (30 ft) or less than 1.5 m (5 ft). For greater distances, an accommodation ladder leading aft or other means should be provided. Side ports should be of adequate size and height to facilitate safe entry and any closure should be designed to not interfere with operation of pilot vessel including its antenna. Provision for over-the-side lighting should be made as well as for the deck landing area. Design features

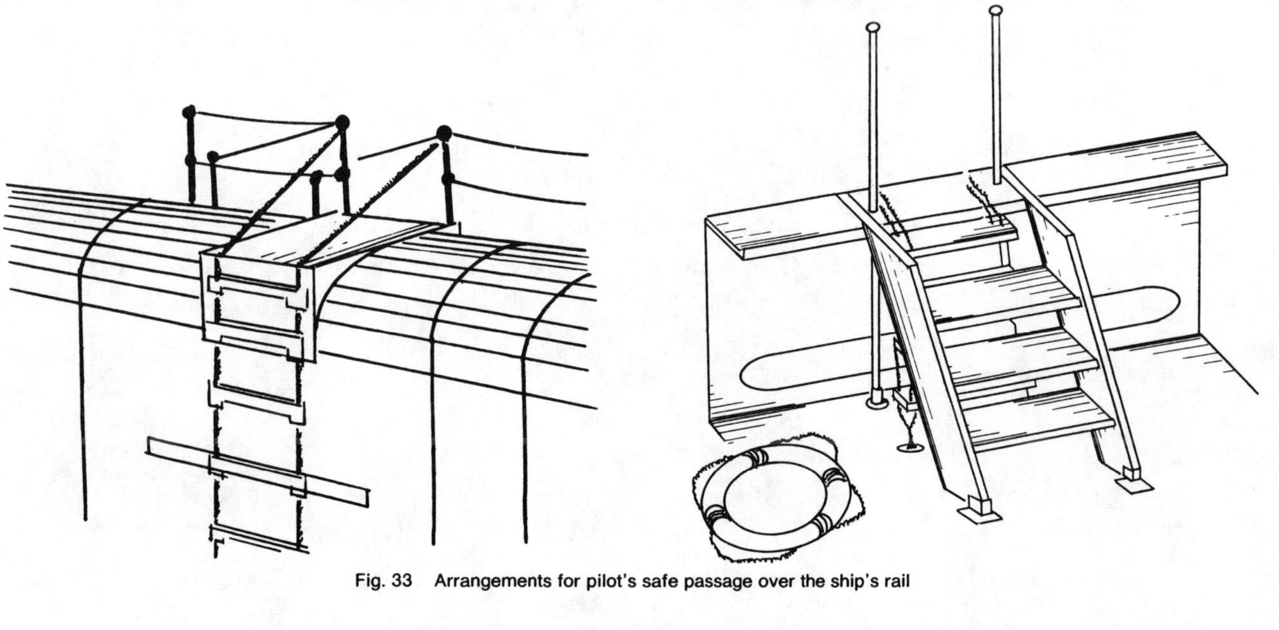

Fig. 33 Arrangements for pilot's safe passage over the ship's rail

which complicate safe boarding include decks projecting beyond the ship side, non-vertical sides, rounded bulwarks, boarding hatches, outward opening doors that get in way of the pilot vessel, boarding hatches too low to permit a quick climb up the ladder to safely clear the pilot vessel, rubbing bands, overboard discharges, and failure to provide clear access to the deck with adequate handholds.

Figs. 32 and 33 depict some relevant features which indicate possible solutions for safe pilot ladder arrangements. A recent development gaining popular acceptance is a pilots' boarding hoist raised by a power winch. Provision is made for this portable unit, both port and starboard, on permanent platform locations. The installation is particularly advantageous on larger ships.

References

American Bureau of Shipping (Annual) Rules for Building and Classing Steel Vessels, ABS, New York.

Cherrix, Charles B. and Eugene L. Coffman (1976), "The Evolution of Shipboard Accommodation and Habitability Standards on U. S. Merchant Ships," *Marine Technology,* July.

Henry, J. J. and H. J. Karsch, (1966) "Container Ships," *Transactions,* SNAME.

Holm, J. T. and J. B. Woodward, (1971) Chapter II *Marine Engineering,* SNAME.

Leeming, J., (1942) *Modern Ship Stowage,* U. S. Government Printing Office, Washington, D. C..

Maritime Administration, (1979) *Standard Specifications for Merchant Ship Construction,* U. S. Department of Commerce.

Price, R. I., (1967) "Fire Protection in Future Passenger Ships," *Marine Technology,* July.

U. S. Coast Guard Rules, Current Code of Federal Regulations, Title 46, Chapter I.

Sub-chapter	Part	Subject
A	1–9	Procedures Applicable to the Public
B	10–16	Merchant Marine Officers & Seamen
C	24–26	Uninspected Vessels
D	30–40	Tank Vessels
E	42–46	Load Lines
F	50–64	Marine Engineering
G	66–69	Documentation & Measurement of Vessels
H	70–89	Passenger Vessels
I	90–109	Cargo & Miscellaneous Vessels
J	110–139	Electrical Engineering
N	146–149	Dangerous Cargoes
O	150–154	Certain Bulk Dangerous Cargoes
P	157	Manning of Vessels
Q	160–164	Specifications
R	166–168	Nautical Schools
T	175–187	Small Passenger Vessels (under 100 gross tons)
U	188–199	Oceanographic Vessels

U. S. Public Health Service (USPHS), (1965) "Handbook on Sanitation of Vessel Construction," Department of Health, Education and Welfare, Pub. No. 393.

CHAPTER X

John W. Boylston

Cargo Handling—Dry Cargo

Section 1
Introduction

1.1 General. Twenty-five years ago, breakbulk cargo handling systems were almost universally employed for dry cargo, except for bulk cargo. The containership, barge carrier, Ro/Ro and others were experimental modes of transportation. The method of handling dry cargo has become so important in the ensuing years that selection of the cargo handling technique often dictates the remaining design characteristics. What has caused this revolution in the design process?

As with all change, the fundamental problem in the 1950's was the rising cost of ocean shipping. Almost everyone in the industry believed that stevedore labor costs were the root problem, and a National Academy of Sciences study (NAS-NRS, 1955)[1] more than confirmed these beliefs. Data derived from that study showed that 47 percent of the shipping dollar was spent loading and unloading the ship, 33 percent spent in domestic movement, while only 20 percent of the shipping dollar was actually spent for the ship voyage. Surely, to reduce shipping costs, cargo handling was the target area.

One obvious solution was to increase the size of the unit moved beyond the normal pallet or piece-good size so that less lifts (cycles) would be required and, thus, less manhours expended in the loading and discharging process.

As the early containership operators were to find out, not only were stevedoring costs reduced, but port time itself was reduced. As most of the ships at that time had the same speed (about 15 knots), this reduced port time allowed their ships to make more trips per year, and thus, they carried more cargo per year than their breakbulk competitors. It was this reduced port time/extra voyage relationship that revolutionized ship operating economics.

1.2 Port Turn-Around Time. To illustrate the importance of port time as related to cargo handling, consider Fig. 1 and two well-known ports, "A" and "B." These ports are 1,500 miles apart, and it is desired to operate a weekly cargo service, such that a departure is made from "A" the same day each week, and similarly, an arrival at "B" is made on the same day each week. This type of liner service is very popular with certain cargo shippers and allows the shipowner to count on regular or repeat customers.

In Table 1, using required port time as a variable, it is interesting to note the dramatic effect that changes in port time have on the required sea speed. All things being equal, sea speed is, of course, directly related to horsepower, fuel consumption and, thus, operating cost.

To illustrate the effect of other operational factors, consider the hypothetical ship *W* which appears to have an unbelievable loading and discharge rate. Actually, as can be seen in Fig. 2, the entire cargo section of this vessel is floated free and then discharged (and reloaded) while the vessel is at sea. Port time only reflects the time necessary to remove the large cargo section and to replace it with another one. However, now factors such as:

- fueling time,
- depth of water for sinkage, and
- capital cost of three cargo sections per ship, become dominant and possibly may cancel the economic effect of the reduced port time.

1.3 Length of Voyage. Another factor in considering the selection of the cargo handling system is the length of the intended voyage. Returning to Fig. 1, if the distance between "A" and "B" were increased to 4500 miles and the voyage frequency extended to approximately 3 weeks, the following would apply:

There is, thus, less incentive in this case to provide an extremely efficient cargo handling system, particularly if a less efficient system would allow an increase in cargo deadweight or a decrease in first cost.

1.4 Commodity Carried and Other Items. Certainly, to ensure the efficient transport of specialized cargos, specially designed ships are required. The design of the ship as influenced by its mission is covered in Chapter I. Within each commercial ship type, however, the relationship of the type or size of the shore or shipboard cargo handling system should be considered carefully.

[1] Complete references are listed at end of chapter.

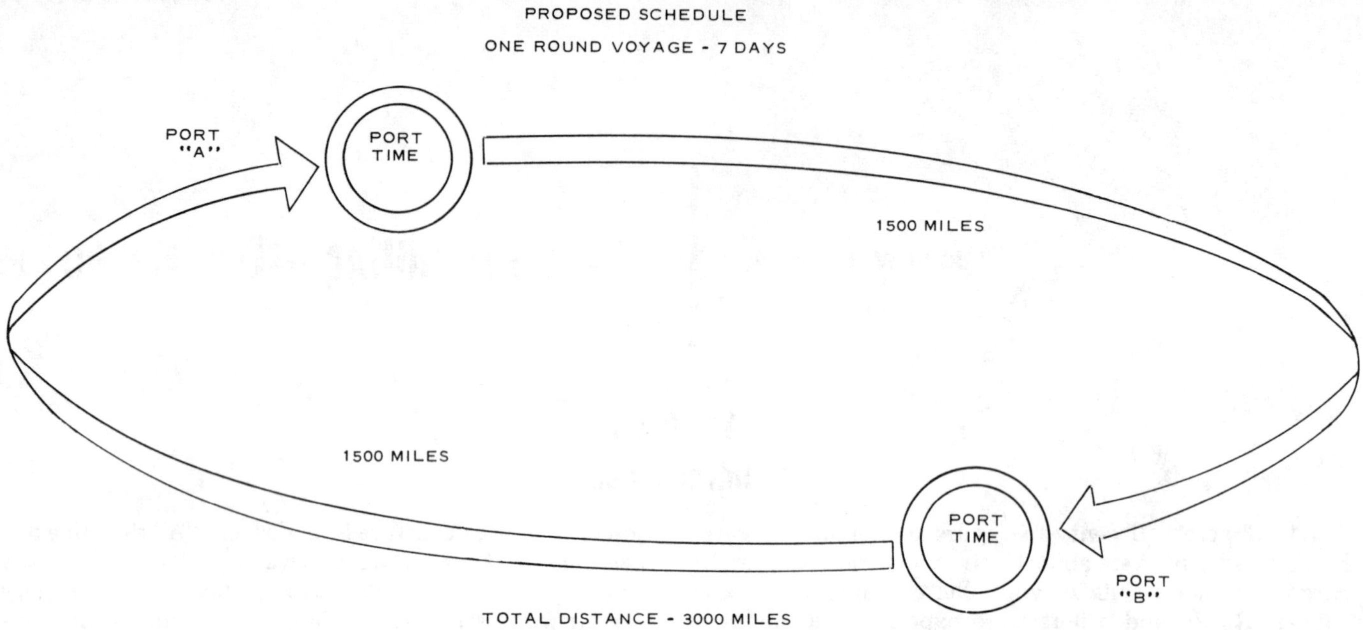

Fig. 1 Pro-forma voyage schedule

In most cases, ships are designed to fit into an existing shorebased loading or berthing system, and often these constraints result in a less than optimum design. It is necessary, however, to consider the whole system of berth, cargo handling system, and ship to ascertain whether changes in one or more components will make a better total system. The needs of the ship operator should be carefully explored and factors such as planned expansion, varying cargo types, possible conversion, and port facilities should all be taken into account.

Alternatively, if the ship design is the first in a particular system, then all of the above considerations take on an increased importance. The selection of the cargo handling system will exert additional influence on ship design, as well as on a number of base parameters directly related to the long-term economic viability of the total system.

Table 1—Variation of Operational Factors with Port Time

Ship	*Loading/ Discharge Time Each Port, hr*	*Total Loading/ Discharge Time, hr*	*Pilotage and Docking, hr*	*Total Port Time, hr*	*Underway Time, hr*	*Required Sea Speed, knots*	*Port Time, percent of Voyage Time*
W	3	6	2	8	160	18.75	5
X	12	24	8	32	136	22.06	19
Y	24	48	8	56	112	26.78	33
Z	36	72	8	80	88	34.09	48

Table 2—Variation of Operational Factors with Length of Voyage

Ship	*Loading/ Discharge Time Each Port, hr*	*Loading/ Discharge Time, hr*	*Pilotage and Docking, hr*	*Total Port Time, hr*	*Underway Time, hr*	*Required Sea Speed, knots*	*Port Time, percent of Voyage Time*
W	3	6	2	8	480	18.75	2
X	12	24	8	32	456	19.73	7
Y	24	48	8	56	432	20.83	11
Z	36	72	8	80	408	22.06	16

In comparison with Table 1, a number of interesting conclusions can be drawn:

- The required sea speed has little variance among the four ships.
- Port time as a percentage of voyage time is reduced, as well as the variance in the percentage among the four ships.
- Since the distance between ports and thus underway time is so much longer, the annual number of port calls is only one-third of the Fig. 2, Table 1 example.

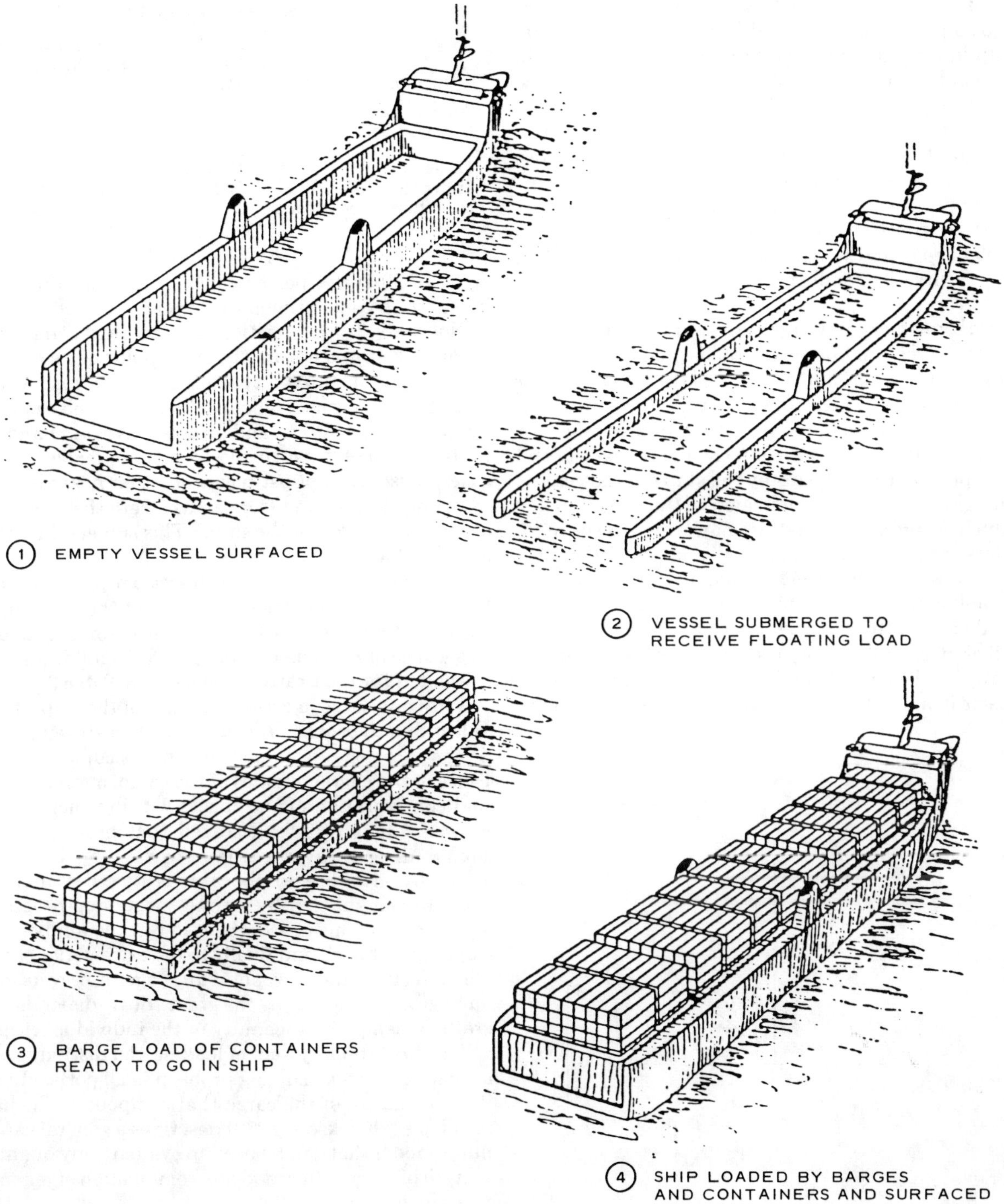

Fig. 2 Advanced cargo discharging scheme

Section 2
The General Cargo Ship

2.1 Mission. There is a worldwide demand for the general cargo ship. A schematic midship section of a general cargo ship in its loading or discharging mode is shown in Fig. 3. General cargo may include items of various sizes, shapes and weights. On the whole, it is predominantly a cubic cargo in contrast to deadweight bulk cargos. The cargo may be heavy machinery, steel shapes and plates, bagged goods, and/or lightweight package goods.

Cargo handling is usually divided into four phases: terminal storage, transfer from storage area to apron cargo hook, hook to hold, and stowage of cargo. Cargo, depending upon its nature, is handled as breakbulk or unitized cargo.

a. Breakbulk Cargo. Cargo arrives at the terminal by railroad cars, trucks and/or barges and is unloaded item by item onto pallets. A forklift truck moves the pallets to their storage positions in the terminal to await transshipment. The pallets are moved to the ship's side by forklift trucks or trailer-towing tractors, slings are attached to the pallets and then the pallets hoisted aboard by the ship's cargo gear.

The slingloads are generally landed in the square of the hatch which covers about one-third of the area of the hold. The pallets or nets are then unloaded, and each item is individually stowed by the longshoreman hold gang, starting from the bulkheads and shell and working toward the hatch square. Thus, two-thirds of the cargo must be man-handled to the wings of the hold. The reverse procedure is followed in discharge. The cargo hook is also employed in snaking heavy items in and out of the wings. Stowage rates on the order of 30 cycles per hr can be expected for each set of booms with conventional cargo gear.

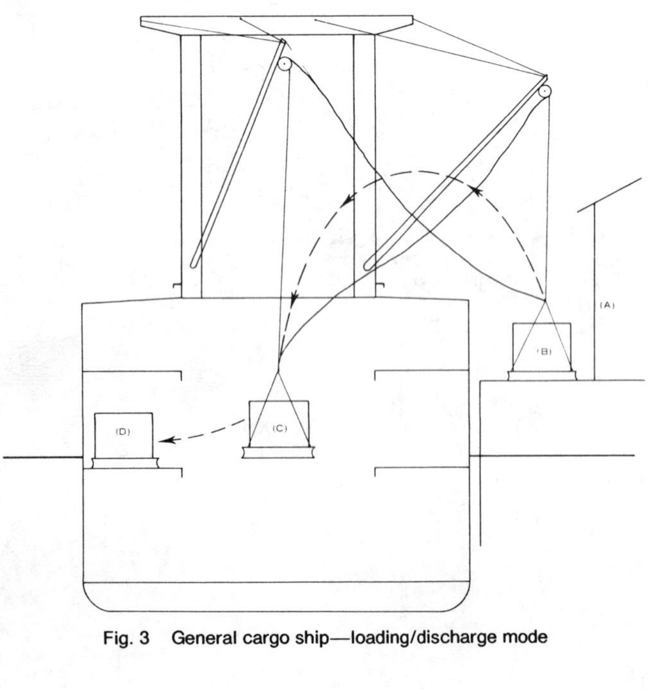

Fig. 3 General cargo ship—loading/discharge mode

Table 3—Stowage Factors

Commodity	Packing	Stowage Factor m^3/metric Ton	(ft^3/ Long Ton)
Alcohol	Drums	1.89	(68)
Caraway Seed	Bags	1.50	(54)
Cement	Sacks	4.18	(150)
Hides (wet)	Bundles	1.23	(44)
Hides (dry)	Bales	5.57	(200)
Structural Steel Beams	Piece	0.67	(24)

Not only is the particular commodity important in breakbulk stowage from the standpoint of its own characteristics, but its proximity to other substances must also be considered. Various labeling codes are used as prescribed by the U.S.C.G., I.C.C. and others to assist in handling and in stowage to segregate explosives, inflammables, corrosives, gases, poisons, combustibles, and other hazardous articles from one another and from areas where they could be hazardous to the crew of the ship. This subject is further covered in Chapter XI.

Each commodity has its own *stowage factor* which is defined as the volume which one ton of the commodity will occupy. Each stowage factor assumes some standard type of package or containers is used, and the following items are presented for illustrative purposes, in Table 3.

With any cargo, a certain amount of the ship's bale cubic is lost due to *broken stowage.* Broken stowage is defined as that portion of a ship's cubic not occupied by cargo because of the characteristics of the cargo, method of stowage, and configuration of the spaces within which the cargo is stowed. It is usually assumed that broken stowage for breakbulk cargo averages 15 percent.

Finally, after breakbulk cargo is stowed, it must be prevented from shifting by a time-consuming process called *dunnaging.* This process uses low-grade lumber to wedge cargo against itself and finally to the extremeties of the hold. The effectiveness of wooden dunnage in preventing cargo damage is suspect as a great deal depends on the craftsmanship and ingenuity of the individual carpenter.

This breakbulk system of loading and discharge has been a common time-consuming problem in all ports. In general, it is the nature of the cargo that compounds the handling problems. Erickson (1970) describes a general cargo computer model that can be used to evaluate any quantity and pattern of cargo inflow and any combination of manning and pier layout.

b. Unitized Cargo. Unitizing is the grouping of quantities of cargo into larger units that are handled as single drafts and stowed as a unit. Unitized cargo can be classed as palletized, containerized or individually packaged or banded cargo not suitable to be handled as breakbulk.

Palletized cargo is breakbulk or large pieces of general

cargo that have been banded or shrink wrapped to a pallet and thus handled as a unit. Broken stowage for palletized cargo can be also 15 percent or more. Containerized cargo on a general cargo ship is usually restricted to *Conex* containers (8 × 8 × 6 ft) and the I.S.O. standard 20 × 8 × 8-ft containers. Large units are not generally carried because of the capacity of ships gear required (in excess of 20 tons) and the normal size of hatch squares.

Large containers, heavy lift, and large general cargo not suitable for below-deck stowage can be stowed on deck and often, in general cargo ship design, this load bearing requirement creates scantlings in excess of those required for longitudinal strength.

2.2 The Burtoning System. This system is the method most commonly used for loading and unloading cargo with a ship's own gear. MacNaught (1955) covers the many types of burtoning gear used and the various methods still used to load and discharge breakbulk cargo in different ports of the world.

Fig. 4 shows one such rig which is called the Ebel mechanical guy arrangement. For purposes of nomenclature, the cargo boom is married to the vertical mast or kingpost by a swivel fitting called the *gooseneck* and associated boom heel fittings. Up and down movement of the boom is accomplished by a topping lift which, in this case, consists of two topping lifts that not only share this load, but help rotate the boom in the horizontal plane toward the centerline (slewing). The outboard purchase from the tip or head of the boom, which again slews the boom horizontally outboard, is called the *vang.* The tackle suspended from the head of the boom is called the *whip,* and in this case it is a *double whip* which denotes the number of parts supporting the block attached to the hook.

As shown in Fig. 4, both whips are married to one hook with one boom permanently spotted over the hatch and the other over the dock, Fig. 3.

In cargo unloading operations, the hatch whip raises the hook load to a height sufficient to clear deck gear, bulwarks, and other obstructions. The pier whip is used to pull the load outboard, while at the same time, the hatch whip pays out rope until the draft reaches a point over the pier. At this time the entire load is taken by the pier boom, and the hatch whip is slack. After the load is lowered onto the pier apron, the common hook is released, and the hoisting winches reeve in the whips in such a way that the hook travels vertically above the pier, and then transversely across the ship, and finally, vertically down into the hatch for another draft. The transfer of cargo from the hold to the pier is carried out by the two winches in one continuous operation, the load remaining in suspension from start to finish. Each whip supports the load while lowering or raising it, but both must take part in the horizontal, transverse movement. Whenever possible, the two winches are operated by one man who becomes so adept that the transfer of cargo is performed rapidly and smoothly. In cargo loading, the burtoning operations just described are carried out in the reverse order. The draft can be landed in the hold in a vertical plane containing the heads of the two booms. The draft can also be landed forward or aft of this plane by manually pushing the load and landing it in the direction of stowage.

In lieu of positioning one boom over the pier, whenever dock facilities permit, the pier whip may be led through a lead block (house fall) attached to the dock structure, in way of the pier landing area. This whip is then married to the hatch whip to permit the burtoning operation. This latter arrangement system offers added flexibility in burtoning when the cargo handling area of the deck is not abreast that of the pier.

When the vangs are used to fix the position of the boom heads, care must be taken in design to locate the pads properly and to ensure a solid foundation for the pads, since severe stresses can result if the angle between the two whips becomes excessive. Of all cargo handling systems, the burtoning system requires the most complete understanding by all concerned if safety is to be maintained.

Port congestion in emerging nations has caused a resurgence in interest in cargo ships with conventional cargo gear, as owners with non-self-sustaining ships must wait long periods for berths fitted with shore cargo handling facilities. Recent developments in burtoning rigging allow systems where light loads are lifted by a single whip at conventional burtoning rates (30 cycles per hr), and then with a four-minute change to a multi-purchase whip, without any re-rigging, heavy loads may be handled at reduced-cycle rates.

2.3 The Swinging Boom Arrangement. Consider one-half

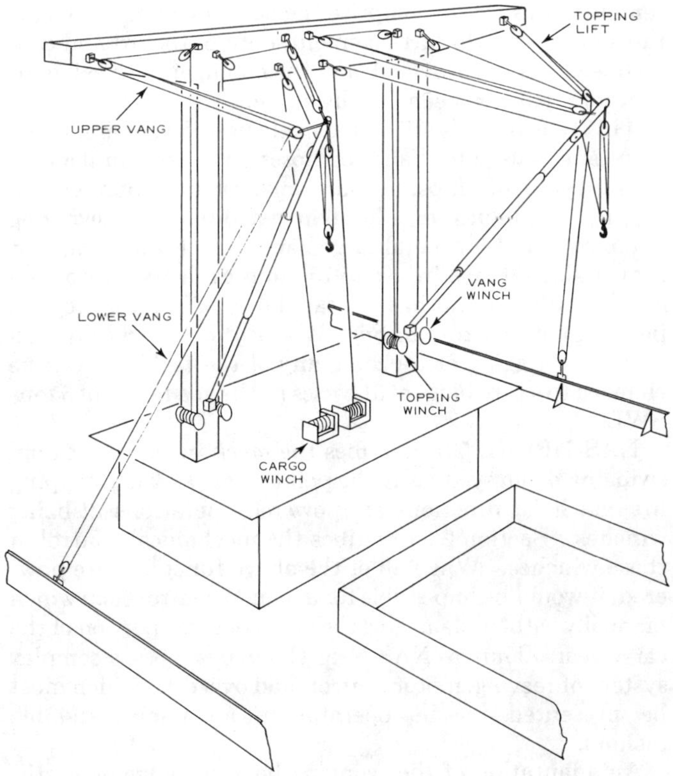

Fig. 4 Ebel mechanical guy or split vang gear rigged for burtoning

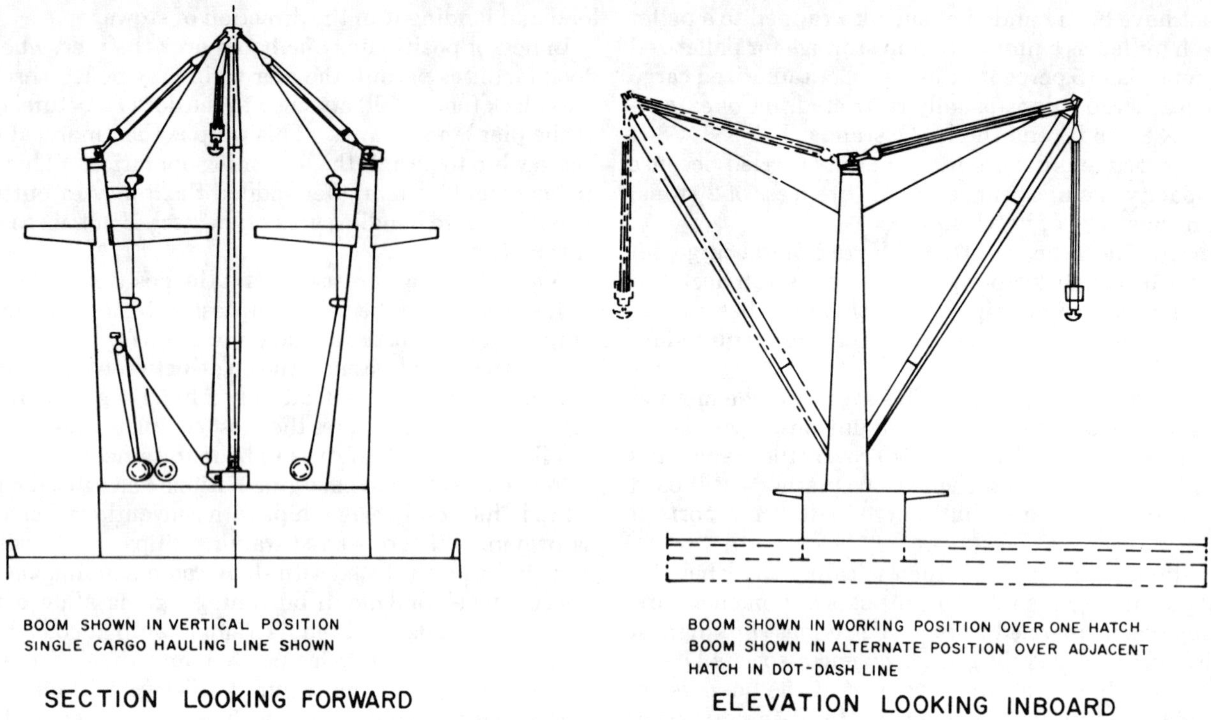

Fig. 5 Stuelcken type heavy-lift

of Fig. 4, with a single topping lift run to the top of one of the kingposts (mast) with the connecting (crosstree) structure removed; this would comprise the basic swinging boom arrangement. In this arrangement, both outboard and centerline vangs are provided, and the boom, once lifted, is either manually or mechanically slewed.

This type of rig is still extensively used on general cargo ships and is used for handling hoses on tankers and stores on all types of ships. Again, modern developments in rigging techniques have brought new life to the swinging boom concept for general cargo handling. Equipment now is available ranging from 5 to 150 tons capacity. As an example, a 50-ton unit has the capability of making 5 cycles per hr at its maximum *Safe Working Load* (SWL), and without rerigging or touching any of the tackle, it can be changed to a capability of 31 cycles per hr at a rating of 3 tons SWL.

NAS-NRC (1959) describes the *mechanization* of dual swinging boom systems to the point where all vang, topping lift and hoist functions are powered operations utilizing winches. Section 2.6 describes the mechanical control of these winches. When all of the above functions are powered, it would be impossible for a man to control each winch manually without damaging the boom or some portion of the cargo gear. Thus, as NAS-NRC (1959) describes, a complex system of reeving, motor controls and overrides which must be presented to the operator in an understandable fashion.

An adaptation of the swinging boom arrangement still used on general cargo ships is the heavy lift boom. Fig. 5 depicts one variation called the *Stuelcken*-type heavy lift cargo gear, which allows one boom and set of kingposts to serve two adjacent hatches. The boom is stepped on the ship's centerline at the midpoint between two free-standing kingposts and is supported at its upper end by twin topping lifts running from rotating heads on the kingposts. Each topping lift is handled from a separate winch that provides quick and accurate positioning of the boom.

The upper block of the cargo falls is supported from a rotating sleeve near the head of the boom. The hauling part of the cargo falls leads to blocks that are connected by a self-positioning universal joint to a pivoting arm at the boom heel. From the heel blocks, the leads travel upwards to a kingpost and thence to the winches. The lower block of the cargo falls is weighted and fitted with a hook or other attachment.

Moving the boom from one hatch to the adjacent hatch forward or aft of the kingposts is accomplished by raising the boom to a near-vertical centerline position. A pendant is then attached to the cargo falls lower block and led to a pad in the area of the hatch toward which the boom is to be moved. Hoisting the lower block will tighten the pendant; this will pull the boom over deadcenter and rotate the sleeve to which the upper cargo falls are attached. The pendant is removed and the boom is ready for use in its new position.

2.4 Cargo Gear Design Example. The analysis and design of mast-boom gear of a certain configuration consists mainly of the determination of the rigging forces needed for the selection of winches, wires and fittings, and of the design of the supporting structure. It is the objective of this section to show:

- How the rigging forces can easily be calculated even for more complicated rigging schemes if a simple vector representation of the forces is used rather than force decompositions in various planes.
- How the analysis and design of the supporting structure can be based on a structural engineering language, such as STRESS (MIT, 1964), STRUDL (MIT, 1967), SAMIS (JPL, 1967), and STARDYNE (CDE, 1968), using the reactions to the rigging force components as loadings.

a. Approach. The approach shown for both steps is applicable to any gear with a determinate rigging scheme and arbitrary support structure. The analysis and design of boom-mast type cargo handling gear have conventionally been based on force decomposition in planes by graphical methods or by trigonometric calculations, and on standard structural analysis methods for the design of the support structure. A thorough and detailed description of this approach is given by MacNaught (1955). However, for the analysis and design of advanced mast-boom arrangements with their more complicated rigging schemes, it becomes necessary to apply methods which are simpler and more adaptable to automation. The suitability for computerization of any new approach is especially important in view of the more complicated kinematics of the arrangements under consideration. For example, the movement could be simulated by a sequence of positions, and all conditions appearing to be critical could be analyzed if computer-adaptable methods for the calculation of the rigging forces and for the structural analysis were available.

As stated, these calculations can be performed in a rather simple way by using a vector representation for the calculation of the rigging forces and a problem-oriented structural analysis system (STRESS language) for the structural design. It is assumed that:

- Dynamic problems can be treated by a pseudo-static analysis in connection with large safety factors which are usually specified.
- The rig arrangement is not structurally redundant.

In the following example from Reuter (1968) the calculation of rigging forces, using three component equations for the determination of axial forces and x-y-z components, is illustrated for a boom-mast gear with upper and lower vangs (Ebel Rig, Fig. 4) which may have a different number of running parts. The utilization of the STRESS system for the analysis of the unstayed supporting structure is described for Ebel Rig kingposts connected by a wide crosstree supporting four booms in either swinging or burtoning operation.

b. The Rigging Force Calculation. It is assumed that the design starts from some form of general layout similar to contract plans. Accompanying specifications will further define hook load capacities (rated loads and test loads) for the considered modes of operation; i.e., swinging, burtoning, etc., required coverages of deck and pier areas, and perhaps allowances for hook cycle times, requirements of standardization; i.e., interchangeability of booms, maximum number of different sizes of wire ropes, blocks, etc., and other guidelines of strong influence upon the final design of the cargo-handling gear. However, before the rig design and analysis are started, the proposed arrangement should be checked against some general ground rules and requirements, such as:

- Compatibility of specified boom length, topping angle and swinging angle with the required outreach.
- Certification rules and other applicable regulations (angle between married falls not to exceed 120 deg).
- Obstructive effects of kingposts and crosstrees on the visibility from the bridge.
- Suitability of gear arrangement for efficient structural implementation (ratio of boom length to mast height, location of kingpost relative to longitudinal deck structure and bulkheads, etc.).

After it has been assured that these ground rules and requirements are satisfied, the rigging forces are calculated for the following purposes:

- Strength analysis of running parts and supporting structure. The selection of the number of running parts, wire sizes and fittings is based on the axial forces in the different components of the rig. It is practical, for the analysis of the supporting structure, to obtain from the rigging force calculation the x-y-z force components in a Cartesian coordinate system, especially if a structural analysis system is used, such as the MIT-STRESS language.
- Checking of kinematics of rig arrangement. The rig must be checked for possible instabilities, such as jackknifing of the boom. This check should include all positions in the various modes of operation, especially those at the boundaries of operational ranges, e.g., maximum swing angle or minimum topping angle.
- Arrangement optimization. The calculation of rigging forces should be performed for configurations of different proportions, having alternate points of attachment to hull, etc. This will permit selection of an arrangement with small rigging forces, requiring minimum (topside) structural weight, and wires of minimum size, etc.

Since it is not always obvious which rig position and which mode of operation will result in the maximum force in a certain rigging component, a large number of rig positions may have to be investigated in order to obtain the maximum force values for all rigging components (vangs, topping lift, boom, cargo whips, etc.). Consequently, a great amount of time is usually spent on such investigations, and they are often complex and voluminous. These characteristics make the computerization of the calculation of rigging forces very attractive, especially in connection with more complicated rigging schemes and a computer-aided analysis of the supporting structure.

After all modes of operation for which rigging forces are to be calculated have been defined, the rig movements of these operational modes are described by a sequence of discrete rig positions. Rigging forces and components are then determined (and printed upon request) for each of these positions and for other critical rig positions which may

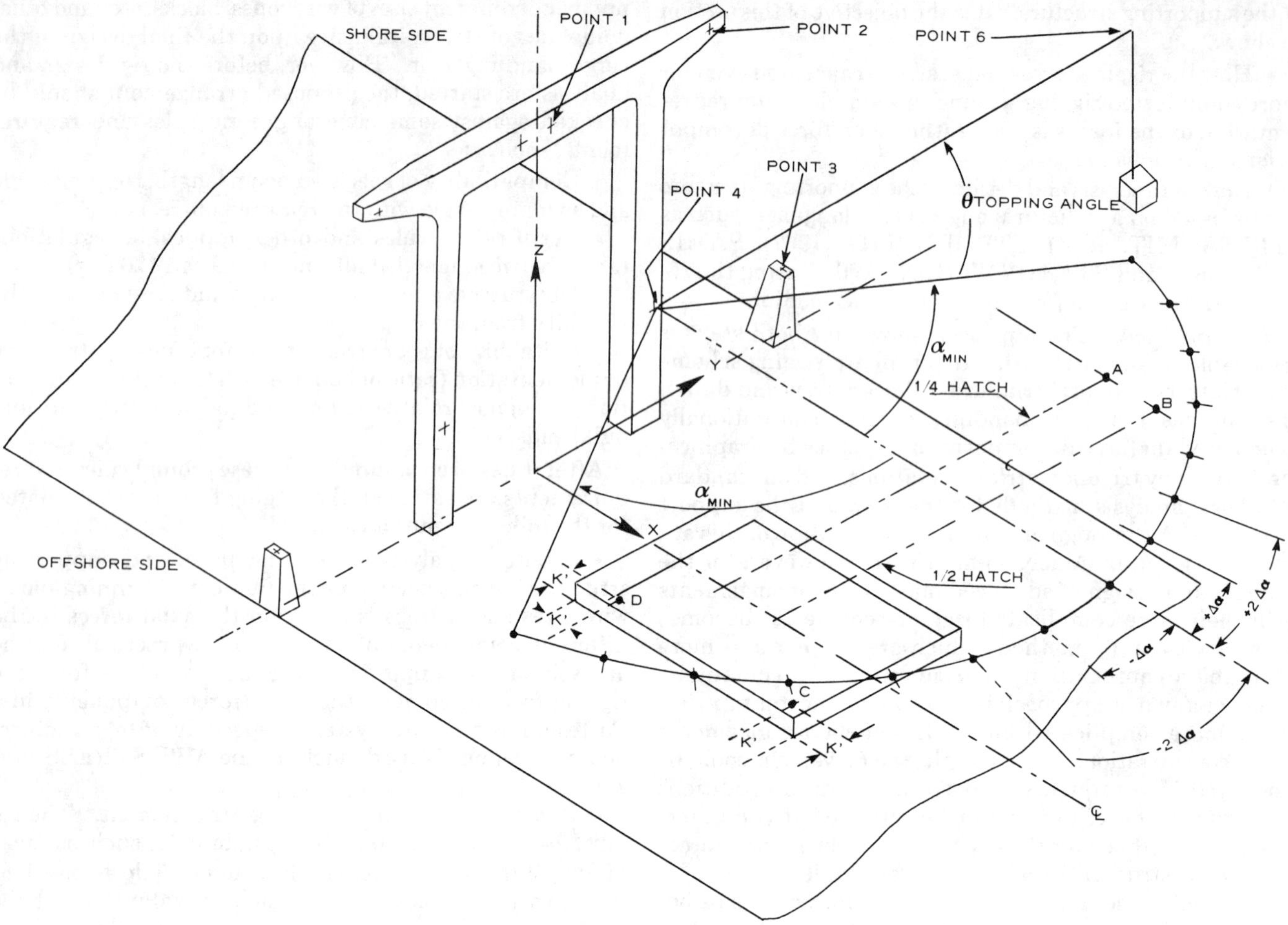

Fig. 6 Swinging arrangement

be specified in addition to the rig movements.

For a certain rig position, the rigging forces and components are calculated in the following way:

1. The force vector in the cargo whip is determined from the known hook load and the weights of the gear components, and from the orientation of the cargo whip in the position under consideration.

2. The cargo whip force is equated to the sum of all other force vectors acting through the head of the boom, e.g., tension in vang(s) plus tension in topping lift(s) plus compression in the boom,

$$\vec{F}_{\substack{\text{cargo}\\\text{whip}}} = \vec{F}_{\substack{\text{boom}\\\text{compression}}} + \vec{F}_{\substack{\text{topping}\\\text{lift}}} + \vec{F}_{\text{vang}} + \vec{F}_{\substack{(\text{additional}\\\text{vang})}} + \vec{F}_{\substack{(\text{additional}\\\text{topping lift})}} + \cdots \quad (1)$$

Since the rig geometry is known for the position under consideration, all unit vectors, e along the forces involved can be calculated, and equation (1) can be written as

$$F_{\substack{\text{cargo}\\\text{whip}}} \cdot \vec{e}_{\substack{\text{cargo}\\\text{whip}}} = F_{\substack{\text{boom}\\\text{compression}}} \cdot \vec{e}_{\substack{\text{boom}\\\text{compression}}} + F_{\substack{\text{topping}\\\text{lift}}} \cdot \vec{e}_{\substack{\text{topping}\\\text{lift}}} + \cdots \quad (2)$$

where $\vec{e}$ is the directional unit vector

In equation (2), all vectors on the right side and all of the left side are now available, allowing us to write three component equations replacing equation (2). Solving the three equations for the x-y-z components yields the rigging force components and thus also the axial force values. Since these are three component equations, only statically determinate gear configurations can be analyzed, i.e., the set of forces on the right side of equation (1) may not contain more than three independent force vectors.

3. After the completion of a rigging force calculation for a certain gear position, the results are compared with previously obtained figures. The maximum values of force in all rigging components (e.g., boom compression, topping lift(s), vang(s), etc.) are retained for final printout of the list of critical forces.

A typical arrangement of a split vang rig (Ebel Rig) is shown in Fig. 4 and described in Section 2.2. One of the important features of this rigging scheme is the vang which is split into an upper and lower vang. Both vangs, however, consist of one continuous running part reeved through the block at the boom head. Therefore, the ratio of the forces in upper and lower vang equals the ratio of the number of running parts in upper and lower vang.

For the calculation of rigging forces, the following modes

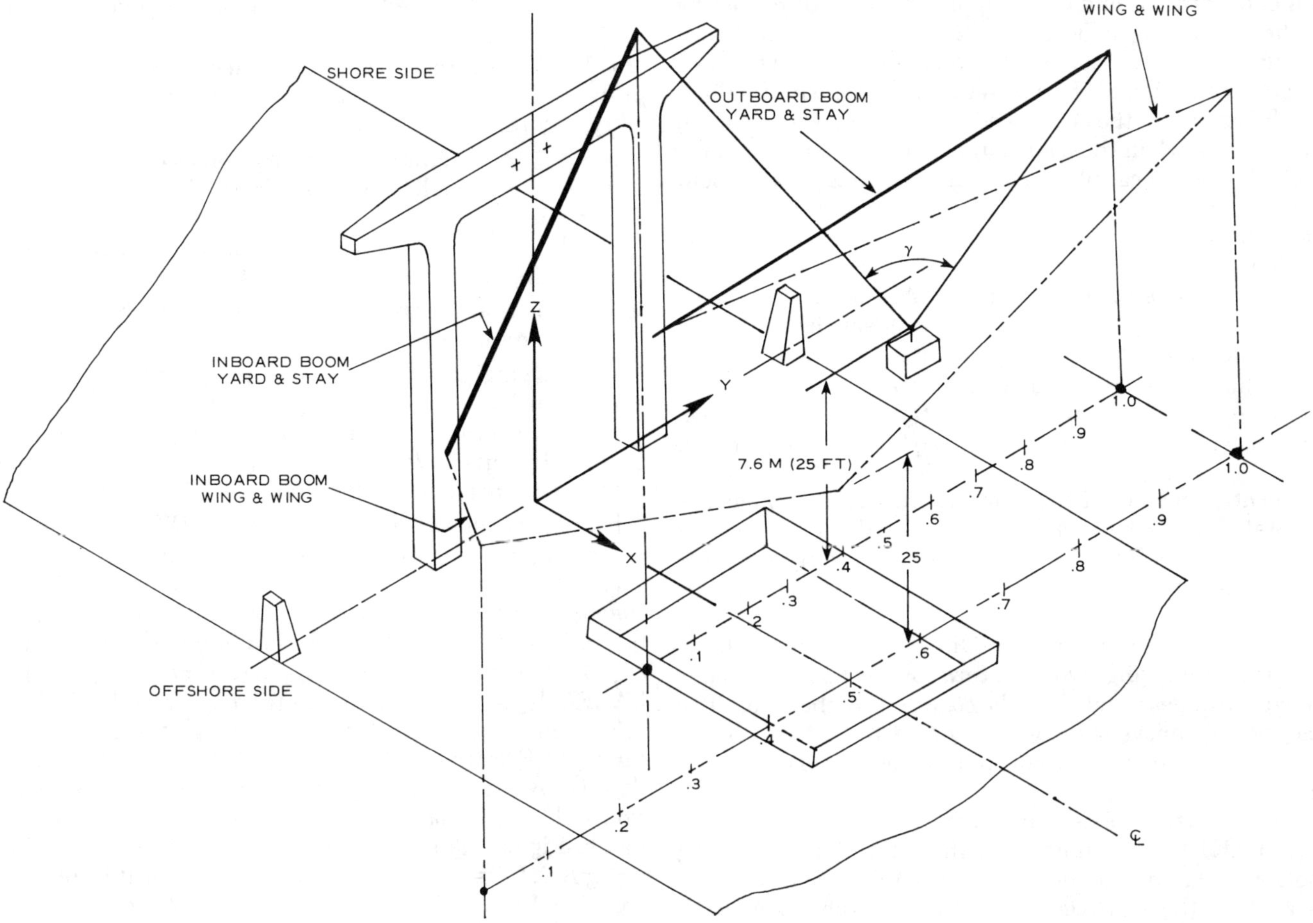

Fig. 7 Burtoning arrangement

of operation are to be considered in this example:

1. A single boom swings at minimum design topping (elevation) angle, handling its rated load and, secondly, at minimum topping angle, handling a test load with maximum coverage of deck and pier.

2. Two booms with their heads over specified points handle a given load at the married falls with a given clearance above deck.

The swinging calculation is concerned only with the swinging operation of the shoreside rig. Results obtained for the shoreside rig are applicable to the offshore rig but to the opposite hand. Wire forces, boom compression, and longitudinal and vertical components on the supporting structure will be alike in magnitude and direction for the two rigs, with the boom in corresponding (reflected) positions. Transverse components will be in opposite directions.

Calculations for swinging the rated load are completed for the boom head in four working positions, designated *A*, *B*, *C*, and *D* (see Fig. 6), and swinging in increments of DELALFA deg at minimum design topping angle THETA 1 through a design range from ALFAMIN (inboard) to ALFAMAX deg (outboard). Calculations for swinging the test load are performed for swinging in increments of DELALFA deg through the same design range, but at minimum test topping THETA 2 which represents the maximum possible area that can be covered by the boom where:

— Maximum shore swing: ALFAMAX (see Fig. 6)
— Maximum offshore swing: ALFAMIN (see Fig. 6)
— Increment in ALFA when swinging: DELALFA
— Minimum design topping angle (rated load) THETA 1
— Minimum test topping angle (test load) THETA 2

The burtoning calculations employ the shoreside and offshore rigs termed, hereinafter, outboard and inboard, respectively. For the yard and stay arrangement, the outboard boom is generally over the shore, while the inboard boom is over the hatch. For the wing-wing arrangements, the outboard boom is over the shore, and the inboard boom is symmetrically placed over a lighter. The cargo falls are married at a given height over the deck, and this height is assumed constant as the load is burtoned between the boom heads, Fig. 7.

In the burtoning calculation for the yard and stay arrangement, the load travels from the hatch to the outboard boom head in equal increments.

The calculation is performed first with the load under the inboard boom. After printing out the axial forces in the elements of the rig and the components of the forces acting on the supporting structure for the inboard and outboard rig, the load is advanced by 10 percent of its travel towards the shore, Fig. 7, and the calculations are repeated. For a certain position of the boom heads, the burtoning calculation is completed when the load is under the outboard boom.

The burtoning calculation for the wing-wing arrangement is similar, except that the calculation begins with the load under the inboard boom head and travels to the center of the hatch, which, for this arrangement, is midway (50 percent of travel) between the boom heads. As the rig is symmetrical about the centerline, the values repeat themselves in reverse order when the load is advanced outboard.

The burtoning calculations are performed first for two wing-wing arrangements ($A_I A_\phi$, $B_I B_\phi$), and then for two yard and stay arrangements ($C_I C_\phi$, $D_I D_\phi$), as shown in Fig. 8.

The arrangement of the gear and the rigging scheme, as illustrated in Fig. 6 (swinging boom) and Fig. 7 (two booms rigged for burtoning), can be described for the rigging force calculation by coordinates in a Cartesian system as follows.

The positive x (longitudinal) direction is assumed to be from the supporting structure towards the hatch, which is served by the gear, and along the centerline of the ship. The positive y (transverse) axis is assumed to be towards the shoreside and along the centerline of the supporting structure.

The z (vertical) axis is positive upward. The topping angle THETA is measured above the horizontal. The swing angle ALFA is measured in the horizontal plane from a line parallel to the ship centerline through the boom heel. Positive swing angles denote a shoreside swing, and negative angles, an offshore swing.

The fixed-rig geometry can thus be described by:

- x-y-z coordinates of topping lift swivel (Point 1, see Fig. 6)
- x-y-z coordinates of upper vang swivel (Point 2)
- x-y-z coordinates of lower vang swivel (Point 3)
- x-y-z coordinates of boom heel (Point 4)
- Length of the boom: BL

  NOTE: x and z coordinates are the same for inboard and outboard rigs, while y coordinates of inboard rig have same magnitude as those for outboard rig, but negative signs due to the symmetry of the ship about its centerline.

The characteristics of the rigging scheme will be defined by:

- Number of parts in cargo fall (swinging): NB
- Number of parts in cargo fall (burtoning): NBB
- Ratio M:N of number of parts in lower vang (M) to number of parts in upper vang (N)
- One plus friction factor per sheave in blocks, fairleads, etc.: FR
- Deadweight affecting topping lift and vang forces:

$TL\ (1V)$ = (half boom weight)
+ (weight of topping lift and vang blocks at boom head)
+ (half weight of topping lift and vang wire ropes)

- Hook load and deadweight affecting cargo lead line pull:

$TL\ (2V)$ = (hook load)
+ (weight of lower cargo block)
+ (weight of cargo falls rope)

For the hook load, three values can be specified: rated load for swinging, test load for swinging and burtoning load.

The positions of the boom heads for which the rigging forces are to be calculated are specified:

- For swinging by x and y coordinates of Points A, B, C, and D (see Fig. 6)
- For burtoning by x and y coordinates for inboard and outboard points A, B, C, and D, representing the positions of the boom heads when burtoning (see Fig. 8).

A maximum allowable angle (MAXGAMMA) between the cargo whips can be specified to limit the axial whip force.

c. *Calculation of Rigging Forces Due to a Swinging Boom of a Split Vang Rig.* The swinging calculations will be performed for the boom head in four working positions (A, B, C, and D) and for swinging in increments of DELALFA deg at minimum design topping (rated load) and at minimum test topping (test load) through the design swing range (ALFAMIN to ALFAMAX). The points A and B are defined by the maximum outreach in the transverse direction and longitudinally by the quarter and half-points of the hatch, Fig. 6. Points C and D can be taken as locations being K ft (K = 4 was used) within the hatch at its offshore coaming, Fig. 6. With the X,Y coordinates of the points J = A,B,C,D, thus obtained from the deck plan, the corresponding coordinates Z_{6J} of the boom head are calculated by:

$$Z_{6J} = 24$$

$$B \cdot sin \left[Arccos \left(\frac{\sqrt{(X_{6J} - X_4)^2 + (Y_{6J} - Y_4)^2}}{BL} \right) \right] \quad (3)$$

NOTE: For points A,B,C,D, X_6, Y_6 are known. Boom length BL is input.

$$\theta = \text{Arccos} \left(\frac{\sqrt{(X_{6J} - X_4)^2 + (Y_{6J} - Y_4)^2}}{BL} \right) \quad (4)$$

From the sketch, read also that the coordinates of the boom head, i.e., point 6, as defined by ALFA and THETA, are computed by

$$X_6 = X_4 + BL \cdot cos\ \theta \cdot cos\ \alpha \quad (5)$$

$$Y_6 = Y_4 + BL \cdot \cos\theta \cdot \sin\alpha \quad (6)$$

$$Z_6 = Z_4 + BL \cdot \sin\theta \quad (7)$$

The axial forces in boom vangs and topping lift may be thought of as caused by a resulting load, acting at the boom head and consisting of the hook load and the deadweights of the gear components. Assuming that the cargo lead line is led along the boom, which is normally the case, then the vertical loads acting at the boom heads should be considered

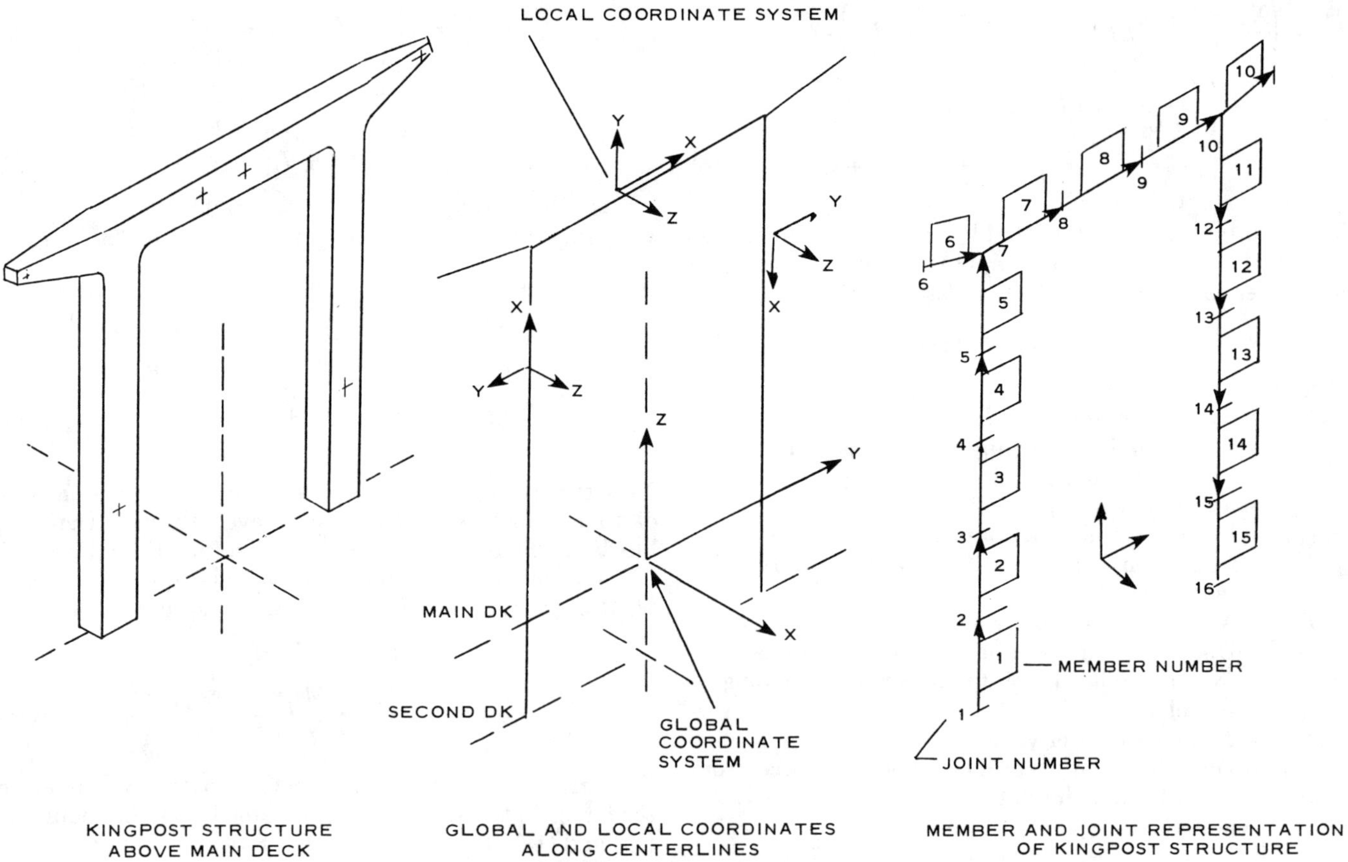

Fig. 8 Model of supporting structure for "stress" analysis

as belonging to either one of the following two groups:

- Loads at boom head affecting topping lift and vang forces:

  Half weight of boom
  + Half weight of topping lift and vang ropes
  + Weight of topping lift and vang blocks at boom head

  Sum = Vertical Load 1 = TL (1V)

- Loads at boom head affecting cargo lead line pull:

  Load on hook
  + Weight of lower hoist block and hook
  + Weight of cargo falls wire rope

  Sum = Vertical Load 2 = TL (2V)

The cargo lead line pull (CLP) is obtained by

$$CLP = FR^{NB} \cdot \frac{TL(2V)}{NB} \tag{8}$$

Where

FR = One plus friction factor per sheave
NB = Number of running parts (sheaves) in cargo fall between load hook and boom head

Thus, the loads $L(1V)$, $L(2V)$ and CLP acting at the boom head are known. To calculate the resultant of the three loads, the unit vectors of their directions are computed and multiplied by $L(1V)$, $L(2V)$ and CLP, respectively. They are added componentwise to yield the resultant load RL:

$$\overrightarrow{L(1V)} = L(1V) \cdot (0{,}0{,}-1)$$
$$\overrightarrow{L(2V)} = L(2V) \cdot (0{,}0{,}-1)$$

To write the CLP as a vector, find first the unit vector of its direction (along the boom, i.e., from point 6 to point 4):

$$\overrightarrow{e_{64}} = \left(\frac{X_4 - X_6}{BL}\right) \cdot (1{,}0{,}0) + \left(\frac{Y_4 - Y_6}{BL}\right) \cdot (0{,}1{,}0) + \left(\frac{Z_4 - Z_6}{BL}\right) \cdot (0{,}0{,}1) \tag{9}$$

Using unit vectors,

$$\vec{i_1} = 1{,}0{,}0$$
$$\vec{i_2} = 0{,}1{,}0$$
$$\vec{i_3} = 0{,}0{,}1$$

and adding $\overrightarrow{L(1V)}$, $\overrightarrow{L(2V)}$, $CLP \cdot \overrightarrow{e_{64}}$, yields the resultant load at the boom head $\overrightarrow{RL}$:

$$\vec{RL} = \left(\frac{X_4 - X_6}{BL} \cdot CLP\right) \cdot \vec{i_1} + \left(\frac{Y_4 - Y_6}{BL} \cdot CLP\right) \cdot \vec{i_2} + \left(\frac{Z_4 - Z_6}{BL} \cdot CLP - L(1V) - L(2V)\right) \cdot \vec{i_3} = AG \cdot \vec{i_1} + BG \cdot \vec{i_2} + CG \cdot \vec{i_3} \quad (10)$$

The resultant load is supported by the boom compression and by the wire forces in the topping lift and in the upper and lower vang. This can be expressed by (Fig. 6)

$$\vec{RL} = F_1 \cdot \vec{e_{46}} + F_2 \cdot \vec{e_{61}} + F_3 \cdot \vec{e_{62}} + F_4 \cdot \vec{e_{63}} \quad (11)$$

Where:

F_1 = Axial boom force.
$\vec{e_{46}}$ = Directional unit vector of compression in boom towards boom head, point 6.
F_2 = Axial force in topping lift.
$\vec{e_{61}}$ = Directional unit vector of topping lift force, from boom head, point 6, towards topping lift swivel, point 1.
F_3 = Axial force in upper vang.
$\vec{e_{62}}$ = Directional unit vector of upper vang force, from boom head, point 6, towards upper vang swivel, point 2.
F_4 = Axial force lower vang.
$\vec{e_{63}}$ = Directional unit vector of lower vang force, from boom head, point 6, towards lower vang swivel, point 3.

The directional unit vectors can be computed directly from the coordinates of the points involved.

$$\vec{e_{46}} = \frac{X_6 - X_4}{BL} \cdot \vec{i_1} + \frac{Y_6 - Y_4}{BL} \cdot \vec{i_2} + \frac{Z_6 - Z_4}{BL} \cdot \vec{i_3} = a_1 \cdot \vec{i_1} + b_1 \cdot \vec{i_2} + c_1 \cdot \vec{i_3} \quad (12)$$

$$\vec{e_{61}} = \frac{X_1 - X_6}{DIS\ 16} \cdot \vec{i_1} + \frac{Y_1 - Y_6}{DIS\ 16} \cdot \vec{i_2} + \frac{Z_1 - Z_6}{DIS\ 16} \cdot \vec{i_3} = a_2 \cdot \vec{i_1} + b_2 \cdot \vec{i_2} + c_2 \cdot \vec{i_3} \quad (13)$$

The distance between points 6 and 1 is

$$DIS\ 16 = \sqrt{(X_1 - X_6)^2 + (Y_1 - Y_6)^2 + (Z_1 - Z_6)^2}$$

$$\vec{e_{62}} = \frac{X_2 - X_6}{DIS\ 26} \cdot \vec{i_1} + \frac{Y_2 - Y_6}{DIS\ 26} \cdot \vec{i_2} + \frac{Z_2 - Z_6}{DIS\ 26} \cdot \vec{i_3} = a_3 \cdot \vec{i_1} + b_3 \cdot \vec{i_2} + c_3 \cdot \vec{i_3} \quad (14)$$

$$DIS\ 26 = \sqrt{(X_2 - X_6)^2 + (Y_2 - Y_6)^2 + (Z_2 - Z_6)^2}$$

$$\vec{e_{63}} = \frac{X_3 - X_6}{DIS\ 36} \cdot \vec{i_1} + \frac{Y_3 - Y_6}{DIS\ 36} \cdot \vec{i_2} + \frac{Z_3 - Z_6}{DIS\ 36} \cdot \vec{i_3} = a_4 \cdot \vec{i_1} + b_4 \cdot \vec{i_2} + c_4 \cdot \vec{i_3} \quad (15)$$

$$DIS\ 36 = \sqrt{(X_3 - X_6)^2 + (Y_3 - Y_6)^2 + (Z_3 - Z_6)^2}$$

In order to obtain relations for F_1, F_2, F_3, F_4, introduce equation (10) and equations (12) through (15) into equation (11):

$$\begin{aligned} AG \cdot \vec{i_1} + BG \cdot \vec{i_2} + CG \cdot \vec{i_3} &= F_1 \cdot a_1 \cdot \vec{i_1} + F_1 \cdot b_1 \cdot \vec{i_2} + F_1 \cdot c_1 \cdot \vec{i_3} \\ &+ F_2 \cdot a_2 \cdot \vec{i_1} + F_2 \cdot b_2 \cdot \vec{i_2} + F_2 \cdot c_2 \cdot \vec{i_3} \\ &+ F_3 \cdot a_3 \cdot \vec{i_1} + F_3 \cdot b_3 \cdot \vec{i_2} + F_3 \cdot c_3 \cdot \vec{i_3} \\ &+ F_4 \cdot a_4 \cdot \vec{i_1} + F_4 \cdot b_4 \cdot \vec{i_2} + F_4 \cdot c_4 \cdot \vec{i_3} \end{aligned} \quad (16)$$

Equation (16) is equivalent to three equations for the components having the orientation of $\vec{i_1}$, $\vec{i_2}$, $\vec{i_3}$:

$$\begin{aligned} AG &= a_1F_1 + a_2F_2 + a_3F_3 + a_4F_4 \\ BG &= b_1F_2 + b_2F_2 + b_3F_3 + b_4F_4 \\ CG &= c_1F_3 + c_2F_2 + c_3F_3 + c_4F_4 \end{aligned} \quad (17)$$

Note that the upper and the lower vang consist of one piece of rope, which is continuously reeved through the vang blocks at the points 2, 6 and 3. Therefore, the ratio F_3/F_4 of forces in the upper and lower vang is equal to the ratio N/M of the number of parts in the upper and lower vang:

$$F_3 : F_4 = N : M$$

$$F_4 = \frac{M}{N} F_3 \quad (18)$$

The arrangement is statically determinate, and therefore, the forces F_1 through F_4 can be obtained from the conditions (17) and (18) of equilibrium only. Since the fourth relation, equation (18), involves F_3 and F_4 only, it appears to be advantageous to reduce equation (17) to a 3 × 3 system by replacing F_4 via equation (18). This yields

$$\begin{aligned} AG &= a_1F_1 + a_2F_3 + \left(a_3 + a_4 \frac{M}{N}\right) F_3 \\ BG &= b_1F_1 + b_2F_2 + \left(b_3 + b_4 \frac{M}{N}\right) F_3 \\ CG &= c_1F_1 + c_2F_2 + \left(c_3 + c_4 \frac{M}{N}\right) F_3 \end{aligned} \quad (19)$$

By using the matrix M

$$M = \begin{Bmatrix} a_1 a_2\ (a_3 + a_4\ M/N) \\ b_1 b_2\ (b_3 + b_4\ M/N) \\ c_1 c_2\ (c_3 + c_4\ M/N) \end{Bmatrix} \quad (20)$$

The equation (20) may be written as

$$M \begin{Bmatrix} F_1 \\ F_2 \\ F_3 \end{Bmatrix} = \begin{Bmatrix} AG \\ BG \\ CG \end{Bmatrix} \quad (21)$$

and provided that M is nonsingular, i.e., that $DET.\ M \neq 0$, obtain the forces F_1, F_2, F_3 by

$$\begin{Bmatrix} F_1 \\ F_2 \\ F_3 \end{Bmatrix} = M^{-1} \begin{Bmatrix} AG \\ BG \\ CG \end{Bmatrix} \tag{22}$$

and F_4 by equation (18). The output of forces and components comprises the following:

$$\begin{array}{lll} \text{Boom Compression:} & \text{Axial} \sim & = F_1 \\ & X\text{-Component} & = F_1 a_1 \\ & Y\text{-} \sim & = F_1 b_1 \\ & Z\text{-} \sim & = F_1 c_1 \end{array} \tag{23}$$

$$\begin{array}{lll} \text{Topping Lift Force:} & \text{Axial} \sim & = -F_2 \\ & X\text{-Component} & = F_2 a_2 \\ & Y\text{-} \sim & = F_2 b_2 \\ & Z\text{-} \sim & = F_2 c_2 \end{array} \tag{24}$$

$$\begin{array}{lll} \text{Upper Vang Force:} & \text{Axial} \sim & = -F_3 \\ & X\text{-Component} & = F_3 a_3 \\ & Y\text{-} \sim & = F_3 b_3 \\ & Z\text{-} \sim & = F_3 c_3 \end{array} \tag{25}$$

$$\begin{array}{lll} \text{Lower Vang Force:} & \text{Axial} \sim & = -F_4 \end{array}$$

The calculation of rigging forces in split vang burtoning gear basically follows the same pattern as the swinging calculation for a certain position of the rig and of the load.

d. Computer Programming of the Rigging Force Calculation. The rigging force calculation, as outlined in the previous sections, can easily be programmed since it comprises straightforward calculations of a linear flow pattern only. As in any other programming task, it is practical to subdivide the problem into functionally independent blocks for which subroutines can be written and checked out individually. For the rigging force calculation, a breakdown has been chosen which employs three calculating and two administrative subroutines:

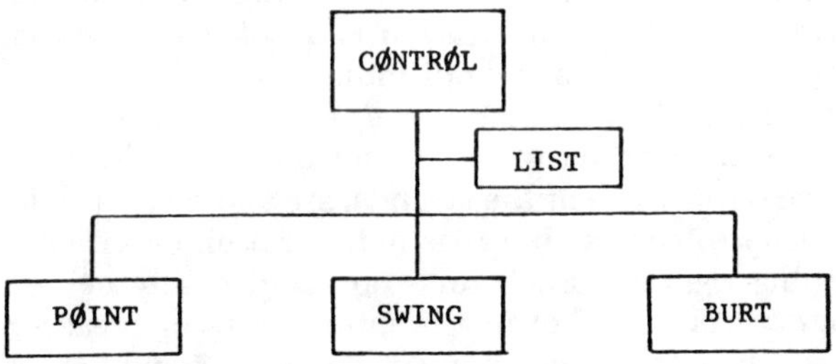

The three calculating routines are direct implementations of the procedures outlined earlier, calculation for swinging boom (subroutines "PØINT" and "SWING") and calculation for burtoning gear (subroutine "BURT").

The user can specify by an *option card* which operational modes he wants to be analyzed. Upon encountering an option card, the main CØNTRØL transfers control to the subroutine indicated on the option card. The subroutine LIST is a search program which maintains a file of maximum axial rigging forces acting in the various rig components during the operational modes under consideration.

e. Analysis and Design of the Supporting Structure by Application of a Problem-Oriented Computer Language. Modern cargo handling gear of the mast-boom type is in most cases supported by an unstayed structure. In addition, deck arrangements often call for two kingposts located athwartships between two hatches, so that each hatch can be served by four booms. The two kingposts are often connected by crosstrees for structural or rig arrangement reasons. Two cross-connected kingposts represent a redundant structure which can efficiently be analyzed and designed by using a problem-oriented structural engineering computer language, such the previously referenced programs STRESS, STRUDL, SAMIS or STARDYNE. Normally, the supporting structure can be adequately modeled as a line element configuration which can be treated by any simple structural engineering language, such as STRESS. A large number of similar structural analysis systems have been developed and are available for computers of much smaller size than the 7094 for which the STRESS system has originally been written.

As an example for the analysis and design of cargo handling gear support structure by the application of problem-oriented languages, this section describes the use of STRESS for the analysis of two unstayed kingposts connected by a crosstree, Fig. 8. The analysis of conventionally stayed masts is treated by Reuter (1968).

The structural design of the rig support must be based on the bending moment envelope due to the support reactions of all operational positions of the rig. Since the support reactions of only a few gear positions generate the maximum bending moments in the various planes, these gear positions must be identified first by inspection or by a search program similar to LIST, Section 2.4.*d*. For the resultant sets of critical loadings, the support structure must be designed for minimum visual obstruction and minimum topside weight, compatible with the design stress limits and the gear arrangement which provides rig stability through the specified ranges of swinging and topping of each boom.

The supporting structure of cargo handling gear with rigidly connected kingposts normally consists of plane frames which could be analyzed by any conventional method. However, if the set of critical loadings is available from a rigging force calculation, as outlined in Section 2.4.*c*, it is practical to analyze and design the rig support by a problem-oriented structural language, such as STRESS. The actual structure is described for that purpose by line elements along the center lines. The geometry of this line element model, together with the cross-sectional properties of each line element, describe the structure. This description can be given relative to the same coordinate system employed for the calculation of the rigging forces and components which now represent the loads.

The output, consisting of stresses or bending moments along the line elements, must be analyzed manually and translated into refined scantlings if deficiencies or redundancies are detected. Thus, the iterative nature of structural design is not overcome, since structural engineering languages, such as STRESS, are strictly analysis systems.

f. Structural Analysis and Design of Two Connected Kingposts Using the M.I.T. STRESS Language. All coding details of the program formulation are covered by the STRESS User Manual (MIT, 1964). Therefore, this section contains only the few pertinent aspects which are peculiar to this application of the STRESS system.

For analysis by the STRESS system, the geometry of the actual kingpost structure is replaced by line elements located on the center lines of the actual scantlings, Fig. 8. Artificial joints and members may be introduced to reflect the off-center line location of a vang swivel or a winch drum mounted on the outside of, for instance, a box-shaped kingpost. Two right-handed, orthogonal, Cartesian coordinate systems are used to describe the geometry of the structure and the cross-sectional properties of the scantlings under consideration. The global coordinate system (Fig. 8) is used for the identification of joint locations, joint loads, joint displacements and joint reactions. Since the orientation of the global system can be chosen arbitrarily, it is made identical to the system used for the rigging force calculation. This will allow the use of the force components, as obtained from the rigging force calculation, directly as input to the STRESS program. A local coordinate system is associated with each member, e.g., the line element between two joints. The local coordinate system is used for the identification of all member properties (cross-sectional area, moment of inertia, section modulus, . . .) and resulting member data (stresses, bending moments, . . .). The orientation of the local system is defined relative to the global system. Joints can be defined arbitrarily along the structural center lines, and a member is defined by a sequence of two joints, Fig. 8.

The STRESS system is based on the stiffness method. In order to maintain proper accuracy, it is therefore advisable to define the members in such a manner that the ratio of their stiffness does not exceed 1:1000. Since stiffness is proportional to l^3/I, members of about the same moment of inertia should not vary in length more than 1:10.

The original version of STRESS has been enabled by an addendum to provide output not only in the form of bending moments but also directly in the form of stresses. The same addition also allows STRESS to provide this output (i.e., combined axial and bending stresses or bending moments), along the member in specified steps rather than at the joints only. These added capabilities greatly improve the usefulness of STRESS for an iterative design approach. It should be mentioned also that all critical loading conditions can be covered by one execution of the program through the use of certain *modification statements* which can modify loads as well as geometry or scantlings. Load modifications are particularly fast and inexpensive, since they do not require changes in the stiffness matrix.

An important question may be raised regarding the convergence of an iterative design and analysis procedure which uses STRESS for the analysis. The experience of most designers shows that only three or four iterations are sufficient for the determination of fully stressed structures. However, the analysis of more complicated structures has shown that it is possible for members, which were actually overdesigned, to appear deficient due to the extra load they picked up because of excessive stiffness. A balanced structure should therefore be defined as the initial configuration by comparison with similar designs or from classification Rules, or by checking the relative stiffness of adjacent members.

The two programs described here for the calculation of the rigging forces and for the structural design of the support are applicable only to a certain gear arrangement. However, similar formulations can be developed for any other determinate rigging scheme. The availability of such a design program enables the designer to respond rapidly to arrangement changes. It also allows him to investigate a larger array of arrangement alternatives from which to select the best candidate for installation.

2.5 Wire Rope, Blocks and Tackle. For general reference material and particular fittings and nomenclature, refer to Chapter IX.

a. Wire Rope Construction and Service Considerations. Cargo handling gear running rigging wire rope is generally preformed of 6-strand construction, with 19 wires per strand, right regular lay, bright (without coating) and with fiber core. Wire rope for standing rigging is usually 6 by 19 construction, right regular lay, galvanized or aluminumized and with fiber or wire rope core. Several advantages are gained in using preformed wire rope for running rigging; i.e., removal of initial torsional stress in wires, easier handling, no tendency of a cut end to unwind which facilitates socketing, better and smoother spooling, and improved conformity to sharp bends around minimum-diameter sheaves. The protrusion of broken wires from regular rope surfaces causes hand injuries and has a nicking effect on other strands, with attendant accelerating rope deterioration, all of which is eliminated by using preformed wire rope. Longer service life results from its use, offsetting the slightly increased initial cost.

A pertinent consideration in the life and performance of running rigging is that the individual wires and strands must move freely; thus, lubrication is absolutely essential for proper performance and optimum service life.

b. Lubrication, Fatigue and Abrasive Wear. Bending fatigue and abrasive wear, combined with bearing pressure on the outer wires of the strands, are two factors which, in addition to corrosion, cause deterioration of wire rope in running rigging. Such ropes are continually being bent around sheaves and drums, eventually causing breaks in the wires. Breaks in the outer wires with no evident excessive wear are usually an indication that the wire rope diameter is disproportionately large for the size of sheave being used, or, conversely, the sheave diameter is too small for the size of wire rope. Heavily loaded ropes on small-diameter sheaves result in high bearing pressures between the rope and the sheave, and between the wires in the rope, causing nicks in the wires where they cross in making up the strands of the rope. This nicking tendency, coupled with repeated bending, eventually breaks the wires and locks the strands into a practically solid unit, preventing the sliding movement which allows each wire to carry its proportion of the load. The outside wires, already nicked, become overloaded when locked and accelerate the failure of the remaining wires.

c. Relation of Wire Rope to Sheave Size. The relation of wire rope and sheave size is especially important in cargo gear running rigging and, in general, it is advisable to use the largest practicable diameter of sheave with the smallest possible size of steel wire rope. In connection with the comparatively small-diameter sheaves used for running rigging, a fiber core is generally use as it provides a more resilient cushion for the strands and does not *cross-nick* the outer wires of adjoining strands. The fiber supplies internal lubrication and contributes to the flexibility and resiliency of the rope. Steel wire cores are acceptable for standing rigging.

d. Lay of a Rope. Reference was made previously to the lay of the rope, an important property affecting its stability under load. When a load is freely suspended from the end of a single wire rope, the strands tend to unwind, making the load spin. To prevent this, a *regular lay* rope is used. The tendency for the strands to unwind is resisted by the tightening of the wires in the strands, which are initially wound in a direction opposite to that of the strands of the rope. Regular lay rope is the most stable and spins less under load than any other in common use and is selected for application where long ropes are employed to handle loads which are detached at the completion of every haul. The term "lay" is also used as a unit of measure, indicating the length of the spiral of the strands along the axis of the rope.

e. Ratio of Block Sheave to Rope Diameter. The main reason for determining the proper ratio of sheave diameter to rope diameter is to reduce the wire bending stress to a minimum. The minimum sheave size for 6 by 19 wire rope is 20 times the rope diameter. Other rope constructions have different requirements. Nothing breaks a wire more quickly than repeated flexing, especially in rapid sequence. However, in the case of wire rope bent around sheaves in a block, the flexing occurs over an arc of about 180 degrees on the sheave. Bending stresses exist only during contact with the sheave and are proportional to the arc of contact and the number of lays involved. For any given rope, the bending stress varies inversely with the diameter of the sheave, and the maximum bending stress occurs at the smallest diameter sheave in the system. Work is done in bending and unbending the rope as it passes around each and every sheave, resulting in an unavoidable loss, which may be reduced somewhat by a proper choice of sheaves and lubrication. If a rope makes contact with a sheave so that less than one lay of the rope is involved, then every wire in the rope is not subject to the maximum bending stress.

The diameter of sheaves should be as large as practicable, but the prevalent use of smaller diameter sheaves, especially for continuously used running gear, usually is dictated by practical considerations and, hence, results in shortened rope life. In installations where reversal of bending occurs in the ropes (i.e., when a rope runs clockwise about one sheave and counterclockwise on a succeeding sheave), it is recommended that sheaves causing reverse bending in the rope be of larger diameter than other sheaves in the system.

f. Sheave Groove. To support a wire rope properly under tension, the sheave groove should have a diameter slightly greater than the rope diameter, Fig. 9. The rope will then be well supported over approximately half its circumference. With a proper sheave groove throat angle, the wire rope will not resist normal rotation of the sheave. When, due to wear, the rope and groove become too small and a new rope is used, the larger diameter rope will not fit properly in the worn groove, and a wedging action of the rope will result. In service, this condition causes difficulty in the rotation of the sheave and unnecessary work is performed by the rope; abrasive wear occurs along two narrow strips, parallel to the rope, instead of the normal wear around the full circumference. The rope is also unbalanced when under tension, causing excessive nicking of wires between the strands in contact, which in turn weakens the rope. Sheaves should be checked and regrooved periodically to eliminate these defects.

If the sheave groove is too large, it will fail to offer support for the wire rope and will result in the rope flattening out under tension, increasing the fatigue in the individual wires, leading to earlier failure of the wires and the rope.

g. Ultimate Strength and Area of 6 by 19 Wire Rope. The wire rope strength values should be obtained from the

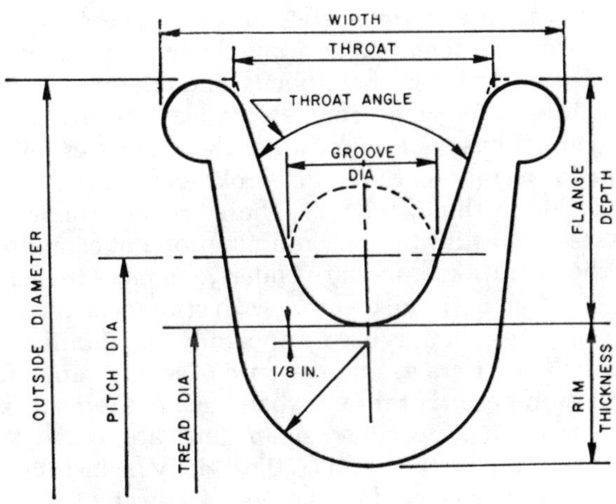

PROPORTIONS IN TERMS OF ROPE DIAMETER, d

	DEPTH	THROAT	RIM THK	NECK	GROOVE DIAMETER
HEAD SHEAVE	2 d	2 d	d (STEEL) 1.125 d	$\frac{d}{2}$	d PLUS TOLERANCE
OTHER SHEAVES	1.5 d	1.625 d	FOR CAST IRON		GIVEN IN TABLE

Fig. 9 Sheave groove nomenclature, proportions and tolerances

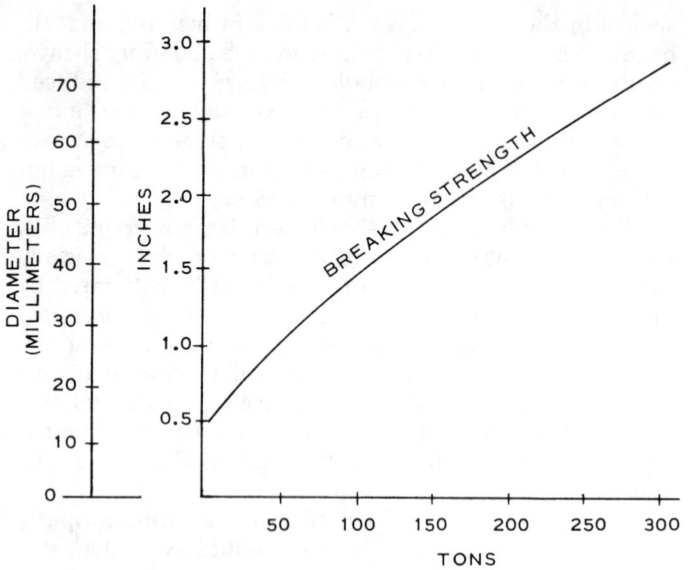

Fig. 10 Properties of 6 × 19 wire rope

manufacturer, since the minimum tensile strengths given vary with the manufacturer or publishing source. The ultimate stress increases for cold-drawn wire rope as the size of the wire is reduced. Since the factor of safety varies with the breaking strength or ultimate stress, it is advisable to use the minimum known values in order to be on the conservative side for design purposes.

The breaking strength for wire rope is higher for stiff rope, such as 6 by 7, and lower for flexible ropes, such as 6 by 37. It is also higher for ropes with a long twist than for ropes with a short twist. The strength of a wire rope is between 80 to 90 percent of the total aggregate strength of all the wire in a wire rope. The strength of the wires in a rope ranges from 414 to 2413 MPa (60,000 to 350,000 psi) depending on the material, the chemical composition and heat treatments used, and the diameter of the wire.

Properties of a certain (6 by 19) wire rope with a hemp core are shown in Fig. 10.

h. Block and Tackle. Chapter IX describes the method of reeving blocks and figuring the *parts of line* for a particular application. When reeving blocks with more than two sheaves, the standing part of the tackle should be reeved through the sheave nearest to the center of the block. The pull of the hauling part will then have less tendency to topple the block, and the chafing of the rope against the edge of the mortises will be reduced.

A block and tackle is a mechanism which permits a comparatively heavy weight to be moved with a light pull on the hauling part. However, in moving the large weight through a distance of, for example, one foot, the pull on the hauling part must travel a distance in feet equal to the number of parts of the purchase when using a simple tackle. With no losses in the system, the work input equals the work output. The input, and the ratio of the actual output L to the input P, is termed the *efficiency* (E) of the system. It should be noted that L must include not only the weight moved but also the weight of the lower block and any other moving parts.

This formula and the data given in Chapter IX are based on idealized conditions and additional losses should be expected as follows:

- friction within the rope due to bending action,
- friction between any ropes in contact,
- friction between the ropes and sheaves.

The idealized losses used in Chapter IX assume a fixed percentage loss per sheave of about 4 percent per sheave for wire rope blocks having bronze-bushed sleeve type bearings, and 2 percent per sheave for blocks fitted with roller bearings. In manila rope systems utilizing blocks with iron bearings, a loss of about 10 percent per sheave should be used.

Detailed formulas for the tension in each part of a block and tackle system for raising, holding and lowering are given in Table 4, which is to be used in conjunction with Fig. 11.

Note that when a load W is being raised, the dynamic pull P will be maximum, since it is working against both gravity and all losses. The value of P will be less when the load W is being lowered, gravity aids in overcoming the total losses, and the dynamic pull P becomes a minimum.

It should be noted that the foregoing information applies to the block and tackle proper. The effect of sheaves, used merely to change the direction of the hauling part or lead line, must be treated separately but similarly.

Overhauling weights are used chiefly to enable the empty hook of a multipurchase cargo falls to be lowered. Such weights are either unit weights added to the cargo hook gear or are built into the lower cargo hoist block in the form of weights bolted to the outer cheek plates.

2.6 Winches. Johnson and Levy (1973) detail recent developments in the uses of winches in general cargo gear. The selection of a winch motive system should be made with careful thought as each type of driving system has advantages and disadvantages. It is hoped that the following gives an objective overview of the types currently available.

a. Steam Winch. The steam winch has the least initial cost of any type now available, can be maintained by engineering personnel, and its performance characteristics are so ideal that it has been the yardstick to evaluate other types of prime mover or transmissions applied to cargo winches. Against these advantages of simplicity and operation, the steam winch requires long, troublesome leads to live and exhaust steam; it has poor efficiency as the engines usually take steam during most of the piston stroke and there is very little expansion in the cylinders. There are also condensation losses and feedwater oil contamination not compatible with the operation of modern boilers. In modern construction, the steam winch, together with other deck auxiliaries powered by steam, is used principally on oil tankers and LNG/LPG carriers as one measure of minimizing oil vapor explosion hazards present when such equipment is operated electrically. Auxiliary steam generators for saturated steam at about 1034 kPa (150 psi) may be installed for their service to avoid contamination of feedwater for the high-pressure boilers of the main propulsion system.

b. Electric Winches. Steam winches have been supplanted in general practice by the electric-motor-driven cargo winch. The electrical systems on shipboard were originally based on the use of direct current; the *mill type* motor, with controls furnishing up to six speeds in the hoisting and lowering direction of rotation, has been successfully used for power cargo winches for many years. Dynamic braking of the motor is employed to afford safe lowering control of loads, augmented by a spring set and electrically released electric motor brake which sets automatically when the master controller is brought to the *off* position or when current fails. The multispeed control provides satisfactory operation characteristics, and the controllers can be located remotely when desired to give operators the best visibility. The foot brake for the winch drum remains as an anachronism to satisfy outmoded safety rules. On more complex winches, a latched foot brake, or a screw and hand-operated brake, may serve a useful purpose in holding a drum which may be declutched when using the winch gypsy heads or an auxiliary drum driven by the same motor.

The demands of hoisting speed have resulted in gradually increased motor ratings for handling general cargo from about 15 kW (20 hp) in early installations to about 37 kW (50 hp) which are generally used at the present time. In burtoning operations, there may not be time in the cycle for the winches to attain full speed before they are decelerated and reversed for lowering. The usual direct-current, *constant voltage* method of motor speed control requires banks of external resistors for inclusion in the motor armature circuit in different combinations. These are determined by contactors which are energized in a control circuit by the winch master controller. There is no regeneration feeding current back into the line when the motor is overhauled, and all power not required by the load at reduced speeds is expended in heating the resistors. Even so, the economy of operation is far superior to that of the steam winch. With the universal adoption of alternating current for modern ships, some applications of multispeed, wound rotor and squirrel cage induction motors have been made for winch operation, but they have not been widely accepted. Their operating characteristics are not as attractive as are those of the direct-current motor, and the problem of providing dynamic braking is a complication. A modified *Ward Leonard* type of control has particular advantages in an arrangement of one or more direct-current generators driven by an alternating-current induction motor. Each generator supplies a winch motor in a variable voltage system. Excitation control of the generator field provides winch motor operating characteristics comparable with, and possibly better than, those of the direct-current, constant voltage system. The wasting of power in external resistance is practically eliminated; the power actually used is proportional to the load requirements, and some amount is recoverable to the ship supply system in regeneration when low-

Table 4—Formulas for Simple Block Systems

n = Total number of sheaves in both blocks
$f = 1 + (1/100) \times$ (percentage total loss per sheave)

Raising	*Holding*	*Lowering*
Last Sheave Fixed—See Fig. 11(a)		
$t_1 = W\left[\frac{f-1}{f^n-1}\right]$	$t_1 = W\left[\frac{1}{n}\right]$	$t_1 = W\left[\frac{(f-1)f^{n-1}}{f^n-1}\right]$
$P = f^n t_1$	$P = t_1$	$P = \frac{t_1}{f^n}$
$\frac{P}{W} = \frac{f^n(f-1)}{f^n-1}$	$\frac{P}{W} = \frac{1}{n}$	$\frac{P}{W} = \frac{f-1}{f(f^n-1)}$
$e = \frac{W}{nP}$	$e = \frac{W}{nP}$	$e = \frac{W}{nP}$
Last Sheave Floating—Fig. 11(b)		
$t_1 = W\left[\frac{f-1}{f^{n+1}-1}\right]$	$t_1 = W\left[\frac{1}{n+1}\right]$	$t_1 = W\left[\frac{f^n(f-1)}{f^{n+1}-1}\right]$
$P = f^n t_1$	$P = t_1$	$P = \frac{t_1}{f^n}$
$\frac{P}{W} = \left[\frac{f^n(f-1)}{f^{n+1}-1}\right]$	$\frac{P}{W} = \frac{1}{n+1}$	$\frac{P}{W} = \frac{f-1}{f^{n+1}-1}$
$e = \frac{W}{(n+1)P}$	$e = \frac{W}{(n+1)P}$	$e = \frac{W}{(n+1)P}$

Note: Tension in any rope part is the geometric mean of the corresponding loads when raising and lowering a hook load. It is not an average.

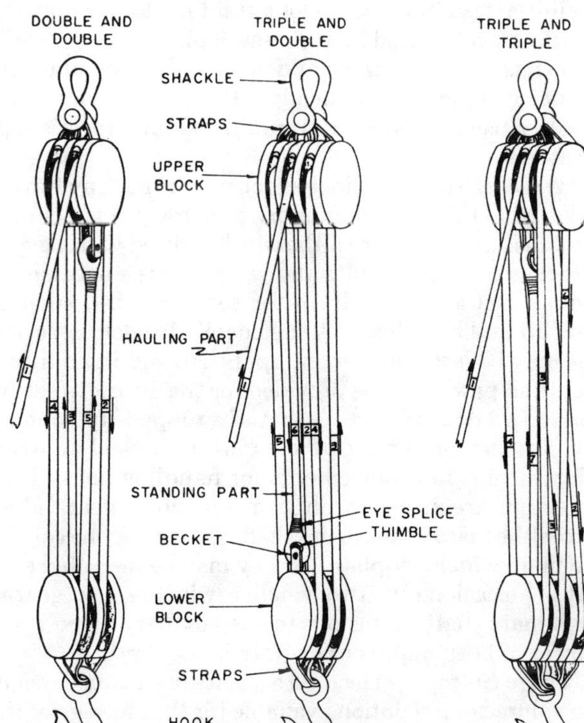

Fig. 11 Reeving tackle blocks

ering loads overhaul the winch motors. The principal objections to the system are in its initial cost per winch and additional maintenance, as basically three major rotating units are involved. These disadvantages may be compensated for by the decreased cost of main generating equipment for the ship supply, which might otherwise include sufficient direct-current capacity for the operation of the deck machinery.

c. Electro-Hydraulic Winches. An alternative that has been used extensively in auxiliaries for naval ships of the United States, and to a greater degree recently in merchant ships, is the electrohydraulic drive, consisting of a *variable stroke* and reversible delivery pump driven by an induction motor and a hydraulic motor geared to the driven unit. The pump is driven continuously in one direction of rotation while the auxiliary is in use, and the speed and direction of rotation of the hydraulic motor is varied by moving the pump stroke control spindle in one direction or the other from a neutral position. This is effected by a control lever or hand wheel, and when in *neutral,* the speed of the hydraulic motor is zero. In this position also, the hydraulic motor is blocked so that the load cannot overhaul the winch. The transmission provides smooth, stepless speed control, regeneration of current when lowering, and protection against excessive overloads by the provision of relief values in the hydraulic system. The hydraulic components may be close-coupled, or the pumping unit may be at any reasonable distance from the hydraulic motor. The electric motor can drive two or more pumps if desired and can be rated for average duty cycles of the winches served. Against the advantages of the hydraulic transmission in operation are its initial cost and the use of a fluid medium requiring tanks, filters, valves and high-pressure piping. Also, while the transmissions are rugged and dependable, the units are somewhat complex mechanically. If repair service is required, it is usually obtained from the manufacturer's representatives.

d. Winch Drums. Winches for wire rope may have a drum keyed to the main shaft or driven through a jaw clutch. The clutch is a convenience if the winch is fitted with a gypsy head for setting topping lifts and vangs. It is a necessity if the winch has a double drum for separate rigs, such as topping lift and hoist for the same boom. In such cases, the brakes may be latched or screw operated, and in addition, ratchets and pawls may be provided for the drums to secure and hold a static load such as that of a topped-up boom.

Jaw clutches or sliding pinions are also used to drive auxiliary drums in arrangements for handling heavy lifts. These drums are driven through an additional gear reduction from the main winch drum shaft. They may be carried on the main winch bedplate, or they may be mounted separately. Occasionally, two smaller winches are geared through main shaft extensions to one auxiliary drum when the power of both motors or engines is required.

e. Drum Storage. The rope capacity of a drum depends on the volume of revolution available for the stowage of the rope. For a given drum, the length of rope that can be wound varies with the diameter of rope to be used. For a given rope, the drum dimensions can be ascertained to suit any required length of rope. Hence, many combinations are possible. In general, there are two cases to be considered: the first is when the drum is filled just flush with the drum flange; the second is when the drum is not completely filled. In any case, the available volume divided by the volume occupied by one foot of the rope, and then multiplied by a suitable factor to correct for the interstitial volume between the ropes, will give the total length of rope that can be spooled upon the drum. Formulas expressing this relation vary only in form, and the factor identified with each size of rope varies slightly due to variation of the rope dimensions by different manufacturers. All such formulas assume uniform rope winding and will not give correct lengths if overwinding or nonuniform winding is used. The amount of tension in the rope will also affect the total length of rope that can be spooled, since all ropes are reduced slightly in diameter and interstitial spaces are reduced by the crushing and flattening action of multilayer spooling. Hence, such formulas serve as useful guides to estimating the capacity

Table 5—Drum Capacities

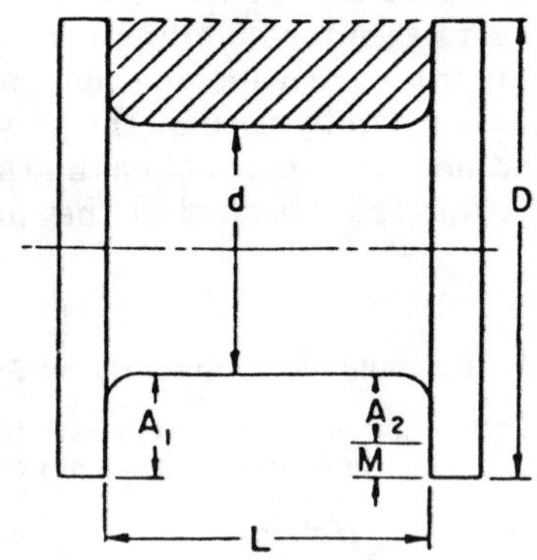

Wire Rope Drum

CASE 1—Drum filled flush with flanges

length of rope = $(A_1 + d)\,KA_1L$, m (ft)

where

- d = diameter of drum, mm (in.)
- D = diameter of flanges, mm (in.)
- L = length of drum, mm (in.)
- A_1 = $\frac{1}{2}(D - d)$, mm (in.)
- K = factor from table below for size rope desired

CASE 2—Drum filled below flanges

length of rope = $(A_2 + d)\,KA_2L$, m (ft)

where

$A_2 = \frac{1}{2}(D - d - 2M)$, mm (in.)

M = clearance, as shown, in millimeters (inches) with symbols having same meaning as above

K Factors

Rope Diameter mm	(in.)	*K* metric $\times 10^{-6}$	(English)
6.35	(1/4)	61.19	(3.29)
12.70	(1/2)	17.20	(0.925)
19.05	(3/4)	7.96	(0.428)
25.40	(1)	4.44	(0.239)
31.75	(1 1/4)	2.82	(0.152)
38.10	(1 1/2)	1.99	(0.107)
44.45	(1 3/4)	1.43	(0.0770)
50.80	(2)	1.11	(0.0597)
57.15	(2 1/4)	0.88	(0.0476)
63.50	(2 1/2)	0.71	(0.0380)

Table 6—Cargo Handling—Nominal Winch Ratings

STEAM WINCHES WITH NET STEAM PRESSURE OF 100 PSI AT THE THROTTLE

	Engine Size Bore & Stroke (cm)	(in.)	Drum Line Pull (Ton)	Drum Line Speed (mpm)	(fpm)
Single gear reduction	12.7 × 17.8	5 × 7	1.45	54.9	180
	15.2 × 20.3	6 × 8	1.81	67.1	220
	20.4 × 20.3	8 × 8	2.54	73.2	240
	21.0 × 25.4	8¼ × 10	3.06	64.0	210
	22.9 × 30.5	9 × 12	4.08	64.0	210
Component geared with mechanical speed change	17.8 × 30.5	7 × 12	2.26	76.2	250 High gear
			4.54	38.1	125 Low gear
	21.0 × 25.4	8¼ × 10	2.72	60.9	200 High gear
			5.44	30.5	100 Low gear
	22.9 × 30.5	9 × 12	3.38	76.2	250 High gear
			9.07	33.5	110 Low gear

ELECTRIC MOTOR DRIVEN WINCHES

Type	Motor	Drum Line Pull (Tons)	Drum Line Speed (mpm)	(fpm)
Cargo	37 kW (50 HP) DC	1.68	88.4	290
		3.38	67.1	220
		6.55	32.0	105 Low gear
		8.84	21.9	72 Separately geared auxiliary drum
Topping	2.2 kW (3 HP) AC/DC	0.445	20.4	67 } For topping boom alone
	3.7 kW (5 HP) AC/DC	0.748	20.4	67 } For topping boom alone
	5 kW (7 HP) AC/DC	1.12	20.4	67 } For topping boom alone
		4.54	*Static* holding load on above three	
	11.1 kW (15 HP) AC	4.54	13.7	45 For topping 10-ton boom under load

of a drum. The formulas in Table 5 are based on an equal number of wraps or turns in each layer, and the error is negligible unless the drum length is small compared to the flange diameter.

f. Topping Winches. In recent years, there has been considerable interest in boom topping winches as an aid to setting booms for burtoning. These are arranged for mounting on mast or kingposts and on the vertical sides of houses, as well as on deck. Generally, these winches are worm geared, powered for topping the empty boom without load on the hoisting hook, and capable of sustaining statically the normal working loads due to operation of the hoist winch. These winches enable the topping line to be stowed on a drum, eliminating the need of handling the topping lifts on gypsy heads and stopping them off on a cleat. A spur-geared winch will enable topping of 10-ton booms under load at reduced speed adequate for topping duty, as an aid in handling the occasional heavier lifts for which the boom is swung. The slow speeds permissible for topping duty enable these winches to be operated without speed control of the driving motors, except as determined by their normal ratings.

g. Nominal Winch Ratings. A summary of nominal winch ratings in U.S. practice is given in Table 6.

h. Location of Winches. The heel block at the base of the cargo boom swings horizontally with the movement of the boom and may increase or decrease the fleet angle to the winch, which is defined as the angle at which the rope enters the drum. A rope entering perpendicular to the drum has a zero fleet angle. This fleet angle is of vital importance, since the wire rope can pile up, fatigue, scrub, and wear while running on and off the winch drum, causing short wire life, jerking of the load, etc. Winch drums must be aligned at right angles to the center of the total movement of the fleet line. Satisfactory performance on smooth drums with proper winch alignment has been obtained with a total fleet angle of 3 degrees; that is, 1.5 degrees each side of the centerline of the drum, or 4-degrees total fleet angle with a grooved drum. Winch operation stations should be located on the centerline of the ship and adjacent to the hatch end. The controls for winches usually are located on top of the resistor houses overlooking the hatch-end coaming. For efficient operation and safety, the operating station should be located high enough so that the winch operator will have the draft in full view at all times.

2.6 Hatch Size. As discussed in Chapter III, hatches should be as large as feasible to minimize the amount of horizontal movement required to stow cargo. The width of hatch opening is usually limited to about 40 percent of the breadth of the ship for general cargo vessels. Excessively wide hatches restrict the open deck for stowing general cargo on deck. For multiport discharging of cargo, deck area often becomes a necessity for the segregation of cargo.

There are many factors that affect the final determination of hatch size and location. The first consideration is the type of cargo to be carried. If any restrictive cargo is contemplated, such as heavy machinery, locomotives, or long structural steel, then the minimum size is dictated by the requirements for handling that particular commodity.

The hatch should be located so that the booms may service it adequately and still have a suitable outreach over the ship's side. Hatches should be located also so that cargo will require a minimum amount of horizontal movement from the square of the hatch to extreme stowage position. The

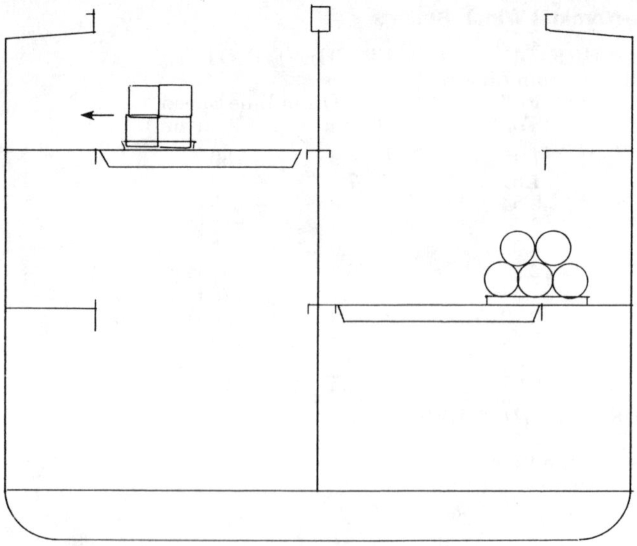

Fig. 12 Multiple hatch ship

distance between the hatch ends depends on the deck space required for winch accommodations and for the fore-and-aft stowage of hatch covers, if they are of the mechanical type.

2.7 Hatch Covers. Cargo hatches traditionally were closed by hatch boards, strongbacks, or steel pontoon covers; also tarpaulins were required. These types of covers created many problems. They were difficult to keep watertight, which resulted in damage to the cargo; removal and replacing of covers was time consuming at the start and close of the day; and the storage of the covers on deck was an obstruction to loading and discharging cargo, as well as a source of injury to crew and longshoremen. Cargo liners are presently equipped with a variety of steel mechanical hatch covers. These covers are composed of one, two, or more sections having multiple hinged, gasketed (if watertight) leaves in each section, the end leaf being hinged to the hatch-end coaming. A number of types of these mechanical hatch covers are described in Chapter IX.

2.8 Improving the Productivity of the General Cargo Ship. Dillon et al (1961) propose some improvements in the traditional general cargo ship with the addition of one or more of the following:

- twin or multiple hatches,
- the sideport and platform elevator,
- revolving cranes in lieu of masts/booms, and
- gantry cranes.

a. Multiple-Hatch Openings. Fig. 12 shows schematically how multiple-hatch openings affect cargo productivity. The twin hatch (which has been extended to triple- and quadruple-hatch openings on some designs) allows cargo to be handled at many levels simultaneously. This provides more flexibility in stowage, particularly when a ship must serve many discharge and loading ports.

Multiple hatches can also be set up fore and aft with the separation strips running transversely rather than longitudinally as shown. This configuration should be considered if the vessel is to continue burtoning as the longitudinal girders will interfere with outboard stows.

The Mariner class (Russo and Sullivan, 1953) has become the basic yardstick for comparison in cargo handling and productivity. In a comparison of cargo handling of the single-hatch Mariner and the triple-hatch Challenger classes, in the same service, it was found that the Challenger class has increased the average productivity rate from the Mariner's 23.3 measurement tons per gang hour to 33.8 tons, an increase of 45 percent. This increase is not entirely due to the hatch configuration, but it also is due in part because of the cargo gear installed in the Challenger class.

b. Sideport—Platform Elevator. Referring to Fig. 13 the use of the sideport in conjunction with multiple internal-hatch openings builds on the idea of increased productivity from the general cargo ship. As outlined in the introduction to this chapter, each step taken in the evolution of cargo ship design has been toward reducing labor cost or increasing productivity. It is not hard to extend Fig. 13 toward the concept of the containership or the ro/ro ship.

Dillon et al (1961) state that while the sideport/elevator could increase potential hatch productivity to 100 long tons per hr, in actuality only 60 tons per hr was achieved due to the actual weight of the cargo involved and less than optimum distribution and scheduling. Thus, in any general cargo design study, the average conditions under which the cargo systems are to be utilized will certainly be less than the maximum capacity of that system. In the general cargo system, there are still simply too many variables.

c. The Revolving Cargo Crane. Fig. 14 shows the application of the revolving deck crane to a type of general cargo ship. Revolving deck cranes have a number of advantages over conventional burtoning gear and have become increasingly popular. The crane is a compact unit which eliminates the resistor houses, kingposts, vangs, and miscellaneous wire rope leads and blocks, providing more usable deck space and greater visibility from the bridge.

Cranes can be put in service or stowed in less time than conventional gear and thus reduce port time. The working load of a crane is limited only by the gears' capacity and the willingness of the longshoremen to handle the load. Greater safety is inherent in the crane over conventional cargo gear where there is a danger of the cargo whips being tightlined, overloading the whips and vangs, or overloading the vangs by improperly locating them at the deck level. The crane has the ability to spot or pick up a draft in the hatchway or pier within the operating radius of the boom head. However, spotting the boom head over the hold or pier is of little value unless the hatch is of sufficient size to land the draft near its point of stowage, or unless port facilities are capable of keeping pace with the cargo hook. Spotting is advantageous in double-ganged hatches and in holds having multiple hatches.

The crane's operating gear and controls can be located within a protecting enclosure. This protection reduces maintenance and provides protection for the operator. During inclement weather, it is common practice for the conventional gear winch operator to spend time in erecting

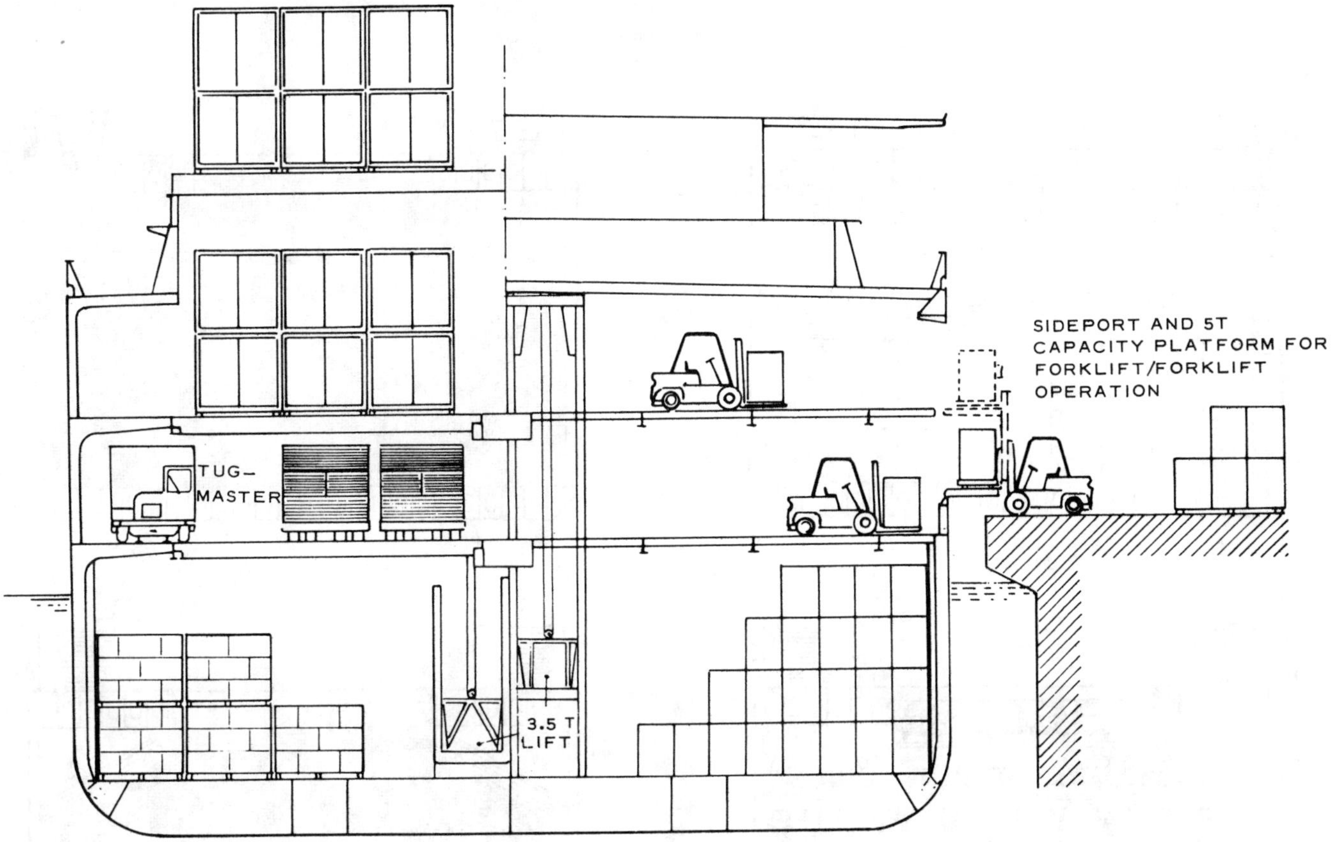

Fig. 13 Sideport and platform elevator

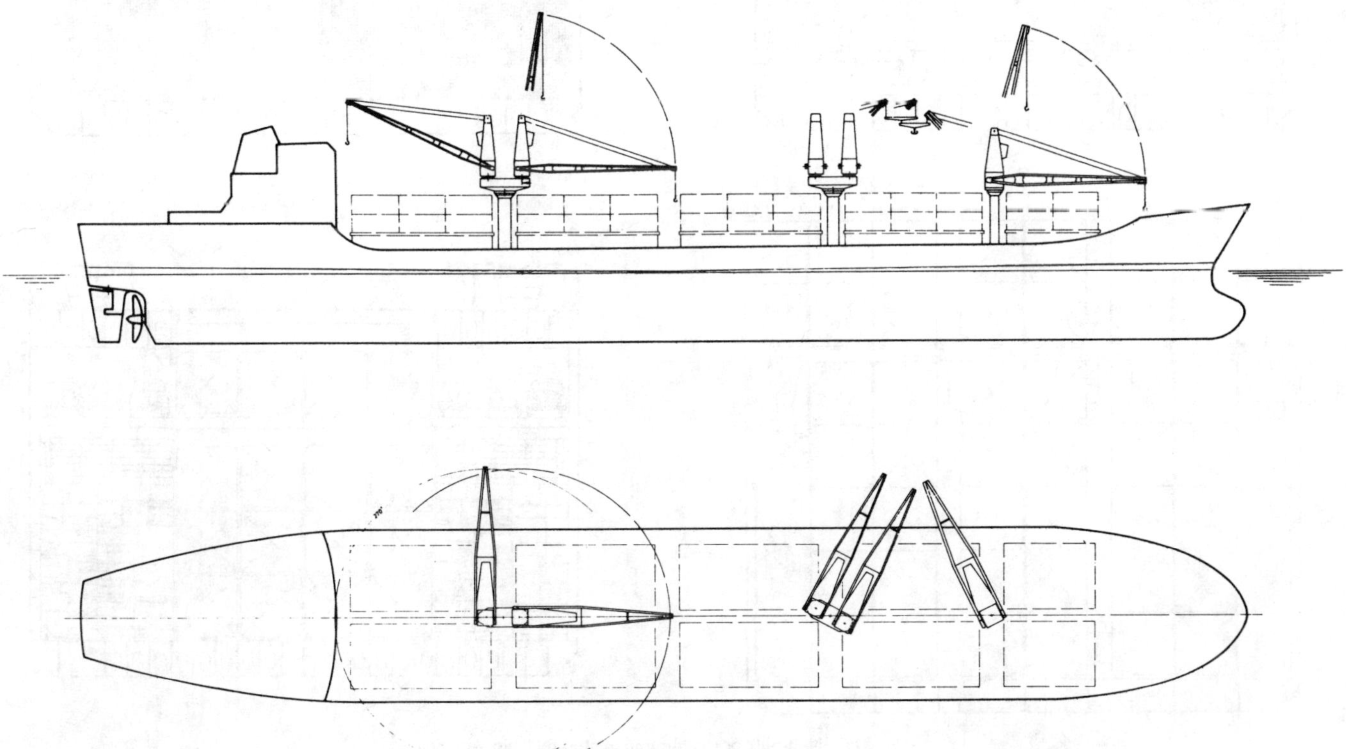

Fig. 14 Neo-bulk on multipurpose ship

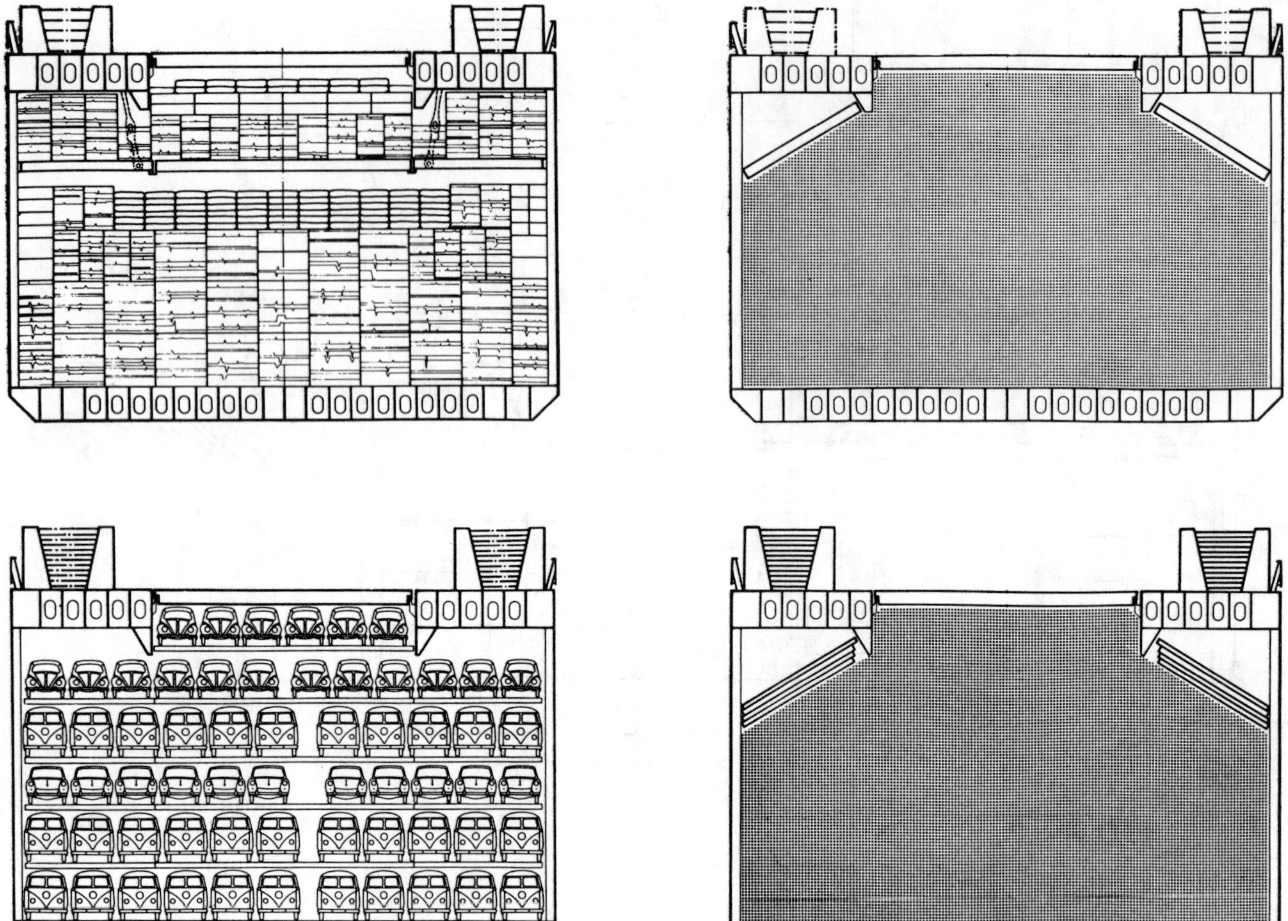

Fig. 15 Neo-bulk or multi-purpose ship-midship sections

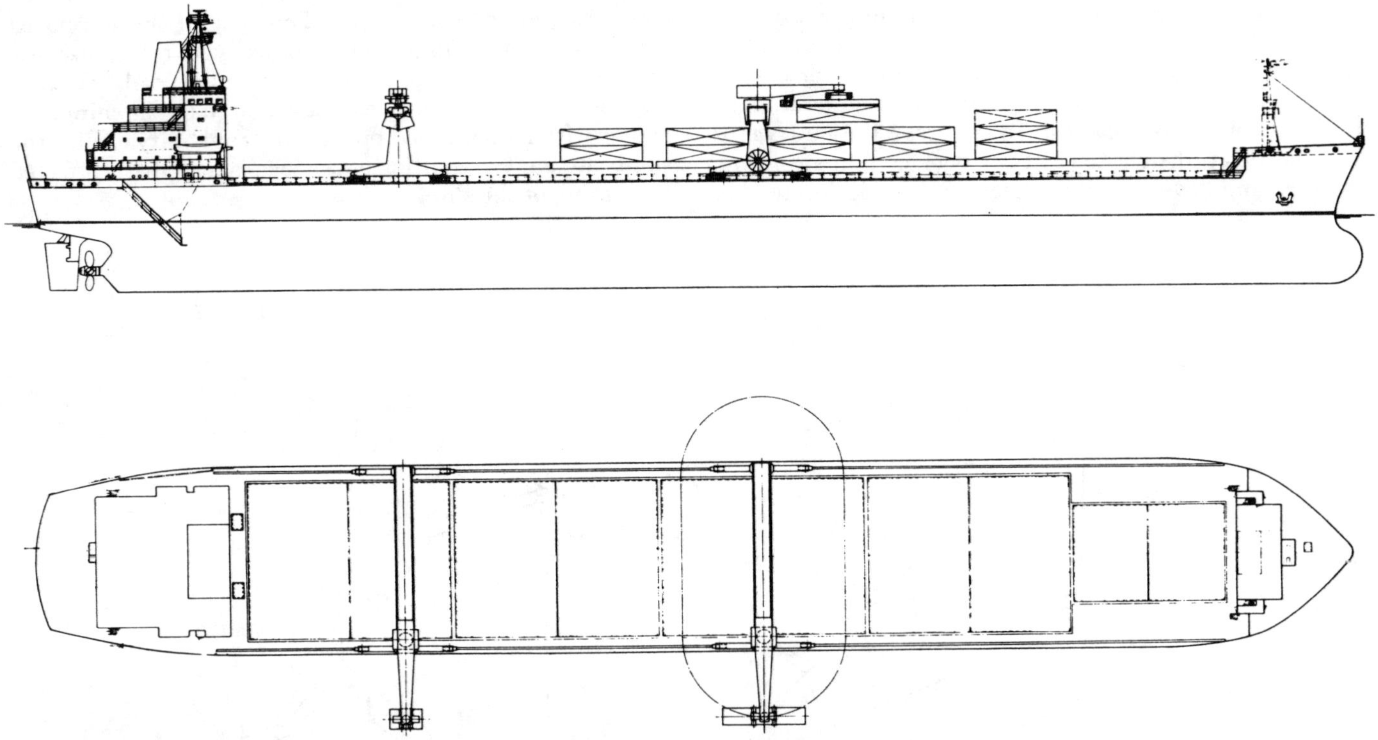

Fig. 16 Multipurpose ship fitted with rotating gantry crane

some sort of weather protection over this station, which reduces the operating time and the amount of cargo handled. The loading or unloading hook cycle for the conventional gear, for loads up to 3 tons, is faster than the crane cycle. However, the ability of the crane to spot and pick up a load from the pier results in approximately equal cargo handling rates. The location of cranes in relation to the hatches should be such that the luffing (topping) and slewing (swinging) of a crane is kept to a minimum for maximum performance.

Note in Fig. 14 that the aftmost cranes are located on common pedestals which can rotate 360 degrees. This allows each crane to serve two adjacent hatches or for the two cranes to be parallel for a heavier lift at one hatch. The midship hatches could be served by two sets of two cranes each to provide a heavy lift capability.

For heavy lifts on light duty cycles, where it is necessary to raise and lower the boom with the load attached, cranes and cargo gear can be reeved for *level luffing*. The level-luffing feature employed maintains the load at a nearly constant elevation during topping of the boom from maximum to minimum radius. The special hoist line reeving system provides the level-luffing feature (Ebel, 1958) and (NAS-NRS, 1958). Some European pierside level-luffing cranes have a pantograph boom which permits the level-luffing action.

Fig. 14 is actually more indicative of what has been called the *neo-bulk* or *multipurpose* ship. Unlike the general cargo ship whose cargos are now carried primarily by the containership, the multi-purpose ship is designed so that it is suitable for features such as:

- bulk (stowage factors from 46–58),
- containers,
- forklift operation or stowage of heavy-wheeled vehicles on tank top (steel coils, rolling stock, structural steel).

To handle this type of cargo, each of the rotating cranes shown in Fig. 14 would have a capacity of 15–20 tons, and the ship would be fitted with hatch lengths to accommodate at least a 40-ft container. In order to carry grain cargos without fitting shifting boards, the split hatches employ a centerline bulkhead.

Another adaptation of the *neo-bulk* or *multipurpose* ship is shown in Fig. 15. In this configuration, a portable deck is used as both an upper 'tween deck for light general cargo, and as an antigrain shifting device in its stowed position. With additional suspended decks, the multipurpose ship even becomes a car carrier. Multipurpose ships are definitely limited when they run in competition with ships designed especially for one commodity. These specialized carriers will be described in succeeding sections.

d. Gantry Cranes. As mentioned, the general cargo ship's cargo is most suitable for containerization, provided both the loading and discharge port have pier and road facilities to accommodate containerized cargo. Dillon et al (1961) proposed the use of *C*-type gantry cranes, described by Kimball and Wudtke (1975), so that a single *C* crane could accommodate 20-ft containers or, since alternate cranes could be arranged with facing openings, two cranes could be married to handle a 40-ft container.

Another gantry-type crane used on multipurpose-type ships is shown in Fig. 16. All gantry cranes move up and down the deck on rails, but in this case, the transverse bridge is fitted with a rotating crane, making the crane suitable for a number of different cargos. In each case, the gantry must have sufficient height to clear the deck container or other stowage and have a positive means of locomotion to overcome list or trim of the ship. The hoist of the shipboard gantry is shorter than that of the other types of loading gear and thus allows smaller pendulation of the load. Greater accuracy in spotting cargo can be achieved by the shipboard gantry with its improved positioning capability. With the shipboard gantry, spotting of the load is not greatly affected by surging of the ship.

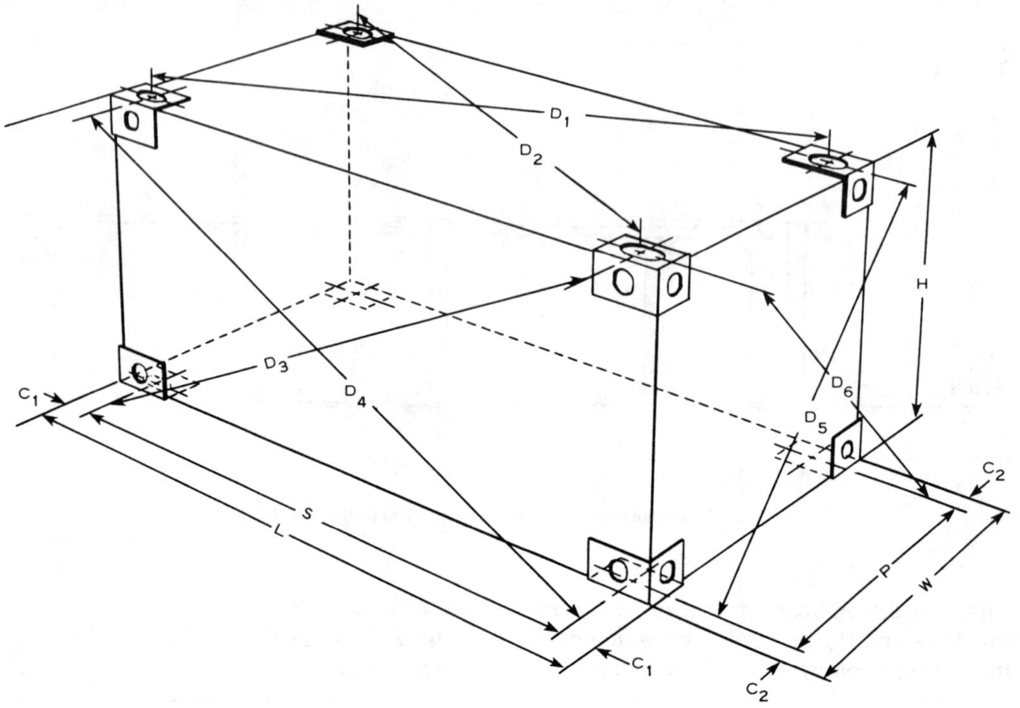

S — Length between centers of apertures in corner fittings
P — Width between centers of apertures in corner fittings
C_1 — Corner fitting measurement $4^{+0}_{-1/4}$ inches ($101.5^{+0}_{-1.5}$ mm)
C_2 — Corner fitting measurement $3\frac{1}{2}{}^{+0}_{-1/4}$ inches ($89^{+0}_{-1.5}$ mm)
L — External length of container
W — External width of container

D — Distance between centers of apertures of diagonally opposite corner fittings resulting in 6 measurements. D_1, D_2, D_3, D_4, D_5 and D_6
K_1 — Difference between D_1 and D_2 or between D_3 and D_4. i.e., $K_1 = D_1 - D_3$ or $K_1 = D_2 - D_1$ or $K_1 = D_3 - D_4$ or $K_1 = D_4 - D_3$
K_2 — Difference between D_1 and D_4. i.e., $K_2 = D_3 - D_6$ or $D_6 - D_3$
H — Overall height

Dimensions in Inches, Millimeters

Freight Container Designation	Length (External)					S			P			K_1 Max		K_2 Max	
	ft	in.	tol.	mm	tol.	ft	in.	mm	ft	in.	mm	in.	mm	in.	mm
1A	40	0	+0 −3/8	12190	+2 −8	39	4	11984	7	5	2258	3/4	19	3/8	10
1B	29	11 1/4	+0 −3/8	9125	+0 −10	29	3 1/4	8918	7	5	2258	5/8	16	3/8	10
1C	19	10 1/2	+0 −1/4	6055	+3 −2	19	2 1/2	5852	7	5	2258	1/2	13	3/8	10
1D	9	9 3/4	+0 −3/16	2990	+1 −4	9	1 3/4	2786	7	5	2258	3/8	10	3/8	10

Width (external) Containers 1A, 1B, 1C, 1D—8 ft$^{+0}_{-3/16}$ in, 2435^{+3}_{-2} mm
Height (external) Containers 1A, 1B, 1C, 1D—8 ft$^{+0}_{-3/16}$ in, 2435^{+3}_{-2} mm

Fig. 17 Dimensions of the standard I.S.O. container

Section 3
Containerships

3.1 General. The containership carries the improvement of the general cargo ship (Section 2.8) one step further by unitizing all of its cargo within containers. The containership system is most suited for finished goods shipment as:

- the individual containers are suited in size to relatively small shipments which are to be expected in the finished goods trade;
- door to door (shipper to consigner) shipping is possible without the integrity of the container being broached. This protects valuable finished goods from the elements, handling damage, and pilferage without expensive crating;
- the time and cost of shipment door to door is reduced which is a requirement of some finished (perishable) goods.

The containership represents one of the types of ships in which the total shipping system must be carefully engineered before operations can start. Thus, this section will attempt to explain the total operation as it applies to the cargo itself.

3.2 Container. The dimensions of the standard ISO (International Organization for Standardization) 10, 20, 30 and 40-ft containers are shown in Fig. 17. Note the rather precise tolerances to which each is built. Unlike the breakbulk carrier, the containership is fitted with cell guides and fittings with similar tolerances in order to ensure that any *standard* container will fit these stowage fittings. Some of the various types of containers used are described briefly as follows:

- *Standard dry container*—comprises the majority of today's containers and simply provides a container with a solid, watertight roof and sides. It is suitable for the carriage of any cargo not requiring temperature control. Like the general cargo ship hold, the dry container is not immune to sweating or cargo generated heat buildup.
- *Ventilated dry container*—dry, insulated container has a small heating and air conditioning unit built in which can maintain a desired temperature (above freezing) and in some cases a desired humidity.
- *Refrigerated container*—dry, insulated container contains a refrigeration unit capable of taking the interior stowed cargo below freezing and holding it there at a desired temperature.
- *Open top*—a dry container with a tarpaulin type roof. Large bulky goods can be loaded through the top by a crane and then the item, even if it extends beyond the top of the container, can be covered by the tarpaulin.
- *Open flat*—structure is more closely related to a large pallet with corner posts so that it can be picked up with container handling gear and so that it can be stacked with other containers. The open flat usually has a much more

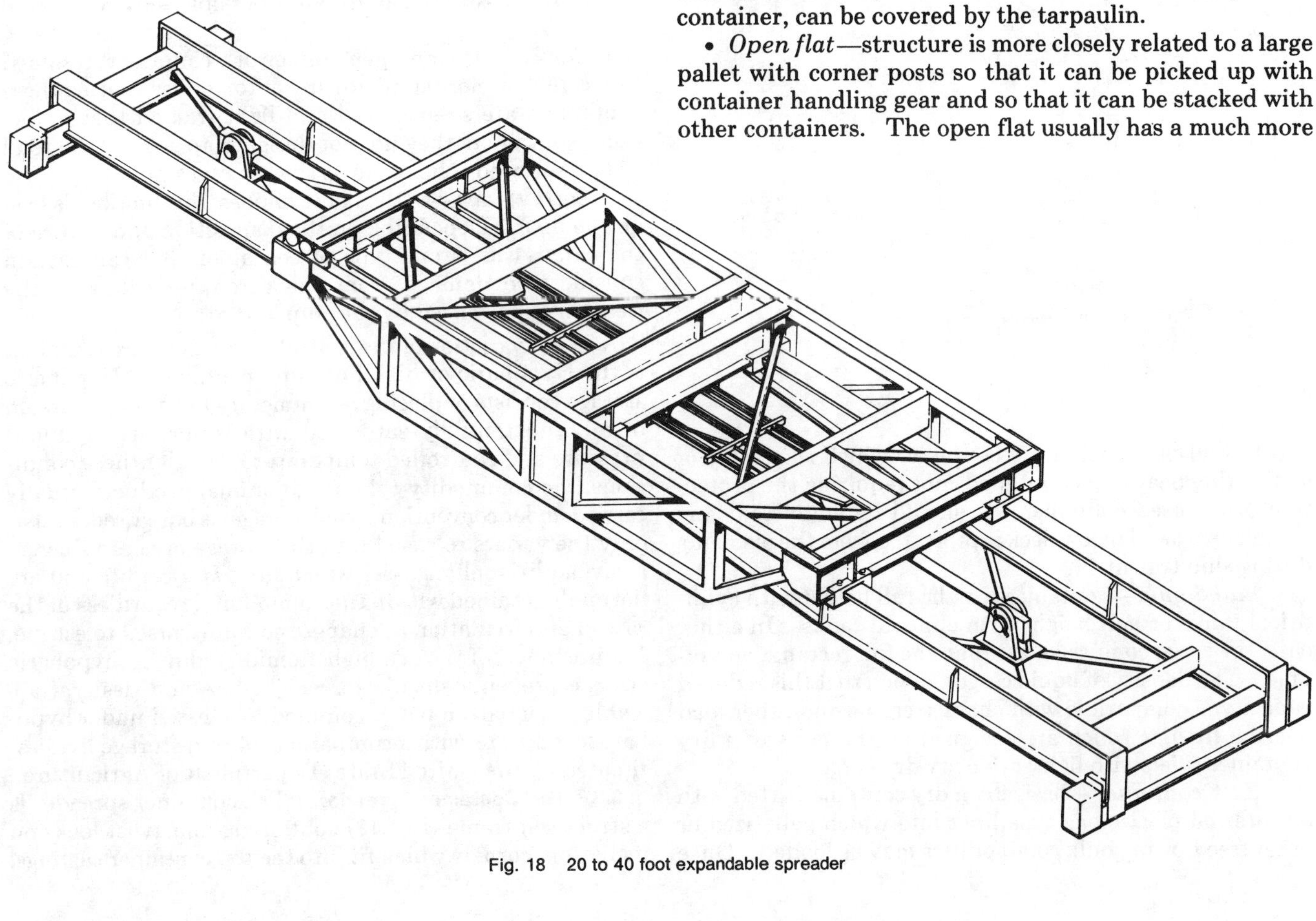

Fig. 18 20 to 40 foot expandable spreader

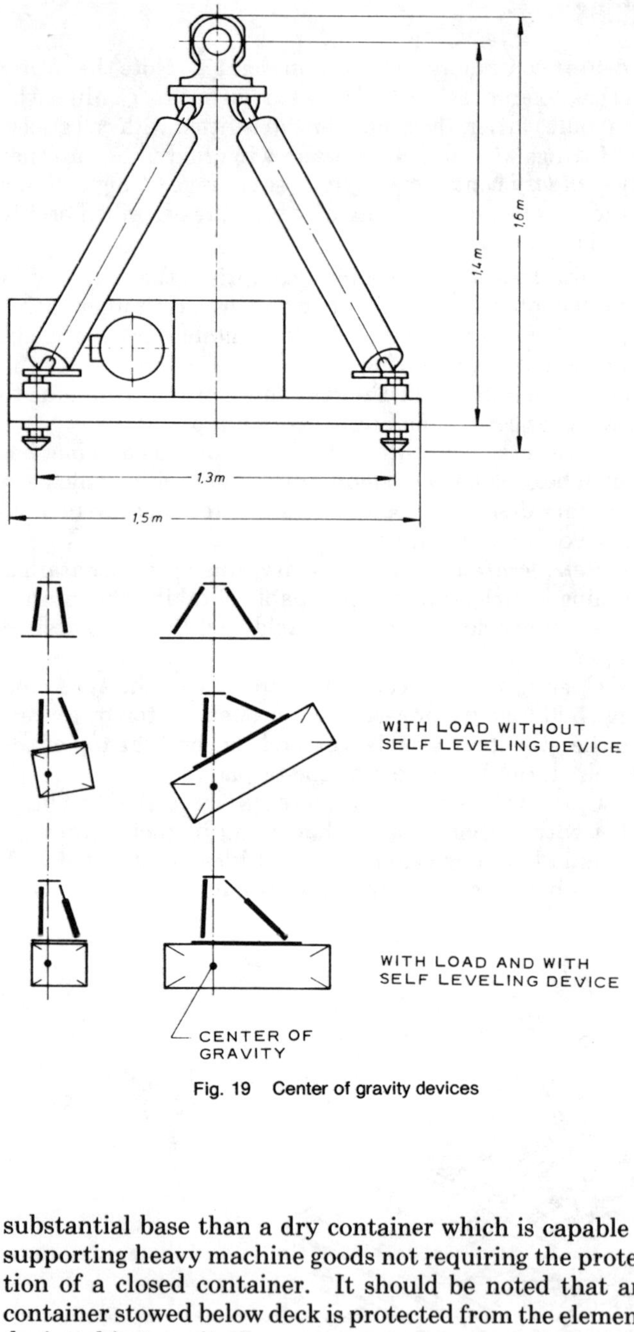

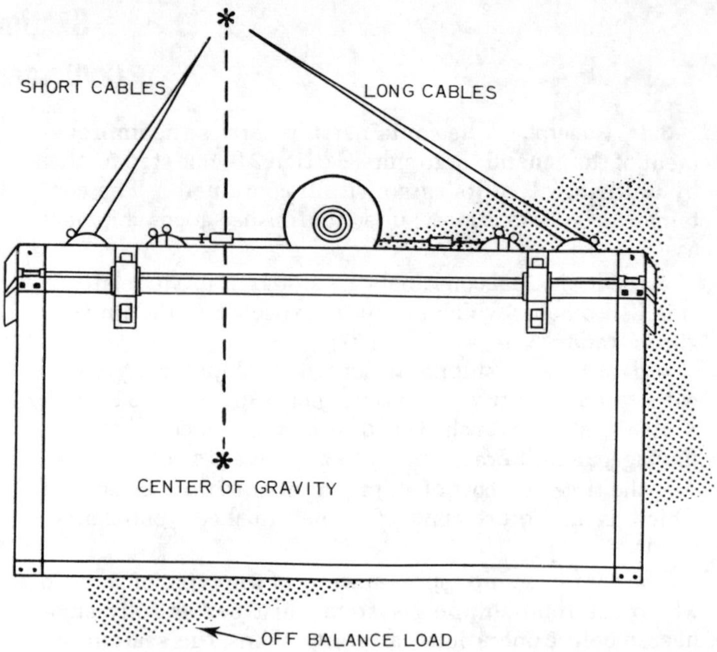

Fig. 19 Center of gravity devices

substantial base than a dry container which is capable of supporting heavy machine goods not requiring the protection of a closed container. It should be noted that any container stowed below deck is protected from the elements during ship transit.

- *Bulk liquid*—containers can be full height with cylindrical tanks or half height with eliptical tanks. In either case, the cargo comprises less than the full rectangular volume. The density of liquid cargos is such that this reduced capacity is compatible with chassis, crane, and other load bearing fittings which are designed on the basis of a dry container filled with lighter density dry cargo.
- *Bulk container*—basically a dry container fitted with an internal plastic bag type liner into which pelletized or other free flowing bulk commodities may be loaded. Once the rear doors are open, discharge can be facilited by tipping the container, dump truck style, up at one end. The liner minimizes costly cleanup and possible environmental problems.
- *Car carrier*—an open framework container designed to accept 4 or more standard size automobiles. Often these containers are so oversized and light framed that unlike other containers they are not designed to go over the road or to be stowed other than at the top of a stack.
- *Cattletainer*—as the name implies, this unit hauls beef on the hoof. Depending on the disposition and nature of the cargo, this is one type of container that can sustain considerable damage from the cargo regardless of the weather experienced by the ship.

A recent variation of the controlled temperature container is the Hypobaric or high humidity container. Hypobaric storage consists of placing a commodity in a flowing stream of air, substantially saturated with water, at a reduced pressure and controlled temperature. Under these conditions, the commodity (plant and animal products usually carried under conventional cold storage) is outgassed, so that only the vapors released into the storage area are flushed away; additionally, gasses which limit storage life and are normally retained within the commodity, regardless of the rate of air circulation or change, now are caused to escape. Maintenance of a very high humidity during hypobaric storage prevents shrinkage, weight loss and desiccation. Table 7 is a partial list of commodities tested under hypobaric conditions with a comparison of cold storage lives estimated by the United States Department of Agriculture.

3.3 The Container Spreader. The container spreader is a structural frame (Fig. 18) containing four twist locks on its bottom corners which fit into the top container castings,

Fig. 17. These twist locks are rotated 90 deg by hydraulic/ electrical or manually activated mechanisms and lock the spreader to the container.

Container spreaders are best lifted directly by individual purchases at all four corners (Fig. 23) to prevent the container from tipping transversely or longitudinally. However, when the spreader must be suspended from a single point, as would be necessary with a Whirley crane, ships' conventional or heavy-lift cargo gear, a leveling mechanism must be placed between the spreader and the hook. Fig. 19 shows how a hydraulic or cable device would be rigged. It can be seen from this figure that for level hoisting it is necessary to have the container center of gravity directly below the hoist point in both planes.

The container spreader shown in Fig. 19 is adjustable longitudinally (expandable) so that it may, in this case, accomodate 20 ft, 30 ft and 40-ft containers.

3.4 The Containership. The concept and preliminary design processes as applied to the containership were developed in Chapter I. Additional developments in the design process are given in those chapters devoted to arrangements, structural configuration, and outfitting. Here, we are concerned with another aspect of the containership design, that is how the ship must be adapted to load and unload its intended cargo and thus perform its mission.

Fig. 20 shows the midship section of a containership with its cell-guide structure, entry guides and deck stowage locations. Fig. 21 shows the configuration of container cell guide fittings, hatch-cover guides and stacking chocks. Pontoon hatch covers contain lifting sockets at the same dimension as the container lifting sockets so that the container spreader can be used to lift both container and hatch cover.

Figs. 20 and 21 also show the hatch cover configuration so that one can be placed on the other transversly and *leap frogged.* Thus, the whole hatch can be made accessible without resetting the shipboard or shore crane longitudinally.

3.5 The Container Terminal. A composite container terminal is shown in Fig. 22. The containership is being loaded and discharged by one shore container crane *A* with containers transported to the storage area by wheeled straddle carriers *B* resembling lumber carriers. The straddle carriers move the containers into open aisles where rail mounted shore gantry cranes *C* stack the containers for storage and/or reshipment.

There are many other methods of loading/unloading, transporting and storing containers as follows.

a. Loading/Unloading Gear

- Shipboard gantry cranes (Ebel, 1958), (SW&S, 1966), (Cushing, 1963), (Fig. 23)
- shore gantry cranes (Fig. 22)
- shore or shipboard Whirley cranes (Fig. 14)
- conventional break-bulk ship cargo gear.

b. Storage Yard Transport

- Straddle carrier (Fig. 22),
- over-the-road container chassis,
- forklift equipped with spreader (certain types of containers have forklift pockets in the lower frame doing away with the spreader equipment)
- wheeled dollies (one to six containers are carried)
- air casters.

c. Storage Yard Stacking and Sorting

- Gantry crane (Fig. 22),
- straddle carrier—referring to Fig. 22 note in this case that an aisle would be required between each individual stack of containers requiring more stowage area per container stack,
- forklift (stack height is limited)
- large parking garage type structures where containers are stacked and accessed through a system of elevators and trolley lifts in a totally mechanized operation. To date, none of these have been built.

Table 7—Hypobaric Storage Life of Commodities

Commodity	*Cold Storage (days)*	*Hypobaric Storage (days)*
Meats		
Beef	12–14	45
Chicken	7–14	21
Pork	5–7	21
Shrimp, fish	3–5	14–21
Ripe Fruit		
Pineapple	9–7	40
Grapefruit	30–40	90–120
Strawberry	5–7	21–28
Cherry, sweet	14	60–90
Vegetables		
Green Pepper	16–18	50
Bean, pole	10–13	30
Onion, green	2–3	15
Cucumber	10–14	41
Corn	4–8	21
Lettuce	14	40–50
Nonripe Fruit		
Tomato (green)	14–21	60–100
Tomato (breaker)	10–12	28–42
Banana	10–14	90–150
Avocado	23–30	90–100
Lime	14–35	60–90
Apples	60–90	300
Pear, Bartlett	45–60	300
Cut Flowers		
Red Ginger	5–7	28–35
Vanda Joaquim	16	41
Carnation	10	91
Chrysanthemum	6–8	21–28
Snapdragon	14	42–56
Gladiolus	5–7	30
Rose	7–14	56
Plants and Cuttings		
Potted Plants:		
Chrysanthemum	7	28–35
Easter Lily	14	42
Nonrooted Cuttings:		
Chrysanthemum	10–28	42–94
Geranium	1–5	21–28
Carnation	90	300
Rooted Cuttings:		
Chrysanthemum	7–14	90

In many cases, the hypobaric storage life given is not the maximum attainable, but rather the storage time at which the test was terminated because of laboratory requirements, or because there is no economic need for longer storage times.

3.6 Computerized Container Operations. The organization and efficient execution of the loading and unloading operations of a containership is a complex task. Careful coordination and planning of both manpower and equipment is required to obtain the most economical results. Because of its complexity, computer-based data processing and handling technology is of great assistance to the human operator in the efficient execution of this task. Recently, under sponsorship of the Maritime Administration (Hydronautics, 1975), certain programs were created and installed on the General Electric Mark III Network for use by U.S. Flag operators in the United States. In addition, communication satellites provide U.S. operators access to the same programs and data banks from many maritime jurisdictions including Puerto Rico, Australia, Hong Kong, Japan, and Western Europe. The only user equipment required is an interactive computer terminal, a telephone, and a GE user's number.

a. Ship Inbound. As the ship approaches the port, the data files listing the containers on board, the ballast, and the fuel conditions upon arrival are accessed and modified, if necessary, by the Cargo Supervisor. Additional files are created which contain sorted data to allow generation of summaries, maps, and lists of containers for unloading, reloading, and those remaining on board. The Cargo Loading Supervisor can:

- obtain the fuel burnoff, bunkering, and ballast conditions from the previous port,
- make changes to the liquid conditions based on projected arrival date,
- obtain container summaries, maps (bay plans), and listings from the previous port.

Based on the expected trade and the containers already on board, each hold is dedicated, according to a loading strategy, to a particular destination, mix, commodities or container size. Using a statistical distribution of container weights, a set of target loading weights for each available container position is generated based on ship stability, draft,

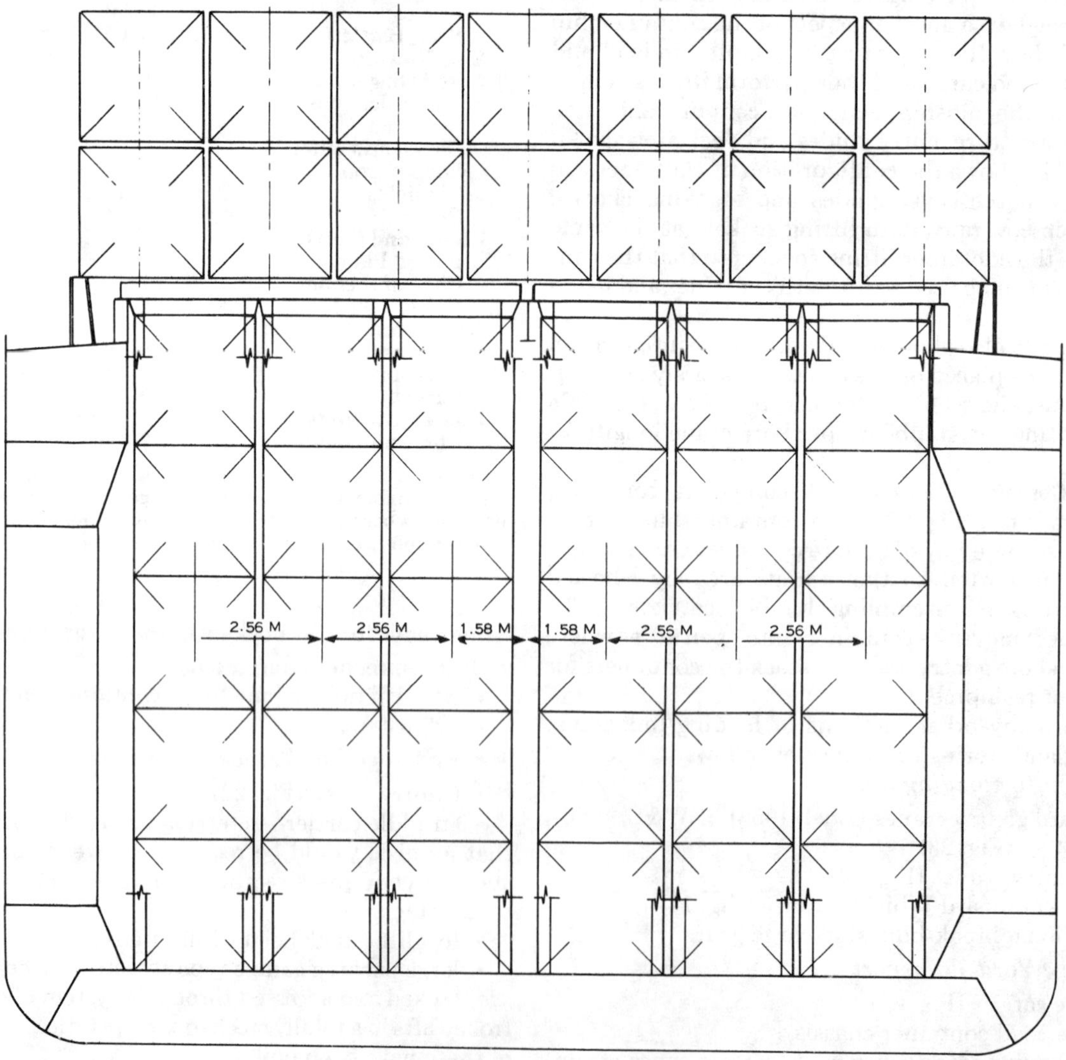

Fig. 20 Midship section—containership

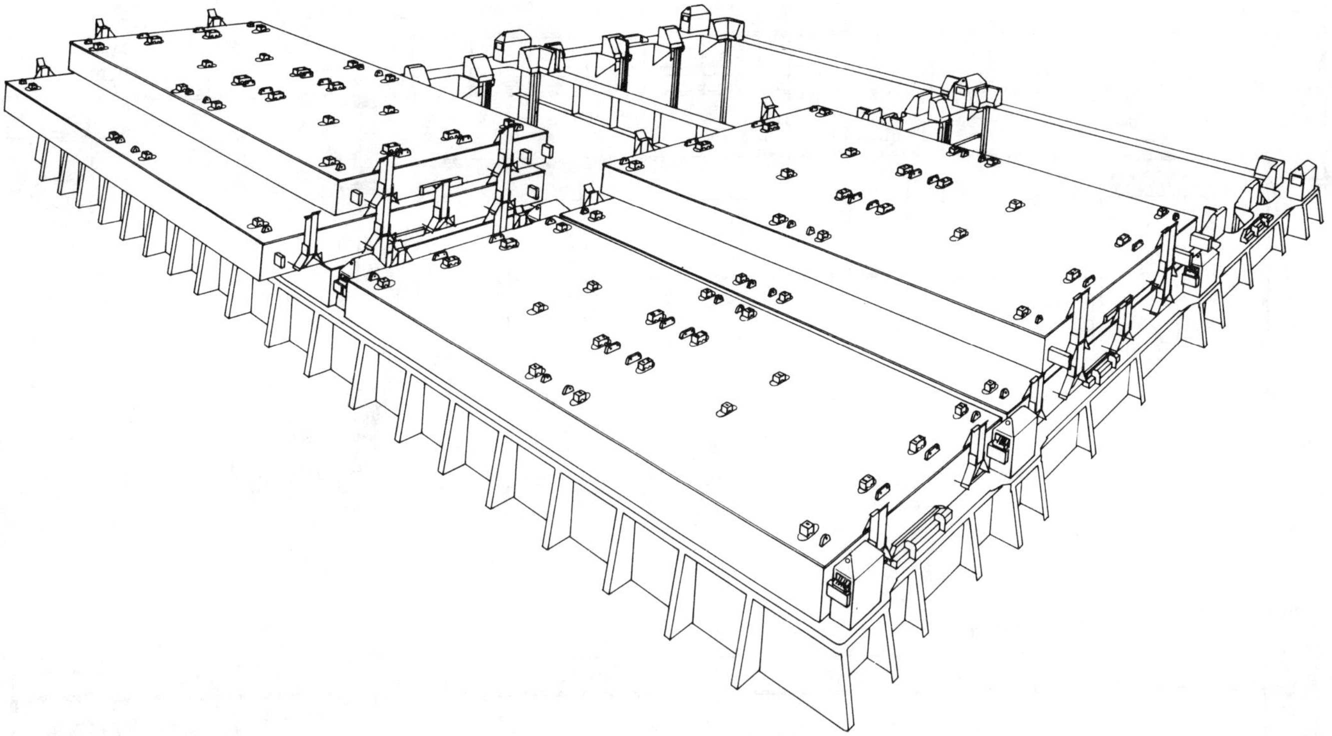

Fig. 21 Lift-away pontoon hatch cover

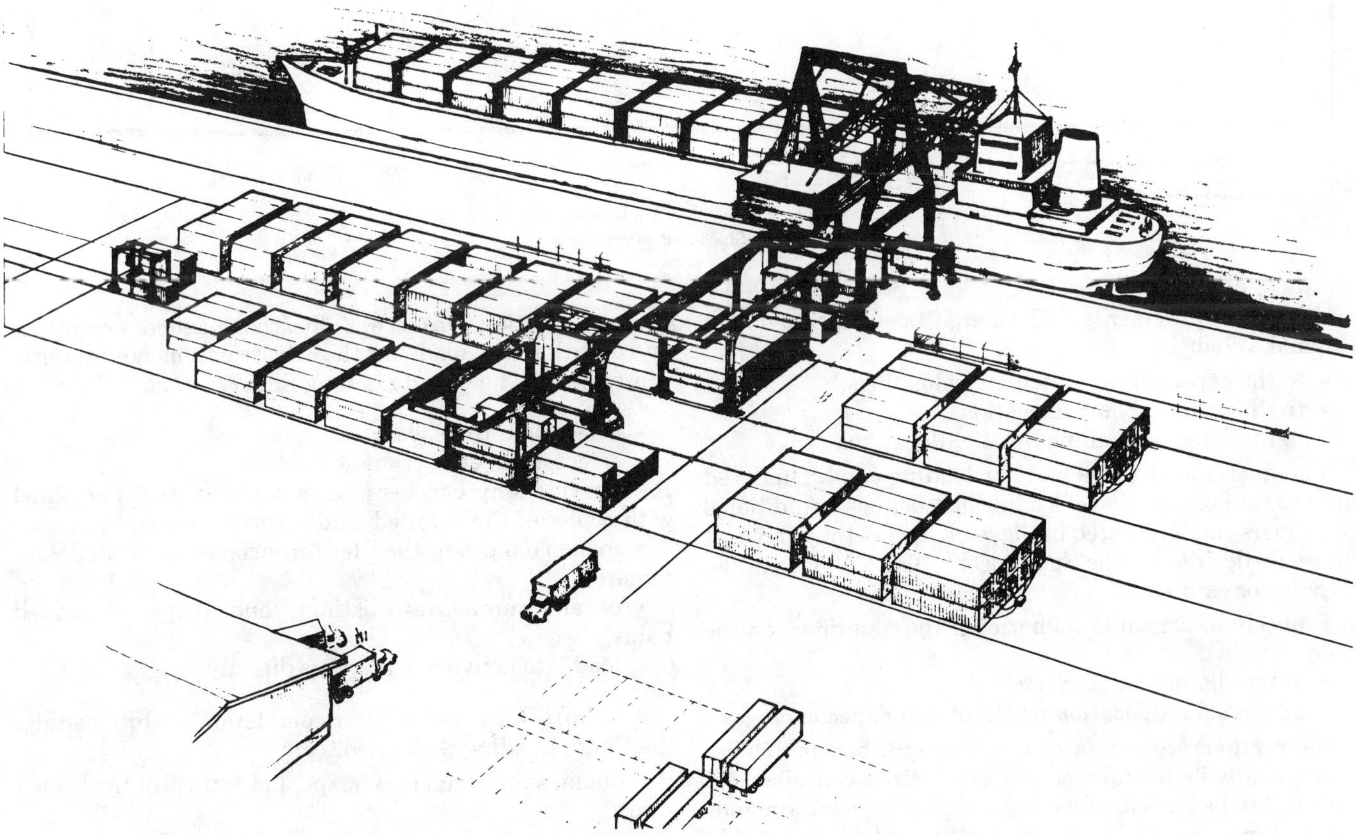

Fig. 22 Composite container terminal

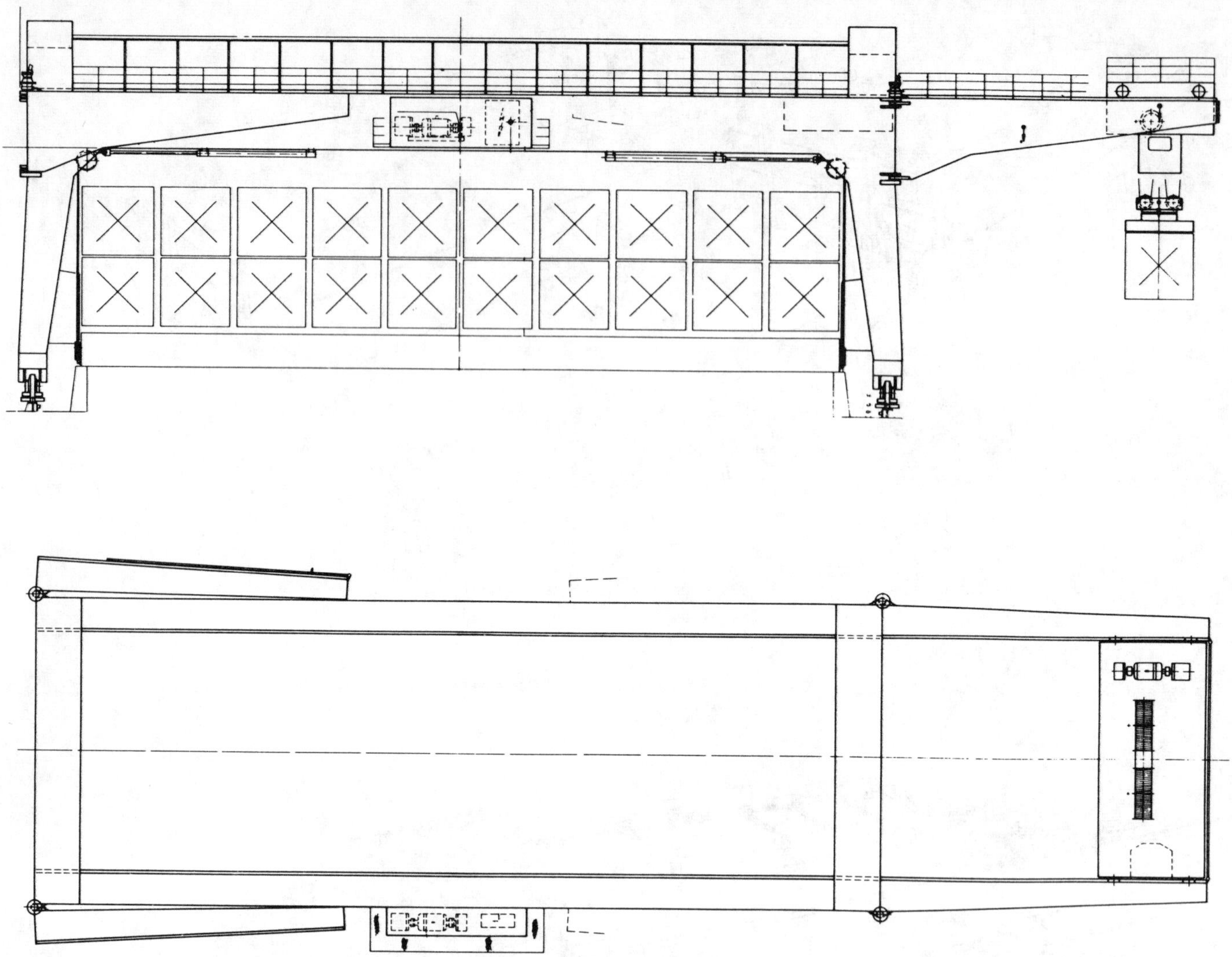

Fig. 23 Shipboard gantry crane

and stress requirements. The Cargo Loading Supervisor can make changes:

- to the expected (or anticipated) loadings,
- to the resulting hold allocations,
- to the selective loading weight allocations.

The dock file of containers for loading can be updated prior to the specification of loading instructions. Additional containers can be entered in the dock files as they arrive at intervals desired by the Supervisor. The Cargo Loading Supervisor can:

- obtain listing and summaries of the containers on the pier.
- obtain listing for each port.
- assign a *fixed location* on the ship for special cargos.

Instructions are generated to achieve the best match between available containers and the desired container for each cell to be loaded. Loading and discharge instructions are determined under control of specific availability and ship attitude constraints. An updating routine is used to verify those instructions unsuccessfully accomplished; exceptions and additions are made for those instructions not properly completed. The Cargo Loading Supervisor can:

- select hold to be worked,
- select number of cranes,
- provide ship checkers, terminal, and yard personnel with copies of the detailed loading instructions,
- make changes to the files for incorrectly located containers,
- obtain summaries, listings, and maps of loaded holds,
- load late arriving containers directly.

b. Ship Outbound. After completion of ship loading, the Cargo Loading Supervisor can:

- obtain ship summaries, maps, and listings of the loaded ship,
- provide the master and/or first officer of the vessel with these as part of the cargo manifest documentation,

• provide the first mate with fuel and ship conditions for the outgoing voyage,

• establish the ship and container files for the vessel's next port of call.

Computerized container operations provide the Cargo Loading Supervisor and terminal personnel with the required documentation and control to assist them in the efficient loading and unloading of containers. Many of the following computer-produced documents can be used directly by loading personnel:

• summary of ship weight, attitude (trim and heel), and stability characteristics,

• summary of containers (number and weight) by hold, length, and destination,

• map of containers in the hold,

• list of containers by hold, identification and special handling, destination, weight, and exact ship location,

• listing of the containers on the dock, location, identification and special handling, weight, and destination,

• hold allocation by container size, destination, and above or below deck location,

• selective loadings with target weights for each available position,

• sequential loading/unloading instructions with container identification, dock and ship location, and ship attitude.

3.7 Loading Rates and Other Study Related Data. For the purpose of designing systems and/or making comparative studies, it is useful for the naval architect to have the cycle time for different types of operations. A cycle is defined as the average time required for the transporting device to acquire a container, to pick the container from the hold, move it out of the hold on to a dock mounted vehicle, release the container, and return to the hold position ready to acquire another container. The designed time based on maximum equipment speeds is often not obtainable because of human interaction in the system and the following cycle times are based on actual experience with 40-ft containers:

• dockside gantry crane or ship mounted gantry crane of modern design—3.0 min,

• shipmounted gantry crane where rotation of load (90 deg) is incorporated in cycle—3.5 min,

• twin-boom rotating cranes, single-point lift with adjustment for off-center weight—10.0 min,

• ships heavy-lift gear, swinging boom with adjustment for off-center weight—15.0 min,

• general cargo ship gear doubled up utilizing one set of

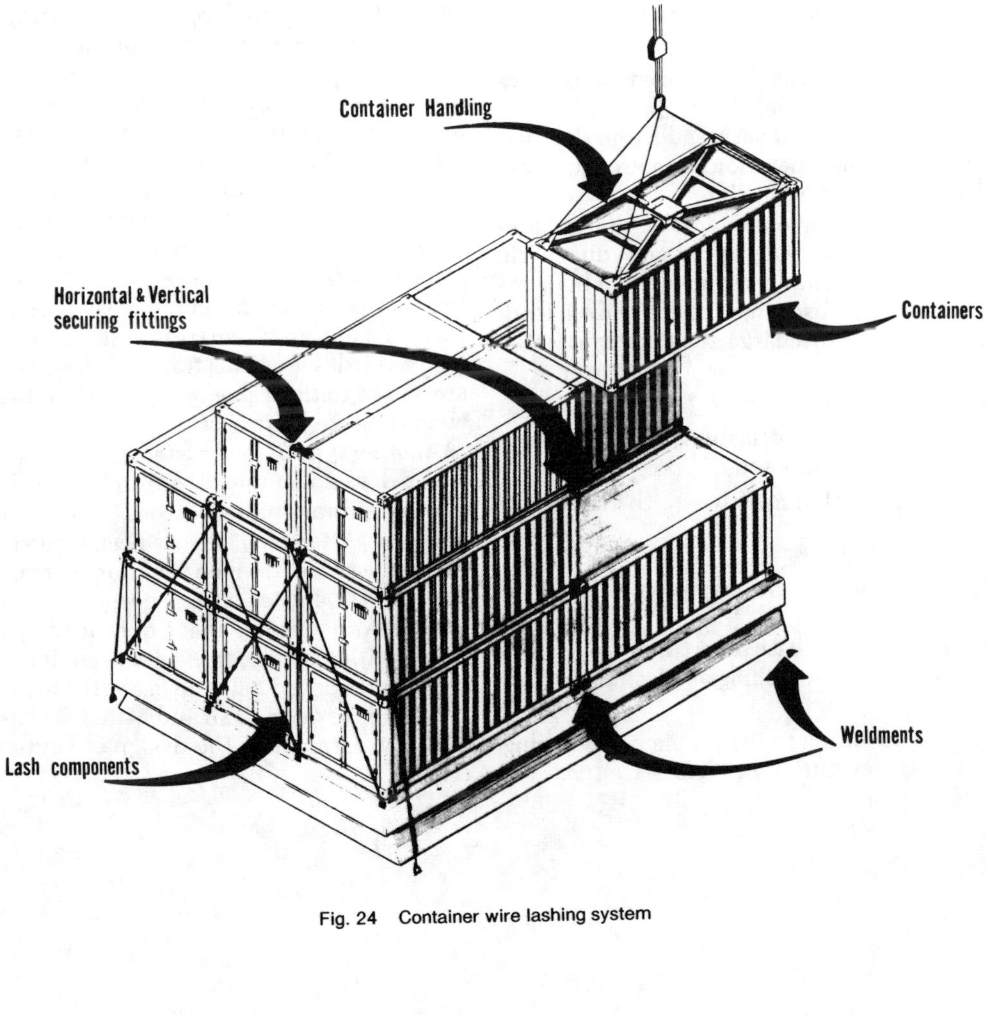

Fig. 24 Container wire lashing system

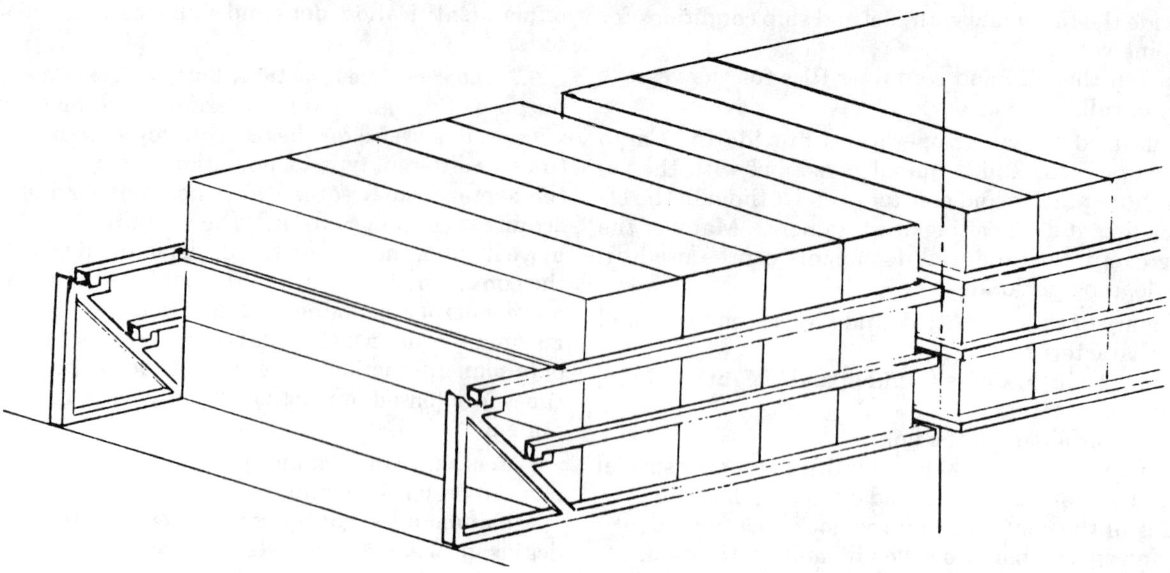

Fig. 25 Rigid structural restraint system

gear at each end of spreader—15.0 min.

For smaller containers (i.e. 20 ft and Conex boxes) better rates can be expected in the last three modes with singling up of gear because of the reduced load.

Referring to Fig. 23 it is also important to understand when making time studies that:

- hatch covers require about the time of one to two cycles depending on fittings and gear use,
- lashing and unlashing times must be added to the total container loading time as this operation cannot completely be carried on while loading or unloading,
- to obtain a cycle where one container is discharged and one full container reloaded in the ship in the same duty cycle (i.e. with a gantry crane 40 containers moved in a hour instead of 20) it is necessary to discharge the entire deck load, remove the hatch cover, and discharge all of the containers from one stack below deck.

3.8 Container Lashing Systems. Fig. 24 shows a typical pattern for an on-deck tensioned lashing system. Each lashing assembly is usually composed of:

- a hook or other fitting that attaches to the container casting,
- a length of wire or solid steel bar,
- a tensioning device (turnbuckle, clamp type tensioner, etc.),
- deck padeye, ring or other base structure,
- shackles to connect hooks to lashing and lashing to deck padeye.

An interesting recent solution to the problem of securing containers has been the evolution of *rigid structural restraint systems.* One such system was achieved by affixing a system of cell guides, similar to those located below decks. Another system avoided the need to travel to the top of each cell before loading or unloading, and also simplified accessibility to the hatch covers and holds, Fig. 25. This system requires that the first tier of on-deck containers be loaded and are restrained from sliding at their bases by the normal hatch cover chocks. After one layer of containers has been loaded, a large frame, covering the first tier is placed on them. The frame is handled with the container crane. This frame has down-facing chocks to assist in positioning itself on the containers. It also engages, at its outboard edges, a kingpost-like (buttress) structure which is welded into the ships structure. After the second frame is placed on that tier this second frame on top of the second tier also engages the same outboard buttress system. It is then possible to load a third tier on this frame. These third tier containers are locked to the frame by a gang-operated set of twist-locks which engage their bases.

Another type of rigid system which has been devised is one which does not employ the frames. This system has a permanent athwartships structure located in the space between hatches. After each tier has been loaded, hinged chocks on the frame swing down on the container corners, thus locking them to the buttress, and permitting the next tier to be loaded. And, finally, above deck loading of containers can be simplified by designing deeper ships wherein more containers are carried below decks with only two-high container stacks above deck. Liu and Mitchell (1977) provide an excellent treatment of the design of container securing systems.

Section 4
Barge Carrying Vessels

4.1 Types. There are two existing types of barge carriers utilized today: The LASH system whereby the barge (lighter) is hoisted on board the ship by a large gantry crane, and the SEABEE type where the barge is floated onto a synchrolift platform, elevated to the proper level, and then rolled along that deck into its stowed position. Fig. 2, at the beginning of this chapter, illustrates a total float-on concept where the ship itself sinks similar to a floating drydock. The barge carrier is best suited to trades where the barge can be utilized at both ends for distribution by river or other inland waterways. Although a point of debate, it would appear that the barge carrier is best suited to high grade bulk shipments in small quantities. Finished products would seem to be better shipped in containerships and large quantities of bulk would be most efficiently shipped in a bulk carrier.

a. LASH System. The general arrangement of the LASH ship is shown in Chapters II and III. Although there are many variations in LASH ships with regard to barge and/or container capacity, only the barge aspect will be discussed here. In referring to the LASH arrangement, it can be seen that the LASH ship is arranged along the lines of a bulk carrier with a single deck and all accomodations forward. Machinery is located just aft of midships with port and starboard stacks to allow the crane access all the way to the cantilevered crane supports on the stern (SW&S, 1969).

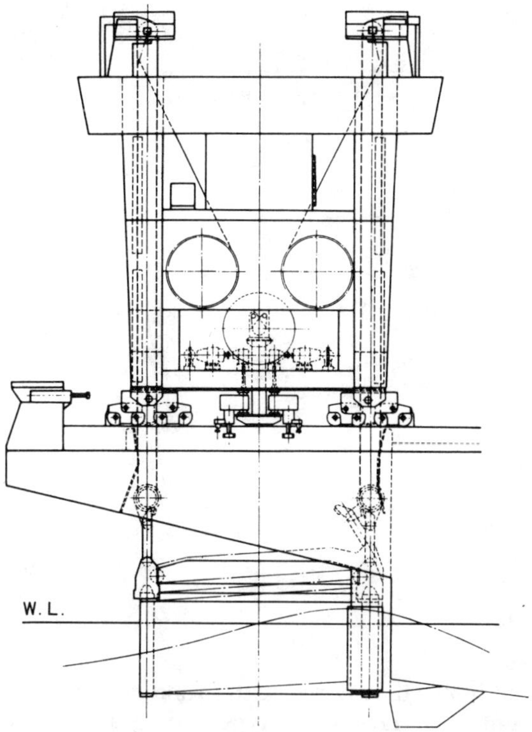

Fig. 26 Stern arrangement of LASH ship

The lighters are brought in between these cantilevered arms and hoisted vertically by a gantry crane of 500 tons capacity, Fig. 26. The crane legs are equipped with guides that line up with the stern and cargo hold guides. These guides ensure that the lighter does not sway while the crane is traveling along the deck. Traveling speed of the crane is 1.02 m/s (200 ft/min) and a typical duty cycle is shown in Fig. 27. With fully loaded barges of 453 tons each, the maximum cargo rate should be about 1500 tons per hour. The barge is then carried down the deck and deposited in a hold on special fittings on the tank top or on top of another barge already stowed. Pontoon hatch covers are raised by the spreader that hoists the barges; these are stowed on adjacent hatch covers during the loading/unloading cycle.

There is also an automatic clamping device that pulls the lighter constantly against the ship's stern even when the ship or lighter is swaying, as well as a swell-compensating device to aid lifting when the lighters are pitching in rough weather. The operation of the self-compensating mechanism is illustrated in Fig. 28. The compensator will respond to the sea movements in 5–7 seconds.

The barge (lighter) itself is shown in Fig. 29. Three small pontoon covers cover this type of barge and can be lifted by a small shoreside crane and stacked on one another until the barge is loaded or unloaded. The barge itself weighs about 80 tons giving each barge, in this case, a deadweight of 373 tons.

The LASH ship configuration carries 73 barges of which 49 are in the holds and 24 are on deck. The resulting cargo deadweight of about 27,000 tons is certainly less than a bulk carrier of similar dimensions and is nearly equal to the deadweight of a containership of similar dimensions.

With the large concentrated weights carried, stability, trim, longitudinal bending moment and vertical shear become major design problems. In these vessels, like the containership, the number of variables becomes difficult to handle with conventional manual loading sheets. Thus many owners have adopted computer controlled systems

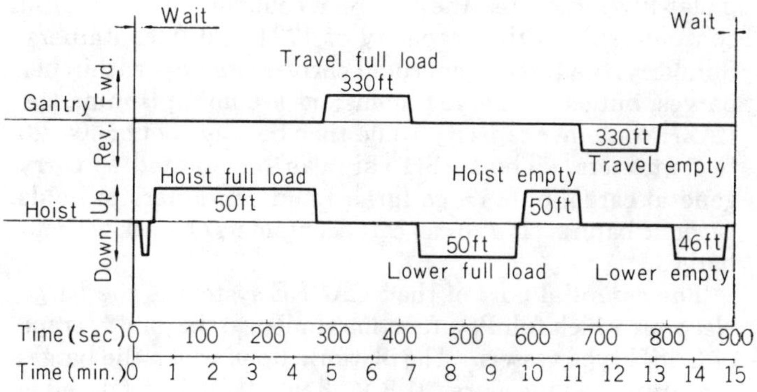

Fig. 27 Duty cycle of LASH lighter gantry crane

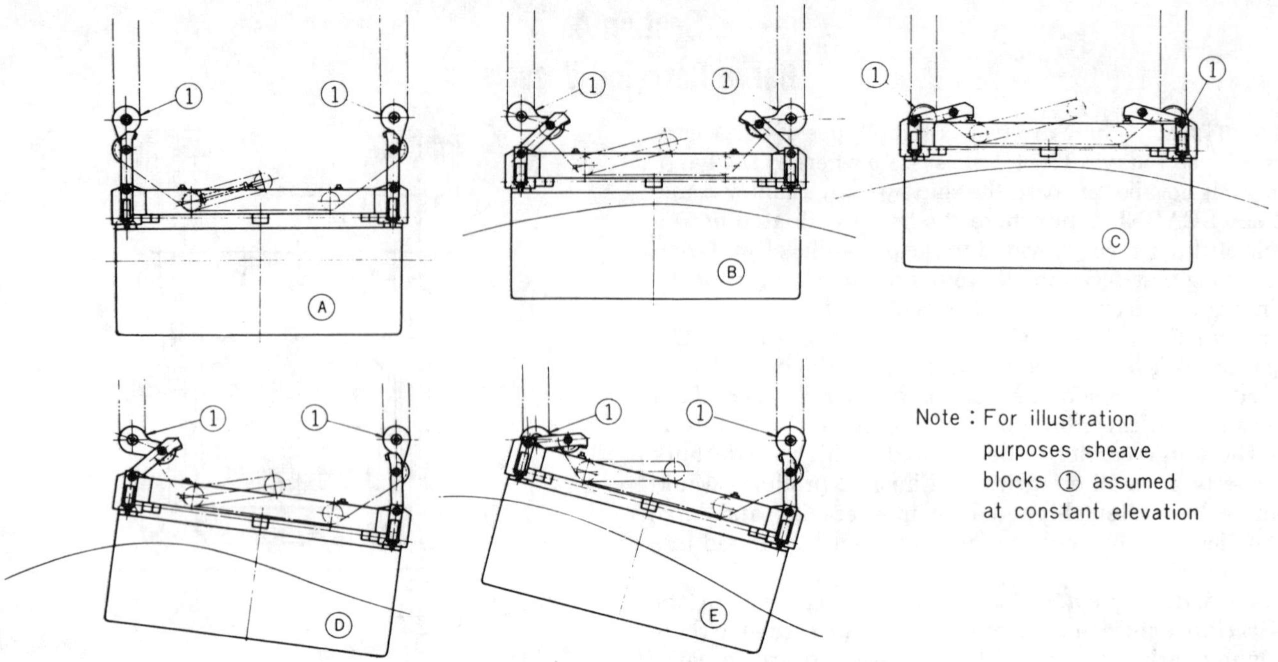

Fig. 28 Illustration of the swell compensating mechanism

which both plan the stowage pattern and maintain inventory control of the barges. This type of program is covered in Section 3.

b. SEABEE System. The SEABEE system is somewhat more flexible than the LASH system in that a larger barge is carried and the lifting and rolling mechanisms that can transport any flat item of appropriate dimensions and weight, Fig. 30. Each barge has a maximum deadweight of about 830 tons and therefore the ship cargo-handling rate approaches 3,000 tons/hr (SW&S, 1972).

The SEABEE could easily be converted to a medium heavy-lift ship transporting small craft and/or wheeled vehicles. It is also possible to convert this type of ship to a roll-on/roll-off operation by the addition of a hinged ramp to the after edge of the elevator. With the barge transporters stowed, a total of 13,569 m^2 (146,055 ft^2) of deck space is available and if fitted with portable "'tween decks, 21,375 m^2 (230,078 ft^2) would be available. Utilizing large pallet-like structures, the SEABEE could also become a full containership with a capacity of 1784—20 ft containers. Similarly, the LASH ship could carry containers within her barges, but as the barge dimensions are not optimum, the LASH container capacity would then be only about 500—20 ft containers. The LASH ship can be adapted to carry general cargo, bulk cargo (grain) and containers in holds *without* barges. It also can carry containers on deck without barges.

The essential part of the SEABEE system is the *barge elevator,* which fulfills a function similar to the gantry crane of LASH type vessels. The platform upon which the barges are supported measures 31.8 × 23 m (104 × 75.5 ft) and is of a box girder construction fitted with a double set of barge support blocks and transporter rails. The elevator can handle double depth barges in one lift, measuring 29.7 × 21 m (97.5 × 70 ft) with a total weight of 2000 tons. However, these larger barges can only be stowed on the upper deck which is free from obstructions. The platform and lifting cables weigh 540 tons.

The elevator is raised and lowered by means of three double-drum hydraulically operated winches located port and starboard on the poop deck, with the winches synchronized to keep the platform level. The elevator is raised and lowered by means of six, six-sheaved pulleys, port and starboard and in the stowed position, the elevator rests on 12 triangular *ears* which are folded out from the mooring area at main deck level. In order to position the barges over the support blocks ready for lifting, two electric winches are arranged in each sponson at upper deck level.

Once the barge has been positioned on the elevator and lifted to the appropriate deck level, it is moved to its intended stowage area by means of the *barge transporter.* This device comprises a steel frame, 29.5m (97 ft) long, which supports 12 sets of four-wheeled, self-propelled bogies driven by electric motors. The barge is raised and lowered by means of eight hydraulic power packs located within the chassis of the transporter. The transporters are controlled from consoles located aft on each deck level or from panels fitted at each barge stowage space, and are moved to the different deck levels by means of the elevator. Electric power is supplied to the drive motors and power pack motors by means of a special loop system which runs on the same rails as the transporter. The transporter is fitted with a limit switch which automatically stops the machine when it contacts a vertical surface. The transporters are stowed in a special garage located forward on the main deck, which is provided with maintenance equipment and spares. This is essential since, if the transporters were to become unserviceable, the ship would be unable to load or discharge barges.

The transporter is thus run under the barge, the hydraulic jacks activated, and the barge driven along a double set of rails to the intended stowage area, where the rotation of the hydraulic power pack motors is reversed and the barge rested on the support blocks in the hold. A typical cargo handling cycle would be:

Operation	Time Required, min
1. lower empty, average distance 13m (42 ft) at 1.2m/min (4 ft/min)	10.5
2. push pair of barges on elevator using tugs. Secure barges in position	8.0
3. Raise elevator	10.5
4. lower make-up rails, move transporter on to elevator and raise hydraulic jacks	3.0
5. remove transporter with barge from elevator at 9.2m/min (30 ft/min), raise make-up rail	6.0
	38.0

During the first three parts of the cycle (totaling 20 min for the main deck), the transporter moves the barge to its stowage position and returns. For a barge located at the forward end on either the main or lower decks, the cycle would be:

Operation	Time required, min
1. travel with loaded barge, 153m (500 ft) at 9.2m/min	17.0
2. lower barge in place	1.0
3. return empty 153m (500 ft.) at 18.3m/min (60 ft/min)	8.5
	26.5

The transporter will theoretically have a 2.5 min wait while the elevator is raised with the next barge, with the exception of barges loaded on the upper deck, where the period is longer.

Thus for the SEABEE it would require 19 cycles to empty a full vessel or load a full cargo of barges, giving a theoretical loading or discharge time of 13 hr. This gives a theoretical cargo handling total of 64 hr for each round trip, although allowing a 50 percent margin for unforseen delays, the total cargo handling time is expected to be about four days.

Each barge deck of the SEABEE can accommodate two barge widths, with the upper deck having a capacity for 14 barges while the main and lower decks are each suitable for 12 units, with each hold stowage area measuring 11.1m (36.54 ft) wide by 5.9m (19.30 ft) high. As the upper deck is free of obstructions, except in the forward area, it is pos-

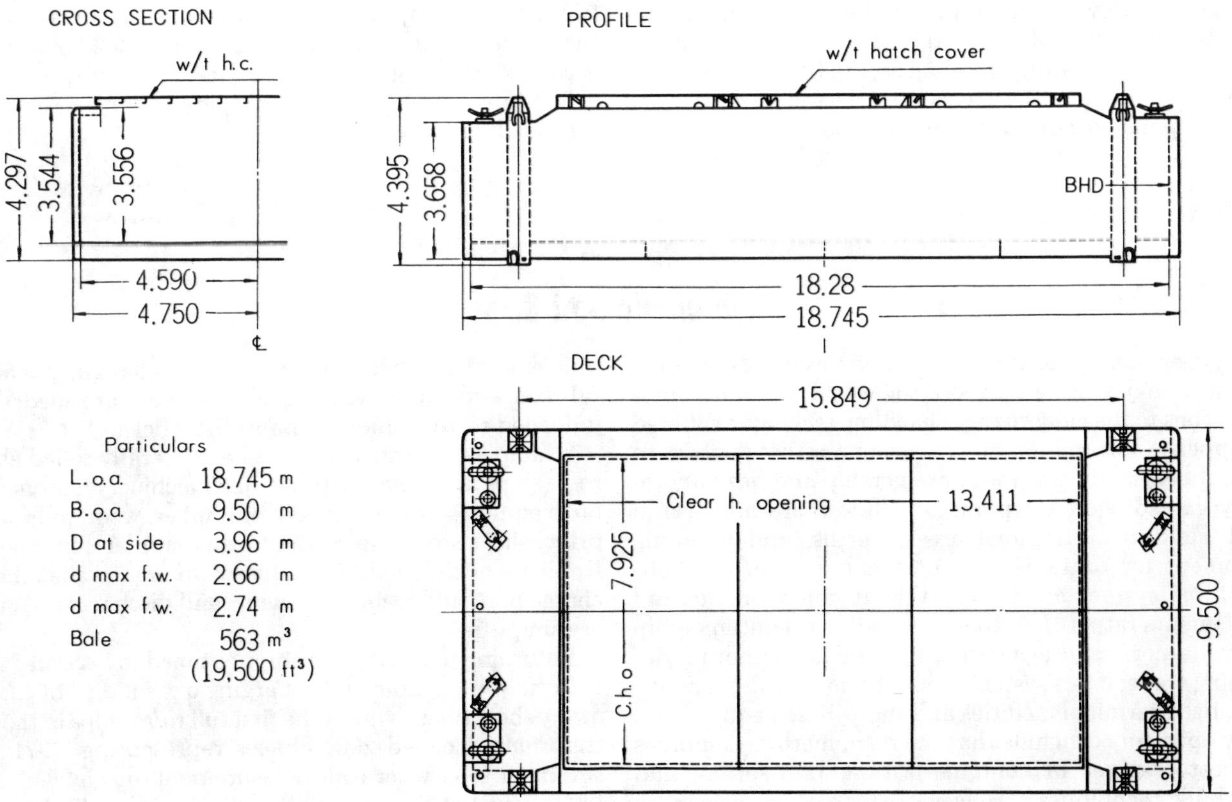

Fig. 29 General arrangement of LASH lighter (pontoon type hatch cover)

Fig. 30 Barges being moved to SEABEE stern elevator

sible to carry double-width barges or up to 10 open barges with cargos of unlimited height (up to the barge stability limit). It is also possible to carry containers on the top of the upper deck standard type barges.

The barges are secured by means of screw-lashing assemblies on the upper deck, while pneumatically-operated screw jacks are lowered from the deck head on the main and lower deck levels to hold the barges securely during transit. Two rows of lashing rings, totaling 68, are fitted around the barge deck itself for securing cargo. Between these rows, a row of combined cargo lashing cleats and pedestal sockets is fitted. Portable pedestals are fitted in these sockets to support up to ten 30-ft containers.

In order to bridge the gap between the forward end of the elevator and the vessel, a hydraulically-operated, hinged ramp is fitted along the after end of the three deck levels. This small ramp, some 1.5m (5 ft) wide, is arranged in four sections on each level. At the service draft of 9.7m (31.83 ft) the lower deck is just above the waterline, while at the scantling draft 11.9m (39 ft), this deck would be below water level. Therefore, the after end is sealed by means of two watertight, vertically sliding covers, which are moved up to the main deck level when working cargo on the lower deck.

The SEABEE barges are considerably larger than the standard LASH lighter, measuring 29.7 × 10.7 m (97.5 × 35 ft) compared with the dimension in Fig. 29 for the LASH lighter. The barges are of double-skin construction with a forward and after collision bulkhead, giving a 27.4 × 10 m (90 × 32 ft) hold space having a depth of 4.4m (14.6 ft) from the tank top to the hatch coaming. The 24.4 × 10 m (80 × 32 ft) hatchway is sealed by means of seven pontoon covers. The barges may be fitted with 'tween decks to give greater cargo capacity and portable container support beams can be provided at either end and at amidships for the stowage of up to sixteen 40-ft containers. It is also possible to carry containers inside the barge. This may not be justified if the containers are stuffed at the port of loading, as unit loads are then merely being carried in other unit loads.

Each barge has a bale capacity of approximately 1,132m^3 (40,000 ft^3) with a LASH barge capacity of 564m^3 (19,600 ft^3) with the majority of the hold space easily accessible through the large hatch opening. If a full load of 38 barges is carried, the cargo deadweight based on a stowage rate of 2.4m^3/ton, would be 17,650 tons, giving an average of 465 tons loading for each barge.

Section 5
Roll-on/Roll-off Ships

5.1 General. The introduction of the roll-on/roll-off system to major trade routes of the world has added new dimensions to the modern cargo handling techniques offered to shippers. The roll-on/roll-off system was first introduced about the same time as the containership and has only recently gained wide acceptance. While there has been a rapid buildup of terminal areas, berths, and handling equipment for containerization at many ports, the roll-on/roll-off system has proved that it complements and supplements rather than competes with the containership and container handling methods. More important for developing trade, it is a system that does not require massive specialized terminal facilities and shore-based equipment. Many operators conclude that the ro/ro method combines the best features of containerization, unitization, and breakbulk techniques. However, these ships also have unsatisfactory features such as wasted space and lashing problems. Recent ro/ro ships have been designed as almost full containerships where the containers are loaded and unloaded ro/ro fashion with forklift trucks.

Examples of cargo that have been literally rolled aboard ro/ro ships are heavy earth-moving machinery, automobiles farm equipment, large pieces of lumber, wood pulp, newsprint, sheet steel, piping, and other similar commodities. Rolling stock is ready for delivery upon arrival at the discharge port, and loading, stowing, and discharge operations are simplified.

Returning to the relationships outlined in Section 1.2, the ro/ro ship is capable of discharging a great deal of cargo in a very short time. One of the first full ro/ro ships in the U.S., the *Comet,* loaded 298 vehicles representing 7,971 measurement tons, where one measurement ton equals 1.13 m^3 (40 ft^3), in 4 hr 55 min. This same cargo was discharged in 2 hr 23 min. The ro/ro ship is volume-limited, rather than

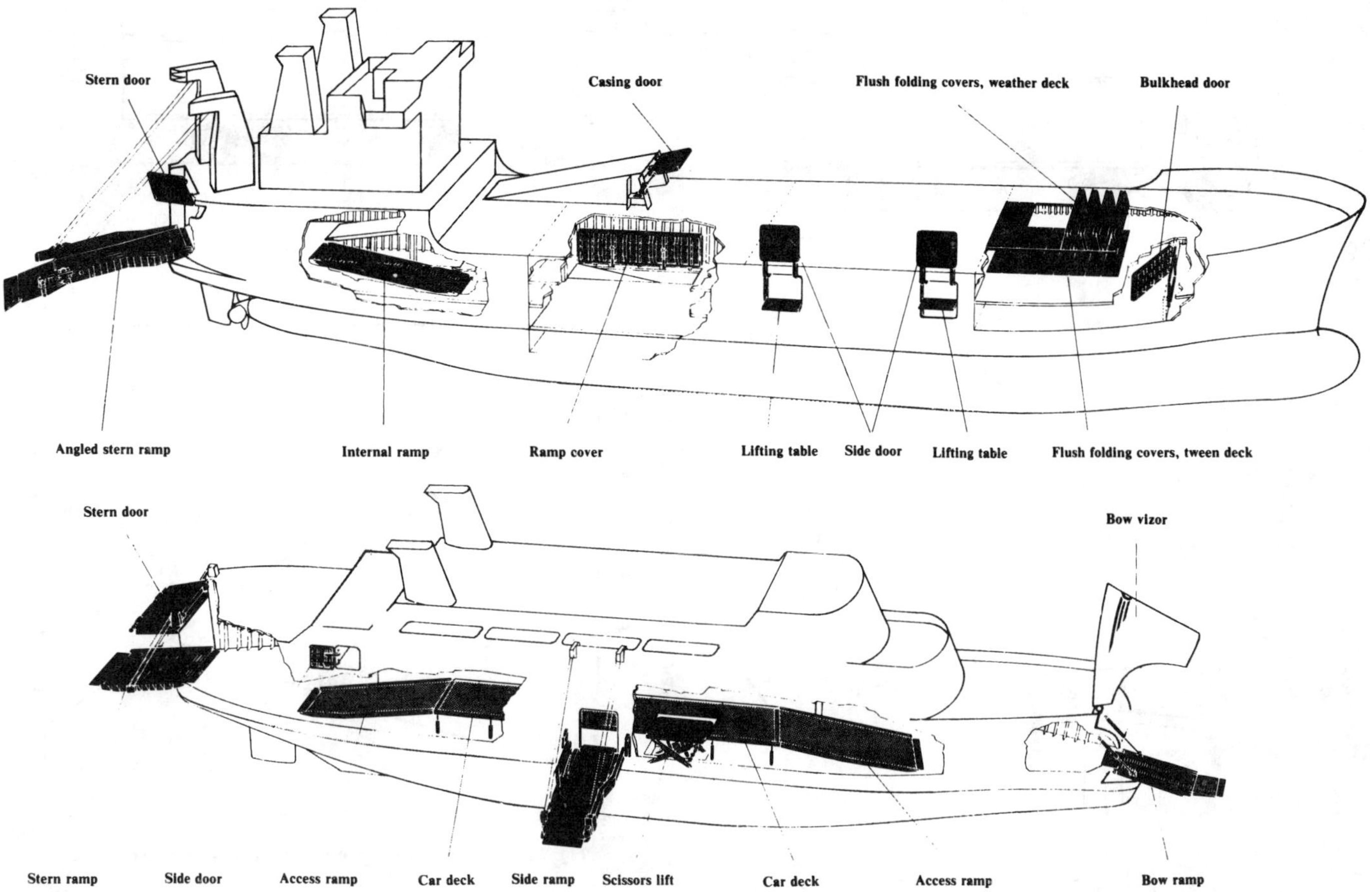

Fig. 31 Typical Navire cargo gear equipment for roll-on/roll-off handling

weight-limited, due to the space required for ramps, access, and underneath the wheeled cargo itself. It is, therefore, usual to discuss cargo stowage in terms of cubic capacity instead of tons.

5.2 Ro/ro Ship Cargo Handling Gear. The ro/ro ship contains some of the most inventive equipment to be seen on any ship. Motive power is largely by hydraulics and electric-powered purchase systems. Fig. 31 shows some of this cargo equipment which is described in the following text.

a. Stern Doors. On some ships with short external ramps, the ramp itself has been used as a watertight door in the closed position. As ramps have become larger, similar to the quarter stern ramp in the upper diagram of Fig. 31, it has been found difficult to limit distortion and always provide a watertight closure. Thus, most modern ro/ro ships use auxiliary watertight doors inside the ramp. This not only does away with the sealing problem, but makes the operating and stowing of the door much more flexible.

On many modern vehicle ferries, the stern ramps are designed of sufficient width to allow three simultaneous lanes of traffic. This could mean two lanes discharging and one lane loading simultaneously, which could reduce the variation of height of ship relative to pier and achieve a very quick turnaround of the vessel.

In some cases where unrestricted wide access is required, multiple doors are fitted, and the dividing pillar of the ship structure is removed or retracted when the ramps are open.

b. Bow Doors. The introduction of bow ramps on ferries created the *drive-through* ships which greatly decreased turnaround time for smaller vessels by dispensing with maneuvering of vehicles and allowing loading and discharging to be carried out simultaneously.

The development of the familiar *bow visor* was necessary at this stage to allow the bow door/ramp principle to be applied to sea-going ships. The original operating principle of raising the visor by hydraulic cylinders is still regularly used today with minor variations in pivoting and linkage arrangements to provide the necessary working clearance for specific shore installations. To prevent damage, the bow door scantlings, supporting structure, and locking devices must be properly developed. Depending upon the relative location of the bow door and the freeboard deck, an additional door in continuation of the collision bulkhead may be required.

c. Side Doors. A typical side door with ramp is shown in Fig. 31. For non- self-sustaining ships, shore ramps may be used in conjunction with watertight side doors. If the trade route of the ro/ro ship is fixed, this type of fixed installation increases ship deadweight, decreases ramp maintenance and usually reduces port time.

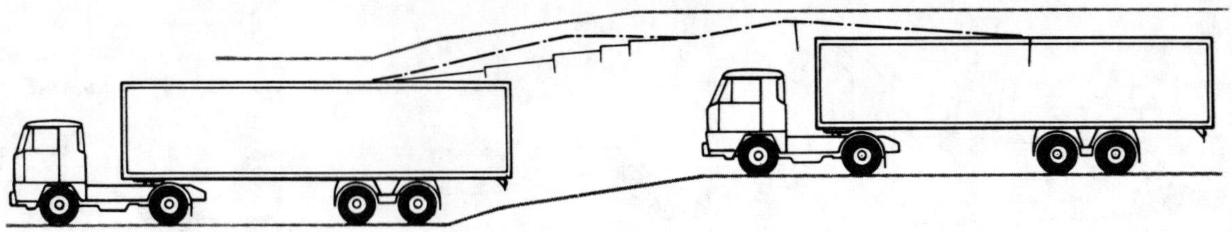

Fig. 32 Vertical shaped envelopes

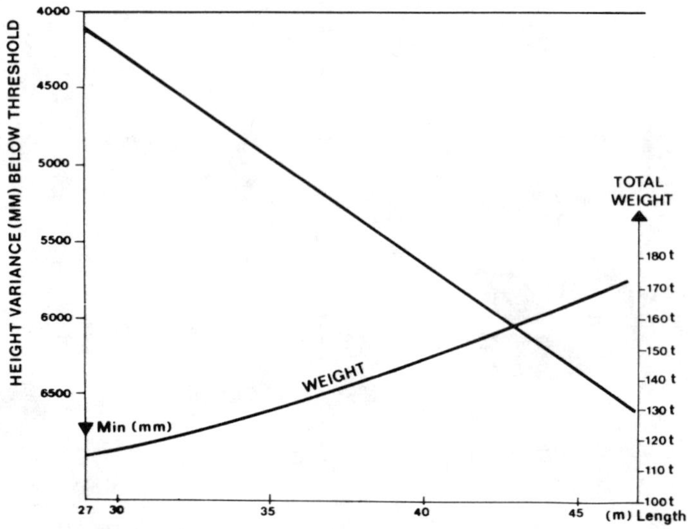

Fig. 34 Height variance, ramp length and weight relationship

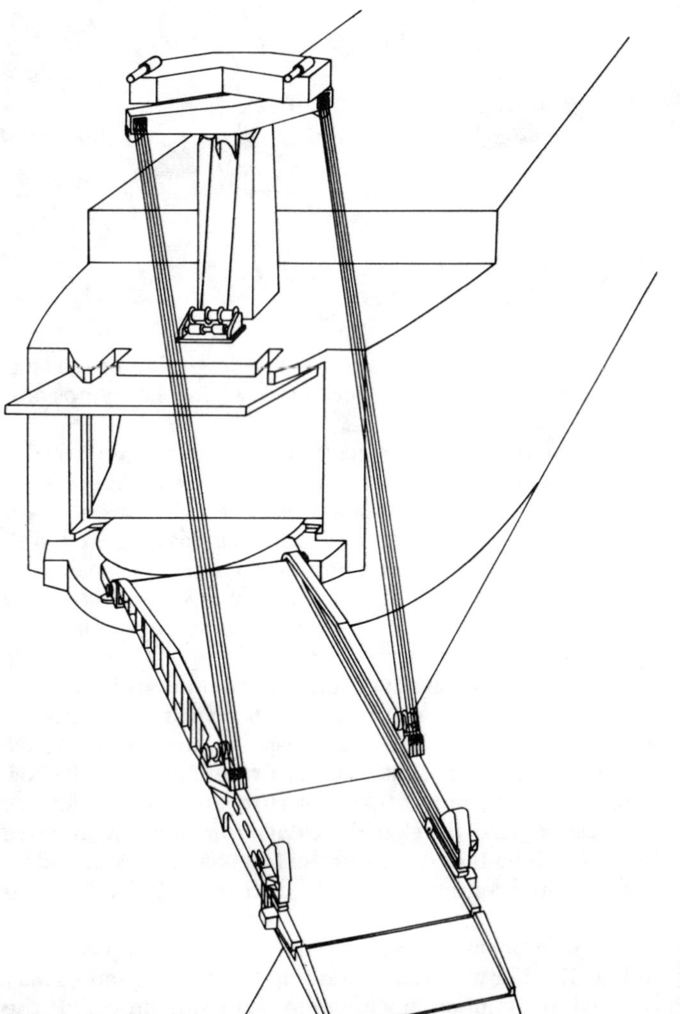

Fig. 33 Slewing stern ramp

Shore ramps are also particularly desirable in ports having great tidal variation where the ramp may be extremely long. By utilizing a shore ramp, the ramp can be stored in its near-operating condition, whereas a ship-mounted unit would have to be rotated more than 90 deg for stowage.

The side door/lifting table shown in Fig. 31 is utilized for pallet cargos if such cargos are carried in addition to ro/ro cargos. Its use is not restricted to ro/ro ships, as it is used on general cargo and multipurpose ships. The operation is similar to the sideporter discussed in Section 2.8.

d. Internal Ramps. Hoistable ramps usually have a gradient between 1 in 6 and 1 in 10, dependent upon the size of the trucks being used, and some of these internal ramps can be very long. This usually means that they need support at a number of points during the hoisting operation and also by articulated links or chains when in the ramp position. In many of the smaller types of ship, these ramps are of a simple one-piece design and are occasionally used as hatch covers when raised into their horizontal position.

Fig. 32 shows the trace of a 40-ft van as it negotiates an internal ramp. Due consideration must be given to structural clearances, not only for internal ramps, but external ramps where the vehicle goes over the doorsill as its load is still coming up the ramp at the ramp angle.

e. External Ramps. Fig. 31 shows a bow ramp, a side ramp and a straight and angled stern ramp. Bow and stern ramps, as shown in the lower diagram of Fig. 31, require a special dock facility. A recent innovation utilizes a mounted intermediate ramp to take up high tidal variations between the normally short, straight stern and bow ramps.

The angled stern ramp is relatively new and is a major factor in the new type of self-sustaining ro/ro ship. The angled ramp allows the ro/ro ship to dock at any conventional pier and to run its traffic pattern more or less in the direction of the normal *string piece* between the pier side and warehouse side. Due to their function, angled ramps are quite long and, thus, can accommodate a wide variance in tidal range. The slewing stern ramp is shown in Fig. 33 and allows the ship to berth either side to a normal pier. Some angled ramps are equipped with a loading sensing device, such that the full weight of the ramp never bears on the pier. Ramp end load could be a problem on some piers. This feature can also limit the heeling effect when heavy

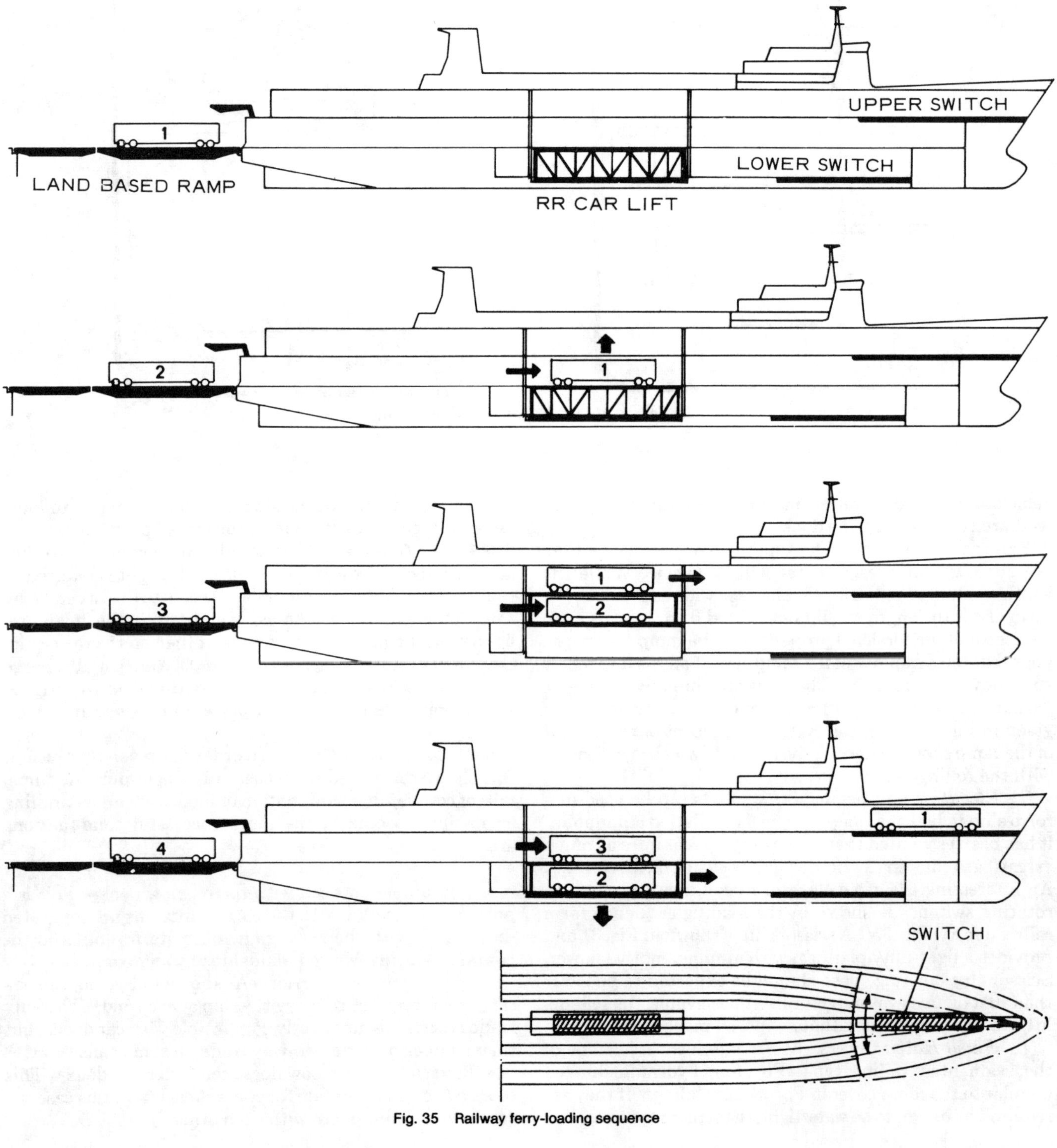

Fig. 35 Railway ferry-loading sequence

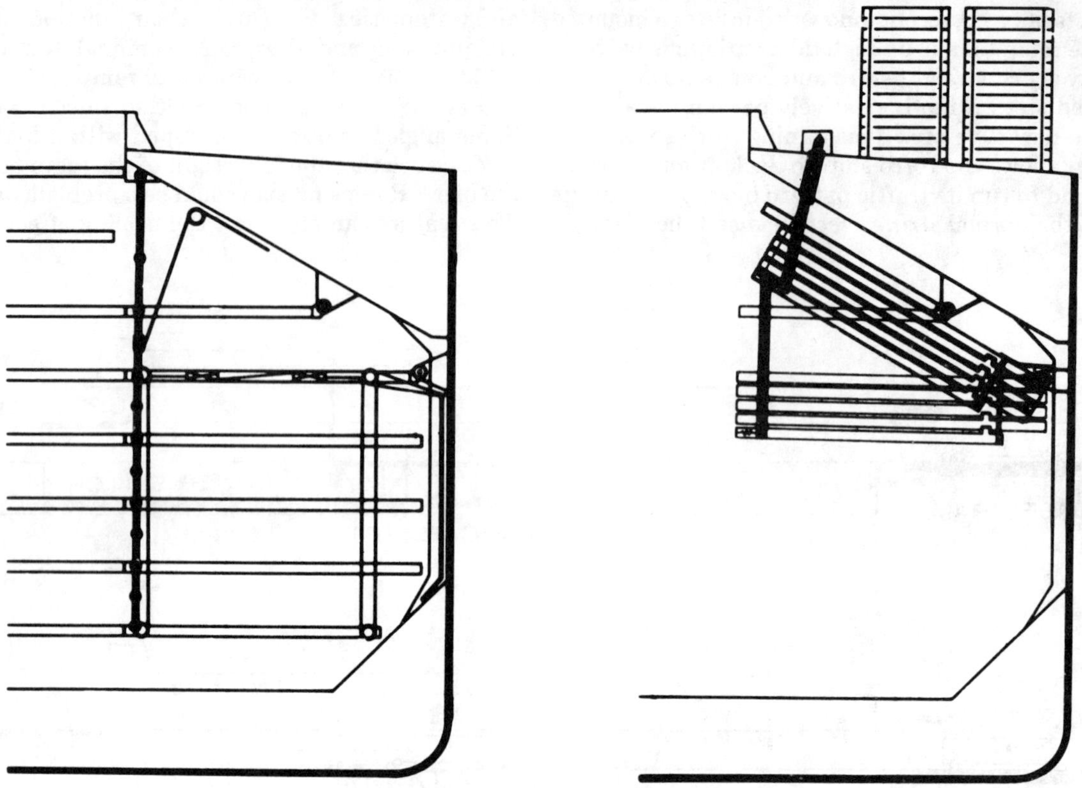

Fig. 36 Hoistable car decks fitted in bulk carrier

vehicles are loaded as the sensing equipment can be adjusted to share the load with the dock.

External ramp design is beyond the scope of this section, but some general design notes will be given. While the width of a ramp (and door) should be as wide as possible, the ratio of width/length should not exceed 0.95. If this ratio is exceeded, the torsional properties of the ramp become so poor that the flexibility of the ramp almost allows it to follow ship movement in heel. The relationship between height variance, required ramp length, and estimated weight is given in Fig. 34. Approximately 50 percent of the weight of the ramp should comprise the top plate which is in contact with the rolling vehicle.

f. Elevators and Scissor Lifts. Lifts can be used in a ro/ro ship in lieu of a ramp. From a ventilation standpoint, it has been estimated that a truck expels 3.5 times as much exhaust gas climbing a ramp, as it does sitting on an elevator. An interesting use of an elevator in conjunction with two rotating switches is shown in the loading sequence for a railroad ferry, Fig. 35. A scissors lift is shown in Fig. 31 and provides a light-duty platform with minimal motive power. Scissors lift hoisting speed should be expected to be lower than that of an elevator, but unlike the elevator, the scissors lift can be stowed flat without exposed machinery.

g. Flush Hatch Covers. Ro/ro hatch covers can be of the segmented rolling type (Fig. 31), hydraulic hinged (similar to the ramp cover in Fig. 31), or sliding. If they are required by design to be watertight, the structure and sealing arrangements must be considered as it is possible to have heavy vehicles transit them in the closed position.

5.3 Car Carriers. The straight car carrier is a special adaptation of the ro/ro, where the deck height is reduced to that required for an automobile. Access to decks can be by ramp or elevating mechanism, and, since the loads are lighter, most equipment moves faster than on the ro/ro ship. Car carriers with capacities up to 4,000 cars can discharge in as little as 20 hours, utilizing two to three ramps. Fig. 36 shows hoistable car decks in place and stowed in a bulk carrier.

5.4 Combination Ships. In order to prevent confusion, this chapter has tried to discuss only single-purpose ships, but, of course, any combination of modes of cargo handling is possible. Some of the more successful combinations are:

- Containers on deck with ro/ro below deck—This is a particularly good mix as the cargo functions are separated on the pier with the ro/ro going off a stern ramp and the containers being worked amidships by shore cranes.
- Bulk carrier/car carrier—As shown in Fig. 36, this arrangement requires refitting the ship at each end. The bulk trade is certainly not overly compatible with car decks, but the mix does provide two-way trade on some runs.
- Pallet/elevator below deck, containers on deck—This too is a good combination for some runs, but in this case the two modes do interfere with each other.

Fig. 37 shows one midship section of a combination ship that almost employs every loading device known: hinged ramp, lift, straddle carrier, trolley, ro/ro, and dolly. Fig. 38 shows a schematic of a similar *everything* ship, combining a barge carrier with containers and ro/ro. To many owners,

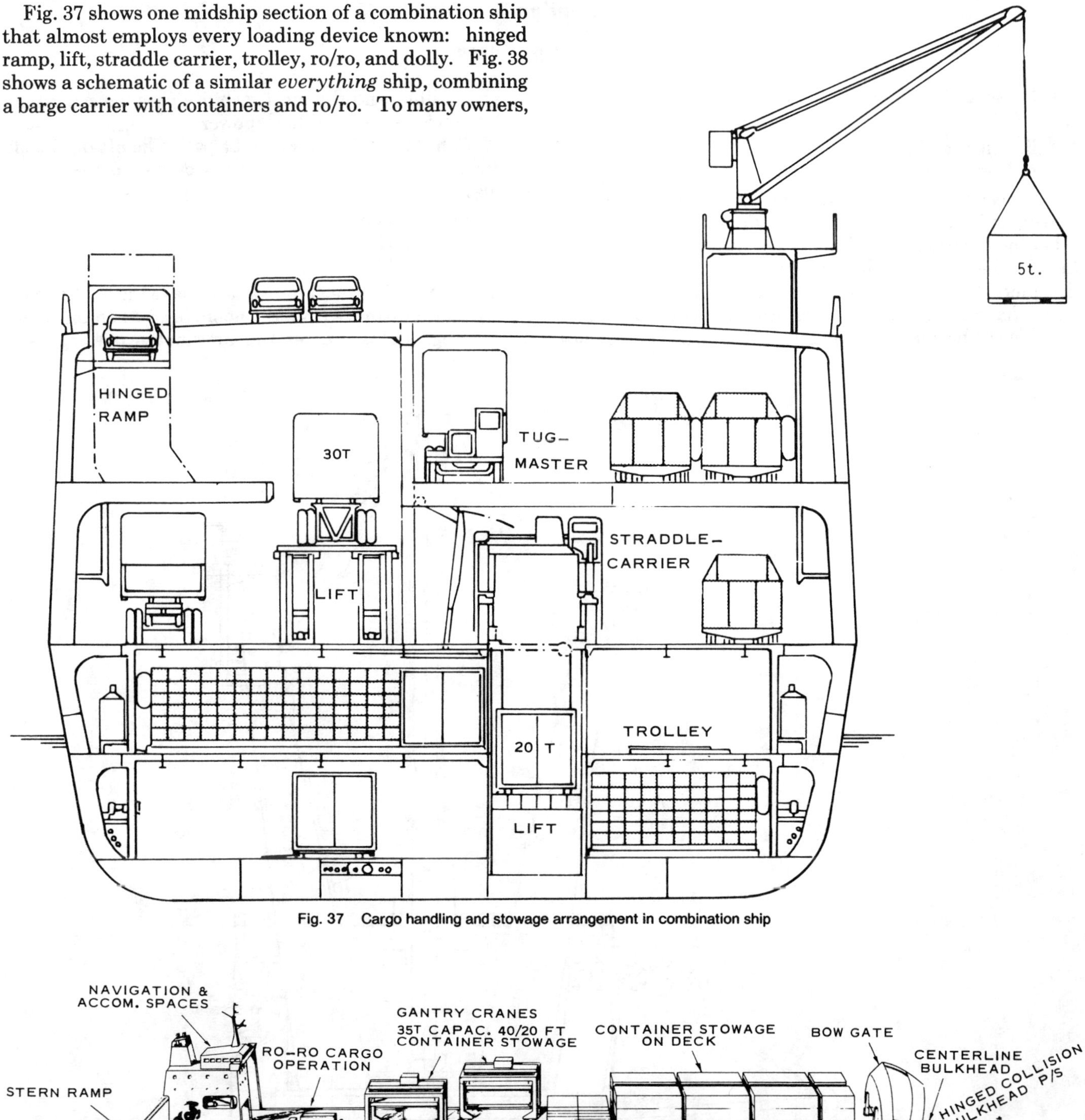

Fig. 37 Cargo handling and stowage arrangement in combination ship

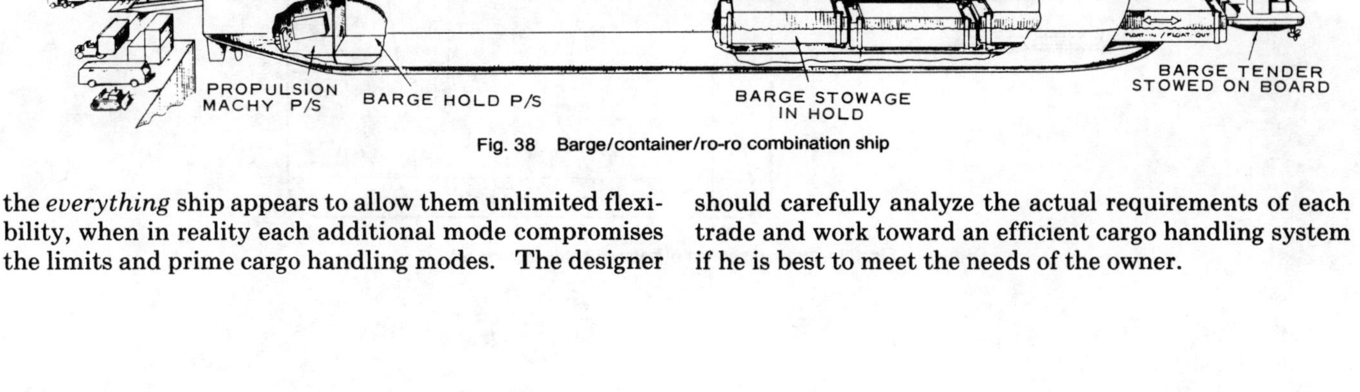

Fig. 38 Barge/container/ro-ro combination ship

the *everything* ship appears to allow them unlimited flexibility, when in reality each additional mode compromises the limits and prime cargo handling modes. The designer should carefully analyze the actual requirements of each trade and work toward an efficient cargo handling system if he is best to meet the needs of the owner.

Section 6
Heavy Lift Ships

6.1 General. There are two basic types of heavy lift ships:

- lift-on/lift-off,
- roll-on/roll-off.

Since each load carried in a heavy lift ship varies from trip to trip, and thus support points, maximum dimensions and other factors usually used to design a cargo system are not known, the heavy lift ship presents a real challenge to the designer. Basically, the designer must, constrained by either a fixed budget, hull size, or other limiting parameter, provide the heaviest lift capability, the longest and widest hatch possible, and the most stability possible consistent with required speed and horsepower.

6.2 Lift-On/Lift-Off Heavy Lift Ships. The lift-on/lift-off heavy lift ship can use many of the cargo rigs previously discussed, such as:

- swinging boom,
- Stuelcken rig (Fig. 39),
- rotating (Whirley) crane.

In each case, it is most desirable to have the load remain at a constant distance from the boom head when the boom is swinging. In many cases, the same motive power is used

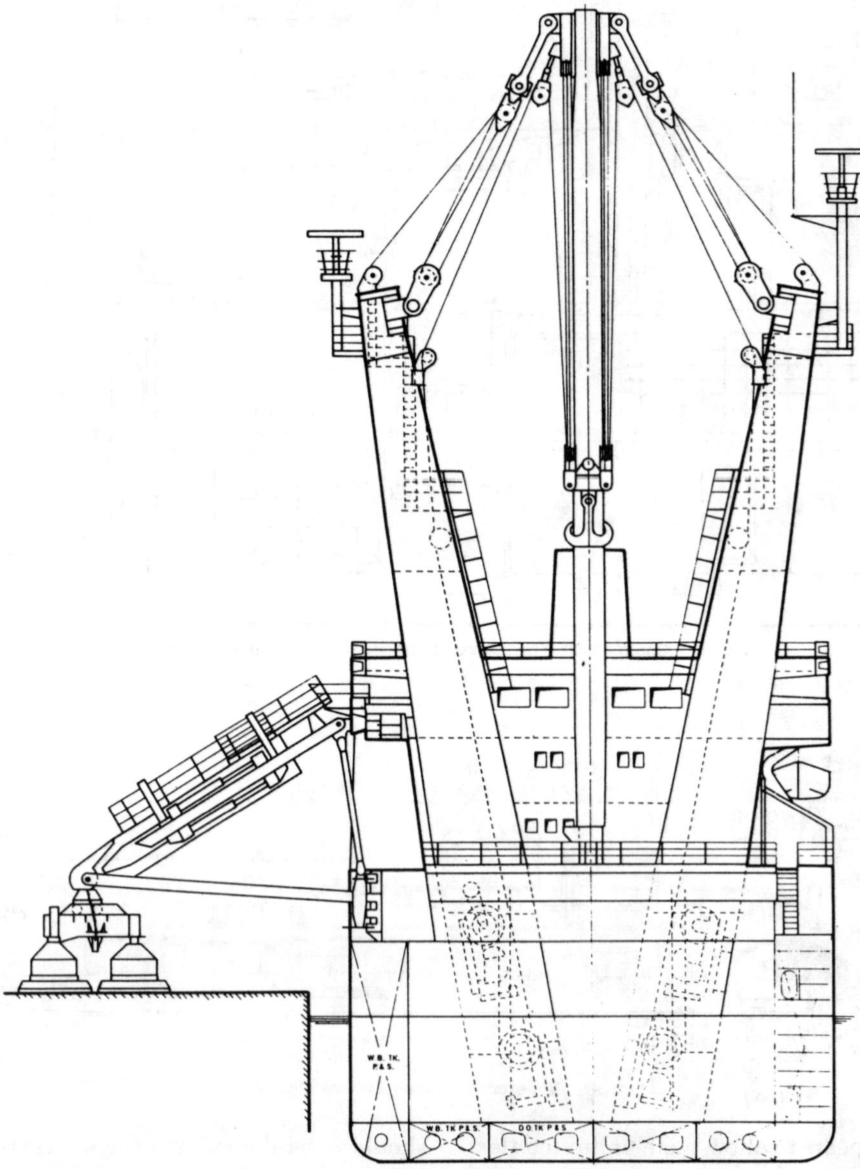

Fig. 39 Midship section of heavy lift ship looking aft

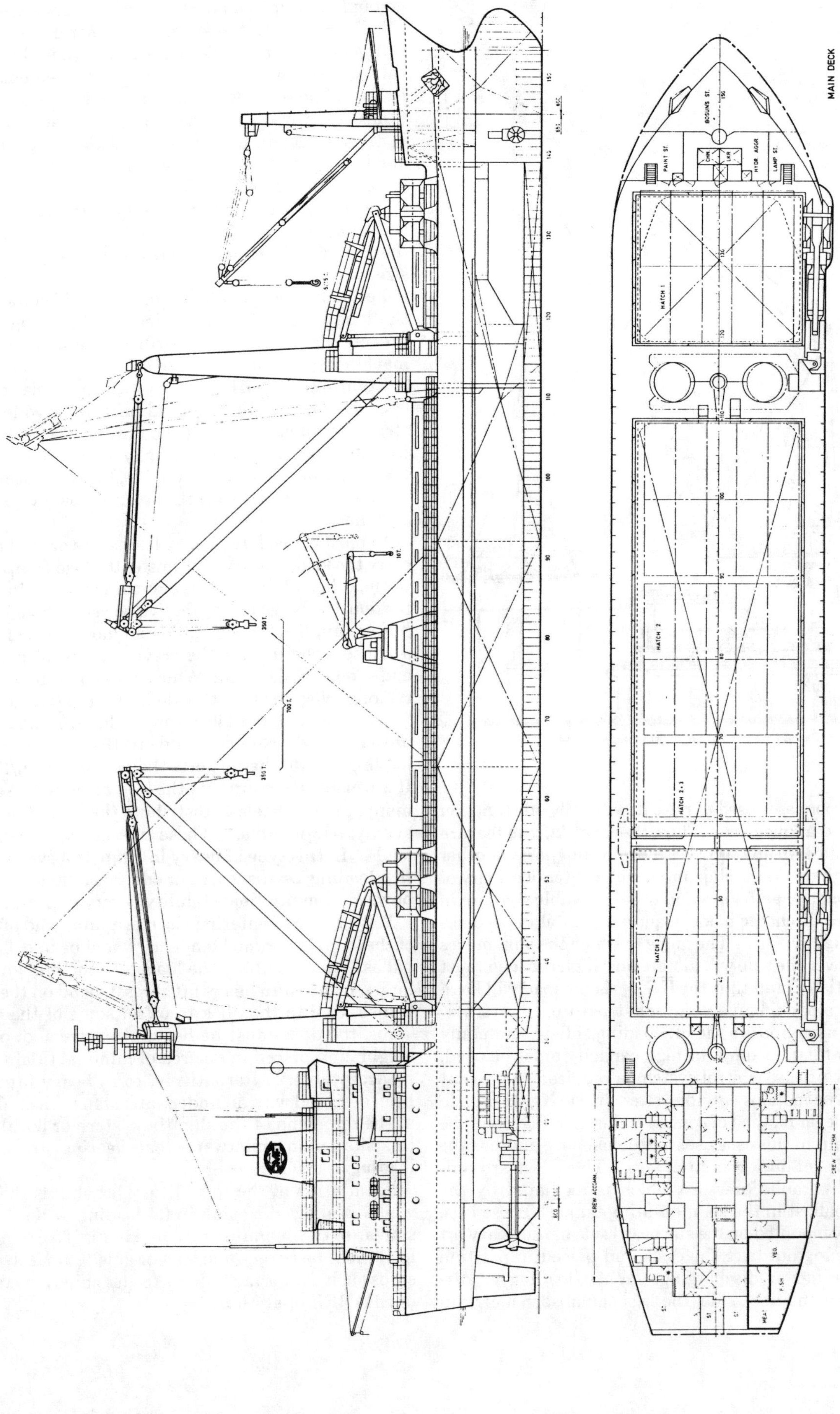

Fig 40 General arrangement of a heavy lift vessel

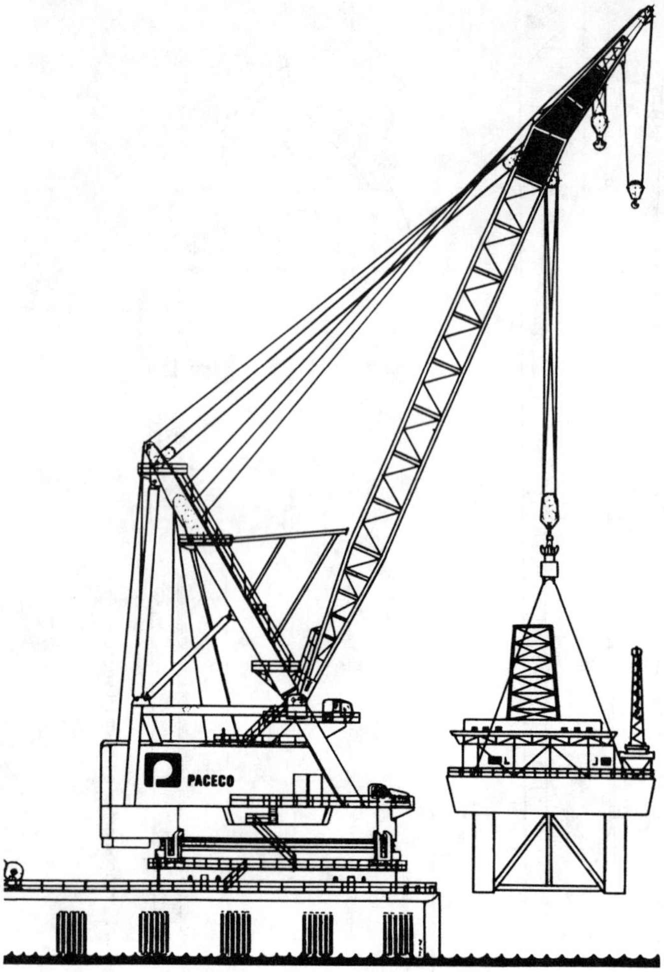

Fig. 41 Rotating on Whirley heavy lift crane

to slew the boom as is used to raise it, with only one function being coupled at one time. Thus, the *level-luffing* feature is usually built into any rig. With heavy kingposts or other structure high up on the ship and a high center of gravity of the heavy load, when booms and other movable gear are in the loading or discharge mode, transverse stability becomes of primary significance. The angle to which the ship moves with a load over the side will dictate how high the load must be lifted initially (and thus the $\overline{KG}$). As the load will have to be set on a dock, barge, or other structure, counterballasting of tanks opposite the off-loading side is commonly used to offset list. Pumps of high capacity must be used, and a rather precise control system is required to prevent pendulum-type motions or undue stress in cargo gear or hull.

A rather clever method of control is shown in Fig. 39. The heavy lift ship in this case uses a Stuelcken rig and has the added feature of dock-mounted *legs*. These legs are controlled by hydraulic pressure which is regulated by the amount of ballast in the compensating tanks. These legs, permanently mounted on the starboard side of the heavy lift ship, are swung out over the dock and placed in position prior to loading or discharging. When loading or discharging over the dock side, the legs can absorb a certain amount of initial load until the ballast tanks are sufficiently filled. By properly setting the controls in the dockside lifting or lowering mode, a *positive* load is always shown in the leg assuring minimal transverse movement of the ship. In the offshore lifting or lowering mode, the dockside ballast tanks are overfilled so that a positive load always remains in the leg, again, minimizing motion. Fig. 39 shows the amount of structure above deck in a heavy lift ship and gives an idea of the height of the $\overline{KG}$.

Each of the Stuelcken rigs on this vessel (Fig. 40 shows two that can work long loads in tandem) has a capacity of 350 tons. Rigs of 500 tons or more are possible, but usually ship size must increase. The type of rig shown in Fig. 40 (whether Stuelcken or rotating boom), in conjunction with the long central hatch, is ideal for handling long, bulky process plant equipment, drill rig equipment, and heavy machinery, such as locomotives.

Rotating or Whirley cranes with capacities to 3000 tons (Fig. 41) are usually mounted on a fairly wide barge-type structure for stability. If the unit is intended to go to sea with its cargo, the crane is usually mounted on the bow with the house located aft. If possible, the main deck is used for stowage of the load with the space below assigned to voids or ballast.

In most cases, large heavy lift barges are not designed to carry the load, but only to transfer the load from one position to the other while the barge remains fixed. In this configuration, the barge has no motive power, although it will have power supplies for the crane and mooring gears, as well as a house structure for the crew. A typical barge with an end-mounted, 3,000-ton Whirley would be about 152 m (500 ft) long by 40 m (130 ft) wide by 12 m (40 ft) depth.

6.3 Ro/ro Heavy Lift Ships. The ro/ro heavy lift ships are a recent development and one that is more limited in its loading and discharge ports than is the lift-on/lift-off ship. It is usually necessary that the load be placed on a dolly with many, many wheels to distribute the weight of the unit as evenly as possible. These vessels are ideal for long loads. In this type of heavy lift ship, transverse stability in the loading or discharge mode is of little consequence, however, longitudinal stability is very important.

With the load entering the stern ramp and after portion of the ship, the normal tendency would be to ballast forward ballast tanks to offset the load. As the hull girder has its limits, many ro/ro heavy lift ships depend on the ramp load transmitted to the dock to offset some of the cargo load. Thus, the dock must be heavily constructed, of a certain height, and placed in an area of minimal tidal variation to be acceptable. Alternatively, ro/ro heavy lift ships have been designed with an underwater step in the hull structure at the ramp end of the ship (bow, stern or both), such that this sits on an underwater load bearing structure during loading or discharge.

A variation of the ro/ro heavy lift ship is the dock ship which is allowed to sink in its loading or discharge mode similar to the operation depicted in Fig. 2. In this case, the load must be barge-mounted or able to float itself. Small ships such as this are used as feeder ships for major LASH or SEABEE operations.

Section 7
Bulk Cargo Handling

7.1 Scope. For the purpose of definition, the following list comprises all types of bulk carriers and their principal differences:

- Tanker—No hatch covers.
- Bulk Carrier—Large hold volume for low density cargos.
- Ore Carrier—Small central hold with high double bottom. High density cargo of 0.34 to 0.51m^3 (12 to 18 ft^3) per ton.
- Ore/Oil Carrier—Virtually a tanker with a double bottom, with a hatch provided on top of the central tank, through which high density ore can be loaded. Ore and oil are never carried simultaneously due to the explosion risk, unless they are separated by cofferdams and proper loading/unloading procedure is observed.
- Oil/Bulk/Ore (OBO)—A bulk carrier in which the structure is reinforced to handle oil and high density cargo. Furthermore, the arrangement of different hold lengths provides various possibilities for loading different cargos according to their densities. Very often alternate spaces such as holds No. 1, 3, 5 and 7 are smaller and are the only holds to be loaded for ore transport. One of the greatest difficulties arising in the design of such ships is the resistance of the ship's structure to hydrostatic effects due to the sloshing of liquid cargos inside the tanks. These difficulties come from increased hold length and breadth, and must be carefully taken into account in ship design especially on the larger vessels of today.

Only dry bulk carriers will be covered in this chapter with the liquid bulk types covered in Chapter XI. Additionally, some of the behavioral characteristics of ore and other bulk cargos, are also addressed in Chapter XI.

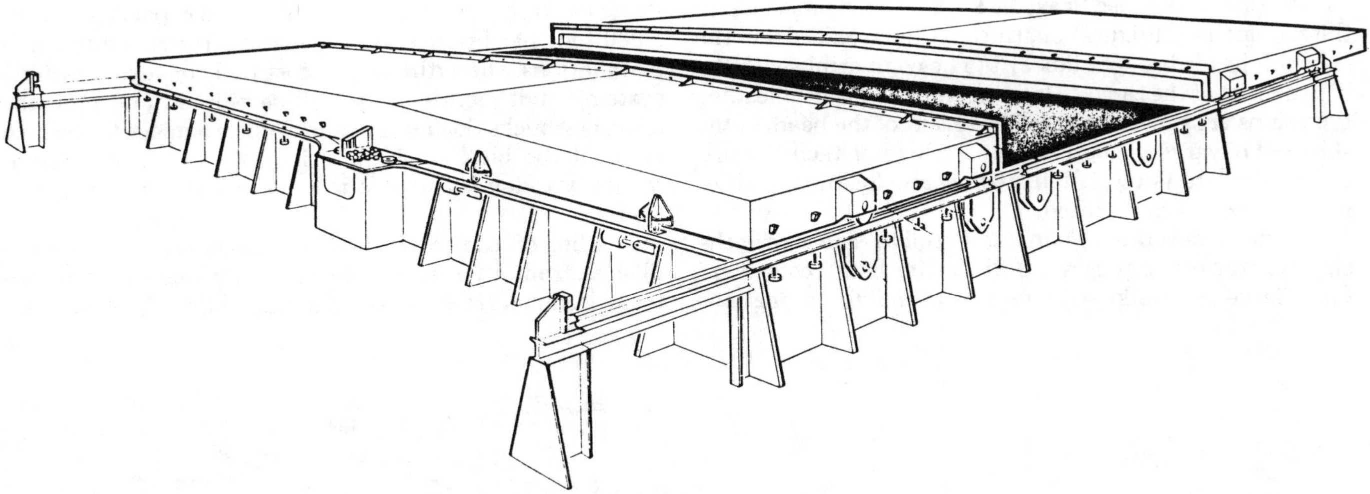

Fig. 42 Rack and pinion driven rolling hatch covers

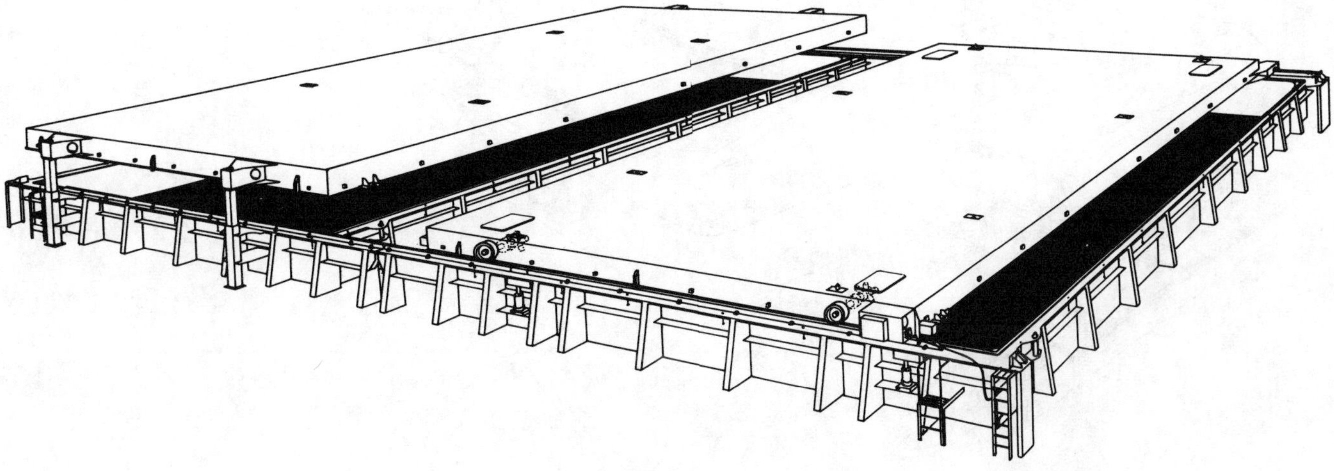

Fig. 43 Piggyback bulk carrier hatch cover

However, in designing any of the above types of vessels, due consideration must be given to the loading and discharging facilities they must serve whether the ships are self-sustaining, partially self-sustaining (discharging only), or non-self sustaining.

The standard bulk carrier still comprises the major amount of the world's tonnage, second only to the tanker. Bulk carriers have remained fairly standard in general configuration but, as with all other ship types, have increased tremendously in size and capacity in recent years.

The configuration of the modern bulk carrier with its canted upper and lower wing tanks allow: 1. A small area for clean up under the square of the hatch once most of the cargo has been discharged as the remaining cargo slides down the canted sides. This also allows discharging gear to *reach* all areas as the tank top breadth is roughly equal to the hatch opening breadth. 2. Stowage free of *shifting boards* or other temporary devices to prevent the load from shifting to one side. The upper wing tank or "topside" tank configuration presents minimum free surface when the bulk cargo is stowed to the top of the hold.

7.2 Ore Carriers and Heavy Bulk—Ship Configuration. In the ore carrier configuration, the tank top is raised and the side bulkheads brought inboard to be compatible with the high density of the cargo. Hatch openings for self-unloading type ships are generally 50 to 70 percent of the beam of the ship and of varying length to suit the loading facilities utilized. Unfortunately, hatch opening and loading facilities have not been standardized.

The hatch covers shown in Fig. 42 roll to each side of the ship on transverse rollers. Bulk carrier hatch covers are fairly large and bulky with weights of individual sections exceeding 50 tons. Driving mechanisms utilized to move these covers are rack and pinion, hydraulic ram and mechanical. Mechanical systems employing chain or cable must either use ship's gear, if the ship is so fitted, or the ship must be fitted with individual tugging winches. Fig. 43 shows another possible arrangement of side rolling or end rolling (longitudinal movement) covers where it is desired that both sections be moved away from the loading or discharging side of the ship. As can be seen in Fig. 43, one section is lifted with hydraulic jacks and the rolling section is rolled underneath. By lowering the first section onto the second both sections can be rolled to either side of the ship. If the bulk carrier has its own cargo gear (revolving or gantry crane) then pontoon or the mechanical covers previously described also may be fitted.

a. Types of Loading Facilities. Ore, coal and other heavy bulk cargos are usually loaded by:

- clamshell bucket,
- chute or spout loaders, or
- conveyor belt loaders.

The clamshell bucket is not generally used in modern installations for loading due to its limited capacity of about 2,000 tons per hr. Its use is generally restricted to small installations where the higher cost of a conveyor or chute system is not justified. As will be pointed out later, the clamshell bucket loader/unloader inflicts damage to the hold areas of the bulk carrier and is less acceptable environmentally today because of the dust created by its operation.

Loading of iron ore on the Great Lakes is usually accomplished from a large number of evenly spaced, gate-controlled chutes from dockside storage bins. The bins are

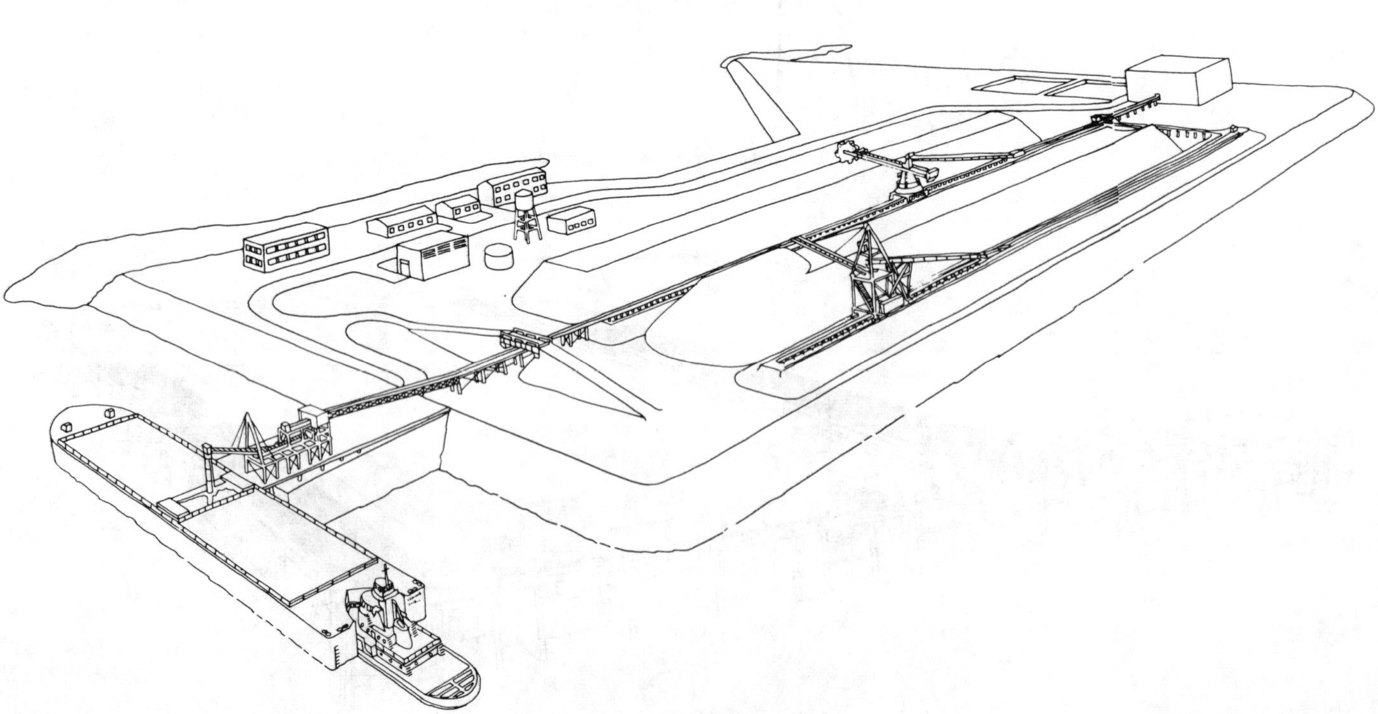

Fig. 44 Single belt conveyor loading system

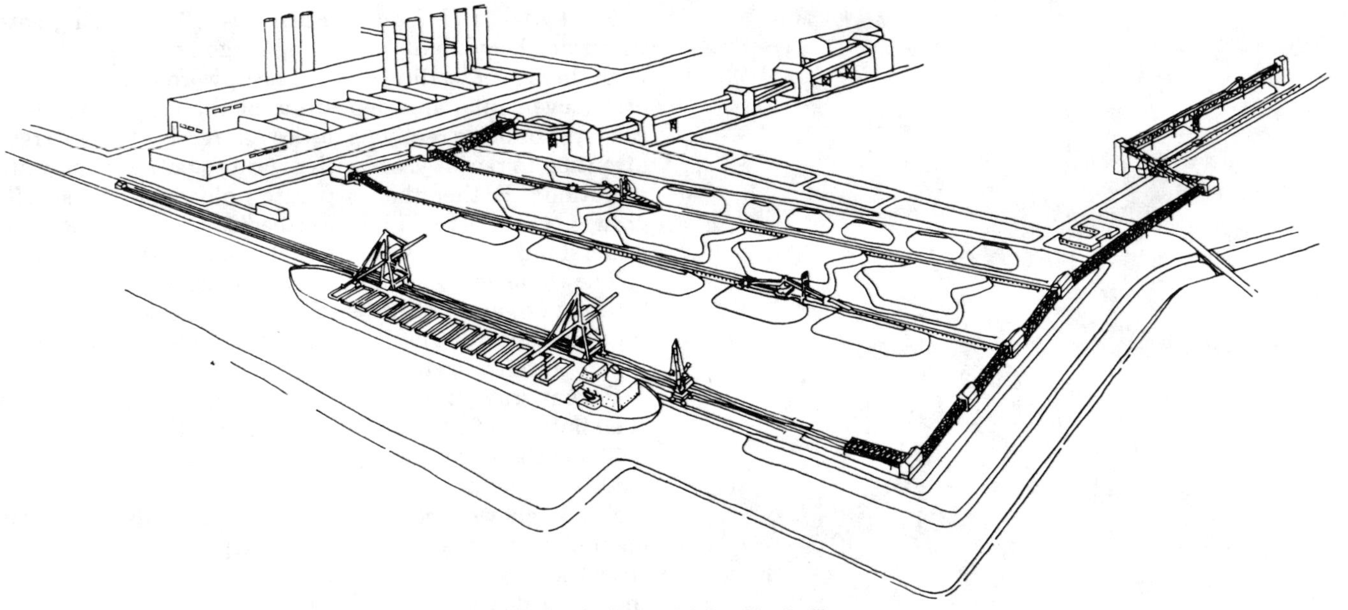

Fig. 45 Steel mill materials handling facility

elevated above the piers so as to provide sufficient chute slope for gravity feed when lowered into the cargo hold. The rate of feeding is controlled by a chute-operated gate. The chutes are lowered for gravity flow and loading of the ore. They can be directed for the proper distribution of the ore within the cargo hold. Ore is brought from the mines in hopper-bottom railroad cars and dumped into the storage bins for ship loading. These bins have capacities that vary from 250 to 400 tons.

For chute or spout loading, the ship's hatch arrangement must match the spacing of the pier loading chutes or spouts. The hatch width must be such as to allow the chutes or spouts to distribute the cargo evenly and to permit the lowering of the chute or spout for free flow of the ore. Average loading rates of 4,000 to 5,000 tons per hr have been achieved by chutes, spouts, and conveyors.

The belt conveyor loader has gained acceptance in the Great Lakes, rivers and offshore as the loading system of the future, largely due to its environmental acceptability. Facilities for loading coal on the Great Lakes of 11,000 tons per hr are in operation and an offshore facility recently completed can presently load at 20,000 tons per hr utilizing two rotating conveyor shiploaders. Fig. 44 shows a single belt system utilized to load offshore and inland barges where the barge is moved under the conveyor head or chute to distribute the cargo longitudinally. Fig. 45 shows a discharge system where the unloaders run on rails up and down the ship length. This type of system is also used for ship loading.

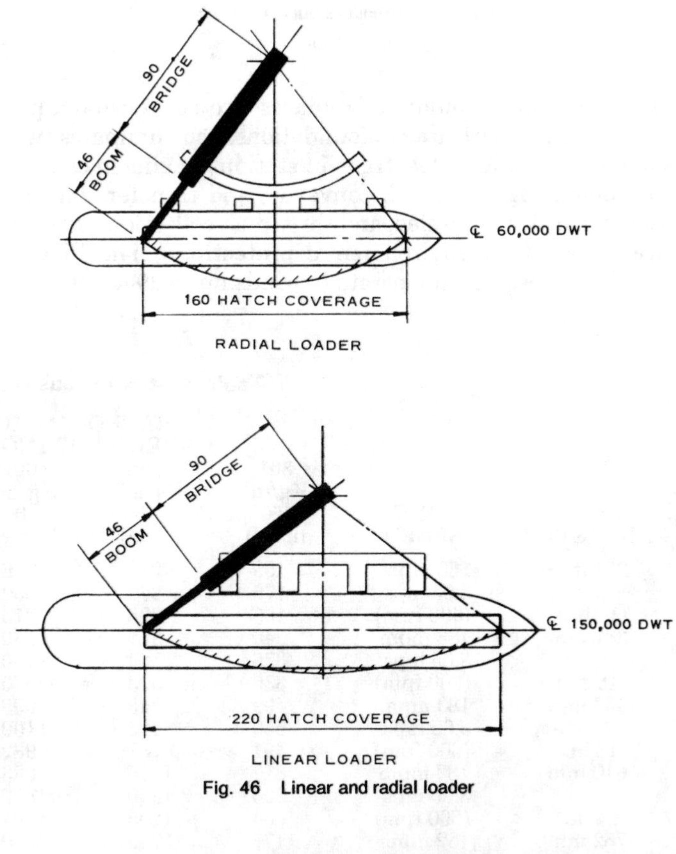

Fig. 46 Linear and radial loader

A recent innovation in loading gear is the Linear Loader. The Linear Loader utilizes a double telescopic, variable-span pivoting bridge supporting a shuttling boom conveyor. The ship side end of the Linear Loader bridge is mounted on a traveling turntable Fig. 46. At the pivot, the same type of turntable, turned upside down, allows the bridge to pivot as well as to slide, with the ship-side end of the bridge moving in a linear path next to the side of the ship, rather than in an arc. The linear concept is said to increase ship coverage to such an extent that one Linear Loader with

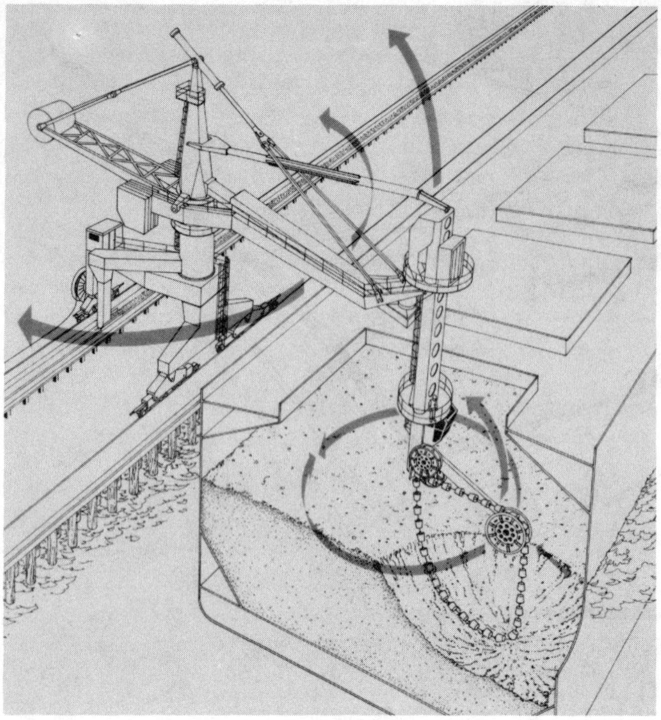

Fig. 47 Continuous bucket unloader

simple marine foundations replaces two conventional pivoting loaders with arc rail foundations, and eliminates two cross conveyors and two transfer stations, saving in capital and operating costs. All conveyors and transfers on the Linear Loader are sealed, and three dust collection systems are installed for environmental protection. The Linear Loader is designed to operate in winds up to 39 knots.

b. Types of Unloading Facilities. There is a greater variety of heavy bulk unloading cargo gear (Durst, 1974) than there is of loading gear. The upward movement, lateral movements, and depositing of the cargo on a shore conveyor or other receiving device has received considerable attention in recent years with regard to both air and water pollution. Rather than separate unloaders into specific groups, a number will be described so that the reader can understand the basic operational points of these devices.

The clamshell bucket fitted to a shoreside or shipboard rotating or gantry crane is still used in some installations, but it is doubtful that large installations will be allowed to utilize this arrangement in the future.

Recent grab bucket installations utilize units with clamshell buckets of a 20 ton capacity which, when operating on a 40-s cycle, have the capacity to discharge at the rate of 1,800 tons per hr.

Fig. 47 shows a continuous bucket unloader where the open buckets on an endless chain deposit the cargo into a closed conveyor system from the bucket head all the way to the shoreside conveyor. The side loading of a covered shore conveyor would give an all-weather, environmentally acceptable discharge system.

Fig. 45 shows how such an unloading system is integrated into the complete shoreside facility, in this case a steel mill. Shoreside *stacker/reclaimers* deposit discharged material from the shoreside conveyor belts usually with a moveable conveyor belt head as previously described for ship loading. Note in Fig. 45 that it is possible to segregate many cargos for shore storage and thus to draw from each of those storage areas as needed.

Capacities for the continuous bucket unloader for various cargo densities are given in Table 8.

Fig. 48 shows the Siwertell ship-unloader which uses a rotating screw encased in a column for vertical movement

Table 8—Continuous Bucket Unloader Capacities

		RATED CAPACITY—FREE DIGGING FREE FLOWING MATERIAL—TONS/HR							
BUCKET DIAMETER	LINE SPEED	801 Kg/m³ (50 lb/ft³)	1201 Kg/m³ (75 lb/ft³)	1602 Kg/m³ (100 lb/ft³)	2002 Kg/m³ (125 lb/ft³)	2403 Kg/m³ (150 lb/ft³)	2803 Kg/m³ (175 lb/ft³)	*	BUCKET UNIT CAPACITY
254 mm	183 mpm	159	239	318	397	476	556	MT	0.0082 m³
	3.05 mps	175	263	350	438	525	613	ST	
(10 in.)	(600 fpm)	156	235	313	391	469	547	LT	(0.29 ft³)
381 mm	183 mpm	340	511	680	851	1021	1191	MT	0.030 m³
	3.05 mps	375	563	750	938	1125	1313	ST	
(15 in.)	(600 fpm)	335	503	670	838	1005	1172	LT	(1.05 ft³)
457 mm	183 mpm	499	748	998	1247	1497	1746	MT	0.052 m³
	3.05 mps	550	825	1100	1375	1650	1925	ST	
(18 in.)	(600 fpm)	491	737	982	1228	1473	1719	LT	(1.82 ft³)
610 mm	183 mpm	817	1225	1633	2041	2449	2858	MT	0.114 m³
	3.05 mps	900	1350	1800	2250	2700	3150	ST	
(24 in.)	(600 fpm)	804	1205	1607	2009	2411	2813	LT	(4.03 ft³)
762 mm	152 mpm	1179	1769	2359	2948	3538	4128	MT	0.248 m³
	2.53 mps	1300	1950	2600	3250	3900	4550	ST	
(30 in.)	(500 fpm)	1161	1741	2321	2902	3482	4063	LT	(8.75 ft³)
914 mm	146 mpm	1588	2381	3175	3969	4763	5557	MT	0.411 m³
	2.43 mps	1750	2625	3500	4375	5250	6125	ST	
(36 in.)	(480 fpm)	1563	2344	3125	3906	4688	5469	LT	(14.5 ft³)

* MT, metric tons; ST, short tons; LT, long tons.

Fig. 48 Rotating screw unloader

of the cargo. Horizontal movement can also be accomplished by this same type of screw installation or an enclosed conveyor. The unusual feature of the Siwertell system is a counter-rotating head unit which reportedly fills the discharge tube to 70—90 percent of its capacity (fill factor). Conventional screw conveyors have been limited to a fill factor of 10–25 percent. Existing installations have achieved rates of 600 tons per hr on heavy bulk and the digging action provided by the counter-rotating head is said to facilitate breaking through hard crusts and breaking up solid banks of material. The amount of mechanical, or hand trimming, of the final cargo left in the hatch should also be reduced.

c. Slurries. The use of slurries for heavy bulk cargo discharge is another step toward minimizing product loss and environmental impact. Generally, the construction of bulk carriers is such that the cargo is loaded dry and then discharged from the top via a pump, after some sort of water jet is used on the cargo to form the slurry.

It is conceivable that cargo could be loaded, transported and discharged from an OBO or tanker as a slurry. The ship designer will have to note, however, the density of the slurry when designing bulkheads and supporting structure. Agitation machinery to keep the slurry homogenous may also be required to be fitted to ships carrying slurries.

7.3 Grain Carrier and Light Bulk. There is a considerable world commerce in the carriage of grain and similar light bulk cargos which has resulted in the evolution of ships specially designed for these services.

a. Grain Stowage Characteristics. Grain exhibits a behavior that is somewhat between general cargo, where the center of gravity of the cargo remains fixed with respect to the ship regardless of sea motion and liquid cargo, where the cargo center always moves in response to sea motion. This has necessitated the imposition of strict stability requirements that are promulgated in international regulations. Fig. 49 shows a comparison between the cross sections of a typical general cargo ship, a tankship, and a bulk carrier to illustrate the effect of ship design differences on bulk cargo stowage.

When a cargo hold on a general cargo ship is loaded with bulk grain or similar cargo, sizeable void spaces remain above the grain that cannot be completely filled under the horizontal deckheads at every deck level. Careful trimming of the cargo, which is required by the regulations, cannot eliminate these voids. Thus there is the possibility that, where the ship experiences sea motion, the grain will shift into these empty spaces and thereby create a heeling moment which may dangerously impair the stability of the vessel, as described by Price and Middleton (1969). Therefore, in order to meet stability requirements it is usually necessary to secure the grain from shifting by strapping it down or overstowing it with other cargo. Another alternative is to minimize the extent of the possible grain shift by constructing temporary longitudinal bulkheads. All of these expedients are expensive and decrease cargo revenue.

A tankship loaded with bulk grain may have even larger void spaces above the grain but, since the ship is designed with at least two longitudinal bulkheads it generally has ample stability to meet the grain regulations without the necessity of trimming or installing special fittings.

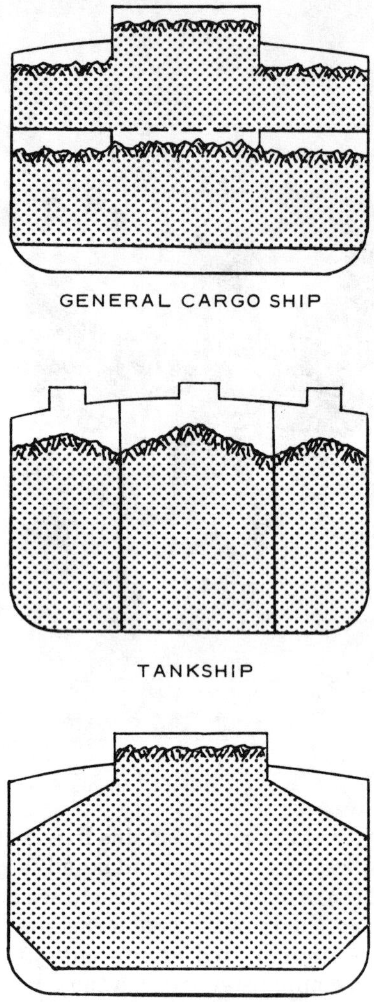

Fig. 49 Grain distribution in various types of ships

The bulk carrier is designed to minimize, where completely filled, the void spaces above the cargo. This is accomplished by eliminating horizontal surfaces above the grain insofar as possible. Instead, such surfaces are sloped at an angle of not less than 30 deg to the horizontal, Fig. 49. Since this angle exceeds the natural angle of repose of the grain, the free-flow of the commodity will completely fill all such spaces. The remaining void spaces, hatchways for example, are minimal and only cause heeling moments that are well within the stability capabilities of the ship. Fittings, and especially longitudinal bulkheads which would interfere with cargo discharge operations, are not needed.

All of the above vessel types are presently required to be approved under the requirements of IMCO Resolution A.264(VIII) Amendment to Chapter VI of the International Convention for the Safety of Life at Sea, 1960 and the subsequent documents (Middleton and Samis, 1970). This Regulation prescribes voids to be assumed under all horizontal surfaces (less than 30 deg to the horizontal) that generate heeling moments due to a transverse shift of grain. The vessel is deemed to be in compliance with the requirements provided that, after such a shift is assumed, she will not exceed a maximum prescribed angle of heel nor have less than a certain amount of residual area between the heeling arm curve and the righting arm curve at any stage of a voyage.

b. Cargo Handling Gear. The light bulk cargo ship can use any of the cargo handling gear described in the previous section. Although there is an infinite variety of gear used, some of the most interesting units developed to date are for the handling of forest products. Like other cargo operations, the forest product industry has started to standardize and to modularize its raw finished products for more efficient handling. Fig. 50 shows some of the handling units used in this industry. The forest products industry also utilizes large barges that carry logs on deck in bulk. When ready for discharge, one wing wall of the barge swings down, the side hull tanks are flooded and the barge lists to that side. An entire barge can be discharged into the water in a matter of minutes.

7.4 The Self-Unloading Vessel. The demand for this style of vessel was one of the consequences of the rapid industrial growth within the Great Lakes Basin. Salt water self-unloading vessels are also growing in number, but still represent an extremely small component of the dry bulk fleet on the oceans of the world.

On the Great Lakes today, the self-unloading vessel is the primary means for transporting dry bulk cargos, Fig. 51. On these vessels, the additional cost of the unloading gear is

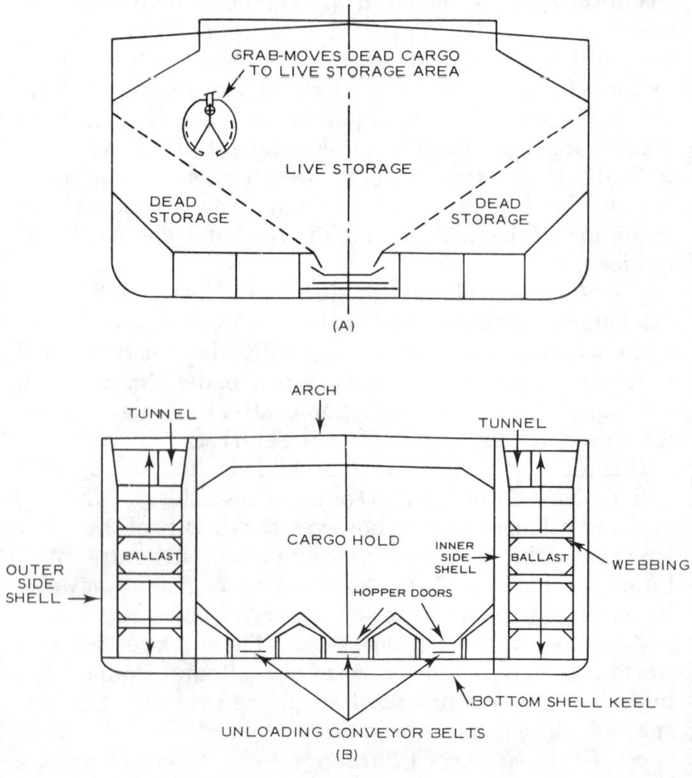

Fig. 51 Self-unloading bulk carriers

Forest Products Handling Gear

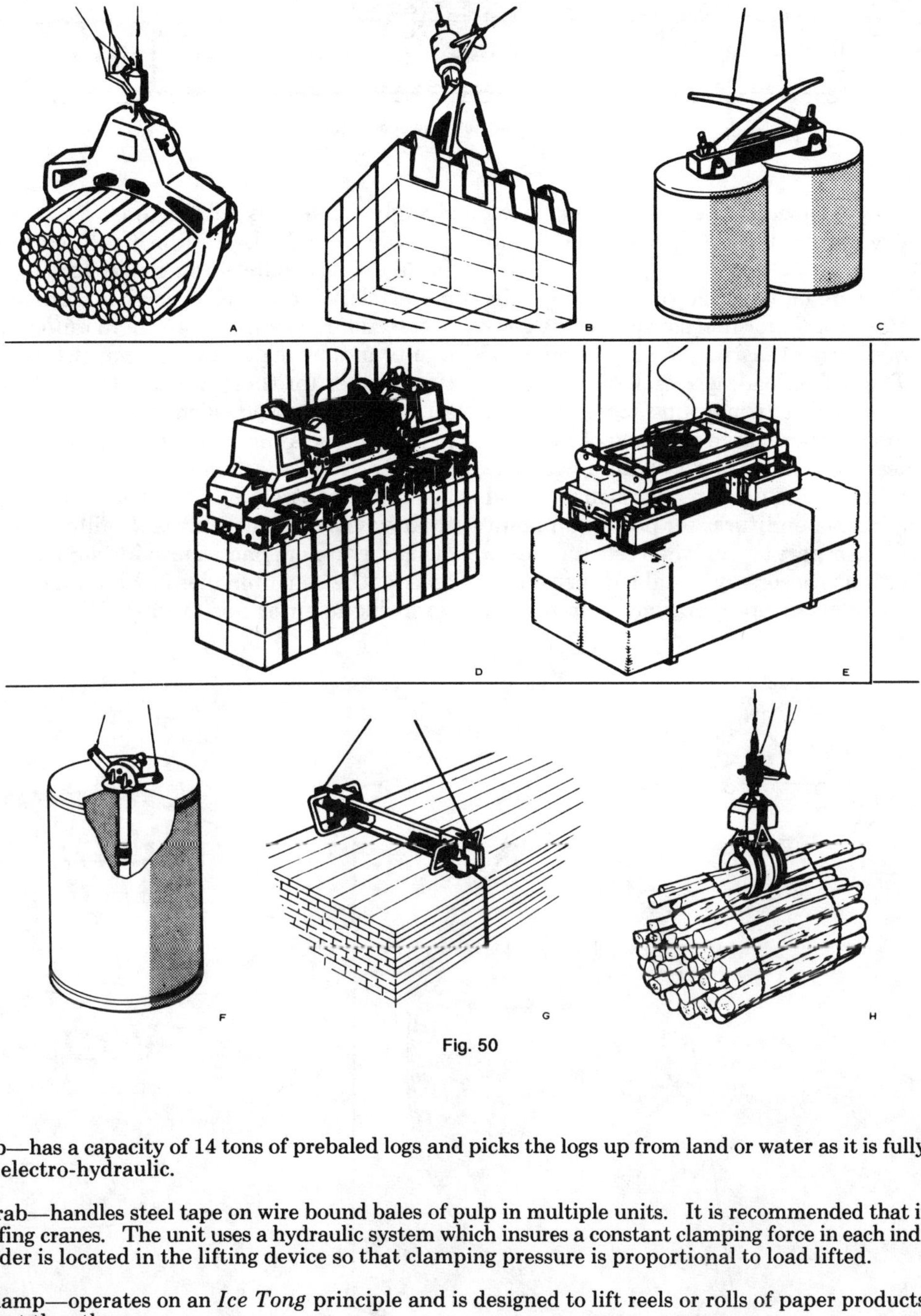

Fig. 50

A Log Grab—has a capacity of 14 tons of prebaled logs and picks the logs up from land or water as it is fully submersible. Actuation is electro-hydraulic.

B Multi-Grab—handles steel tape on wire bound bales of pulp in multiple units. It is recommended that it only be used with level luffing cranes. The unit uses a hydraulic system which insures a constant clamping force in each individual clamp. The oil cylinder is located in the lifting device so that clamping pressure is proportional to load lifted.

C Probe Clamp—operates on an *Ice Tong* principle and is designed to lift reels or rolls of paper products by clamping one roll against the other.

D Bale Clamp—utilizes clamp heads which are positioned on the steel straps of the bales by magnetic means. The heads then clamp under the band forcing tension in the band. This same principle is used to lift stacks of plywood, E finished lumber, G and any other commodity fastened with wire banding.

F Core Probe—expands within an individual roll of paper with the expansion action actuated by the two lifting arms. Again expansion force is proportional to weight lifted and so damage to the inner (core) layers of paper is minimized.

H Log Grapple—is a variation of A and grabs only the top logs of a banded pile.

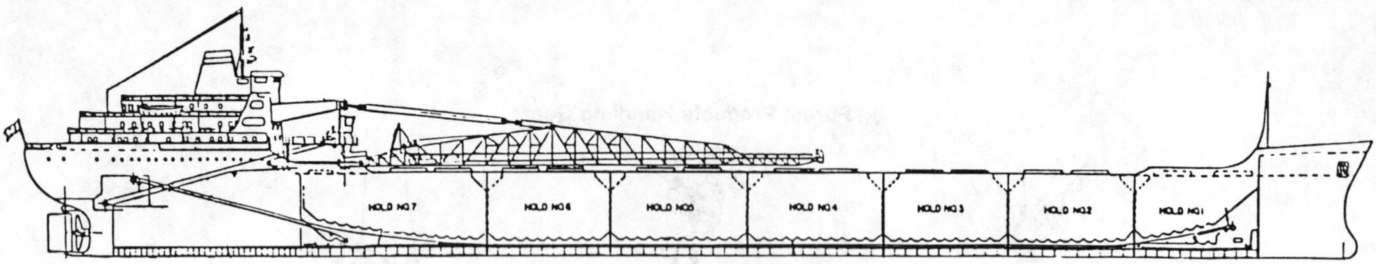

Fig. 52 Inclined belt conveyor system

reimbursed to the owners through the vessel's increased flexibility and reduced port time (Young, 1967) and (Vaughn, 1977).

Vast quantities of iron ore move to the various steel mills around the Lakes. Coal is a major fuel source for the generation of electric power, as well as a major energy source of the steel industry. The self-unloading vessel, with its high discharge capability and short turnaround time, offers a solution to the handling of these large volumes of raw material. Shipping lines can handle more volume with the same size of vessel and receiving ports realize a substantial reduction in their capital expenditures for port equipment and a simplification of their port layout when self-unloaders are used. Shore labor is not necessary to unload the vessels and the time-consuming clean-up operation, common to straight bulkers, is eliminated.

Aside from the Lakes, during the early 1950s many of the world's steel producers started to look offshore for their raw materials. Unlike the North American steel mills, they invested large sums of capital in unloading rigs at all their plants and, by taking this action, dictated the use of the box type bulker for their trades. Up until this time, virtually all dry bulk vessels, although small in size, were self-unloading vessels that used grabs located on deck.

Since the only major movement of raw materials was toward these steel mills, the world shipping community built their vessels to suit the steel mill requirements and their other customers had to provide similar facilities. For this reason, the self-unloader has not made great strides in attempting to enter this market.

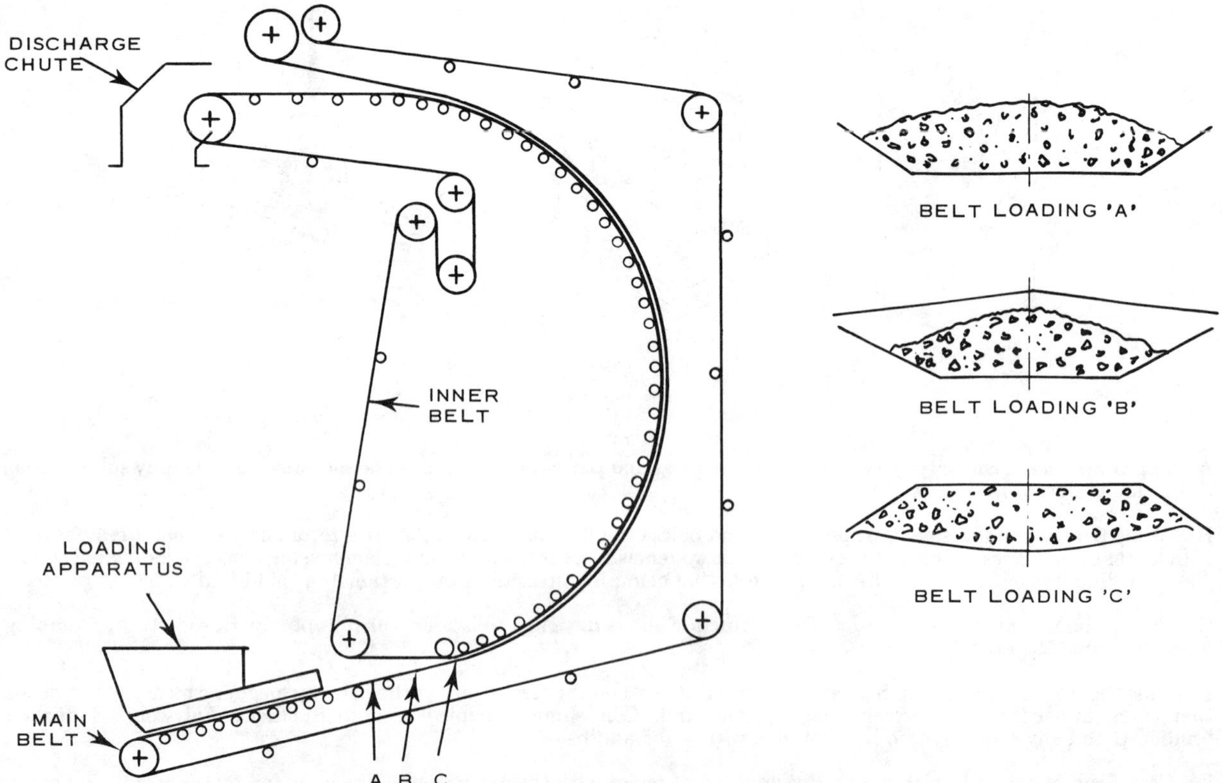

Fig. 53 Schematic of loop belt system. Sections A, B, and C, show how the material is sealed between the two belts during the lift.

Today's oceangoing self-discharging vessels range in size up to 155,000 dwt and are actively engaged in moving a variety of bulk cargos across ocean trade routes and in the coastal distribution trades. Cargos hauled in these vessels include grain, coal, iron, ore, phosphate, gypsum, salt, and aragonite sand.

a. The Tunnel Area. The basic design of virtually all high capacity self-discharging vessels incorporates the use of gravity-fed belt conveyors. As gravity is used to load ships, why not unload by gravity? The bottom of the hold is hoppered and fitted with hydraulically operated gates. These gates are sequentially opened in a controlled manner in order to allow the cargo to be discharged onto the conveyor belts located beneath the hoppered area of the vessel, Fig. 51.

The hold belt conveyors move the cargo to an area either just aft of the collision bulkhead, or forward of the engine room bulkhead, so that the material can be collected and then elevated to the desired deck level for discharge from the vessel.

In recent years a considerable amount of time has been spent in the redesign of gates and the reduction in space requirements for the tunnel conveyors. Improvement has been made in the gates, skirting has been eliminated on many vessels, and the headroom requirements for the unloading equipment have been reduced. Fig. 51B shows the midship section of the *James L. Barker* and the amount of dead storage associated with this system. As a grab bucket or other means must be used to move this dead storage cargo to the belt, discharge time is extended and rate of discharge slowed. In utilizing the three belt system all cargo is gravity-fed to the belts without any extra handling.

b. The Elevating Phase. The elevating of cargo from the hold conveyor to the discharge boom, or shuttle, has always been a problem area to the designers of equipment for self-unloading vessels. It is in this area that substantial improvements have been made recently by eliminating the *bucket elevator system.*

An obvious means of eliminating the bucket elevator was to use an inclined belt conveyor system similar to most land installations handling high tonnages, but is was soon found that space which is available on land is not available aboard ships, Fig. 52. As can be seen, the inclined belt system has to penetrate the engine room space, requires transfer points, and affects the layout of accommodations. Cargo space is lost in the vessel design and the layout and selection of propulsion equipment for the vessel is seriously affected. Vessels using this system were built primarily for the iron ore and stone trades and, therefore, cubic space was of little significance.

Another possible solution was the use of rotary wheel elevators. Two such rotary wheel elevators have been installed in the *M.V. Stewart J. Cort* and the tug barge *M.V. Presque Isle.* These vessels were completed in 1972 and 1973 respectively, and utilized a rotary elevator mounted on centerline in the engine room. The *Cort's* self-unloading system empties the ship at a rate of 6,000 to 20,000 tons per hr when the cargo is a free-flowing material of 50 mm (2 in.) lump size and below. Main components of the system consist of metering feed gates, a single 3m (10-ft) wide steel-cord conveyor belt running almost the length of the ship, an 18m (60-ft) diam wheel elevator and a 27.5m (98-ft) wide transverse boom conveyor. Material is fed from the cargo hold to the tunnel conveyor by means of 105 metering gates. All gates discharge simultaneously, allowing unloading in a manner that induces the least strain on the hull and simplified ballasting.

The conveyor transports the material to the unloading boom via the wheel elevator. The belt can move at speeds to 300 m/min (1000 fpm). The rotary elevator is similar to an undershot waterwheel except that it is supported by two body idlers and has no spokes or axle. The wheel is not powered. It rotates as a result of belt friction (the belt is wrapped 210-deg around its circumference.

After the belt leaves the wheel, it wraps around two drive pulleys and returns to the hold along the same route used during the elevating process, but separated from the wheel by idlers. The conveyor contacts the wheel at its lowest point. Material on the belt is trapped in one of a number of compartments (similar to paddles on a waterwheel) until it reaches the highest point on the wheel. From there it falls by gravity to a hopper where it is transferred to the transverse boom for discharge over the side.

Another problem associated with the use of an inclined belt is the fact that it is limited to an angle of inclination of roughly 18 deg. Once this angle has been exceeded, the material tends to roll or slide back and off the conveying belt. If a belt conveyor could be made to climb at an angle of 30 to 45 deg, much of the problem could be resolved and the belt system, while it would still affect accommodations, could be designed to enter the engine room area above the main engine and gear box.

The *retainer belt system* is a steep angle conveyor and was developed for use in ships. In this system, an additional heavy conveyor belt is laid down atop the material transported on the regular conveyor system as the belt conveyor enters the steep sloped portion of its conveying run. This second belt traps the material between itself and the standard belt and holds it there until the two belts are separated at the discharge point.

Conveyors of this design were installed in two vessels which have entered service since 1969. The *S.S. Quetico* of Canada Steamship Lines has three retainer belts with each belt designed to elevate 2,000 tons per hr of iron ore pellets, or 1,400 tons per hr of coal. The second vessel, the *M.V. Phosphore Conveyor,* a salt water self-unloader, incorporates a retainer belt system to elevate cargo at a rate of 3,000 tons per hr of ore and 2,000 tons per hr of coal. This attempt to solve the elevating problem was not sufficient to please the owners of the vessels, or the shipyards. The cubic problem was reduced, but that improvement was offset by the problems of accommodation layout and potential maintenance. The heavy belt was difficult to repair, if damaged, and extremely expensive to replace.

In 1967, major research work was undertaken to solve the space problems associated with belt conveyor systems. The industry was asking for a belt conveyor which could climb vertically. This is the *Loop Belt Elevator*.

The Loop Belt system uses the same components as a standard belt conveyor and, since the material comes in contact with nothing but the rubber of the conveyor belts, it naturally requires less maintenance than the old bucket elevator system. This system uses two endless conveyor belts, one overlying the other in the curved area so that the material is trapped between the belts until the desired elevation is reached. The loading and discharging of this conveyor is identical to that of a standard belt conveyor. The difference between this system and the inclined belt is the basic fact that it requires little horizontal space and is capable of reversing the flow of material without the need for high maintenance transfer points. This system appeared to be ideal for the ships.

One of the American vessels to use the Loop Belt system is the *M.V. James L. Barker,* the largest on the Lakes. This vessel has a capacity of 54,000 tons of ore pellets or 47,000 tons of coal and is capable of discharging pellets at the rate of 9,000 tons per hr through a 76m (250-ft) boom able to be swung out to either side of the ship.

c. Conveyor Discharge Booms. Once the material has been elevated to the main deck, a luffing, slewing boom, approximately 76m (250 ft) in length is usually used to discharge the cargo from the vessel. This boom permits the vessel to discharge its cargo directly to the customer's wharf, into a shore based receiving hopper for distribution inland by conveyor, or to transfer its cargo directly to the holds of another vessel. All self-discharging Lake vessels are equipped with a boom of this type.

However, some vessels are equipped with a shorter boom, which is only 30–40m (100–160 ft) in length. These booms are to be found on salt water self-unloaders and are usually totally enclosed for dust-free operation. Vessels equipped with this type of boom are unable to stockpile the cargo for the receiving customer and, therefore, forfeit some of the operational flexibility of the Lakes type self-unloader.

Boom design has changed. They are now fabricated of high strength structural tubing and are triangular in cross section. The reason for this is to reduce the weight of these structures and, therefore, add to the cargo carrying capability of the vessel. The only other major change in the design of these booms is the use of hydraulic motors to drive both luffing and slewing.

A variation in discharge equipment is found in the use of a shuttle conveyor. This type of system involves a conveyor mounted athwartship in the vessel that is capable of being extended or traversed laterally on the vessel in order to allow the discharge of the cargo from a fixed point on the vessel. This type of discharge reduces the flexibility of the discharge system further since it eliminates the slewing capability offered by even the short boom system.

The self-unloading vessel is a more expensive ship than the standard bulk carrier. The advantages of the concept can only be realized if the additional cost of the vessel is compared with the savings in capital and operating costs at all the various discharge terminals. Such comparisons have frequently demonstrated, when considered in conjunction with the economics of specific trade routes and the additional flexibility offered, that the total transportation cost can be reduced.

References

CDE (1968), "MRI/STARDYNE, Structural Analysis System, User's Manual," Data Centers Division of CDE, February.

Cragin, J. Q. (1970), "Ship Response Instrumentation Aboard the Container Vessel *S.S. Boston,*" Teledyne Materials Research Co., Technical Report No. E-1214(c).

Cushing, Charles R. (1963), "Shipboard Gantry Cranes," ASME, New York, October.

Cushing, Charles R. (1969), "Protection and Securing of Deck-Stowed Containers," Marine Technology, July.

Dillon, E. S., F. G. Ebel and A. R. Goobeck (1961), "Ship Design for Improved Cargo Handling," SNAME, Chesapeake Section, October.

Durst, Walter (1974), "The Continuous Discharge of Bulk Cargos," *Shipping World and Shipbuilder,* April.

Ebel, F. G. (1953), "Notes on Cargo Handling," SNAME, Philadelphia Section, January.

Ebel, F. G. (1958), "An Analysis of Shipboard Cargo Cranes," *Trans.* SNAME.

Erichsen, Stian (1970), "Simulation of Receiving, Storing and Loading General Cargo," The University of Michigan, Department of Naval Architecture and Marine Engineering, May.

J. J. Henry, Co. Inc. (1970), "Guidelines for Deck Stowage of Containers," prepared for U. S. Department of Commerce, Maritime Administration, Contract MA-4832, J. J. Henry, Co. Inc., July.

Hopper, A. G., P. H. Judd, and G. Williams (1964), "Cargo Handling and Its Effect on Dry Cargo Ship Design," INA.

Hydronautics, Inc. (1975), "Automated Sequencing of Containers for Containers Ship Loading and Unloading."

Johnston, W. H., Monroe Levy (1973), "Automation of Cargo Booms," SNAME, Gulf Section, February.

JPL (1967), "User's Manual for SAMIS (Structural Analysis and Matrix Interpretive Systems)," Technical Memorandum TM-33-305, Jet Propulsion Laboratory.

Kimball, P. B., D. J. Wudtke (1975), "A Shipboard Gantry Crane—Design and Performance," SNAME, New York Section, April.

Lewis, E. V. (1974), "Long Term Applications of St. Denis/Pierson Technique to Ship Design," Seakeeping 1953—1973, SNAME.

Liu, King-Tao, and R. Keith Michell (1977), "Design of On-Deck Container Securing Systems," *Marine Technology,* April.

MacNaught, D. F. (1955), "Cargo Handling Arrangements," *Design and Construction of Steel Merchant Ships,* edited by D. Arnott, SNAME.

Middleton, E. H., and S. F. Samis (1970), "Rules and Regulations for Bulk Grain Cargos: Guidance for Preparing Documents for Approval under 1969 Equivalent to Subchapter M," *Marine Technology,* January.

MIT, "STRESS: 1964 A User's Manual," The Department of Civil Engineering, Massachusetts Institute of Technology, The MIT Press.

MIT (1967), "ICES-STRUDL-1: The Structural Design Language," Engineering User's Manual, The Department of Civil Engineering, Massachusetts Institute of Technology.

NAS-NRC (1955), "The *S.S. Warrior,*" An Analysis of an Export Transportation System from Shipper to Consignee, Publication 339.

NAS-NRC (1958), "Recent Research in Maritime Transportation," Publication 592.

NAS-NRC (1959), "Maritime Transportation of Unitized Cargo," A Comparative Economic Analysis of Break-Bulk and Unit Load Systems, Publication 745.

NAS-NRC (1964), "San Francisco Port Study," Publication 1140.

Price, R. I., and E. R. Middleton (1969), "Stability and the Stowage of Bulk Grain in Ships—Activities of the Intergovernmental Maritime Consultative Organization," *Marine Technology,* October.

Reuter, Wolfgang (1968), "A Method for the Analysis and Design of Mast-Boom Type Cargo Handling Gear," SNAME, Los Angeles Metropolitan Section, September.

Russo, V. L., and E. K. Sullivan (1953), "Design of the Mariner-Type Ship," *Trans.* SNAME.

SW&S (1966), "Container Handling System Developments," *Shipping World and Shipbuilder,* October.

SW&S (1969), "Acadia Forest—First Ship to Operate on Lash Inland Water and Ocean Transport System," *Shipping World and Shipbuilder,* December.

SW&S (1972), "Doctor Lykes—First of the Seabee Type Barge Carrying Vessels," *Shipping World and Shipbuilder,* September.

UCLA (1954), "An Engineering Analysis of Cargo Handling—II, The Field Study of 1953," Report 55-2, Department of Engineering, University of California at Los Angeles, November.

UCLA (1956), "An Engineering Analysis of Cargo Handling—V, Simulation of Cargo Handling Systems," Report 56-37, Department of Engineering, University of California at Los Angeles, September.

Vaughan, W. R. (1977), "Update on Self-Unloading Dry Bulk Vessels," IME/SNAME, Montreal Section, January.

Vosslers, G. Swaan, and H. Rijken (1960), "Experiments with Series 60 Models in Waves," *Trans.* SNAME.

Young, J. C. (1967), "Development of Self-Unloading Bulk Cargo Ships," IME/SNAME, Montreal Section, March.

CHAPTER XI

Robert I. Price

Design for Transport of Liquid and Hazardous Cargos

Section 1
Introduction

1.1 Historical Background. We can be confident that among the first problems seafaring men had to solve before venturing any distance from shore was the transportation of liquid cargo, that is, a supply of fresh water. Across the centuries the retention of an adequate amount of drinkable fresh water at sea has held prime importance. As water craft progressed from canoe and coracle to merchant vessels to men-of-war, the variety of liquids carried increased. Bladders, gourds, jugs, amphorae, casks, and tuns were among the many forms of containers utilized. The ancient craft of cooper had to do with preparation of these containers, and one of the concerns of a ship's carpenter was the staunchness of barrels and kegs aboard.

This transport of liquids in individual, portable quantities and containers continued unchanged until the late 19th century. Several ships were designed after 1863 endeavoring to improve liquid transport, by use of the hull as container. The first oil tanker, *Gluckauf,* in 1886 with her engine room aft was, in many respects, save for her suit of sails, similar to the tankers of today. She was dedicated exclusively to the carriage of oil, some 2,300 tons, and reportedly served successfully until she ran aground in 1893 on Fire Island. In the course of nearly a century since, oil tankers have grown to vessels carrying over 200 times that quantity, as described and illustrated in Chapter II.

1.2 Hazards Inherent in Liquid Cargos. In the late 1950's the successful experience in the transport of petroleum products in tankers together with the growth of chemical products, many derived from petroleum, brought in the movement of bulk cargos of considerable variety, including the initial experimentation with the transport of cryogenic products such as liquefied natural gas (LNG) and liquefied petroleum gas (LPG). Much of this early activity was conducted in barges plying the Mississippi and Ohio River systems.

In 1961, a hopper barge of minimum reserve buoyancy carrying about 1,000 tons of chlorine in four independent tanks, sank in the vicinity of Natchez, Mississippi, and was lost in the silt of the river bottom. Recovery of the barge and its cargo to eliminate a potential substantial hazard to the general populace cost $4 million. From this incident it was clear that a new system of regulation was necessary.

For future shipments of substances as hazardous as chlorine, three classes of barge construction were set out in U. S. Coast Guard (USCG) Regulations. Emphasis was placed on ability of the vessel to survive damage it might suffer in normal service and still retain its cargo after damage. Existing craft were required either to be modified or taken out of the hazardous cargo trade. At the same time, the National Academy of Sciences was requested to evolve a classification system that would identify and rank the hazards in the properties of new chemicals for water transport.

Coincident with the development of these construction controls for barges, there began a significant sea trade from American petro-chemical plants to Europe and Japan. There was concern that the ships utilized for this service, all of foreign registry, would not have the requisite safeguards to assure containment of the cargo after damage. In 1964, a program was begun requiring that plans of the cargo containment area, related piping systems, and electric systems of chemical ships which operate from United States ports be submitted for technical review by the USCG. By 1967 the workload imposed by world-wide construction under the conditions of this Letter of Compliance Program was overcoming the capability of the USCG technical staff. The solution seen was an international system of regulations under which the Government of registry would enforce requirements for construction. The United States requested the Intergovernmental Maritime Consultative Organization (IMCO) to create a subcommittee for this purpose.

The IMCO Subcommittee on Ship Design and Equipment was authorized and initially assigned the program of developing a system of regulations appropriate to the transport of bulk chemicals, including shipment in large portable containers and tanks. Incorporating United States contributions relative to the classification of hazardous substances and the development of the related construction program, IMCO eventually adopted a Code for the construction of bulk chemical carriers and another Code for those transporting liquefied gases. These Codes are based on the principle of countering increased risk with increased defense by relating particular cargos to certain classes and levels of vessel construction and containment.

With respect to oil cargos, in March 1967 the Liberian tanker *Torrey Canyon* ran aground on Seven Stones Reef off the south coast of England. Had the ship and all hands been lost without trace, the world today would not know the ship's name. But between the grounding and the salvage efforts, a good deal of the nearly 120,000-ton cargo of crude oil was released in an area dependent on fishing and tourism. This incident focused attention on the subject of pollution of the sea leading to a number of new international treaties and the revision of existing agreements. The "International Convention for the Prevention of Pollution from Ships, 1973" incorporated many requirements affecting ship construction, especially of tank ships. The USCG has translated these provisions into detailed domestic regulations in implementing the Ports and Waterways Safety Act (P.L. 92-340,86 STAT 424).

The Winter of 1976–7 was remarkable for a great many accidents off the coast of the United States involving oil tankers. In a burst of national outrage, new restrictive regulations were proposed.

In February 1978 at another international meeting under the auspices of IMCO a number of new provisions were adopted as protocols to the 1973 Convention and the 1974 Safety of Life at Sea (SOLAS) Convention.

It is extremely difficult to provide a full description of regulations applied to construction and operation of tank ships especially those carrying oil. Many of these requirements may be retroactive, requiring refit of existing ships; others apply to new construction under different time scales for implementation.

A thorough job of research is recommended before undertaking design work of this nature.

Section 2
Cargo—Variety and Characteristics

2.1 Scope. Cargos capable of transport in a liquid or semi-liquid state are now of an incredible variety with widely differing properties. Table 1 is a listing of some typical bulk cargos now being carried. This section will describe a number of cargos in limited detail to illustrate the necessity of thorough investigation of the intended use of the vessel before undertaking the design process. The physical properties of the cargo and the regulatory limitations imposed because of safety or environmental aspects often dominate the design and determine the method of cargo handling used. If the owner is to have any latitude in the use of the ship, detailed knowledge of the cargos to be carried is absolutely essential.

2.2 Liquids. Many of the cargos carried in tank-type vessels are not strictly liquid by definition. Among those that can, however, be categorized as liquids are oil, petroleum products and a good many chemicals; also liquefied gases and cryogenics, wine, juices, and vegetable and animal oils.

a. Oil. Crude oil is the most common liquid cargo, the commodity upon which the economies of both technically advanced and developing countries depend. The world's oil reserves are located in remote, undeveloped areas, such as Saudi Arabia or Alaska or beneath the sea. To find, extract, and transport crude oil to refineries nearly always involves sea transport.

Physical and chemical properties of crude oil vary considerably depending upon its origin. Crudes are composed of a number of hydrocarbon products each with a different flash point and vapor pressure. The result is a mixture capable of being flammable at moderate to low temperatures. The degree of flammability is usually measured in terms of flash point and Reid vapor pressure, as defined in Section 2.6b. Crudes generally have a flash point below 26.7°C (80°F) and a Reid vapor pressure from about 42 to 84 kPa (6 to 12 psi), although there are crudes which have higher and lower flammability. Crudes may therefore be classed according to the criteria of the USCG "Rules and Regulations for Tank Vessels" (CG-123) as either Flammable Liquids (Grade A, B, or C) or as Combustible Liquids (Grade D or E).

Crude oil is an impure substance. Besides differing as to hydrocarbon composition, crude oil also varies in the quality of the contaminants. The contaminants encourage corrosion of a steel container and create amounts of residue which must be removed periodically.

Refineries are designed to deal with certain crudes and are unable to handle others. For example, in addition to the difference in volatility, *sour* or *sweet* crude oils are so named for the presence or absence of sulphur. Hence the cargo destination will often be determined by the requisite refinery capacity, or by the manipulating and exchanging of oil quantities within the oil industry intended to bring together oil and refining capability.

Crude oils include paraffins and asphaltic elements which cling significantly to the interior surfaces of the tanks after the ship is emptied. Webs, longitudinals, and stiffeners within the tank further accentuate this tendency. Crudes also vary in the temperature at which they are loaded, but normally loading at ambient temperature can be expected.

The standard temperature for the determination of specific gravity is about 15.5°C/(60°F). A correction must therefore be made for departures from that value because a small difference can represent a variation of several tons in the loading of the ship. A system of measure for the ratio of weight to volume is the API scale, which derives from an arbitrary formula:

Table 1—Bulk Liquid Cargos

Cargo	Specific gravity at 15.5°C (60°F)	Specific heat	Approximate melting point deg C (deg F)	Flash point deg C (deg F)	Recommended pumping temperature deg C (deg F)
Vegetable Fats					
coconut oil	0.926	0.56	18 (64.4)	215.5 (419.9)	48 (118.4)
palm kernel oil	0.926		23.5 (74.3)	203.5 (393.3)	27 (80.6)
Crude Semidrying Vegetable Oils					
corn oil (maize)	0.923		−12 (10.4)	255 (491.0)	27 (80.6)
cotton seed oil	0.919		0 (32)	252 (485.6)	35 (95)
Nondrying Vegetable Oils					
peanut oil	0.921		−1 (30.2)	282 (539.6)	24 (75.2)
castor oil	0.964		−14 (6.8)	229.5 (445.1)	24 (75.2)
Drying Vegetable Oils					
linseed oil	0.936		−11.5 (11.3)	207 (404.6)	27 (80.6)
Animal Fat—Tallow	0.915	0.53	31.5 (88.7)	265 (509)	52 (125.6)
Glycerine	1.2647	0.58	20 (68)	160 (320)	32 (89.6)
Cashew Shell Liquid	0.959				29 (84.2)
Latex	0.944				ambient
Creosote	1.0821			74 (165.2)	52 (125.6)
Bitumens	1.05				42 (107.6)
Lube Oils	0.895	0.40	−7 (19.4)	313 (595.4)	
Fuel Oils (Diesel)	0.8708	0.56		65.5 (149.9) min	
Grade E Liquids, USCG				65.5 (149.9) min	
Trichlorethylene	1.46		50.5 (122.9)	none	
Molasses	1.44	0.50			49 (120.2)
Alkane				110 (230)	
Aqueous Ammonia	0.90				
Ethylene Glycol	1.019–1.073	0.54			
Tetrachlorethylene	1.63	0.22			
Methylchlorethylene	1.59	0.198			
Polyxyethylene	1.125–1.204				
Creosol (cresylic acid)	1.034–1.054	0.479–0.499			

$$\mathrm{API}_{\mathrm{deg}} = \frac{141.5}{\mathrm{sg\ at\ 15.5°C}} - 131.5$$

The relationship between API degrees and specific gravity, weight and volume appears in Table 2.

Table 3 illustrates the range of properties in crudes from different sources.

b. Petroleum Products. Petroleum oils are characterized as either *black*, or *white* (clean). Ships carrying these substances are frequently so designated. Black oils include crude oil, (previously discussed), furnace oil, and fuel oil. Tar and asphalt also come under this heading having certain of the difficulties of handling associated with crude oil but without the high vapor pressure. White oil includes benzene, kerosene, and gasoline. The white oils are generally more volatile than the black, except for crude oil from whence all products come and which contains all the volatile elements before processing.

The refined products have a wide spread of physical properties, Table 4, ranging as they do from gasolines through kerosene, and jet fuel to lubricating oil. Some commercial grades of gasoline even have properties that are varied with the time of year and section of the country where the product is to be used, sometimes by adjusting the vapor pressure.

Although the vapor pressure of some refined products is too low to form vapor—air mixtures at ordinary room temperature, foam mist may form during loading, thus producing an explosive mixture.

A cargo of special concern is jet fuel which has a wide boiling range. For example, the flash point of JP-4 may be lower than 17.8°C (64°F) as compared to commercial aviation kerosene at 49°C (120°F) or higher.

c. Chemicals. Liquid chemicals being transported by sea may be elements, mixtures, or compounds. These cargos may be the final form or may be feedstock for an essential ingredient of a process industry. (A ship may deliver substances which are in a raw material state and, tramp-like,

Table 2—Gravity, Weight and Volume Conversions for Petroleum Products

Gravity, degrees API	Specific gravity	kg/bbl	lb/gal	lb/bbl	bbl/ton
0	1.0760	170.65	8.962	376.40	5.86
10	1.0000	158.73	8.328	349.78	6.30
15	0.9659	153.37	8.044	337.85	6.52
20	0.9340	148.15	7.778	326.68	6.75
30	0.8762	139.08	7.296	306.43	7.19
40	0.8251	130.89	6.870	288.54	7.64
50	0.7796	123.6	6.491	272.62	8.09
60	0.7389	117.23	6.151	258.34	8.53
70	0.7022	111.36	5.845	245.49	8.98
80	0.6690	106.04	5.569	233.90	9.43
85	0.6536	103.63	5.440	228.48	9.65
90	0.6388	101.32	5.317	223.31	9.87
100	0.6112	96.90	5.087	213.65	10.32

Table 3—Variation Among Crude Oils by Origin

Loading Terminal	Crude Name	Gravity API	Sp. Gr. at 60°F	ft³/ton at 60°F	bbl/ton at 60°F	Viscosity (centistokes) 70°F	Viscosity (centistokes) 100°F	Pour point °F	Wax % wt	H_2S (as loaded) ppm	RVP (approx) lb	BP Cargo Heating Code No Heat	BP Cargo Heating Code Heat
WEST AFRICA													
Forcados (Nigeria)	Forcados Blend	30.5	0.8735	41.17	7.332	8.5	5.4	5	7	nil	4–7	N	—
Brass River	Brass River	43.1	8.8105	44.38	7.904	3.3	—	−25	10.5	nil	7–10	N	—
Dyero (Congo)	Emeraude Marine	23.7	0.912	39.43	7.023	203	—	45	—	30	<4	N	—
VENEZUELA													
Puerto Miranda Bajo Grande	Boscan	10.1	0.9995	35.97	6.407	—	19.400	60	2.3	nil	<4	—	H
Puerto Miranda & Bach-aquero	Lagotreco	27.3	0.891	40.36	7.188	34.8	17.5	−55	5.9	nil	<4	N	—
Bachaquero & Cardon	Lagocinco	35.6	0.847	42.46	7.562	8.8	5.2	5	—	nil	4–7	N	—
PERSIAN GULF													
Kharg Island	Iranian Light	33.6	0.857	41.96	7.474	11	6.1	0	7.5	70. max	7–10	N	—
Bahrgan	Iranian Nowruz	18.3	0.9445	38.07	6.780	917	270	−15	3.5	nil	<4	—	H
Umm Said	Qatar	41.4	0.8185	43.94	7.826	3.7	2.6	5	6.6	200–400	7–10	N	—
INDONESIA													
Dumai (Suma-tra)	Mingi	35.3	0.8485	42.38	7.549	—	17.6	95	ca 30	nil	<4	—	H
Lotong (Brunei)	Seria Light	38.8	0.8305	43.30	7.712	—	2.13	60	7.5	nil	4–7	N	—
Dumai (Suma-tra)	Duria	22.8	0.917	39.21	6.984	ca	98.4	35.55	17.5	nil	4	—	H

depart with a cargo of a finished or a more advanced stage of composition.) Many of these substances are derived from petroleum (Fig. 1) others from coal (Fig. 2). Others are caustics, acids, or products of polymerization. The uses to which these substances may be put include fertilizers, plastics, fuels, solvents, disinfectants, detergents, pharmaceuticals, and insecticides.

Table 4—Representative Weights of Petroleum Products

Product	kg/bbl	lb/gal	lb/bbl	bbl/ton
	Basis: 42 gal/bbl			
Aviation gasoline	112.36	5.90	248	8.90
Motor gasoline	117.65	6.17	259	8.50
Natural gasoline	117.65	6.17	259	8.50
Kerosine	129.03	6.76	284	7.75
Gas oil, diesel oil, distillate fuel oils	134.05	7.05	296	7.46
Residual fuel oils	150.15	7.88	332	6.66
Lubricating oils	142.86	7.50	315	7.00
Unspecified oils	142.86	7.50	315	7.00
Benzol	140.06	7.36	309	7.14
Mineral spirits	116.01	6.10	256	8.62
Liquefied petroleum gas	86.21	4.52	190	11.60
Paraffin wax	127.06	6.67	280	7.87
Petrolatum	127.06	6.67	280	7.87
Grease	158.73	8.33	350	6.30
Asphalt and road oil	165.02	8.67	364	6.06
Petroleum coke	181.82	9.55	401	5.50

Approximate or representative figures used by the Bureau of Mines in converting international trade data; should be used only for rough estimating. When API or specific gravity is known, table entitled "Gravity, Weight and Volume Conversions for Petroleum Products" should be used.

Not all chemicals are hazardous. Many of them can be transported safely with very little special attention. However, it is those substances which present a wider range of hazard than simply combustibility or flammability for which there are special requirements when transported in bulk.

Table 5 shows the relation between various chemicals.

d. Liquefied Gases and Cryogenics. A liquefied gas is arbitrarily defined as a product having a vapor pressure exceeding 275 kPa (40 psi) absolute at a temperature of 37.8°C (100°F) corresponding to a Reid vapor pressure of 275 kPa (40 psia). Among this class of product regulated by the IMCO "Code for the Construction and Equipment of Ships Carrying Liquefied Gases in Bulk" are liquefied petroleum gas (LPG), liquefied natural gas (LNG), and anhydrous ammonia. LPG, primarily hydrocarbons such as propane, butane, propylene, butylene or mixtures of these, is either transported at ambient temperature in pressure tanks independent of the hull or at reduced temperature approximately equivalent to the boiling point of the cargo at atmospheric pressure with the pressure approximately equivalent to atmospheric. Various combinations of higher pressure and temperature may also be employed.

The most extreme cryogenic cargos that have been shipped are liquid oxygen and liquid hydrogen. The liquid oxygen is transported at a temperature of −183°C (−297°F), liquid hydrogen at −253°C (−423°F). One liter of liquid oxygen weighs 1.15 kg (2.53 lb); one liter of liquid hydrogen weighs 0.072 kg (0.159 lb). These substances were moved in specially designed barges in support of the U. S. space program.

Another liquefied gas used on board ship from time to time is liquid nitrogen. It is not a cargo itself but a supportive element used to inert the atmosphere in void spaces surrounding the cargo tanks or as *padding* in a tank to prevent a cargo from coming in contact with air. A further use of liquid nitrogen is for fast freezing of foodstuffs stored in containers for shipment.

e. Wine, Juices, Vegetable and Animal Oils. Ships also transport in bulk such substances as wine, whisky, olive oil, and coconut oil. Some fats and oils are used in the manufacture of margarine and soap. The principal problem associated with the transport of these substances is the preservation of their quality and purity. When intended to be used as foodstuffs, there are special requirements imposed in Title 21, Code of Federal Regulations, by the Food and Drug Administration regarding the coatings which may be used on the interior of the container.

2.3 Bulk Semi-Liquids and Granular Substances. A number of substances exhibit a semi-liquid behavior although they are actually solid matter. A significant characteristic of these materials is the *angle of repose.* This is the angle between the horizontal and the side of the cone

Table 5—Principal Uses of Selected Chemicals

Commodity	1	2	3	4	5	6	7	8	9	10	Remarks
Asphalt and Bitumen	Fung	P	F	EB		R					
Acetone Cyanohydrin	I									C	
Acrylonitrile	Fum					RPF		P T		C	Acrilan Dynel and Orlon
Ammonia (Aqueous)			LE		Phot Phar	RPF	SD	T	I	FC	Nitric acid Rayon ceramics
Benzene	I Fum Fung	SRP VL					DP		D	C	Nylon intermediates
Carbon Disulphide	I Fum	SVL			Phar	R		T			Matches, Viscose Rayon Cellophane
Caustic soda					Phot	R	SD	TP		C	Rayon
Chlorosulphonic acid			P							C	Saccharin
Cresols	D Fum	RS APV	LE		Phot Phar	P			ID	C	Flotation agents
Formaldehyde	DI Fung	RV A	E		Phar	RP		TP	D	C	Reducing agent
Furfural	HI Fung	SA R	L	B	F	P			D	C	Refining metals
Hydrochloric acid				E	F					C	Ore reduction
TEL and TML		F									Anti-knock additive
Nitric acid	HI Fung		E	E	Phar				D	F	Ore Float'n
Oleum			E						D	C	
Phenol	H Fung	RSP			Phar	RP	D				Nylon (Tyres)
Phosphoric acid				E	F Phar	P	P	T		CF	Animal feeds
Phosphorus			FE	E			SD			CF	Animal feeds
Styrene monomer		R		E		P				C	Polystyrene
Sulphur	Fung		E		Phar	R		P	D	C	Sulphuric acid
Sulphuric acid		P	E	E	Phar	P	D	T	D	*FC	Rayon
Coal Tar Naptha		SRA PL	EF	E		R		TP	I	C	Alkyd Resins Floor coverings
Napthalene	Fung I	RS	EL	E	Phar	P	DPS	T	D	C	Solvents Tetralin and Decalin
Methanol	I Fung	S	F		Phot Phar	R		T	D	C	Anti-Freeze

*Calcium superphosphate and Ammonium sulphate

[1] Fungicides (Fung) Insecticides (I) Herbicides (H) Fumigants (Fum) Disinfectants (D)
[2] Paints (P) Solvents (S) Resins (R) Varnishes (V) Adhesives (A) Lacquers (L)
[3] Fuels (F) Lubricants (L) Explosives (E) Poison gas (P)
[4] Engineering (E) Building (B)
[5] Photographic (Phot) Pharmaceutical (Phar) Food (F)
[6] Rubber (R) Plastics (P) Fibres (F)
[7] Soaps (S) Detergents (D) Polishes (P)
[8] Textile industries (T) Paper industries (P)
[9] Dyestuffs (D) Inks (I)
[10] Fertilizers (F) Chemical and petroleum industries (C)

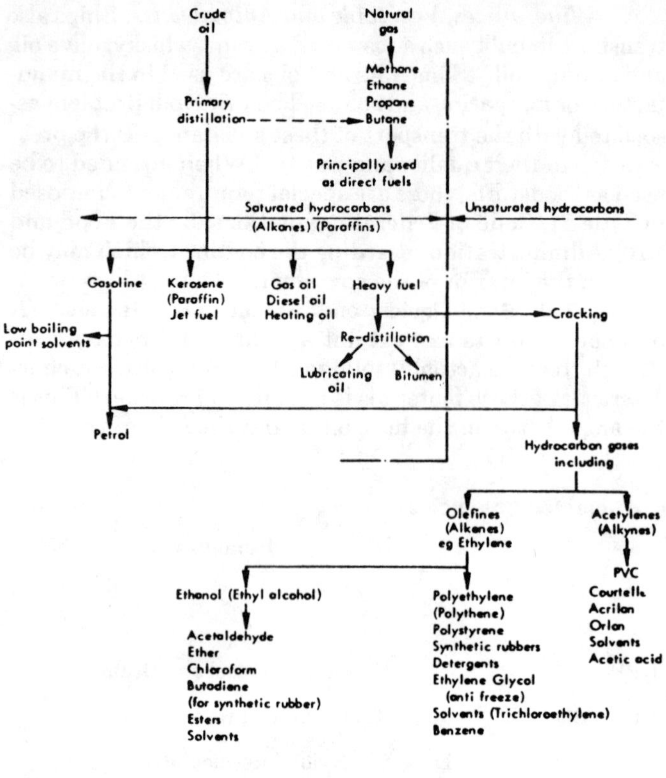

Fig. 1 Petroleum derivatives

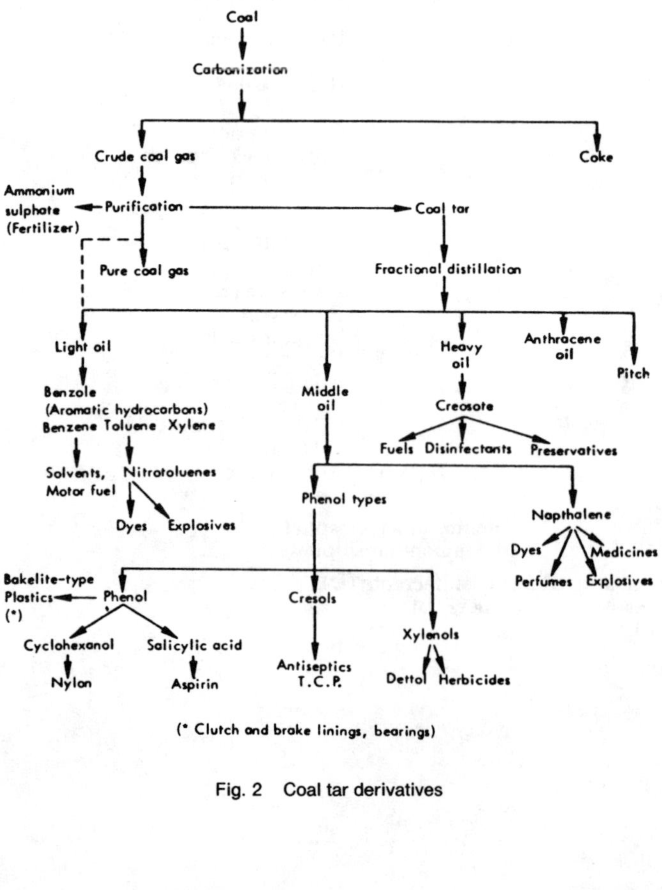

Fig. 2 Coal tar derivatives

formed when the material is poured onto a flat surface. A low angle of repose indicates a bulk cargo susceptible of transverse movement aboard a ship in a seaway owing to shearing of the surface layers, somewhat like a landslide.

Although handling systems for loading and off-loading these bulk cargos are described in more detail in Chapter X, their behavioral characteristics in terms of liquidity and potential hazardous properties are covered herein.

a. Grain. Grain is surely the most basic of commodities. It must have been a cargo very early in marine trade. (Grain vessels are mentioned in the commentaries of Julius Caesar). Grain can be shipped in bags, but it is far easier, faster, and cheaper to load and unload it in bulk. However, bulk or loose grain has long been recognized as a problem cargo requiring special precautions to prevent shifting. It must also be kept dry as it will swell and expand when wet, developing considerable pressure on the boundaries containing it.

Loss or hazarding of several vessels transporting grain in bulk subsequent to the 1960 Safety of Life at Sea Convention led to a research program in depth. This uncovered a fallacy in the International Grain Regulations which supposed that a sinkage or settling of the grain occurred and which had to be compensated by the installation of a feeder constructed in the hatch trunk.

The regulations presumed the holds to be trimmed full, a condition desired but found not to be attained in practice. Consequently, substantial void spaces remained under the horizontal surfaces clear of the hatchways with the result that the feeders did not perform as anticipated and, in fact, aggravated the condition once the ship acquired a list by feeding grain to the low side.

The International Grain Regulations have since been revised. It is now necessary to carry out a stability calculation assuming a certain void beneath the deck round about the hatch in a conventional cargo ship to ascertain that there is sufficient residual dynamic stability in the event of an adverse shift occurring.

Specific details as to the requisite strength of grain fittings required to diminish the potential angle of heel due to shifting, along with particulars regarding cleanliness, infestation, structural integrity, sheathing of hot bulkheads, loading in deep tanks and wing tanks, trimming of holds and hatches, securing of 'tween decks and hatch covers are published by the National Cargo Bureau (1978).

Table 6 shows the variation among grains.

b. Ore and Ore Concentrates. There is an IMCO "Code of Safe Practice For Bulk Cargos" for storage of these cargos which are capable of presenting three forms of hazard:

1. Structural damage because of either improper weight distribution between holds or excessive weight on decks or inner bottom,

2. Adverse stability at sea, which is either excessive stability with the potential for structural damage and subsequent shifting or insufficient stability because of the transverse shifting of the cargo surface layers with certain cargos considered *dry*, or the transverse shifting of the entire cargo in the case of those cargos considered *wet*.

3. Under certain conditions, some of these cargos are also given to spontaneous heating.

Table 6—Data on Principal Grains Loaded at United States Ports

	kg/liter	lb/ bushel	angle of repose, deg
Barley	0.645	50	46
Corn	0.710	55	21
Linseed	0.645	50	21
Oats	0.516	40	21
Rice	0.773	60	20
Rye	0.750	58	32
Safflower Seed	0.530	41	28
Sorghum (Milo)	0.735	57	31
Soybeans	0.722	56	22
Wheat	0.800	62	23

The test weight of a particular grain is the actual weight in lb of a U. S. (Winchester) Bushel which is a unit of volume (dry measure) equalling 35.238 L, 2,150.42 in.^{3} or 1.2445 ft^3. The test weights are average figures based on information obtained from grain loading ports. Particular grain cargos may vary from the figures shown.

Substances considered *dry* are gaged for their tendency to shift by their internal frictional properties as indicated by the natural angle of repose. An angle of repose exceeding 35 degrees does not generally require any special measures except for appropriate trimming athwartships. Where, however, the angle of repose is 35 degrees or less, there is a potential for shifting of the surface layers and it may be necessary to fit shifting boards of sufficient strength to restrain the cargo having regard for its density.

Cargos considered *wet* are those which retain moisture in free suspension between the individual particles. If this moisture level is exceeded, the cargo may flow or shift with the vibration or motion of the ship. Concentrates of iron ore, lead ore, zinc ore, and copper are among the substances which exhibit this tendency.

It is necessary to determine the actual moisture content of the material to be transported and to compare it with the safe transportable moisture limit, which is 90 percent of the flow point as determined by lab vibrator tests.

When it is intended to transport a cargo having a propensity for shifting, either because the angle of repose is less than 35 degrees or because of an excessive moisture content, transportation may still be authorized if the intact stability of the ship is found adequate after allowing for a cargo shift occurring, provided the angle of heel is not excessive and the residual dynamic stability is still sufficient. The significant characteristics of some ores and concentrates appear in Table 7.

c. Fish and Crab. Whole fish when carried in bulk in a vessel's hold behave like liquid. They either have no angle of repose or a very small one. When first introduced into the hold still wet from the sea, they definitely act like a liquid. As they dry off, depending on the type of fish, the depth of the hold, the temperature of the water and the hold, they do develop a small resistance to sliding and the cargo will have a small angle of repose. The longer they are in the hold, however, and as decomposition starts to set in, oil exudes through the skin and they eventually return to a zero angle of repose. For this reason designers of ships carrying bulk quantities of such fish as anchovies, herring, hake, salmon, and trash fish commonly treat holds partially filled with such cargo as if it were liquid. The Master must be instructed in the proper loading procedures so that the ship does not suffer from excessively reduced stability due to free surface not only in the ship's tanks but in the cargo holds.

Some ships carry fish under refrigeration in chilled seawater or brine. Such cargos are part fish and part liquid. When the holds are in a partially filled condition they must be treated as carrying a liquid cargo since the fish are below the surface of the liquid and full free surface effects prevail. However, unlike conventional liquid tankers which very seldom empty or fill a tank during the progress of a voyage, fishing vessels must fill their tanks as the fish come aboard. Fishing vessel designers must take these conditions into account and so arrange the tanks that loss of stability due to free surface is minimized.

Crab fishermen conventionally put the catch immediately into tanks to keep the crab alive until delivered to the pro-

Table 7—Ore Concentrates and Similar Materials

Concentrate Type	Origin Country (District or Mine) Port(s) of Loading	Angle of Repose Range, deg	Transportable moisture limit, percent	Average Moisture Content at origin percent	Stowage Factor m^3/ton	Stowage Factor ft^3/ton	Summary of particle size distribution
Apatite (28 percent phosphorus)	U.S.S.R. (Khibiny) Murmansk	37	—	0.5 to 1.5	0.20	7	60:0.088 mm (0.0035 in); 40 percent < 0.088 mm (0.0035 in).
Blende	Algeria (—) Nemours	40 to 45	11.4	7.5 to 10	0.46	17	75 percent < 100 microns.
Copper Ore (25 percent copper)	Norway (Folldal VerkA/S Tverrfjellet) Trondheim	34	10.5	5 to 9	0.40	14	Fine granulated; 97 percent < 20 mesh dark yellow/ green
Copper Ore	Peru (—) Callao	60	9.5	8	0.44	16	Fine powder; liable to spontaneous combustion

cessing plant. Water in the tanks must be constantly replenished in order to keep the crab from dying. To prevent the crabs injuring themselves it is desirable to make the tanks as large as possible. To reduce free surface, the tanks are generally fitted with relatively small hatches with high coamings. Replenishment water is introduced at the bottom of the tank and overflows through pipes set in at the top of the hatch coamings. This insures that the water level remains up in the hatches and free surface is negligible. Care must be taken to investigate the vessel's stability during loading and unloading the tanks when free surface of the whole tanks occurs. Piping must be so arranged that in the event of pump failure or electrical power failure the water level cannot drop below the hatch coamings.

d. Slurries. An industrial technique relatively recently applied to shipboard loading is the slurry system. Commodities such as iron ore, coal, bauxite, salt, and phosphate rock may be mixed with water to permit them to be transported by pipeline for loading or discharging a ship. The ability to use this method for a particular material depends upon the size-range of the particles, specific gravity, any chemical limitations associated with wetting it, environmental requirements, and the location of the loading and discharge ports. The technique requires the ship to be fitted to permit removal of the excess water within the holds after the slurry has been loaded. Water must continue to be removed throughout the voyage until the material reaches its natural moisture content. The abrasive nature of the slurry necessitates careful selection of the materials of the piping system including the filtration process.

The economics of this method may, for some materials, be enhanced by recirculating the drainage fluid back to the shore facility.

Discharging the cargo at destination may be conducted by grab bucket or by returning the cargo to the slurry state with special equipment available at the discharge port.

e. Bulk Sugar. This is an interesting cargo which ranges in stowage from 1.06 to 1.2 m^3/ton (37 to 42 ft^3/ton). Sugar readily absorbs moisture which may cause it to solidify in the surface layers with subsequent difficulty in discharging the cargo. This is especially true of raw, brown, crystallized sugar. White crystallized sugar is inclined to flow and must be trimmed to prevent shifting. Additionally, fermenting sugar can give off gases which are harmful to humans and only a moderate amount of salt water can cause fermentation.

2.4 Hazardous Liquids in Drums and Containers. Liquids are still carried aboard in packages ranging from portable

Table 8—Various Restrictions on the Shipment of Package Lots of Hazardous Materials

(1)	(2)	(3)	(4)	(5) Packaging		(6) Maximum net quantity in one package		(7) Water shipments		
				(a)	(b)	(a)	(b)	(a)	(b)	(c)
*/ W/ A	Hazardous materials descriptions and proper shipping names	Hazard class	Label(s) required (if not excepted)	Exceptions	Specific requirements	Passenger carrying aircraft or railcar	Cargo only aircraft	Cargo vessel	Passenger vessel	Other requirements
	Accumulator, pressurized (*pneumatic or hydraulic*), *containing nonflammable gas*	Nonflammable gas	Nonflammable gas	173.306		No limit	No limit	1.2	1.2	
	Acetal	Flammable liquid	Flammable liquid	173.118	173.119	1 quart	10 gallons	1.3	4	
	Acetaldehyde (*ethyl aldehyde*)	Flammable liquid	Flammable liquid	None	173.119	Forbidden	10 gallons	1.3	5	
A	Acetaldehyde ammonia	ORM-A	None	173.505	173.510	No limit	No limit			
*	Acetic acid (*aqueous solution*)	Corrosive material	Corrosive	173.244	173.245	1 quart	10 gallons	1.2	1.2	Stow separate from nitric acid or oxidizing materials.
	Acetic acid, glacial	Corrosive material	Corrosive	173.244	173.245	1 quart	10 gallons	1.2	1.2	Stow separate from nitric acid or oxidizing materials. Segregation same as for flammable liquids

Table 8—Various Restrictions on the Shipment of Package Lots of Hazardous Materials (Continued)

(1)	(2)	(3)	(4)	(5)		(6)		(7)		
				Packaging		Maximum net quantity in one package		Water shipments		
				(a)	(b)	(a)	(b)	(a)	(b)	(c)
*/ W/ A	Hazardous materials descriptions and proper shipping names	Hazard class	Label(s) required (if not excepted)	Exceptions	Specific requirements	Passenger carrying aircraft or railcar	Cargo only aircraft	Cargo vessel	Passenger vessel	Other requirements
	Acetic anhydride	Corrosive material	Corrosive	173.244	173.245	1 quart	1 gallon	1.2	1.2	
	Acetone	Flammable liquid	Flammable liquid*	173.118	173.119	1 quart	10 gallons	1.3	4	
	Acetone cyanohydrin	Poison B	Poison	None	173.346	Forbidden	55 gallons	1	5	Shade from radiant heat. Stow away from corrosive materials
	Acetone oil	Flammable liquid	Flammable liquid	173.118	173.119	1 quart	10 gallons	1.2	1	
	Acetonitrile	Flammable liquid	Flammable liquid	173.118	173.119	1 quart	10 gallons	1	4	Shade from radiant heat
	Acetyl benzoyl peroxide, solid	Forbidden								
	Acetyl benzoyl peroxide solution, not over *40% peroxide*	Organic peroxide	Organic peroxide	None	173.222	Forbidden	1 quart	1.2	1	
	Acetyl bromide	Corrosive material	Corrosive	173.244	173.247	1 quart	1 gallon	1	1	Keep dry. Glass carboys not permitted on passenger vessels
	Acetyl chloride	Flammable liquid	Flammable liquid	173.244	173.247	1 quart	1 gallon	1	1	Stow away from alcohols. Keep cool and dry. Separate longitudinally by an intervening complete compartment or hold from explosives
	Acetylene	Flammable gas	Flammable gas	None	173.303	Forbidden	300 pounds	1	1	Shade from radiant heat

In columns 7(a) and (b), water transport limitations are stated by number as follows:

(1) means the material may be stowed "on deck" subject to the requirements of § 176.63(b) of this subchapter. When both "on deck" and "under deck" are authorized, "under deck" should be used if it is available.

(2) means the material may be stowed "under deck" in a compartment or hold subject to the requirements of § 176.63(c). When both "on deck" and "under deck" are authorized, "under deck" should be used if it is available.

(3) means the material may be stowed "under deck away from heat" in a ventilated compartment or hold subject to the requirements of § 176.63(d) of this subchapter.

(4) means the material is authorized to be transported in only the limited quantities specified in the CFR section listed in Column (5) and is subject to the stowage requirements specified for a cargo vessel for the same material.

(5) means the material is forbidden and may not be offered or accepted for transportation.

(6) means the material is authorized to be transported in a magazine subject to the requirements of §§ 176.135 through 176.144 of this subchapter.

tanks mounted on freight container frames to through drums, carboys, cylinders, and bottles. These liquid parcels may be loaded individually, palletized, packed in boxes, stowed in freight containers or even in barges to be carried aboard specially designed vessels. When these cargos are hazardous materials; i.e., flammable, combustible, corrosive, toxic, etc., they must meet package specifications, label and placarding standards, stowage, cargo segregation, and manifesting requirements. These regulatory requirements appear in the IMCO Dangerous Goods Code and in Department of Transportation regulations in 49 CFR 171 through 179. Table 8, a sample from 49 CFR, shows the variety of stowage segregation and separation requirements to be considered. In some cases, stowage below decks is prohibited. In others, proximity to certain products is forbidden. These regulatory requirements are extensive, complex, and susceptible of change—usually of minor details.

While the emphasis is on the containment required of individual parcels, there are some basic factors which relate to ship design. These include fire and smoke sensing, fire extinguishing and water flushing and water removal capability (in case of spillage in the holds), ventilation of holds and capability to seal the hold, location of vent supply and exhaust fittings (so as not to unnecessarily restrict where certain cargos can be placed and to insure that flammable or toxic vapors are not discharged in the vicinity of living quarters), facilitation of dunnaging, securing and blocking out the stowage. When on-deck stowage is likely to be required, the vessel design should provide protection from boarding seas and the rapid freeing of same.

IMCO is currently studying the carriage of packaged hazardous materials aboard vessels looking towards the eventual amendment of SOLAS 1974 to include specific vessel construction requirements. Under consideration are increased fire-main system capabilities, specified minimum ventilation levels, and requirements for design and installation of fire detection systems.

The idea of an upper limit on the total quantity of individually packaged hazardous cargo that may be carried on board a vessel before reaching a risk comparable to bulk shipment has not been fully addressed. The USCG has imposed an upper limit of four tanks of 300 tons each for chlorine barges, and has prohibited any bulk shipment of certain carcinogens and nitroglycerine. In the future, with novel vessel configurations, increased quantity of hazardous cargo on board a single vessel, new appreciation of existing hazardous properties, and newly identified hazards (i.e., mutagenic, teratogenic, etc.), vessel design considerations may become extremely sensitive to the properties and quantity of cargos carried in package form.

Table 8 drawn from Title 49 CFR is indicative of various restrictions on the shipment of package lots of hazardous materials. Although certain of the restrictions do not apply to water shipments, they are shown to make the reader aware of the problems to be overcome in intermodal transport.

2.5 Properties in Bulk Transport. The three factors which have the strongest influence upon the way that the available volume of the vessel is utilized are the specific gravity of the cargo, the temperature under which it is to be transported, and the pressure under which it is retained en route. This Section will elaborate the impact of those three factors.

a. Specific Gravity. Of all the physical properties a liquid cargo may possess, its specific gravity has greatest influence on the shape of its container and on the design of the transporting ship or barge. The most extensive experience with bulk liquid cargos has been with those petroleum products having specific gravities a little less than that of water. With a cargo specific gravity near unity, there is little difficulty designing a vessel for efficient distribution and arrangement of the cargo volume and deadweight. Very simply, volumes not used for cargo need only provide the buoyancy to support, with an adequate margin of safety, the weight of the hull and outfit. In this respect, the modern crude oil tanker is similar in principle to a simple tank barge in which the rake compartments at either end provide the buoyancy.

However, the general range of specific gravity in liquid bulk cargos today runs from 0.4 to 2.0. (Slurries and concentrates are heavier, and the *exotics* such as liquid hydrogen are lighter.) At these extremes, there are certain complications in arranging the vessel.

With a very light product, a conventional ship hull form will have an excess of buoyancy, even when most efficient use is made of the space within the hull. To bring the conventional ship form down to its loadline marks under these circumstances, more volume must be gained outside the hull and tanks may be extended or added above the weather deck. Extending or adding tanks vertically raises the center of gravity and increases windage area, both factors affecting seaworthiness.

Ballasting may be more difficult to resolve for low specific gravity cargo carriers than for conventional ships. Even if ballast water is compatible with the cargo, tanks for such cargos are often independent of the hull and are not ordinarily designed for the weight of salt water. It is also difficult in such a ship, which needs the maximum cargo volume, to find space for separate ballast tanks.

Unmanned barges—having no propulsion, control, or accommodations spaces—have relatively more usable volume. Relatively large, long containers for light cargo can be fitted. However, in supporting long independent containers within an extremely flexible barge hull, serious attention needs to be given to the distribution of saddle loads into the hull and the effect of saddle reactions on the tanks.

With a heavy product, a large portion of the conventional hull is required to keep the vessel afloat. Cargo tanks must be relatively smaller. Free surface and surge effects should be minimized in such tanks since the effects of having partial loads are magnified in proportion to the increased cargo density.

However, partial loading is probably inevitable if there is to be flexibility in operation of the vessel. Handling a variety of cargos over a certain range in specific gravity can be accomplished safely if the initial design adequately considers the effects on stability and strength of the highest anticipated specific gravity. The actual tank structure can be designed for a full head of a somewhat lower nominal specific gravity; e.g., tanks could be designed for a full head of 1.3 specific gravity cargo which would load the vessel to its marks. This would establish the maximum load in tons to be carried in each tank. Then a partial load of 1.8 specific gravity could be carried to the same maximum load/tank limitation providing stability and dynamics had been checked in the initial design and assuming that both cargos were at the same pressure.

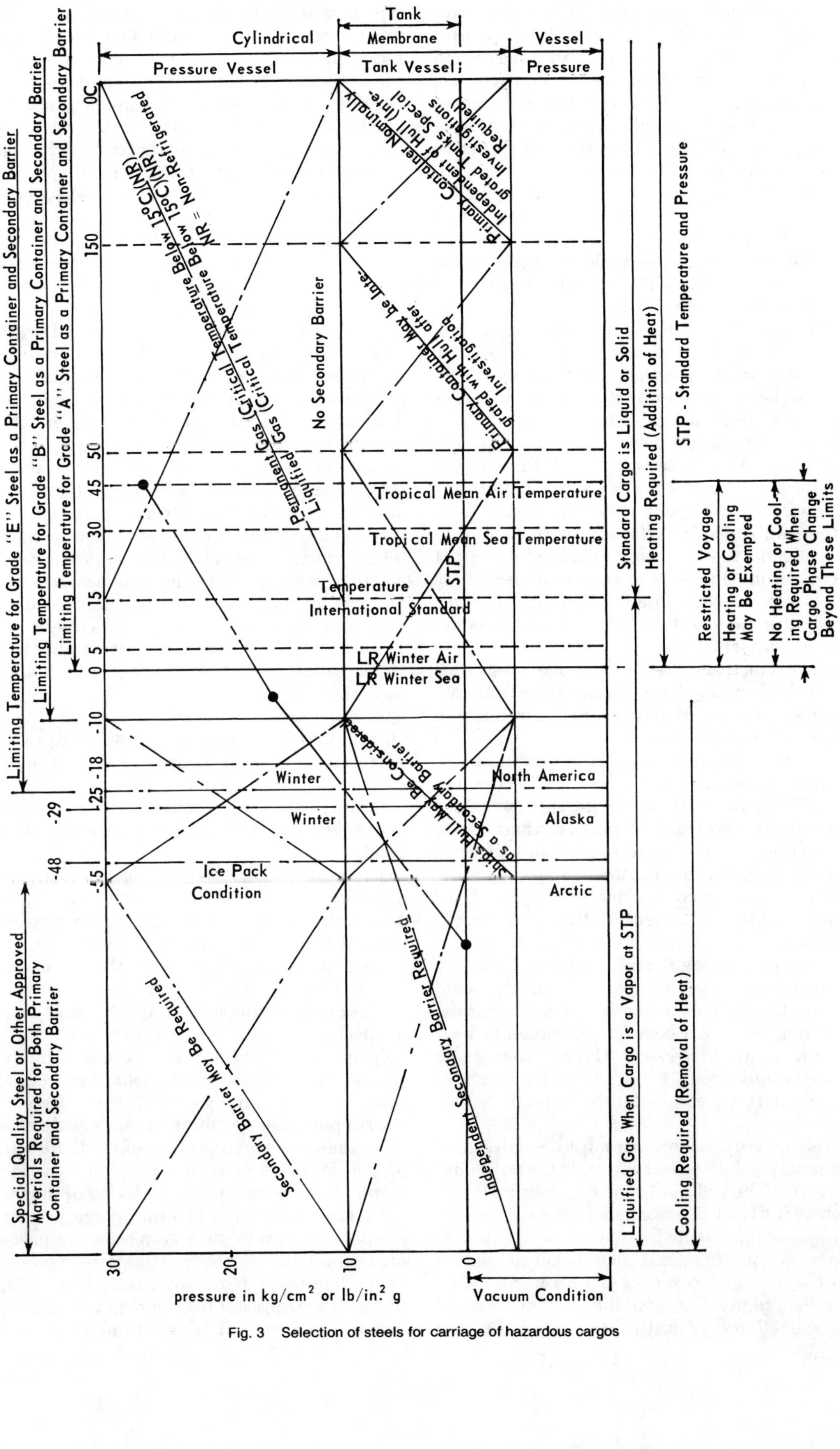

Fig. 3 Selection of steels for carriage of hazardous cargos

The large non-revenue bearing volumes in a carrier for heavy liquid cargo challenge the designer and the operator to find cargo for the return voyage so the ship may avoid unprofitable sailing in ballast. Heavy chemical carriers may use the wing spaces for the installation of tanks for a return cargo. Separate tankage is usually provided because of difficulty finding a compatible return cargo which can be loaded into the main cargo tanks. Separate tankage generally means separate piping systems as well. It is possible to fill cargo tanks of a high specific gravity cargo vessel with salt water for ballasting purposes, but salt water may be incompatible with the cargo and is usually not beneficial to the life of container or piping. The alternative is special ballast tanks outside the cargo spaces.

The ship transporting a dense liquid cargo has a high degree of permeability in the void or ballast spaces. Without attention to its compartmentation, it is more susceptible to loss of buoyancy and/or stability if the hull is breached by accident or structural failure.

An additional complication when the cargo's specific gravity departs from unity is testing for tightness of the completed container. It is no longer a straightforward matter of using water for the time-honored hydro-static test. To fill a container intended for light cargo with water may overstress the container or vessel, while to half-fill it may not test the upper portion. In a heavy cargo container, filling with water does not test the container bottom, and overfilling to a head sufficient to test the bottom of the container may overstress the upper portion.

For cargos such as methane, special atmosphere tests such as with freon or helium may be necessary to indicate acceptable limits of porosity. Water testing is not likely to show excess porosity of the welds.

b. Temperature. Liquid bulk cargos now range in temperature as transported from +149°C (300°F) (molten sulphur) to −253°C (−423°F) (liquid hydrogen). When the cargo temperature is elevated or depressed, thermal expansion, contraction, and temperature gradients must be anticipated in the design of the container and its supports. The extent of expansion or contraction is directly related to the dimensions of the container. With long tanks, the amount of movement from thermal effects can be considerable, calling for careful engineering and installation to avoid unintended constraints. The point of fixation should be centrally located to reduce the amount of thermal movement. Piping and other connections should be near the fixed point to minimize stresses. Hot and cold cargos also have to be isolated from the hull to eliminate sharp thermal gradients and unacceptable temperatures in the hull structure.

With very cold cargo, the container must be so designed and use of material carefully selected to avoid the possibility of brittle fracture. The USCG Marine Engineering Regulations (Subchapter F) include information on materials and special toughness requirements for weldments, test procedures, chemical composition, and mechanical properties according to the anticipated service temperatures. The design must also minimize the possibility of minor leakage because of the likelihood of brittle fracture of the ship structure if chilled by the cold cargo. Toward this end, present concepts and requirements emphasize provision of both a primary and secondary cargo container. The use of the *double container* further complicates design and fabrication. Relief from the requirement of a full secondary barrier has been permitted when the primary containers are designed as pressure vessels because of the higher standard of quality control and greater capability of analysis.

When cargos are transported at elevated or depressed temperatures, tanks must be insulated. A careful thermal analysis, taking account of the extreme environmental thermal conditions, is required to insure adequacy of insulation systems. To be confident that the system will remain effective in normal service, factors such as vapor protection for cold cargo, mechanical protection, and weather resistance must be considered as well. In addition to proper design, competent installation and adequate maintenance are essential to satisfactory performance of an insulation system. The insulation system, besides being effective in normal service, must not itself create a hazard. This necessitates that the insulation be non-combustible or, alternatively, that it be provided with fire exposure protection consisting of a steel jacket or equally effective material. (The latter is readily accomplished by vessel configuration in some cases.) The required relief valve capacity is also related to the insulation provided. If the insulation system can be expected to remain intact to protect the tank during external fire exposure, the relief valve capacities may be reduced.

A *catch basin* beneath such a container to prevent even the slightest contact of the cargo with the hull is still essential.

c. Pressure. Obviously, it is possible to package substances under pressure, at atmospheric pressure, or under vacuum. No liquid cargos are being transported under vacuum, although some cryogenic substances are carried in containers which resemble vacuum bottles, having an evacuated annular space between the inner and outer shell.

A container for cargo at elevated pressure is customarily a *pressure vessel* with clean design details, quality materials and workmanship, and with relief valves to assure that the structural limits are not approached within a respectable margin of safety. Piping and other attachments must be of similar strength.

The shape of pressure containers must be considered in connection with the fact that they are being supported by a flexible foundation and subject to acceleration forces from the motion of the sea. Long containers consequently have to withstand bending as well as hoop stress. In analyzing the support of long containers, one should recognize that if more than two supports are involved there is the possibility of considerable increase in saddle reactions and tank bending moments due to the flexing of the foundation.

Containers designed for atmospheric pressure at ambient temperature require suitable venting arrangements. There are containers designed for atmospheric pressure which are dependent upon depressed temperature, where the cargo presents a significant relationship between vapor pressure and temperature. It is usually necessary to insulate the container or to provide a refrigeration system (or both) to hold the temperature, and thereby the pressure, within limits. Relief valves are also required to safeguard against rupture if the temperature control fails.

Indicative of the way in which regulatory requirements are imposed on cargos carried at various combinations of temperature and pressure is Fig. 3. As will be explained later, most of the conditions illustrated which control the material specifications of the container and secondary barrier where required are drawn from the IMCO Code for carriage of liquefied gases.

2.6 Safety and Environment. The nature of hazard presented by a bulk liquid cargo depends on the chemical properties of the substance together with the physical state in which it is contained for transport. The quantity released, the time rate of release and the location and circumstances of the causative event all have a bearing in an accident on how far-reaching the outcome. The risks in transporting and handling hazardous substances must be anticipated by appropriate design countermeasures to minimize the likelihood of accidents and the consequences of such accidents as do occur.

a. Toxicity. The extent of personnel and public hazard a toxic cargo may present is difficult to define. In deciding how dangerous a chemical is, some questions arise:

- Does it cause illness in every case?
- If not, what percentage of exposed persons are affected?
- What concentration can be tolerated without harmful effects?
- Are the effects cumulative?
- What degree of contact? Inhalation? Touching? Tasting?
- Is it volatile? Water soluble? Are the vapors heavy or light?
- Has it a characteristic odor or color?

The classification of a substance as toxic is generally the result of tests conducted upon laboratory animals. For small packaged quantities 49 CFR distinguishes among the required packing and stowage of *poisons* as Class A—, Class B, and Irritating Materials, (or Class C). Products of these classes are not generally allowed bulk shipment.

For bulk transport, the toxicity recommendations were developed for the USCG by the National Academy of Science (NAS) Committee on Toxicology using threshold limit values as published by The American Conference of Governmental Industrial Hygienists. The threshold limits pertain to the safe concentration based on an eight hour working day with continuous exposure for industrial workers.

b. Flammability. Liquids are categorized as either flammable or combustible as follows:

Flammable refers to a liquid which gives off flammable vapors (as determined by flash point from an open cup tester) at or below a temperature of 26.8°C (80°F). They are further categorized by grades as follows:

1. Grade A—Having a Reid vapor pressure of 98 kPa (14 lb) or more.
2. Grade B—Having a Reid vapor pressure under 98 kPa (14 psi) and over 59.5 kPa (8.5 psi).
3. Grade C—Having a Reid vapor pressure of 59.5 kPa (8.5 psi) or less.

Combustible liquids are those having a flash point above 26.8°C (80°F), are further categorized as follows:

4. Grade D—Having a flash point below 65.6°C (150°F) and above 26.8°C (80°F).
5. Grade E—Having a flash point of 65.6°C (150°F) or more.

The Reid vapor pressure is measured at a temperature of 31.8°C (100°F) expressed in kPa (psi) absolute, according to the "Reid Method" as per ASTM Standard D-323 "Method of Test for Vapor Pressure of Petroleum Products."

Flammability is fairly well established on the basis of vapor pressure and flash point from experience with petroleum. There are many chemicals whose flammability is of an order not exceeding that of petroleum products, but there are others which are much more susceptible to burning because of wide flammable limits, low ignition temperatures, etc.

If the vapor of a flammable chemical is heavy, there is a risk of it gathering in low points within the hull and finding a source of ignition. While light vapors tend to rise and be dissipated, it is possible to have temperature-inversion effects accompanying the release of a large amount of light vapor so that it could remain at ground level for some time. Also, flammable mixtures of light vapors with air may not dissipate rapidly.

Some flammable substances provide their own oxygen and when ignited are difficult to extinguish by smothering techniques which exclude air. Others; e.g., alcohols, are water soluble, dissolving ordinary foam solutions.

c. Corrosivity. If the cargo is corrosive or caustic, the materials for the container and piping and the use of internal coatings are important considerations in assuring the retention of the cargo and the life of the container. Rubber (usually neoprene) has been used extensively. Newer coatings such as epoxy and plastics have been used shoreside and are now being employed at sea. Because of the flexing action of a vessel, special attention must be paid to resiliency in selecting coating materials.

In addition, it is important to recognize the ability of the coating to survive the cyclic operations of tank washing and ballasting if conducted in the tank. Occasionally it is not the cargo but other stages in the operating pattern which cause the corrosive condition to arise. For example, sulfuric acid, when carried concentrated, is considerably less corrosive than dilute sulfuric acid which is formed in the process of changing the cargo. Petroleum carriers are not free from corrosive problems. This has been a long-standing problem which affects ships differently depending on whether they are in the white or black trade. The alternate wetting and drying of upper surfaces of tanks carrying the white oils tends to concentrate the corrosion there. On the other hand, the buildup of sludge and water residues in the bottom of black oil tanks is responsible for corrosion occurring in the

(Continued on page 491)

Table 9—Incompatible Substances

CARGO GROUPS \ REACTIVE GROUP	1 Non oxidizing mineral acids	2 Sulfuric Acid	3 Nitric Acid	4 Organic Acid	5 Caustic	6 Ammonia	7 Aliphatic Amines	8 Alkanolamines	9 Aromatic Amines	10 Organic Anhydrides	11 Isocyanates*	12 Olefins	13 Vinyl Acetate	14 Acrylates	15 Substituted Allyls	16 Oxidizers*	17 Alkylene Oxides	18 Epichlorohydrin	19 Ketones	20 Aldehydes	21 Unsaturated Aldehydes/Ketones	22 Alcohols	23 Glycols and Glycol Ethers	24 Phenols and Cresols	25 Misc. Water Solutions
1 Non oxidizing Mineral Acids		x			x	x	x	x	x	x			x				x	x				x			
2 Sulfuric Acid	x		x	x	x	x	x	x	x	x		x	x	x	x		x	x	x	x	x	x	x	x	x
3 Nitric Acid		x			x	x	x	x	x	x		x	x	x	x		x	x		x	x			x	
4 Organic Acid		x			x	x	x	x																	
5 Caustic	x	x	x	x						x										x	x	x	x	x	
6 Ammonia	x	x	x	x						x							x			x	x				
7 Aliphatic Amines	x	x	x	x						x			x	x	x		x	x	x	x	x	x	x	x	
8 Alkanolamines	x	x	x	x						x			x	x	x		x	x		x	x				
9 Aromatic Amines	x	x	x							x										x	x				
10 Organic Anhydrides	x	x	x		x	x	x	x	x																
11 Isocyanates*																									
12 Olefins		x	x																						
13 Vinyl Acetate	x	x	x			x	x	x																	
14 Acrylates		x	x				x	x																	
15 Substituted Allyls		x	x				x	x																	
16 Oxidizers*																									
17 Alkylene Oxides	x	x	x			x		x																	
18 Epichlorohydrin	x	x	x				x	x																	
19 Ketones		x					x																		
20 Aldehydes		x	x		x	x	x	x	x																
21 Unsaturated Aldehydes/Ketones	x	x	x		x	x	x	x	x																
22 Alcohols		x			x		x																		
23 Glycols & Glyco Ethers		x			x		x																		
24 Phenols and Cresols		x	x		x		x																		
25 Misc. Water Solutions		x																							
31 Paraffins**																									
32 Aromatic hydrocarbons**																									

Table 9—Incompatible Substances (Continued)

	1	2	3	4	5	6	7	8	9	10	11	12	13	14	15	16	17	18	19	20	21	22	23	24	25
33 Misc hydrocarbon mixtures**																									
34 Esters		x																							
35 Vinyl halides			x																						
36 Halogenated hydrocarbons		x																							
37 Nitriles		x																							
38 Carbon disulfide							x	x																	
39 Amides**																									
40 Sulfolane**																									
41 Ethers		x	x																						
42 Nitroaromatics																									
43 Nitroparaffins					x		x																		
44 Edible oils																									

* To be determined

** Not incompatible—included for completeness

Table 10—Ratings of Chemicals

	Column: I	II	III	IV	V	VI	VII	VIII	IX	X
	Fire	Health			Water Pollution			Reactivity		
CHEMICALS		Vapor Irritant	Liquid or Solid Irritant	Poisons	Human Toxicity	Aquatic Toxicity	Aesthetic Effect	Other Chemicals	Water	Self-Reaction
Acetaldehyde	4	3	1	2	2	3	2	2	0	1
Acetic Acid (Glacial)	2	2	3	2	1	2	2	2	0	0
Acetic Anhydride	2	3	3	3	1	2[a]	2	3	2	0
Acetone	3	1	0	0	1	1	1	2	0	1
Acetonitrile	3*	1	1	3	2	1	1	2	0	0
Acetone Cyanohydrin	1*	1[b]	2	4	4	3[a]	3	2[b]	3[b]	0
Acrolein (Inhibited)	3	4	3	4	4	3	3	3	0	3
Acrylic Acid (Inhibited)	2	3	3	2	1	3[a]	2	2	0	3
Acrylonitrile	3*	3	1	3	4	3	2	3	0	3
Adiponitrile	1*	1	1	3	3	2	3	2	0	0
Allyl Alcohol	3	3	2	3	2	3	2	2	0	1
Allyl Chloride	3*	3	2	3	2	1[a]	2	2	0	1
Aluminum Triethyl	4*	NA	4	3	0	2[a]	3	4	4	0
n-Aminoethyl Ethanolamine	1*	1	3	1	1	1[a]	3	3	0	0
Ammonia, Anhydrous	1*	4	2	2	2	2	2	3	2	0
Ammonia, Aqua (28%)	1*	2	2	2	2	3	2	3	0	0
Amyl Acetate, iso-	3	1	0	1	1	3	2	1	0	0
Amyl Acetate, primary	3	2	1	1	1	3	2	1	0	0
Amyl Alcohol, n-	3	1	0	2	2	2	2	2	0	0
Aniline	1*	1	1	3	3	2	4	3	0	0
Asphalt (Typical)	1	1	2[c]	1	0	1	4	0	0	0
Benzene	3	1	1	3	1	3	2	1	0	0

Table 11—Outline of Rating System for Health, Water Pollution, and Reactivity

	I	II	III	IV	V	VI	VII	VIII	IX	X
		HEALTH			WATER POLLUTION			REACTIVITY		
	FIRE	Vapor Irritant	Liquid or Solid Irritant	Human Poisons	Human Toxicity	Aquatic Toxicity	Aesthetic Effect	Other Chemicals	Water	Self-reaction
Grade 0	No hazard	No effect	No effect	No effect	Nontoxic; LD_{10} 15 g/kg	Acute threshold limits above 10,000 ppm	No significant pollution; gases and odorless liquids	Inactive; may be attacked by Grade 4	No reaction	No reaction
Grade 1	Closed cup: Flash point above 140°F	Slight effect	Causes skin smarting	Slightly toxic	Practically nontoxic: LD_{10} 5–15 g/kg	Threshold limits 1,000 to 10,000 ppm	Mild-odored light oils and soluble chemicals	React only with Grade 4	Mild reaction: unlikely to be hazard-ous	Mild self-reaction under some conditions
Grade 2	Flash point 100 to 140°F	Moderate irritation; temporary effect	First-degree burns, short exposure	Intermediate toxicity	Slightly toxic; LD_{10} 0.5 to 5 g/kg	Threshold limits 100 to 1,000 ppm	Mild-odored oils; B.P. 150–400°F	React with Grades 3 and 1	Moderate reaction	Will undergo self-reaction if con-taminated; do not require stabilizer
Grade 3	Flash point below 100°F; B.P. above 100°F	Irritating; cannot be tolerated	Second-degree burns, few minutes exposure	Moderately toxic	Moderately toxic; LD_{10} 50–500 mg/kg	Threshold limits 1 to 100 ppm	Light-colored high-boiling oils; odorous water-soluble compounds	React with each other and with those of Grades 2 and 4	More vigor-ous re-action; may be hazardous	Vigorous self-reaction; requires stabilizer
Grade 4	Flash point below 100°F; B.P. below 100°F	Severe effect; may do permanent injury	Second- and third-degree burns	Severely toxic	Toxic chemicals; LD_{50} 50 mg/kg	Threshold limits below 1 ppm	Heavy oils, colored or malodorous	React with each other and all other grades	Vigorous reaction; likely to be hazardous	Self-oxidizing chemical; capable of explosion or detonation

These definitions were employed as a broad scheme in rating chemicals, but were not always followed rigidly. Ratings were shifted upward or downward to reflect other properties not shown in this brief outline.

(*Continued from page 487*)

bottoms of those ships. Additional treatment of this subject may be found in Chapter XIV.

d. Reactivity. Reactivity refers to chemical properties of the cargo which may cause it to decompose, react with itself (*polymerize* for example) react with other cargos, or react with structural materials. The extent of the ensuing reaction depends upon the substances involved and can generate large or small amounts of heat, or merely contamination of the principal cargo. As an extreme example, liquid oxygen (LOX) is not flammable, but will make most things burn rapidly, and can cause tremendous reactions with a great many other substances.

It is essential to know what materials will make a chemical cargo unstable. If the cargo reacts violently in content with a common substance such as water, more care is obviously required in its handling. Some combinations of substances can reach a state at which only a small amount of a third substance (or catalyst) is required to produce a violent reaction.

Table 9 shows combinations of substances which are incompatible.

e. Personnel Safety. The individuals most likely to come in contact with the cargo are members of the ship's crew going about their routine duties. Having regard for the nature of the substances to be carried, the designer must anticipate situations which may affect crew members. Piping must be insulated. Vent systems must be oriented to preclude noxious vapors entering the accommodations. The taking of ullage readings by dipstick or tape may not be permissible calling for closed gaging. If periodic entry into the tanks is required, there must be an ability to provide a gas-free, safe-for-entry atmosphere. Special clothing, breathing apparatus, emergency rinse-off stations, and changing rooms may have to be provided.

f. Ship Safety. Accidental large scale release of some liquid cargos may menace the ship itself and all its crew. Casualties of this order include fire, explosion, flooding, stability loss, and brittle fracture structural failure due to the release of cryogenic cargo. These may be brought about by external injury caused by collision, stranding or heavy sea conditions, or by internal failure of cargo containment or of the integrity of critical interior systems.

Against the contingency of fire, it should be recognized that there are hazardous substances for which special extinguishing agents are required. There are some cargos once ignited for which the prospect of extinguishment is extremely poor. It is therefore especially necessary to minimize possible leakage, and to make provision for protecting the crew until extinguishment or abandonment can be carried out. The hazard of fire or explosion is not restricted to the laden condition but may be present at other points in the transport cycle. For example, an oil tanker in an empty, gaseous state may confine an explosive mixture of enormous volume. The preparation of the tanks for other cargos or in connection with the ballasting process includes the risk of explosion. One recently adopted measure requires the superstructure to be insulated with structural fire protection of A-60 Class across the front facing the cargo area to serve as a thermal shield behind which the crew might retreat. These requirements appear in 46 CFR 32.56, and in SOLAS 74, Chapter II-2, Part E.

g. Port and Environmental Safety. The transport of hazardous and polluting substances holds a potential for consequences far beyond the environs of the ship. The great quantities contained in a single ship, recent marine accidents, plus sensational news reporting have raised the spectre of holocaust, of poisoning of organisms in the food chain, or of aesthetic degradation of the shore. Notwithstanding extreme and oftentimes unsupportable assertions, there are some bases for these fears, such as the explosion of the *Sansinena* in December 1976 and the oil spilled on the Brittany Coast by the *Amoco Cadiz* in March 1978.

Virtually every port of any size contains large amounts of flammable and/or hazardous cargo in fixed storage tanks. Of course, the ships calling may contain significant quantities. The expansion of the port areas has frequently brought about close proximity of port activities to residential areas or industrial activity. An accident of a relatively low order in the port may therefore have consequences of considerable proportions.

Heavy penalties are levied for polluting and the longstanding practice of disposing of bilge waste and tank washings at sea is now highly restricted if permitted at all. Oil released on the water persists for a long time and there is considerable controversy as to what actually becomes of it.

The accidental release of chemical substances is less readily detected but the impact upon the civilian community can be severe, for instance, in a river in proximity to a water intake. Those substances which are water soluble are incapable of being recovered. Some heavy chemicals may tend to accumulate in pockets on the bottom. The ability to recover them depends upon the ease with which they can be located and the depth at which they are found. Some liquids are not water soluble and, on warming, convert to gaseous form, threatening persons and property downwind.

Environmental impacts may even occur from well intended safety measures. For example, inert gas systems in scrubbing sulphur out of stack gases form acid products which must be discharged overboard in the port. Over the long term a negative effect may be felt in that location.

In addition to water pollution is the effect on air quality. Some ports are situated within topographic boundaries causing high incidence of smog affecting the respiration of persons living in the area. Concerns have been raised over the emissions from ship operations at terminals, as adding to the ambient level. Local jurisdictions are therefore likely to impose limitations on the ordinary operations of vessels. At the present time, vapor recovery systems are being investigated for possible application in the ship-to-shore and ship-to-ship transfer of hydrocarbons and chemicals through research sponsored jointly by the USCG and the Environmental Protection Agency (EPA).

h. Hazard Ratings. Toxicity, flammability, corrosivity, and reactivity discussed earlier are individual dimensions of risk. The implications of release of a substance upon environment or public health and the physical state of transport are other dimensions of risk. The possible combinations are enormous. NAS was requested by the USCG in the early 1960's to develop a method by which to compare the risk presented on a composite rational basis. Such a system helps guard against disproportionate regulations and constraints. This NAS system was strongly motivated by concern for public safety as shown by the four broad hazard categories employed:

1. Fire hazard,
2. health hazard—whether a vapor irritant, a liquid or solid irritant or a poison,
3. water pollution hazard—the distinction being between human toxicity, aquatic toxicity or aesthetic impact, and
4. reactivity hazard—distinguishing between reaction with other chemicals with water or with itself.

The NAS system when introduced for international attention through IMCO, influenced the construction and design restrictions developed in the Codes for Gas Ships and for Chemical Carriers.

A second hazard-rating procedure, closely resembling NAS was later drawn up by IMCO on the basis of the potential effect upon the environment if a substance is spilled or released to the sea. Called the *GESAMP* (Group of Experts on the Scientific Aspects of Marine Pollution) System, the hazard profiles were applied to determine the acceptability of operational discharges such as those associated with the washing of tanks. The concerns of GESAMP are reflected by the attention given to bio-accumulation; damage to species through the food chain; hazard to humans by drinking the water, eating fish or exposure to the tainted water or atmosphere; and degradation of environmental amenities such as beaches, fishing, and other recreational areas.

Tables 10 and 11 are excerpted from the NAS system.

Section 3
Transport of Liquid Cargos

3.1 Transport System Elements. It is important that the naval architect recognize certain considerations which determine whether the vessel will be efficient in the trade for which intended. Two major elements are the economic factors governing the cargo, and the operating cycle of the ship when engaged in physical movement of the cargo.

a. Economy of Scale. Surface increases as the square and volume as the cube. Hence the cargo deadweight increases faster than the hull light weight. Further, it requires about the same personnel to move a large ship from point to point as a small one. These considerations of efficiency have promoted the growth of large bulk carriers. As an indicator of the economies to be realized, the Required Freight Rate (RFR), which is the ability to move a ton of cargo without profit, is often used. See Chapter I.

However, there are some practical factors which militate against a continued upward growth in the size of vessels. It is necessary to examine the origin and destination of the substance being transported, and as well the route that will be followed between. Vessels of great draft are obviously limited in the ports they can visit and sometimes in the waterways they can traverse. It was the stimulus of the closing of the Suez Canal in 1956, precluding the use of the relatively shallow Mediterranean Sea as a passage and forcing the much longer distances round the Cape of Good Hope, which provoked the idea of moving great quantities per ship the long distance from the Persian Gulf to northern Europe. At the destination, faced with the limited draft of most northern European ports, it was necessary to transfer into smaller vessels to complete the distribution process.

Relatively few ports in the United States have the depth of channel necessary to large ships. Efforts to promote deep water ports have had limited success for the delivery of crude oil to the U. S. Gulf Coast. Hence, before determining that a vessel is of an efficient size, it is necessary to examine all of the limitations which relate to the transport from origin to destination.

b. Product Purity. Successful transportation of multiple products in a chemical tanker or product carrier requires not only that great pains be taken to separate and segregate the cargos but that the individual substances be properly treated in shipment. Many products, especially chemicals, pharmaceuticals, lubricating oils, and foodstuffs, require great care to insure their quality and purity. If contaminated or altered they may become worthless.

It is therefore necessary to know the tolerance of a product to insure its marketability is not impaired by exposure to improper temperature levels, moisture, surfaces of piping, hoses, or tanks, or traces of a previous cargo. To reduce the risk of losing the cargo by tainting, it may be possible to load the product adjacent to another commodity in the same chemical *family* so that any mixing of the two will only cause a downgrading of the premium substance. Where a hazardous reaction may develop from mixing with another cargo, provision of cofferdams may be essential or, alternatively, the fitting of independent tanks. Where exposure to moisture is deleterious a dehumidifying unit can be fitted to dry out the tank for loading or to remove moisture which may enter through the venting system en route. Particular attention must be given to the interior coatings not only in the tanks but in all valves, pumps, casings, seals and gaskets which may come in contact with the cargo. In some cases, the seals and gaskets may be dissolved by the substance.

Where the purity of the product is so critical that it cannot be satisfied in any other way, it may be necessary to consider a dedicated vessel or one for which the range of acceptable products is extremely limited, or the installation of dedicated independent tanks, pumps, and piping. As an ex-

Table 12 Relation of Cargo to Ship Design and Operation

CARGO PARTICULARS	Containment	Segregation from waterway	Segregation from air	Segregation from other cargo	Segregation from ignition sources	Segregation from accommodations	Material selection	Pumping system	Tank venting	Tank gaging	Cargo sampling	Vapor space padding	Vapor space inerting	Fire protection	Cargo handling space ventilation	Cargo heating	Cargo cooling	Thermal insulation	Temperature controls	Electric components	Tank cleaning	First aid	Emergency rescue equipment	Personnel protection	Gas detection equipment	Safety procedures	Monitoring system
Specific gravity	x							x		x																x	
Vapor pressure	x				x			x	x	x	x		x	x	x											x	x
Vapor density	x				x			x	x	x	x		x	x	x											x	x
Viscosity								x		x						x					x						
Flash point	x				x	x			x	x	x		x	x	x				x	x	x			x		x	
Boiling point	x			x	x				x							x	x	x	x								
Pour point								x		x						x					x						
Freezing point								x								x	x	x	x		x						
Autoignition point	x				x	x		x					x	x				x	x								
Explosive concentration limits					x	x				x	x		x	x	x						x				x	x	x
Color	x						x				x	x				x			x		x						
Odor						x			x												x					x	
Water solubility		x												x												x	x
Vapor irritant	x					x			x	x	x				x						x	x	x	x	x	x	
Liquid irritant	x					x		x		x	x										x	x	x	x	x	x	
Absorptive poison	x					x		x	x	x	x				x						x	x	x	x	x	x	
Threshold limit value	x					x		x	x	x	x				x						x	x	x	x	x	x	
Water pollution	x	x																			x					x	
toxicity—human																					x					x	
Water pollution		x															x	x	x			x	x	x		x	x
toxicity—aquatic																					x	x	x	x		x	
Self-reactivity	x					x	x							x							x	x	x	x		x	
Reactivity with water	x	x												x							x	x	x	x		x	x
Reactivity with other	x			x			x					x		x							x						x
substances																					x	x	x	x		x	
Pyroforicity	x		x				x		x	x	x		x	x	x									x		x	
Reactivity with air (purity control)	x		x				x		x	x	x	x						x	x								
Corrosivity	x						x		x					x													
Fire extinguishing media														x													
Loading temperature																x											
Polymerization	x						x														x						x
Requires liquefaction	x						x	x	x	x	x						x	x	x		x					x	

treme example a product which, once carried, neutralizes or eliminates the possibility of carrying any other substance because its traces can never be completely eliminated, is tetraethyl lead.

c. Flexibility. For a crude oil carrier the run is from the producing area to the refinery and back again empty with the ship in ballast, delivering a full one-grade cargo. As a rule a gas carrier is in much the same situation. However, ships in the white oil trade or chemical carriers are expected to be able to carry many different grades, many different products and to deliver quantities of each to a number of different locations.

The range of products and their quantities, and the ability to deliver them in parcels without impairing trim, stability, or stress condition of the ship require careful analysis of the owner's prospects. While it might appear the easiest solution to design for the most extreme cargos that could conceivably be carried, there are restrictions in the way of such a proposal. Furthermore, the cost of construction bears a relationship to the degree of hazard of the cargo since the more dangerous the substance, the greater the number of safeguards which must be incorporated. If there is only a limited potential for the transport of extreme risk substances, a design which endeavors to achieve the greatest range of cargos will call for many systems and control measures which will have limited use. It is therefore important to analyze the intended service of the ship with emphasis upon particulars of the cargo. Not only must consideration be given the compatibility of those substances and their individual requirements for purity and quality, but for the complex of systems of inerting, temperature control, venting, gaging, and so on. These relationships are indicated in Table 12.

d. Maintenance. The more complex the ship and its systems the more difficult it is to maintain. The designer must make provisions for upkeep. The great number of systems along with gages and alarms that must be in operating condition to insure the successful transport of bulk liquid cargo necessitates a carefully developed maintenance program. The network of the entire operation must be studied for its critical paths to insure that a loss of any one system will not rob the operator of the ability to receive, transport, or transfer the cargo. The ability to test periodically and replace instruments should be considered in connection with their selection and installation. The availability of spare parts for critical systems must be evaluated. Specially selected materials and coatings must

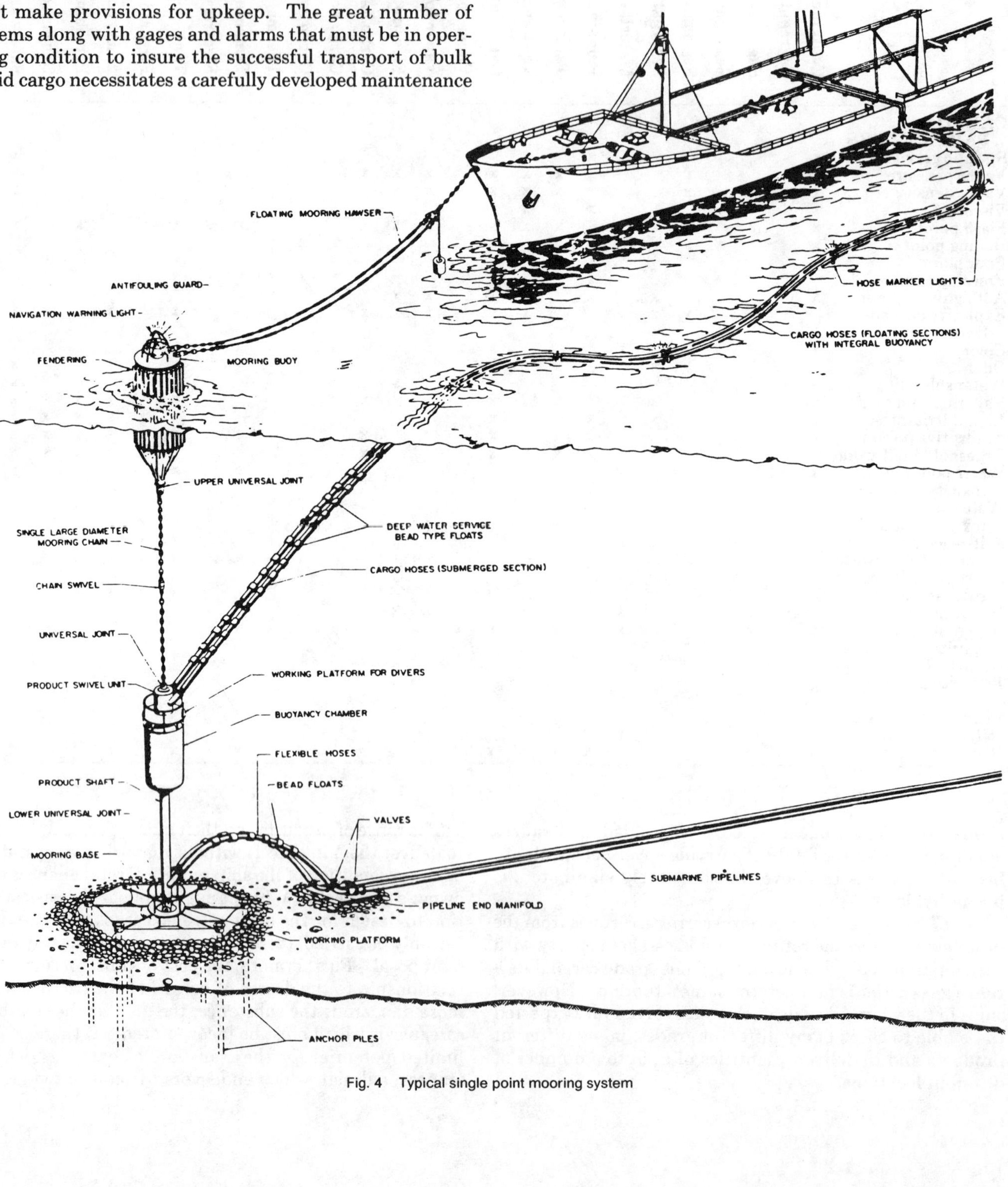

Fig. 4 Typical single point mooring system

be identified in sufficient detail to prevent replacement or repair at a later time in the life of the ship with materials which do not conform to original specifications.

There are new considerations in doing repair work. The marine chemist has an even more critical job in assuring that the atmosphere is not only safe to enter but additionally, that performing *hot work* in a space where chemicals have been carried will not release toxic products despite being apparently *gas free*. The quality of welding repairs may be adversely affected by the presence of small quantities of certain chemicals.

e. Delivery Terminals. The naval architect cannot consider his task of designing the ship a success unless assured that it can safely deliver its cargo at the intended destination. Public concern over the importation of hazardous substances, opposition to dredging as a means of gaining entree for ships, and questions of coastal land use especially the impact on pristine areas for any new development, have made it difficult to know that the community will accept the arriving ship and its cargo, or that it will be possible to move the material onward to its next stage in the distribution sequence, whether by reshipment, pipeline, or other mode. Local, state and Federal agencies concerned with any new proposal are legion. The procedure for gaining acceptance is protracted and frequently indecisive.

In the face of these constraints, the naval architect may well find it a requirement to solve the landing of the cargo at locations remote from the usual services available in a port. Or it may be necessary actually to design a reception facility in an offshore location at which the product may be stored pending transshipment in smaller vessels, transferred ashore by pipeline, or as in the case of LNG, changed from the liquid state by regasification and pipelined ashore as gas. The growing use of single point moorings is an attempt to get around the administrative delays imposed in the process of approving a new facility. The costs associated with this factor are part of the total solution and allowance must be made for both the delay and the expense. These effects may well influence the overall plan for the movement of the cargo and alter the parameters governing the design of the ship.

Offshore facilities for storage, transshipment or conversion of hazardous cargos may take a number of forms. Floating vessels weathervaning around an anchored column, built of steel or concrete—of ship shape or even toroid—have been built. Other concepts suggest creation of ship-formed multi-cellular concrete structures, either ballasted and floating or bottom supported. Fig. 4 illustrates the single-point mooring concept.

3.2 Operating Cycle. The designer must be able to perceive the entire sequence of operations in delivering the cargo to be able to specify and provide the ship and its personnel with the proper means and the capacity to execute each required step safely.

a. Loading. In loading the ship, the ship's officers must satisfy a number of conditions, some of which are at cross purposes. It is necessary to:

- maximize the cargo lift,
- satisfy the loadline limitations,
- place the vessel in proper trim,
- minimize the turn-around by loading at an efficient rate,
- avoid over-pressuring the tanks,
- avoid the creation of hazardous stress conditions to the hull girder,
- dispose of any ballast water aboard,
- satisfy the damage stability conditions.
- If the vessel is transporting more than one product, provide a suitable segregation of the cargos,
- avoid contamination of the cargo,
- load in such a fashion as to permit discharging at different ports without causing adverse effects on the overall stowage as the separate parcels are offloaded,
- avoid pollution, and
- eliminate fire and explosion hazards.

This problem is the Master's to solve, but the conditions are generally so complex that preparations by the naval architect are necessary to assist him.

b. Laden Passage. The principal concern during the laden passage is to insure that the substance being transported remains in a fluid state to retain the ability to be pumped off at the destination. Most substances present no problem as they are normally liquid at atmospheric pressure and within the ambient temperature range. When, however, owing either to its viscosity, melting point, pour point, freezing point, vapor pressure, or specific gravity, the substance is not liquid at ambient conditions, provision must be made for additional heating or cooling enroute. The potential heat loss or gain must be evaluated to establish the type of and size of system necessary. Detecting sensors must be fitted that will insure that the cargo is kept within the intended thermal range and that any pressure limitations are not exceeded.

Some substances must be shielded from exposure to the atmosphere to prevent oxidization. As mentioned earlier, nitrogen is sometimes used as a *padding* to prevent contact with air. Where such systems are fitted it is essential to have a means of assuring that the desired atmosphere is created and preserved. The initial tank atmosphere may be provided by a shoreside source. This reduces the size of the shipboard system which is only used thereafter to make up for any loss or reduction in the quality of the padding.

The heating system and its controls require special attention in connection with chemical and similar products. When the cargo is toxic it may not be passed through heat exchangers located in the engine room or other space interior to the ship and its accommodations. The pressure relationship should insure that any leakage is from the heating medium into the cargo and also that the heating medium is compatible with the cargo involved.

In the case of certain cargos an adverse interaction with the metal used in the heating coils is possible if the local temperature at the common surface is not closely controlled. With some cargos for example–molten sulphur, the heat transfer to the cargo is a problem. If heated beyond its transport temperature, sulphur becomes more viscous, leading to the formation of an overheated layer in proximity to the heat exchange surface.

With cargos transported in a refrigerated state, provision

must be made for a conditioning plant and transfer system. The capacity of this plant must be determined on the basis of a heat loss study allowing for conditions of operation and delay the ship may encounter. It is also necessary to allow for a back-up system in the event of failure of the primary unit.

Certain products are unstable as shipped and may deteriorate or self-react if provision is not made to retard the process. This will not only affect the marketability of the product but may lead to the creation of a substantial amount of heat. Polymerization is a form of reaction of this sort and is associated with such a product as styrene monomer. The substance is used in synthetic rubber. If polymerization occurs, the heat generated speeds the reaction and the product becomes increasingly viscous until it reaches a solid state. To avoid this condition an inhibitor is applied after loading. During the voyage it may become necessary to add additional inhibitor and periodic testing must be carried out to determine when and how much of the inhibitor to add.

In addition to sensing the conditions of the cargo, it is also important to have the ability to detect the atmosphere in void spaces, if any, surrounding the cargo tanks.

Steam remains the most common method of heating cargos although hot oil systems have been used in some cases, such as for asphalt barges, and hot water and other systems may also be employed. In most cases, heating is done at or near the bottom of the tank, using continuous 38 mm (1$^3/_8$ in.) steel pipe or gilled cast iron heating elements. This is basically an inefficient system because of formation of condensate in the coils limiting the amount of effective heat transfer surface to a small part of the installed surface. The materials employed are also basically unsuitable, except in special cases, owing to the high weight of the cast iron and the susceptibility to corrosion of the steel coils.

Systems consisting of a series of grids, each with its own riser and downcomer and designed for maximim efficiency of each heating unit, are considerably better. In addition to reducing radically the surface required for a particular level of heating, this arrangement permits individual sections to be cut when not needed due to low heating demand. Helical coils are also used, either permanently installed or as portable coils to supplement the grid system. When cargos require heating levels higher than those for which the permanently installed system was designed, portable helical coils can readily be fitted. Both grids and helical coils are normally aluminum-brass for cargo tanks and steel for fuel oil tanks.

Also, heating coil surface has been greatly reduced by careful programming of heating during a voyage rather than by maintenance of the required unloading temperature throughout the voyage.

c. Discharging. Discharging cargo resembles the loading operation in that there are again a number of competing objectives. It is necessary to:

- Minimize the turn-around by discharging at an efficient rate,
- avoid contamination of the cargo,
- avoid pollution of both water and air,
- avoid creation of hazardous stress conditions,
- minimize the amount of residual cargo, that is, insofar as practicable, to discharge the full amount of cargo.

To carry out all of these functions satisfactorily, the discharge operation must be pre-planned, taking advantage of the trimming and heeling which will occur to avoid the loss of suction by the cargo pumps and to promote as complete a process of stripping in the final stages as may be achieved.

It may also be necessary to apply heat in the final stages in order to minimize the amount of sediment remaining.

In the case of chemical carriers and product carriers, great care must be exercised in lining up the system. The use of deep-well pumps applied directly to the cargo tank, one pump per tank or group of tanks, is one mode by which contamination is avoided. In any event, it is important that the individual piping systems be coded in color or other marking schemes to reduce the possibility of personnel error.

d. Ballast Passage. Ballasting may commence during the discharge of the cargo in order to achieve the various objectives mentioned under *discharging* and to avert having the vessel in an extremely light state alongside of the pier.

1. Taking on Ballast. Where segregated ballast spaces are fitted, because the piping systems are independent, there is no difficulty in conducting this process coincident with the discharging as long as structural difficulties are avoided and draft limits are not exceeded.

It is necessary to ballast for immersion of the hull, propeller, and rudder in order that the ship may be capable of being handled on the return voyage. The distribution of ballast is also governed by the Master's intentions during the ballast passage as well as the duration and sea conditions expected. Tanks may be intended to be cleaned to remove residue accumulated or for entry for inspection.

The problems associated with ballasting arise from the standpoint of pollution. Introduction of salt water into a recently emptied cargo tank will drive the vapor in the tank out to atmosphere. In some port areas this may be objectionable because of air quality degradation. Depending on the nature of the cargo, th introduction of salt water leads to a potential for water pollution in the disposition of the water now contaminated by the tank residues.

2. Discharging Ballast. Dirty ballast may no longer be discharged freely into the sea. If the cargo previously carried may pollute the marine environment, it may not be possible to discharge any dirty ballast at all. At the present time it is possible to discharge a very small amount of oil when carrying out the procedure known as *load on top* in the course of which separation of salt water from the oil is gradually carried out. In addition to the air quality problem, the ballasting process endangers the ship itself if the cargo previously carried is a volatile one. The vapors and gases forced out of the compartment are extremely dangerous and great care must be exercised to avoid the possibility of igniting the gases.

3. Tank Cleaning. During the ballast passage some or all of the tanks may be cleaned. Except in the case of ships whose cargo tanks are exclusively dedicated to the carriage of a particular cargo whose purity must be preserved or the materials of which cannot be exposed to salt water, it is eventually necessary to clean the tanks to remove accumulated contaminants, to permit a change to another cargo, and to permit inspection and repair. The crude oil carrier is difficult to clean because of the large amounts of contaminants combined with the explosive potential of the vapor in the residue. As a safety measure the atmosphere in the tank may be inerted before the process begins. This action has been adopted relatively recently because of unexplained explosions which occurred several years ago in tank ships engaged in this operation. Alternatively, the hydrocarbon level in the tank may be reduced by blowing in fresh air to bring the tank below the explosive limit before cleaning begins.

Tank cleaning is accomplished by washing the tank with high pressure water streams directed at the tank surface by portable rotating nozzles lowered through a special tank-cleaning opening in the deck and suspended and lowered by its water supply hose. Salt water at high pressure 29 kPa (200 psi) is supplied through the deck fire line by a tank-cleaning pump in the machinery space. A tank-cleaning heater is also provided to heat the water if desired. Either hot or cold water is used depending primarily on the cargo last carried in the tank. The nozzles are lowered and raised by hand and must be repositioned a number of times depending on the depth of the tank. Fixed nozzles are being supplied on some ships in specialized services, particularly where tanks are coated so that the scale problem does not exist and where, therefore the effective range of the nozzle is increased without any increase in water pressure. A final washdown with hoses is often necessary to reach inaccessible corners and undersides of structure which cannot be reached effectively by the stream from the nozzles.

Oily water and scale are removed through the tank stripping line either by the stripping pump or by eductors located in the pump room which are powered by water from the cargo pumps. The eductors are preferable because handling of scale by the stripping pump can result in damage to the pump. The accumulation of oil and water is gathered and separated. It was once common practice to pump the accumulation overboard and this practice, termed operational pollution, has been responsible for most of the oil dumped in the sea much more than due to tanker accidents. The procedure is now prohibited; instead, the separation process returns to the sea clean water and retains the slops on board. The slops are either discharged ashore or are

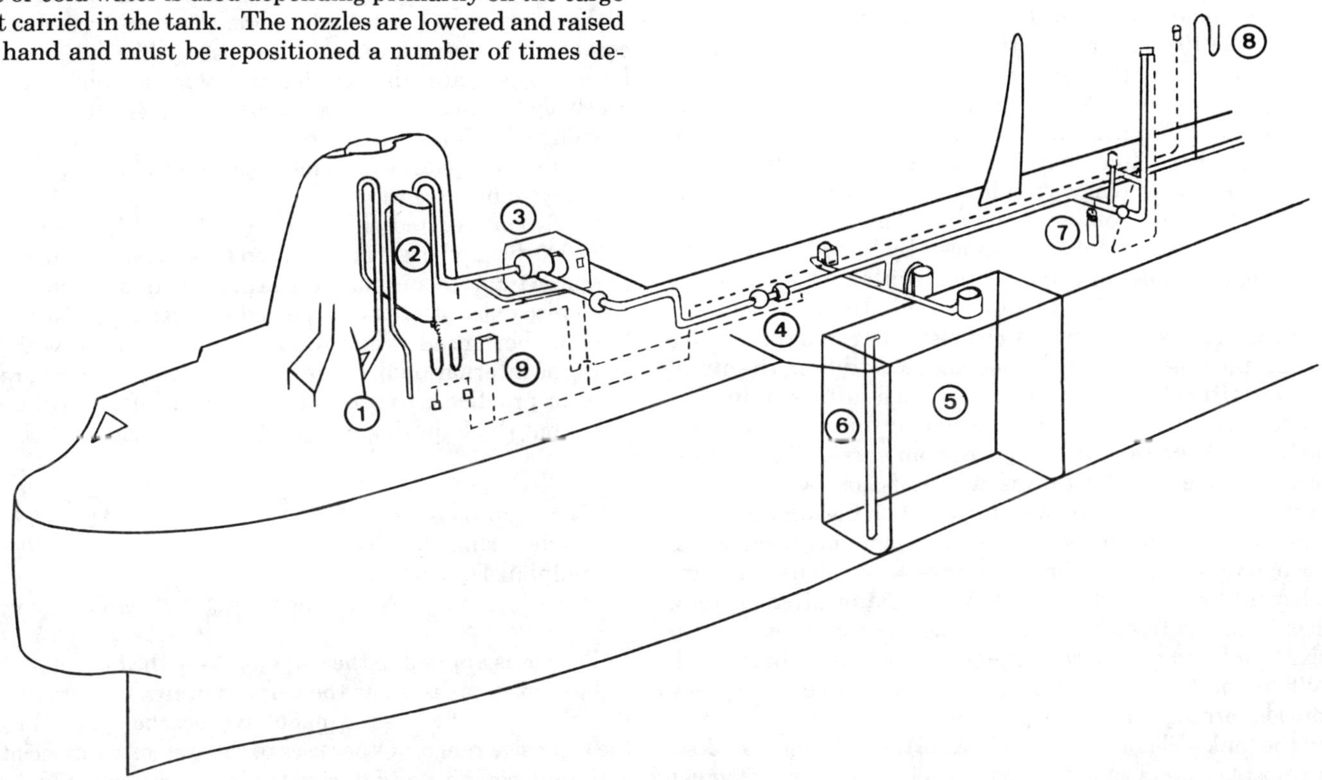

Combustion gases are drawn from the boiler uptakes at (1) and led through a water scrubber (2), drawn by a high-capacity fan (3), which forces the washed, cooled gas along a main header through a system of check, control, and safety valves (4). From the main header, the inert gas enters branch lines into the tank trunks to the cargo tank (5). Each tank has a purge pipe extending from the top of the tank to within a foot of the bottom (6). When the purge pipe is opened at the top it permits a complete circulation of the inert gas within the tank and when the oxygen level has been sufficiently reduced—ideally to about 5 percent instead of 21 percent of air—the purge pipe is closed. Thereafter the system remains under slight pressure, limited by an oil seal safety arrangement in the line at (7) which relieves to atmosphere. Gaging and alarms are installed on the bridge (8) and the engine room (9). The scrubbing tower, in some designs, is built into the top of the engine room.

Fig. 5 Inert gas installation

added on the top of the next load of cargo lending the name *load on top* to this practice. Once a tank has been cleaned and stripped of residue, it is ready for the new, different cargo. Clean sea ballast may also be taken in permitting the vessel on loading the next cargo to discharge the clean ballast water into the harbor at the loading facility without causing pollution.

A new method the industry is experimenting with in the interest of reducing water pollution is called *crude washing* wherein the oil itself is used as the solvent to speed removal of the residues on the interior surfaces of the tank. The oil is more effective than water in removing the clingage from the tank interior surfaces and therefore produces a tank which, following discharge, contains less oil residue. The washing is conducted in port during the final stages of discharge prolonging that phase of operation, whereas water washing is carried out on the return voyage. Because of the generating of vapor in the tank, inerting is always required during crude oil washing.

The virtues of crude oil washing in preventing operational pollution have been recognized and international requirements are in process to require crude oil washing capability to be fitted in new crude oil tankers of 20,000 dwt or greater. In existing tankers crude oil washing is among three alternatives one of which must be fitted in a phase-in program under the 1978 protocol to the IMCO 1973 Convention for Prevention of Pollution.

e. Inspection. At infrequent periods personnel must enter a tank for inspection or repairs involving burning or welding. From time to time it will be necessary for men to enter the tank to remove residues which accumulate and do not drain. This combination of scale and rust along with contaminants can only be removed by hand methods, requiring personnel to enter the tank with buckets. Before persons are permitted to enter a tank, the tank must be certified *gas free* by a marine chemist. To reach the gas-free stage, the tank must be cleaned and, in addition, the atmosphere after cleaning must be changed a sufficient number of times to insure that no explosive or toxic condition remains and the tank is safe for human entry. The tank is ventilated either by canvas windsails, or by steam- or water-driven exhaust blowers mounted on the tank-cleaning openings. Windsails are no longer commonly used as the primary means of ventilation but they are still used to supplement blowers and to maintain ventilation after the tank has been gas-freed. Systems using the cargo mains with exhaust blowers located in the pump room have been used. Where inert gas systems are provided for cargo tanks, they are also arranged for use in gas-freeing. Since all of the scale in the tank will not be removed by the tank-cleaning process, explosive vapors will continue to evolve and the tank must be frequently checked to ensure that it remains gas-free and safe for entry and work.

It is particularly important in the case of chemical tankers that attention be given to the problem of cleaning between cargos. The interior design of the tanks to insure effective draining, the provision of cleaning openings, stripping pumps to remove the wash water, and the ability to remove accumulated vapors all have to be considered. When dealing with chemicals, fresh water washing systems must be utilized because salt residues are unacceptable. This means there must not only be a sufficient quantity of fresh water but a suitable system involving a water heater and a pump and means of drying out the tanks afterwards.

f. Inerting. Inerting systems are fitted to prevent explosive conditions, reduce corrosion, detect leakage, or reduce heat transfer. Gases used for the purpose include nitrogen (either from charged cylinders or a nitrogen generating plant), carbon dioxide, argon and helium, or exhaust gases drawn from the stack or from an inert gas generator and scrubbed to remove particulate matter and acid-forming elements.

In chemical carriers the void spaces surrounding the tanks are often kept inerted. The presence of an inert atmosphere is a safeguard against fire or explosion should cargos of a flammable nature escape to the voids. It is also essential in some applications to preclude moisture that would enter if the voids were permitted to *breathe* freely. Moisture combined with small quantities of cargos like sulfur could form acids, or cause the deterioration of insulation. The gas is usually force-circulated past sensing elements to give early warning of leakage. The same procedure may be applied to the annular spaces within insulating assemblies associated with cryogenic cargos.

Newly constructed oil tankers (both crude and product of 20,000 dwt and over) are required to be capable of inerting the cargo tanks to minimize the explosive state which would otherwise exist after the cargo is discharged. Existing oil tankers, both crude and product, are to be fitted with inerting as internationally agreed under a phase-in program. Present practice is to use stack gas which, after scrubbing, has about the following composition percentages:

Oxygen (O_2)	2½–4½
Nitrogen (N_2)	About 77
Carbon Dioxide (CO_2)	12–14½
Sulphur Dioxides (SO_2)	0.02–0.03
And some particles of water vapor	

The gas is applied to the cargo tanks with vents closed at a slight over-pressure as the cargo is drawn off. Tank atmosphere thus becomes a mix of hydrocarbon vapor below the explosive range. Other uses of the system are associated with tank cleaning and during the laden voyage. Fig. 5 explains the operation as applied to an oil tank ship (King, 1971).[1]

---

[1] Complete references are listed at end of chapter.

Section 4
Design Requirements

4.1 Hull Arrangement and Cargo Containment. Because of the concerns over safety of the ports and of the environment, ships transporting oil and hazardous substances in bulk are subject to a tremendous amount of regulation. In recognition of the international character of the trade, the USCG effort has been to develop detailed international regulations through IMCO and to avoid unilateral requirements peculiar to U. S. operations. It is then the responsibility of the government of registry to insure compliance of the ship's design with the international criteria. This program has generally succeeded and the major shipbuilding countries virtually all conform to the same construction requirements. However, when the vessel enters service, differences develop according to differing levels of attention to maintenance by the nations of registry.

The following discussion draws heavily upon the IMCO criteria which are broadly stated in concept. This is usual and necessary in international standards and criteria. When translated into national requirements a greater degree of detail can be expected. The reader is therefore advised to pursue the particulars in the case of U. S. construction through a study of the relevant USCG Regulations.

Common to the codes and regulations governing construction of oil, gas, and chemical carriers are procedures in the design stage in which certain damage conditions are assumed against which criteria for the determination of the survival of the ship must be met, and limits on the possible release of the cargo satisfied. These criteria of the International Loadline Convention of 1966 (ILLC, 1966) apply to all ships. An oil tanker is also governed by the "International Convention for the Prevention of Pollution from Ships, 1973" (ICPPS, 1973). In the case of a chemical carrier, the IMCO Chemical Code must be met and, in the case of a gas carrier, the Gas Carrier Code. Table 13 lists the extent of damage assumed in these respective instances. Table 14 specifies the location of the assumed damage with particular emphasis on the flooding, whether externally or internally caused, of the most vulnerable space in any ship—the engine room. Table 15 stipulates the survival capability requirements and Table 16 the limitations on the

Table 13—Extent of Damage

		ILLC, 1966 (5)	ICPPS, 1973 (5)	IMCO CHEMICAL CODE	IMCO GAS CODE (5)
SIDE DAMAGE (1)	Longitudinal	Two-compartment standard (2)	$\frac{1}{3} L^{2/3}$ or 14.5 m	$\frac{1}{3} L^{2/3}$ or 14.5 m	$\frac{1}{3} L^{2/3}$ or 14.5 m
	Transverse	$B/5$	$B/5$ or 11.5 m	$B/5$ or 11.5 m	$B/5$ or 11.5 m
	Vertical	From baseline upward without limit	From baseline upward without limit	From baseline upward without limit	From baseline upward without limit
	Minor/local	None required	None required	760mm (3)	760mm(4)
BOTTOM DAMAGE (1)	Forward 0.3 L	None required	$\frac{1}{3} L^{2/3}$ or 14.5m	$L/10$	$\frac{1}{3} L^{2/3}$ or 14.5m
	Aft 0.7 L	None required	$L/10$ or 5m	$L/10$ or 5m	$L/10$ or 5m
	Forward 0.3 L	None required	$B/6$ or 10m	$B/6$ or 10m	$B/6$ or 10m
	Aft 0.7 L	None required	5m	5m	$B/6$ or 5m
	Vertical	None required	$B/15$ or 6m	$B/15$ or 6m	$B/15$ or 2m

(1) Where two values are given, the lesser extent applies for each specific case on a particular vessel.

(2) ILLC (1966) does not specify longitudinal extent of damage. However, for the assignment of a Type "A" freeboard, a two-compartment damage stability standard within the cargo length of the ship is required. The minimum bulkhead spacing is assumed to be $(3.05 + 0.03L)$ or 10.7 m, whichever is less.

(3) See tank location requirements in Table 16.

(4) See survival requirements in Table 15 and tank location requirements in Table 16.

(5) If lesser extent of damage than the maximum specified would result in a more severe condition, such damage must be considered.

(6) Length (L) means 96 percent of the total length in meters on a waterline at 85 percent of the least molded depth measured from the top of the keel, or the length from the fore side of the stem to the axis of the rudder stock on that waterline, if that be greater. In ships designed with a rake of keel the waterline on which this length is measured shall be parallel to the designed waterline.

(7) Breadth (B) means the maximum breadth of the ship, measured in meters amidships to the molded line of the frame in a ship with a metal shell and to the outer surface of the hull in a ship of any other material.

Table 14—Damage Locations

	ILLC (1966)
$L < 225$m:	Two-compartment damage within the cargo length; engine room not considered.
$L > 226$m:	Two-compartment damage within the cargo length and one-compartment damage for the engine room (engine room considered separately as a single flooded compartment).
	ICPPS (1973)
$L < 100$	The administration may allow relaxations from paragraph (3) of Regulation 25 (General Damage Stability Requirements).
$L < 150$m:	Damage assumed between transverse bulkheads; engine room not considered.
150m $< L <$ 225m:	Damage assumed anywhere in the ship's length, except the engine room; engine room considered separately as a single flooded compartment.
$L > 225$m:	Damage assumed anywhere in the ship's length, including the engine room.
	IMCO GAS CODE
Type IG:	Damage assumed anywhere in ship's length, including engine room. Minor damage assumed to occur anywhere in the cargo length.
Type IIG, $L < 150$m:	Damage assumed anywhere in cargo length; engine room considered separately as a single flooded compartment. Minor damage assumed to occur anywhere in the cargo length.
Type IIG, $L > 150$m:	Damage assumed anywhere in ship's length, including engine room. Minor damage assumed to occur anywhere in the cargo length.
Type IIPG, $L < 150$m:	Damage assumed between transverse bulkheads[1], including engine room. Minor damage assumed to occur anywhere in the cargo length.
Type IIPG, $L > 150$m:	No definition for a Type IIPG ship over 150m.
Type IIIG, $L < 125$m:	Damage assumed between transverse bulkheads[1], engine room not considered. The ability to survive flooding of the engine room should be considered by the administration. Minor damage assumed to occur anywhere in the cargo length except on a transverse bulkhead.
Type IIIG, $L > 125$m:	Damage assumed between transverse bulkheads[1], including engine room. Minor damage assumed to occur anywhere in cargo length except on a transverse bulkhead.
	IMCO CHEMICAL CODE
Type I:	Damage assumed anywhere in ship's length, including the engine room.
Type II, $L < 150$m:	Damage assumed anywhere in cargo length; engine room considered separately as a single flooded compartment.
Type II, $L > 150$m:	Damage assumed anywhere in ship's length; including the engine room.
Type III, $L < 125$m:	Damage assumed anywhere in cargo length; engine room not considered. The administration will determine the ship's ability to survive flooding of the engine room.
Type III, $L > 125$m:	Damage assumed anywhere in cargo length; engine room considered separately as a single flooded compartment.

[1] Spaced greater than the assumed longitudinal extent of damage

Note: For type definition of gas ships see Table 17 and for chemical carriers, Table 18.

placement of tanks in proximity to the shell.

The assumed damage conditions are rather severe and are intended to include a significant portion of the potential scope of damages which might be experienced. The permeability of spaces flooded is also defined. Depending upon the character of the cargo and its hazard, higher risks must meet with greater safeguards.

Differences among these criteria as applied to oil tankers, gas carriers and chemical ships are the consequence of the decade extending from the development of the International Loadline Conference of 1966 to the adoption of the Gas Carrier Code during the course of which international concern intensified with respect to environmental safety. In addition, an attempt has been made to address the economics of the operation and to avoid very sharp and unproven changes in standards between existing practice and future construction. More severe requirements for gas carriers and chemical ships were possible because the most significant of the international requirements were under draft in advance of any extensive development of those vessels, whereas the oil tanker was a longstanding ship type. With the accidents which have overtaken oil tankers, more severe requirements have been raised. With the passage of time a closer alignment of these criteria will likely occur. Tables included indicate their present condition or statement.

a. IMCO Gas Ship Code. The Gas Carrier Code adopted by IMCO in 1975 is directed toward liquefied gases which do not generally present a water pollution hazard, but rather the hazards of flammability, toxicity, and either depressed temperature or pressurized stowage. Ships subject to the Code are expected to survive the normal effects of flooding following certain assumed levels of hull damage externally caused and, in addition, in order to protect ship and environment, the cargo tanks are required to be protected from penetration in the event of minor damage to the ship as for example, from handling alongside by tugs, and having as well a degree of protection from damage in the event of collision or stranding. To this end, the cargo tanks are required to be situated within the hull a certain minimum distance from the shell. The damage assumed and the proximity of the cargo tanks to the side and bottom are dependent upon the hazard inherent in the product.

A graduation is made in three levels, Type IG, Type IIG and Type IIIG with the Type IG representing the greatest overall hazard. Hence, a ship intended for the transport of such a product is required to sustain and survive the greatest extent of hull damage with the least likelihood of releasing the cargo. A list of products as against the type required is shown in Table 17. When it is proposed to carry more than one product, the requirements for ship survival are those appropriate to the product having the highest ship type requirement but the required placement of the cargo tanks is governed by that of the specific product.

In addition to those considerations mentioned previously concerning the arrangement of the ship and its cargo area, the Code goes into considerable detail with respect to the segregation of the cargo from accommodations, service spaces, and other areas of the ship. However, the portion of the Code that is developed in the most profound detail pertains to cargo containment. This Section develops in great detail four different forms of container;

1. integral tanks,
2. membrane tanks,
3. semi-membrane tanks, and
4. independent tanks.

In each instance the Code develops the required particulars of the design loads, the required structural analysis, the allowable stresses and corrosion allowance plus supports, provision for secondary barriers, insulation and choice of materials, construction and testing, and requirements for stress relief.

The compound structure and insulation details of example forms of LNG tank designs were given earlier in Chapter II.

Table 15—Survival Capability Requirements

	ILLC (1966)	ICPPS (1973)	IMCO CHEMICAL CODE	IMCO GAS CODE
Maximum angle of heel after damage, degrees	15, or 17 if no deck edge immersion	25, or 30 if no deck edge immersion	15, or 17 if no deck edge immersion (1)	30 (2)
Residual stability after damage	The damaged righting arm curve must be positive for 20 deg beyond the final equilibrium angle with a maximum righting arm of at least 100mm within this range.			
Stability upright after flooding	$GM \geq$ 50mm	NONE	NONE	See Note (3)
Minor/local damage	NONE	NONE	NONE	30 deg, but also (4)

(1) $L < 150$m: The Administration can accept maximum angles of heel after damage up to 25 degrees, provided it is positively shown that a lesser angle is not attainable. $L > 150$m: 15 or 17 deg if no deck edge immersion.

(2) At the final angle of heel, the emergency power supply must be capable of operating, and lifesaving appliances must be capable of operating from at least the low side.

(3) Equalization arrangements requiring mechanical aids are not to be considered for the purpose of reducing an angle of heel to meet survival requirements; and if used, positive stability is to be maintained during all stages of equalization.

(4) Under local damage to a depth of 760 mm (30 in.) the angle of heel attained must not be such as to prohibit the use of the ballast system, the restoration of propulsion capability at reduced speed, nor the restoration of steering engine power at reduced speed.

Table 16—Tank Location Requirements

	ILLC (1966)
	None
	ICPPS (1973)
	None
	IMCO CHEMICAL CODE
Type I:	$B/5$ or 11.5m, whichever is less, from the side shell at the summer load line; and $B/15$ or 6m, whichever is less, from baseline nowhere closer than 760mm to shell plating
Type II:	$B/15$ or 6m, whichever is less, from baseline; and 760mm inboard from ship's side at right angles to the summer load line
Type III:	None
	IMCO GAS CODE
Type IG:	$B/5$ or 11.5, whichever is less, from the side shell at the summer load line; and $B/15$ or 2m, whichever is less, from baseline nowhere closer than 760mm to shell plating
Type IIG:	$B/15$ or 2m, whichever is less, from baseline; and nowhere closer than 760mm to shell plating
Type IIPG:	Same as Type IIG
Type IIIG:	Same as Type IIG

b. Chemical Carrier Code. This Code, adopted by the IMCO Assembly in 1971, applies to bulk cargos of dangerous chemical substances other than petroleum or similar flammable products; 1, having significant fire hazards exceeding those of petroleum products, and 2, having significant hazards in addition to or other than flammability. The Code is limited to products which are liquid at normal temperatures. Products which fall within the Code are listed in Table 18 and products which have been determined *not* to fall within the scope of the Code are in Table 19. Essentially the products included are those which have higher hazard ratings according to the scheme discussed earlier developed by NAS. The Code is an extraordinarily complex document which attempts to elaborate broad principles applicable to a wide variety of substances and then goes into detail with respect to certain individual substances in order to reflect

Table 17—Products to Which IMCO Gas Code Applies, Including Ship-Type Requirements

PRODUCT	SHIP TYPE	PRODUCT	SHIP TYPE
Acetaldehyde	IIG/IIPG	Ethylene oxide	IG
Ammonia, anhydrous	IIG/IIPG	Methane (LNG)	IIG
Butadiene	IIG/IIPG	Methyl acetylene—propadiene mixture	IIG/IIPG
Butane	IIG/IIPG		
Butane/propane mixtures	IIG/IIPG	Methyl bromide	IG
Butylenes	IIG/IIPG	Methyl chloride	IIG/IIPG
Chlorine*	IG	Nitrogen	IIIG
Dimethylamine	IIG/IIPG	Propane	IIG/IIPG
		Propylene	IIG/IIPG
Ethane	IIG	Refrigerant gases (dichlorodifluoromethane, etc.)	IIIG
Ethylamine	IIG/IIPG		
Ethyl chloride	IIG/IIPG	Sulphur dioxide	IG
Ethylene	IIG	Vinyl chloride	IIG/IIPG

* Certain special requirements specifically for the carriage of chlorine are presently in a developmental stage.

Table 18—Chemicals to which IMCO Chemical Code applies, including ship type requirements

PRODUCT	SHIP TYPE	PRODUCT	SHIP TYPE
Acetic acid	III	Ethylene dichloride	II
Acetic anhydride	II	2-Ethylhexyl acrylate	III
Acetone cyanohydrin	II	Ethyl methacrylate	
Acetonitrile	II	Formaldehyde, 37% aqueous solution	III
Acrylic acid	III	Formic Acid	III
Acrylonitrile	II	Furfural	III
Adiponitrile	III	Hydrochloric acid	III
Allyl alcohol	II	Isoprene	III
Allyl chloride	II	Iso propyl amine	II
Aminoethylethanolamine	III	Mesityl oxide	III
Ammonia aqua less than 28%	III	Methyl acrylate	II
Aniline	II	2-Methyl, 5 ethyl pyridine	III
Benzene	III	Methyl methacrylate	II
Benzyl chloride	II	Methyl styrene	III
iso-Butyl acrylate	II	Monoethanolamine	III
n-Butyl acrylate	II	Mono ethyl amine	II
n-Butyl ether	III	Monoisopropanol-0; amine	III
Butyl methacrylate	III	Mono-nitrobenzene	II
iso-Butyraldehyde	III	Morpholine	III
n-Butyraldehyde	III	M.F. anti-knock compounds	II
Camphor oil	III	Naphthalene, molten	III
Carbolic acid	II	Nitric acid, 70% & over	II
Carbon disulfide	II	2-Nitropropane	III
Carbon tetrachloride	III	o- and p-Nitrotoluene	II
Caustic potash	III	Oleum	II
Caustic soda	III	Paraldehyde	III
Chlorobenzene	III	Pentachloroethane	III
2-Chloroethanol (ethylene chlorohydrin)	II	Phenol	II
Chloroform	III	Phosphoric acid	III
Chlorohydrins, crude	II	Phosphorus	I
Chloroprene	II	Phthalic anhydride	III
Chloro sulfonic acid	I	B-Propiolactone	II
Coal tar naphtha	III	Propionic acid	III
Creosote	III	Propionic anhydride	III
Cresols (mixed isomers)	III	Propylamine	II
Crotonaldehyde	II	Propylene oxide	II
Cyclohexanone	III	Pyridine	III
Cyclohexylamine	III	Sodium Hypochlorite (15% solution)	III
Decyl acrylate	III	Sodium hydrosulphide (45% solution)	III
Dibutylamine	III	Styrene monomer	III
o-Dichlorobenzene	III	Sulphuric acid (spent)	III
1,1-Dichloroethane	III	Sulfur liquid	III
Dichloroethyl ether	II	Sulfuric acid	III
Dichloromethane (methylene chloride)	III	Tetrachloroethane	III
1,1-Dichloropropane	II	Tetrahydrofuran	III
1,2 Dichloropropane	II	Toluene diisocyanate	II
1,3 Dichloropropene	II	Tricresyl phosphate (containing at least 1% ortho-)	II
Diethanolamine	III	Triethanolamine	III
Diethylamine	III	Triethylamine	II
Diethylaminoethanol (diethylethanolamine)		Triethylenetetramine	III
Diethylenetriamine	III	Urea ammonium solution (containing aqua ammonia)	III
Diisopropanolamine	III	n- and iso-Valeraldehyde	III
Dimethyl ethanolamine	III	Vinyl acetate	III
Dimethylamine (40% Aq)	III	Vinyl ethyl ether	II
Dimethylformamide	III	Vinylidene chloride	II
1-4 Dioxane	II	Vinyl toluene	III
Diisopropylamine	II		
Epichlorohydrin	II		
Ethyl acrylate	II		
Ethyl ether	II		
2 Ethyl 3 propyl acrolein	III		
Ethylene cyanohydrin	III		
Ethylene diamine	II		
Ethylene dibromide	II		

properly the required safeguards appropriate to those substances. Indicative of the specific Code requirements are the following which apply to Propylene oxide:

"4.7.1 Propylene oxide transported under provisions of this section should be acetylene free.

Table 19—Chemicals to which IMCO Chemical Code has been determined not to apply

Acetone	p-Cymene (isopropyl voluene)
Amyl acetate—iso	n-Decanol
Amyl acetate—n	iso-Decanol
Amyl acetate—sec	Decyl alcohol—n
Amyl acohol—n	Diacetone alcohol
Amyl alcohol—p, iso	Di iso butylene
Amyl alcohol—sec, n	Di iso butyl ketone
Amyl alcohol—sec, iso	Dibutyl phthalate
Amyl alcohol—tert	Di iso butyl phthalate
Butyl acetate—iso	Diethylbenzene
Butyl acetate—n	Diethylene glycol
Butyl acetate—sec	Diethylene glycol monobutyl ether
Butyl alcohol—iso, n, sec, tert	Diethylene glycol—monoethyl ether
Butyl benzyl phthalate	Diethylene glycol monomethyl ether
Cumene (isopropyl benzene)	Nonyl alcohol
Cyclohexane	Nonyl phenol
Cyclohexanol	iso-Octanol
Dipentene	Parafin wax
Dipropylene glycol	Pentane—n, iso
Dipropylene glycol monomethyl ether	Pentene—n, iso
Dodecyl benzene	Iso-Phorone
2-Ethoxyethyl acetate	Petrolatum
2-Ethoxyethanol	Petroleum naphtha
Ethyl acetate	Perchloroethylene
Ethyl alcohol	Proprionaldehyde
Ethylene glycol	Propyl acetate–iso, n
Ethylene glycol methyl butyl ether	Propyl alcohol—iso, n
Ethylene glycol monobutyl ether acetate	Propylene glycol
	Propylene glycol monomethyl ether
Ethylene glycol monomethyl ether	Propylene tetramer
2-Ethyl hexanol	Propylene trimer
Furfuryl alcohol	Solvent naphtha
Glycerine	Tetrahydronaphthalene
Heptane—n	Toluene
n-Heptanol	Trichloroethane—alpha, beta, 1.1.1
Heptene (mixed isomers)	Tridecanol
Hexane—n	Triethyl benzene
1-Hexanol	Triethylene glycol
Hexylene glycol	Tripropylene glycol
Methyl alcohol	Tripropylene glycol monomethyl ether
Methyl acetate	Turpentine
Methylamyl acetate	Wine
Methyl ethyl Ketone	Xylenes
Methyl isobutyl ketone	
Molasses	

NOTE: The list may be used as a guide in considering bulk carriage of products whose hazards have not yet been evaluated. Although the products listed above fall outside the scope of the Code, attention of Administrations is drawn to the fact that some safety precautions may be needed for their safe transportation. Accordingly Administrations should prescribe appropriate safety requirements. USCG requirements may differ also.

4.7.2 No other product should be transported in tanks certified for propylene oxide except that the Administration may approve subsequent transportation of other products and return to propylene oxide service if tanks, piping and auxiliary equipment are satisfactorily cleaned.

4.7.3 All valves, flanges, fittings and accessory equipment should be of a type suitable for use with propylene oxides and should be constructed of steel or stainless steel, or other materials acceptable to the Administration. Impurities of copper, magnesium and other acetylides-forming metals should be kept to a minimum. The chemical composition of all material used should be submitted to the Administration for approval prior to fabrication. Discs or disc faces, seats and other wearing parts of valves should be made of stainless steel containing not less than 11 percent chromium. Mercury, silver, aluminum, magnesium, copper and their alloys should not be used for any valves, gages, thermometers, etc. All packing and gaskets should be constructed of materials which do not react spontaneously with or lower the auto-ignition temperature of the propylene oxides.

4.7.4 Pressure rating of valves, fittings and accessories should be not less than the maximum pressure for which the cargo tank is designed or the shut-off head of the cargo pump, whichever is greater. Threaded joints in the cargo liquid and vapor lines are prohibited.

4.7.5 Filling and discharge piping should extend to within 100 mm (4 in.) of the tank bottom or any sump pit.

4.7.6 Suitable means should be provided to return vapors to the shore during cargo transfer. For this purpose, a valved connection should be provided to a vapor return line to shore.

4.7.7 Tanks carrying propylene oxide should be vented independently of tanks carrying other products.

4.7.8 Manifolds for mounting multiple safety relief valves may be fitted with acceptable interlocking shut-off valves so arranged that at all times the required relief valve capacity will be available to relieve internal pressure. The valving arrangements should be such that no vapor will escape even if the "out-of-service" relief valve is removed.

4.7.9 Enclosed spaces in which cargo tanks are located should be:

(a) inerted by injection of a suitable inert gas or well ventilated and monitored, or

(b) if an inerting system is not installed, be fitted with forced ventilation of such capacity to provide a complete change of air every three minutes and arranged in such a manner that any vapors lost into the space will be removed. The ventilation system should be in operation at all times during cargo transfer.

4.7.10 All ventilation machinery should be of non-sparking construction.

4.7.11 In case air should be allowed to enter the cargo pump or piping system during cargo transfer, vapor should not be discharged to the atmosphere.

4.7.12 Prior to disconnecting shore lines, the pressure in liquid and vapor lines should be relieved through suitable valves installed at the loading header. Liquid and vapor from these lines should not be discharged to the atmosphere.

4.7.13 Propylene oxide may be carried in gravity type tanks when carried at pressures less than 1.45 kPa (10 psig). Tanks should be designed for the maximum pressure expected to be encountered during loading, storing and discharging cargo.

4.7.14 Cargo tanks with a design pressure less than 1.31 kPa (9 psig) require a cooling system to maintain the propylene oxide below the boiling temperature at the pressure at which it is carried. The cooling system may not be required if it can be demonstrated that the propylene oxide can always be maintained below its boiling temperature at the pressure at which it is carried.

4.7.15 (a) Any cooling system should maintain the liquid temperature below 40°C (104°F) or below the boiling temperature, whichever is less. At least two complete cooling plants, automatically regulated by temperature variations within the tanks should be provided, each to be complete with the necessary auxiliaries for proper operation. The control system should also be capable of being manually operated. An alarm should be provided to indicate malfunctioning of the temperature controls. The capacity of each cooling system should be sufficient to maintain the temperature of the liquid cargo at or below the design temperature of the system.

(b) An alternate arrangement may consist of three cooling plants, any two of which should be sufficient to maintain the liquid temperature at or below the design temperature.

(c) Cooling systems requiring compression of propylene oxide are prohibited.

4.7.16 Pressure relief valve settings should not be less than 0.44 kPa (3 psig) for gravity tanks.

4.7.17 When propylene oxide is carried, piping systems in propylene oxide service should not be used for any other product and should be completely separate from all other systems. The piping system should be designed so that no cross connection may be made either through accident or design.

4.7.18 Filling density should not exceed 80 percent for non-refrigerated pressure vessels.

4.7.19 The cargo should be shipped under a suitable protective padding, such as nitrogen gas. Original charging of the gas pad at the loading facility is not adequate. Additional gas should be provided to maintain pad gas concentration. Any padding gas selected should be at least 98.0 percent pure and free of reactive materials.

4.7.20 Prior to, during, and after loading, if necessary, the cargo tank vapor space should be tested to ensure that oxygen content is 2 percent or less.

4.7.21 A water spray extinguishing system should be provided in the area where loading and unloading operations are conducted. The capacity and arrangement should be such as to blanket effectively the area in way of the loading manifold and exposed deck piping for propylene oxide. The rate of discharge and the arrangement of piping and nozzles should be such as to give a uniform distribution over the entire area protected. Additionally, means should be provided for local and remote manual operation. The arrangement should ensure that any spilled cargo is washed away. A water hose with pressure to the nozzle, when atmospheric temperatures permit, should be connected ready for immediate use during filling and discharge operations and any spillage of propylene oxide should immediately be washed away. The water spray extinguishing system should provide a uniform spray over the area of application of 0.175 $L/m^2/s$ (0.5 $gal/ft^2/s$).

4.7.22 A remote operational, quick closing shut-off valve should be provided at each cargo hose connection used in cargo transfer. Such valves should be of the fail-closed (closed on loss of power) type and be capable of local manual operation. The operating time for such valves should be such as to avoid excessive pressures in the piping on both ship and shore."

Like the Gas Code, the procedure is to set out degrees of physical protection in terms of the extent of damage a ship should be able to survive coupled to the separation of the cargo containment with respect to the ship's boundaries as a means of preventing the release of the cargo in the event of accident. The more severe requirements are imposed against the more hazardous substances in three gradations, Types I, II and III.

In addition, a limit has been established on the quantity of cargo which would be required to be carried in a Type I ship as not to exceed 1,250 m^3 (44,144 ft^3) in any one tank, and in the case of a cargo which would be required to be carried in a Type II ship, that it not exceed 3,000 m^3 (105,930 ft^3) in any one tank.

In ships which carry a multitude of products, it is the cargo with the greatest hazard which sets the requirements for the hull type. However, the location of the individual cargo containment is only applied to the cargo itself and not to the entire package. It is therefore possible to make certain combinations as long as the damage criteria assumed for collision and grounding can be imposed upon the design and still have it meet the final state required for survival, (Fig. 6).

The Code develops in detail additional requirements on the segregation of cargo, on arrangement of cargo pump rooms, cargo piping and hose, tank vent systems, cargo temperature control and control of vapor space in cargo tanks, pump and pipeline identification, gaging, etc.

By way of exhibiting the interrelation between other particulars and ship design and operational features, Figure 7 may be consulted for an appreciation of the intricacies of this Code.

c. Oil Tankers. In the case of oil tankers, the assumed damage is applied in connection with a formula for the size limitations of cargo tanks. It is intended to restrict the amount of oil which may escape in the event of accidental breaching of the hull from collision or stranding.

The cargo tanks are required to be so sized and arranged that the hypothetical oil outflow from the assumed damage imposed anywhere in the length of the ship does not exceed 30,000 m^3 (1,059,440 ft^3) or 400 times the cube root of the deadweight, whichever is the greater, subject to a maximum of 40,000 m^3 (1,412,590 ft^3). The volume of any one wing cargo oil tank is not permitted to exceed 75 percent of this

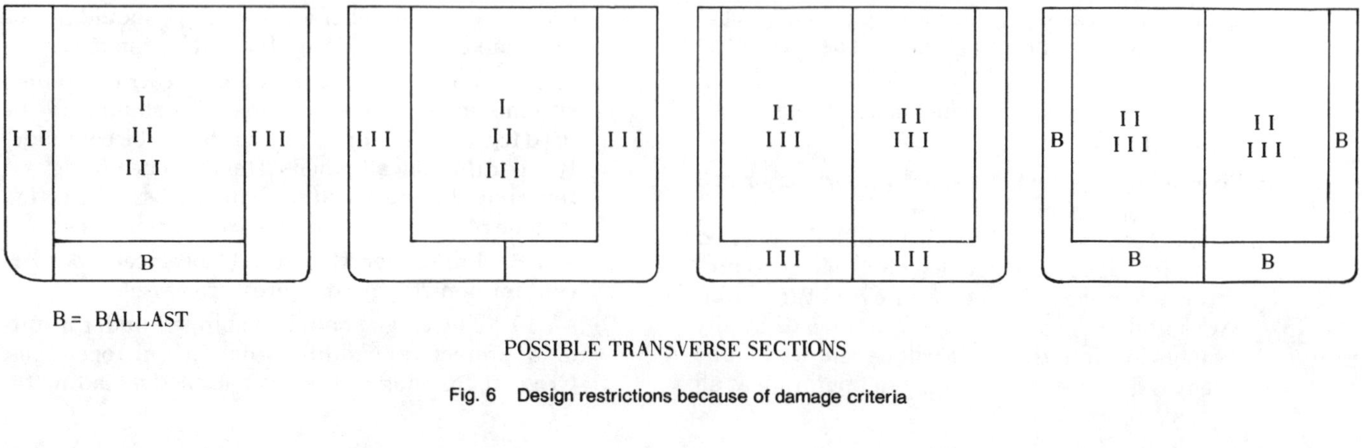

Fig. 6 Design restrictions because of damage criteria

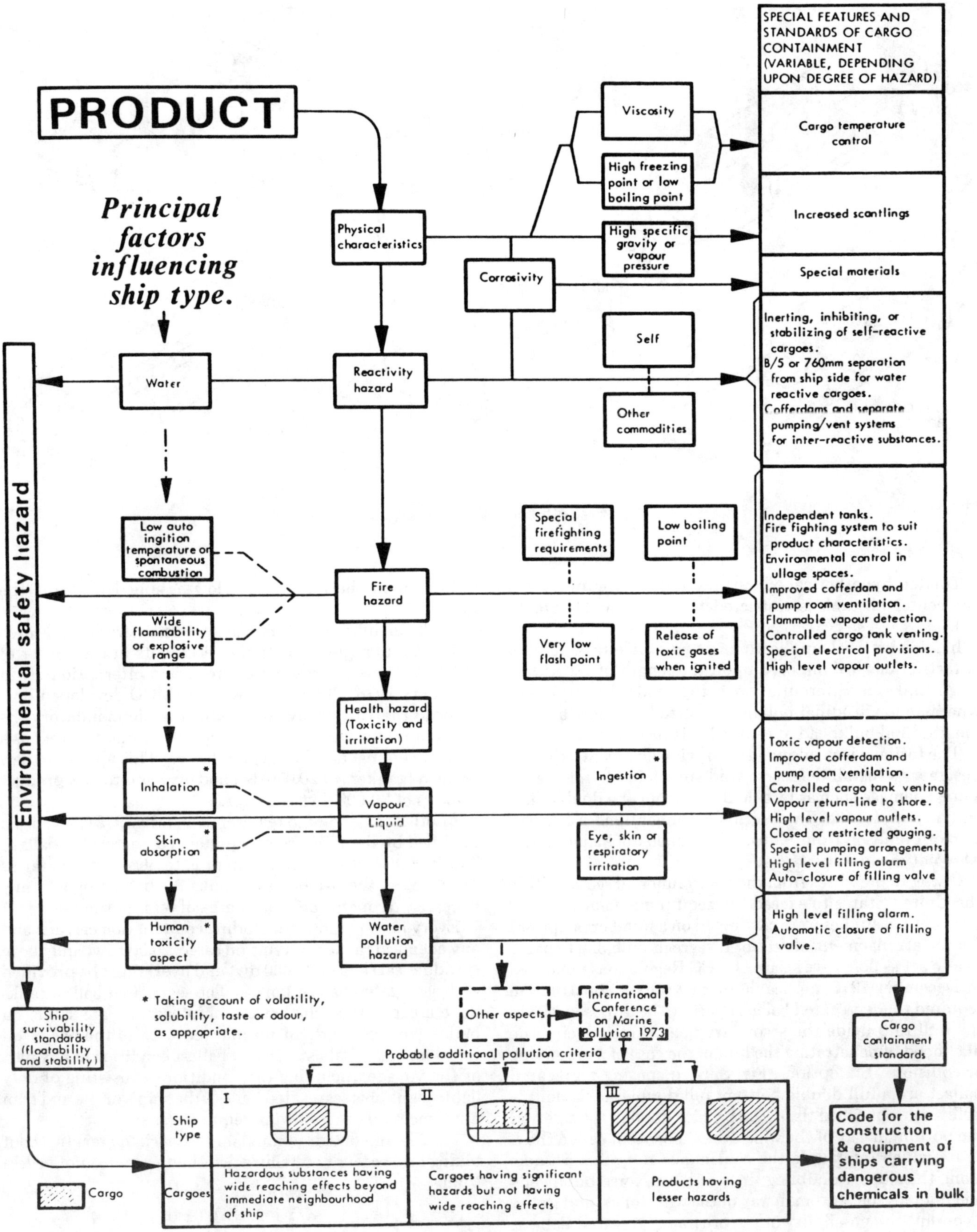

Fig. 7 Influence of physical and chemical characteristics of cargos on ship design

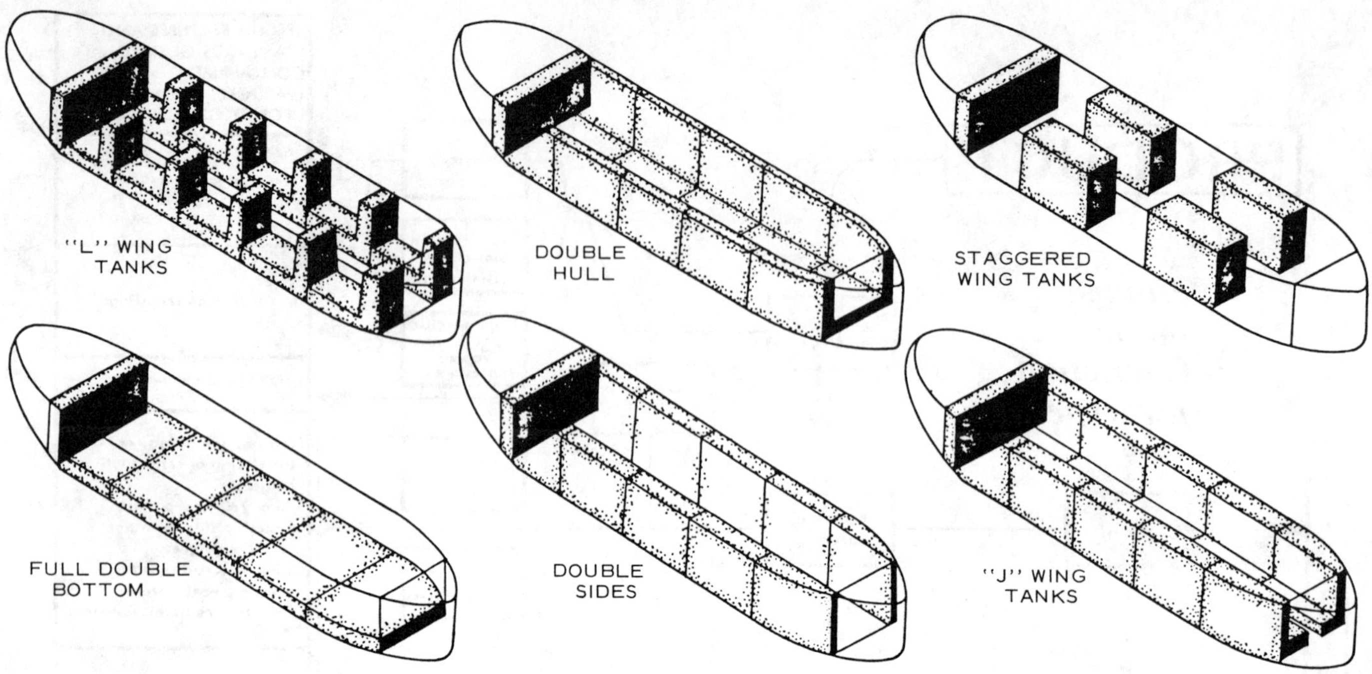

Fig. 8 Segregated ballast tanker study—ship configuration

calculated hypothetical oil outflow and the volume of any one center oil tank is restricted to 50,000 m^3 (1,765,730 ft^3).

In addition, with application to small tankers, a number of restrictions are imposed upon the length of individual cargo tanks, a differentiation being made for situations where no longitudinal bulkhead is fitted or where a single longitudinal bulkhead at centerline is provided.

The formula was intended to provide an incentive to the designer to install defensive void spaces in way of the assumed damage such as by a double bottom or a double side. Initially, under the "International Convention for Prevention of Pollution from Ships, 1973," such defensive spaces were not mandatory.

Owing to pressures from the environmental sector within the United States to exceed the requirements of the 1973 Convention, studies were conducted on a number of options for the arrangement of required segregated ballast spaces to also act as defensive space. USCG Regulations published in December 1976 applicable to new tanker construction required the segregated ballast spaces to be disposed within the hull to provide the secondary benefit of reducing the likelihood of penetrating the hull in the case of a grounding or collision. Six options, Fig. 8, for disposing segregated ballast are a full double bottom; full double sides; double hull; a concept called the *J Tank* which doubles the side and the bottom in way of the wing tanks; a concept called the *L Tank* which interrupts the continuous run of skin tanks along the side by doubling the wing transverse bulkheads and doubles the bottom in way of the wing tanks; and lastly, to permit the possibility of retrofitting segregated ballast, the alternating of wing tanks, empty and full, together with reduction in the allowed outflow assumed in sizing the tanks.

Under an initiative of the Carter Administration in March 1977, stimulated by the tanker accidents in the winter of 1976–77, these proposals were reviewed internationally in February 1978. The result was an IMCO developed protocol to the 1973 Convention which made mandatory the concept that segregated ballast spaces be placed to act as a defense against accidental oil release. This applies to new crude oil tankers of 20,000 dwt and over and to new product tankers of 30,000 dwt and over.

Another major design requirement of the 1973 Convention modified by the 1978 protocol pertains to segregated ballast. The purpose of this requirement is to eliminate, so far as practicable, the use of cargo tanks for ballasting and any necessity of operational discharge of oil into the sea.

Every new oil tanker, including combination carriers and any chemical tanker carrying oil cargo, 20,000 dwt and above (product carriers of 30,000 dwt and over) must be provided with segregated ballast tanks. The segregated ballast tanks have a capacity such that the ship can operate safely on ballast voyages without normal recourse to the use of oil tanks for water ballast. In any ballast condition at any part of the voyage, including the conditions consisting of only lightweight plus segregated ballast, the ship's drafts and trim must meet all the following requirements:

1. The molded draft amidships (dm) in meters (without taking into account any ship's deformation) must not be less than:

$$dm = 2.0 + 0.02\,L$$

where L is 96 percent of the total length on a waterline at 85 percent of the least molded depth measured from the top of the keel, or the length from the fore side of the stem to the axis of the rudder stock on that waterline, whichever is greater.

2. The drafts at the forward and after perpendiculars must correspond to those determined by the draft amidships (dm) in association with the trim by the stern of not greater than 0.015 L.

3. In any case, the draft at the after perpendicular must be sufficient to obtain full immersion of the propeller or propellers.

Ballast water may be carried in oil tanks only in severe weather conditions when the safety of the ship demands carrying additional ballast water in oil tanks. This additional ballast water must be processed and discharged in accordance with the operational discharge requirements of the Convention.

Seagoing vessels of less than 150 gross tons must retain on board any oily mixtures or transfer them to a reception facility. (Clean ballast and segregated ballast may be discharged overboard.) Seagoing vessels of 150 gross tons or more must discharge oil mixtures overboard in accordance with the criteria outlined below, or retain the oily mixture on board, or transfer the oily mixture to a reception facility. The use of chemicals to treat an oily mixture to circumvent the discharge requirements is not allowed.

An oily mixture from a cargo tank may be discharged into the sea if a tank vessel complies with all of the following:

1. Is more than 50 nautical miles from the nearest land,
2. is proceeding en route,
3. is discharging at an instantaneous rate of oil content not exceeding 60 liters per nautical mile,
4. does not discharge a total quantity of more than 1/15,000 for an existing vessel or 1/30,000 for a new vessel of the total quantity of cargo of which the discharge formed a part, and
5. has in operation the required oil discharge monitoring and control system.

An oily mixture from a machinery space bilge, except cargo pump rooms, may be discharged into the sea, unless combined with an oily cargo mixture, if the tank vessel complies with all of the following:

1. Is more than 12 nautical miles from the nearest land,
2. is proceeding en route,
3. is discharging an effluent with an oil content of less than 100 parts per million, and

Table 20—Classification of Barges Carrying Dangerous Cargos

<table>
<tr><th rowspan="3">BARGE CLASSIFICATION AND TYPICAL CARGOS / DESIGN FEATURES</th><th>Maximum Preventive Measures</th><th>Substantial Preventive Measures</th><th>Moderate Preventive Measures</th></tr>
<tr><th>TYPE I</th><th>TYPE II</th><th>TYPE III</th></tr>
<tr><td>Chlorine, Liquid Oxygen, Liquid Hydrogen, Molten Phosphorus Ethylene Oxide</td><td>Liquefied Petroleum Gases, Anhydrous Ammonia</td><td>Hydrochloric Acid, Sulfuric Acid, Sodium Hydroxide</td></tr>
<tr><td>Bow Rake</td><td colspan="3">Suitable for Design Speed of Tow</td></tr>
<tr><td>Open Hopper Coamings</td><td colspan="3">916mm (36 in.) FWD—610mm (24 in.) MIDLENGTH—457mm (18 in.) AFT</td></tr>
<tr><td>Breakwater Height</td><td colspan="3">916mm (36 in.)</td></tr>
<tr><td rowspan="2">Subdivision Stability</td><td colspan="2">Remain Afloat and Upright with Hopper Flooded to Deck Height</td><td>Remain Afloat and Upright Under Operating Conditions</td></tr>
<tr><td>Remain Afloat and Upright After Holing in way of Transverse Bulkhead</td><td>Remain Afloat and Upright After Holing not in way of Transverse Bulkhead</td><td>No Damaged Stability Requirement</td></tr>
<tr><td rowspan="2">Hull Structure</td><td colspan="3">CLASSIFICATION SOCIETY STANDARD</td></tr>
<tr><td colspan="2">Additional Strength Based on Grounding Study</td><td>No Additional Requirement</td></tr>
<tr><td rowspan="2">Cargo Tank Collision Protection</td><td colspan="2">Independent 1.3 g Longitudinal Collision Chock</td><td>Independent 1.0 g Longitudinal Collision Chock</td></tr>
<tr><td>7.5m (25 ft) from Headlog
1.22m (4 ft) from Side</td><td>7.5m 23 ft from Headlog
0.92 m (3 ft) from Side</td><td>No Distance from Headlog Specified 0.61m (2 ft) from Side Existing Regulations</td></tr>
<tr><td>Independent Cargo Tank Strength</td><td colspan="2">Sufficient to Withstand Loading Due to Barge Grounding Hull Deflection</td><td>Sufficient to Withstand Loading due to Stillwater Barge Hull Deflection</td></tr>
</table>

4. has in operation the required oil discharge monitoring and control system or the required oily water separating equipment.

Oil-water separating and filtering equipment will be required on new and existing tank vessels. These devices will be used for oily bilge water and ballast water from oil fuel tanks. All discharges of effluent from the cargo spaces of a tank vessel will be required to go through a monitoring and control system which will ensure that any oil discharge is automatically stopped when the oil content of the effluent exceeds that permitted by the discharge criteria. The monitoring and control system must be fitted with a recording device to provide a continuous permanent record of the oil content of the effluent. All of this equipment is essential in practicing the improved LOT system for shipboard handling cargo oil. New tank vessels will not be allowed to put ballast water in oil fuel tanks.

New tank vessels of less than 70,000 dwt and all existing tank vessels must have at least one slop tank. A new vessel of 70,000 dwt or more must have two slop tanks. A slop tank must have the capacity to retain slop from tank washings, oil residues, and dirty ballast residues, but the total capacity may not be less than three percent of the oil capacity of the vessel except two percent of the oil capacity of the vessel will be accepted if there is the required amount of segregated ballast space; or eductors that use water in addition to the washing waters are not fitted. Each slop tank must be designed with a separate inlet and outlet. Slop tanks may be used to carry cargo on the loaded leg of a voyage, since they are not required for treating oily mixtures during that time.

A tank vessel of 400 gross tons or more must have a tank that receives and holds oily residue resulting from purification of fuel and lubricating oil and oil leakages in machinery spaces. This sludge tank must have an adequate capacity determined by type of machinery installed on the vessel and the maximum fuel oil capacity. Each oily residue tank must facilitate cleaning and transfer of residue to a reception facility.

d. Barges. The concept of relating the hazard of the cargo to defensive construction of the vessel was first implemented in the United States for barges carrying hazardous substances. All facets of construction and design for these craft were eventually assembled into a body of regulations called Subchapter 0 of 46 CFR.

Barges are classed as Type I, II or III, according to the order of hazard as follows:

- Type I—Designed to carry substances requiring maximum preventive measures to avoid uncontrolled release of the cargo to water or atmosphere.
- Type II—Designed to carry substances requiring substantial preventive measures to avoid uncontrolled release to the atmosphere or waterways which substances may cause local or temporary pollution but no long lasting hazard.
- Type III—Designed to carry substances of a sufficient hazard to require moderate safeguards against release.

The constraints imposed affect the strength of the vessel and its component parts, the arrangement of the hull with respect to the buoyancy chambers of rake and wing spaces and hopper, if any, and the subdivision of the hull as well as the location of the cargo containment. The damage assumed in a barge is to provide for the abuse such craft receive in service; i.e., intentional grounding, and striking lock walls or other barges in fleeting. It is not intended to cope with a collision with an ocean going vessel.

For Type I and II barges in inland river service, the strength must be shown as adequate to endure an assumed grounding condition in which the forward rake bulkhead is resting upon a pinnacle at the water surface. The contribution to the strength and stiffness of the hull by independent tanks may be considered under certain conditions.

Cargo tanks for products to be carried in Types I and II barge hulls are required to be separated from the side and headlog. Barges in the Type III category correspond to the garden variety tank barge with the cargo carried in *skin* tanks. A summary of the requirements for all three classes appears in Table 20.

With respect to the container for the cargo, Subchapter 0 stipulates requirements comparable to those later adopted in the IMCO Gas Carrier Code and Chemical Ship Code concerning the provision of secondary barriers to protect the hull, stress relief, welding details, relief valve settings, and so on. Table 21 is an excerpt from USCG regulations setting out the relation between the cargo and the barge class required along with other limitations.

With the increased concern over environment, the regulations with respect to barges transporting oil and petroleum products subsequently came under review. At the present time the regulations are being studied but it is expected that requirements for barges carrying oil will be comparable in many respects to those corresponding to cargo intended for a Type II barge.

4.2 Piping Systems and Pumps. Tanker cargo systems, except in the case of certain specialized ships, are basically of two types, piped or free-flow, or combinations of the two. Except in the case of deep-well and submersible pump systems, one or more cargo pump rooms are provided, the number being determined primarily by the number of grades of cargo the vessel must carry simultaneously, as this determines the number of pumps and the size and complexity of the system. The cargo pump drives normally are located in the machinery space and the pumps are driven by shafting which penetrates the bulkhead through a liquid-and-gas-tight stuffing box. Deep-well pump drives are located on deck and pump drives using low temperature steam are located in the pump room. The main cargo pumps are of the centrifugal type, although both gear and reciprocating types are used where sizes are small or where the nature of the product dictates such a choice. The pump room also contains the stripping pumps which are normally steam reciprocating on the larger ships although steam or motor driven gear pumps are used also.

Piped systems are most used and consist of a number of large size mains, one for each tank grouping, with branches to each individual tank in the group. Fig. 9 is a typical diagram of a piped system, showing the main cargo oil piping in the pump room. Fig. 10 is a typical diagram of a piped

Table 21—Summary of Minimum Requirements

Cargo identification: Name	Cargo identification: Pressure	Temp.	Hull type	Cargo segregation Tank	Tanks: Type	Tanks: Vent	Tanks: Gaging device	Cargo transfer: Piping class	Cargo transfer: Control	Environmental control: Cargo tanks	Environmental control: Cargo handling space	Fire protection required	Special requirements (Section)	Electrical hazard group-class	Temp. control install.	Tank internal inspect. period—years
Acetaldehyde	Press.	Amb.	II	INA 2ii	Ind. Pressure	SR	Restr.	II	P-1	Inerted	Vent F	Yes	151.55-1(h)	I-C	NA	G
Acetic acid	Atmos.	Amb.	III	1i 2ii	Integral Gravity	Open	Open	II	G-1	NR	Vent N	Yes	151.55-l(g)	I	NA	G
Acetic Anhydride	Atmos.	Amb.	III	ii 2ii	Integral Gravity	PV	Restr.	II	G-1	NR	Vent F	Yes	151.55-l(g)	I	NA	G
Acetone cyanohydrin	Atmos.	Amb.	I	1ii 2i	Integral Gravity	PV	Closed	I	G-1	NR	Vent F	Yes	151.50-5	I	NA	G
Acetonitrile	Atmos.	Amb.	III	1i 2ii	Integral Gravity	PV	Restr.	II	G-1	NR	Vent F	Yes	No	I-D	NA	G
Acrylonitrile	Atmos.	Amb.	II	1ii 2ii	Integral Gravity	PV	Closed	II	G-1	NR	Vent F	Yes	151.55-1(e)	I-D	NA	G
Adiponitrile	Atmos.	Amb.	II	1ii 2i	Integral Gravity	PV	Open	II	G-1	NR	Vent F	Yes	No	I	NA	G
Allyl alcohol	Atmos.	Amb.	I	1ii 2ii	Integral Gravity	PV	Closed	I	G-1	NR	Vent F	Yes	151.50-5	I	NA	G
Allyl chloride	Atmos.	Amb.	I	1ii 2ii	Integral Gravity	PV	Closed	I	G-1	NR	Vent F	Yes	151.50-5	I-D	NA	G
Aminoethyl-ethanolami	Atmos.	Amb.	III	1i 2i	Integral Gravity	Open	Open	II	G-1	NR	Vent N	Yes	151.55-1(b)		NA	G
Ammonia, anhydrous	Press.	Amb.	II	1NA 2ii	Ind. Pressure	8R 250 p.s.i.	Restr.	II	P-2	NR	Vent F	No	151.50-30 151.50–32	I-D	NA	8
Ammonia, anhydrous	Atmos.	Low	II	1NA 2ii	Ind. Gravity	PV	Restr.	II-L	G-2	NR	Vent F	No	151.50-30 151.50-32	I-D	151.40-1(b)-(1)	

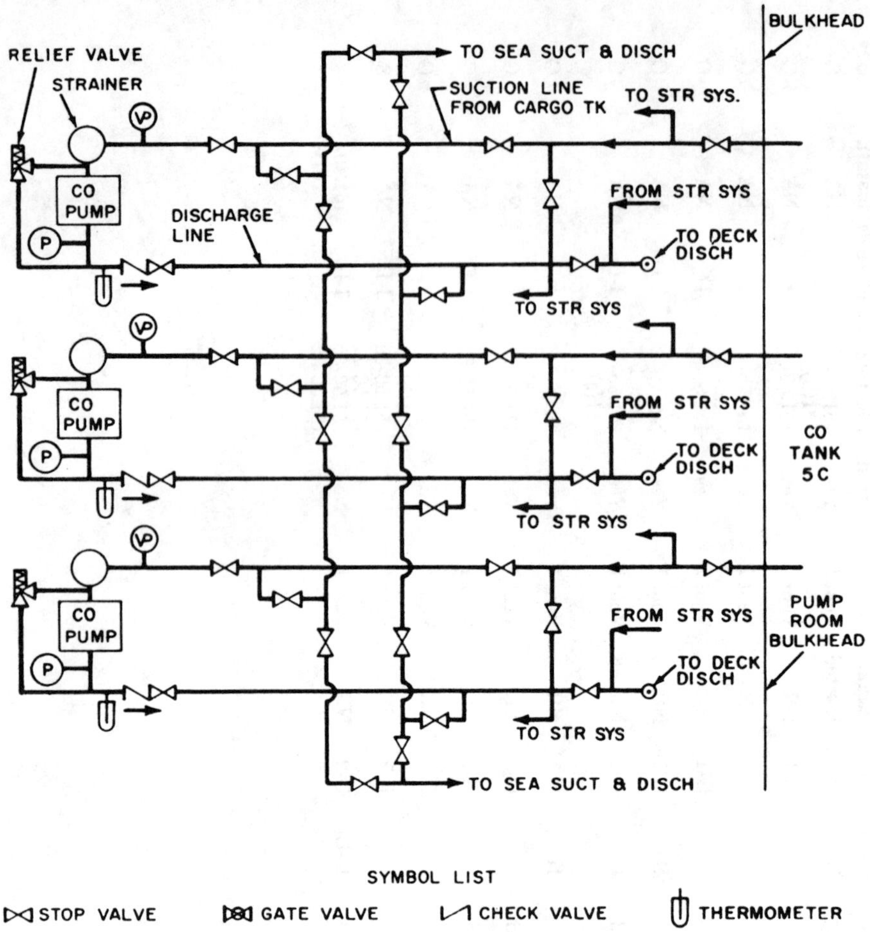

GENERAL NOTES:

1. ALL SYSTEMS BASED ON SAME TANK ARRANGEMENT
2. ALL SYSTEMS BASED ON CARRYING THREE GRADES OF CARGO
3. GRADE 1 CARRIED IN TANKS 1P, 1C, 1S, 2P, 2S
GRADE 2 CARRIED IN TANKS 2C, 3C, 4C,
GRADE 3 CARRIED IN TANKS 4P, 4S, 5P, 5C, 5S
THIS PROVIDES APPROXIMATELY EQUAL PARCEL SIZE
4. GRADES 1 AND 2 CROSS-CONNECTED ON CARGO MAINS
GRADES 2 AND 3 CROSS-CONNECTED ON CARGO MAINS
STRIPPING LINES NOT CROSS-CONNECTED
5. CLEAN BALLAST SYSTEM NOT SHOWN FOR SIMPLICITY
6. PUMPROOM BILGE PIPING NOT SHOWN FOR SIMPLICITY
7. LOADING DROPS TO CO TANKS NOT SHOWN

Fig. 9 Piped system. Main cargo-oil piping in pump room

system (with sluice valves), showing the piping in tanks. Fig. 11 is a typical diagram of a piped system, showing the piping on deck. The branches terminate in suction valves and bellmouths. The mains are cross-connected in the pump room on both the suction and discharge sides of the cargo pumps. A sea cross-connection is also provided. Where a ship is designed to handle varying package sizes or multiple grades of cargo, cross-connections and block valves are also provided between mains in the cargo tanks. When loading, cargo is distributed to the tanks through the suction lines, using either the cross-connections in the pump room or drops from the deck discharge lines to the suction mains. Tank valves are deck-operated either through reach rods which may be manually, pneumatically, or hydraulically operated, or by hydraulic operators located directly on the valves. Pneumatically or hydraulically assisted valve operation was originally introduced because of the increasing difficulty in opening and closing manually the larger valves, particularly those of about 400 mm (16-in.) diam size and over. While some ships now have power assistance for all valves because the entire system is remotely controlled, it is more common to have a combination of valve operators to suit the sizes involved. Even on remotely operated cargo systems, it is often desirable to limit such operation to those valves which must be manipulated during loading and discharging and to have manual operation for those valves which can be set up prior to starting and which do not need resetting during the operation.

In full free-flow systems, all piping in the cargo tanks is eliminated except for suction connections from the pumps to the after tanks. Individual tank bulkheads are fitted with sluice valves to allow the cargo to flow by gravity to the pump suctions. Although suction lines are eliminated, it is still necessary to have drop lines from the deck for loading. Full free-flow systems are normally not practical from the standpoint of trim control, speed of discharge, and limitation of the ship to a single grade of cargo. Modified free-flow systems are practical and have been employed. These systems are provided with bulkhead sluice valves to dispense with branch piping but have a suction main to each group of tanks. This arrangement provides better trim control and allows some cargo flexibility. Valve operation is similar to operation with conventional piped systems.

As tankers have increased in size without allowing a corresponding increase in discharge time, pump suction conditions have become an increasing problem. *Increased* tanker size not only has meant longer main suction pipes with resultant increased friction losses, but it has meant larger structure in the tanks. This has forced the piping to be carried higher in the ship, thus increasing the static lift of the pumps. Attempts to reduce friction losses have resulted in increases in suction pipe sizes to as large as 710 mm

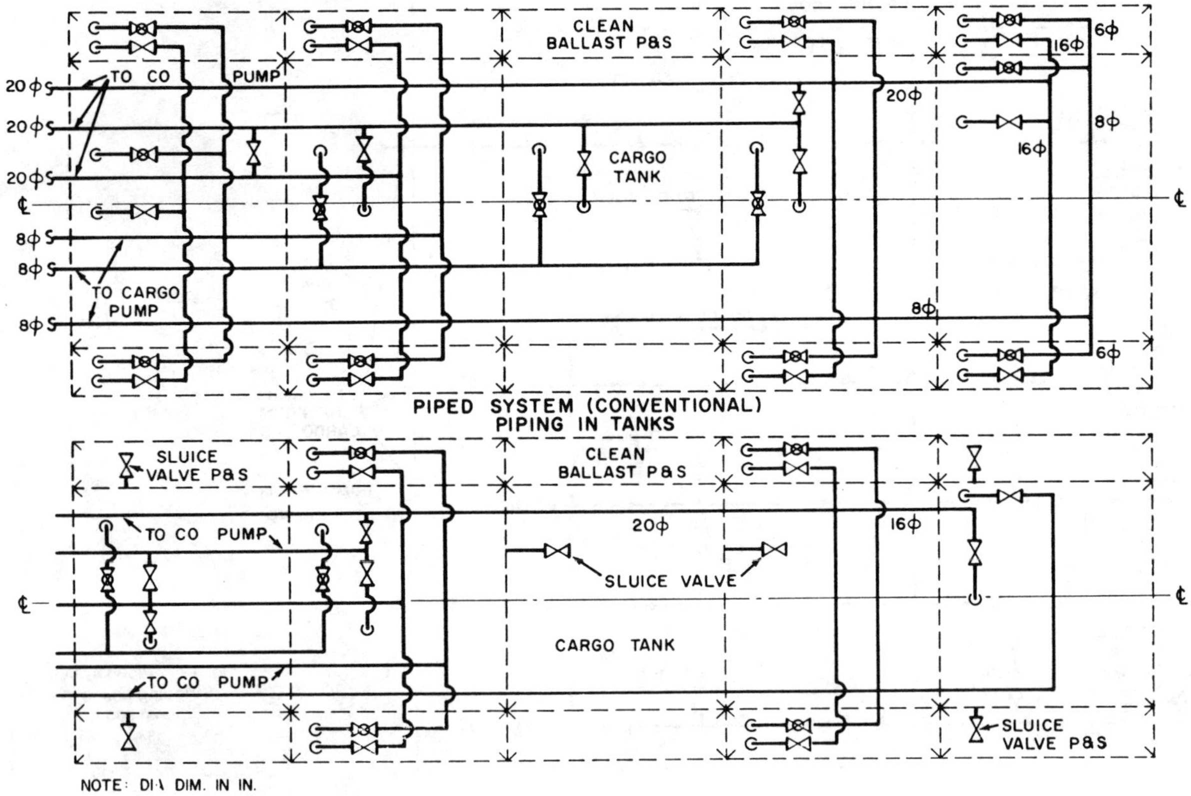

Fig. 10 Piped system (with sluice valves). Piping in tanks

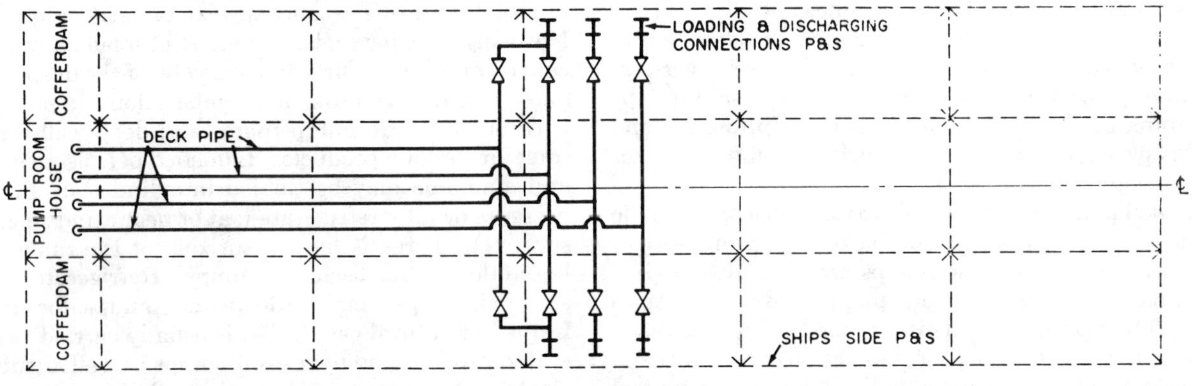

Fig. 11 Piped system. Piping on deck

(28 in.) in diam, but little can be done to decrease the height of the suction mains. Attempts to reduce the friction losses to a minimum and, particularly, to reduce the height of the cargo pumps have resulted in limited use of duct systems. A duct system is essentially a piped system using a large rectangular suction duct in lieu of the multiple suction mains of the conventional system. Either sluice valves or branch lines are provided. This system successfully reduces losses but restricts the ship to single grades of cargo where sluice valves are used, or to a limited number of compatible grades where branches are used. In any case, since the large vessels on which a duct is used are single grade ships this is not a handicap.

Piped, free-flow, and duct systems require stripping systems to prevent excessive amounts of cargo being left in the vessel because of a limited suction capability of the centrifugal-type main cargo pumps. Except for reduced size of pumps and lines, the usual stripping system is similar to the piped main cargo system. Fig. 12 is a typical diagram of a piped stripping piping system in the pump room. In addition to stripping the tanks of cargo left by the main pumps, the stripping pumps are also used to strip the piping

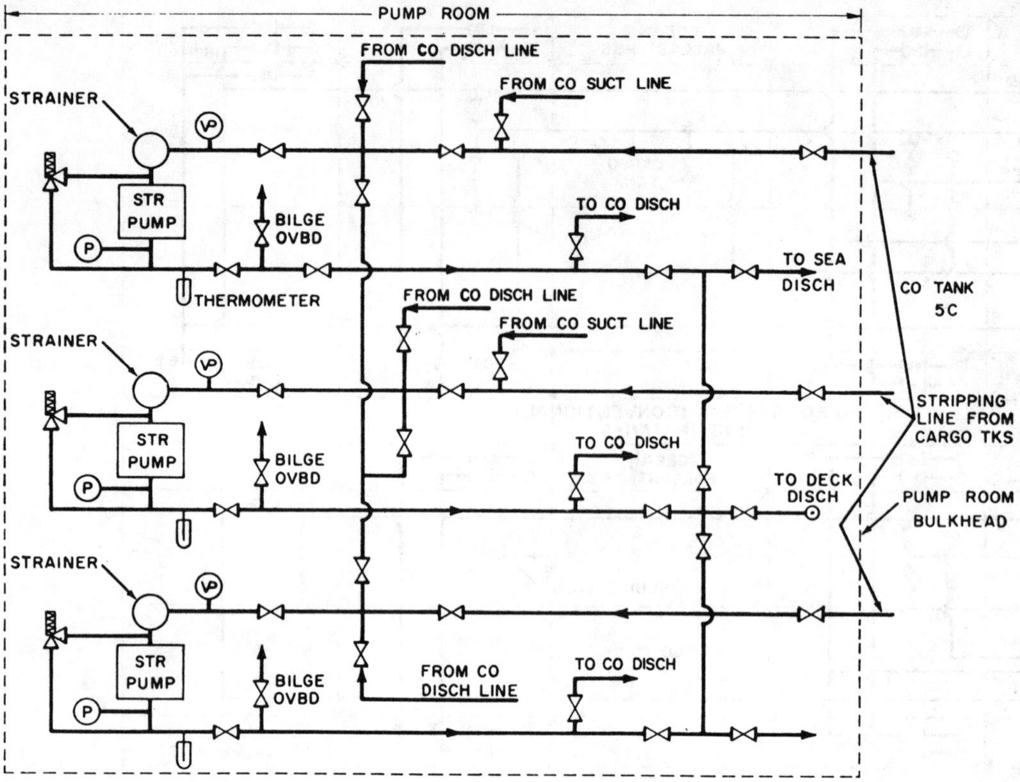

Fig. 12 Piped system. Stripping piping in pump room

of the main system. Use of a separate stripper discharge permits draining of main discharge lines and discharge of maximum cargo from the ship. The latest development in stripping systems is the vacuum type, which uses a vacuum pump to improve the suction capacity of the main pumps and eliminates the need for a separate stripping system. With this system it is necessary to provide a pump or eductor system for pump room bilge service.

Deep-well pumps were originally used to discharge single small tanks on multigrade or specialized ships and were of low capacity, but now these pumps are used also in sizes comparable to conventional cargo pumps to discharge large vessels. They are used with either piped or free-flow systems and have the advantage of elimination of conventional pump rooms and the associated piping. Fig. 13 is a typical diagram of a deep-well pump piping system in tanks. Fig. 14 is a typical diagram of a deep-well pump piping system on deck. Small pump houses may be required for the protection of the pump drives which may be electric, steam, or hydraulic, depending on the nature of the cargo and the main propulsion machinery. Suction problems are considerably reduced and, if self-priming pumps are used, stripping systems are eliminated except for small air-driven pumps used to drain the deep-well pumps for product changes and gas-freeing. Total power requirements are also reduced since elimination of the pump room piping reduces friction losses, and a deep-well pump requires a smaller total head than a conventional pump for the same rail pressure.

Chemical tankers are similar to conventional tankers except for tank size and special material requirements. Normally, they have a large number of small tanks and small capacity pumps. The pumps may be of the deep-well type, one for each cargo tank, or a combination of a piped system with one or more pump rooms and deep-well pumps for some single-tank products. Liquefied petroleum gas (LPG) and ammonia may be carried in cylindrical tanks under pressure or fully refrigerated, at or near atmospheric pressure. Discharge is by pressurizing of the tanks to force liquid flow to the discharge pumps. Refrigerated LPG and ammonia ships employ deep-well pumps for discharge. Liquefied natural gas (LNG) is usually carried as a refrigerated product and deep-well pumps as well as submerged electric-driven pumps are used for discharge.

4.3 Electrical Systems. In addition to adhering to good marine practice in the installation of wiring and the selection of equipment, particular attention must be given to areas of the ship in way of the cargo spaces where the risk of fire or explosion may be met. Such zones include open decks within three meters of a cargo tank outlet, gas or vapor outlet, cargo pipe flange, cargo valve, or entrances or ventilation openings to cargo pump rooms and compressor rooms. These places call for certified safe equipment and through runs of cable. More stringent requirements are imposed in the case of cargo pump rooms and cargo compressor rooms.

A distinction is drawn between those spaces which must be regarded as dangerous and those which are safe, and in the dangerous category, between those which can be inerted and those which cannot.

Fig. 15 serves to explain the treatment accorded electrical equipment in association with various spaces. Certain differences will be found in the specific application to oil tankers, chemical ships, and gas ships under their respective regulations, but the premises employed generally conform to that which is graphically depicted.

4.4 Instrumentation and Alarms. For oil cargos other than grade A, except where inert gas systems are provided, open loading is permitted and sounding may be done through an open hatch or through ullage connections. For grade A cargos and for others using closed loading systems, it is necessary to have a means of sounding without opening a hatch or an ullage connection. Conventionally, this has been a guided float attached to a calibrated tape which is read directly on deck through a sight glass. These have the advantage of providing an accurate reading throughout the depth of the tank, but have the disadvantage of being subject to malfunction due to the corrosive nature of petroleum vapors, so newer types of indicators are being introduced.

Ultrasonic probes are generally limited to reading definite fixed levels rather than reading the entire range of the tank. They are used therefore primarily in connection with remote shutoff of pumps and closing of valves. Other systems using differential liquid and gas pressures in a tube, or tubes, within the tank are also used. These have the advantage of reading the entire range of the tank. Electronic or electrical liquid-level gages used in cargo tanks containing flammable or combustible products are required to be intrinsically safe in order to prevent a possible source of vapor ignition within the tank.

Float systems are often used for ballast tanks and where open loading is permitted because of the greater ease and speed in reading of liquid levels in comparison to manual systems.

Completely enclosed gaging devices with remote reading features are required in most chemical carriers and gas ships. The respective codes recognize four different classes of tank gages with the more restrictive forms required to be fitted for the carriage of the more dangerous substances. The codes also specify the installation of gages and thermometers

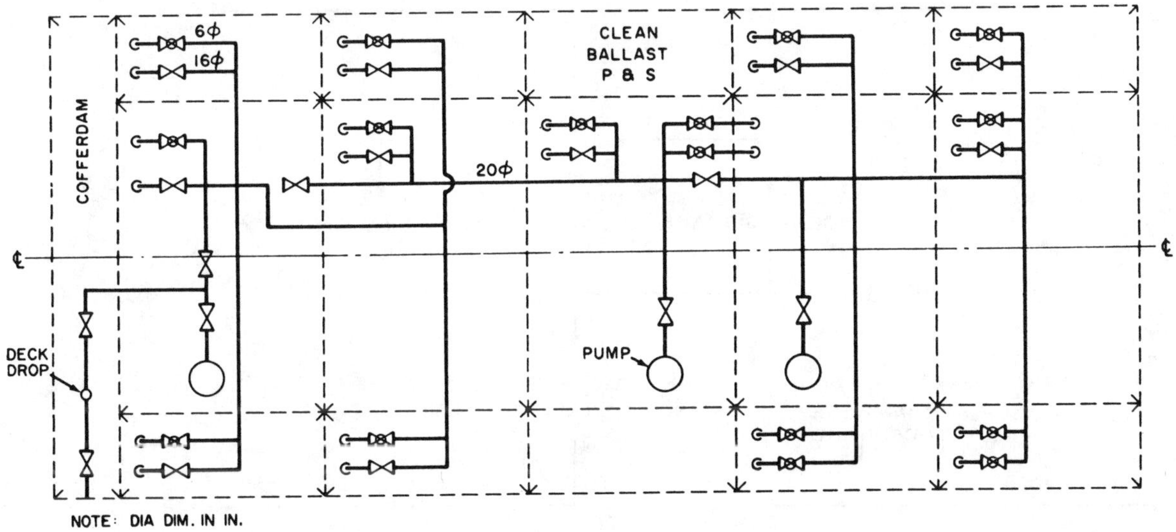

Fig. 13 Deep-well pump system. Piping in tanks

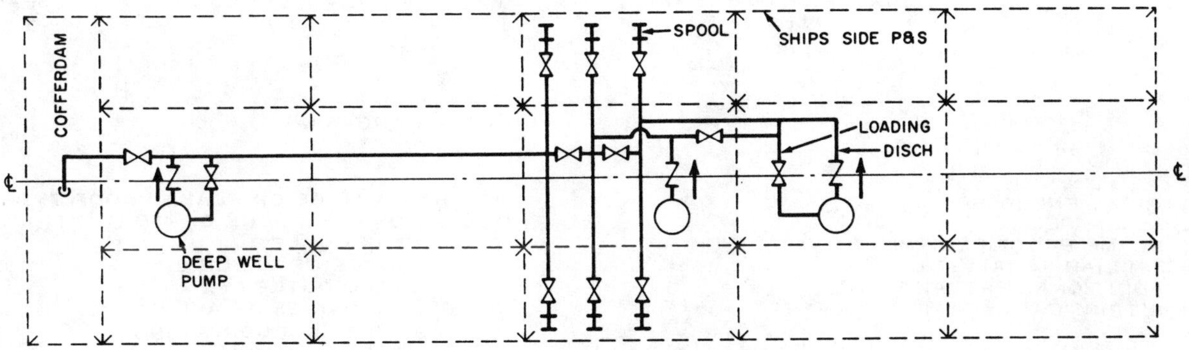

Fig. 14 Deep-well pump system. Piping on deck

for the control of pressure and temperature in the cargo spaces, for the installation of gas detection equipment in various critical spaces including the hold spaces and inter-barrier spaces. In addition, audible and visual alarms are required as safeguards against the release of gas and as indicators of the liquid level being exceeded during loading. Portable sets of detection equipment and instruments for the measurement of oxygen levels in inert atmospheres are other elements required to be carried on board.

4.5 Venting and Emission Control. The functions of venting are to relieve pressure and prevent the formation of a vacuum where such pressure differential could impair the structure of the container. These conditions may occur during the diurnal variations of temperature changes, the weather, and during loading or discharging.

For venting purposes, tanker cargos are classified into grades in accordance with their flash point and Reid vapor pressures, and the type of venting system is selected for the highest grade of cargo for which the ship is designed. The vent system must be designed to permit release of the air or vapor in the tank on loading without over-pressurizing the tank and also to permit entry of air into the tank on discharging to prevent creation of a vacuum. Provision is made also to control vapors evolved during the tanker voyage. The largest venting requirement is for loading, and piping is sized for that service.

Vent systems are either of the closed type for grade A oil cargos or of the standpipe type for other grades, although cargos which could legally qualify for standpipe systems are also sometimes fitted with closed systems where the cargo has a high sulphur content, because of the noxious odors such cargos release. Of course, the fitting of inert gas systems is compelling systems of the closed type.

The closed system consists of a single vertical riser, fitted with a flame arrester for each group of tanks, with branch connections to each tank of the group. To permit carriage

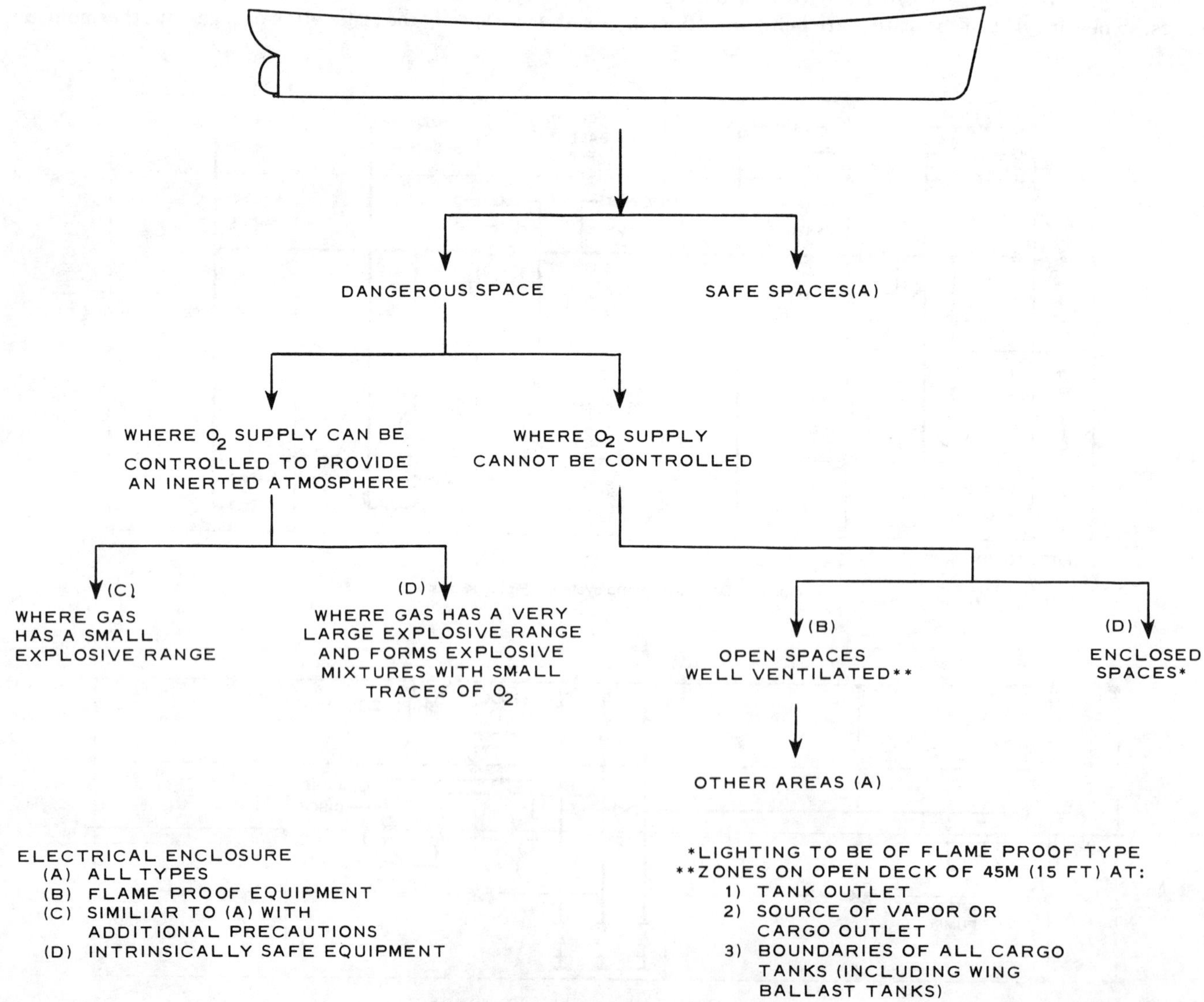

Fig. 15 Simple division of spaces for electrical equipment

of multiple grades of cargo without danger of contamination through the vent system, tanks served by different pumps are not connected to the same vent riser. Branch lines should not be fitted with stop valves. A spill or emergency overfill flange valve is fitted at the base of the riser to prevent overpressurization of the tank due to excessive filling during loading or due to plugging of the vent line. Pressure-vacuum, *P-V* valves are not required by the Rules but are fitted to control emission of vapors during the voyage. An individual valve may be provided for each tank or one valve may be provided for a number of tanks within a group. Where these P-V valves are fitted so that vapor or air must pass through them during loading or discharging operations, they must be fitted with a manual means of opening and must be kept fully opened during loading and discharge. These valves may be either spring-or weight-loaded.

Standpipes may be fitted either directly to the tank access hatch or to the tank, the former being more common since it is the most convenient location. The pipe may terminate in a pressure-vacuum valve or may be fitted with a flame arrester and a closing device at the upper end, and with a small P-V valve for use during the voyage. It is more common to terminate in a P-V valve, in which case the flame arrester is part of the valve.

Ballast tanks are fitted with gooseneck vents terminating in a ball-check valve.

Venting and emission control is a good deal more complex in the case of chemical carriers and gas carriers. In the case of gas carriers the determination of the size of pressure relief valves is varied according to a formula which takes account of the location of tanks whether in holds or above decks, whether insulated or not, and whether membrane or independent. In the case of the Chemical Carrier Code, pressure relief valve settings are varied according to the cargo. Stipulations also are made for the height of and location of the vent exits above the weather deck. Great care must be exercised in designing the vent piping system to insure that the liquid is effectively drained back into the tank and may not spray out endangering personnel in the case of toxic substances or endangering the ship through brittle fracture in the case of refrigerated substances.

4.6 Fire Prevention and Firefighting. Owing to the hazard of flammability present in most of the bulk liquid cargos, great attention has to be paid to preventing the possibility of explosive conditions. The regulations specify separation of the accommodation areas, and the machinery spaces and other ignition sources from the cargo and its ventilation systems. As have been mentioned, various means of inerting void spaces to prevent the accumulation of explosive vapor are required. To provide fire shields within the ship, the 1974 Safety of Life at Sea Convention includes provisions for structural fire protection in certain important boundaries. The Safety Convention also specifies for all ships the minimum conditions for fire main and fire pumps. However, other requirements, depending upon the service of the vessel, may make it essential that the pump capacity be increased. For example, the Gas Carrier Code requires a water spray system to cool any exposed cargo tank domes, or exposed on-deck storage vessels containing flammable or toxic products. In addition, the Gas Carrier Code requires the carriage of dry chemical powder fire extinguishing systems. It includes details as to the size and arrangement of such systems.

In the case of an oil tanker, the USCG Regulations stipulate provision of a deck foam system for the protection of all cargo tank spaces.

In chemical ships, the problem is a good deal more complex. A chemical may dissolve the extinguishing agent; for example, alcohols are soluble in water and are unaffected by ordinary foam solutions. A special foam which will affect alcohol has been developed, but it is extremely expensive and difficult to apply. Hence it can be expected that the chemical carrier will have a great many extinguishing systems installed.

4.7 Visibility from the Bridge. In some applications, particularly those where light cargos are involved, the cargo tanks protruding above the main deck create a problem of visibility from the bridge, a condition also encountered in the containership. Since, like the containership, the LNG vessel is likely to be a fast ship and in respect of its dangerous cargo, it is important that the designer recognize the need of a clear view forward of the ship to anticipate and to avoid obstacles in the ship's path, particularly in the approach waters to a harbor. To encourage other vessels to give a ship carrying dangerous commodities a wide berth, some owners identify these vessels with special coloration.

4.8 Special Conditions. LNG is the only cargo for which arrangements have been agreed for the consumption of the boil-off gas to be used in driving the ship. The details of this arrangement are spelled out in the Gas Carrier Code and stipulate the location of the lines carrying the gas which are required to be of a double wall construction. The annular space is to be pressurized with inert gas and have suitable alarms to indicate any loss of pressure. The Code describes in detail the required shut-down features, ventilation systems and inerting requirements to insure that the fuel is prevented from creating any hazard whatever.

Some thought needs to be given to the provisions for salvage of the cargo and/or the ship in the event the vessel is damaged or sunk. The loss of a cargo as deadly as tetraethyl lead, for example, in a geographic area associated with tourism or commercial fishing, would lead to a demand for the removal of the sunken vessel and/or its cargo.

An LNG ship in difficulties might be obliged to jettison a portion of its cargo in the absence of vessels into which the cargo could be lightered. These and other problems will arise as part of the contingency planning demanded in the process of approving entry of a hazardous cargo vessel to certain port areas. It is therefore necessary that the naval architect carry out a *fault-tree* analysis in order to ascertain conditions which might occur and measures that may be taken to counteract them.

References

Benkert, William M. and Douglas H. Williams (1974) "The Impact of the 1973 IMCO Convention on the Maritime Industry," *Marine Technology,* January.

Crighton L. J. and I. E. Telfer (1973), "Segregated Ballast Tankers," RINA Symposium on Marine Pollution, February.

Evans, Herbert (1966), "Design and Development of the Saturn Marine System," *Marine Technology,* July.

Farrell, T. R. (1974), "Chemical Tankers—The Quiet Evolution," RINA, October.

Kime, J. W., R. E. Johnson and W. D. Rabe (1976), "Damage Stability Requirements for Tankships, Chemical Ships, and Gas Ships," *Marine Technology,* April.

Kime, J. W., R. J. Lakey and T. R. Dickey (1977), "A Review of the IMCO Code for Gas Ships," SNAME STAR.

King, G. A. B. (1971), *Tanker Practice,* Stanford Maritime Press, London.

National Cargo Bureau (1978), "General Information for Grain Loading."

Navaz, M. Z. (1971), "The Carriage by Sea of Hazardous Cargos Requiring Environmental Control," Institute of Marine Engineers, January.

Neal, Jr., Walter G. (1976), "Chemical Tankers—Design Concepts of Operation," Spring Meeting SNAME, June.

Price, Robert I., R. D. Goldbach, A. F. Dyer and R. J. Wheeler (1967) "Sending Chemicals to Sea," N.Y. Metropolitan Section, SNAME, March.

Steinman, G. C., and T. E. Carman (1966), "Design Considerations for Barges Transporting Hazardous Commodities," *Marine Technology,* July.

Thomas, W. DuBarry (1974), "Whither the LNG Ship," RINA, October.

CHAPTER XII

William H. Hunley
Norman W. Lemley

Ship Maneuvering, Navigation, and Motion Control

Section 1
Maneuvering Systems

1.1 Definitions. A surface ship *maneuver* can be broadly defined as an acceleration in a horizontal plane which may be either voluntary or involuntary. A voluntary acceleration results from a controlled force applied by a propeller, rudder, tug, or line while an involuntary acceleration normally stems from wind forces, current forces, or wave action. A ship has adequate maneuvering control when it can apply controlled forces of sufficient magnitude to overcome the involuntary forces encountered plus the additional force necessary to move the ship along a selected path at a desired rate of speed.

Mooring is a term applied to the operation of anchoring a vessel in a harbor, securing to a mooring buoy, or securing to a wharf or quay by means of chains or ropes. Mooring means to make a vessel fast to a buoy, quay or wharf, or by anchoring but, in addition the term is also used as a noun to describe the hardware involved in the securing. A moored vessel need not necessarily be truly stationary but may, in some types of moors, by free to swing around a single anchor on a chain or around a buoy to which the vessel is secured. Similarly, a moored ship may be free to rise and fall with the tide or to take on oscillatory motions in response to the action of waves on the hull. In this respect a moored vessel is restricted to a limited amount of movement and is restrained only to the extent necessary to keep that movement within well defined bounds.

1.2 Environmental Forces. The forces acting on a moored ship include those imposed by the environment and those opposing forces that are applied to keep the ship stationary or in a desired position and attitude. These forces are involved when mooring to piers and quays, mooring to buoys, anchoring, or when dynamically positioning a ship. The environmentally imposed forces result from wind, current, tidal action, and wave action.

Of primary concern in holding a ship stationary is opposing forces resulting from wind and current action. Tidal forces, aside from tidal currents, are essentially irresistible forces; the resulting vertical movement of the ship must be allowed for and cannot be opposed. Similarly, the wave forces acting on a stationary ship are essentially oscillatory in nature; they can be minimized by reorienting the ship or by detuning its natural response but little can be done to oppose the oscillatory ship motions resulting from wave action. However, a mooring or dynamic positioning system must be designed to adapt to wave imposed motions. Thus, the brief discussion herein of the environmental constraints involved in a mooring or a dynamic positioning system design will be limited to wind and current forces, not waves nor tides.

The literature is replete with dissertations on wind and current forces acting upon floating bodies. However, when a designer is faced with the task of calculating the forces and moments acting on a ship while moored or while dynamically positioned, he is hard pressed to sift out those which give the most reasonable answers. The method given below for wind force and moment calculation and that given for current force calculation are presented in simplified form for design use and appear to fit reasonably well the majority of the published data.

a. Wind Force Calculations. Air moving across a water surface has a varying velocity with altitude due to the interaction between the surface and the air mass. Saunders (1957)[1] evaluated several published velocity distributions and came to the conclusion that the most consistent average of various data was one where the velocity varied as the fifth root of the height above the surface. This is expressed as:

$$\frac{V}{V_0} = \left(\frac{h}{h_0}\right)^{0.2} \qquad (1)$$

where V is the airstream velocity at any height h and V_0 is a measured airstream velocity at some standard height h_0 above the surface, Fig. 1.

Saunders (1957) also concluded that the best average of wind drag force data was expressed by:

[1] Complete references are listed at end of chapter.

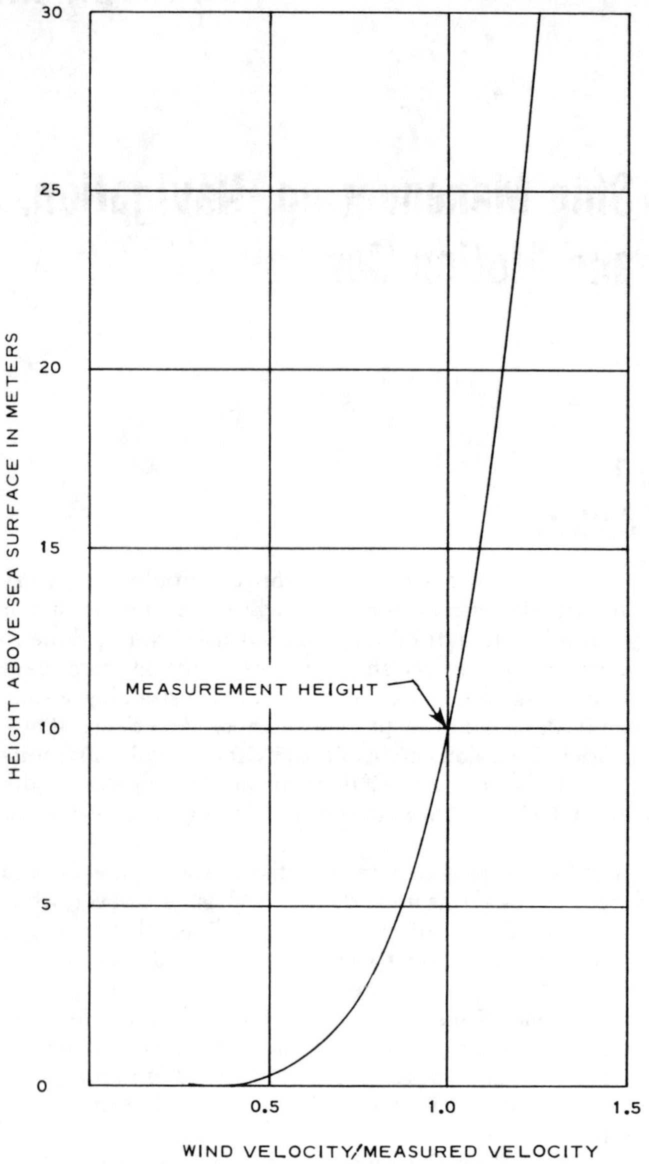

Fig. 1 Variation of wind velocity with height above water

$$F = 0.004 A V_\omega^2 \tag{2}$$

where:

F is the drag force in lb
A is the projected area in ft^2
V_ω is the airstream velocity in knots

When this equation is expressed in the terms:

$$F = C_D \frac{\rho}{2} A v^2 \tag{3}$$

where:

ρ is the mass density of air
v is the wind velocity
A is the area with all terms in consistent dimensions, then:

$C_D = 1.185$, a constant drag coefficient.

One standard used for wind velocity measurements over ground is a height h_0 of 10 m. However, since in many areas where calculations will be required the velocity may be measured at a different height, h_0 will be kept as a variable. To calculate the wind force on a projected strip of a ship of length l and height $h_2 - h_1$, the following expression is derived:

$$F = klV_0^2 \int_{h_1}^{h_2} \left(\frac{h}{h_0}\right)^{0.4} dh$$

$$F = \frac{klV_0^2}{1.4 h_0^{0.4}} (h_2^{1.4} - h_1^{1.4}) \tag{4}$$

When calculating wind forces and moments for any mooring or dynamic positioning application it is usually desirable to determine the force distribution with the wind at various angles to the ship and also the yawing moment that the wind can apply as a function of wind angle. The first step is to employ equation (4) to calculate the effective wind force that acts on the ship when the wind is bow on (angle of attack $\alpha = 0$ deg) and when the wind is abeam ($\alpha = 90$ deg). The abovewater areas to be included are all of those structures that the wind will impinge upon. Judgement can be used to estimate those elements that will always be in the lee of other elements and these can be neglected. However, in calculating bow-on drag forces it should not be assumed that lower parts of a midship deckhouse will be in the lee of a forecastle deckhouse, for example, since the wind will close in abaft the forecastle and apply its full force amidships. Similarly, smaller structures such as port and starboard kingposts should both be included and allowance should be added for rigging and antennas.

When making the profile force calculations, the longitudinal location of each element should be noted and the moment from the midship section determined. These moments can then be summed and divided by the transverse force factor to obtain the center of lateral resistance. A typical cargo ship wind force calculation for a one knot wind acting bow-on and beam-on is given in Table 1. Here it has been assumed that the wind velocity, V_0, was measured at a height of 10 meters above the surface. In this case the dimensions are in meters and the unit forces are expressed in Newtons. The typical ship used is that depicted in Fig. 10 of Chapter III.

Once the side force for a beam-on wind, the axial force for a bow-on wind, and the center of lateral resistance have been calculated, the force pattern for wind action at various angles of attack can be derived. Both model and full scale tests have demonstrated that this is not a simple sine-cosine relationship but is a complex pattern resembling the lift, drag, and center of pressure characteristics of an airfoil.

As the wind angle of attack off the bow, α, increases from zero, there develops both an increase in the axial force and an increase in the side force, with the latter analogous to the lift on an airfoil. The center of pressure of the side force remains at about the quarter length of the hull from the bow

Table 1—Forces for a One-Knot Wind on Beam and on Bow of Sample Ship

Above-water Region	h_2 (Dimensions meters)	h_1	l	b	LCP from ⊗	$\frac{F_T}{V_0^2}$ N	$\frac{F_L}{V_0^2}$ N
Forecastle to Fr 45	10.27	0.00	31.88	24.63	73.58	45.34	35.02
Bulwark Fr 45 to ⊗	5.73	0.00	58.28	24.63	29.14	36.6.	—
Kingpost 1 (× 2)	28.44	9.57	2.19	2.19	58.00	10.16	10.16
Kingpost 2 (× 2)	24.63	9.57	2.19	2.19	35.30	7.79	7.79
Winch house 2	9.57	5.73	8.47	9.02	35.30	5.59	5.95
Kingpost 3 (× 2)	24.08	8.75	2.19	2.19	14.23	7.79	7.79
Winch house 3	9.02	4.94	7.65	9.02	14.23	5.17	6.10
Bulwark ⊗ to stern	5.46	0.00	92.17	24.63	−46.09	54.05	—
Kingpost 4 (× 2)	26.79	5.27	2.19	2.19	− 2.19	10.72	10.72
Kingpost 5 (× 2)	26.79	5.27	2.19	2.19	−25.73	10.72	10.72
Bulwark to Upper Dk	6.55	5.46	27.89	24.63	−18.59	4.78	4.22
Upper Dk to Boat Dk	9.30	6.55	25.18	24.63	−17.50	12.05	11.79
Boat Dk to Bridge Dk	11.77	9.30	24.08	20.51	−17.22	11.63	9.91
Bridge Dk to Nav Bridge	14.23	11.77	19.96	20.51	−15.33	10.49	10.78
Nav Bridge to House Top	16.95	14.23	13.41	19.94	−12.04	8.33	6.80
Stack	21.88	14.23	6.28	4.39	−19.14	11.65	8.14
Kingpost 6 (× 2)	28.99	8.47	1.65	1.65	−46.51	8.21	8.21
Winch house 6	8.47	5.21	8.75	9.02	−48.13	4.69	4.84
Kingpost 7 (× 2)	24.08	8.75	2.19	2.19	−64.83	7.79	7.79
Winch house 7	8.47	5.21	7.65	9.02	−66.75	4.10	4.84
After Deckhouse	7.65	5.21	9.30	11.49	−84.80	3.64	4.50
			Totals				
			F_{L_0}/V_0^2				176.07N
			F_{T_0}/V_0^2			281.30N	
Center of lateral resistance			LCR		0.64 m fwd ⊗		

(where: b = breadth of strip
LCP = longitudinal center of pressure)

while the side force, or lift force, increases almost linearly with increasing angle of attack. When the angle of attack reaches about 30 deg, a condition similar to a stall takes place. The lift force drops rapidly and the center of pressure moves back to the center of lateral resistance as the angle of attack increases from 30 to 90 deg. At an angle of attack of 90 deg, in other words a beam wind, the athwartship drag equals the calculated side force and acts at the center of lateral resistance. An inverse variation occurs as the angle of attack varies from 90 to 180 deg with the center of lateral resistance acting as the midpoint for the center of pressure movement.

Using the results of a calculation such as that shown for a sample ship in Table 1 to get the bow-on force, F_{L_0}, the beam force, F_{T_0}, and the center of lateral resistance, LCR from ⊗, the multipliers in Table 2 can be applied to derive the force distribution for any wind velocity at a series of angles of attack. The results of this calculation for a one-knot wind acting on the sample ship are given in Table 3. It can be noted that the angle of application of the resultant force differs from the angle of attack of the wind, Fig. 2.

The technique described above can be considered as reasonably valid for calculating wind drag on a generally ship-shaped form. For a much more detailed analyses and for an extensive list of references on this subject the reader is referred to a report by Altmann (1971).

b. Current Force Calculations. The movement of the body of water in which a ship is moored or dynamically positioned applies a downstream force to the hull that must be counteracted by the mooring or the dynamic positioning system to the extent that the ship does not move outside its established boundaries. When secured to a single-point mooring buoy, when swinging on a single anchor, or when being positioned dynamically, the boundaries are often established to allow the ship to move around to a heading where the current has a minimum effect, and thus the mooring or positioning system can be designed for this ad-

Table 2—Flow Force and Centers Multipliers

Flow Angle, deg α	$\frac{F_L}{F_{L_0}}$	$\frac{F_T}{F_{T_0}}$	$\frac{LCP \text{ fwd} ⊗ - LCR \text{ fwd} ⊗}{\frac{L}{2} - LCR \text{ fwd} ⊗}$
0	1.000	0	—
10	1.087	0.174	0.500
20	1.321	0.342	0.500
30	1.500	0.500	0.500
40	1.400	0.643	0.451
50	1.125	0.766	0.415
60	0.750	0.866	0.325
70	0.375	0.940	0.225
80	0.100	0.985	0.133
90	0	1.000	0
100	−0.100	0.985	−0.133
110	−0.375	0.940	−0.225
120	−0.750	0.866	−0.325
130	−1.125	0.766	−0.415
140	−1.400	0.643	−0.451
150	−1.500	0.050	−0.500
160	−1.321	0.342	−0.500
170	−1.087	0.174	−0.500
180	−1.000	0	—

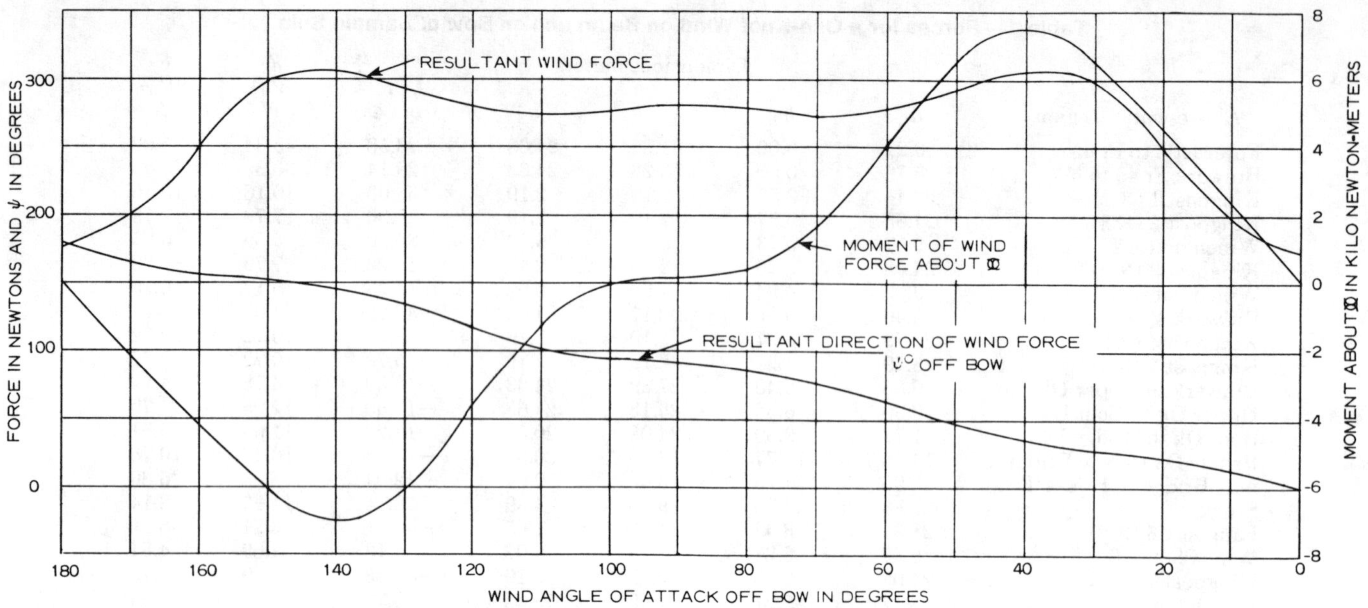

Fig. 2 Distribution of wind force factors for a one-knot-wind

aptation to minimal environmental forces. However, when moored to a pier, quay, or lock wall there probably will be no means of minimizing the current effect and the mooring hardware will be required to withstand the full brunt of the current force upon the hull.

As in the calculation of wind effects on the abovewater structure, the current forces must be calculated for any direction of flow relative to the hull and similarly the center of application of these forces must be determined. Again, there is a wealth of literature published on the subject (Altmann, 1971) but it is difficult to extract a simple means of calculating the effects of slow moving currents on a hull

Table 3—Wind Forces and Moments for the Sample Ship in a One-Knot Wind

Wind Angle, deg α	Resultant Force N	Resultant Angle, deg Ψ	Moment about ℄ $N-m$
0	176	0	0
10	197	14.3	2350
20	251	22.5	4431
30	299	28.0	6477
40	305	36.3	7516
50	292	47.4	6494
60	277	61.6	4067
70	272	76.0	1669
80	277	86.4	395
90	281	90.0	180
100	277	93.6	−36
110	272	104.0	−1298
120	277	118.4	−3668
130	292	132.6	−6075
140	305	143.7	−7118
150	299	152.0	−6153
160	251	157.5	−4209
170	197	165.7	−2137
180	176	180.0	0

over a range of angles of attack.

A significant factor in the longitudinal force exerted on a stationary ship by a current is the drag of locked propellers which often can exceed the drag of the hull itself. Additionally, current speeds are frequently so low that the Reynolds Number, based either upon length or beam of the ship, is in a range where transitional flow occurs with a correspondingly wide variation in the drag coefficients. Thus, a simple answer to the calculation of current forces is difficult to come by but there is no guarantee that a much more complex calculation method will be any more reliable. Therefore, the calculation procedure outlined below, while admittedly simplified to the extreme, is probably as good as can be obtained within the current state of the art.

First, it will be assumed that the maximum section area of the underwater body of the hull, A_X, can be obtained and that the profile area, A_Y, is equal to the product of the mean underbody length and the draft. The resistance of a ship form to a bow-on current is extremely low. An average of some 60 ships from the SNAME Resistance Data sheets gives an average drag coefficient of 0.088 with a standard deviation of 0.039. Although there are numerous refinements that can be applied, it is doubtful whether they significantly improve the accuracy of simply using this average value. Similarly, if the spread of data for the beam-on resistance is examined it will be found that an athwarthships drag coefficient of 0.50 is a consistent average for determining the transverse resistance of a ship hull to a current.

Thus, for a ship type hull, without locked propellers, the axial resistance to a bow-on current can be calculated as:

$$F_{L0} = \frac{\rho}{2} \times (0.088) \times C_X \times B \times T \times v^2 \quad (5)$$

with all terms in consistent units;

the transverse resistance to a beam-on current can be calculated as:

$$F_{T_0} = \frac{\rho}{2} \times (0.50) \times L \times T \times \nu^2 \qquad (6)$$

where:

ρ = mass density of water
C_X = maximum section coefficient
L = mean underbody length
T = mean draft
ν = current velocity

The high axial resistance of locked propellers has been mentioned previously and this must be added to the axial resistance of the ship for other than dynamic positioning situations. A reasonable value is:

$$F_{L_0}/\text{Prop} = \frac{\rho}{2}(0.50)d^2\nu^2 \qquad (7)$$

where:

d = propeller diameter in consistent units

If the underwater profile of the ship is such that the center of its area is at a point other than amidship, the position of this lateral center of resistance, *LCR,* should be calculated.

According to the majority of available data the axial, or longitudinal resistance variation with the angle of attack of the current is strictly a function of the product of the bow-on current resistance, F_{L_0}, and the cosine of the attack angle. The athwartships resistance, on the other hand, is proportional to the 1.5 power of the sine of the attack angle, multiplied by the beam-on current resistance, F_{T_0}.

As with the wind resistance, the center of pressure is at the quarter-point from 0 to about 30 deg angle of attack of the current off the centerline. The center of pressure then moves rapidly back to the center of lateral resistance as the angle of attack reaches 90 deg. A similar movement of the center of pressure is associated with current angles off the stern. Thus, the center of pressure variation with angle of attack given for wind forces in Table 2 is also applicable to the centers of underwater current forces. These force and centers relationships are shown in Fig. 3.

The sample ship for wind force and moment calculations has a maximum underwater section area at load draft of 256 m^2 (2,755 ft^2). The underwater profile area is 1,862 m^2 (20,038 ft^2) with the center of lateral resistance located 1.4 m (4.7 ft) or 0.8 percent of the length forward of amidships. The propeller has a diameter of 6.7 m (22.0 ft). This gives a bow-on current force, F_{L_0}, of 6,109 N (1,375 lb) per knot squared and a beam-on current force, F_{T_0}, of 126,350 N (28,405 lb) per knot squared. Using the force and moment arm distributions of Fig. 3, the resultant forces and moments of Fig. 4 are obtained.

Ships are often moored in relatively shallow water and when a beam current acts on the hull its velocity is altered as it flows from one side of the ship to the other. The restricted passage between the hull and the floor of the body of water causes increased flow velocities under the ship bottom and around the bow and stern resulting in a greater athwartships force being applied to the ship. This increase in force is usually expressed as a drag coefficient plotted against the ratio of water depth, D, to ship draft, T. This drag coefficient augmentation in shallow water is shown in Fig. 5 as obtained from a large number of model tests; it should be applied instead of C_D = 0.50 whenever the D/T value is less than 8 to all current force and moment calculations.

c. Combination of Environmental Forces and Moments. For any mooring or dynamic positioning calculation it is expedient to establish a set of environmental conditions in which the system is expected to exert the controlling forces

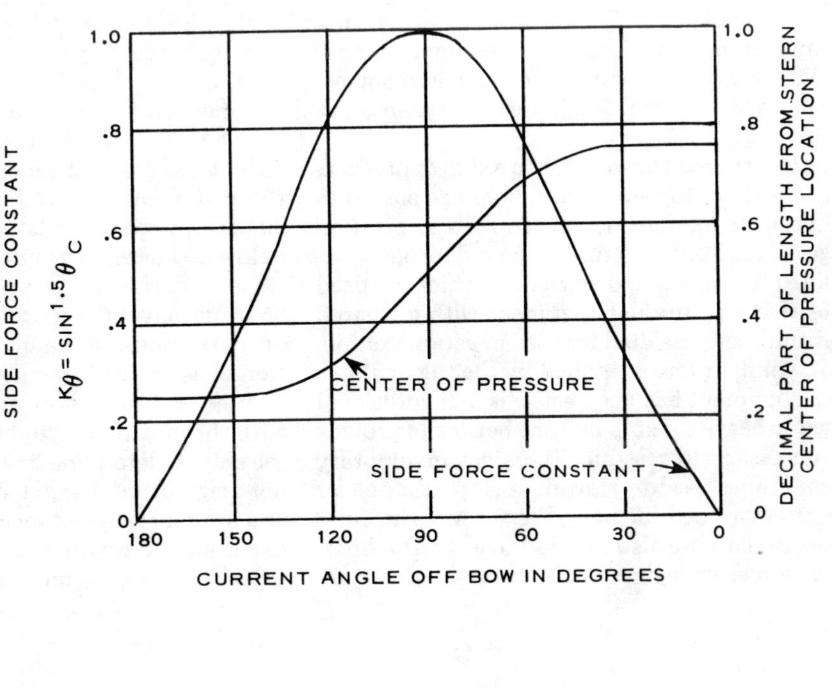

Fig. 3 Current force and center relationships

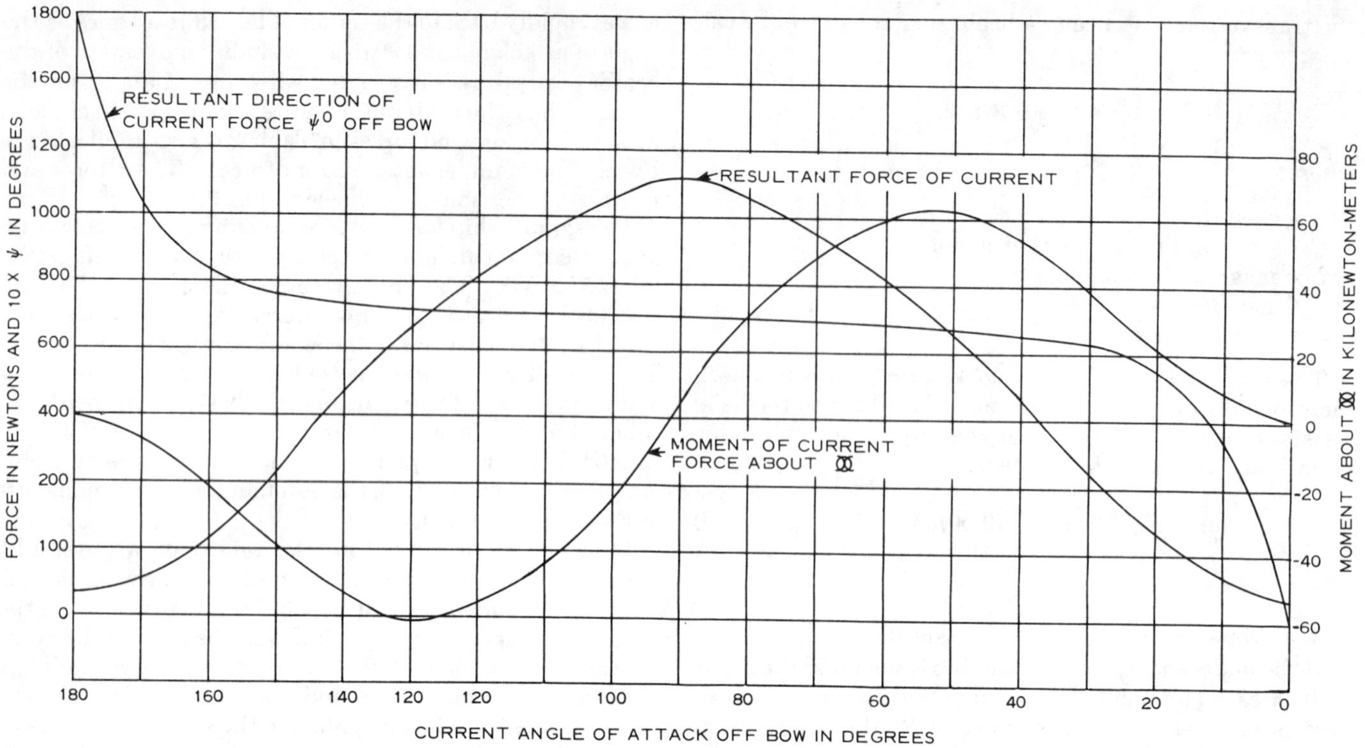

Fig. 4 Distribution of current force factors for a one-tenth knot current

and moments required to oppose the environmental forces and moments. These environmental specifications will generally be formulated as a wind velocity measured at a specified height, a current velocity in a specified depth of water, and, in some cases, a limiting range of directions over which the wind and current may be applied. It then becomes necessary to determine what will be the most critical combination of these forces and moments insofar as the mooring or positioning system design is concerned.

For each combination that appears critical, the forces and moments can be calculated. These will be resolved into longitudinal and transverse forces and moments about amidships and the wind forces, current forces, and moments can then be summed to give a single force, moment, and force direction.

1.3 Mooring to Piers and Quays. When a ship approaches a berth, with or without tug assistance, lines are passed to establish the positive motion control required to position the ship alongside the fixed structure without damage. The first lines ashore are the bow and stern lines which are used to provide positive longitudinal alignment with a desired berth location and check residual forward or astern motion, due to current, wind, or forces applied by the tug or ship's screws. After alignment has been established, additional lines are passed to heave the ship into the berth and provide the security necessary to restrain it against involuntary adverse forces. Lines used to restrain a ship at a pier or quay are referred to as mooring lines, Fig. 6. Mooring lines which tend fore and aft are also referred to as *spring* lines, and resist surge force, while lines tending at right angles are referred to as *breast* lines and resist sway and yaw forces. These lines may also be referred to as *bow, stern, waist,* or *quarter* lines depending on their location on the ship. Mooring lines are connected to bollards on the pier and extend through chocks on the ship to bitts, winches, or traction units. Bow and stern lines used during the ship's approach to the berth usually pass through roller chocks to capstans or gypsy heads at the bow and stern, Fig. 6. Fairleaders are used to lead mooring lines around obstructions and provide alignment with the gypsy heads.

a. Design Considerations. Mooring line design is usually based on providing an arrangement to restrain the ship against all but the most adverse forces that may be experienced in the life of the ship. The rationale considers that a ship would be unduly penalized in terms of added weight and cost if the mooring system was designed to resist these atypical adverse forces. Based on available studies, 90 percent of the winds experienced at commercial ports are below 35 knots, discounting gusts of less than 5 min duration. Mooring design criteria using wind forces exceeding 35 knots may reflect port conditions on a specific trade route, or more conservative design criteria. When the same rationale is applied to current forces, maximum current velocities of 2 to 3 knots are assumed to act simultaneously with the maximum wind force, parallel or perpendicular to the ship to determine line loads for design. Wave forces are not a significant design factor in typical mooring calculations and are generally not included unless the ship is committed to a trade route with mooring sites characteristically exposed to ocean waves. A mooring system is usually designed with a reserve strength which is a function of conservative safety factors and the availability of fittings to which additional lines may be attached. It is prudent for a moored ship to leave its berth and go to sea when unusual adverse weather is approaching. If the ship cannot leave its berth, additional lines may be provided by the ship or shore facility to supplement the ship's mooring system.

Bow and stern lines are sized to restrain the ship against the dynamic forces generated by the ship's momentum during its approach to the pier or quay, or the mooring loads generated by wind and current. The force due to approach velocity is principally a function of the ship's mass times the square of its velocity. For design purposes, velocity is based on prudent seamanship, usually less than one-half knot for large ships. It would not be practical to attempt a design based on resisting the excessive forces that may be generated by careless seamanship. The mooring lines used to warp the ship broadside into the berth should be sufficiently strong to perform safely the warping operation when there is an adverse wind force of 15 to 25 knots, or the maximum wind force in which the ship can be safely maneuvered alongside a fixed structure.

b. Arrangement of Lines. An efficient mooring line arrangement uses the minimum number of lines in a simple arrangement which can be consistently repeated. If practical considerations such as risk and line size are ignored, it

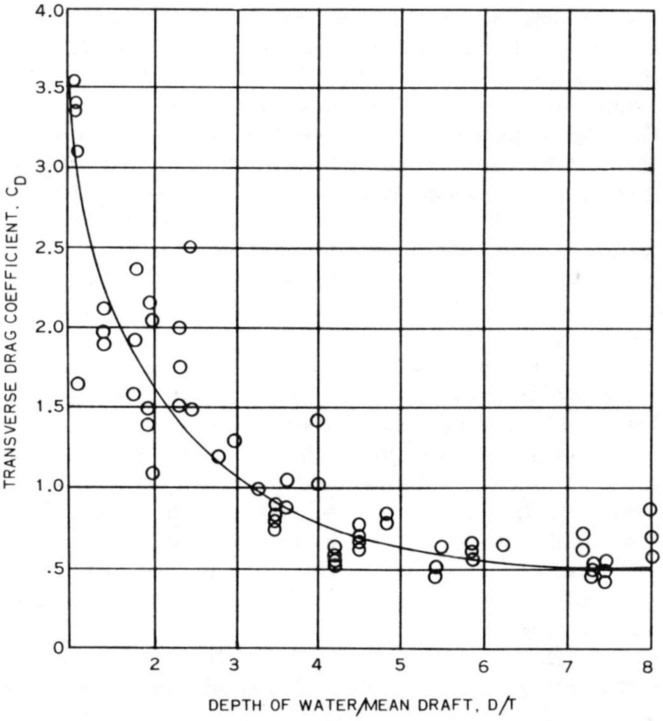

Fig. 5 Variation of drag coefficient in various water depths

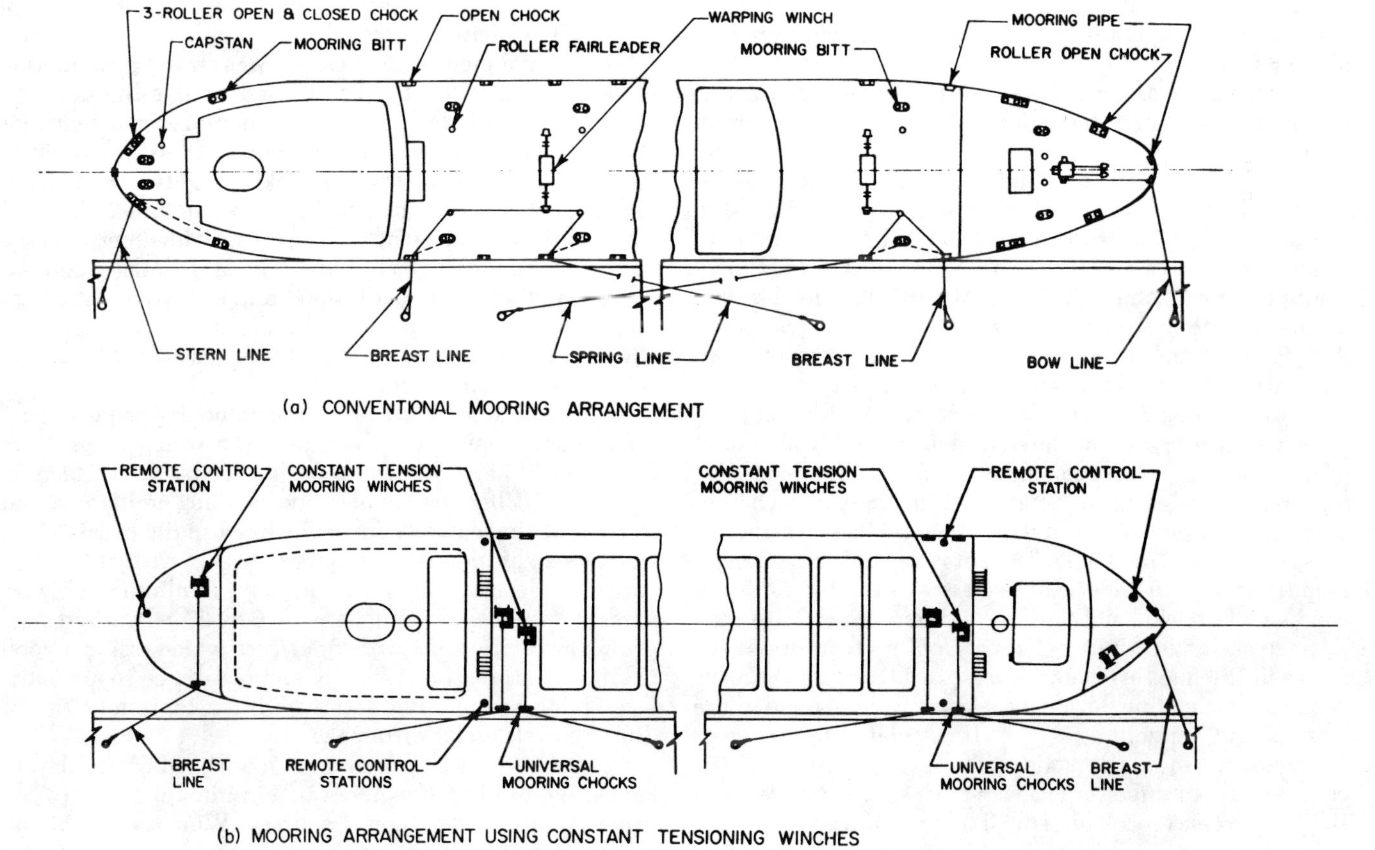

Fig. 6 Arrangements of mooring lines

is possible to moor a ship with a single spring line at the bow and stern. To reduce the risk of catastrophic damage due to a casualty in one of the lines it is common practice to use a minimum of four lines, two at the bow and two at the stern. On large ships where large mooring loads are involved the size and number of mooring lines are governed by the limiting strength of available lines and existing mooring fittings. As the number of lines increases it is proportionately more difficult to locate each line to provide maximum effectiveness. Mooring lines are located at the bow and stern when possible to provide the largest moment arm to resist eccentrically applied environmental forces.

Ideally, for maximum line efficiency, all lines should be horizontal from the ship to the pier fittings; lines acting in the same direction should be of equal length and size to share the load equally; breast lines should be at 90 deg to the ship centerline and spring lines should make the smallest possible angles with the ship's centerline. However, in practice, lines are usually attached on the weather deck of the ship well above the pier, reducing the capability of the line to resist horizontal forces; the point of exit for breast lines typically does not line up with the pier fittings, hence breast lines are not at right angles to the ship; lines acting in the same direction are of unequal length and, assuming equal elasticity and size, the shorter line will carry a larger share of the load; mooring fittings on the pier vary in strength and spacing from port to port; mooring lines which are not controlled by winches or traction units with tensioning features cannot be equally tensioned when handled manually or with power assist such as capstans, gypsies, or winches without tensioning features.

c. Sizing the Mooring Line. Although the forces acting on a moored ship and the angles of the mooring lines cannot be accurately predicted, an assumed mooring line arrangement is used to select lines and equipment. The arrangement may never be exactly duplicated in the life of the ship unless the ship uses dedicated pier facilities. A number of design approaches have successfully provided approximate solutions. For example, in NAVFAC (1968) model test data are used to determine the wind and current forces; in the ABS Rules (Annual), empirical equations are used to select lines; also the U. S. Navy uses a mathematical approach. In a typical mooring design in which ABS Rules are not applied the acting forces are calculated or determined from model tests, a mooring arrangement selected, and line loads determined considering the length and elasticity of each line. The ship and pier are assumed to be rigid bodies attached by springs (mooring lines). A tentative mooring arrangement is developed and the line stresses calculated for beam wind and head wind and current conditions which have been selected as design criteria. For each of these conditions the forces in the lines are determined considering the relative position of each line, its length, elasticity, and angle in the vertical and horizontal planes. If calculated line stresses are grossly unequal the mooring line arrangement is inefficient and the number or location of lines are changed and the line stresses recalculated. The procedure is repeated until an arrangement is achieved in which the lines are approximately equally stressed.

The acting forces due to current and wind can be approximated by employing the calculations described in Section 1.2. The mooring arrangement layout assumes a typical pier configuration. Fittings are assumed to be spaced 5 to 23 m (15 to 75 ft) apart, 2 to 3 m (8 to 10 ft) above the water. The ship is separated from the pier by fenders or camels allowing 2 to 4 m (6 to 12 ft) between the midbody and the pier or a distance consistent with the size and overhang of the specific ship and the type of berthing facilities to be used.

d. Mooring Lines. Synthetic lines have replaced ropes made from natural fibers for all mooring line applications because of their superior strength, durability, and reduced weight. While natural fiber ropes require special drying and storage to prevent dry rot and mildew, synthetic lines rarely mildew and never rot. However, the lines must be stowed in areas where temperatures are moderate and out of direct sunlight to prevent deterioration.

Of the three types of synthetic fiber material used for mooring lines, nylon is the strongest and most elastic providing high energy absorption capacity. Elongation can be reduced by line construction technique for applications where excessive elongation is not desired. Polyester has the least elongation of the synthetic lines, is second in strength to nylon and is the heaviest and most abrasion resistant. Polyester or nylon are most commonly used. Polypropylene is the lightest of the three, the least expensive, with less strength than the two others, sensitive to sunlight unless protected by a special coating, and is subject to fiber damage due to a low melting point.

Traditional mooring line using three strand construction is being superseded by plaited and double-braided line construction. Plaited lines use a balanced construction, four strands with a left hand lay and four right lay, to provide a very flexible line which resists kinking and twisting. Double-braided construction provides a braided sheath around a braided core. The loads are equally divided between the two providing a very flexible line of greater strength and less elongation than plaited or three strand line. Synthetic lines are used with capstans, gypsy heads, or traction machines when power is required for line handling, usually lines above 127 mm (5 in.) in circumference.

When wire rope is selected it is commonly used with wire rope winches which stow the wire on the winch drum. Wire rope size is based on the mooring loads, safety factors, strength of bollard on the pier, line handling limitations, and the size of the winch drum. The line usually consists of a fiber, steel, or independent wire rope core with at least six wire strands, each strand containing a minimum of twelve wires. A typical mooring wire is a 6 by 37 galvanized steel pre-formed wire rope with fiber core which offers a good balance of strength, flexibility and resistance to abrasion. Combining wire and synthetic rope in a moor is not efficient due to the elasticity difference.

e. Deck Layout. A practical mooring equipment layout keeps the number of winches to a minimum and they are driven by a common power source. Winches or traction units are positioned so that they can be used to tension lines on either the port or starboard side. Winches should be positioned so that the operator can have direct visual contact with the officer in charge of the mooring operation. Remote winch controls can be provided to place the operator in the least hazardous location and most advantageous position to observe pierside line handling. At least one chock is installed at each mooring station to provide a fairlead for mooring lines between the ship and shore. Chocks should be arranged to accommodate Panama Canal requirements and the additional lines that may be required for extreme weather conditions or special handling such as required for drydocking. One bitt should be provided for each mooring line. Fairleads are located to accommodate lines from either side of the ship to suit the required lead to gypsy heads or capstans. Control stations must not be located in the bight of the line or in the direct line of pull to avoid being in way of a snapback which occurs when a tensioned line breaks.

f. Mooring Fittings. Fittings are sized to suit the line size and strength requirement of the selected synthetic or wire mooring line. The contact surface of each fitting should have a radius of curvature equal to or exceeding the minimum recommended bending radius for the selected line size to prevent rapid reduction in the useful life of the line. The following fittings, used in a mooring system, are illustrated in Chapter IX:

- *Mooring bitts* consist of two vertical hollow steel cylinders rigidly attached to a base. Their function is to secure the shipboard end of the mooring line. A padeye with shackle is frequently attached to the base to make fast a stopper line, which is used to secure temporarily the mooring line when transferring the tensioned line from a gypsy or capstan. Mooring bitts are located a minimum of 1.8 to 2.4 m (6 to 8 ft) from chocks to allow space for stoppers. They should be oriented parallel to the tensioned line as close as practical so as to minimize the loss in tension when the line is transferred from the power source to the bitt. The bitt

Fig. 7 Hydraulic automatic mooring winch

Fig. 8 Automatic traction winch for use with synthetic lines

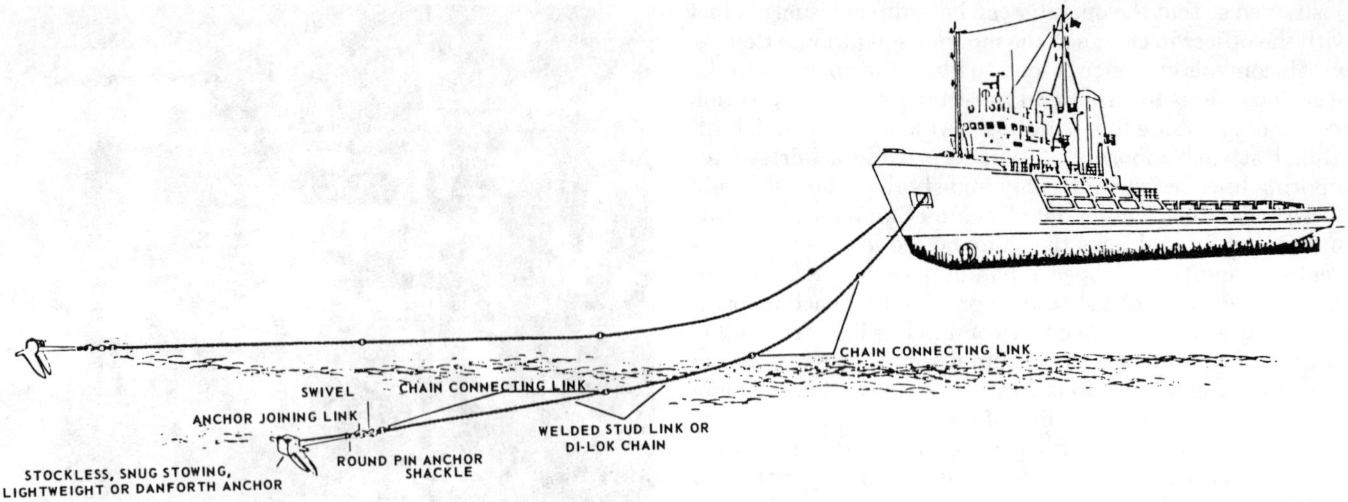

Fig. 9 Offshore workboat with two anchors set

location should not require a sharp bend around the chock.

• *Chocks* are either closed, open, ring, roller, or special. Roller chocks are normally provided at the bow or stern to reduce frictional resistance on the running lines used to warp the ship into a pier. Roller chocks consist of a cast steel frame and base in which are mounted two or three vertical rollers. When chocks are located in way of bulwarks, the chock consists of a heavy ring welded into the bulwark. In way of open rails, chocks can be either closed chocks, which are steel castings with a large oval opening through which the eye splice on the end of the mooring line can be passed, or open chocks which are similar in configuration except that an opening is provided on the upper side through which a line can be inserted. Special chocks have been developed for use with constant-tension mooring winches where the line is constantly being payed in or out with continual changes in direction. These chocks are particularly advantageous on Great Lakes ships, which normally dock without tugs, to facilitate moving the ship along the pier, in and out of canal locks, and to adjust for rapid changes in draft through the use of constant-tensioning mooring winches. The chock consists of a pair of sheaves, between which the mooring line passes, mounted eccentrically on a circular swivel plate so that the pull of the mooring line will cause the swivel plate to turn as the lead of the line changes. The line thus rides constantly on the lower sheave when under load.

• *Roller fairleads* are mounted on deck when a direct lead from a warping head to a chock cannot be obtained. The fairlead consists of a single roller mounted on a cantilever pin, forward or aft of the warping head, so that a lead may be obtained in any direction.

• *Fenders.* When a ship is moored alongside a pier or wharf, it is usually breasted off by a series of floating fenders or camels placed between the ship and the pier to protect padeyes, scupper lips, and other fittings extended beyond the shell plating. Smaller craft coming alongside a ship are breasted off by hanging fenders. Hanging fenders are usually provided for small ships which are nested at a pier, such as fishing ships. Large bow fenders, or puddings, are provided on tugs to prevent damage when pushing against a ship's plating.

g. Capstans and Winches. Capstans, warping winches, or gypsy heads on cargo winches or anchor windlasses are provided for heaving in mooring lines. Gypsy heads or capstans driven by cargo winch or anchor windlass machinery are used where possible for economy of cost and deck area. Steam, direct-current electric, and electrohydraulic drives are designed to give speed variation from creeping to a maximum of about 43 m/min (140 ft/min). Alternating-current electric drives usually have two speeds; full and quarter. Capstan machinery is frequently placed below deck, with controls conveniently mounted on deck. Mooring winches and capstans rotate generally in one direction only, and the line is allowed to slip on the head to pay out; however, some are made reversible. A friction brake generally is fitted so that the gypsy head or capstan head may be braked and used as a bitt for snubbing a line, in which case the shaft and head must be designed to withstand the breaking strength of the largest line to be used. The gypsy head or capstan must be designed to suit the size and material of the furnished lines.

• *Constant-tension.* Ships that change draft rapidly when loading or unloading, such as tankers, colliers, and ore carriers; ships which must moor alongside a pier where there is a large rise and fall of tides; and ships which often traverse canals, frequently are fitted with constant-tension mooring winches and wire rope mooring lines. The use of constant-tension mooring winches, Fig. 7, allows a reduction in the number of personnel required for the mooring operation, better control of the ship while mooring, elimination of line tending during loading and unloading, and improved safety to personnel by eliminating manhandling or line tending under tension. These winches may have steam, electric, or electrohydraulic drives, keep tension constantly on the lines, and haul in or pay out lines as the tension varies above or below the limits for which the controls are set. The wire

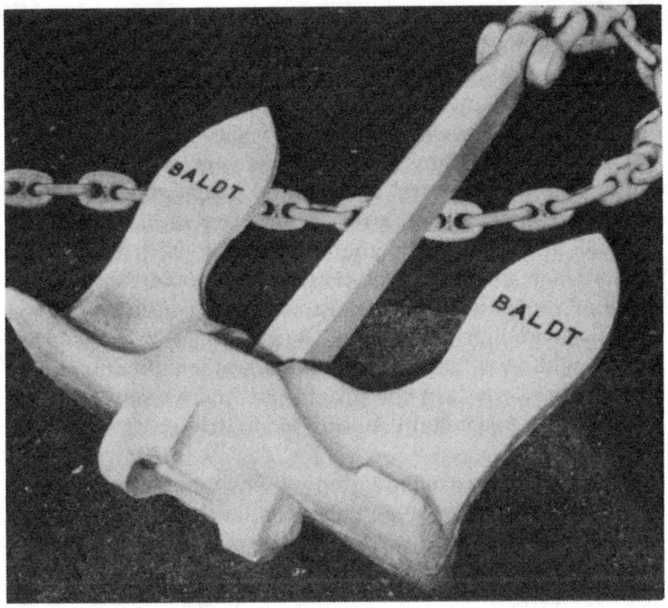

Fig. 10 Stockless anchor

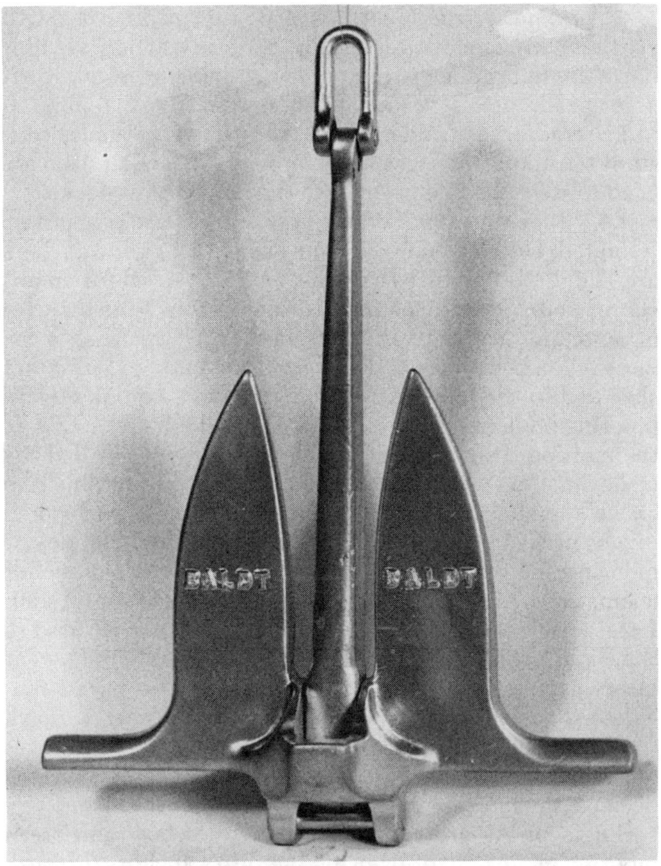

Fig. 12 Snug stowing anchor

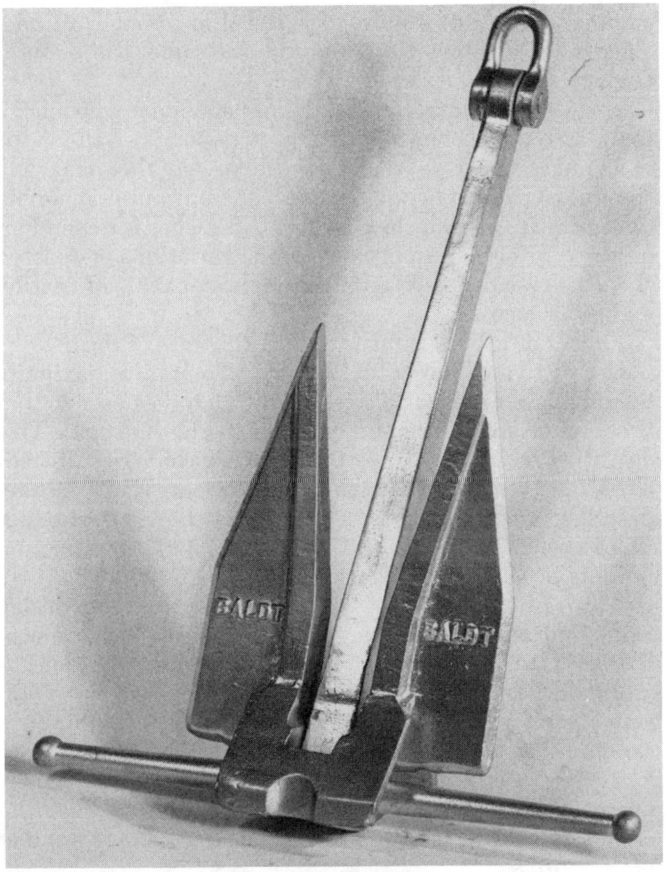

Fig. 11 Danforth anchor

Fig. 13 Mushroom anchor

rope mooring lines, stowed on the winch drums, are led through roller chocks and secured to bollards on the pier. Self-aligning fairlead chocks are used to prevent chafing of the mooring lines. Mooring lines for use with constant-tension mooring winches must have greater strength than normal mooring hawsers, since the single line must carry the expected maximum load. For ease in handling, these lines should not exceed 32 mm ($1\frac{1}{4}$ in.) in diameter.

• *Automatic-traction winches* are designed for use with synthetic rope mooring lines, Fig. 8. The traction winch is normally hydraulically powered to provide infinitely variable speed. It can accommodate line sizes to 250 mm (10 in.) circumference and performs the same function as constant tension mooring winches for wire rope. Fairleads are used to stow the line in rope stowage bins below decks.

1.4 Buoy Mooring. Where anchorage space in a harbor is limited, mooring buoys fixed in position by several anchors are installed to reduce the swing circle required for an anchored ship. Provisions are made on the top of the buoy for attachment of a chain or line from the ship mooring to the buoy. Anchor chain or lines sized for buoy mooring are shackled to the buoy fitting or attached to a quick release mooring hook provided on the top of some buoys. One of the methods used to send a chain to the buoy is called the trolley method. A line is passed through the centerline bow chock and shackled to the ring or link on the buoy by men in a boat used to perform the work at the buoy. A buoy or mooring shackle is attached to the end link of an anchor chain, which has been detached from the anchor swivel shot. Large shackles are placed over the buoy line and secured to the chain a short distance from the mooring shackles and at intervals of every three or four links as it is payed out. These shackles act as trolleys for the chain running down the buoy line. In moderate wind and current conditions mooring lines are used instead of chain by some captains because the mooring line is easier and safer to handle.

Single buoys are sometimes used in a bow and stern mooring arrangement by using the ship's anchors forward. The ship is maneuvered to drop two forward anchors spread well apart before backing down to make up the stern mooring to the buoy. The ship is moved forward while paying out the buoy line to provide a reasonable scope of line to the buoy and the anchor chains are tensioned up. Multi-buoy mooring berths are provided to hold the ship in position for offshore cargo handling. A minimum of four and as many as eight mooring buoys may be provided to secure the ship.

1.5 Shallow-Water Anchoring. An anchor system provides a simple reliable means of safely holding the ship in a relatively fixed position in shallow water without the use of the ship's main or secondary power source. The anchor system may also be used to assist in stopping or controlling the ship in an emergency to avoid grounding or a collision. In adverse circumstances the safety of the crew, the ship, and its cargo may be totally dependent on the holding power of the anchor system.

The system is designed to hold the ship against the combined forces of wind, current, and waves considered to be representative of typical maximum adverse weather conditions. When extremely adverse weather conditions are anticipated at the anchorage the ship could weigh anchor and go to sea to ride out the storm. However, if the ship is anchored and the weather unexpectedly deteriorates, the ship can turn its propeller to relieve the load on the anchor.

Methods used to determine the load on an anchor system are usually simplifications of more rigid mathematical approaches and should always tend to be conservative because of the large number of variables which cannot be well defined. The holding power of the anchor is a function of the anchor design and bottom soil characteristics, and varies from place to place as the bottom varies.

The anchor system consists of an anchor shackled to a chain which is engaged by a chain sprocket driven by a windlass; this may be powered electrically, hydraulically, or by steam. When the anchor is retrieved it is normally stowed in an anchor pocket or against a bolster surrounding the hawsepipe with the chain stowed in a chain locker. When the anchor is deployed, the holding power of the anchor should be slightly more than the applied maximum load and slightly less than the proof load for the anchor. The proof load for the chain should be slightly greater than either.

Two anchors are usually deployed when the swing circle of a ship using a single anchor is too large for the available anchorage. The radius can be reduced by using more than one anchor. The anchors are placed well apart so that a line drawn between the two anchors is perpendicular to the current, see Fig. 9. In some cases, with ample room at the anchorage, it is desirable to restrain the ship's bow when adverse conditions would cause the ship to horse around. The bridle or hammerlock moor is used under these conditions.

a. Anchor Characteristics. The capability of an anchor to dig into the bottom when dragged depends on the shape of its flukes, the angle between the flukes and the shank, and the fluke tripping arrangement. The optimum angle between the flukes and the shank of the anchor is dependent on the type of bottom because of the variations in density, shearing resistance, internal and external angle of friction and other factors.

Anchor efficiency is measured by the ratio of the holding power of the anchor to its weight. To obtain maximum holding power the load is applied through the chain so that the anchor shank remains parallel to the bottom. This condition is obtained by using an anchor chain of sufficient length and weight for a given depth to assure that the chain describes a catenary parallel to the bottom at the anchor shackle when the pull of the chain is equal to the maximum holding power of the anchor. The pull on the anchor will then not lift the anchor shank when the maximum design load is applied. If a catenary is not obtained and the anchor shank is raised, an upward force is applied, lifting the anchor and reducing its holding power. Holding power to weight ratios are normally given for anchors embedded in a firm sand bottom with the shank parallel to the bottom.

b. Types of Anchors. Ship's *bower* anchors are usually one of two general types—*stock* anchors or *stockless* anchors. In each of these categories there are a large number of variations.

• The stockless anchors, Fig. 10, of cast or forged steel, generally are used for ships' anchors. The relatively short shank has a knob at the lower end which is loosely retained in a socket in the anchor crown, allowing the flukes to swing about 45 degrees to either side of the shank. The crown is broad to resist rotation when dragged, and is shaped so as to force the fluke points downward for engagement in the bottom. The stockless anchor has good holding-power-to-weight ratio and is the easiest type of anchor to stow with its shank in a hawsepipe. The commercial stockless anchor has flukes shaped in symmetrical rounded points. The weight of the crown is not less than 60 percent of the total anchor weight. The Navy stockless anchor is similar to the commercial anchor except that the flukes are wider, with outward turning points, giving added holding power. The weight of the crown is 73 percent of the total anchor weight.

• Stock anchors are characterized by a transverse bar, or stock which orients the flukes in the proper position to dig in one fluke when the anchor is dragged along the bottom. Incorporation of the stock in the anchor head, as in the *Danforth* anchor, Fig. 11, allows stowage of a stock type anchor with its shank in a hawsepipe. The Danforth anchor is generally similar to the stockless anchor, except that the fluke motion is limited to about 30 deg each side of the shank axis. The stock forms the hinge pin for the flukes and stabilizes the anchor against rotation when dragged over the bottom. The Danforth anchor digs in readily, and its holding-power-to-weight ratio in favorable bottom far exceeds that of a stockless anchor. The narrow crown allows it to be stowed snug against the shell, with the crown in the hawsepipe opening.

The *snug stowing* anchor, Fig. 12, has stocklike projections made integral with the crown which give it good stability against rotation when dragged, and it develops holding power comparable to the stockless anchor. The small crown nests into the hawsepipe, which has no shell bolster, allowing the flukes to lie snugly against the shell.

• A *stern anchor* is used to keep a ship from swinging with the current in a river or channel and with the tide when anchored in ports which have a restricted anchorage. It may be housed in a hawsepipe so located that the anchor will clear the rudder and the propeller when being dropped or heaved in. The requirements for the design of a stern hawsepipe are generally similar to those of the bow hawsepipes. The chain is led from the hawsepipe around a wildcat on either a capstan or a deck winch and then to the chain locker.

• The *stream anchor* and chain or stream wire rope, when provided, may be used as a stern anchor. The stream anchor usually is stowed on deck and put overboard by a crane or davit, in which case a large, closed chock will serve to guide the anchor chain. The hawsepipe stowage is preferable for a ship which uses the stern anchor frequently under service conditions.

• *Kedge anchors* are usually stock anchors and are carried by small coastal or inland vessels. Deep sea anchors are stowed on deck and handled by a davit.

• *Mushroom anchors,* Fig. 13, so named for their shape, are used on light-ships, canal barges, and for anchoring permanent mooring buoys.

c. Anchor Chains. The bower anchor chains required by classification society rules may be stud-link chains of normal, high, or extra high-strength steel. Normal strength stud-link chain is made of medium steel with cast iron studs pressed into the links to retain the shape of the link under load and to prevent kinking. Anchor chain is manufactured in 15-fathom *shots* (27.4 m or 90 ft in length). Normal strength chain has an enlarged link at each end of the shot to accommodate the joining shackles used to link up the chain. Each shot is made with an odd number of links to assure that the joining shackles will pass over the wildcat in the same relative position. The first shot is sometimes made 30 fathoms long to allow anchoring in shallow water without passing a joining shackle over the wildcat. An outboard swivel shot, Fig. 10, consisting of a swivel separated by three to seven links from the bending shackle, is usually provided to allow rotation of the anchor without twisting the chain.

Stud-link chains also are made of cast steel and of drop-forged alloy steel, rated as high strength and extra high strength respectively. The studs form integral parts of the links, which are of uniform size, including the joining links. This uniformity assures a better fit of chain to the wildcat, reducing chances of the chain jumping the wildcat, as frequently happens with enlarged links and joining shackles. The classification society rules permit a reduction in the chain size for high strength steel on the basis of the increased strength of the chain. This allows a marked weight saving in the ship, but sharply decreases the holding power of the ground tackle for a given scope of chain. Navy practice is to select the chain size on the basis of the chain weight necessary to form a catenary which will ensure a horizontal pull at the anchor for the prescribed chain length under maximum design anchoring conditions.

d. Determination of Anchor and Chain Sizes. Ships are normally equipped with a minimum of two bower anchors and anchor chains, each set sized to hold the ship when subjected to the maximum design load, which is a function of the intended ship's service and applicable regulations. The anchor chain is usually much heavier than would be required to withstand the tensile load alone. The chain hangs in a catenary between the hawsepipe and the bottom, and must be selected so that the lower end of the catenary will be horizontal when the tension at the anchor shackle is equal to the maximum holding power of the anchor.

Anchor size is based on anchor holding power in a firm sand bottom and the forces acting on the ship under specified design conditions of wind and current. Calculations should consider the ship at light and full load draft. The condition giving the greater load is used to determine the anchor size.

When a ship rides on a single anchor it will describe a characteristic horsing motion when disturbed by high winds. This horsing motion orients the ship at an angle to the wind and can subject the anchor and chain to larger loads than possible when the ship lies head to the wind. The maximum longitudinal load occurs when the wind is approximately 30 degrees off the bow.

Chain and anchor sizes may be determined by calculation or classification society rules. To calculate the size of the anchors and chains required, the wind load and current load on the ship must first be determined for the most severe conditions under which the ship will be expected to remain at anchor. The steady wind and current loads are calculated

Table 4—Anchor Holding Power

Type of Anchor	Holding Power/Weight Ratio
Mushroom	2.5
Standard Navy Stockless	7.0
Navy Mark 2 Stockless	8.0–9.0
Commercial Stockless	5.0
Danforth	16
Balanced Fluke	9.0

by the methods described in Section 1.2.

The anchor system also is subjected to dynamic loads as the ship surges, sways and yaws in adverse weather and is buffeted by wave action. A Z-factor multiple is applied to approximate the anticipated force increase required in a static analysis. This factor, based on Navy experience, uses a factor of 1.25–1.50 for small ships with fine lines; 1.50–1.70 for larger more fully shaped ships; 1.75–2.00 for cargo ships and auxiliaries with more blunt shapes.

After selecting the Z-factor the horizontal component of the anchor line tension becomes:

$$H = ZF_R$$

where:

F_R = total wind and current load
H = total resistance corrected for dynamic forces

Required anchor weight (in air), $W = H \div H/W$

where H/W is the holding power-to-weight ratio for the selected anchor according to Table 4.

Chain size and scope are determined to suit the anchor size and the depth in which the ship will anchor, using the following catenary equations:

$$a = \frac{H}{0.87\omega} \text{ or } \omega = \frac{H}{0.87a} \tag{8}$$

$$a = \frac{s^2}{2y} - \frac{y}{2} \tag{9}$$

$$\frac{H}{0.87\omega} = \frac{s^2}{2y} - \frac{y}{2} \tag{10}$$

$$s = \left(\frac{2yH}{0.87\omega} + y^2\right)^{1/2} \tag{11}$$

where:

H = holding power of anchor (pounds)
a = catenary parameter, determined from equation (9) using assumed scope (180 fathoms)
0.87 = constant to convert chain weight in air to weight in water
s = scope of chain, fathoms
y = water depth + height of hawsepipe above water
ω = chain weight (in air) per unit length

Determine chain weight, using equation (8). If calculated weight falls between chain sizes, use next larger chain size. Determine required scope of this size chain by using equation (11). The static load safety factor for the selected chain may be determined by:

$$S.F. = \frac{\text{Ultimate strength}}{H + s(0.87W)} \tag{12}$$

A length of chain to reach from the hawsepipe, through the windlass to the chain locker, must be added to the calculated scope. Since chain is manufactured in 15-fathom shots, the length of chain must be the next full multiple of 15 fathoms.

The sizes of anchors and chains also may be determined by reference to tables provided by American Bureau of Shipping (Annual). The required sizes, based on experience, are tabulated according to an equipment number determined from characteristic dimensions of the ship by use of a formula given in the Rules. When the equipment of a ship, including anchors, chains, tow lines, windlass, and hawsepipe is in accordance with the Rules or with requirements which have been approved for the particular service to which the ship is limited, the letter E is placed after the symbols of classification in the *Record*. The Rule tables are based on commercial stockless anchors, but the American Bureau of Shipping, upon request by the owners, will consider use of special types of anchors, and will allow reduction in anchor weight up to 25 percent for anchors of superior holding power.

e. Anchor Handling Arrangements. The anchor handling arrangements are determined by the type of anchor and windlass to be used. The most common arrangement utilizes a stockless anchor and a horizontal windless wherein the wildcats are mounted on a horizontal shaft, Fig. 14. In laying out this arrangement, the centerline of the deck opening and deck bolster of each hawsepipe is placed about in line with the centerline of a wildcat so that the anchor chain will lead fairly to it. For very wide ships such as large tankers, and for ships with large bulbs at the bow, it is necessary to split the windlass and set each wildcat at an angle to the centerline in order to obtain proper leads to the hawsepipe. The windlass is located so that the chain lead from the hawsepipe to the wildcat will allow ample space for fitting a chain stopper and a devil's claw, and so that the chain openings in the windlass bedplate will approximately plumb the chain locker. A chain stopper, made of cast steel, is usually placed between the wildcat and the hawsepipe and bolted to a plate foundation on the deck at such a height that the anchor chain will lead fairly through the stopper. A hinged tongue is fitted in the chain stopper which may be dropped between two links of the chain in order to prevent the chain from running out when the windlass brake is released. This stopper also is used for holding the chain when the ship is riding at anchor. Each anchor chain contacts about 120 deg of the circumference of the wildcat before passing through an opening in the windlass bedplate and into the chain locker by way of a chain pipe which is fitted in line with this opening.

Large passenger ships and naval ships generally are equipped with a vertical windlass where each wildcat is mounted on a vertical shaft, Fig. 15. Passenger ships usually have two wildcats and three hawsepipes. One hawsepipe is placed on the centerline and is used for stowing the spare bower anchor. The anchor chains normally are connected

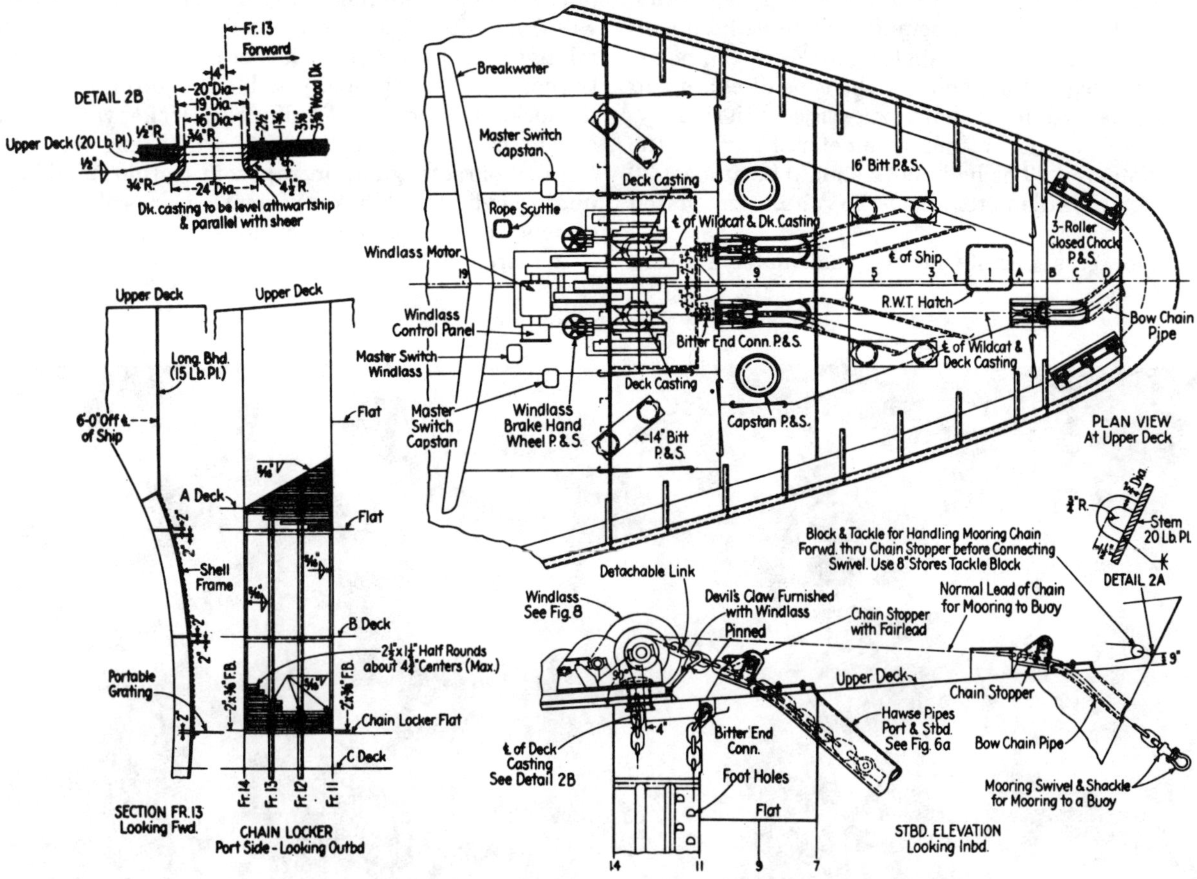

Fig. 14 Anchor handling arrangements, horizontal shaft windlass

to the anchors in the port and starboard hawsepipes; however, the anchor chain is disconnected from one of these anchors, and connected to the spare bower anchor when it is to be used. The wildcats are fitted relatively close to the deck to reduce the bending moment in the shaft and to minimize the height of the hawsepipe deck bolster needed to lead the chains to the centers of the wildcats. If the distance between the wildcats and the hawsepipes is so great that the chains strike the deck, it is customary to fit a strip of plating, or the equivalent, to prevent chafing of the deck. The anchor chains contact about 180 deg of the circumference of the wildcats before passing over the deck bolsters leading into the chain pipes.

The chain pipe may be made of pipe or welded steel plate. The inside diameter usually is made about 6.5 times the diameter of the chain wire. The lower edge of the pipe is fitted with a half-round or a casting forming a small bellmouth, Fig. 14, to prevent the chain from catching as it is drawn out of the chain locker when the anchor is dropped. The lead to the chain locker should be as straight as practicable and nearly vertical. A cover of steel plate can be fitted to each chain pipe opening in the windlass bedplate to keep seawater from entering in rough weather. This cover can be made in halves, neatly fitted around a link in the chain for easy removal before windlass is operated. Canvas covers over wood plugs may be used also.

The chain locker should be sized to hold the total volume of chain, plus an allowance of 1.5 to 1.8 m (5 to 6 ft) from the top of the stowed chain to the bottom of the bellmouth on the chain pipe for clearance. In a circular chain locker, the top of the chain pile will be nearly flat, while in a rectangular locker the pile will be humped. This should be considered in establishing the height of the locker above the chain pile. The drainage sump below the chain locker is in addition to the volume of the locker itself.

Existing equations for the volume of the stowed chain are:

$$V = 0.0150Ld^2 \text{ for U. S. Navy Die-Lock chain} \quad (13)$$

$$V = 0.0313Ld^2 \text{ for commercial stud-link chain} \quad (14)$$

where:

V = chain volume
L = chain length
d = chain wire diameter

all expressed in the same linear dimensions.

The above expressions determine the volume required

without regard to the geometry of the space. Chain entering a locker through an overhead opening (bellmouth) tends to form a cone shaped pile. The angle at the apex can be assumed to be 60 deg but depends on the type of chain and frictional factors. An ideal chain locker is self-tiering, cylindrical in shape with an optimum diameter of 24 times the chain wire diameter. The bellmouth should be at least 30 times the chain wire diameter above the volume calculated for the chain.

The chain locker is generally located immediately forward of the forepeak bulkhead, Fig. 14. The hull configuration will frequently dictate the shape of the chain locker, but relative proportions of the locker must be set to assure that the chain will stow properly without requiring a man to enter the locker to guide it. A cylindrical locker with a diameter of $24d$ is optimum to provide self-tiering of the chain. Lockers of rectangular or trapezoid section should approximate a cylindrical locker as nearly as possible in the ratio of depth to horizontal dimensions.

In merchant ships, the chain locker is usually divided by

Fig. 15 Deck arrangement for vertical shaft windlasses and capstans

a centerline bulkhead to within 1 m (3 ft) of the top, stiffened with steel pipe split in half and welded to the bulkhead. The stiffening of the chain locker should be outside the locker if practicable. If not, the stiffeners should be bulb angles, or if inverted angles or channels are used, their bosoms should be filled with wood, or the chain locker should be lined with half-rounded battens inside the stiffeners to prevent the chain from wedging under the flanges.

Access to the chain locker is by way of a hinged manhole located either in the deck over, or in a boundary bulkhead just below, the deck beams. Semicircular cuts are made in the centerline bulkhead, in line with the access opening, to serve as toe and hand holds instead of a ladder—which would be subject to damage by the chain.

The bottom of the chain locker generally is covered with a layer of cement and sand having a thickness of about 50 mm (2 in.). The upper surface is sloped in order to direct any water brought in by the chain into a scupper fitted with a strainer. The chain rests on a grating of steel bars placed on top of the cement and sand or on a perforated plate supported about 750 mm (30 in.) above the bottom of the locker. For access and cleaning a watertight manhole is fitted to the space below the perforated plate.

The inboard or *bitter end* of the anchor chain is fastened by a large shackle to a pad at the top of the chain locker, Fig. 14, or to a heavy plate bracket at the bottom of the locker and, in small ships, to the center keelson. In some cases, the bitter ends of the port and the starboard anchor chains are fastened together with a connecting link or shackle through a hole in the centerline bulkhead. The plating in way of this hole is doubled and the hole is bounded with double half-rounds. This arrangement eases the shock caused by the chain being stopped suddenly, as by a fixed attachment, should the chain jump the wildcat or run out without braking control.

f. Anchor Handling Machinery. The ABS Classification Rules require a windlass of sufficient power and suitable for the specified anchor chain. The ship specifications usually require a windlass of sufficient power to lift each anchor and 60 fathoms of chain at an average speed of not less than 0.15 m/s (0.5 ft/s). Also, it must be able to lower each anchor under power by reversing the machinery and by gravity under the control of a hand-operated friction brake.

The standard horsepower equation is used to calculate the required horsepower for the windlass:

$$\text{Power} = \frac{\text{LOAD} \times \text{hoisting speed}}{\text{efficiency factors}} \tag{15}$$

The load is the immersed weight of the anchor and chain used. Efficiency is determined by estimating the efficiency of the individual parts of the system and calculating an overall efficiency. The following is a brief list of efficiencies generally used for anchor handling system components:

Electric motor under 75 kW (100 hp)	95 percent
Electric motors over 75 kW (100 hp)	98 percent
Chain in hawsepipe and along deck	50–85 percent (varies with pipe geometry and surface roughness)
Mechanical, through gearing	80 percent
Hydraulic system (piping and valves)	95 percent
Hydraulic pumps	95 percent

Various types of windlasses and other types of anchor handling machinery are described in the following:

• *Horizontal-Shaft Windlass.* The steam driven, spur-geared, horizontal-shaft windlass with a two-cylinder engine, all mounted on the same bedplate, is the type used on many tankers, Fig. 14. The two wildcats are mounted on the horizontal shaft, which is rotated by a large spur gear. Each wildcat is located independently on the shaft when its anchor is to be hoisted or lowered by power. The speed is controlled by the engine throttle. When an anchor is to be dropped by gravity, the hand-operated friction brake is first set up on the wildcat. The wildcat is then disengaged from the shaft by withdrawing the locking blocks, after which the brake is released and the anchor is free to fall. When enough chain is out, the brake is set up quickly to prevent burning the brake linings, and the vessel rides on the brake until the chain stopper is adjusted.

A windlass of this type usually is fitted with two gypsy heads for use in handling the bow mooring lines. The engine may be installed on the deck below, in which case power is transmitted to the windlass through bevel gears and vertical shafting. This protected engine location provides access at all times for overhaul. A windlass of this type is readily kept in commission; however, the maintenance of the comparatively long leads of steam and exhaust pipes is a disadvantage. These pipes must be kept drained in cold weather to prevent freezing. Also, time is needed to warm up the engine before using the windlass.

The horizontal-shaft windlass, similar to the steam winch except that it is powered by an electric motor, is the most common type fitted on merchant ships. The wildcat and gypsy head speeds are varied by an electric controller which gives multiple speeds with direct current. Full and quarter speeds usually are provided with alternating current. Electric motor-driven windlasses are provided on most modern oceangoing ships, except in cases where there is danger of spark-initiated explosions or fires because of the nature of the cargo carried.

• *Vertical Shaft Windlass.* This, with horizontal wildcats, Fig. 15, is driven through spur and worm gears by electric motors located on the deck below. Capstan heads, driven by the windlass machinery, usually are fitted on each side outboard of the wildcats for handling the forward mooring lines. The handwheels for controlling the brakes and the power are located above deck. The operation of this windlass is similar to the horizontal shaft windlass; i.e., hand-operated friction brakes, locking and unlocking wildcats, etc. Vertical windlasses are usually fitted on naval ships because the greater wrap of the chain around wildcats gives better control of the chains when dropping the anchor, and the machinery is more easily maintained at sea.

Fig. 16 Deep sea anchor winch

A vessel equipped with alternating current may use electrohydraulic machinery to drive the windlass at varying speeds. This machinery customarily is located on the deck below the windlass. The power is supplied by a constant-speed electric motor driving a variable-stroke hydraulic pump (A-end) which is piped to a hydraulic motor (B-end), which drives the windlass through shafting and gearing. This power plant may be used with either the horizontal or the vertical shaft windlass. The speed and direction of rotation of the wildcats and the gypsy heads or capstans are regulated by varying the stroke and reversing the discharge of the A-end pump. The handwheels for controlling the A-end pump and the friction brakes are located above deck. The windlass is operated in the same manner as those previously described. The electrohydraulic drive has all the advantages of the direct electric drive, plus a very fine speed control, which is helpful when housing the anchors. Also, a relief valve in the hydraulic system provides a better means of controlling overloads on the windlass than the overload relays of a direct electric drive. This drive is more complicated than the direct electric drive and ordinarily is more costly.

• *Deep Sea Anchor Winch.* This usually consists of a multisheave traction unit, mounted on deck, and a separate single or double drum storage winch, capable of spooling about 6,000 to 15,000 m (20,000 to 50,000 ft) of anchor wire rope, located below. The anchor rope is pulled in by the traction unit, and spooled on the storage winch, Fig. 16, at a lower tension to prevent the rope from cutting through the many layers on the drum and jamming or damaging the rope. Frequently, both the traction unit and the storage winch are powered by a single hydraulic plant, and controlled so as to maintain the proper tension in the cable between the traction unit and the storage winch. The anchor rope usually is tapered to decrease its weight and cost.

g. Anchor Storage. Stockless bower anchors usually are stowed in anchor pockets, Fig. 17, or with the shank stowed in a hawsepipe. However, inland river vessels, because of low freeboard, stow them on *billboards*. Stowage in a hawsepipe is accomplished by heaving in slowly with the anchor windlass, until the shank is fully housed in the hawsepipe, and the flukes are thrown against the shell plating by the tripping lugs on the anchor crown bearing against the shell bolster. The anchor is secured against movement caused by ship motion by setting the chain stopper, hooking a devil's claw into the chain, and setting up on the turnbuckle. The anchor may be secured also by a wire rope or chain through the anchor shackle and made fast to a bitt or pad eye.

The basic requirements for hawsepipes are stated in classification society rules. Generally they are made of cast steel in one or more pieces and welded in place. The shell bolster of the one-piece hawsepipe is so formed, Fig. 18, that it may be cut readily by chipping or grinding to fit closely against the shell surface before welding. The deck plating is fitted around and welded neatly to the deck bolster. The hawsepipes ordinarily are circular in cross section, with the thickness increased where the chain bears. The diameter must be large enough to house the anchor shank (not less

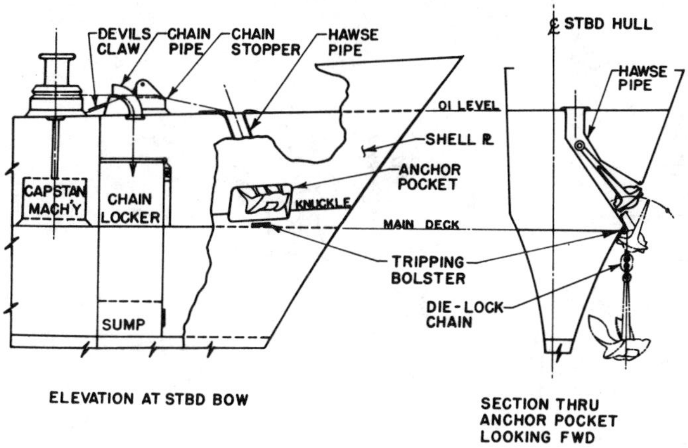

Fig. 17 Anchor handling arrangement with anchor pocket

than nine times the diameter of the normal strength chain wire), and the length must be great enough to keep the anchor shackle below the deck bolsters when the flukes are bearing on the shell plating. The axis of the hawsepipe should be not more than 45 deg from the vertical in the fore-and-aft plane, as a greater angle makes it difficult to pull the anchor shank into it. The shell bolster of the hawsepipe must be large enough, and so located, that the flukes will clear the forefoot of the ship when the anchor is being heaved in. This bolster is shaped to provide an easy entrance for the anchor shank and an easy lead for the chain. In cases where it is impractical to make the bolster large enough to keep the anchor flukes clear of the forefoot, the shell plating is bossed out in way of the hawsepipe to provide this clearance, and the shell plating is made thicker in the area which may be struck by the anchor. This condition is common with bulbous bows. A heavy insert plate is fitted to the shell plating and to the deck plating in way of each hawsepipe.

When the ship is to be fitted with high-strength alloy steel chains, the chafing surfaces of the hawsepipe should be clad with about 6 mm (1/4 in.) of hard weld metal to prevent the hard chain from cutting through the softer cast steel of the hawsepipe.

The shell bolster of the hawsepipe has been replaced by an anchor pocket, Fig. 17, on some seagoing ships and on many vessels on the rivers and Great Lakes. These anchor pockets are formed by recessing the shell plating enough to allow the anchor crown to stow inside the line of the shell surface. The anchor shank is stowed in the hawsepipe as usual, while the anchor head and flukes rest on a curved shelf or billboard. The bolster or lip on the end of the hawepipe at the top of the anchor pocket is only large enough to provide an easy lead for the anchor chain.

A half-model of the bow of the ship is usually made to assure that the anchor stowage is arranged so that the anchor can clear the bottom of the ship without catching and will stow properly in the hawsepipe, whether the flukes come up turned inboard or outboard.

1.6 Deep Water Mooring. Many special ships and craft are required to moor in very deep water to perform their designed functions. The station-keeping precision required and the water depth determine the type of mooring array to be used. A three-legged single moor may be used when positioning precision is not critical and some swinging of the craft is permissible. Submarine rescue ships, which must moor in water depths to 1,000 ft with a precision of 5 percent of depth, are equipped for a self-laid four-legged array. Each leg consists of a lightweight type anchor, secured by stud-link chain to a spring buoy, to which the ship is moored by synthetic hawsers. The hawsers are adjusted by winches on the ship to maintain tension in the moor and to adjust for changes in wind and current load.

Craft which must be moored with greater precision, such as oil drilling platforms, usually require mooring arrays of six or eight legs to maintain position with the required predision.

a. Anchors. Generally, lightweight anchors are used for deep sea moors to develop the greatest possible holding power while keeping the weight of the mooring leg to a minimum.

b. Chains and cable. Mooring legs for depths too great for use of chain cables are made up of chain and wire rope or fiber rope. Several shots of chain are connected next to the anchor to provide the weight necessary to hold the anchor shank parallel to the bottom. The remainder of the leg is made up of rope. If wire rope is used, it is tappered from the lower end upward by joining successively larger diameter sections to compensate for the increasing weight of wire rope supported.

c. Buoys. Buoys are used to support the riser legs of the moor and to provide spring in the array to reduce shocks due to wave action and motion of the ship in the moor. Vertical heave loads are reduced or eliminated by the horizontal hawsers connecting the ship to the buoys, and surge loads are smoothed out by the bobbing action of the buoys. Intermediate buoys, supporting the moor leg at intervals with risers, may be required to keep the wire rope catenary from dropping vertically to the bottom, subjecting the rope to damage by chafing, and reducing the horizontal spread and thus the tautness of the moor.

e. Efficiency. The efficiency of a deep sea moor, that is, the ability to hold a ship within a very restricted range of movement, decreases rapidly as the water depth increases. Similarly, the cost of the mooring equipment escalates with depth. For drill rigs, mooring in more than 600 m (2,000 ft) of water is impractical due to the high cost of providing buoyant cables with adequate strength and the minimum elasticity required. Between 500 and 750 m (1,640 and 2,460 ft) the relative costs of deep sea mooring and dynamic positioning should be assessed.

1.7 Dynamic Positioning. The term dynamic positioning was coined in 1959 to define a means of holding a ship in a relatively fixed position with respect to the ocean floor without using anchors. During preparations for the Mohole Experimental Drilling Project, when it was estimated that mooring of the drilling ship *Cuss I* in 3,700 m water depths would cost on the order of $1.5 million, it was evident that some other means would have to be found to maintain position using propulsors instead of anchors.

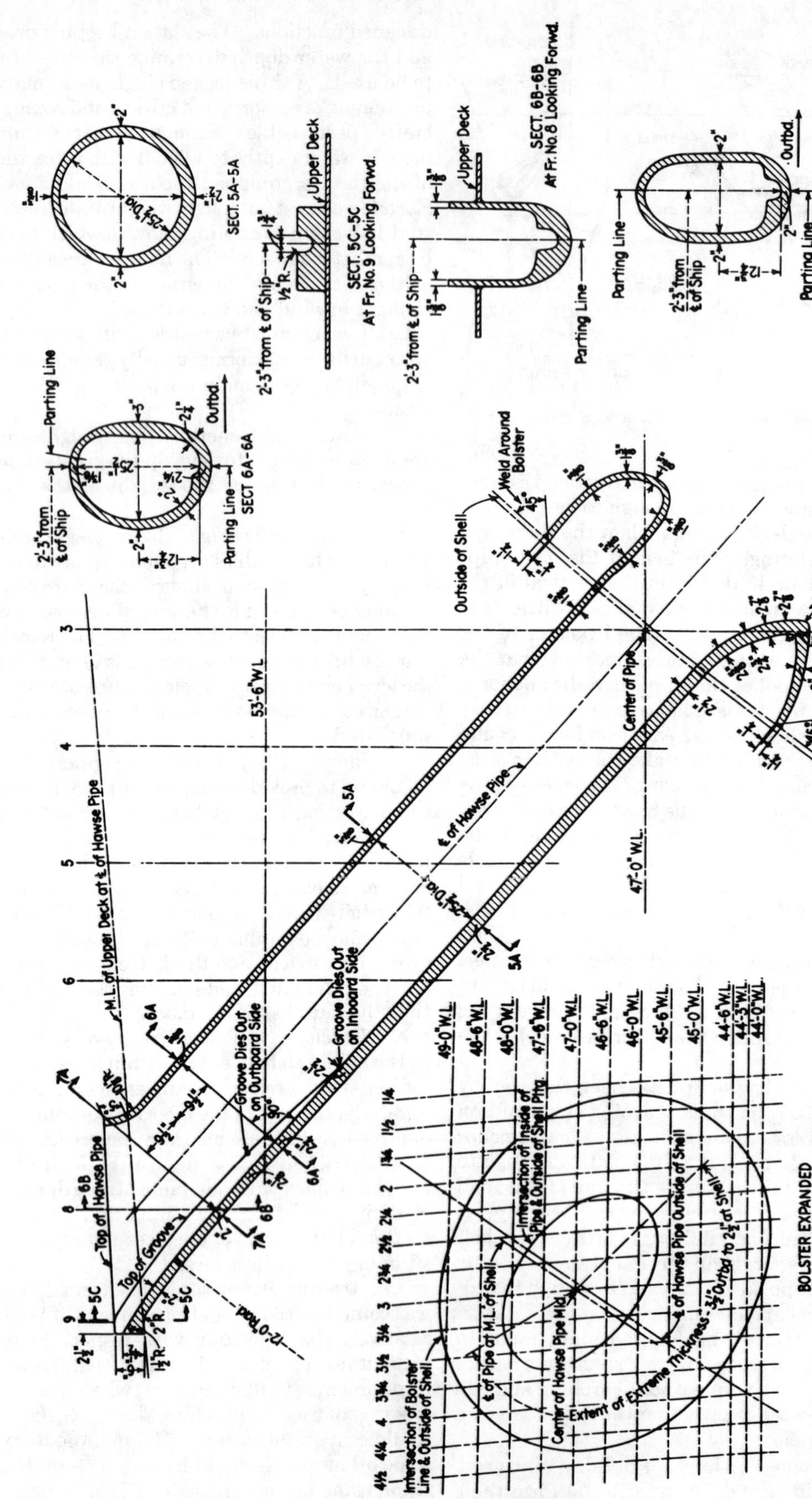

Fig. 18 One piece cast steel hawspipe

As mentioned in Chapter II, the *Cuss I* was a converted Navy YFNB with a hull identical to that of the ocean construction platform *Seacon.* In fact the dynamic positioning propulsors initially chosen were the same as those later selected for the *Seacon,* that is three rotating blade propellers as shown in Fig. 19. This planned installation was voided because of the excessive cost of the propulsors and an unacceptable delivery time. Instead, four right-angle drive steering propellers were procured and installed in the first dynamically positioned ship used for ocean drilling. The diesel-driven 200 horsepower units were mounted with the diesel base on the main deck, two at the stern, port and starboard and two forward just abaft the ramps leading up to the forecastle deck (Taggart, 1961).

a. Control System Requirements. The function of the dynamic positioning system was to hold the ship in a fixed position and head it into the prevailing swell regardless of the direction of wind and current. This dictated that a combined thrust of all propulsors be resolvable into a single vector of any direction or magnitude, into a couple which could rotate the ship in either direction, or both concurrently. If possible, these resolved forces and moments should utilize the total thrust available from each of the propulsors.

The optimum combinations of thrust reactions for several types of ship motion are illustrated in Fig. 20. An optimum control system should be able to direct the propeller thrusts in these ways with variable magnitudes. There were, however, several other functions for which the *Cuss I* control system design had to provide.

First, the diesel engines which drove the steering screws could not be started instantaneously to provide thrust of a given magnitude at the flick of a switch. For proper operation, the engines had to be turning over, at least at idling speed, continuously. Unless the control system was to be complicated by the unnecessary addition of a clutching arrangement, this meant that the propellers would always be rotating and delivering some minimal thrust. To keep the ship from moving while idling, it was necessary that the idling thrusts of the four screws be directed so that they opposed one another, with a resultant zero thrust and zero turning couple.

A second consideration involved the effect of propeller race on equipment suspended from the center well of *Cuss*

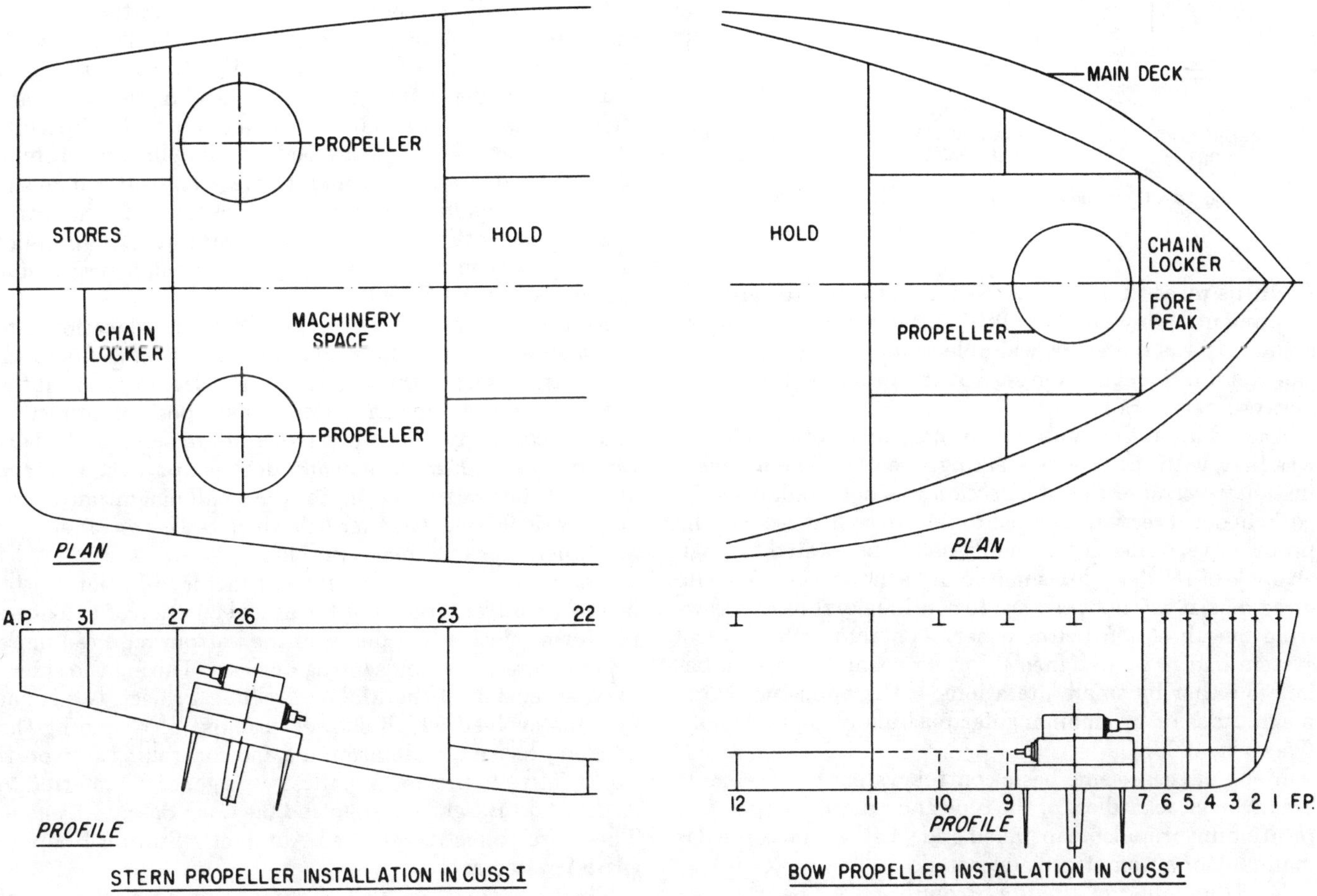

Fig. 19 Proposed arrangement of Voith-Schnedier propellers in CUSS I

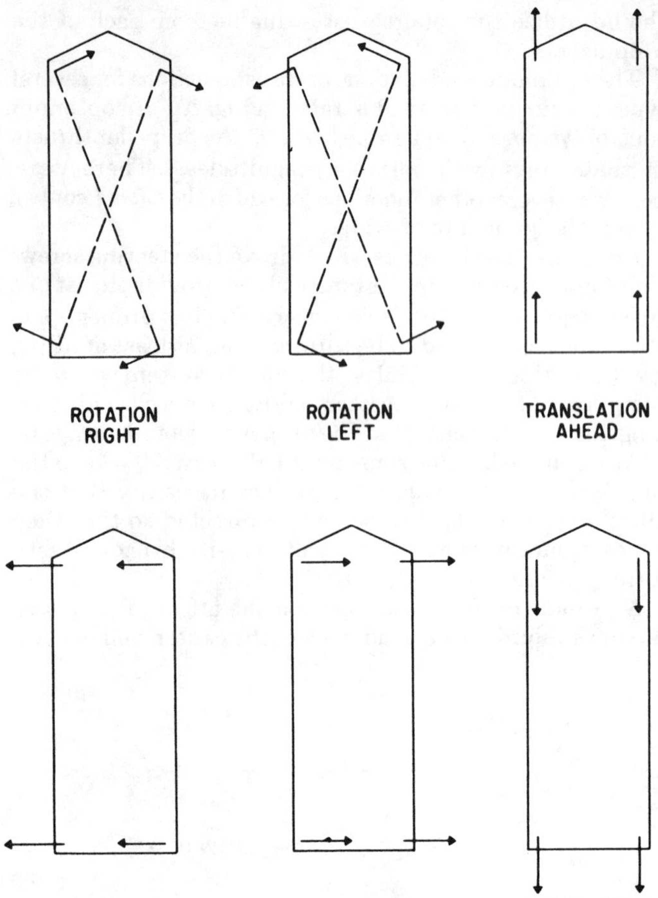

Fig. 20 Thrust reactions required for various ship motions

I. If the races from four screws were directed toward this equipment, many handling difficulties and failures might result. It was therefore desirable to limit the screw vertical-axis rotational motion so that the propeller races were always directed outboard.

A third limitation imposed on the control system design was that, with the type of steering screws to be employed, instant reversal of thrust direction was not available. To go from one direction of thrust to the opposite direction, the propeller vertical-shaft assembly had to be rotated through an angle of 180 deg. For small changes in thrust in opposite directions, it was not practical to continually flip the screws from one direction to the other. Therefore the control system had to be designed to provide small thrust imbalances either by small alterations of the opposing thrust magnitudes or by small angular position changes or both (Taggart, 1969).

Many of these same basic control system requirements are involved regardless of the type and number of dynamic positioning propulsors installed and whether the control is manually or automatically operated.

b. Dynamic Positioning Propulsors. The rotating blade propeller and the steering screw have already been mentioned as candidate propulsors for dynamic positioning systems. Other alternatives are air screws, hydraulic jets, fixed screws trained in orthogonal directions, propellers in tunnels selectively disposed along the ship length, and vortex propellers. Steering propellers in flow accelerating nozzles are also used and propellers in tunnels may be either of the reversing type or may be controllable-reversible-pitch. The selection by the designer may be on the basis of simplicity, reliability, efficiency, cost, flexibility, operational considerations, or the powering system available. A number of these factors will be discussed in greater detail in Section 1.9.

c. Ship Location and Heading Inputs. The system designed for the *Cuss I* operations in 1961 necessarily used a crude combination of sensors to determine the heading and location of the ship. A gyrocompass repeater indicated the ship heading to the pilot which he then corrected manually with a steering wheel. The general practice was to seek a heading relative to waves and swell that gave minimum roll and then to maintain that gyro heading. Location was determined from a group of taut line buoys anchored to the bottom in a somewhat random pattern. Each submerged buoy was fitted with a sonar transponder and a surface float supporting a radar reflector. The ship was manually controlled into a fixed position with respect to the surface buoy indications on the radar screen.

It is interesting to note that a human operator has great difficulty controlling both heading and the translational position of a dynamically positioned platform at the same time. His tendency is first to make a heading correction and then to make the translational correction. The difficulty of this process can be mitigated by employing an autopilot of one type or another to operate the propulsor controls to correct the heading automatically, leaving only the translational correction to the human operator. An autopilot control has been incorporated in most dynamic positioning systems since the *Cuss I.*

Another means of sensing translational dislocation is to run a wire line over the side to an anchor on the bottom. Either the angles of the line with the horizontal plane at the bottom or at the ship can be sensed and a position correction made accordingly. This technique worked on the Shell Oil Company ship *Eureka* in water depths up to about 500 m (1,500 ft) but in greater depths the small magnitude of the angular signals and the time delay in detecting an off-station movement became unacceptable.

The autopilot was a highly satisfactory solution to the heading control aspect but the actual position of a floating platform relative to a point on the bottom required more sophisticated position sensing devices. During the short-lived extension of the Mohole Drilling Project, two sonar systems evolved which showed promise of providing the needed precision in determining and controlling ship position relative to the bottom; these were generally referred to as the Short Baseline System and the Long Baseline System. These are collectively called Automatic Station Keeping (ASK) systems.

The Long Baseline ASK System is used primarily when precise positioning of the ship is required over a large working area while operating in water depths over 1,000 m (3,050 ft). Four or more transponders are dropped to the ocean floor at a horizontal distance from the ship on the order of 70 percent of the water depth. From an interrogation transducer on the bottom of the ship a sound pulse is sent out. The replies from the transponders are received by corresponding transducers on the ship bottom and the slant range to each transponder is calculated. The processed signals can give a computed ship position with an accuracy of approximately 12 m (36 ft) in 5000-m (15,000-ft) water depths or about one-quarter of a percent of depth.

In the Short Baseline ASK System a single beacon is installed on the ocean floor; its signal is received by a series of transducers mounted at precisely spaced positions on the bottom of the platform and the position relative to the beacon is determined by difference in arrival times of the signals at the transducers. The accuracy of this system is on the order of 45 meters in 5,000 or about one percent of depth.

Systems are also designed to accept the input from anemometers and wave height measuring systems. The computed position and direction changes from the combination of direction and position error plus estimated force reactions are then converted into control signals for the dynamic positioning propulsors. The result is to maintain the position and orientation of the platform on the surface relative to a point on the bottom within a very small circle of error.

d. The Deep Sea Drilling Project. The next major advance in dynamic positioning development was the first expedition of the *Glomar Challenger* in the Gulf of Mexico in 1968 which launched the Deep Sea Drilling Project (DSDP). The *Challenger* was fitted with two main screws, two stern lateral thrusters, and two bow lateral thrusters, all of which were fixed pitch propellers driven by variable-speed DC motors. The drilling equipment was also powered by DC traction motors so that the entire DC power plant could be interchanged between main propulsion, dynamic positioning, and drilling (Graham et al, 1970).

As far as dynamic positioning capability is concerned, the greatest step forward was taken when the *Glomar Challenger* first achieved hole reentry; that is, after tripping out the drill string, a new bit was attached and the ship was held in position while the drill string was lowered and the drill guided back into the same hole in the ocean floor. This is now a routine operation that has been achieved in water depths as great as 5500 meters (18,000 ft).

1.8 Ship Steering Systems. Steering is a special case of maneuvering involved in keeping an underway ship on a desired constant heading. Other definitions of steering are used in the literature in special cases such as keeping the ship center of gravity moving along a straight line. In such cases (e.g., following a meridian of longitude) steering would involve changing the ship heading as the ship center of gravity went east or west of the desired path.

The officer in charge on the bridge of the ship, serves as the navigator and calls out orders to a helmsman. The helmsman controls a wheel (the "helm") which transmits orders for rudder angle to the steering machinery aft. Fig. 21 shows the system in block diagram form. Elements of the steering system, as described here are the rudder and steering gear. Information displays and controls are described in Section 2.2.

a. Types of Rudders. Some common rudder types, bearings arrangements, and the associated terminology are shown in Fig. 22. It will be noted that unbalanced rudders have no area forward of the rudderstock or pintles, semi-balanced rudders have area forward for part of the rudder height (or span, as it is sometimes called) and balanced rudders have area forward for the full span.

b. Rudder Area and Shape. Regulatory bodies generally do not specify steering, maneuvering, or stopping capability for ships. Also the American Bureau of Shipping (ABS) and the Lloyd's Register of Shipping Rules do not require any specific rudder type or area. Det norske Veritas issues the most demanding requirements in its 1972 Rules; for rudders working directly behind a propeller, this classification society calls for rudder area to be at least the size of:

$$A = \frac{TL}{100}\left(1 + 25\left(\frac{B}{L}\right)^2\right) \tag{16}$$

Where the symbols have the following meanings, in any consistent set of units:

A = Rudder area
T = Ship draft
B = Ship beam
L = Ship length

Usual practice is to compare a new design with a known successful design, and to use the same shape and general arrangement of rudder. Rudder area would also be taken as about the same percentage of LT. For merchant ships, the area of the rudder is usually about 2 percent of the product LT for ships 120 m (400 ft) long and over; for smaller ships it may increase to about 3 percent for a 30-m (100-ft) ship. Ships requiring special maneuverability will have more rudder area; seagoing tugs may have rudder areas of 4 percent, and harbor tugboats may have rudder areas of 6 to 8 percent of LT. When operational requirements are stringent, or if usual practice is not considered to be a reliable guide, there are two basic design methods available. The first is analytic, and would involve using a theoretical approach. The second method, more reliable, would involve model tests; these can be either with a freely maneuvering remote controlled model, or else with certain restricted model measurements followed by computer prediction of trajectories.

c. Rudder Angle Effects. Most ship rudders operate to 35 deg right or left. Tugs and Great Lakes vessels often have rudders operating to 45 deg for turning at zero or low speed. The choice of the rudder angle ordered by the officer on watch is basically a matter of experience in a particular maneuvering situation. In general, the use of large, rather than small, rudder angles has the following effects:

- The ship turns in a smaller (tighter) space

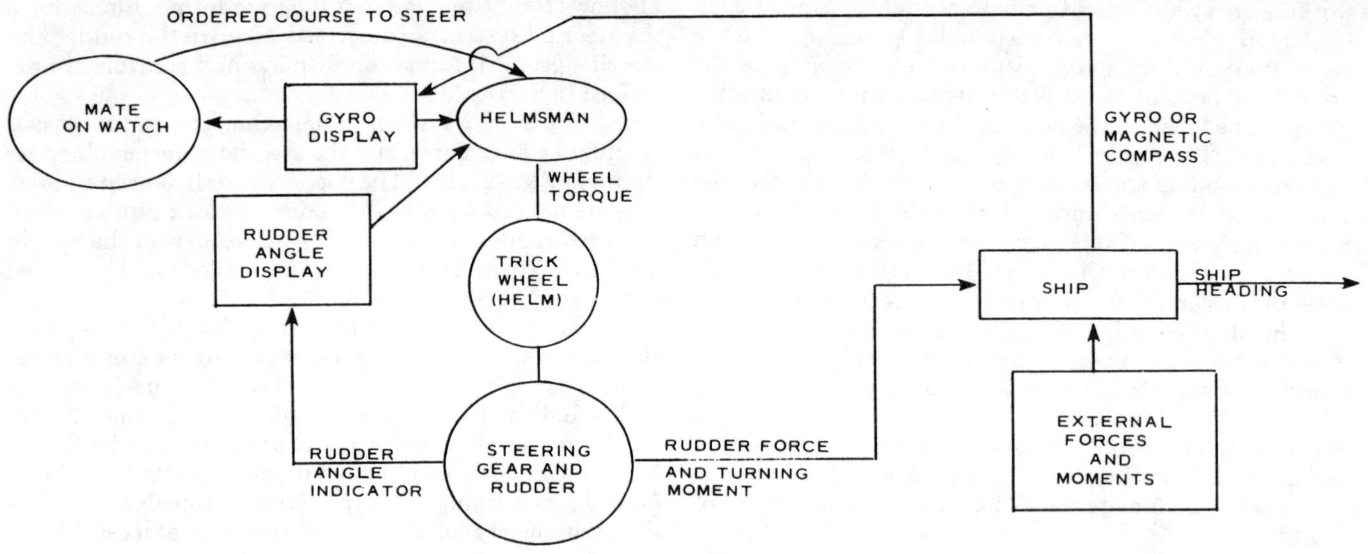

Fig. 21 Block diagram of steering and control system elements

- The speed loss is greater, as is time to regain speed
- Heel angle in the turn is greater. (This may be significant for container ships or others with loads which have to be secured.)

Use of large rudder angles is the fastest way to change heading; that is, the tighter turn more than compensates for the speed reduction. Typical data relationships of turning rate and ship speed are shown in Fig. 23. The size of the

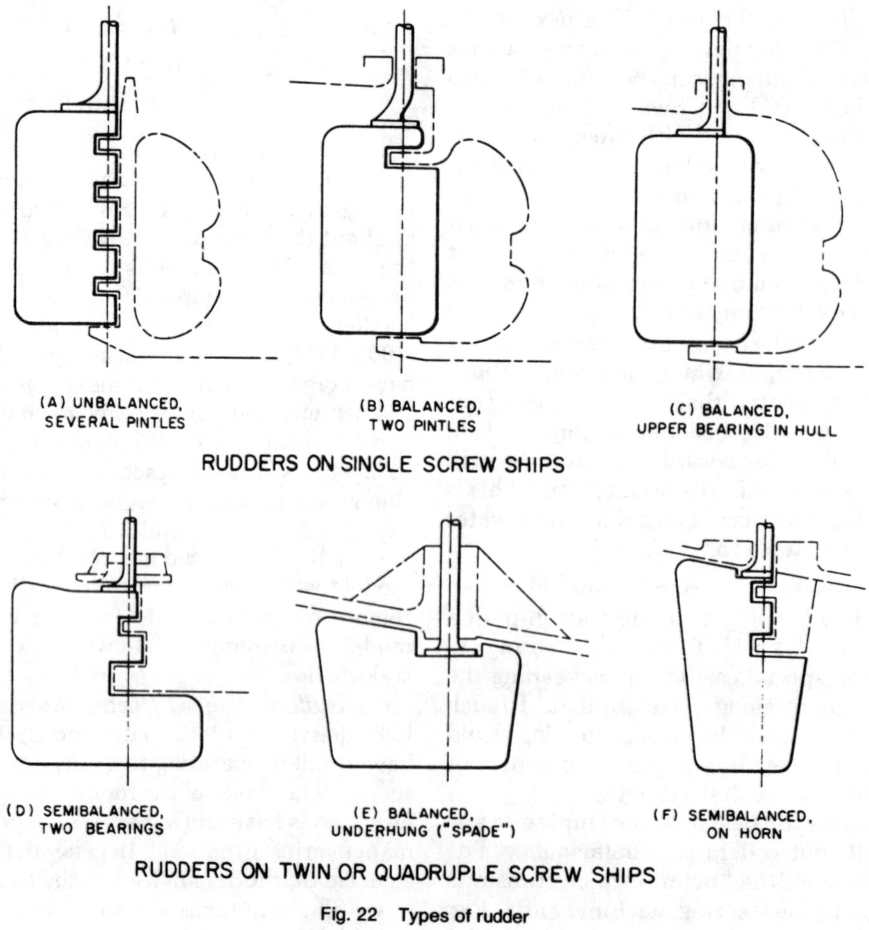

Fig. 22 Types of rudder

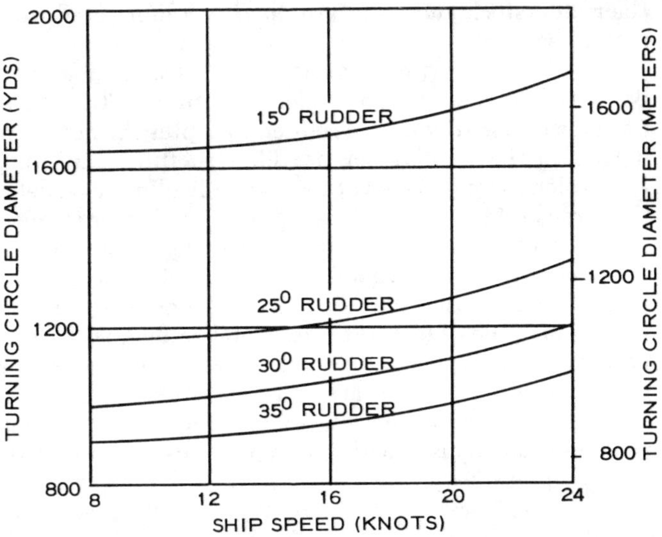

Fig. 23 Representative curves of turning circle diameter for a 152 m (500 ft) ship for various speeds and rudder angles

ship turning circle is practically constant in the lower range of ship speeds. At about Froude Numbers of 0.2 and higher, the turning circles become larger.

In shallow water the flow around a ship changes, as may be seen from increased draft and different surface wave patterns and trim. The ship tends to steer in a straight line a little better than in deep water, but turning circles usually become larger. For practical purposes, shallow water may be taken as existing when the under-keel clearance becomes less than 25 percent of ship draft (where D/T in Section 1.2 becomes less than 1.25).

d. Rudder Forces. The various rudders shown in Fig. 22 have the common requirements that:

- strength be adequate for the dynamic loads in a turn;
- the bearings for transverse loads and for vertical loads (weight and flooding water in the rudder) be of adequate capacity and low friction;
- deflections under loading conditions should not cause binding;
- there be adequate clearance to allow for normal wear in bearings, preferably for the life of the ship, but at least for several years of operation; and
- the maintenance requirements should be simple.

Strength calculations involve the transverse force approximately normal to the rudder, and the torque, or twisting moment, required to turn the rudder against the pressure of the water flowing past it, all at the highest designed speed, plus propeller race effects where applicable. In addition, rudders withstand similar, but probably greater, forces resulting from the impact of seas in heavy weather. In practice, rudders are designed only for the normal turning forces anticipated, with margins of strength for heavy weather as found necessary by experience. These margins are greater than for most shipboard structures, because failure of a rudder, particularly in a single-screw ship, may mean the loss of the ship. Rigorous calculation of forces is complex, requiring estimating flow velocity as affected by the boundary layer and propeller race, geometric shape factors, streamline shape factors, gap between hull and top of rudder, true angle of attack of flow as affected by the ship turning, Reynolds Number, and seaway effects. Detailed information is given by Mandel (1967) and by Taplin (1960). For an understanding of the magnitude of forces on rudders, the usual basic dynamic flow formula may be applied:

$$F_R = C_R \frac{\rho}{2} A \nu^2 \qquad (17)$$

where the symbols have the following meaning:

F_R = total force
C_R = total non-dimensional force coefficient for the angle of attack
ρ = mass density of the fluid
A = total moveable rudder area
ν = fluid flow velocity over the rudder

A simple approximate formula for rudder force at 35 deg rudder with ship speed in knots can be derived using the above formula, and estimating $C_R = 1.05$ (a reasonable value for 35 deg rudder with drift angle). Also the speed over the rudder can be estimated as 1.17 times ship speed, to allow for propeller race effects.

when: V_K is the ship speed in knots:

$$F_R = 196\, A V_K{}^2 \text{ Newtons when } A \text{ is in m}^2 \qquad (18)$$

$$F_R = 4.08\, A V_K{}^2 \text{ lb when } A \text{ is in ft}^2 \qquad (19)$$

e. Rudder Bending Moment and Torque. The above formulas provide the total magnitude of rudder force. The point of application in the vertical direction is generally taken at the centroid of the rudder area, for use in calculating vertical (spanwise) bending moments and bearing reactions. In the horizontal (chordwise) direction of the rudder, strip theory is applied. In each strip, rudder force times its lever arm forward or aft of the rudderstock is summed algebraically to determine rudder torque. There is no generally accepted procedure for this. In the formerly used Jossel flat plate method (Pollard and Dudebout, 1892), the fore-and-aft center of pressure on a rudder such as in Fig. 22(B) and 22(C) at various rudder angles would be:

$$\chi = (0.195 + 0.305 \sin\alpha)\, b \qquad (20)$$

where χ = center of pressure aft of leading edge
α = rudder angle or angle of inflow to the rudder
b = rudder chord (i.e., fore-and-aft dimension of the rudder or elemental strip).

On that basis, at 35 deg rudder, the center of pressure would be $0.37b$ aft of the leading edge. In more modern methods, using aerodynamic data, the chordwise center of pressure aft of the leading edge of a strip varies from about $0.15b$ at small rudder angles to about $0.27b$ at large angles. Essentially rudder torque calculation is not yet an exact science. It is closely tied in to rudder balance, which is defined as rudder area forward of the stock divided by total rudder area

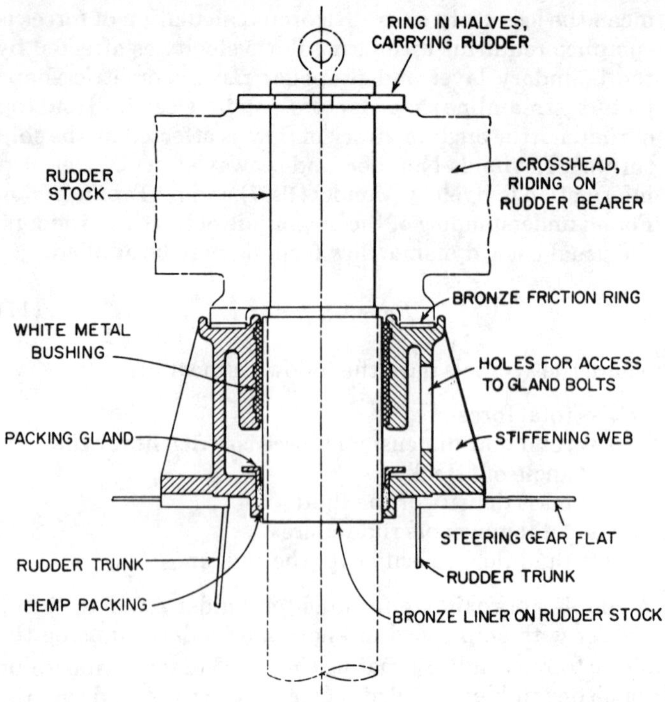

Fig. 24 Rudder bearer

times 100 percent. Rudder balance in new designs is generally taken the same as on old designs which have worked well, usually in the 15 to 20 percent range. For astern operation the ship speed is usually a fraction of ahead speed and the rudder is not acted upon by the propeller race so that rudder force is comparatively low. Lever arms for rudder torque in astern operation are generally larger than for ahead operation. The calculation procedure is basically the same as for ahead operation, with whatever allowances individual designers may make.

f. Rudder Structure and Supports. Steel is the predominant rudder material. Classification society rules are generally used for determining scantlings of rudder castings, forgings, and weldments, stress relief requirements for large weldments, plating thickness, spacing of internal diaphragms, plus other structural details.

Commonly used bearings involve pintles or rudderstocks having bronze sleeves operating in grease-lubricated bearings. Bearings are usually made of reinforced laminated phenolics or white metal. Where friction reduction can be counted upon to reduce steering gear size, it is sometimes worth installing anti-friction bearings. Such bearings would degrade rapidly in sea water, so their principal application is for a protected inboard rudderstock bearing, where they can take both radial (side force) and axial (vertical) loads. Anti-friction bearings at the hull or outboard require carefully designed and installed seals, so that bearings can be kept in oil or grease, with no sea water intrusion. Fig. 24 shows a common design, in which vertical rudder loads are carried via the rudderstock to the cross head, and then through a lubricated bronze friction ring to the rudder trunk and shell structure. Transverse loads are transmitted via the rudderstock and its bronze liner to a lubricated white metal bushing.

Fig. 25 shows typical construction for a semibalanced double-plate rudder on a single-screw ship. The tapered fit between the rudder and rudderstock plus the nut at the bottom of the rudderstock provide structural continuity. An erosion plate is shown at the rudder leading edge, below the fixed rudder horn. This plate, of thicker or stronger steel than normally used, is for resisting cavitation impingement from the propeller tail cone, which would be a short distance upstream. The external rudder stop, shown at the top of the rudder, provides metal-to-metal contact to prevent wild swinging of the rudder after damage to internal mechanisms. The external stop is set at one or two degrees beyond the last internal stop. The rudder horn, where provided, usually is a casting, and provides support for the

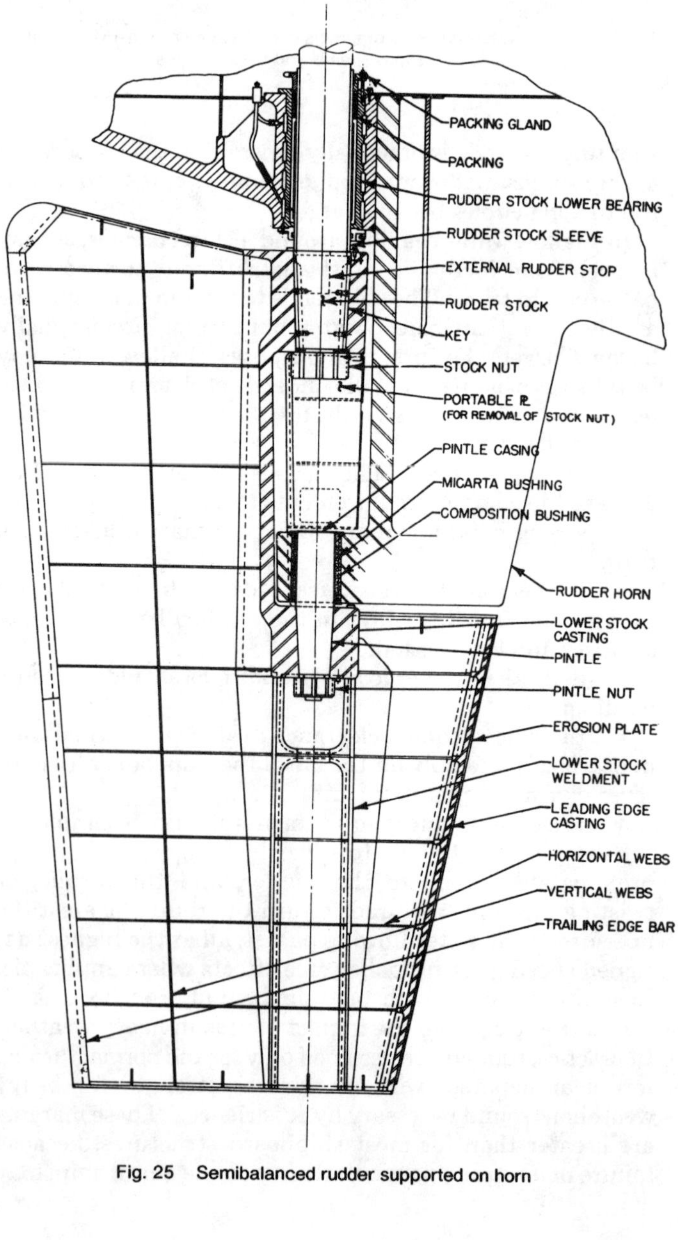

Fig. 25 Semibalanced rudder supported on horn

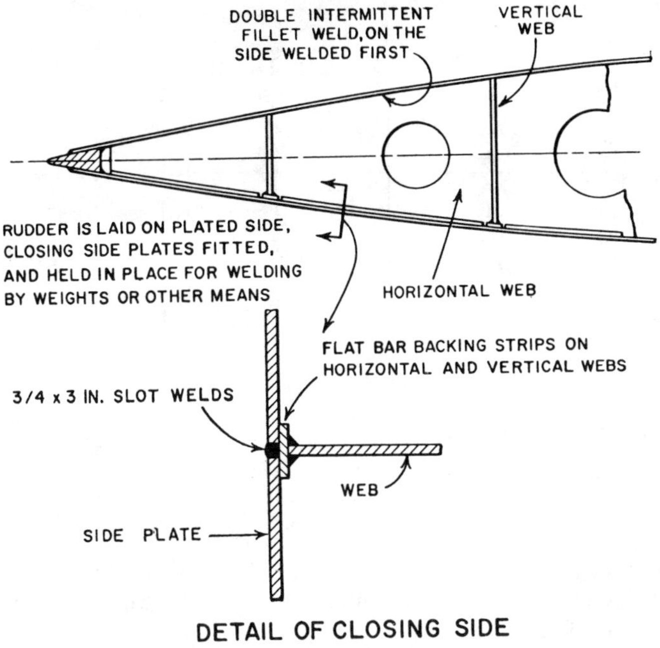

Fig. 26 Attachment of closing side of double plate rudder

pintle. The structure shown in Fig. 25 is carefully designed to permit disassembly in a drydock. Portable plates are shown below the pintle nut and stock nut. The supporting structure is suitably reinforced, and in some cases the securing nut can be used to jack the rudder off the tapered fit. The keys at the lower end of the rudderstock transmit torque.

The closing side of the rudder plating must be welded entirely from the outside, and approved practice for that is shown in Fig. 26. In the rudders shown in Fig. 22 (A), (B), (C), (D), and (F) the rudderstock scantlings are governed by torsional stress, using the standard formula

$$S_t = \frac{TC}{J} \qquad (21)$$

where:

S_t = torsional stress
T = torque
C = rudderstock radius
J = polar moment of inertia of rudderstock cross-sectional area. For solid circular sections

$$J = \frac{\pi\,(\text{diameter})^4}{32}$$

In the spade rudder shown in Fig. 22(E), bending stress predominates, because:

- The hydrodynamic force has a large lever arm for bending (approximately half the rudder height), but only a small lever for torsion (a few percent of the fore-and-aft dimension).
- Stress is inversely proportional to sectional inertia. For torsion, polar inertia is used, which is double the inertia for flexure. In such cases the bending stress must be computed and properly combined with torsional stress. The bending stress is computed

$$S_f = \frac{MC}{I} \qquad (22)$$

where

S_f = bending stress
M = bending moment
C = radius of rudderstock
I = moment of inertia of rudderstock cross sectional area about its diameter. For solid circular sections

$$I = \frac{\pi\,(\text{diameter})^4}{64}$$

The two stresses are then combined as follows:

$$S_c = \frac{S_f}{2} + \left[\left(\frac{S_f}{2}\right)^2 + (S_t)^2\right]^{1/2} \qquad (23)$$

where

S_c = combined stress.

g. Streamlined Contour for Rudders. After rudderstock diameter and associated casting thicknesses are settled, the rudder streamline shape is chosen so as to surround the structure smoothly. Bulges in way of the rudderstock are acceptable for low speed but at high speed, discontinuities are avoided in order to prevent cavitation and early flow separation. Many streamlined shapes have been used. A particularly useful shape, because of the great amount of associated wind tunnel data, is the National Advisory Committee for Aeronautics (NACA) 4-digit symmetrical series, also known as NACA OOXX. The XX digits represent the maximum thickness as a percentage of the fore-and-aft length. Thus, for an NACS 0015 section, thickness is 15 percent of the length. The offsets as decimal parts of the maximum thickness are shown in Fig. 27. It will be noted that the maximum thickness occurs 30 percent aft of the leading edge, and that there is a definite sharp corner thickness at the trailing edge. The rudder surface is then determined by selecting a streamline contour at the top (say NACA 0022) and another at the bottom (say NACA 0010), and connecting like-numbered stations by straight lines.

h. Steering Gear. In the earliest known ships, the rudder was put over directly by hand power, as with a tiller. When rudder area and ship speed increased, the human operator needed more and more leverage or mechanical advantage. Mechanical devices were developed, such as pulleys, windlasses, and worm gear-quadrants. These could be used for rotating the rudderstock, all powered by one or more sailors. Within the past century, steering gear driven by steam or electrical power became feasible and also became required by further increased rudder torque. Seagoing ships now have steering gear motors in the capacity range of 15 to 150 kW (20 to 200 hp). Some small ships and boats still use human powered mechanisms or an electromechanical type power drive. Nickerson and Smith (1971)

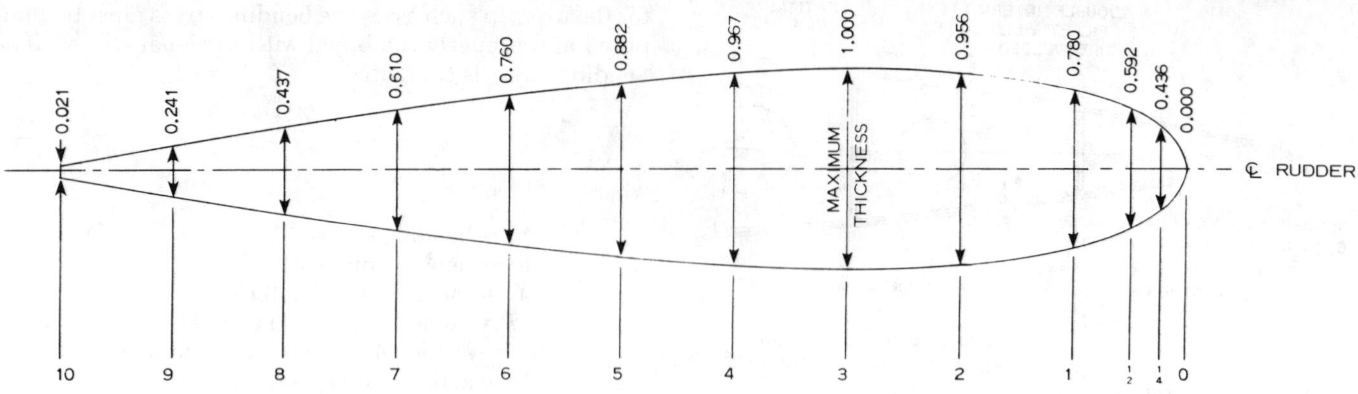

Fig. 27 Offsets for double-plate rudders, NACA 4-digit series

provide more coverage on steering gear development and detailed design.

The most common type of power steering gear for seagoing ships is the electrohydraulic drive. There are many variations, so that a typical unit will be described. It has two major sub-divisions, called the power unit and the ram unit. The power unit consists of a continuously running electric motor coupled to a hydraulic oil pump of the variable delivery type. These pumps are relatively small, about half the size of the driving electric motor. They have the important capability of varying the amount of delivery from 0 to 100 percent capacity and reversing the flow direction by simply moving one actuator. Moving the actuator (often called "putting the pump on stroke") provides a smooth, readily controllable supply of oil to the ram unit. In a basic arrangement, a double-acting ram is connected to the rudderstock tiller by means of either a link mechanism or a Rapson-slide mechanism. The arrangement shown in Fig. 28 has two double-acting rams and two independent power units, which is common practice for large ships. In that way the ship can have some steering control even if one ram unit is disabled (e.g., from leaking seals) or one power unit is disabled (e.g., from worn bearings). This capability may eliminate classification society or regulatory body requirements for auxiliary steering gear. Other variants of this basic type are:

- Positive displacement pumps, such as the screw type, plus flow control valves can be used in the power unit.
- The hydraulic oil pipes and valves can be arranged so that both pumps operate to double the flow rate. That scheme is sometimes used for maneuvering in a harbor, where fast rudder action is desired, although only one pumping unit is used for normal steering at sea.
- The swiveling cylinder type is common in tugs. Instead of cylinders fixed to the deck, as in Fig. 28, it uses a double-acting cylinder with a piston rod protruding from one end. The cylinder end is attached to the deck by a pinned connection, free to rotate. The other end, the piston rod end, is attached by a pinned connection to the tiller. This is a relatively inexpensive steering gear, based on using standard off-the-shelf cylinders. Because the hydraulic cylinder swivels, it is of course necessary to supply oil by means of flexible hoses.
- A rotary vane type electrohydraulic steering gear, Fig. 29, consists of a limited travel vane actuator connected directly to the rudderstock, powered and controlled in the same manner as ram type gear. The rotary vane type gear is lighter and requires significantly less deck area than the ram type gear for equivalent capacity. It is, however, difficult to gain access to the vanes in case of an emergency which would require holding the rudder in place.

The general application of electro-mechanical gear is in

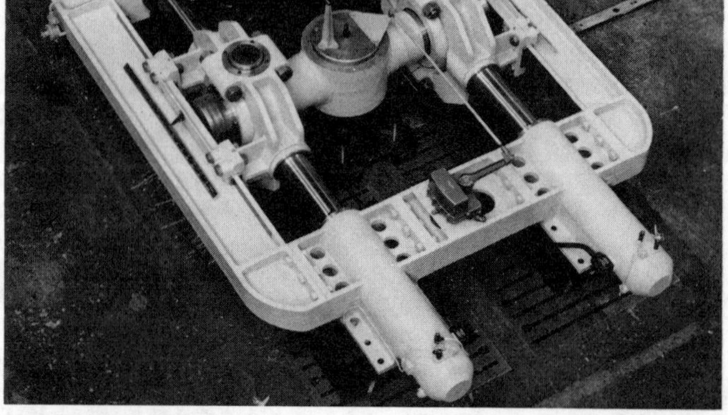

This steering gear has two 34.5 MPa pumping units and changeover valves. Cylinders, pumping units, and Rapson slide crossheads are mounted in a unitized frame for ease of installation.

Fig. 28 Electro-hydraulic steering gear

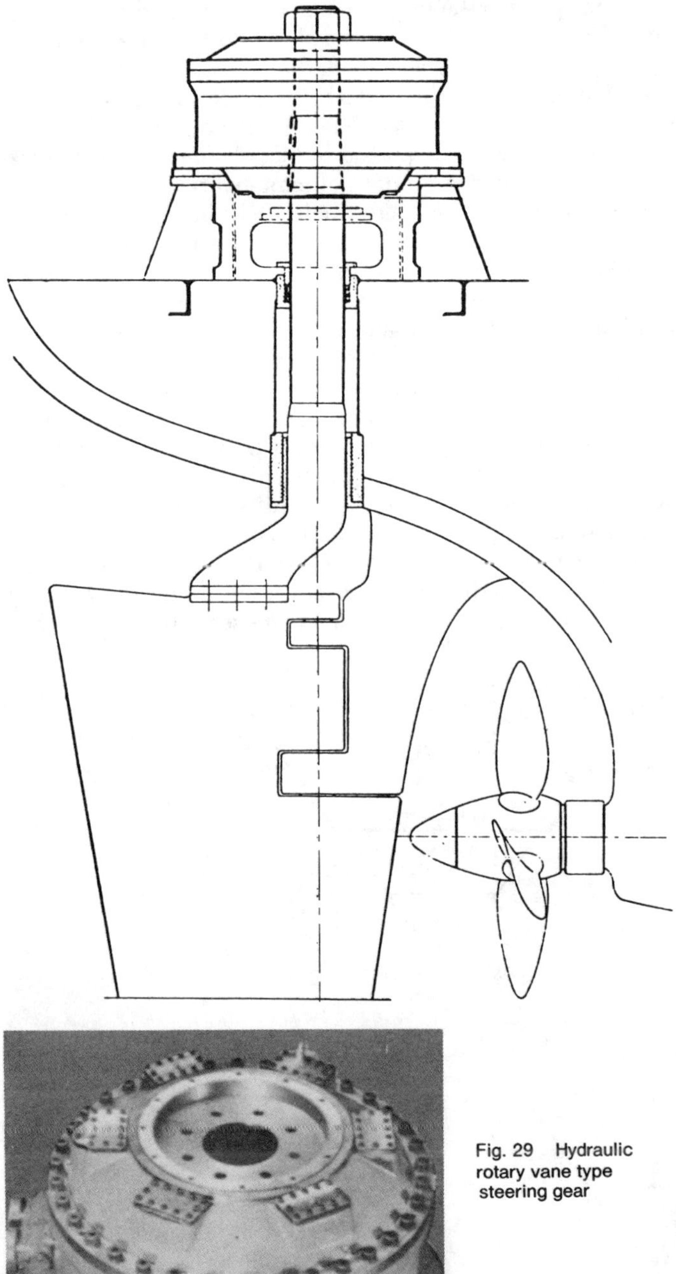

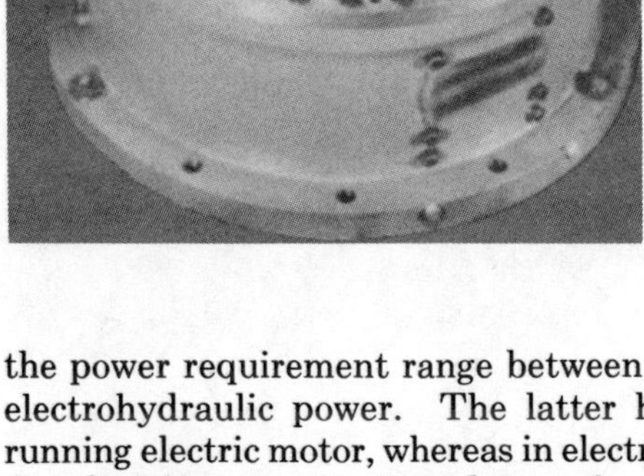

Fig. 29 Hydraulic rotary vane type steering gear

the power requirement range between human power and electrohydraulic power. The latter has a continuously running electric motor, whereas in electro-mechanical gears the electric motor starts and stops for each rudder movement. This starting, stopping, and reversing, particularly under load, requires special attention to motor heat loss and to controller contacts. Typically, the motor drives gears or a drum. Linkage to the tiller is then made by several methods, such as chain, wire rope, or rods.

The drum type steering engine is fitted on many ships under 100 m (300 ft) in length. The drum, which is grooved for the steering chain, is driven through spur gears by a reversible electric motor. Steering engines of this type customarily are located in or near the engine room for convenience in servicing. Several turns of a short link chain are wrapped around the drum and the ends are carried by sheaves down each side of the ship to the rudder quadrant.

i. Classification Society and Regulatory Body Requirements. There has been renewed study of the traditional steering gear information displays, and personnel training. Due to recent accidents, it is probable that new regulations, expected around 1981, will call for more redundancy. For new design, therefore, the up-to-date applicable rules should be checked. A few requirements that have been in use for years are provided for reference and, as a check-off list, the following are given:

- The main steering gear for ships over 76.2 m (250 ft) in length and in ships requiring 229 mm (9 in.) or larger upper rudderstock diameter must be power operated.
- Steering gears may be hand-operated on vessels less than 76.2 m (250 ft) in length when the ABS Rule diameter of the upper rudderstock does not exceed 229 mm (9 in.). Larger ships must be fitted with power-driven steering gears.
- During sea trials, the main gear must be capable of moving the rudder from 35 deg on one side to 35 deg on the other side while underway at maximum continuous rated shaft rpm, and take no more than 28 seconds between 35 deg on one side and 30 deg on the other (American Bureau of Shipping, Annual).
- An auxiliary steering gear must also be provided except when the main gear consists of duplicate power-operated units, or is electrohydraulic with two independent pumping plants, on separate power circuits, either of which can operate the gear. The auxiliary gear, when fitted, must be capable of moving the rudder from 15 deg on one side to 15 deg on the other side in not more than 60 seconds when the ship is going ahead at half speed, or 7 knots, whichever is greater. The auxiliary steering gear may be a small, independent, manually powered ram. However, power-operated auxiliary gear is required when the upper rudderstock diameter exceeds 229 mm (9 in.) in passenger ships and 356 mm (14 in.) in cargo ships (Coast Guard, 1977).
- The United States Coast Guard now requires, by Title 46 CFR 58.30, that ship maneuvering data be posted on the bridge.

1.9 Maneuvering Propulsion Devices. The use of some type of maneuvering device will be required during the major portion of the life of a merchant ship. When at sea, courses will have to be set and maintained and, as ports are approached changes in course will occur with ever increasing frequency. Under conditions such as these, the rudder and propeller, in most instances, provide sufficient control. However, the dimensions of ships are increasing more and more each year. This growth, coupled with the static nature of port area, channel, canal lock, and turning basin dimensions, has led to a situation wherein ships are operating in relatively more confined areas. When operating in these confined areas, ship speeds are considerably less than normal sea speeds.

The force-producing capabilities of a rudder are dependent upon the velocity of flow over its surface. As ship and propeller speeds decrease, the control effectiveness also decreases. Thus, when ships are operating in congested and confined areas, where the requirements for rapid and positive maneuvering control are at their maximum, the effectiveness of their control device becomes less and less. Eventually, at very slow speeds, independent maneuvering control is, at best, marginal. At times such as these, reliance must be placed on some other form of maneuvering assistance.

An external source of assistance is available in the form of tugboats. One or more of these craft can be attached to the ship, and commands to push or pull are given with whistles or by the use of small portable radios. In most instances, the ship is maneuvered into the desired position in a routine fashion. Occasionally, though, space restrictions will not permit the use of a tug at the desired location, or communications between the ship and the tugs will break down, resulting in the ship striking a pier and causing extensive damage to the hull, the propeller, or the pier.

In an attempt to provide ships with a greater degree of independent control, and to reduce the need for costly external assistance, a number of maneuvering assistance devices have been developed which can be installed within the hull of a ship. Many of these devices can be used for most applications, whether the ship is a merchant vessel or one with a highly specialized mission. Others have limitations which for one reason or another preclude their use in various situations. These are referred to here as *maneuvering propulsion devices* (*MPDs*). Numerous types of maneuvering propulsion devices exist, covering a broad range of powers, and having varying thrust producing capabilities. They can be considered to fall within two general classifications: those with fixed thrust directions and those capable of providing a trainable thrust.

a. Fixed Thrust MPDs. Fixed thrust direction MPDs, Figs. 30 and 31 are characterized by one of various types of motivating devices located in a transverse tunnel through

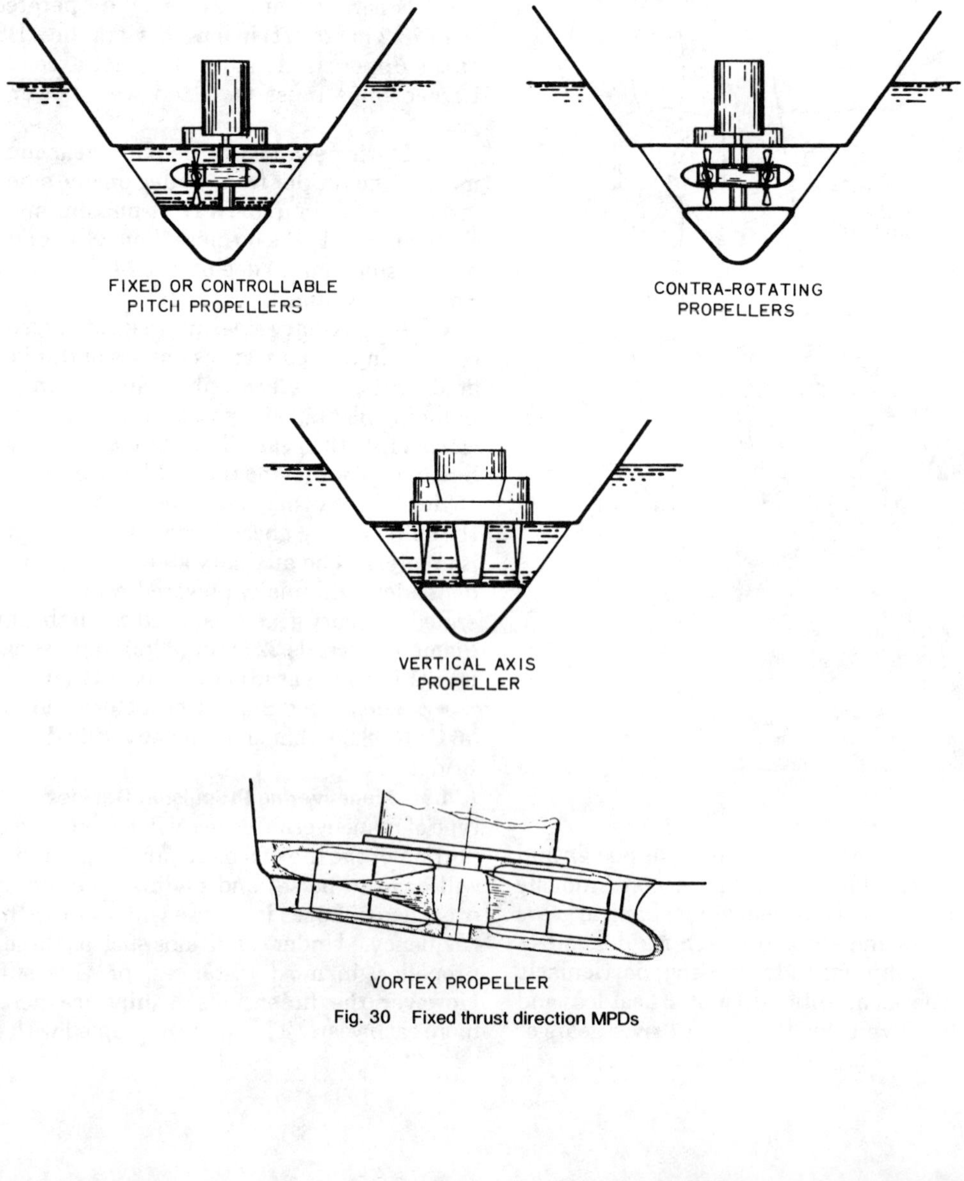

Fig. 30 Fixed thrust direction MPDs

Fig. 31 Cross-tunnel bow thruster with controllable-pitch propeller

the ship's hull. They may be either fixed pitch, controllable pitch, contra-rotating or vertical axis propellers, or, in some instances, centrifugal pumps. Fixed and controllable pitch screw propeller installations are, by far, the most common, either or both types being available through numerous manufacturers in the U. S. and abroad. Two manufacturers provide contrarotating propeller tunnel thrusters and one a vertical axis propeller type installation (English, 1962). Fixed axis thruster installations using centrifugal pumps are even less common; usually they involve the addition of a tunnel and the utilization of existing pumping equipment. However, a recently invented device called the Vortex Propeller shows considerably improved performance over the basic centrifugal pump (Taggart, 1975). Fixed axis thrusters can be located at either the bow or stern; although in most instances the bow is used. The tunnel is wholly contained within the hull and is usually located in an area of the ship which does not interfere with the carrying of cargo.

Location of the device in the bow has many attributes.

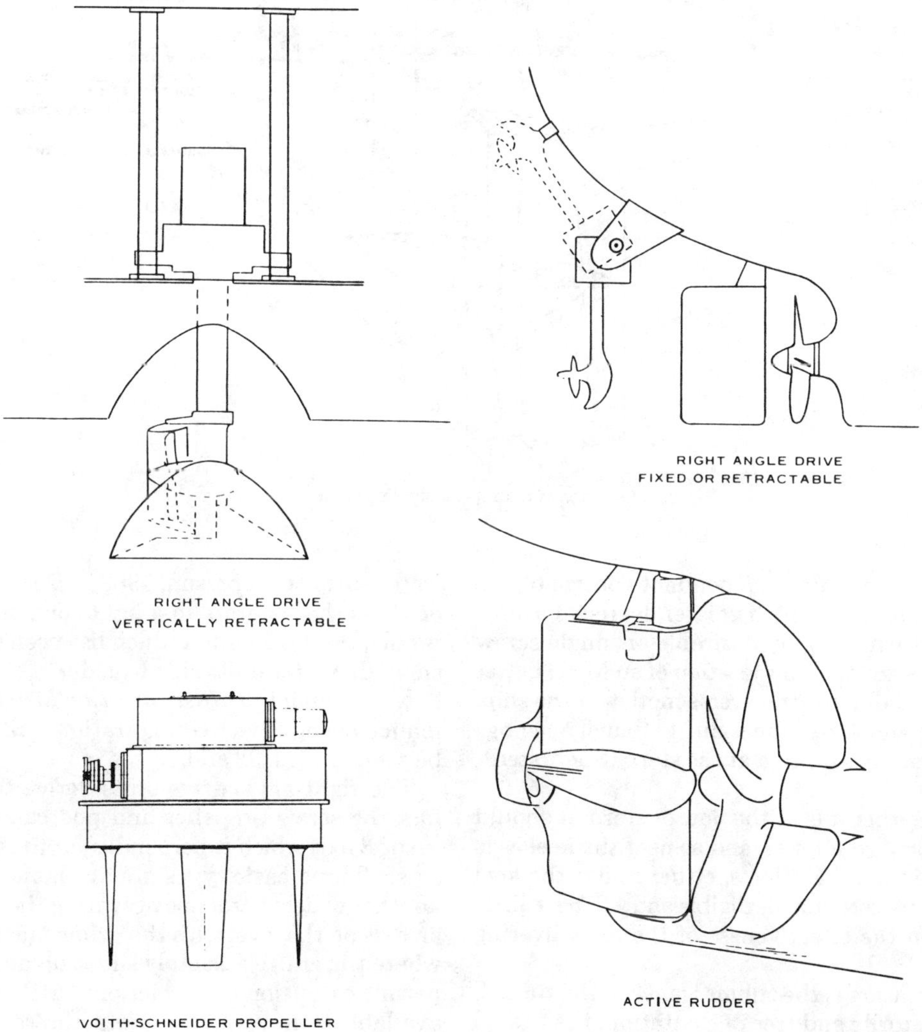

Fig. 32 Trainable thrust MPDs

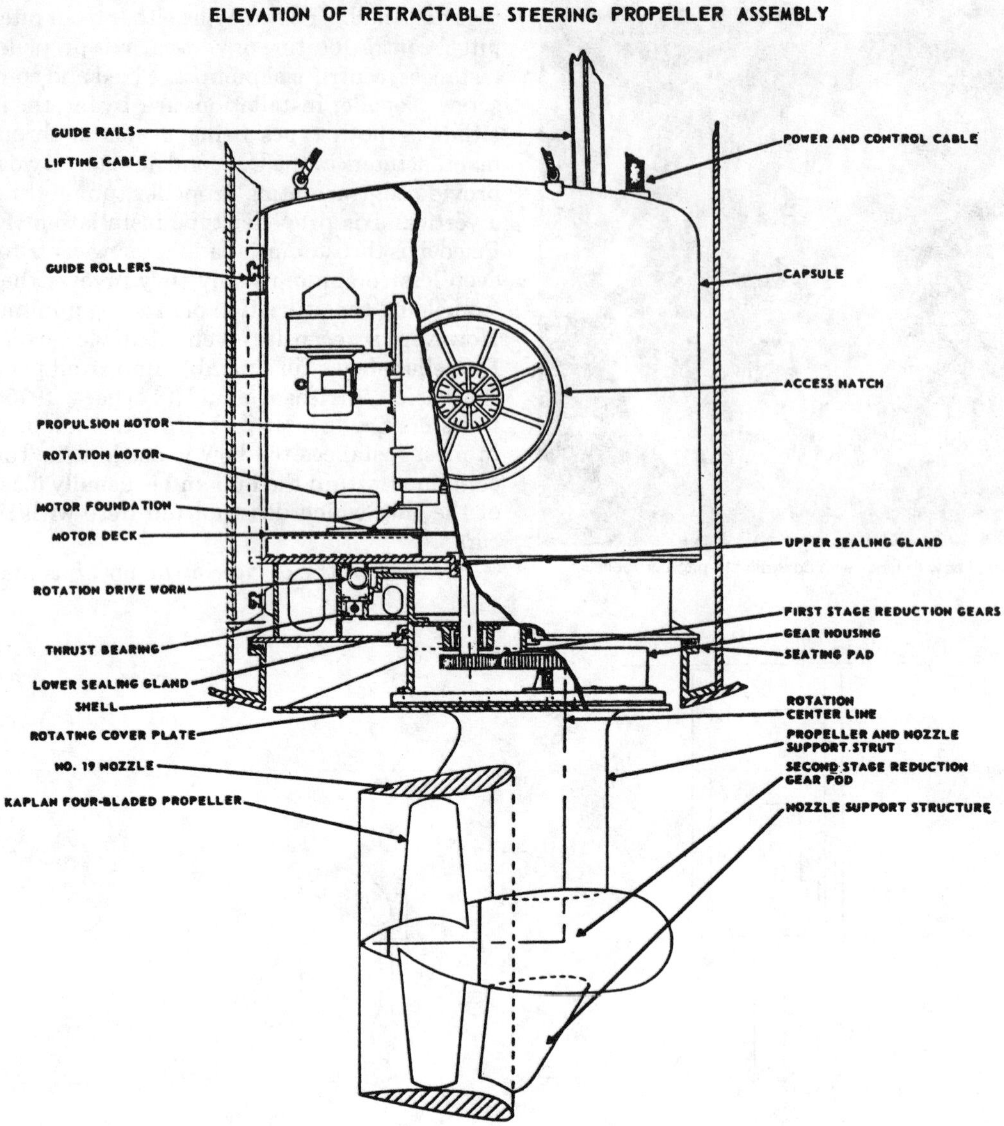

Fig. 33 Elevation of retractable steering propeller assembly

First and foremost, it permits independent control of the bow. Also, it provides for a much greater degree of astern steering control, which is highly desirable on single screw ships. Other factors are that bow location of an MPD better facilitates control of sidling or transverse motion of the ship, and apparently, the added resistance due to tunnel openings is less than if the device is located at the stern (Suehrstedt, 1960).

Regardless of whether it is at the bow or stern, it should be located as far forward (or aft), and as near the keel as is possible. Any recirculating effects, either under the keel or around the bow, are certainly negligible and will have little or no bearing upon the effectiveness of the maneuvering device.

Another consideration is the submergence of the tunnel. To operate satisfactorily and free of cavitation, the tunnel centerline should be located at least one diameter below the water surface (Pehrsson, 1960). For shallow draft vessels or ships that operate in a light condition, this may pose a problem. One solution which has been used is to install two small diameter units side by side.

b. Trainable Thrust Direction MPDs. Trainable thrust maneuvering device configurations are quite varied, as can be seen in Figs. 32 and 33.

The right-angle drive units derive their name from the fact the screw propeller and pod can be rotated about a vertical axis which is perpendicular to the propeller rotation axis. Three basic types are available, one which is fixed, another which retracts by pivoting the vertical shaft about an axis on the level with the prime mover, and still another where the entire assembly slides up and down on shafts to permit extension or retraction. MPDs of these types are available from various manufacturers.

Open propellers or ducted propellers can be used on

Fig. 34 A range of sizes of vertical axis type rotating blade propellers

right-angle drive MPDs. Flow accelerating nozzles are extremely effective in increasing available thrust under the high slip conditions at which maneuvering propulsion devices are primarily used.

The *active rudder* comprises a submersible AC motor, housed in a nacelle, mounted in the rudder assembly. The motor shaft is fitted with propeller and, as a rule, a flow accelerating nozzle is provided. Trainable thrust is obtained by moving the rudder in the desired direction and by reversing the motor rotation. In ships with the common type of steering gear, the angularity of thrust application is limited to plus and minus 35 to 45 deg. Often, however, if an active rudder is to be installed, the steering mechanism will be modified such that angles of 70 to 90 deg on either side of amidships are attainable, with appropriate interlocks to prevent having the rudder thrown over to these angles when traveling at high speeds. Until recently, variable thrust magnitudes were obtained by varying the frequency of the generator supplying power to the rudder motor. This, of course, necessitated a governor control on the power supply for the generator, and often precluded the use of the generator for other purposes than operating the maneuvering device. Now, however, active rudders are available with controllable-pitch propellers. The entire assembly of the device is the same with the exception that the motor shaft is hollow so as to accommodate part of the pitch control mechanism. Thus, the motor can be driven at a constant speed from 50 or 60 Hz generating sources, alleviating the disadvantages of a variable frequency system.

One of the unique and interesting propulsion devices which supplies a controllable directional thrust is the vertical axis propeller. The term refers to those propellers having a vertical axis of rotation with a series of vertical blades disposed in a circle around the rotational axes to which are also imparted some form of rotation on their own axes, as shown in Fig. 34. Directional thrust is obtained by varying the phase angle between the blades as the rotor turns. The amount of thrust, depending upon the type of vertical axis propeller, can be controlled by varying the speed of the rotor or, in a particular case, by a series of linkages which alleviate the need for variable rotor speed.

This device is unique in that, in the sense used here, the device is the main propulsion system as well as a maneuvering device. Vessels having this type of propulsion system usually must be designed specifically for this purpose, with hull shape and the desired hydrodynamic characteristics being taken into consideration.

c. Other MPDs. Many other forms of maneuvering propulsion devices have been suggested and used from time to time. A few of these are worthy of consideration and shall be discussed briefly:

• *Steering Nozzle.* One type of maneuvering device integrated with the propulsion system is the steering nozzle. It is in essence a duct which can be rotated around the propeller plane. The duct can be so designed that at slow speed or bollard pull conditions, the advantages of a flow accelerating nozzle are obtained, in addition to the superior turning forces which can be obtained by diverting the entire jet stream. This type of system can be superior to a nozzle-propeller and rudder system of equal size and power. Its primary disadvantage lies in the vulnerability of the nozzle to damage, possibly to the extent that the propeller might not be able to turn. At high speeds and when going astern, this type of system becomes less advantageous, because of increased drag and loss of propulsive efficiency.

• *Stopping Devices.* The ability to stop or decelerate a ship is often a great problem in the area of ship control and maneuvering. Basically, this involves applying a force in a direction opposite to that of the ship's motion such as is usually done by reversing the propeller, by increasing the drag in the direction of motion, or by a combination of the two. The use of the ship's propellers for stopping is as old as steam and their usefulness is familiar to the operator and naval architect. In general, it is felt that there is, in many instances, room for improvement.

A paper by Jaeger (1963), describes this problem and suggests the use of hydraulically actuated flaps in the forebody of the ship. The author concluded from his tests that:

• Ahead reach can be reduced by as much as 50 to 60 percent;

• the flaps do not disturb the coursekeeping qualities; and,

• the flaps can assist in steering the ship, but only to a small degree.

Another type of device which increases the stopping ability of a vessel, but which also has merits as a turning device is what might be called a clamshell rudder. It has the same thrust producing characteristics as the Kitchen rudder (Saunders, 1957), and was once put into use on a Japanese Coast Guard vessel (*Marine Engineering/Log,* 1964). Tests were conducted wherein the vessel was decelerated from full speed (8 knots) to 2 knots. With just the propeller reversed the ship reached 2 knots in $8\frac{1}{2}$ ship lengths, however, when using the clamshell rudder the ship stopped in 2 to 3 ship lengths. When used as a stern side-thruster, the rudder is opened and turned in the desired direction. The propeller slip stream is then directed in a somewhat similar manner to that mentioned for the steering nozzle. One obvious disadvantage that can be foreseen with this type of device on large merchant vessels is the magnitude of the forces involved and the concomitant structural design problems.

Other devices such as the air screw, and what might be called exotic variations on the hydraulic jet are often mentioned. Usually these devices suffer from one or more of the following disadvantages: high cost, inefficiency, complexity, and susceptibility to damage.

One other system warrants mention with regard to ship maneuverability: controllable-pitch main propellers. The advantages result from the fact that a fixed-pitch propeller is designed to operate most efficiently at one load and speed condition whereas the controllable-pitch propeller can operate efficiently at any load and speed condition. Further attributes are its ability to use full power astern by merely reversing the blade pitch, the ability to adjust and rapidly reverse the pitch, and the fact that this control is directly in the hands of the person conning the ship.

d. Commercial Ship Operational Limitations. The fact that merchant vessels operate in waterways and harbors where draft limitations exist will in most instances preclude the use of certain MPDs. In merchant ship design, it is the practice to configure the hull for the maximum usable draft for the minimum depth of waterway which the ship will encounter. Obviously, any maneuvering device which extends below the bottom of the ship would be subject to damage. Additionally, there would be many instances when such an installation could not be used, due to limited depths of water. Occasions may arise when retractable right angle drive MPDs may be justified on the basis of providing the ship with emergency take-home capability.

Ship drafts also can have an effect on the size and location of an MPD. On shallow draft vessels, or vessels which often operate in a light condition, due consideration will have to be given to maintaining proper MPD submergence for effective operation. For tunnel thrusters, this means a limitation on the diameter, and possibly the location of the tunnel.

Another operational limitation is that of the maximum size MPD for a given application. The selection of MPDs to maneuver a merchant ship in 60-knot winds or against very strong currents is out of the question. Economics will dictate, in most instances, an MPD size which will provide maneuvering control under what might be called normal operating conditions in confined waters.

e. Commercial Ship Physical Limitations. In many instances the excess generator capacity available may influence the determination of the power of an MPD installation. If, for one reason or another, a choice must be made between an MPD which can be powered within existing generator capabilities and the addition of another generator to supply additional power to a larger MPD, economics will undoubtedly favor the former approach. Generally, though, another alternative is available in the form of other types of prime movers. Nevertheless, if the excess generator power falls within the MPD power range being investigated, it should be one of the specific MPD sizes considered.

When an MPD is to be installed on an existing ship, certain modifications to the hull, and possibly to the machinery, will be involved. Occasionally, these modifications may not be economically justifiable. One example has been given above. Another may be encountered with an active rudder installation. Depending upon the particular configuration, it may be necessary to bore out the rudder stock to accommodate the active rudder power cables in addition to the modifications necessary to the rudder itself. Still further, it may be desirable to modify the steering system so as to obtain greater angular travel to fully utilize the potential of the active rudder. In addition to the steering gear modifications, it would also be necessary to include an interlock system to prevent the rudder from traveling to these large angles when underway at high speeds. On new vessels these factors become of less importance as they can be accounted for in the design stages.

The use of fixed right angle drive MPDs aft on merchant vessels will usually be precluded on the basis of problems encountered from having the device in the propeller race, or near to it. Ship resistance will increase and, from past experience, mechanical problems may be encountered due to windmilling of the propeller, or if the propeller is locked, due to high resistance, when traveling at high speeds. These problems can be alleviated by the use of a retraction system such as that shown earlier. This, however, will necessitate a modification to the ship, and the incorporation of the necessary retraction equipment.

Ships with large amounts of pumping capacity have an alternative approach available in the form of a centrifugal pump type MPD. The pumping power and capacity available may be of such a magnitude as to provide a satisfactory increase in maneuverability in spite of the inherent inefficiency of hydraulic jet maneuvering devices. If existing pumping capacity can be used, this may be the most economical solution.

f. Types of MPDs Applicable to Merchant Ships. From the previous discussion, it would appear that only certain types of MPDs will be applicable to merchant ships. Draft limitations will dictate that the MPD be of such a type as to be above the baseline of the ship when in operation. This then leaves all of the tunnel type thrusters and the active rudder as MPDs which will be generally applicable to merchant vessels.

It should be pointed out that the foregoing will not always be the case. Often ships with highly specialized missions such as holding station in the open ocean will use MPDs which extend below the hull. In these instances the need for such an MPD will override the limitations imposed by ship draft and in some instances the economics of extensive modifications.

g. Factors Involved in MPD Selection. When selecting a maneuvering device for a given application many factors must be considered. In general, these fall into three categories; i.e., technical, practical, and economic. The technical factors include maneuvering requirements, external forces and moments applied to the ship, by the environment, forces and moments exerted by rudders and maneuvering propulsion devices, and ship response to external and ship generated forces. For additional consideration are such practical factors as the physical characteristics of MPDs, powering and control, added resistance when underway, maintenance and repair, and regulatory requirements. The economics of MPD application must be treated in terms of an analysis of costs related to ship maneuvering, an analysis of MPD costs, and the determination of the economic feasibility of using maneuvering devices. For analytical techniques see Hawkins et al (1965).

One primary consideration is that of developing the maximum possible thrust within limitations of the size of the unit and the power required to drive it. A figure of merit for the comparison of various types of MDPs was suggested by Hoyt (1962). This comparison has as its basis a nondimensional relation between propulsive thrust at dead pull, delivered shaft power, and swept area of the propulsive unit. This figure of merit, designated "C" is derived as follows:

$$C = \frac{T^{3/2}}{P\sqrt{\rho A}}$$

where:

T = thrust at bollard pull
P = delivered shaft power
A = swept area
ρ = mass density of fluid

with all of the above being in consistent units.

Fig. 35 is a plot of these relationships showing contours

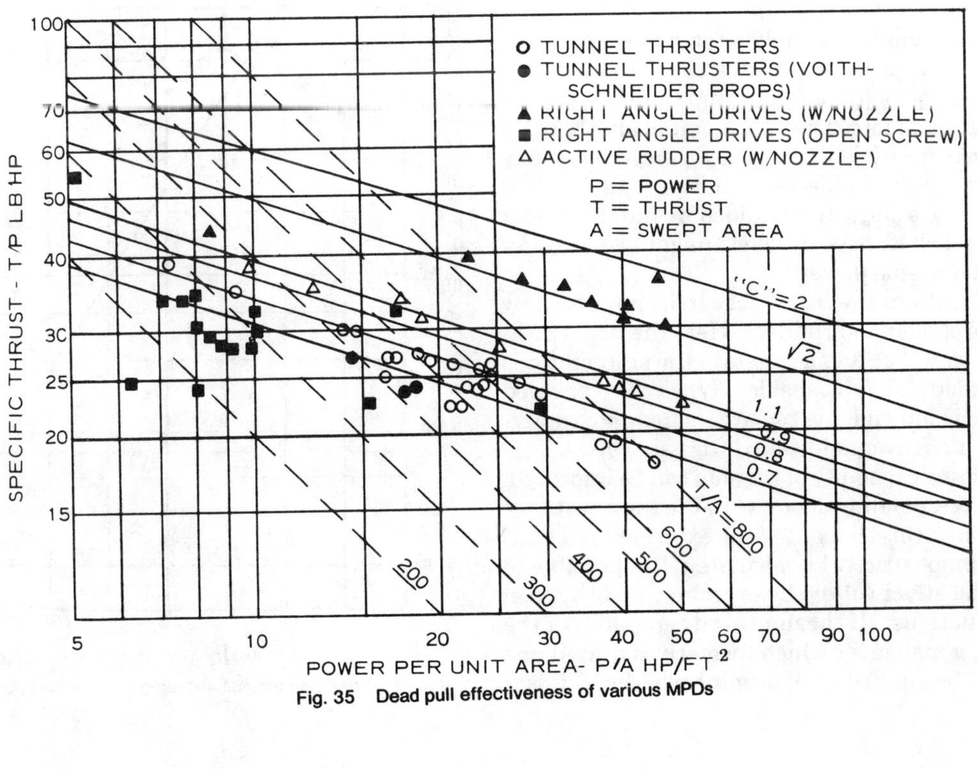

Fig. 35 Dead pull effectiveness of various MPDs

of C and of thrust per unit area plotted against thrust in lb/hp as an ordinate and power per unit area in hp/ft^2 as the abscissa. The T/A contours are in lb/ft^2 and the C contours are non-dimensional.

For an ideal open water propeller in dead pull operation the theoretical value of C is $\sqrt{2}$. For a propeller in a flow accelerating nozzle, the comparable value of C is 2. Both of these contours are plotted and, in addition, model and full scale test results for several different types of MPDs are also shown. These curves provide a reasonably good basis for estimating the performance of various propulsors under the constraints that are imposed in ship installations.

1.10 Maneuvering Assistance from Tugs. It may have been noted in Table 2 of Chapter I that tugboats were categorized as "service vessels" whereas towboats and integrated tug/barges were categorized as "commercial vessels." Both tugboats and towboats can, of course, perform services to other vessels; however, when they are working with a group of barges that are carrying cargo, these vessels are essentially engaged as a part of a commercial system. They are basically a floating power plant used to propel the cargo-carrying barges.

The tug or towboat may be designed as a service vessel only, as a part of a cargo movement system, or it may be designed to fulfill both functions. In this chapter however, the primary concern is with the use of tugs to supplement or to replace the basic maneuvering capability of a ship. Most frequently this occurs during berthing operations in a port but occasionally tug assistance will be required in salvage operations and towing a disabled ship back to port.

a. Tug Assistance in Port. The forces and moments exerted by a tug on a maneuvering ship are hardly calculable but needless to say they are an important factor in maneuvering in confined waters. Some discussion of what a tug can do is therefore in order.

A tugboat has the ability to apply a force to a ship in almost any desired direction. However, a single tug can only apply this force at one point at a time and requires time to move the application of force from one point to another. In addition, when maneuvering in confined waters, the tug may well be restricted by fixed objects from complete freedom of movement. Thus, a single tug, unaided by the ship under tow, cannot control all maneuvers which might be necessary even though it has ample power.

Under most situations the tug effect can be improved by use of the ship propeller and rudder. Here the tug is positioned in such a way that it can generate the ship motions of which the unaided ship is incapable. When the combined capabilities of the ship and one tug are inadequate to perform a required maneuver, additional tugs are called.

If the maneuvering capability of the ship can be improved, there will be a corresponding decrease in the tug assistance required. Although this element of the external forces applied to a ship cannot strictly be calculated, it is possible to estimate what the effect will be by examination of typical maneuvering situations. If the improved capability of the ship is assumed, a maneuver which formerly required one or more tugs can be studied to determine whether a lesser

Table 5—Tug Services Used by One Shipping Line

Operations	Tugs per Operation	Number of Tugs	Percent Operations	Cumulative Operations
105	1	105	40.8	40.8
128	2	256	49.8	90.6
19	3	57	7.4	98.0
2	4	8	0.8	98.8
3	5	15	1.2	100.0
257		441	100.0	

number will be as effective.

The tug services used by one shipping line over an extended period were examined. Several different types of ships made up the sample with the average being about the size of a C-2 cargo ship. A total of 257 operations were involved, including docking and undocking, using a total of 441 tugs, Table 5.

Assuming that the wind velocities in the harbors in which the operation in Table 5 took place had the cumulative distribution curve shown in Fig. 36 which represents a year-round average for the East Coast of the U. S., then the average wind velocity can be obtained for each type of operation.

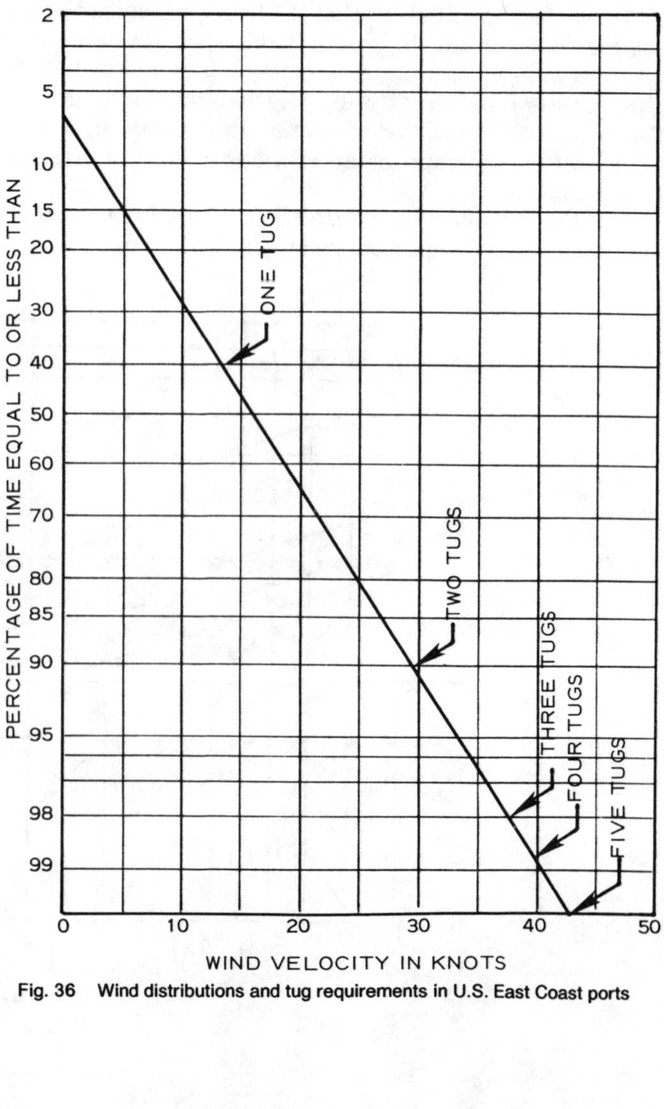

Fig. 36 Wind distributions and tug requirements in U.S. East Coast ports

Table 6—Total Force for the Period

Tugs per Operation	V_W Knots	Side Force kN	Side Force (Kips)	Force/Tug kN	Force/Tug (Kips)	Force/Period kN	Force/Period (Kips)
1	13.5	42	(9)	42	(9)	4,422	(994)
2	29.9	206	(46)	103	(23)	26,391	(5,933)
3	37.6	327	(74)	109	(24)	6,214	(1,397)
4	39.7	366	(82)	91	(21)	730	(164)
5	43.4	435	(98)	87	(20)	1,308	(294)
			Average	86.6	(19.5)		
			Total			39,064	(8782)

Using the wind resistance curve for the ship, the side force due to beam winds of these velocities can then be calculated. If each tug is assumed to contribute its share to holding a ship broadside to these winds, an average force contribution per tug can be derived. Now if the force per tug is multiplied by the total number of tugs involved in each operation over the sample period, a total force for the period is obtained, Table 6. These postulated tug requirements are also plotted on Fig. 36.

If the average ship of this group were to be fitted with a 600 hp MPD which had a calculated thrust of 66.7 kN (15 kips), it would contribute a total force of 17,148 kN (3,855 kips) in 257 maneuvering operations. This would leave a balance of 21,916 kN (4,927 kips) to be contributed by tugs. At an average contribution of 86.6 kN (19.5 kips) per tug, a total of 253 tugs would be required over the period as compared with 441 tugs previously required. The tug cost with an MPD could then be estimated at 253/441 = 57.4 percent of the previous cost. Therefore, the savings due to the installation of the MPD on this particular ship would be 42.6 percent.

The general concept of evaluation can be extended to other sizes of ships and other weather conditions. It does, however, rely on the assumption of an average value of say 89 kN (20 kips) usable thrust per tug derived above which seems a reasonable value. It also assumes that the cumulative distribution of wind velocity in the ports of interest has been derived and that the broadside wind force constant has been calculated.

b. Emergency Towing Operations. Traditionally ships have assisted each other by towing to safety when the need arose. Proof that towing assistance is traditional at sea is exemplified by the 132 specific towing signals that appear in the *International Code of Signals.* The capability to provide towing assistance, however, has been greatly diminished in the last century with the rapid increase in the size of ships. The ability to jury rig a tow between two ships in the 3,000 to 7,000-dwt range still exists but that same capability cannot always be extrapolated to ships of the 30,000 ton to VLCC size. If these ships are to be provided with a basic towing capability, specific equipment designed and installed for the purpose of meeting the tow requirements of the ship must be provided.

Towing gear aboard a ship is used in three situations: emergency towing, debeaching, and ocean towing. The needs of each of these situations has to be understood for the equipment to be properly designed. Of the three situations cited, emergency towing has the lowest strength requirements. This gear should be designed to the tow load required to move the ship through the water one knot faster than the highest current that may be encountered. The debeaching strength requirement is based on resources available to pull the ship off the beach. Ship salvors will utilize available hard points and often create points of adequate strength (bitts tied together, chain looped around deck house, etc.). Invariably the towing pad will be used as one of the hard points. From the strength standpoint the ocean towing requirement is often the most severe. Since ships are usually designed to go faster than tugs the maximum tow speed is not limited by the ship but rather by the tug. Ocean tug bollard pulls range from 222 kN (50 kips) up to 1,957 kN (440 kips) with the desirable tow speeds being between 5 and 10 knots. The high towing loads are often accommodated by utilizing the ship's own ground tackle which is usually much larger than is needed for the towing requirement. When the ground tackle is inadequate in size or configuration, or the towing is to be performed from the stern, a separate tow pad must be installed to provide a hard point on the ship for the towing hawser. To fit a ship out with a reliable emergency towing system that can meet the three situations of emergency towing, debeaching, and ocean towing, the following design criteria should be used to select the components:

- The working load of the tow pad should be the ship's total resistance at five knots with screws locked.
- The towing system should be fitted to both bow and stern of the ship.
- The towing rig should be designed to take tow cable fairleads from dead ahead or astern to athwartship.
- The tow hawser should be 1.25 times the ship's length with a minimum length of 90 fathoms.
- A large radius should be provided wherever the towing hawser or chafing pendant contacts the ship so as not to decrease the strength of the towing gear through bending or chafing.
- The towing equipment should be designed, configured, and located so that it can be rigged and deployed with no power available.

Section 2
Navigation and Control Systems

2.1 Navigation System Elements. From a navigational point of view, the safety of a ship underway depends on the ability of personnel to determine accurately the position of the ship, to keep on a desired course, to determine the presence of other vessels in the area, and to make any such vessels aware of their own presence. To accomplish these objectives, the ship must be provided with essential navigating equipment.

International convention establishes the minimum navigation equipment standards which require a ship to be equipped with the following: a marine radar system, a radio direction finder, a gyrocompass, an echo-depth sounding device, a radiotelephone or radiotelegraph, and the necessary nautical publications as appropriate for the intended voyage. Certain items of navigation equipment are specified in detail by governmental rules and regulations or by international agreement. For most of these items, the primary purpose of establishing requirements is to assure that equipment such as navigation lights and sound signals will be uniform and recognizable by all ships. As an example, U. S. regulations [33 CFR 164.35(b)] require all vessels operating in U. S. navigable waters to have an illuminated magnetic compass.

a. Navigation Lights. The lights required by the International Regulations for Preventing Collisions at Sea, 1972 (72 Colregs) include the masthead and range lights, side lights, stern light, anchor lights, and not-under-command lights, as well as special lights which are prescribed to identify ships engaged in unusual operations which affect their maneuverability. Ships which require these special lights include towing vessels, dredges, cable ships, underwater survey ships, pilot vessels, and fishing vessels, as well as certain military vessels.

The types and shapes of the lights, together with their general location and spacing, are indicated in the above regulations; strict adherence to the specified details is necessary to avoid liability in case of a casualty. Within the requirements of these Rules, some variation in the exact location of these lights is permissible, and this matter should have the designer's attention at an early stage, so that proper locations can be worked out in conjunction with the proposed arrangement of masts, stacks, and other topside gear.

b. Sound Signals. The 72 Colregs require that a ship carry equipment for making certain prescribed sound signals when maneuvering with respect to another ship, and for use in fog or other conditions of restricted visibility. All vessels of 12 m (39.4 ft) or more in length should be provided with a whistle and a bell. Usually the whistle is located such that the sound will not be intercepted by any structure. It is often located on the forward side of the stack at a level above the tops of all houses. However, on bridge-aft tankers and bulk carriers, a location on the span between the forward king posts is preferable. The bell, which varies in size from a minimum mouth diam of 200 mm (7.87 in.) on vessels from 12 to 20 m (65.6 ft) in length to a mouth diameter of not less than 300 mm (11.8 in.) for vessels of more than 20 m in length, should be located on the open deck near the bow. It shall be made of corrosion-resistant material and designed to give a clear tone. Additionally, a vessel of 100 m (328 ft) or more in length shall also be provided with a gong, the tone and sound of which cannot be confused with that of the bell.

c. Other Signals. All ships of over 150 gross tons, when engaged on international voyages, shall have on board an efficient daylight signalling lamp which shall not be solely dependent upon the ship's main source of electrical power.

d. Depth Sounding Gear. By international agreement, all new ships of 500 gross tons and upwards, when engaged on international voyages, are required to be fitted with an echo depth sounding device. Electronic depth-sounding apparatus consists of a visual indicator on the bridge and a recording unit on the bridge or in the chart room. The transducer is located near the keel. It is essential that this transducer be accessible for service and maintenance and located to avoid interference from water turbulence or entrapped air.

In addition, seagoing ships carry an electronic or a mechanical deep sea sounding machine and a variety of hand lead lines for determining the depths of water. For the proper use of the hand lead lines, small platforms extending beyond the bulwark are necessary. These should be located port and starboard just below the bridge. The sounding machine requires a small boom, normally at about the same location which can be rigged outboard over the bulwark to take the lead and depth-recording apparatus clear of the side of the ship.

Aside from the conventional use of sounding gear to determine clearance depth of water, is its use as an aid to navigation. Various locations of ocean bottom have been extensively plotted and charted and systems are available to determine a vessel's location by interpretation of bottom soundings.

e. Compasses. All ships of 1,600 gross tons and upwards, when engaged on international voyages, are required to be fitted with a gyrocompass in addition to the magnetic compass required for all vessels in U. S. waters, foreign or domestic.

The number and types of compasses carried vary somewhat for different types of ships. A typical installation on a seagoing cargo ship would provide a magnetic steering compass in the wheelhouse and a magnetic standard compass fitted in the open on a suitable stand on top of the wheelhouse. The wheelhouse compass can be dispensed with if the standard compass has a reflector projecting into the wheelhouse. At one time an additional magnetic compass was customary aft at the emergency steering station.

The master gyrocompass and its power supply and controls are located in a small room usually containing no other equipment. The master gyrocompass controls a steering repeater compass in the wheelhouse, another at the emergency steering station, and a bearing repeater compass on each wing of the bridge. The master gyrocompass can also provide the control for other useful navigational aids, such as automatic steering control, course recorder, radio direction finder and radar. Location of the gyrocompass should be as close to the center of pitch and roll as possible to minimize maintenance costs. However, it is generally located in a room near the bridge as a convenience to the ship's officers.

f. Radio Direction Finder. A radio direction finder is also required by international agreement on vessels of 1,600 gross tons and over. Additionally, all new ships of 1,600 gross tons and upwards, when engaged on international voyages, shall be fitted with radio equipment for homing on the international radio distress frequency.

Practical considerations dictate the need for a radio direction finder antenna system which has a minimum of interference from large metal structures or from other antennas. The instrument is usually located in the chart room and is useful as an aid in determining position, as well as an aid in locating ships or aircraft capable of emitting a homing signal.

g. Radar. All ships of 1,600 gross tons and upwards are required to be fitted with a marine radar system of a type approved by the government of the State under whose flag they are registered. Facilities for plotting radar readings shall be provided on the bridge in those ships. Radar has proved its usefulness as a means of avoiding collisions and as an aid in piloting. It has become an essential item of navigating equipment.

The radar indicator unit is located in the wheelhouse and the radar antenna is supported by a mast, the location of which requires consideration along with the mast supporting the navigation lights. The antennas should be located on a rigid platform in such a manner that there are no shadows or dead zones. The radar receiver should be located in close proximity to the radar mast in order to minimize the length of wave guide tubing transmitting the signal to the console unit.

In addition to a vessel's conventional radar installation, an electronic relative motion analyzer is available called a *collision radar.* It serves as a computer-aided system to assist in correctly interpreting radar data to avoid collisions. Courses and speeds of nearby vessels are evaluated to determine if collision danger exists or will exist.

There are several collision-avoidance systems commercially available for doing automatically some of what the deck officer has been doing. The navigator and watch officer can readily conn a ship in the open ocean in the presence of one or two other ships. Their task can become difficult, however, when there are several ships within radar range and when there is land nearby (or navigational channel restrictions). The collision-avoidance system takes the ship radar, gyro, and speed inputs, plus previously prepared navigational displays, and computes and displays information such as:

- Given the present ship speeds and headings, which ship presents the most immediate collision hazard?
- If one's own ship changes course and heading by a trial value, what would then be the most dangerous ship?

The collision-avoidance systems have digital computers which permit rapid evaluation of trial changes of course and heading, and may have automatic alarms. Also, given that the digital computer is aboard, it can be readily used to adapt the autopilot to the speed and prevailing seaway conditions. This adaptive autopilot feature, essentially changing gains, does have promise of increasing speed made good by a few percent. Additional features of such systems are alarms which sound when the system picks up a target within a set limit, and the ability to compute the most economical course from the ship's present position to a desired port. This latter feature is available with pre-programmed routings for a number of ports.

h. Other Aids to Navigation. These aids consist of Doppler sonar, satellite communication receivers, Loran, and Omega. The Doppler sonars consist of keel mounted transducers which send down pulsed sonic beams and convert the return signals into displays of ship speed, ahead and transverse. The read-outs have display arrows showing direction of bow and stern movement. These permit a quick visual and quantitative display of translational and rotational movement relative to the bottom. This is of particular use for large ships which make up to sea buoys or come to a pier, because in the very low speed range (less than a knot over ground) precise docking maneuvers are very difficult to judge.

Satellite navigation signals require a special shipboard antenna and computer. These are commercially available, and use a satellite signal available roughly every hour and a half. The result is the ability to have latitude and longitude in the open seas about every hour and a half precise to within one or two nautical miles. Loran C provides position to a few hundred yards when close to shore-based stations.

The Loran set is useful in determining an accurate position electronically and its use is becoming commonplace on almost all merchant ships. It is generally located in the chart room. A Loran set is particularly sensitive to vibration and its location in the chart room should be selected with care so as to minimize the effects of vibration and thus reduce maintenance costs.

Loran-A has been in common use by marine commerce for electronic position determination since World War II. Areas of Loran-A coverage included coastal North America, shipping lanes of the North Atlantic, and selected areas in European waters and in the Pacific. This system has been gradually phased out with most service scheduled for termination by the end of 1979.

Loran-C, a more accurate navigation system, replaced Loran-A in North American coastal waters including all of the U. S. Coastal Confluence Zone by 1979. Loran-C receivers have come into wide use on merchant ships as the old Loran-A service was phased out. Loran-C receivers are located in the chart room or, more commonly, in the pilot house since no equipment expertise is required to operate them.

The Omega receiver set is useful in determining position fixing information electronically and provides accuracy equivalent to that obtained from Loran-A. The system is partially in operation now and covers most of the Northern Hemisphere. When the last station is completed and coverage areas and accuracy validated, it is expected that most of the open seas areas around the globe will be covered with the eight stations. This will occur in the early 1980's. The receiver set may be located in the chart room, but is commonly found in the pilot house.

i. Nautical Publications. All ships are required to carry as appropriate for the intended voyage, adequate and up-to-date

- charts,
- sailing directions,
- lists of lights,
- notices to mariners,
- tide tables, and
- all other nautical publications necessary for the intended voyage.

A chart table and adequate stowage space for these various publications must be provided in the bridge area.

2.2 Steering Control Systems. As depicted earlier in Fig. 21 the purpose of the steering control system is to convey the commands of the officer in charge on the bridge to the steering machinery system in such a manner that it will cause the ship to change heading in response to the command. Steering control systems may range in complexity from a man physically turning a tiller to swing the rudder to an automatic system wherein the pilot sets a desired heading and that heading is automatically maintained without further human input.

a. Basic Steering Control Elements. There are two fundamentally different types of control exercised in steering systems that can be designated as follow-up controls and non-follow-up controls. The former is the most prevalent but the latter deserves some mention since occasionally it is employed in certain types of steering systems. An example of a non-follow-up control would be one where the helmsman throws a switch that starts the steering power element turning in a given direction. In response, the rudder would turn in that direction until the switch was cut off manually or by travel limit stops. The disadvantages of such a system are obvious since the helmsman must always be observing what the changing rudder angle is as well as how the ship is responding. Non-follow-up control is used primarily in emergency steering systems where these disadvantages are less important than the need to steer the ship. With a follow-up control, the helmsman sets a desired rudder angle. The powering element is then actuated only for the length of time necessary for the rudder to reach that angle and then the power cuts off automatically.

In ship steering control systems, where powerful machinery and heavy equipment must be rotated in response to helm orders, it is customary to cushion the shock of starting and stopping the swing of a rudder. Some systems accomplish this with the use of mechanical, hydraulic, or pneumatic springs but more often the cushioning is a function of the control system. In such systems the controls are arranged so that the rotating torque application is eased off as the rudder begins and ends its swing.

Another basic element of steering control systems is the need for complete assurance that they will always be able to work under any conditions when the ship is underway. This has led to a plethora of emergency control systems and redundant elements that sometimes, because of their complexity, cause more casualties than they avoid. Redundance often starts with dual controls on the bridge and dual control transmission systems leading from the bridge to the steering engine room. An after steering station is generally provided from which the steering machinery can be directly controlled based upon voice communication from the bridge. In the case of power failure, hand pumps for hydraulic steering systems are provided; mechanical geared drives are installed for use with electric motor or steam driven systems. Some older ships are still fitted with rudderstock extensions to a quadrant on the deck above that can be turned by winches or capstans.

Unfortunately, redundant elements are sometimes placed in series rather than in parallel which renders them useless. In other cases, once an emergency rig is cut in and then the power comes back on, critical machinery components can be destroyed. For these and many other reasons a designer should carefully analyze the complete steering control, powering, and emergency system before placing orders for assemblies of off-the-shelf components that may be totally incompatible.

b. Automatic Steering Control. The idea of tying in the steering controls of a ship with the compass heading is far from new. When Elmer Sperry presented his first paper to SNAME on automatic steering (Sperry, 1922) he indicated development work had been underway since 1913. Yet it was many years before the "Iron Mike" was installed aboard ship and many more years before it moved from an out-of-the-way corner of the bridge to become the main steering control stand.

When this transition to center stage did occur, there was some trepidation that a needed redundant steering system was being eliminated. Thus, in the late 1960s there was felt to be a requirement not only to have both a manual and automatic mode of operation but also to have dual control elements in the steering stand connected to dual powering elements in the steering engine room. The particular system to be described here is of this type; although it is not the most modern equipment available, it does have the major characteristics of those combination manual and automatic steering control systems that are in use today.

The basic control element is the steering stand located on the bridge. This unit may be operated in either a manual or an automatic mode and provision is made for operating with either a port or a starboard set of components. This redundancy is intended to increase the reliability of the system.

Fig. 37 shows the steering stand mechanical elements. One steering wheel, one step motor, and one indicator are provided but from these elements there are two control potentiometers driven. From there the system splits into two sides, each of which have separate magnetic amplifiers,

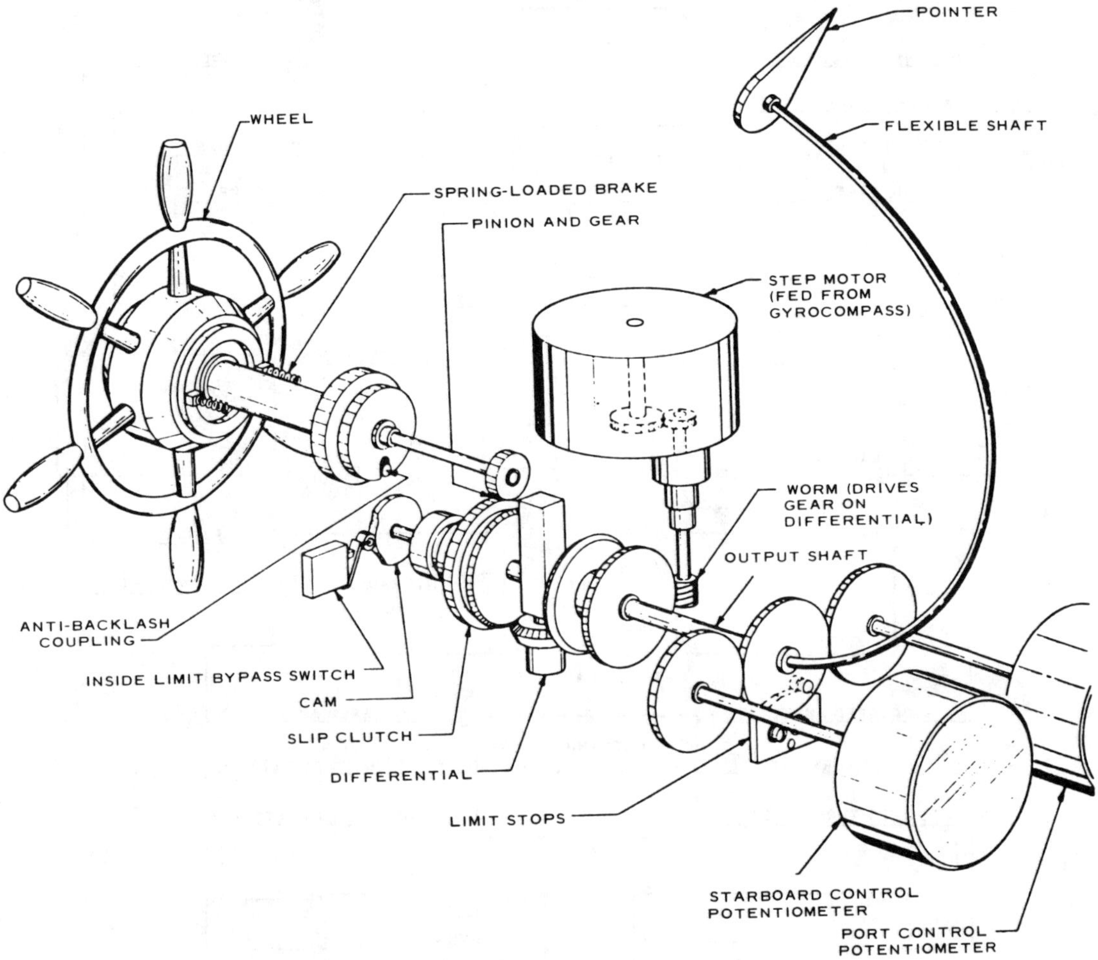

Fig. 37 Steering stand mechanical elements

Wheatstone Bridge circuits, adjustment potentiometers, differential relays, and power units in the steering engine room consisting of hydraulic pump, solenoid valves, control cylinder, follow-up rack, and follow-up potentiometer. These power units are mounted port and starboard on the steering engine. The piston rods of the two control cylinders are connected together with another rack so that when either the port or starboard side of the dual control is activated this control rack is driven from side to side for a distance equivalent to the ordered rudder angle. It is this control rack which actuates the steering engine.

Steering power is obtained from two 60 kW (80 hp) motor-driven positive displacement hydraulic pumps, one on the port side of the steering engine and one on the starboard side. When the port side of the dual control system is operating the port hydraulic pump motor is running and when the switchover to starboard control takes place the port motor stops and the starboard pump and motor are activated. The pumps are both connected hydaulically to port and starboard rams which are linked to rotate the tiller. A block diagram of the entire arrangement is shown in Fig. 38 (Taggart, 1970).

2.3 Total Maneuvering Control from Bridge. Although bridge control of steering systems has been standard practice for many years, the trend toward bridge control of main propulsion is a relatively recent development. Typically, on larger ships, it was always the practice to transmit propulsion maneuvering orders by voice or mechanical communication from the bridge to engine room personnel who would then, in turn, effect any ordered changes in propulsive thrust by manipulation of the machinery controls.

With the advent of diesel engines, diesel- and turbo-electric plants, and controllable-reversible-propellers, it became a simple matter to transfer the control manipulative functions to the bridge. Deck officers soon adapted to this added responsibility because it gave them a better feeling of direct control over the maneuvering of the ship in tight situations. As a result, bridge controls for other types of machinery plants, such as geared turbines, were soon developed and are now becoming standard installations on the major of new vessels. When maneuvering propulsion devices (MPDs) were added for improved ship maneuverability, the controls for these devices were installed on the bridge instead of in the engine room.

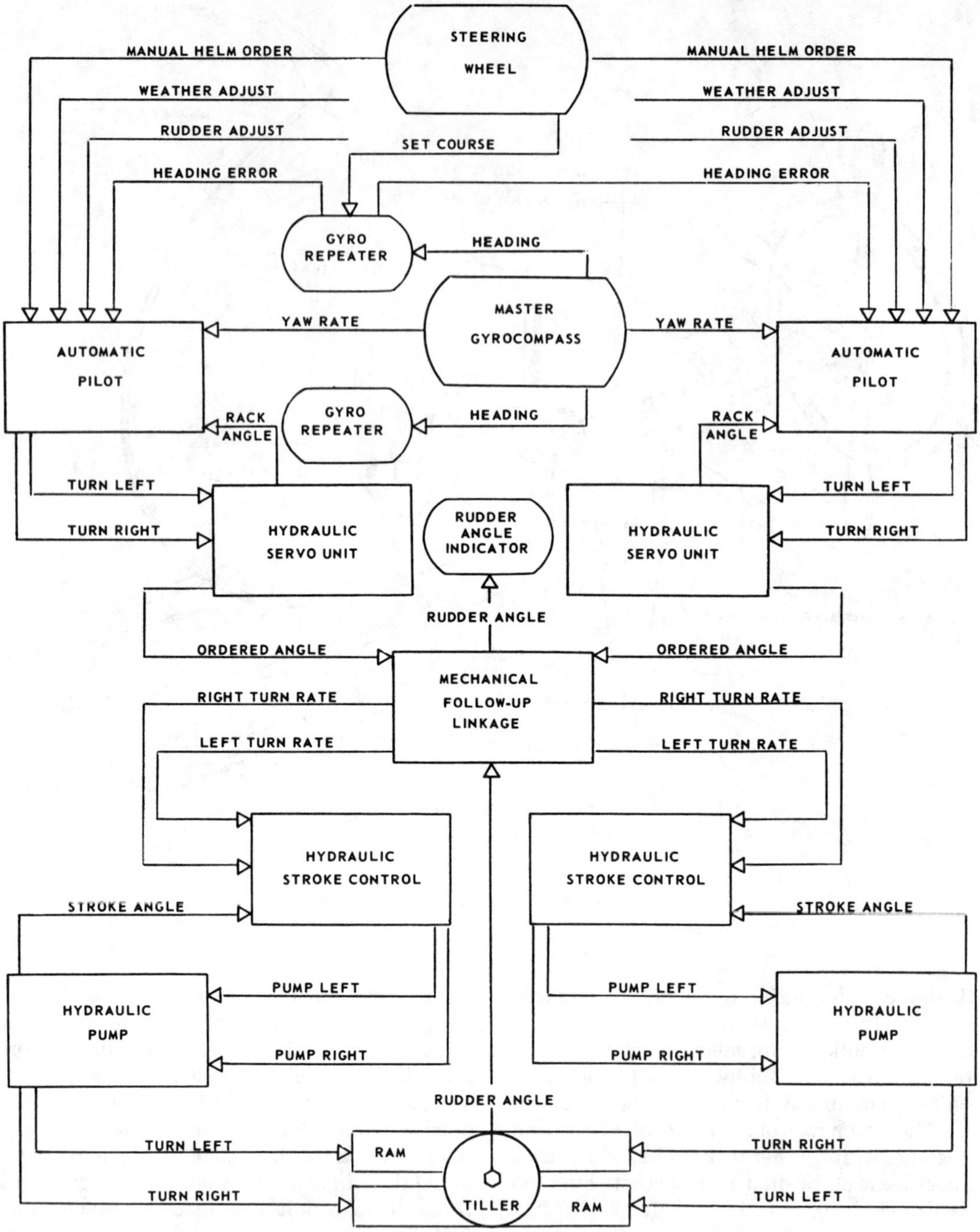

Fig. 38 Functional block diagram of dual steering system

The transition from remote control to bridge control was not without its trauma. Deck officers often failed to understand that the compressed air supply for reversing diesel engines was not unlimited and that steam pressures and temperatures did not always react in a kindly manner to rapid manipulations of the throttle. However, both improved training of deck officers and subtle adaptations of machinery control systems have been such that, in most cases, workable methods of bridge control for all operating situations have evolved.

Given control over steering, main propulsion, and a maneuvering propulsion device, a deck officer can develop skills in ship maneuvering control that may often permit doing away with tug services in docking and undocking situations. Cruise ships, operating in the Caribbean, are often equipped with twin controllable-pitch propellers, twin rudders, and

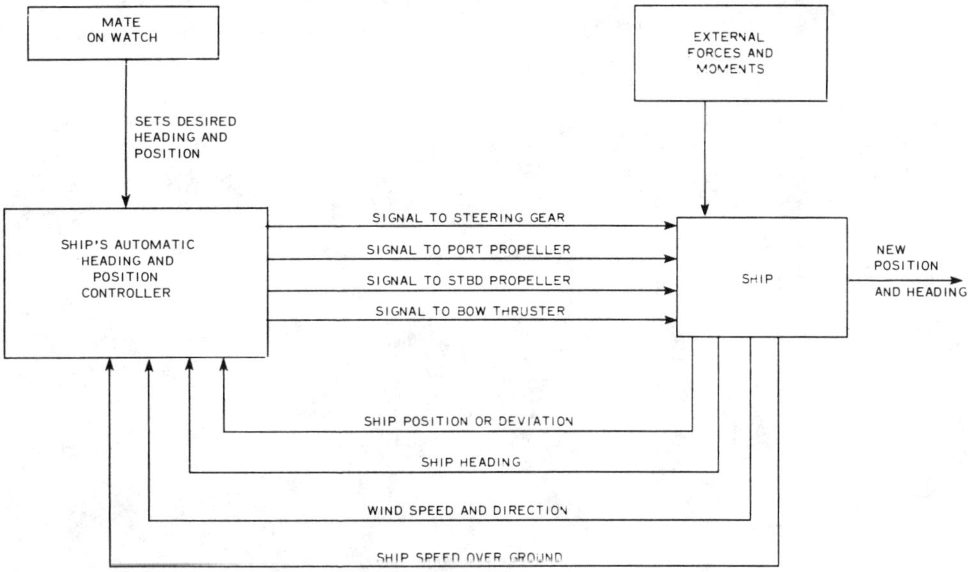

Fig. 39 Block diagram for automatic position-keeping system for twin-screw ship with bow thruster

a bow thruster. With this combination it is possible to translate athwartships into and away from a pier.

Suppose, for example, translation to port is required. Ahead pitch is applied to the port propeller and astern pitch to the starboard propeller so that there is no net fore-and-aft thrust but a turning moment to the right is applied. The bow thruster is then controlled to thrust to port and right rudder is applied thus exerting a thrust to port at both the bow and stern. Any unbalanced turning moment can be corrected by propeller pitch changes as can any imbalance

Fig. 40 Bridge console with steering, main propulsion, and thruster control

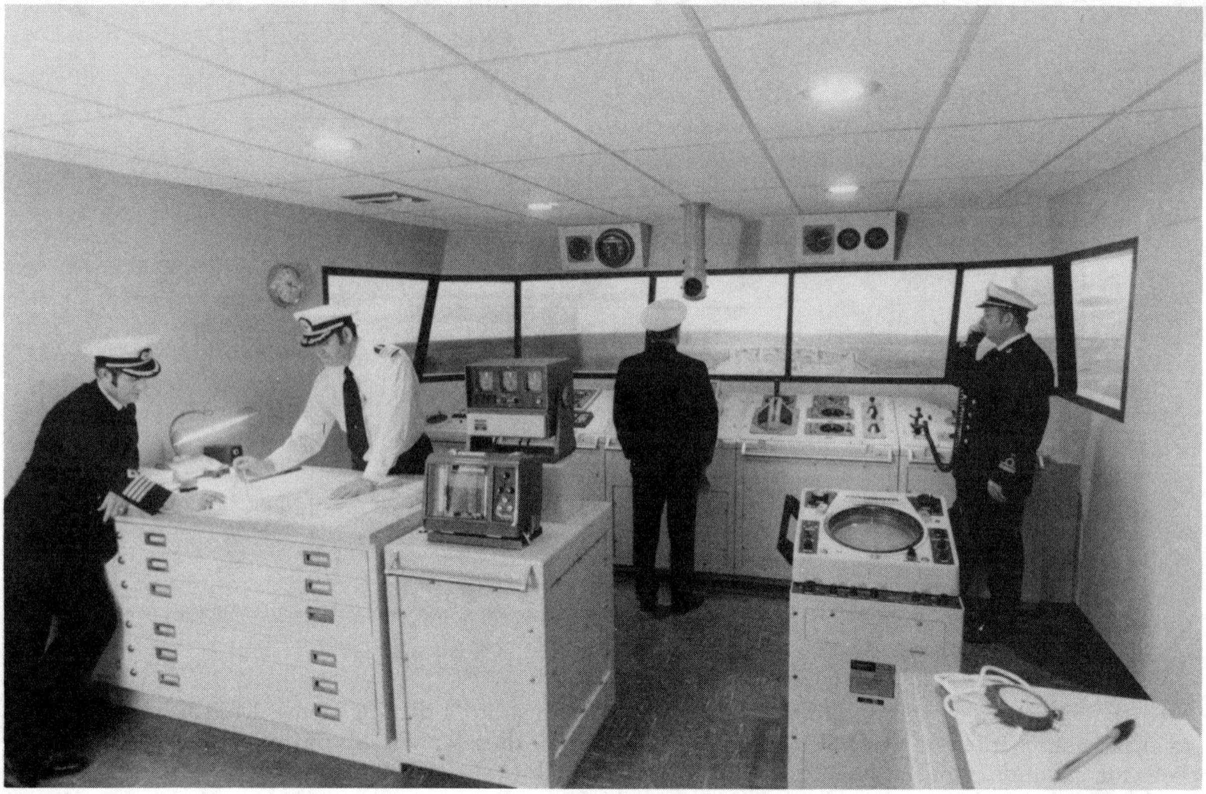

Fig. 41 Bridge arrangement

in the fore-and-aft thrust. As a result the ship will move to port without any change of heading or ahead or astern movement.

With a vessel so equipped, dynamic positioning is possible, provided that the imposed environmental forces are not too large. Fore-and-aft translation and rotation can be accomplished using similar control manipulations. Fig. 39 is a block diagram of a control system that is capable of duplicating the above maneuver with any twin-screw ship fitted with a bow thruster.

2.4 Bridge Systems Design. An optimum bridge design is one which will fully integrate the watch officer with his equipment. The person in charge of the vessel's navigation must receive and process information from various sources. Review of marine casualties indicates that the lack of needed information, or inadequately supplied information, is a factor in many of these accidents.

The design of a bridge with the person in charge in mind should help solve this problem. All vessels differ in size and purpose and a single optimum design suitable for all vessels is highly improbable. Locating essential equipment in close proximity to the work area of the watch officer is of prime concern. The location and type of available nevigational equipment must be integrated with the available electronic relative motion analyzers used to avoid collisions. The vessel's internal communication system becomes the focal point of activity during emergency situations and must be considered in the design.

Limiting the working area so that the watch officer is within three to five steps of all the essential equipment, the chart table, and the bridge windows is realistic and beneficial. Particular vessel designs may present unique bridge design problems but keeping the watch officers duties in mind will result in the optimum bridge design for that vessel. Figs. 40 and 41 depict one such bridge design, incorporating built-in consoles.

Section 3
Ship Motion Control Systems

3.1 Acceptable Roll Motion Characteristics. The determination of whether or not a stabilization system could be installed in a projected new ship, and the selection of the most appropriate system, involves consideration of the following factors:

1. The level of operational performance (expressed in terms of acceptable roll motion characteristics without appreciable performance degradation) which the ship is required to attain.

2. The comparative levels of operational performance

of the unstabilized ship and of the ship with each of several sizes and types of stabilization systems installed.

3. The cost and total impact of each stabilization system on the ship design.

The required operational performance of a ship can, to some extent, be stated in terms of its required capability to carry out each of the elements of a mission under specified environmental conditions. Those capabilities where roll motion is a critical factor (e.g., the ability to transit from point A to point B within a specified time without damaging the cargo) have normally been stated in rather general terms. However, for the purpose of determining the need for roll stabilization and in selecting and sizing the appropriate stabilizer, these capabilities must be stated in statistically definable quantities. For example, Cox and Lloyd (1977) state that it is necessary to establish the maximum permissible percentage of roll occurrences which can exceed a specified limiting roll angle for certain critical ship operations, and they give the following as an example:

> *"No more than X percent of roll angles shall exceed ±Y degrees in a sea state Z at the worst heading at a ship speed of V knots."*

Once the acceptable roll motion characteristics without appreciable performance degradation have been established, the next step is to compare the acceptable roll motions with predictions of the unstabilized roll motions of the ship being designed. Assuming that the unstabilized roll motions are greater than the acceptable roll motions, the ship designer must then investigate the effectiveness (or roll reduction) of various alternative roll stabilization systems in order to provide an amount of stabilization in consonance with the acceptable roll motions and the constraints of the particular ship design.

3.2 Types of Roll Stabilization Systems. The decision on the type of stabilization system should be made in the very early stages of the ship design, since this decision will affect many important aspects of the design; i.e., ship arrangement, ship displacement, power required for propulsion and/or ship service, stability, and acquisition and operating costs. The final selection of type of stabilization system should be made on the basis of trade-off studies between quantitative estimates of stabilizer effectiveness (roll reduction), reliability, costs, weight, added drag, and required space and power. Miller et al (1974), Cox and Lloyd (1977), Barr and Ankudinov (1977), McCallum (1976), Conolly (1969), and Lloyd (1972) provide parametric relationships for estimating the effectiveness of various roll stabilization systems. It is emphasized, however, that these parametric calculations are intended for use only in the early stages of ship design during the trade-off studies mentioned above. As soon as practicable, specific calculations and/or model tests, as described by Cox and Lloyd (1977), Barr and Ankudinov (1977) Zarnick and Diskin (1972), and Lloyd (1975) should be conducted for a definitive determination that a particular design will satisfy the roll motion requirements.

In the early stages of ship design, there are key design parameters that should be generally considered in determining the candidate stabilization systems for the trade-off studies mentioned above. For ships with low initial metacentric height, $\overline{GM}$, anti-roll tanks should not be considered due to the tank free-surface motions and the ensuing reduction of $\overline{GM}$. In ship designs where there are severe restrictions on available space, anti-roll tanks, particularly passive free-surface tanks, may not be feasible systems. When roll stabilization is critical only at low speeds (less than 10–12 knots), active fin stabilizers should not be considered. When roll stabilization is critical at both low speeds and high speeds, it may be desirable to consider combinations of systems, bilge keels and active fin stabilizers, or passive anti-roll tanks and active fin stabilizers. Generalized information on the effectiveness of the roll stabilization systems described in the following sections is as follows:

a. Bilge Keels. These keels increase hull roll damping and hence reduce roll motions, and are particularly effective at low speeds, where bare hull damping is very small. Roll motions with and without bilge keels can be estimated using appropriate roll damping ratios.

b. Anti-Roll Tanks. Tanks are effective at almost all ship speeds and headings, but, are most effective at low ship speeds. There are certain speeds and headings (usually aft quartering seas), however, where passive free surface tanks may destabilize, that is, increase the ship's natural roll motions. The control system of controlled-passive and active tanks can be designed to avoid these anomalies.

c. Active Fin Stabilizers. Active fins are effective in all sea states at all headings but only at higher ship speeds (greater than 10–12 knots). Their effectiveness is reduced with reduced speed since the fin moment due to fin lift is a function of the ship speed. At zero speed the fins make only a small contribution to the passive damping of the ship. The effectiveness of active fins has traditionally been determined primarily by specifying a static heel angle (usually 5 degrees) that the fins must produce at a prescribed ship speed (Miller et al, 1974). As Cox and Lloyd (1977) point out, however, the *static heel angle* measure of effectiveness is not related to a specific roll reduction requirement and ignores the fin system in its dynamic mode of operation; thus, it can result in a larger and more expensive fin system than actually required.

Cox and Lloyd (1977) provide a thorough assessment of the state-of-the art in predicting the effectiveness of active fin stabilizers. They emphasize that the actual performance of active fins can be significantly less than that predicted by "simple roll theory," which neglects the adverse effects on performance of the hull boundary layer, interference between fins and bilge keels, and fin-induced yaw and sway motions. Lloyd (1972, 1975, 1976) has developed a more advanced theory which accounts for these fin performance degradation effects in assessing fin effectiveness; his theory is considered adequate for design purposes.

3.3 Bilge Keel Design. It has been recognized since the 19th century that the rolling motions of a ship are large because the hydrodynamic damping in this degree-of-freedom is small. W. Froude in 1865 recommended that bilge keels be fitted in order to increase the roll damping and thus reduce the roll motions. The damping moment generated by the bilge keels is due to a component supplied by the pressure resistance of the bilge keel itself and to a component created by the change in the pressure distribution on the hull. On a hull without bilge keels, the roll damping is provided by the dissipation of energy in surface waves, in viscous flow around the hull, and by surface tension. (The latter component is not important for the full-scale case.) The addition of bilge keels greatly increases the energy dissipation due to viscous flow. The controlling coefficient in bilge keel sizing is the ratio of the actual damping to the critical damping of the linear system, and is a function of the ship characteristics and the size of the bilge keels. Methods of estimating the relationship between the ship characteristics, bilge keel size, and damping ratio are given by Miller et al (1974) and Cox and Lloyd (1977). The damping coefficient is also a function of the roll amplitude and tends to increase with amplitude. Bilge keel size will usually be limited by considerations of vulnerability and added drag.

Cox and Lloyd (1977) note that the damping moment attributed to bilge keels varies considerably with girthwise position and that the best results are obtained with the bilge keels about one quarter of the way around the girth from the waterline toward the keel. They also note that adverse effects on the ship can be produced if the bilge keels are too long and that the optimum length with respect to effect on rolling is when the bilge keels are limited in length to about one-third of the ship length from the bow.

Bilge keels have been shown to be effective even in the highest sea states and can be installed with only minor deleterious effects on a normal ship. The only impact is an increase in the ship's total resistance due to the additional wetted surface. Since this may be expected to be small except on very high speed ships, the fitting of bilge keels is a customary and routine procedure.

3.4 Passive Anti-Roll Tank Design. The concept behind passive stabilization, as described by Miller et al (1974), is a simple one. Most ships have very little roll damping. As a result, the energy which the waves impart to the ship is exhibited in large roll motions. The magnitude of these motions must be sufficient so that the energy dissipated equals the energy imparted by the waves. When a passive stabilizer tank system is placed on board a ship, it is a dynamic system which has a resonance. This resonance is not so pronounced as the ship's resonance, since the stabilizer is much better damped. Therefore, the stabilizer is chosen so that its natural resonance in roll is close to that of the ship. The tuning of the stabilizer provides excellent dynamic coupling between the ship and stabilizer. This means that the roll energy of the ship can be efficiently transferred to the stabilizer, which, because of its high damping, converts this energy into heat.

There are two important points. First, the stabilizer drains roll energy from the ship and thus greatly reduces the roll motion (particularly at the ship's roll resonance). Second, the stabilizer does not work unless the ship is already rolling. In other words, a passive stabilizer cannot eliminate roll motion entirely. Rather, it reduces the potential motion that is already there.

From the above discussion, it is clear that a passive roll stabilizer can be any resonant system which couples well with roll. It must also be big enough so that it can absorb a significant amount of the ship's roll energy. Several general realizations of such passive systems have been invented, mostly during the latter part of the last century. The passive stabilizer system predominantly in use today is the anti-roll tank.

Although passive anti-roll tanks require large volumes of liquids in order to absorb a significant amount of roll energy, it is relatively easy to develop a resonant tank system. Consider the difficulty one has when walking when carrying a full cup of liquid. The liquid sloshes from side-to-side in a well defined resonant mode. Tanks of water which run the full beam of a ship, or nearly so, also have such a resonance. By proper installation of sufficient structure within this tank, an adequate amount of damping is provided.

There are two general types of passive anti-roll tanks in current usage: the free-surface tank and the U-tube. The respective typical general arrangements of these two are shown in Fig. 42a and Fig. 42b. Free-surface tanks are totally passive. The U-tube tank, which is commonly referred to as a controlled-passive tank, is discussed in Section 3.5 further on.

One of the most important steps in the design of a passive anti-roll tank is the determination of the tank size required to produce the desired roll reductions. Procedures for estimating the size of free-surface tanks have been developed by Miller et al (1974), Cox and Lloyd (1977), and McCallum (1976). A design procedure for sizing U-tube tanks has been developed by Webster (1976).

Miller et al (1974) point out that there are a number of factors which must be addressed in the design of a passive tank system (either free-surface type or U-tube type) that are related to the size of the tank; they are as follows:

a. Tuning. The tank should be tuned to a natural frequency near the ship's resonant roll frequency. Experience has shown that this frequency should be 6 to 10 percent higher than the ship's resonant roll frequency.

b. $\overline{GM}$ Loss. The tank should have sufficient size so that it can absorb a significant amount of energy from the ship's roll motion. A theoretical analysis of this problem shows that the pertinent size parameter is the ratio of the free surface loss of the stabilizer to the metacentric height of the ship with the fluid in the stabilizer tank but frozen in position (no free surface loss). This ratio $\delta\overline{GM}/\overline{GM}_{uncorrected}$ should be in the range of 0.15 to 0.30.

c. Damping. The equivalent linear damping ratio for the tank sloshing should be in the range from 0.2 to 0.6 (typically a damping ratio of 0.2 for active tanks and 0.4 to 0.6 for passive tanks). The damping of this slosh motion comes about from the drag of the fluid as it passes by the structure within the tank and when it enters and exits from the wing tanks. All of these losses are quadratic in nature and this causes the damping ratio of the tank to depend on the amplitude of roll motion as well as the actual structural configuration.

d. Capacity. The volume of water in the tank must be sufficient so that the tank does not saturate in a low sea state, Fig. 43. The requirement for free surface loss usually

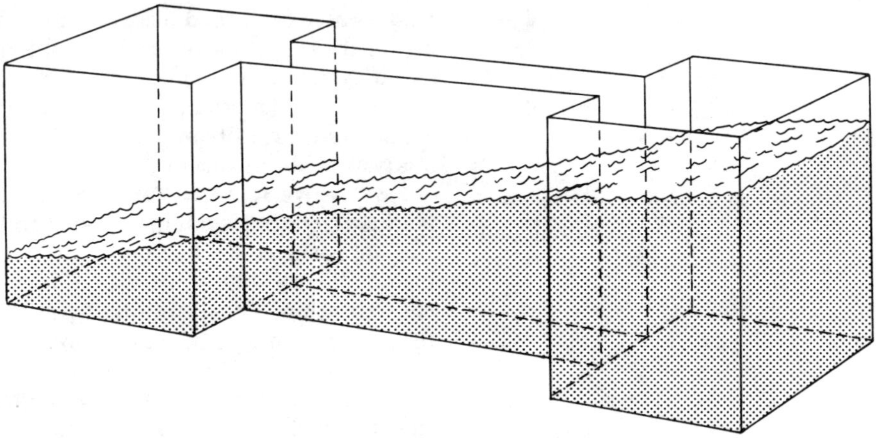

Fig. 42a Free surface passive anti-roll tank

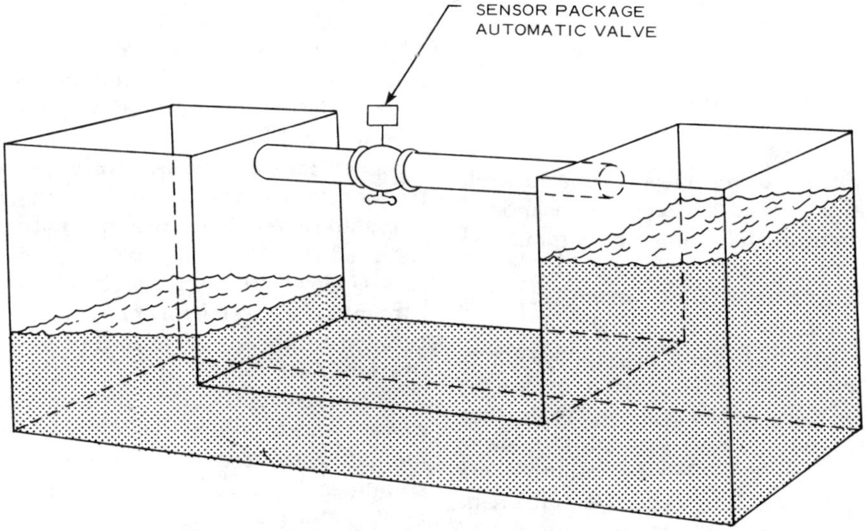

Fig. 42b U-tube passive anti-roll tank

dictates the plan-form of the tank; e.g., rectangular, C-shaped, I-shaped, and the requirement for capacity then dictates the height of the tank. A good rule of thumb appears to be that the tank should have sufficient height so that the tank must be rolled 10 to 15 deg before either the fluid hits the tank top (saturation) on the low side or the bottom runs dry on the high side. If the tank is a given height, then the greatest tank capacity in this sense occurs when the tank is about one-half full.

e. Location. Experience has shown that a passive tank system can be located almost anywhere in the midlength of the ship, preferably no further forward than $0.25L$ forward of amidships or aft of $0.35L$ aft of amidships. For some ship types, the vertical location of the tanks can be important. Generally speaking, the higher the tank, the more effective it will be because the moment generated by the transverse acceleration of the fluid acts to reduce motions when the tank is located above the center of gravity. However, it is usually impracticable to mount a tank high in a ship, since this is a most useful area. As a rule of thumb, if the ratio of $\overline{GM}$ to beam is 0.1 or less, it makes little difference where the tank is located vertically. For ratios of $\overline{GM}/B$ greater than 0.2, a tank located in the bilges may lose 50 percent of its effectiveness or more.

Almost any liquid can be, and has been used in tank systems. The only requirement is that it remains a liquid. Ordinary residual oils can become too viscous if not heated. In larger tanks material as viscous as Navy Special Fuel Oil has been used successfully. Since in tank systems the weight which moves back-and-forth across the ship is a liquid, it can be easily disposed of. For instance, if a tank containing water is installed high in the ship, it can readily be dumped into the ocean. This gives the ship additional transverse static stability for emergencies. An alternate arrangement is to provide a void tank low in the ship (usually directly below the stabilizer tank) into which the water (or fuel) can be dumped in an emergency.

The major impacts of a passive anti-roll tank on a ship design are engendered by the volume and weight required and the effects of the reduction in stability attributable to

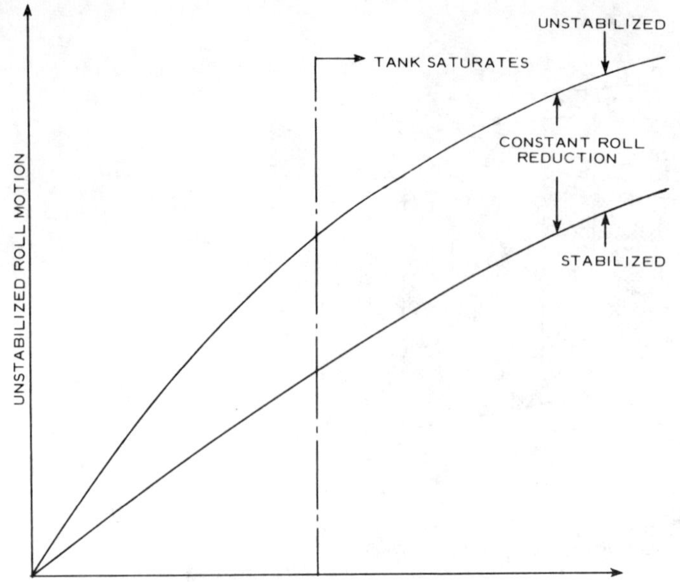

Fig. 43 The effect of saturation

the tank free-surface. There are only a few U-tube anti-roll tank installations, however, there are literally thousands of free-surface anti-roll tank installations in commercial ships.

3.5 Controlled-Passive Anti-Roll Tank Design. In the U-tube configuration, one must provide a path for the air above the liquid in one wing tank to move to the space above the liquid in the opposite wing tank. To provide continuity, the air flow (volume rate) must be the same in this path as in the water flow path (volume rate) through the lower crossover. If this air path is valved, then the amount of air flow (and thus the amount of water flow) can be controlled. Completely closing the valve virtually prevents motion of the tank water from side-to-side. In other words, closing of the valve turns off the tank. This might be a very important damage control feature, since closing a valve can be accomplished more quickly than dumping a tank.

It should be pointed out that well designed controlled-passive tanks or active tanks (described in Section 3.6) have slosh periods which are considerably below that which would be desirable for a good passive tank. As a result, faulure of any of the electronic or mechanical components will lead to a tank system which will not stabilize effectively or may even worsen the roll motion. This characteristic also needs to be considered carefully in performance of the overall systems analysis of stabilization.

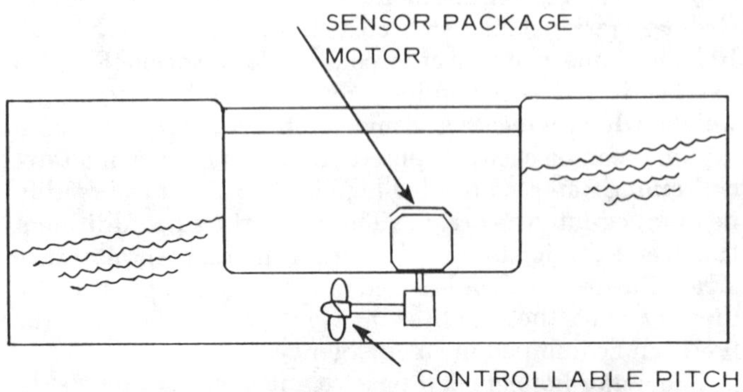

Fig. 44 Active anti-roll tank

3.6 Active Anti-Roll Tank Design. It is possible to use feedback control systems in anti-roll tanks in a fashion similar to the control systems used in active-fin roll stabilizers. These tank systems are invariably of the U-tube configuration. The motion of the ship is sensed, this information is processed, and action is taken to change some features of the tank system accordingly. If the action is such that energy is put into (or extended from) the tank liquid, then it is called a fully-active tank system, usually referred to simply as an active anti-roll tank. Fig. 44 shows the typical arrangement of this type of roll stabilization system. As illustrated, an active anti-roll tank system is basically the same configuration as a U-tube tank. Due to the complexity of fully-active systems, information about the pump and control system cannot be generalized. However, it is possible to define the required differences in tank geometry. The major differences between an active tank and those designed for pure passive stabilization is that the natural frequency of the tank with the air valve open is 30 to 40 percent greater than the ship's natural frequency and it is required that the tank have an equivalent linear damping ratio somewhat lower than a good passive tank. As a result, care must be taken in order to avoid any superfluous structure within the tank itself. The design of the tank configuration itself, follows along exactly as a U-tube design.

For active tanks, it is necessary to design the control system and to select the control system gains to provide maximum roll reduction and to avoid control system instabilities. A detailed discussion of control systems is given by Webster (1976).

Of the three different types of tank systems, the active anti-roll tank offers the best performance in roll motion reduction. However, this system is more costly and requires more complex components. For instance, a typical arrangement includes a controllable-pitch propeller pump connected to a motor. The pitch of the propeller is varied by hydraulic actuators commanded by the automatic control system. In a well designed system, during part of the roll cycle, power is extracted from the tank, and in other parts of the cycle, power is supplied to the tank. In such a well designed system, the average power required is near zero and usually negative (meaning that a net amount of power must be extracted from the tank). However, the instantaneous power required (either into or out of the tank) is usually large. A typical 6,000 ton ship may require an 1,865 kW (2,500 hp) pump for this purpose with an average net hp out of the tank of about 100. It is not surprising that the system extracts energy from the tank! It is this energy that the tank has extracted from the ship roll motion which must be dissipated. The pump system provides the means of energy dissipation.

3.7 Active Fin Roll Stabilizers. In an active-fin roll stabilization system, flow over the fin caused by the forward speed of the ship generates roll moments which oppose the heave excitation roll moments in response to the command of a control system. The roll motions are reduced by the resulting dissipation of energy. The fin angle is controlled by a system which may sense the roll motions, velocity, acceleration, and in some cases the lift on the fin. The fin is actuated by a hydraulic system which in most cases can change the fin angle from stop to stop in two seconds or less. The fin system may consist of a simple fin or a fin with a trailing edge flap, or articulated fin which, for a given planform area, can develop greater lift than a simple fin. However, care must be taken in the selection of material and in the design of the flap to ensure against corrosion and other maintenance problems. A retractable, flapped active fin stabilizer has the following main components.

a. Fins. The fin planform is the shape of the fin in plan view and is defined by a span and mean chord in terms of the *geometric aspect ratio,* which is equal to span divided by mean chord. The tip of the fin is the outermost end, and the root is that section closest to the hull.

The fin is connected to the shaft (or stock) by means of either a tapered key or a stock nut. The fin is fabricated from steel plate, similar to a rudder. The plating is normally 6 to 9 mm (1/4 to 3/8 in.) thick depending on fin size and plate strength. Plating support is provided by means of webs, intercostals, and leading and trailing edge castings. The fin is watertight and is normally filled and drained with anti-corrosive coatings prior to installation. Another popular preservative is foamed-in-place polyurethane; this has the additional advantage of preventing ingress of sea water, due to the cellular composition of polyurethane.

b. Fin Shafts. Fin shafts or stocks are normally steel forgings or centrifugal castings, usually with hollowed centers. Use of high-strength steel is encouraged, since stock diameter can then be kept to a minimum. This, in turn, reduces the size of bearings and castings in addition to the weight, and provides for a narrow root section. The stock diameter and casting thickness at the root of the fin determine the maximum width of the fin at the root. This is usually around 15 percent of the root chord; in fact, from hydrodynamic considerations, a 15 percent ratio is nominally accepted as being a satisfactory compromise between delayed cavitation at the leading edge and adequate lift characteristics.

c. Fin Shaft Bearings. These bearings are described as either *outboard* or *inboard bearings.* Normally outboard bearings are sleeve stave-type bearings and inboard are anti-friction roller bearings. Stave bearings are preferred for the outboard application because they are easier to maintain, are less susceptible to damage from sea water and contaminants, and are more resistant to leakage in conjunction with a conventional stuffing tube. However, stave bearings have higher frictional torque coefficients and this results in higher power requirements than roller bearings.

d. Hydraulic Power Plant and Acutators. The power unit provides all the hydraulic power required by the fin system. Each power unit assembly is mounted on a fabricated bedplate and is driven by a squirrel cage marine-type electric motor. Both power unit and hydraulic pump are normally resiliently mounted, to decrease structureborne noise and vibration.

The hydraulic pump is normally an axial piston variable-delivery type and is connected to the electric motor by means of a flexible coupling. The hydraulic pump delivers to the fin tilting mechanism hydraulic power which is governed by a hydraulic control system.

A constant-delivery vane servo pump is driven from the other end of the electric motor. When stabilizing, this supplies fluid at low pressure.

Hydraulic power is applied to the fin tilting mechanism which applies torque to the tiller that is passed through the fin shaft. The tilting mechanism may be either a Rapson slide or double acting hydraulic cylinders. Most modern fin systems employ the latter, since the cylinders are lighter, easier to manufacture and normally off-the-shelf items. On the other hand, the Rapson slide units are custom built, heavier, and more cumbersome. The tiller or yoke is normally a split steel casting or forging.

In designing the cylinders, care should be taken to provide enough stroke. This is normally equivalent to ±30 deg tiller rotation, with an additional allowance for positive stops. These are incorporated into a standard cylinder design which has a cushioned stop, using hydraulic oil for this purpose, prior to bottoming out.

The major impacts of an active-fin roll-stabilization system on the ship design are the added space, weight, and power required for the fin control and activation system. However, the space and weight requirements for active fin stabilizers are much less than those for anti-roll tanks. Because of the large space required, fin installations in space-limited ships are generally not retractable. There is also a small increase in the ship's total resistance due to the additional wetted surface of the fins. Because of their proven effectiveness, there are a large number of active fin stabilizer installations on naval vessels and on cruise ships.

References

Altman, Ronald (1971) "Forces on Ships Moored in Protected Waters," Hydronautics, Incorporated, Technical Report 7096-1, July.

American Bureau of Shipping (Annual) "*Rules for Building and Classing Steel Vessels.*"

Barr, R. A. and V. Ankudinov (1977) "Ship Rolling, Its Prediction and Reduction Using Roll Stabilization," *Marine Technology,* SNAME, January.

Conolly, J. E. (1969) "Rolling and Its Stabilization by Active Fins," *Transactions,* RINA.

Cox, G. G. and A. R. J. M. Lloyd (1977) "Hydrodynamic Design Basis for Navy Ship Roll Motion Stabilization," *Transactions,* SNAME.

English, J. W. (1962) "The Design and Performance of Lateral Thrust Units for Ships," paper presented to RINA, London, October.

Global Marine Development Inc. (1976) "The Glomar

Explorer; Deep Ocean Working Vessel Technical Description and Specification,"

Graham, J. R., M. N. A. Peterson, and A. R. McLerran (1970) "Deep Sea Drilling Project—Prelude to a Decade of Ocean Exploration," *Marine Technology Society Journal,* Sept.–Oct.

Hawkins, Seth, Robert Taggart, Edgar D. Hoyt (1965) "The Use of Maneuvering Propulsion Devices on Merchant Ships," Robert Taggart Inc. Report RT-8518, January.

Hoyt, Edgar D. (1969) "Note on the Bollard Pull of Marine Propulsion Devices," *Naval Engineers Journal,* November.

Jaeger, H. (1963) "The Braking of Large Vessels," *International Shipbuilding Progress,* August.

Lloyd, A. R. J. M. (1972) "The Hydrodynamic Performance of Roll Stabilizer Fins," Third Ship Control Systems Symposium, Bath University, England, September.

Lloyd, A. R. J. M. (1975) "Roll Stabilizer Fins—A Design Method," *Transactions,* RINA.

Lloyd, A. R. J. M. (1976) "Roll Stabilization Fins: Interference at Non-Zero Frequencies," RINA, Paper W3.

Mandel, Philip (1967) "Ship Maneuvering and Control," *Principles of Naval Architecture,* Edited by John P. Comstock, SNAME, New York.

Marine Engineering/Log (1964) "Ship Can 'Slam on the Brakes' with New Rudder," November.

McCallum, D. N. (1976) "Passive Anti-Roll Tanks, Design Considerations," Thirteenth Annual Technical Symposium, Association of Scientists and Engineers of the Naval Air And Sea Systems Commands, Department of the Navy, Washington, D. C.

Miller, E. R., J. J. Slager, and W. C. Webster (1974) "Development of a Technical Practice for Roll Stabilization System Selection," NAVSEC Report 6136-74-280, October.

Naval Facilities Engineering Command (1968) "Design Manual Harbor and Coastal Facilities," July.

Nickerson, A. M. Jr. and I. W. Smith (1971) "Hull Machinery," *Marine Engineering,* Edited by R. L. Harrington, SNAME, New York.

Pehrsson, L. (1960) "Model Tests on Bow-Jet Screw Propellers," DTMB Report 1461, October.

Pollard, J. and A. Dudebout (1892) "*Theorie du Navire,*" Paris.

Saunders, H. E. (1957) "*Hydrodynamics in Ship Design,*" Vol. II, SNAME, New York.

Sperry, Elmer A. (1922) "Automatic Steering," *Transactions,* SNAME, Vol. 30.

Suehrstedt, R. H. (1960) "Maneuvering Devices in Great Lakes Vessels," Great Lakes Section, SNAME, October.

Taggart, Robert (1961) "Positioning the Drilling Ship over the Mohole," Chesapeake Section, SNAME, February.

Taggart, Robert (1969) "Dynamic Positioning of Surface Ships," *Topics in Ocean Engineering,* Gulf Publishing Co., Houston.

Taggart, Robert (1970) "Anomalous Behaviour of Merchant Ship Steering Systems," *Marine Technology,* SNAME, New York, April.

Taggart, Robert (1975) "Vortex Propeller," Patent No. 3,882,674, May.

Taplan, A. (1960) "Notes on Rudder Design Practice," First Symposium on Ship Maneuverability." DTMB Report 1461, October.

U. S. Coast Guard (1977) "Marine Engineering Regulations," CG 115, August.

Webster, W. C. (1976) "Analysis of the Control of Activated Anti-Roll Tanks," *Transactions,* SNAME, New York.

Zarnick, E. E. and J. A. Diskin (1972) "Modelling Techniques for the Evaluation of Anti-Roll Tank Devices," Third Ship Control Systems Symposium, Bath University, England, September.

CHAPTER XIII

Victor R. Schellenberg

Control of the Ship's Interior Environment

Section 1
Introduction

1.1 General. The interior environment aboard ship is controlled to provide an atmosphere which is agreeable to the operating personnel, machinery and equipment, cargo and ship stores, and passengers. The environment may be modified by means of ventilation, heating, cooling, and dehumidification or by any combination of these means. Additionally, the elements of noise and vibration must be controlled. The means of controlling these environmental factors must be as reliable, simple, and maintenance free as practical, consistent with the desired results.

Section 2
Ventilation Systems

2.1 General Description and Definition. Ventilation is the process used to provide fresh outside air to various spaces within the ship. The air is distributed by means of a duct network and suitable weather openings in the ship's envelope. The type of ventilation used depends upon the nature of the space and the service of the ship. The fresh air may be supplied by natural draft or mechanical means and is provided for the removal of heat, noxious or explosive vapors, and to assure an adequate supply of oxygen to personnel. The quantity of air required for each space ventilated is determined by heat transfer or empirical calculations.

2.2 Types of Systems. There are two basic types of ventilation systems; natural and mechanical. In the natural ventilation system, air movement is created by the difference in temperature and density of inside and outside air and the trimming of cowls or scoops toward the wind. A typical system is shown in Fig. 1. In the mechanical ventilation system, a fan or similar device is used to propel air through the system. A typical system is shown in Fig. 2.

A typical system consists of a suitably weather-protected inlet or outlet, a duct to the area served and distribution branch ducts with terminals designed to suit the supply or exhaust air function required. The mechanical ventilation system may be of the mechanical-supply/natural-exhaust type, the mechanical-exhaust/natural-supply type, or the mechanical-supply/mechanical-exhaust type, depending upon the location of the fan or fans within the system. In general, the mechanical-supply/natural-exhaust type system will maintain a slight positive pressure within the spaces served. The natural-supply/mechanical-exhaust type system will maintain a slight negative pressure within the spaces served; this type of system is used in spaces such as galleys, toilets, and pantries where a positive pressure might dispel the heat and odors into adjacent spaces. The mechanical-supply/mechanical-exhaust type system may produce either a slight positive or negative pressure within the spaces served depending upon the relative ratings of the supply and exhaust fans.

Special consideration must be given to ventilation systems which serve spaces having the potential for containing hazardous vapors. Examples of such spaces are the cargo

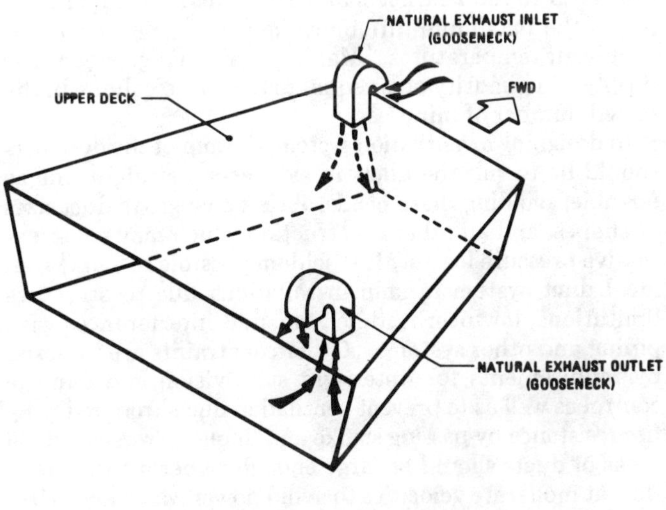

Fig. 1 Natural ventilation system

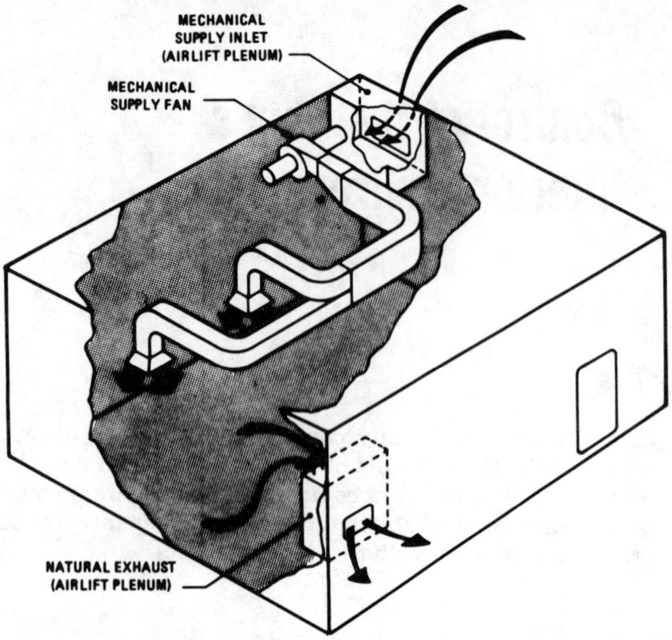

Fig. 2 Mechanical supply ventilation system

pump room on tank vessels, the compressor room on liquefied gas carriers, and passageways in way of the cargo tanks on liquefied gas carriers. The relative densities of air and the vapor to be handled determine the locations of the supply and exhaust terminals. A typical system serving the cargo pump room on a tank vessel is shown in Fig. 3. For a description of the systems employed for the inerting and ventilating of liquid cargo tanks, see Chapter XI.

2.3 Design Criteria. The application of the various types of ventilation systems and the associated design criteria for representative spaces aboard ship is indicated in Table 1. The word "Yes" under "Natural" means that a separate ventilator to the weather shall be provided. Temperature rise is the maximum difference between exhaust air and supply air temperatures. Maximum air change means the supply of a quantity of air equal to the space volume in the stated number of minutes.

In designing a ventilation system, the aim of the designers should be to run the ducts in as nearly a straight line as possible, avoiding sharp bends, abrupt changes in duct sizes or shapes, and all other construction which may cause excessive pressure losses. It is seldom possible to achieve an ideal duct system in ship installations due to structure limitations, low overhead heights, and interferences with piping and other systems. Other constraints are imposed by requirements for watertight subdivision and damage control as well as to prevent ventilation ducts from reducing fire resistance by passing smoke and flame. Cross-sectional areas of ducts should be large enough to permit the air to flow at moderate velocities to avoid power waste and to reduce noise. The following maximum duct velocities represent good practice; in areas where quiet operation is essential, 10 m/s (33 ft/s), and in areas where quiet operation is not essential 18 m/s (59 ft/s).

Sometimes obstructions are found during the installation of the ductwork and about 10 percent of the total fan pressure should be reserved in the system for these additional losses.

a. Design of Duct Fittings. Good design can contribute much to the economy of system operation. Abrupt changes in duct sizes or in direction of airflow are always to

Table 1—Ventilation Criteria

Type Space	Mechanical Sup.	Mechanical Exh.	Natural Sup.	Natural Exh.	Temp. Rise °C(°F)	Max. Air Change In Minutes
Air Cond. Equip. and Fan Rooms	Either		Either		8(15)	
Baggage		Yes			8(15)	15
Battery Room		Yes	Yes			2
Bonded Stores		Yes			8(15)	10
Bosn's Stores	Yes			Yes		20
Butcher Shop		Yes			6(10)	4
Canvas Room	Yes			Yes		6
Clean. Gear Locker		Yes				4
CO_2 Room		Yes	Yes		8(15)	6
Deck Lockers			Yes	Yes	8(15)	
Dry Cargo	Yes			Yes		30
Holds & Deep Tanks	Yes			Yes		20
Elec. & Eng. Workshops	Yes				6(10)	6
Elev. Machy. Room		Yes	Yes		8(15)	6
Eng. Stores	Yes				11(20)	20
Emer. Gen. Room	Yes			Yes	8(15)	10
Galley and Galley Pantry		Yes	Yes		6(10)	1
Garbage Room		Yes			6(10)	
Gyro Room		Yes			6(10)	4
Hawser Room	Yes			Yes		20
Lateral Thruster Comp't.	Yes			Yes	8(15)	6
Laundries—Service		Yes			6(10)	4
Linen Room Clean	Yes				6(10)	5
Linen Room Soiled		Yes			6(10)	5
Linen Lockers, Daily Service				Yes		15
Machine Shop	Yes				6(10)	6
Main Mach. Space	Yes	Yes				
Oil Skin Lockers		Yes				4
Paint Rooms Deck & Engr's.		Yes	Yes		6(10)	6
Refrigerated Cargo	Yes			Yes		60
Shaft Alley		Yes	Yes			10
Steering Gear Room	Yes			Yes	8(15)	6
Stores-Ships Dry Daily Galley & Bulk Food	Yes				4(7)	4
Sewage Tank Space	Yes	Yes			8(15)	6
Toilet & Showers		Yes			6(10)	4
Wheelhouse						10 (infiltration)

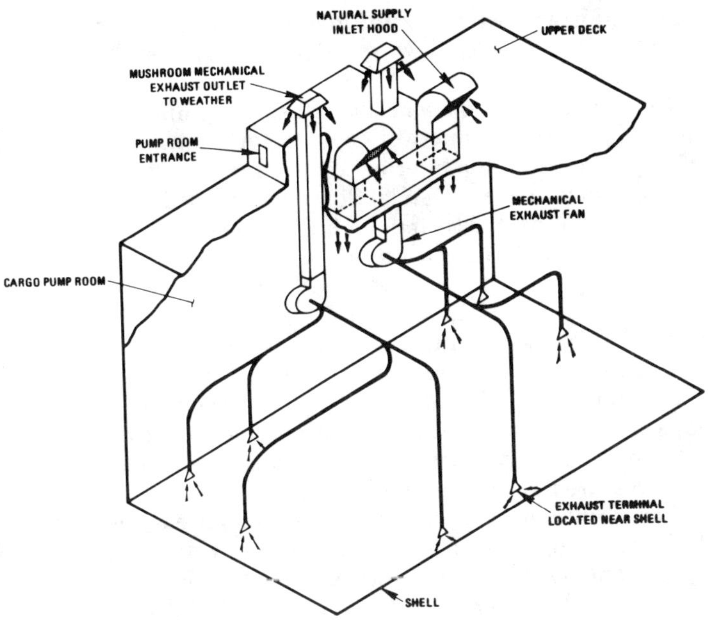

Fig. 3 Cargo pump room ventilation

be avoided. Abrupt expansions are particularly bad, since as much as five percent of the total available fan pressure can be lost in a single expansion.

Elbows should be designed for smooth airflow. It is desirable to make the throat radius of the elbow equal to the dimension of the duct in the plane of the bend. Where this is impossible and a shorter radius is required, splitters should be added in these elbows. Where conditions require square elbows, curved turning vanes should be provided.

When smaller ducts are taken off the main supply duct to serve individual spaces, a division of the main duct is the most desirable method and is based on the proportional division of air; e.g., the air quantity is divided in proportion to the area of the duct. Where headroom and structural requirements do not allow this arrangement, and for small air quantities, branch takeoffs may be used; e.g., small ducts connected into the main duct at a 30-degree angle from the direction of airflow in the main duct. It should be recognized, however, that these branch takeoffs create a loss in the main duct. A series of such branches may generate sufficient loss in the system to require an increase in fan horsepower. See Market (1971)[1] for a typical ventilation system pressure loss calculation.

b. System Balance. All systems should be checked for proper air delivery after installation is complete. Usual requirements are a minimum of 90 percent of design air quantity to each space, and 80 percent of design air quantity at any terminal where more than one terminal serves a space. During the system design stage, the pressure loss through each flow path must be established to ensure adequate fan pressure to achieve the design flow rates. This process will identify the pressure loss in each branch circuit at the design flow rate and the individual circuit having the greatest pressure loss.

Some easily accessible means of controlling air flow in each branch, such as an orifice or adjustable damper, should be provided in the design of the system especially when ducts are placed above the ceiling. This will save considerable money and time when balancing the system. System balance is achieved by adjusting the position of the balancing dampers or changing orifice sizes such that the measured flow rates are within the tolerances noted. Care must be exercised in adjusting the flow in any one circuit since to do so will result in changing the flows in all other circuits of the system.

c. Construction Details. Ducts may be constructed of galvanized sheet steel in order to withstand corrosion and vibration, or ducts may be constructed of aluminum in order to save topside weight. If aluminum is used, special attention must be given to compliance with the U.S. Coast Guard fire-protection requirements. Built-in trunk construction may be used when the minimum dimension is 230 mm (9 in.) using adjacent bulkheads and similar structures for one or more sides. These trunks and all ducts exposed to the weather are built of not less than 32 mm (1/8 in.) plate and made watertight. Vertical and horizontal ducts in general cargo holds are usually constructed of 65 mm (1/4 in.) and 48 mm (3/16 in.) plates respectively. See Table 2 for a standard of ventilation duct gages.

Because of headroom requirements, most ducts are rectangular; round ducts being used only in the smaller sizes. Circular or oval ducts are used when passing through beams, girders, and other strength members. Usually a heavy section of ductwork is welded into the penetrated structure where structural compensation is required. Handholes, access holes, and portable sections are provided to permit cleaning, painting, and inspection.

Ducts passing over electrical equipment are made watertight. Flanged connections are provided for making all ducts portable, and flanged coamings are provided where ducts penetrate bulkheads, decks, and other structures. Ducts are made with either riveted, welded, or hook seams (Pittsburgh lock seams) and are airtight. Slip joints may

Table 2—Standard of Ventilation Duct Gages

The minimum thickness of the material shall be determined by the diameter for round ducts or the maximum dimension for rectangular ducts as follows:

All vertical exposed ducts	No. 16 USSG
Horizontal or concealed vertical ducts over 750 mm (30 in.)	No. 16 USSG
Horizontal or concealed vertical ducts 460 mm–750 mm (18½–30 in.)	No. 18 USSG
Horizontal or concealed vertical ducts 310 mm–450 mm (12½–18 in.)	No. 20 USSG
Horizontal or concealed vertical ducts 150 mm–300 mm (6–12 in.)	No. 22 USSG
Horizontal or concealed vertical ducts less than 150 mm (6 in.)	No. 24 USSG

All ducts in machinery spaces shall not be less than No. 16 USSG in thickness.

---

[1] Complete references are listed at end of chapter.

be used for joining duct sections (they are especially useful in spiral wound duct installations), but good workmanship is essential to prevent leakage.

Galleys, bakeries, and food handling spaces present a special problem of heat, smoke, and odor removal. Mechanical exhaust is provided to exhaust the air from hoods over heat producing equipment. The mechanical supply should equal about 50 percent of the exhaust quantity. Supply air is discharged directly into the work areas. The remaining 50 percent is supplied by indraft from surrounding spaces. Supply air to the galley is usually preheated in winter and steam convectors or electric grids are often used for heating spaces such as toilets and lavatories. The heating system is discussed in Section 3 of this chapter.

To maintain habitable temperatures in machinery spaces and for the removal of fumes, spot cooling with large quantities of air at high terminal velocities, 13–15 m/s (42–50 ft/s), is used at operating stations and other strategic locations. Exhaust terminals should be located in the vicinity of heat-producing equipment and on the upper levels of the machinery space. Occasionally, natural exhaust is used. The exhaust air is discharged into the outer casing of the smokestack, the rest being utilized for forced draft blowers supplying it to the boilers of steam plants or to the intake air ducts of gas turbine and diesel plants for combustion.

Generally, the supply air to machinery spaces is unheated

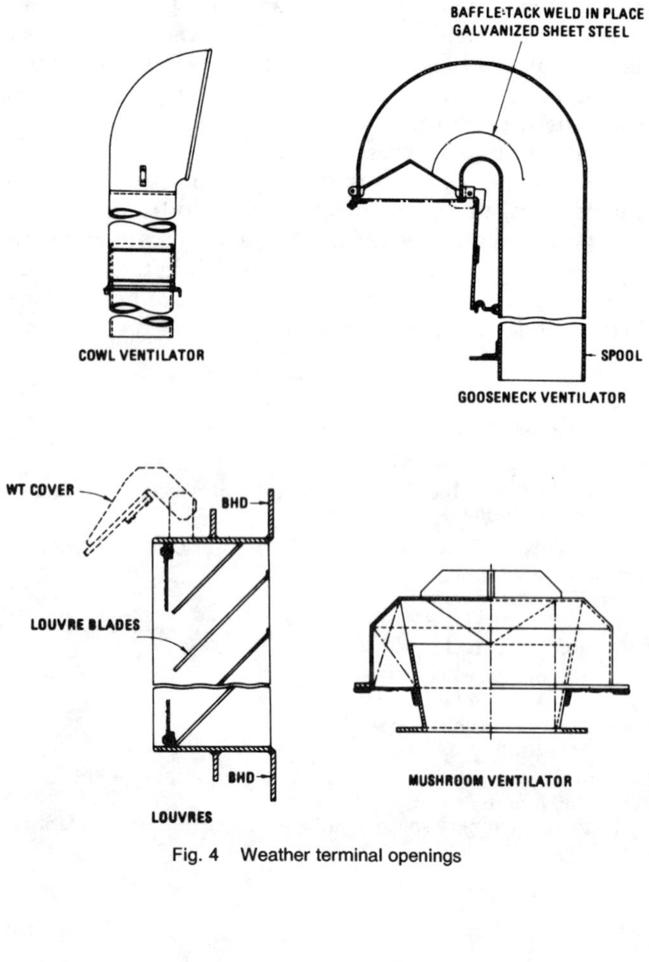

Fig. 4 Weather terminal openings

and comfortable winter conditions are maintained by terminal volume dampers, two-speed fans, or by-pass ducts to admit outside air.

2.4 Ventilation Components. Various components are included within the ductwork system for the transmission of air to and from the interior compartments of the ship.

a. Types of Fans. Axial flow fans are used widely because of compactness and high efficiency and are well adapted for ventilation of cargo spaces, machinery spaces, and other places where noise is not a significant consideration. Centrifugal fans are used for ventilation where quiet operation is desired and also for galleys, battery room exhaust, and areas where explosive vapors are removed, where the motor is not to be located in the air stream. Propeller fans are used in bulkhead installations and sometimes in a cowl for machinery space supply and exhaust systems where the pressure required is small.

For maintenance, fans must be located for easy accessibility to the motors. Many motors are provided with two-speed controls to permit reduction in supply air during cold weather. Motors are selected for 40 °C (104°F) ambient temperature except when located where high temperatures prevail, in which case they may be selected for 50 °C (122°F) or 65 °C (149°F) ambient temperatures. Axial and propeller fans generally are provided with waterproof or totally enclosed motors.

b. Weather—Terminal Openings include cowls, goosenecks, mushrooms, louvres and airlift boxes on deck or in bulkheads, Fig. 4.

All of these devices are fitted with wire mesh for ratproofing. They are made either watertight or weathertight depending upon their location and on the space served. Weather terminals should be so located that exhaust air or stack gases do not contaminate supply air. On tank vessels, the supply terminals must be located outside the hazardous area as defined by the U.S. Coast Guard. Cowls usually are provided with portable covers.

c. Interior Terminals vary in type depending upon the application. High velocity directional terminals are used in galleys, pantries, laundries, machinery spaces, and similar heat-producing spaces where spot cooling is desired. Slotted outlets may be used in front of galley hoods and switchboards. Supply registers and ceiling or wall type diffusers are used for ventilated living spaces, with terminal velocities so as to provide diffusion and throw, without objectionable air movement and air noise in the space.

Care must be exercised to insure that moving parts of terminals are rugged and rattleproof.

Terminals for holds and storerooms are merely open-ended ducts or a cut in the side of the duct fitted with wire-mesh screens for ratproofing. High velocity ducts may require expanding cones to reduce the terminal loss.

d. Exhaust Terminals are located close to heat sources and are usually an open-ended duct covered with wire-mesh or with grilles where appearance is important. Exhaust-inlet velocities should be about 5–8 m/s (17–25 ft/s) in living spaces and up to about 10 m/s (33 ft/s) in other spaces. Guards made of round bars are usually fitted in way of terminals in cargo holds to prevent damage by cargo.

e. Air Filters located in the inlet are a highly desirable item in supply ventilation systems. Viscous-coated metallic filters or manual roll renewable type air filters are satisfactory for marine use. Filters without the viscous-coating are also used in galley exhaust systems as grease filters and are cleaned easily by washing. Also, each filter is usually provided with a permanently installed air filter gage to indicate dirt loading. Filters apt to become laden with dirt and grease are potential fire hazards and should be accessible for regular cleaning.

f. Dampers are sometimes used to control the volume of air delivered at terminals. They must be of rugged construction and rattleproof. Manually operated dampers must be provided on passenger ships at the weather opening in all ventilating systems to shut off the passage of air in the event of fire, except in exhaust ducts from film lockers and projection rooms. When ducts pass through main fire-zone bulkheads, automatic fire dampers are required which will operate by melting a fusible link at 74 °C (165°F). Automatic dampers are also required in exhaust ducts over potential sources of fire in galleys and are designed to operate by melting a fusible link. Dampers are designed to close against the anticipated draft in the duct. The U.S. Coast Guard requires also that all electrical ventilation systems be provided with remote control means for stopping the motors in case of fire or other emergency.

Section 3
Air Conditioning Systems

3.1 General Description and Definition. Air conditioning systems serve to modify the outside fresh air to improve the ship's interior environment. This is accomplished by heating, cooling, dehumidifying, and contaminant removal processes. These processes may be used singularly or in combination to achieve the required environmental conditions.

Air conditioning is used almost exclusively for living spaces such as staterooms, messrooms, offices, lounges, and other public areas. However, many items of electronic equipment must also be maintained at a controlled temperature and humidity. Conditioning of the air is accomplished by a cooling medium of chilled water or freon and a heating medium of steam, hot water or electricity and is designed to permit simultaneous heating and cooling as may be necessary to satisfy specified design conditions. Dehumidification or humidity control may be accomplished by cooling or through the use of desiccant dryers. Contaminant removal is accomplished by means of filtering, adsorption, electrostatic charging or absorption.

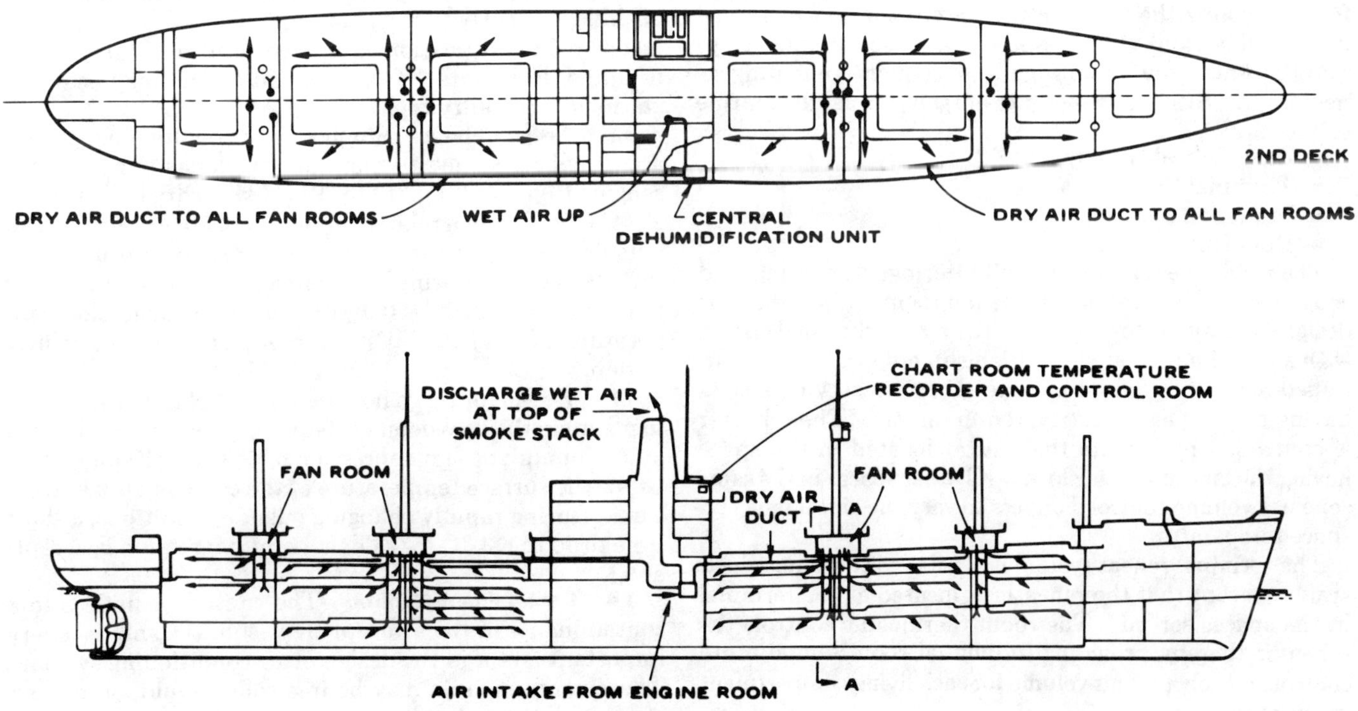

Fig. 5 Cargo ventilation and dehumidification

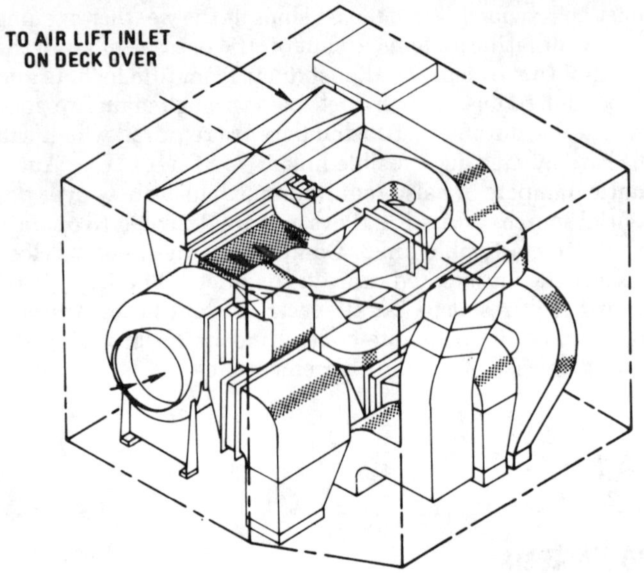

Fig. 6 Minimum volume fan room

The conditioned air is distributed to the spaces served by the same network of ducts used for space ventilation systems as described in Section 2.

3.2 Types of Systems. Centralized air conditioning systems may be generally categorized by function such as: Systems which provide a combination of heating and cooling, systems which provide cooling only, systems which provide heating only, and systems which provide dehumidification only.

a. Combination Heating and Cooling Systems. The combination heating and cooling systems provide the means for controlling the temperature within the spaces served during all seasons. Such systems are also capable of humidity control by cooling and subsequent reheating as necessary. Systems which are in general use in marine service are:

- Zone Reheat System;
- Terminal Reheat system;
- Induction System;
- Dual Duct System.

The zone reheat system has all filtering, preheating, and cooling equipment located in the fan room. This system is designed to mix return air from the air conditioned spaces with a minimum of outside replenishment air. The conditioned air is distributed to zone reheaters serving spaces having more or less identical requirements. The reheater is controlled by a room thermostat located in the space having the largest heating load. All other spaces in the same zone use volume control dampers to vary the air volume and space temperature.

The terminal reheat system is similar to the zone reheat system except that the reheater is located at the terminal in the spaces served. The room thermostat controls the reheater thereby providing individual room temperature control and constant air volume for each living compartment in the ship.

The induction system provides for filtering, cooling and reheating (as necessary) of outside air. The air is delivered by a fan to individual ceiling or bulkhead mounted room units. The primary air is discharged through nozzles creating a region of low pressure behind a secondary coil in the room unit. Secondary air (room air) is induced across the coil where it is heated or cooled to maintain the desired temperature.

The dual duct systems provide for preheating the combined replenishment and outside air and discharge of this air to two parallel supply duct systems. The air is cooled in one duct and heated in the other. The cold air and warm air ducts are routed to the individual spaces and terminate in an overhead mixing box. The mixing box contains a damper which can be regulated to vary the quantity of cold or warm air to satisfy the room load.

b. Cooling Systems. The previously noted combination heating and cooling systems are designed to permit securing the cooling unit during the peak of the heating season. At such times certain spaces within the ship which are subjected to high internal heat load such as galleys or enclosed operating stations within machinery spaces may require cooling to maintain habitable conditions. The environment within these spaces may be maintained through the use of self-contained air conditioning units.

c. Dehumidification. One or more complete, automatically operated dehumidification systems may be provided for preventing moisture damage to or condensation of moisture on cargo and internal structures of all those portions of holds suitable for carrying dry cargo.

Mechanical dry air supply and natural exhaust systems are fitted in each hold as well as means for recirculating the air in each hold, Fig. 5. Recirculation dampers or valves are provided to control supply, exhaust, recirculated, and dry air independently to each hold.

The dehumidification equipment utilizes solid granular or liquid desiccants. The desiccant carrier may be of the stationary or rotary type.

3.3 Design Criteria. Shipboard air conditioning systems are usually designed to maintain inside air temperatures ranging from 24 °C (76°F) to 29 °C (85°F) dry bulb, usually 27 °C (80°F), and a relative humidity of 50 percent with an outside air temperature of 35 °C (95°F) dry bulb and 28 °C (82°F) wet bulb during the summer season and inside air temperature of 21 °C (70°F) dry bulb with an outside temperature of −18 °C (0°F) dry bulb during the winter season.

The criteria for cargo hold dehumidification is normally to maintain the dew-point of the atmosphere within the hold at a minimum design depression of 6° C (10°F) dew-point below the surface temperature of the cargo or ship's structure. During rapidly changing outside conditions, a short time drop to a 3 °C (5°F) dew-point depression is acceptable.

3.4 Contaminant Control. The control of undesirable contaminants in the atmosphere within the ship is a very important factor in the design of air conditioning systems. These contaminants may be in a solid, liquid, or gaseous state and range in size from submicroscopic to macroscopic.

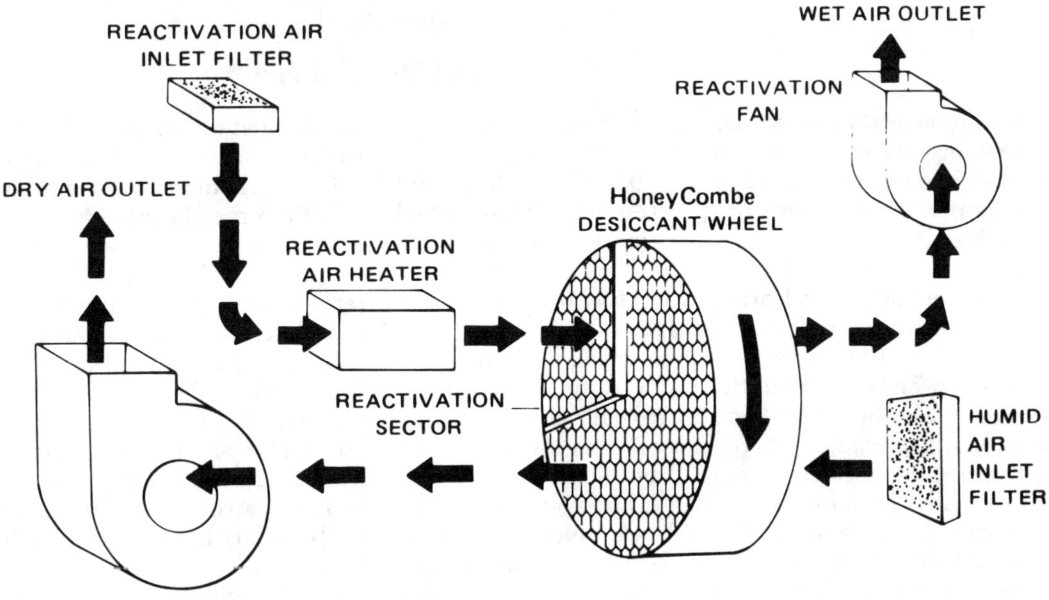

Fig. 7 Cargo hold dehumidification unit

The bulk of these undesirable contaminants can be removed by conventional mechanical means. Passing air through filter media of metallic, fibrous, or plastic foam composition removes contaminants by impingement and/or straining with the efficiency of removal dependent upon the size of openings in the filter media. Passing air through cooling coils removes contaminants by condensation and to some extent by absorption. Passing air through beds of silica gel, activated carbon, or activated alumina removes contaminants by adsorption. Passing air through electrostatic precipitators removes contaminants by electrically charging the contaminant and then attracting the contaminant to an oppositely charged plate.

Many contaminants, however, cannot be removed by mechanical methods, used either separately or in combination, and must be removed by absorption. By passing the air through beds of solid absorbents, such as calcium chloride, lithium hydroxide, or causing the air to come into intimate contact with liquid absorbents such as ethylene glycol, lithium chloride, lithium bromide, or monoethanolamine, various contaminants can be removed by a chemical process. The various types of dehumidifiers, scrubbers, and burners available remove contaminants by utilizing a chemical change.

Submersible craft present unique problems in contaminant control due to the absence of a supply of fresh air. The elimination of contaminants from the atmosphere by absorption techniques is of paramount importance to the safety, health and welfare of the crew. As they develop and improve, undersea craft techniques will be employed more widely in surface ships. These include replenishment of oxygen by oxygen generators, removal of carbon monoxide, hydrogen, and hydrocarbons by burners, and removal of carbon dioxide by scrubbers. Of major importance is the careful selection of construction materials to eliminate or minimize the generation of atmospheric contaminants.

3.5 Air Conditioning Components. The components employed within air conditioning systems, such as fans, ductwork, terminals and dampers are the same as described in Section 1, Ventilation Systems. The preheaters, reheaters, cooling coils, and refrigeration units are commercially available units similar to those employed in shoreside installations.

When these units are located in the fan room, space constraints may challenge the most experienced designer. Frequently, these spaces are too small for a properly engineered installation. Fig. 6 illustrates the problem of fan rooms with minimum volume.

The principle of a representative dehumidification unit for cargo hold service is shown in Fig. 7. A typical design is based on inlet air having 21 g of moisture per kg of dry air (150 grains per lb) with the dehumidified outlet air having no more than 7 g per kg (50 grains per lb) of dry air. The dehumidification unit consists of the vapor removal media, the dry air and wet air fans, and reactivation heater. This equipment should be constructed to the best marine standards and designed to facilitate access to all working parts.

Section 4
Acoustical Habitability

4.1 General Considerations. The human ear has certain interesting characteristics. The normal hearing range for a young person extends from about 20 to 15,000 Hertz (Hz) (cycles per second), with the greatest sensitivity around 1,000 Hz. Aside from the noise attributes of loudness and annoyance, there is the factor of physical tolerance to noise, that is, the noise sound-pressure levels which the ear can stand without discomfort or damage.

The effect of noise on the human being with regard to hearing loss and communication has been studied and design criteria established through extensive habitability research in naval ship design. Airborne noise levels generally should not exceed the decibel values in Table 3.

The effect of noise annoyance is, however, not as well defined. The wide range of noise levels which various persons find disturbing makes this aspect of noise control more subjective and difficult to define. Factors which influence a person's reaction to noise include interest of the listener in the sound, whether the noise is unnecessary and could be avoided, the degree to which the listener can disregard the noise, the activity with which the noise interferes (sleeping, reading, recreation, working), the character of the listener.

For the more noisy spaces aboard ship, it has been determined that people can readily adjust to various environments and actually consider them normal, once they are conditioned to accepting them, provided the environment includes no hostile sounds. How well the naval architect can handle the problem of acoustical habitability depends largely upon how well he can control the magnitude of sound levels and how much he can shape the noise spectrum in any of the various ship's spaces.

The reduction of noise within certain spaces may produce counter-productive results. For example, staterooms normally receive a high degree of isolation from passageway noise, however, the resultant number and location of general alarm bells must be carefully reviewed to assure audibility within all staterooms. Also, the enclosing of machinery control stations has prompted some reaction from operating personnel that not hearing the machinery has degraded their effectiveness.

4.2 Sources of Noise Generation. The major sources of noise generation aboard ship may be categorized into flow generated noise and mechanically generated noise. Each of these general categories contain several elements each of which must be considered by the designer to preclude objectionable noise conditions aboard an operational ship.

Table 3—Permissible Airborne Noise Levels in Decibels

	Octave Band Frequencies (Hz)							
	20–75	75–150	150–300	300–600	600–1200	1200–2400	2400–4800	4800–10000
Living Spaces	72	66	60	55	52	50	48	47
Passageways	75	69	64	59	57	55	53	52
Machinery Spaces	90	90	90	90	85	85	85	85

Flow generated noise is produced by a fluid in motion. The fluid may be either a liquid or gas and may be either within the ship envelope or external to it.

a. Noise Generated by Ship Moving Through the Water. Flow of water around the hull of a ship is almost completely turbulent, particularly in the bow and stern areas. The turbulent water flow path produces pressure fluctuations which tend to drive the hull plating into vibration. The resultant noise may be transmitted within the hull either as airborne noise or as structureborne noise. Discontinuities of the hull such as sonar domes, sea chests, and shaft struts function to increase the turbulence within the boundary layer thereby tending to be sources of external flow noise.

b. Propeller Generated Noise. Several different types of noise may be generated by the ship's propeller. The two types of propeller noise associated with fluid flow include cavitation and vortex shedding. When a ship's propeller is rotated at high speed cavities can form and collapse radiating a loud and continuous noise. Also vortices are shed from the trailing edges of the propeller blades. If the frequency of this shedding corresponds with a resonant frequency of the propeller blade the blade vibration will radiate a loud ringing noise.

c. Fluid Flow Noise. The noise sources within piping and duct systems are similar to those produced by the ship moving through the water and the propeller. The flow of fluid through a piping system may produce noise due to turbulence, cavitation and vortex shedding. In piping systems the noise may be intensified by the organ pipe effect of the pipe or duct. Restrictions or obstructions in the fluid flow path which increase the velocity are prime sources of cavitation and turbulence generated noise. Dampers and splitters within ventilation ducts may also produce noise due to vortex shedding.

Mechanically generated noise usually originates in rotating and reciprocating machinery. The sources of such noise may be the result of improper balancing, excessive tolerance between mating parts such as gears or the result of loose or worn parts. The characteristics of the machine and the noise produced may provide an indication of the problem area.

4.3 Design Implications. The acoustical aspects of controlling the ship's interior environment must be included at the beginning of the design process. In many cases, it is extremely difficult and expensive to correct a noise problem whereas the impact of precluding the problem at the early design stage would be minimal.

When addressing the acoustic or noise considerations both the noise source and the transmission path must be reviewed. Control of noise at the source may be accomplished by improved dynamic balance of rotating machinery, limiting velocity within fluid systems, avoiding turbulence within fluid systems, improved tolerance between mating parts, and application of suppression material to the noise source. In addition, the noise level within spaces such as staterooms may be controlled by locating them as remote as practical from spaces having a higher noise level. For example, it would not be prudent to locate a stateroom adjacent to a fan room.

Noise may be transmitted from the source to other areas via the structure as vibration, as airborne noise or as fluidborne noise. Structureborne noise usually originates at machinery or equipment foundations. Transmission of noise from a machine to the supporting structure may be reduced by mounting the unit on resilient mounts or distributed isolation material (DIM). Where such type mountings are used, isolation of the connecting piping or ductwork must also be accomplished. When resilient mounts are employed, care must be taken to assure the natural frequency of the mount does not coincide with the exciting frequency of the vibration source.

Airborne noise may be reduced by locating the offending unit within a space lined with sound absorbent material or, as in the case within the engine room, as remote as practical from the operating station.

Fluidborne noise may best be controlled by eliminating the source where practical. If a large pressure drop is induced by a single orifice in a piping system, consideration should be given to a multiple-step orifice thereby reducing the velocity through each step. If the source of the noise may not be eliminated as in the case of ventilation fans, absorbent material may be installed in the ductwork or, in some cases, the ductwork must be increased in mass to reduce the amplitude of response of the ductwork to the vibratory excitation.

With respect to naval vessels, acoustical considerations may extend well beyond the element of habitability due to the nature of the ship's function. The methods employed aboard submarines, for example, for controlling noise are several orders of magnitude more extensive than those normally encountered in merchant ship practice. Additional information relative to methods of noise control are discussed in various papers of the SNAME Ship Vibration Symposium, October 1978.

Section 5
Vibrational Habitability

5.1 General Considerations. On board ships equipped with machinery developing power for propulsion and auxiliary purposes, personnel are subjected to vibration. The vibration may cause annoyance, physiological damage to body organs, psychological disturbance of the crew or damage to shipboard equipment.

The effect of vibration upon equipment installed aboard ship has been one of the major factors resulting in premature failures of equipment which has previously proved satisfactory in land-based installations. The equipment supplier must consider the vibration aspect of the shipboard environment in the design and construction of marine hardware.

A magnitude of vibration which can do no harm to equipment or structure can, however, be a great nuisance to the crew. The degree of human perception to vibration in the frequency range of 30 to 4800 cycles per minute is a function of the amplitude of the vibration. Frequencies below 0.5 Hz may cause motion sickness. While it is difficult to define precisely acceptable limits of intensities of vibration, Fig. 8 and Fig. 9 identify vibration zones which may be used as a guide in determining general acceptability. Zone A defines that area within which a high probability of vibration difficulties exist; Zone C defines that area within which no vibration difficulties are anticipated; Zone B defines that area within which the subjective nature of vibration does not permit a reasonable assurance of acceptability.

5.2 Design Implications. Recent design practices have resulted in an increase in the flexibility of the hull structure with an attendant lowering of the natural frequency. The

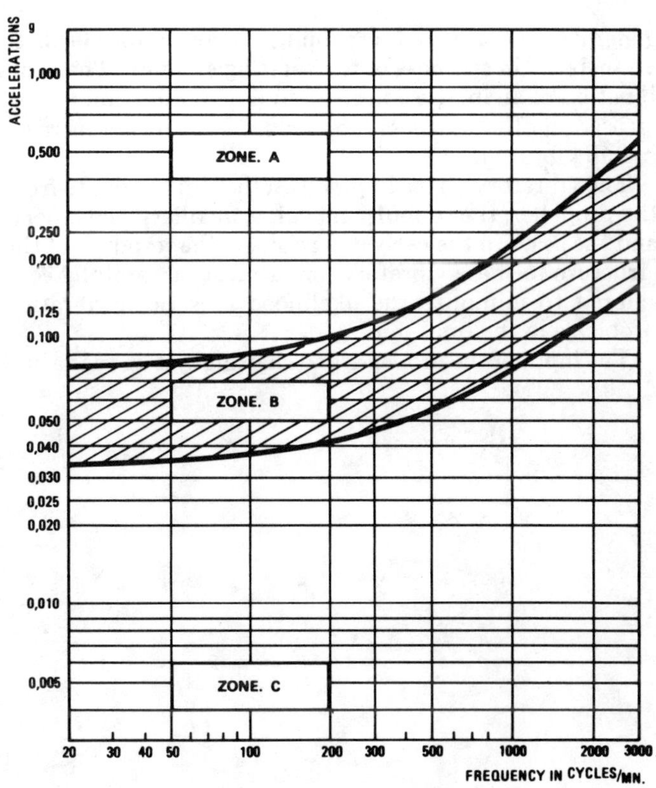

Fig. 8 Vertical vibration criteria

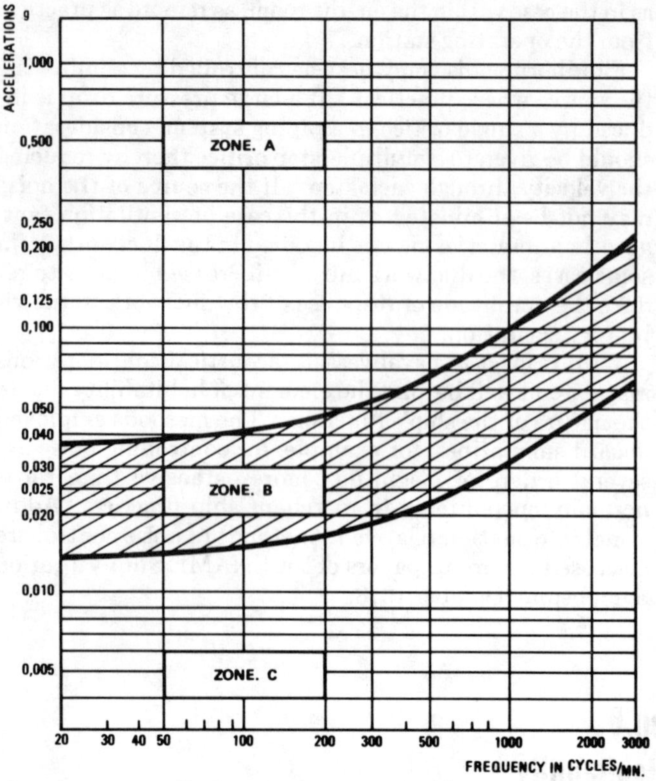

Fig. 9 Horizontal vibration criteria

general trend of increased propulsion horsepower leads to an increase in stiffness of the shafting system. The combination of an increase in structural flexibility and the increased shafting stiffness can produce a significant increase in hull vibration.

The vibratory forces imposed on the ship originate from the propeller, the propulsion engine, auxiliary machinery, and the sea. In the early design stage, the response of the structure to these vibratory forces must be carefully considered to minimize the likelihood of serious vibration problems in the final ship.

The input of the propeller and shafting system to the hull is influenced by propeller geometry, including number of blades, the shape of the afterbody underwater hull, the clearance between the propeller and the hull and rudder, the shafting system alignment, the structure supporting the thrust bearing, and the location of thrust bearing and line shaft bearings. The interactions between the propeller itself and the hull, including the flow of water into the propeller, may be analyzed by model testing. The shafting system may be evaluated using developed analytical techniques (Lewis 1967).

The propulsion machinery, whether of the rotating or reciprocating type, imposes forces on the ship's hull as does reduction gearing. The engine and gear manufacturers should be consulted in the design of the supporting structure to assure compatibility.

The excitations generated by the auxiliary machinery units are generally of the same character as those generated by the propulsion machinery. Since the vibratory forces imposed on the structure are usually less severe than for the propulsion machinery, empirical methods are usually employed to limit the resultant vibration.

The sea itself imposes transient forces on the ship's hull. These forces may result from either sea-swell or rough seas. In either case the vibration inducing effect may be altered by varying such factors as ship's speed, heading, or draft and trim.

The factors which affect the vibration characteristics of a given ship must be addressed at the very inception of the design process. The correction of vibration problems identified during vessel sea trial may be both extremely costly and delay entry of the ship into service. Further insight on vibration control techniques is given in many of the papers presented at the SNAME Ship Vibration Symposium, October 1978.

References

Lewis, Frank M. (1967), "Vibration of Ships," *Principles of Naval Architecture,* SNAME.

Markert, John W. (1971) "Environmental Control," *Marine Engineering,* SNAME.

CHAPTER XIV

Raymond P. Devoluy
David T. Bloodgood

Hull Preservation

Section 1
Introduction

1.1 Purpose. The purpose of hull preservation measures is primarily to protect against corrosion, but it also includes maintaining good appearance, protecting cargo, minimizing abrasion damage, and slowing the rate of underwater hull fouling. These measures involve both the design of the preservation system to be provided during the construction of the ship, and the development of a suitable preventive maintenance program when the ship enters service. Both of these aspects of hull preservation will be discussed.

1.2 Initiation of a Hull Preservation Program. A new ship is constructed, blasted with abrasive and coated under conditions more suitable for surface preparation and coating than it is likely to encounter at any other time during its service life.

Maintenance planning should begin during design and be incorporated into new construction. The prudent owner, at the time of construction, should establish a tentative hull preservation and maintenance program based on planned service and cargos.

The selection of corrosion control measures, which will include surface preparation, coating materials, metals and cathodic protection should be on a life cycle cost basis.

Section 2
Objectives of a Hull Preservation and Maintenance Program

2.1 Definition. The objectives of a hull maintenance program are corrosion protection to maintain appearance, to protect cargo, and to increase the life of the ship and its components; and fouling prevention to reduce hull roughness to promote more efficient operation and fuel savings. Preventive maintenance may be defined as the judicious expenditure of funds with the purpose of restricting further compulsory expenditure for maintenance and repair.

2.2 Vessel Appearance and Protection. Traditionally, shipowners take pride in the appearance of their vessels. In the past, appearance and cleanliness were maintained by frequent repainting. Modern, high performance coatings maintain their appearance for longer time periods, particularly when applied over primers that arrest corrosion. The boottop, topsides, deck, and superstructure can be kept in good condition with minor maintenance. The prevention of cargo contamination, particularly of valuable liquid cargos such as chemicals or edibles, is of great economic importance to both the shipper and the shipowner.

Damage from abrasion and other mechanical causes can be reduced by the use of high performance coating systems using coating materials such as inorganic zinc, epoxies, vinyls and reinforced coatings. In the underwater area of a vessel the abrasion resistant coatings help to limit mechanical damage and, with the aid of cathodic protection, keep corrosion of the affected areas (Fig. 1) to a minimum.

2.3 Motivations for Using Better Materials and Improved Maintenance Procedures. Classification societies such as the American Bureau of Shipping and Lloyds Register of Shipping, recognizing the effectiveness of improved coatings in reducing corrosion, allow lowered requirements for plate and frame thicknesses in most areas of cargo ships except the bottom and side shell below the maximum load line. Therefore, the owner of a new ship can evaluate the cost of a good coating system and material selection against the initial savings in steel costs, and enjoy greater revenues due to the increased cargo carrying capacity of the lighter ship.

ABS does not define or name specific coatings but does refer to "an effective method of protection against corrosion," to the application of special protective coatings, and to cases where "other effective methods adopted as a means of corrosion control." The decision of which coating to use is left to the shipowner because, if he does not choose wisely,

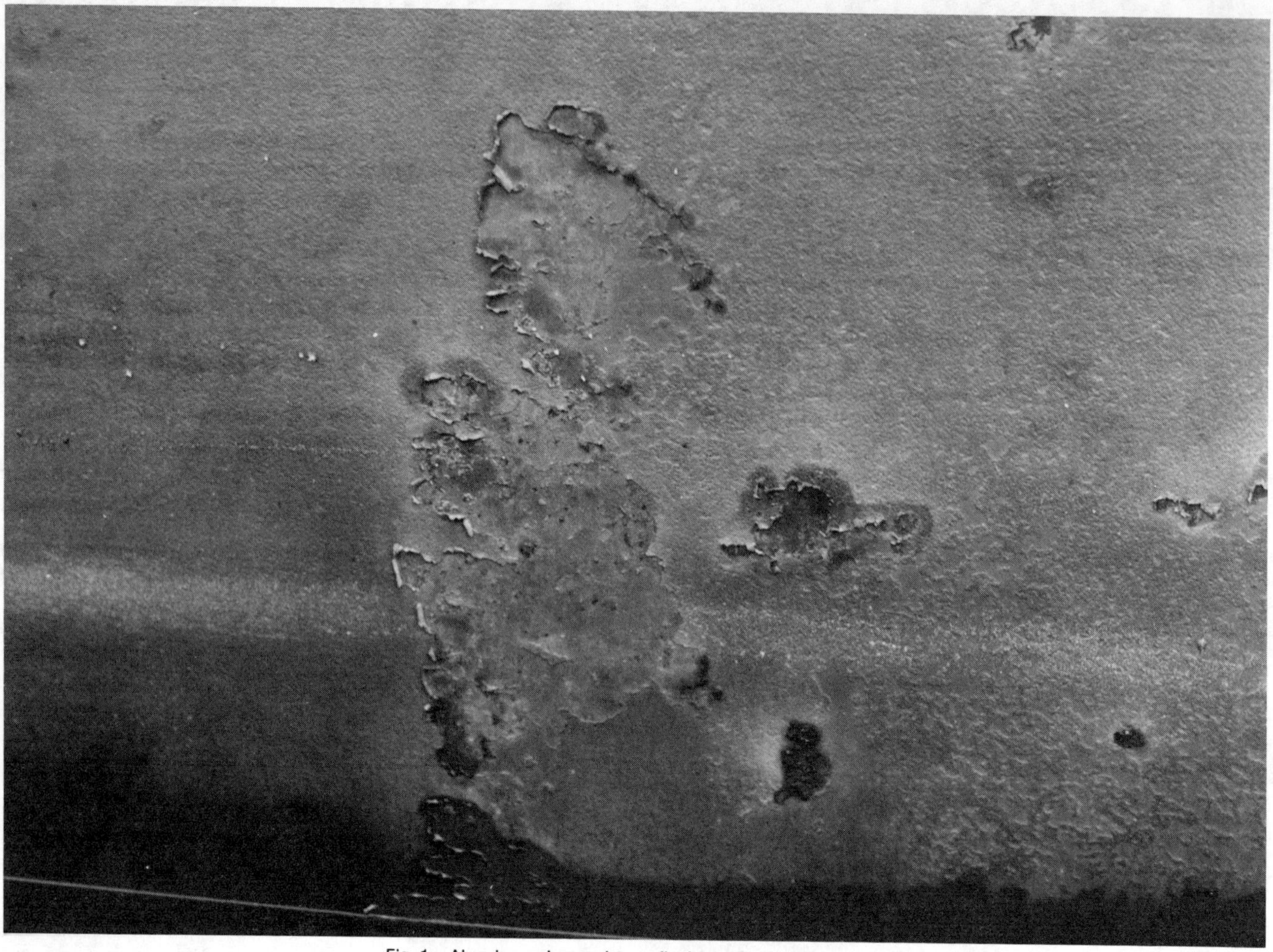

Fig. 1 Abrasion and corrosion on flat bottom of coastwise tanker

and the plating corrodes below the accepted minimum thickness, it will have to be replaced at his expense.

ABS and the U. S. Coast Guard now permit two-year regular drydock inspections and special surveys at four-year intervals. Furthermore, technically trained divers inspect welds and plates by ultrasonic techniques. They also use underwater TV cameras to show the condition of propellers and other appendages to shipowner representatives. All this provides strong motives for the use of better corrosion and fouling protective measures.

The above incentives, combined with the increased cost of ship time out of service, smaller crews on mechanized ships, and increase in the deleterious effect of roughness at higher speeds, are swelling the demand for better materials, greater adherence to preparation and coating procedures and improved maintenance.

Section 3
Corrosion

3.1 Character of Corrosion. The study of corrosion considers reactions between a metal and its environment. It also concerns the suppression of corrosion by changing the characteristics of metals. The great progress in the battle against corrosion has been based on a better understanding of the natural forces concerned and on the development of the scientific principles upon which effective control of corrosion must be based. Corrosion aboard a ship covers the broad spectrum of corrosion technology because most types of corrosion are present. Uhlig (1948)[1] provides a

[1] Complete references are listed at end of chapter.

comprehensive text on corrosion and the International Nickel Company book, "Corrosion in Action."

3.2 Types of Corrosion. The most common causes and types of corrosion found on a ship are:

- *Galvanic corrosion* occurring when two metals of different potentials are in metallic contact in an electrolyte such as salt water. The acceleration of the attack on the less noble metal may be observed. The farther apart the metals are in the galvanic series the greater the rate of corrosion.
- *Pitting corrosion* is localized corrosion at an accelerated rate which is characterized by deep penetrations into the metal. This type of pitting is prevelant on the underwater hull and in tanks that carry liquid cargos or ballast. The pitted areas, if not repaired, can lead to performations of the steel, or the weak spots may become focal points for corrosion fatigue or stress corrosion failure.
- *Anaerobic corrosion* is caused by the sulphate reducing bacteria that are present in many harbors.

Other forces that cause metal breakdown are cavitation erosion and impingement. These forces can cause severe pitting on struts, rudders, etc.

3.3 Corrosion Prevention Measures. Useful corrosion measures for ships are organic and inorganic coatings, metallic coatings, corrosion resistant metals and alloys, cathodic protection, linings and laminates.

- *Organic coatings* prevent corrosion in accordance with their electroyltic resistance, or by inhibitive pigments and, in the case of metal fillers such as zinc, by galvanic protection (Bacon et al, 1948), (Brown, 1959), (Brown, 1961).
- *Inorganic coatings,* such as the inorganic zinc coatings, are used extensively as the base or prime coat on many areas of marine structures for corrosion prevention. These protect the metal by the same process as the organic zinc-rich materials.
- *Metallic coatings,* including hot dip galvanizing, and flame sprayed aluminum and zinc, are used as protective measures. Hot dip galvanizing is used extensively on sheet metal, piping, etc., whereas the flame sprayed coatings, being

Fig. 2 Underwater hull showing heavy fouling

more expensive to apply, are used for specialized services such as steam valves and high heat exhausts.

• *Corrosion resistant metals and alloys* are used extensively for ship piping (copper nickel), bolts, nuts, fittings, fasteners, and tank linings for carriage of chemicals and other critical cargos (stainless steel and stainless clad). Aluminum alloys are used for hulls and superstructures, mostly on specialized craft where light weight is required, and on navy vessels.

• *Cathodic protection* is used to provide corrosion protection usually as an adjunct to coatings, for underwater hull areas and tanks subjected to salt water. There are two basic types of cathodic protection—impressed current and sacrificial anodes. Impressed current systems are used to protect the underwater hull of a ship by superimposing on the hull an impressed current provided by a remote power source through a small number of inert anodes. Galvanic or sacrificial anodes are used in tanks and for partial underwater hull protection such as the stern area of a ship and the rudder. These anodes are made of either aluminum, magnesium or zinc. The design of cathodic protection systems has been discussed in many technical papers. SNAME T&R Bulletin R-21 (1976) provides a thorough discussion on the subject.

• *Linings and laminates* are thick coatings used to protect a metal surface that is unavailable for maintenance, such as the stern tube section of a propeller shaft, the interior of piping carrying highly corrosive materials, or to prevent chemical attack in tanks carrying very corrosive cargos. Elastomeric or plastic linings may be applied as precured sheets or spray depending on the nature of the polymer. Similarly laminates can be produced by various techniques. As shipowners strive for greater flexibility in accepting a variety of chemical cargos for deep tanks, it is possible that linings and laminates will find greater acceptance.

Section 4
Fouling

4.1 General. Fouling of a ship's bottom causes increased hull roughness which will be reflected in loss of speed and/or increased power requirements necessitating higher fuel consumption. Fouling can also cause breakdown of coating systems which lead to corrosion and provide an additional source of hull roughness, Fig. 2.

4.2 Prevention of Marine Fouling. Fouling prevention is being studied with increased vigor as various marine research groups (Lageveen-Van Kuyk, 1967) are finding that even minor fouling attachment will cause increases in frictional resistance and thus increase fuel consumption. Data obtained from well instrumented ship trials, run before and after various service periods, show that even modest algae attachment can cause significant increase in hull friction.

Another reason for wanting better and longer-life antifouling coatings is the desire by shipowners and operators to extend periods between drydockings. This has become economically important as ships increase in size; shipyard services become more expensive and loss of operating time can cause significant losses in revenue.

The better fouling coatings available on today's market will keep a ship's bottom clean for a minimum of two years under most circumstances. The development of longer life antifouling systems has been made possible through a better understanding of how toxics are released (Van London, 1963) and the development of polymer-type coatings.

However, the development of new antifoulings and new toxics has been seriously hampered by environmental restrictions imposed by cognizant government agencies in the interest of environmental protection.

Section 5
Preservation Design

5.1 Definition. Preservation design is more than the selection of materials with proven performance and ease of maintenance; it calls for continuous welds in exterior areas and wet spaces (Thayer, 1966); the elimination of crevices, cracks, and small drain holes; the use of flat bars and bulb angles instead of plain angles and T-bars. All these reduce edge corrosion necessitating frequent maintenance. Rough welds must be ground and all weld splatter removed to eliminate common causes of premature coating failure.

Preservation design includes providing cargo deep tanks with flush sides and with heating coils installed externally. It involves using the principles of corrosion engineering for the judicious selection of metals and alloys rather than dependence upon traditional concepts which have survived by chance. The design stage is the time for assessing the potential rewards obtainable through the use of new coatings and procedures, especially for surfaces of limited accessibility such as the keel plates.

5.2 Surface Preparation. For new construction the initial surface preparation normally will comprise the automatic sand blasting of plates and structural shapes, as shown in Fig. 3. The blast cleaning may be immediately followed by

Fig. 3 Plate leaving dryer and entering 8-wheel Wheelabrator blasting cabinet

Fig. 4 Double-nozzle spray assembly which oscillates across entire upper surface of blasted plate

Fig. 5 Coated plate leaving drying oven

Fig. 6 Fabricated section in blasting and coating building

Fig. 7 Abrasive blasting under controlled conditions

a preconstruction primer applied to a dry film thickness of 20–25 micron or μm (0.9-1 mil), Figs. 4 and 5.

a. Surface Preparation Standards. Surface preparation standards are provided by the specifications of the Steel Structures Painting Council, National Association of Corrosion Engineers and others. The most widely used in the marine industry are the Steel Structures Painting Council Specifications. Pictorial Standards present photographic descriptions of the desired degree of surface preparation and are used as an adjunct to the specifications. The Pictorial Surface Preparation Standards for Painting Steel Surfaces that display conditions for new steel are available through the Steel Structures Painting Council (1967). Pictorial Standards for aged or coated steel surfaces are available in SNAME T&R Bulletin 4-9 (1969). An excellent adjunct to the above publications for those who must select Surface Preparation equipment is the "Catalog of Existing Small Tools for Surface Preparation and Support Equipment for Blasters and Painters" (MarAd, 1977) prepared under the auspices of the National Shipbuilding Research Program.

5.3 Quality Control and Inspection. The surfaces to be coated during construction of a ship are readily available and subject to greater scrutiny and inspection than they will be in a repair situation. Surface preparation and coating work are inspected as work progresses and each area of a ship is subject to inspection more than one time during construction, Figs. 6 and 7.

Inspection is the responsibility of a number of parties, including the owner or owner's representative, coating vendor, and the yard. The number of inspecting parties will vary with the requirements of the ship specification. The inspection requirements for any job should be specific in scope.

5.4 Basis for Agreement on Preservation Systems. There have been many improvements in the materials and methods for preserving ships. Potential suppliers will continue their research as long as there is a reasonable chance of a return on their investment from large volume new construction orders. Naval architects and shipowners spend much time critically analyzing test and service data, and supplementing these with ship inspections before preparing specifications for new vessels. Shipbuilders' technical departments are also an excellent source of information on application costs, health hazards, and feasibility of using special equipment. It is most important that performance and other technical considerations prevail in the selection of materials for new ships.

Section 6
Selection of Preservation and Maintenance Systems

6.1 Basic Selection Criteria. Coating materials should be selected on a life cycle cost basis. The optimum coating for each situation should be chosen. As the cost of a coating material is normally less than 25 percent of the total applied cost, including surface preparation, the economics of using materials with the best performance capabilities should be the first consideration. Information on coating systems are available. The "Coating Systems Guide for Exterior Surfaces of Steel Vessels," SNAME T&R Bulletin 4-15 (1979) presents data on surface preparation, application and performance data for generic types of coatings, and recognizes that variations in properties will exist due to individual manufacturer's research and formulation capabilities.

Criteria for selecting coatings and other preservation methods for various types of marine exposure follow.

6.2 Continuous Immersion (Ship Bottoms). It has been shown that electrolytic resistance is the most important property of an underwater anticorrosive coating. With the advent of full hull impressed current cathodic systems it is necessary to have anticorrosive systems that are not alkali sensitive.

The ship bottom anticorrosive is also burdened with preventing copper ions, released by the toxic cuprous oxide in the antifouling paint, from migrating to the steel surface. Such penetration by copper ions can cause pitting of the steel and may result in the inactivation of the antifouling paint.

Other important properties of underwater coating systems are adhesion, abrasion resistance, resistance to the effects of cathodic protection, and compatibility with drydock environments and schedules.

Cathodic protection is an important part of the answer to the bottom abrasion-corrosion problem, but the coatings used on cathodically protected ships must have good electrolytic and alkaline resistance. The applied potential should be limited to the minimum necessary for the protection of the steel if damage to coatings is to be avoided.

Hull metal loss on a ship bottom is due largely to electrochemical corrosion rather than to mechanical removal of metal by abrasion. However, both forms of metal removal are illustrated in Fig. 8. Hence, the necessary use of cathodic protection to supplement abrasion resistant coatings is endorsed. Coatings based on vinyls, coal tar epoxies, epoxies, and chlorinated rubber are suitable for use with controlled cathodic protection.

The effect of hull roughness on the performance of ships is well documented in the work of the Norwegian Ship Research Institute (SFI) (Matzow-Sorenson, 1963). Hull roughness associated with new ships is in the order of 75-125 μm (3–5 mil) and this initial roughness increases at about 100 μm/yr (4 mil/yr) on ships in poorer condition and increases at about half this rate even on the best maintained vessels.

The roughness of 75-125 μm on new ships is made up of two components. First there is the original surface profile of the steel which is reflected to a considerable extent in the paint films subsequently applied. In addition, during paint application, there is a superimposed roughness caused by spray patterns, runs, sags, and inclusions of foreign particulate matter. Under the best possible conditions the surface is left with this unavoidable initial roughness of approximately 100 μm (4 mil).

Ships in service show a progressive increase in roughness caused by mechanical damage to the coating system and the accompanying corrosion. On poorly maintained vessels the increase in hull roughness can amount to some 125 μm/yr (5 mil/yr) which can be translated into an effective increase of 15 percent in shaft horsepower if speed is to be kept.

Roughness due to fouling is the most widely recognized cause of roughness which when present will significantly affect performance. Since the dimensions of roughness associated with fouling are much larger than those of physical roughness it is obvious that complete control of fouling is essential. Considering the SFI data it is apparent that even microfouling associated with slime films has to be eliminated if adverse effects on performance are to be avoided. Practical observations from ships in service have shown speed losses of over three knots on a 15.5-knot tanker caused by fouling.

6.3 Zinc-Rich Primers. The use of zinc-rich primers, particularly inorganic zinc silicates have been one of the recent technological developments that have had a positive impact on preservation of metals. This impact, although not dramatic, has had the effect of preserving scarce materials, eliminating the need for replacement of existing structures, reducing the cost of steel structures, and pro-

Fig. 8 Abrasion and corrosion on flat bottom

viding new structures with a substantial increase in life expectancy.

Some of the advantages of inorganic zinc coatings are:

• The coating is unaffected by weathering—sunlight, rain, dew, ultraviolet and wide changes in temperature or bacteria and fungus. Since it does not chalk or dissipate itself as a result of the above causes, as is usual with an organic coating, the inorganic coating remains intact and with essentially the same thickness over many years in the weather.

• Due to the strong permanent bonding of the coating to the steel the base coat forms a permanent primer which does not undercut or allow underfilm corrosion. Where topcoats are applied, this property cannot be over-emphasized since most failure of organic films in a corrosive atmosphere is due to undercutting of the coating at breaks or underfilm corrosion through the coating. With the inorganic base, this cannot take place—thus multiplying the effective life of the organic film many times.

• The cured film is hard, metallic, and abrasion resistant.

• The coatings do not shrink on curing as do organic coatings. They wet the steel surface well and, because of these properties, completely follow the configuration of the surface over which they are applied. This is a great advantage in coating rough, pitted surfaces.

Because of the above properties the inorganic zinc silicates are the predominant primer coating used on new construction for the exterior of ships from the ballast waterline up, including superstructure and all deck appurtenances and structures. This has reduced surface preparation for maintenance since corrosion is arrested and normally only topcoats have to be renewed.

Other areas of use for the zinc coatings are cargo tanks (cargos to be carried must be defined), interior wet spaces, and to some degree on underwater hulls and ballast tanks.

6.4 Coating Requirements—Ship Areas Other Than Underwater Hull. The alternate immersion, or so-called boottop area, is one that is subject to heavy abrasion, alternate immersion, and exposure to the weather, sunlight, and harbor pollution. Top-coating over a zinc-rich primer is normally accomplished with a chlorinated rubber, epoxy, or vinyl system as each of these have the required properties for the service conditions expected. Some shipowners, particularly in tanker service, will overcoat the system with antifouling coatings to preclude heavy grass growth in these areas because of the extended times the ships will be in deep draft service. These antifouling materials will be of such a nature that they can retain their properties with extended periods of exposure to the atmosphere.

The topside will normally be coated the same as the boottop area except for a change in color in accordance with owner's preference. The service is less severe than in the boottop area. However, it is still subject to heavy wave action, salt spray and continuous moisture.

The weather decks on a ship are subject to a variety of conditions. All ship's deck coating systems must have good abrasion resistance and impact resistance, as well as resistance to water, continuous moisture, thermal shock, and air pollution.

The decks of tankers and chemical carriers, in addition, are susceptible to the spills of cargos; therefore, the coatings are required to have resistance to chemicals, oils, and solvents. The coating materials that exhibit the best all around properties are chlorinated rubber, epoxy, vinyl, and urethane coating systems. These, when applied over a zinc-rich primer, can meet the criteria. For safety purposes a non-skid surface is usually required. This may be applied to the overall deck or just in walkways and work areas. The non-skid additive is normally incorporated into the last coat applied.

The superstructure is an area subject to salt spray, salt atmosphere, air pollutants in port areas, and sunlight. The combination of these effects is very deleterious to the life and appearance of a coating system. For most owners the appearance of these areas is of primary importance, therefore topcoats with good gloss and color retention predominate. Resistance to dirt retention and cleanability are important assets.

Most ships constructed in the last twenty years will have a coat of zinc-rich primer on the steel to reduce corrosion. Deck structures, including the house, have many sharp edges, corners, cutouts, hand welding, and attachments. These are all places where it is difficult to apply a proper film and unsightly rust bleed often occurs. The zinc-rich primers have been very helpful in reducing these effects. Topcoated with chlorinated rubbers, epoxies, vinyls, vinyl acrylics, and high grade alkyds such as the silicone alkyd, it is possible to maintain structures with good corrosion protection and appearance.

6.5 Maintenance. Because of quick turn arounds, reduced crews, and larger ships, most maintenance of exterior coatings is done in shipyards or by companies that specialize in surface preparation and coating application. Hence, the increase in the use of more sophisticated coatings which have a longer life expectancy should be measured against the overall life cycle economics.

6.6 Living Spaces. The joiner bulkheads in living spaces are faced with a decorative veneer, usually of a melamine type. These are considered as permanent finishes and will need no further treatment during the life of the ship unless they sustain damage. The decks are covered with various decking materials depending upon area or service. Crew spaces will normally have vinyl tile with terrazzo finishes in wet spaces; galleys will have quarry tile; some spaces will be carpeted. These spaces are more of a housekeeping problem rather than a problem for a corrosion engineer.

6.7 Internal Tank Coatings and Linings. One of the largest and probably the most critical ship spaces requiring coatings or linings are the tank spaces. All ships have ballast tanks. Cargo tanks vary in size, construction, and cargo carrying capability. Coatings in tanks are to prevent corrosion and protect cargos from contamination. For cargo tanks the resistance of a particular coating to a cargo must be considered to some extent in relation to the cycle of cargos already carried.

The investigation by practical tests, of the interaction with coating materials of all possible permutations of cargos which are normally compatible with the tank coatings, would require extremely lengthy trials. In general, sufficient information is not readily available from shipowners to indicate whether one type of cargo will more likely be carried than another, and the impression is that any cargo is liable to follow any other, in a particular tank; sequences are generally completely unpredictable. Cycling tests by paint manufacturers are generally restricted to particular sequences which are known to exist, or to have existed, for a particular ship's tank. Perhaps the most frequent sequence in smaller tankers without segregated ballast tanks is the alternation between chemical cargos and ballast, with a drying-out period between, and this is the basis for sequential exposure of a number of paint manufacturers' products. Of particular importance is the need for allowing sufficient time to elapse between cargos which are of borderline compatibility, or which may temporarily soften coatings, to allow the paint to recover. On a short voyage of a few days, a cargo which is considered safe for that period, if followed by a similar one in the same tank, can cause cumulative damage unless forced ventilation is carried out, for periods on the order of 30 hr between the cargos.

a. Ballast Tanks. Any tank intended solely for the carriage of salt water ballast must be protected if steel renewals are to be avoided long before the vessel has reached its planned economic life. Protection of the steel can be accomplished by coatings alone or coatings in conjunction with galvanic anodes. The most commonly used coating for these tanks has been an epoxy. However, if the ballast cycle is such that the tanks are only in ballast approximately 50 percent of the time, and are dry the balance of the time, an inorganic zinc coating can be used successfully.

Interest is being shown, at this time, in coatings that do not require such high degrees of surface preparation as the epoxies and zincs. These coatings are based on corrosion inhibitors incorporated into wax, petrolatum, wool grease, and coal tar type resins. These are soft film formers but give dense films with high water impermeability. These may become more popular as many of the older ships convert tanks, which had previously carried cargo as well as ballast, into dedicated ballast tanks. The cost for such a coating will be considerably less than the epoxy zinc coatings because of the relaxed surface preparation requirements.

b. Cargo/Ballast Tanks. In clean product carriers, the tanks that carry both cargo and ballast are fully coated. For similar service, on ships carrying exclusively crude oils, the tanks are coated on the bottom and on the top including horizontal webs where water can lie. One of the reasons for the partial coating is economics, as many of the crude carriers are extremely large and the cost for coating the total surface area would be prohibitive.

Another reason for concern in coating the overhead in tanks carrying cargo is the requirement now for inert gas systems where flue gas is blown into the tanks to prevent explosions and fires. The flue gases contain acidic compounds that are deleterious to the steel on the tank overheads.

The coating materials used to coat the internals of cargo and cargo/ballast tanks are normally epoxy, epoxy-urethane, or inorganic zinc coatings. These coatings are used in clean product carriers, with the epoxy type being used in crude carriers, and for any ship having an inert gas system. Before selection of a coating or coating system it is always wise to check with the manufacturer and cite the product list that is planned for carriage.

c. Chemical Tanks. Cargo tanks for the carriage of liquid chemicals and solvents are among the most difficult areas to protect from corrosion, and from possible contamination of the cargos. Use of the tanks for ballast water on return voyages increases the possibility of corrosion and degradation of both the cargo and the tank coating. For the carriage of many extremely aggressive liquid chemicals required to be transported in bulk today, the coatings themselves must have a high resistance to chemical attack and, due to their chemical inertness, the repair of such coatings is therefore made more difficult. If the ship is to be available for all possible charters, the avoidance of damage to cargo-tank coatings, or the speedy remedial treatment of accidentally damaged areas is of the utmost importance. Added to this is the restricted time allowed for repairs.

Complete information on the cargos to be carried, impurities, etc., should be fully known before making any selection of coating materials. More than one type of coating will normally be required to carry successfully a wide variety of cargos. Both solid stainless steel and stainless clad mild steel are used as tank lining materials. The high cost of the solid stainless steel must be justified by a guarantee of long term charters for cargos requiring this material. The clad steel, being much less expensive, has wider usage. However in the event of pitting corrosion, penetration to the underlying mild steel can lead to catastrophic failure. For this reason with certain cargos the solid stainless can be justified. Ships with stainless tanks, either solid or clad, will normally have a number of coated tanks to handle the less severe cargos. This also helps to reduce the cost of the ship.

Ballast water is not usually carried in stainless steel tanks. Therefore, other provisions for ballast, such as double-bottom tanks, peak tanks or cofferdams will have to be capable of carrying ballast water.

The requirements for liquid cargo deep tanks on cargo ships will be similar to that for chemical carriers. There is one added criterion: since many of the cargos are edible products, the tank lining materials must have Food and Drug Administration approval.

d. Miscellaneous Interior Spaces. Machinery space bilges continue to be a problem, not because there are no coatings available for the service required, but because the conditions for application are extremely poor. On new construction a base coat of inorganic zinc is normally used. This will be followed by an epoxy or chlorinated rubber.

Fresh water tanks, which include potable and distilled water tanks, are still generally coated with zinc dust paint. However, the epoxy coatings, which are longer lasting, are being applied in many of the newer ships.

6.8 Evaluation Techniques. The marine industry has a strong economic motivation to use better coatings for new ships, but is often held back by time-honored testing procedures that require extensive time periods to complete. Even then the results are often misleading because the controlling application or service factor may have been omitted for reasons beyond the investigator's control.

The more resistant the coating, the longer it takes to break down in normal test. There is an urgent need for the adoption of a suitable simulated service test, wherein detection of failure is accelerated without changing the relative rate of coating breakdown. Devoluy et al (1967) review some simulated service techniques that have been tried for the evaluation of ship bottom coatings. For example, several investigators were able to predict the order of corrosion breakdown between several ship bottom anticorrosives long before the failures were determined by visual observations. They measured decrease in electrolytic resistance versus time for the coatings applied to panels immersed in flowing seawater.

It is important to correlate laboratory test methods with ship service, and this can best be done by running coating systems of known performance through the laboratory procedures. Test panels coated under a variety of controlled conditions, and then mounted on racks fastened to the appropriate area of the ship, can be run concurrently with the laboratory evaluation to provide a second check.

Section 7
Planned Maintenance Programs

7.1 Examination of the Alternatives for a Planned Maintenance Program. The choice of when and where maintenance work will be done are important considerations. The answers will vary according to the type of ship and her trade. There are alternatives available to the ship owner. Assuming no emergency the questions to be answered are:

- What is the availability of the ship with the least interference to cargo carriage?
- Where and when is the climate likely to be most suitable for cleaning and recoating work?
- Who will best be able to perform the maintenance work at the least overall cost, including ship's time?

Table 2—Time Underway, Loading and Unloading, and at Shipyard

	Percent of Time		
	Underway	Loading and Unloading	At Shipyard
General Cargo Ships	(1) 74–50	25–49	1
Long Haul Tankers	90–82	8–16	2
Short Haul Tankers	76–60	22–38	2

Note: (1) Most general cargo ship operators reported about 50 percent of the time underway. It is expected that container ships will be underway approximately 80 percent of the time.

Table 1—Availability of Surface to be Recoated

	Shipyard	Loading Pier	Underway
Bottom to Ballast Line	X	1	—
Upper Boottopping and Topsides	X	X	—
Weather Decks	X	—	X
Superstructure	X	X	X
Dry Cargo Spaces	X	—	—
Liquid Cargo Tanks	X	—	2
Living Areas	X	X	X
Miscellaneous Tanks and Voids	X	2	2

Key: "X" Feasible; practicality depends on the ship and her trade.

"1" Removal of light fouling by underwater cleaning is useful.

"2" Preliminary cleaning and gasfreeing on way to shipyard.

The question as to where work can feasibly be accomplished may best be answered by Table 1.

It is immediately apparent that all areas can be maintained at a shipyard or repair yard. Unfortunately, repairs requiring burning and welding are not compatible with repainting, and even well-organized repair yards may not be able to cover all the areas that need recoating during a normal availability period. Very few shipowners can afford the loss of ship time just for maintenance painting; therefore, planned maintenance schedules are very important.

Table 2 may help to pinpoint maintenance alternatives by indicating how several types of ships spend their time.

It is evident from Table 2 that time spent at shipyards for planned maintenance is small compared to the time a ship is in active service. It is therefore very important, when selecting coating systems, that the planned maintenance schedule be realistic and take optimum advantage of the qualities of the coating materials that are applied.

References

Bacon, R. C., J. J. Smith, and F. M. Rugg (1948), "Electrolytic Resistance in Evaluating Protective Merit of Coatings on Metals," *Industrial Engineering Chemistry,* Vol. 40, January.

Brown, J. R. (1959), "Evaluation of Protective Coatings for Ship Bottoms," *Corrosion,* Vol. 15, June.

Brown, J. R. (1961), "Laboratory Tests Give Rapid Evaluation of Ship Bottom Coatings," *Corrosion,* Vol. 17, July.

Devoluy, R. P., L. J. Nowacki, and F. W. Fink (1967), "Simulated Service Evaluation and Marine Coatings," *Marine Technology,* Vol. 4, No. 2, April.

Lageveen-Van Kuyk, H. J. (1967), "Cost Relations of the Treatments of Ship Hulls and the Fuel Consumption of Ships," Report 93C, Netherlands Ship Research Center TNO, March.

MarAd (1977), "Catalog of Existing Small Tools for Surface Preparation and Support Equipment for Painters and Blasters," National Shipbuilding Research Program, Maritime Administration, May.

Matzow-Sorenson, B. (1963), "Measurements of Surface Roughness of Ship Bottoms Can Show the Way to Great Savings," Shipftknisk Forskninginstitutt-Nytt, September.

Steel Structures Painting Council (1967), "Pictorial Surface Preparation Standards for Painting Steel Structures," SSPC-VIA-1-67T.

Thayer, S. W. (1966), "A Shipowner's Views on Preservative Coatings," National Paint, Varnish and Lacquer Association, Marine Coatings Conference, May.

Uhlig, H. H. (1948), *Corrosion Handbook,* John Wiley & Sons, New York.

Van London, A. M. (1963), "A Study of Ship Bottom Paints, in Particular Pertaining to the Behavior and Action of Antifouling Paints," Report 54C, Paint Research Institute TNO, Netherlands Ship Research Center, September.

CHAPTER XV

Nicholas Bachko
Ludwig Hoffmann

Shipbuilding Costing and Contract Arrangements

Section 1
Introduction

1.1 Purpose. Naval architects and engineers in structured organizations are frequently excluded from participating in the contracting and financing arrangements of vessel construction. This exclusion is most unfortunate and it is anticipated that this section will assist naval architects and engineers in contributing to these arrangements.

Far too often the approach to vessel selection and financing involves lawyers, accountants, financial planners, and ship construction people working independently of each other without the continuing interchange of ideas that is so essential during the planning stage. The widest knowledge of any proposed vessel construction and the fullest participation in the mission aspects of new vessels provide the best climate for producing the most effective design and construction.

If continuing interaction of the interested parties is not feasible, the next best thing is to have those entering the field of vessel construction understand fully (*1*) the various contributions of the lawyers, the accountants, the financial planners, and the operators to the design, and (*2*) the documents and instruments developed over the years to ensure the continuum of events which must occur timely to produce the desired vessel within the desired time at an acceptable price. With the intellect considered to be the prerequisite of such understanding comes the patience to accept a less perfect alternative when it is more important to see the project move onward;

1.2 Scope. Accordingly, this section is proceeding on a format which is intended first as a narrative of a typical ship design genesis, its subsequent contracting and construction, and its delivery and operating inception. After this over-simplified case history approach, a more detailed discussion of the legal and financial aspects and impacts, including the documents usually involved in such transactions will be presented. Where deemed appropriate, a sampling of alternative approaches will be included, all to show the reader that the kinds of agreements which can be made between a purchaser and a builder or between a financier and a purchaser/borrower are limited only by the law of the land and the ingenuity of the parties dealing in the matter—the assumption being in all cases that the objective is the best product at the least all-inclusive cost to the owner.

Also, where the narrative leads to identifiable problem areas, sufficient analysis will be outlined to permit understanding and insight into the course of these problems so that the naval architect or marine engineer might be better prepared to avoid controversial approaches in preparing ship construction documents for the clients or principals.

A significant number of naval architects, engineers and others are directly involved in United States shipbuilding and shipping business practices which differ in many respects from those in other shipbuilding countries. Although this chapter discusses international costing and contracting arrangements to some extent it is primarily concerned with U. S. practice.

To the extent considered necessary, reference is therefore frequently made to specific United States rules, organizations and operating procedures. At the same time, the general discussions apply equally to all international shipbuilding and shipping.

When dealing specifically with the United States government, those procurements made directly by the U. S. Navy and the U. S. Coast Guard are generally so identified; when the term *government aid* is used it generally means the government is not procuring the vessel but that the owner, buyer, or purchaser has arranged a *construction differential subsidy* (CDS) for the shipbuilder and/or obtained a *government insured mortgage* or other benefits available under the Merchant Marine Act.

1.3 Definitions. In discussions of shipbuilding costing and contract arrangements, a number of terms are used that have specific connotations in this aspect of the shipbuilding process. For the purposes of this chapter the following definitions can be considered to apply.

- *Architect of Contract:* A term borrowed from the legal profession to indicate the person or entity that authored the contract document.
- *Builder:* In this text, *builder, contractor,* and *shipyard* are used synonymously. It is the entity that signs the construction contract and undertakes to physically build the vessel. The various forms are used as these terms are encountered in invitations, contracts and specifications.
- *Owner:* In this chapter, this term is used to identify the *buyer* of a vessel to be constructed. In the parlance of MarAd contracts the term *purchaser* is usually substituted

for *buyer*. Primarily, the intent is to name the party who selects the design and causes the initiation of the contract to build. It is recognized that in leveraged lease situations the owner of record of the constructed vessel may be someone or some group having only a financial interest. In such cases *owner* as used herein is the *charterer*.

• *Naval Architect:* Anyone having decision authority over the design of the vessel to be constructed or reconstructed. *In-house naval architects* are those on the wage payroll of the shipowner or entity contracting for a vessel. *Outside* or *Contract* naval architects are those persons whose business is the design and engineering of vessels, and who contract with owners or shipyards to perform their services for a fee.

• *Design Agent:* A term used interchangeably with an outside or contract naval architect. It has come into use as shipyard designs have become prevalent. Shipyards are frequently design agents. They do employ naval architects, and those who work at design are usually in the engineering department or the planning department.

• *Reps:* Abbreviation for representatives; as, for instance, owner's *reps* are the inspectors and plan approvers working on-site during ship construction.

• *Lead Ship:* The first vessel built to a new set of plans and specifications. It is not necessarily the first vessel delivered because under circumstances of the order being alloted to two or more yards, the first vessel in one of the other yards may be delivered first. This occurs because of better production methods, or because of unforeseen delays in the lead yard.

• *Following Ships:* Ships built to the same plans and specifications whether in the same yard as the lead ship, or in other yards, are *following ships*.

• *Berth Term:* This refers to dry cargo liner operations utilizing publicly issued schedules of port calls.

• *Cease and determine:* A phrase used in Maritime contracts to indicate a full unconditional stop action plus an inventory of the financial position as of that moment.

Section 2
Genesis and Framework of a Typical Ship Construction Program

2.1 How the Decision to Build Vessels Germinates. Except sometimes in the world of chartering (where new ventures generated by the ability to secure long term charters lead to ship contracts), the impetus for new ship construction usually comes from established companies already operating some tonnage and sufficiently well indoctrinated in the every day workings of their business to be planning a continued future, utilizing newer and better ships. Entrepreneurs who can develop a bona fide commitment to charter a vessel for a sufficiently long term, can and do build new vessels with the help of the banking community's eagerness to accept a bona fide charter commitment as collateral for the credit required.

The climate for speculative ship building varies with the bullishness of the world economy and the presence or absence of over-tonnaging of the kind of vessels best suited to a trade or service. When the movement of foreign-source crude oil appeared to be on a steady growth curve, entrepreneurs and oil companies worked out deal upon deal. When the embargo occurred in 1973, confidence was shaken and continued deliveries of previously contracted vessels built up a world-wide stockpile of unemployable tonnage. This over-tonnaging cast a pall on speculative construction which tended to limit risk-taking in new construction.

Accordingly, this narrative will use the berth term, established operator's vessel construction program because it illustrates a wider spectrum of contractual possibilities involving the architect. The entrepreneurial vessel built for chartering out is most likely to be one more of a class built for someone else and modifications or changes are considered items to be sacrificed in the interest of timely signing of the contract. The banking community feels most secure with concrete examples described in the documents and would rather know the ship is a sister ship to one identified in a previous deal than cope with new specifications.

In the commercial world, the initiation of planning for new vessels is a decision by a commercial company's top management that new vessels, particularly those employing innovation and improved reliability, will improve the company's profitability. If the company is a subsidized company, or if the company has certain other statutory arrangements with the government, a contractual obligation to build one or more vessels may exist, and may be the impetus to start management's vessel replacement planning.

Vessels have always been costly enough structures that their acquisition is and continues to be a priority item for action by the company's board of directors. Prior to management's submitting a proposal to the board, an indeterminate development period or stage must be anticipated. During this period, many ideas will be forthcoming from almost every participant in the business. Management usually begins with suggestions for characteristics which it hopes will improve earnings. These are most likely to be simply stated and will include, as a rule, desires for higher speeds and greater capacities. Some owners, however, tend to emphasize the goal of low-cost carriage of cargo which relates to a low acquisition cost per-ton or per container.

2.2 Design Data, Origin, and Infusion into the Accepted Design. Suggestions for the selection of a ship design in this period come also from shipyards, customers, vessel component vendors, interested friends, seagoing personnel, shore staffs, agents, etc. The company's outside naval architect or the company's in-house vessel replacement department culls all these suggestions and molds what is good into a concept which also considers naval architectural practices, shipyard availability and limitations (present day term for this is *producibility*), initial cost impacts, rules and regulations, etc. From this work a conceptual design or designs is produced with a table of characteristics and a precise description of the salient features.

It has not been without precedent to have a design competition at this stage. Design competitions were common in shore-side construction but only rarely used in modern shipbuilding. Owners will consider a design competition if they are not committed to a favorite architect, if the workload for architects is so limited all are actively seeking the owner's work, and if choosing from several designs suits the owner's posture at that particular time. For those owners contemplating building under construction differential subsidy (CDS), the federal government may effectively influence the choice of design to enable participation in multi-ship programs.

2.3 Shipyard Developed Design. Since the advent of the negotiated contract in the United States and in recognition of long standing tradition in other countries, owners will nowadays tend to progress from in-house conceptual outlines to a shipyard developed design, either by-passing the outside naval architect entirely, or using him as a consultant to evaluate a shipyard design and to suggest to the owner ways to incorporate owner's preferred features in the shipyard concept. Thus, the owner may benefit from a proven design and, hopefully, at a favorable price.

If the construction is to be built with government aid, the owner begins informal discussions with the government's staff to acquaint them with a preliminary economic analysis of the proposed service, the conceptual design and to insure that the design meets with the standards currently advocated by the government.

One should be mindful of the fact that up to this point minimal expenditures have been made. Most of the work has been done in-house by the owner's staff and if anything has been spent outside, it has been on travel to various shipyards and, perhaps, 2,000 or 3,000 man-hours of engineering effort by an outside architect. These kinds of expenditures fall into the category of day-to-day expenses.

2.4 Government's Financial Involvement in the Ship Design. At some point in time, top management, armed with a suitable draft proposal, sketches, plans, and economic analyses, presents the new construction proposal to its board. Assuming the construction is to proceed, the board will in due course authorize the further development of the plans, subject to refined estimates, financing and contracts satisfying some particular set of conditions or the management's best judgement. When so authorized, the owner's staff undertakes to have contract plans and specifications developed either by an outside naval architect, by the shipyard whose design was acceptable to management, or by the owner's own organization.

Simultaneously, if the vessel is eligible to receive government aid, an application is prepared by the owner to be submitted to the government. This application will incorporate a rather complete set of plans and data from which the government staff is able to understand the vessel's concept, and make predictions as to cost, and as to delivery. Such applications are usually conditional on the owner's board of directors accepting the available financial aid rate and contract conditions, if any.

During the course of time in which government is considering the application, the owner will be editing and re-editing the plans and specifications; determining the estimated domestic cost; cost to construct in another country, if applicable; source of equity; plan approval and inspection estimated; loan arrangements, etc.

2.5 Invitation to Bid, or Request for Proposal. If the vessel is to be bid or negotiated, at the earliest date that the design is set and contract plans and specifications are ready, an *Invitation to Bid* or *Request for Proposal* (RFP) is sent to a list of prospective builders describing briefly the vessel to be built and the conditions under which it is to be built. Such prospective builders as are interested respond by posting a cash bond to insure return of all bidding material received by them and by requesting the full set of plans and specifications prepared by the owner and his naval architect, if one has been employed. It could well be that the owner's list of builders in deference to the location of his business may not include all the builders. The government may, however, insist upon an expanded list of bidders.

The Invitation to Bid specifies the date on which sealed construction bids are to be opened and sets out the conditions under which bids are determined to be responsive. Similarly, the Request for Proposal sets forth the conditions under which designs and construction proposals are to be submitted.

The bidding period is usually chosen by the owner upon consultation with selected shipyards and the government, if it is to be involved; it is set at the minimum practical period, usually 60 to 90 days, in which a shipyard can secure quotations and establish its proper selling price.

The contractor is guided by his own order book, his potential opportunities with other customers, his problems in the yard as to availability of trades, market conditions, etc., to determine whether he is in a position to compete for the new work. If he considers that the new work will not conform to his own production schedule, he makes no attempt to bid, but returns the bidding material and repossesses his bond.

If he deems the work a practical possibility, he begins the process of estimating. This process is excellently set forth by Mack-Forlist and Goldbach (1976).[1]

Contracts may be of the fixed price or escalating type wherein the bid price may or may not be adjusted to reduce the risk to the shipbuilder of the escalating costs of labor and material. Responding to a bid request for an escalating type of contract is a costly and time-consuming process. Not all bidders or proposers can complete their version of the bidding processes in the time allotted. Any party can ask for an extension of time. Unless there are circumstances which prevent such an extension, it is the practice to grant any requests for reasonable extensions.

Bid or RFP responses, although requested in a set form,

[1] Complete references are listed at end of chapter.

usually vary from "no bid" to careful compliance with the bid form; others will include explanations, exclusions, differing times of completion, or other non-conforming items. Shipyards may thus become non-responsive legally.

The true low bidder is rarely discoverable by casual inspection. Analysis is necessary to establish those bids which are non-responsive and then to establish the low bidder from the remaining competitors. The government, if involved, or the owner, or both, mutually may then announce the low bidder; or, reject any or all bids, a condition generally written into the Invitation to Bid or Request for Proposal.

2.6 Award of Contract. If the low bid price is viable and reasonable, an award may be announced. The proforma contract which was included in the Invitation to Bid, or RFP, for general information, is finalized and perfected after receipt of bids, or RFP, including the amount of government support. At this juncture, the owner's management is in a position to calculate the required equity for its program, the contract price, the estimated delivered price, the amount of construction loans, fees, legal costs, plan approval, engineering, inspection, etc. A rather close estimate of the capital cost and the required cash flow is now available. The management is in a position to either use the authority previously granted by its board of directors, or refer back to the board for final approval. At this stage, any or all of the parties can find themselves in the position of having the venture frustrated by failure to obtain their respective boards' approval. Sometimes there are penalties involved in failure of the shipyard to conclude a contract, but the action is taken after deliberation and with full awareness of the penalties, monetary or otherwise.

In recent years, at least one shipyard refused to build the vessel on which it was low bidder, and more than one owner has forsaken construction after seeing the bids. On occasion, government has been dissatisfied with the results of bidding.

It is not unheard of to see the venture salvaged through negotiation after taking bids. Usually it is accomplished by changing the scope of work, or by changing conditions of contracts, guarantees, etc. At these times, the value of an independent, outside naval architect can be immense because his guidance to the owner is objective and professional; whereas the owner is more concerned with a business decision. If the contract is executed, a significant milestone in the vessel's construction is reached.

2.7 Shipbuilding Construction Contract Administration. The contract signing initiates a series of important events which are vital to the naval architect. Among these are the preparation of:

- Principal event schedule
- Plan lists and working drawings
- Production schedules
- Progress payment schedule
- Procedure and distribution memoranda
- Organization of plan approval
- Organization of inspection

From a planning standpoint, these are informative days and much of the success of the relationship between the shipyard and owner can be traced to the foundation built in those days immediately after contract signing. One of the incentives for negotiating rather than bidding a construction contract is that the relationship of mutuality of interests is enhanced as compared with the relationship between owner and a low bidder, particularly if the low bidder pared to the bone to be sure of continued work to keep a nucleus organization intact. Contract and specification interpretations are more apt to become important. It then becomes even more important to establish clear cut and unambiguous procedures, schedules, communication processes, etc., with particular preference to time periods mutually agreed upon to perform the functions, as for instance, the period of time to do plan approvals.

The example of plan approval is good to use as an illustration of how the relationship may work in practice because plan approval is the origin of a large part of the disputes generated in the course of completing the contract. This comes about because the production of the plans often places a strain on the shipyard. No one can expect the various yards to be fully staffed in the engineering sections for a peak production work-load at all times. The plans require concentrated effort at the outset of a new contract, whereas the production of the physical vessel stretches out for a long time, particularly if the contract is for multiple vessels, or is a forerunner of repeat orders. Thus, the job of getting the working drawings for the lead vessel is under pressure of time constraints. Frequently, portions of this plan work will be contracted out to naval architectural firms by the shipyard to meet schedule deadlines. Similarly, the use of unproven lead ship plans frequently forms the basis for claims with follow-ship contracts.

The most effective plan approval relationship is one permitting peer-to-peer personal contact of the owner's representatives with the shipyard's drafting division. If the functions are performed in different physical locations, much of the time usually allotted to plan approval is lost in the mails, or in transportation. Rigidly structured lines of communication also delay the resolution of valid comments because much understanding is lost in the transmission of information through several layers of supervision.

But, notwithstanding the foregoing, it is not likely that shipyards will, as a rule, welcome wide-open access to the drawing rooms by the owner's plan approval personnel. It is even less likely that a sub-contractor for working drawings, who has probably agreed to a set price for his service, will permit such access to his drawing room personnel. Accordingly, it behooves both parties to arrive at a formal memorandum on procedures establishing a standard routine for the preparation and distribution of all documents related to building and delivering the vessels contracted. Each class of documents should be identified; the number of copies to be sent to each named participant; the authority in each instance; the time for action on such submittals as require action; and the first, second and subsequent responses with time limits for each. All these things should be set down in procedure memoranda which list the various roles of the contractor and purchaser and may be used to guide the development of mutually agreed upon contracting procedures.

Two basic approaches to plan approval cover most instances:

1. Owner organizes a staff and carries out plan approval on site and/or at his place of business.
2. Owner contracts these services with a naval architectural firm, or some outside entity (as, for instance, another owner constructing vessels in the same yard).

Each of these approaches has its own advantages. The first permits the owner positive and direct control, easier adaptability to shipyard location, better record access (particularly post delivery), and an operation tuned to owner's desires. The contracted type of service presumes a broader spectrum of talent available for difficult problems and avoidance of labor impacts on the day-to-day operations of the owner's existing fleet.

Because the contracted method of plan approval is usually performed by a naval architectural firm, executed drawings are checked by marine electricians, piping by marine piping draftsmen, hull drawings by those knowledgeable in scantlings and stresses, etc. Although the billing rates may be higher than the owner could otherwise incur if he had his own cadre of plan approvers, he will not be faced with severance pay expenses or possible union involvement in reassignment when the plan approval activity is completed.

A detailed cost analysis is required for in-house versus contracted plan approval to see which approach will be less costly. Experience indicates there is no consistency of position. There is, however, a general preference by outside naval architectural firms to split plan approval work permitting certain selected drawings to be dealt with on site, while requiring the more important ones to be reviewed in the home office. Sometimes all first approvals are done at the home office, and all second and subsequent re-approvals are done at the shipyard. A prospective owner must consider this, particularly with the trend today of allowing something on the order of twenty days for first approvals (a very short period of time). A substantial part of this can be lost in transmission.

Because the shipyard has responsibility for the American Bureau of Shipping, the United States Coast Guard, and other regulatory body approvals, the submittals to everybody are sometimes simultaneous, which complicates approval by the owner who may find his actions being nulled by changes invoked by others. These cases generally are resolved in the procedures of resubmission and reapproval.

Proper planning of the approval function requires early submittal by the shipyard of the listing of working drawings (the plan schedule), and production schedule, as well as the principal event schedule. These will identify the level of effort required throughout the working drawing production and also, to the knowledgeable vessel replacement executive, establish the manpower needed for inspection of the physical construction, including testing of components at vendors' plants, and other travel assignments. This is important because it is quite within the bounds of reason that plan approval personnel will also be inspectors of the physical work in the areas of expertise they possess. In any event, the number of men, their assignments, and their transfer or separation, can be closely predicted from a work flow chart developed from the plan schedule and the principal event schedule.

Having stated the foregoing, the point can now be made that a builder who assumes an affirmative and amicable approach with the owner can effectively reduce the contention and the time of approval by establishing an open drafting room and encouraging give and take in the peer exchanges on approaches to settling interpretations of the specifications and bidding plans. A reasonable oversight climate, if tactfully practised by the owner's representative, can prevent backing and filling by draftsmen whose time is lost if the approach they are pursuing is doomed to disapproval. The owner must be ever alert to discourage his reps from meddling, indecisive waffling, or passing the buck. It is a delicate balance because ten thousand successful instances of cooperation can be forgotten over one instance causing disruption.

There is simply no substitute for an amicable professionalism in the relationship between owner and builder. Otherwise, both parties expend disproportionate time and effort in complaint correspondence and legal matters, whereas the time is better spent respectively in building vessels and in operating them efficiently. This is more easily attained under negotiated contracts and when there is a long history of customer/contractor relationship.

2.8 Progress Payment Schedules. Reverting to the narrative of events, the progress payment schedule will be spelled out in the contract. It is important both to the builder and to the buyer that major expenditures for both labor and materials be made early and the builder compensated before escalation takes its toll. With vessel prices escalating as they have, it is usual nowadays to see contracts requiring periodic payments twice monthly. Because a good deal of outside procurement is undertaken at the outset of a new building contract, the shipyard progress payment curve is heavy on the front end. Also, as the delivery date approaches, more labor and premium time are spent to finish. Accordingly, the progress payment curve is a *bath tub curve.* In series construction, the curve for each vessel is plotted against time, and a composite curve of aggregate semi-monthly payments can be predicted.

Actual billings will be made by the shipyard on the basis of completed *points* in the case of most United States shipbuilding contracts. In the usual approach to determining the status of completion, each shipyard generally assigns weighted points to the various sub-divisions, or crafts, including engineering, lofting, and all the production departments. These weighted points are the total manhour selling price and material costs, plus surcharges to cover overhead. The accountant will, as a rule, equate the contract price of the vessel to 10,000 weighted points, which total may include direct application of points for major vendors.

At any given time, the shipyard accountants, utilizing work tickets and material assignments, can produce a substantially accurate summation of completed points. If the other parties accept these completion figures, the progress billings are made accordingly. This affirmation process requires a judgment by owner's representative on-site as to the percentage completion. He is not under pressure to be absolutely precise because the process is self-correcting as the construction work continues. In U. S. yards there is also a nominal holdback, of say two percent during progress payments, part of which is paid upon ship completion and the balance after the warranty period.

International shipbuilding practice, other than in the United States, is quite likely to be different. In a majority of cases, the vessel is constructed under some form of national credit assistance or guarantee. These financing arrangements usually involve a twenty percent equity payment by the purchaser, with eighty percent financing through banks having a construction loan arrangement with the shipyard. The national government usually has a guarantee relationship with those banks so as to foster shipbuilding.

In such cases, the traditional approach most widely encountered is one requiring the purchaser to pay a portion of his equity at contract signing, followed by payments at launching and at delivery. A formula often used is ten percent of the contract price at contract signing, and five percent again at launching, and at delivery. By delivery time, the shipbuilder has taken down the eighty percent of contract price from the banks and the exchange of documents at delivery essentially transfers the bank mortgages from the shipyard to the purchaser. In matter of form, these are always newly written mortgages which are drawn to suit the flag of registry and the banks. As a matter of information, the usual term of these mortgages has settled around eight years, but, of course, terms and conditions change to suit the financial climate and national policy.

The take-down from the banks by the builder is in accordance with local custom, but is always based on a formula equating certain principal events to a percentage of completion.

Thus, the principal event schedule for such construction is quite important to cash flow projection. Additionally, certain public relations efforts are contrived, as a rule, to take advantage of keel layings, launchings, etc. It is quite normal in the United States to have the launching and christening combined, and to make this an occasion for improving relationships with political figures, union leaders, press, and customers. In many other countries, the same weight is not given to these events, and the gala might be saved for christening, which may take place at time of delivery (or even later) rather than at launching. One can foresee a decline and passing of the launching ceremony in the future because of the disruptive impact of the ceremony on production work in the shipyard, as well as the exposure to liability for claims which can be expected when crowds gather in heavy construction areas.

Section 3
General Aspects of Contracts

3.1 The Ship Construction Contract. Regardless of who is the architect of the Ship Construction Contract, be it builder or purchaser, it will include by reference and by direct instruction, additional prepared documents, such as contract plans, guidance plans, specifications, rules and regulations applicable to the construction of such vessels, and standard codes, statutes, general provision, and legal precedent. Additionally, the contract is subject to the jurisdiction of the law of the land which in itself is a matter of such impact that it is usual to specify in the contract which law applies. Frequently, the language of the contract can take an unexpected meaning when viewed from the position of the Standard Contract Code in one country, versus the code in another country. Such items as Statutes of Limitations vary from country to country (as well as from state to state within the United States). Also, there may be a distinction in interpreting whether the ship contract is a contract or a sales agreement (i.e., Sale of Goods Contract appears to be beyond legal redress much sooner than is the case for a construction contract). Similarly, time begins to toll at different thresholds for different contracts, thus the Statutes of Limitations begin to run from the occurrence of different events.

One can appreciate also the wide difference in litigation costs for both parties if the contract has been executed with a clause naming the applicable law. Otherwise, the parties are likely to resort to litigation in whatever location is to their own advantage, thus precipitating countersuits in other locations, all of which greatly increases the cost of dispute settlement.

As a minimum, all parties to a Ship Construction Contract have the right to expect:

- that the contract documents can be drawn with sufficient expertise to describe the vessel to be built in sufficient detail so that both buyer and builder agree on the end product with complete confidence;
- that the price stated in the contract is the delivery price, and the delivery price will be that price, plus only such extra costs as are explicitly permitted by the contract;
- that the vessel will be delivered on the day stated, unless modified by time extensions explicitly permitted by the contract; and
- that all perils which could impact on the foregoing are identified and the curative action and procedures in the event such risks obtain are stated.

These basic elements of the contract must be stated without ambiguity. It is the elimination of ambiguities which concerns the naval architect most. If he fails to be clear and precise, and if he fails to be consistent in the plans and specifications which are included in the contract, then he has failed to provide the owner who has engaged him with the expertise required to protect the owner's interest.

The rule of thumb is that he who writes the contract suffers the burden of any ambiguity. If the financial climate is good, and if the relationship between the buyer and the builder is amicable, most minor ambiguities are settled in day-to-day compromises. If the builder is in a financially disadvantageous contract, then even an item such as identifying a component by name and model becomes a serious problem should the manufacturer discontinue the model, or cease his business. Builder will invariably claim a price increase in such event, to which he may be entitled.

These things are equally important in foreign contracts, and the problem may be compounded, particularly if American standards and interpretations are depended upon to describe the vessel to be constructed in another country. Under these circumstances, it behooves the naval architect to be certain that contract compliance is viewed in the same light and with equal importance by the builder as it would be in the United States. Frequently, local contract law is entirely different and compliance in a legal sense may allow rather broad interpretation—a fact known to the builder, but not necessarily to the buyer. Obviously, under these circumstances, buyer will be disappointed and frustrated.

Furthermore, it does no good to specify American components in another country's contract if local law limits the content of the manufactured vessel to substantially local flag sources. In fact, such a contract will be fraught with ambiguities arising out of mandated changes of vessel components. In the business of buying vessels constructed in other countries, the owner soon discovers that the usual solution to coping with components is to supply the component built by a domestic licensee in the country where the ship is being constructed.

It is, ultimately, the long term of the shipbuilding contract that precipitates the adversities. In a high priced contract extending over two or more years, particularly in an escalating market, the time factor gives opportunity for prices to change, profitability to erode, managements and management strategies to change, etc. Thus a contract entered into in good faith can later create administrative difficulties.

A United States contract may be a private contract (no government aid); a CDS contract, or it may be one not eligible for CDS but yet eligible for Title XI Mortgage Insurance. Additionally, it may be a contract directly with a federal, state, or local government. It is incumbent upon the naval architect to be familiar with these types of contractual arrangements in detail so that he may advise his client correctly and knowledgeably relative to the scope of design and engineering functions the client will need.

In the private contract, it is quite likely that the builder will either have in hand, or will prepare, the specification and design. The owner's naval architect, in such case, acts as a review agent whose value to the owner is in insuring that the owner will be delivered the vessel to which he has agreed.

The naval architect must be thoroughly familiar with government assistance programs and all government design and engineering requirements which are caveats to the approval of such assistance.

As to the details in any construction contract, the naval architect can be very helpful to his client and his client's lawyers. In the international contract, for instance, local customs and usages may be much better known by the local naval architect who can interpret the accepted (and, therefore, most likely to occur) practice for his client. The client may also be totally unfamiliar with the regulatory bodies having jurisdiction over the construction and thus be disappointed in the results, unless the contract especially covers those areas where higher quality is important to the client. A similar area of competence for the naval architect is the arbitration clause. The naval architect may be more familiar from past experience in the procedures, and in the forum named in the contract.

It is, however, in the matter of pricing that the naval architect must have thorough competence. This is true in all shipbuilding contracts. A naval architect with international operations is doubly valuable to his client in the course of a United States CDS contract because he is best situated to provide substantive documentation of the comparable low cost foreign price estimate for purposes of his client's efforts to obtain a true differential to establish the price net of subsidy. If the naval architect has had recent experience in the foreign locality in which a client is building vessels, he may also further his client's interests by advising him on the currency to be preferred for the contract price; on export permits, documentation and fees; and on other delivery matters, including renegotiation.

As to this last, it is not unusual in times of building way shortages to see an owner with a perfectly good contract forced to renegotiate to secure delivery of his vessel. A contract is an understanding between the parties and can be mutually amended at any time, including the eve of delivery.

Similar factors are equally important in private contracts because the owner's board of directors is entitled to make its decisions based on accurate information submitted to the board by the officers of the naval architect. Overruns of time and money resulting from owner induced changes are disastrous developments in the eyes of the board of directors and everyone else involved. Overruns do not enhance the credibility of the staff which is forced to correct the price estimates; even if the causes could not be foreseen.

Additional cost can impact upon the owner, particularly when competitive bids are taken, because of the location of the builder. To the owner, all expenses incurred up to the moment the vessel arrives at its first loading berth, are part of the capital cost of the vessel; accordingly, a construction contract in a yard distant from the owner's regular service is not advantageous, unless the price differential or other factors cover all out-of-pocket expenses of positioning the vessel at a regular berth in the owner's service. Obviously, a possibility exists for a one-way charter, or other hire which could help offset the positioning costs, but, as a rule, the owner prefers to get a delivered vessel into his intended deployment as quickly as possible.

Before leaving the topic of contracts, one clause which invites some discussion is the so-called liquidated damage clause. Should the builder fail to deliver a vessel on the date of delivery stated in the contract, the presumption is that the owner suffers damage in that his business is disrupted. If the builder has been unable to document the reasons for the delay with sufficient proof to have been granted excused delays, those days not excused are multiplied by the dollar amount stipulated in the contract to arrive at the payment to be made to owner by builder. The contract identifies this dollar amount per diem as liquidated damages and not as a penalty.

Even as early as at contract signing, the owner can make a rather accurate calculation of the interest lost on his equity and the interest obligated on his construction funds. In a financial climate of high interest and big prices, interest can be $10,000 to $20,000 per day. It is illogical to stipulate damages of $3,000 or $4,000 per day in such a case. However, the liquidated damages are frequently stipulated at something less than actual out-of-pocket interest costs. Proposed liquidated damages are usually agreed to by contractor and owner prior to contract signing.

3.2 Non-performance of Contract. There are grave risks attendant in every shipbuilding contract, both for the builder and for the owner (and, where an interest appears, for government). In these days of large, complicated vessels, the costs are so high that a single contract which goes wrong may well precipitate a financial crisis, or bankruptcy, for one or the other of the parties.

The greatest risk is the one of non-performance. Non-delivery, cancellation and default are the most devastating events. In these cases, a party to the contract is unable to conduct the business in which he is engaged as planned.

Once the event of non-performance is upon the parties, it behooves both to move quickly and in accord with the procedures and remedies set forth in the contract to minimize the damage and permit recovery. It is as this point approaches that the naval architect's advice to his client is valuable. If he has been the inspection agent also, he should have been able to identify the crisis long before the other party has given notice. Moreover, he is in a position to broadly inventory delivered components and the progress status of non-delivered items, as well as the status of the work. His client will make his own decisions, but both the client and the client's legal advisors must rely on the best up-date of facts from a reliable and competent technical source.

It is a widely respected rule that in situations of one party damaging another in business events, both parties must thereafter at all times act in a way to minimize the damage. When working together to minimize the effects of a cancellation or failure to deliver, both will look to each other and their advisors to generate ideas for ameliorating the effects. Frequently, the naval architect is helpful in that he is aware of alternative uses for the engines, or the hull, etc.

At this juncture, it is necessary to comment on the government's obligations, if any, in the event of cancellation or default. The contract will specify a procedure for builder to give notice in event of default to the owner, and to the government. If the MarAd Subsidy Board is involved, it must within a time period (usually 15 days) be given written notice to undertake one of the following courses of action:

- Assume the payments required under the contract to be paid by the owner, or
- Elect to complete certain vessels and to optionally terminate other vessels in the contract, or
- Elect to optionally terminate all contract work.

Optional termination is a cease and determine action. The contractor must carry out prescribed functions under the control of the Maritime Subsidy Board to transfer title to the work (including all purchased items) to the Board and the purchaser, or to either, and to liquidate or preserve for future disposition the work (and the materials). This includes settling all claims of subcontractors, and other parties damaged, to the extent authorized by the Board.

Termination is expensive and the amounts payable include the cost of settling and paying all claims, plus a prescribed rate of profit (unless it can be determined that builder would have sustained a loss if the contract had been completed, in which event a profit would not be payable).

In the event of default by the contractor (failure to proceed on the contract with due diligence, bankruptcy, etc.), the owner and the Board, where their interests appear (provided the builder fails to cure the default timely) occupy the yard to perform the work and complete the vessels; or to sell the uncompleted vessels. These rights are stipulated to be in addition to, and not in substitution of, any rights the damaged parties would have in law or in equity upon experiencing the events of default.

3.3 Contract Work Changes. One aspect of contracting for ships not heretofore discussed is the matter of cost change estimates and adjustments. In days gone by, cost changes in U. S. yards were an unending source of disputes arising in part because government proforma contracts were basically *Rice Doctrine* contracts. The owner and the government could demand a change in scope and the builder was contractually bound to go forward with the work. The builder then had to rely on an administrative procedure to determine the cost of the work.

A predetermined profit was to be added to the adjudicated cost. The same administrative procedure determined the number of days of extension for the vessel. The builder did not have a contractual right to refuse the work, and he was not entitled to consideration for impacts on other contracts or other work. This is the essence of the Rice Doctrine.

Under the U. S. Merchant Marine Act of 1970, the builder does have a right to refuse the change work if he has valid reasons, including a failure of the owner and the contractor to agree on what constitutes a proper cost of the changed work. In this respect, the contract provides for an equitable adjustment to the contract price, thus opening the door to inclusion of the dollar impacts upon unchanged work and the contractor's other business. This is essentially the transfer of the philosophy of the Armed Services contract to the shipyard's commercial work. Accordingly, if the contractor wants to do the change work, he will negotiate with the owner amicably to minimize the consequential costs of doing the change.

If, from his own business requirements, the contractor finds the change to be disruptive, he may claim substantial amounts over and beyond the actual costs, thus forcing the owner into a position of dropping the item. In cases where the work must be done to meet regulatory body rules, as an example, the owner may order the work to be done even though the price has not been agreed upon; but, in any event, the final agreed price is an equitable adjustment, permitting consideration of all aspects of cost. This latter situation is described under the "Essential Change" clause of the 1970 Act contracts. These procedures substantially follow the rights of the owner and shipbuilder as expressed in non-U. S. shipbuilding contracts.

An owner may, if he elects, accept the vessel without the work changed, and then do the work after delivery. This would be based on a trade-off analysis involving comparison of out-of-pocket monies in each instance.

Contracts which do not involve the U. S. Government are less rigid as to the construction of the change clauses. In other countries, the parties generally rely on clauses which stipulate mutual agreement on cost and time impacts of changes prior to any work being undertaken. Essentially, these are amended contracts. In cases where work must be undertaken in any event, such as regulatory body work, provisions are generally incorporated in the contract for arbitration. In other words, the change becomes a dispute which must be arbitrated.

3.4 Disputes. Disputes in the U. S. are frequently agreed to be remanded to an arbitration as codified by a reçognized arbitration association. It is also possible for disputes in U. S. CDS contracts to be remanded to an arbitrator or arbitration panels. Both parties to the contract must agree to be bound by the arbitration, or in lieu of thereof, proceed to the Courts for recourse. In the United States, it is usual to refer to the Commercial Arbitration Rules of the American Arbitration Association.

It is important to the parties to the contract to understand the rules of arbitration, to understand the forum for arbitration, and to amend the contract to the extent mutually agreeable to limit the expense and inconvenience to both parties. Unless the owner guards against it, the builder will desire that the arbitration be at a location convenient to him, which is excellent if it is also convenient to the owner. Usually in construction in the United States this is not a problem unless the owner is on one coast and builder on another. In non-U. S. construction, it is a more difficult situation if arbitration is at the site of the builder. Frequently, in foreign building contracts, arbitration is stipulated in a neutral country such as Switzerland, or in a recognized arbitration capital such as London.

If the non-U. S. construction contract permits either party to sue rather than arbitrate, the contract must be carefully negotiated; otherwise, lawsuits in the builder's country may tend to advantage the builder who at a minimum need not underwrite substantial travel expense and two sets of corresponding lawyers. Furthermore, local law may not permit a non-national citizen the same rights in Court. In at least one shipbuilding country, for example, a party to the contract has no rights in the Courts unless the contract was preregistered at a cost for registry determined as a percentage of the total value.

The practical aspects of the non-national contract disputes procedure may be so overwhelming in cost impacts as to effectively undermine the disputes procedures in the contract, a fact which may be more apparent to one of the parties than to the other.

By and large, the greatest impetus to settlement of disputes is the high cost in dollars and time of permitting the dispute to be settled by others. To begin with, once the dispute leaves the hands of the supervisors of both the shipyard and the owner, knowledge of the facts grows increasingly less important than knowledge of the law or other extraneous impacts. At this time, the professional advice of a naval architect in suggesting alternative disposition of a problem may eliminate the cause of a dispute, or, if the work must be done, provide owner with the facts whereby he can make a decision to do the work after delivery.

The aim of every owner should be to associate himself with specifications and plans sufficiently explicit to permit a zero change construction effort. Additionally, one cannot escape the fact that the faster one can have a vessel built, the less the opportunity for escalation and dispute. The benefits of high productivity are substantial. The owner is most always advantaged by early delivery and it is quite obvious on the face of it that builder overheads are reduced below those used in the original estimate. The cash flow is at higher levels giving the builder more cash to manage, and reducing the owner's construction interest payments. All the impacts of labor contracts and regulatory body changes are mitigated because the impending rules are visible relative to the vessel construction deliveries.

Unfortunately, however, shipyard productivity in many yards is not generally improving on a broad front. One of the identifiable reasons comes under the heading of social legislation. And, it is not the traditional agencies, such as the U.S. Coast Guard and the American Bureau of Shipping, Lloyd's, the Veritas bodies, etc., but the environmental groups and the safety of work groups that create new constraints. For example, in the United States, the Occupational Safety and Health Act (OSHA) impedes production work by requiring procedures and approaches which delay the work and add significantly to its cost. Shipbuilding has always been hazardous work, and shipyards have always fostered safety programs, but some work has always been treated as involving hazards, even after all practicable precautions have been taken. OSHA and other social legislation directly affects ship construction costs as well as ship repair costs aboard an operating vessel.

3.5 Other Naval Architectural Roles. Contracting arrangements for vessels are very intricate and very complex. Whereas every vessel must perform its design function (or the function for which it is converted), the action impetus for contracting may be as remote from operation, as, for instance, a financial investment. Profitability may be measured in part against the financing expertise for construction rather than the voyage returns in the life of the vessel.

A naval architectural consulting organization would be advised to be informed and knowledgeable in changing financial aspects of vessel construction including all government assistance programs, tax shelter and other tax aids, equity financing, debt structuring, and various current chartering agreements. It should be familiar with documentation and its possibilities insofar as a client has a free choice as to the flag of operation. In this respect, it must have the acumen to advise on, and to facilitate, both direct construction and off-the-balance-sheet acquisition or contracting. If possible, it should be sufficiently well versed in maritime law to assist the owner in contracting in various shipbuilding countries with a view to reducing the owner's risks and assuring him of timely delivery. This facet of his expertise requires an understanding of marine insurance and international finance in that the risk of late delivery can be ameliorated sometimes by insurance and sometimes by purchasing futures in currency.

The foregoing account of the fields of expertise needed to be a well-rounded naval architectural organization is by no means all inclusive, but it suffices to say that in addition to the foregoing, the firm must know the engineering and technical aspects of its business so that it can produce specifications and plans for designs which will do the job for the owner. It must know how the work should be done by the shipyard, how to inspect it, how the trials should be conducted, and how the efficient operation of the vessel's plant should be conducted.

To the uninitiated, the naval architect's work is narrowly defined and his product is thought to be plans and specifications. To those who know the intricacies of ship construction, the naval architect's really important work may well be in the general counseling areas previously mentioned.

3.6 Escalation. The post World War II era of U. S. Government-aided ship construction provided an opportunity to experiment with differences between fixed price contracts and adjusted price contracts. Prices were not escalating madly, and labor at the shipyards was not overly militant. The adjusted price inquiries were based on using rather simplified indices provided by the Bureau of Labor Statistics (BLS). Accordingly, the bidding was rather stylized in that the fixed price bid was directly related to the adjusted price contracts being the latter price increased by the BLS trend, multiplied by the years of construction. On balance, the contractor offered the prices, one with the risk of escalation on himself, and one with the risk on the owner. Even with such a relatively straightforward approach, the adjusted price contract was unpopular, and the preference for the fixed price contract was so overwhelming that all the contracts in the late fifties and the decade of the sixties were fixed price contracts, which for owners, are considered preferable. In fact, until recently, MarAd insisted upon fixed price contracts.

The experience with double digit rates of inflation in the early seventies forced the shipyards to insist on escalation clauses in commercial contracts. Those yards building vessels for the U. S. Department of Defense had already been using some escalation clauses on government work. MarAd was reluctant to return to these escalated contracts because of the opportunity for disputes and the difficulty of meeting budgets. Nevertheless, the drafting of fair and reasonable escalation clauses began and today the escalated contract is much more a likelihood than a fixed price contract.

The general nature of escalation clauses is to identify the categories of costs which may be subject to inflation and to establish a fair way to adjust the price of the vessel to satisfy both parties. For these purposes, it is important to establish the content of labor, materials, overhead and profit in the base price of the vessel. Whatever proportion of the price is determined to be man-hours, the escalation should be limited to the amount of these man-hours yet to be expended at the expiration of the labor contract held by the shipyard when the contract is signed. Thus the theory is that up to the expiration of the labor contract, the shipyard is fully informed on the total cost of its labor force. Accordingly, the labor escalation can be determined to start at a certain date within the life of the contract. The escalation for labor is then usually determined by using the BLS, or other appropriate index to establish a rate of increase, and this rate is applied to the part of the total contract price attributable to unforeseen labor increase. This is treated by the time applicable and the total is determined by the rate multiplied by the hours so affected.

However, it should be pointed out that the BLS indices reflect only the labor *earnings;* they do not reflect labor *cost*. The cost of labor to the shipyard includes fringe benefits, and social security contributions which frequently change more drastically then the rate of pay. Thus, the shipyard may still be trapped by rising costs over which it has no control or recourse.

Materials escalation constitutes the major exposure for added cost. Most major ship component suppliers today have a long lead time on orders and have come to insist on orders being placed on a price at delivery basis. This places the shipyard in a position of relying on quotations at price setting time, which may be totally inadequate at delivery.

The clauses on materials again are usually based on the BLS Shipbuilding Material Index, which is a weighted index composed of three material indices. The contract may require the shipyard to order all large components at the outset of the work, rather than at the times which might otherwise be more prudent. In such an event, the risk of heavier than usual commitments in event of cancellation and the risk of shutting off desirable changes must be recognized.

The matter of overhead escalation is not clear. In U. S. shipbuilding work, the contractor establishes an overhead rate which is applied to the man-hour labor rate, and this overhead rate is used throughout the contract. Shipyards are chafing since overhead is escalating because of tax and energy increases, management labor increases, safety and other regulatory costs (OSHA), etc. Generally, the escalation clauses do not mention overhead, probably because it is felt overhead is reflected in the labor escalation. In the future, more detailed commercial escalation theory will treat overhead also in a manner now recognized in the newer U. S. Navy shipbuilding contracts.

Profit is never included in commercial escalation clauses. This exclusion is considered a disincentive necessary to encourage productivity and early delivery.

Escalation clauses are negotiated for each contract with

some degree of business risk ensuing for both the contractor and owner.

In a contract with escalation clauses, the owner must be vigilant in monitoring the final estimated delivery price, particularly in consideration of changing operating and market conditions. The naval architect should be of valuable assistance to his client in managing the ship cost, leaving to the owner the analysis of whether or not the escalated price will permit him to conduct his business in a viable way. If the owner determines that accepting delivery at a higher than anticipated price is tantamount to destroying his profitability, the naval architect will be needed to help the owner in determining the most advantageous course of action.

Section 4
U. S. Government and Shipbuilding Contracts

4.1 Government Assistance to the Maritime Industry. Since a significant number of readers of this section are engaged in U.S. shipbuilding activities, it is necessary to discuss the peculiarities of the U. S. Government-maritime industry relationships. Here the term "government" does not apply to U. S. Navy, U. S. Coast Guard (USCG) or other federal and state government procurements.

Government's interest, and its administrative acts and instruments, pertaining to ship construction arise out of the Merchant Marine Act of 1936, as amended. It is important to quote the Declaration of Policy as written into the 1936 Act because the activities of MarAd will be directed to achieving this policy within the confines of the statutes bearing on ship construction:

> "Title I—Declaration of Policy Section 101. It is necessary for the national defense and development of its foreign and domestic commerce that the United States shall have a merchant marine (a) sufficient to carry its domestic water-borne commerce and a substantial portion of the water-borne export and import foreign commerce of the United States and to provide shipping service essential for maintaining the flow of such domestic and foreign water-borne commerce at all times, (b) capable of serving as a naval and military auxiliary in time of war or national emergency, (c) owned and operated under the United States flag by citizens of the United States insofar as may be practicable, (d) composed of the best-equipped, safest, and most suitable types of vessels, constructed in the United States and manned with a trained and efficient citizen personnel, and (e) supplemented by efficient facilities for shipbuilding and ship repair. It is hereby declared to be the policy of the United States to foster the development and encourage the maintenance of such a merchant marine."

The significant government financial aid programs in which naval architects and marine engineers are usually involved are:

- *Construction Differential Subsidy* provides for government to pay the shipyard the difference between the reasonable price of constructing the ship in the selected U. S. yard and the cost of constructing the same ship in a recognized competitive foreign shipbuilding center. Thus, the prospective shipowner is placed on parity with his non United States competitors who have access to non United States shipyards.
- *Title XI Mortgage Insurance* provides for U. S. Government guarantees to the private lending organization for repayment of the mortgage loan (and interest) granted to a prospective shipowner. This guarantee results in favorable financial terms for the owner.
- *Title VII Procurement* involves 100 percent government funds and is generally invoked in times of national emergency.
- *Capital Construction Fund* (CCF) permits a qualified U. S. operator who has an established CCF to draw from this fund the equity required to finance his investment in a new vessel thus allowing him to use deferred tax dollars for this purpose.

There are other financial aid programs for the shipowner for which reference should be made to the Merchant Marine Act of 1936, as amended.

The contract between the Maritime Subsidy Board and the contractor (builder) begins with a preamble certifying the findings that the vessel construction conforms to Title V requirements and conditions. The total price is stated, plus the cost of national defense features, if any. The amount of subsidy to be paid to the contractor by the Subsidy Board is stated in amount and as an equivalent percentage of the total price. A stipulation is included that the remainder of the price (over and above the subsidy) is to be paid by the purchaser (owner) as required by the construction contract. Other items covered are the "Buy American" provisions, the functions and rights of the Board, the right to direct national defense feature changes, value engineering, rights of purchaser and Board to engineering and design data, hold harmless clauses for actions arising out of contractor's actions, total loss clauses, limitation of liability, patent infringement, optional termination by the Board, default of purchaser, default of contractor, fees, assignment of claims, fair labor clauses, renegotiation, disputes, etc.

The contract between the Maritime Subsidy Board and the purchaser begins also with the preamble mentioned above. The full price is stated and the subsidy rate. In this contract, the Board agrees to pay the subsidy at the same rate for any changes determined to be eligible, and, further, an amount at the same rate for a total stipulated amount determined to be the sum required for design work, plan approval and inspection, including travel, communication, and other office expenses directly related to the construction. This contract also authorizes the purchaser to withdraw his portion of the vessel cost from his CCF (less any trade-in allowance). If the purchaser is also taking Title XI loans, this contract stipulates such proceeds shall be deposited in the statutory fund that the owner is using for the construction.

Other clauses cover responsibilities of the owner after delivery and acceptance of the vessel (for the statutory life of the vessel). One of these clauses commits the owner to carry insurance covering total loss to the full extent of his investment at delivery. Government also requires the owner of a CDS vessel to agree to insure government's invested funds on the vessel should the need arise. MarAd agrees to pay for the extra premium.

Another clause binds the owner to American flag documentation for twenty-five (25) years and for a longer period if Title XI funds are still unpaid at the end of 25 years. Also, the owner agrees to repay subsidy whenever the vessel aided by CDS is used in domestic service. The pay back formula is based on a relationship between revenues derived in the domestic portion of the voyage and revenues for the total voyage.

Should the vessel include national defense features, the owner agrees not to alter such features without approval (unless he actually improves the features). Furthermore, if he makes use of national defense features, he agrees to pay the depreciated foreign cost of such features.

The owner is bound not to sell the vessel nor lease, nor charter it for a period of ten years or longer without government's prior approval. Owner is responsible for all the agreements and cannot assign the obligations without MarAd approval.

The contract between contractor and purchaser is structured to describe the vessel to be built, deal with the financial aspects of the work, cover the events of non-performance and the relief available, and preserve the commitments written into the other two contracts. The remedies for the situations that obtain in contractual non-performance or impacts on work are discussed elsewhere in this chapter.

The Title XI contract, if not also a CDS contract, is substantially simpler. Government is far less involved in a Title XI contract than a CDS program. Evidence of this is indicated by the fact that one contract, not three exists and that one is between owner and shipyard. Government's interests are protected by a clause in the contract and through a review of plans and specifications submitted with the Title XI application. Certain key working plans are transmitted to government during the construction period, but government approval is not a prerequisite for initiating work.

The naval architect's work is somewhat lessened by the lighter load of plan and data submittals and resubmittals; but, if the naval architect is also the owner's approval and inspection agent, his field work may be somewhat increased for purposes of certifying percentage of completion for progress payments.

As soon as an owner files an application for aid in construction, MarAd begins detailed reviews of submittals to establish consonance with the policy objectives and statutory fiat. Both the CDS Application and the Title XI Application are designed to assist MarAd in making the necessary determination as to citizenship, ability and experience of an applicant to meet the financing requirements, suitability of the vessel for the intended operation and to meet the requirements of the foreign commerce of the United States. MarAd further conforms with the 1936 Act by submitting ship design plans and specifications to the Navy Department for examination and suggestion. It bears repeating that MarAd's activities relative to approvals must be faithful to the 1936 Act, as amended; therefore, the functions will faithfully track the language of the Act. Accordingly, an owner applying for CDS or Title XI serves his own ends best by cooperating with MarAd's data requests fully and timely.

The applications also provide for further flow of detailed information, particularly as to the design of the vessel. Much technical review is carried out in this period and is very helpful to the owner in that the exchange of comments produces a better design.

When deemed appropriate, MarAd arranges for a valuation of the vessel under an assumption that the same vessel is to be built at the same time in a foreign low-cost shipbuilding center. This requires a determination of what foreign country qualifies as the low-cost shipbuilding center for the type and size vessel under construction, after which an estimated price is established by the Maritime Subsidy Board. This price is extremely important as it is the basis for calculating the construction subsidy to be given.

Needless to say, this method of determining the actual foreign price is imperfect because the only way one could establish the true price is to bid the vessel in foreign yards under a covenant to place the construction irrevocably with the low bidder. In the absence of such an understanding, a foreign estimate, however professionally drawn, may be somewhat different from the price as determined in the market place. Nevertheless, this is the best system available and it works well enough as proven by past experience. An owner knowledgeable in foreign contracting may submit his own report of the estimated price and MarAd will give it consideration as one of the data sources from which MarAd determines the foreign price.

As the contracting time approaches, MarAd's calendar of events requires staff recommendations to approve the design to be submitted to the Maritime Subsidy Board. When so indicated, the Board will approve the design and price. This Board action specifically sets the conditions of approval and the scope, i.e., approval of CDS at a determined rate for a set number of vessels, approval of Title XI commitment, etc., including approved estimated amounts for engineering the design and plan approval and inspection. It is not unusual for the Board to set the manning and limit subsidy to work to accommodate only the approved manning.

From MarAd's standpoint, the next step is contracting. This has different aspects depending on whether it is a CDS program, a Title XI, a CDS with Title XI, etc. Involved in a CDS program will be a MarAd contract with the purchaser, a subsidy contract with the shipyard, and a construction contract between the shipyard and the purchaser. If the owner is using existing vessels as his equity, a trade-in agreement is necessary and a use agreement will be required if purchaser intends to operate the vessels until the new construction vessels are delivered. These five basic documents are drafted by the government and are subject to modification on a special provision basis. Where applicable, general provisions are standard and made part of the contract by reference. If any article of the General Provisions is unacceptable, amended versions when agreed upon are stipulated in the Special Provisions.

It is obvious from the foregoing discussion that there is extensive involvement of MarAd in the contracting process. However, it should be emphasized that MarAd is not a party to the ship construction contract even though a construction differential subsidy is being provided. This has been the case since 1970 and it has resulted in much less government involvement in the construction contract, where most of the troublesome issues are left to the two principal parties, that is, the owner and the contractor.

For general business reasons and for facility in handling construction funds, it is to the advantage of the owner to have a construction reserve fund. If the owner has not previously applied for, and had been granted an agreement for, such a fund he will have to make another application to MarAd at the time of construction. Further discussion of the details of financing are included elsewhere in this section.

4.2 Maritime Administration Construction. MarAd frequently acts as an agent for federal departments or agencies.

In this role, MarAd approximates the activity of the naval architect in that it may design the vessel, write the specifications, invite bids or negotiate the construction cost, perform the plan approval and inspection, accept delivery, and then transfer the vessel to the sister agency or federal department. This function is rather limited, but not often appreciated by casual observers because of the absence of publicity.

Examples of this function of MarAd are various research vessels for different federal functionaries including NASA, the National Ocean Survey, and others. Other vessels which may be built in this way are the training vessels for the federal and state maritime academies and the Alaska Indian tribe service and replenishment vessel (feeder ship).

Additionally, MarAd contracts from time to time for itself, thus assuming the role of owner. Large scale contracts for new merchant vessels are made from time to time as the political climate may dictate under Title VII of the 1936 Act, as amended. Use of Title VII procurement generally obtains in times of national emergency. The war-built C-type vessels were, for the most part, built under Title VII, and sold to private owners after the war under the Ship Sales Act. Also, the construction of the *Mariner* fleet was carried out under the Title VII. Again, except for those *Mariner* vessels delivered to the Department of Defense for use as tracking vessels, and for other military purposes, the remaining ones were finally made available to private owners under the Ship Sales Act.

MarAd also has built experimental vessels for research purposes under this Title. The nuclear ship *Savannah* and the hydrofoil ship *Denison* are good examples of this function.

4.3 U. S. Department of Defense Contracts. There are a great many differences in contract terms and practice between commercial and U. S. Department of Defense ship construction. As previously mentioned, the basic difference in the past was a matter of contract philosophy with commercial work being done under a Rice Doctrine contract, whereas defense work was under an equitable adjustment contract. This difference has been virtually eliminated because the 1970 Act provided for equitable adjustments on commercial contracts. However, disputes arising in commercial work now have direct access to the Courts, whereas disputes in defense work must pass through an administrative procedure involving the Armed Forces Board of Contract Appeals prior to establishing standing in the Courts. This administrative procedure is a time consuming process.

Whereas MarAd and the owner are not prohibited from furnishing material to the contractor for use in constructing the commercial vessel or vessels under contract, they almost never do because of the inevitable complications and the probability of disputes and claims. DOD, on the other hand almost always provides a substantial portion of the finished vessel, particularly the weapon systems, guidance controls, communications, electronic early warning devices, and other long lead or multiple procured equipment and material. Faulty government furnished equipment (and its late delivery) is a major cause of the increasing number of contract changes, hence the shipbuilder should purchase these to the extent possible. He would thus have responsibility for acquisition and performance.

The foregoing general statements are meant to invite the naval architect's attention to the fact that DOD contracts are complicated and rigidly restricted by the Armed Services Procurement Regulations (ASPR). These regulations are formidable in scope and legalistic in detail. Anyone having dealings contractually with DOD should refer to the Code of Federal Regulations, Title 32, and gain access to a service updating the regulations on a current basis and giving synopses of the cases determined by the Armed Forces Board of Contract Appeals (AFBCA).

It is of interest to note that all government contracts have been subject to renegotiation under the Renegotiation Act of 1951. This statute technically expired September 30, 1976, and the sitting Congress is debating a number of bills to revive and extend the powers of the Renegotiation Board. It will be important for anyone using this book to verify the situations in respect to renegotiation for contracts executed under ASPR, Public Law 87-553, Federal Procurement Regulations, and such parallel regulations adopted by the U. S. Coast Guard, and other governmental agencies.

4.4 U. S. Coast Guard Construction. This agency maintains its own design capability and as a rule, prepares its own concept designs. The federal procurement regulations apply. However, since the Coast Guard is not an agency of the Department of Defense, its procurement is not limited by the *Buy American Act*. Accordingly, it is possible that solicitation of the Coast Guard for shipbuilding bids may be seen by the political sector as an opportunity to stimulate foreign trade, leading to the submission of bids from outside the country. Although a recent bid for icebreakers indicated a foreign shipyard to be the apparent low bidder, no award was ultimately made when a full appraisal uncovered that the submission did not include a number of the systems which were considered integral to the contract by American bidders, and the foreign bid was eventually withdrawn. Considering the long life expected of Coast Guard cutters, and the consequent problems of the procurement of parts from foreign sources over the life of the ships, it is problematic that this agency will build outside the United States.

Because the USCG may in time of war become an element in the Department of Defense operating within the Navy, in the construction of its own vessels the standards employed are similar to or are identical with naval construction.

The commercial regulations which the Coast Guard publishes applicable to merchant construction as a regulatory agency are therefore not employed in the construction of its own vessels.

4.5 Private Construction Contracts. Having discussed American shipbuilding from the government aided construction view, a few special words are fitting as to private construction in non-U. S. yards, the American-flag documented foreign built vessel, and unaided construction in U. S. shipyards.

Private construction in any shipyard is a matter of mutually agreeable contract documents. These are generally oversimplified, providing a brief description of the vessel and establishing the class to which built, the price, the payments to be made, and incorporating a financing commitment for the portion not paid during construction. If the vessel is to be American-flag documented, the contract must provide for materials and workmanship to U. S. Coast Guard and other U. S. regulations, and the filing of citizenship compliance documents as they become timely. Such requirements when ordered into a vessel under construction in a non-U. S. shipyard may require importation of U. S. built material capable of meeting these regulations. This could impact upon local laws limiting the vessel's content as to imported goods.

Non-U. S. contracts generally specify the currency in which the contract is payable. In times of inflation and monetary instability, the professional naval architect acting for his client may do well to encourage his principal to hedge against these financing costs by buying the currency advantageously to fix the contract cost (at least for the equity portion). However, in some countries this may not be permitted.

A substantial tonnage of construction which is totally unaided by government financing programs takes place in U. S. shipyards. A variety of reasons is involved in making a decision to construct vessels in a U. S. shipyard at the full price without any benefit of aid. Usually it is a time constraint which causes a decision to build using one hundred-cent dollars. In the pre-embargo tanker tonnage build-up, a number of tankers were ordered in U. S. yards under private contracts even though going through the time consuming application route seemed certain of achieving a CDS contract.

Small vessels and specialty vessels are built in large numbers in the United States without aid. Drilling rigs and their work and service boats are in evidence at this time in almost every small facility, and also in many of the largest facilities. The price differential on these types of vessels is not as great as it is on the conventional large vessels, primarily because the productivity advantage in steel erection held by non-U. S. shipbuilding centers is not a factor in small vessels.

Occasionally a desire to avoid the restrictions which attend taking government aid, causes an owner to build independently. And some services such as the domestic service (Jones Act), by statute are eligible only for Title XI assistance.

Vessels built in U. S. shipyards without government aid may be documented under foreign flag. Such vessels can be re-documented under the American flag, but current statutes limit their access to certain government impelled cargo for a period of time upon return to the American flag.

Section 5
Additional Elements of the Contracting Process

5.1 Estimates. It is important to a naval architect to be familiar with shipyard techniques of price estimating so as to give his client good factual estimates prior to the bid or negotiations (for purposes of filing applications). It is also important for him to guide his client in negotiation for purpose of obtaining the best price, or to advise his client that further negotiation is unproductive. He should have alternative strategies to suggest in such a case.

The owner/client usually relies on information in the market place to be generally knowledgeable as to the price of vessels but he does not usually require in-house competence on precise shipyard estimating procedures. The naval architect should have such competence, however, because from time to time the shipyard will also be his client. He must also be able to help the owner in cost estimating of changes to the contract.

Earlier in this chapter, reference is made to a paper presented at the SNAME 1976 Annual Meeting by D. M.

Mack-Forlist and Richard A. Goldbach entitled, "Bid Preparation in Shipbuilding." Readers are urged to refer to this paper for a full understanding of the shipyard estimating procedure. However, a brief, general discussion of this procedure is included here for the reader's convenience.

The shipyard's contract bid estimate is a large scale computation based on the work and materials as depicted in the plans and specifications and the conditions for doing such work at that particular site and at that particular time. It begins usually with a competent staff listing all material and applying the standards of that builder for units of labor per unit of material.

Such summation, if done with reasonable accuracy, will produce a total of man-hours, which, by inspection, in the light of experience, can be judged as to general acceptability. A review of the plans and specifications to identify any flagrant omissions is standard procedure in the checks of man-hours.

The material costs are generally priced out by requesting vendor quotations on the basis of projected delivery times.

The raw ship cost thus established in the early development of the contract bid estimate is compared against bids and recent construction of ships most closely approximating the ship to be built. The vessels' costs are adjusted by pricing the differences and applying these to tune the comparison. Where costs and return labor are used (when available) a more meaningful comparison can be developed. If the comparison gives credence to the detailed cost developed in the first instance, confidence in the bid estimate will obtain. Otherwise, checking and rechecking will be needed to be certain that the lack of correlation is valid.

A large number of factors are then applied to the bid estimate, an important one being changed conditions of construction, which requires a strong background of experience. The learning curve, together with the experience in labor stability, must be applied to the follow-on ships in order to produce a realistic overall, or average cost. The known impacts, such as increases in labor rates in existing contracts, must also be applied.

The shipyard's management uses the foregoing and other such data as inputs for deciding on the price to be bid, or the price to be negotiated. They also consider the market conditions, including the availability of alternative work; prospects; changing productivity trends; changing overhead burdens; availability of craft and labor.

These factors will influence the management in determining whether to bid at all, whether to bid on part or all of the desired order, what profit level to seek, etc. Assuming the decision is to build, then a price will be established but it is important to understand that the estimate, while a necessary step in arriving at the price, is not a *sine-qua-non.* In fact, the price could well be below the estimate if maintaining a nucleus of skilled workers is more important than making a profit on a particular contract.

5.2 Award Procedures. The determination of the price of the proposed new vessel, whether arrived at by bidding or by negotiation, is an important event in the vessel acquisition continuum. If the vessel is to be government aided and American flag documented, the elapsed time between setting the price and signing the contract may be prolonged because of the structured formalities, availability of financing, changed conditions, budgets and authorizations, etc. This can be a busy period for the owner and his naval architect. If owner finds the price acceptable and is assured of his own financing, he may yet be quite concerned as to the non-U. S. shipyard cost determination for purposes of establishing government support, and direct his naval architect to provide a supporting study attesting to a parity price in a non-U. S. low cost shipbuilding center. If government's determination varies from owner's estimate of non-U. S. price, the matter has to be amicably settled or the financing adjusted to suit any off-budget CDS determination; or the projected vessel must be redesigned and renegotiated to suit the facts of life. No rigid rules exist as to renegotiation of the low bid, or the originally negotiated price, but it is generally conceded that owner and contractor may alter the scope of work by mutual agreement to permit the price of the work to fit the parameters of financing stipulated as requisite conditions to the contract signing. If such adjustments cannot be worked out, the owner must either start afresh, or choose an alternative approach. The alternatives are usually radical changes in philosophy, including different approaches to service, such as feeder ship concepts, giving up services or ports of service, chartering in lieu of owning or changing flag documentation.

One method of temporizing is to convert existing owned tonnage or newly acquired tonnage. In such cases, the naval architect must begin work on an ad hoc basis in order to regain time lost on the aborted design. Conversions are intended to be short term expediencies to permit recouping finances for building new tonnage. Nevertheless, many owners find the conversion method so viable as to continue the cycle over and over.

Conversion compounds the practical problems for the owner in documentation and financing because all the caveats agreed to at original construction still apply where the interest appears and to these caveats must be added all the waivers and new commitments for the conversion work. Accordingly, the U. S. owner and his naval architect must be careful, in selecting vessels for conversion, that they do not overlook the restrictions; as for instance, the original vessel may have been built with CDS funds, whereas the converted vessel is to trade under the Jones Act. In such a case, the depreciated subsidy would have to be paid to the government if approved, creating an elevated capital cost (and perhaps make the intended service unprofitable). Similarly, if the unions covering the manning are to be different on the finished vessel, then the labor contracts applying to the existing vessels must not preclude such a change.

If government subsidy support is not a condition of the new construction, the award is generally made rather quickly. Even if there are some unsettled financial items, a contract, or letter of intent conditional on obtaining certain waivers or approvals, can be executed forthwith; thus permitting timely and orderly progress with the work.

As to non-U. S. shipyard contracts, the customs and practices vary from nation to nation. An award may be very informal and consist of little more than an exchange of correspondence, or it can be ritualistic and formal requiring notaries and registration.

5.3 Plans and Specifications. From the standpoint of costing and contractual arrangements, it is important that the plans and specifications be clear, concise, and accurate in communicating to all parties exactly what the owner expects in the delivered vessel. It may come to pass that the naval architect describes an important item differently in various sections of the specifications, or differently in various bidding documents. Such anomalies are a recurring source of disputes, particularly when the owner provides the specifications and plans, and subsequently, the builder experiences overruns in buying the materials and in doing the work.

In the past, naval architects and owners have relied on general clauses to enforce requirements intended to improve the quality of the work. However, the value of the general statements has been eroded because of the trend toward the philosophy that a builder may follow his own way of doing work, except when the specifications and plans specifically require something different.

Similarly, the former practice of reserving large segments of plan details for development by the builder after contract signing has been a source of trouble, in that the developed work might differ from bidding documents and allotted estimated costs. In such cases, the builder frequently cited the bidding plans as proof of owner's intentions. The comparison of the developed work to the bidding plans was used as the basis for a claim for the benefit of the builder to the extent builder could make a case for a change.

To avoid such confrontations, some owners and naval architects have resorted to labeling all bidding plans as guidance plans for the purpose of assisting the builders in preparation of the working plans, which, when approved by owner and regulatory bodies, become the true description of the vessel to be built.

This discussion simply reinforces the premise that an owner and his naval architect must avoid every source of ambiguity that can be identified. The most effective way to do that is to purchase an already completed vessel, but the opportunity to do so is quite limited. Speculative building in an inflationary period is rarely practiced by shipyards. Speculative building by entrepreneurs for resale is sometimes encountered in the tanker segment, but not often on the dry cargo side.

An effective approach is to build under contract to a set of specifications and plans developed in advance by the builder. The burden of righting the ambiguities is then on the builder. Assuming the prospective owner is involved with a bona fide replacement program, and assuming he is properly informed at the outset as to the details of the vessel, he will be less inclined to raising disputes than is the builder who may be irritated by overruns. There may be other sources of disputes, but those generated by ambiguities and their resolution will be decreased. When plans and specifications prepared by prospective shipbuilders have been issued for competitive bids, it has been historically demonstrated that the low bidder is not always the designing shipyard. It has been stated that the designer knew more about the non-obvious high cost items than the final low bidder.

Probably the greatest number of disputes (and the costliest) arise out of plan approval.

5.4 Plan Approval. The widest industry participation in the course of a ship construction contract comes during plan approval and inspection of the work. Plan approval is the review of the shipyard and vendor furnished plans by the owner's representative to assure compliance to specifications and conformance to good marine practice. At this time, owner, builder, government (if involved), naval architect and all regulatory bodies must interface in timely fashion to avoid work disruption. To this end, a plan schedule is agreed upon and an administrative and correspondence instruction prepared and distributed. This latter sets down a procedure for submittals of the first (original) plans; time elapsed to review and comment; resubmittal; time elapsed to return with comment and/or approval. It is usual to allow two or three weeks for initial plan review and one week for resubmittals. Obviously, the time lost in mail systems can be defeating and accordingly, the owner/naval architect must utilize a system which provides for private delivery or arrange for the plan approval function to be performed at the builder's plant. Even if the initial approval is conducted at the owner's or naval architect's place of business, reapprovals are best handled at the builder's plant.

One can see that even the best predetermined plan approval system can be affected by surges in work load, considering that the modern vessel is likely to require 700 to 1,000 shipyard working plans (including technical and scientific plans) and perhaps two or three times that many vendor plans. The owner does not review all plans in detail and normally concentrates on so-called key or principal plans. If owner does not guard against the one and two day overruns on plan approval return times, the builder may claim that the owner delayed the orderly progress of the work, or may defend non-conforming actions on the basis he proceeded on his submitted course to avoid delay. This is a very critical segment of the owner's contract administration.

The same problems occur with the plans submitted to government, to the U. S. Coast Guard, American Bureau of Shipping, U. S. Public Health, etc. In these instances, builder cannot look to owner for redress, but he can claim that events beyond his control affected his delivery.

The owner's inspection force is frequently the same staff as the owner's plan approval on-site staff. This is a most satisfactory arrangement because of the technical competence and manpower loading, considering the heavy plan approval load comes before the physical work begins. Properly chosen for competence and tact, these men will head off many potential disputes by discussion and agreement at on-site supervisory level. Difficulties may generate by conflicting test schedules and off-hour attendance demands. Many tests are scheduled at vendor plants, often inconveniently planned and in varied geographical locations. Inspection items not resolved on the spot are listed in non-compliance reports and if not dealt with prior to delivery, are incorporated into a schedule of outstanding work items affixed to the delivery documents for completion in the guarantee period. Obviously, this list of items cannot include such work as would render the vessel unsafe.

The plan approval function is complicated when vessels are built outside the United States for American owners. Customs and practices are different and the language barrier increases the communication burden. By and large, the effort parallels the U. S. practice. More and more, international shipbuilding organizations depend on English as the business language. In any event, large shipbuilding programs for American owners have demonstrated that satisfactory procedures have evolved.

Section 6
U. S. Regulatory Bodies and Construction

6.1 The U. S. Coast Guard. Every vessel engaged in foreign or domestic commerce is required by various laws and treaties (which are also national law) to be constructed in ways leading to the issuance of certain documents without which it cannot ply the seas under that particular flag. The Coast Guard is the instrumentality of the U. S. Government charged with carrying out these duties.

All of the responsibilities on behalf of the U. S. Government are assigned to the Coast Guard which *in turn* is empowered to and has delegated certain of these functions to *classification societies,* the principal of which is the American Bureau of Shipping. The Coast Guard has also delegated functions to other organizations in the marine field such as the National Cargo Bureau and the International Cargo Gear Bureau. However, the responsibility of the government is vested in the Coast Guard and delegation to these other agencies does not remove the basic responsibility.

The Coast Guard is more a law enforcement agency than a regulatory body. However, most of the federal laws that the Coast Guard enforces are, to a large extent, implemented by federal regulation. These regulations give order, consistency and interpretation to the many utterances of Congresses which have been emitted over the past century and a half in the aftermath of maritime disasters. For the purpose of discussion within the Coast Guard, the terms *law enforcement* and *regulatory* are used interchangeably.

The stated objective of most all the laws enforced by the Coast Guard is the protection of United States citizens. Most of these laws (marine safety laws) were enacted to protect U. S. citizens from the consequences of incidents in the marine environment. Of particular concern in early years were the passengers on steam vessels. Later this interest was expanded to include crew members; the waterways and other vessels; port facilities, port populations and other private property; and, finally, the marine environment itself. Only minimal reference is made within these statutory mandates to provide protection for the capital investment of vessel owners, i.e. the ship.

U. S. law also reflects international treaty, therefore, the Coast Guard has taken an active role internationally in both developing and establishing international safety standards, as well as enforcing those standards on U. S. vessels as part of domestic regulation and on foreign vessels within the *control* provisions of the accords. Where treaty and bilateral accords are silent, the Coast Guard must enforce domestic law upon both U. S. and foreign vessels. The Coast Guard represents the U. S. Government in IMCO in an effort to standardize the various safety criteria among governments.

The Coast Guard has approached its marine safety responsibilities in several ways, tracking, of course, the statutory authority provided by the Congress. The Coast Guard seeks, within this large body of marine safety authority, to prevent accidents from happening by developing, establishing and enforcing minimal safety standards applicable to commercial vessels of the United States. These standards are addressed to a wide range of commercial marine activities which include the traditional deep water oceangoing vessels, commuter ferries, tank barges and offshore structures. And the depth, as well as the scope of regulation varies. Some U. S. water-craft are *inspected,* meaning they are fully regulated: vessel design construction standards; vessel maintenance standards; manning and crew qualification standards are thus established and enforced. A host of other commercial vessels are *uninspected,* meaning that they are not fully regulated. These vessels are required to comply with minimal safety equipment standards and some with certain manning and crew qualification standards. Virtually all U. S. vessels must comply with established operating requirements, pollution prevention standards, and accident reporting requirements.

Other Coast Guard marine safety responsibilities entail ensuring safe and expeditious transit through our ports and waterways, search and rescue operations, recreational boating safety programs, and pollution response.

6.2 Classification Societies. Of all the regulatory bodies the classification societies generally have the largest influence on the construction of vessels whether registered in the U. S. or other nations. Classification societies establish and administer standards, called Rules, for the design, construction, and periodic survey of ships and other marine structures. Classification provides evidence to underwriters, governmental bodies, charters, financial institutions and other interested parties that an owner has exercised due diligence in making his vessel structurally and mechanically fit for its intended service. As such, and since the major societies are recognized as impartial and authoritative, virtually all sizeable vessels are classed.

Under Section 25 of the Merchant Marine Act of 1920,

classification for all United States owned vessels, and all functions of a classification society required by all departments, boards, bureaus and agencies are to be American Bureau of Shipping. Section 1104 (b) (6) of the Merchant Marine Act clearly holds that vessels built under the Title XI program must be Class A-1, American Bureau of Shipping, or meet other standards acceptable to the Secretary of Commerce.

There are reasons to believe that notwithstanding past experience to the contrary, all classification societies today are moving toward some conformity in rules and regulations. One particular impetus arises out of standardization upon the metric system. Under the sponsorship of the International Association of Classification Societies (IACS) and the proliferation of IMCO ship design and safety rules, the Rule books are beginning to be standardized to some degree. This simplifies the development of design for international construction.

Nevertheless, the naval architect should be cognizant of the existence of the American Bureau of Shipping, Lloyd's Register of Shipping, Det norske Veritas, Bureau Veritas, Germanischer Lloyd, Nippon Kaiji Kyokai, and the Register of Shipping of the USSR. There are other societies as well and some now in the process of formation. At present, the naval architect might have contact with: the China Corporation Register of Shipping (Taiwan); Czechoslavik Register of Shipping; Hellenic Register of Shipping; Registro Italiano Navale; Korean Register of Shipping; and Polski Regestr Stratkow.

6.3 Interaction between Regulatory Bodies. In the past, the Coast Guard's rule making relative to vessel construction was largely delegated. The Rules of the American Bureau of Shipping and codes of Institute of Electrical Engineers were adopted by reference. While the USCG still accepts ABS and IEEE rules, inconsistencies are beginning to appear because of changes in laws and international treaties provoked by concern for the environment, requiring the Coast Guard to publish them as regulations. As required by the Administrative Procedures Act, these rules are drafted and promulgated to the interested public by publishing in the Federal Register. Hearings are set for discussion of changes advanced by individuals or industry associations. Final rules are then adopted. As in all cases of having more than one oversight body of rules, occasions do arise when ABS Rules, for instance, differ from newly promulgated USCG rules. If the differences are not self defeating, the owner can live with the discrepancy; but, if the difference is in the time scheduling area, owners must use the means available to amend the orders so they coincide. For instance, if one set of rules would permit a period of twice the duration of the other set of rules between drydockings, a correlation is obviously needed. The Coast Guard works closely with the industry, the American Bureau of Shipping, and other groups to minimize problems of this nature.

Such other activities as performing compliance work under the Environmental Protection Act are being added on a continual basis. The Coast Guard some time ago was given the admeasurement function for U. S.-flag vessels. Priorly, this had been accomplished by U. S. Customs Bureau. Another area of jurisdiction is documentation. Vessels built all over the world which are intended to be documented under the U. S. flag, must comply with the Coast Guard regulations when constructed or must undergo extensive re-work when application is made for American flag documentation. To facilitate compliance to United States standards, if the construction is at a foreign site, the Coast Guard cooperates with applicants to provide inspection teams in these foreign shipyards. It is becoming a substantial level of effort for the Coast Guard and of great importance to the owner.

6.4 Other Governmental Regulations. Aside from classification societies and the Coast Guard, many other Federal bodies have an impact on new vessel construction and ship repairs. Additionally, there is the U. S. Public Health Service (Department of Health, Education & Welfare) whose inspection for compliance is required to obtain the Certificates of Deratization Exemption and of Sanitary Construction. For communications, one must comply with the requirements of the Federal Communications Commission (FCC). The regulations for safety and health applicable to longshoring were promulgated under Public Law 85-742, August 23, 1958 (Longshoremen's and Harbor Workers' Compensation Act). These regulations are now part of the Occupational Safety and Health Administration (OSHA) and impinge directly on the vessel construction and in matters not under the control of the Coast Guard.

If the vessel is to be powered with a nuclear plant, then the Nuclear Regulatory Commission (supplants Atomic Energy Commission) takes a very important oversight and compliance role. Every aspect and every phase of the reactor program comes under their purview, including the construction and operation of a prototype plant ashore, plus the shipboard installation, crew training, fueling facilities, etc.

Just as we have federal regulation affecting vessels plying international waters, we also have state law impacts and other country regulations. Many coastal states are making efforts to protect their seashores and citizens, and some have sanctioned rules affecting vessels even more stringent than the federal rules.

Many countries in addition to the United States have local laws concerning safety as, for instance, the British Factories Act, Dock Regulations. Most countries have rules as to pollution with some requiring connections to shoreside facilities unless the vessel is designed to effectively cease overboard disposal of any sewage effluent. Other types of regulations requiring special fittings, cleats, bitts, fairleads, lighting, communications, etc., exist in such restricted passages as the Panama Canal and the Suez Canal.

Section 7
Financing

7.1 Private Financing. Once it is understood that interest cost is just as much a cost of the vessel as is the steel in the hull, the naval architect can deal in a broader arena than if he confines himself to the purely technical aspects.

In private construction in all shipbuilding countries, the emphasis in financing is on ingenuity. The options available to an entrepreneur attempting to build a vessel will change as often as tax laws change, or government assistance and the market for transportation changes. Nevertheless, it is predictable that, except for isolated instances, privately contracted vessels meant for the international common carrier trades will be technically off-balance-sheet contracts which in the usual form, are sale and lease-back arrangements.

The over-simplified description of the process is that an entrepreneur obtains a long term charter or contract of affreightment. He then uses this document as collateral with a financial institution to initiate the venture. A usual result is that the financial institution promotes an owning company which is title holder of the vessel to be constructed. This owning company bareboat charters the vessel to the entrepreneur who found the deal at a rate which pays off the bond holders and provides a profit. The contract of affreightment, or charter, is assigned to the owning company should the entrepreneur fail to operate successfully. The two direct assets are the charter and the residual value of the vessel. The indirect asset, and the one which fathers the owning company, is the value of the tax shelter available through prepayment of construction interest, etc. Of course, if government aid is also available, the deal is that much more attractive.

Obviously, the charter rate has to be sufficient to cover the capital costs and the ongoing operating costs. A long-term time charter, accordingly, must include appropriate escalation or the operator will be forced out of business in an inflationary climate.

Properly structured, this leveraged lease approach can be utilized under government aided financing also, but an established company of standing and experience will tend to use an owned vessel rather than a leased vessel because it is more advantageous. Assuming that an established U. S. operator has a *capital construction fund,* then it is quite likely that he has made sufficient qualified deposits of pre-tax dollars to this fund to cover the equity needed to build a new vessel. If not, he may trade-in existing tonnage, or otherwise secure funds for deposit to this account.

7.2 Government Aided Financing. If an established U. S. operator is building in the United States for foreign commerce, he has an opportunity to apply for construction differential subsidy which, if granted, under today's laws permits him to acquire the U. S. built vessel at the determined foreign cost. Whatever the price of the vessel is to the owner under CDS, he is obligated to present 25 percent of the owner's cost as his equity. He is free to arrange for 75 percent of his cost through financing.

Usually the U. S. owner will also apply for government (Title XI) insured financing for the financed amount. To this end, he will make application for Title XI at the onset of the contracting so as to be certain of satisfying all administrative procedures in time to get his money. It is usual to use a financial house to act for the owner to cope with the extensive work of a Title XI bond issue. The printing costs alone may exceed $100,000. The costs of making the bond issue, including legal costs and other fees, as well as the interest during the construction period are added to the insurable portion of the contract price to establish the total of the Title XI bond issue. Accordingly, the total ship price cost at delivery includes this override, as well as escalation and cost changes.

When received, the Title XI receipts are deposited in the owner's capital construction fund. Because the early progress payments are made from the equity portion, there is, as a rule, ample time to accomplish the Title XI work.

If an owner is not eligible for CDS, or has other reasons for not applying (as, for instance, if he is planning to use the vessel in domestic trade), he may apply for Title XI financing only. In such case, his equity contribution must be $12\frac{1}{2}$ percent of the contract price. Otherwise, the events as herein described apply.

7.3 Foreign Financing. Vessels built in other countries are financed usually in accordance with the arrangements made by the government having jurisdiction. That is to say, the governments of shipbuilding nations regard export shipbuilding as a desirable part of their gross national product and, from time to time, make concessions to buyers to encourage orders from abroad. Tax incentives and accelerated depreciation are such concessions. Also, direct grants to shipbuilders are sometimes made to permit lower contract prices.

Recently some progress appears to be evident in shipbuilding centers outside the United States to supplant the guarantees on shipbuilding loans with guarantees of first class banks. This is noted only to emphasize the continual changing climate in shipbuilding.

7.4 Special Aspects of Nuclear Vessels. In closing this chapter, it is wished to invite attention to the special aspects of nuclear powered vessels relative to vessel financing which stem from the unusually long period of time currently required to build a nuclear powered vessel. Because of the cost of nuclear power, we are confined to thinking in terms of relatively large vessels in order to use this energy source efficiently. Such vessels, if conventionally powered, can be built in two or three years from contracting time in the U. S. and in one or two years in some foreign yards. For the same size nuclear vessel, at this time, one must think in terms of seven years, which includes prototype operation, and qualifying for approvals.

Furthermore, the established operator who is concerned with offering a reliable service is concerned with fleet obsolescence. He will be thinking in terms of six, seven or more vessels, and not one test vehicle. Accordingly, if he finds that a conventional fleet of six or seven vessels will insure a regular service (say, weekly sailings), then he must contemplate one more vessel if he decides to use nuclear power because of the long refueling time required at a special refueling facility. Even if time can be reduced to three weeks per fueling and even if he has paid the substantial extra cost to obtain exotic rods and re-shuffling capability, he is still faced with a cumulative 18 or 21 weeks' outage at three to four-year intervals. If his frequency is weekly sailings, this is untenable and he must arrange for a picket substitute vessel to fill the weekly slot when the regular vessel is refueling.

These two considerations, i.e., the long lead time for delivery and the capital cost of a substitute vessel, combine to militate against the nuclear vessel. With interest rates in double digits, high nuclear plant operating costs, four or more years of construction interest, plus a proportionate charge for the substitute vessel could very well result in a delivered price totally beyond the ability of the trade to support the nuclear vessel profitably. Seven years is a long time; the management of the shipping company making the decision to construct such vessels may have long departed, and the nature of the business may have changed so drastically as to have rendered both the price and the design unattractive.

References

Mack-Forlist, D. and R. A. Goldbach, (1976), "Bid Preparation in Shipbuilding," *Trans.* SNAME.

CHAPTER XVI

Harold G. Acker
Francis G. Bartlett

Ship Construction

Section 1
Introduction

1.1 Scope. After the award of a contract, the shipbuilder will determine in detail the construction procedures and methods by which the ship or ships will be built. Invariably these procedures will be developments of the general scheme of construction decided upon previously during the bidding and precontract period. In this connection, this Chapter highlights pertinent aspects of shipyard planning and scheduling and describes some of the more modern shipyard facilities and construction practices, including the installation of machinery and outfit. The role of computer-aids is described, particularly as related to planning lofting, and fabrication.

1.2 Trends. As a part of industry's desire to introduce assembly line methods, the trend in ship construction is toward the increased use of mechanized equipment and larger structural assembly units to permit more efficient use of the work force and to shorten the time on shipways. In this respect some machinery and outfitting items, some in modular form, are incorporated into structural assembly units whenever practicable, and this, of course, requires that these items and related plan work be available sooner than would normally be the case. These developments have resulted in a trend toward fewer shipways or building sites and have necessitated more rigid controls over the production operation.

Trends toward larger and special purpose ships have resulted in ships becoming more complicated to build. Lead times for design development and for manufacture of components increase with the amount of sophistication, and this tends to complicate the planning effort. New materials, such as higher strength steels and special paint coatings are being used more frequently. In addition, some engineering, and also some computer work. is often subcontracted by a shipyard to an independent organization, which may not be fully acquainted with fabrication details or with fabrication procedures peculiar to the shipyard. All of these trends add to the complexity of building a ship and point to the importance of early and precise planning.

Section 2
Modern Shipyard Facilities

2.1 General. There has been a notable attempt in new and modernized yards to achieve an assembly line type of flow of materials to assembly areas and an orderly flow of assemblies to the final building site. Due primarily to the unpredictability of the market, it is difficult to achieve a balance between introducing expensive assembly line equipment to suit one type of ship and, at the same time, maintaining the yard's capability to construct different types of ships and other marine structures. It is not unusual to have several different types of ships building in a yard at the same time, and with the building periods ranging from a few months to several years. The justification for large expenditures is more precise in multi-ship programs. Nevertheless, the general scheme is to adopt some method whereby an orderly flow of material will be achieved.

The trend toward larger assembly areas and fewer building sites has led to the building site being incorporated more fully into the overall scheme of material flow.

2.2 Building Sites. The building site may be a conventional sloping shipway, a building basin, or a ground level assembly area where the ship is completed for launch. In modern yards, it is significant that the building basin or ground level assembly area has been adopted in lieu of the traditional shipway. Some modernized yards have retained a few shipways to maintain some flexibility for constructing different types of ships at the same time.

a. The Building Basin. A graving dock type of building basin is generally preferred for constructing the very large ships. The principal advantages of the basin over the sloping way are that the headroom limitation for cranes is minimized, the ship can be built without the inconvenience of declivity corrections, and the launching operation is eliminated. In the larger basins, more than one moderately sized ship can be built at the same time. Quite often construction of the stern of a second ship may be well underway in the basin before the first ship is floated out. This is done to shorten the time between floatouts and to maintain a more even labor force. In order to make basins even more flexible, some basins have been provided with a portable gate to seal off one end of the basin while the other end is flooded.

b. The Ground Level Building Site. This requires a special launching facility. Unique methods of launching have been developed where the ship is transferred to a special floating drydock facility and then undocked in the usual manner. One method is to transfer sideways the completed ship to the drydock facility, Fig. 1. This facility is held in place by ballasting it down onto a supporting structure while the transfer is made. After the transfer, the drydock is de-ballasted, floated clear of the supporting structure and then ballasted again to permit the ship to float free. Another method is to transfer end-on the forward half of a ship to one portion of a floating drydock and the after half of the ship to another portion of the drydock. After the two drydock portions are brought together and aligned, the two halves of the ship are welded together.

2.3 Steel Storage. In a typical modern shipyard there is a compact plate and shape storage area from which steel is fed, via an automatic roller conveyor system, through automatic blasting and paint priming facilities into a large enclosed fabrication shop where assembly units over 200 tons can be constructed.

Plate storage areas as well as plate handling areas within some fabrication shops are usually serviced by multiple-magnet gantry or bridge cranes having lifting capacities up to 20 tons. When non-magnetic plates, such as aluminum, are lifted, suction cups instead of the magnetic heads are used. In fact the air suction, or vacuum technique of plate handling is used exclusively in some facilities for handling large bottom and side plates in barge construction.

2.4 Steel Fabrication and Assembly. One of the first notable examples of an assembly line type of assembly system was the one at Gotaverken's Arendal yard in Sweden where material flows into a single assembly line. Ship sections, complete with some outfitting, are assembled under cover and then joined to previously completed sections. The completed portion of the ship is systematically moved out of the covered area and into the building basin as new sections are added. This operation continues until the completed ship is entirely in the basin ready for floatout. In another yard, the completed sections are moved out of a covered shop onto a sloping way, and the ship, when completed, is launched in the conventional manner.

In contrast to Arendal's assembly system, Ingalls' Pascagoula yard, Fig. 2, features several parallel assembly flow lines, each beginning at the fabrication shops and extending through the subassembly area to the module assembly area. These structural modules are essentially complete sections of the ship, including most outfit. The modules are then moved to the final ship assembly area, Fig. 3.

The assembly systems begin with plates and shapes being moved into a fabrication shop, Fig. 4, from a steel storage area. The first step in many fabrication operations is the cutting of plates. The numerically controlled (N/C) burning machine, Fig. 5, is one of the most versatile of steel cutting machines. It can cut automatically duplicate pieces and make mirror-image pieces as a first step in flat-plate fabrication.

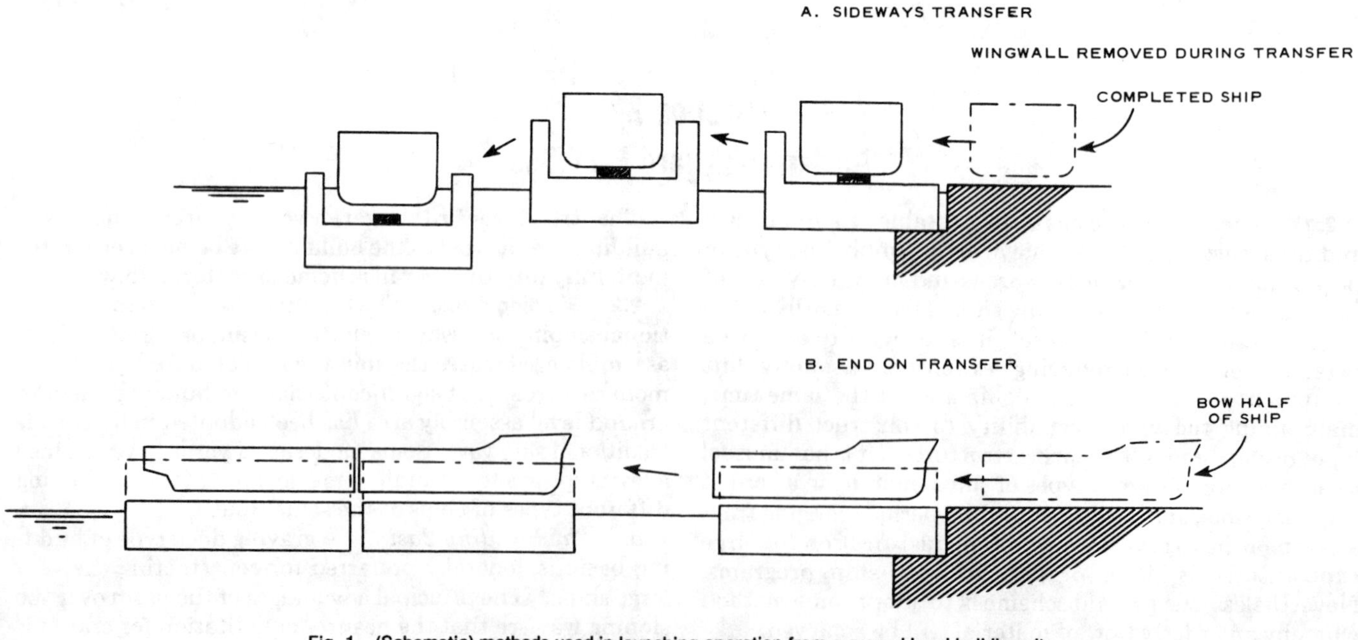

Fig. 1 (Schematic) methods used in launching operation from a ground level building site

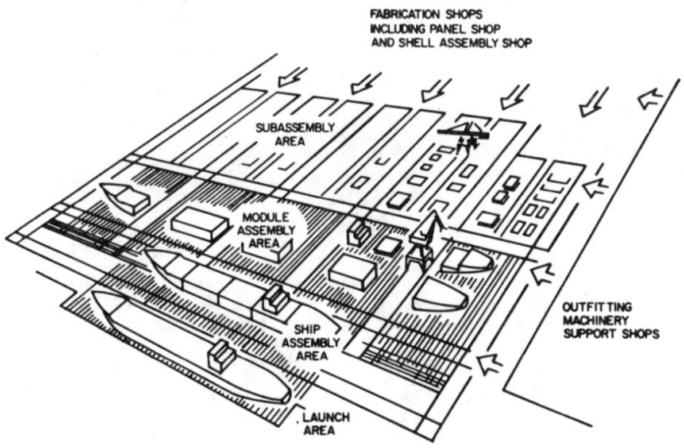

Fig. 2 Principal elements in assembly lineflow at Ingalls Pascagoula

Fig. 3 The Ingalls Pascagoula yard building the DD963 class destroyers

Some shipyards fabricate stiffener angle and T shapes in the depth range of 200 to 460 mm (8 to 18 in.) rather than purchase mill shapes and remove one of the flanges. In some cases the work is moved past fixed automatic welding machines which can weld several shapes at a time. Shapes over 460 mm deep must be fabricated in any case. In connection with angle shapes, some European mills can furnish rolled shapes up to 500 mm (19.7 in.) deep and fabricated shapes, consisting of a rolled angle butt welded to a flat plate, up to 1000 mm (39.4 in.) deep.

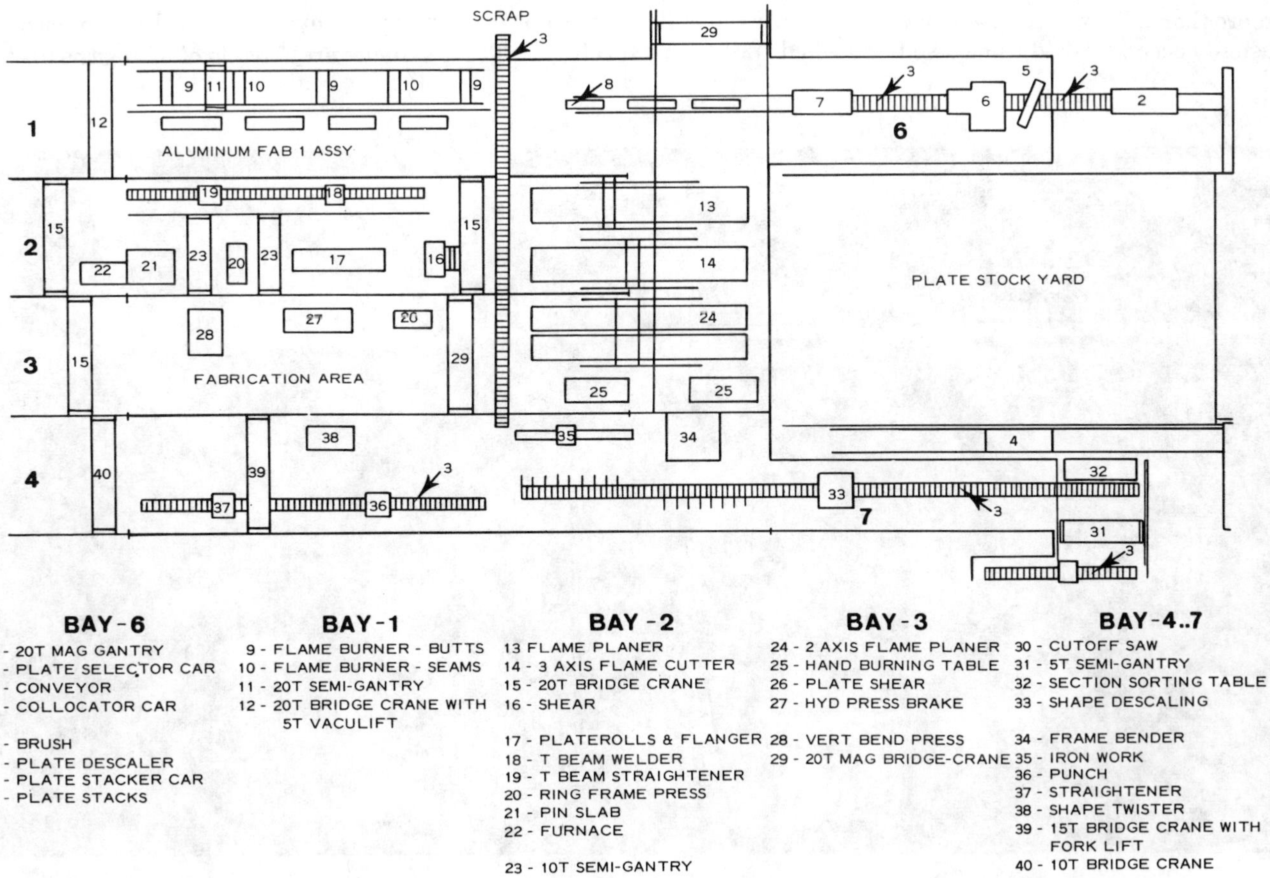

Fig. 4 Typical steel fabrication shop

Many shipyards now have a panel line system included in their fabrication area. This system provides for welding together flat plate panels, attaching stiffeners with automatic welding, and moving the panels along a line where webs and other steel members may be attached. A typical panel line function diagram is shown in Fig. 6. Movement of all plates and stiffeners is completely automatic from location 2 through location 5. Outfit items, if added, are installed at locations 6 and 7. The weight of an assembly unit may be over 200 tons.

Assemblies with curved surfaces are assembled on specially prepared forms or on pin jigs.

a. Painting Facilities. Another unique facility installed by several shipyards is the large blast and paint facility which can accommodate large units such as those coming off of a panel line. In some installations, the whole assembly can be rotated while it is being blasted.

b. Buffer Area. As the completed assembly units accumulate from the panel line, the paint facility and other assembly areas, space must be found to store these units until needed. This space is sometimes called a *buffer area,* i.e. an area which can absorb any overflow of units. An ideal area, of course, would be one alongside a building basin where a large gantry crane would extend over the area alongside the basin as well as over the basin, thus making it possible to transfer the largest and heaviest units from the side of the basin as well as from the end.

2.5 Materials Handling. Movement of heavy units, some of more than 200 tons, from a panel line or painting facility is usually accomplished using a self-propelled trailer type transporter employing a hydraulic mechanism within the vehicle. These vehicles are capable of lifting the units un-

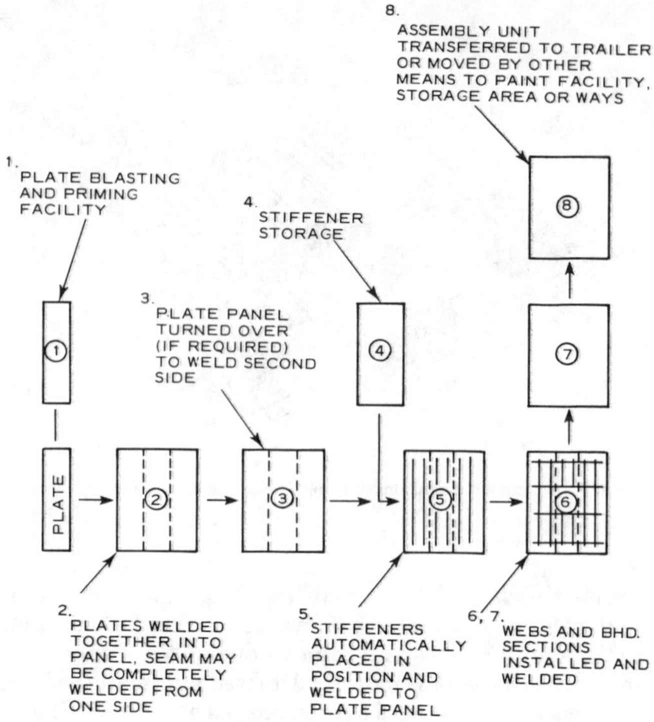

Fig. 6 Typical panel line functions

Fig. 5 Tape-controlled automatic burning machine cutting mirror image web frames

aided. For light weight transfers, mobile cranes, straddle trucks, and forklift trucks are used wherever practicable. Inside shops, crane capabilities often dictate the method of construction.

The most noticeable of material handling facilities are the giant gantry cranes recently installed in a few U. S. yards. In one yard, a 1200 ton crane over 60 m (200 ft) high was installed to handle large aluminum spheres for LNG carriers. In another yard a 900 ton crane, with a span of about 165 m (540 ft) was installed to service not only a building basin but also a large assembly area at the side of the basin. In some cases the cranes extend out over the water to permit handling waterborne loads.

Other yards employ two or four revolving cranes in tandem to lift heavy units. There are certain advantages in having several cranes capable of servicing a large area such as a building basin, but the lifting capacity along the side of the basin is limited. In some yards the revolving cranes can be made to travel along two different sets of rails, one set at right angles to the other, by rotating the crane trucks. At outfitting piers, revolving cranes, even those having a small lifting capacity, must have very long booms, some 60 m (200 ft) long, to reach all parts of the larger ships.

Floating cranes having lifting capacities of 500 tons or more are available in most large ports, and a few shipyards, especially those also engaged in repair work, have their own floating cranes.

Perhaps the most unique transfers are made when whole ships are transferred from the building site to a launching platform or special drydock, or when large units or portions of ships are moved horizontally. This is usually done by synchronized hydraulic equipment. The action principle is simply that of a hydraulic jack; it is not unlike that for installing and withdrawing propellers using hydraulic nuts as described in Section 14.4. The two basic actions are illustrated schematically in Fig. 7. Devices are provided to prevent the load from breaking away while the load is being moved; in this respect the difference between starting and moving friction becomes important. Fig. 7 shows a single unit. Actually, several synchronized units are used together to move ships and other large structures.

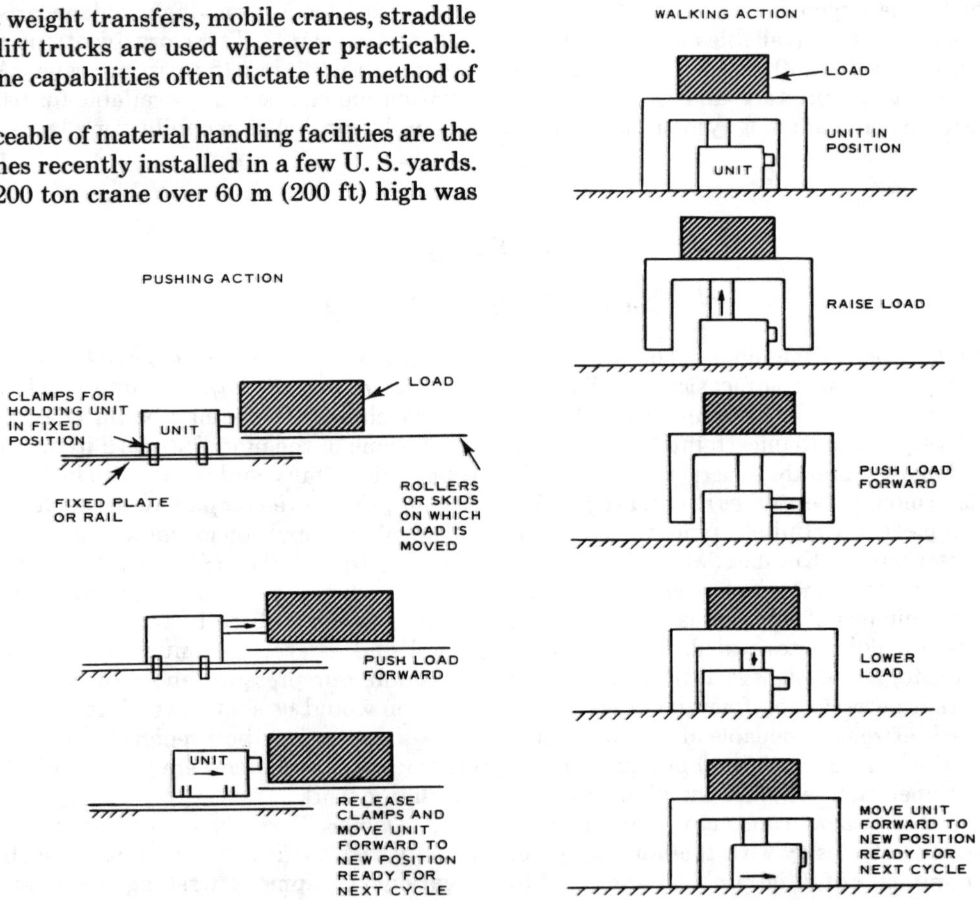

Fig. 7 Schematic steps for moving heavy loads using hydraulic equipment

2.6 Computer Aids. Computers are being used to assist in defining a ship's lines and the geometry of much of the structure without actually lofting the work. N/C tapes can be produced directly from lofting information to guide the cutting torches on the burning machines. Thus, the character of lofting has undergone a marked change as will be discussed in Section 4.

Computer aids are used in a number of ways in engineering and in fabrication. Naval architecture calculations are now done by computers using the same data base information as used in the computer lofting system. In some instances N/C tapes are used to guide frame bending and pipe bending machines.

Modern telecommunications can provide facilities through a telephone hook-up for using large capacity computers on a rental basis through remote terminals. In addition, small in-house or mini computers have been found economically attractive for doing many routine calculations which do not require the large capacity computers. Not only are computers (hardware) available on a rental basis, but computer programs (software) are also available from many sources, including proprietary and governmental. Organizations controlling programming systems have service groups capable of resolving problems arising from the use of their programs. These organizations also generate new programs that fit into the existing systems. Thus, computer information can be made available for use without having a complete in-house capability or a large staff.

Section 3
Planning and Scheduling

3.1 Preliminary Planning. Preliminary planning is done at the time of bidding and before contract signing. The first step is to determine dates for such key events as keel laying, launching, and delivery. Due to uncertainties of the final design, the material market, and the general labor situation, it may be desirable to modify these dates in order to provide a margin of time in meeting definite commitments. This is a management prerogative. Key dates are usually shown on a shipway schedule chart; these dates become increasingly critical when the number of shipways is decreased (just one or two shipways or building sites available). Estimating the shipway schedule for new designs is difficult, especially if development work is necessary of if new facilities are required. Experience is extremely valuable in this connection because there is seldom time for any in-depth analysis.

For any efficient operation, whether the plans are prepared in the shipyard or by a subcontractor, the engineers and draftsmen must work closely with the building yard. The only way to minimize potential problem areas and to ensure a maximum degree of success is to consider each step in the construction and scheduling process, no matter how trivial the step may appear. This applies to machinery installation and outfitting as well. It is imperative that a design engineer be familiar with all of the pertinent factors affecting production, such as maximum size and weight of plates, subassemblies, and erection units. But most important is the time and effort needed to plan construction at the very beginning of the design stage.

In order to provide some flexibility in design and construction, specifications are often written to give the contractor reasonable options, such as to use a casting, a forging, or a weldment for a stern frame or to use either radiography or ultrasonic testing for weld inspection. In other cases, the specification will describe the construction method or system preferred by the contractor.

a. Purchase of Working Plans. When a series of ships is built in different shipyards, one yard usually has the option of purchasing working plans from the other, with no obligation on either party. However, it is difficult for one yard to build efficiently a ship to another yard's plans because of the physical constraints peculiar to each yard. Normally, there are enough differences in desired fabrication and erection procedures to require more than just a small amount of plan revision, and allowances must be made for this contingency. The same may be said when working to plans prepared by a design agent unless the design agent has worked closely with the building yard. At best, the results are not likely to be very satisfactory.

b. Purchase of Outfitting and Machinery. Outfitting and machinery items may be purchased in several ways. First, some of the more standard items such as rigging fittings, valves, fans, and pumps can be bought off-the-shelf. Second, the more complex items such as boilers, steering gears, main propulsion machinery, and loading instruments require a technical specification to be prepared by the engineering department. For example, a steering gear specification would include hydrodynamic rudder torques for ahead and astern, maximum rudder angle, rudder rate, maximum ram pressure, and rudder stock size. The specification would be sent to vendors for quotations. Third, certain items may be prepared and installed by outside contractors. Such items are joiner work, floor covering, and insulation work.

c. Owner Furnished Equipment. The owner will usually wish to furnish certain items such as radio and navigational equipment, washing machines and dryers, mattresses and linens, galley equipment and dishes, and many of the spare parts. In naval vessels, essentially all of the weaponry and detection equipment is government furnished. In some instances, the government furnished equipment has been under development at the time of contract signing, and this made planning and scheduling more difficult.

d. Special Facilities. Shipyards have been called upon to construct unusual types of ships such as LNG carriers and semi-submersible drilling rigs. In some instances completely new facilities had to be built. Notable among these new facilities are those required for the construction and handling of the large tanks and spheres for some of the LNG carriers. The 850-ton spheres for one ship design had to be moved out of the construction site by a specially designed transporter to a barge, built specifically for the job, and then transported 900 miles to the shipyard where a new 1200-ton crane had been installed to handle the spheres.

In addition to new facilities, new methods for constructing and installing unusual structures and equipment may have to be developed, such as for nuclear powered ships or for various LNG containment and insulation systems. Some new methods may require rigid temperature and humidity controls. Intensive training and control programs are usually required when integrating these new methods with normal shipyard practices.

3.2 Engineering and Design. The engineering department is called upon to aid in preliminary planning due principally to the many rules and regulations which must be satisfied and to the complexity of both yard and shipboard equipment. It is important to note that new regulations, such as those of Intergovernmental Maritime Consultative Organization (IMCO) and the U. S. Coast Guard (USCG), have greatly increased the amount of engineering work necessary for both the preliminary and final designs. Engineering work is started at the earliest possible time, even before schedules are prepared.

The extent and thoroughness of engineering work done on contract plans and specifications by the owner or the owner's design agent varies considerably, and this can markedly influence the engineering effort required by the building yard. The building yard is often required to make an independent weight estimate and to conduct basic design studies which will enable the drawing room to develop working plans and the purchasing department to proceed with ordering component parts.

The shipyard drafting department prepares detail working plans and bills of material. Plan schedules are prepared which list the plans to be developed and scheduled dates for start and completion, approval, and yard issue. The lead time required to order materials will often determine the start date of drawings. Whenever material, including steel, is in short supply, or whenever there is a likhood of a long delivery period, orders are often placed before plan approval. This, of course, is done at the shipyard's risk.

Design changes are inevitable as the design develops and during construction. These changes can affect schedules, procurement, and even the ship's delivery date. Changes affecting the ship's weight, center of gravity, or cubic must be reviewed to determine their influence on the ship's characteristics. The effect of such changes and related costs must be acceptable to all interested parties before the changes are authorized. However, most contracts require that the shipbuilder, without prior agreement on cost, put in hand those changes, identified as *Essential Changes*, which result from applicable rules and regulations which were not in effect at the time of contract signing. Any addition or change by either the owner or builder should obviously be brought forward at the earliest possible time.

3.3 Production Planning. A schematic diagram as shown in Fig. 8 is useful for providing the various departments with a guide for the general division and sequence of work. For a ship with machinery amidships, the sequence of assembly usually starts amidships and proceeds toward the after end and then forward. When machinery is aft, the first sections are set aft to provide early availability of machinery space.

Scheduling methods are unique to each yard and generally reflect practices developed from experience. An overall schedule which is most useful to both management and production departments is one which highlights major tasks and events, and which shows the sequence of work and the relation of the various tasks to each other and to the whole project. Such a schedule can be prepared using the simple principles employed in network flow techniques.

The basic principle in network flow is the task-to-task relationship. That is, task C cannot start until its two prerequisite tasks A and B are completed. There is, of course, the usual task-to-time relationship for each task. These principles have always been employed in one form or another throughout industry, but the computer has now made it possible to utilize to the fullest these principles in network form.

a. Network Flow Scheduling Technique. This technique is often used for controlling large, complex, and possibly non-repetitive projects such as the Polaris submarine project where the nuclear-powered submarine, the missile, and the inertial guidance system were planned and scheduled simultaneously. Examples of the technique are *PERT* (*Program Evaluation and Review Technique*) and *CPM* (*Critical Path Method*), both of which provide a means of representing graphically the different operations that make up a project. These networks can be revised to show the effects of adjustments to a schedule necessitated by changes in design, delays, etc. It is also possible to treat the network statistically in order to obtain an idea of the probable longest and shortest times for completion of a project.

Fig. 9a shows a CPM network for a small portion of an overall ship schedule. The tasks are diagrammed in network form, an arrow representing each task. The length of line representing each task has no time significance. The total length of each path through the network can be calculated and the longest path is the Critical Path for this particular portion of the network.

The amount of spare time (that portion of allotted time not actually needed) connected with tasks not on the Critical Path is called *float*. These tasks may be accomplished any time within their respective time ranges without delaying the project.

b. Practical Aspects of Network Flow. It is important to keep networks as simple as practicable and to eliminate all relatively unimportant tasks; otherwise, the network will be too cumbersome to be of significant value.

If it is found necessary to reduce the total time along the Critical Path, overtime, rescheduling, or subcontracting

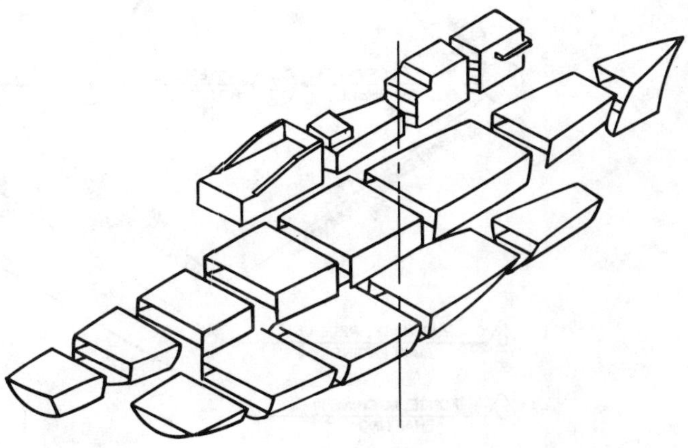

Fig. 8 Schematic diagram for scheduling sequence of work

work may be employed. Due to the branching effect of the network, it is usually more economical to reduce the time allotted for earlier tasks.

Overlapping of tasks is always a difficult problem to handle. In some instances, there will be several overlapping steps between the start of a drawing room activity for a particular job and the completion of the job. For example, in the case of ventilation systems, one portion of a job might be drawn and fabricated before the drawing room activity for the remaining portion of the same job has been completed.

As revisions are made to a schedule, a new Critical Path may be created. It is important to point out that there will be several paths, perhaps in different areas, which will be considered critical. In fact, each department will have its own critical tasks.

Abbreviated networks are useful for investigating selected areas of interest and for simplifying the more complex networks for management use. Occasionally, it may be necessary to blend one network with networks of other contracts to aid in maintaining a fairly constant work force or to make multiple purchases for several contracts.

3.4 Gantt Chart. This type of chart is particularly useful for management in general and for the less complicated jobs such as for shipyard work in particular. The Gantt type chart is a multiple-bar chart showing the main activities with accompanying key dates laid out along a time base. Fig. 9b shows a Gantt Chart for the same portion of the schedule as shown in Fig. 9a. Since this chart must show the relationship between various activities, the same principles as employed in network flow are used in its development.

3.5 Materials and Component Parts. The scheduling of purchases and deliveries of materials and component parts requires great care to ensure a flow of needed material to the yard without overstocking. For some items, factory or preinstallation tests are required.

It is general practice to purchase semifinished items such as large castings, forgings, piping, and switchboards. In addition, many shipyards sublet joiner work, floor covering, insulation, and other outfitting items. Important components such as boilers, main propulsion units, and auxiliaries may be subcontracted as complete units. This procedure requires early settlement with the vendor of components and agreement on dates for delivery. Further, it is customary for the shipbuilder to keep in close contact with the progress of work in the vendor's plant.

Procurement on some items includes design and development by the vendor. In these cases, the level of responsibility should be stated clearly. This is particularly important if subsequent changes are found necessary in either the vendor item or associated ship items. There is also an interdependence of shipyard plan work required by the

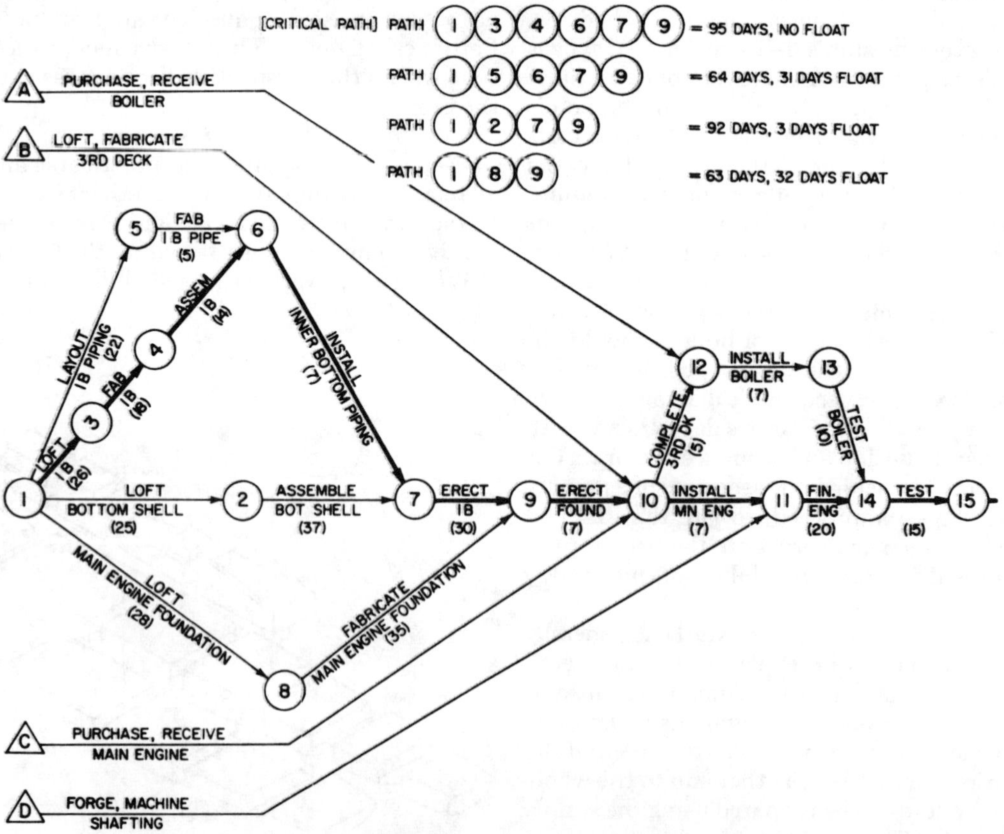

Fig. 9a Scheduling network for critical path method

vendor and of vendor information required by the yard.

3.6 Estimated Labor Load. Labor-load curves are prepared for various departments, and this is done at an early stage for the information of the employment office and department heads. Such an early prediction can be approximate only, but revised schedules can be made and issued periodically. It is desirable to furnish the department heads with regular statements of labor hours expended for comparison with labor hours budgeted.

3.7 Production Control. Because there are many different parts of a ship being worked on simultaneously, there is a need to monitor the progress of work in order that management will know what is actually happening. In some yards, this is done by a production control group. Since ships are rarely built without deviations from planned schedules, this group must be able to anticipate, or at least pinpoint, problems as they develop so that management can take remedial action. As accurate basic production information is accumulated, better first estimates and fewer problem areas can be expected.

In some instances, the information source for the progress of work has been cost accounting procedures. Through additional programming of a computerized accounting system it has been possible to monitor the expenditures associated with the progress of some elements of work and from this extract work progress data.

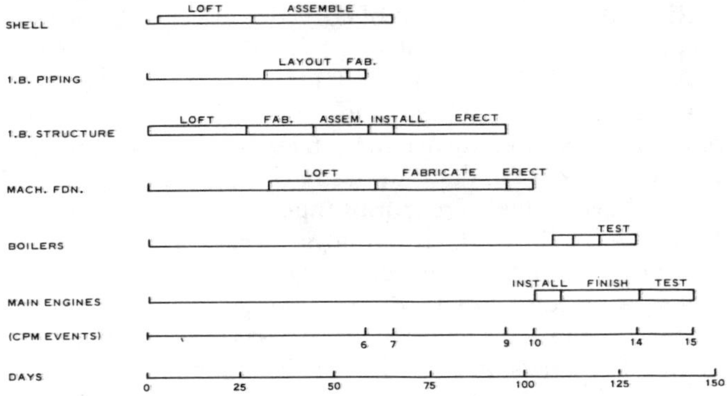

Fig. 9b Gantt type scheduling chart

3.8 Preoutfitting. When preoutfitting of machinery and outfit is desired, additional planning and tighter controls are needed for production. This is covered in Section 12.

Section 4
Lofting

4.1 General. Lofting work has historically been associated with laying down a full size body plan and making full size templates. However, as a result of advances in optics and electronic computing, lofting, especially in the larger shipyards, is now done either by optical transfer techniques or by computer-aided methods. While optical lofting commonly uses $\frac{1}{10}$-size body plans and lines, computerized lofting can be programmed for any desired scale.

Offsets are usually expressed in millimeters (4991 mm for example) or in feet-inches-eighths of an inch (16′-4½″ would be 16-4-4 for example). Offsets in English units are sometimes given in sixteenths rather than eighths of an inch.

4.2 Full Size Body Plan and Templates. The full-size mold loft had a large, well-lighted, clear floor having a smooth wood surface. The fairing of the lines from preliminary lines and offsets was done using the full-size beam and depth of the ship but having the length contracted to $\frac{1}{4}$-size due to space limitations. After the lines had been faired, the body plan was laid down, the frame lines were run in, and the final mold loft offsets were picked off.

Templates were made of different materials such as basswood, plywood, Masonite and heavy waxed paper. On the templates were marked necessary information such as plate edges, frame lines and various check lines to ensure that accurate dimensions were maintained. This information was transferred to the plate by scribing or center-punching. Templates were usually stored only for the duration of the contract.

4.3 Optical Detailing. Optical lofting or detailing consists of drawing structural members $\frac{1}{10}$ full size, photographing the drawing on a glass negative to about $\frac{1}{140}$ full size and then projecting the negative image to full size on a steel plate.

The $\frac{1}{10}$ scale drawings, or simplified forms thereof, are sometimes used directly as templates for optically controlled burning machines, the expansion to full size being handled within the burning machine mechanism.

Optical lofting requires a much smaller working space than that required for full scale lofting, but the work is more exacting because of the small scale used. The high degree of accuracy requires the use of magnifying glasses, special vernier scales,and other precise drawing instruments.

The body plan is drawn with fine-line pens on a smooth, flat table, usually an aluminum plate painted white, or on drafting film having a high degree of dimensional stability. In order to minimize the effects of climatic changes, most lofting areas are air conditioned.

After the body plan has been prepared, $\frac{1}{10}$ scale pencil-line laydowns of decks, bulkheads, shell plating, etc. are made on drafting film preparatory to making the ink drawings for photographing. The ink drawings, most of which are tracings of portions of the laydowns, are photographed on a glass negative which can then be used to project layout information onto the plate, Fig. 10. The projected image appears as white lines about 3 mm ($\frac{1}{8}$ in.) wide on the steel plate.

Simplified development of curved plates can be accomplished through the use of a $\frac{1}{10}$ scale jig.

Numerically Controlled (N/C) tapes can be prepared by using $\frac{1}{10}$ scale information and a digitizer. A digitizer is a manually operated device for converting points along any line into numerical data for use in the preparation of N/C tapes. This method of preparing tapes is much more arduous than the method which employs direct computer-aided lofting through numerical detailing.

4.4 Computer-aided Lofting. The use of computers to define a ship's lines and structural parts developed rapidly during the 1960s and 1970s. During this period, various combinations of optical lofting and computerized lofting were used to define the parts, particularly in connection with the development of tapes for N/C burning machines.

Computerized lofting requires the use of various items of computer *hardware* to handle the programming or *software*, which consists of program modules tailored to produce the output information. The hardware consists of a large capacity computer and associated equipment including cathode ray tube displays, computer driven drafting machines, and a mechanism for producing magnetic or punched tapes.

4.5 Computer Software Systems. These systems cover all phases of lofting from the development of ship's lines to the production of N/C tapes in addition to a variety of supplementary design and fabrication information. The systems are normally made up of several related programs, all tied in with a common data base. The program or module names usually relate to the work they do such as fairing of lines, parts programming, or naval architectural calculations.

Computer systems for lofting require personnel with special talents. A combination of two vitally important skills is essential, namely an in-depth knowledge of the particular phase of shipbuilding concerned, and a thorough familiarity with the techniques of manipulating the programming to obtain the required results. This work encompasses so much minute coding detail that only continuous work in such areas can ensure reasonable levels of productivity. Even though attempts have been made in some systems to simplify the programming so that less skilled personnel can be used to model portions of certain programs, the overall personnel requirements in both of the above mentioned areas become increasingly critical to the success of the computer applications as the programs expand their coverage.

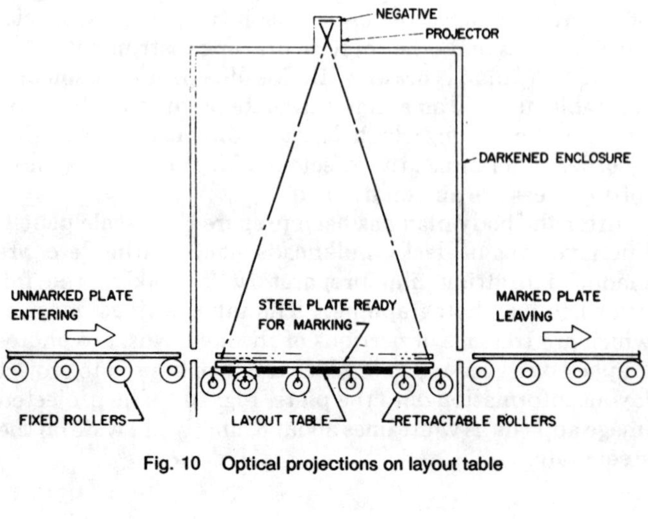

Fig. 10 Optical projections on layout table

a. Fairing of Lines. The programming for this module fairs the lines of the vessel from offsets taken from preliminary lines and generates a data base. Normally 20 stations are used and the computer fairs the lines between the stations. It should be pointed out that it is highly desirable that the preliminary lines be accurate, otherwise the points on the stations may have to be altered which involves a time-consuming analysis and conversion operation by highly skilled personnel.

The ends of the ship are very difficult to fair, and it has been found expedient to use conventional lofting techniques to fair the surfaces in the bow and stern regions.

If only a preliminary type of lines were desired, the computer could produce lines from very rough input, but the output lines would not be considered satisfactory for accurate hull surface development.

Basically, a fairing program generally adapts in mathematical form the techniques of a loftsman by using elastic beam theory to create an exact representation of the loftsman's spline. In fact the mathematical development of *spline theory* was a result of this analytical treatment. The resulting mathematical curves are faired using an iteration process. If necessary, the position of the station data points may have to be moved slightly by a skilled operator using the programming to ease the strain, or curvature, in the beam (spline) a minimum amount.

Before computer fairing begins, the controlling boundary of the hull as required by the program must be determined. These bounds usually involve bow and stern ending points, side and bottom tangents, and deck lines. Fairing the remainder of the hull surface is done iteratively. After each step in the fairing process, a drawing of stations and diagonals can be requested by the user to ensure that the program and the method of using it are functioning properly. Fairing programs may have the capability of allowing the user to select fairing planes at any position and angle to the centerline plane of the hull. On completion of fairing, a lines drawing can be prepared. Construction frames can be positioned and their shape stored in the data base for future use along with the lines data. The final frame and waterline offsets are then routinely tabulated as part of the computer output.

One of the advantages of a computerized lines program is that a variety of hulls can be developed from one basic hull. That is, the size and proportions of the basic hull can be changed easily, thus enabling comparisons to be made. This is valuable in research and in preliminary design studies.

b. Body Plan. To define traces of decks, longitudinals, girders, and seam sight edges, the draftsman must specify a number of points on the associated trace along the length of the ship. The computer program can then fit a curve through the points and store the resulting trace in the data base. Final offsets for these items are routinely tabulated. From this information, a shell expansion plan for most of the ship's length can be drawn to the desired scale, usually 1:50 ~ (1/4 in. ≐ 1 ft).

c. Parts Programming. This primarily involves preparing N/C tapes to cut out selected structural parts such as web frames, shell plates, and deck plates. The shape or configuration of the part to be cut is usually made to suit some simple mathematical form if at all possible, except of course for a molded shell line, which would come from the data base. For example, a cutout in a web plate for a longitudinal may be made using straight lines and arcs of circles, Fig. 11. The radius centers in this case must be precisely determined to provide a smooth curve but this can be done very easily using internal computer program capabilities. The same mathematical routine can be followed for cutouts for different size longitudinals. These cutouts are described in subroutine form and stored in the computer for repeatable insertion in the development of the N/C tapes. Subroutines for all kinds of repeatable cutouts including access holes can be developed and stored in like manner.

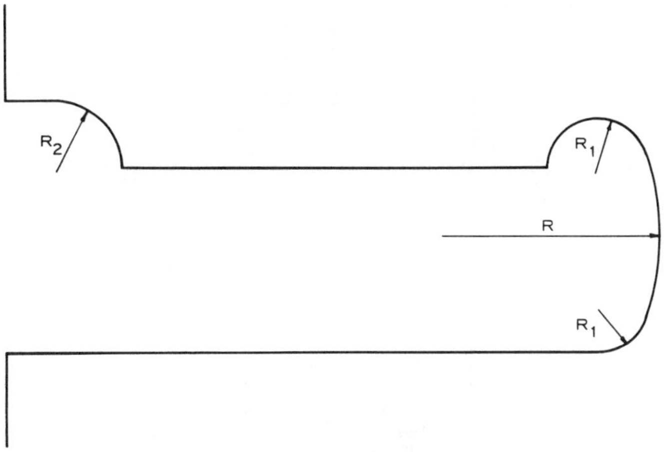

Fig. 11 Typical cutout developed as subroutine

Fig. 12 shows an innerbottom floor plate with typical cutouts. Key reference lines can be stamped on automatically for checking alignment and dimensions. When the cut is made, the cutting torch usually travels continuously in one general rotation for outside cuts. Kerf compensation, Fig. 23, can be accommodated for either clockwise or counterclockwise rotations. The N/C tape can direct the burning machines to cut several similar or mirror image pieces at one time. It should be emphasized that the location of the cuts relative to the molded lines must be exact, especially for internal structures where the plating may be on either side of the molded line; an error here could result in the piece being either too short or too long.

The fairing of lines and the development of N/C tapes can be subcontracted so that the smaller yards can make use of automatic burning machines without requiring a large computer staff.

d. Shell Plate Development. Shell plates are defined by specifying their boundary seams and butts. The shell plate development program automatically develops each plate into a flat plane so the plate can be cut from a flat plate. Reference lines such as frame lines and longitudinal traces as well as the roll line are also included and information for making roll set templates is provided. Many of the methods used to develop the shape of the plates are simply those previously used by loftsmen and consist of a progressive geometric development. When the shape of the plate involves sharp compound curvatures, the accuracy of the development is decreased, the chance for error becomes greater and, as a result, extra stock is usually added.

The computer can also supply information for setting the height of pin jigs which are required for fabricating curved plate assemblies.

e. Nesting. For merchant ship work, it has been found convenient to lay out the pieces using N/C drawings and then to arrange the pieces by hand to suit the plate size, Fig.

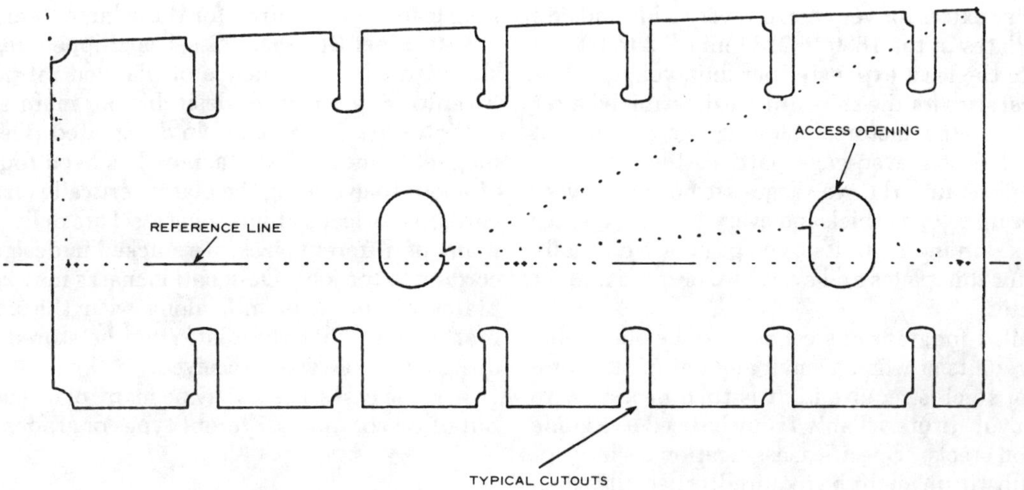

Fig. 12 Inner bottom floor as drawn by computer drafting machine

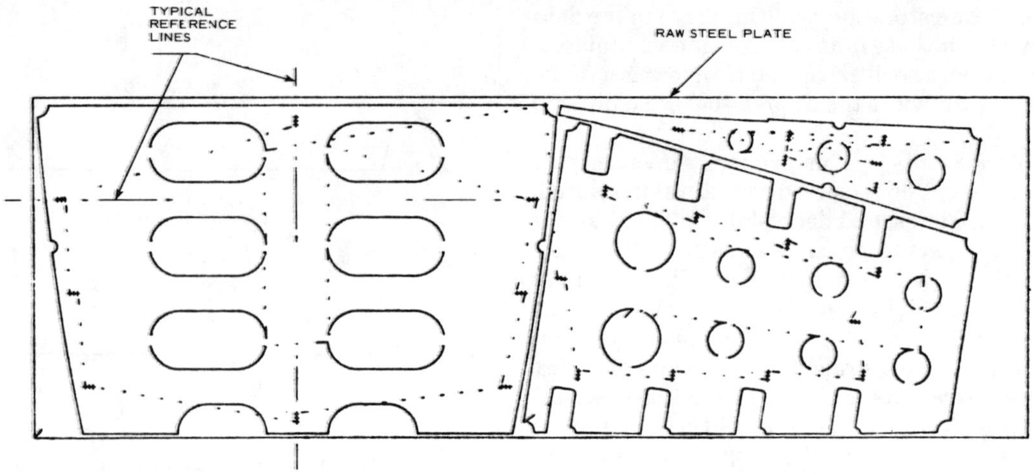

Fig. 13 Nesting of pieces

13. Having done this, each piece can be programmed for N/C burning. Connecting tabs can be programmed into the burning operation to prevent accidental interferences by loose pieces. Compensation for kerf is also provided. In addition, wastage and percentage of steel utilized can be calculated and included in the output.

The use of computer graphics using a cathode ray tube (CRT) picture has been recently employed in some cases to speed the nesting process by replacing the manual positioning of N/C parts with a quick response, variable positioning capability of scaled images of the parts on the outline of the raw steel pictured on a CRT.

Section 5
Steel Ordering and Storage

5.1 Ordering of Steel. Mill orders are prepared by the drawing room, the plate sizes being lifted from plans or plating models or obtained directly from computer printouts. Each piece of steel is given an identifying mark on the plan and on the bill of material. Plates to be severely hot-formed are ordered somewhat thicker than the required size to allow for the thinning down that occurs during the forming process.

There are price extras for very narrow, very wide and odd gage plates. Plates in the 1830 to 2286 mm (72 to 90-in.) width range are the least expensive per unit weight. The most economical sizes for the shipyard are determined after considering all related factors including the number of welded seams. Selected even-gage plates cost less per unit weight. In English units, the even gages are normally every $\frac{1}{32}$ in. for plates up to $\frac{1}{2}$ in. thick and every $\frac{1}{16}$ in. for plates over $\frac{1}{2}$ in. thick. In metric units, even gages are normally every one mm for thin plates and every two or more mm for the thicker plates.

The exact called-for grade of steel may not be obtainable for some ship repairs or when a small amount of steel or a special grade of steel is required. It is then necessary to obtain a proper substitute. Table 1 may be used as a guide for classification steels. Specific classification society approval of the substitute might be required, especially when ASTM and other non-ship grades are involved.

There has been little effort to increase the amount of standardization of plate or shape sizes beyond that normally adopted because such increases have been considered to be of little, if any, economic value. This view seems to have been confirmed by a MarAd sponsored study "Improved Design Process," Bath Iron Works (1977)[1].

Design work on structures incorporating large castings or forgings, such as a stern frame, is started early due to the long lead time required for these large items.

5.2 Steel Storage. Steel is shipped regularly to the shipyard in the sequence of planned fabrication. Large inventories are neither desirable nor maintained.

Plates are flat stacked in designated piles by multiple-magnet cranes. Flat stacking has been found to be more efficient than storing the plates vertically on edge. In some yards, as carloads of incoming steel are delivered, the plates, many of different sizes, are stacked in designated piles according to the job. Designation marks may be placed on the plates at the steel mill along with the approval stamp markings so that the plates can be stored in the correct designated pile when received.

In some cases, there may be many plates of one thickness but of two or more different types or grades. This poses a

[1] Complete list of references is at end of chapter.

Table 1—Appropriate Grade Relationships between Classification Society Steels

Class. Society	Year	Ordinary Strength Hull Steel Designations[1,4]				
ABS	1979	A (<½″)	A (>½″)	B	D or DS[2]	CS, E[3]
ABS	Prior to 1974	A	R Before '72	B	C, D	CS, E
BV	1977	A	A	B	D	E
DNV	1977	NVA	NVA	NVW	NVD	NVE
GL	1973	A	A	B	D	E
LR	1978	A	A	B	D	E
NKK	1978	KA	KA	KB	KD	KE
RINa	1977	A	A	B	D	E
IACS	1978	A	A	B	D	E

1. Any grade to the right may be substituted for any grade to its left.
2. DS to 1⅜ in.
3. Any grade in the extreme right hand column may replace any ordinary strength hull steel including ABS CN, GL C, DNV, NVC, NKK KC, RINa DDS.
4. Although these are the various society grades, one cannot be necessarily substituted for another for any particular structural item as at present the various societies have different requirements for the application of these grades to hull structural items.

ABS	=	American Bureau of Shipping	(USA)
BV	=	Bureau Veritas	(France)
DNV	=	Det norske Veritas	(Norway)
GL	=	Germanischer Lloyd's	(Germany)
LR	=	Lloyd's Register	(United Kingdom)
NKK	=	Nippon Kaiji Kyokai	(Japan)
RINa	=	Registro Italiano Navale	(Italy)
IACS	=	International Association of Classification Societies	(International)

problem in steel segregation not only in storage but also throughout the construction period. Therefore, mill orders should distinguish between the grades of steels, and identifying markings should be made clear. Sometimes, different colored preconstruction priming paints are used to identify the various grades of steel.

Clad plates must be specially handled and stored because the cladding is very thin and can be easily damaged and because foreign matter must be prevented from becoming imbedded in the cladding. Clad plates are usually received with a paper or plastic covering on the cladding. Stacked plates are usually separated by wood strips. Obviously, plate grabs with serrated teeth should not be used for handling these plates.

Section 6
Steel Cutting and Forming

6.1 Cutting. Several methods of cutting and gouging are used in shipbuilding to cut steel, to prepare bevels for welded joints, to back gouge welded joints, and to remove defects.

a. Oxygen Cutting. The various types of hull steel used in ships, including the higher strength and low alloy steels, can be cut by the oxygen cutting (burning) method. Oxygen and fuel gas, such as acetylene, usually are piped through distribution mains to various parts of the yard. There are many different types of machines used for multiple and automatic cutting operations, both for straight line and contour work.

One machine for straight line cutting is the flame planer. With the cutting heads attached, the bridge arm simply moves linearly along the line of travel. Each cutting head may mount three torches for making double bevels.

Machines for contour cutting are of the cross carriage type or the pantograph type, both types can mount several cutting heads. In the cross carriage type, as used in N/C burning machines, the bridge arm moves only along the line of travel and the cutting heads move sideways along the arm. In the pantograph type, the cutting heads are located firmly on the bridge arm, which can move sideways as well as along the line of travel. Movement of the arm is directed by a tracing device employing either an electric eye or a magnetic tracer.

b. Plasma Cutting. In plasma-arc cutting, also used in N/C burning machines, the normal cutting speed for a 12 mm (½-in.) plate is about three times the speed of oxygen cutting. However, the speed advantage decreases as the plate thickness increases. The plasma-arc nozzle is bulky and is, therefore, used primarily for machine cutting. Substantial reduction in the amount of fumes and noise result when cutting is done over a water surface. Consequently, large cutting tables have water basins immediately below them, the water level being less than 25 mm (1 in.) below the bottom surface of the plate being cut.

The quality of the cut is superior to that obtained with oxygen cutting. Also, less heat is transferred to the plate by the plasma process by virtue of the jet-like stream of gas removing the molten metal, thus, there is less tendency for the plate to distort during the cutting operation, especially if cutting is done only along one edge of the plate.

c. Air Carbon Arc. This operation is useful for back gouging and for excavating defective areas. The steel is melted locally by the intense heat of the carbon arc, and the high pressure air introduced around the carbon rod blows away the molten metal. Thus, the temperature of the adjacent surrounding material is much lower than would be the case with oxygen gouging, and this is one reason why carbon arc gouging is particularly useful in repairing castings. In addition, generally the excavation will show clearly the extent of a crack, if one is present, without requiring further mechanical gouging.

6.2 Cold Forming. Cold forming is used to produce plating of desired configuration. Excessive straining can reduce notch toughness properties in the direction normal to the forming, i.e. in the lengthwise direction of a rolled tube. The effect of cold forming bilge plates (outside fibre strain up to 1 percent elongation) should be of little consequence. However, cold forming a tube from thick plate, such as a heavy mast tube (outside fibre strain greater than say 4 percent elongation) might be significant in a highly stressed area.

a. Rolling. Rolled plates with shape in only one direction, such as bilge plates in the parallel middle-body, are shaped in bending rolls consisting of a large-diameter top roll and two smaller bottom rolls. If the housing at one end is demountable, complete circles may be rolled. A small amount of lengthwise set may be rolled into a plate by packing the large roll at the center with thin metal or wood strips. Normally about 150 mm (6 in.) of the lengthwise edges of the plate will be flat because the rolls cannot shape these edges. Sometimes it may be desirable to shape the edges in a press prior to rolling.

More force is needed for rolling high tensile steel (HTS) plates than for rolling the same thickness ordinary strength mild steel (MS) plates. Thus, the rolling capacity limits are affected by the strength of the material as well as by plate thickness and length. The equivalent thickness relationship is:

$$t_{\text{HTS}} = t_{\text{MS}} \sqrt{\frac{\text{yield point MS}}{\text{yield point HTS}}}$$

The equivalent thickness for ABS grade H36 steels would be 80 percent of the mild steel thickness for the same roll force.

b. Pressing. Some plates may be shaped entirely by a hydraulic press, sometimes referred to as a keel bender, provided the curvature and width are not excessive. Some yards have large press capacities and can shape heavy plates without furnacing.

Cold forming of shapes is sometimes performed in a hydraulic press with both horizontal and vertical rams operating independently. Some presses are N/C tape controlled. Channel shapes and I beams which are split through the web or have one flange removed are ordinarily straightened, or bent to the desired shape, by cold bending. One such frame bending machine consists of a horizontal reciprocating ram, with an adjustable stroke, located between two fixed supports.

Another method of bending angles and Tee shapes features a clamping device which can shrink (compress) or stretch the free edge of the web portion of the shape, Fig. 14. Shrinking the free edge bends the shape one way and stretching bends it the other. The work is moved intermittently past the clamp which is held in place and actuated by a press.

6.3 Hot Forming. Where curved plates or shapes cannot be cold formed by mechanical means, they may be hot formed by furnacing (to a red heat) or, in some cases, may be formed by line heating.

a. Furnacing. The floor area in front of the furnaces is made of perforated cast-iron blocks. The perforation holes are used to place drift pins and dogs for holding down guide bars and heated members.

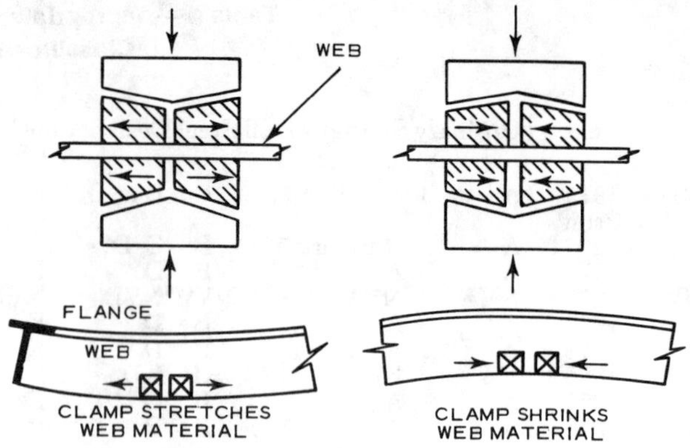

Fig. 14 Frame bending clamp

In shaping plates where there is considerable curvature in two directions, it is necessary to form the plate on a firm bed or jig constructed of plate sections and stiffeners and held securely on the furnace floor. If the bed is convex, the heated plate is forced down over the bed by hammers and dogs. If the bed is concave, a heavy iron ball may be suspended by an overhead crane, with a tripping device, and made to fall repeatedly on the plate to force it down onto the bed.

It should be noted that any heat-treated plates used in construction should be formed cold, or formed by careful line heating as noted below, if possible. If furnace-heated above the temperature which would affect the material, the plate's physical properties would be impaired and the furnaced plate would require a reheat-treatment.

In shaping frames, a heavy steel flat bar guide is bent to conform to the desired curve and bevel of the member to be shaped and held in place by drift pins and dogs. The heated frame member is placed against the guide bar and jacked into place as rapidly as possible.

b. Line Heating. In this process, a combination of line heating and quenching is used to shape plates. Quenching is accomplished by applying a water spray immediately behind the heating torch. The operator of the heating and quenching apparatus follows line patterns predetermined by calculation and experience. Compound curvatures can be achieved by this method with repeated heating and quenching. In some cases bulbous bows have been shaped by this method. Higher strength steels are more difficult to shape by this method than ordinary strength steels because of the higher yield strength of the material.

The temperature for line heating or any type of flame forming of ordinary and higher strength carbon steels is about 650°C (1200°F). For quenched-and-tempered steel, the temperature should not exceed the tempering temperature. In any case, special permission must be obtained from regulatory agencies before using this method to form quenched-and-tempered steels.

Section 7
Fabrication and Erection

7.1 Design for Production. Design for production has come to mean design for an orderly flow of assembly units and a very short erection period before launch because most shipyards now rely on just one or two building basins or building sites instead of several building ways. Therefore, design must be heavily oriented toward yard production. It follows that many constraints will be imposed on design by the requirements of yard facilities and production planning, and compromises in many areas will have to be made.

a. Constraints. Reference is often made to optimizing the design. From the practical point of view, however, the many constraints imposed by the shipyard characteristics and facilities will determine to a large extent the design which will be the most efficient for the yard when considering other factors such as availability of plans and vendor items, schedule of other ships, and time necessary for development work.

Some of the usual physical constraints are as follows:

- Maximum plate width for automatic plate blasting and priming facility.
- Longest plate length and largest panel size for panel line.
- Headroom in shops and painting facilities.
- Crane and transporting facilities in shops, on ship ways and outfitting pier.
- Cutting and welding equipment in shops and on ship ways.
- Environmental requirements involving access, ventilation, staging, etc.
- Machinery and outfitting requirements and incorporation with assembly and erection.
- Available labor force for the various trades.

Many of the above physical constraints would influence, if not determine, bulkhead spacing, web spacing, stiffener spacing, subassembly size and weight, locations of erection butts and seams and type of structural details. The degree of difficulty in fitting and welding, especially of erection units, is also taken into account.

A reliable set of contract plans and specifications is essential to orderly procurement and design development, otherwise several additional engineering studies must be undertaken with the result that further unforeseen compromises and delays may be imposed on the builder. Additional constraints would be applied when working with plans prepared by another yard, as often happens when a series of ships is built in several shipyards.

b. Design Details. Some design features conducive to ease of construction are:

- Flat surfaces instead of curved surfaces; single curvature rather than compound curvature.
- Flat bottom instead of deadrise.
- Flat sheer and camber.
- Weld details that permit machine welding or require no back gouging.
- Assembly procedures that minimize the amount of overhead and vertical welding.
- Drain holes in stiffeners that permit continuous fillet welds instead of drain holes that extend down to the plate surface.
- Features that permit selected machinery and outfit items to be installed in subassembly rather than on ship.
- Standard structural and welding details that best suit the particular yard.
- Provisions for good access for all trades.
- Simplification through reduction in number of pieces to be cut and handled.

A shipyard will generally have both standard structural details and standard welding details. Standard cutouts, standard brackets, standard drain holes, and standard framing connections, if repeated many times in construction, can be particularly helpful as regards providing proper access and clearances so the welder (and also the painter) can do a good job.

Because the fatigue strength of the higher strength steels such as the ABS H36 steels is little different from that of ordinary strength steels, greater attention to critical details should be given when the higher strength materials are used.

Even though several recent studies, such as that of the Ship Structure Committee (1977), have been made in connection with structural details, each shipyard must develop its own details to suit the particular ship. For example, there are several ways of making *oil and water stops* at bulkhead intersections as shown in Fig. 15. The one best suited for the job should be selected.

A feature frequently used to facilitate ship erection is a *shelf* type of connection, used in lieu of a butt welded connection. The shelf is a horizontal member on which a large structural unit can be placed and may take various forms as indicated in Fig. 16.

Another example relates to *scallop cutouts.* Scallops should be large enough, say a minimum radius of 40 mm (1.5 in.), so a proper fillet weld can be carried around the end of the scallop. However, the trend is away from the widespread use of scallops, and scallops are not recommended in stiffening members in way of completed plate butts; rather, it is recommended that the butt weld reinforcement be removed in way of the stiffener. In any case, scallops should not be cut near the end of brackets.

In connection with locating and cutting *holes,* it is desirable to set up a procedure to ensure proper layout of all major holes in decks and bulkheads. On most drawings, it is not possible to locate many of these holes with sufficient accuracy to serve the best purpose for the system under consideration as well as to provide proper clearance for other systems, which may only be in the development stage. In many structural members, holes or cuts not shown on plans should be approved by the engineering department before

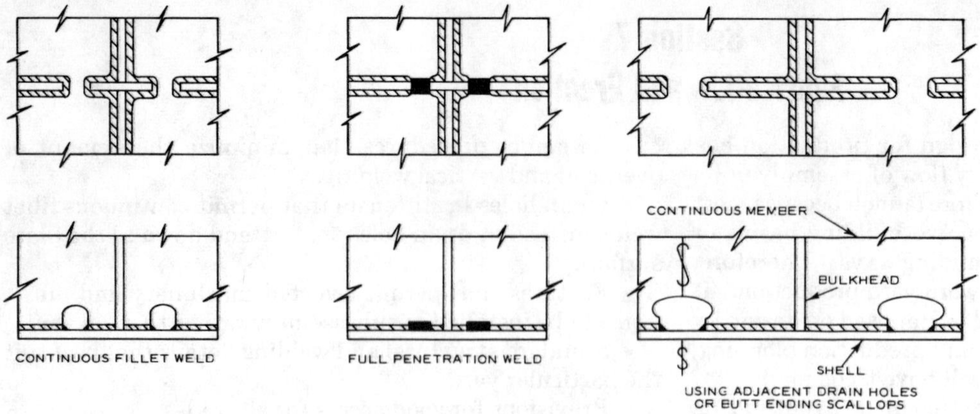

Fig. 15 Oil and water stops

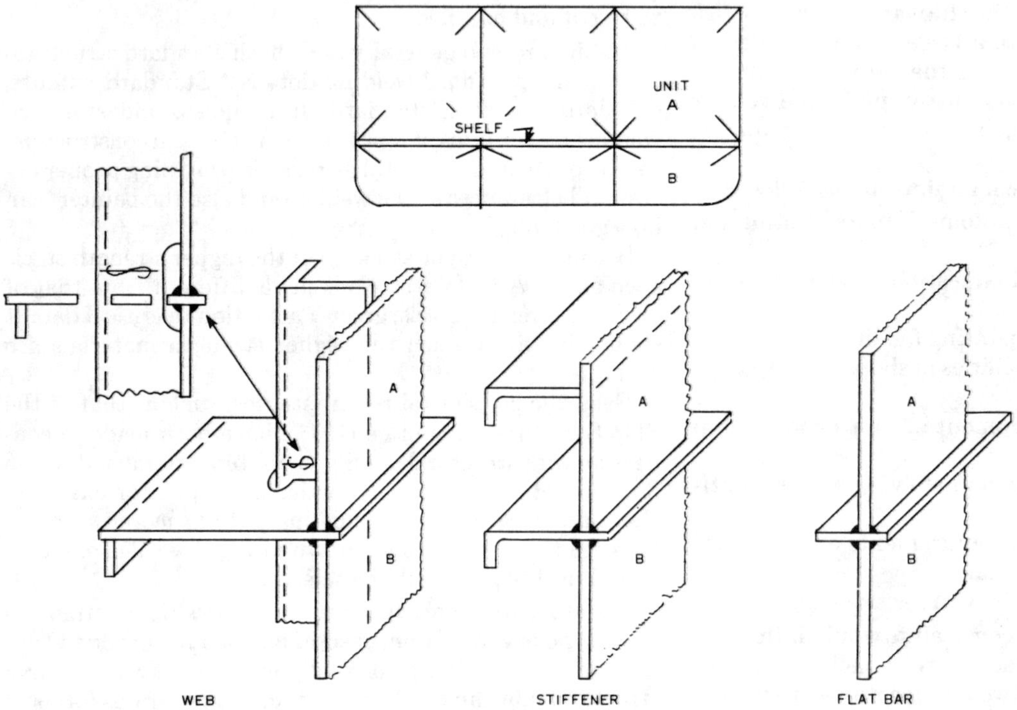

Fig. 16 Typical shelf type of connections

they are cut. Often it is found convenient to have a checker, who is in constant touch with the engineering department, handle approval of the hole layouts. Quite often, large elongated holes, for such items as ventilation ducts, have to be cut in structural girders or beams, and it is important that these holes be checked and approved for any required reinforcement before they are cut.

7.2 Access. Access for construction has three main facets. One has to do with providing necessary access for constructing the ship. Another has to do with meeting required safety standards, such as those imposed by the Occupational Safety and Health Administration (OSHA) and by state governments. The third has to do with design, which should attempt to include features which would help meet the requirements of the other two.

a. Access to Compartments. Several openings are normally cut in the bottom shell for access into the bottom structure during construction. These openings are also used for ventilation and cleaning purposes. In some cases, a portion of the bilge or bottom plating is left off. Large openings are often cut in bulkheads and in the side shell, especially in way of the machinery space, for easy access and for installing machinery and outfit items. Sometimes bulkhead openings are large enough to allow fork lift trucks to pass through. The final closing of any of these openings requires an approved welding sequence. Typical sequences are given in Section 10.

The shipyard design office, in many cases, provides permanent structural openings which serve the dual purpose of meeting the ship requirements as well as providing access during construction. An example of this is ladder openings cut in horizontal webs. If the ships permanent ladders were installed at subassembly so they could be used during construction, an added benefit would be realized.

When an assembly is built upside down, any ship's access openings will be near the overhead of the assembly.

Fig. 17 Mock-up of complex foundation

Therefore, the design should include additional openings near the bottom of the upside down assembly, if practicable, for access during construction.

b. Access for Working. The design office plays an important role in providing proper access for the trades and especially for fitting, welding, and painting. It should not be forgotten that a welder's equipment and protective clothing are rather bulky.

The joining together of large assemblies by automatic welding procedures, such as submerged arc or electroslag, requires a clear run for the welding machines. Where submerged arc is used to weld butts or seams in bottom shell or deck plates, openings must be cut in any vertical member in the path of the machine. These openings are quite large and are usually required to be filled in with insert plates.

In the case of electroslag or electrogas welding of a vertical shell butt, small cutouts must be cut in any stiffening member crossing the butt so that the sliding shoe, cables and hoses can pass through the member. Generally it is not necessary to fill in these cutouts unless the stiffening member is small.

c. Mock-ups and Models. Full size mock-ups of critical areas can be used to determine feasible accessibility arrangements, mainly for structure, ventilation, and piping. Fig. 17 shows a mock-up made to determine accessibility as well as the exact fabricating sequence for a complex foundation. Some specifications require that mock-ups of certain spaces be approved before the detail design is started.

Scale models in many cases can be used to serve the same purpose when accessibility is, perhaps, less critical.

7.3 Staging and Ventilation. The federal regulations put forward by OSHA along with applicable state regulations must be adhered to when a ship is being built or repaired. Although these regulations are extensive, only two pertinent items will be discussed here, staging and ventilation.

a. Staging. There are specific requirements for staging, scaffolding, ladders, backrails, gangways, etc. For example, backrails are required when staging is more than 1.5 m (5 ft) above a solid surface or at any distance above water, otherwise safety belts and lifelines are required. In some cases safety nets might be employed.

Common practice is to provide staging planks rough sawn to at least 50 × 250 mm (2 × 10 in.) and made of spruce, fir or long-leaf yellow pine. Staging walkways should be a minimum of two planks wide.

Both ramps and elevators are often built into access staging towers, offering the advantage of ease in carrying tools and light material. Some towers have double stairways, one for upgoing and one for downgoing traffic.

In order to avoid erecting elaborate staging or using cumbersome scaffolding, special mechanical lifts are often used. The mobile hydraulic, articulated tower type of machine can place a man almost anywhere on the outside of the ship provided there is room below for the tower to maneuver. These towers can also be placed on a barge so that the outside of the ship can be serviced without hanging staging over the side.

There are numerous patented staging structures, and each would require acceptance by regulatory agencies before use. One device for welding and painting on a flat side is the cage lift or portable air platform. The vertical positioning is regulated by the workman in the cage. Sometimes, a structural part of the ship, such as a longitudinal girder or stiffener, will be made larger than normal so that it can act as a walkway during construction. Interior staging is usually held up by wood or pipe supports. Steel bars or shapes may be bolted to uprights to support the planking. Overhead clips welded to the structure are normally left in place if they do not interfere with the operation of the ship, and they may be helpful for future overhead work or inspection. Movable platforms mounted on rollers or, where practicable, hydraulic lifts or towers may be used.

Unique stagings are often required for specialized work. Movable welding towers developed for use on the outside of large tanks protect the welder and welding equipment from wind and moisture. Fig. 18 shows a highly maneuverable lift used for construction inside a large cargo tank. In some tanks, the rather delicate insulated tank bottoms, such as in some LNG ships, must be specially protected against damage from staging feet or from the wheels and support feet of movable lifts.

b. Ventilation. Strict rules govern air quality in working spaces, especially in confined areas where fumes and smoke from welding or painting exist. In confined spaces, there must be more than one access except where the structure or arrangement of the ship makes this provision impracticable. The access openings are also used for ventilation. Where the working space cannot be ventilated adequately, the workers must wear respirators, either a filter cartridge respirator or an air line respirator, depending on the conditions. Requirements are somewhat more severe when welding is done with inert gas or low hydrogen electrodes or where metals of toxic significance such as galvanized or lead based materials are cut or welded.

Fig. 18 Hydraulic lift inside tank during erection of steel structure on LNG carrier

Air is monitored in confined spaces to detect toxic or explosive atmospheres. Dangerous atmospheres may be removed by inerting in some cases or by ventilating in others. Where an explosive atmosphere exists, special precautions are taken to prevent any work, such as hot work, being done in the vicinity that could trigger an explosion. As an example, if a tank contained fuel oil, no welding or hot work would be permitted on the tank's fill or vent lines or on any other critical place in the system.

7.4 Assembly Units. Large areas are required for constructing assembly units. Flat platens and assembly panel lines, as described in Section 2.4, are used for flat work. Plate jigs or pin jigs are used for curved plate units.

While the normal assembly procedure is first to weld framing members to plate panels and then weld the deep web frames, the alternate procedure of first welding the web frames is occasionally used.

a. Platens and Jigs. Flat platens are made in a number of ways but the essential feature is that they be true and on a firm, horizontal base, especially when large assemblies are constructed.

The most difficult assembly work is with curved surface units built in jigs, either plate jigs or pin jigs. Most pin jigs have adjustable pins. The construction of an accurate jig is difficult and requires a firm base on which to construct the jig. One thing that creates difficulty is the orientation of the jig. The frame lines, waterlines, and buttocks are often skewed so that the unit will be approximately horizontal when being assembled. Since the outside of the assembly plating is against the jig, corrections have to be made for plate thickness because the molded surface is invariably on the upper side of the plate where the framing is attached. Computer programs have been developed for determining the jig's surface points from a given horizontal flat base.

b. Unit Plans. Unit plans are sometimes developed to facilitate work on a unit and to utilize the less skilled yard personnel. Since these plans give only that information needed to construct, outfit and check the unit, the workmen need not be concerned with the relation of one unit to other units or to the ship. Occasionally, a plan may be drawn to the orientation of the unit during assembly; for example, a deck unit may be drawn upside down. As regards alignment with other units, dimensional control, as discussed in Section 8, must be relied upon to ensure proper erection and joining on the ways. This control must be clearly understood and adhered to.

Fig. 19 is a simplified unit plan to illustrate the type of information required. Normally all principal dimensions are given from three master reference planes. In practice, several sheets or plans may be necessary to describe a complete unit. Standardization of details makes it possible to refer to a standard details plan rather than show all details on each unit plan. If extra stock is to be added, this should

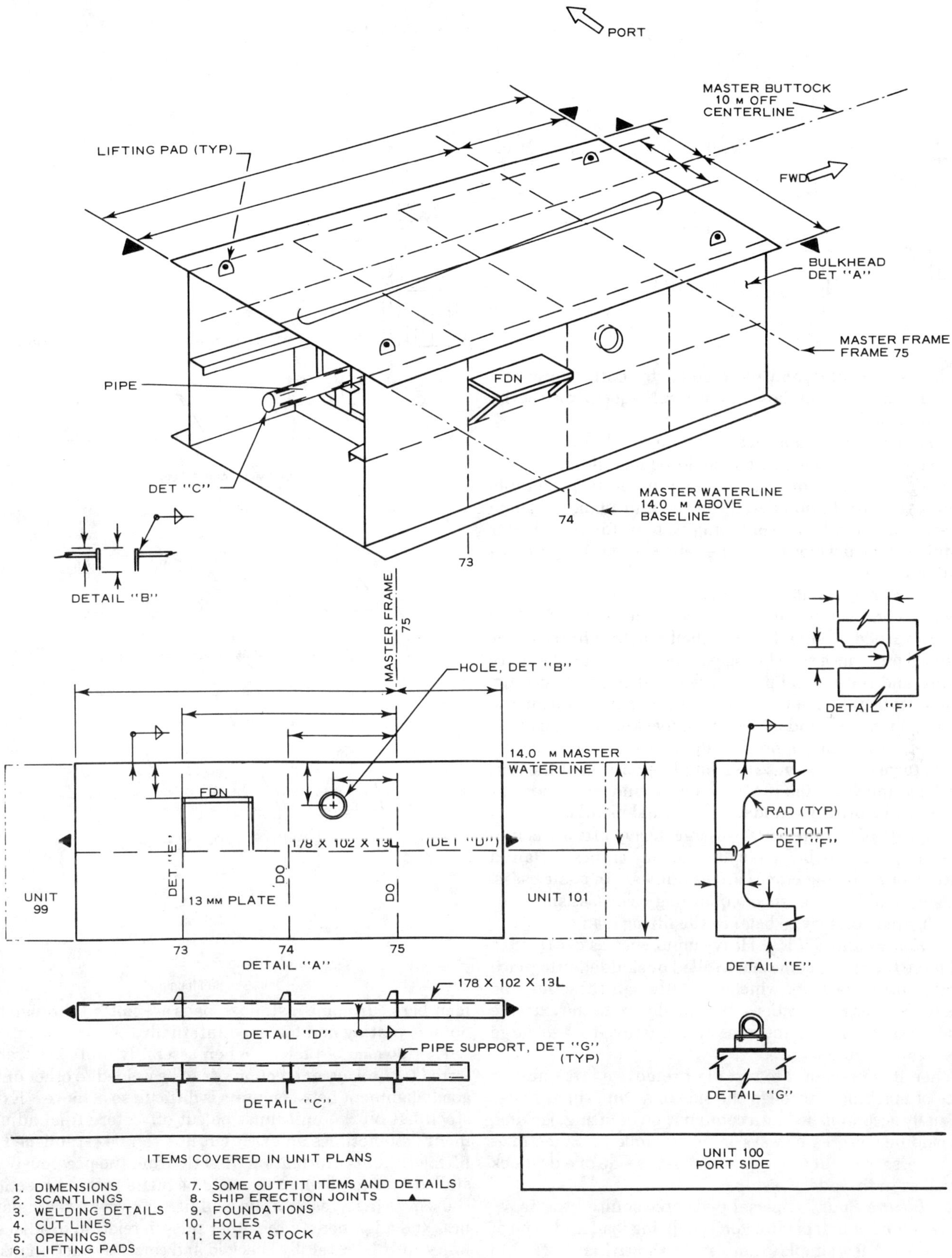

Fig. 19 Simplified illustration of unit plan

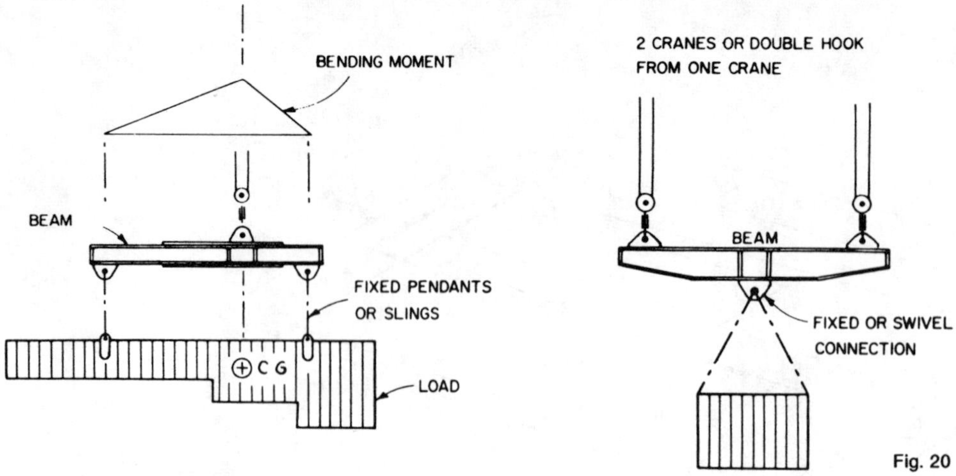

Fig. 20 Equalizing beams to balance lifting loads

be clearly indicated and understood by both those who construct the unit and those who will subsequently erect the unit on the ship.

Too much information on a unit plan could be confusing. For example, if an outfit item could be installed easily by referring to a layout or system plan, it might be preferable to omit the installation details on the unit plan. In some instances a separate plan of piping systems for a particular module will be developed. Experience of the shipyard is a reliable guide.

7.5 Handling of Units. Some units, when they leave the panel line or assembly area, must be turned over (if a deck unit) or turned on side (if a side shell unit). This requires technical assistance as to lifting pad design, center of gravity of unit, and the general procedure for lifting and turning. Care must be taken not to deform the assembly during the turning operation, and added strongbacking is sometimes required. In addition, proper supports must be provided for the turned unit because the flat plate portion of the unit may be as much as 10m (30 ft) above ground level and support must be provided under the internal members.

a. Multiple Crane Lifts. Large, heavy lifts are sometimes required to be made by pairing cranes. Paired gantries or revolving cranes are examples. In some cases, single or double spreaders or equalizing beams, as shown in Fig. 20, are necessary to balance the lifting load.

b. Rollers and Skids. Heavy units, such as deckhouses and large tanks, are sometimes rolled or skidded into place using multiple jacks or winches. Although these may be considered special operations, they do illustrate the extreme efforts taken, in the interest of economy, to use large preassembled units.

When it is necessary to install or remove a large unit or piece of machinery through the side of a ship, the transfer is usually accomplished by a combination of lifting, jacking, and skidding on greased ways or on steel rollers. Sometimes it is necessary to cut large openings in the side of a drydock as well as in the side of a ship.

c. Lifting Pads. Special pads are required for heavy lifts, their design depending on the lifting load and type of structure. A few simple examples are shown in Fig. 21. On heavy lifts, the pads should be backed up by stiffening

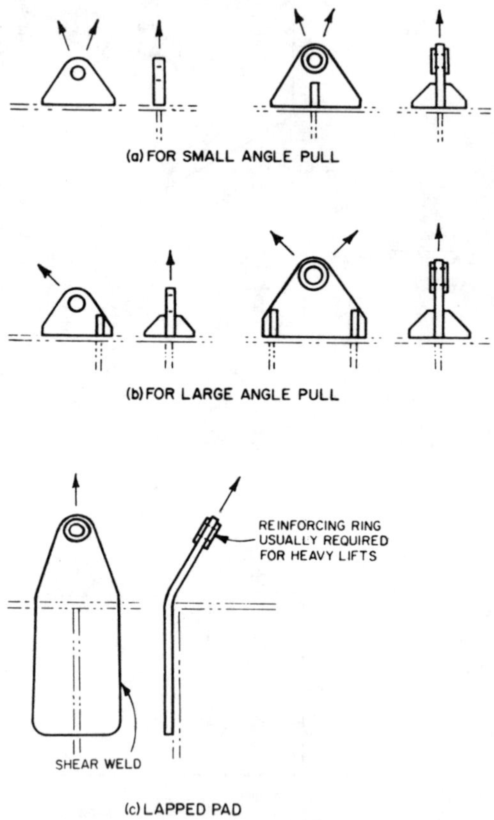

Fig. 21 Simple lifting pads

members. In some instances, pads are slotted through the outside plating and fastened internally.

7.6 Joining of Units. When assembly units are transferred to the ways or erection site to be welded to other units, good alignment between units is difficult to achieve. Extra stock left on the unit must be cut off before final adjustment. Sometimes an exact cut line can be established by lifting the cut information directly from the previously installed unit to enable the cut to be made prior to erection, thus ensuring a reasonably good fit. However, the squareness and alignment of the other sides in relation to adjacent units must be carefully checked and corrective action taken if necessary.

If heavy lift equipment is available, large units made up of several smaller units welded together can be lifted in place. In some yards these large units are simply moved horizontally to be joined to other units, such as at Ingalls' Pascagoula yard where complete transverse sections are joined together to form a completed ship.

The largest units to be joined together are those for lengthening or jumboizing a ship. Here a center section or a whole new forebody might be fastened to the remaining portion of the original ship in a drydock. In the case where a forebody section is joined to a stern section, the connection joint will usually require fairing plates to join the sections or a cutting back of the plate seams and longitudinal stiffener welds on one of the sections so as to provide some flexibility in alignment.

Some newly constructed ships are built in two sections, and the two joined together in a drydock or afloat. Sometimes, where two sections of a new ship were joined afloat (Belch, 1976) a watertight working cofferdam was placed under the ship so that the two halves could be fitted and welded afloat. Large aligning devices fastened to the ship sections were used to help position the two sections after the sections had been properly ballasted.

7.7 Compartment Testing. Testing may be conducted to ensure both watertightness and structural integrity. The tests made on more or less conventional structures are described here; the subject of containment for specialized cargos of different specific gravities or other special characteristics are covered more completely in Chapter XI.

a. Hydrostatic Tests. These tests are conducted on boundaries of tanks, chain lockers, and sometimes on tunnels. Specified water test heads are usually to a few feet above the top of the tank or to the top of the overflow.

Cargo oil tanks should be tested before launch, but in large tankers where the building ways would not support the weight of the test water, and where a water test is specified, it is customary to test the tightness of the lower portion of the tank before launching by filling the tanks about one-third full, and then to test the remaining portion of the tank when the ship is afloat. Test water should be removed from clad tanks as soon as possible in order to prevent possible contamination or pitting of the clad surface. Rudders are usually tested with a nominal air pressure of 35 to 70 kPa (5 to 10 psi) rather than with a water head.

An air hull tightness test may be accepted as an alternative to hydrostatic testing of deep tanks and double bottoms intended for water ballast, and of cargo tanks other than those adjacent to cofferdams, pump rooms, machinery, or water ballast tanks.

b. Hull Tightness Tests. A compartment may be put under air pressure of about 14 kPa (2 psi). All fillet welded boundary connections and erection joints are to be examined under air test by use of a suitable leak detection solution.

In tanks where special paint coatings are applied, it is permitted to apply coatings before hydrostatic tests provided all welds are surveyed before coating, and all fillet welded boundary welds and erection joints are air tested before coating. In some instances the tank is given an air test before coating is applied to the boundary welds, and then, after the final coating has been applied, the tank is given a hydrostatic test.

A hose test is given to ordinary watertight compartment bulkheads, decks, side shell, deckhouses, etc., where it is impractical or unnecessary to conduct a hydrostatic test. A fire hose is held at about 3 m (10 ft) with 207 kPa (30 psi) pressure. In some cases an air test could be used.

c. Hydropneumatic Tests. For independent tanks designed to carry cargo of a density markedly greater or less than that of sea water, special consideration is given to a combination of hydrostatic and air tests (hydropneumatic tests) so that stresses approximate, as far as practicable, the maximum design stresses without exceeding 90 percent of yield stress of the tank or its supports. For example, the 36.5 m (120 ft) diameter aluminum spheres for one design of LNG carriers were tested by filling the tank half full of fresh water and then applying 214 kPa (31 psi) of air pressure above the water for the strength test (Veliotis, 1977). The air pressure was reduced to 172 kPa (25 psi) and the tank inspected for leaks, the upper portion being inspected with a soap solution. The large rectangular aluminum tanks for another design of LNG carriers were tested in similar manner but pressurized to a lesser degree.

d. Low Temperature Cargos. Classification society rules require all primary containers, insulation, and cargo-handling equipment to be tested under service conditions with the cargo at minimum service temperature.

Section 8
Dimensional Control

8.1 Dimensional Control and Tolerances. The broad subjects of dimensional control and tolerances are, to a large degree, related. Each is involved with:

- Structural acceptability to satisfy regulatory agencies and any special strength requirements
- Operational acceptability to satisfy an owner's special requirements, if any
- Fabrication needs of the shipyard to facilitate construction with a view to reducing costs.

Dimensional control is primarily concerned with both assembly and fabrication methods and practices. In connection with the last item above, the shipyard must decide on how much effort, if any, should be expended on improving dimensional control. For instance, the yard may wish to reduce certain dimensional or fit-up tolerances in one operation in order to improve the efficiency of a following operation. Some kind of analysis of cost-effectiveness would be required. Dimensional control would be expected to be much more exacting in the case of a 3-dimensional block assembly, such as an innerbottom unit, than for a 2-dimensional assembly such as a bottom shell unit.

While proposals have been made favoring standard tolerances, it is reasonable to state that tolerance *limits* should be based on technical acceptability and experience. Any guidelines for tolerance limits should come from classification societies based on input from research, industry, and service experience. The matter of acceptable tolerances is similar to acceptable welds in that much would depend on the stress condition, the location in the ship and the judgment of the classification surveyor.

8.2 Molded Lines and Measurements. The key to all lofting and N/C development is the relation of the various structural members to the molded lines. This relationship should be based on the intended fabrication procedure and sequence rather than on historical precedent. The inside of the shell plate is normally on the molded surface. Occasionally, the molded line may be somewhere between the plate surfaces as for rudders, stern frames or even for decks where a smooth outside or top surface is desired and where plates of different thicknesses are used. Generally, the molded line for a bulkhead made of various thickness plates will be on the smooth side and also on the side of the plate where stiffeners are attached, but this is not always the case. It may depend on whether the plate panels are turned over in the panel line, or on which way the plates are lapped or in which direction the flange of a stiffener faces. Therefore, a well established molded line base is the first order of business in the preparation of plans for lofting, but first, knowledge of how the ship is to be constructed is essential.

In cases where bulkhead stiffeners are not on the smooth side and where the stiffeners cross plate seams, the increment of increasing plate thickness is normally only about 1.5 mm (1/16 in.) and should cause no serious welding problems although it is necessary to remove the crown of the seam weld in way of the stiffener or to notch the stiffener in way of the seam. In order to avoid these inconveniences in panel line work, it is desirable to arrange the stiffening members parallel to plate seams where practicable.

Measurements are taken from main structural members or from reference frame lines, waterlines and buttocks marked on a plate or assembly. Measurements are sometimes taken from reference marks on platens and jigs on which the assemblies are built. The reliability of the measurement depends on the past history of the plate or assembly. For example, if a reference line were marked on a plate and then the plate welded into a plate panel, the true location of the marked line would be questionable. A reference line marked after final cut would be more reliable. From a practical point of view, measurements in the field made from main structural members, such as a bulkhead, are often preferred. However, the relation of the member to the molded line must be known because most dimensions on plans are given from molded lines and not from structure.

Satisfactory measurements may be accomplished by a number of methods, but the techniques must be proper. The most common tool is the steel tape. New steel tapes can be checked against a standard tape for accuracy. The determination of squareness requires very accurate calculations and measurements. A check for squareness may be made using the familiar 3-4-5 triangulation or using widely spaced trammel points with metal scribes. Piano wire is satisfactory for aligning over short distances, but the sag in the wire must be taken into account. The water tube method for determining level planes is a reliable method. The transit telescope and *laser* (Todd, 1974) are employed for various aligning jobs such as for placing keel blocks and aligning shafting. The laser is considered more convenient for some work because the laser spot can be readily and accurately picked up, especially under poor lighting conditions. An unusual application is described in Belch (1976) where the transverse mating surfaces of the forward and after sections of a newly built tanker were trimmed based on lines established by a laser using a right angle pentagonal prism. Care must be taken when using lasers emitting more than 1 mW/cm^2 because of possible radiation damage to the eye. Other sophisticated measuring and aligning methods, such as *photogrammetry* (Todd, 1976) are being investigated as to possible applications in shipbuilding. On the spheres for an LNG carrier, the shape and volume were surveyed internally by photogrammetry using 400 targets and nine cameras, each target appearing on four photographic plates (Veliotis, 1977). In summary, it is reasonable to say that the accuracy of any measurement will depend as much on the techniques employed and quality control as on the method of measurement.

An example of simplifying measurements is shown in Fig. 22. The location of the stiffener butt in *A* can be checked more easily than that in *B*. The exact cut line would depend upon the welding detail. Of course, this presupposes that the plating edges in both cases are straight and true. If they are not, then it would be necessary to make a slight correction to the stiffener cuts.

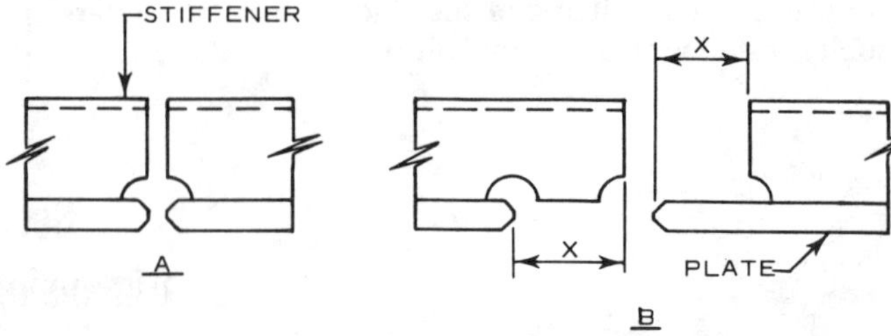

Fig. 22 Location of stiffener butt

When cutting plate edges or measuring plates, the kerf allowance (the distance from center of cutting torch to cut edge of the plate, Fig. 23) must be taken into account. Normally the allowance is 2 to 3 mm (1/16 to 1/8 in.) depending on the torch, the type of cutting and the plate thickness. In

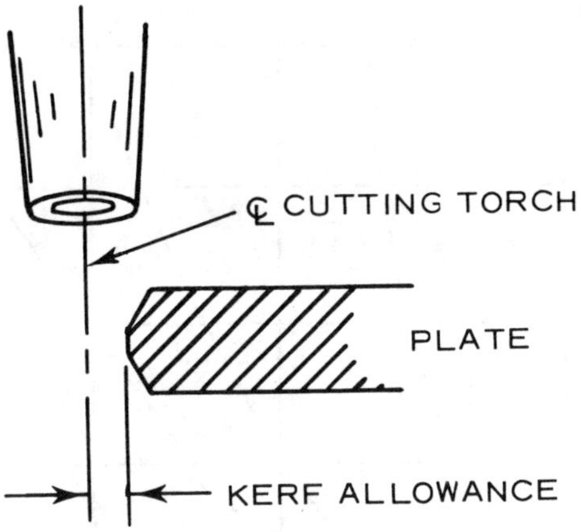

Fig. 23 Kerf allowance

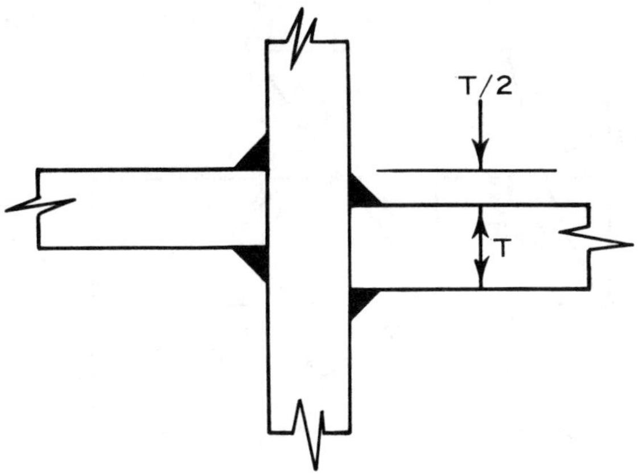

Fig. 24 Plate alignment

some machine operations, this allowance can be taken care of automatically.

8.3 Alignment of Structure. One of the problems in assembly work is to ensure proper alignment of one assembly with another. Here again 3-dimensional alignment is much more difficult to achieve than 2-dimensional alignment. In some cases, as for framing members, it is possible to leave the members loose for a short distance at the ends so they can be properly fitted when the assemblies are tied in. Unfortunately, it is not practicable to do this for many major bulkheads and webs.

Another problem is that of aligning structures on opposite sides of a bulkhead where there is no penetration of the bulkhead. Classification society rules generally state that the alignment should be within one-half the plate thickness, Fig. 24. On relatively thin plating, such as most bulkhead plating, it is possible to see the outline of the fillet welds made on the opposite side of the plate, thus, making it possible to locate accurately the opposite-side member. Ultrasonics or predetermined reference marks may also be used to locate welded members on the opposite side of the plate. To use ultrasonics, the member must be welded because the ultrasonic beam will pick up only the welds connecting the member and not the member itself.

When aligning units that are not square, corrections may be made on the job to an individual unit, but, if this problem is prevalent among several adjacent units, the cumulative effect could be substantial and could even affect the alignment of other non-adjacent units. It is rather difficult to check the squareness of a large assembly because the plate edge preparation usually makes it necessary to mark very carefully a corner reference point close to the corner of the assembly. The measuring tapes should be accurate and care should be taken that temperature variations along the tape and throughout the assembly are small, i.e. the sun should not be shining on one side of the assembly.

a. Curved Plate Assembly. Not only is a curved plate assembly the most difficult to construct but it is also the most difficult to check for dimensional accuracy. If extra stock on the plating and main stiffening members is added for cutting in on the ship and if the ends of the framing are left loose, more than moderate accuracy should not be necessary or expected. Advance planning should take these factors into account.

b. Shipways. Whether the building site incorporates launching ways or whether the ship is built in a flat basin, the bottom of the ship will distort somewhat as structural weight is added because the foundation and blocking are inherently flexible to various degrees. Heavy weight-boxes are often placed on the bottom units to help hold them down and align them with other bottom units. After the bottom is secure, the additional distortion due to adding more structure is not significant to the strength of the ship. The keel line can be kept reasonably straight by proper erection and welding sequence and by adding additional blocking under heavy concentrated weights. Sometimes cribs are kept slightly high at first to allow for settlement when heavy concentrated loads are added.

8.4 Distortion Due to Temperature Differentials. The heating or cooling of a portion of a structure will cause distortion and, if the heating or cooling is off center, the structure will bend. The action of the sun's rays on the deck or on one side of the ship can cause the ends of the ship to deflect several inches. Bending is caused by the heated deck or side wanting to expand and the remainder of the ship resisting the expansion. Even when the ship is afloat after launching, it may deflect several inches as a result of these thermal gradients.

When measuring the keel line or when taking strain gage measurements on the structure, it is customary to take these measurements at night to minimize the diurnal temperature variation effects.

8.5 Welding Shrinkages. When welding is applied to a plate, the plate material will shrink slightly. If shrinkage data based on experience are not available, the estimates given in Table 2 may be used. The figures in this table have appeared in many specifications and guidelines, but considerable variation from these estimates can be expected. Shrinkage allowances can be programmed into computer input for neat cutting of plates. Of course, if the final cut

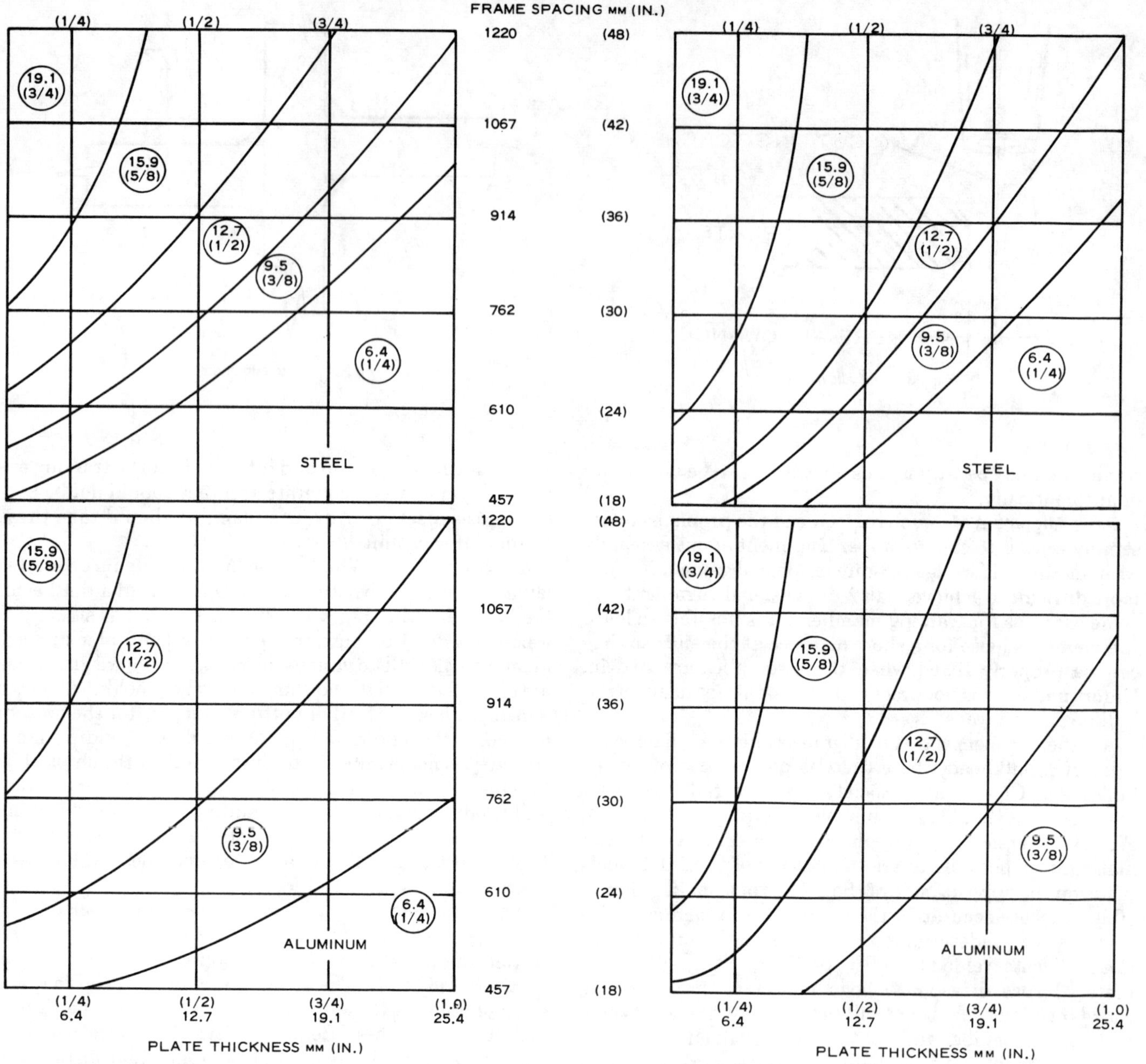

(A) PRIMARY STRUCTURE

SHELL, UPPER STRENGTH DECK, MAIN LONGITUDINAL STRENGTH MEMBERS INCLUDING TANK TOP WITHIN THE MIDSHIP 3/5 LENGTH, BULWARKS AND EXTERIOR SUPERSTRUCTURE BULKHEADS. FOR TRANSVERSELY FRAMED SHIPS, REDUCE TOLERANCE BY 3.2 MM (1/8 IN.) FOR ABOVE MEMBERS EXCEPT BULWARKS AND EXTERIOR SUPERSTRUCTURE BULKHEADS.

(B) SECONDARY STRUCTURE

BULKHEADS FORMING BOUNDARY OF LIVING SPACE, DECKS WITHIN HULL AND SUPERSTRUCTURE IN WAY OF LIVING SPACES, DECKS EXPOSED TO WEATHER, MAIN TRANSVERSE BULKHEADS, INNER BOTTOM GIRDERS. FOR OTHER STRUCTURAL BULKHEADS AND DECKS, TOLERANCE MAY BE INCREASED 3.2 MM (1/8 IN.).

Fig. 25 Permissible unfairness in welded structures as given in U. S. Navy Specifications 0900-000-1001

is made after welding, no shrinkage allowance is necessary.

8.6 Unfairness of Plating. This subject as regards acceptable unfairness has been handled over the years primarily by good judgment. Since plate unfairness is often unavoidable, consideration must be given to the structural importance of the part and the past experience with similar structures. Although Fig. 25 was developed for military

Table 2—Weld Shrinkage Allowances

BUTT WELDS	
Transverse	1.6 to 2.4 mm ($^1/_{16}$ to $^3/_{32}$ in.) for all thicknesses
Longitudinal	
Over 12.5 mm ($^1/_2$ in.) thick	0.8 mm ($^1/_{32}$ in.) in 3 m (10 ft)
9.5 to 12.5 mm ($^3/_8$ to $^1/_2$ in.) thick	0.8 to 1.6 mm in 3 m ($^1/_{32}$ to $^1/_{16}$ in. in 10 ft)
6.5 to 9.5 mm ($^1/_4$ to $^3/_8$ in.) thick	1.6 to 2.4 mm in 3 m ($^1/_{16}$ to $^3/_{32}$ in. in 10 ft)
6.5 mm ($^1/_4$ in.) and thinner	1.6 to 3.2 mm in 3 m ($^1/_{16}$ to $^1/_8$ in. in 10 ft)
FILLET WELDS	
Tucking Allowance[1]	
Over 12.5 mm ($^1/_2$ in.) thick	No allowance
9.5 to 12.5 mm ($^3/_8$ to $^1/_2$ in.) thick	0.4 mm ($^1/_{64}$ in.) each stiffener
6.5 to 9.5 mm ($^1/_4$ to $^3/_8$ in.) thick	0.8 mm ($^1/_{32}$ in.) each stiffener
6.5 mm ($^1/_4$ in.) and thinner	1.6 mm ($^1/_{16}$ in.) each stiffener

1. Tucking allowance was developed for use on flat plates with continuously welded stiffeners. Intermittent welding will result in about one half the tucking tabulated above.

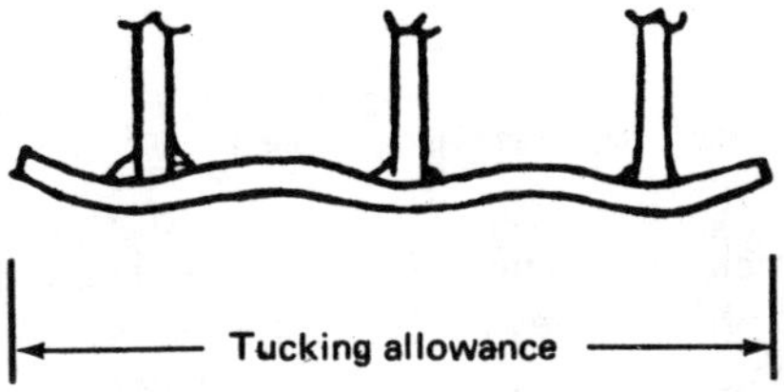

vessels it may be used for guidance in commercial ship construction.

Methods for straightening plate panels are given in Section 10.7 of this Chapter.

Section 9
Surface Preparation and Painting

9.1 General. Surface preparation and painting are major operations and must be integrated with the sequence of fabrication, assembly, and erection. They also must be considered in connection with tank testing as indicated in Section 7 of this chapter. Details of coating systems and applications may be found in Chapter XIV.

9.2 Abrasive Blasting and Priming. Plates and shapes are generally blasted and primed at the shipyard before fabrication. This is usually done in automatic blasting and priming facilities. In cases when fabrication schedules are tight, some steel may be blasted and primed before it is shipped to the shipyard. Blasting removes mill scale and cleans the plate. The primer protects the steel during fabrication and provides a surface on which final coats of paint can be applied without a major cleaning operation after erection.

Blasting on the ship and on some assembly units is done with portable equipment, and often it is done under shelters which provide various degrees of controlled ambient conditions. When humid conditions exist inside of tanks, dehumidifiers are sometimes required to keep the blasted steel dry until paint can be applied. Abrasive blasting, of course, interferes with other work being done in the immediate vicinity. Clean-up inside a tank is a major chore, and holes are often cut in the bottom of the ship to remove the grit.

Some shipyards have large blasting and painting facilities which can handle completed assembly units. These facilities are normally used when finish coatings are applied.

9.3 Finish Coatings. It is generally found more economical to apply internal finish coatings after assembly, such as in a large painting facility, than after ship erection. Nevertheless, touch-up work would be necessary in way of damaged areas, erection welds, and patched access holes.

Where a multi-coat system is applied, a complete finish coat might be applied after a tank is completed and tested.

Some special coating systems require dehumidifiers and explosion-proof equipment. In addition, there may be critical maximum and minimum drying times for each coat of a multi-coat system, thus establishing a range of time within which the overcoats should be scheduled.

The exterior finish coatings are usually applied just before sea trials or delivery, the underwater coatings including antifouling are, of course, applied prior to launch or float-out. Sometimes finish bottom coats are applied in drydock, if drydocking is scheduled at about the time of trials or delivery. Finish topside painting is usually done as late as possible when the ship is relatively clean.

9.4 Reduction of Scantlings Due to Corrosion Control. Classification societies will permit a substantial reduction in certain scantlings if the steel is effectively coated with a special protective coating. If such a reduction were desired and approved, the tanks would receive special coatings; otherwise the tanks might be left uncoated. Where reduced scantlings are used, the structural plans show two plate thicknesses, one, the actual reduced thickness and the other,

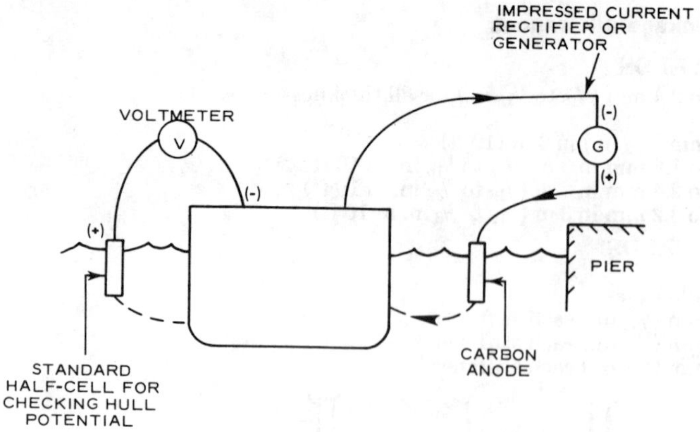

Fig. 26 Schematic of impressed current system

the unreduced Rule thickness. The latter thickness is given to aid in subsequent ship surveys. Maintenance of the coating to avoid corrosion and early replacement becomes the owner's responsibility.

9.5 Impressed Current Protection. After launching and while the ship is lying at the outfitting pier, it is customary to protect the underwater hull of the ship from accelerated electrolytic corrosion by an impressed current system. The more efficient the underwater paint is in isolating the steel hull from seawater, the more active will be the corrosion at damaged paint areas and holidays.

The impressed current system consists of a low voltage power source with its positive side connected to an anode placed in the water near the ship (cathode) and its negative side connected to the ship, Fig. 26. The most commonly used anode is the carbon anode, which wastes away very slowly, although a steel plate may be used. Normally several anodes are placed in the water along the pier.

When the impressed current raises the potential of the steel hull to about −0.8 volts relative to a standard copper-copper sulfate *half-cell*, the hull is considered reasonably well protected. Higher potentials will not help. The potential is checked at various places around the ship to assure overall coverage.

When a ship receives electric current from a shore source, the ship should be properly grounded so that the current can flow back to shore via the ground. If the ship is not grounded, electrolytic corrosion can occur at bare or thinly painted places where the current leaves the ship on its way through the water back to shore.

When welding is done on the ship with DC, the ground should be made from the ship directly back to the DC generator and not to the pier. If convenient, the DC generator could be placed on the ship. With AC welding, the influence of welding on electrolytic corrosion is far less significant.

Section 10
Hull Steel Welding

10.1 Design. This section discusses the application to ship construction of the welding processes and inspection described in Chapter VIII. Design for welding encompasses those things which facilitate welding and which promote efficiency. Four obvious factors are downhand welding, good fit-up, good access, and the high deposition rate associated with machine welding. Three other factors which can markedly affect welding progress are one-side welding, priming paints, and standard details.

a. One-Side and One-Pass Welding. One-side welding with submerged arc has been developed so that reliable seam welds can be made in panel line production. Because the weld quality is sometimes inconsistent, necessitating repairs on the opposite or root side, some yards turn over the plate panels and weld the back side. An advantage in turning over a plate panel is that it presents a flat surface on which to place webs and stiffening members when different thickness plates are used.

• *Tape backing,* such as a fiber glass tape or a ceramic backing held on with tape has made it possible to weld many butt welds in the field from one side without back gouging. Ceramic backing has been particularly useful. Automatic or semiautomatic welding can be used to complete a joint after one or two root passes have sealed the tape-backed joint to prevent burn-through. The root passes are usually made with flux-core wire, inert gas (MIG) welding.

• *Electroslag and electrogas* welding requires no back gouging or second weld and has been used extensively for welding vertical butts in flat side shells. However, the butts in the sheerstrake and other areas where preservation of toughness is of concern, should not be welded by this process unless it can be demonstrated that the required notch toughness properties can be met.

• *Consumable guide tubes* can be used to make short vertical welds using the electroslag process and by constructing a dam, normally made of heavy copper, around the joint to be welded. This method has been used to weld vertical butts of slab longitudinals and butts in the web plates of tee longitudinals. In some cases, holes have been cut in the deck to permit the butt weld of a slab longitudinal to be completely welded from above the deck. After welding, the weld portion protruding above the deck was flushed off.

b. Welding Over Priming Paints. A dry paint film thickness less than about 18 microns will normally permit manual fillet welds to be made satisfactorily on the primed surface. However, when automatic submerged arc fillet welds are made simultaneously on both sides of a stiffener, as in panel line work, it is generally necessary to remove the paint film in way of the welds by grinding or other means. The U. S. Navy has rather restrictive requirements for making fillet welds on painted surfaces.

Priming paint should be removed from groove or beveled joints. Critical areas are sometimes masked off with masking tape, but then the taped area must be cleaned before welding or painting.

c. Welding Details. Welding details are determined primarily by the type of work and by yard facilities. There are many variations of so-called *standard* welding details and each variation will require classification society approval. Fig. 27 gives examples of typical *butt weld* joint details for various methods of welding:

- *For manual welding,* plates up to 6 mm (1/4 in.) can be welded without beveling the plate edges. For plates over 6 mm, a 60 degree included angle bevel is normally used and back gouging is required to remove slag from the root. The 1/3–2/3 weld is used primarily in the field where the 1/3 weld is overhead and the 2/3 weld is downhand. When a butt weld is made from one side, using either a backing bar on a tape backing, the minimum root opening is 6 mm (1/4 in.) and the included angle may be reduced to 45 degrees. The included angle may be reduced further if the root opening is increased. This further reduction in included angle could be beneficial when the plates are over about 40 mm (1 1/2 in.) thick. For example, for a 50 mm (2 in.) plate, a 20 degree included angle is usually satisfactory for a 12 mm (1/2 in.) root opening; the amount of weld metal required would be reduced by about 20 percent and the angular distortion would also be reduced.
- *For single-arc machine welding,* plates between 15 and 20 mm (5/8 and 3/4 in.) in thickness require a bevel on one side, and for plates over about 20 mm, a bevel on both sides.
- *For 3-arc machine welding,* as used in panel line work, the joint details shown are for plates that are turned over and also for one-side welding. There are many different details for making one-side welding, each having been developed after much experimentation to suit a particular yard's needs.
- *For electroslag and electrogas welding,* the gap is large and the fit-up requirement less exact. Consumable guide electroslag welding requires essentially the same joint details.
- *For inert gas (MIG) welding,* satisfactory welds can be obtained using an included angle of 45 degrees rather than the normal 60 degrees.
- *For different thickness plates* at butt welds, no special detail is required unless the difference in plate thickness is greater than about 3 mm (1/8 in.) in which case the final butt welded joint contour should have a taper of about 2.5:1. In some highly stressed locations such as at strength deck and bottom shell amidships, the taper may be required to be 3:1. No taper is required for fore- and -aft seam welds or at places where the stress across the weld is relatively low.

d. Fillet welds are sized generally from requirements of classification societies. When automatic deep penetration fillet welds are used, such as in panel line welding of longitudinals, the weld size may be reduced slightly. For example, a 6.5 mm deep penetration weld is considered equivalent to an 8 mm manual weld. For large fillet welds having many weld passes, it is advantageous to have a gap, say 2 mm (1/16 in.), between the members being welded. This gap allows the weld to shrink slightly so as to reduce the chance of cracking during cooling and to reduce the probability of lamellar tearing.

e. Low Hydrogen Electrodes. It should be noted that low hydrogen electrodes are more difficult to store and handle than ordinary mild steel electrodes and generally require more exact fit-up and joint preparation. These electrodes must be kept dry. When they are first removed from their hermetically sealed containers, they are placed in hot ovens, at about 120° to 230°C (250° to 450°F). If the electrodes are not used within a few hours on the job, they are baked for about one hour at 260° to 425°C (500° to 800°F) to ensure that the permissible moisture content is not exceeded. Portable ovens are used on the job during inclement weather conditions to help keep the electrodes from picking up moisture.

10.2 Higher Strength Hull Steels. Carbon-manganese steels such as ABS H36 with yield point about 345 MPa (50,000 psi) would be welded in a manner similar to ordinary strength mild steel except that low hydrogen electrodes must be used and some preheat might be required for structures under conditions of high restraint.

a. Low Alloy Steels. When welding quenched and tempered low alloy steels with yield point of 345 to 690 MPa (50,000 to 100,000 psi), low hydrogen electrodes are used and special precautions, as explained in Chapter VIII, are required as regards preheat and welding heat input control in order to meet notch toughness requirements, especially in the heat-affected zone (*HAZ*).

Since low heat input is required, special care must be taken when using submerged arc or other automatic processes, and many more weld passes may be required than for the same size weld in ordinary strength steel.

When arc gouging is used to prepare a joint, a hard surface film might be formed, and this film must be removed by grinding.

Tempering passes are often used in critical areas to reduce the tendency for cracking at the toe of the finished weld. These passes are made adjacent to the toe pass and after the toe pass is made. Fig. 28 shows the procedure for a butt weld. The procedure for a fillet weld is similar.

When ordinary strength steels are welded to high strength steels, low hydrogen electrodes should be used, but not necessarily to match the strength of the higher strength material. Under conditions of restraint, the lower strength weld deposit would be preferable in order to provide increased ductility.

10.3 Low Temperature Service. Low temperature service aboard ship is encountered both in vessels carrying liquefied gases and those carrying conventional refrigerated cargo.

a. Liquefied Gases. For low temperature shipboard service, the USCG and classification societies have rather severe notch toughness requirements based on the IMCO Gas Code as outlined in Chapter VIII. These requirements specify the same notch toughness for the base metal, the weld metal, and the (*HAZ*) heat-affected zone regardless of plate orientation. Since toughness properties transverse to the direction of plate rolling are not as good as those

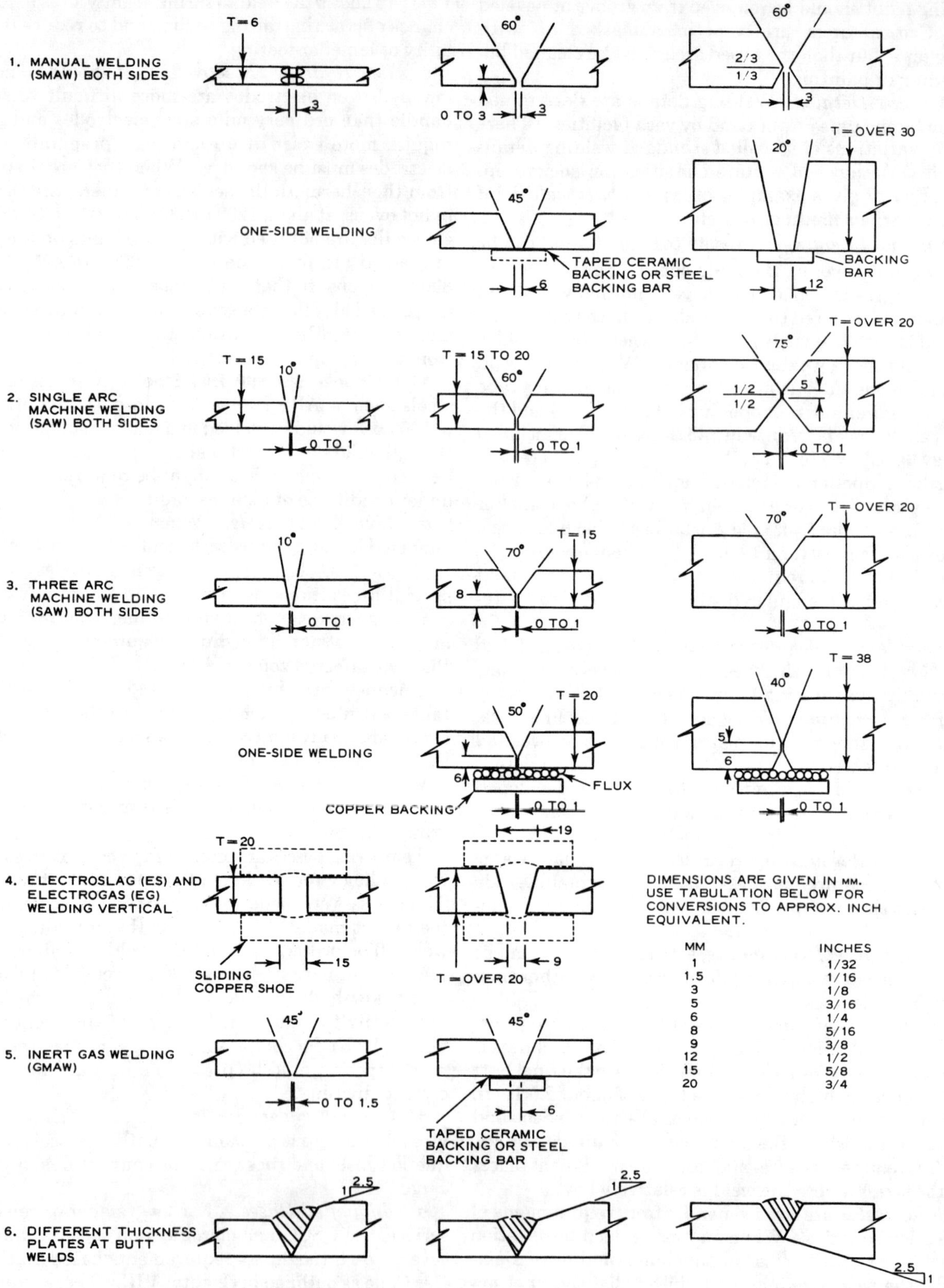

DIMENSIONS ARE GIVEN IN MM. USE TABULATION BELOW FOR CONVERSIONS TO APPROX. INCH EQUIVALENT.

MM	INCHES
1	1/32
1.5	1/16
3	1/8
5	3/16
6	1/4
8	5/16
9	3/8
12	1/2
15	5/8
20	3/4

Fig. 27 Typical butt weld joints details

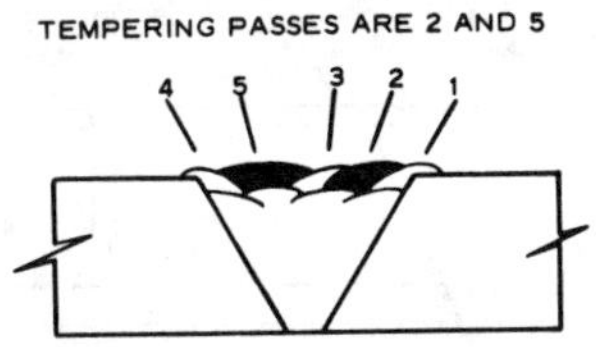

Fig. 28 Tempering passes

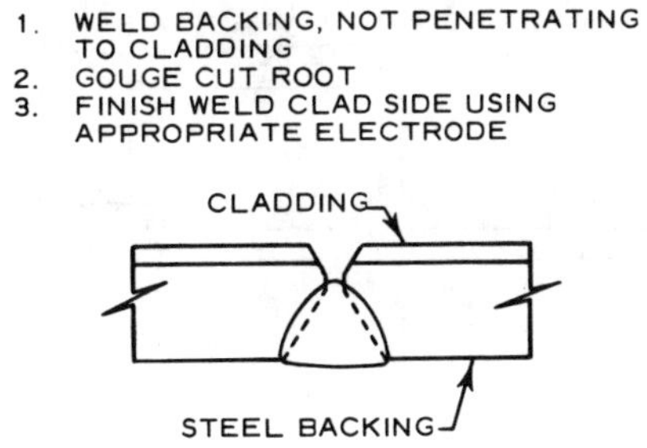

Fig. 29 Typical clapplate butt weld

parallel to the direction of rolling, the transverse properties are the critical ones. The best grades of ship steel (E, CS, and EH) are acceptable down to only about −25°C (−13°F) when applied to this low temperature service. Below that temperature, special steels are required.

For service temperature below −25°C down to −55°C (−67°F), steel manufacturers can produce acceptable carbon-manganese steels but the welding of these steels remains a problem. It is now not possible to meet the notch toughness requirements for the weld and HAZ using normal high-heat input production welding procedures, especially at the lower temperatures. To meet the requirements, welding procedures using controlled rates of heat input and interpass temperature control are necessary.

For service temperatures below −55°C, nickel steels, austenitic (stainless) steels and aluminum are required, and all have special welding procedures.

Approximate service témperatures for three main products are:

	°C	(°F)
Ammonia	−33	(−28)
Liquefied petroleum gas (LPG)	−46	(−50)
Liquefied natural gas (LNG)	−162	(−260)

b. Refrigerated Cargos. Hull steel in way of refrigèrated spaces may be cooled well below normal temperatures. In addition, the local cooling of a deck, or other structure, can cause high tensile drumheading stresses. Because of these two factors, more notch tough steel, as given in Table 6 in Chapter VIII are required in way of refrigerated cargos. However, no special welding procedures, such as for liquefied gases, are requred.

10.4 Clad Steel. Extraordinary care must be taken in handling, cutting, and welding clad plating because the cladding is very thin, normally 1.5 to 3 mm ($^1/_{16}$ to $^1/_8$ in.) thick, and is easily damaged.

Special welding precautions are required to prevent contamination of the clad surface welds by iron pickup from the steel backing. Referring to Fig. 29, normal practice is to weld the steel backing first with ordinary steel electrodes, but being careful not to penetrate the cladding, especially if submerged arc welding is used. The clad side is then welded, usually with gas metal arc. Undercutting should be avoided because it could destroy the effectiveness of the cladding.

Samples of surface weld are examined to ensure that iron pickup is not excessive. If it is excessive, the surface weld must be removed, usually by grinding, and a new layer of clad deposited, and multiple layer welds may be necessary to meet requirements.

10.5 Welding Sequence. The welding sequence should be aimed at minimizing overall distortion and facilitating construction. It is also aimed at minimizing the chance of cracking during the welding process in areas of high restraint. The welding sequence should be simple and practical so that an assembly line type of production can be easily accomplished and that a large number of welders can be put to work at one time.

a. Overall Sequence. As illustrated in Fig. 30, there are just two basic rules to follow for an overall sequence:

- First, tie-in plates which are relatively free to draw together.
- Second, do not weld across an unwelded butt or seam.

The same reasoning may be applied when welding internal webs and framing to plating. Referring to Fig. 31, the plate butt is welded first, then the butts in the stiffening members, and finally the fillet weld attachment is completed. Generally, the fillet weld attachment made prior to welding the butt joints is kept back from these joints by at least 300 mm (12 in.).

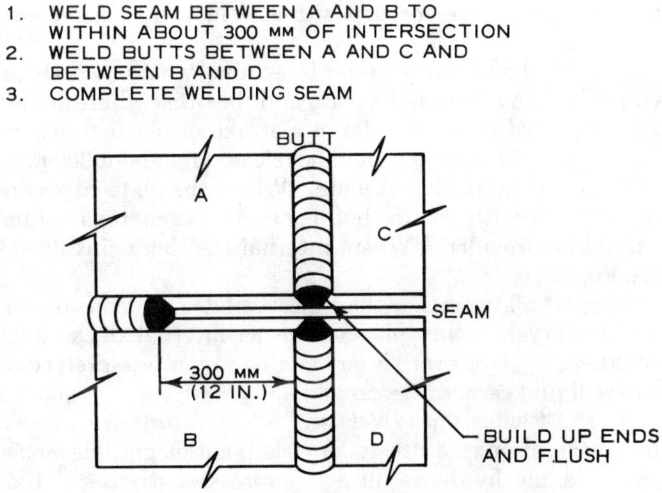

Fig. 30 Welding sequence at intersection of butt and seam

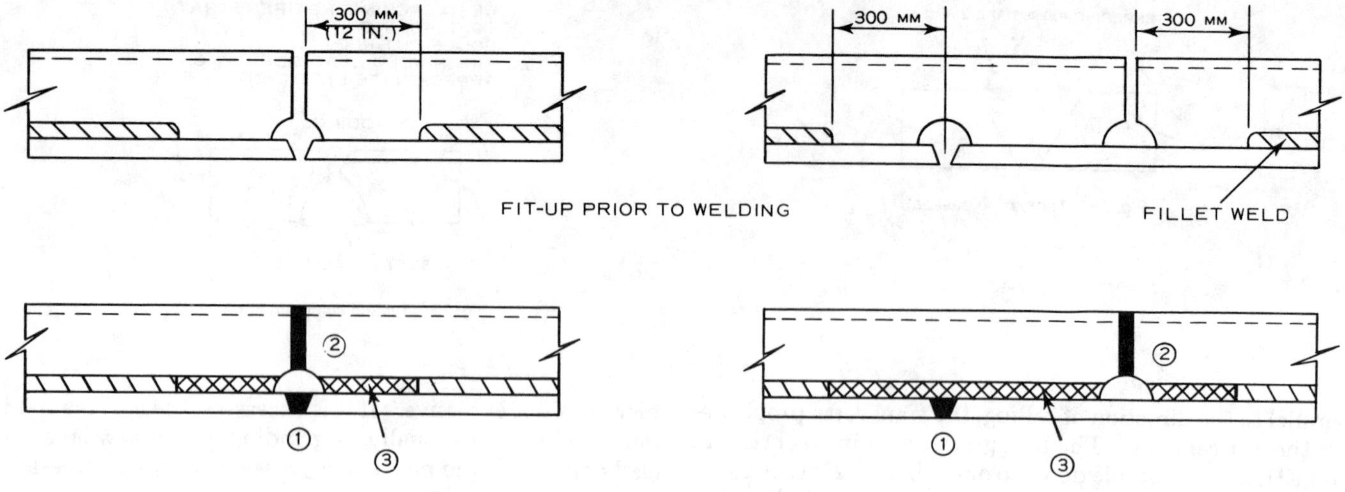

1. WELD PLATE BUTT
2. WELD STIFFENER BUTT
3. COMPLETE FILLET WELD ATTACHMENT

Fig. 31 Sequence for welding stiffener in way of buttplate or seam

When large panel sections are welded together on the shipways, the sequence of joining them is essentially the same as that shown in Fig. 31.

When a portion of deck, shell plate, or plate assembly is removed for access during construction or for repair of a damaged plate, the replacement is welded under restrained conditions. A typical sequence for replacement is shown in Fig. 32. The seams should be cut back approximately 300 mm (12 in.) to reduce restraint when the butts are being welded and to eliminate weld starts and stops at the weld intersections. Any abutting frames or webs of the original intact structure should also be released for about 300 mm clear of the joint opening.

b. Closure Plates. Fig. 33 shows several methods of closing small openings. Generally, a backstep, cascade welding procedure for depositing weld beads is recommended.

Fig. 34 shows two insert plates, one where two sides coincide with a butt and a seam and the other where an insert is placed in the middle of a strake of plating. Here again, the weld grooves should be released for about 300 mm (12 in.) beyond the insert plate. Where the plate edges do not coincide with existing butts or seams, a generous radius should be provided. Where internal framing is involved, see Fig. 32.

Lapped plates rather than insert plates are considered satisfactory in many places such as internal decks and bulkheads. However, insert plates are often preferred where liquid cargoes are carried.

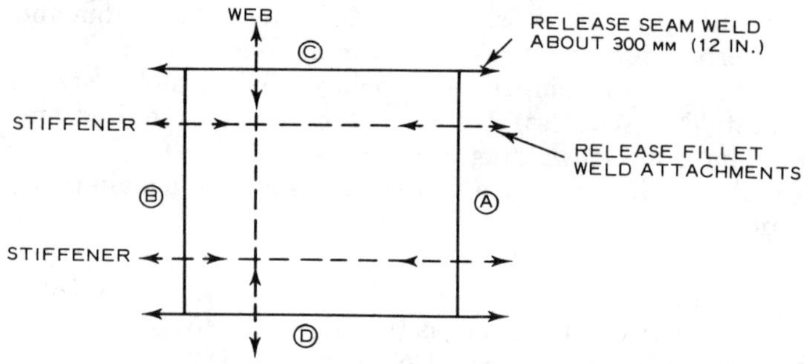

1. RELEASE SEAM WELDS FOR ABOUT 300 MM (12 IN.) AT EACH CORNER OF INSERT PLATE. RELEASE WEB PLATES AND STIFFENERS FOR ABOUT 300 MM ON EACH SIDE OF UNWELDED BUTT AND SEAM JOINTS.
2. WELD BUTT (A).
3. WELD BUTT (B).
4. WELD UNWELDED FRAMING IN WAY OF BUTTS (A) AND (B) FOLLOWING PROCEDURE SHOWN IN FIG. 31.
5. WELD SEAM (C) INCLUDING RELEASE LENGTHS.
6. WELD SEAM (D) LIKEWISE.
7. WELD UNWELDED FRAMING IN WAY OF SEAMS (C) AND (D) FOLLOWING PROCEDURE SHOWN IN FIG. 31.

NOTE: IF JUST THE PLATE IS TO BE REPLACED, THE SEQUENCE WOULD FIRST REQUIRE ALL INTERNAL FRAMING TO BE WELDED TO NEW PLATE TO WITHIN 300 MM OF THE PLATE EDGE. THE REMAINING SEQUENCE WOULD BE THE SAME AS ABOVE.

Fig. 32 General sequence for welding a large insert plate panel

10.6 Distortion Due to Welding. Weld distortion is caused by tensile, shrinkage stresses of yield point magnitude along the weld and by the resulting compressive stresses in the adjacent plate or structure. If the compressive stresses are high, natural buckling waves may develop, especially in light plating, Fig. 35A. Shrinkage also takes place across the weld, and this shrinkage can be substantial.

If the tensile weld stresses are not exerted at the neutral axis of assembly or structure, the assembly will bend.

Distortion can occur if there is too much unbalanced welding or burning along one edge of a long narrow plate,

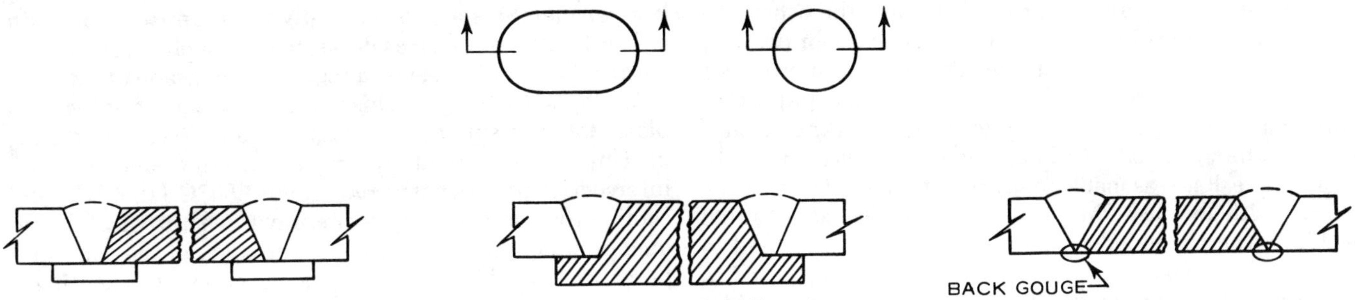

Fig. 33 Closure for small openings

Fig. 35C. Ideally, of course, welding should start along the midline of assembled units and progress upward and downward, but this is not always practicable. The greater the inertia of the structure to resist bending, the smaller will be the distortion. This is one reason why ships built with large, deep assembly units fastened together on the building ways will invariably have straighter keel lines than ships built with shallow assembly units, where the ends of the ship might rise off the keel blocks.

Welding details and procedures can affect distortion. Balanced and concentric welds will have less local distortion than unbalanced and eccentric welds. For a given weld size, the greater the number of weld passes used, the greater will be the distortion. In this respect, automatic welding will usually cause less distortion than manual welding. It follows that the smallest practicable fillet welds (to minimize the washboard and bowing effect, Fig. 35B) and the smallest groove angles should be used.

When two plates or assemblies are being welded together

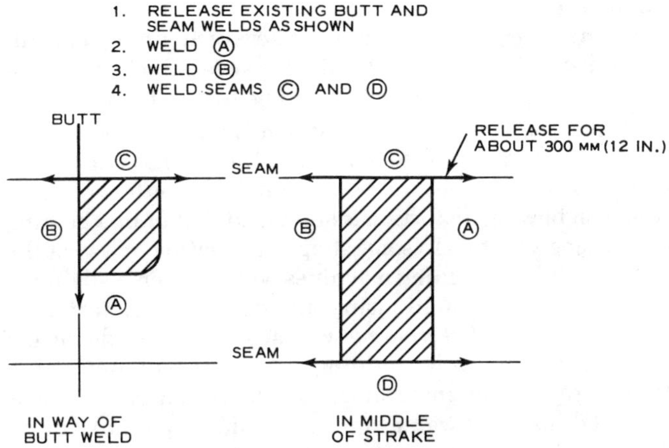

Fig. 34 Insert closure plates

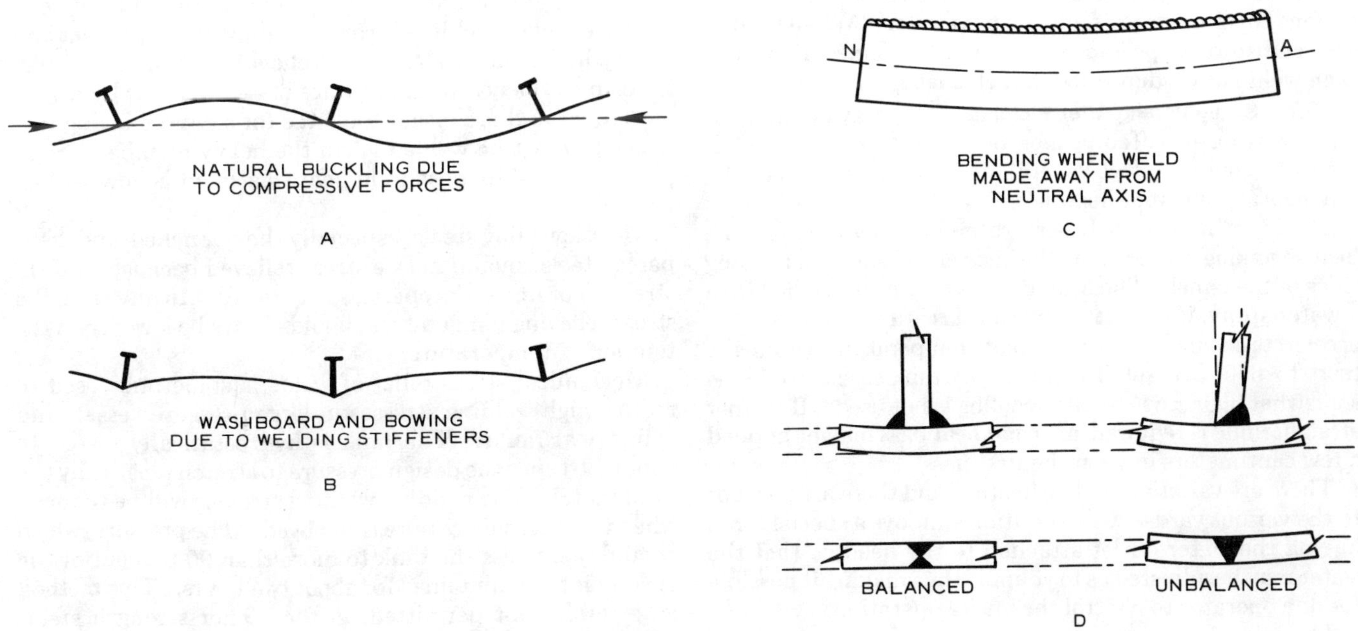

Fig. 35 Typical local weld distortion

starting at one end and progressing toward the other, the unwelded portion of the joint may tend to close or possibly to open slightly, depending on the heating pattern produced in the plating adjacent to the welding. This is important when making long butt or seam welds such as the vertical butts in the side shell. Although experience is the only reliable guide, it is reasonable to say that if the weld were made slowly with low heat input, as in multiple pass manual welding, the unwelded portion of the joint would tend to close. Thus, block welding of such joints is often used to counteract this tendency. If the weld were made rapidly with high heat input, as with submerged arc or electroslag, the unwelded portion of the joint would tend to remain fairly stable. With submerged arc, tack welds help to maintain proper gap alignment. With electroslag or electrogas, stiffening members help hold the alignment since no tack welds can be used.

Strongbacks, clamps, hold-down dogs or anything that will resist distortion will be helpful. However, the heavier the members, the more difficult it will be to resist distortion.

Presetting can be useful. When the preset member is welded, it is expected to deform to the desired shape. In connection with fillet welds, the cumulative effect of these welds on bowing distortion is similar to that of line heating on shaping a plate. In some shops, the plating in way of the automatic fillet welding machines will be given a slight set for the purpose of minimizing the cumulative distortion.

Intermittent fillet welding will cause much less distortion than continuous fillet welding. Intermittent welding is frequently used on internal framing but is not permitted or not used inside of water tanks or liquid cargo tanks or on weather structure where unsightly rust stains from the unwelded faying surfaces would result.

Peening as a means of controlling distortion can be used, but with limitations. Peening should not be used on single pass welds, on root passes of multiple pass welds, or on the surface cover passes of multiple pass welds. When used to correct distortion, peening should be done immediately after each weld pass is deposited and cleaned.

10.7 Straightening After Welding. One way of straightening plating or stiffening members is by torch heating followed by cooling. Two general methods are used, namely, line heating and spot heating.

a. Line Heating. In straightening a plate panel, line heat shrinking is used along the back of stiffeners on the long sides of the panel. The heated areas may be quenched with a water spray, if permitted by the classification society, to accelerate cooling. Maximum plate temperature is normally held to 650°C (1200°F) but, in certain cases, could be somewhat higher when water-cooling is not used. If further straightening is required, a second heat pass may be applied a few centimeters in from the first pass.

There are variations to line heating, and these are peculiar to the various yards. One variation employs a special torch having the water outlet attached to the head so that the water may be adjusted as to volume, thus making it possible for one operator to control the entire operation.

The use of a water spray is not recommended for the heavier members and is generally not permitted for main strength members such as decks, bottom shell, and side shell within the middle 3/5 length regardless of grade of steel.

b. Spot Heating. This method consists of lining off a plate panel in squares of about 150 mm (6 in.) on a side, heating of spots about 40 mm (1½ in.) in diameter at the intersections to a cherry red, about 650°C (1200°F), and cooling the spots with a water spray if permitted. The sequence of heating is to start near the edges of the panel and progress toward the center. In some cases, spot heating is used in conjunction with line heating.

c. Higher Strength Steels. Flame straightening on the higher strength steels should be kept to an absolute minimum. For main strength members, fairing may include jacking or strong-backing the plate and may also include releasing all restraint and then cold straightening.

On some *low alloy quenched-and-tempered steels,* flame straightening is not permitted. Refairing should be done by releasing adjacent joints or by making new joints, fairing by strong-backing and then rewelding the joints. Tests reported (SSC, 1974) indicate that moderate heating, not to exceed the steel's tempering temperature, can be used with caution. In these tests, a water spray was used to help control the heating and to speed the operation. However, special permission should be obtained from regulatory agencies before adopting this low-heat method on quenched-and-tempered steels.

d. Framing. Framing may be straightened by area heating or line heating along the edge requiring shrinking. The general method used in line heating may be employed when shaping plates and girders.

10.8 Stress Relief of Weldments. A post-weld heat treatment is occasionally given to heavy, medium steel weldments or castings where large amounts of weld metal are deposited. The treatment temperature is about 625°C (1150°F) held for one hr per 25 mm (per in.) of thickness. However, when welding is carefully done using preheat and low hydrogen electrodes, stress relieving is usually omitted. Since most heavy weldments are made with low hydrogen electrodes and are either too large for stress relieving furnaces or must be welded when the heavy members are in place on the ship, post-weld heat treatment is now seldom used.

Most low alloy steels, especially the quenched-and-tempered steels, should not be stress relieved because a loss in strength or impact properties may result. In any case, the stress relieving temperature should always be lower than the tempering temperature.

Mechanical stress relief of welds is sometimes used to reduce high welding stresses in large pressure vessels and cylindrical tanks. The tank is hydrostatically tested to about 1.5 times the design pressure to stretch plastically the weld metal. The residual welding stresses will be reduced when the internal pressure is removed. The pressure, which should not stress the tank to more than 90 percent of the yield point, is maintained for about two hours. This method is generally not permitted on the higher strength steels where the yield point is greater than 80 percent of the ultimate tensile strength. Application is covered in USCG regulations, Subchapter F (see Table 1, Chapter XVIII).

10.9 Preheat. Preheating, by reducing the cooling rate during welding, reduces the concentration of shrinkage stresses and helps to prevent reduction in impact properties in the weld. Preheating in cold weather, say below freezing, is usually specified, particularly when welding thick plates. Preheat temperatures normally are from 60 to 90°C (125 to 200°F).

When an insert is welded in a heavy plate, a minimum amount of preheat should be used because, when the structure cools, the plate shrinkage from the preheated areas adjacent to the weld will be added to the weld shrinkage and, thus, increase the tendency toward shrinkage cracking.

When welding some of the higher strength steels, especially the quenched and tempered steels, the same preheats required for the main welds should be used for tack welds and other miscellaneous welds.

Preheating may be accomplished in several ways. Torch heat is the simplest to apply for small jobs. For the larger jobs, strip heaters are used. Strip heaters are tack welded to the plating or casting and the current regulated to produce the required temperature. They are used on heavy weldments where welding is done over a long period of time such as with fabricated stern frames.

10.10 Inspection of Hull Welding. The most important type of weld inspection to a shipyard is the non-destructive, subsurface examination of full penetration welds. This type of inspection has an obvious psychological value as well as a practical value.

The main purpose of subsurface inspection is to maintain welding quality at a high level, and not to eliminate all defects. In fracture mechanics it is recognized that a certain number of minor defects are inevitable and that perfect welds are impossible to achieve. Studies dealing with the probable adverse effects on the performance of a structure indicate that many types of so-called minor defects would have little or no adverse effect and that, in many cases, it would be better not to try to remove the minor defects. New fracture mechanics techniques using such tests as the *Dynamic Tear* test (*DT*) and the *Crack Opening Displacement* test (*COD*) are now being applied to varying degrees in designs of LNG tank structures, pipelines, pressure vessels, etc. and will undoubtedly have an influence on the acceptability criteria for non-destructive inspection in the future by introducing a more realistic fitness for purpose assessment.

Classification societies necessarily have rather general requirements for guidance of their inspection personnel regarding subsurface inspection. Acceptability of welds is governed largely by the particular type of weld defect and the location of the weld in the ship. A given defect may be considered undesirable in a highly stressed area amidship, say at a hatch corner, but may be considered quite acceptable in another part of the ship, say in the side shell. This requires experience and sound judgment on the part of the inspector.

Subsurface inspection is usually done on a random, spot-check basis and is accomplished by *Radiographic Testing* (*RT*) or by *Ultrasonic Testing* (*UT*) as described in Chapter VIII. Inspection is normally confined to the most highly stressed areas within the midship 3/5 length and concentrated on transverse butts and weld intersections in the vicinity of the gunwale, bilge, and hatch corners.

Seam welds are considered relatively unimportant because the primary ship stresses are parallel to the weld rather than across the weld and because the failure record shows that, where major deck or shell fractures started in welded joints, essentially all fractures started in transverse butt welds and not in seam welds.

The number of locations inspected on a ship is generally 200 to 500 with more in special purpose ships. Some increase in inspection can be expected when higher strength steels are used. The extent of inspection for some offshore drilling rigs may be several times greater than that for normal ship construction.

Increasing use is being made of UT and it might replace radiography almost completely in the inspection of hull welds. UT is especially valuable in detecting laminations which are parallel to the surface. RT on the other hand, can readily pick up cracks oriented perpendicular to the plate surface but will have difficulty picking up a flat discontinuity, such as a lamination, parallel to the plate surface.

One advantage of UT is that it can be carried out with much less interference with production than can RT. For this reason it is used extensively in panel line production to inspect automatically welded seams to check the quality of weld and to ensure that the welding equipment is operating properly.

10.11 Types of Welding Discontinuities. The most common type of discontinuities that may require repair are: cracks; incomplete fusion, particularly at root; slag inclusions; porosity; and undercut.

- *Cracks* may occur in the weld metal for a number of reasons such as failure to back gouge, excessive shrinkage when root opening is too large and cracked tack welds not removed.
- *Incomplete fusion* may be caused by poor fit-up, a poor root condition or by improper welding techniques. Back gouging can remove areas of incomplete fusion at or near the root of the joint.
- *Slag inclusions* are nonmetallic and are usually associated with incomplete fusion between weld passes or between a weld pass and the base plate. Small, widely scattered slag inclusions are generally not considered harmful.
- *Porosity* is caused by gas pockets or blow holes. To minimize porosity when welding with submerged arc or low hydrogen electrodes, it is important that the joint be clean and that the moisture content of the electrode covering be kept low. Fine, widely scattered porosity is generally not considered harmful.
- *Undercut* is a small groove melted into the surface of the base plate adjacent to the toe of a weld, primarily a fillet weld. It can also occur along the sidewalls of a groove joint and, thus, may lead to incomplete fusion in the interior of a weld. A slight amount of surface undercut is almost al-

ways present, and an undercut of about 1 mm (1/32 in.) is normally acceptable for both merchant and naval work.

10.12 Lamellar Tearing. Lamellar tearing is a form of cracking which has only recently been observed and studied. It is associated with strains created by welding shrinkage which tends to pull the plate fibres apart in the thickness direction, the weakest direction of the plate, Fig. 36. The elongated, fibrous type of inclusions are believed to be one of the prime causes of the problem. The tears usually develop during welding of highly restrained thick plates. Most notable are those tears which have occurred at joints connecting large tubular members in drilling rigs.

Lamellar tears have been reported in full penetration joints similar to those which are commonly used in ships and which have given excellent performance over the years. To this extent, the explanation for the recently reported tearing failures is somewhat uncertain. Nevertheless, several things have been suggested to minimize the chance of lamellar tearing in joints which would experience high, through-thickness tensile stresses:

- Use a rather expensive low sulphur steel made to reduce the number of inclusions and to eliminate the elongated inclusions. So far, there does not seem to be a need for this special type of steel in normal ship construction.
- Use fillet welds instead of full penetration welds to spread out the welding forces.
- Use a buttering weld on the plate surface prior to welding the members together.
- Use as low a strength weld metal as practicable when welding higher strength steels to allow more stretching in the weld metal and less in the base plate.

If lamellar tearing were thought to be a potential problem in an angle joint, Fig. 37A would be preferred to Fig. 37B, and Fig. 37C would be preferred to Fig. 37D.

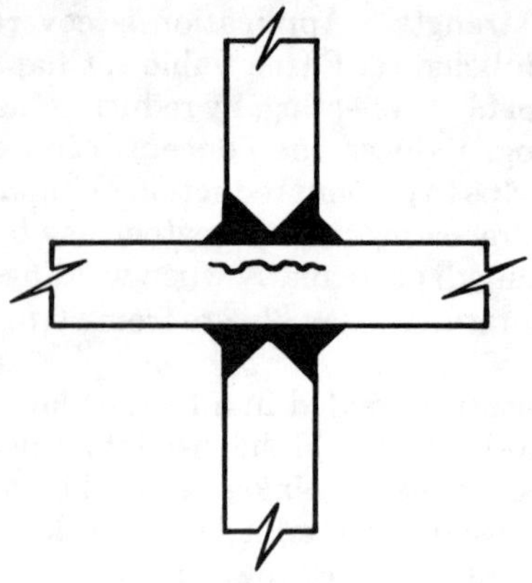

Fig. 36 Lamellar tear

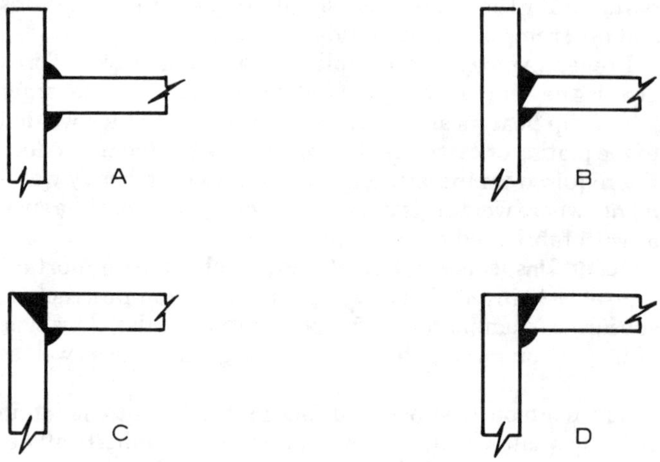

Fig. 37 Corner weld details

10.13 Repair of Defects. Defects that require repair during construction include: damaged or cracked plates; large root opening butt welds; large root opening fillet welds; plate laminations; and plate scars and edge imperfections.

a. Damaged or Cracked Plate. If a crack is found in a plate while the ship is afloat, a hole should be drilled in the plate slightly beyond the visible crack to prevent the crack from spreading. If high tensile bending moment stresses are present, they should be reduced by ballasting or shifting loads where practicable. The crack can then be repaired by excavating to sound metal and welding with an approved procedure.

The recommendations for *emergency* repairs in the case of cracks in low alloy steels given in SSC (1969) include using a stainless steel electrode (E-310) which is relatively easy to handle and produces a tough weld. Where high-tensile stresses exist, the report also recommends welding overlay beads of weld about 400 mm (15 in.) long perpendicular to the crack and just beyond the drilled hole, Fig. 38. This overlay not only produces a tough material which would resist any tendency for the crack to propagate further but also introduces compressive stresses in the material at the end of the crack.

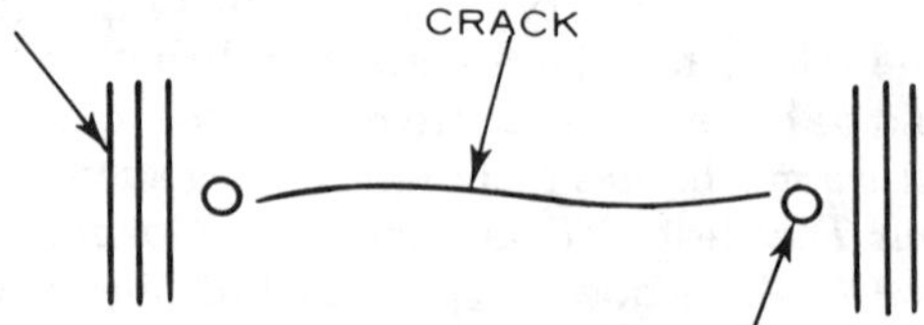

Fig. 38 Overlay beads

Where a section of plating in a main strength member, such as a main deck or shell, must be removed due to cracking or damage, no more than about 1 m (3 ft) of plating need be removed unless, of course, the damage is extensive. For other members, such as internal structure, 0.5 m should be sufficient unless the plating is exceptionally thick. A typical sequence for repair is shown in Fig. 32 and an approved welding procedure must be followed. Under restrained conditions, the smaller the repair, the more important is the procedure. On the other hand, the repair to a plate edge where there is no restraining structure in the immediate vicinity of the repair is relatively simple.

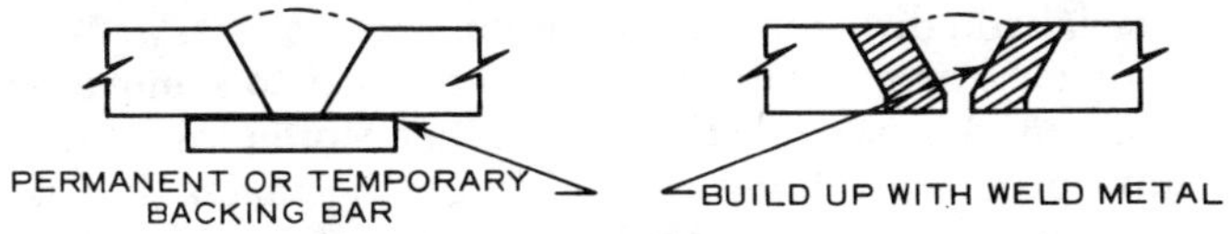

Fig. 39 Correction for excessive butt weld root opening

Fig. 40 Fair in plate edge

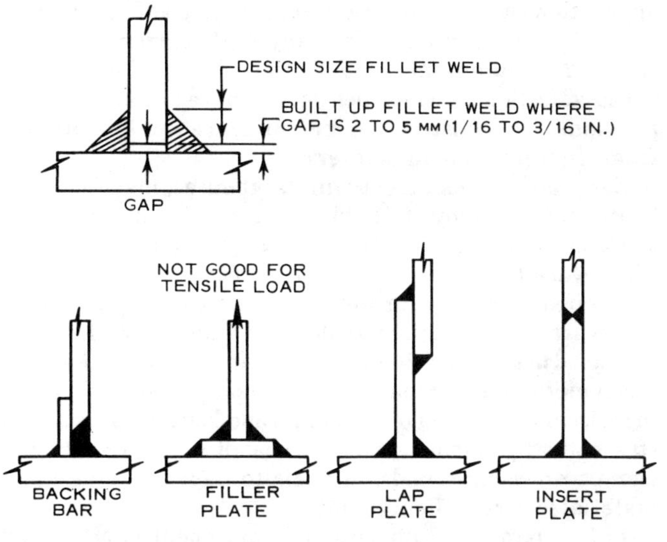

Fig. 41 Correction for excessive fillet weld root opening

b. Large Root Opening, Butt Weld. Where the root opening is too large, it may be built-up on one or both edges as indicated in Fig. 39. The plate edge or edges should be built-up by welding to within the specified root opening before the joint welding is started. With adequate supervision, there seems to be no technical reason why the build-up cannot extend well beyond the normal limits of about one half the plate thickness provided the properties of the weld are equal to those required of the plate. Large build-ups of several inches and more have often been made to castings and forgings.

A permanent or temporary backing strip is sometimes used where the root opening is excessive. Where the gap is considered too large for building-up with weld metal, an insert plate is required. Here, as in the case of the damaged plate mentioned above, an insert of 1m (3 ft) for main strength members and 0.5 m (18 in.) for other members should be sufficient. In the highly stressed areas of the main hull, the rolling direction of the insert plate should be the same as that of the original plate.

c. Large Root Opening, Fillet Weld. Where the root opening exceeds about 2 mm ($^1/_{16}$ in.), the fillet weld is required to be increased as shown in Fig. 41. When the opening exceeds about 5 mm ($^3/_{16}$ in.), other correction methods must be employed as indicated in Fig. 41.

d. Plate Laminations. Laminations occasionally occur in rolled plates and are oriented parallel to the plate surface near mid-thickness. Laminations may be any size from a few square inches upward. Small laminations are not considered harmful unless they appear along a plate edge which is to be welded. It is reasonable to assume that laminations near the middle of a plate would not be harmful unless a high, tensile load at that particular place were applied in the thickness direction, a loading situation which does not often occur. Plate edge laminations should be removed by excavating and the cavity welded or, alternatively, should be removed by cropping and a small insert plate welded in.

Plates generally are not ordered with steel mill inspection for laminations, although they may be at extra cost. If an area of plate is required to be checked, this may be done with ultrasonic inspection.

e. Plate Scars and Edge Imperfections. Deep scars in the plate surface must be repaired by welding and flushing off. Shallow scars, say less than 3 mm ($^1/_8$ in.) deep for thick plate, should be faired in by grinding and not be repaired by welding. Many deep scars are created when removing erection clips with a maul, the clip weld metal actually pulling out a piece of the plate surface. Removing clips by chipping or by air carbon arc and grinding is preferable. Temporary clips and lifting pads are often cut off just above the plate surface without further treatment. These clips in the interior of the ship are usually left in place if they do not interfere with the operations of the ship.

Plate edge imperfections, such as notches at the top of the sheer strake or at the edge of a face plate should not be repaired by welding unless the imperfection is exceptionally deep. The imperfection should be removed and faired in by grinding or by flame cutting followed by grinding, Fig. 40.

Section 11
Aluminum Hull Construction

11.1 General. Aluminum has been used in some superstructures and deckhouses to save topside weight, especially in naval vessels. It is also used in the construction of small boats, barges, outfit items such as ladders, and in applications which make use of its non-magnetic properties. The large primary tanks in many LNG carriers are constructed of aluminum to handle the cold liquid cargo.

Classification societies such as ABS have formulated rules for aluminum construction and guidance can be obtained from aluminum manufacturers.

Early application, mostly in deckhouses, made use of a heat-treatable alloy 6061, which has good corrosion resistance but poor weldability characteristics and, as a result, was invariably riveted. The newer 5000 alloy series was developed primarily for welding. The various aluminum alloys are discussed in more detail in Chapter VIII.

11.2 Handling and Storage. Aluminum must be handled much more carefully than steel and should be stored indoors in racks made of wood or aluminum and kept separated from steel and other metals. It must be handled with smooth grip clamps or vacuum pads. Repair to damaged aluminum material is very difficult to effect.

11.3 Forming. Conventional equipment is employed to form aluminum plates, shapes and tubing. Where severe forming is necessary, the softer tempers of aluminum alloys are used. A moderate amount of heat, generally 200° to 260°C (400° to 500°F) may be applied but rigid control of temperatures is required. Special approval is usually required for this operation.

Since the surface of aluminum is relatively soft and easily damaged, tools and equipment for bending and forming must be smooth and free from dirt. Sometimes a tool surface is covered with a paper sheet. For some of the harder tempers, allowances must be made for springback as in the case of spring steel.

11.4 Extruded Shapes. Aluminum has the advantage of having available many extruded shape configurations. In addition, a die can be made at reasonable cost to produce shapes to meet special requirements.

11.5 Welding. Joint edges may be prepared by mechanical means such as saws, millers, and routers and by plasma-arc cutting, especially for the thicker materials. Plasma-arc has also been developed so that it can now be used for bevelling and back gouging. Band sawed edges are normally required for normal aluminum ship fabrication. Machining would be required for U or J grooves.

Shearing is not recommended for preparing plate edges. The roughness of sheared edges may entrap oil or dirt, which must be removed. If an abrasive wheel is used, it must be of a type suitable for removing rough edges of aluminum and must be used lightly so as to avoid embedding abrasive material and dirt into the aluminum surface.

Solvents or mechanical means are used to remove oil, grease, markings, oxide films, and other contaminants from joints prior to welding. Care should be taken that degreasing chemicals or machining lubricants do not collect in crevices such as at faying surfaces between plate and backing bars and lap welds. Power driven stainless steel wire brushes are frequently used in cleaning operations. Welding should be done before oxide surfaces can again form. Welding should not be performed on anodically-treated aluminum except where the surface oxide is removed from the joint areas.

Welding is done mainly with a rather bulky gas metal-arc welding gun, and care must be taken in the design to provide proper access for welding.

Welding sequences used for steel generally apply to aluminum alloys.

11.6 Distortion. Fairing by heating or flame shrinking is not recommended for correcting distortion in main strength members, but, if done, should be carried out only with special approval of regulatory agencies. For the 5000 series alloys, heating and cooling through the sensitized range of 65° to 200°C (150° to 400°F) should be done as rapidly as practicable. There is always the danger that the material will be damaged by overheating if heating is carelessly applied. A plate buckle may be reduced by laying down weld beads in the middle of the panel to drumhead the plating. The use of strongbacks to help force the plating is usually necessary in such cases. Guidance on permissible aluminum plate unfairness is given in Fig. 25 in Section 8.6.

Distortion, which is much more of a problem in aluminum than in steel due primarily to the high coefficient of expansion, may be minimized by tacking together several plates with their stiffeners attached before welding the assembly. The high coefficient of expansion also causes fit-up problems in that the clearances at unwelded joints can be markedly affected by changes in ambient temperatures and by the sun's rays.

11.7 Electrolytic Corrosion. When aluminum is combined with other materials, such as steel, by bolting or riveting, it is necessary to provide insulation between the dissimilar materials in order to avoid electrolytic corrosion. Repair can be expensive if the insulation breaks down. The explosion bonded (steel to aluminum) connections offer certain advantages in weather areas as regards corrosion compared to bolted or riveted connections.

Section 12
Preoutfitting

12.1 General. *Preoutfitting* as used in this chapter refers to the installation of both outfit and machinery items in large structural assembly units prior to these units being erected in the ship. The term also covers the installation of other elements such as piping, ventilation, and electrical cable in subassemblies where access is simplified if the in-

Fig. 42a Underside of deck with ventilation system elements being installed

stallation is made before erecting the subassembly in the ship. Decisions as to the extent to which preoutfitting is carried out depend largely on the time available for early planning, development work, and procurement of equipment. A major effort in this direction is usually not justified for single ship contracts but has proved worthwhile in recent multiship contracts.

As preoutfitting techniques developed during World War II, it was found to be efficient in many instances to carry the process one step further by pre-assembling machinery, piping, and electrical elements into *modules* that could then be mounted in the ship assemblies. Thus, the module concept evolved during the same period as preoutfitting. The two terms are not, however, always necessarily interrelated since modules may be installed after as well as before assembly erection in the ship. In some cases the mechanism and piping in the modules could be cleaned and tested before installation. The ship foundations for most of these modules were relatively easy to install because they required little exact alignment work, and, in recent years, the widespread use of epoxy resin in lieu of steel chocks has greatly simplified the installations.

Transverse bulkheads and underdeck areas are good candidates for preoutfitting both of complete modules and of piping, ventilation, and electrical system components because installation can be made on the flat rather than vertically or overhead on the ship, Fig. 42. The ultimate in preoutfitting is a completely finished deckhouse placed on board with only attachment welds and electrical and piping connections to be made up.

Certain cable splicing techniques, as noted in Section 14.7b. are now approved by regulatory agencies for joining cables at assembly interfaces, thus avoiding the necessity of pulling long lengths of cable after the hull is completely assembled.

The use of computers in outfitting has been limited

Fig. 42b Tank top module with piping system elements installed

mainly to the manipulation of scheduling networks and to the conventional accounting procedures for man-hours and materials. More extensive use is anticipated with the refinement of various computer programming systems. Interest is evident in computer-aided pipe design and manufacture which relates to preoutfitting of piping systems. As soon as costs associated with the development and application of such precision programming are reduced and errors brought under control, an extended usage of computers in this area can be anticipated.

12.2 Planning. Increased preoutfitting increases scheduling, design, and construction problems. First, there must be assurance that the outfitting and machinery items will be available when needed. This usually means early delivery. Second, many items will require protection against damage during and after the erection period. In some cases it may be more profitable not to install the item than to provide protection. Third, there must be a high degree of accuracy in system layout and close coordination among various drafting rooms, and there must be a carefully planned sequence of installation in order to avoid installation interferences, i.e. to avoid having to remove or relocate some item in order to be able to install another. Further complications may result if some of the work involves outside contractors. If corrective action must be taken after erection, additional staging, special lifting gear, and repainting might be required, and delay is likely.

When a major preoutfitting effort is to be made, special preparations must be made to coordinate the efforts of the various yard crafts as well as those of the drawing rooms. Regardless of the method or system adopted, this preparation should be done by experienced personnel from the crafts and should include the determination of manpower requirements and the time required to do the work. In some cases unit plans, as described in Section 7.4, are prepared to aid in assemblying outfit into structural units.

One method of coordinating the various crafts is through a *work package system* (Goldbach, 1973) (Bath Iron Works, 1978). Work packages are a form of work order that also serve as an aid to management. The work package comprises all of the paperwork, or documentation, that is associated with creating a specific element of the ship, transporting that element, or incorporating it into the ship itself. This documentation can include schedules, drawings, material lists, personnel assignments, or any other items of information that are required to aid in the efficient accomplishment of the work element covered by that particular work package.

There may be a work package for manufacturing an outfit item and another for installing that item in a structural unit. There may also be work packages for other divisions of work such as for construction services and testing. Work packages can take many forms and can also apply to steel fabrication and assembly work as well as to outfit. Computers are invaluable in scheduling these packages. A brief description of several interdependent work packages is given below to illustrate how a complex problem can be simplified by planning.

A *Manufacturing* work package may designate work to be performed in one of the shops such as the pipe shop, electrical shop, or fabricating shop. It defines by drawings, dimensions, material lists, and specifications the unit to be manufactured and gives the start and completion schedules for the production of some of the specific items covered.

An *Installation* work package may designate the work to be done in installing such items as steering gear units, cable runs, piping runs, ventilation ducts and fans, and machinery modules. Machinery modules may consist of several components with their piping and wiring secured to a base plate.

A *Construction Service* work package may apply to the services needed to accomplish the work required in the above work packages. It may also apply to work to be done in moving material and work to be done in providing access, temporary lights, and temporary supports.

A *Materials Package* may be made up as needed for accomplishing the work described in the work packages.

The materials are usually assembled in a warehouse or storage area and then delivered to the job site or work station. In some cases the work may move progressively through several work stations.

Section 13
General Outfitting

13.1 Scheduling. Generally, planning and scheduling for outfitting is more detailed than for hull construction or for machinery installation. This scheduling includes that of tests for rigging, lifeboats, deck gear, and other equipment outside of the machinery space. An additional complication is that much of the outfitting work in the shipyard is done by vendors or outside contractors.

13.2 Subcontractor's Work. Examples of items frequently installed on a subcontract basis are elevators, heat insulation and pipe covering, some floor covering, and special electronic equipment. Important galley equipment also may be installed by the vendor.

Some yards subcontract all joiner work, including installation as well as fabrication. This procedure is advantageous when the demand load for work is extremely variable in its nature. It is advisable to give the subcontractors and vendors as much lead time as possible to perform their tasks so that accelerated and costly work schedules will not be required as the delivery date approaches. Although the subcontractor may assume responsibility for protecting some of his installations, the yard must be alert to provide good working conditions without undue interference from other trades and to protect the completed installation from damage.

13.3 Living Spaces. The steps involved in the progress of outfitting work in living quarters should follow some definite order, based on experience, to avoid confusion and possible renewal of outfitting work completed previously. The usual steps following the approximate completion of steel work are given below:

- Make pipe and electrical penetrations
- Fair buckled plating
- Install air ports or windows
- Perform hose testing
- Fit insulation, wiring, ventilation ducts, piping, and joiner furring
- Fit joiner partitions and ceilings
- Install plumbing fixtures
- Lay floor covering
- Install built-in furniture
- Fit joiner trim and doors
- Install wiring drops and fixtures
- Complete painting
- Fit carpets and drapes
- Install portable furniture

Where there is a considerable duplication of certain types of staterooms, as in passenger ships, it is usual to construct a sample stateroom in order to standarize such items as built-in-furniture, ceiling heights, shower stalls and bathroom layouts, as well as associated electrical, piping, ventilation, and joiner work. This project is started at an early date with the cooperation of an industrial design agent, if one has been engaged by the owner or yard. Perhaps the greatest advantage of preparing a sample stateroom is that early approval by the owner can be obtained.

13.4 Right of Way in Spaces. It is desirable to set up a procedure for locating and cutting openings, especially in a ship of complex design. It is also important to control the occupation of the spaces by the outfitting trades in the larger and more complicated type of vessel. Regardless of the system used, authorization should be given only after it is determined that the work to be performed will not interfere with subsequent work to be performed by other trades. If planning and scheduling is carefully done, the authorization may be given automatically when orders are given for work to be done in a compartment or in accordance with a work package. These orders are often given in the form of computer printouts.

13.5 Work Lists. It is desirable, because of the nature of outfitting work, to prepare and issue work lists during its progress. In addition to critical items to be completed, the work list should include items pertaining to safety, inspection and approval, and the scheduling of tests for the various systems. As the work nears completion, unsatisfactory items also should be noted on the work lists.

Inspectors and owner's representatives should be approached in connection with any additional or unsatisfactory work on which action is required in a particular space before final painting in the area is started. Early inclusion of such items usually will ensure prompt correction, thus avoiding a long list of work to be performed just prior to acceptance of the ship. By following this procedure, it is found that items involving changes in cost under the contract or questions of design can be noted and settled quickly.

13.6 Underwater Protection. Most ships are protected from underwater corrosion for the duration of the outfitting period by an impressed current system, such as that described in Section 9.5, supplied and installed by the yard and removed before delivery.

13.7 Completion of the Spaces. As the outfitting work nears completion, the work effort is concentrated in selected areas to facilitate completion of the work in accordance with the compartment completion schedule. After acceptance of the space, rooms are locked and the keys are kept under close supervision and are usually kept in a key locker when not in use. For some rooms the contractor may supply temporary padlocks.

Many of the spaces must be used during ship trials, and a certain amount of cleaning up after trials is required. Frequently, considerable effort is expended by the yard to protect painted areas and floor covering during the trials.

In some cases, certain easily removed fixtures are not installed until the ship is delivered. As regards storerooms and stowage of spare parts, only that equipment necessary for possible use should be put aboard before trials. In the post-trial period, all equipment supplied by the builder should be placed in the proper spaces and the various items checked off by the personnel of the ship.

Section 14
Machinery Installations

14.1 Installation Priorities. Priority in early machinery layouts during the design stage is given to components of the main propulsion system because of their influence on the arrangement of the hull structure and on the location of associated auxiliary machinery and piping. This high priority of propulsion components, most of which are vendor supplied, continues through the procurement and installation stages. The shafting arrangement and location of main propulsion units have to be closely integrated with the location of the web frames, stanchions, machinery flats, and the configuration of the stern frame or the location of shaft struts. These arrangements also take into account installation procedures and maintenance operations such as withdrawing condenser tubes and shafting. For example, a transverse web may have to be relocated in order to withdraw the condenser tubes or the design of the condenser may have to be modified. The machinery arrangements must, of course, be dovetailed with the design of large structural assembly units when any preoutfitting, as with machinery modules, is planned.

Preoutfitting adds an element of constraint in that compromises may have to be made when machinery or main foundations are to be installed on structural assembly units. For instance, a relocation of a machinery component or of an erection butt, or both, might be required so that a component, possibly with its own subbase, could be completely installed on a structural unit. Such relocations should be worked out before detailed structural plans are developed. The same may be said for piping, cable runs and electrical equipment, and even for main foundations such as those for main gears, circulating water pumps, and boilers. For example, quite often the entire top foundation plating for one or two boilers will be incorporated in one structural assembly unit.

a. Accessibility. Proper access for installation, protection and removal of main machinery is developed in the early design and may involve leaving off part of the structure, cutting temporary holes in the side of the ship or providing temporary protective covering. Lifting gear and overhead trolley tracks, which may have to be portable, are also considered at this time.

A common problem is that dealing with the arrangement of piping and the sequence for installing or removing sections of large pipe and valves without disturbing adjacent machinery or piping. Two particular problem areas are the pump room in tankers, and the area around the main condenser and main circulating water pumps, including the scoop if one is provided, particularly in ships with machinery aft.

Another problem area is the space required for removal and replacement of the tailshaft or for stowing a spare tailshaft, if one is required. The trend is away from carrying a spare below decks, if at all. Nevertheless, provision must be made for withdrawing the tailshaft, generally through the side shell of the engine room. This assumes that the tailshaft must be withdrawn inboard, as is generally the case for single-screw ships.

These problems of piping installation in tanker pump rooms and withdrawal of the tailshaft in all types of ships require full consideration in the early design stage. The problems are not complicated if recognized in advance.

It is important to realize that allowances should be made for some structural deviations from the design dimensions; in this respect, machinery and outfitting layouts should not call for unreasonably close clearances between equipment and ship structure. If close clearances are necessary, special attention should be directed to all interested parties, especially to yard supervision, for adherence to the details of the design layout.

b. Models and Mock-ups. A model of a complete machinery space, or a portion thereof, could be of value, especially for a multiship program, provided the model work sufficiently precedes the plan development. Generally, small scale models have not been successful in highlighting general interferences. However, a model of a limited area could be helpful when considering an unusually tight arrangement or a complex design by showing interferences and inaccessible spaces.

In some cases, full-size mock-ups of critical areas have been useful in determining feasible accessibility arrangements and in demonstrating how the actual fabrication or assembly can be accomplished. An example is given in Fig. 17, Section 7.

14.2 Foundations. Foundations are discussed here from the point of view of machinery installation and alignment. The functions and design of the foundations are covered in Chapter VII.

a. Subbase. In many cases, the vendor supplies the machinery unit as a module already mounted on a firm subbase or bedplate to ensure proper alignment of parts. The subbases are fastened to associated ship foundations by direct bolting or by bolting through chocks (liners). Examples of these machinery items are turbo-generators, motor-driven pumps and gas turbine propulsion units.

The gear case for main reduction gears may be considered a subbase even though the gears are often dissassembled at the manufacturer's plant and then reassembled in the shipyard's machine shop and in the ship. The more parts that can be assembled in the shop, the less will be the chance of damage to the gears during final ship assembly.

b. Reduction Gears and Thrust Bearing. The trend toward higher power, larger diameter (stiffer) shafting and larger and more flexible ships has led to increased concern over the effect of distortions of the gear and thrust bearing foundations and innerbottom on the alignment of shafting. In this respect, gear and thrust bearing foundations are made as rugged as practicable to minimize distortions in the shafting system that could lead to excessive vibration or improper meshing of gears. Maintaining a nearly equal load under service conditions on the forward and aft gear bearings provides proper load distribution on the gear and pinion teeth. This is accomplished by proper alignment of shafting.

Thermal expansion presents a problem in alignment in that it must be handled without causing excessive distortion and overstressing of the machinery units. The primary thermal growth affecting alignment of the gears is the vertical expansion of the gear foundation and the gear case. This growth is compensated for in the initial cold alignment of the shafting system as explained in Section 14.3f. Vertical and athwartship turbine and gear growth is compensated for in the initial turbine and gear alignment. Fore and aft turbine thermal expansion is compensated for by attaching the unit to flexplates which are fastened at one end to the main foundation or to the subbase.

c. Diesels. Because slow-speed main propulsion diesels are large and heavy, they are generally dismantled at the manufacturer's plant and reassembled in the ship. The engine bedplate is leveled and aligned with the shaft line by means of jacking screws. Allowance is usually provided for some settling as engine weight is increased during engine assembly. The engine may be set slightly high initially so that, on final alignment, the probability would be that the engine would be lowered, rather than raised, by the jacking screws, it being much easier to lower the engine than to raise it. Final alignment with shafting is accomplished when the ship is afloat.

Bottom and side chocks are fitted to keep the engine from moving on the ship's foundation. Fitted bolts are installed at the thrust end of the foundation and clearance bolts are fitted at the other end to allow for expansion.

Medium speed diesels are usually light enough so they can be completely assembled at the manufacturer's plant and installed intact into the ship at the shipyard. Reduction gears are usually installed separately.

d. Gas Turbines. In the case of gas turbine foundations, the subbase is normally mounted on a 3-point support to prevent movements of the ship's structure from distorting the subbase and possibly causing machinery misalignment. In one case, the GT unit was designed in modular fashion so that it could be shipped with its subbase in a normal 40 ft container fitted out for proper storing and handling of the unit at the shipyard. Reduction gears are installed separately.

Since gas turbines require frequent overhauls, the modular package greatly facilitates this work. However, because of these frequent overhauls, the scheme for access and handling the units should be carefully developed.

e. Bolts and Chocks. Foundation bolts for main machinery are made from high tensile bar stock such as AISI 1045 and are normally tightened with a torque wrench. Holes for fitted bolts are reamed. Holding-down bolts, or studs, in the thrust foundation are all fitted as are about 20 percent of the bolts in the main gear foundation. To prevent bending of the bolts, the foundation plate surface at the bottom end of the bolt, as well as at the top, may be required to be spot faced.

Chocks are usually 12 to 50 mm (0.5 to 2 in.) in thickness and may be made of steel or, as discussed below, made of a poured-in-place epoxy resin. There are normally two or more bolts per chock.

Steel chocks are often tapered to facilitate installation. The taper is provided by either machine tapering the heavy top plate of the foundation or by welding tapered pads to a flat foundation plate, Fig. 43. On some main gear foundations, the top plate is tapered both forward and aft from the centerline of the gear.

After the steel chocks are fitted and tack welded in place, bolt holes are drilled using undersized holes in the bedplate for a guide. Undersized temporary bolts are used for initial positioning and to hold the unit firmly on the foundation while drilling and reaming for the fitted bolts.

f. Poured-in-Place Resin Chocks. The use of epoxy resin for chocking, in lieu of fitted steel chocks, eliminates the need for machining foundation plates and the time-consuming task of milling and fitting chocks. When resin chocks are used, a taper would not be of any benefit. Resin chocks are used primarily in auxiliary and deck machinery mountings, and, in some cases, have been used under main reduction gears.

The procedure for installing resin chocks is to align the machinery or subbase to the foundation with jacking screws or spacer blocks and install the hold-down bolts slightly torqued. A dam is then built around each set of bolts to

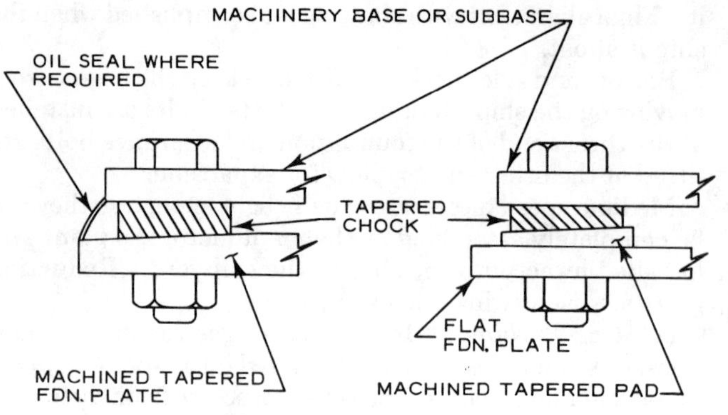

Fig. 43 Tapered steel chocks

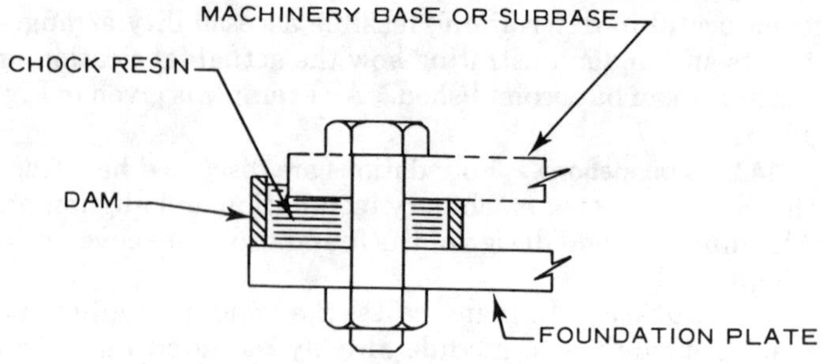

Fig. 44 Poured-in-place resin chock

receive the poured resin. The dam is built up about 12 mm (0.5 in.) above the bottom of the footing to ensure good resin contact at the upper surface of the chock, Fig. 44. In cold weather, the area must be heated. To facilitate machinery removal in case of repair, a thin liquid paraffin coating is applied to bolts and certain bearing surfaces to prevent adhesion. After the resin has hardened (12 to 48 hours, depending on the temperature) the dams are removed and the bolts are fully torqued.

14.3 Shafting. The manufacture and installation of shafting involves several extraordinary procedures which are described in this section. In addition, it should be mentioned that unusual new designs such as the hollow shaft torque tube (thin-walled cylinder made of rolled plate and fastened to end forgings) and the extra long stern tube, which extends all the way back to the main strut barrel (eliminating the intermediate strut), would obviously necessitate significant changes to normal installation practices.

a. Balancing. Propulsion shafting sections, for most merchant ships are solid and require no balancing treatment. However, hollow shafting, as used primarily in naval ships, does require balancing. This is normally done, after the first machining cuts are made, by placing the shaft on horizontal rails to determine any unbalance. Then the lathe centers are changed if found necessary. After further machining, balancing checks are made until the required balance is achieved.

b. Cold-Rolling. In order to improve fatigue characteristics at the critical areas of a tailshaft, the shaft surface at these areas is usually cold-rolled. Cold-rolling, which introduces compressive stresses over the surface, requires special shop equipment and is normally applied for about 1.5 m (5 ft) in way of the aft end of the tailshaft bearing and part way down the shaft taper. Rolling is accomplished by turning the shaft at a surface speed of about 20 m per min (60 ft per min) and applying the cold-working with a small roller, about 230 mm (9 in.) in diam, which is moved down the shaft about 2 mm ($^1/_{16}$ in.) per revolution. After rolling, the shaft is finish machined, removing as little material as possible, usually less than 0.4 mm (.015 in.), so as not to lose the benefits of the cold-rolling. If shaft struts are used, cold-rolling is generally applied at the forward end of the main strut bearing as well as the aft end.

c. Protective Coatings. When shafting is exposed to sea water, the shafting forward of the main strut is sometimes protected by a coating of synthetic rubber or a glass reinforced epoxy plastic. These coatings require rather exacting application procedures (NAVSHIPS, 1964). The epoxy plastic coating is usually preferred because of its relative ease of application. Briefly, application of the epoxy plastic to the straight portion of the shaft consists of cleaning the shaft surface, by a light abrasive blast if necessary, and then applying the resin and hardeners along with wrappings of glass fiber tape. Work must be done fairly rapidly because of the limited pot-life of the epoxy before it begins to gel.

The sealing of the coating to the ends of bronze sleeves and around flanged joints is rather difficult. At the junction of the sleeve, the end of the sleeve is recessed and tapered as indicated in Fig. 45. Several coats of epoxy without the glass tape are used for fairing-in, and then the ending is finished off following the method used for the straight section. The flanged couplings are coated with several coats of epoxy without the glass tape.

d. Tailshaft Bearing Arrangements. The selection of an oil-lubricated or water-lubricated tailshaft bearing will affect not only the design of shafting and stern frame or struts but also the shop work and ship installation.

If a water-lubricated bearing is used, a bronze sleeve is shrunk onto the shaft and the bearing is made up of lignum vitae wood, phenolic, or rubber staves which are dovetailed into a heavy bronze bushing.

If an oil-lubricated bearing is used, no sleeve is required, the bearing is usually babbitt (resilient plastic bearings have recently been developed) and there must be effective seals at each end of the stern tube to keep out water and to retain the lubricating oil which is under pressure. In the case of babbitt tailshaft bearings, the babbitt is normally centrifugally cast in a ductile iron or steel shell (bush) to ensure a good bond. This shell is machined slightly larger than the bore in the stern frame to provide an interference fit.

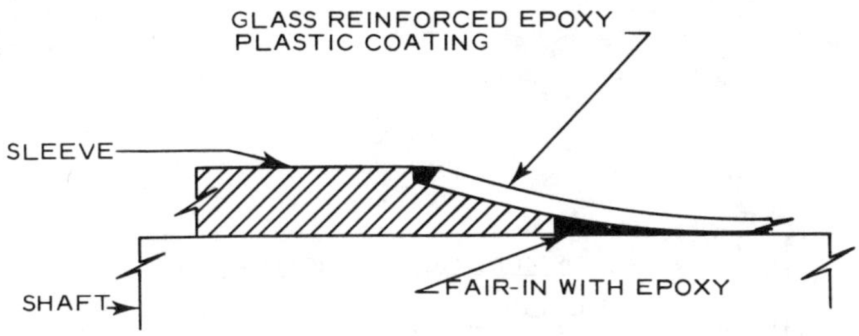

Fig. 45 Plastic coating on shaft

More exacting installation is required for oil-lubricated than for water-lubricated bearings as there is no provision for wear-down in the oil-lubricated bearing arrangement. The success of the oil lubricated bearing depends primarily on an initial satisfactory alignment and on the reliability of the seals, and the oil lubrication system must be thoroughly cleaned and flushed to remove all foreign matter before use.

Recent designs have attempted to deal with tail-shaft installation and servicing problems. Some stern bearing designs for some single-screw ships feature an arrangement whereby oil-lubricated bearings and associated seals can be installed, inspected, or replaced without disturbing the propeller or shafting and without drydocking the ship. One design incorporates a conventional journal type bearing while another design incorporates spherical roller bearings. In each case, the bossing structure must be enlarged in way of the bearing to provide the clearance necessary for installing and servicing the bearings from inside the ship.

In some recent designs of twin-screw ships, the normal intermediate strut has been eliminated and the stern tube is extended aft to the main strut. In a few cases, the main strut and tube bearings are lubricated by circulating oil through this tube.

e. Coupling Bolts. Fitted coupling bolts are prepared slightly larger than the anticipated final size so that the shipboard installation will ensure a proper fit of bolt to reamed hole as required during the assembly. Each bolt is marked to identify it with its corresponding marked hole in the coupling.

Two normal types of fitted bolts are used: One is a hex head bolt with only a slight body taper of about 1:100 on diam; the other is a tapered body bolt (no head) with a large taper of about 1:15 on diam. Bolt holes, whenever practicable, are final bored and reamed when two mating sections of shafting are assembled in the machine shop. Otherwise, this fitting must be done under less ideal conditions aboard ship. When a single section of shafting is replaced, the holes must be rebored and re-reamed aboard ship.

A new type of bolt is one which has a head but no taper. This bolt can be mechanically stretched through an internal device using hydraulic pressure. The bolt, when stretched, can be inserted in the straight hole in the coupling and the nut installed. When the internal pressure is released allowing the bolt to contract in length, the body of the bolt swells so that it fills the hole with a tight fit.

f. Shaft Alignment. Although both vertical and horizontal alignment must be considered, only vertical alignment is discussed in this section except to say that whenever a bearing is moved, the horizontal alignment must be rechecked.

The basic procedure for determining vertical alignment of shafting and gears is to calculate, for the *cold* condition at time of aligning, a shape of shaft line and corresponding bearing loads that will produce, in the *hot* operating condition, essentially equal loads on the main gear bearings and reasonably equal loads on line bearings. Negative bearing reactions should be avoided. These calculations are done iteratively, usually on a computer, using influence numbers. Influence numbers represent the change in reaction at the various bearings due to raising or lowering a given bearing one unit (say one mm).

The basic steps in the calculation are as follows:

1. Calculate the static vertical reactions at all bearings in cold condition, with all bearings in a straight line (along centerline of tailshaft boring).
2. Calculate or estimate the amount the two main gear bearings will rise due to thermal expansion of the gear case in the hot operating condition. This rise may be about 0.8 mm ($\frac{1}{32}$ in.).
3. Estimate an amount of wear at the aft end of the tailshaft bearing if bearing is wood or rubber. If an oil-lubricated bearing is used, this wear factor need not be considered.
4. If there is a wide range of operating drafts, calculate the relative innerbottom deflections in way of the shaft bearings between the alignment draft condition and the extreme operating drafts. This is important for a stiff shaft system.
5. Using the influence numbers, calculate the respective bearing displacements from the initial straight line that will give a shafting alignment which fulfils the requirements for the hot operating condition.
6. Considering the bearing displacements found in paragraphs 2, 3 and 4 above, calculate the cold condition for the draft at the time of alignment. This calculation will give the desired cold bearing loads and corresponding shape of shafting, Fig. 46A.

It is important to remember that these calculated reactions are necessarily approximate due not only to the imprecise nature of some of the input such as thermal expansion and innerbottom deflections but also to seaway loadings and diurnal temperature effects.

If a shaft is known to be bent slightly, the shaft is usually first rotated at 90 degree intervals and a mean line or position used.

The boring of the tailshaft bearing is delayed as long as practicable so that as much of the stern structure as possible can be erected and welded. Shaft alignment sightings are ideally made at night so as to minimize the distortion of the hull due to diurnal temperature effects. Once the boring for the tailshaft has been completed, all bearing heights, including those for the main reduction gear, are referred to

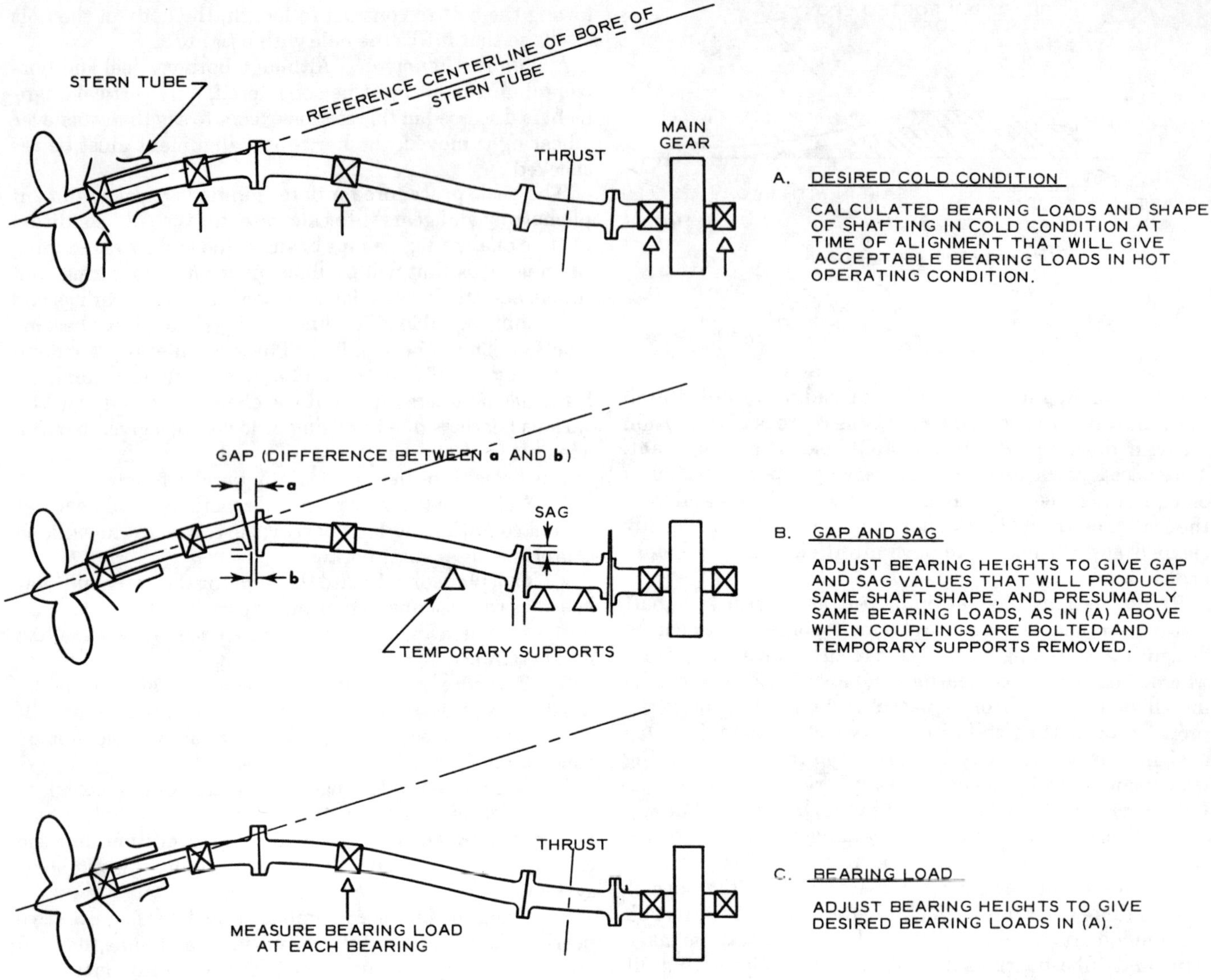

Fig. 46 Schematic diagram for shaft alignment

the straight line through the centerline of bore, see Fig. 46A. When bearings are chocked, the thickness of the chocks are adjusted so that the bearing lies at the same slope as the shafting at that point.

Two methods presently used for establishing the desired alignment, always with the ship afloat, are the *gap and sag method* and the *bearing load measurement* method.

• *Gap and Sag Method.* In this method, Fig. 46B, the desired bearing loads are obtained indirectly by establishing the desired shape of shaft line in terms of offsets from the straight line.

Starting from the calculated desired shape of shaft line in the cold condition, mating flanges are then imagined to be free (unbolted) and their deflections (gap and sag in Fig. 46B) are calculated. If temporary supports are required, these supports are included in the calculations. With the actual unbolted shafting approximately aligned, the gap and sag values are measured. If they are not in agreement with calculated values, the bearings are moved up or down until agreement is reached. The mating flanges are then drawn together and bolted.

• *Bearing Load Method.* In this method, Fig. 46C, the desired bearing loads are obtained directly by measuring the actual bearing reactions.

With the bolted shafting approximately aligned by gap and sag or optical sighting, the bearing reactions are then measured by a calibrated hydraulic jack or with electronic load cell and jack (bearing foundations are often designed with an extension to provide a jack foundation) and the deviations from desired reactions noted. These deviations are then corrected by adjusting several selected bearings. The amount of adjustment at those bearings in terms of bearing loads is obtained by using the above mentioned influence numbers.

When determining the relation between the jacking lift load and the actual shaft lift, it is found that more load change is required to lift the shaft a given amount than to lower the shaft by the same amount. This change is assumed to be due to friction in the shafting and jacking systems. Therefore, a mean line relationship, between ascending and descending loads vs vertical movement plots, is normally adopted.

Initial alignment is sometimes made by the gap and sag method and corrections, as necessary, made by the bearing load method, which is considered more accurate than the gap and sag method.

14.4 Propellers. Propellers are supplied by propeller manufacturers who cast, machine, finish and balance propellers. A static balance is considered quite adequate for merchant work. To provide a tight final fit, the propeller hub is bored to be slightly smaller than a specially prepared plug gage, which is representative of the tailshaft taper. The end of the tailshaft taper is made to suit a corresponding ring gage. These two gages ensure interchangeability of shafts and propellers, and this is especially valuable in a multiship program. The final fitting of the propeller to the shaft is done at the shipyard just prior to installation on the ship.

The permissible static unbalance is based on a maximum (centrifugal) force at rated rpm. The maximum force is usually taken as about 1 percent of the propeller weight. This information is normally given on the propeller plan in terms of an equivalent moment. When correcting for unbalance, metal should not be removed from the vicinity of the leading or trailing edges.

When installed on the ship, the propeller is forced beyond the snug-fit position by a predetermined amount (assuming the temperature of the propeller and shaft is moderate) in order to ensure a tight fit. This is accomplished by slogging the propeller nut or but hydraulic means. Since the propeller material has a larger coefficient of expansion than does the shaft material, a correction is made when installation is made in cold or hot weather as directed by the engineering department. When the temperature is high, the propeller is pushed onto the shaft slightly less than the basic moderate temperature amount whereas, when the temperature is low, the propeller is pushed slightly more than that amount.

New methods developed for installing and removing propellers have facilitated this work, especially as regards the larger propellers. These new methods have in common the use of hydraulic pressure to force the propeller onto or off of the tailshaft. The correction for temperature, as mentioned in the foregoing is accomplished by varying the oil pressure when hydraulic nuts are used (lower pressure when temperature is high, and higher pressure when temperature is low). Schematically, the general method is outlined in Fig. 47.

One modification to this general method involves backing off the propeller nut and then injecting oil under pressure between the hub and tailshaft so as to expand the hub slightly and thus release the tight fit and free the propeller.

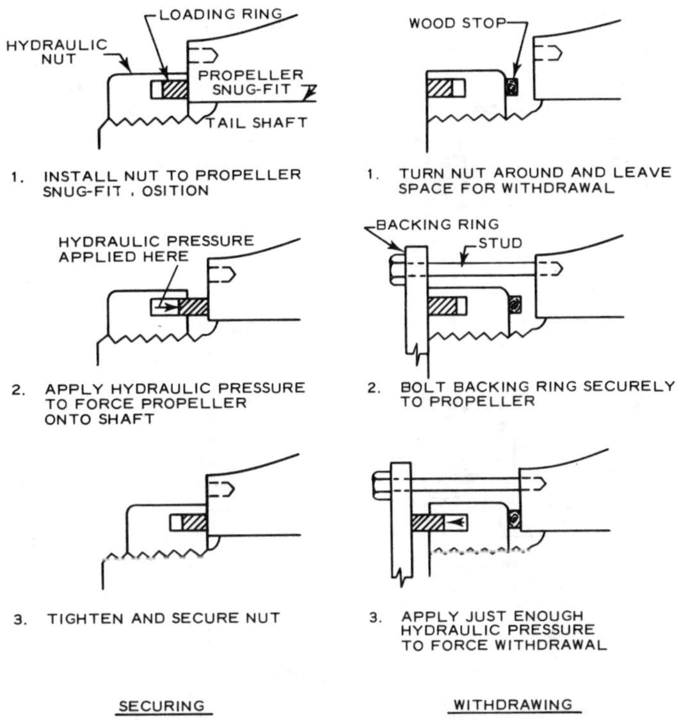

Fig. 47 Schematic steps for securing or withdrawing propeller using hydraulic pressure

Another modification involves injecting oil under pressure in like manner to expand the hub slightly in order to provide better control of the installation operation. Then when the propeller is in position, the release of oil pressure allows the hub to shrink onto the shaft. These modified methods have generally been associated with keyless propellers.

Keyless propellers rely on friction between the hub surface and the tailshaft surface to hold the propeller tightly in place, and, in this respect, require carefully controlled installation procedures as recommended by the equipment manufacturer. Also, except for grooves for the hydraulic oil, there are no large recesses within the propeller hub similar to those usually provided for keyed propellers.

Controllable pitch propellers have a special bolted connection to the tail shaft and, therefore, do not have the tapered connection typical of fixed pitch propellers.

On destroyers and some other high-speed ships, the propellers project below the base line, and this requires special attention in handling the propellers and in drydocking the ships.

14.5 Boilers. Boilers are assembled in the shipyard or local vendor's shop and installed in the ship as one unit if handling facilities will permit, otherwise, the boilers must be assembled in the ship. The main pressure elements may be pre-assembled and such items as economizer and superheater elements, casing, and brickwork installed after the boilers are in place in the ship. The advantages of erecting the boilers as completely as practicable before placing them aboard ship are obvious. In any event, care must be taken to protect the internal and external surfaces

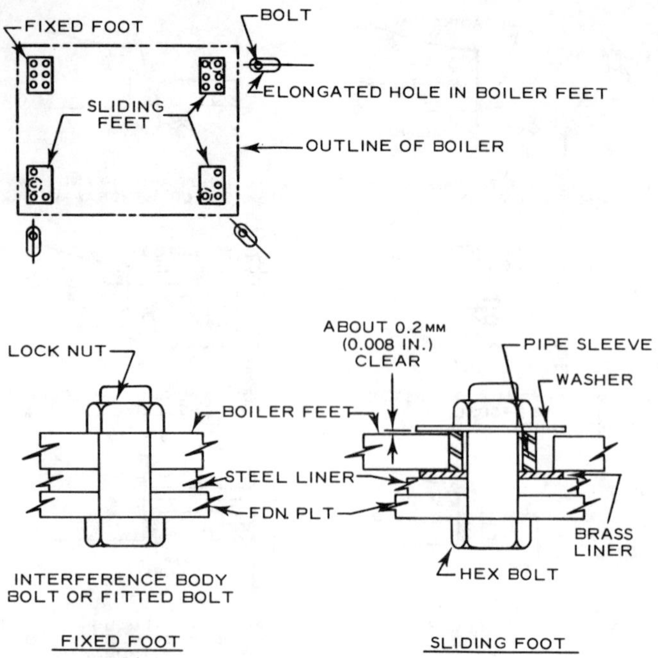

Fig. 48 Typical arrangement of boiler feet

of the pressure elements and gas outlets until such time as the uptakes and stack are in position.

There is no need for special alignment in connection with boiler installations except that provisions must be made for thermal expansion at the boiler feet (saddles) locations when the boiler is in hot operating condition. Expansion may be in the order of 10 mm (½ in.). Assuming a four point support, one foot will be bolted down hard to the foundation, while the other three feet will be bolted so as to permit the feet to slide, Fig. 48. Holes for the fixed-foot bolts are reamed in place. The foundation bolts in way of the sliding feet must be accurately located with respect to the elongated holes in the boiler feet to enable the feet to slide clear of the bolts. The pipe sleeves prevent the bolts from tightening down on the sliding feet. A brass liner or a lubricant such as molybdenum disulfide is normally used to facilitate the sliding action.

In some cases, the pipe sleeves shown in Fig. 48 are not used. Instead, the bolts are tightened only slightly so that the boiler foot can slide.

14.6 Piping. Large and important piping systems such as for main steam, main circulating water, and for cargo handling on tankers and liquid product carriers are drawn in detail by the drafting room. Particular attention is given to pipe supports, and take-down joints are strategically located to facilitate pipe removal and valve repair. Piping requirements for low temperature product carriers, such as for LNG, are covered in Chapter XI and in the IMCO Gas Code.

Great care is taken to ensure proper application of Class I (or Group I) piping which includes all piping intended for high pressures or temperatures, such as for main steam, and for piping for lethal gases and liquids. Each weld is detailed on a plan or sketch and is usually given a specific reference number. Accurate welded joint records are kept of such information as welder's name, heat treatment, and inspection.

Subsurface inspection, where required, is by radiography (RT). Where RT is not practicable, inspection is by magnetic particle or dye penetrant. For butt welds where access to the interior of the pipe is not possible, the joints are radiographed across the pipe diameter using at least three exposure locations to cover the entire butt. Normally, butt welds in Cr-Mo piping 62 mm (2½ in.) in diam and over are radiographed. For mild steel piping over 9.5 mm (⅜ in.) thick, the butt welds are radiographed. For thinner mild steel piping over 76 mm (3 in.) in diam), butt weld inspection is by magnetic particle, as are welds of many other connections such as pipe flanges.

Pipes are bent hot or cold. Cr-Mo pipe bent hot must be packed with sand to prevent buckling and must be normalized before welding. If the pipe is bent cold, to a minimum radius equal to five diameters, no sand packing is needed but the pipe must be normalized before welding. Welding is done with preheat. After welding and inspection, the pipe is stress relieved if the wall thickness exceeds 12.5 mm (½ in.). Mild steel pipe is handled in a somewhat similar manner and stress relieved if wall thickness exceeds 19 mm (¾ in.).

When pipe connections are made aboard ship or when minor alignment adjustments to the piping must be made aboard ship, local heating is applied to the pipe with resistance coils and the temperature cycle recorded using portable equipment. After a pipe is realigned, the pipe connections should fit without forcing.

a. Main Steam Piping. Main steam piping diagrams and layouts must consider not only installation and maintenance factors but also the stresses and reaction forces created by thermal movements of the piping. Computer programs have been developed to make pipe stress analysis to aid in establishing support points and also to determine reaction forces and moments on machinery units such as turbines and boilers. These forces and moments must not exceed limits set by the machinery manufacturers. In some cases, large pipe loops or 3-dimensional bends are required to avoid excessive stresses or reactions.

Main steam piping is usually made of seamless low alloy steel. Where temperatures are above 455°C (850°F), the most widely used alloy contains 0.5 percent molybdenum (Mo) and 1.25 percent chromium (Cr).

Gaskets between flanges are generally made of thin stainless steel strips spiral wound with insulating filler in between successive layers.

There are three types of supports to carry the weight of main steam piping: rod hangers, variable force spring hangers, and constant force spring hangers. Fig. 49 shows typical spring hanger arrangements where vertical movement at a support point occurs as a result of thermal expansion of the piping.

When variable force spring hangers are used, it is necessary to make an initial adjustment when the pipe is installed

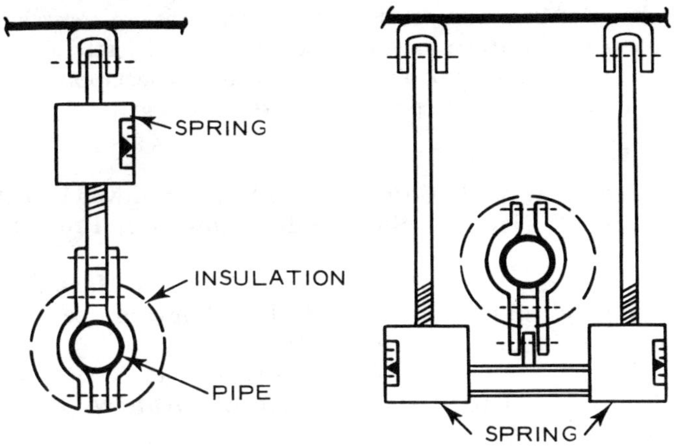

Fig. 49 Typical spring hangar pipe support

cold and a final adjustment when the pipe is in hot operating condition. Final adjustment is normally accomplished during ship trials by adjusting the spring force to the value determined by the pipe stress analysis for the system.

Horizontal sway braces are often used to resist dynamic forces applied to piping due to rolling and pitching or due to vibrations transmitted to pipe anchor points. Sway braces may be of the turnbuckle-rod type, the preloaded spring type or the hydraulically damped type.

At fixed anchor points, the pipe anchor brackets are separated from the anchor foundation by insulating material to reduce heat transfer from the pipe to the hull structure.

b. Insulation Work. This work is normally done by subcontractors. Insulation on main steam piping is usually of sufficient thickness as to reduce the outer surface temperature to 14°C (25°F) above the ambient temperature of the space. The insulation is manufactured in either molded segmented form or block form and is held in place by wire or metal bands. When the insulation is over about 75 mm (3 in.) thick, two layers of insulation may be used, and the joints of the two layers are staggered. The joints are sealed with insulating cement. Where molded sections cannot be used, such as around valves, the insulation may be in the form of blankets. The insulation may be covered with a cloth lagging using a lagging adhesive, or it may be covered by hard finish insulating cement. Sheet metal covering is applied at places where mechanical damage to the insulation would be likely.

Diesel or gas turbine exhaust piping and boiler uptakes are insulated and lagged in a manner similar to that for main steam piping. In the case of diesels and gas turbines, the insulation also helps to reduce the noise level radiated from the exhaust gas ducts and pipes.

c. Main Circulation Water Piping. As with piping layouts in tanker pump rooms, main circulating water piping layouts require special attention because this piping is located at the innerbottom level in a very congested area of the ship, and the pipes, valves, and pumps are large and heavy.

Vessels having higher service speeds, say above 16 knots, may have a scoop in lieu of pumped circulation of seawater. A standby circulation pump is installed for use at lower maneuvering speeds and in port. The scoop is a difficult structure to fabricate and install because of the acute angle intersection between the scoop and the innerbottom and the shell.

In order to allow for thermal expansion of piping and for thermal movement of the condenser, bellows type expansion joints are inserted in the piping near the condenser and at other points, such as around the circulating water pumps. These bellows are normally made of molded synthetic rubber. Care must be taken not to use these joints to correct installation misalignment or to support any load.

14.7 Electrical Installations. The number and size of electrical installations aboard ships have increased due primarily to new developments in the areas of mechanized engine room control, mechanized cargo handling, and sophisticated electronic systems. The installation of major items of equipment such as generators, switchboards, and associated wiring must be in accordance with the requirements of regulatory agencies and also the requirements of either IEEE Standard No. 45, or Publication No. 92, International Electrotechnical Commission, depending upon the classification of the ship.

The gyroscope system, underwater log, depth sounder, loran, radar, and other electronic systems are installed as indicated by manufacturers' requirements.

Items such as switchboards and electric generators are not placed in position until there is no longer danger of damage from falling objects, dirt, and water. Some equipment is provided with a waterproof covering and even with internal heating elements which may be turned on to keep the equipment dry when not in use.

Major generating and motorized units are mounted on subbases which, in turn, are fastened to their corresponding ship foundations. For minor electrical installations, the electrical shops prepare foundations for many items of electrical equipment. When completed, these foundation brackets and frames are stored for subsequent installation in the ship.

The work of installing the smaller foundations such as those for connection boxes, control panels, and power panels, as well as cable supports and stuffing tubes, can be done at various stages of construction. This work is done before thermal or other insulation is installed.

a. Wiring. Many cable runs are laid out by yard electricians from schematics drawn in the drafting room.

All cable ends must be fitted with connectors, and ends of the insulation must be sealed to keep out moisture. Where mineral insulated, metal shielded cable is used, the end must be sealed immediately after cutting. Where several wiring installations are duplicated, wiring mock-ups can be made in the shop and the above-mentioned detail work can be done under shop working conditions.

The minimum bend radius for armored cable is eight cable diameters and six diameters for other cables. The armor on electric cable, when used, must be continuous, except at cable splices as discussed in Section 14.7b, and must be grounded to the hull.

Single cables may be supported by clips welded to structure. Cables in groups are supported by metal cable hangers and usually arranged so as to permit painting the surrounding structure whenever practicable.

Typical hanger arrangements are described by Melvin (1971). This Reference also describes typical cable penetration details in way of plain or insulated decks and bulkheads.

The preparation of stuffing tubes and multiple cable transits for locations where cables pass through watertight boundaries may be done in the shops before installation in the ship. Similarly, those lighting and power panels that are unique to the ship may be fabricated in the electrical shops; other standard sizes are available from commercial suppliers.

Cables must be protected from damage in way of places such as hatches, tank tops, and open decks. Protection is usually provided by removable metal coverings, which must be grounded to the hull. Horizontal pipes used for cable protection are to be provided with drain holes. Cable ways are to be kept as far from heat dissipating equipment as possible; where this is impracticable, shielding and insulation must be installed for protection of the cables. Cables are not to be run behind insulation, through refrigerated spaces, or in bilges.

b. Cable Splices. Splicing methods have been developed so that splice connections can now be made at assembly interfaces with regulatory agency approval. However, the location of the splice and the method of protection requires specific approval.

The typical splice is made using an approved splice kit and requires several steps. First, the ends of the conductor cables are joined together by a butt connector which is crimped around the ends of the conductor being joined. Crimping is done by a special one-cycle compression tool. Second, an insulating sleeve of heat shrink tube is shrunk onto each conductor splice by blowing hot air on the tube. There may be several conductor splices in a cable joint. Third, the armored covering at each end of the splice joint is electrically connected by a wire provided in the kit. It is not necessary to replace the armor in way of the splices. And lastly, a heat shrink tube is shrunk over the whole joint, overlapping the armor at each end to provide watertight integrity.

References

Bath Iron Works Corporation (1977), "Improved Design Process," National Shipbuilding Research Program, April.

Bath Iron Works Corporation (1978), "A Manual on Planning and Production Control for Shipyard Use," National Shipbuilding Research Program.

Belch, A. Ross (1976), "Construction of Very Large Tankers in Two Parts," *The Naval Architect,* RINA, July.

Goldbach, R. D. (1973), "Applications of Preoutfitting During Construction of Ammunition Ships AE 32–35," *Marine Technology,* January.

Melvin, B. (1971), "Electric Plants," *Marine Engineering,* edited by Roy L. Harrington, SNAME, New York.

Naval Ship Systems Command (1964), "Instruction Manual, Glass Reinforced Plastic Coatings for Propeller Shafting," NAVSHIPS 250-634-4, January.

Ship Structure Committee (1969), "Recommended Emergency Welding Procedure for Temporary Repairs of Ships," Report No. 196.

Ship Structure Comittee (1974), "Flame Straightening Quenched and Tempered Steels in Ship Construction," Report No. 247.

Ship Structure Committee (1977), "Review of Ship Structural Details," Report No. 266.

Todd Shipyards Corporation (1974), "Shipbuilding Alignment with Lasers," National Shipbuilding Research Program, April.

Todd Shipyards Corporation (1976), "Photogrammetry in Shipbuilding," National Shipbuilding Research Program, July.

Veliotis, P. T. (1977), "A Solution to the Series Production of Aluminum LNG Spheres," *Trans.* SNAME.

CHAPTER XVII

Clyde M. Leavitt

Launching

Section 1
Launching Methods

1.1 General. The launching of a vessel is a critical event in the building process and one that is potentially hazardous if the movement of the large, yet fragile, mass that is supported on a comparatively frail structure is not properly planned and executed. Configuration of design and selection of materials for suitability and strength and the control of, and making use of, the natural forces of gravity, buoyancy, water resistance and friction is, in essence, the art of ship launching. A properly engineered launching can be accomplished safely and efficiently. The development of this engineering plan is one of the most important tasks facing the shipyard naval architect.

At the time of receipt of a request for a proposal, preliminary calculations are made to verify that the vessel can be constructed and launched, in one or more pieces, using facilities existing or suitably modified. Following contract award, further calculations are prepared to exactly locate the vessel on the building slip, determine forces acting on the ground ways, cradle and ship, and to provide a basis for the design and delineation of the launching arrangement.

Consideration must be given to building slip and ground way capacity from standpoint of length, breadth, vertical clearance and local and overall load carrying capacity. Light weight of vessel is obtained from specifications, or is estimated, and is used as a basis for estimated weight in launching condition. Longitudinal, vertical, and transverse position of vessel's center of gravity is estimated. Availability of suitable existing sliding and ground ways is verified or proportions of new ways are calculated to provide suitable unit pressures on the launching grease. Method of release is determined. Preliminary Bonjean curves are made, if not available, and drafts of vessel afloat are determined. Assuming a tide height conservatively less than high water springs, maximum way end and pivoting pressures are determined as is possible tipping and drop off. Satisfactory transverse stability is verified and means of checking the launched vessel are determined. Following contract award, the preliminary calculations are reviewed and refined as a basis for the final studies, calculations, and plans.

1.2 Launch Methods. There are two methods of launching vessels. In the traditional method the vessel is supported by a cradle which rests on inclined, or inclinable, ways. At the interface between cradle and ways there is launching grease or some other means of reducing friction. Release of the cradle lets the vessel and cradle slide into the water. The vessel may enter the water stern first, bow first, or sideways.

In the other method, the vessel is supported by a translation system cradle and/or by blocking and shoring which rest on a platform which may be fixed or movable. If the platform is movable, it is lowered, possibly with trimming, until the vessel is afloat. With a fixed platform, water is caused to rise around the vessel until it floats.

The launching characteristics are logically grouped in Table 1 without regard for launching mode while the check marks in Table 2 indicate possible problems with certain parameters for the eleven launching modes shown. Side launching is shown to be superior to end launching. Launching from a trimmable submersible platform promises a minimum of difficulties. End haul and side haul marine railway and heavy lift crane launchings are attractive for relatively small vessels.

a. End Launching. The vessel is ordinarily launched from the building position. The vessel's weight is transferred from the keel block, bilge crib and shoring building supports to the launching cradle. The cradle is supported on one or more ground ways which extend longitudinally under water. Attachments between sliding and ground ways prevent movement of the former. Restraining attachments may be burn off sole plates or triggers, release of which allows the cradle-supported vessel to slide down the ground ways under the influence of gravity and enter the water. As the stern acquires buoyancy, it lifts, and when this happens the ship pivots about its fore poppet. Checking of the sternward motion may be required to prevent grounding of the vessel on an opposite bank or shoal.

b. Side Launching. If the vessel is supported by a translation system, on arrival in the launching position the vessel's weight is transferred to the sliding ways or sleds resting on ground ways perpendicular to the vessel's centerline. If the vessel is built in the launching position, its weight is transferred as described above for end launching.

c. Floating Drydock. Varying deformations due to

Table 1—Launching Characteristics

ENVIRONMENT
- Astronomical tide range and frequency
- River stage
- Wind tide
- Current
- Slack water
- Wind velocity and directions
- Way end submergence
- Water depth at way ends
- Water breadth off ground ways
- Water depths outshore of ground ways
- Traffic off way ends
- Water depth for poppet removal
- Water depth over submersible platform
- Chain drag path size and surface

GROUND & SLIDING WAY INTERFACE
- Base coat(s)
- Slip coat
- Grease irons
- Grease protection
- Grease salvage and reuse
- Outshore ground way cleaning
- Grease reaction to temperature and pressure
- Tetrafluoroethylene
- Water casters
- Rollers, spherical and cylindrical
- Wheels
- Side launch dagger starting shores
- Dog shores
- Burn off plate release
- Mechanical trigger release
- Tumble shores
- Starting rams
- Emergency starting arrangements for stuck vessel

INSTRUMENTATION
- Recording
- Indicating
- Between ship and ground
- On ground ways
- On cradle
- Aboard ship
- On submersible platform

LAUNCHING EVENTS
- Cradle wedged up
- Vessel partly supported by cradle
- Other means of support removed
- Vessel entirely supported by cradle
- Dog shores removed
- Vessel released by triggers or burn off
- High way end pressure
- Pivoting for end launch or possible tipping for side launch
- Drop off or float off ground ways
- Start of checking
- Vessel afloat
- Cradle removed all or in part
- Vessel secured at berth

VESSEL PARAMETERS
- Length
- Breadth
- Depth
- Deadrise
- Bottom framing arrangement and spacing
- Bottom local strength in way of cradle
- Sonar dome, propeller(s), rudder(s) below base line
- Launching drafts and trim
- Weight and weight distribution
- Transverse stability
- Asymmetry of hull
- Water resistance
- Structural strength and watertight integrity
- Overhang inshore of inshore poppet
- Shape of forefoot
- Vessel location on ways
- Bottom height above ways
- Sail area

CRADLE
- Sliding ways
- Wedges and ram rails
- Wedge riders
- Packing
- End poppets
- Fore poppet, crushing strip or rocker
- Spreader ties
- Sliding way end connections
- Side launch sleds
- Steel structure and bracing
- Reuse of cradle or sleds
- Positive or negative buoyancy
- Bow buoyancy pontoons
- Stern buoyancy pontoon(s)

RIGGING
- Cradle tricing, backstays
- Wedge heel securing wires
- Cradle hauling line
- Sled securing lines
- Drag cables
- Drag cable tricing
- Mooring and heaving lines
- New or existing rigging
- Bunting for ship and platform decoration

GROUND WAYS
- Straight
- Declivity
- Cambered
- Radius
- Transverse inclination
- Spread of end launching ways
- Side launch tilting ways
- Side launch rocker or pivoting ways
- Pivoting area and way end reinforcement
- Ribbands
- Material. Wood, concrete, steel plate or crane rail
- Way spacing
- Fixed or portable
- Existing or new
- Spur shores
- Number of ground ways. One, two, three, or four
- Outboard balancing ways
- Treatment of submerged wood to resist marine borer damage
- Non-trimmable submersible platform
- Trimmable submersible platform
- Crushing packing outshore
- Outshore throat
- Depth over way ends

COMMUNICATIONS
- Prelaunch between control station, sponsor's platform, dog shores, releasing gear, vessel to be launched and tug boats. Postlaunch between launched vessel, shore and tug boats.

CHECKING
- Chain cable drags
- Concrete block or other drags
- Mask(s)
- Anchor(s)
- Breaking stops
- Friction-braked wire
- Water's edge chain piles to stop whipping and for final checking
- Turning vessel by drags on one side
- Turning vessel by rudder
- Mask effect of propeller(s) and rudders
- Drag cable release at vessel
- Emergency checking
- Resilient fenders for side launch
- Soft mud bank for end launch

changes in solar radiation, air and water temperatures and varying loads, as well as the absence of a horizontal reference plane, makes a floating drydock unsuitable for the new building of vessels of any great size or weight.

d. Ground Supported and Floating Platform. The platform, which is essentially a floating drydock, is ballasted to rest on a horizontal underwater pile-supported grid, or in another version, has one side or one end of the platform temporarily connected to, and aligned with, the shore. The vessel is moved onto the platform, the inshore wing walls of the platform being temporarily removed if the vessel is moved transversely. Alignment between shore side and platform rails, tracks or ways is maintained by the relative fixity of platform deck and shore. If only side or end support is provided, the platform must be progressively dewatered as the vessel is moved on to it to keep the platform deck horizontal and aligned vertically. Even if the platform has the overall support afforded by a grid, progressive dewatering may be required to reduce piling loads if a heavy vessel is being moved onto the platform. With the vessel centered on the platform, any removed wing walls are reinstalled and ballast tanks are dewatered to float the platform clear of any submerged or above water supports. The platform is moved to a dredged area of enough depth for it to operate in a conventional floating drydock mode and submerge sufficiently to float the vessel being launched.

e. Ground Supported Trimmable Platform Lift. The platform is supported on both sides and can be raised and lowered vertically by means of steel wire ropes and winches or chain cables operated by hydraulic rams or windlasses with wildcats. Winch or windlass controls are such that the platform can be moved in a horizontal position or can be lowered to a trimmed position to approximate the afloat trim of the vessel being launched. The platform and surrounding

Table 2—Characteristics with Possible Significant Effects on Launching

CHARACTERISTICS	LAUNCHING METHODS										
	END LAUNCHING	SIDE LAUNCHING	FLOATING DRYDOCK	GROUND SUPPORTED AND FLOATING PLATFORM	GROUND SUPPORTED TRIMMABLE PLATFORM LIFT	GROUND SUPPORTED NONTRIMMABLE PLATFORM LIFT	GRAVING DOCK	MARINE RAILWAY	FOUR BAR LINKAGE GROUND SUPPORTED PLATFORM	PIGGY-BACK LAUNCHING	HEAVY LIFT CRANE LAUNCHING
VESSEL WEIGHT	✓		✓	✓	✓	✓		✓		✓	✓
VESSEL LENGTH	✓				✓	✓	✓			✓	✓
VESSEL BREADTH	✓		✓	✓			✓			✓	
VESSEL DEPTH	✓										
VESSEL LAUNCHING DRAFT	✓	✓					✓		✓		
VESSEL TRIM		✓				✓	✓	✓	✓		
UNDER KEEL PROJECTIONS	✓						✓			✓	
LAUNCH PREPARATION ELAPSED TIME	✓	✓							✓		
LAUNCHING ENGINEERING	✓	✓									
CRADLE STRENGTH	✓										
STARTING AFTER RELEASE	✓										
VESSEL BOTTOM LOCAL DAMAGE	✓	✓				✓	✓				
GROUND WAY LOCAL DAMAGE	✓										
HULL GIRDER BENDING DAMAGE	✓										
VESSEL CHECKING	✓										
VESSEL RETRIEVAL SUITABILITY	✓	✓							✓	✓	
OUTSHORE GROUND WAY LENGTH	✓										
LAUNCH BASIN BREADTH	✓										
DREDGED DEPTH	✓	✓	✓	✓							
WAY END TIPPING	✓										
KNUCKLE BLOCK PIVOTING						✓	✓				
SLEWING ON WAYS	✓	✓									
TRANSVERSE STABILITY		✓				✓	✓				
NONDISPERSAL OF PAINT AND WELDING FUMES			✓	✓			✓				
EXPENDABLE MATERIALS	✓	✓								✓	

shipyard decking are provided with a translation system for longitudinal and/or transverse movement onto or off the platform.

f. Ground Supported Non-trimmable Platform Lift. The platform is identical to the design just described except that the platform remains horizontal with no trimming.

g. Graving Dock. The vessel is supported during construction by keel, side and bilge blocks, and shores stepped on the level, reinforced concrete floor of the graving dock. The vessel is launched by controlled flooding of the dock.

h. Marine Railway. Both end haul and side haul marine railways can be used for new building launching if the vessel is built over the railway or if it can be moved onto the railway. Launchings can be controlled or run free.

i. Four Bar Linkage Ground Supported Platform. A platform at ground level and parallel to the shore is supported by vertical, or nearly vertical, struts pinned at their upper ends to platform inboard and outboard edges and at their lower ends to foundations on the basin bottom. After the vessel has been moved onto the platform, mechanical triggers restraining the platform are released and the four bar linkage system results in the upper ends of the struts describing arcs so that the platform swings outshore and downward to deposit the vessel in the water.

j. Piggy-Back Launching. The deck of a barge-like vessel can be used as a building berth or launching slip for another vessel. If the former, both vessels can be launched as one unit. In the case of the latter, the vessel to be launched, if an offshore jacket for example, may have been built ashore and then moved horizontally onto the deck of the supporting vessel. Depending on relative sizes, configurations and weights of the two units, the supporting vessel can be ballasted and trimmed for a float off or sliding end launching or can be ballasted and heeled for a sliding side launching.

k. Heavy-Lift Crane Launchings. One or more heavy-lift cranes with the necessary outfit of spreader beams, steel wire rope slings and lifting eyes on, or bottom straps for, the vessel to be launched provide a means of launching relatively small vessels.

Section 2
Ground Ways

2.1 General. Ground ways support the cradle which supports the vessel to be launched. The ways may be steel-reinforced concrete, built up of structural timbers stud-bolted together or welded steel plates forming box beams if launching lubricants are to be used, or in the case of wheeled cradles, may consist of crane rails. Composite construction may be used. Wooden ground ways may have steel plate ribbands and the concrete ribbands of steel-reinforced concrete ground ways may be sheathed with structural steel shapes. Ground ways may include outshore ways which are permanently submerged in locations where the tidal range is small, or submerged at high water when the tidal range or variation in river stage is large, and inshore ground ways where only the outshore ends may, on occasion, be submerged. Ground ways may be taken up and stored following a launching or may be left in place ready for the next launching. There is a limit to the lengths of wedges which can be manually driven to remove vertical slack from the cradle. Consequently sliding way width is limited and this, in turn, limits ground way width.

2.2 End Launch Ground Ways. Ways may be straight or cambered to a circular arc of large radius. If two ground ways are used, there may be a transverse inclination of each way forming a shallow V tending to keep the cradle centered. Viewed from above, each ground way is straight and there may be a slight spread between the ground ways, going outshore, to reduce the probability of the cradle binding between the ribbands if vessel and cradle should slew due to wind and/or current. Ribbands, usually on the outboard sides of the ground ways, maintain the alignment of the cradle. Due to clearance between ground way ribbands and cradle as well as possible divergence of the ground ways, they should be wider than the sliding ways if maximum bearing area is to be obtained at the possibly critical pivoting and ground way end areas.

The length of costly outshore ground ways can be reduced by the use of high declivity and/or cambered ways. A high declivity straight way results in large forces on releasing arrangements. These forces are reduced if cambered ways are used as the declivity under the vessel's center of gravity at time of release need only be large enough to ensure starting. If the inshore end of the cradle may drop, rather than float, off the ground ways, and the vessel's forefoot projects inshore of the inshore end of the fore poppet, then outshore cross bracing between the ground ways must be omitted to provide a throat of sufficient longitudinal extent into which the forefoot of the launched vessel can drop. Three or four ground ways can be used for large and/or heavy vessels. Vessels can also be launched on a single centerline way with relatively narrow balancing ways on each side to provide stability until the vessel is afloat. Ground ways can be perpendicular to the shore line, or at an angle, depending on water width and depth offshore.

2.3 Side Launch Ground Ways. Ways are usually straight but may be cambered, the latter being more likely if float off, rather than tipping, launchings are planned. The top surfaces of straight ground ways lie in a plane inclined at a considerably steeper angle to the horizontal than usual for end launching ways. In plan view, the ground ways are parallel to each other and perpendicular to the shore line. Unlike end launching ways, side launching ways may have ribbands on each side or may have no ribbands at all if the cradle has butter-boards instead of sleds. The outshore ends of the ground ways may be above water so that the cradle and vessel tip about the way ends rather than floating off the ground ways as would be the case if the ways extended sufficiently far under water.

a. Side Launch Tilting Ground Ways. Tilting ways are a variation of conventional ground ways where the outshore way ends are above water. Underneath each ground way and at the edge of shipyard decking or other outshore support, fulcrum blocking is provided. Each fulcrum supports a length of ground way, more than half the length of the ground way being inshore of the fulcrum. Each ground way length is restrained against transverse or longitudinal movement but is free to rotate or tip about the fulcrum. As the vessel's center of gravity passes the fulcrum line, tipping of vessel, cradle and outshore way sections commences. Tilting ways act as fenders to prevent the vessel striking the near shore on the return roll, distribute ground way end support over the underside of the cradle and protect the vessel's bottom from high way-end pressure if one end of the vessel leads the other during launching.

b. Side Launch Rocker Ground Ways. Rocker or pivoting ground ways can be fabricated from structural steel, each with a trunnion at its midlength. The ways are initially horizontal and are aligned with horizontal translation ways. The vessel to be launched is supported on simple sliding ways and is moved athwartships onto the ground ways and is stopped with the center of gravity just inboard, or just outboard, of the trunnions. With the center of gravity inshore, the inshore ends of the rocker ways are elevated by one or more cranes. As soon as the vessel's center of gravity passes the trunnion centerlines, rotation of the ways and vessel about the trunnions commences and continues until the outboard ends of the pivoting ways strike ground-supported chocks. If the center of gravity is outboard of the trunnion centerlines, launching is effected by releasing mechanical triggers at the inshore ends of the rocker ways or by simultaneously snatching away shores under the outshore ends of the ways. Double the normal side launching declivity can be obtained, trigger loads are low, and the launching system is largely an extension of the translation system.

2.4 Design Data. General design data for end and side-launch ways are given by the following:

a. End Launch Straight Ways.

Slope . . . . 0.045–0.065 to 1.00
Transverse Inclination . . . . 0.020 to 1.00
Total Spread Going Outshore . . . . 1 to 2400
Sliding Way—Ribband Clearance
Each side . . . . 25 mm (1 in.)
Ground Way Center
to Center Separation . . . . One third vessel's breadth

Table 3—Slopes versus Weights

Weight Tons	Way Slope	Keel Slope
10,000	0.052 to 1.00	0.047 to 1.00
13,000	0.046 to 1.00	0.044 to 1.00
23,000	0.045 to 1.00	0.042 to 1.00

Way slopes are the slopes of the tangents to the ground ways under the vessels' centers of gravity. Typical ground way radii are 12,000 to 23,000 meters (40,000 to 75,000 ft) (Andrews, 1967).

b. End Launch Cambered Ways. Typical way and keel slopes for various launching weights are shown in Table 3.

c. End-Launch and Side-Launch Cambered Ways. Rise of arc above the chord joining any two points can be found from Keith (1939):

$$b = R - \tfrac{1}{2}\sqrt{4R^2 - C^2} \tag{1}$$

and approximately (Robb, 1952)[1]

$$b = \frac{C^2}{8R}$$

also:

$$R = \frac{1}{2b}\left(\frac{C^2}{4} + b^2\right)$$

where:

b = rise of arc above chord
R = radius of ground ways
C = chord length between the two points.

Slope of ways at any point:

$$\theta = \frac{x}{\sqrt{R^2 - x^2}} = sin^{-1}\frac{x}{R} \tag{2}$$

Height of any point above ground way ends:

$$y = \sqrt{R^2 - x^2} - \sqrt{R^2 - x_0{}^2} \tag{3}$$

where:

θ = slope of tangent to ground way
y = vertical distance from ground way ends to any point on ways
x = horizontal distance from origin where slope is zero to any point on ways
x_0 = horizontal distance from origin where slope is zero to ground way ends.

also

$$y = b - R + \sqrt{R^2 - x^2} \tag{4}$$

If a parabola of the form $x^2 = k(b - y)$ is calculated so that it passes through the ground way ends and a point on the ground ways about under the vessel's center of gravity, it will be found to conform closely to the arc of the cambered ways.

d. Side Launch Straight Way Inclination. Typical ground way inclination is 0.065–0.165 to 1.00

e. Ground Way Sizes. Typical sizes of ground ways are:

Maximum nominal width one way . . . . 2.5 m (8 ft).
Width and thickness of timbers
. . . . 290 to 380 mm (11.5 to 15.5 in.).
Length of timbers . . . . 9 to 11 m (30 to 35 ft).

---

[1] Complete references are listed at end of chapter.

Section 3
Ground Way and Sliding Way Interface and Launching Lubricants

3.1 Requirements. Some friction reducing coating, mechanism, or condition must exist between the top surface of the ground ways and the bottom surface of the sliding ways so that the cradle and vessel can move down the inclined ways and into the water under the influence of gravity. Free or uncontrolled launching friction reducers include slippery substances, slippery surfaces, and mechanical rollers.

3.2 Slippery Substances. Slippery substances include proprietary launching greases, skid grease, and even ripe bananas. For barges built on sloping river banks, wet mud may be an adequate lubricant, particularly if a bulldozer is available to assist gravity.

A conventional launching grease system consists of a mineral-base grease base coat and a lime-soap slip coat. A specially formulated slip coat, containing finely divided zinc as ballast, can be manually applied to outshore ground ways which are always submerged due to great underwater length and/or a small tidal range.

Two or more grease irons are required at each sliding way section to keep the weight of the cradle off the grease until shortly before launching. The grease irons bear on the wood of the ground ways and grooves must therefore be provided in the base coat. This can be accomplished by the use of wooden strips or by cutting away the base coat. Good clearance should be provided to avoid any tendency of grease irons to jam.

The coefficient of sliding friction of a satisfactory grease system may vary from 0.015 to 0.030. The word "system" is used advisedly because laboratory tests of grease friction are meaningless for actual ship launchings. Grease friction varies with unit pressure on the grease, grease temperature, planeness or fairness of ground ways, alignment of ground ways, support and possible settling of ground ways, uniformity of cradle wedging up, pressure between ribband and sliding ways due to slewing of ship, base coat smoothness and resistance of base coat to squeezing out and/or melting, the speed of the vessel as it moves down the ways, presence or absence of lubrication on outshore ground ways, marine growths, sediment or debris on outshore ways and finally the lubricating qualities of the base coat working in conjunction with the slip coat.

a. Starting Grease Friction. Tables 4 and 5 show typical coefficients of starting or initial friction for launching grease systems at various unit pressures and temperatures. The least friction is shown to be for the highest unit pressure and highest temperature. High unit pressures and temperatures, if maintained for too long a time, may cause squeezing out of the grease and softening and even melting of the base coat. Therefore prolonged high unit pressures and temperatures should be avoided. For a vessel to start when released, the tangent of the angle of inclination of the ground ways under the vessel's center of gravity must be greater than the starting coefficient of friction. If tumble shores are used, then there is additional resistance to starting and the apparent coefficient of starting friction will be higher than its true value. Equation (47) in Section 16.2 shows calculation of coefficient of starting friction.

b. Sliding Grease Friction. Coefficients of sliding friction can be expected to vary from 0.015 to 0.030 with a maximum range of 0.010 to 0.050. Calculation of coefficient of sliding friction is shown by equation (48).

c. Grease Pressures. Average initial pressures on launching grease can vary from 16 tons/m^2 (1.5 tons/ft^2) to 33 tons/m^2 (3.0 tons/ft^2) with 22 tons/m^2 (2.0 tons/ft^2) being an acceptable average value. High way-end pressures may necessitate wider ways which will reduce initial grease pressure. Construction delays may require the warm weather launching of a vessel scheduled to be launched in cold weather with the result that launching grease may squeeze out and melt if the cradle and grease system have been designed for a cold weather launching. Grease pressures tend to decrease with increasing declivity as shown by Table 6. Subject to avoidance of damage to the vessel's bottom and/or ground ways, maximum way-end pressures can be taken as twice the average initial pressure. Much higher pressures have been successfully used. Maximum pivoting pressure can be taken as four times the average initial pressure. High pivoting pressures may cause smoking or ignition of the launching grease or may cause the fore poppet to stick on the ways.

Table 4—Starting Friction Coefficients—Metric Units

Temperature Grease °C	Grease Pressure in Metric Tons/Square Meter		
	20.0	30.0	40.0
0	0.0183	0.0120	0.0093
5	0.0171	0.0110	0.0089
10	0.0159	0.0101	0.0085
15	0.0143	0.0092	0.0078
20	0.0125	0.0084	0.0070
25	0.0112	0.0076	0.0062
30	0.0100	0.0068	0.0058
35	0.0093	0.0063	0.0055

Table 5—Starting Friction Coefficients—English Units

Temperature Grease °F.	Grease Pressure in Long Tons/Square Foot		
	1.750	2.625	3.500
30	0.0193	0.0128	0.0097
40	0.0182	0.0115	0.0092
50	0.0168	0.0105	0.0086
60	0.0151	0.0096	0.0080
70	0.0133	0.0089	0.0073
80	0.0119	0.0080	0.0065
90	0.0106	0.0072	0.0060
100	0.0099	0.0066	0.0057

Table 6—Grease Pressures vs. Declivities

Declivity	0.045	0.050	0.055	0.060	0.065	0.070	0.075
Tons/m^2	27.1	25.0	22.9	20.8	18.6	16.5	14.4
Tons/ft^2	2.48	2.28	2.09	1.90	1.70	1.51	1.32

3.3 Slippery Surfaces. Tetrafluoroethylene has a sliding coefficient of friction of 0.04 to 0.12 at normal temperatures and unit pressures of 50 to 1400 tons/m^2 (4.6 to 128.0 tons/ft^2). A sliding-coefficient of friction of 0.10 can be used as a round figure without regard to temperature, load, sliding speed or maintenance. The starting coefficient of friction is equal to, or a little less than the sliding coefficient of friction. The 2.4 mm (3/32 in.) fluorocarbon resin is bonded to a steel ground way backing plate and can accept some misalignment and dirt and grit embedment. The mating sliding surface can be fluorocarbon resin or stainless steel plate.

3.4 Steel Roller Systems. The coefficient of rolling friction of ship-launching steel balls is 0.025 with a somewhat lower coefficient of starting friction, Andrews (1967). The coefficient of rolling friction of caged steel rollers, working between steel-sheathed fixed and moving ways, varies from 0.015 to 0.025. The starting coefficient of friction for roller dollies is 0.075 with a value of 0.025 for rolling.

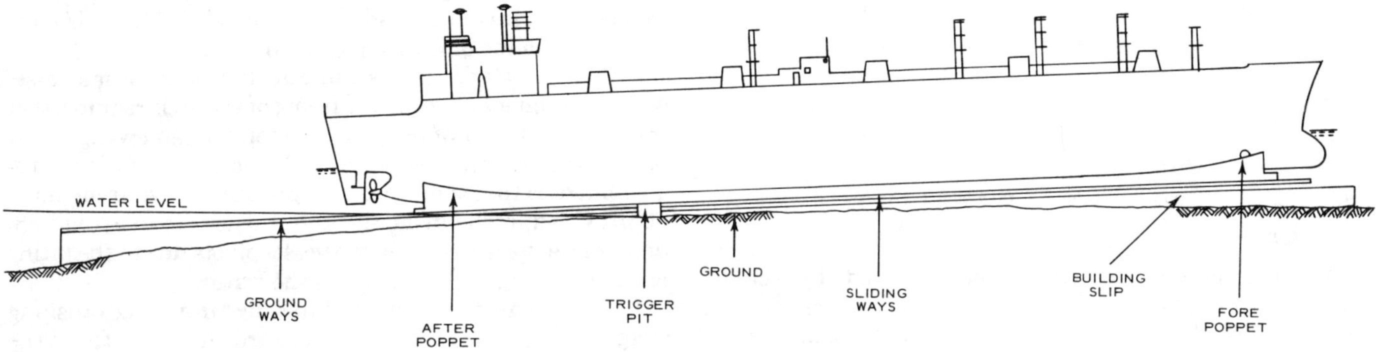

Fig. 1 Launching arrangements

Section 4
End Launch Cradle

4.1 Cradle Structure. The cradle rests on the grease or other friction reducer on the ground ways and supports the ship immediately before, and during its travel down the ground ways. The cradle is composed of sliding ways, wedges, wedge riders, packing, fore and after poppets, sub-poppets and the necessary hardware, fittings and rigging to attach the cradle to the vessel, Fig. 1 and 2.

Sliding Ways. The sliding ways can be built up from 290 by 290 mm (11.5 by 11.5 in.) timbers, 9 m (30 ft) in length, of Douglas fir, long leaf yellow pine or other timber with similar properties. Sizes of timbers can vary according to size of vessel and material available. Wood should have good resistance to crushing and should be light enough so that the cradle with all its hardware and fastenings will float. Sliding way sections can be from one to five or more timbers wide to suit the weight of the vessel but nothing will be gained by making sliding ways wider than ground ways. Bottom surfaces should be smooth and plane. The top and bottom transverse edges at the ends of each sliding way section should be rounded and the outboard lower end of the outboard section should be chamfered to avoid any tendency to gouge the base coat or wipe it clean of slip coat. Sliding way fastenings can be bolts, washers, and nuts. Fastenings on the outboard sides of the sliding ways must clear the groundway ribbands, assuming a complete absence of both base and slip coats.

End Connections. Sliding way sections are joined together by connecting links and stud bolts, Fig. 3. For a burn-off plate release, the grease friction can be conservatively neglected in designing sliding way joints so that the longitudinal tensile force on one sliding way at its inshore

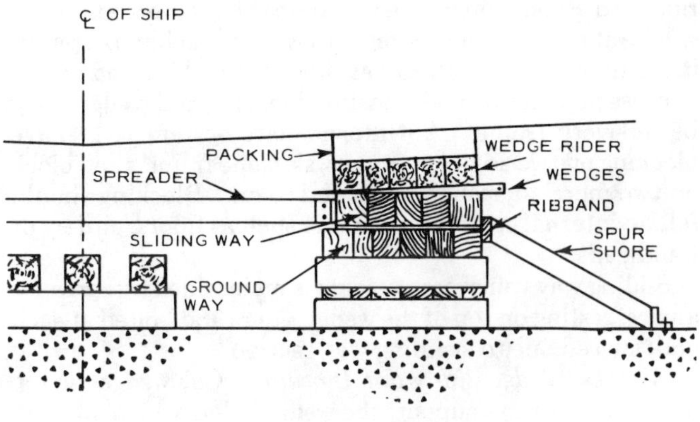

Fig. 2 Transverse section amidships

end is one half the weight of the vessel times the sine of the slope of the ways. This force reduces linearly to zero at the outshore end of the cradle. The force tending to pull the cradle apart can thus be determined for any joint.

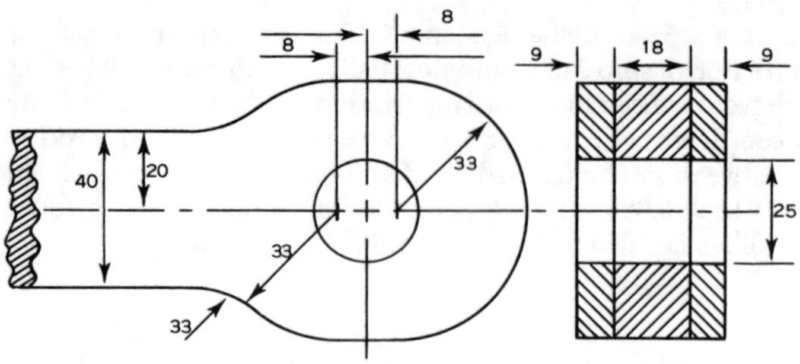

ULTIMATE STRENGTHS FOR 50 MM (2.0 IN.) DIAMETER PIN

SHEAR LOADING	KILOGRAMS	POUNDS	LINK t / PIN d	SIDE PLATE t / PIN d
SINGLE	68,400	150,800	0.393	0.393
DOUBLE	123,120	271,440	0.707	0.355

Fig. 3 Sliding way connecting link end

When triggers, located near the outshore end of the cradle are used for release, the sliding way section end connections inshore of the triggers are in compression before launching. The sliding way section ends touch each other and the need for sliding way joint strength is reduced.

Blocking between top of wedge rider and underside of hull in the flat bottom area may consist of single blocks cut from 290 by 290 mm (11.5 by 11.5 in.) timber and of length greater than the width of the wedge rider. The top surfaces of the wedge riders will have transverse inclination due to the taper of the wedges and due also to the transverse inclination of the groundways, if such exists. If there is little or no deadrise, the transverse blocks on top of the wedge rider will be tapered to suit the sum of the transverse inclinations and thus will facilitate taking slack out of the blocking before wedging up. Where there is deadrise, single block packing can be replaced by packing units in the form of tapered marrying blocks.

For a vessel with transverse bottom frames spaced at 750 to 900 mm (30 to 36 in.), transverse blocks above the wedge rider and accompanying wedge groups below the wedge rider can be at every second frame. In way of machinery spaces, if machinery is to be aboard at time of launching and in way of areas of high way end pressure, blocking and wedges may be at every frame. An intermediate system is to have blocking on two adjacent frames, skip one frame, then block on two more adjacent frames and so on. Blocking should fall on internal strength members such as floors, girders or bulkheads.

Sliding way transverse spacing is maintained by spreader angles, resting on top of the wedge riders and bolted at each end, spaced one to a sliding way section.

Wedges, Wedge Riders and Blocking. Oak wedges on top of the sliding ways support the wedge riders which, in turn, support the blocking and poppets which support the vessel. Wedge lengths are greater than sliding way widths. Wedges can be in groups of two or three underneath each block on top of wedge rider. Wedge riders are similar to sliding ways but are somewhat thinner, i.e. if sliding ways are built up from 290 by 290 mm (11.5 in.) timbers, then 230 by 290 mm (9.0 by 11.5 in.) timbers might be used for wedge riders. In determining number of wedges to use in any one group, account must be taken of the compressive strength of the material of sliding ways and wedge rider perpendicular to the grain. There must be sufficient space between adjacent wedges in any one group so that the ram or maul will hit only one wedge at a time when ramming up.

If narrow sliding ways are used, it may be advantageous to make the wedge riders 150 to 300 mm (6 to 12 in.) wider than the sliding ways. Transverse stability of the port and starboard sides of the cradle is improved and there is more bearing area for blocking and poppet heels.

Fore and After Poppets and Sub-Poppets. For a vessel of conventional hull form, the fining of the ends requires that the vertical height of packing between hull and wedge rider be progressively increased towards the ends. A second requirement is that some means of pressure distribution must be provided at the inshore end of the cradle to avoid high unit grease pressures when the vessel pivots due to the lifting of the outshore end as it enters the water.

Pivoting can be accommodated by providing crushing strip packing over roughly the forward three fourths of the fore poppet. Crushing strips are arranged in vertical rows, the row spacing being greater towards the forward end of the fore poppet. Crushing strips are in horizontal layers, separated by stage planking or thick plywood. The numbers of layers forward are greater than farther aft. The volume not occupied by crushing packing is packed solid and in side elevation appears as a right trapezoid, the vertical line forming the after end of the solid packing, the opposite side marking the start of the crushing packing sloping downward as it goes forward. In the early stages of pivoting the crushing packing has not crushed sufficiently to offer significant resistance and therefore practically all of the fore poppet load, which is at or near its maximum, is taken on the solid packing at the after end of the fore poppet. This packing should have as broad a base as possible for optimum distribution of load.

The fore poppet wedge rider is of usual form and oak wedges are installed continuously with just enough spacing for efficient driving. Sufficient spreader angles must be provided to connect port and starboard wedge riders, or port and starboard sliding ways, to resist the tendency of these members to spread during pivoting.

Structure connecting the top of fore poppet packing to the ship can be one or more plate belly bands or slings, the upper ends of the plates being bracketed and bolted to the timber structure at the top of the packing. The space between shell and belly band is filled with light-weight concrete or is packed with wood with concrete poured in to fill any voids. Red template paper can be used to protect the bottom paint.

Belly bands and concrete filling are heavy and buoyancy in the form of empty oil drums can be provided below the belly bands for flotation and to reduce the tendency of the fore poppet to jack knife when it has been removed from under the ship after launching. In lieu of belly bands, the fore poppet may be fastened to the shell by welded brackets. Bolted or other suitable connections must be provided to permit removal of the fore poppet along with the rest of the cradle following launching.

Instead of crushing strip packing, the rocker fore poppet offers considerable advantages. The wedge rider is built up until there is enough depth of solid timber for a cylindrical surface with transverse horizontal axis and a radius of say 15 m (50 ft) for a medium size vessel to be sawn out. A mating part of timber construction is built up and the top of this upper part of the fore poppet, the rocker, is connected to the hull by belly bands or by welded brackets. The mating, cylindrical sliding surfaces should be true and smooth, separated by a 3 mm (1/8 in.) plate to avoid any tendency to lock, lubricated with launching slip coat. Advantages of the rocker fore poppet include avoidance of unequal unit loads associated with the crushing strip fore poppet, maximum capacity, reliability and durability. A crushing strip fore poppet may require rebuilding after use. A rocker fore poppet will work regardless of the angle pivoted through while a crushing strip fore poppet is, or should be, designed for only one maximum pivoting angle.

If a barge-type hull or vessel with a short entrance and low deadrise is to be launched, it may be necessary to construct the inshore end of the vessel at a greater height above the ground ways than would be usual to provide sufficient vertical height for the necessary number of layers of crushing strips. Vertical height can be saved if plywood sheets are used instead of stage planking to separate crushing strip layers. If the pivoting angle is large, a rocker fore poppet will require less height than a crushing strip fore poppet. A rocker fore poppet would be particularly advantageous when launching a ship with its weight carried on a single centerline ground way. If the bottom of the vessel is dead flat in way of the fore poppet, depth can be saved if the wedge rider is eliminated and wedges are fitted between the top of the fore poppet and the bottom shell plating.

Between the after end of the fore poppet and the flat bottom area, vertical struts or poppets are erected on top of the wedge rider. In side elevation these struts are perpendicular to the wedge rider. In end view they are canted inboard to avoid excessive height. Poppets are made up of timbers as used to construct the sliding ways or poppets may be of composite construction, partly structural steel with wood bearing against the shell to provide a measure of compressibility. The structure just described, together with associated wedge rider and sliding way is called the fore sub-poppet.

After poppets and sub-poppets support the vessel aft of the flat bottom area where blocking consists simply of transverse timbers on top of the wedge rider. After poppets are similar to fore sub-poppets except that in side elevation they should be perpendicular to the horizontal so as to better resist the tendency for their heads to be forced aft due to the rising of the buttock line of the vessel's hull. Poppets may require to be spaced more closely than every second frame.

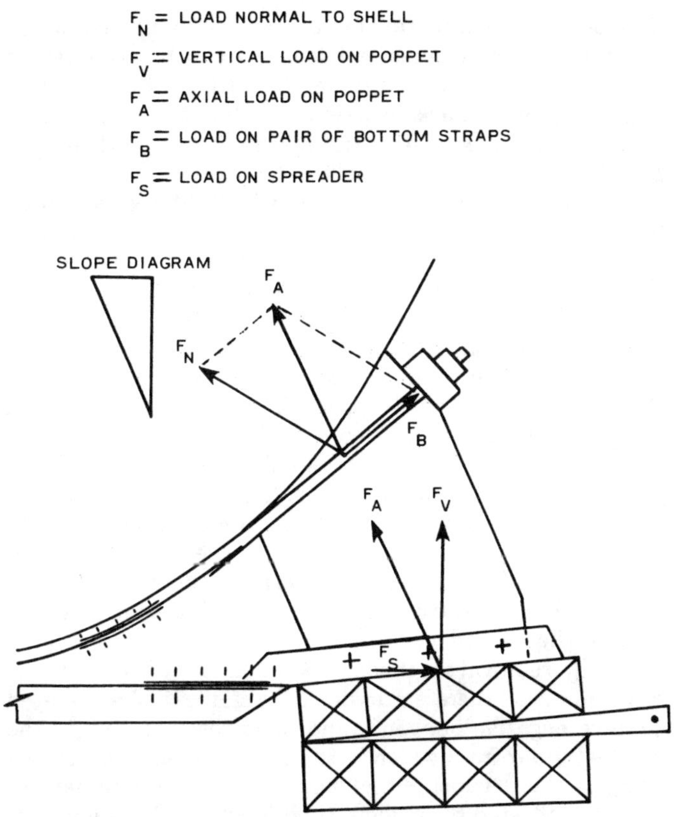

Fig. 4 Socketed poppet with bottom straps

Unlike the remainder of the cradle, the after poppets and sub-poppets are ballasted to sink, rather than float, and are usually removed after the cradle has been hauled.

Loading of bottom straps and spreader angles where fitted, poppets, wedge rider, wedges, sliding way, grease and ground way are determined by vector analysis, Figs. 4 and 5. Bearing strength of wood loaded normal to the grain may be critical and it may be found that additional wedges are necessary to prevent local overloading of sliding way and wedge rider. Adequacy of strength of bottom straps connecting poppet heads and spreader angles joining poppet heels must be verified as adequate to carry the loads resulting from that part of the vessel's weight to be carried by the poppets.

Where allowable vertical loads are much greater than actual vertical loads, spreader angles can be omitted at every second or third poppet. Where spreaders are omitted, angle clips or other suitable means must be provided to fasten poppet heels to wedge riders.

4.2 Rigging. Tricing rigging resists relative transverse moment of the ship and cradle during pivoting and after the vessel is afloat and, depending on amount of deadrise, tends to hold the cradle up under the bottom, supplementing the effect of cradle buoyancy. As the vessel enters the water there is a considerable drag on poppets, wedge riders, sliding ways and spreaders. Diagonal back stays are therefore run from the gunwale right aft to the sliding ways and after poppet. To secure favorable leads without too much obliquity, forward ends of backstays can be attached to forward end of after poppet and well forward on sliding ways. Failure of backstays could result in the after portion of the cradle jack knifing downward, wrecking itself and possibly damaging the vessel's bottom.

A cradle hauling line consists of a bight of steel wire rope with the ends shackled to the forward ends of the sliding ways. At time of launching, the bight of line is hauled up out of the way. After the vessel is afloat, the wire is dropped over a pile cluster, or secured to a buoy, the cradle tricing lines are released and tugs pull the vessel astern off the cradle which is prevented from moving with the vessel by the hauling line. The after poppet and sub-poppets remain triced in place to be removed with the assistance of a crane after the vessel has been berthed alongside a pier.

Another means of cradle removal is to secure the inshore ends of the sliding ways to the shore by means of lines which take up shortly after the cradle has left the ground ways. If backstays and tricing are released in a timely manner, the cradle is snatched from under the vessel as soon as it has cleared the ground ways.

Cradle removal methods depend on facilities available. For example, if there is a drydock, the cradle can be ballasted to have negative buoyancy and be released to sink to the drydock floor from which it can be recovered after dewatering of the drydock. Launching cradles for submarines may be ballasted throughout their lengths to have negative buoyancy.

During and following a launching, all components of the cradle should remain together. This facilitates the recovery of the cradle which may be used again in its entirety or in part. All parts which might become separated should therefore be lashed, clipped or dogged together.

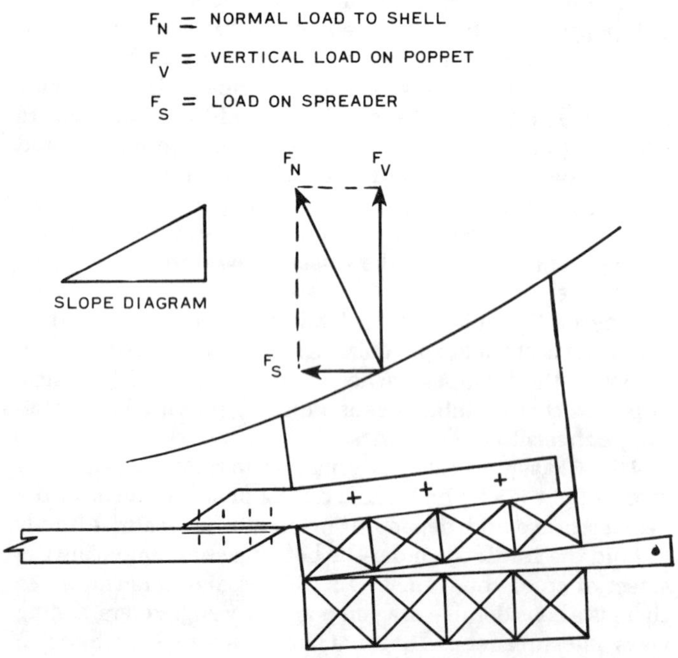

Fig. 5 Clipped poppet without bottom straps

4.3 Design Data. The following data are provided for use in end launch cradle design:

- Length of typical cradle equals 0.80 times vessel's length between perpendiculars with ten percent overhang at each end.
- Taper of oak wedges can be from 0.03 to 0.06 to one with greater tapers being used for lighter vessels.
- Sliding way timbers can be 290 to 390 mm (11.5 to 15.5 in.) square and 9 to 11 m (30 to 36 ft) long to make sliding way sections up to about 2.4 m (8 ft) maximum width.
- Vertical distance from top of groundway at centerline of underside of vessel's bottom over groundway can be 750 to 850 mm (30 to 33 in.).
- Keel can be 1.5 m (5 ft) or more above building slip deck at amidships.
- Rocker fore poppet.

$$b = R - \tfrac{1}{2}\sqrt{4R^2 - C^2} \tag{5}$$

where:

b = drop of arc below chord
R = rocker radius
C = rocker chord length

For a vessel of medium size the radius can be 15 to 18 m (49 to 59 ft), (Keith, 1939). For a 15 m radius and a rocker length of 7.5 m (25 ft), the drop of arc below the chord is 0.48 m (1.57 ft).

Table 7—Poppet and Blocking Working Loads

Size	Loading	Tons
290 × 290 mm (11.5 × 11.5 in.)	Parallel to Grain	85
290 × 290 mm (11.5 × 11.5 in.)	Perpendicular to Grain	14

The specific gravity of Douglas fir, coast type, 12 percent moisture content, is 0.48.

Table 8—Wedge and Tumble Shore Strength

Size	Loading	Tons
290 × 290 mm (11.5 × 11.5 in.)	Parallel to Grain	80
290 × 290 mm (11.5 × 11.5 in.)	Perpendicular to Grain	22

- Table 7 shows working loads for Douglas Fir, coast type, or southern yellow pine poppets and blocking. Factors of safety are about four for long-time loading and six for short-time loading such as exists between start of blocking removal and launching. Factors of safety are based on proportional limit.
- Table 8 shows working loads for white oak and red oak which may be used for wedges or tumble shores. Factors of safety are about four for long-time loading and six for short-time loading such as exists between start of wedging up and launching. Factors of safety are based on proportional limit.
- Table 9 shows depths of wedge butts for sliding way widths and wedge lengths shown:

Table 9—Cradle Wedges

Way Width mm (in.)	Wedge Length mm (in.)	Wedge Butt mm (in.)
290 (11.5)	650 (25.5)	56 (2.2)
585 (23.0)	940 (37.0)	71 (2.8)
875 (34.5)	1230 (48.5)	89 (3.5)
1170 (46.0)	1525 (60.0)	114 (4.5)
1460 (57.5)	1815 (71.5)	127 (5.0)

All wedges are 13 mm (0.5 in.) deep at their points. Wedge lengths are about 360 mm (14 in.) longer than the width of ways with which they are used. Wedges are about 150 mm (6 in.) wide, the vertical surfaces not being dressed. Taper of wedges is 1 to 16. If transverse inclination of ground ways is 1 to 48, the transverse inclination of the wedge rider will be 1 to 12.

• *Bolted Fastenings.* Allowable total lateral loads in pounds on both ends of one bolt with steel plates or angles at each end and bolt length to diameter ratio of six or greater are as follows in metric and (English) units (Forest Products Laboratory, 1974):

Load Parallel to Grain = $3.10\,D^2$ ($4400D^2$)
Load Normal to Grain = $1.35\,D^2$ ($1900D^2$)

where

Load = kg (lb)
D = bolt diameter mm (in.)

Locations of bolt holes in wood should be as in Fig. 6. The center to center distance along the grain between bolts acting parallel to the grain should be at least 4 times the bolt diameter. When the joint is in tension, the bolt nearest the end of the timber should be at least 7 times the bolt diameter from the end of the timber. When the joint is in compression, the bolt nearest the end of the timber should be at least 4 times the bolt diameter from the end of the timber. For bolts bearing parallel to the grain, the distance from the edge of a timber to the center of a bolt should be at least 1.5 times the bolt diameter. For bolts bearing perpendicular to the grain, the distance between the edge towards which the bolt pressure is acting and the center of a bolt nearest this edge should be at least 4 times the bolt diameter. The minimum center to center spacing of bolts in the across the grain direction for loads acting through metal side plates need only be sufficient to permit tightening of the nuts.

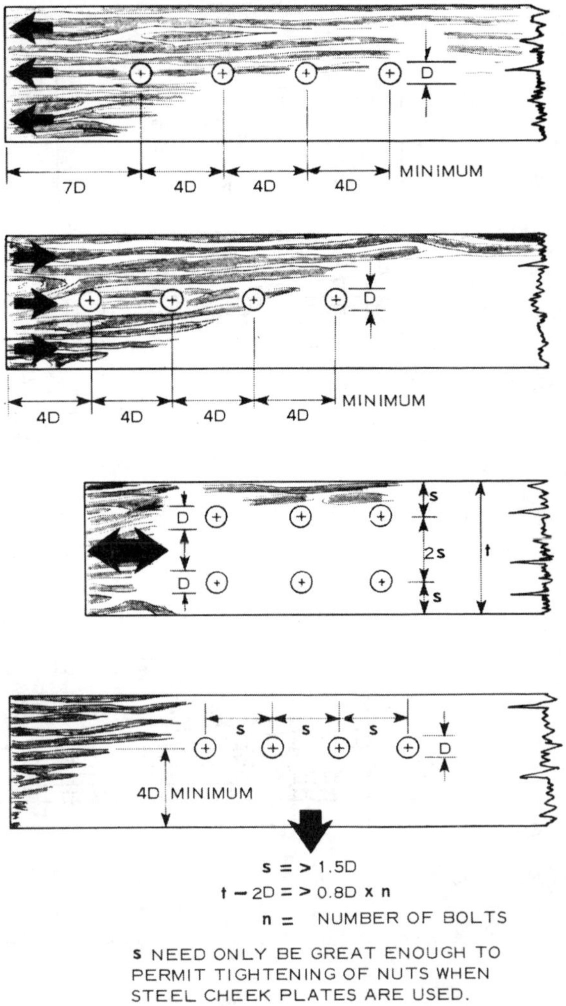

Fig. 6 Bolt locations in wood timbers

Section 5
Side Launch Cradle

5.1 Cradle Structure. The cradle transmits the weight of the vessel onto the launching grease or other friction reducer and consists of transverse sleds or longitudinally disposed sliding ways, wedges, wedge riders, packing and the necessary hardware and fittings. Fore and after poppets as required for end launching are not necessary.

• *Sleds* are in the form of a right angle triangle, the hypotenuse resting on the ground ways with one side horizontal to take wedges and wedge rider Fig. 7. Construction can be all timber or can be welded structural steel with hypotenuse and horizontal sides being built up of heavy timbers through-bolted as for end-launch sliding ways. If all-steel construction is used it should be of double plate cellular form with vertical webs between the side plates. A watertight and buoyant design is desirable to facilitate retrieval of the sleds after launching. Sled inshore ends can have trigger fittings.

The simplest type of sleds are those used for launching from rocker ground ways trunnioned at their mid-lengths. Such sleds consist simply of sliding ways, wedges, wedge

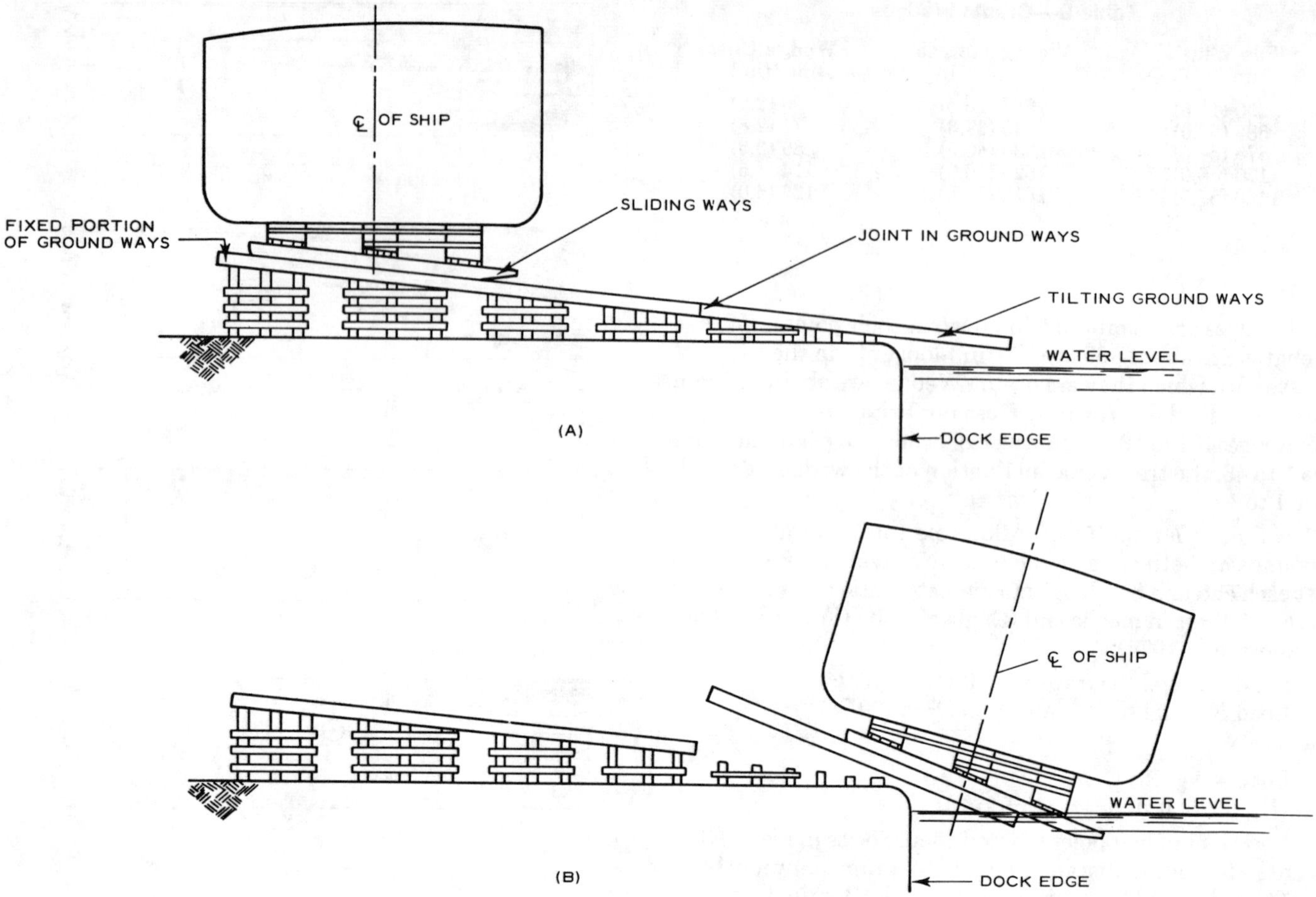

Fig. 7 Timber sleds and tilting ground ways

rider and packing between wedge rider and vessel's bottom. For some conditions, wedging arrangements can be omitted and the sleds consist only of sliding ways and packing.

Sleds of the triangular type, or of the simple type described in the foregoing are well-suited to a ship-manufacturing facility where vessels are moved into, rather than built in, the launching position.

• *Longitudinal Sliding Ways.* Sliding ways or butter-boards as in Fig. 8 are placed parallel to the keel and perpendicular to the ground ways on which they rest. The butter-boards span two or more ground ways which are arranged in groups and the butter-board ends project beyond the ground ways bounding each group. The butter-boards are heavy timbers as would be used for ground or end launch sliding way construction.

A butter-board cradle is not restrained by ground way ribbands as is a sled-type cradle and can therefore tolerate some slewing of the vessel if both ends do not start at the same instant when released or if one end of the vessel tends to lag during movement down the ways.

• *Wedges, Wedge Riders and Blocking.* Launching sled tops carry wedges, wedge riders, and blocking to support the vessel. Oak wedges are fitted between sled tops and wedge riders. In the case of wide sleds, wedges may be driven from both sides of the sleds, the undersides of the wedge riders being shaped to a shallow Vee to suit the wedge slopes or, alternatively, the wedges can be married. Construction of wedge riders, wedges and blocking can be about as described for an end launching cradle.

Sleds are usually all of the same size. Accordingly sleds

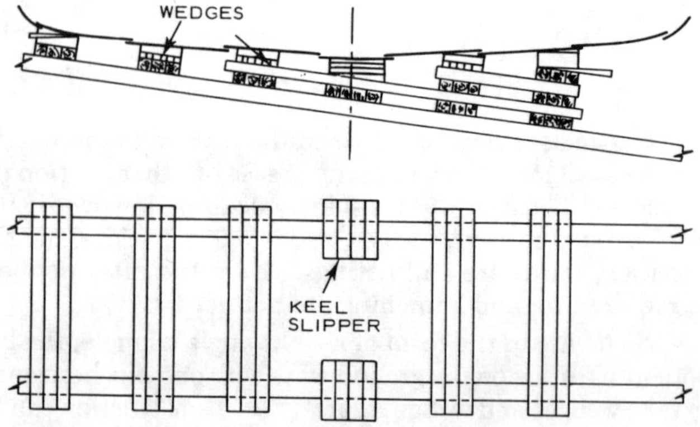

Fig. 8 Butter-boards and packing

installed under the ends of a vessel of ship form are lòaded only at their mid-lengths where they support the keel. Additional loading can be obtained by shores in the plane of each sled, extending from near each sled end to chocks welded to the vessel's shell, port and starboard. Solid packing built up of horizontal heavy timbers is practically useless for distributing load over the length of a sled under the end of a fine-lined vessel. If no shores are fitted, the grease friction under the end sleds will be slightly increased as will be the loading of the sleds towards amidships. A possibility is to build sleds in two or three lengths, the shorter sleds being used under the ends of vessels where there is only keel bearing.

Butter-boards carry heavy timber cribbing with tie timbers parallel to the ground ways to resist any tendency of the cribbing to topple. Crushing packing may be necessary to reduce pressures on the vessel's bottom as the fore and aft sliding ways and cribbing are unable to provide the support given by triangular sleds as the vessel tips about the ground way ends. Oak wedges are installed between cribbing and blocks which bear against the vessel's shell. It is essential that there be free access to the heels of the oak wedges so that the blocking can be tightened as much as possible prior to removal of blocking and shoring. Accordingly, wedges at outboard and inboard cribbing can be driven inboard, all other wedges being driven fore and aft.

The butter-board type cradle has the advantage that butter-boards and blocking need be installed only on centerline at the ends of a vessel of fine-lined ship form. Additional butter-boards can be provided under heavy parts of the vessel. Accordingly a more uniform grease loading is obtained and the center of gravity of the grease-bearing area can be located under the vessel's center of gravity.

To facilitate removal from the water following a launching, butter-boards, cribbing and wedges should be wired together in groups of components.

5.2 Rigging. For sled launchings from inclined ground ways or for launchings from pivoting, trunnioned ground ways, practically no rigging is required. However, sleds or sliding way sections can be secured to the ground by wire ropes so that sleds or sliding ways will not be carried away by the current. These mooring lines also tend to ensure that sleds or sliding ways will come out from under the vessel. The wedge rider should be wired or otherwise flexibly secured to the top of each sled or sliding way and all wedges should have wires rove through holes, one near the heel of each wedge. Blocking can be dogged, wired, or clipped to the wedge riders.

For butter-board cradles similar rigging can be used. This includes mooring wires attaching butter-board units to the shore plus sufficient dogs, wires, and clips to hold the components of each butter-board unit together, but not in their initial relative locations, as each unit is drawn by its mooring wire from beneath the sideways moving vessel.

Section 6
Platform Launch Blocking and Cradle

6.1 Definition. If a vessel is launched from a fixed platform such as a graving dock floor, it will have been built in that position and will float off the supporting blocking when the dock is flooded to the necessary depth. For moving platforms which are at times partly or entirely ground supported or which are always ground supported, it is probable that the vessel will not have been built on the platform and that when in launching position the vessel will be supported by a translation system cradle used to move the vessel transversely or longitudinally onto the platform. Depending on the resistance to submergence damage of the translation system mechanical and electrical components, portions of the translation system cradle may be removed before launching and replaced by conventional fixed blocking and shoring.

6.2 Fixed Platform Blocking. Launching is from the keel and side blocks and bilge cribs which supported the vessel during construction. Fore and afters should be provided to stabilize keel and side blocking. If considerable trim is expected at float off, crushing packing can be installed at the after keel blocks or a rocker can be fitted, as described for fore poppets, to uniformly distribute the knuckle load with minimum risk of damage to vessel and blocking. In the case of a vessel with propellers and/or sonar dome projecting below the base line, or a submarine with keel blocking heights increasing towards the ends of the vessel, it may be necessary to move the vessel just launched in a sideways direction until it is clear of the keel and side blocking and can then be moved longitudinally to clear the basin.

6.3 Moving Platform Blocking. If the vessel has been moved onto the platform by means of a translation system consisting of ground ways, sliding ways, wedge riders and packing or rails, wheeled cars, strong beams and packing, or some other system, the platform can be submerged and the vessel floated off the translation cradle. A drydock type floating platform may list inadverdently or be trimmed intentionally during submergence and shores or other preventers must be installed to prevent any transverse or longitudinal movement of the cradle. The discussion under Section 6.2, relative to moving a vessel sideways to clear fixed platform blocking, applies to moving platform blocking as well.

Section 7
Releasing and Starting

7.1 Requirements. Launching a cradle supported vessel from inclined ground ways requires that after all blocking, cribbing and shoring tending to restrain movement down the ways has been removed, there remains a connection between cradle and ground ways, or between vessel and ground, that can be released to permit the cradle and vessel to move down the ways and into the water. In case the vessel fails to move when released, some means of pushing on the cradle or vessel is provided to augment the gravity component acting along the ways.

7.2 Types of Releasing and Starting Gear. These include burn-off sole plates, mechanical triggers, hydraulic triggers, timber lever and rope triggers, dog shores as triggers and preventers, starting rams, side-launch starting shores, and side-launch trunnioned pivoting ground ways.

a. Burn Off Sole Plates. For end launchings a burn off sole plate connects the underside of the inshore sliding way section to the upper surface of the groundway inshore section. Each sole plate, and its fastenings to timber sliding and ground ways, must withstand one half the component of weight of ship and cradle along the ways, grease friction being conservatively neglected. On the sliding ways the sole plate can be welded to side sheathing and centerline plates which may be carried outshore to reinforce the inshore sliding way joint. Cast iron ogee or large thick plate washers can be used on underside of ground ways at bolts. Sole plates can have transverse bars welded to them and buried in grooves cut in the wood of ground or sliding ways to increase bearing area of wood in compression, thus reducing the loading on the bolts.

Each sole plate should have a transverse row of drilled holes, each segment being numbered from outboard towards the plate centerline. Each sole plate is cut by two burners working towards the plate centerline. Each numbered plate segment of each plate on port and starboard ways is burnt on a word of command so that the four burners keep together.

Immediately before and during burn off, the down ways component of ship and cradle weight is hanging, as it were, on the ground way heads. Due to the vertical distance between each sole plate and mid-depth of ground ways, there is a moment tending to lift the ground way heads. This should be resisted by hold down bolts or steel wire rope lashings securing the ground way heads to the building slip head house. In the absence of a head house, hold down fastenings can be secured to pilings or to heavy concrete blocks resting on the ground.

The suitability of burn off sole plate release depends on the strength of the inshore sliding way joints. Unless special sheathing and bearing chocks are used, joint strength depends on the bearing strength of wood under bolts and connecting link bolts. When the cost of reinforced sliding way joints and burn off sole plates becomes large, then triggers should be considered as a means of release.

Burn off release can be used for side launching small vessels where sleds, rather than butter board sliding ways,

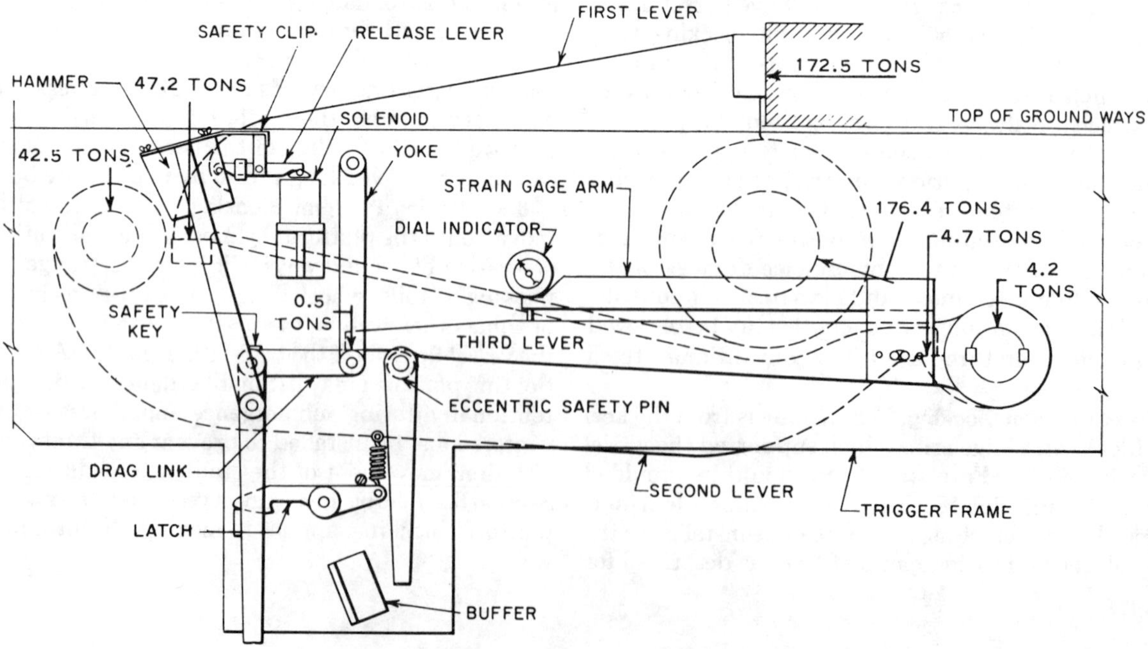

Fig. 9 170 ton mechanical trigger in holding position

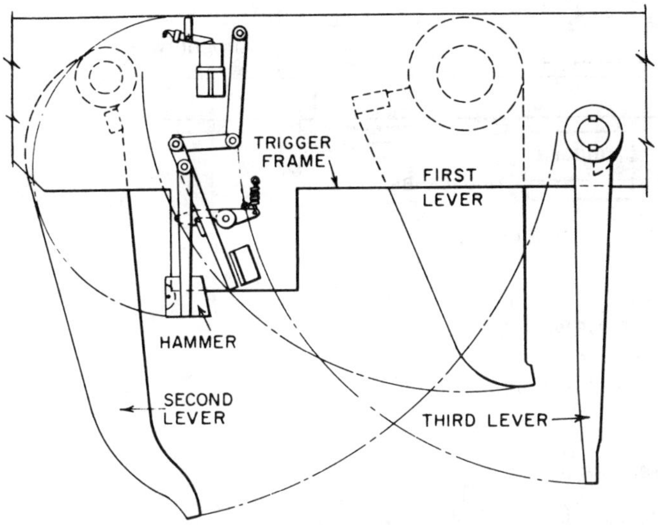

Fig. 10 170 ton mechanical trigger in released position

are used and not more than two burn-off plates are required. Sleds with burn-off plates should be adjacent to amidships.

b. Mechanical Triggers. Mechanical triggers, Fig. 9 and 10, can be located about three quarters of the way outshore from the inshore end of the cradle and offer the great advantage that sliding ways inshore of the triggers are, up to the instant of release, in compression. The possibility of the sliding ways pulling apart is thus eliminated. Each trigger can consist of three levers mounted in a structural steel frame. The short arm of the first lever projects above the top of the ground way and engages in a cast or fabricated steel shoe on the underside of a sliding way section. The first lever has a long arm which engages with the short arm of a second lever and so on until only a small restraining force is required at the long arm of the third lever.

For very heavy vessels, triggers can be used in two or three pairs. Simultaneous release can be by solenoids, wired in a common electrical circuit. If only one pair of triggers is used, the triggers can be connected by an athwartships shaft, manual rotation of which through a thirty degree angle releases both triggers. Each trigger can be fitted with a dial indicator to read lever deflection, a function of load. A build up of trigger load as blocking and shoring are removed indicates a *liveliness* of vessel and cradle and suggests that the vessel will start when released.

Side launchings in which the vessel is sled supported can also use mechanical triggers, the inshore end of each sled being connected to the supporting ground way by means of a mechanical trigger. As there is one trigger for each sled, the load per trigger is reduced in spite of the large declivity associated with a side launching. Each trigger release lever can be connected to a horizontal shaft or pipe, just above ground level and parallel to centerline of vessel. A weight suspended by a wire rope and sheave system puts the longitudinal shaft in tension. The pull is resisted by a burn-off link or mechanical trigger. Burning the plate or releasing the trigger permits longitudinal movement of the shaft which releases all the sled triggers. For a side launching it is essential that all triggers release simultaneously.

c. Hydraulic Triggers. Hydraulic triggers are similar to mechanical triggers except that the second and third levers are replaced by a hydraulic ram, the piston of which bears on and supports the long arm of the first, and only, lever. Trigger load can be shown by a calibrated hydraulic gage. Hydraulic pressure is released to release the trigger. Compared with mechanical triggers, hydraulic triggers are unreliable as hydraulic fluid leakage can result in a trigger failure and premature release of a vessel.

d. Timber Lever and Rope Triggers. Structural timbers can be used to construct simple and reliable mechanical triggers of considerably less capacity than the three-lever mechanical triggers described in the foregoing. For the end-launching trigger; Fig. 11, a heavy timber or structural steel chock bolted to the outboard side of the ground ways acts as a fulcrum. The trigger is a structural timber, horizontal and perpendicular to the ground ways with its heel bearing on the ground way chock. Just outboard of the chock, the outshore end of a timber spur shore bears on the inshore face of the trigger timber. The spur shore inshore end bears against, and prevents outshore movement of, a heavy timber or structural steel sliding way chock. A fiber rope lashing, passing over a chopping block, restrains the outboard end of the trigger timber from outshore movement. Trigger capacity can be increased if bearing surfaces are steel shod to distribute the load over wood loaded perpendicular to the grain.

Similar triggers can be used for side launching sleds. For side launchings on butter-boards, the cradle units are unable to take trigger thrust and the inshore ends of triggers must therefore bear against the forward and after deadwoods or temporary chocks or other suitable bearing locations on the underside of the vessel, Fig. 12. For large vessels, several triggers will be required. Simultaneous cutting of the

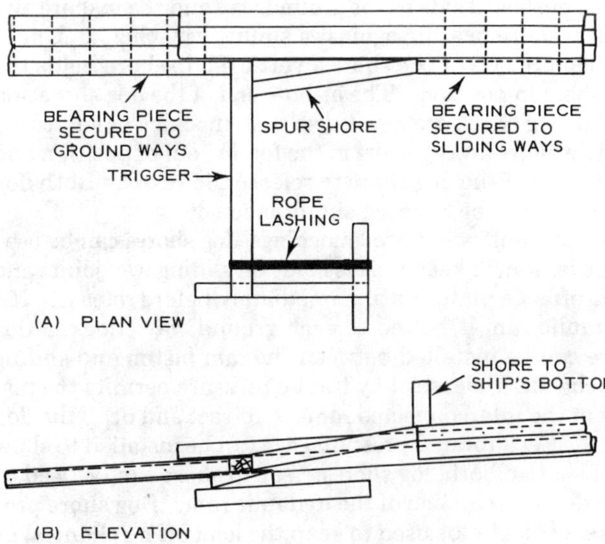

Fig. 11 Timber and rope trigger

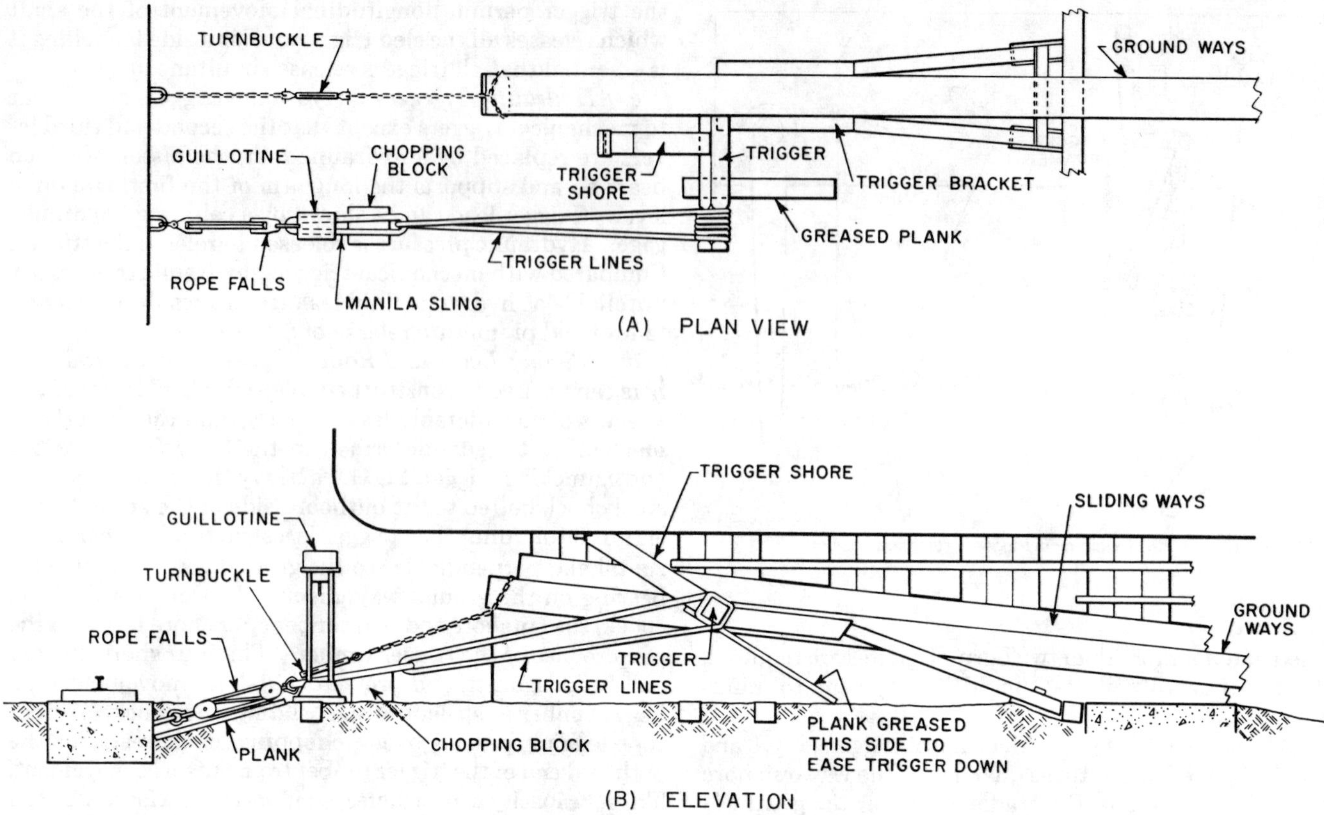

Fig. 12 Side launch trigger

trigger beam lashings can be done by guillotines which are electrically, pneumatically or hydraulically released. If compressed air is used, the lines should be cleared of condensed water and should be of equal length to ensure that each unit operates at the same time.

e. Dog Shores. Dog or dagger shores can be used as triggers, the outshore end of each shore bearing on a chock on the outboard side of the ground way and the inshore end of each shore bearing against a sliding way chock. Under the inshore end of the shore a vertical trip shore holds the dog shore in position. The inshore end of the dog shore and the mating chock are beveled so that knocking or snatching the trip shore away results in the forcing out of position and the falling of the dog shore to release the vessel. Both dog shores should be released simultaneously.

For burn-off sole plate launchings, dog shores can be used as preventers to keep tensile loads off sliding way joints and burn-off sole plates until immediately before release. If a hydraulic ram is bolted to each ground way chock, a dog shore can be installed between the ram piston and sliding way chock. Release of hydraulic pressure permits the pistons of the interconnected rams to retract and drop the dog shores. A calibrated pressure gage can be installed to show total load on both dog shores. A trip shore can be used at each dog shore in lieu of the hydaulic ram. Dog shore preventers can also be used to keep the load off mechanical or hydraulic triggers until just before release.

f. Starting Rams. Launching grease system poor quality or deterioration, unequal wedging of the cradle, minor settling of the ground ways; and other similar defects, acting singly or in combination, may be sufficient to prevent starting but not enough to stop sliding once started. It is therefore prudent to provide ground way head starting rams ready to push on an end launching cradle or side launching sleds if the vessel fails to start when released (Robb, 1952). The less the declivity, the greater the probable need for starting rams. Depending upon the type of releasing arrangements, it may be possible to have ram pistons pushing against the inshore ends of the end launching cradle or side launching sleds immediately before and during release.

g. Side Launch Starting Shores. To avoid high pressures on the bottom of a side launched vessel, it is important that the vessel's centerline be parallel to the line of the ground way ends as the vessel enters the water. This requires the simultaneous starting and continued sliding of both ends of the vessel when released. If one end of a vessel sticks, or tends to stick, when released, there may be a tendency for rotation about a vertical axis, at or near the center of gravity, with the result that initially the sticking end of the vessel may tend to move inshore. If at the sticking end of the vessel, an athwartships shore extends from the ground to a point on the deadwood or lower shell, the outshore reaction of the shore in resisting the inshore motion of the end of the vessel, produces an outshore push to overcome the reluctance of the sticking end. Depending on the vessel's size one or more starting shores should be installed at each end of the vessel.

Starting shores are particularly well suited to butter-board launchings where neither the butter-boards nor launching blocking and cribbing is suitable for taking the thrust of starting rams.

h. Side Launch Trunnioned Pivoting Ground Ways. When the ground way holding gear, shores, or mechanical triggers are released, the vessel's weight outshore of the trunnions causes rotation of ways and vessel about the centerline of the trunnions through an angle which can, if desired, be considerably steeper than that normally used for side launching. The vessel and sliding ways move down the inclined ground ways and enter the water. The loading of the shores or triggers securing the ground ways against pivoting is small compared with the loading of triggers for a side launching on fixed ground ways. The absence of conventional triggers, the steep slope of the trunnioned ground ways and the shock as the outshore ends of the pivoting ground ways come to a stop, tends to result in a high degree of starting reliability.

7.3 Design Data. For a burn-off sole plate of medium ship steel, 19 mm (0.75 in.) thick with 19 mm holes drilled at 50 mm (2 in.) centers, the strength of one segment 32 mm (1.25 in.) wide is as follows:

	Load, N (lb)	Stress, MPa (psi)
Yield	183,200 (41,190)	302.9 (43,935)
Ultimate	273,010 (61,375)	451.4 (65,465)

The sole-plate segment breaking strength can be used to estimate strength of other segments with different thicknesses and widths.

Starting ram capacity can be determined by:

$$F = K \times W \times \sin\theta \tag{6}$$

where:

F = Total force capacity of both starting rams
K = A coefficient of 0.5 to 1.0. Larger coefficients are appropriate for smaller way inclinations.
W = Weight of ship and cradle
θ = Slope of ground ways.

Section 8
Checking

8.1 Requirements for Checking. As a result of measures to ensure starting and avoid sticking on the ways, an end-launched vessel may have considerable velocity, and kinetic energy, when the cradle leaves the ground ways. If the run before the outshore end of vessel enters shoal water or strikes a wharf or vessel on the opposite side of the stream is limited, then some means of absorbing or transforming the kinetic energy of the vessel just launched, or modifying its track and/or heading, may be required.

Vessels side-launched into a narrow slip or graving dock have great water resistance due to their sideways motion and may or may not require checking.

8.2 Types of Checking. Means of checking involving work done by water resistance include outshore (stern) masks, outshore pontoons which may be needed to reduce ground way end pressures, inshore (bow) pontoons which may be needed to reduce drop off, outshore end of vessel built close to the water, and conversion of translation energy into rotational energy by slewing the vessel.

Checking by means other than water resistance include chain and concrete block drags, anchors, fiber rope breaking stops, friction brakes, and soft mud banks. Pneumatic fenders and resilient camels are a means of checking side-launched vessels.

Undesired checking can be reduced by installing false rake-type *bows* on vessels with square or very blunt outshore ends, by delaying entry of the vessel into the water by means of a long run, building the vessel with less than ground way declivity to raise the stern, or by launching at slack low water.

a. Masks. For a vessel of ship form, launched stern first, the mask can be at the trailing edge of the rudder, or rudders, or can be in two pieces at the outshore end of the cradle. The area should be large and perpendicular to the direction of travel. Construction can be of timber or welded steel. Masks have the following disadvantages: resistance varies as the product of the immersed area and square of velocity. Thus resistance is at a maximum before the vessel has left the ways (when a high velocity is wanted) and is very low shortly before the vessel has come to rest (when effective checking may be most wanted). An end launching cradle has great water resistance and this resistance is reduced if the cradle is in the turbulent wake of the mask.

A propeller which has been secured to prevent rotation has a considerable masking effect and it may be desirable to launch with locked propellers. A vessel with twin rudders can have each rudder locked in the hardover outboard position.

b. Pontoons. An outshore pontoon may be required to reduce outshore ground way end pressure and inshore pontoons may be required to limit drop off. If the outshore ends of such pontoons are box-shaped and perpendicular to the direction of travel, mask-type resistance will result. Resistance can be reduced if the lower outshore part of the pontoon is cut away by an inclined plane to produce a scow-like bow.

c. Outshore End of Vessel Close to Water. If the vessel is built with a declivity greater than that of the ground ways, and as far outshore as possible, the maximum velocity will be less due to the advanced start of water resistance and less distance slid. Less checking will be required and distance run will tend to be reduced.

d. Slewing. If a vessel, immediately after the cradle clears the ground ways, can be subjected to a checking force with an athwartships component at one end of the vessel, the vessel will turn, i.e. rotate about a vertical axis, the kinetic energy of linear translation being transformed into the kinetic energy of rotation. Rotation about a vertical axis means that the ends of the vessel experience the great resistance associated with sideways movement through the water and that sternward movement of the vessel will cease during the latter part of checking. Turning of the vessel can be done by putting the rudder over as soon as the cradle has cleared the ways, although this is an ineffective means of slewing the vessel. A more common way of causing slewing is by the use of chain or concrete block drags or other suitable means of checking, secured to steel wire rope cables which are fastened to the vessel aft and are on one side only.

If a ship-shaped vessel is launched stern first and is light forward and deep aft with a large sail area forward, and slewing is intended to swing the bow to port, a wind blowing on the port side with a strong athwartships component can reduce or prevent slewing with the result that the vessel's heading will not change and the chain drags may travel farther than expected.

e. Chain Drags. A very satisfactory method of checking the motion of an end launched vessel is by chain drags, attached to the vessel by steel wire cables, which take up as the cradle leaves the ground ways. The chain drags can be located well inshore so they come to rest at or near the water's edge. Chain drags can be on both sides of the vessel or only on one side. In the case of the latter disposition, total drag weight can be reduced by one third or more due to absorption of energy by slewing of the vessel. In addition, the number of chain drags and drag cables can be reduced and a chain drag path need be prepared on only one side of the vessel. If a larger slewing moment is desired, the drag initial positions may be at some distance from the vessel's side, shipyard arrangements permitting, so that the initial lever arm is several times the vessel's breadth rather than about one half the vessel's breadth as in the case if the chain drag path is close alongside and parallel to the building slip. Chain drag paths may be completely on dry land or may extend into the water or may be completely in the water. Chain drags are easier to retrieve if they stop clear of the water or in shallow water. Checking reliability is reduced if the drag path is extensively or totally submerged as there may be undetected debris on the bottom or bottom irregularities may exist that would cause the drags to hang up. In addition the resistance of drags on a muddy bottom may be less than that of drags on soft dry earth. Each chain drag that may become submerged by design or accident should be buoyed.

Chain drags are customarily arranged in U-shaped piles, the open end of the U being outshore. A steel wire rope choker is passed around the inshore end of the drag in a round turn and choker ends are shackled to the inshore end of the steel wire rope drag cable. The other end of the wire is secured to the vessel's shell by a quick release device, operable from the vessel's deck. If slewing assisted checking is desired, the ship ends of the drag cables are secured to the vessel as far outshore as is practicable and on only one side. If slewing is not wanted, then drag cables are attached to both sides of the inshore end of the vessel's hull.

Chain drag cable brackets should be welded to sufficiently heavy shell plating in way of decks, stringers or frames where this can be arranged. If plating and/or framing of the ship's side appears to be weak, it may be necessary to fit internal headers in way of the chain drag cable brackets. These headers may or may not have to be removed following launching. Chain drag cable brackets should be kept clear of riveted crack arrestors as well as areas of the ship's side to receive insulation.

The purpose of the U-shaped arrangement is to ensure a gradual acceleration of the chain pile's mass as the bight of the chain pulls through the open end of the U. The complete mass of chain is not in motion until the drag cable inshore end has moved about twice the length of the pile. Gradual acceleration of chain piles reduces loads on drag cables, cable end fittings and drag chokers. All drags can be of the same weight and as large as practicable considering strength of drag cables and width of chain drag path. Drag take up produces negligible shock on the vessel and there is no need to have a progressive increase in chain drag weight in the sequence of chain drag take up.

If the width available for the chain drag path is limited, one large U-shaped pile may be replaced by two small U-shaped piles of the same total weight. One drag cable is used for both piles and is connected to the outshore pile by a choker. A steel wire rope pendant extends from the outshore pile to the choker of the inshore pile.

Instead of disposing two small chain piles in tandem, to save longitudinal space one chain pile may be located on top of the other. The upper pile must have a shorter choker than the lower pile so that the top pile will have started, and possibly completed, its movement before the bottom pile starts.

Chain drags can be used for checking side-launched vessels. If maximum effect is required as soon as possible, no U-shaped pile pull through should be provided and the full mass of each chain clump should start moving as the sleds or sliding ways clear the ground way ends.

f. Concrete Blocks. Concrete blocks can be filled with plate punching or iron ore aggregates for maximum density, and can have structural angle shod edges. Blocks of cubic form cannot be expected to have the desirable characteristics of stability, burrowing and high resistance exhibited by chain drags in softened earth. Such blocks can be used for letting go after a vessel's cradle has cleared the ways. Blocks of rectangular, shallow form, to prevent tumbling, or steel boxes filled with plate punchings or other heavy ballast, can be used on prepared sliding surfaces for checking side-launched vessels.

g. Anchors. An anchor being dragged at fairly high speed cannot be expected to have a uniform or predictable resistance. In addition, there is the hazard that the anchor may perform its intended function, that is, dig in and hold, in which case drag cable or fittings can be expected to part unless the vessel is moving at very slow speed. Anchors, suitably buoyed, should be provided ready to let go in case wind or current tends to drift the vessel just launched onto the launching ways or into some other hazardous situation.

h. Fiber Rope Breaking Stops. In lieu of checking by means of chain drags, the work done by the parting in succession of fiber rope breaking stops can be used. As with chain drags, checking arrangements can be on one or on both sides of the vessel. The vessel can be partially checked by slewing. A ground chain is laid down alongside the vessel. The inshore end of the chain is shackled to a buried anchor. Parallel to the ground chain, and secured to it by many breaking stops is the ship chain. Part of the ship chain is triced along the vessel's side and one end of the chain is secured to the vessel, forward or aft, with a quick release fitting. As the vessel moves outshore, the ship's side tricing stops part in succession and as the cradle clears the ground ways the inshore breaking stop connecting the ground and ship chain cables stretches and parts. As the remaining rope stops break in succession, the catenary of the ship chain is momentarily flattened as each stop stretches and breaks. The energy absorbed by the repeated tightening and slacking of the ship chain may be greater than that absorbed by the rope stops. This method of checking can be considered when a suitable chain drag path is not available.

i. Friction Brakes. One or more wire ropes are secured to the vessel and led through spring or hydraulic loaded friction brakes which are firmly secured to the ground. No chain drag path is required but objections to this method include heat generated by friction between the brake shoes and the wire ropes and damage to the wires.

j. Soft Mud Banks. Grounding on a soft bottom may be an acceptable method of checking a vessel if hull and underwater appendages are not liable to damage and if possible damage to bottom paint and choking of sea chests with mud are not considerations.

k. Resilient Fenders. Final motion of a vessel side launched into a basin can be checked by means of pneumatic fenders or resilient camels. The basin must be configured so that the fenders can be compressed between the vessel's side and the nearly vertical opposite side of the basin.

8.3 Design Data for End-Launch Checking Calculations. The data supplied below are primarily empirical formulas from which can be derived resistive force and energy absorption values that can be utilized in launching calculations. Nondimensional equations have been used where practicable and correct results can be obtained if consistent systems of units are employed. Otherwise dimensions are given in both SI and English units.

The hydrodynamic resistance of a stern mask or buoyancy pontoon or bow pontoons is obtained from the equation (Saunders, 1957):

$$R = \frac{\rho b h v^2 C_D}{2} \tag{7}$$

R = resistance
ρ = mass density of water
b = breadth of mask
h = draft of mask
v = velocity of ship motion
C_D = drag coefficient varying as follows:

b/h	C_D
1	1.16
5	1.20
10	1.50

The hydrodynamic resistance of an end launched vessel, with its propellers locked, plus the cradle can be derived from Robb (1952):

$$R = \rho(0.63\, A_c + 0.004\, A_{ws})v^2/2 \tag{8}$$

where:

A_c = Projected area of submerged portion of cradle, propellers, and hull
A_{ws} = Wetted surface of submerged portion of hull and sides of cradle

It can be seen that the above coefficients are a pressure drag coefficient and a frictional resistance coefficient respectively that are average values for all depths of immersion of the hull and cradle.

Frictional resistance developed by various types of drag devices such as chains, anchors, and brakes varies widely. A rough estimate of the energy absorbed by chain drags, or the work required as a function of launching weight, can be obtained from:

$$E_{CD} = 24.3\,(W - 2500 \pm 360)\text{ kJ} \tag{9}$$

$$= 8.00\,(W - 2500 \pm 360)\text{ ton-ft} \tag{10}$$

where:

E_{CD} = absorbed energy which is the sum of the products of chain drag weights times distance slid
W = launching weight of vessel

This formula is for a ship-form vessel, end launched, with drag cables on one side only and secured aft so that checking is partially by slewing. The spread of energy values is attributable to differences in tide height, grease temperature and pressure, alignment of ground ways, smoothness of base coat, and wind strength and direction acting on the vessel during slewing. As a more accurate means of determining chain drag resistance, the following coefficients of friction can be used:

Surface	Drag/ Weight
Broken stone and gravel (running)	0.93
Earth (starting)	0.55
Earth (running)	0.75
Cinders on hard clay (starting)	0.40
Cinders on hard clay (running)	0.70
Sand on concrete (running)	0.70
Smooth concrete or wood (running)	0.16

The dragging resistance of a Navy type stockless anchor can be taken as 6.5 times the anchor weight; the buried resistance of a similar anchor is 15 times the anchor weight. The resistance of friction brake cast steel shoes and steel wire rope cables can be based on a coefficient of friction of 0.25.

When fiber ropes are used either as breaking stops or to secure the vessel to a fixed or dragging anchor the rope will absorb energy from the point where it becomes taut to the point where it breaks. The breaking strength, in a dynamic situation, is approximately 80 percent of the ultimate strength determined from static test. The energy absorbed in elongating the rope is roughly one-third of the product of the elongation and the breaking strength as indicated below:

$$E_{FR} = KF_sL_R\left(\frac{e_r}{3}\right) \qquad (11)$$

where:

E_{FR} = fiber rope energy absorption
K = applied load/breaking strength
F_s = breaking strength of fiber rope
L_R = unstretched length of rope
e_r = ratio of elongation to unstretched length (0.15 for wet manila rope)

Thus the work done, or energy absorbed, in snubbing with wet manila rope stressed to 70 percent of its breaking strength would be:

$$E_{FR} = 0.7\,F_s\,L_R\,(0.15/3) = 0.035\,F_s\,L_R \qquad (12)$$

The absorbed energy obtained is in the same units of force times distance in which F_s and L_R are expressed.

Section 9
End Launch Calculations

9.1 Final Calculations. The following calculations provide a basis for developing the arrangement and details delineated on the launching drawing:

a. Weight and Center of Gravity. This estimate should be made in a systematic manner in accordance with a standard weight-grouping system, (See Chapter I). Systems, machinery, equipments, and outfitting items not to be aboard at launching are deducted. Weights of temporary offices and storerooms, temporary lighting and ventilation, staging, ladders, ballast, raw materials, shipyard machines and tools, welding cable, air hoses, the vessel's share of checking arrangements, launching cradle, mooring lines and launching crew are added.

The longitudinal and vertical centers of gravity are determined for the vessel in launching condition. If uncertainty exists as to the location of the longitudinal center of gravity, it is conservative to assume that it is somewhat out-shore of the estimated location in determining way end pressures and somewhat inshore of the estimated location in determining drop off. Ordinarily a vessel of normal ship form will have a large metacentric radius when launched, and if topside outfitting is not far advanced, will have ample stability. The vertical center of gravity should be estimated in a conservatively high location. In the case of asymmetrical vessels, such as drilling platforms launched in two halves, or ship-shaped vessels with asymmetrical machinery arrangements, say two boilers on one side and one boiler on the other, or submarines with little stability, then the transverse center of gravity becomes of importance and ballasting or some adjustment of weight may be required to ensure that the vessel launched floats with positive stability and little or no list.

b. Vessel Location on Building Slip. The vessel's weight and longitudinal center of gravity in launching condition being known, the extent and location of the cradle under the vessel can be determined to support the vessel properly and avoid excessively low or high grease pressures. The cradle will have burn-off sole plate or trigger connections which must mate with the corresponding ground way fittings. Assuming that the inshore ground ways are made in standard lengths, different possible locations of the vessel on the building slip will vary by increments equal to the ground way section lengths.

c. Condition Afloat. The condition afloat is based on the vessel's launching weight and position of the center of gravity. The cradle, except for the after poppets, should have slight positive buoyancy and its weight is ignored. Buoyant effects of any bow and/or stern pontoons must be considered. The drafts at the perpendiculars and at the ends of the cradle can be found from the hydrostatic curves or by trial and error from the Bonjean curves. The draft to the underside of the cradle at its forward end is calculated and, according to the depth of water over the ground way ends, determines if the vessel will float off the ways or if the cradle will drop off the way ends. Clearance of forefoot and stem, the sides of the vessel forward, and the keel forward should be verified, using a calculated sternward velocity. The metacentric height of the vessel should be calculated and determined to be adequate.

d. Buoyancy During Launching. It is necessary to assume a certain tide height and depth of water over the way ends at time of launching. Water level assumed can be based on the predicted tide height as determined from the tide tables for the date and time of launching. Launching should be at slack water. An unfavorable wind or river stage and high barometric pressure can be assumed to reduce the actual water level below its predicted level.

The vessel is assumed to move down the ways and displacements and locations of the longitudinal center of buoyancy are determined by means of Bonjean curves as intersections of the vessel's base line and certain displacement stations enter the water. If displacements and centers of buoyancy are to be determined using Simpson's First Rule, the number of stations in the water must be a multiple of two. Centers of buoyancy are located relative to the intersection of base line and water. For each subsequent

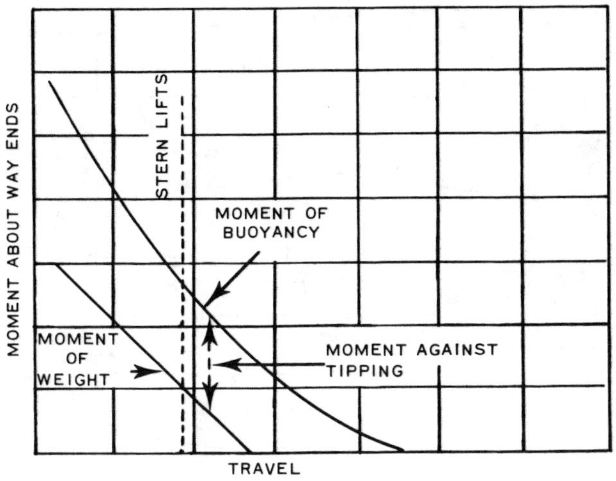

Fig. 13 Tipping curves

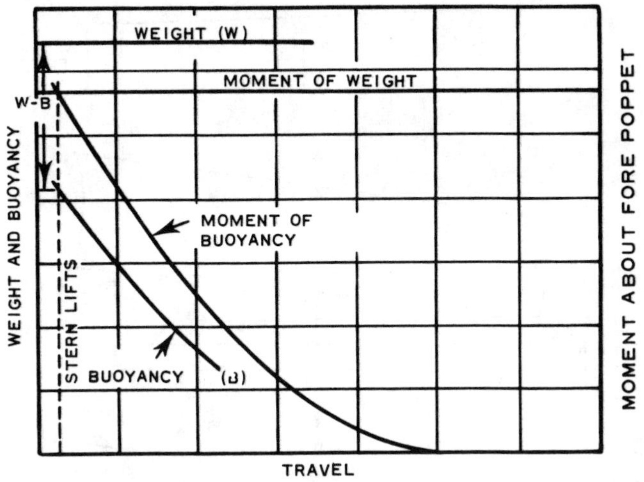

Fig. 14 Pivoting curves

calculation, the vessel is assumed to have moved two displacement stations. As for the condition of the vessel afloat, the displacement of the cradle can usually be disregarded. If hull form equations are in computer storage, the computer can be used to determine buoyancy and moments at successive positions of the ship moving down the ways.

e. Way-End Pressure and Moment Against Tipping. After the center of gravity of the vessel passes the ground way ends, there is a moment tending to cause the inshore end of the vessel and cradle to lift off the ground ways, that is, tip. This moment is resisted by the moment of buoyancy about the way ends. Moments of Buoyancy and Moments of Weight can be plotted as ordinates on an abscissa with scales of distance slid and stations entering the water, Fig. 13. To avoid the unacceptable condition of tipping, the moment of buoyancy must always be greater than the moment of weight. The difference between the moments is the Moment Against Tipping.

After the vessel enters the water, it is acted on by the buoyant force and the force of gravity. The resultant will be the difference of these two forces and represents the support of the ground ways. If the resultant is within the middle third of the length of sliding and ground ways in contact, the load can be represented by a trapezoid where the ordinates are weight per unit length of way or force per unit area of grease. If the resultant is outshore of the middle third of the length in contact, then the load is represented by a triangle with base equal to three times the distance from the resultant to the way ends.

For any distance traveled, or station entering the water, the loading of the ground ways can be determined. This is the loading on the ship's bottom but, in the absence of continuous longitudinal packing, is carried into the hull by means of relatively widely spaced packing blocks. The loading of the vessel's bottom structure being known, calculations can be made to determine if reinforcement is necessary to prevent damage due to high way-end pressure.

f. Pivoting. Lifting of the outshore end of the vessel, or pivoting, occurs when the moment of buoyancy about the fore poppet equals the moment of weight about the fore poppet, Fig. 14. The latter moment is, of course, constant. With a crushing packing fore poppet, initial pivoting is about a point near the after end of the poppet because the only slightly compressed crushing strips are unable to offer significant resistance. As crushing progresses, the pivoting point moves forward and if the pivoting angle is large, may approach the forward end of the fore poppet. With a rocker fore poppet, pivoting is about the midlength of the fore poppet. In launching calculations the extent of the cradle is known but the length of the fore poppet may be unknown. Pivoting can be assumed to be about the estimated midlength or the forward end of the fore poppet. At the instant of pivoting, the load on the fore poppet is a maximum and is equal to the vessel's weight minus its buoyancy.

Even if vessel and cradle will drop off the ground way ends, it is convenient to assume that the ground ways extend sufficiently far under water for float off. During pivoting the moment of buoyancy about the fore poppet must equal the moment of weight about the fore poppet. For any distance slid, the draft at the fore poppet is known. It can be assumed that the path of the intersection of the vessel's after perpendicular and base line at the start of pivoting travels along a straight line to the point of floatoff. The draft at the after perpendicular was previously determined for condition of vessel afloat.

Lifting of the stern results in a lowering of the bow and it must be verified that the forefoot of the vessel will not come in contact with the building slip or dismantled keel blocking.

As the vessel's stern enters the water there is a rise in the water level over and beyond the outshore ground ways. In addition, the relative motion of the sloping stern and water tends to produce a lifting action. As a result, pivoting may be somewhat earlier than indicated by calculations for an assumed static condition. On the other hand, if flow of

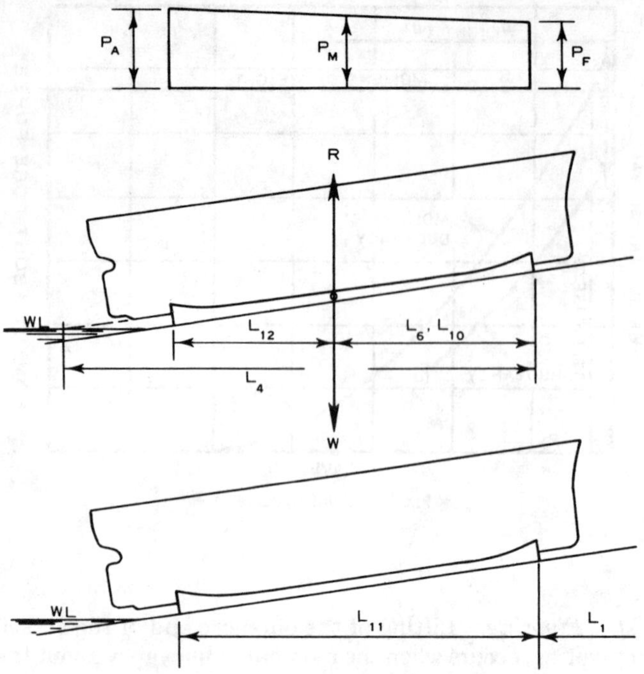

Fig. 15 Vessel before entering water

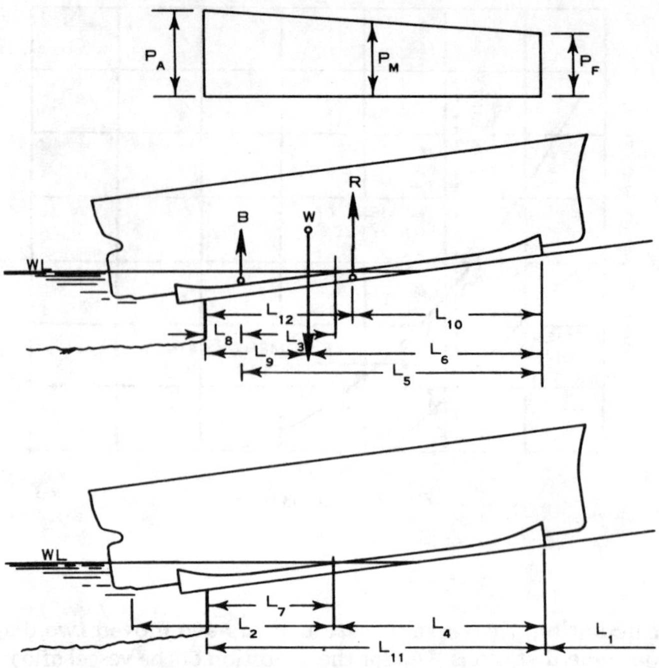

Fig. 16 Vessel with centers of buoyancy and weight inshore of ground ways ends before pivoting

Table 10—General Data

Launching weight ....
Length between perpendiculars ....
Station spacing (*S*) ....

Center of gravity forward/aft of amidships ....
Center of gravity to forward end of fore poppet (L_6)(L_{10} initial) ....
Center of gravity to ground way ends (L_9 initial) ....

Moment of weight about forward end of fore poppet (M_1) ....
Effective width of one way (L_{13}) ....
Length of sliding ways (L_{11} initial) ....

Effective areas of both sliding ways ....
Forward end of cradle to fore perpendicular ....
After end of cradle to after perpendicular ....
Distance slid to point where *A.P.* at *B.L.* enters water ....
Intersection *B.L.* projected and W.L. to forward end of fore poppet (L_4 initial) ....

Declivity of ground ways ....

Declivity of base line ....
Height of base line above ground ways at fore perpendicular ....
Height of base line above ground ways at amidships ....
Height of base line above ground ways at after perpendicular ....

Height of tide above mean low water ....
Depth of water over ground way ends ....
Specific gravity of water (*s.g.*) ....

water near the water's edge and perpendicular to the ground ways is restricted, there may be a local reduction in water level with the result that dynamic pivoting may be delayed beyond static pivoting.

If drop-off occurs, the static drop is the difference between the draft to the underside of the sliding ways at the forward end of the cradle and depth of water over the ground way ends. The dynamic drop can be taken as twice the static drop.

g. Strength. Hull girder stresses may be high in vessels which are long in relation to depth or which have large strength deck openings or discontinuities in longitudinal strength members. High stresses are also possible if longitudinal strength members have large temporary openings or are not complete as regards structure or welding.

As an end-launched vessel moves down the ways, the hull girder is subjected to hogging when the outshore end of the cradle is beyond the ground way ends and before pivoting has occurred. Calculation of the hogging stresses is done

Table 11—Launching Calculation Symbols

DISTANCES

L_1 Distance slid by vessel
L_2 Immersed hull length
L_3 Intersection of base line and water to center of buoyancy
L_4 Intersection of base line and water to forward end of fore poppet
L_5 Center of buoyancy to forward end of fore poppet
L_6 Center of gravity to forward end of fore poppet (constant)
L_7 Length of immersed ways in contact to intersection of base line and water
L_8 Center of buoyancy to ground way ends
L_9 Center of gravity to ground way ends
L_{10} Resultant force R (support of ways) to forward end of fore poppet
L_{11} Length of ways in contact
L_{12} Resultant force R (support of ways) to outshore end of sliding or ground ways, whichever is least
L_{13} Effective width of one way

FORCES

W Vessel launching weight (constant)
B Displacement or buoyancy
R Support of ways

PRESSURES

P_M Average unit pressure on ways
P_A Unit pressure at outshore end of ways in contact
P_F Unit pressure at inshore end of ways in contact

MOMENTS

M_1 Moment of weight about forward end of fore poppet (constant)
M_2 Moment of buoyancy about forward end of fore poppet
M_3 Anti-pivoting moment
M_4 Moment of buoyancy about ground way ends
M_5 Moment of weight about ground way ends
M_6 Anti-tipping moment

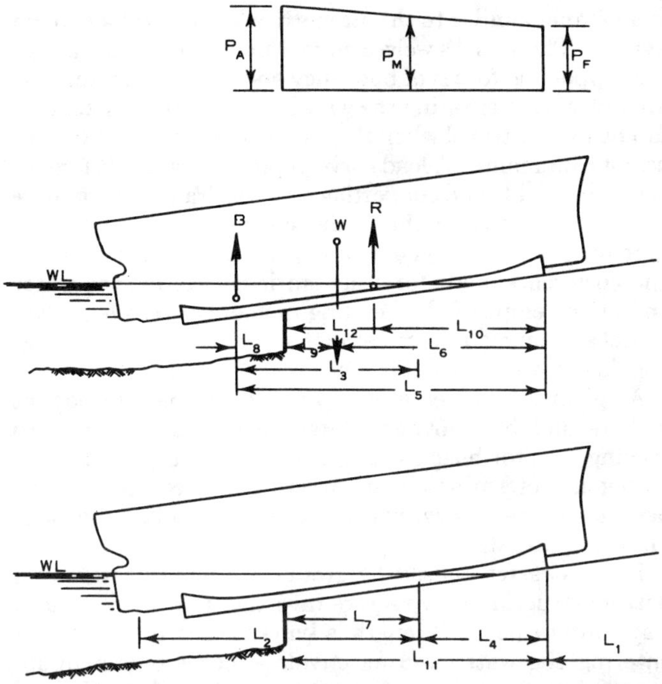

Fig. 17 Vessel with center of gravity inshore, and center of buoyancy out shore, of ground way ends before pivoting

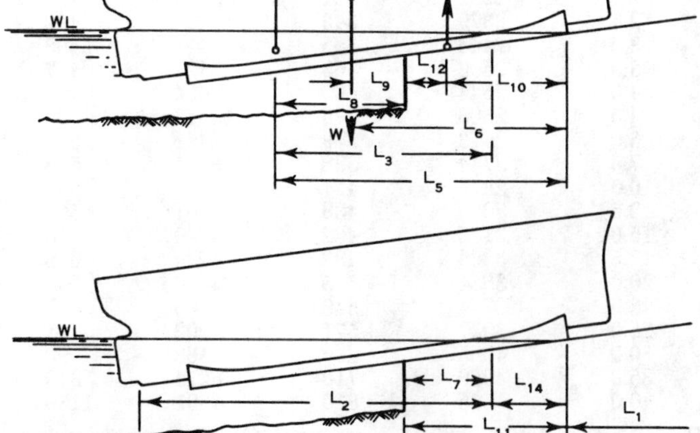

Fig. 18 Vessel with centers of gravity and buoyancy outshore of ground way ends before pivoting

in a manner similar to the strength calculation of a ship on a wave. The vessel's weight curve can be determined as can the supporting forces of buoyancy and ground way support for any station entering the water. The initial calculation can be for the travel when the anti-tipping moment is at or near its minimum. A load curve, representing the difference between weights and supporting forces is drawn. This curve is integrated once to obtain the shear curve and a second time for the bending moment curve. Similar calculations and curves are made for other stations entering the water and an envelope of the bending moment curves is drawn. Vessel section moduli are calculated for various stations and hogging stresses are determined.

At pivoting the vessel is supported by the fore poppet inshore and by buoyancy outshore and is subjected to sagging. As for hogging, the load curve at the start of pivoting is determined and integrated twice to give the bending moment curve which is used to calculate the maximum sagging stress.

Local stresses caused by ground way-end pressure on bottom structure may require investigation. The ground way support per unit length is known for various stations entering the water and an envelope curve of maximum way-end pressures is drawn on the vessel's profile. Ground way support is transmitted to the vessel via launching grease, sliding ways, wedges, wedge riders, and packing blocks or poppet timbers. Accordingly, the loading in way of each block or timber can be determined.

h. Pre-pivoting, Pivoting and Vessel Afloat Calculations. Tables 10 and 11 and Fig. 15 to 21 present a systematic method of tabulating data and performing calculations for displacement, support of ground ways, grease pressures, and anti-pivoting and anti-tipping moments.

Table 12—Crushing Strip Resistance

Percent Crushing	Spruce kPa	Spruce psi	Douglas Fir kPa	Douglas Fir psi
0.0	0	0	0	0
0.5	97	14	862	125
1.0	1276	185	1724	250
1.5	2068	300	2586	375
2.0	2282	331	3289	477
2.5	2379	345	3392	492
3.0	2448	355	3413	495
3.5	2503	363	3427	497
4.0	2544	369	3434	498
4.5	2586	375	3447	500
5.0	2613	379	3461	502
7.5	2744	398	3571	518
10.0	2875	417	3709	538
12.5	3020	438	3861	560
15.0	3185	462	4040	586
17.5	3365	488	4240	615
20.0	3551	515	4440	644
22.5	3744	543	4668	677
25.0	3937	571	4902	711
30.0	4385	636	5468	793
35.0	4950	718	6274	910
40.0	5585	810	7791	1130

Data are for 67 mm (2.625 in.) square flat grain crushing strips. Proportional limit for spruce is at 1.20 percent crushing with 1712 kPa (258 psi) resistance. The proportional limit for Douglas fir is at 1.80 percent crushing with 3082 kPa (450 psi) resistance.

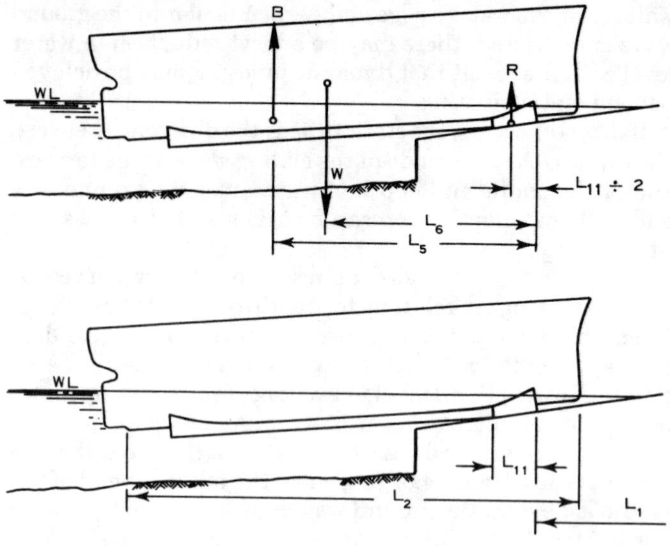

Fig. 20 Vessel pivoting

The following curves should be plotted:
1. Weight
2. Buoyancy
3. Moment of weight about after end of groundways
4. Moment of buoyancy about after end of ground ways
5. Moment of weight about fore poppet
6. Moment of buoyancy about fore poppet

i. Fore Poppet Calculations. Table 12 and Fig. 22 provide a method of fore poppet crushing strip design and calculation of distribution of loading and fore poppet support during pivoting. Crushing strip arrangements for fore poppets of various capacities are shown in Fig. 23.

During the initial stage of pivoting when the cradle aft of the fore poppet has just started to lift from the ground ways, the angle between base line of ship and ground ways will be but little changed from the initial angle. There will be practically parallel crushing of the fore poppet. The crushing will be relatively small and most of the load will be transmitted to the ground ways through the solid packing at the after end of the fore poppet. In side elevation the solid packing should be trapezoidal in shape with the shorter of the two parallel sides at the top so as to distribute the load over as great a length of ground ways as possible.

As pivoting continues, the drafts of the ship forward and aft can be determined for any stage of pivoting and consequently the angle between base line of ship and ground ways can be determined. The load on the fore poppet is also known for any pivoting position. If a point of zero crushing is assumed at the top of the fore poppet packing the actual and percent crushing for the assumed condition can be determined for each vertical row of crushing strips. The crushing resistance of spruce crushing packing is known and thus the resistance of each vertical row of crushing strips can be calculated. The sum of these resistances equals the supporting force of the fore poppet for the assumed condi-

Fig. 19—Pre-Pivoting Launching Calculation

Station entering water ....

Distance slid, L_1 ....

Immersed hull length, L_2 ....

Intersection base line and water to forward end of fore poppet, L_4 ....

Length of immersed ways in contact to intersection base line & water, L_7 ....

Center of gravity to ground way ends, L_9 ....

Length of ways in contact, L_{11} ....

STATION NUMBERS	DRAFTS	AREAS	SIMPSON'S MULTIPLIERS	FUNCTIONS OF AREAS	WATER & BASE LINE LEVERS	FUNCTIONS OF MOMENTS
			Σf(A) =		Σf(M) =	

$B = s.g \times S \times \Sigma fA/3$ metric tons $= s.g \times S \times \Sigma fA/108$ long tons (13)

where: S is station spacing in meters.
A is area in square meters.

where: S is station spacing in feet
A is area in square feet

$$L_3 = S \times \Sigma fM/\Sigma fA \quad (14)$$

L_3 L_4 ________ Add L_5	$M_5 = L_9 \times W =$
$M_1 = L_6 \times W =$	M_4 M_5 ________ Subtract M_6
$M_2 = L_5 \times B =$	W B ________ Subtract R
M_1 M_2 ________ Subtract M_3	$L_{10} = M_3/R =$
L_3 L_7 ________ Subtract L_8	L_4 L_7 ________ Add L_{11} L_{12} ________ Subtract L
$M_4 = L_8 \times B =$	$L_{11}/3$
$P_M = R/(2 \times L_{11} \times L_{13}) =$	
If R is within the middle third of L_{11} $P_A = (1.0 - 1.5\,L_{12}/L_{11})4P_M =$ $P_F = (1.5\,L_{12}/L_{11} - 0.5)4P_M =$	
If R is aft of the middle third of L_{11} $P_F = 0$ $P_A = R/3L_{12}L_{13} =$	
If R is forward of the middle third of L_{11} $P_A = 0$ $P_F = R/3L_{10}L_{13} =$	

Fig. 21—Pivoting and Ship Afloat Launching Calculation

Station entering water ....

Distance slid, L_1 ....

Immersed hull length, L_2 ....

STATION NUMBERS	DRAFTS	AREAS	SIMPSON'S MULTIPLIERS	FUNCTIONS OF AREAS	FORE POPPET LEVERS	FUNCTIONS OF MOMENTS
			ΣfA		ΣfM	

$B = s.g. \times S \times \Sigma fA/3$ metric tons

Where: S is station spacing in meters
A is area in sq. m.

$= s.g. \times S \times \Sigma fA/108$ long tons (15)

Where: S is station spacing in feet
A is area in sq. ft.

$$L_5 = S \times \Sigma fM/\Sigma fA \quad (16)$$

$$M_1 = (L_6 - L_{11}/2)W \quad (17)$$

$$M_2 = (L_5 - L_{11}/2)W \quad (18)$$

$W =$

$B =$ ____________ *subtract*

$R =$

tion. This must equal the load on the fore poppet. If the resistance and loads are not equal, the assumed point of zero crushing must be moved forward or aft and repeated trials made until approximate equality is obtained between the resistance of the fore poppet and the load on the fore poppet.

The foregoing calculations are repeated for several stages of pivoting and for each stage the resistances of the various vertical rows of crushing strips are examined and plotted to reveal any irregularities. The curves of resistance should be as uniform as possible and any abrupt humps or hollows will indicate the need for respacing of the vertical rows or changes to the number of crushing strips in one or more vertical rows.

If changes are made to the crushing strip or solid packing arrangement, the calculations are repeated as necessary. Failure to equalize the pressures under the fore poppet may result in burning of the inshore launching grease or, in an extreme case, could result in the ship sticking on the ways following pivoting.

Fig. 23 illustrates crushing strips for standard fore poppets and shows crushing strips arranged in up to eight horizontal layers and eighteen vertical rows for eighteen different arrangements. In the diagram, which is not to scale, the black squares represent the crushing strip ends. The solid packing is trapezoidal shaped and is at the left, being shown to be 1.79 m (5.88 ft) long at the top. The solid packing for a single effective width of 305 mm (12 in.) can accommodate a load of about 170 tons at the start of pivoting. As pivoting progresses, there is crushing at, and rotation around, the forward upper corner of the solid packing. Bearing pressures at the after end of the solid packing diminish as the pivoting angle increases and therefore great lengths of solid packing do not increase and load carrying capacity of the fore poppet except in the early stages of pivoting.

Determination of which of the eighteen crushing strip arrangements to use is based on the calculation of acceptable unit grease pressures for an arrangement selected according to experience and/or intuition. Many vertical rows of crushing strips are advantageous in distributing pivoting pressure after pivoting is well advanced. Unless there are enough horizontal layers of crushing strips, the percent crushing of the forward (inshore) strips may be so large that vertical row crushing resistance and local unit grease pressure is too high. On the other hand, too many layers will not result in enough vertical row resistance and too much of the pivoting load may be carried by the solid packing aft of the crushing strips.

As an example, Arrangement H shows five horizontal layers with eight crushing strips in the top layer and four crushing strips in the bottom layer. There are four complete vertical rows, each with five crushing strips. Crushing strips in bottom layer 5 rest on the wedge rider. The dashed line below the type designation H and sloping downward and to the right (forward), defines the after boundary of the crushing strips and the forward boundary of the solid packing. If calculations for different stages of pivoting show that there are not enough vertical rows of crushing strips to support the pivoting load, then Arrangement K with eleven vertical rows and six horizontal layers can be tried.

j. Pre-Pivoting Drafts on Cambered Ways. The following formulas permit the calculation of perpendicular drafts for any distance slid before pivoting on cambered ways.

$$H_{FP} = h - B \times x + \frac{x}{2R_c}(C - x) + S_{PP}\left(x + \frac{x}{R_c}\right) \quad (19)$$

$$H_{AP} = h - B \times x + \frac{x}{2R_c}(C - x) - (L_{PP} - S_{PP})\left(\alpha + \frac{x}{R_c}\right) \tag{20}$$

where:

H_{FP} = height of keel at fore perpendicular above water level.

H_{AP} = height of keel at aft perpendicular above water level.

h = initial height of ground ways at forward end of fore poppet above water level.

B = slope of ground way chord from forward end of fore poppet to ground way ends.

x = distance slid by vessel measured along chord.

C = length of chord from inshore end of fore poppet to ground way ends.

R_c = radius of ground way camber.

S_{PP} = distance from inshore end of fore poppet to fore perpendicular.

α = initial slope of vessel's keel

L_{PP} = length of vessel between perpendiculars.

Fig. 22—Calculation of Fore Poppet Loading During Pivoting

Length between perpendiculars . . . .
Number of layers of crushing strips . . . .
Number of rows of crushing strips . . . .
Separation of first and last rows of crushing strips . . . .
Length of one crushing strip . . . .
Width of one crushing strip . . . .
Area of one crushing strip . . . .
Distance slid to start of pivoting . . . .
Distance slid to float off . . . .

DISTANCE SLID	DRAFT AT A.P.	DRAFT AT F.P.	TRIM IN L.B.P.	TRIM IN L.B.P. BEFORE PIVOTING	CHANGE OF TRIM IN L.B.P.	SLOPE RATIO	FORE POPPET TOTAL LOAD	FORE POPPET LOAD ONE SIDE

Distance slid . . . .
Trial number . . . .
Load on one side of fore poppet . . . .
Thickness of one crushing strip . . . .
Change in slope, due to pivoting, between base line and ground ways . . . .

1	2	3	4	5	6	7	8	9	10
STRIP VERTICAL ROW NO.	NUMBER OF LAYERS	TOTAL STRIP THICKNESS	DISTANCE FROM ROW NO.	AMOUNT CRUSHED	PERCENT CRUSHED	RESISTANCE	RESISTANCE EACH ROW	STRIP SPACING	ROW PRESSURE
TOTAL									
ACTUAL LOAD ON ONE SIDE									
PERCENT DIFFERENCE									

COLUMN	HOW OBTAINED
5	Multiply column 4 by change in slope, due to pivoting, between base line and ground ways.
6	Multiply column 5 by 100 and divide the product by column 3.
7	Reference to table showing unit resistance to crushing of crushing strips.
8	Multiply column 7 by the area of one strip.

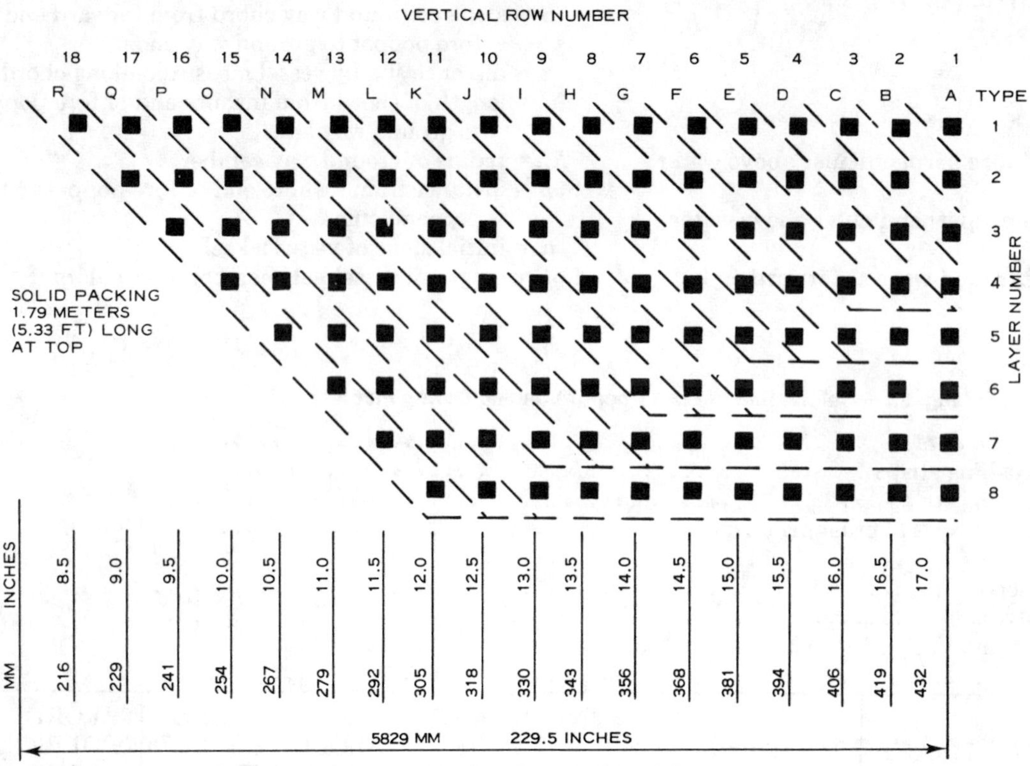

TYPE	TONS RANGE
A	0 - 10
B	10 - 19
C	19 - 29
D	29 - 39
E	39 - 49
F	49 - 58
G	58 - 68
H	68 - 78
I	78 - 88
J	88 - 97
K	97 - 107
L	107 - 117
M	117 - 126
N	126 - 136
O	136 - 146
P	146 - 156
Q	156 - 165
R	165 - 175

Fig. 23 Fore poppet crushing strips

k. Drop-off Calculations. The elapsed time from drop off to the farthest descent of the bow (twice the static drop) can be taken as half the pitching period of the vessel launched. The distance of the forward end of the underside of the fore poppet below and beyond the ground way ends can accordingly be determined.

l. Sliding Speed. Speed of sliding before there is appreciable water resistance can be determined from the following formula:

$$V = \sqrt{2 \times g \times S \times \cos\theta\,(\tan\theta - f_s)} \qquad (21)$$

where:

V = velocity
g = acceleration of gravity
S = distance slid along ways
θ = slope of ground ways
f_s = coefficient of sliding friction

and:

$$a = g \times \cos\theta\,(\tan\theta - f_s)$$

where:

a = acceleration of vessel.

m. Velocity by Energy Analysis. Loss of Potential Energy = Gain in Kinetic Energy + Work Done Against Friction + Work Done by Water Resistance, Fig. 24 (Andrews, 1967).

With cradle and ground ways in contact (Robb, 1952):

$$\frac{W}{2g}(V_2^2 - V_1^2) = W \times S \times \sin\theta - f_s(W - \Delta)S \times \cos\theta - R \times S - \Delta \times S \times \sin\theta \qquad (22)$$

where:

W = weight of vessel and cradle
V_2 = velocity at end of distance slid increment
V_1 = velocity at start of distance slid increment
S = incremental distance slid along ways
Δ = displacement of vessel
R = water resistance

As the vessel slides down the ways, velocities are calculated based on assumed value of coefficient of sliding friction and, after the stern enters the water, on calculated values of water resistance. Incremental distance can conveniently be the station spacing.

After cradle leaves the ground ways:

$$\frac{W}{2g}(V_2^2 - V_1^2) = -R \times S \qquad (23)$$

When the first drag has started:

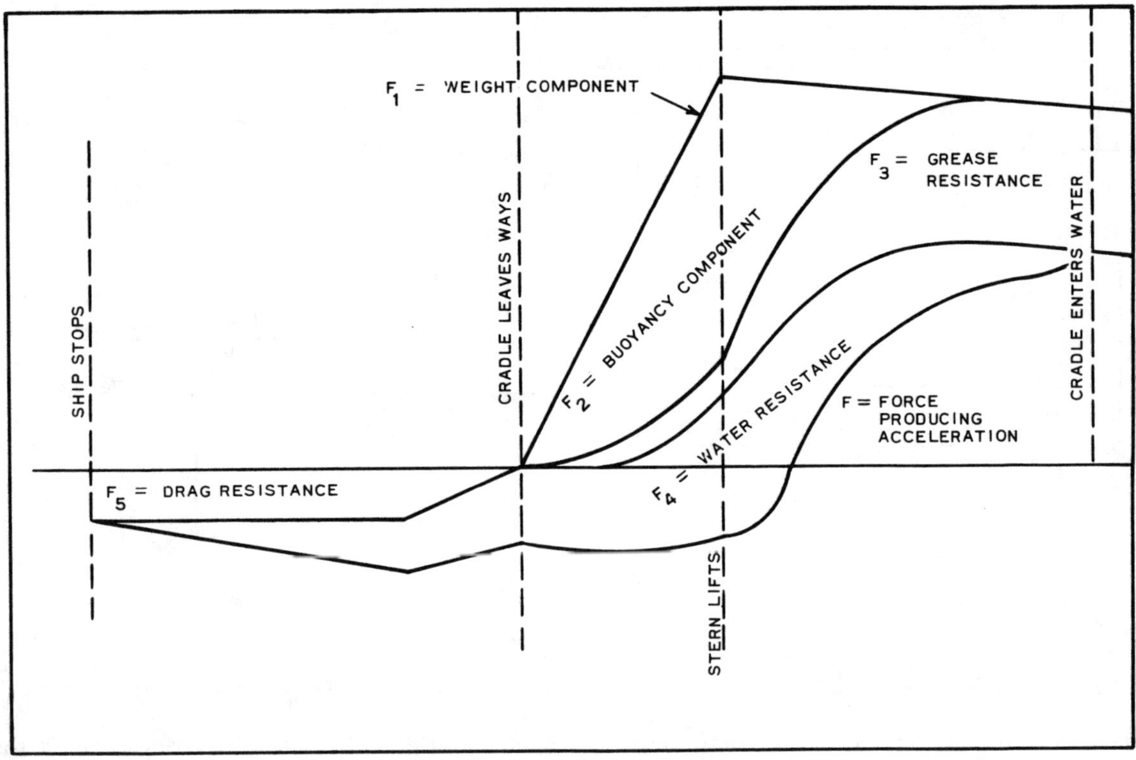

Fig. 24 Force diagram for cambered ways

$$\frac{W}{2g}(V_2^2 - V_1^2) + \frac{W_{D1} \times V_2^2}{2g} = -f_D \times W_{D1} \times S - R \times S \quad (24)$$

where:

W_{D1} = weight of the first drag(s)
f_D = coefficient of drag friction

When the second drag has started:

$$\frac{(W + W_{D1})}{2g}(V_2^2 - V_1^2) + \frac{W_{D2} \times V_2^2}{2g} = -f_D(W_{D1} + W_{D2})S - R \times S \quad (25)$$

where:

W_{D2} = weight of the second drag(s).

Section 10
Side Launch Calculations

10.1 Preliminary and Final Calculations. The end launching calculation material is generally applicable. Calculation of the vertical center of gravity is of importance for narrow vessels in which the topside outfitting may be well advanced at time of launching. The longitudinal center of gravity should be located over the center of gravity of the load-carrying grease. The purpose is to reduce the tendency for non-uniform sliding of bow and stern and slewing on the ways. Measures to adjust the position of the longitudinal center of gravity include omission of structure, machinery and equipments from the vessel to be launched, provision of temporary ballast, adjustment of bow and stern overhangs, widening of existing ground ways, installation of additional ground ways, and changes in sled lengths or widths to change grease bearing area. In the case of a butter-board launching as opposed to a sled launching, the number, width and spacing of the butter-boards can be adjusted to give the desired longitudinal center of gravity of the load-carrying grease.

If the ground ways extend sufficiently far under water, the vessel floats off the sleds and there are no difficulties with high bottom pressures during tipping about the way ends, possible capsizing at the end of the outboard roll, or striking the way ends or wharf edge at the end of the inboard roll.

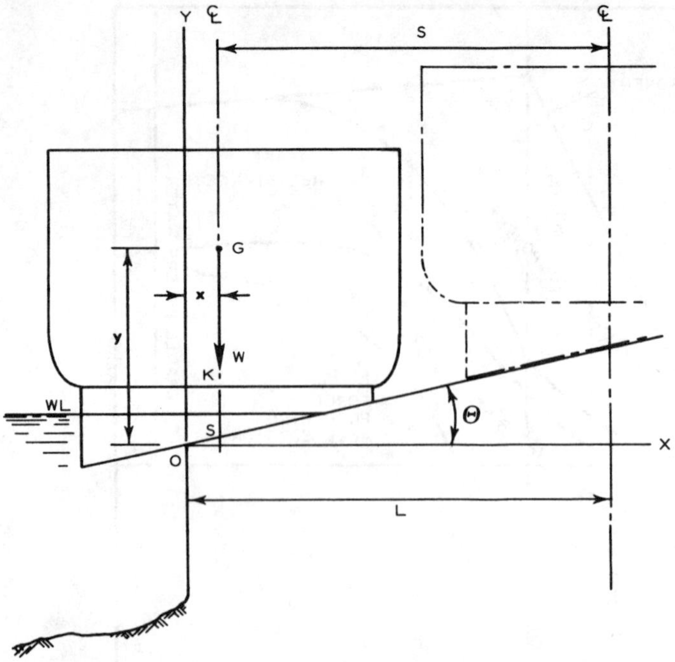

Fig. 25 Phase 1—sliding

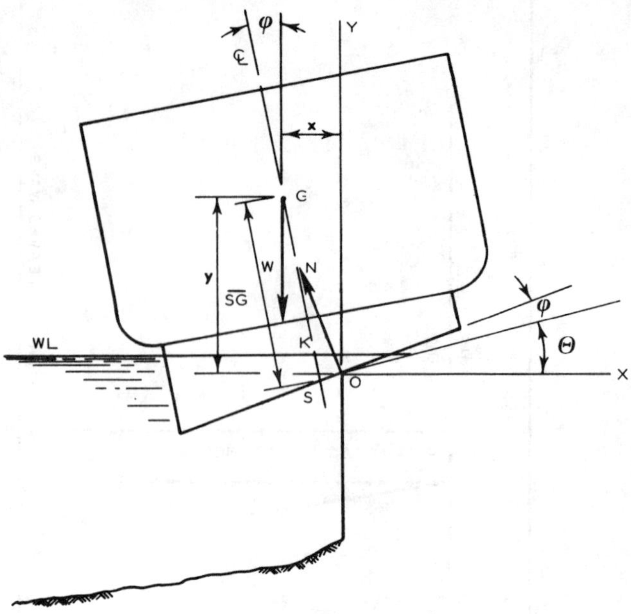

Fig. 26 Phase 2—tipping

Long underwater ground ways, or some compromise of submerged length, may be desirable for very large and heavy vessels.

The following applies to launchings where straight ground ways terminate at or near the water's edge and the sled-supported vessel tips about the ground way ends.

10.2 Calculation of Travel, Velocities, Accelerations, Heel Angles and Resultant Forces. Vessel movement is divided into the following four phases (Semyonov-Tyan-Shansky, undated):

Phase 1. Sliding. Linear translation of the vessel parallel to and down the ground ways from release to start of tipping about the ground way ends.

Phase 2. Tipping. Start of tipping until the hull first touches the water.

Phase 3. Tipping and Immersion. From when hull first touches the water until sleds leave the way ends, resulting in the end of tipping.

Phase 4. Dropping. Bodily dropping after sleds leave the way ends and tipping ceases. Heaving, rolling and translation until the vessel's motion ceases.

a. Phase 1—Sliding. Referring to Fig. 25, velocity along the ways, as for end launchings, is:

$$V = \sqrt{2 \times g \times S \times cos\theta\ (tan\theta - f_s)} \tag{21}$$

or:

$$V = \sqrt{2 \times g \times (L - x) \times cos\theta\ (tan\theta - f_s)}$$

where:

V = velocity along ways = $\dot{s}$
S = distance slid along way
L = initial distance of vessel's center of gravity from way ends.
x = distance of vessel's center of gravity from way ends.

b. Phase 2—Tipping. Referring to Fig. 26 the following assumptions are made for this and following phases:

$$sin\theta = tan\theta = \theta$$

$$sin\phi = tan\phi = \phi$$

$$cos\theta = cos\phi = 1$$

where:

ϕ = inclination of vessel about an axis through its center of gravity.

The origin is at the top surface of the outshore end of the ground ways.

The following formulas apply to Phase 2:

$$x = \dot{s}t = vt \tag{26}$$

where

x = horizontal distance of vessel's center of gravity outshore of origin.
S = distance slid from end of Phase 1.
t = elapsed time from end of Phase 1.
$\dot{s}$ = v = velocity at end of Phase 1.

$$y = x\ (\theta + \phi) - \overline{SG} \tag{27}$$

where:

y = vertical distance of vessel's center of gravity from origin.
$\overline{SG}$ = height of vessel's center of gravity, measured along centerline, above undersides of sleds.

$$\phi = \frac{gk_\phi}{2(\dot{s})^2}\left(\frac{\dot{s}t}{k_\phi} - arc\ tan\ \frac{\dot{s}t}{k_\phi}\right) \tag{28}$$

where:

ϕ = heel angle of vessel

$k_\phi = \sqrt{\frac{I_\phi g}{W}}$ = radius of gyration of vessel and sleds about a longitudinal axis through the vessel's center of gravity.

$\dot{s}$ = vessel's velocity at end of Phase 1.

I_ϕ = moment of inertia of mass of vessel and sleds about a longitudinal axis through the vessel's center of gravity.

W = weight of vessel and sleds.

t = elapsed time from end of Phase 1.

$$\dot{\phi} = \frac{g}{2\dot{s}} \frac{\left(\frac{\dot{s}t}{k_\phi}\right)^2}{1 + \left(\frac{\dot{s}t}{k_\phi}\right)^2} \tag{29}$$

where:

$\dot{\phi}$ = angular velocity.

$$\ddot{\phi} = \frac{g}{k_\phi} \frac{\frac{\dot{s}t}{k_\phi}}{\left[1 + \left(\frac{\dot{s}t}{k_\phi}\right)^2\right]^2} \tag{30}$$

where:

$\ddot{\phi}$ = angular acceleration.

$$N = \frac{W}{\left[1 + \left(\frac{\dot{s}t}{k_\phi}\right)^2\right]^2} \tag{31}$$

where:

N = ground way end support during tipping.

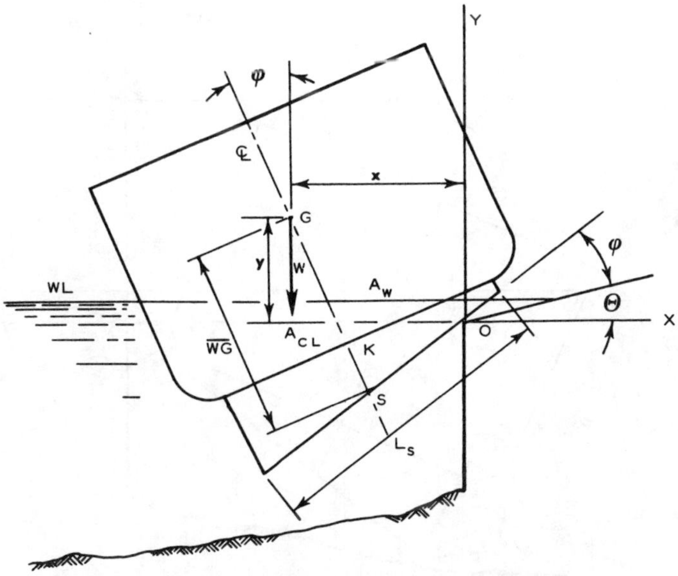

Fig. 27 Phase 3—tipping and immersion

c. *Phase 3—Tipping and Immersion.* Referring to Fig. 27, the equations of motion can be derived from those of Phase 2 if account is taken of water pressure on the immersed portion of the vessel. The effect of buoyancy and water resistance is to reduce downward velocity. The maximum angle of heel is of special interest as capsizing may result or the sled outshore ends may strike bottom.

The following formulas apply to Phase 3.

$$\dot{x}_m = \frac{\dot{x}_2}{2}\left[1 + \frac{e^{-n(x-x_2)}}{1+k}\right] \tag{32}$$

where:

$\dot{x}_m$ = mean horizontal velocity.

x_2 = horizontal distance traveled during Phase 2.

$\dot{x}_2 \approx \dot{s}$.

$n = \frac{C_x \times \rho \times g \times A_{CL}}{4 \times W\,(1+k)}$

$k = \frac{\lambda_x g}{W}$

C_x = a non-dimensional resistance coefficient = 1.00.

$\lambda_x = \frac{2}{3}\rho\,\nabla\,\frac{A_{CL}}{A_W}$ = coefficient of added mass for horizontal motion component.

A_W = waterplane area.

A_{CL} = area of centerline plane below water.

e = 2.718 = natural logarithm base.

∇ = displacement volume.

ρ = mass density of water.

W = weight of vessel and sleds.

At the start of Phase 3.

$$t = 0$$
$$x = x_2$$
$$\dot{x} = \dot{x}_2$$
$$A_{CL} = 0$$
$$\lambda_y = 0$$

where:

$$x = x_2 + \dot{x}_m t \tag{33}$$

x = horizontal distance traveled from start of Phase 2.

x_2 = horizontal distance traveled during Phase 2.

x_m = horizontal distance traveled during Phase 3.

t = elapsed time during Phase 3.

$\dot{x}_m$ = mean horizontal component of velocity during Phase 3.

$$\dot{\phi} = \frac{g}{2\,\dot{x}_m} \frac{(1-a_1)\tau^2 - a_2}{\left(1 + \frac{gb}{2\dot{x}_m}\right)\tau_1^2 + 1} \tag{34}$$

where:

$\dot{\phi}$ = angular velocity during Phase 3.

$\dot{x}_m$ = mean horizontal component of velocity during Phase 3.

$$a_1 = \frac{\rho \times g \times \nabla}{2 \times W}$$

∇ = displacement volume.

$$\tau = \frac{x}{k_\phi} = \frac{x_2}{k_\phi} + \frac{\dot{x}_m t}{k_\phi}$$

$k_\phi = \sqrt{\dfrac{I_\phi g}{W}}$ = radius of gyration of vessel and sleds about a longitudinal axis through the vessel's center of gravity.

x_2 = horizontal distance traveled during Phase 2.

x_m = horizontal distance traveled during Phase 3.

$$a_2 = \frac{x_2{}^2}{k_\phi{}^2} \frac{\gamma\nabla}{2W} \left[\frac{2\, C_y\, \dot{x}_m \dot{\phi}_2}{g} - 1\right] \quad (35)$$

where:

C_y = a nondimensional resistance coefficient = 1.00.

$$b = \left[\frac{C_y\, \rho\, \nabla + 4\, \lambda_y}{W}\right] \dot{x}_m$$

$\lambda_y = \dfrac{1}{2} \rho\, \nabla \dfrac{A_W}{A_{CL}}$ = coefficient of added mass for vertical component of motion.

$$\phi = \frac{g\, k_\phi}{2\, \dot{x}_m{}^2} \left\{\left[1 - \frac{1}{2}(a_1 + b\dot{\phi}\right] \tau - \left[1 - \frac{1}{2}(a_1 + b\dot{\phi} - a_2)\right] arc\ tan\ \tau\right\} + \phi_2 \left(1 - \frac{\dot{s}_1{}^2}{\dot{x}_m{}^2}\right) \quad (36)$$

Phase 3 motion is calculated as follows. Using equation $y = x\,(\theta + \phi) - \overline{SG}$ for one value of x and several equal incremental values of ϕ, the attitude of the vessel is determined, it being remembered that sleds and ground way ends remain in contact. Values of A_W, A_{CL} and ∇ are calculated and a set of curves is plotted for the value of x with A_W, A_{CL} and ∇ as ordinates and ϕ as abscissa. Similar calculations and plots of A_W, A_{CL} and ∇ are made for increasing values of and angles of ϕ.

At the start of Phase 3, $A_W = A_{CL} = \nabla = 0$ and $\dot{x}_m = \dot{S}$. ϕ is calculated for the first assumed value of x. Values of A_w, A_{cl} and ∇ are taken from the plotted curves for x and ϕ and substituted in formulae for b, a_1, a_2, λ_x and λ_y. Further approximations can be similarly made for the first value of x. Values of $\dot{x}_m$, $\dot{\phi}$ and ϕ can be determined similarly for other values of x and corresponding times.

At the end of Phase 3

$$x_3 = \frac{L_s}{2} + \overline{GS}\, \phi_3 \quad (37)$$

where

x_3 = distance of center of gravity outshore of ground way ends.

L_s = length of sled bottom sliding surface.

$\overline{GS}$ = vertical distance from center of gravity to sled underside.

ϕ_3 = heel at end of Phase 3

d. Phase 4—Dropping. As illustrated in Fig. 28, the sleds clear the ground way ends and the vessel moves transversely while heaving and rolling.

The following formulas apply to Phase 4.

$$w_y = \sqrt{\frac{\rho \times g \times A_W}{\dfrac{W}{g} + \lambda_y}} \quad (38)$$

where:

w_y = heaving frequency $= \dfrac{2\pi}{T_y}$

W = launching weight $= \rho \times g \times \nabla = \Delta$

λ_y = coefficient of added mass for vertical motion $= \dfrac{\rho \times \nabla \times A_W}{2 \times A_{CL}}$

$$T_y = \frac{2\pi}{w_y} = 2\pi \sqrt{\frac{\dfrac{W}{g} + \lambda_y}{\rho \times g \times A_W}} \quad (39)$$

where:

T_y = heaving period

$$y_2 = \sqrt{(y_3 - y_0)^2 + \left(\frac{\dot{y}_3}{w_y}\right)^2} \quad (40)$$

where:

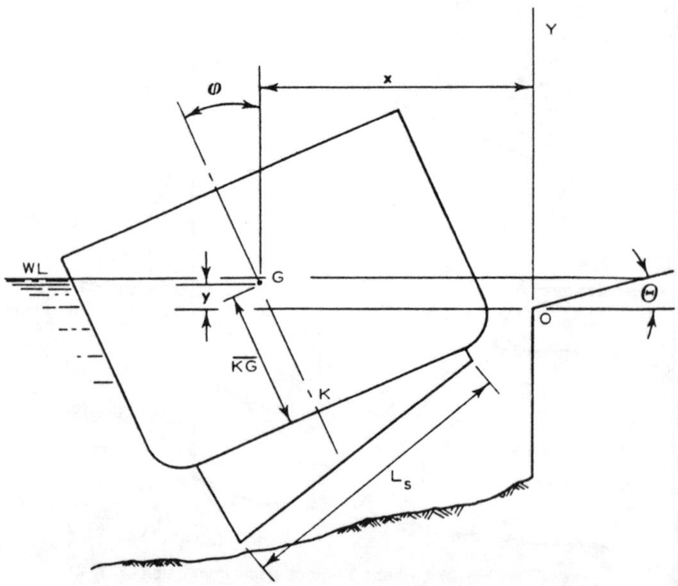

Fig. 28 Phase 4—dropping

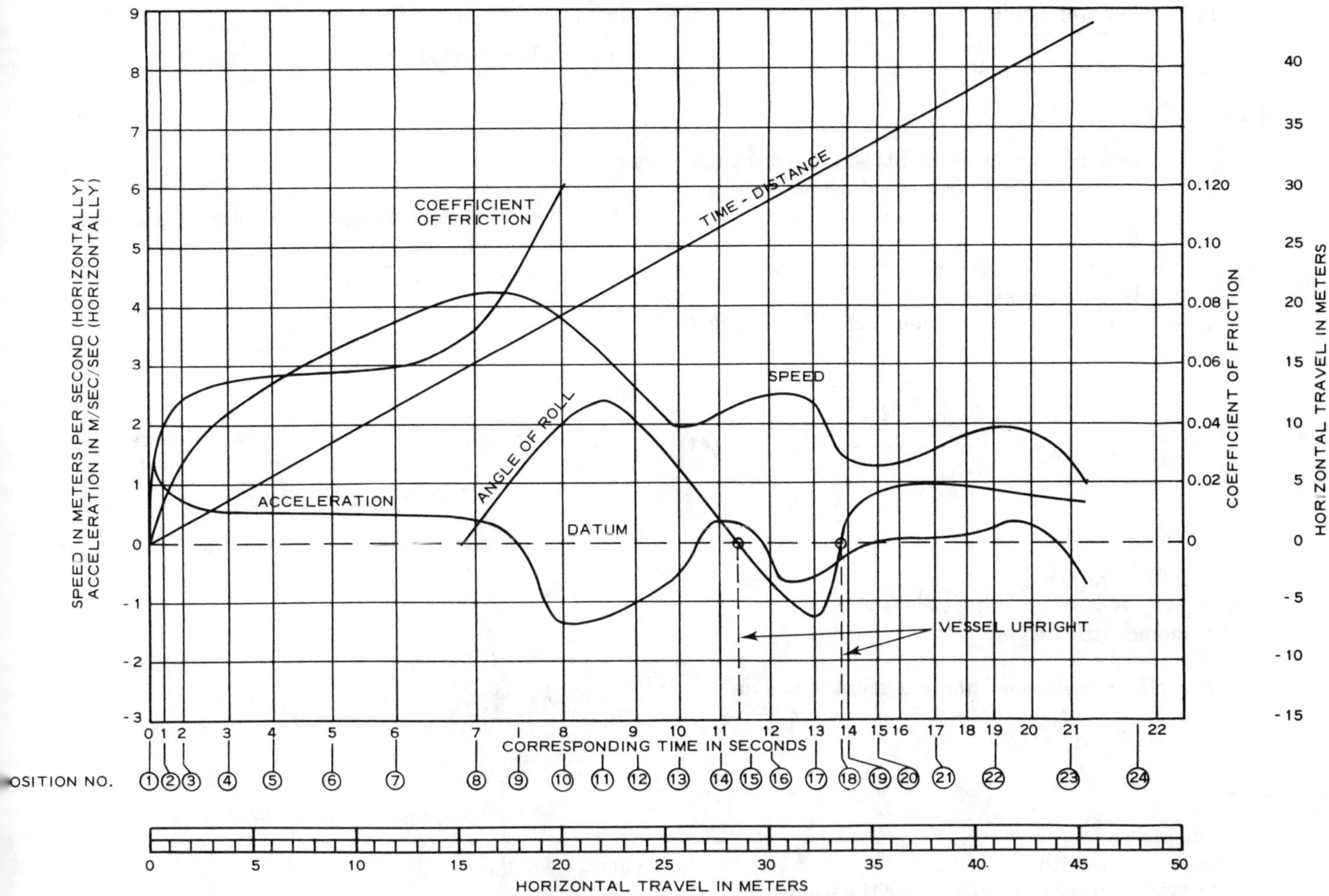

Fig. 29 Observed data

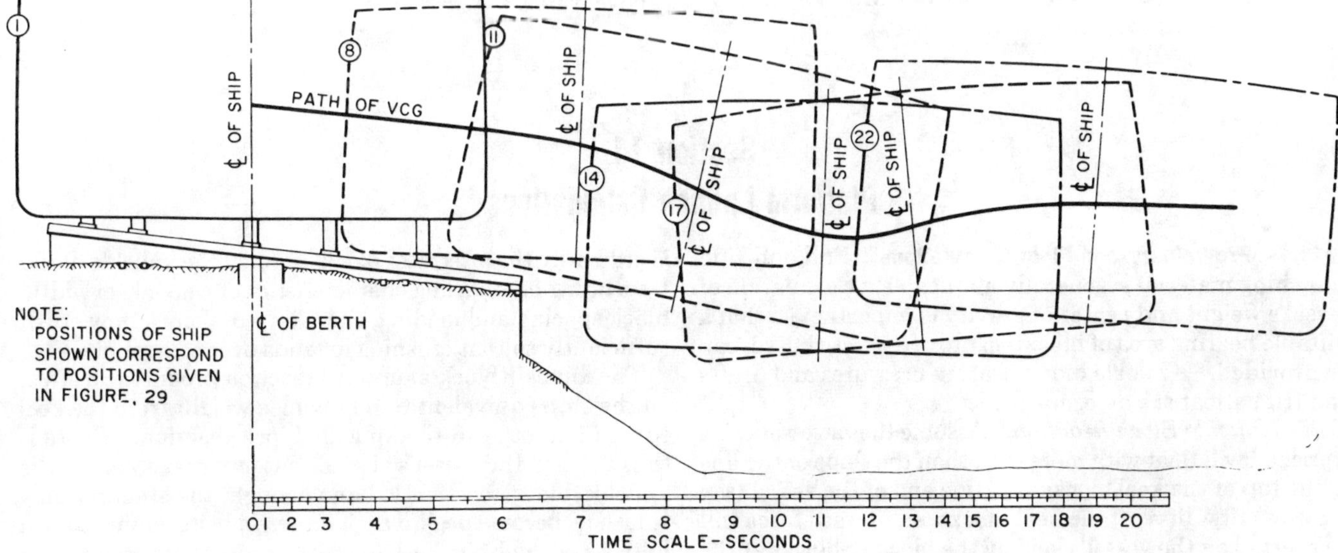

Fig. 30 Observed trajectory

y_a = heave amplitude

$$y_{max.} = 2\sqrt{(y_3 - y_0)^2 + \left(\frac{\dot{y}_3}{w_y}\right)^2} + y_3$$

where:

$y_{max.}$ = maximum downward displacement of vessel's center of gravity below waterline.
y_3 = ordinate of vessel's center of gravity at end of Phase 3.
y_0 = ordinate of vessel's center of gravity with launched vessel at rest.
$\dot{y}_3$ = vertical velocity component at end of Phase 3.

$$\omega_\phi = \sqrt{\frac{W\,\overline{GM}}{\frac{Wk_\phi^2}{g} + \lambda_\phi}} \tag{41}$$

where:

ω_ϕ = rolling frequency
$W = \rho \times g \times \nabla$ = launching weight = Δ
$\overline{GM}$ = metacentric height

λ_ϕ = coefficient of added mass moment of inertia

$$\approx \frac{\Delta}{12g}(B^2 + 4Z^2)$$

where:

Δ = displacement
B = vessel's breadth
Z = vessel's center of gravity above the waterline.

$$\mathrm{T}_\phi = \frac{2\pi}{\omega_\phi} = 2\pi\sqrt{\frac{\frac{Wk_\phi^2}{g} + \lambda_\phi}{W\,\overline{GM}}} \tag{42}$$

where:

T_ϕ = rolling period

$$\phi_{max.} = \sqrt{\phi_3^2 + \frac{\dot{\phi}_3^2}{\omega_\phi^2}} \tag{43}$$

where:

$\phi_{max.}$ = maximum heel angle at $\frac{T_\phi}{2}$ seconds after start of Phase 4.
ϕ_3 = heel at end of Phase 3.

$$x_{max.o.b.\phi} = x_3 + \frac{1}{k_4} ln\left(k\dot{x}_3\frac{T_\phi}{2} + 1\right) \tag{44}$$

where:

$x_{max.o.b.\phi}$ = horizontal distance of vessel's center of gravity from way ends at time of maximum heel outboard.

$\dot{x}_3$ = horizontal component of velocity at end of Phase 3.
T_ϕ = rolling period

$$k_4 = \frac{C_y\frac{\rho}{2}A_{CL}}{\frac{W}{g} + \lambda_y}$$

$$x_{max.i.b.\phi} = x_3 + \frac{1}{k_4} ln\,(k\dot{x}_3T_\phi + 1) \tag{45}$$

where:

$x_{max.i.b.\phi}$ = horizontal distance of vessel's center of gravity from way ends at time of maximum heel inboard.

Figs. 29 and 30 show observed data and trajectory for a typical side launching with ground way ends terminating above water.

Section 11
Platform Launch Calculations

11.1 Preliminary and Final Calculations. End and side launching material is generally applicable. Calculation of vessel's weight and center of gravity is important so that a suitable bearing area of blocking, properly distributed, can be provided. Knuckle block bearing pressures and drafts and trim afloat are determined.

a. Knuckle Block Reaction. Assume that a vessel being launched will float with more trim than the slope of the line of the top of the keel blocks. As one end of the vessel (say the bow) first lifts off the keel blocks, an upward force will be exerted on the vessel's keel by the block or blocks at the other end of the line of keel blocks. The end block, which is ordinarily the aftermost block, is called the knuckle block. Depending on crushing characteristics of wood, keel width, block spacing and loading, the aftermost block may crush sufficiently so that crushing extends over several blocks.

The knuckle block(s) upward reaction produces an effect on the ship equivalent to removing a weight from the keel equal in amount to the knuckle block reaction. There is thus a rise in the vessel's virtual center of gravity and if the knuckle block reaction is large enough, the stability may vanish or become negative. The vessel being launched will initially be unable to heel due to the support of the outboard blocking. However, as the launching, or undocking, progresses and one end of the vessel lifts off the keel blocks, clearance between the lower turn of the bilge and outboard blocking will increase so that a significant list can occur and persist until the knuckle block reaction has diminished sufficiently for the metacentric height to have become positive.

The knuckle block pressure can be determined as follows. The moment of the vessel's weight about the knuckle keel block is calculated. This constant value is equal to the weight of the vessel multiplied by the lever from the knuckle block to the vessel's center of gravity (longitudinal center of buoyancy in the afloat condition). The moment of buoyancy about the knuckle block and displacement are calculated for the displacement afloat and with the vessel trimmed to the slope of the keel blocks. Other moments of buoyancy and displacement are determined for waterlines parallel to and above and below the first waterline. With moment as the ordinate and displacement as the abscissa, the moment of weight is plotted as a straight horizontal line. A second plot is made to show the curve of moments of buoyancy for various displacements.

The intersection of the moment of weight line and moment of buoyancy curve gives the vessel's displacement at the instant of one end of the vessel lifting off the keel blocks. The difference between this displacement and displacement afloat is the knuckle block reaction. Knuckle block reactions for conditions as the launching progresses to floatoff can be determined from the plot.

High knuckle block reactions and excessive loss of stability due to knuckle block reactions can be reduced by reducing the angle between the underside of the keel and the top of the keel blocks at the moment of one end of the vessel lifting off one end of the keel blocks. In some cases the top of the keel block line can be inclined relative to the platform, the platform itself may be inclinable or the vessel can be ballasted to reduce trim. The optimum condition would be one in which the inclination of the line of the top of the keel blocks at floatoff equaled the trim of the keel with the vessel afloat.

b. Transverse Stability. The upward knuckle block reaction produces a rise in the vessel's virtual center of gravity equal to (Andrews, 1967):

$$\overline{GG}_v = \frac{\overline{KG} \times R}{\Delta - R} \tag{46}$$

where:

$\overline{GG}_v$ = rise in virtual center of gravity.
$\overline{KG}$ = center of gravity above keel with vessel afloat.
R = knuckle block reaction.
Δ = displacement of vessel afloat.

c. Blocking. Vessels launched from platforms may be supported entirely on the cradle used for moving the vessel horizontally onto the platform. In case of need, the cradle support can be supplemented by blocking and cribbing on centerline and outboard or conventional blocking may provide all of the support.

The weight and longitudinal center of gravity being known, as is the location and longitudinal extent of blocking or other supports, the load distribution on the blocking can be calculated and will usually be trapezoidal with vertical scale of weight per unit length. In the case of a long overhang with no blocking, the load distribution may be triangular. Distributions of loading on blocking can be determined more closely if a weight curve is available for the vessel being launched.

d. Design Data. The following general design information is presented:

• Maximum average block loading for keel and side blocking is 23.6 MPa (20 tons per sq ft) and for knuckle block(s) is 57.5 MPa (32 tons per sq ft).

• Maximum trim difference between keel of vessel afloat and top of keel blocks is 0.012 Lpp/MTC (0.030 Lpp/MTI) but should not exceed 0.01.

where:

MTC = moment to trim one centimeter
MTI = moment to trim one inch
L_{pp} = length between perpendiculars

• Keel height above dock floor should be 1.8 m (6 ft).

Section 12
Launching Tests

12.1 Purpose. Model and full scale test data are required on which to base calculations for future and completed launchings. From model tests the motions of both end-launched and side-launched vessels can be determined by launching a properly ballasted model into a water-filled model launching basin.

Full scale tests include the following:

• *Launch Grease Temperature and Pressure Resistance.* This can best be observed at an actual launching. For pre-launch evaluation of different makes of grease, or evaluation of the effect of different pressures and temperatures, a section of ground way can be greased and loaded by means of a sliding way section placed on the grease and ballasted to give the desired unit pressure on the grease. Tests should be performed in an enclosure where temperature can be controlled.

• *Launch Grease Coefficients of Starting and Sliding Friction.* Coefficients can best be determined by analysis of launching data or by reference to reported data.

• *Burn-off Release Strength.* The resultant force down the ways can be determined if the number of broken segments and the ultimate strength of one segment are known.

A test specimen representing a segment of burn off plate, the segment being bounded on either side by a drilled hole, can be tested in a tensile testing machine.

• *Drag Resistance.* Chain or other drags can be pulled over broken up ground, or other surface, by a crane or locomotive, the resistance being measured by a dynamometer.

• *Wood Crushing Resistance.* Resistance of fore poppet or other crushing packing can be measured in a testing machine. Test specimen widths and thicknesses should be to scale.

• *Current Velocity.* Launchings should be at slack water, if such exists, to avoid slewing on the ways of end-launched vessels and to facilitate the post-launch handling of end, side and platform-launched vessels. Launching basin water movement can be averaged by means of a buoyed line of length equal to the maximum draft of the vessel to be launched, the line having a weight at its lower end and baffles along its length. Repeated observations of the buoy will indicate the interval between slack water and high water or other reference event. Wind, river stage, and point on tidal cycle should be considered.

• *Tidal Height.* Actual tide heights and times before launching should be observed or recorded by a recording gage for comparison with predicted heights and times based on tide tables.

• *Water Depths.* Availability of sufficient water depth to avoid the possibility of the vessel or its cradle touching the ground should be verified sufficiently far in advance of the launching to permit any necessary dredging.

Section 13
Instrumentation and Equipment

13.1 Purpose. The instrumentation and equipments described can be used at end launchings, and as applicable at side launchings, to provide technical data for monitoring of the launching operation as well as to provide a data base for the launching of future vessels.

13.2 Installed Devices. The following devices may be installed on the ground or on ways, cradle or in the ship.

• *Distance-time Recorder.* For analysis of the launching it is desirable that distance slid-time data be recorded from release until the vessel is at rest. A chronograph is a suitable instrument for recording these data. The chronograph can consist of a steel drum with a single layer of piano wire wound on it. One end of the wire is secured to the inshore end of the cradle, the chronograph being temporarily secured to the inshore end of a ground way. As the vessel moves down the ways the drum rotates as the wire unwinds. The drum shaft is reduction geared to a circular horizontal platen with blank paper chart. A pen near the perimeter of the chart draws a curved line which is deflected at one second intervals by a synchronous motor. The pen trace can be spiral, rather than circular, so that after the chart has rotated a full turn, the pen trace will not run over itself. Drum diameter and reduction gearing ratio being known, the travel of the launching cradle for any angular rotation of the chart can be calculated. Differentiation of the plotted distance-time curve gives the velocity curve and differentiation of the velocity curve gives the acceleration curve. In lieu of the circular paper chart with the spiral jogged pen line around its perimeter, a strip chart can be used. For a launching in which the final motion is checked by slewing, the latter part of the chronograph data may require adjustment as the outshore end of the vessel may be stopped while the inshore end is swinging and unreeling the chronograph wire. The paper chart can be marked to record events such as drop off, start of chain drag movement and so on or a separate pen can draw a line which is jogged when a contact is made to record an event.

• *Visual Observations.* If a chronograph is unavailable, vertical marks a known distance apart can be painted on the vessel's side, the passage of the marks past a fixed point near the water's edge being timed by one or more observers or recorded by a motion picture camera. After the vessel is apparently at rest, a visual signal can be made from the vessel for two or more shore observers to sight through transits and obtain the bearings of bow and stern at the waterline, masts or other suitable reference points. Transit locations being known, the vessel's final position relative to the building slip and its prelaunching position can be plotted and the distance slid measured for comparison with the chronograph distance-time record.

• *Tide Gages.* These can include a recording eight-day tide gage and painted board indicating gages, one of which should be a master gage with another similar gage being temporarily installed at the launching site.

• *Cameras.* Motion picture and/or still cameras can be used to provide photographic records before, during and after the launching. Motion picture records are particularly valuable for the recording of vessel motion during side launching of models and full size vessels.

• *Creep Gages.* A creep gage consists of a graduated scale and a pointer, one component being fixed to the ground way ribband and the other to the sliding way. Four or more creep gages can be installed on each side of end launching ways to show the local movement or creep of the sliding ways relative to the ground ways as the blocking and shoring are removed.

• *Vessel Creep and Settling Gages.* On each side of the vessel at amidships and just above the turn of the bilge a paper or sheet metal target is installed. Each target has two intersecting scales, one parallel to the keel and one perpendicular to the keel. On each side of the vessel a ground observer sights through a transit initially fixed on each target's scale intersection. Each line of sight should be perpendicular, insofar as practicable, to the vessel's centerline plane. As blocking and shoring are removed, the vessel's bodily creep outshore and bodily settling can be observed and recorded. If one side of the vessel settles more than the other side, a possibility of unequal wedging up or settling of the ground ways exists. Similar installations can be made at bow and stern.

• *Strain Gages.* Strain gages, electrical recording or indicating or mechanical indicating, can be installed on the main deck near amidships or elsewhere to measure the elastic deformations occuring during blocking removal, hogging, sagging and with the vessel afloat.

• *Hog and Sag Wire.* To the extent permitted by superstructure and deckhouses, a piano wire can be stretched the length of the vessel, above the weather deck, one end of the wire being fixed and the other end passing over a sheave and attached to a suspended weight. At amidships a vertical batten is fixed so that relative movement of batten and wire will indicate hull deflection during blocking removal, hogging, sagging and with the vessel afloat.

• *Accelerometer.* A recording accelerometer can be used to record acceleration. Pivoting requires that a correction be made as does checking of the vessel's motion by slewing. In lieu of, or as a check on the accelerometer, an oil-damped pendulum can be suspended to show by its angular displacement in a plane paralled to centerline the acceleration to at least the start of pivoting. For side launchings two recording accelerometers can be installed at the calculated position of the vessel's center of gravity. One accelerometer would be oriented to record accelerations athwartships and parallel to the base plane. The other accelerometer would record accelerations perpendicular to the base plane.

• *Pivoting Indicator.* The elapsed time from release to start of pivoting should be observed and recorded. Pivoting can be indicated by movement of the bubble in a spirit level as the outshore end of the vessel lifts, or by a contact maker on the fore poppet completing an electrical circuit to actuate a visual or audible signal on deck. The contact maker can be operated by the initial crushing of a crushing strip fore poppet or by the initial rotation of a rocker fore poppet.

• *Crushing Indicators.* For crushing strip fore poppets, a telescoping crushing indicator can be installed at each corner between the wedge rider and the solid packing above the crushing strips. Each indicator is marked just before release. Fore poppet crushing causes telescoping of the indicators which can be read when the fore poppet is placed ashore after hauling and partial dismantling of the cradle.

Similar devices can be used to determine bodily settling of a vessel at any point or points under the flat or approximately flat bottom where it is desired to know the total settling from start of blocking removal to launching.

Crushing indicators can be used for platform launchings to record maximum amount of knuckle block crushing.

• *Roll Recorder.* A gyroscopic roll recorder that will plot angle of vessel's inclination against elapsed time can be used at side launchings. The data obtained can be compared with motion picture and model test data.

• *Trim Change Indicator.* For a platform launching, it is useful to know the instant that one end of the vessel lifts from the keel blocking. Assuming that there is no change in the angle, if any, between water surface and base plane of the vessel as the two surfaces move away from each other, liftoff at one end will be accompanied by a change in trim which can be noted by observing a carpenter's water U tube disposed longitudinally on the weather deck of the vessel. A similar water U tube can be disposed athwartships to indicate amount of, and change in, heel.

An alternate, and for some types of platform a more reliable method, is to extend a vertical piano wire from platform deck to the vessel's deck at bow and stern. The lower end of each wire is fastened to the platform. The upper end of each wire is fastened to a visual and/or audible indicator which operates when the vessel's end moves upwards relative to the fixed wire. In lieu of this arrangement, an electrical contact maker can be fitted at end keel blocks. As the vessel lifts off the keel blocks at one end or the other, or at both ends, one or both contact makers operate to actuate visual and/or audible signals.

• *Pressure Cells.* Pressure cells, transducers and recorders can be used to record outshore way-end pressures.

Section 14
Launch Observations

14.1 Introduction. Launching observations may commence several days before the launching and continue until the vessel has been secured at the outfitting pier. Observations may indicate some failure in operations or material that may be correctable for that particular launching and also provide data which, properly interpreted, may enhance the reliability and reduce the cost of future launchings.

14.2 End and Side Launch Observer Assignments. Five observers, with whatever assistants are considered necessary, with the following stations and observation assignments can take the data for a typical burn-off sole-plate end launching. End-launch observer duties can be modified as required for launchings that will take place at yards that customarily side-launch vessels.

a. Way Head Observer. Record number of collapsible keel blocks, number of grease irons, number of tumble shores, make of base coat and slip coat, number of times base coat was previously used, use of slip coat outshore. Record inshore creep gage readings. Operate chronograph, record elapsed times from release to pivoting and to ship at rest. Record burn-off sole plate release time and record number of segments broken. Record condition of slip and base coats following launching and location of pivoting if visible on launching grease.

b. Port Ground Observer. Record number of port bilge cribs, total number of port cradle wedges, number of port fore poppet wedges, direction and velocity of current off way ends at release, wind direction and velocity at release, weather, creep gage readings, elapsed times to outshore and inshore ends of cradle entering water, maximum settling of port side of vessel. Record start and completion times of following events: Wedge driving rallies, ram rail removal, shore removal, bilge crib removal and keel block removal.

c. Starboard Ground Observer. Record number of starboard bilge cribs, total number of starboard cradle wedges, number of starboard fore poppet wedges, air temperature under vessel, launching grease temperature, maximum settling of starboard side of vessel, creep gage readings, dog shore force gage readings. Record elapsed times from release to chain drags starting and stopping, distances chain drags moved and height of tide at release. Record start and completion times of following events: grease iron removal, grease protection strip removal, wedge driving rallies, ram rail removal, shore removal, bilge crib removal, keel block removal, and dropping of dog shores. It is assumed that chain drags are only on the starboard side of the vessel.

d. Ship Deflection and Draft Observer. Record hog and sag wire batten readings before rallies, after removal of shores, after removal of bilge cribs, after removal of keel blocks, at maximum hogging before pivoting, at maximum sagging during pivoting, and after vessel is afloat. Record drafts and density and temperature of water at outfitting berth.

e. Ship Pivoting and Log Observer. Record following elapsed times from release: pivoting by bell indicator and visual sight, vessel dead in water. Record tug names. Record clock times for following events: last chain drag cable slipped, tugs fast to vessel, vessel's bow at cradle hauling pile cluster, hauling line dropped over pile cluster, start and completion of release of cradle tricing lines and backstays, vessel hauled clear of cradle, vessel's bow enters outfitting slip, first mooring line to pier, start and completion of release of aft poppet and aft sub-poppet tricing lines and backstays and vessel secured in final position. Record estimated weight of tools, welding machines, welding cables, air hoses, temporary lighting and ventilation, staging and ladders, raw materials, etc. Record fore poppet crushing indicator readings following retrieval of fore poppet.

f. Other Observers. Additional observers will be required if accelerometer, roll recorder (for side launchings) and strain gages are installed. If release is by triggers, rather than by burn-off sole plates, then two trigger observers will be required to record and plot forces acting on triggers.

14.3 Platform Launch Observations. The simplicity of a platform launching is reflected by the relatively few observations which are required to assure that the launching plan is being followed. Observations include the following with applicability as indicated.

- Monitoring of movement of vessel onto platform and of deballasting of platform to support vessel's weight. (Movement of vessel is not usually involved in a graving dock launching.) Platform deballasting may or may not be required in the case of a buoyancy-supported platform.
- Monitoring of deballasting of platform to float free of any ground support. (Applies only to buoyancy-supported platforms). Recording of air and water temperatures and extent of solar radiation.
- Monitoring of ballasting as platform is submerged. (Applies only to buoyancy-supported platforms).
- Monitoring of submergence of platform as indicated by draft marks and alignment of platform as indicated by optical sights and, for a buoyancy-supported platform, as shown by indicating or recording strain gages.
- Monitoring of deflection of vessel's hull girder by means of optical sights.
- Observation of first lifting of one end of keel off keel blocks.
- Observation of lifting of both ends of keel off keel blocks.

Section 15
Launch Preparations, Crew and Schedule

15.1 Introduction. This section describes the preparations and personnel required for cradle installation and the actual launching itself and also presents a launching event schedule. End launchings are principally considered in view of their relatively greater complexity.

15.2 Launching Preparations. Inshore ground ways are installed under the vessel, aligned, blocked, and spur shored, unless permanently installed ground ways are already in place. Ways are greased and grease irons are placed in grooves cut in the base coat. Brush base coated sliding ways are skidded into place to rest on the grease irons. Wedge riders and wedges are installed as are amidships blocking and end poppets. Top surfaces of blocking and poppet heads must conform to the contours of the shell plating and there should be a minimum of slack to be removed during wedging up. On the other hand, there must be enough clearance so that deformation of the hull due to solar radiation or settling will not pinch one or more grease irons between ground and sliding ways and make grease iron removal a difficult and time consuming operation.

Shortly before launching, the grease irons are hauled and wedges are then driven to refusal by ram crews using battering rams. With all of the vertical slack removed from the cradle, which now carries part of the vessel's weight, ram rails, shores, bilge cribs, and keel blocks are expeditously removed. The cradle now carries all of the vessel's weight, unless tumble shores are installed. It now remains only to drop the dog shores and then release the vessel by burning the way head sole plates or releasing the triggers to let the vessel go.

In the case of a side launching on trapezoidal-shaped sleds and if a vessel translation system exists, the ground ways can be greased and the launching sleds installed before the vessel is moved into launching position. This is in contrast to the laborious piece-by-piece assembly of an end-launching cradle.

15.3 Launching Crew. Installation of the cradle, wedging up and removal of shores, cribbing, and blocking is the responsibility of the Carpenter Department. Launching personnel are divided into ram crews which drive the wedges in their assigned sections of cradle and remove vessel supporting arrangement in accordance with the schedule.

Ship checking arrangements are the responsibility of the Rigging Department. Following are responsibilities of departments who have personnel aboard the vessel during the actual launching: Riggers to handle lines and release cradle rigging, Tank Testers to inspect compartments, Outside Machinists to inspect sea valves and the Safety Department for first aid.

After the vessel is afloat, the launching master aboard the vessel directs tug operations, cradle hauling, aft poppet dropping and final securing of the vessel.

Section 16
Post Launch Calculations

16.1 Purpose. Following a launching, calculations are made and analyzed and recorded data studied with the following aims:

- Verification of satisfactory launching arrangement design, material performance, scheduling and launching operations
- Identification of any unusual results that are desirable for future launchings or that should be avoided in future on account of their undesirability
- Determination of performance quality of any innovative launching design features or procedures
- Acquisition of data useful for future launchings for comparison with data from former launchings, data assumed for prelaunch calculations and data obtained during model launching tests
- Accurate determination of the weight and longitudinal center of gravity of the launched vessel, and hog or sag of the hull girder afloat.

The foregoing principally applies to end launching but with some modifications can be applied to side launchings. For platform launchings the principal post launching calculations are those for displacement, longitudinal center of gravity and hog or sag.

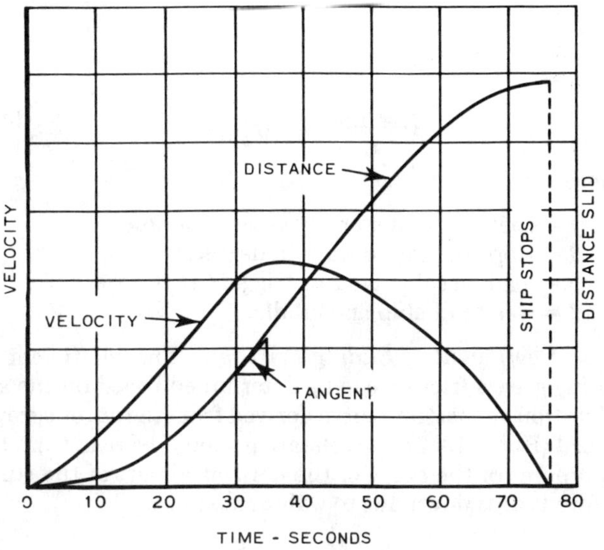

Fig. 31 Distance and velocity—time curves

16.2 Calculations. These include calculation and plotting of time-distance and time-velocity curves, Fig. 31, and time-acceleration curve, Fig. 32, calculation and plotting of displacement and longitudinal center of gravity, and calculation and plotting of displacement curve versus distance slid for actual depth of water over the ground way ends. They also include determination of way-end and pivoting pressures and distribution and distance slid to start of pivoting and floatoff or drop off, determination of distances slid by chain drags, and calculation of coefficients of initial or starting grease friction and sliding grease friction and preparation of force diagram.

a. Velocity and Acceleration. The time-distance curve slopes are determined graphically for equal intervals of time and are plotted as a curve of velocity versus time. Similarly, the time-velocity curve slopes are determined and are plotted as a curve of acceleration versus time. The second graphical differentiation is not very precise and there may be considerable scatter in calculated acceleration points. If precise acceleration data are required, gyro-stabilized recording accelerometer(s) should be used. As the velocity of the vessel is initially and finally zero, the tangents at the ends of the time-distance curve are horizontal. Maximum velocity is at the point of inflection of the curve. At the point of inflection of the time-velocity curve the acceleration curve crosses the base line and becomes negative. The area between the positive acceleration curve and the base line must equal the area between the negative acceleration curve and the base line.

b. Coefficient of Starting Friction. The coefficient of starting or initial grease friction can be determined if the strength of the broken segments of sole plates, or the load on the triggers at release, is known (Robb, 1952). The following formula is used.

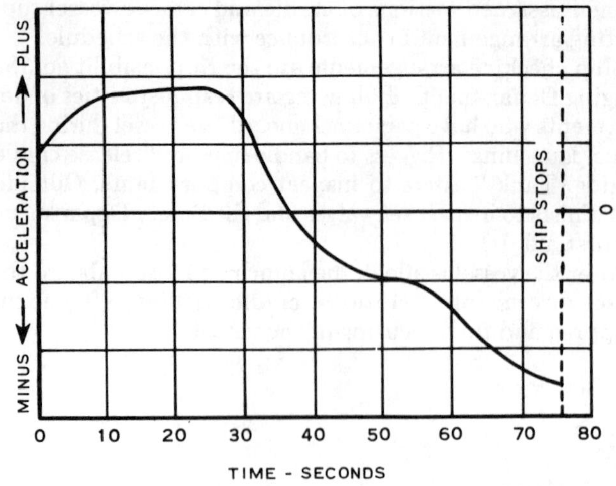

Fig. 32 Acceleration—time curve

$$f_I = tan\theta - \frac{F_T}{W\,cos\theta} \qquad (47)$$

where:

f_I = coefficient of starting grease friction
θ = slope of ground ways in degrees
F_T = load on sole plates or triggers at release
W = weight of ship and cradle.

c. Coefficient of Sliding Friction. The coefficient of sliding grease friction can be determined based on the acceleration versus time curve derived from the chronograph record (Robb, 1952). Acceleration should be read from the high point of the curve or the horizontal part of the curve before it dips down due to water resistance.

$$f_s = tan\theta - \frac{a}{g \times cos\theta} \qquad (48)$$

where:

f_s = coefficient of sliding grease friction
a = acceleration of ship under influence of resultant of component of weight down the ways and grease friction force up the ways
g = acceleration due to gravity.

d. Starting Acceleration. Acceleration at the instant of release cannot be determined from the time-distance data. However, by equating the product of mass of the vessel and acceleration to component of weight down the ways minus the starting grease friction force up the ways, the starting acceleration can be calculated.

$$a_i = \frac{g \times F_T}{W} \qquad (49)$$

where:

a_i = initial acceleration.

e. Force Diagram. The following forces make up the force diagram. Refer to Fig. 24, Section 9 End Launch Calculations.

$F_1 = W\,sin\theta$ = component of weight down ways
$F_2 = B\,sin\theta$ = component of buoyancy up ways
$F_3 = f_s\,(W - B)$ = component of grease sliding friction force up ways
$F_4 = \rho\,(0.63A_C + 0.004A_{WS})v^2/2$ = retarding water resistance
$F_5 = f_D W_D$ = checking resistance.

where:

W = weight of vessel.
B = buoyancy.
f_s = coefficient of sliding grease friction.
f_D = coefficient of chain drag friction.
W_D = weight of chain drags.

All of the above forces can be calculated with at least reasonable accuracy except possibly the water resistance. However, the curve bounding the top of the water resistance area can be determined. The curve defining the lower edge of this area can be obtained by multiplying the known accelerations by the vessel's mass. Effect of added mass due to hull and cradle entrained water can be allowed for or can conservatively be ignored. Distances slid to cradle entering the water, stern entering the water, start of pivoting, floatoff or drop off, chain drags starting and fully effective, and chain drags and vessel stopped are known or can be determined. A realistic force diagram can be thus drawn.

Table 13—Typical Schedule of Principal Events Immediately Preceding an End Launching.

CLOCK TIME	EVENT
10:30 to Noon	Pull grease irons and place in racks. Remove No. 3 row of shores and install ram rails.
Noon to 12:30	Lunch period.
12:30	All personnel stand by for first wedge rally.
12:40 to 12:50	First rally. Four blows on each wedge.
12:50 to 13:00	Second rally. Four blows on each wedge.
13:00 to 13:05	Rest period.
13:05 to 13:15	Third rally. Four blows on each wedge.
13:15 to 13:25	Fourth rally. Two blows on each wedge.
13:25 to 13:35	Fifth rally. One blow on each wedge. Remove ram rails.
13:35 to 13:45	Ram Crew No. 3 embark to cross river to assist with cradle hauling. Ram Crew Nos. 6 and 7 remove first row of shores, starting at first white shore aft and working forward. Ram Crew Nos. 4 and 5 remove second row of shores, starting at first white shore aft and working forward.
13:45 to 13:55	Ram Crew Nos. 8 and 9 remove first row of shores, starting at first red shore amidships and working forward. Ram Crew Nos. 1 and 2 remove second row of shores, starting at first red shore amidships and working forward.
13:55 to 14:05	Ram Crew Nos. 6 and 7 remove fourth row of shores, starting at first white shore aft and working forward. Ram Crew Nos. 4 and 5 remove fourth row of shores, starting at first red shore amidships and working forward.
14:05 to 14:10	Rest period.
14:10 to 14:30	Ram Crew Nos. 1, 2, 4 and 5 remove bilge cribs on signal for each crib. Start amidships and work

	forward. Ram Crew Nos. 6, 7, 8, 9 and 10 remove bilge cribs on signal for each crib. Start amidships and work aft.
14:30 to 14:40	Ram Crew Nos. 1, 2, 4 and 5 remove marked keel blocks and stack on port side of keel track.
14:40 to 14:55	Ram Crew Nos. 6, 7, 8, 9 and 10 remove remaining keel blocks starting at block marked for each crew.
15:00	Release by mechanical triggers.

References

Andrews, Harrison B. (1967), "Launching," *Principles of Naval Architecture,* edited by John P. Comstock, SNAME.

Forest Products Laboratory (1974), *Wood Handbook,* U. S. Department of Agriculture, U. S. Government Printing Office.

Keith, Henry H. W. (1939), "Launching," *Principles of Naval Architecture,* edited by Henry E. Rossell and Lawrence B. Chapman, SNAME.

Robb, Andrew McCance (1952), *Theory of Naval Architecture,* Charles Griffin and Co., Ltd., London.

Saunders, Harold E. (1957), *Hydrodynamics in Ship Design,* SNAME, New York.

Semyonov-Tyan-Shansky, V. (date unknown) *Statics and Dynamics of the Ship,* Peace Publishers, Moscow (Four Continent Book Corporation, New York).

CHAPTER XVIII

Robert L. Jack

Trials and Preparation For Delivery

Section 1
Testing

1.1 Introduction. A few weeks before delivery, after many months of planning, procuring tons of materials, and expending thousands of man-hours involving many trades, the shipbuilding contractor is faced with one of the most critical periods of construction. For it is during this relatively short span of time that the many diverse activities associated with final testing and delivery must be systematically scheduled so all contract requirements can be completed in a timely manner. A few loose ends at this juncture can result in costly delays in delivery.

It is the intent of this chapter to discuss those activities that are associated with the delivery process and the fulfillment of guarantee obligations and to suggest procedures that have been found helpful by some shipbuilders and owners. There are a number of references to other material where more detailed information and instruction may be found.

For obvious reasons, in the design of a ship's structure and its machinery systems, reliability is a primary consideration. In keeping with this philosophy, the testing during all stages of construction should be extensive and thorough. In addition to the owner, there are several regulatory bodies which also must witness the testing procedures and be assured that their numerous requirements are being met.

1.2 Regulatory Body Requirements. The American Bureau of Shipping (ABS) is a ship classification society which performs the primary service of certifying the soundness and seaworthiness of merchant ships and other marine structures. ABS establishes standards known as *Rules* for the design, construction, and periodic survey of vessels, as discussed in Chapter VII. By applying these internationally accepted Rules, ABS classes ships, that is, assures that ships are fit for their intended service.

The men who work for a classification society are collectively called *surveyors*. Individually they are naval architects, marine engineers, metallurgists, computer specialists, and men with experience as seagoing engineers. They represent all the essential skills, knowledge and technology required by a ship classification society. Because shipping is an international business, ABS technical and field surveyors, numbering about 640, are stationed around the world at major seaports and in shipbuilding, material, and machinery production centers. Surveyors stationed in the shipyard during ship construction are responsible for insuring that the Bureau's requirements in regard to ship design and construction, as well as testing procedures, may be found in its "Rules for Building and Classing Steel Vessels," published annually. In addition, there are more than 20 other publications covering special ship designs and areas of interest such as "Requirements for Certification of the Construction and Survey of Cargo Gear on Merchant Vessels."

The safety features of the ship for the protection of its personnel are primarily of concern to the U. S. Coast Guard. These would include the vessel's structure, the lifeboats and davits, general alarm and fire fighting systems, as well as all major mechanical and electrical components and systems of the vessel. As discussed in Section 6.1 of Chapter XV, all of the responsibilites on behalf of the U. S. Government are assigned to the Coast Guard which has delegated certain of these functions to the American Bureau of Shipping. The installation and test requirements of the Coast Guard may be found in the "Code of Federal Regulations (CFR), Title 46—Shipping" Chapter I—"Coast Guard, Department of Transportation." Chapter I comprises eight volumes containing 17 subchapters, the titles of which are listed in Table 1.

The Federal Communications Commission (FCC) must approve the installation and witness the testing of the radio communication equipment. FCC's regulations in this regard may be found in "Code of Federal Regulations (CFR), Title 47—Telecommunications," Part 83—"Stations on Shipboard in the Maritime Services."

Compliance with sanitation and ratproofing standards is under the authority of the U. S. Public Health Service, whose requirements are set out in USPHS Publication No. 393, "Handbook on Sanitation of Vessel Construction." Included are regulations regarding fresh water storage and piping systems, evaporator performance, sanitary drains, food storage and preparation spaces, as well as standards for ratproof construction.

Certification that the construction and tests meet the standards of each of the above regulatory bodies is a requirement for all ships built in the United States. The lack of such certification is generally justification for the purchaser to refuse to accept delivery of the vessel. In other

Table 1—Subchapters of Chapter I of Code of Federal Regulations Title 46—Shipping

Sub-chapter	*Parts*	*Title*
A	1–9	**Procedures Applicable to the Public** Organization, functions, and methods of the USCG re inspections and investigations.
B	10–16	**Merchant Marine Officers and Seamen** Licensing of officers and certification of seamen.
C	24–26	**Uninspected Vessels** Requirements for equipment and operation of small motor vessels under 15 gross tons.
D	30–40	**Tank Vessels** Requirements, inspection, and certification of tank vessels. Lifesaving & fire-fighting equipment.
E	42–46	**Load Lines** Load line requirements for foreign and domestic voyages.
F	50–64	**Marine Engineering** Requirements, inspection, & certification of boilers, pressure vessels, piping systems, and machinery equipment.
G	66–69	**Documentation and Measurement of Vessels** Provisions for documentation and measurement.
H	70–8-	**Passenger Vessels** Design, construction, inspection, and certification. Provisions for subdivision, stability, lifesaving and fire protection equipment.
I	90–105	**Cargo and Miscellaneous Vessels** Design, construction, inspection, and certification. Special provisions for nuclear vessels.
J	110–113	**Electrical Engineering** Requirements for electrical systems, including emergency lighting and power, communications, and alarms.
N	146–148	**Dangerous Cargoes** Regulations for storage or transportation of explosives, combustibles, and other dangerous substances.
O	151–154	**Certain Bulk Dangerous Cargoes** Unmanned barges.
P	157	**Manning of Vessels** Manning requirements
Q	160–164	**Specifications** Requirements for lifesaving, electrical, and engineering equipment.
R	166–168	**Nautical Schools** Designation and approval of nautical school ships.
T	175–187	**Small Passenger Vessels** Design, construction, inspection, certification, operation, manning, and licensing of vessels under 100 gross tons.
U	188–196	**Oceangraphic Vessels** Design, construction, inspection, and certification.

countries, there are similar regulatory agencies whose requirements are mandatory for vessels built there. Additional regulatory requirements will be discussed where pertinent in later sections of this chapter.

1.3 Published Testing Procedures. The Society of Naval Architects and Marine Engineers (SNAME) published the "Code on Shop and Installation Tests of Merchant Marine Propulsion and Auxiliary Equipment," Technical and Research (T&R) Bulletin No. 3-8, in 1960. The stated objective of this code is to provide the industry with a standard test procedure for machinery and thereby

- Simplify the preparation of ship's specifications and provide a uniform basis for bids.
- Reduce the uncertainty of the manufacturer and the shipbuilder as to test obligations.
- Remove from current shop and ship test practices many items which can no longer be considered to add to the quality of the vessel.

The U. S. Maritime Administration (MarAd) developed and published in 1964 its "Supplementary Procedure for Testing Machinery," which included test requirements for equipment and systems not covered by SNAME's T&R Bulletin No. 3-8, such as ventilation and air conditioning, boat winches, mechanical hatch covers, propellers, bow thrusters and stabilizers. It also added to SNAME's requirements for a number of tests, including a short circuit test for electrical circuit breakers. A year later, in 1965, "Addendum No. 1 to Maritime Administration Supplementary Procedure for Testing Machinery, Central Control Systems Testing" was issued to cover the testing of the remote control and mechanized operation systems that were beginning to be installed at that time on the majority of new ships.

Since the time of the publication of these directives, there have been significant changes in machinery installations aboard ships, including higher powers, more sophisticated mechanization, as well as radical new kinds of cargo and types of cargo handling systems. These procedures, therefore, are sorely in need of revision and supplementation. As this publication goes to press, SNAME Panel M-19 is in the process of revising and updating Bulletin No. 3-8 as well as incorporating the pertinent portions of MarAd's Supplementary Procedure and Addendum No. 1 thereto. Additional testing procedure information will follow.

1.4 Test Schedule and Test Memoranda. Early in the construction stage, a test schedule is developed listing all components and systems subject to test. Most yards divide these tests into three groups of numerical sequences, the 100 series for mechanical and piping systems, 200 series for electrical and electronic equipment, and 300 series for hull tests. Altogether, the total number exceeds 100 tests, with many having multiple subdivisions. The hydrostatic pressure test for piping, for example, generally involves more than 50 independent systems. Table 2 is an abbreviated example of a test schedule which illustrates the manner used in breaking down the extensive testing program into finite and manageable increments.

Estimated dates are established for each item to provide a logical sequence of testing and to ensure that the dates for trials and delivery can be met. This schedule is a very useful instrument for management, serving as a production gage with multiple bench marks for the determination of construction progress.

Table 2—Example of Test Schedule

Test No.	*Title*	*Date Scheduled*	*Date Completed*
100	Piping Systems Pressure Test		
101 to 115	Engine Room Pumps & Systems		
116	Main Shafting & Gear Alignment		
117	Steam Generating Plant		
118	Forced Draft Blowers		
119	Distilling Plant & Pumps		
120	Sewage Treatment Plant		
121	A/C Refrigeration Plant		
122	Calibration of Gages & Therms.		
123	Air Compressors & System		
124	Combustion Controls		
125	Centralized Throttle Control		
(Plus approximately 25 more tests in this series)			
201	Ship's Service Generators		
202	Emergency Diesel Generator		
203	Meter Calibration		
204	Circuit Breakers		
205	Transformers		
206	Insulation Resistance		
207	Lighting System		
208	Commissary Equipment		
209	Echo Depth Sounder		
210	Radio Equipment		
211	Interior Communications		
(Plus approximately 20 more tests in this series)			
300	Hull, Tank & Compartment Testing		
301	Cargo Hatch Covers		
302	Ventilation & A/C		
303	Accommodation Ladder & Winch		
304	Life Boats & Davits		
305	Cargo Gear		
306	Hydraulic Doors		
307	Anchor Windlass		
308	Mooring Winches		
(Plus approximately 15 more tests in this series)			

A test memorandum or procedure, is prepared for the test of each component or system based on the requirements of the specifications and those of the regulatory bodies, as well as incorporating the recommendations of the manufacturer and the desires of the owner. These memoranda include such information as:

- List of applicable drawings and documents.
- Basic design data of equipment or system.
- Pre-test inspection procedure.
- Detailed step by step procedure for conducting the test.
- Tabulation of data to be recorded together with blank data sheets.
- Post-test inspection procedure.

After approval by all concerned, these agenda should become the undisputable authority for the manner of conducting the tests.

1.5 Piping, Electrical, and Component Tests. As the piping installations are completed, the systems are cleaned and flushed and then hydrostatically tested prior to the application of lagging. The test pressures are generally 150 percent of the working pressure or 207 kPa (30 psi) for suction piping. Working pressure is the pressure at which the relief valve is set, or the highest pressure obtainable in service, for instance, the shutoff pressure of a centrifugal pump. All equipment in the system such as heaters, coolers, strainers, etc. which are subjected to the system pressure should be tested with the system. A detailed description of the recommended procedure for testing piping systems can be found in Section 2.40 of MarAd's 1964 "Supplementary Procedure for Testing Machinery."

Similarly, all electrical cable must be tested after installation for continuity of circuits and resistance of insulation.

As the installations of the various machinery and electrical systems are completed, testing is carried out in accordance with the test memoranda described above in Section 1.4. A typical memorandum would include the following steps:

- Pretest inspection of installation for proper foundations, accessibility for maintenance, and damage.
- Record cold insulation resistance of motor and controller.
- Determine actual setting values of all safety and control devices.
- Demonstrate system in all modes of control.
- Operate system at design capacity, if possible, for extended period, recording all pertinent data. Check unit for deficiencies such as vibration, overheating of bearings, and capacity.
- At conclusion of run, record hot insulation resistance of motor and controller.
- Post-test inspection, including disassembly if operation suggests damage or defective parts.

1.6 Dock Trials. After all pertinent component and system tests have been completed, the main propulsion system is subjected to a low power test while the ship is restrained at the dock. For a steam-turbine-propelled vessel, a spin test will have been performed wherein the shaft is uncoupled, and the turbine operated at no load to set the overspeed governor and to check out the turbine for vibration. After recoupling the shaft, the propulsion system is run under load at the maximum power attainable at the dock. This power level is usually restricted by the strength of the moorings or that of the pier itself. The risk of damaging propellers from debris and ingestion of mud through suction sea chests should be considered. Also, the wash from the propeller may cause scouring around the pilings of the pier. For these reasons, care should be taken in establishing the maximum rpm for the dock trials.

Since it is obviously desirable to conduct the dock trials at powers as high as possible, all yards devise means to exploit any natural advantages available to them. Some yards situated on rivers can use the river current as a counter action, and tugs on the offshore side of the ship have been used on occasion. Factors unique to each construction site establish the power limitation which varies widely from yard to yard. At one facility, it is possible to develop thrusts as high as 1,779,000 N (400,000 lb) which translate into powers of about 16,000 shp. This is unusual, and the maximum power levels at the dock for most yards is less than 5,000 shp.

The dock trials are a major step in the testing program, being the culmination of component testing and an essential preliminary for sea trials. They should always be considered to be, and conducted as, an integrated plant test and in no way a substitute for the thorough prerequisite testing of individual components and systems.

The dock trials provide an excellent opportunity to demonstrate under safe but realistic operating conditions the performance of various mechanized control and safety devices for the boilers and main engine. The remote controls for the throttle valve in both the engine room and bridge consoles can be adjusted and calibrated over the lower rpm ranges. The boiler combustion controls likewise can be given an initial adjustment, and by operating first one boiler and then the other, the range can be extended significantly.

All safety devices activated by such abnormalities as high and low boiler water level, high and low steam pressure, flame-out, and forced-draft fan failure can be simulated and tested. With a few exceptions, all remote temperature and pressure indicators can be calibrated. On vessels so equipped, the dock trial provides an excellent opportunity to check out, adjust, and calibrate controllable-reversible-pitch propellers.

If the dock trials are comprehensive and properly performed, few adjustments other than final calibration of the throttle and combustion controls will need to be made when the ship goes on sea trials. Otherwise, many costly hours can be wasted at the beginning of the sea trials checking out control devices and instrumentation that could have been more efficiently and economically serviced at the dock.

Section 2
Stability Test

2.1 Requirements and Objective. The International Convention for Safety of Life at Sea, 1974, the International Convention on Load Lines, 1966, and the U. S. Coast Guard require that the first ship of every class be subjected to a stability test, or inclining experiment, to determine the light displacement and center of gravity. If the following ships prove to be *sisterly,* the data from the first is considered applicable to all others. If any ship is later modified, another inclining may be required if the effect on total weight or center of gravity is in doubt.

The regulations for these tests are stated in the Code of Federal Regulations, Title 46 "Shipping" Parts 31, 74, 93, 100, and 179. Supplementing these requirements, the U. S. Coast Guard has prepared Navigation and Vessel Inspection Circular No. 1-67, "Stability Test—Preparations and Procedures." As the title indicates, the purpose of this circular is to minimize the unnecessary and costly delays due to inadequate preparation and improper procedures. Similar information may be found in Inter-Governmental Maritime Consultative Organization (IMCO) Resolution A-267.

An excellent explanation of the elementary principles of stability, as well as the basic theory of the inclining experiment, can be found in Moore (1967).[1] Also included is a brief description of the various steps in performing the test and some guidance for the preparation of the test report.

2.2 Condition of the Vessel for Inclining. The vessel should be complete or nearly complete as defined by the estimated list of weights to complete, to deduct, and to relocate. Acceptability of the condition of the vessel depends on the accuracy with which each such weight and its center of gravity can be determined, and the expected cumulative effect on the test results.

The vessel should be clean, with all loose gear, scaffolding, scrap, and debris removed. Bilges should be dry and decks free of liquid. Boilers, wet machinery, and piping should have liquids at operating level. No work should be in progress aboard ship during the test, and there should be a minimum number of personnel on board.

U. S. Coast Guard Circular No. 1-67 specifies in detail the requirements that must be met in regard to the condition of the ship and its tanks. Since the test is generally conducted just prior to sea trials, slack tanks containing fuel oil and fresh water are unavoidable, but there are restrictions as to the number and configuration of such tanks. Full tanks must be pressed up, 100 percent full, with no pockets or voids caused by inadequate venting. Empty tanks must have a final stripping using portable pumps.

The test may be aborted due to adverse weather conditions. Strong gusty winds can cause irregular heeling moments, and rough water makes it difficult to determine the drafts accurately. Excessive amounts of snow or standing water may be cause for cancellation of the test.

The total test weight should be sufficient to produce an inclination of not less than one degree for merchant ships. The weights should be compact and of such a configuration that the center of gravity can be readily determined. The weights and handling equipment should be such as to permit rapid movement of weights in order to minimize inaccuracies due to changing weather or current conditions.

The U. S. Coast Guard requires that the inclination be measured by three pendulums installed in areas protected from the wind. The pendulums should be as long as practicable, and in no case should the deflection at maximum inclination be less than 153 mm (6 in.). This corresponds to about 9 m (30 ft) for one degree of inclination.

---

[1] Complete references are listed at end of chapter.

2.3 Conduct of Test and Preparation of Report. The stability test must be conducted under the supervision of the U. S. Coast Guard and in accordance with Circular No. 1-67, which provides detailed instructions for the movement of weights and the recording of data. In preparing the report, the first step is to calculate the displacement, longitudinal center of gravity, and vertical center of gravity of the ship as inclined, following the principles as outlined by Moore (1967). These values must then be corrected for the weights to be added, to be removed, and to be relocated to bring the ship to the standard" *light ship* condition. The effect of these weight changes on the center of gravity are calculated by the summation of the moments of these weights about a common reference plane.

The Coast Guard has developed a booklet CG-993 (Forms for Stability Test Report), complete with instructions for the preparation of the stability report from the recording of raw data to the final determination of weight and centers of gravity of the prescribed light ship condition. *Light ship* is defined as the ship complete in every respect with water in boilers at steaming level and liquids in machinery and piping, but with all tanks and bunkers empty and no passengers, crew, cargo, stores or baggage on board.

2.4 Trim and Stability Booklet. Having determined the light ship characteristics, the displacement, trim and stability can be calculated for any condition of loading by employing the basic principle of the summation of moments. This tedious task, which must be performed for the departure and arrival condition of each voyage, is simplified by the *Trim and Stability Booklets* prepared by the shipbuilding contractor.

In addition to calculations of typical ship loading conditions, these booklets contain tables listing capacities, centers of gravity, and weights of each tank when containing various liquids. Included are helpful charts, graphs, and tables such as hydrostatic characteristics and deadweight scales, minimum metacentric height requirements, and a draft diagram for estimating hydrostatic properties. Generally, blank forms are furnished with instructions and examples for making either an approximate or detailed analysis of the effect of any cargo loading distribution. Some ships are equipped with special instrumentation or computing aids to calculate hydrostatic stability.

On container ships, this chore is usually accomplished by computer ashore as the containers are loaded aboard ship. Many large ships have computers aboard that given instant values of not only drafts and stability but also of structural bending moments and sheer at the transverse bulkheads in response to inputs of tank ullages. An obligation of the shipbuilder is to furnish the computer manufacturer with the information needed to produce an instrument which recognizes the characteristics of the particular ship.

2.5 Certificate of Deadweight and Stability Letter. For all ships which do not require inclining, deadweight surveys must be made to verify that such ships are identical to the inclined sister ship in regard to light ship displacement and trim. These surveys are conducted in much the same manner as the stability test with the exception of the movement of weights and reading of pendulums.

The ship must be in a nearly finished condition and clean of extraneous material. Drafts are carefully determined at multiple stations and a complete inventory is made of weights to be added, weights to be removed, and those to be relocated. From these data, the true light ship weight can be determined. These data and calculations constitute the *Certificate of Deadweight* which is one of the documents required to be furnished by the shipbuilder at delivery.

After witnessing and approving either the stability test or the deadweight survey, the Coast Guard issues a *Stability Letter* stating that on the basis of such test, the ship is approved for operation when loaded in accordance with the limitations imposed by the approved Stability Booklet. The Stability Letter is a required delivery document and must be framed and permanently mounted in the pilot house.

Section 3
Drydocking

3.1 General. Ship design must include considerations of drydocking, especially for naval vessels and large merchant ships, and special internal structure is often provided to resist docking loads. The ship's position on the dock is determined in advance, and the vessel must be ballasted and trimmed to suit the operation. The complete drydocking operation is the responsibility of the dockmaster who is aided by the ship's docking plan which gives all necessary dimensions, location of appendages and drains, irregularities in the keel line, and offsets for the construction of side piers, if required.

3.2 Pre-trial Drydocking. Most construction specifications require the ship to be drydocked not more than 20 days prior to sea trials and again before delivery if in the water more than four months after sea trials. On the first ship of a class, it is essential that the bottom be clean and uniformly coated, and the propeller free of all fouling, if meaningful correlation between the standardization trials and model tests is to be achieved. While the 20 day limitation is adequate for most yards, particularly those located in colder regions or on fresh water, this restriction may be too liberal for ships outfitted in warm salt water, such as along the Gulf of Mexico where the fouling rate is amazingly high. For vessels built in such areas, it is advisable to drydock immediately before trials, especially those ships that are to be utilized for standardization and fuel economy trials.

In areas of low fouling rates, pretrial drydocking is not absolutely essential for follow-on ships of a class. Furthermore, drydocking after trials has the advantage of permitting an inspection for damage to the bottom or propeller that may have been incurred during trials. It also affords an opportunity to correct underwater deficiencies detected during trials, such as faulty fathometer transducers and leaking stern bearing seals.

3.3 Types of Drydocks. There are four general types of drydocks for handling large vessels:

- Graving docks
- Continuous wing wall floating drydocks
- Sectional floating drydocks
- Marine elevator drydocks.

In many modern shipyards there are also building basins and launching systems that serve the dual function of transferring the newly constructed ship from land to a waterborne condition but also serve as a drydock prior to delivery when the ship is again lifted from the water to provide access for final delivery preparations. These dual purpose types of drydocks have been described both in Chapter XVI and in Chapter XVII.

Section 4
Sea Trials

4.1 Introduction. When the vessel is substantially complete, and after successful completion of dock trials and other tests, to be reasonably sure that all machinery installed will perform satisfactorily, trials will be conducted in the open sea and deep water to demonstrate designed power, speed, and rpm capabilities. Sea trials can be classified into three broad categories as follows:

a. Standardization Trials. These are speed trials which establish the relationship of speed to power and propeller rpm, thus providing a measure of performance of the ship and its propulsion system; they also provide a comparison with model test predictions. Standardization trials are usually conducted on only the first vessel of a multiple ship contract.

b. Economy and Endurance Trials. These trials consist of carefully conducted runs of several hours duration at sea for the accurate determination of power and fuel consumption. Since they require special and sophisticated instrumentation, economy trials are usually conducted on only the first of a class of ships. For those vessels that follow, endurance trials of about four hours duration are conducted to verify that the power plant can perform for that period at full power and design conditions without difficulty.

c. Maneuvering Trials and Special Tests. These include miscellaneous tests to demonstrate the maneuvering capability of the ship. Included in this category are:

- Emergency reversals from ahead to astern and vice versa to establish stopping times and distances as well as the ability of the machinery system to withstand unusual operating demands
- Steering test (hard over to hard over) to prove the capability of the steering gear
- Turning circles, Z-maneuver, and spiral maneuver to demonstrate the responsiveness of the ship to the rudder
- Astern operation (30 minutes) to demonstrate capability for sustained operation astern
- Anchor windlass test to prove adequacy of the anchor handling equipment.

Special ship designs require special tests or trials to demonstrate their unique capabilities. A test of the constant-tension towing winch is often required for seagoing tugs, and harbor tugs may be given a bollard strain test. Fishing boats must prove their ability in handling nets and other fishing gear.

The material of this section has been developed by the most part for trials of a steam turbine propelled merchant ship. However, the principles are equally applicable to gas turbine or diesel propelled vessels.

All sea trials, regardless of scope, should be conducted in accordance with SNAME's Technical and Research (T&R) Code C-2, "Code for Sea Trials," a comprehensive treatise on all facets of the subject, from early planning to the preparation of the final report. Some helpful suggestions regarding the conduct of trials may also be found in a SNAME Marine Technology paper, "Sea Trials: Some Recommended Practices" (Jack, 1973).

For most cargo ships and tankers, this optimum trial date has been found to be about two weeks before delivery. The tests should be scheduled so that they are not compromised by unfinished work: they should be followed by a scheduled construction period that will provide time for the correction of normal trial deficiencies. All testing of components and systems associated with the trials should be complete, as well as the calibration of all instrumentation. Habitability systems, such as heating, air conditioning, plumbing and commissary, should be fully operational if not completely tested. Work items that can be deferred until after trials include the installation of furniture, floor coverings, draperies, and the final cleaning and painting, in fact, cost considerations may dictate this.

4.2 Pre-Trial Preparation. Too much emphasis cannot be placed on the importance of advanced planning for successful trials, particularly those of the first ship of a class. It is generally inadequate preparation in this area rather than poor performance of equipment that produces questionable or unacceptable results. For a fossil fueled, steam turbine ship, these preparations should include the following:

a. Fuel Rate Correction Factors. In the determination of the weight of fuel burned per horsepower-hour, the rate as measured must be corrected for off-design conditions of excessive steam pressure and temperature, condenser vacuum, generator load, evaporator output, propeller rpm, and domestic steam usage. The fuel to be burned on trial must be analyzed and the heating value corrected to standard fuel heat content, density, and chemical composition. Detailed instructions for the calculation and use of these correction factors are presented in SNAME's Technical and Research Bulletin No. 3-17, "Recommended Practices for Correcting Steam Power Plant Plant Trial Performance." Curves for these correction factors should be prepared and submitted to the purchaser for review and approval 60 days prior to trials.

b. Special Instrumentation. The first ship of a class requires special instrumentation, including a torsionmeter and shaft revolution counters for horsepower determination, two calibrated fuel meters for measuring fuel consumed, thermocouples for measuring superheater outlet steam temperature, and an absolute pressure gage for the determination of main condenser vacuum. Depending upon the arrangement of the fuel-oil piping system, a diesel-propelled ship may require a dozen or more calibrated fuel meters in order to determine the total net amount of fuel consumed by the main engines and generators. Radiometric navigational equipment, such as *Raydist,* is now universally used for conducting standardization trials and for the plotting of turning circles and ahead and astern reaches during quick reversals. The shipbuilder must obtain much of the above instrumentation from outside sources, and the torsionmeter must be tailored to fit the propeller shaft. Requisitions for these items must therefore be prepared well in advance of the trials.

c. Pre-Trial Conference. Several weeks before trials of a first-of-class ship, a conference should be held between the shipbuilder and purchaser to review the proposed trial agenda and to agree on the many details associated with the manner of conducting the various events. This can be done by correspondence, but direct discussion between knowledgeable parties can accomplish far more with less effort. There are many questions that should be resolved such as what instrumentation is to be used for various measurements, how fuel samples are to be taken, the time intervals for recording data, and where and in what water depths the standardization trials are to be conducted. All verbal agreements should be documented by minutes or memoranda.

d. Data Recording. Well organized data sheets can greatly simplify the tedious task of collecting engine room data. It is desirable that readings from every gage, thermometer, and indicator be recorded during the economy or endurance trials. To facilitate this recording, the items on the data sheets should be grouped, not by systems, but according to physical location of the instruments in the engine room. Each instrument should be temporarily numbered, with the numbers appearing on the data sheets. Finally, each person recording data should become familiar with the instruments of his responsibility prior to the beginning of the trials. Special instructions should be given regarding the reading of critical instruments such as fuel meters, thermocouples and the torsionmeter.

e. Fuel Analysis. A preliminary analysis of the fuel should be made from a representative sample taken during the loading process. The resulting values of heat content, density, and chemical composition should be used for preliminary calculations of fuel rate attained during trials. Final calculations should be made using the results of the analysis of the sample taken during the trials of the fuel as burned.

f. Instrument Calibration. All temperature indicators should be calibrated prior to or during dock trials. Pressure gages should be calibrated using a dead weight testing stand or certified gages. The torsionmeter should be calibrated within 24 hours prior to trials. In far too many cases, the potential value of the trials is never realized because the data are suspect and the instrument readings are of questionable reliability.

Sea trials represent a very expensive effort on the part of the shipbuilder. At no other time during the life of the ship is there an opportunity to test the equipment to its maximum capability and to have the performance fully documented by a technical staff using special instrumentation. Every effort should be made, therefore, to ensure that the results are as complete and accurate as possible. Proper pretrial preparations are the first and most essential steps toward this goal.

4.3 Builder's Trials. Builder's trials are unofficial tests conducted at sea prior to official trials, their principal purpose being to provide the shipbuilder with assurance that the vessel will achieve contract performance requirements when subjected to official trials. Builder's trials are usually not specified as a contractual obligation but are conducted at the discretion of the contractor. However, it is generally accepted good practice, especially for first-of-class ships, that they be conducted to the extent possible. Some factors that influence this decision are:

- Drydocking arrangements
- Distance from shipyard to deep water
- Anticipated problems
- Fuel rate bonus/penalty clauses.

Since some shipyards do not have drydocking facilities, the ship must sometimes be taken to another location for the required predelivery drydocking. This trip affords an excellent opportunity for the shipbuilder to shakedown the propulsion system and to perform special tests without the expense of the full complement of the official trials.

At other shipyards, and particularly those remote from deep water, the shipbuilders are generally reluctant to assume the additional expense of an unspecified builder's trial unless it is to verify the correction of some serious deficiency or perhaps to ensure an acceptable fuel rate if there is a significant bonus/penalty clause in the contract. But in most instances, the shipbuilders merely allocate time at the beginning of the official trials for their special testing and trust that no serious problems will arise.

4.4 Trial Events. A typical official sea trial agenda for the first ship of a class would include the following events:

- Compass and radio direction finder calibration
- Pretrial shaft drag for torsionmeter zero*
- Standardization trials*
- Turning circles*
- Z-maneuver*

- Turbine water rate test*
- Economy trials—4 hours*
- Ahead steering test
- Emergency steering test
- Post-trial shaft drag for torsionmeter zero*
- Crash ahead from astern
- Crash astern from ahead
- Astern run—30 minutes
- Astern steering test
- Boiler overload test
- Mechanized control demonstration
- Anchor handling test.

The foregoing events marked by an asterisk would not normally be repeated on follow-on ships. A four hour endurance trial without fuel rate determination would be conducted in lieu of the economy trial. For diesel ships, the propeller shaft must be jacked in both directions, in lieu of being dragged, to determine the torsionmeter zero. This is usually performed at the yard immediately before departure for trials and again upon returning. Detailed information in regard to all of these events may be found in SNAME's T&R Code C-2.

Liquefied natural gas carriers must be subjected to two sets of trials. The first are conventional trials using residual oil for fuel. The second trials prove the capability of the plant to operate with LNG boil-off as fuel. Some shipbuilding contracts provide that the LNG trials be the responsibility of the purchaser after delivery. However, a most significant trial requirement for this type ship is the demonstration of the LNG containment and handling features. These gas trials can only be carried out where there is an available supply of LNG in significant quantity which, in effect, means after the vessel has been turned over to the owner. Final delivery, in the contractual sense, cannot take place until after successful completion of the gas trials.

4.5 Trial Displacement. It is important that the drafts during standardization trials be the same as those of the model test if there is to be meaningful correlation between model and ship. This is no problem for a tanker as the ship can be ballasted to almost any desired condition. But for a cargo ship, the trial displacement is often limited to that obtainable by ballasting the available tankage, resulting in light drafts with severe trim by the stern.

However, many cargo ships have been standardized at deeper drafts than can be obtained by tankage alone. Normal dry cargo holds can be filled with water and temporary solid ballast, such as steel ingots, can be used. Although frequently the standardization trials must be conducted with abnormal loading conditions, correlation with model tests is possible, only if the model test results include performance data and curves at the anticipated trial drafts. If model tests were not originally carried out at equivalent ship trial drafts, the model can be retested.

4.6 Maneuvering Data Requirements. By Resolution A.209(VII) of October 12, 1971, the Inter-governmental Maritime Consultative Organization (IMCO) adopted a recommendation on information to be included in maneuvering booklets available aboard large ships and ships carrying dangerous chemicals. The U. S. Coast Guard requirements in this regard, for tankships of 1,600 gross tons or over, may be found in Title 46, Code of Federal Regulations, paragraph 35.20, which requires that the following maneuvering information be prominently displayed in the pilot house on a fact sheet:

- Turning circle diagrams, port and starboard, showing time and distance of advance and transfer for 90 degrees at full and half speed
- Time and distance to stop from full and half speed
- Table of speeds at which an installed auxiliary device, such as a bow thruster, is effective.

This regulation further states that the above information, required for both fully loaded and ballast conditions, may be determined by:

- Trial observations
- Model tests
- Analytical calculations
- Simulation
- Information from a similar vessel.

If this information is to be developed by trial observation, trials will need to be extended considerably beyond the standard agenda and will require at least another full day at sea. On normal tanker trials, all tests are conducted with the ship ballasted to the full-load displacement and the turning circles and crash stops are performed at full power only. Additional trials would therefore be necessary to obtain the following information required by the Coast Guard:

- Turning circles and stopping distance from half speed at full displacement
- Data for the above items at ballast draft.

The Coast Guard extended these requirements to general cargo ships by 46 CFR 97.19-1 and to any vessel of 1,600 or more tons operating in or on navigable waters of the United States by (CFR Title 33) 164.01. As pointed out in the foregoing operation a dry cargo ship on trials at any draft condition other than the extremely light displacement obtainable by ballasting available tankage is usually impracticable. To develop the desired maneuvering data, it will be necessary to conduct such tests after the vessel is in service and laden to the desired displacements, to depend upon model test results, or to calculate results at the required conditions from basic ship response characteristics obtained at other conditions.

4.7 Special Instrumentation. The fuel economy and standardization runs constitute the most significant tests of any trial agenda. The first is a measure of the efficiency of the power plant, and the second is indicative of how well the ship's performance compares with model predictions and determines whether the ship meets contract or design speed requirements. Together they delineate the economic capability of the vessel, or the time and cost of carrying cargo from point A to point B. The purchaser's concern with the ship's performance in these areas is evidenced by the fact that many contracts include a bonus/penalty clause for fuel economy and, in some cases, for speed also. These tests involve the accurate measurement of horsepower, fuel consumption, and ship's position or distance traveled.

a. Torsionmeter. Except for ships with electric drive, the accurate determination of horsepower is not easily accomplished. The accepted procedure is to measure the twist in the propeller shaft and calculate the torque by applying a factor of torsional modulus of elasticity. Some contracts require that this modulus be determined by twisting the shaft before installation aboard ship and measuring the deflection and torque as described in SNAME's T&R Code C-2. Otherwise, the use of the generally accepted value of 82,000 MPa (11.9 × 10^6 psi) is permissible.

The twist of the shaft is measured by a torsionmeter, a unique instrument that is described in detail in T&R Code C-2. There have been any number of devices conceived to measure and transmit the twist in a rotating shaft, but the one that has found universal acceptance based on its reliability and accuracy for the past 40 years is the variable mutual inductance type. This instrument does not give an automatic readout but must be manually manipulated. It also must be calibrated immediately before and after trials.

Since there is always some residual torque in a propeller shaft at rest, it is impossible to set the torsionmeter to read zero torque when it is installed. However, if the ship is allowed to glide slowly through the water without power, the driving force of the propeller moving through the water is approximately equal to the inertia of freely spinning turbine and gears, and the torque in the shaft is nearly zero. This procedure is performed ahead and stern, and the average of the two sets of torsionmeter readings is considered to be the value of zero shaft torque. If this value is positive, the amount must be subtracted from all torsionmeter readings of the trial, and added if the value is negative.

In spite of these complications, the instrument has proven to be reliable, and the potential accuracy with an experienced operator is well within one percent, provided the seas are not so rough as to be an adverse factor.

b. Fuel Meters. To measure the fuel consumption, two fuel meters, previously calibrated using fuel of approximately the same temperature and viscosity as that to be used on trials, are temporarily installed in the fuel oil service system. Each should be provided with an individual bypass to permit servicing or removal in the event of malfunction during trials. The high viscosity residual fuels normally used on steam propelled vessels do not lend themselves to accurate measurement by the commercial fluid meters available, and the failure of one meter during trials is not unusual.

Should the corrected individual readings from each of the two meters diverge during the economy run, the meters must be recalibrated after trials to bring the data into closer agreement. Unfortunately, this procedure is not always successful, and the true fuel consumption on some trials has remained in doubt because of the unreliability of the fuel meters. The industry is badly in need of reliable instruments of laboratory quality suitable for this rugged service.

In the case of diesel ships, it is possible to use dual fuel tanks with quick changeover valves for the measurement of the fuel consumed. This technique is particularly adaptable to smaller craft such as tugs and fishing vessels.

c. Radiometric Tracking Systems. All standardization and maneuvering trials on first-of-a-class vessels are now routinely conducted with the aid of radiometric equipment such as that provided by Raydist, Loran, or Decca and described in SNAME's T&R Code C-2. A more detailed explanation of these systems and associated hardware can be found in Hastings and Comstock (1969). This equipment has not only greatly improved the accuracy of the standardization trials, but has afforded a means of plotting the course, heading and speed of the ship during turning circles and crash stop maneuvers with a precision previously impossible.

The Raydist system consists of two shore-based transmitter stations and a receiver with a strip recorder aboard the ship which indicates the distance in *lanes* from each of the transmitters. Once the receiver is properly calibrated in relation to the distance to the transmitters and *locked in*, the position of the ship can be determined and/or recorded for the duration of the trials. In addition to the lane counters, the instrument, with proper auxiliary connections to the ship's gyro and revolution counter, can also simultaneously record ship's heading, shaft revolutions and elapsed time.

Fig. 1 is a reproduction of a portion of a radiometric strip chart recorded during a trial. Knowing that each lane is equivalent to 45.18 m (148 ft), the following information can be developed from the chart at time instant "X − X":

Red station distance	45.18 × 1670.3 = 75,465 m(247,589 ft)
Green station distance	45.18 × 2060.6 = 93,097 m(305,442 ft)
Ship's heading	062.7 deg
Propeller speed	41 rpm

Fig. 2 is a typical turning circle plot developed from radiometric data, and Fig. 3 is a similar plot of a crash astern maneuver. As can be seen, not only is the ship's position accurately traced, but the heading and elapsed time are also clearly indicated.

For standardization trials, a course is selected in line with one of the two transmitters, and the speed runs are made directly to and from that station over the same area of water. This simplifies the data processing as the distance traveled is readily determined from the lane count of the one station. The lane count from the other station must be used to position the ship properly and to maintain its alignment with the predetermined track.

4.8 Vibration Analysis. With the advent of higher powers and unique ship designs of recent years, vibration has become a matter of serious concern for both the designer and builder. As a first step toward establishing acceptable limits of vibration as well as to improve the predictability of this problem, SNAME through the cooperation of MarAd has promoted a program of collecting vibration data during the sea trials of at least one ship of each class built under the subsidy program.

The data are gathered under a uniform procedure utilizing a twelve channel recording oscillograph with transducers installed in similar if not identical locations of each ship tested. These transducers are located so as to measure three

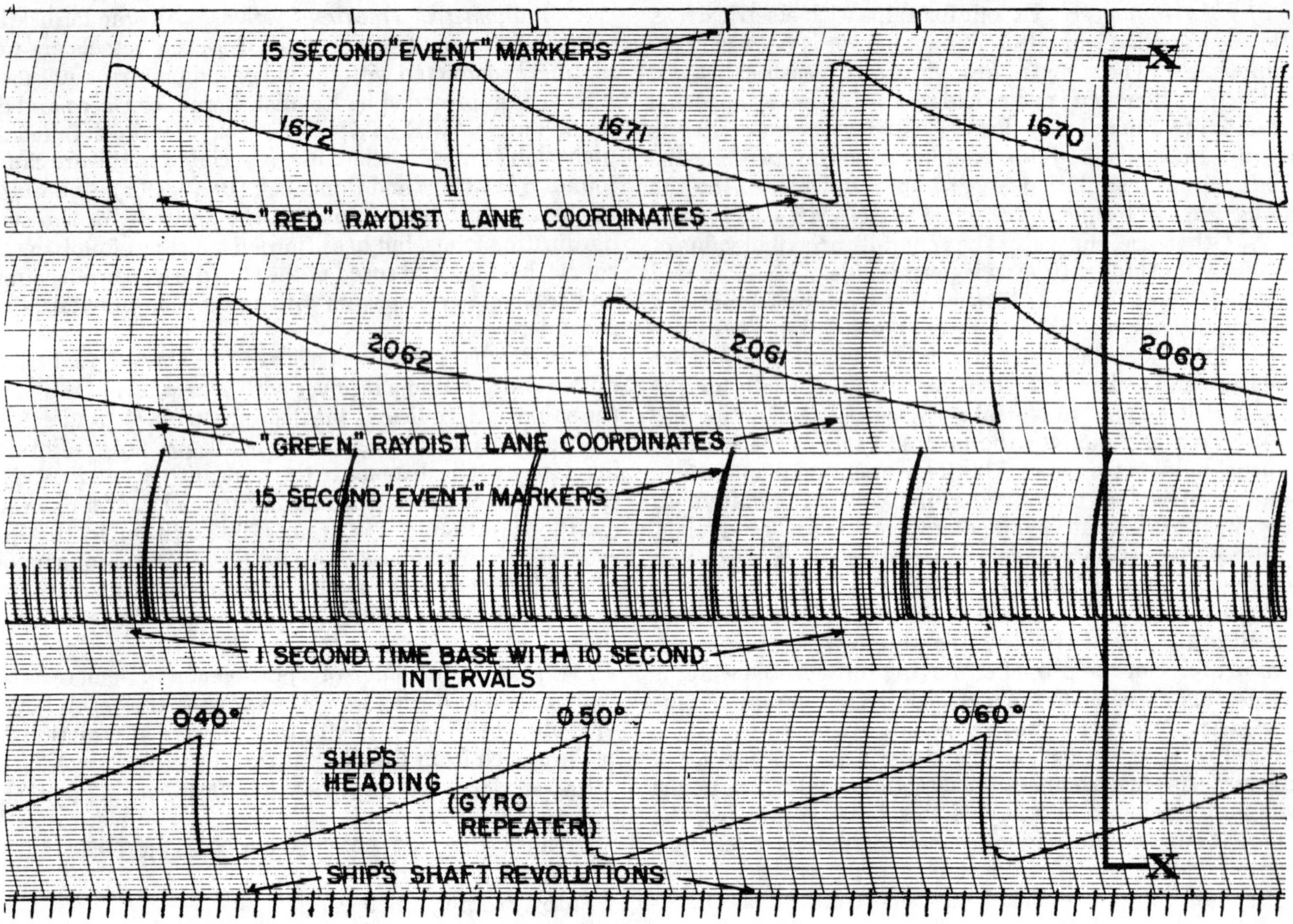

Fig. 1 Typical radiometric record chart

dimensional vibration of the hull structure in the area of the steering gear foundation and of the main thrust bearing and longitudinal vibration of the main reduction gears and high pressure and low pressure turbines. Readings are taken at a range of powers from "half" to "full" in five rpm increments.

The results of these vibration surveys for a number of ships have been summarized and published by SNAME as "Vibration Data Sheets." The data for five ships of 1976 are included in booklet "D-14" and five ships for 1977 are in booklet "D-15."

A full description of the procedure for the installation of equipment and conducting the test, together with the standard format for tabulating the data, is presented in SNAME Code C-1, "Code for Shipboard Vibration Measurements." Similar information in regard to local vibration can be found in SNAME Code C-4, "Code for Shipboard Local Structures and Machinery Vibration Measurements." Acceptable vibration limits have been established for principal machinery components, and this information is contained in SNAME Code C-5, "Acceptable Vibration of Marine Steam and Gas Turbine Main and Auxiliary Machinery Plants."

4.9 Trial Conferences. It is often desirable to convene an impromptu conference of the interested parties during sea trials, particularly if trouble develops or if the data are suspect. In such cases, every effort should be made to reach an agreement in regard to what corrective measures should be taken, whether the trials should be extended or other alternatives. If the trials must be aborted, as much meaningful data as possible should be salvaged, and perhaps special tests should be conducted to aid in taking proper corrective action or in conducting retrials.

While still at sea, a post-trial conference should always be held at the conclusion of all testing to review the results and to affirm that no further demonstrations are desired by the purchaser or regulatory bodies. During this conference, the shipbuilder should be made aware of all operational deficiencies observed during trials so that an opportunity is afforded for corrections to be made and proven before returning to the yard. The shipbuilder should also be advised as to the machinery components that may be required to be opened for a post-trial inspection. The following is a typical list of such equipment routinely examined:

- Boilers—fire and water sides
- Air heaters

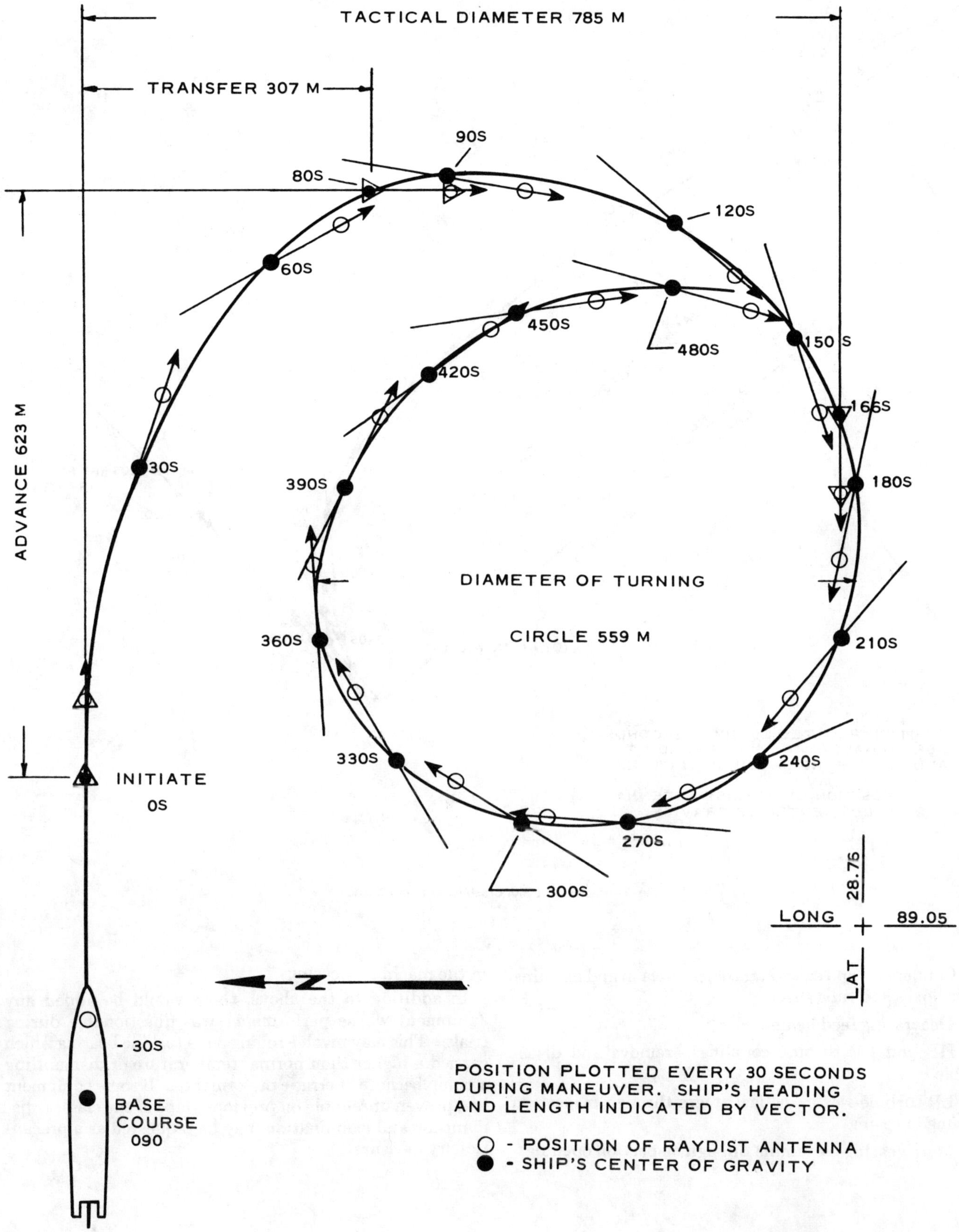

Fig. 2 Radiometric plot right turning circle

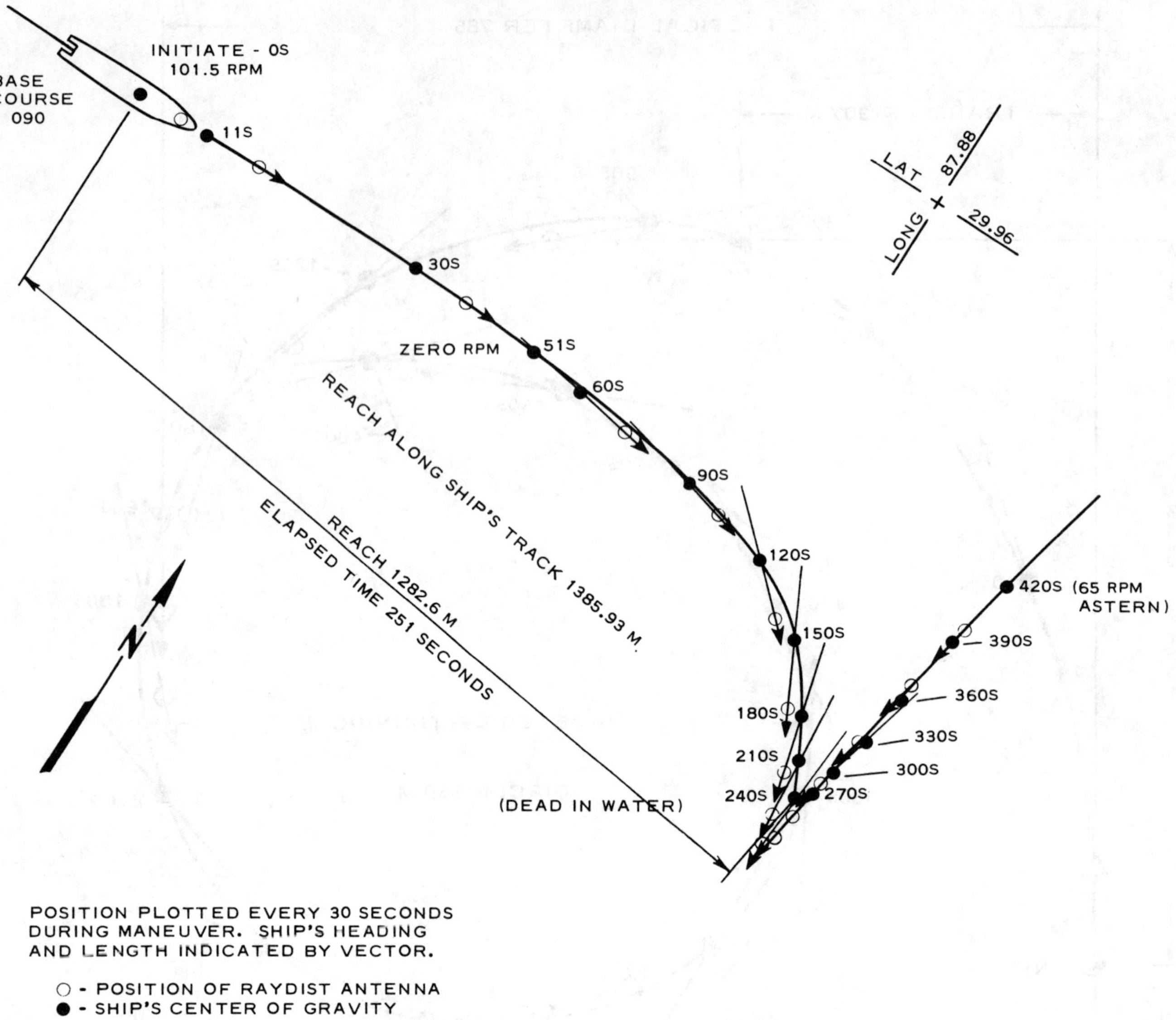

Fig. 3 Radiometric Plot Crash Astern from Ahead

- Condensers and salt water coolers—clean and examine water boxes and tube sheets
- Deaerating feed heater
- HP and LP turbine couplings—remove and disassemble
- LP turbine—examine last stages through inspection opening of casing
- Main gearing—examine through inspection openings, in the machinery casings.

In addition to the above, there would be added any equipment whose performance was questionable during trials. This may involve rolling out a turbine bearing which carried a higher than normal temperature or dismantling a pump because of erratic performance. If some equipment had proven unreliable on previous ships of the classes, dismantling and examination may be requested as a precautionary measure.

Section 5
Delivery

5.1 Deficiency Items. Throughout the course of construction, but particularly during the latter stages when testing is underway and when compartments are being completed and inspected, the purchaser and the regulatory bodies will develop their lists of incomplete or unsatisfactory work. In view of the possible magnitude of this tabulation which continues to grow until delivery, it is important that the purchaser's list be developed in an orderly fashion to avoid duplication of items and to facilitate ready identification. It is general practice to subdivide the items into four categories of hull, outfit, machinery, and electrical, numbering each item consecutively as it is added. Items should not be permitted to accumulate but should be forwarded to the contractor at frequent intervals, perhaps on a daily basis as delivery approaches. The purchaser, in turn, should be informed of all regulatory body deficiencies to avoid duplication.

It is in the shipbuilder's interest to act promptly on all items as they are submitted to keep the number to a minimum. If any item is considered not to be a contractual obligation, the purchaser should be so advised in order that the issue may be resolved. As items are corrected, they should be inspected promptly by the purchaser and, if acceptable, should be signed off by initialing the shipbuilder's work sheet. As items are eliminated, it is often desirable for the purchaser to revise the list, including only the outstanding items, but under no circumstances should the items be renumbered. It is obvious that in the resolution of the deficiency items, there must be open discourse and mutual cooperation between the shipbuilder and purchaser if serious last minute disputes regarding delivery exceptions are to be avoided.

5.2 Allowance List Inventory. With the increased sophistication of equipment installations, the tedious task of taking inventory of the myriad spare parts and tools has become ever more time-consuming. The specifications generally require that the shipbuilder prepare, for the purchaser's approval, a complete *allowance list* of not only spares and tools but also all portable and readily removable equipment on board at the time of delivery. The inventory of this material is taken jointly by representatives of the purchaser and shipbuilder and, in order for it to be completed by the delivery date, it is often necessary to begin as early as three months in advance.

The specifications generally require the spares to be stowed in specially constructed and padlocked boxes that are placed in locations convenient to the equipment. It is advisable to inventory these boxes first, since they can then be secured by seal to prevent unauthorized removal of contents. Inventory of portable and loose gear that cannot be stowed in a secured space should be deferred until immediately before delivery. All oil tanks should be sounded and all drums of oil, grease, etc. should be jointly inventoried just prior to delivery to determine quantities for which the shipbuilder must be reimbursed if in excess of requirements.

5.3 Delivery Documents. Typical construction specifications include the stipulation that the ship as delivered shall comply with all applicable laws of the United States and the requirements of the various regulatory bodies and rules noted in the individual specification sections. It is further stated that all necessary certifications and/or documents indicating approval and compliance in construction shall be obtained by the contractor. The scope and significance of this requirement is illustrated by the typical list of documents shown in Table 3 that must be prepared or acquired by the contractor for presentation to the purchaser at the time of delivery.

At the time of delivery, the purchaser will desire that all documents be available without any exceptions or limitations attached. But, for several reasons including administrative procedures of the regulatory bodies, it is seldom possible to deliver a complete set of final certificates with the ship, and temporary letters are customarily issued pending receipt of the official documents. The purchaser, however, must be provided with adequate documentation to prove that there are no citations against the vessel that would compromise its operation or jeopardize its insurance coverage.

5.4 Delivery Date. About a week before the scheduled delivery date, a joint survey of the vessel should be made by representatives of the purchaser and the shipbuilder to identify the remaining unfinished work and to determine the status of the outstanding deficiency items. With this information and under normal circumstances, agreement as to a firm delivery date should be possible. However, there are a number of factors other than the cold statistics of construction progress that enter into the delivery process at this point.

The purchaser would normally desire that the ship be complete in every respect, including all testing, cleaning and painting, as well as the correction of all deficiency items and Allowance List shortages. But this position is often compromised by the economic advantage of placing the vessel in service as quickly as possible. This is particularly true if there are large cargo commitments, and the ship is needed to fill a gap in sailing schedules. In such cases, many of the less significant deficiencies may be waived, and others will be deferred for correction until after delivery. The shipbuilder encounters little difficulty in effecting delivery of a ship under these circumstances.

On the other hand, if the purchaser has little immediate need for the ship, every deficiency item tends to become justification for refusing to accept delivery. The language of most contracts provides for this eventuality, however, stating that when the work is complete or substantially complete and the required tests have been passed, the vessel shall be delivered by the contractor and accepted by the purchaser with not less than five days' written notice to the purchaser. *Substantially complete* is further defined as complete except for minor items not affecting the commercial utility of the vessel. It is not surprising that argu-

Table 3—List of Delivery Documents

Title of Document	*Issued By*	*Signers*
Inter Class Certificate	ABS	Attending Surveyor
Provisional Load Line Certificate	ABS	Attending Surveyor
Register of Cargo Gear	ABS	Attending Surveyor
Cargo Ship Safety Construction Certificate	ABS	Attending Surveyor
Port Bower Anchor Certificate	ABS	Attending Surveyor at manufacturer's plant
Starboard Bower Anchor Certificate	ABS	Attending Surveyor at manufacturer's plant
Chain and Link Certificate	ABS	Attending Surveyor at manufacturer's plant
Main Boiler Certificate	ABS	Attending Surveyor at manufacturer's plant
Main Turbine & Reduction Gear Certificates	ABS	Attending Surveyor at manufacturer's plant
Turbo-Generator Certificate	ABS	Attending Surveyor at manufacturer's plant
Propeller Report	ABS	Attending Surveyor at manufacturer's plant
Tailshaft and Lineshaft Report	ABS	Attending Surveyor at manufacturer's plant
Certificate of Inspection	USCG	Inspection Officer
Stability Letter	USCG	Inspection Officer
Cargo Ship Safety Equipment Certificate	USCG	Commandant, USCG
*Designation of Home Port	USCG	Documentation Officer
*Application and Notice of Award of Official No. & Signal Letters	USCG	Documentation Officer
Certificate of Admeasurement	USCG	Office of Admeasurement
Panama Canal Tonnage Certificate	USCG	Office of Admeasurement
Suez Canal Tonnage Certificate	USCG	Office of Admeasurement
Safety Radiotelegraphy	FCC/USCG	FCC
Sanitary Construction Certificate	USPHS	USPHS
Deratization Exemption Certificate	USPHS	USPHS
Certificate of Deadweight	Contractor	USCG
Inclining Experiment Report	Contractor	USCG
Master Carpenter Certificate	Contractor	Contractor
Compass Deviation Cards	Contractor	Compass Adjuster
Soundings of Fuel Oil & Water Tanks	Contractor	Purchaser
List of Instruction Books	Contractor	Purchaser
List of Plans	Contractor	Purchaser
Operating Manual	Contractor	Purchaser
Lab Report on Water Tanks	Contractor	Purchaser
Trim & Stability Booklet	Contractor	Purchaser
CO_2 Fire Equipment Certificate	Contractor	Vendor
RDF Calibration Record	Contractor	Vendor
Receipt of Safe Combinations	Contractor	Captain of Vessel
Receipt of Ship's Keys	Contractor	Captain of Vessel
Certificate of Delivery of Vessel and Certificate of Completion	Purchaser	Purchaser & Contractor
Powers of Attorney	Purchaser/ Contractor	Company Executives

* Application must be made by the Master of the vessel.

ments often develop over this wording as to what constitutes a *minor item.*

5.5 Predelivery Conference. A day or two before delivery, a work session attended by representatives of the purchaser, shipbuilder, and regulatory bodies should be scheduled to establish the conditions for delivery of the vessel. A typical agenda would include the following:

• Review the outstanding deficencies from the regulatory bodies and verify that there are no items that would preclude placing the vessel in service, nor any shortages of critical Allowance List material.

• Review the list of documents to identify those certificates which will not be available at delivery and verify that there will be no exceptions that would place any restriction on operation or insurance coverage. The regulatory body representatives could be excused at this juncture.

• Review all outstanding items of deficiency. Disposition should be made in one of the following categories:

1. Items satisfactorily completed or cancelled by the purchaser.
2. Items acknowledged to be the shipbuilder's responsibility to correct but which will not be completed at the time of delivery. These items should be renumbered consecutively and recorded as Exhibit "A" to the Certificate of Delivery and Completion of the vessel.
3. Allowance List items missing on the inventory should be recorded as Exhibit "B" to the Certificate of Delivery and Completion.
4. Items which will be held for service experience during the guarantee period.

It is hoped that all items can be resolved by the above dispositions but there is always the possibility of an impasse preventing agreement on certain items. In order to facilitate delivery, any items in dispute can be made a matter of record by statement or special exhibit attached to the Certification of Delivery. These disagreements can then be resolved at the end of the guarantee period, or if necessary, by arbitration.

After review of and concurrence with the language of the Certificate of Delivery and Completion of Vessel, the time and place for delivery can be established.

5.6 Delivery. If all matters of the agenda for the predelivery conference have been resolved, there should remain little to be done at delivery other than the formal signing and distribution of the delivery documents. This generally consists of

- Delivery by the shipbuilder of the various certificates and documents listed in Table 3. After a check off, the purchaser and shipbuilder will sign a Receipted Distribution List of Documents certifying that the available documents have been delivered.
- Signing by the purchaser for onboard plans.
- Signing by the purchaser for onboard instruction books.
- Signing by the Master for the receipt of ship's keys and safe combinations.
- Signing by the shipbuilder and purchaser of Certificate of Delivery and Completion of Vessel with Exhibits of Exceptions attached. A final copy of the Exhibits should have been given to the shipbuilder in advance for verification that all listed deficiency items are acceptable as shipbuilder's responsibility.

Section 6
Guarantee Settlement

6.1 Contract Provisions. Most construction contracts for U. S. built ships provide for a guarantee period of six months from the date of delivery. Some contracts have extended this period to a year for certain specific major machinery components such as main turbine, gears, generator sets, boilers, steering gear, and anchor windlasses. Recently, there has been a trend toward a one year guarantee for the entire ship, which is the general practice in other countries.

Although the guarantee provisions may vary to some degree, the contract language generally provides that:

- The shipbuilder is responsible for the correction to the requirements of the plans and specifications for any weakness, deficiency, failure, deterioration in workmanship or material, or any failure of equipment, machinery, or material furnished by the shipbuilder to function as prescribed and intended.
- The shipbuilder's responsibility is limited to actual repair or replacement costs at straight-time commercial shipyard or ship repair-yard rates.
- The shipbuilder's responsibility is limited to the repair or replacement of the item of deficient machinery or equipment and does not extend to consequential damages beyond such machinery and equipment. However, the degree of builder's liability for consequential damages is today a matter for serious concern as a result of court decisions extending product liability concepts to shipyard contracts.
- The repair work may be carried out at the shipyard of the builder if practicable. The shipbuilder may, with the concurrence of the owner, have the work performed at another yard, or the owner may place the work with a shipyard or ship repair yard in any port he chooses.
- The shipbuilder may have an engineer aboard the vessel for the guarantee period.
- Should the vessel be taken out of service because of guarantee deficiencies, the guarantee period will be extended by an amount equal to such idle time.
- The shipbuilder shall be advised of any guarantee deficiency and given an opportunity for inspection prior to repair or correction.
- No items shall be added to the guarantee list after the expiration of the guarantee period. The complete list must be forwarded promptly to the shipbuilder.
- For the determination of underwater guarantee deficiencies, the owner is permitted an extension of six months beyond the guarantee period for drydocking the vessel. The shipbuilder's responsibility is limited to those items which arise during the guarantee period.
- If guarantee deficiencies are discovered while on drydock, the shipbuilder shall be responsible for drydocking charges resulting from any additional lay days in excess of those required for owner's normal maintenance.
- If a special drydocking is required for the correction of guarantee deficiencies, the shipbuilder shall be responsible for all drydocking costs.
- The shipbuilder shall transfer and assign to the owner any vendor guarantees extending beyond the ship's guarantee period.

6.2 Repair Procedures. The above legal framework allows broad latitude in establishing the actual procedure to be followed in fulfilling the guarantee responsibilities. This is essential since the requirements will be found to be different for each combination of ship type, trade route, shipbuilder and owner. Also, the quick turn-arounds of the modern container carriers and tankers dictate different ground rules than those to be followed for the break-bulk general cargo ships which are in port for days at a time. So with each contract, the owner and shipbuilder must establish a *modus operandi* for the effective and efficient correction of guarantee deficiencies and yet provide protection for their individual interests.

Although the contract may permit the return of the vessel to the shipbuilder's yard, this option is rarely exercised except in unusual cases where major alterations are required or where the shipbuilder's expertise is not readily available elsewhere. But the normal method of correcting nearly all guarantee deficiencies is the same as that used by the owner for voyage repairs; i.e., by engaging local ship repair organizations for pier side corrections.

While the shipbuilder may contract for the repairs, he seldom exercises this prerogative as by so doing he tacitly assumes responsibility not only for the deficiency but also for the correction. Evidence may develop during the process of the repair that the deficiency was not due to inferior workmanship or material, but to maloperation or improper maintenance by the owner. The general trend, therefore, has been for the owner to contract for all repair work, both the normal voyage repairs and items that the owner considers to be of a guarantee nature.

By contract, the shipbuilder must be advised of all such items considered to be guarantee deficiencies before any corrective work is undertaken. This is to allow the builder an opportunity for inspection as well as to bring in vendors if the equipment or material is covered by warranty. In order not to preclude the possibility of making future claims for any items as guarantee responsibilities, many owners routinely report all pending repairs to the shipbuilder and reserve the actual designation of guarantee items until the time of the guarantee survey.

It is in the shipbuilder's interest to have a representative present during all guarantee repairs but his presence should not be construed as his tacit acceptance that all such repairs fall into the guarantee category. The owner should impress upon his repair contractors the importance of properly documenting all repairs for which reimbursement will be requested as a guarantee obligation. This would include the taking of photographs and obtaining laboratory reports when appropriate, as well as retaining all replaced parts for examination. If a vendor is involved, an invitation should be issued for a representative to be present. The invoices should be explicit as to what was found, what corrections were made and should also include a detailed breakdown of charges for labor and material with a separate item for premium pay for overtime.

The quick turn-arounds of the modern commercial ship impose a severe handicap in accomplishing any voyage repairs and complicate the already delicate problem of arriving at equitable settlements for corrective costs of guarantee deficiencies. Because of the brief time in port, much of the repair work must be done on an overtime basis. The contracts generally provide that the shipbuilder's liability is limited to the cost of correction at straight time rates, any premium pay for overtime being the responsibility of the owner. This is in keeping with the prevailing philosophy that while the shipbuilder must correct all guarantee deficiencies, the owner must make the vessel available for him to do so.

The problem becomes more involved due to the lack of time to complete even minor repairs during one port stay, requiring repeated opening and closing of equipment and multiple travel time for mechanics and vendor representatives. It is not unusual for the work to be done twice, once as a temporary expedient and again as a final correction. A repair may well result in a cost several times what it would have been had the work been carried through to completion at the outset. In such instances, the shipbuilder is usually relieved of all costs except those associated with the final repair.

6.3 Preparation for Guarantee Survey. As soon as practicable after the expiration of the guarantee period, usually at the first availability of the ship, a *Guarantee Survey* should be scheduled to resolve all outstanding construction matters of the contract. At this conference, an effort should be made to reach agreement regarding the responsibility for the correction of all deficiency items which became apparent during the guarantee period as well as the determination of the status of all exceptions taken at the time of delivery, Exhibits A and B of the delivery document.

The owner can alleviate this task to a great degree by giving proper care to the preparation of the list of deficiencies and to the cataloging of the reference material such as repair invoices, reports by surveyors and laboratories, and photographs. If the list is lengthy, it is often helpful to group the items in categories of hull, outfit, machinery, and electrical as was done during the construction period. This will permit the conservation of the valuable time of the ship's officers whose testimony is often invaluable in determining the cause of a deficiency. It will also assist in locating items in the list when the numbers are not known.

The list should be limited to those items which the owner sincerely considers to be the responsibility of the shipbuilder. Included would be items not yet repaired together with those which have been corrected but which the owner is seeking reimbursement for the repair costs.

6.4 Guarantee Survey Conference. The representatives of the owner and shipbuilder should be knowledgeable with regard to the items of the survey and must have full authority to make binding agreements for their companies. This is essential since the resolution of the many complicated items, which are seldom incisive, inevitably involves a series of compromises. If either of the parties cannot negotiate in this manner, there can be little progress in arriving at settlements.

The status of these representatives within their organizations will vary, but in many cases, the owner will be represented by the Construction Representative who served during the construction of the ship. He may be assisted by the Superintendent Engineer or Port Engineer who supervised the voyage repairs and the ship's officers who would have detailed background information on the various deficiency items. The shipbuilder will usually be represented by a guarantee specialist, often the Guarantee Engineer who has ridden the ship and is familiar with the repair items. On a first-of-class ship or if there are major items involved, the shipbuilder may also have a senior technical officer of the company present. In addition, the contractor will also have representatives from the suppliers of the principal machinery components.

Intelligent and knowledgeable people should be able to

resolve most of the controversial issues. Failure to reach agreement may result in litigation, a long and expensive process that often costs more to all concerned than can be recovered.

The spokesman for the owner should serve as chairman of the conference, presenting the items and the justification for their consideration as guarantee responsibilities. Items involving reimbursement for repairs already made should be documented with a breakdown of costs that are subject to audit by the shipbuilder to verify that the repair did not exceed the original specifications or that premium pay for overtime was not included. In regard to items involving repairs yet to be made, agreement should be reached as to how the repair is to be made, who is to put the work in hand, and how payment is to be accomplished.

Not all deficiencies discovered during the guarantee period are shipbuilder's responsibility to correct. In addition to items resulting from maloperation or improper maintenance, there are included in this category items of inferior workmanship or items of a cosmetic nature not deleterious to performance. The guarantee language of most contracts is directed toward the correction of deteriorating material, or equipment failing to function as intended. The guarantee period is not intended to provide the owner an extended opportunity to inspect the ship for construction flaws; if a ship is accepted at delivery with weld splatter, smeared paint, badly fitted joiner work or mismatched floor tile, no recourse is usually provided for correction of such items under the guarantee provisions. Similarly, omissions or missing materials must be made a matter of record before delivery.

Inevitably, there will be a number of items of the guarantee list which will require inspection before an agreement can be reached as to disposition. Such items should be set aside until the complete list has been reviewed, after which an inspection tour of the ship can be made. The inspection items should be regrouped according to geographical location aboard ship in order to avoid unnecessary backtracking.

6.5 Consequential Damage. Many agreements involving compromise must be reached in the settlement of guarantee responsibilities, but none are more difficult than those involving the interpretation of the consequential damage clause of the contract. By this clause, the shipbuilder's responsibility is limited to the correction of any deficient machinery or equipment and any deficiency in workmanship or material furnished by the contractor in performing the contract work. This clause specifically excludes responsibility for any damage or claims extending beyond such machinery or equipment.

For example suppose, during the guarantee period, a ship develops a malfunction of the steering gear while negotiating a narrow river channel with the result that the vessel strikes a bridge support, collapsing the bridge with loss of life and incurring severe damage to the ship. By the consequential damage clause, the shipbuilder's responsibility would be limited to the correction of the defective steering mechanism, but he could not be held responsible for the repair of the ship or the loss of its earning capability, nor would he be liable for any claims resulting from loss of life, repair of the bridge, or economic loss to the community.

But not all examples are so straightforward, the difficulty being in defining the limits of the "machinery or equipment" and "workmanship or material." Consider the case of a boom that falls while handling cargo as the result of a faulty winch controller. The casualty results in damage to cargo as well as to the ship itself, including the boom in question. Is the "machinery or equipment" limited to the winch controller or does it also include the winch, boom, and running gear? Convincing arguments can be made for either position, but in the past, most settlements have included the repair of any equipment required to provide a completely operative unit. In this instance, the contractor would be responsible for the repair of the entire unit of cargo gear including the boom.

In view of the magnitude of claims that could conceivably arise from a marine disaster resulting from the maloperation of some vital component, it is essential that the shipbuilder's liability be limited as provided by this contract clause. Otherwise, the cost of construction would become prohibitive as provision for this contingency would have to come from shipbuilder reserves. It is inconceivable that underwriters would provide this type of coverage since there has been a reluctance to issue insurance to cover even the limited guarantee responsibilities for which the shipbuilder is now liable.

6.6 Guarantee Survey Report. Ideally, at the conclusion of the Guarantee Survey Conference, agreements will have been reached in regard to responsibility for all items of the survey list. These agreements should be documented by a report that can be prepared by either party, but the shipbuilder usually assumes this task. The report should include only those items which involve some future obligation of the shipbuilder such as making repairs, reimbursing the owner for repairs previously accomplished, or furnishing shortages of Allowance List material as listed on Appendix B of the delivery document.

Several years ago, MarAd developed a group indexing code for the preparation of guarantee survey reports to provide an expedient filing reference for the individual items. Since then, several shipyards have adopted this system in preparing their own reports. It is believed that by this system the reports will serve a broader purpose for all parties with particular reference to the recognition of recurring deficiencies and the improvement of future designs. It also facilitates locating a particular item when the number is not known and prevents duplicating items.

The system utilizes three sets of numbers. The first set indicates the group number which classifies the item into one of 16 subject divisions. The second number is the item number as listed on the owner's guarantee survey list. The third number is a new number assigned consecutively in the Guarantee Survey Report.

For example:

(1) (2) (3)
No. 6 – 117 – 41

- The 6 indicates Group 6—Main Machinery

• The 117 indicates item 117 of the guarantee survey list

• The 41 indicates item 41 of the Guarantee Survey Report

Table 4—Group Index for Items of Guarantee Survey Reports

Group	*Item*
1.	Material Engineering and Miscellaneous Deficiencies
2.	Structural
3.	Arrangements
4.	Hull Equipment and Fittings
5.	Heating, Ventilating, and Hull Piping
6.	Main Machinery
7.	Boilers
8.	Piping
9.	Auxiliary Machinery Equipment
10.	Refrigeration and Air Conditioning
11.	Deck Machinery
12.	Electronic Systems
13.	Electric Power Distribution
14.	Electrical Machinery
15.	Lighting and Interior Communication
16.	Centralized Control and Monitoring Systems

Table 4 is a tabulation of the 16 group numbers of the MarAd indexing system.

After endorsement by both parties, the Guarantee Survey Report becomes the official document delineating the outstanding obligations of the shipbuilder under the guarantee provisions of the contract. Once these responsibilities are accomplished to the owner's satisfaction, the construction contract, unless complicated by claims or litigation, is complete.

References

American Bureau of Shipping (1975), *"Requirements for Certification of the Construction and survey of Cargo Gear on Merchant Vessels."*

American Bureau of Shipping (Annual), *"Rules for Building and Classing Steel Vessels."*

Hastings, C. E. and A. L. Comstock (1969), "Pinpoint Positioning of Surface Vessels Beyond Line of Sight," Hastings-Raydist, Inc.

IMCO (1973), "Code of Practice Concerning the Accuracy of Stability Information for Fishing Vessels," Resolution A.267 (VIII).

Jack, R. L. (1973), "Sea Trials: Some Recommended Practices," *Marine Technology,* October.

MarAd (1964), "Supplementary Procedure for Testing Machinery," Maritime Administration.

MarAd (1965), "Addendum No. 1 to Supplementary Procedure for Testing Machinery, Central Control System Testing," Maritime Administration.

Moore, Charles S. (1967), "Intact Stability," *Principles of Naval Architecture,* edited by J. P. Comstock, SNAME, New York.

SNAME (1960), "Code on Installation and Shop Tests," T&R Bulletin No. 3-8.

SNAME (1968), "Recommended Practices for Correcting Steam Power Plant Performance," T&R Bulletin No. 3-17.

SNAME (1975), "Code for Shipboard Vibration Measurements," T&R Code C-1.

SNAME (1973), "Code for Sea Trials," T&R Code C-2.

SNAME (1976), "Code for Local Structure and Machinery Vibration Measurements," T&R Code C-4.

SNAME (1976), "Acceptable Vibration of Main Steam and Gas Turbine Main and Auxiliary Machinery Plants," T&R Code C-5.

U. S. Coast Guard, "Forms for Stability Test Report," CG-993, Latest Revision.

U. S. Coast Guard (1967), "Stability Test—Preparation and Procedures," Navigation and Vessel Inspection Circular No. 1-67.

U. S. Government, "Code of Federal Regulations, Title 33—Navigation, Navigable Waters," U. S. Government Printing Office.

U. S. Government, "Code of Federal Regulations, Title 46—Shipping," U. S. Government Printing Office.

U. S. Government, "Code of Federal Regulations, Title 47—Telecommunications," U. S. Government Printing Office.

U. S. Public Health Service, "Handbook on Sanitation of Vessel Construction," Department of Health Education and Welfare.

Glossary of Terms

Abaft. Aft of; toward the stern from a designated location.

Accommodation ladder. A portable inclined ladder hinged to a platform attached to the edge of a ship, or at the sill of a shell entry port, and which can be positioned to provide access between ship and small craft or shore.

Administration. As used in international conventions, "the Government of the State whose flag the ship is flying".

Aft. Toward, at, or near the stern.

Afterbody. That portion of a ship's hull abaft amidships.

After peak. The compartment in the stern, abaft the aftermost watertight bulkhead.

After perpendicular. (See **length between perpendiculars.**)

Air port. A hinged glass window generally circular, in the ship's side or deckhouse, for light and ventilation; also called porthole, portlight or side scuttle.

Amidships. In the vicinity of the midlength of a ship as distinguished from the ends. Technically it is exactly half way between the forward and the after perpendiculars. A helm order to indicate the rudder is to be put on the centerline.

Anaerobic corrosion. Corrosion of ship's structure by the action of sulphate reducing bacteria present in some harbors.

Anchor. A heavy forging or casting so shaped as to grip the sea bottom, and by means of a cable or rope, hold a ship or other floating structure in a desired position regardless of wind and current.

Appendages. The portions of a vessel extending beyond the main hull outline including such items as rudder, shafting, struts, bossings, and bilge keels.

Assemble. To fit together small parts, in making a large section or part.

Athwartship. Across the ship, at right angles to the fore-and-aft centerline.

Auxiliary machinery. All machinery other than that required for main propulsion.

Bale cubic. The cubic capacity of a cargo hold measured to the inside of the frames or cargo battens.

Ballast. Any solid or liquid weight placed in a ship to increase the draft, to change the trim, or to regulate the stability.

Ballast tank. Watertight compartment to hold water ballast.

Baseline. A fore-and-aft reference line at the upper surface of the flat plate keel at the centerline for flush shell plated vessels, or the thickness of the garboard strake above that level for ships having lap seam shell plating. Vertical dimensions are measured from a horizontal plane through the baseline, often called the molded baseline.

Battens. (See **cargo battens and hatch battens.**)

Beam, cant. Deck beams aft radially disposed and extending from the transom beam to the cant frame heads at the deck edge.

Beam, deck. An athwartship horizontal structural member, usually a rolled shape, supporting a deck or flat.

Beam knee. (See **knee, beam.**)

Beam, molded. The maximum breadth of the hull measured between the inboard surfaces of the side shell plating of flush-plated ships, or between the inboard surfaces of the inside strakes of lap seam-plated vessels.

Beam, transom. (See **transom beam.**)

Bearding line. The intersection of the inside surface of the shell plating and the stem or sternpost.

Berth. Where a ship is docked or tied up; a place to sleep aboard; a bunk or bed.

Between decks. (See **'tween decks.**)

Bevel. The angle between the flanges of a frame or other member. (When greater than a right angle, open bevel, when less, closed or shut; also, to chamfer).

Bilge. Intersection of bottom and side. May be rounded or angular as in a chine form hull. The lower parts of holds, tanks and machinery spaces where bilge water may accumulate.

Bilge and ballast system. A system of piping generally located in the holds or lower compartments of a ship and connected to pumps. This system is used for pumping overboard accumulations of water in holds and compartments, and also for filling ballast tanks.

Bilge bracket. A vertical transverse flat plate welded or riveted to the tank top or margin plate and to the frame in the area of the bilge.

Bilge keel. A long longitudinal fin fitted at the turn of the bilge to reduce rolling. Commonly it consists of plating attached to the shell plating by welding or by angles.

Bilge strake. Course of shell plates at the bilge.

Billboard. A sloping plate above the rake or gunwale of a vessel having low freeboard, on which an anchor is stowed and from which it can be released for use.

Binnacle. A stand or box for holding and illuminating a compass.

Bitter end. The inboard end of a ship's anchoring cable which is secured in the chain locker.

Bitt, mooring. Short metal column (usually two) extending up from a base plate attached to the deck for the purpose of securing and belaying wire ropes, hawsers, etc. used to secure a ship to a pier or tugboat. Also called a bollard.

Bitumastic. An elastic bituminous cement used in place of paint to protect steel.

Bleeder. A small cock, valve, or plug to drain off small quantities of fluids from a container or system.

Block coefficient. The ratio of the underwater volume of a ship to the volume of a rectangular block, the dimensions of which are the effective length, draft and beam. The relationship is expressed as a decimal.

Body plan. A drawing consisting of two half transverse elevations or end views of a ship, both having a common vertical centerline, so that the right-hand side represents the ship as seen from ahead, and the left-hand side as seen from astern. On the body plan appear the forms of the various cross sections, the curvature of the deck lines at the side, and the projections, as straight lines of the waterlines, the buttock lines, and the diagonal lines.

Bollard. (See **bitt, mooring.**)

Bolster. (See **hawsepipe.**)

Bonjean curves. A set of curves, each of which represents a plot of the cumulative area of a station on the lines plan, from the base line to any point above it.

Booby hatch. An access hatch in a weather deck, protected by a hood from sea and weather. Also called **companionway.**

Boom. A long round spar hinged at its lower end, usually to a mast, and supported by a wire rope or tackle from aloft to the upper end of the boom. Cargo, stores, etc, are lifted by tackle leading from the upper end of the boom.

Boom crutch. A term applied to a light structure built up from a deck to support the free end of a boom when it is not in use. Also called **boom rest**.

Boom table. A stout small platform usually attached to a mast to support the hinged heel bearings of booms and to provide proper working clearances when a number of booms are installed on or around one mast. Also called a **mast table**.

Boottop, or boottopping. The surface of the outside plating between light and load waterlines.

Bosom piece. A short piece of angle riveted inside a butt joint of two angles to form a strap.

Bossing or boss. The curved swelling outboard portion of the ship's shell plating that surrounds and supports the propeller shaft.

Bossing plate. Steel plate covering the bulged portion of hull where the propeller shaft passes outboard.

Bow. The forward end of a ship.

Bow line. The intersection of the molded hull surface forward of amidships with any vertical longitudinal plane not on the centerline. (See **Buttock**.)

Bracket. A plate used to connect rigidly two or more structural parts, such as deck beam to frame, or bulkhead stiffener to the deck or tank top (usually triangular in shape).

Breadth, molded. (See **beam, molded**.)

Break. The end of a partial superstructure such as a poop, bridge or forecastle where it drops to the deck below.

Breakwater. Inclined bulwark-like structure on a weather deck to deflect overboard water coming over the bow or over the gunwale and moving aft.

Breasthook. A triangular plate bracket joining port and starboard side stringers at the stem.

Bridge, flying. The platform forming the top of the pilothouse.

Bridge house. A term applied to an erection fitted on the upper or superstructure deck of a ship. The officers' quarters, staterooms and accommodations are usually located in the bridge house and the pilothouse is located above it.

Bridge, navigating. The conning station or command post of a ship.

Broken stowage. The spaces between and around cargo packages, including dunnage, and spaces not usable because of structural interferences.

Brow. A water shed over an air port; a small inclined ramp to allow passage of trucks over a hatch coaming or bulkhead door sills, etc.

Buckler. A portable cover secured over the deck opening of the hawsepipes and the chain pipes to restrict the flow of water through the openings.

Building basin. A structure essentially similar to a graving dock, in which one or more ships or parts of ships may be built at one time; no launching operation is required, the ship being floated by flooding the basin.

Bulk cargo. Cargo made up of commodities such as oil, coal, ore, grain, etc, and not shipped in bags or containers.

Bulkhead. A term applied to the vertical partition walls which subdivide the interior of a ship into compartments or rooms. The various types of bulkheads are distinguished by their location, use, kind of material or method of fabrication, such as forepeak, longitudinal, transverse, watertight, wire mesh, pilaster, etc. Bulkheads which contribute to the strength of a vessel are called strength bulkheads, those which are essential to the watertight subdivision are watertight or oiltight bulkheads, and gastight bulkheads serve to prevent the passage of gas or fumes.

Bulkhead, after peak. A term applied to the first main transverse bulkhead forward of the sternpost. This bulkhead forms the forward boundary of the after peak tank.

Bulkhead, collision or forepeak. The foremost main transverse watertight bulkhead. It extends from the bottom of the hold to the freeboard deck and it is designed to keep water out of the forward hold in case of bow collision damage.

Bulkhead deck. The bulkhead deck is the uppermost deck up to which the transverse watertight bulkheads and shell are carried.

Bulkhead, screen. A term applied to a light nonwatertight transverse bulkhead fitted in some Great Lakes ore carriers. Its greater flexibility allows it to survive the effects of the unloading machinery.

Bulkhead, swash. (See **swash bulkhead**.)

Bulwark. Fore-and-aft vertical plating immediately above the upper edge of the sheer strake.

Bunk. A berth or bed, usually built in.

Butt. The end joint between two plates or other members which meet end to end.

Buttock. The intersection of the molded surface abaft amidships with any vertical longitudinal plane not on the centerline. (See **Bow line**.)

Butt strap. A strap that overlaps the butt between two plates, serving as a connecting strength strap between the butted ends of the plating.

Calk or caulk. To fill seams in a wood deck with oakum and pay them with pitch, marine glue, etc. To drive or hammer the adjoining edges of metal together to stop or prevent leaks.

Camber. The rise or crown of a deck, athwartship; also called round of beam.

Camel. A fender to keep a vessel away from a pier or quay to prevent damage to the hull or pier; usually a floating body with massive padding of rope, tires, etc.

Cant frame. A frame not square to the centerline at the counter of the ship and connected at the upper end to the cant beams. (See **beams, cant**.)

Capacity plan. A plan outlining the spaces available for cargo, fuel, fresh water, water ballast, etc, and containing cubic or weight capacity lists for such spaces and a scale showing deadweight capacities at varying drafts and displacements.

Capstan. A warping head with a vertical axis used for handling mooring and other lines. It may have at its base a wildcat for handling anchor chain.

Cargo battens. Strips of wood fitted inside the frames to keep cargo away from hull steelwork; also called sparring.

Cargo port. Opening in a ship's side for loading and unloading cargo.

Casing, engine and boiler. Bulkheads enclosing a large opening between the weather deck and the engine and boiler rooms. This provides space for the boiler uptakes, access to these rooms, and permits installing or removing large propulsion units such as boilers or turbines.

Cathodic protection. Protection of a ship's hull against corrosion by superimposing on the hull an impressed current provided by a remote power source through a small number of inert anodes. Also accomplished by fitting aluminum, magnesium or zinc anodes in tanks or the underwater portion of a ship which waste away by galvanic action.

Ceiling, hold and tanktop. A covering usually of wood, placed over the tank top for its protection.

Ceiling, joiner work. The overhead finished surface in quarters, etc.

Center girder. A vertical plate on the ship's centerline between the flat keel and inner bottom or rider plate, extending the length of the ship. Also called center vertical keel, CVK, or center keelson.

Centerline. The middle line of the ship, extending from stem to stern at any level.

Chafing plate. Bent plate for minimizing chafing of ropes, as at hatches.

Chain locker. A compartment for the stowage of anchor chain.

Chain pipe. Pipe for passage of chain from windlass to chain locker.

Chain riveting. Two or more rows of rivets so arranged that the rivets in one row are abreast those in the adjacent row; also see zig-zag riveting.

Chain stopper. A device used to secure the chain cable when riding at anchor, thereby relieving the strain on the windlass, and also for securing the anchor in the housed position in the hawsepipe. (See **devil's claw**.)

Chamfer. To cut off the sharp edge of a 90 deg corner. To trim to an acute angle.

Chock. A heavy smooth-surfaced fitting usually located near the edge of the weather deck through which wire ropes or fiber hawsers may be led, usually to piers. One of several pieces of metal precisely fitted between machinery units and their foundations to assure alignment, also made by pouring plastic material in place. A small piece of plate fitted to one side of a plated structure opposite the landing of a structural member on the other side.

Chock, boat. A cradle or support for a lifeboat.

Cleat. A fitting having two arms or horns around which ropes may be made fast. Clips at intervals on the horizontal stiffeners of hatch coamings; wedges are driven between the clips and the hatch battens.

Clip. A short length of angle to attach or connect structural parts.

Coaming, hatch. The vertical plating bounding a hatch for the purpose of stiffening the edges of the opening and resisting entry of water below.

Cofferdam. Narrow void space between two bulkheads or floors that prevents leakage between the adjoining compartments.

Collision bulkhead. (See **bulkhead, collision**.)

Companionway. An access hatchway in a deck, with a ladder leading below, generally for the crew's use.

Compartmentation. The subdividing of the hull by transverse watertight bulkheads so that the ship may remain afloat under certain assumed conditions of flooding.

Counter. (See **Fantail**.)

Cowl. A hood-shaped top or end of a natural ventilation trunk that may be rotated in direction to cause wind to blow air into or out of the trunk.

Crown. (See **camber**.)

Crow's nest. An elevated lookout station, usually attached to forward side of foremast.

Davit. A crane arm for handling lifeboats, anchors, stores, etc.

Dead cover. A metal cover to close an airport from within the ship in case of breakage of the glass. Also called deadlight.

Dead flat. The portion of a ship's structure that has the same transverse shape as the midship frame. (See **parallel middle-body**.)

Deadlight or fixed light. A term applied to a portlight that does not open.

Deadrise. Athwartship rise of the bottom from the keel to the bilge.

Deadweight. The carrying capacity of a ship at any draft and water density. Includes weight of cargo, fuel, lubricating oil, fresh water in tanks, stores, passengers and baggage, crew and their effects.

Deck. A platform in a ship corresponding to a floor in a building. It is the plating, planking, or covering of any tier of beams either in the hull or superstructure of a ship.

Deck beam. (See **beam**.)

Deck, bulkhead. (See **bulkhead deck**.)

Deck, freeboard. Deck to which freeboard is measured; the uppermost continuous deck having permanent means of closing all weather openings.

Deck height. The vertical distance between the molded lines of two adjacent decks.

Deckhouse. An enclosed erection on or above the weather deck that does not extend from side to side of the ship.

Deck machinery. A term applied to steering gear, capstans, windlasses, winches, and miscellaneous machinery located on the decks of a ship.

Deck, platform. A lower deck, usually in the cargo space, which does not contribute to the longitudinal strength of the ship.

Deck, shelter. (See **shelter deck**.)

Deck stringer. The strake of deck plating that runs along the outboard edge of a deck.

Deck, tonnage. The tonnage deck constitutes the upper boundary of the internal volume of the measurable portions of the ship as defined by the tonnage regulations.

Deck, weather. Uppermost continuous deck with no overhead protection.

Declivity. Inclination of shipways on which a ship slides during launching.

Deep tanks. Tanks extending from the bottom or inner bottom up to or higher than the lowest deck. They are often fitted with hatches so that they also may be used for dry cargo in lieu of fuel oil, ballast water, or liquid cargo.

Depth, molded. The vertical distance from the molded baseline to the top of the freeboard deck beam at side, measured at midlength of the ship.

Derrick. A device for hoisting and lowering heavy weights, cargo, stores, etc.

Devil's claw. A turnbuckle device having two heavy claws designed to fit over a link in the anchor chain for the purpose of securing the anchor hard up in stowed position in the hawsepipe.

Displacement, light. The weight of the ship complete including hull, machinery, outfit, equipment and liquids in machinery.

Displacement, loaded. The displacement of a ship when floating at her greatest allowable draft. It is equal to the weight of water displaced and is the sum of the "light displacement" and the "deadweight."

Dog. A small metal fitting used to hold doors, hatch covers, manhole covers, etc, closed.

Dog shore. A strut extending diagonally from a chock on a ground launch way to a similar chock on the mating sliding way. Dog shores keep the component of ship weight along the ways off triggers or way head burn-off plates and load sliding ways inshore of the dog shores in compression, rather than tension, until dog shores are dropped immediately before launching. Also called **Dagger Shore.**

Double bottom. Compartments at the bottom of a ship between inner bottom and the shell plating, used for ballast water, fresh water, fuel oil, etc.

Doubling (doubler) plate. A plate fitted outside or inside of and faying (touching) against another to give extra local strength or stiffness.

Draft. The depth of the ship below the waterline measured vertically to the lowest part of the hull, propellers, or other reference point. When measured to the lowest projecting portion of the vessel, it is called the **extreme draft,** when measured at the bow, it is called **forward draft,** and when measured at the stern, the **after draft,** the average of the forward draft and the after draft is the **mean draft,** and the mean draft when in full load condition is the **load draft.** Also, in cargo handling, the unit of cargo being hoisted on or off the ship by the cargo gear at one particular hoist.

Draft marks. The numbers on each side of a ship at the bow and stern, and sometimes amidships, to indicate the distance from the lower edge of the number to the bottom of the keel or other fixed reference point. The numbers are 6 in. high and spaced 12 in. bottom to bottom vertically in English units.

Drag. The designed excess of draft aft over that forward when fore and aft drafts are measured from the designed waterline.

Dunnage. Cushioning, loose material placed under or among cargo in the holds to prevent cargo motion or chafing.

Dutchman. A piece of steel fitted or driven into an opening to cover up open joints or crevices usually caused by poor workmanship.

Dynamic positioning. A means of holding a ship in a relatively fixed position with respect to the ocean floor without using anchors, accomplished by two or more propulsive devices controlled by inputs from sonic instruments on the sea bottom and on the ship, by gyrocompass, by satellite navigation or other means.

Effective length. The length used for speed-power calculations and the coefficients therefor. It is determined from the sectional area curve by excluding any abrupt tailing off at the after end of the curve such as often occurs with single-screw, cruiser sternships. In multiscrew normal vessels, it is usually the load waterline length, but in single-screw ships with either cruiser or fantail sterns, it is usually the length from the forward perpendicular to about the middle of the propeller aperture.

Ensign staff. A flagstaff at the stern.

Entrance. That portion of a ship's body forward of the parallel middlebody or the point at which the slope of the sectional area curve is zero.

Erect. To hoist into place and bolt up on the ways fabricated parts of a ship's hull, preparatory to riveting or welding.

Escape trunk. A vertical trunk fitted with a ladder to permit personnel to escape being trapped. Usually provided from the after end of the shaft tunnel to topside spaces.

Even keel. A ship is said to be on an even keel when the keel is horizontal.

Expansion trunk or tank. A trunk extending above a space which is used for the stowage of liquid cargo. The surface of the cargo liquid is kept sufficiently high in the trunk to permit expansion without risk of excessive strain on the hull or of overflowing, and to allow contraction of the liquid without increase of free surface.

Fabricate. To process hull material in the shops prior to assembly or erection. In hull work, fabrication consists of shearing, shaping, punching, drilling, countersinking, scarfing, rabbeting, beveling, and welding.

Face plate. Generally a narrow stiffening plate fitted along the inner edge of web frames, stringers, etc to form the flange of the member.

Fair. To smooth or fair up a ship's lines; eliminating irregularities; to assemble the parts of a ship so that they will be fair, i.e., without kinks, bumps, or waves; to bring rivet holes into alignment.

Fairlead or fairleader. A fitting or device used to preserve or to change the direction of a rope so that it will be delivered on a straight lead to a sheave or drum.

Fairwater. A term applied to plating fitted around the ends of shaft tubes and strut barrels, and shaped to streamline the parts, thus eliminating abrupt changes in the waterflow. Also applied to any casting or plating fitted to the hull for the purpose of preserving a smooth flow of water.

Fall. The rope used with blocks to make up a tackle. The end secured to the block is called the **standing part;** the opposite end, the **hauling part.**

Fantail. The overhanging stern section of ships with round or elliptical after endings to uppermost decks and which extend well abaft the after perpendicular. Also called counter.

Fathom. A measure of length, equivalent to 6 linear feet, used for depths of water and lengths of anchor chain (see shot).

Fathometer. A device to measure the depth of water, by timing the travel of a sound wave from the ship to the ocean bottom and return.

Faying surface. The surface between two adjoining parts.

Fender. The term applied to devices built into or hung over the sides to prevent the shell plating from rubbing or chafing against other ships or piers; a permanent hardwood or steel structure which runs fore and aft on the outside above the waterline and is firmly secured to the hull; wood spars, bundles of rope, used automobile tires, woven cane, or covered cork hung over the sides by lines when permanent fenders are not fitted.

Fidley. The top of engine and boiler room casings on the weather deck. A partially raised deck over the engine and boiler casings, usually around the smokestack.

Fixed light or deadlight. Circular nonopening window with glass in side of ship, door, skylight cover, etc.

Flange. The part of a plate or shape bent at right angles to the main part; to bend over to form an angle.

Flare. The spreading out of the hull form from the central vertical plane, with increasing rapidity as it rises from the waterline to the rail; usually in the forebody. Also a night distress signal.

Flat. A partial deck, usually without camber or sheer.

Flexplate. A plate, at one end of a foundation, which permits a piece of machinery mounted on the foundation to expand or contract freely.

Floodable length. The length of ship which may be flooded without sinking below her safety or margin line. The floodable length of a vessel varies from point to point throughout her length and is usually greatest amidships and least near the quarter length.

Floor. Vertical transverse plate immediately above the bottom shell plating, often located at every frame, extending from bilge to bilge.

Flush deck ship. A ship constructed with an upper deck extending throughout her entire length without a break or a superstructure such as forecastle, bridge or poop.

Fore. A term used in indicating portions or that part of a ship at or adjacent to the bow. Also applied to that portion and parts of the ship lying between amidships and the stem; as, forebody, forehold, and foremast.

Fore-and-aft. In line with the length of the ship; longitudinal.

Forebody. That portion of the ship's body forward of amidships.

Forecastle. A superstructure fitted at the extreme forward end of the upper deck.

Forefoot. The lower end of a ship's stem which curves to meet the keel.

Forepeak. The watertight compartment at the extreme forward end. The forward trimming tank.

Forward. In the direction of the stem.

Forward or fore perpendicular. (See **length between perpendiculars.)**

Foundation. The structural supports for the boilers, main engines or turbines and reduction gears are called main foundations. Supports for machinery space auxiliary machinery are called auxiliary foundations. Deck machinery supports are called, for example, **steering engine foundation, winch foundation,** etc.

Frame. A term used to designate one of the transverse members that make up the riblike part of the skeleton of a ship. The frames act as stiffeners, holding the outside plating in shape and maintaining the transverse form of the ship. (See also **Longitudinal.**)

Frame spacing. The fore-and-aft distance, heel to heel, of adjacent transverse frames.

Freeboard. The distance from the waterline to the upper surface of the freeboard deck at side.

Freeboard deck. (See **deck, freeboard.**)

Freeing port. An opening in the lower portion of a bulwark, which allows deck water to drain overboard. Some freeing ports have hinged gates which allow water to drain overboard but which swing shut to prevent seawater flowing inboard.

Froude number. A non-dimensional number indicating the relation between a vessel's length and its speed, expressed as

$$F_n = \frac{V}{\sqrt{gL}}$$

where
Fn is the Froude number
V is the speed
g is the acceleration due to gravity
L is the length of the vessel all in consistant units.

Full scantling ship. A ship designed with scantlings and weather deck closing arrangements qualifying the ship for minimum freeboard, measured from the uppermost continuous deck, according to the International Load Line Convention.

Funnel. (See **smokestack.**)

Furnaced plate. A plate that requires heating in order to shape it.

Galley. A cookroom or kitchen on a ship.

Gangway. A passageway, side shell opening, or ladderway used for boarding a ship.

Gantry crane. A hoisting device, usually travelling on rails, having the lifting hook suspended from a car which is movable horizontally in a direction transverse to the rails.

Garboard strake. The strake of bottom shell plating adjacent to the keel plate.

Gasket. Flexible material used to pack joints in machinery, piping, doors, hatches, etc, to prevent leakage of liquids or gases.

Girder. A continuous member running fore-and-aft under a deck for the purpose of supporting the deck beams and deck. The girder is generally supported by widely spaced pillars. Also, the vertical fore-and-aft plate members on the bottom of single or double bottom ships.

Girth. Any expanded length, such as the length of a frame from gunwale to gunwale.

Gooseneck, or Pacific iron. A swivel fitting on the end of a boom for connecting it to the mast or mast table. It permits the boom to rotate laterally and to be peaked to any angle. Also, a ventilation terminal in the weather, of rectangular cross section and consisting of a 180 degree bend with the opening facing down; usually fitted with screen and hinged cover.

Grain cubic. The cubic capacity of a cargo hold measured to the shell plating rather than to the inside of the frames or cargo battens.

Graving dock. A structure for taking a ship out of water, consisting of an excavation in the shoreline to a depth at least equal to the draft of ships to be handled, closed at the water side end by a movable gate, and provided with large capacity pumps for removing water; blocks support the ship when the dock is pumped out.

Grommet. A soft ring used under a nut or bolthead to maintain watertightness; also an eye fitted into canvas.

Gross tonnage. (See **tonnage, gross.**)

Ground level building site. A facility wherein the ship is built without sloping building ways, the base line being parallel to the water surface; means of physically moving the completed ship, or large components, to the water are required.

Ground tackle. A general term for anchors, cables, wire ropes, etc, used in anchoring a ship to the bottom.

Gudgeon. Bosses or lugs on sternpost drilled for the pins (pintles) on which the rudder hinges.

Gunwale bar. (See **stringer bar.**)

Gusset plate. A bracket plate lying in a horizontal, or nearly horizontal plane.

Gypsy head. A cylinderlike fitting on the end of winch or windlass shafts. Fiber line or wire rope is hauled or slacked by winding a few turns around it, the free end being held taut manually as it rotates.

Halyard. Light lines used in hoisting signals, flags, etc.

Hatch (hatchway). An opening in a deck through which cargo and stores are loaded or unloaded.

Hatch battens. Flat bars which are wedged against hatch coamings to secure tarpaulins.

Hatch beam. Portable beam across a hatch to support hatch covers.

Hatch, booby. (See **booby hatch.**)

Hatch coaming. (See **coaming, hatch.**)

Hawsepipe. Tube through which anchor chain is led overboard from the windlass wildcat on deck through the ship's side. Bolsters form rounded endings at the deck and shell to avoid sharp edges. Stockless anchors are usually stowed in the hawsepipe.

Head. Toilet; believed to be derived from 'Ship's head', when a small platform outside the bulwarks near the bow was the only semblance of sanitary facilities.

Headlog. In river craft of rectangular shape, the member at the extreme end between the rake shell plating and the deck. Usually a vertical plate of considerable thickness owing to its susceptibility to damage in service.

Heel. The inclination of a ship to one side. (See **list.**) Also the corner of an angle, bulb angle or channel, commonly used in reference to the molded line.

Hogging. Straining of the ship that tends to make the bow and stern lower than the middle portion. (See **sagging.**)

Holds. The large spaces below deck for the stowage of cargo; the lowermost cargo compartments.

Horn, rudder. A heavy casting or weldment projecting down from the hull immediately abaft the propeller, to support the gudgeon fitted to take the single pintle of a semi-balanced rudder.

House, deck. (See **deckhouse.**)

Hull. The structural body of a ship, including shell plating, framing, decks, bulkheads, etc.

Hull girder. That part of the hull structural material effective in the longitudinal strength of the ship as a whole, which may be treated as analagous to a girder.

Inboard. Inside the ship; toward the centerline.

Inner bottom. Plating forming the top of the double bottom; also called tank top.

Intercostal. Made in separate parts: between floors, frames or beams, etc; the opposite of continuous.

Jack staff. A flagstaff at the bow.

Jacob's ladder. A portable ladder with flexible or articulated sides used for access between deck and small craft or piers.

Joggle. To offset a plate or shape to avoid the use of liners.

Keel. The principal fore-and-aft component of a ship's framing, located along the centerline of the bottom and connected to the stem and stern frames. Floors or bottom transverses are attached to the keel.

Keel, bilge. (See **bilge keel.**)

Keel, center vertical. The vertical, centerline web of the keel.

Keel, flat plate. The horizontal, centerline, bottom shell strake constituting the lower flange of the keel.

Keel blocks. Heavy wood or concrete blocks on which ship rests during construction.

Keelson, side. Fore-and-aft vertical plate member located above the bottom shell on each side of the center vertical keel and some distance therefrom.

Kerf. The material removed by the cutting device, such as a burning torch, in preparing a structural member; width of the kerf must be known in programming automatic cutting.

Kingpost. A strong vertical post used instead of a mast to support a boom and rigging to form a derrick: also called **samson post.**

Knee, beam. Bracket connecting a deck beam and frame.

Knot. A unit of speed, equaling one nautical mile per hour; the international nautical mile is 1852 m (6076.1 ft.).

Knuckle. An abrupt change in direction of the plating, frames, keel, deck, or other structure of a ship.

Ladder, accommodation. (See **accommodation ladder.**)

Ladder, Jacob's. (See **Jacob's ladder.**)

Lap. A joint in which one part overlaps the other.

Laying off. The development of the lines of ship's form on the mold-loft floor and making templates therefrom: also called **laying down.**

Length, effective. (See **effective length.**)

Length, overall. The extreme length of a ship measured from the foremost point of the stem to the aftermost part of the stern.

Length between perpendiculars. The length of a ship between the forward and after perpendiculars. The forward perpendicular is a vertical line at the intersection of the fore side of the stern and the summer load waterline. The after perpendicular is a vertical line at the intersection of the summer load line and the after side of the rudder post or sternpost, or the centerline of the rudder stock if there is no rudder post or sternpost.

Lifeboat. A boat carried by a ship for use in emergency.

Life raft. A very buoyant raft, usually of inflatable material, designed to hold people abandoning ship.

Lightening hole. A hole cut in a structural member to reduce its weight.

Lightship weight. (See **displacement, light.**)

Limber hole. A small hole or slot in a frame or plate for the purpose of preventing water or oil from collecting; a drain hole.

Liner. A flat or tapered strip placed under a plate or shape to bring it in line with another part that it overlaps; a filler.

Lines (plan). The plans that show the shape or form of the ship. From the lines drawn full size on the mold-loft floor, templates are made for the various parts of the hull. (See also **Molded Lines.**)

Line shafting. Sections of main propulsion shafting between the machinery and the tail shaft.

List. If the centerline plane of a ship is not vertical, as when there is more weight on one side than on the other, she is said to list, or to heel.

Liverpool head. A metal device fitted to the top of a smoke pipe, as from a galley, to minimize inflow of water and to promote the draft in the pipe. Sometimes mechanically activated by wind.

Load on top. Oil separated from water in the residue from cleaning cargo tanks is loaded on top of the vessel's next cargo, rather than pumping the residue overboard without separation. Slop tanks are provided in which the separation takes place.

Load waterline. The line on the **lines plan** of a ship, representing the intersection of the ship's form with the plane of the water surface when the ship is floating at the **summer freeboard** draft or at the designed draft.

Lofting. The process of developing the size and shape of components of the ship from the designed lines; traditionally, making templates using full scale lines laid down on the floor of the mold loft; today, largely performed at small scale using photographic or computer methods.

Longitudinals. Fore-and-aft structural shape or plate members attached to the underside of decks, flats, or to the inner bottom, or on the inboard side of the shell plating, in association with widely spaced transverses, in the longitudinal framing system.

Louver. An opening to the weather in a ventilation system, fitted with a series of overlapping vanes at about 45 degrees to the vertical intended to minimize the admission of rain or spray to the opening.

Manhole. A round or oval hole cut in decks, tanks, etc, for the purpose of providing access.

Margin angle. Angle connecting margin plate to shell.

Margin bracket. A bracket connecting a side frame to the margin plate at the bilge; sometimes called **bilge bracket.**

Margin line. A line, not less than 3 in. below the top of the bulkhead deck at side, defining the highest permissible location on the side of the ship of any damage waterplane in the final condition of sinkage, trim and heel.

Margin plate. The outboard strake of the inner bottom. When the margin plate is turned down at the bilge it forms the outboard boundary of the double bottom, connecting the inner bottom to the shell plating at the bilge.

Marine railway. An arrangement of tracks from shore to a sufficient depth of water to permit a vessel to be placed on the moveable carriage, which may then be drawn up the track in order to give access to the portion of the vessel below its waterline. Generally limited to comparatively small vessels.

Mast. A tall vertical or raked structure, usually of circular section, located on the centerline of a ship and used to carry navigation lights, radio antennae and, usually, cargo booms. (See **kingpost.**)

Mast step. A term applied to the foundation on which a mast is erected.

Mast table. (See **boom table.**)

Maximum section coefficient, C_X. The ratio of the area of the maximum vertical transverse cross section of the underwater body of a ship to the product of the waterline beam and the draft at that section.

Messroom. Dining room for officers or crew.

Metacenter. The center of buoyancy of a listed ship is not on the vertical centerline plane. The intersection of a vertical line drawn through the center of buoyancy of a slightly listed vessel intersects the centerline plane at a point called the metacenter.

Metacentric height. The distance from the metacenter to the center of gravity of a ship. If the center of gravity is below the metacenter the vessel is stable.

Midship. (See **amidships.**)

Midship section. A drawing showing a typical cross section of the hull and superstructure at or near amidships, and giving the scantlings of the principal structural members.

Mock-up. A three dimensional full size replica of the shape of a portion of a vessel, used where the geometry makes fabrication of steel members from conventional templates difficult, or to avoid interferences by laying out components in three dimensions.

Mold loft. A floor space used for laying down (laying off) the full-size lines of a ship and for making templates to lay out the hull structural components.

Molded lines. Lines defining the geometry of a hull as a surface without thickness; structural members are related to molded lines according to standard practice (unless otherwise shown on drawings) e.g. the inside surface of flush shell plating is on the molded line, also the underside of deck plating.

Moon pool. A large opening through the deck and bottom of a drill ship at about amidships to accomodate the major drilling operations.

Mooring. Securing a ship at a pier or elsewhere by several lines or cables so as to limit her movement.

Mooring ring. A round or oval casting inserted in the bulwark plating through which the mooring lines, or hawsers, are passed.

Mushroom. A cover permanently fitted above a ventilator located in the weather, usually round and of larger diameter than the ventilator.

Nautical mile. (See **knot.**)

Net tonnage. (See **tonnage, net.**)

Network flow. A technique for relating tasks, in the process of scheduling ship production; the principles lend themselves to computer utilization.

OBO. Abbreviation for a vessel designed to carry oil, bulk cargos or ore cargos.

Offsets. A term used for the coordinates of a ship's form, deck heights, etc.

Oil stop or water stop. A detail in the construction of the intersection of members, one or both of which are oiltight, to permit continuous welding of the boundry of the tight member; usually a small semicircular cut in the edge of the "thru" member near the intersection; in a lapped seam, a slot in one member, filled with weld metal.

One compartment subdivision. A standard of subdivision of a ship by bulkheads, which will result in the ship remaining afloat with any one compartment flooded, under specified conditions as to permeability of the compartment and the draft of the ship before flooding of the compartment.

Outboard. Abreast or away from the centerline towards the side; outside the hull.

Overhang. That portion of a ship's bow or stern clear of the water which projects beyond the forward and after perpendiculars.

Pad eye. A fitting having one or more eyes integral with a base to provide means of securing blocks, wire rope or fiber line.

Padding. An inert gas, usually nitrogen, introduced into a cargo tank above the liquid cargo to prevent the cargo coming in contact with air.

Panel line. A production line where individual plates, framing members, webs, etc. are successively welded together to form an assembly unit which may include some items of outfit.

Panting. The pulsation in and out of the bow and stern plating as the ship alternately rises and plunges deep into the water. May also occur abreast the propellers of a multiscrew ship.

Panting frames. The frames in the forward and after portions of the hull framing, to prevent panting action of the shell plating.

Paragraph ship. A vessel designed to measure slightly less than the gross tonnage marking the boundary for application of provisions of a convention, treaty, law or regulation. The word "paragraph" is thought to be used in the sense that tonnage boundaries are, at times, cited as provisions of a specific paragraph of a treaty, law or regulation.

Parallel middlebody. The amidship portion of a ship within which the contour of the underwater hull form is unchanged.

Peak. (See **after peak, forepeak.**)

Period of roll. The time occupied in performing one complete double oscillation or roll of a ship as from port to starboard and back to port.

Pillar. Vertical member or column giving support to a deck girder, flat or similar structure: also called **stanchion.**

Pintles. The pins or bolts that hinge the rudder to the gudgeons on the sternpost or rudder post.

Planking. Wood covering for decks, etc.

Plasma-arc cutting. A process employing an extremely high-temperature, high-velocity constricted arc between an electrode within a torch and the metal to be cut. The intense heat melts the metal which is continuously removed by a jet-like stream of gas issuing from the torch.

Platform. (See **Deck, platform,** and **Flat.**)

Plating, joggled. The edges of the plating are joggled to avoid the use of liners between the plating and the framing.

Poop. A superstructure fitted at the after end of the upper deck.

Port, cargo. An opening in the side plating provided with a watertight cover or door and used for loading and unloading.

Porthole, portlight. (See **air port.**)

Port side. The left-hand side of a ship when looking forward. Opposite to starboard.

Propeller. A revolving screw-like device that drives the ship through the water, consisting of two or more blades; sometimes called a screw or wheel.

Propeller boss. (See **bossing.**)

Propeller post. (See **stern frame.**)

Quarters. Living or sleeping rooms.

Quenching. In steelmaking, an operation consisting of heating the material to a certain temperature and holding at that temperature to obtain desired crystalline structure, and then rapidly cooling it in a suitable medium, such as water or oil. Quenching is often followed by tempering.

Rabbet. A groove, depression, or offset in a member into which the end or edge of another member is fitted, generally so that the two surfaces are flush. A rabbet in the stem or stern frame would take the ends or edges of the shell plating, resulting in a flush surface.

Rail. The rounded member at the upper edge of the bulwark, or the horizontal pipes or chains forming a fencelike railing fitted instead of a bulwark.

Rake. A term applied to the fore-and-aft inclination from the vertical, of the mast, smokestack, stem, etc. In river, and some ocean, barges, the end portion of the hull, in which the bottom rises from the midship portion to meet the deck at the headlog.

Reefer. Colloquial abbreviation for "refrigerator" or "refrigerated".

Reverse frame. A bar forming the top member of an open floor, attached to the underside of the inner bottom.

Ribband. A fore-and-aft wooden strip or heavy batten used to align the transverse frames and keep them in a fair line. Any similar batten for fairing a ship's structure. Also, a layer of insulation on the boundaries of decks in insulated spaces. Also, steel plating or heavy planking bolted to the outboard sides of ground launching ways and projecting above the sliding surface sufficiently to keep the sliding ways in approximate alignment.

Rider plate. A continuous flat plate attached to the top or bottom of a girder.

Rigging. Chains, wire ropes, fiber lines and associated fittings and accessories used to support masts and booms used for handling cargo and stores and for other purposes.

Rise of floor. (See **deadrise.**)

Roll. To impart curvature to a plate. Also the transverse angular motion of the ship in waves. (See **period of roll.**)

Ro-Ro or Ro/Ro. Abbreviation for a vessel designed to carry vehicles, so arranged that the vehicles may be loaded and unloaded by being rolled on or off on their own and/or auxiliary wheels, via ramps fitted in the sides, bow or stern of the vessel.

Round of beam. (See **camber.**)

Rudder. A device used to steer a ship. The most common type consists of a vertical metal area, hinged at the forward edge to the sternpost or rudder post.

Rudder pintle. (See **Pintles.**)

Rudder post. (See **sternpost.**)

Rudder stock. A vertical shaft that connects the rudder to the steering engine.

Rudder stop. Lag on stern frame or a stout bracket on deck at each side of the quadrant, to limit the swing of the rudder to approximately 37 deg port or starboard. A rudder angle of 35 deg is the maximum usually used at sea (45 deg on inland waterways vessels).

Run. That part of a ship's body aft of the parallel middle body or the point at which the slope of the sectional area curve is zero.

Sagging. Straining of the ship that tends to make the middle portion lower than the bow and stern. (See **hogging.**)

Samson post. (See **kingpost.**)

Scantlings. The dimensions of a ship's frames, girders, plating, etc.

Scantling Draft. The maximum draft at which a vessel complies with the governing strength requirements. Usually used when the Scantling Draft is less than the geometrical draft corresponding to the freeboard calculated according to the Load Line Convention.

Scarf. A connection made between two pieces by tapering their ends so that they fit together in a joint of the same breadth and depth as the pieces connected. It is used on bar keels, stem and stern frames, and other parts.

Screen bulkhead. (See **bulkhead, screen.**)

Scuppers. Drains from decks to carry off accumulations of rainwater, condensation or seawater. Scuppers are located in the gutters or waterways, on open decks, and connect to pipes usually leading overboard, and, in corners of enclosed decks to the bilge.

Scuttle. A small circular or oval opening fitted in decks to provide access. When used as escape scuttles and fitted with means whereby the covers can be opened quickly to permit exit, they are called quick-acting scuttles. Sometimes used to refer to an air port.

Scuttlebutt. A container for drinking water. A drinking fountain. Colloquially, rumors heard at the drinking fountain.

Sea chest. An enclosure, attached to the inside of the underwater shell and open to the sea, fitted with a portable strainer plate. A sea valve and piping connected to the sea chest passes sea water into the ship for cooling, fire or sanitary purposes. Compressed air or steam connections may be provided to remove ice or other obstructions.

Seam. Fore-and-aft joint of shell plating, deck and tank top plating, or a lengthwise edge joint of any plating.

Seam strap. A strip of plate serving as a connecting strap between the edges of flush plating. Strap connections at the ends of plates are called butt straps.

Sections. A general term referring to structural bars, rolled or extruded in any cross section, such as angles, channels, bulbs, Tees, H and I bars (or beams). Sometimes called profiles. Also the intersections with the hull of transverse planes perpendicular to the centerline plane of the ship.

Shaft strut. (See **strut.**)

Shaft tunnel, shaft alley. A watertight enclosure for the propeller shafting large enough to walk in, extending aft from the engine room to provide access and protection to the shafting in way of holds.

Shape. A rolled bar of constant cross section such as an angle, bulb angle, channel, etc; also to impart curvature to a plate or other member.

Sheer. The longitudinal curve of a vessel's decks in a vertical plane, the usual reference being to the ship's side; in the case of a deck having a camber, its centerline sheer may also be given in offsets. Due to sheer, a vessel's deck height above the baseline is higher at the ends than amidships.

Sheer strake. The course of shell plating at strength deck level.

Shell expansion. A plan showing the seams and butts, thickness and associated welding on riveting of all plates comprising the shell plating, framing, etc.

Shell landings. Points on the frames where the edges of shell plates are located.

Shell plating. The plates forming the outer side and bottom skin of the hull.

Shelter deck. Formerly, a term applied to a superstructure deck fitted continuous from stem to stern and fitted with at least one tonnage opening.

Shelter deck ship. A ship having two or more complete decks, with scantlings and weather deck closing arrangements based upon the uppermost deck being a superstructure deck and the second deck being the freeboard deck, in contrast to the "Full Scantling Ship".

Shifting boards. Portable bulkhead members, generally constructed of wood planking and fitted fore and aft in cargo holds when carrying grain or other cargo that might shift to one side when the ship is rolling in a seaway. Temporary closing appliances for tonnage openings in superstructure end bulkheads.

Shift of butts. The arrangement of the butts in structural plating members whereby the butts of adjacent members are located a specified distance from one another.

Shore. A brace or prop used for support while building a ship. (See **dogshore.**)

Shot. A length of anchor chain equivalent to 15 fathoms or 90 feet.

Shroud. One of the principal members of the standing rigging, consisting of wire rope which extends from the mast head to the ship's side, affording lateral support for a mast.

SI Units. The system of units now being used internationally is the Systeme International d'Unites (SI). The conversion table (see next page) provides conversion factors for U.S. Customary units, and MKS units that differ from SI units, that are used in this volume. Proper use of significant figures and rounding-off techniques should be given due consideration when using the conversion factors.

Sight edge. The visible edge of shell plating as seen from outside the hull.

Skeg. A deep, vertical, finlike projection on the bottom of a vessel near the stern, installed to support the lower edge of the rudder, to support the propeller shaft for singlescrew ships, and for the

CONVERSION TABLE

Quantity	U.S. Customary or MKS Unit	Multiple by	To Obtain SI Unit	SI Units
Area	ft^2	9.290 304(10^{-2})	m^2	$meters^2$
	in^2	6.451 600(10^2)	mm^2	$millimeters^2$
Bending, Torque	lb-ft	1.355 818	Nm	newton-meters
	lb-in.	1.129 848(10^{-1})	Nm	newton-meters
	kg-m	9.806 650*	Nm	newton-meters
Density	lb/ft^3	1.601 846(10^1)	kg/m^3	$kilograms/meter^3$
	$lb/in.^3$	2.767 990(10^4)	kg/m^3	$kilograms/meter^3$
Electrical	circular mils	5.067 075(10^{-10})	m^2	$meter^2$
Energy, Work	BTU	1.055 056(10^3)	J	joules
	ft-lb	1.355 818	J	joules
	in.-lb	1.129 848(10^{-1})	J	joules
	kcal	4.186 800(10^3)*	J	joules
	kg-m	9.806 650*	J	joules
Force	lb	4.448 222	N	newtons
	Ton(long)	9.964 017(10^3)	N	newtons
	kg	9.806 650	N	newtons
	ton(metric)	9.806 650(10^3)*	N	newtons
Length	fathom	1.828 800*	m	meters
	ft	3.048(10^{-1})*	m	meters
	in.	25.4(10^1)*	mm	millimeters
Mass	lb	4.535 924(10^{-1})	kg	kilograms
	ton(long)	1.016 047(10^3)	kg	kilograms
	ton(metric)	1.000(10^3)	kg	kilograms
Plane angle	degree	$\pi/180$	rad	radians
Power	hp	7.456 999(10^2)	W	watts
	hp(metric)	7.354 99(10^2)	W	watts
Pressure, Stress	ksi	6.894 757(10^6)	N/m^2(or Pa)	$newtons/meter^2$(pascals)
	psi	6.894 757(10^3)	N/m^2(or Pa)	$newtons/meter^2$(pascals)
		6.894 757(10^{-3})	N/mm^2	$newtons/millimeter^2$
		6.894 757(10^{-2})	bar(or $10^5 N/m^2$)	bar(10^5 $newtons/meter^2$)
	$ton(long)/in.^2$	1.544 426(10^7)	N/m^2(or Pa)	$newtons/meter^2$(pascals)
		1.544 426(10^1)	N/mm^2	$newtons/millimeter^2$
	kg/cm^2	9.806 650(10^4)*	N/m^2(or Pa)	$newtons/meter^2$(pascals)
		9.806 650(10^{-1})	bar(or $10^5 N/m^2$)	bar(10 $newtons/meter^2$)
	kg/mm^2	9.806 650(10^6)*	N/m^2(or Pa)	$newtons/meter^2$(pascals)
Temperature	°F	(°F − 32)1.8*	°C	degrees Celsius
Velocity	knot(kn)	0.514 444	m/sec	meters/second
Volume	ft^3	2.831 685(10^{-2})	m^3	$meter^3$
	gallon(liquid)	3.785 412(10^{-3})	m^3	$meter^3$
	$in.^3$	1.638 706(10^4)	mm^3	liters
	$in.^3$	1.638 706(10^4)	mm^3	$millimeter^3$

*Exact value

SI UNIT PREFIXES

Multiplication Factor	Prefix	Symbol	Pronunciation (USA) (1)	Meaning (in USA)	In Other Countries
1 000 000 000 000 000 000 = 10^{18}	exa (2)	E	ex' a (*a* as in *a*bout)	One quintillion times	trillion
1 000 000 000 000 000 = 10^{15}	peta (2)	P	as in *pet*al	One quadrillion times	thousand billion
1 000 000 000 000 = 10^{12}	tera	T	as in *terra*ce	One trillion times	billion
1 000 000 000 = 10^9	giga	G	jig' a (*a* as in *a*bout)	One billion times	milliard
1 000 000 = 10^6	mega	M	as in *mega*phone	One million times	
1 000 = 10^3	kilo	k	as in *kilo*watt	One thousand times	
100 = 10^2	hecto	h (4)	heck' toe	One hundred times	
10 = 10	deka	da (4)	deck' a (*a* as in *a*bout)	Ten times	
0.1 = 10^{-1}	deci	d (4)	as in *deci*mal	One tenth of	
0.01 = 10^{-2}	centi	c (4)	as in *senti*ment	One hundredth of	
0.001 = 10^{-3}	milli	m	as in *mili*tary	One thousandth of	
0.000 001 = 10^{-6}	micro	μ(5)	as in *micro*phone	One millionth of	
0.000 000 001 = 10^{-9}	nano	n	nan' oh (*an* as in *ant*)	One billionth of	milliardth
0.000 000 000 001 = 10^{-12}	pico	p	peek' oh	One trillionth of	billionth
0.000 000 000 000 001 = 10^{-15}	femto	f	fem' toe (*fem* as in *fem*inine)	One quadrillionth of	thousand billionth
0.000 000 000 000 000 001 = 10^{-18}	atto	a	as in an*ato*my	One quintillionth of	trillionth

support of the vessel in dry dock: also used on barges to minimize erratic steering in seaway.

Skids. A skeleton framework used to hold structural assemblies above ground to facilitate riveting or welding.

Skin tank. A tank for liquid cargo or ballast one or more of whose boundaries is the side or bottom plating of the hull.

Skylight. A framework fitted over a deck opening and having covers with glass inserted for the admission of light and air to the compartment below.

Slamming. Heavy impact resulting from a vessel's bottom forward making sudden contact with the sea surface after having risen on a wave. Similar action results from rapid immersion of the bow in vessels with large flare.

Smokestack. A chimney through which combustion products are lead from propulsion and auxiliary machinery to the weather; also called a funnel.

Sounding tube. A pipe leading to the bottom of a bilge, double bottom, deep tank, drainwell, hold or other compartment, used to guide a sounding line, tape or rod to determine the depth and nature of any liquid.

Spar deck. An anachronism used on the Great Lakes to indicate the weather or upper deck; so used because the term Main Deck is applied to the narrow plating forming the top of the usual side tanks abreast the cargo spaces on typical Great Lakes bulk carriers.

Spectacle frame. A large casting extending outboard from the main hull and furnishing support for the ends of the propeller shafts in a multiscrew ship. The shell plating (bossing) encloses the shafts and is attached at its after end to the spectacle frame. Used in place of shaft struts.

Spring bearing. Bearings to support the line shafting.

Springing. A vibration of the complete vessel induced by wave forces in conjunction with the ship's elastic properties. More pronounced in ships having a high length-to-depth ratio.

Squatting. The increase in trim by the stern assumed by a ship when underway over that existing when at rest.

Stability. The tendency of a ship to remain upright or the ability to return to her normal upright position when heeled by the action of waves, wind, etc.

Staging. Horizontal surfaces giving shipbuilding personnel access to the work, either inside or outside the hull; may be supported by temporary members attached to subassemblies or by wood or metal uprights, usually portable.

Stanchion. Vertical column supporting decks, flats, girders, etc; also called a pillar. Rail stanchions are vertical metal columns on which fence-like rails are mounted. (See **rail.**)

Standing rigging. Fixed rigging supporting the masts such as shrouds and stays. Does not include running rigging such as boom topping lifts, vangs and cargo falls.

Starboard side. The right-hand side of a ship when looking forward. Opposite to port.

Stays. Fixed wire ropes leading forward from aloft on a mast to the deck to prevent the mast from bending aft. Backstays lead from aloft to the deck edge well abaft the position of the mast. Preventer stays lead to any point on the deck to provide additional mast support when handling very heavy loads with boom tackle.

Steady bearing. (See **Spring bearing.**)

Stealer. A single wide plate that is butt-connected to two narrow plates, usually near the ends of a ship.

Steering gear. A term applied to the steering wheels, leads, steering engine, and fittings by which the rudder is turned. Usually applied to the steering engine.

Stem. The bow frame forming the apex of the intersection of the forward sides of a ship. It is rigidly connected at its lower end to the keel and may be a heavy flat bar or of rounded plate construction.

Step, mast. (See **mast step.**)

Stern. After end of ship.

Stern, clearwater. A stern with a "shoeless" stern frame.

Stern, cruiser. A spoon-shaped stern used on most merchant ships, designed to give maximum immersed length.

Stern, transom. A square-ended stern used to provide additional hull volume and deck space aft and(or) to decrease resistance in some high speed ships.

Stern frame. Large casting, forging or weldment attached to after end of the keel. Incorporates the rudder gudgeons and in single-screw ships includes the propeller post.

Sternpost. Sometimes, the vertical part of the stern frame to which the rudder is attached.

Stern tube. The watertight tube enclosing and supporting the tail-shaft. It consists of a cast-iron or cast-steel cylinder fitted with bearing surface within which the tailshaft, enclosed in a sleeve, rotates.

Stiff, stiffness. A vessel is said to be stiff if she has an abnormally large metacentric height. Such a ship may have a short period of roll and therefore will roll uncomfortably. The opposite of **tender.**

Stiffener. An angle, T-bar, channel, built-up section, etc. used to stiffen plating of a bulkhead, etc.

Stopwater. (See **Oilstop.**)

Stow. To put away. To stow cargo in a hold.

Stowage factor. The volume of a given type of cargo per unit of its weight.

Strain. The deformation resulting from a stress, measured by the ratio of the change to the total value of the dimension in which the change occurred.

Strake. A course, or row, of shell, deck, bulkhead, or other plating.

Strength deck. The deck that is designed as the uppermost part of the main hull longitudinal strength girder. The bottom shell plating forms the lowermost part of this girder.

Stress. The force per unit section area producing deformation in a body.

Stringer. A term applied to a fore-and-aft girder running along the side of a ship at the shell and also to the outboard strake of plating on any deck. The side pieces of a ladder or staircase into which the treads and risers are fastened.

Stringer bar. The angle connecting the deck plating to the shell plating or to the inside of the frames. The strength deck stringer bar is usually called the gunwale bar.

Stringer plate. (See **deck stringer.**)

Strongback. A piece of plate or a special tool used to align the edges of plates to be welded together. Also a steel bar such as a channel, one or more of which may be used to secure a door in closed position, in addition to the dogs around its edge.

Strut. Outboard column-like support or vee-arranged supports for the propeller shaft, used on some ships with more than one propeller instead of bossings.

Stuelcken gear. A design of heavy lift cargo gear consisting of a pair of freestanding masts, flared, with a boom between them on the centerline capable of use either abaft or forward of the masts.

Submerged arc welding. A fusion welding process in which a machine feeds electrode and a flux in granular form as the machine moves over the joint; the flux covers the arc, submerging it, and is melted to a brittle slag covering the deposited metal.

Superstructure. A decked-over structure above the upper deck, the outboard sides of which are formed by the shell plating as distinguished from a deckhouse that does not extend outboard to the ship's sides. (See **shelter deck,** also **poop, bridge** and **forecastle.**)

Swash bulkhead, swash plate. Longitudinal or transverse nontight bulkheads fitted in a tank to decrease the swashing action of the liquid contents as a ship rolls and pitches at sea. Their function is greatest when the tanks are partially filled. Without them the unrestricted action of the liquid against the sides of the tank might be severe. A plate serving this purpose but not extending to the bottom of the tank is called a swash plate.

Tailshaft. The aftermost section of the propulsion shafting in the stern tube in single screw ships and in the struts of multiple screw ships, to which the propeller is fitted.

Tank, ballast. (See **ballast tank.**)

Tank, peak. (See **after peak, forepeak.**)

Tank, settling. Fuel oil tanks used for separating entrained water from the oil. The oil is allowed to stand for a few hours until the water has settled to the bottom, when the latter is drained or pumped off.

Tank top. (See **inner bottom.**)

Tank, trimming. A tank located near the ends of a ship. Seawater (or fuel oil) is carried in such tanks as necessary to change trim.

Tank, wing. Tanks located well outboard adjacent to the side shell plating, often consisting of a continuation of the double bottom up the sides to a deck or flat.

Tarpaulin. A pliable waterproof cloth cover secured over nonwatertight hatch covers.

Telegraph. An apparatus, either electrical or mechanical, for transmitting orders, as from a ship's bridge to the engine room, steering gear room, or elsewhere about the ship.

Telemotor. A device for operating the control valves of the steering engine from the pilothouse, either by fluid pressure or by electricity.

Tempering. After quenching, the material is reheated to a predetermined temperature below the critical range and then cooled. In steelmaking this is done to relieve stresses set up by quenching and to restore ductility.

Template. Wood or paper full-size patterns to be placed on materials to indicate the size and location of rivet holes, plate edges, etc; also to indicate the curvature to which frames, for example, are to be bent.

Tender. A vessel is said to be tender if she has an abnormally small metacentric height. Such a ship may have a long period of roll but may list excessively in a strong wind and may be dangerous if a hold is flooded following a collision. The opposite of stiff. Also a small general utility boat carried aboard ship.

Test head. The head or height of the column of water which will give a prescribed pressure on the vertical or horizontal sides of a compartment or tank in order to test its tightness, or strength.

Thrust recess. A small compartment at the after end of the main engine room at forward end of shaft tunnel, designed to contain and give access to the thrust shaft and thrust bearing.

Tie plate. A fore-and-aft course of plating attached to deck beams under a wood deck for strength purposes.

Tiller. An arm, attached to rudder stock, which turns the rudder.

Tonnage, gross. Under vessel measurement rules of various nations, the Panama Canal and the Suez Canal, a measure of the internal volume of spaces within a vessel in which 1 ton is equivalent to 2.83 m^3 or 100 cu ft. Under the International Convention on Tonnage Measurement of Ships, 1969, (ICTM), a standardized numerical value that is a logarithmic function of spaces within a vessel There is no definition of a ton under ICTM because the value per unit of volume is greater on a vessel of large volume than on a vessel of small volume. Gross tonnage according to the national and canal rules generally includes spaces bounded by the under surface of the uppermost complete deck, the side frames, and the floor frames or the inner bottom if it rests on the floors or if the double bottom is for water ballast, plus closed-in space in deck structures available for cargo or stores or for the berthing or accommodation of passengers or crew. Rules vary greatly as to exclusion or inclusion of various spaces. Gross tonnage according to ICTM is $GT = K_1 V$ in which V is the total molded volume of all enclosed spaces of the ship in m^3 and K_1 is $0.2 + 0.02 \log_{10} V$.

Tonnage mark. A distinctive symbol placed slightly abaft amidships each side indicating the maximum draft to which a vessel may be loaded if exemption of certain spaces in the uppermost 'tween decks is to be maintained under IMCO Resolution A.48(III) of 18 October 1963. The tonnage mark is a horizontal line 380 mm long and 25 mm wide identified by an inverted equilateral triangle 300 mm long each side with its apex resting on the midpoint of the upper edge of the line.

Tonnage, net. Net tonnage according to national and canal rules is derived from gross tonnage by deducting an allowance for the propelling machinery space and certain other spaces. Net tonnage according to ICTM is a logarithmic function of the volume of cargo space, the draft to-depth ratio, the number of passengers to be accommodated, and the gross tonnage.

Tonnage opening. A device to circumvent the tonnage measurement rules, consisting of a closure just short of meeting the definition of a watertight closing appliance, but nevertheless accepted in the U.S. as a weathertight closing appliance for purposes of the Load Line regulations.

Topping lift. A wire rope or tackle extending from the head of a boom to a mast, or to the ship's structure, for the purpose of supporting the weight of the boom and its loads, and permitting the boom to be raised or lowered.

Topside Tank. A tank usually used for ballast in bulk carriers, formed by the deck outboard of the hatches, the upper shell plating and a logitudinal bulkhead sloping around 45 deg.

Tramp ship. A general breakbulk cargo ship that has no set trade route or schedule.

Transformation temperature. The temperature above which the ferrite form of iron in shipbuilding and other steel is transformed to the austenite form, and below which the ferrite form recurs. The microstructure of the steel is changed upon passing through this temperature.

Transom, stern. (See **Stern, transom.**)

Transom beam. The aftermost transverse deck beam. Abaft and connected to it are the cant beams.

Transom frame. The aftermost transverse side frame, abaft which are the cant frames.

Transverse. A deep member supporting longitudinal frames of bottom or side shell or longitudinal deck beams. At right angles to the fore-and-aft centerline.

Trim. The difference between the draft forward and the draft aft. If the draft forward is the greater, the vessel is said to "trim by the head." If the draft aft is the greater, she is "trimming by the stern." To trim a ship is to adjust the location of cargo, fuel, etc. so as to result in the desired drafts forward and aft.

Tripping bracket. Flat bars or plates fitted at various points on deck girders, stiffeners, or beams as reinforcements to prevent their free flanges from tripping.

Trunk. A vertical or inclined space or passage formed by bulkheads or casings, extending one or more deck heights, around openings in the decks, through which access can be obtained and cargo, stores, etc. handled, or ventilation provided without disturbing or interfering with the contents or arrangements of the adjoining spaces.

Tumble home. Inboard slope of a ship's side, usually above the designed waterline.

Tumble shore. A strut, perpendicular to the plane of the launch ways, between the bottom of a vessel about to be launched and the ground or building slip deck. Used to keep some of ship weight off the launching grease until sliding starts. Motion of the vessel causes the shore to fall or tumble to the ground. Typically, one or more tumble shores might be installed under the outshore end of the keel of a ship to be end launched.

Tunnel, shaft. (See **shaft tunnel.**)

'Tween decks. The space between any two adjacent decks.

Twist lock. A mechanical locking device at the corner of a cargo container. Rotating the movable part about a vertical axis locks the device and container to a mating component on an above-deck stowage or another container. Reversing the motion effects unlocking.

Uptake. A casing connecting a boiler or gas turbine combustion product outlet with the base of the inner casing of the smokestack.

Vang. Wire rope or tackle secured to the end of a cargo boom, the lower end being secured to the deck, top of bulwark, or to a special post at the ship's side. Used to swing the boom and hold it in a desired position.

Ventilator, cowl. (See **cowl.**)

VLCC tanker. A very large crude oil carrier, over about 150,000 DWT.

Waterline. The line of the water's edge when the ship is afloat; technically, the intersection of any horizontal plane with the molded form.

Waterway. A narrow gutter along the edge of a deck for drainage.

Weather deck. (See **deck, weather.**)

Web frame. A built-up frame to provide extra strength, usually consisting of a web plate flanged or otherwise stiffened on its edge, spaced several frame spaces apart, with the smaller, regular frames in between.

Well. Space in the bottom of a ship to which bilge water drains so that it may be pumped overboard; space between partial superstructures.

Whip. In cargo handling, the wire leading to the hook by which the draft of cargo is being hoisted.

Wildcat. A special type of cog-like windlass drum whose faces are formed to fit the links of the anchor chain. The rotating wildcat causes the chain to be slacked off when lowering the anchor, or hauled in when raising it.

Winch. A machine, usually steam or electric, used primarily for hoisting and lowering cargo but also for other purposes.

Windlass. The machine used to hoist and lower anchors.

Zig-Zag riveting. Two or more rows of rivets spaced so that the rivets of one row are offset; also see **chain riveting.**

Index

This page intentionally left blank

This page intentionally left blank

This page intentionally left blank

This page intentionally left blank